Semiconductor
Characterization
Present Status and Future Needs

Semiconductor Characterization
Present Status and Future Needs

EDITORS

W. M. Bullis
Materials & Metrology, Sunnyvale, CA 94087

D. G. Seiler
NIST, Gaithersburg, MD 20899

A. C. Diebold
SEMATECH, Austin, TX 78741

American Institute of Physics   **Woodbury, New York**

©1996 by American Institute of Physics
Printed in the United States of America.

AIP Press
American Institute of Physics
500 Sunnyside Boulevard
Woodbury, NY 11797-2999

Library of Congress Cataloging-in-Publication Data
Semiconductor characterization : present status and future needs / editors, W. M. Bullis, D. G. Seiler, A. C. Diebold.
    p. cm.
  Includes index.
  ISBN 1-56396-503-8
  1. Semiconductors--Characterization--Congresses.
  2. Semiconductors--Design and construction--Congresses.
  I. Bullis, W. Murray, 1930– . II. Seiler, David G. III. Diebold, A. C. (Alain C.).
TK7871.85.S4457 1996   95-41074
621.3815'2--dc20   CIP

10 9 8 7 6 5 4 3 2 1

FOREWORD

Semiconductors and their applications are one of the greatest scientific and technological breakthroughs of this century. Their influence has far reaching effects on society in general and on our daily lives. Consider the impact of microelectronic components used in computers, entertainment equipment, automotive electronics, medical instrumentation, telecommunications, space technology, television, radio, and manufacturing technologies. The semiconductor technology evolution that is producing transistors, rectifiers, photoelectric cells, lasers, detectors, magnetometers, solar cells, microprocessors, and high density memories will continue to affect us all. Almost every factory, hospital, office, bank, school, or household contains transistors, microprocessors, and other semiconductor devices.

One of the most spectacular and important applications of semiconductors is the area of information technology. It is because microelectronics is being used to enhance our computing and reasoning powers that we are even able to talk about the information revolution of today. These developments were possible because of the miniaturization of the transistor dimensions which allowed the construction of compact systems with tremendous computing power and memory. Miniaturization, in turn, is possible because of the perfection of "planar" fabrication techniques that allow "integration" of circuits and thus the production of devices containing many millions of elements per square centimeter. Circuit integration has gone through several major phases – from IC (integrated circuit) to LSI (large-scale integration) to VLSI (very large-scale integration) to ULSI (ultra-large-scale integration) to GSI (giga scale integration) and beyond. The cornerstone of this technology is silicon. Meeting the demands for these large-scale, complex, ICs will continue to require technological advances in materials, processing, circuit design, characterization, testing, and standards.

Compound semiconductors are also being used in a variety of structures for light-emitting diodes, laser diodes, far-infrared detectors, and microwave devices. In addition to being used in their bulk and natural forms, these semiconductor materials are used in artificially created structures such as superlattices and heterostructures. Various compounds are mixed to produce structures in which properties like the bandgap have been engineered to have specific values. Such structures and devices have unique applications in wireless communications, optical communications, visible light sources, and imaging, all of which are critical for information technologies. They also enable functions and enhanced performance that cannot be equaled by silicon-based technologies.

Semiconductor characterization is a key enabler in the development of semiconductor technology and in improving semiconductor manufacturing. Materials and devices must continually meet more stringent requirements as the density and performance of semiconductor devices increases. The purity, perfection, and cleanliness required of the materials; the ultra-small dimensions of the devices; and the device properties themselves require measurements to a higher precision and with a resolution and sensitivity that pushes these techniques to their very limit. In addition, new techniques associated with analysis of process chemicals, control of process steps, and characterization of packages are critically needed. Because the materials and devices themselves exhibit a rich variety of properties, an increasingly wide range of measurement techniques has evolved to meet industry's needs.

As integrated circuit manufacturing plant costs approach $2 billion, financial and economic concerns are becoming of enormous importance to chipmakers, along with the scientific and technological issues. We believe that the challenges facing the semiconductor world are dramatic and dynamic, and that science and technology should be used to create fundamental changes in manufacturing technology. It is important for industry, government, and academia to cooperate on research and development, and the characterization and modeling of semiconductors are increasingly becoming a crucial part of semiconductor manufacturing. This book provides a concise and effective portrayal of industry characterization needs and the problems that must be addressed by industry, government, and academia to continue the dramatic progress in semiconductor technology.

Craig Barrett, Executive Vice President
  and Chief Operating Officer
Intel

Arati Prabhakar, Director
NIST

August 1995

v

CONTENTS

SILICON WAFERS, GATE DIELECTRICS, AND PROCESS SIMULATION

INTERCONNECTS AND FAILURE ANALYSIS

CRITICAL ANALYTICAL METHODS

ELECTRICAL METHODS

CONTAMINATION FREE MANUFACTURING

X-RAY METHODS

SCANNING PROBE MICROSCOPIES

NUCLEAR METHODS

ELECTRON AND ION BEAM TECHNIQUES (INCLUDING SIMS)

OPTICAL AND INFRARED TECHNIQUES

CHEMICAL METHODS

IN SITU, REAL TIME DIAGNOSIS, ANALYSIS, AND CONTROL

FRONTIERS IN COMPOUND SEMICONDUCTORS

APPENDICES

PREFACE

The International Workshop on Semiconductor Characterization: Present Status and Future Needs was held at the National Institute of Standards and Technology from January 30 to February 2, 1995. This comprehensive, "world-class" Workshop was dedicated to summarizing major issues and giving critical reviews of important semiconductor characterization techniques that are useful to the semiconductor industry. Because of the increasing importance of in-line and in-situ characterization methods, the Workshop placed a strong emphasis on these methods.

Specific goals of the Workshop were: (1) to provide a forum in which measurements and technical issues of current and future interest to the semiconductor industry could be reviewed, discussed, critiqued, and summarized; (2) to demonstrate and review important applications for diagnostics, manufacturing, and in-situ monitoring and control in real-time environments; (3) to provide a silicon integrated circuit process– and materials-based view of requirements for off-line, in-line, and in-situ analysis/metrology; (4) to focus attention on the critical and unique requirements related to compound semiconductor materials and devices; and (5) to act as an important stimulus for new progress in the field by providing new perspectives.

Sponsors of the Workshop were: The Advanced Research Projects Agency, SEMATECH, the National Institute of Standards and Technology, the Army Research Office, the U.S. Department of Energy, the National Science Foundation, Semiconductor Equipment and Materials International (SEMI), the Manufacturing Science and Technology Division of the American Vacuum Society, and the Working Group on Electronic Materials of the Committee on Civilian Industrial Technologies.

The Workshop brought together over 280 scientists and engineers concerned with research, development, manufacturing, diagnostics, and other aspects of the characterization of semiconductor materials, processes, and devices. Key, knowledgeable people in the semiconductor field addressed the unique characterization requirements of both silicon IC development and manufacturing and compound semiconductor materials, devices, and manufacturing. Sessions on silicon ICs were based on the technology drivers in the National Technology Roadmap for Semiconductors. Additional sessions covered technology trends and future requirements for compound semiconductor applications. Also

highlighted were recent developments in characterization, including in situ, in-fab, and off-line analysis methods.

The Workshop opened with introductory remarks by Arati Prabhakar, Director of the National Institute for Standards and Technology, on "Redefining the Possibilities: New Horizons in the Semiconductor World" and a Plenary Lecture by Craig Barrett, Chief Operating Officer of Intel Corporation, on "Status and Needs of the Semiconductor Industry." These talks provided a larger context for the detailed discussions of semiconductor characterization issues in the Workshop program which consisted of formal invited presentation sessions, poster sessions for contributed papers, and panel sessions.

The invited papers provided up-to-date reviews of the major issues and characterization techniques for semiconductor device research, development, and manufacturing. Poster papers, which were presented in four groups, one on each day of the Workshop, emphasized new developments and improvements in characterization technology. Three panel sessions, closing each of the first three days of the Workshop, were organized by SEMI to provide for multiple inputs and interactive discussion, with emphasis on the measurement equipment supplier perspective, on important issues related to the topics of the invited paper sessions. A special evening rump session explored the potential and promise of synchrotron x-ray metrology for TXRF and other analytical techniques.

The program was organized by a program committee chaired by David G. Seiler, National Institute of Standards and Technology. Other members of the committee were: David E. Aspnes, North Carolina State University; Ray Balcerak, Advanced Research Projects Agency; W. Murray Bullis, Materials & Metrology; Alain Diebold, SEMATECH, Inc.; Wolfgang Jantz, Fraunhofer-Institut für Angewandte Festkörperphysik; Alan Jung, SEMI; Sanjiv Kamath, Hughes Research Laboratories; Stephen S. Laderman, Hewlett Packard; Bob McDonald, Intel Corporation; William T. Oosterhuis, U.S. Department of Energy; Abbas Ourmazd, AT&T Bell Labs; Paul S. Peercy, Sandia National Laboratories; Fred H. Pollak, Brooklyn College of SUNY; John Prater, Army Research Office; Tom Remmel, Motorola; Linton G. Salmon, National Science Foundation; Tom J. Shaffner, Texas Instruments; Richard A. Singer, Institute for Defense Analyses; and William E.

Tennant, Rockwell International Science Center. A NIST Advisory Committee consisting of Paul M. Amirtharaj, Frank F. Oettinger, Robert I. Scace, and James R. Whetstone also assisted in the organization of the Workshop.

Dirk Bartelink, Hewlett-Packard; James Freedman, SRC; Len Feldman, AT&T Bell Labs; Stephanie Butler, Texas Instruments; Anne Testoni, Digital Equipment Corporation; P. B. Ghate, Texas Instruments; Paul Ho, University of Texas; Richard Brundle, Brundle Associates; Richard S. Hockett, Charles Evans & Associates; David E. Aspnes, North Carolina State University; C. Pickering, Defence Research Agency; John Prater, Army Research Office; Fred H. Pollak, Brooklyn College of SUNY; John Parsey, Motorola, Inc.; and Jerry Woodall, Purdue University served as chairs of the invited paper sessions. Mike Fossey, ADE Corporation; Robert I. Scace, NIST; and John C. Bean, AT&T Bell Labs chaired the panel sessions.

An evaluation survey was conducted following the Workshop. Respondents indicated that the Workshop had been meaningful and relevant for them because it (1) gave them insights and priorities rather than simple listings, (2) provided useful insights of the capabilities of different characterization techniques, (3) gave perspectives on industrial metrology requirements, (4) explored critical needs and issues in semiconductor metrology research, (5) was relevant to technologists, and (6) brought together the international metrology community at NIST, a logical place to hold the Workshop. There was considerable interest in holding another similar workshop in two to three years in order to cover new emerging techniques in semiconductor characterization this important field develops further.

This proceedings volume is intended to serve as a base-line reference for the characterization of semiconductors for the next decade. It begins with a specially prepared paper that describes the business and manufacturing motivations for the application of analytical technology and metrology by the semiconductor industry. This paper provides a comprehensive introduction to the field of semiconductor characterization and metrology, and hence to this volume.

The remainder of the volume is organized along the lines of the Workshop program. The papers are grouped in seven major sections under the topics of the invited paper sessions: Drivers for Silicon Process Development and Manufacturing; Metrology Requirements for beyond 0.35-μm Geometries; Silicon Materials, Gate Dielectrics, and Process Simulation; Interconnects and Failure Analysis; Critical Analytical Methods; In-Situ, Real-Time Diagnosis, Analysis, and Control; and Frontiers in Compound Semiconductors. The section on Critical Analytical Methods is further divided into eight sub-sections according to category of method. Brief summaries of the panel and rump sessions and the after-dinner remarks on Historical Perspective on Semiconductor Characterization are included as appendices.

The Editors thank the members of the Program Committee, the session chairs, and many of the invited speakers for their assistance in reviewing the manuscripts submitted for publication in this volume. They also thank the authors for their diligence in producing the camera-ready copy for their papers and in responding to reviewer comments and suggestions. Finally, they especially thank E. Jane Walters and Tammy Clark of the NIST staff, whose expert assistance greatly facilitated the planning and conduct of the Workshop and the preparation of the proceedings volume. Their tireless efforts helped contribute both to a successful Workshop and to the timely publication of this volume.

W. Murray Bullis
David G. Seiler
Alain C. Diebold
July 1995

ACKNOWLEDGMENTS

Grateful acknowledgment is made to the following for permission to reprint copyright material:

Butterworth-Hinemann for figures taken from "Characterization of Integrated Circuit Packaging Materials" (Materials Science Series) by T. M. Moore and R. G. McKenna (1993).

John Wiley & Sons, Ltd. for figures taken from "Energy Calibration of X-Ray Photoelectron Spectrometers: Results of an Interlaboratory Comparison to Evaluate a Proposed Calibration Procedure," Surf. Interface, 23, 121 (March, 1995).

Elsevier Science B.V. for figures taken from "Elemental Binding Energies for X-Ray Photoelectron Spectroscopy," Applied Surface Science 89, 141 (June, 1995).

Academic Press for figures taken from "Superlattices and Microstructures 15" by R. M. Feenstra, D. A. Collins, and T.C. McGill (1994).

Business and Manufacturing Motivations for the Development of Analytical Technology and Metrology for Semiconductors

T. J. Shaffner
Texas Instruments, Inc., Dallas, TX 75265

A. C. Diebold
SEMATECH, Austin, TX 78741

R. C. McDonald
Intel Corporation, Santa Clara, CA 95052

D. G. Seiler
*National Institute of Standards and Technology
Gaithersburg, MD 20899*

W. M. Bullis
Materials & Metrology, Sunnyvale, CA 94087

Semiconductor characterization is an indispensable enabler of all modern microelectronics and optoelectronic circuits, and is in the critical path for maintaining the steady decline in cost-per-function of silicon integrated circuit technology. It is also driving new developments in compound semiconductor materials and devices (III-V and II-VI). In this overview, we present a perspective on measurement technology relative to business and economic challenges of the semiconductor industry, and illustrate the key role metrology plays in modern process development and manufacturing. We also describe how new characterization techniques evolve for semiconductor applications.

INTRODUCTION

Semiconductors form the backbone of all modern microelectronics and optoelectronic devices. Semiconductor characterization is an integral and indispensable component of these technologies, providing necessary infrastructure for the marketplace, and for continued advances in the R&D laboratory and manufacturing line.

A good description of materials characterization considers it as an integral part of process development and manufacturing. The Office of Science and Technology Policy 1993 fiscal year program booklet describes it as a wide range of interdisciplinary activities that determine the structure, composition, properties, and performance of materials, and the relationships among these (1). Characterization measurements address quality assurance of incoming materials, wafer screening methods, control and monitoring of equipment and manufacturing processes, diagnostic and failure analysis, and end device performance in light of intended design and function.

The semiconductor industry refers to measurements used in process control as *metrology*. Engineers often use the word to describe procedures, such as critical dimension (CD) measurements, which routinely monitor lithography processes inside the clean room. Others generalize it to all in-line measurements. According to the dictionary, *metrology is the science of measurement* (2). It should also be noted that metrology includes all aspects both theoretical and practical with reference to measurements, whatever their uncertainty, and in whatever fields of science or technology they occur (3). The term *characterization* is also sometimes used interchangeably with *metrology*.

In this paper, we provide a perspective on measurement technology for the semiconductor industry, and illustrate its changing role in modern process development and manufacturing. The aim is to provide an overview and introduce some current analytical methods relative to key industry needs. We highlight business and economic challenges, and give examples of critical metrology requirements essential to timely delivery of more and better device functions at lower cost.

SILICON

Business and Economic Challenges

Owens recently described key elements that drive the semiconductor industry under a theme entitled *Keeping the Productivity Engine on Track* (4). We borrow Figure 1 from his presentation, which outlines recent history and future challenges for competitive semiconductor manufacturing. The exponential decrease in cost-per-function with time continues to be driven by four trends:

1) reduction in feature size,
2) conversion to larger diameter wafers,
3) yield improvements, and
4) other improvements in equipment productivity.

Despite the increasing difficulty in maintaining the historical slope of the reduction in feature size and conversion to larger wafers, this trend is expected to continue. Yield improvements are a recent major contributor to cost reduction, and high yields (>90%) are now expected for mature manufacturing facilities for even the most complex million-plus transistor ICs. High yields are enabled through process control, defect reduction, and improved equipment reliability. But yield cannot be increased much further. In short, smaller feature size, 300 mm wafer fabrication facilities (fabs), and yield enhancement remain critical cost savers, but are insufficient for maintaining the historical reduction in cost-per-function.

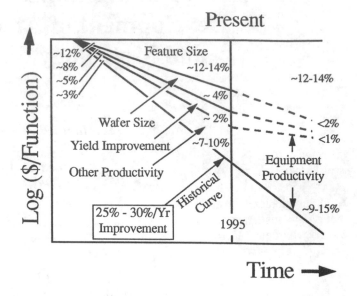

FIGURE 1. Keeping the Productivity Engine on Track
The semi-logarithmic decline in cost-per-function with time continues to be driven by four trends: scaling of feature size, conversion to larger wafers, yield improvements, and equipment productivity improvements (4). Metrology is an important enabler for the latter.

Improvement in equipment productivity (9-15%) therefore, is the best candidate for significant advancements over the next decade, and a last viable avenue for maintaining the cumulative 25-30%-per-year historical drive. Owens points out that factors impacting *other productivity* are those intrinsic to the equipment itself, and those resulting from logistics during use. Intrinsic elements include theoretical and attainable throughput, cost of consumables, and design for overall tool effectiveness. Logistics includes floor layout, scheduled and unscheduled downtime, test wafer occupancy, and product quality.

Metrology must break the status quo image of being a *non value-added* activity. For in situ real time control, it can improve equipment effectiveness by reducing process time, down time, and use of test wafers, while maintaining product quality. In some cases, it is advantageous to replace metrology tools requiring test wafers with others that measure product wafers nonintrusively. In many wafer fabs, defect metrology is a key element in maintaining high yield. This includes particle and defect detection and characterization on patterned wafers, particle sensors, and electrical yield testing.

Obviously, other issues impact yield and yield ramping. Huff and Goodall discuss silicon wafer materials issues that directly affect the ability to manufacture high yielding chips (5). Scharfetter reviews development needs for Technology Computer Automated Design (TCAD) tools which reduce the need for experimentation (6).

Rose describes the rapidly increasing costs of metrology, both in absolute terms and as a fraction of overall factory costs (7). These come from the tool itself, floor space overhead, and automation development. Automation alone is nearly one quarter of the overall metrology installation cost, and maintenance is an ongoing burden. Most emphasis is on defect, thin film, and lithography critical dimension measurements.

Rose's data show that capital costs per wafer associated with metrology in a new factory startup have increased 6.4 times over the last four generations of technology (from 1.0 μm to 0.35 μm). In the same period, capital for the entire factory increased 5.1 times. He concludes that the metrology fraction of the overall factory is increasing 10% faster per technology generation. In addition, Barrett stresses the importance of characterization and modeling as avenues for reducing the expense of building and operating a new factory (8). Metrology cost reduction is now a recognized challenge, and will impact future productivity improvements.

Metrology Alignment with Product Maturity

We thus find ourselves at a crossroads in this changing environment, where clear distinctions need to be made regarding the role of metrology in the overall semiconductor process flow, from beginning to end. Figure 2 illustrates the relative importance of analytical laboratory measurements, in-line and in situ measurement sensors, and the off-line failure analysis laboratory in the various manufacturing cycles. Poor or inadequate metrology in the early phases can result in excessive failure analysis activity, and can even cause the metrology demand to rise during the end use phase.

For a product or process to move successfully into manufacturing from R&D, it must be measured thoroughly by analytical instruments and replicated as closely as possible. Undetected hidden variables are likely to delay the ramp to high yields later on. Early measurements require heavy use of test wafers and multi-parameter analysis to identify and control key operables. High overhead comes from wafers lost to destructive testing and the high cost of fundamental measurements.

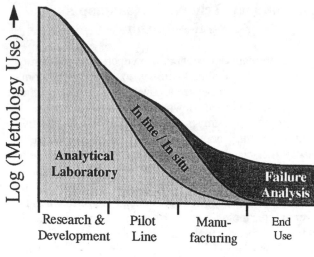

FIGURE 2. Metrology Alignment with Product Maturity Analytical laboratory (both R&D and wafer fab) measurements are in greatest demand in the R&D and pilot line phases of product development. Manufacturing benefits from in-line and in situ measurements and device failure analysis.

In high yield manufacturing, metrology requirements switch from fundamental measurement to verification of repetitive execution. In situ and in-line sensors are evolving for process control, and today reach the manufacturing floor through vendors who integrate them into new equipment. Sensors add value by detecting excursions early on that might persist through the wafer fab with cumulative line yield loss.

The simple depiction of Figure 2 is at the root of common misconceptions regarding materials and device characterization. The laboratory analyst is often dismayed that manufacturing shows no interest in the latest and newest sophisticated technique. In fact, some manufacturing managers will attest to an ultimate goal of eliminating measurement tools altogether. Ideally, a finely tuned mature and stable process does not deviate from plan and requires no mid-course correction.

Analytical laboratory researchers sensitive to this perspective focus their efforts on technologies that are less mature. In contrast, process development engineers sometimes stretch existing metrology too thin by pushing sophisticated instruments into a regime where they are not very effective or technically applicable. Metrology has the potential to establish a common denominator for transitions between all of the process maturity phases, and thereby reduce operational barriers.

3

National Technology Roadmap for Semiconductors

International economic competition is pressing the U.S. semiconductor industry to promote precompetitive cooperation between government, industry, and universities. A unified viewpoint fosters efficient use of government and university resources and effective planning for future activities. In 1992, the Semiconductor Industry Association (SIA) published a consensus view of the requirements for the manufacture of future silicon based integrated circuits in a document known as the *1992 SIA Roadmap*. Here, key industry, government, and university technologists describe a strategy for meeting materials, process, tool, and factory requirements for future IC manufacturing. The document was updated two years later as the *National Technology Roadmap for Semiconductors (NTRS)* (9).

The NTRS embodies a detailed treatment of technical requirements at each IC technology node, which is defined in terms of the smallest feature size of the circuit. Categories addressed include the following: Design and Test; Process Integration; Devices and Structures; Environment; Safety and Health; Lithography; Interconnect; Materials and Bulk Processes; Assembly and Packaging; and Factory Integration.

Broad infrastructure activities are termed *Crosscutting Technologies*, of which metrology is one prime example. The metrology requirements associated with NTRS have been further defined in a *Metrology Roadmap Supplement* developed through SEMATECH in 1994 (10), and are also discussed in a more recent paper (11).

NTRS and the SEMATECH supplement are impacting improvements in volume manufacturing, which in turn impacts semiconductor market share of the U.S. Related metrology activities address wafers-measured-per-hour and nondestructive and noninvasive analysis. These metrology tools are beginning to mature in concert with factory production equipment development and pilot line certification.

The NTRS confirmed what SEMATECH had already taught industry in the U.S., that basic technology used in manufacturing semiconductors is generic. In a constrained economic environment, coordinated planning through roadmaps helps make more intelligent use of resources. Alignment between university, industry, and government activities in building the required metrology infrastructure for semiconductors is a good example.

One avenue for addressing the broad range of metrology problems outlined in the NTRS is the National Semiconductor Metrology Program (NSMP) established at the National Institute of Standards and Technology (NIST) in early 1994, upon urging of the SIA. The NSMP matrix manages a spectrum of projects that draws on multi-disciplinary metrology expertise available at NIST to address key roadmap issues.

Although initially scoped for competitive analysis in the U.S., the NTRS is sufficiently generic to guide the broader international silicon community as well. It is providing a time-line and likely specifications for factory equipment and process procedures of the future.

Critical Dimensions: Example of Evolving Metrology

The critical dimensions (CDs) of features on lithography masks are supplied by vendors, and wafers are routinely measured in-line to ensure etch relief control in the factory. Historically, the optical microscope was the first workhorse tool for this application, providing repeatable measurements in the micron regime. The scanning electron microscope (SEM) has since replaced optical microscopy for modern applications requiring submicron imaging resolution, although it has not been as effective for insulating photomask substrates. Nevertheless, the SEM has not solved the metrology problem for two reasons.

First, accurate dimension measurements can be no better than the edge definition of the feature in question. Planar vertical edges are ideal, but seldom found in practice. Therefore, an arbitrary definition of the edge is made and an average value computed from a series of measurements taken at different locations down the length of a line. Edge quality is an even more serious problem in the fabrication of universal standards for submicron metrology, because these need to be made and calibrated with much smaller uncertainty than targeted by the user.

Second, the classical resolution of an SEM is not necessarily a valid measure of performance for features patterned in thick layers, such as photoresist or polysilicon. Interactions between the incident electron beam and specimen degrade resolution for dimensional measurements to values larger than the probe diameter, which at low accelerating voltages is in the nanometer range. This is considered inadequate for CD measurements below 0.2 μm unless broadening effects are taken into account. Monte Carlo simulations of finite electron beam scattering within the specimen hold promise for extending performance, but this requires detailed knowledge about the shape of the target feature, as well as adjacent structures (12).

We desire universal reference standards to calibrate optical and SEM metrology tools, but there is little if any commonality in the materials, edge structure, or neighboring features in the manufacture of integrated circuits. It is not realistic to count on a single reference for accurate calibration of a metrology instrument used for virtually all specimens. Future flexibility may come through computer modeling of the feature's shape and composition as well as nearest neighbor influences.

Critical dimension control is based on the need to maintain a repeatable process. The purpose of the measurement is to ensure that values fall within an envelope defined by previous performance. In new equipment startup, the SEM is at times calibrated to a known pitch based on an accepted reference standard. A more robust substitute reference is then used to track repeatability of the measurement in volume manufacturing. Others go to the trouble of cross sectioning select product wafers for direct SEM imaging of the feature's profile, but this is a costly approach for routine applications.

Griffith (13) and others propose using scanning probe microscopy to qualify standards as well as perform routine measurements. Atomic force microscopy is a likely candidate for this purpose because it can be used on insulating (e.g., SiO_2) as well as conductive materials, and is nondestructive. However, it is inherently slow and subject to uncertainties arising from the shape of the tip and shank, specimen edge variations, and the inability to access hidden corners. Active topics of research include investigations of probe tip stability and shape, and a variety of operational mode configurations.

We conclude there is no universal remedy for dimensional metrology in the submicron regime. Each approach has limitations that lead to uncertainties larger than those tolerated by the semiconductor industry. The dilemma for the fabrication and calibration of dimensional standards is even worse. Presently, we can only strive to make the edges as smooth and vertical as possible, and minimize uncertainties in dimensional metrology instrumentation.

New reference artifacts are being developed that are less sensitive to the particular metrology instrument used for CD, overlay, and feature placement (14). The pattern is amenable to measurement by the scanning probe microscope, and has a cross section profile and dimensions that can be certified to the accuracy and precision required by NTRS roadmap projections. The artifact is fabricated using preferential planar etching of crystalline materials, an oxide barrier implantation process, and selective removal of the underlying substrate. This architecture allows multiple dimensional certification by footprint capacitance, top surface atomic layer counting, and average profile width extraction by electrical techniques.

In Situ and In-line Measurements

Metrology tools applied in-line to manufacturing, and in situ inside commercial instruments are initially developed as scientific prototypes in the R&D environment. These then evolve into stable user friendly modules, suitable for introduction into the off-line wafer fab analytical laboratory. Further maturity brings durability and reproducibility through robust design adaptable to the wafer fab. Specifications and standard test methods developed by organizations such as SEMI and ASTM, and the availability of applicable certified reference materials such as NIST *Standard Reference Materials* (SRM®), facilitate successful transfer through these three stages. Training in relevant materials science as well as procedures of calibration are also important.

In situ measurements are not common in volume manufacturing in the semiconductor industry today. The NTRS predicts that required factory equipment efficiency improvements will drive future incorporation of in situ metrology. As fab managers realize the impact of a measurement tool on throughput, they will drive equipment vendors to incorporate such in situ sensors. Ellipsometry is an example of a metrology that has progressed from R&D laboratory to an in situ process control tool. It was facilitated by inexpensive hardware and computer algorithms capable of providing thickness and refractive index data at one second intervals (15). One quarter of a century elapsed between the first R&D application of ellipsometry to semiconductor dielectric films and the demonstration of an in situ sensor. Development is typically a little shorter for other metrology tools, ranging from one to two decades. The timetable proposed by the NTRS will require significant shortening of this transition sequence.

COMPOUND SEMICONDUCTORS

The NTRS established groundwork for the creation of a metrology roadmap for silicon, which was developed through SEMATECH (10). A comparable top level roadmap for compound semiconductors does not yet exist, mostly because the materials systems are diverse and highly complex relative to silicon. Applications in the U.S. have been mostly defense related, being coordinated and executed through government funding. Also, profit margins in this country are not expected to be adequate to support a roadmap effort comparable to the NTRS. Nevertheless, segments of the industry do have defined strategies, and it is proving desirable to achieve some level of consensus in planning for future compound semiconductor technology (16).

III-V Microwave and Optoelectronic Semiconductors

The dominant material for microwave and radio frequency (RF) applications is GaAs, even though it represents only 1% of the overall semiconductor market (17). GaAs lags silicon by about ten years in terms of wafer size and scaleup for manufacturing. Unlike the situation for silicon, the industry in the U.S. remains driven by military applications, particularly through the ARPA Microwave Monolithic Integrated Circuit (MIMIC) program.

The contract proposal funding process has provided focus, or we might say an evolving roadmap, for the microwave community in the U.S. This continues today, but the exploding wireless market is drawing microwave devices into the commercial sector, where high frequency power, amplifier, and transmit-receive modules are in great demand. This is placing substantial pressure on producing GaAs material and RF microwave devices in a cost effective manner.

The microwave industry is thus in transition from dominance by high performance, low volume defense applications, to high volume, acceptable performance commercial applications. Wafer fabs are critically dependent on the existing silicon infrastructure for manufacturing equipment. This leads to tailoring of silicon oriented tools for specialized GaAs applications. Typically, investment in custom equipment for III-Vs does not directly benefit the silicon community. Today, pressing technical issues include limited wafer size, complex package interconnects, heat dissipation, and the development of immature and unexplored materials.

In 1994, the Optoelectronics Industry Development Association (OIDA) published a projection of technology developments required to achieve eight-fold growth over the next 20 years ($50B to $400B) (18). It targets markets for displays, communications, optical storage, and image sensors. The growth is broadly based, depending on materials that efficiently generate, detect, modulate, and guide light.

Materials parameters generic to III-V compound semiconductors include bandgap energy, band offsets, interface diffusion and roughness, index of refraction and index dispersion, optical absorption, minority and majority carrier lifetime and mobility, and defects that influence lifetime or trapping of charge. Photoluminescence spectroscopy is a particularly well suited technique for such measurements, because it directly probes the band gap, density of states, and electrical behavior of optically active materials. It is a workhorse of the industry but, at the liquid helium temperatures conventionally used, is limited mostly to R&D and pilot line applications.

Other contactless tools include ellipsometry, light scattering, optical reflectometry, and photoreflectance, which are applied to heterostructures grown by molecular beam epitaxy (MBE) and other advanced thin film techniques, e.g., metalorganic continuous vapor deposition (MOCVD) and metalorganic-MBE (MOMBE). These optical measurements are by nature non-invasive, and are quickly finding niches as in situ process sensors that optimize film thickness, composition, interface quality, and uniformity during growth.

Off-line metrology is also expanding for III-V materials applications. Screening of liquid encapsulated Czochralski (LEC) GaAs wafers by room temperature photoluminescence and infrared transmission mapping exposes stress patterns frozen into the crystal during cooling. These correlate with wafer warpage, breakage, and yield loss. X-ray rocking curves quantify layer thickness, composition, and interface abruptness. Commercial X-ray and photoluminescence instruments are streamlined for high throughput 4-inch (100 mm) wafer mapping.

Maturity of these methods parallels, and in some cases surpasses, those applied by the silicon industry. The complexity of binary and ternary III-V alloys prolongs development in the R&D and pilot phases of Figure 2 relative to silicon, so it is not surprising that novel sensors and unproven diagnostics find fertile ground for application. We must also recall that the critical feature size (gate length) of compound semiconductor microwave devices has always progressed well in advance of silicon.

II-VI Infrared Detector Materials

Infrared detectors and imagers based on II-VI compound semiconductors are key enablers in a variety of military and space applications. Today, focal plane arrays vary in shape from 2x240 linear arrays to full 480x640 two dimensional imagers. The current U.S. military market is close to $75M, with projections for growth to $168M by the year 2000.

HgCdTe is the material of choice for both scanning and staring focal plane arrays. The materials technology evolved from small, bulk grown ingots to large epitaxial wafers prepared in special reactors. The complexity of ternary alloy systems and the stringent demands of large element detector arrays combine to increase the difficulty in working with II-VI compounds by at least an order of magnitude, relative to GaAs.

Material quality has reached the level of maturity required for two dimensional focal plane arrays based on epitaxially grown HgCdTe wafers as large as 30 cm^2. Recent advances in surface passivation, liquid phase epitaxial (LPE) growth, and device processing have resolved many of the major materials problems encountered in the past. The principal limitation is now control of crystal point defects and impurities, especially for devices operating at 78 K or lower temperatures.

Microdefects are sparsely distributed and extremely small, typically being several to hundreds of nanometers across. It is difficult to even locate these by microprobe techniques and weak signals from such small points add to the analysis dilemma. Methods of X-ray diffraction and topography work reasonably well, because they sense minute strain fields that extend far beyond the defect center, and also simultaneously detect hundreds or thousands that are distributed throughout the sampling volume. Combined with transmission electron microscopy and etch pit density, these are workhorse tools for understanding the origin and control of point defects. New scanning probe methods are also being exploited with some success.

Other techniques measure the composition variation of alloys like HgCdTe, CdZnTe, HgTe, and ZnS with high precision (<0.01 mole fraction), and the distribution of Te precipitates. X-ray photoelectron spectroscopy contributes to understanding the complex chemistry of surface passivation films. As with III-Vs, we find growing acceptance of in situ growth monitoring by ellipsometer and light scattering methods.

Common material impurities are Al, Cu, Fe and C. Secondary ion mass spectrometry quantifies these at levels above 10^{16} atoms/cm^2, but is inadequate at the lower levels responsible for type-inversion, dark current, and electrical anomalies during manufacturing. There is critical demand for smaller analytical probes with higher sensitivity to resolve such problems. Incomplete understanding of physical and chemical defects continues to inhibit progress with basic II-VI materials issues.

Finally, we point out that a recent paper reports results of an extensive industrial survey of the importance of characterization measurements for HgCdTe materials, processes, and devices (19). Also, Tennant recently provided a thorough overview of II-VI metrology (20).

Other Thrusts

Compound semiconductors are candidates for high temperature electronics applications. Although diamond offers high thermal stability and mechanical hardness for power transistors, difficulty in growing thin films with suitable lattice matched substrates is making it less attractive to low cost commercial markets. Active research with compounds such as SiC and GaN is in progress. GaN has added advantages for light emitting display applications. The characterization tools outlined above are also applicable to these developments.

Nanoelectronics is a revolutionary digital IC technology directed at continued downsizing of minimum feature size below 0.1 μm. Priority is given to eliminating long interconnects, because these are the most difficult to scale down. The approach leads to an architecture in which resonant electron quantum tunneling effects dominate, and each active element is connected only to its nearest neighbor cells (*cellular automaton*). Material systems are typically heterostructures composed of lattice matched and doped GaAs/AlGaAs and related compounds.

Quantum structures are patterned with nanoscale resolution, which challenges even the most sophisticated small spot diagnostic probes available today. Our best solutions for imaging surface detail and measuring the electrical properties of a single cell are based on scanning probe methodology. Current-voltage characteristics from a single quantum dot have been demonstrated, but are extremely difficult to reproduce. We can expect these new probes to establish high resolution capabilities applicable to future silicon circuits.

LIMITS AND TRENDS

The progressive shrinking of ULSI circuits into the submicron regime and the emergence of quantum structures on the nanometer scale increase the challenge for characterization specialists. One reason is because the inspection of defects typically involves features which are an order of magnitude smaller than the minimum geometry. We can ultimately attain atomic resolution with modern scanning probes, but there is a fundamental compromise that exists between X-Y spatial resolution, depth of analysis, and technique sensitivity. This originates from Poisson statistics that govern the measurement process, and is why we must almost always use more than one technique to successfully diagnose a complex problem.

When the probe radius, depth of analysis or contaminant level is reduced, the volume of analysis ultimately includes only several species atoms of interest. In the extreme, the volume analyzed contains no species atoms at all. The analyst is now confronted with a sampling problem in selecting which atoms are appropriate for the analysis. Single atom detection cannot be considered independently of the sampling problem. This issue is becoming more visible, as we realize a related problem is looming not just in characterization, but also in the fabrication, of lightly doped 3-D shallow junctions of nanometer dimensions, where only a few, or even single, dopant atoms may someday be required.

We sometimes think characterization technology will naturally mature to resolve such dilemmas, and that turnkey diagnostic tools will eventually become available at a reasonable cost. It is evident in Figure 3, however, that our progress in this regard has been anything but steady. As much as we would like to invent new methods to achieve submicron characterization goals, it is more likely they will spring from revitalization of older techniques, or from fields out of the ULSI or compound semiconductor mainstream. Many of the microscopes and microprobes in use today grew from unrelated backgrounds in geology, medicine, nuclear physics, and polymer science.

Nevertheless, it is evident from the figure that subsequent development of a new tool does, in fact,

respond to a driving force for the smallest possible analytical spot, once relevance of the method for semiconductor applications is established. For example, the focused probe of Auger electron spectroscopy progressed from 200 μm in the 1970s to the 15 nm spot attainable today. A similar drive into the sub-parts-per-trillion regime for the analysis of low level material impurities is also taking place. These are *evolutionary* trends, where directed funding and personnel can be expected to attain advertised goals.

The random appearance of new techniques is why cutting edge characterization organizations seek to function as a collection center for new ideas, inventions, and measurement science and technology. The buildup of such infrastructure, and the merging of disciplines provides fertile ground for this *revolutionary* aspect of technique advancement. The scanning tunneling microscope, for example, was not conceived in response to ULSI downsizing, but rather evolved from a marriage of diverse technologies, including piezo-manipulators, ultra-high vacuum science, vibration damping, computer automation, and others. These breakthroughs are what we all wish for, but cannot as easily attain through directed milestone driven programs.

One typically predicts where we will be five or ten years from now, based on extrapolation of known technology and historic trends. We might guess which *evolutionary* and *revolutionary* changes to expect, but can predict with certainty only that we will be partially correct, at best. However, we do know that characterization technology provides a technical foundation for the fabrication tools of tomorrow. Optical microscope lens science is found in modern ultraviolet steppers, e-beam lithography evolved from the scanning electron microscope, focused ion beams preceded implanters, and it is likely that scanning probe fabrication approaches will be found in future wafer fabs. In our daily activities, we must continue to weigh the strengths and weaknesses of characterization techniques not only for the problem at hand, but also with an eye toward the future engineering and scientific foundations of our industry.

ACKNOWLEDGMENTS

The authors gratefully acknowledge beneficial comments and insights from Herbert Bennett, Michael Cresswell, Judson French, Robert Larrabee, and Robert Scace at NIST, and Randal Goodall, Howard Huff, and Lumdas Saraf at SEMATECH.

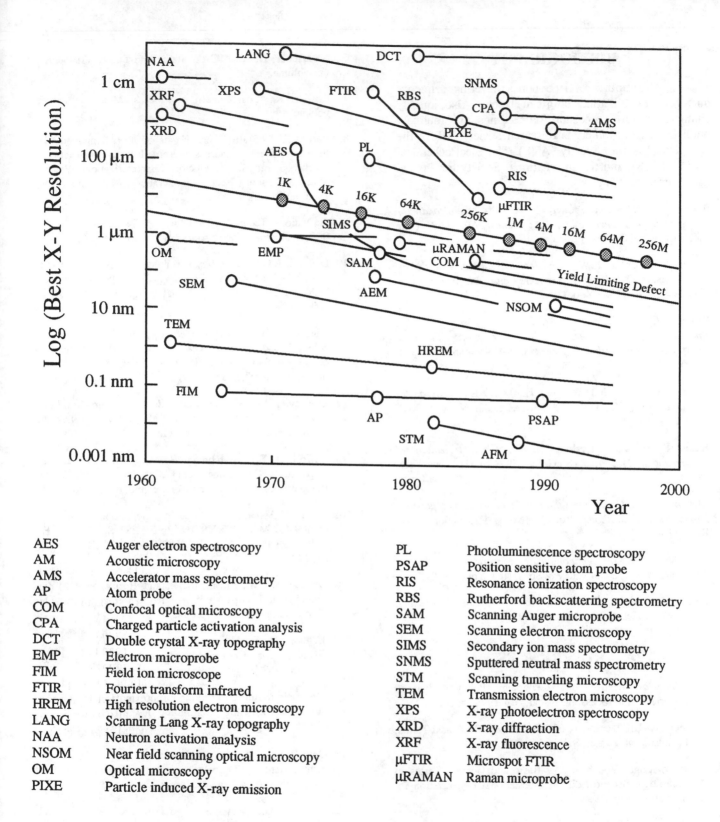

FIGURE 3. History of Analytical Techniques for Semiconductor Applications

The revitalization of existing techniques and the development of new methods follow an erratic history relative to the steady decline of minimum DRAM geometry (1K through 256M). Lines illustrate evolutionary changes, while circles show the introduction of revolutionary new techniques for semiconductor applications.

AES	Auger electron spectroscopy	PL	Photoluminescence spectroscopy
AM	Acoustic microscopy	PSAP	Position sensitive atom probe
AMS	Accelerator mass spectrometry	RIS	Resonance ionization spectroscopy
AP	Atom probe	RBS	Rutherford backscattering spectrometry
COM	Confocal optical microscopy	SAM	Scanning Auger microprobe
CPA	Charged particle activation analysis	SEM	Scanning electron microscopy
DCT	Double crystal X-ray topography	SIMS	Secondary ion mass spectrometry
EMP	Electron microprobe	SNMS	Sputtered neutral mass spectrometry
FIM	Field ion microscope	STM	Scanning tunneling microscopy
FTIR	Fourier transform infrared	TEM	Transmission electron microscopy
HREM	High resolution electron microscopy	XPS	X-ray photoelectron spectroscopy
LANG	Scanning Lang X-ray topography	XRD	X-ray diffraction
NAA	Neutron activation analysis	XRF	X-ray fluorescence
NSOM	Near field scanning optical microscopy	μFTIR	Microspot FTIR
OM	Optical microscopy	μRAMAN	Raman microprobe
PIXE	Particle induced X-ray emission		

BIBLIOGRAPHY

A sampling of fine books covering semiconductor characterization is given below. Also, quality characterization symposium proceedings are routinely published by the Materials Research Society (MRS), the American Vacuum Society (AVS), The Electrochemical Society (ECS), and the International Society for Optical Engineering (SPIE).

Barnes, P. A. and Rozgonyi G. A., eds., *Semiconductor Characterization Techniques*, Proceedings Vol. 78-3, (The Electrochemical Society, Princeton, 1978).

Blood, P. and Orton, J. W., *The Electrical Characterization of Semiconductors: Majority Carriers and Electron States*, (Academic Press, New York, 1992).

Brundle, C. R., Evans, C. A., Jr., and Wilson, S., eds., *Encyclopedia of Materials Characterization: Surfaces, Interfaces, Thin Films*, (Butterworth-Heinemann, Boston, 1992).

Grasserbauer, M. and Werner, H. W., eds., *Analysis of Microelectronic Materials and Devices*, (John Wiley & Sons, New York, 1991).

Kane, P. F. and Larrabee, G. B., *Characterization of Semiconductor Materials*, (McGraw-Hill Book Co., New York, 1970).

McGuire, G. E., *Characterization of Semiconductor Materials, Principles and Methods*, Vol. 1, (Noyes Publications, Park Ridge, 1989)

Schroder, D. K., *Semiconductor Material and Device Characterization*, (John Wiley & Sons, New York, 1990).

Strausser, Y., ed., *Characterization in Silicon Processing*, (Butterworth-Heinemann, Boston, 1993).

REFERENCES

1. Office of Science and Technology Policy, Advanced Materials and Processing: Fiscal Year 1993 Program.

2. McGraw-Hill Dictionary of Scientific and Technical Terms, Fifth Edition, Parker, S. P., ed., 1994.

3. International Vocabulary of Basic and General Terms in Metrology, Second Edition, International Organization for Standardization, 1993.

4. Owens, J., SEMATECH, unpublished results communicated to the authors.

5. Huff, H. and Goodall, R. K., "Silicon Materials and Metrology: Critical Concepts for Optimal IC Performance in the Gigabit Era", this volume.

6. Scharfetter, D. L., "TCAD Status and Future Prospects", this volume.

7. Rose, D. R., "Cost Benefit of Process Metrology", this volume.

8. Barrett, C., "Status and Needs of the Semiconductor Industry", Plenary address presented at the International Workshop on Semiconductor Characterization: Present Status and Future Needs, Gaithersburg, MD, January 30, 1995, unpublished.

9. The National Technology Roadmap for Semiconductors, The Semiconductor Industry Association, San Jose, CA, 1994.

10. Diebold, A.C., "Metrology Roadmap: A Supplement to the National Technology Roadmap for Semiconductors", SEMATECH Technology Transfer Document #94102578 A-TR, 1994.

11. Diebold, A. C., "Critical Metrology and Analytical Technology Based on the Process and Materials Requirements of the 1994 National Technology Roadmap for Semiconductors", this volume.

12. Lowney, J. R., "Use of Monte Carlo Modeling for Interpreting Scanning Electron Microscope Linewidth Measurements", to be published in *Scanning*, 17, 1995.

13. Griffith, J. E., Marchman, H. M., Miller, G. L., and Hopkins, L. C., "Dimensional Metrology with Scanning Probe Microscopes", *J. Vac. Sci. Technol.*, 13, 1100, 1995.

14. NIST - SANDIA U. S. patent application Serial No. 08/409,467 MONOCRYSTALLINE TEST STRUCTURES, AND USE FOR CALIBRATING INSTRUMENTS, March 23, 1995.

15. Duncan, W. M., Bevan, M. J., and Henck, S. A., "Real Time Spectral Ellipsometry Applied to Semiconductor Thin Film Diagnostics", this volume.

16. Bennett, H. S., "Summary Report of the Workshop on Planning for Compound Semiconductor Technology, NISTIR 5702 (August 1995).

17. Electronic Materials Working Group Conference, Civilian Industrial Technology Committee of the National Science and Technology Council, technical report to be published.

18. Optoelectronic Technology Roadmap - Conclusions & Recommendations, Optoelectronics Industry Development Association, 1994.

19. Seiler, D. G., Mayo, S., and Lowney, J. R., "HgCdTe Characterization Measurements: Current Practice and Future Needs", *Semicond. Sci. Technol.*, 8, 753, 1993.

20. Tennant, W.E., "Role of Diagnostics, Characterization, and Modeling in IR Detector Technology Development: HgCdTe - A Paradigm", this volume.

DRIVERS FOR SILICON PROCESS DEVELOPMENT AND MANUFACTURING

Physical Limits on Gigascale Integration (GSI)

James D. Meindl

Joseph M. Pettit Chair Professor of Microelectronics
Georgia Institute of Technology
Atlanta, Georgia 30332

Future opportunities for gigascale integration (GSI) will be governed by a hierarchy of theoretical and practical limits whose levels can be codified as: 1) fundamental, 2) material, 3) device, 4) circuit, and 5) system.   Theoretical system limits indicate that in rank order, the most serious technological problems impeding progress towards GSI are related to: 1) interconnections, 2) heat removal, and 3) transistors. Practical limits on the number of transistors per chip can be extended only if the manufacturing cost per function of GSI continues to decline, which will require, among other advances, enormous progress in semiconductor characterization.

Introduction

The sustained rate of advance of microelectronics over the past four decades is utterly without precedent in technological history [1].   Consequently, the question of how much longer this advance can continue must be explored. The central thesis of this paper is that future opportunities for microelectronics to achieve GSI will be governed by a hierarchy of theoretical and practical limits.   The levels of this hierarchy can be codified as: 1) fundamental, 2) material, 3) device, 4) circuit, and 5) system [2].   Theoretical limits are elucidated by plotting the average power transfer during a binary switching-transition (P) versus the corresponding transition time (t_d). Practical limits are compactly summarized by a graph of the number of transistors per chip (N) versus calendar year (Y). The singular metric which describes the promise of a technology for GSI is the chip performance index (CPI), defined as the quotient of the number of transistors per chip and the associated power delay product or CPI = N/Pt_d. The challenge which GSI presents to the discipline of semiconductor characterization is how to meet the metrological needs of a low cost silicon chip containing many billions of transistors and associated interconnections with feature sizes less than 100 nm.

Theoretical Limits

Theoretical limits are informed by the laws of physics in conjunction with technological invention. In particular, at the first level of the hierarchy, fundamental limits are independent of the properties of specific materials.   The three most important fundamental limits on GSI are derived from thermodynamics, quantum mechanics, and electromagnetic theory.   The thermal energy of an electron gas requires that the power-delay product Pt_d of a switching transition be greater than about 4 kT, where k is Boltzmann's constant, and T is absolute temperature. The Heisenberg uncertainty principle of quantum mechanics requires that the average power transfer during a measurable energy change associated with a switching transition of duration t_d be greater than h/t_d^2 where h is Planck's constant.   These two fundamental limits forbid switching transitions to the left of their loci in Figure 1.   The fundamental limit based on electromagnetic theory simply dictates that the propagation velocity of a high speed pulse on a long distance interconnect or transmission line must be less than the speed of light in free space. As illustrated in Figure 2, this limit prohibits operation to the left of the $L=c_o\tau$ locus for any interconnect regardless of the materials or structure used for its implementation.

13

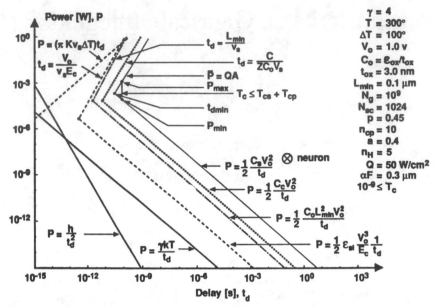

FIGURE 1. Average power transfer versus delay time for a binary switching transition.

A fourth fundamental limit which enters the picture especially for low power/voltage design is derived from Fermi-Dirac quantum statistics. This "carrier density swing" limit applies to all common semiconductor materials and requires a Fermi energy level change of 60 meV in order to produce a one decade change in carrier density.

A final interesting opportunity related to fundamental limits is based upon the second law of thermodynamics which can be stated as follows: There is no way in which you can make the entropy of a thermodynamic system and its environment decrease [3]. Entropy change dS can be expressed as $dQ/T = dS \geq 0$ where dQ is the heat added to the system and T is its absolute temperature. In a computational process, it is only those steps that discard information and therefore increase entropy (dS>0) which have a lower limit on energy consumption or heat generation (dQ>0) imposed by the second law of thermodynamics [4]. Consequently, the intriguing prospect of inventing reversible adiabatic computational technology offers the possibility of reducing power dissipation to levels below those imposed by fundamental limits on conventional logically irreversible processes.

At the second level of the hierarchy, material limits are independent of the geometrical configuration and dimensions of particular device structures. For distances greater than several mean free path lengths, the principal properties of Si and GaAs which determine their key material limits are carrier mobility,

carrier saturation velocity (v_s), breakdown field strength (E_c), and thermal conductivity. The switching or storage energy limit on Si is given by $Pt_d = \frac{1}{2} \varepsilon_{Si} V_o^3/E_c$ for a potential swing V_o. The material limit on carrier transit time is $t_d/V_o = 1/v_s E_c$ and bears only a 33% larger value for Si than for GaAs. A third key material limit is the carrier transit time (t_d) of a generic device per unit of heat removal (P), or $t_d/P = 1/\pi K v_s T$ where K is thermal conductivity and T is the absolute temperature difference between the generic device and a constant temperature heat sink in contact with the bottom of the die containing the device. GaAs suffers about a 300% larger t_d/P than Si. The switching energy, transit time and heat removal limits for Si forbid operation to the left of their loci in Figure 1. The primary interconnect material limit for GSI is defined by substituting a polymer with a relative dielectric constant $\varepsilon_r \cong 2$ as the insulator replacing free space in the fundamental speed of light limit, as illustrated in Figure 2 for the L=vt locus.

The critical device limit on GSI is the allowable minimum effective channel length L_{min} of the Si MOSFET. To achieve L_{min}, both gate oxide thickness t_{ox} and source/drain junction depth y_j should be as small as possible. In addition, the doping profile of the channel region must be optimized. Considering several MOSFET structures illustrated in Figure 3, solutions of the two-dimensional and three-dimensional Poisson equation are coupled to the one-dimensional drift-diffusion equation in the subthreshold region of operation. These analyses

14

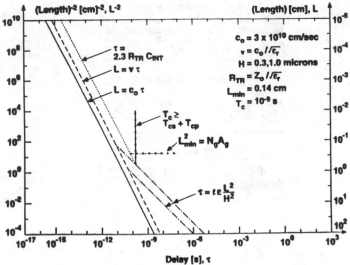

FIGURE 2. $(1/L)^2$ vs τ for all levels of the hierarchy. System limits are imposed by global interconnect response time designated by $T_c \geq T_{cs} + T_{cp}$ and length designated by $L^2_{min} = N_g A_g$.

provide a description of threshold voltage V_T as a function of MOSFET structure, materials and applied voltages as illustrated in Figure 4, which indicates the feasibility of $L_{min} \cong 0.05$ microns for shallow junction retrograde profile bulk MOSFETs and $L_{min} \cong 0.025$ microns for dual gate or delta SOI MOSFETS capable of room temperature operation [5,6]. Assuming a conservative value of $L_{min} = 0.1$ microns, MOSFET switching energy and transit time are plotted in Figure 1 indicating a third forbidden region of operation to the left of the $Pt_d = \frac{1}{2} C_o L^2_{min} V^2_o$ and $t_d = L_{min}/v_s$ loci. For Si MOSFET channel lengths less than about 0.1 microns, in addition to the classical carrier transport mechanisms of drift and diffusion, quasi-ballistic effects which give rise to velocity overshoot must be taken into account in calculating drain current and transconductance characteristics [7]. In addition, the effects of random dopant distribution in the channel must be taken into account [8].

The key device limit on interconnects is represented by the response time of a canonical distributed resistance-capacitance network driven by an ideal voltage source. The 0 to 90% response time of such a network is $\tau = RC$ where R and C are the total resistance and capacitance, respectively of the interconnect. Neglecting fringing effects, for an interconnect of length L, $\tau \cong (rH_r)(\epsilon H_\epsilon)L^2$ where (r/H_r) is the conductor sheet resistance in ohms/square and (ϵ/H_ϵ) is the sheet capacitance in farads/cm^2. Figure 2 illustrates this limit for equal metal and insulator thicknesses $H_r = H_\epsilon = 0.3\mu m$ and $1.0\mu m$. A third forbidden zone is evident. No polymer-copper

interconnect with a thickness H smaller than 0.3, $1.0\mu m$ can operate to the left of the 0.3 1.0mm locus which represents a contour of constant distributed resistance-capacitance product.

Circuit limits occupy the fourth level of the hierarchy. Due to its relatively large operating margins, scalability, low static power drain, and circuit flexibility, the CMOS logic family is the most promising for GSI. To maintain signal quantization and consequently, virtually error-free operation throughout a large computing system, the ultimate limit on the transfer characteristic of a static CMOS logic gate is an incremental gain greater than unity in absolute value for equal input and output voltage levels [2]. This is the first generic circuit limit. For a CMOS gate, the power-delay product $Pt_d = \frac{1}{2} C_c V^2_o$, where C_c is the load capacitance which is charged and discharged each cycle and V_o is the logic level swing. As the switching rate of a gate increases, eventually it is limited by either intrinsic circuit delay or heat removal capability represented by the package cooling coefficient $Q[w/cm^2]$. The power-delay or switching energy limit $Pt_d = \frac{1}{2} C_c V^2_o$ and the intrinsic switching speed limit $t_d = C_c/ZC_o v_s$ are illustrated in Figure 1. Once again, operation to the left of these loci is forbidden. A fourth generic circuit limit is imposed by the output stage of a logic gate driving a long distance interconnect that extends, for example, between opposite corners of a chip. The approximate response time of such a circuit is given by $\tau \cong (2.3R_{tr} + R_{int})C_{int}$ where R_{tr} is the driver transistor output resistance and $R_{int} = r_{int}L$ and $C_{int} = c_{int}L$ are the total resistance and

15

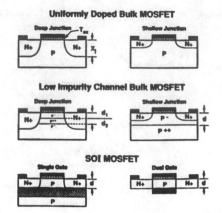

FIGURE 3. MOSFET device structures

capacitance, respectively, of an interconnect of length L [9]. To prevent excessive delay due to wiring resistance, the circuit should be designed so that $R_{int} < 2.3 R_{tr}$ giving $\tau \cong 2.3 R_{tr} C_{int} = 2.3 R_{tr} c_{int} L$ where c_{int} is the capacitance per unit length. This interconnect response time limit is illustrated in Figure 2 for a line characteristic impedance $Z_o \cong [\mu_o/\varepsilon_o \varepsilon_r]^{1/2}$ where μ_o and ε_o are the permeability and permittivity of free space respectively, and $\varepsilon_r = 2.0$.

System limits are the most numerous and nebulous ones in the hierarchy. In addition, because they occupy the highest level of the hierarchy they are necessarily the most restrictive of all limits. Consequently, they must be most carefully considered. The foremost generic system limit is imposed by the architecture of a chip. To simply illustrate this limit,

assume a random logic network consisting of one million gates, $N_g = 10^6$, each of area A in a macrocell configuration described by a Rent's exponent p = 0.45. Using Rent's empirical rule as a foundation, the average interconnect length and therefore the average load capacitance of a gate in the random logic network can be calculated [10]. This in turn enables calculation of the second generic system limit which is imposed by the average power-delay product $Pt_d = \frac{1}{2} C_s V_o^2$ of a gate in the random logic network (whose load capacitance C_s is larger than the circuit level value C_c). The third generic system limit requires that the heat generated due to power dissipation be removed by the chip package according to the relationship $\overline{P} \leq QA$ where in this instance, $\overline{P}$ is the total chip power consumption, Q[watts/cm^2] is the cooling coefficient of

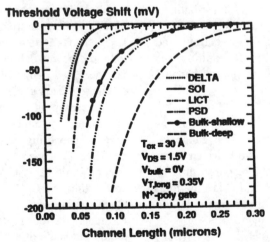

FIGURE 4. Threshold voltage shift versus channel length for the MOSFET device structures illustrated in Figure 2. Device parameters at 300K are: DELTA: $N_A = 5 \times 10^{18}$ cm^{-3}, d = 109 Å; SOI: $N_A = 5 \times 10^{18}$ cm^{-3}, d = 55 Å; LICT: $N_A = 10^{16}$ cm^{-3}, $N_A^+ = 5 \times 10^{18}$ cm^{-3}, d = 218 Å, Junction depth = 50 Å; PSD: $N_A = 10^{16}$ cm^{-3}, N_A^{++} 5×10^{18} cm^{-3}, $N_A^+ = 5 \times 10^{17}$ cm^{-3}, d = 218 Å, Junction depth = 500 Å; Shallow-junction bulk MOSFET: $N_A = 8.6 \times 10^{17}$ cm^{-3}, Junction depth = 50 Å; Deep-junction bulk, MOSFET: $N_A = 8.6 \times 10^{17}$ cm^{-3}, Junction depth = 1000 Å.

the chip package and A is the chip area. Operation to the left of the system level $Pt_d = \frac{1}{2}C_sV_o^2$ and $\overline{P} = QA$ loci in Figure 1 is forbidden. The final generic system limit requires that the clock period T_c must be sufficiently large to accommodate the maximum clock skew T_{cs} of the macrocell and the critical logic path delay T_{cp} assuming, e.g. that it includes 10 average gate delays (i.e., $n_{cp} = 10$). This system timing limit can be expressed as $T_c \geq T_{cs} + T_{cp}$ and forbids operation to the <u>right</u> of this locus in Figure 1. Consequently, the allowable design space for the 10^6 gate random logic macrocell is the hatched triangle where t_{dmin} represents the highest speed design, P_{min} the lowest power design, and P_{max} the most mature technology design that are included within the allowable design space.

At the system level, $(1/L)^2$ vs τ plane limits focus on the longest interconnects since typically they impose the most stringent demands on performance. As illustrated in Figure 2, the response time of the longest global interconnect, i.e. a logic signal path, of length 2L is designated by $T_c \geq T_{cs} + T_{cp}$ since its delay τ must be included in T_{cp}. The actual length of its path is designated $L^2_{min} = N_gA_g$, since a smaller area than L^2_{min} could not accommodate the required number of logic gates N_g using the prescribed technology which requires a gate area A_g. The forbidden zone of operation of the longest interconnects lies external to the small triangle, two of whose sides are defined by the preceding limits. Exploration of various architectures for a one billion gate system ($N_g = 10^9$) operating with a one nanosecond clock cycle time

$(T_c=10^{-9}s)$ in both the P vs t_d and $(1/L)^2$ vs τ planes suggests that the limitations which must be overcome to achieve GSI can be ranked by degree of difficulty as follows:
(1) Long distance interconnect limitations associated with critical logic path delay and to a lesser extent with clock skew,
(2) Heat removal limitations which must be addressed by improved packaging and reduced power dissipation, and
(3) Transistor and specifically MOSFET limitations.

Practical Limits

Practical limits must first, of course, comply with the theoretical limits and then more urgently take into account manufacturing costs and markets. The most compact presentation of the totality of practical limits is simply a plot of the number of transistors per chip N versus calendar year Y for past, present, and projected products of major economic consequence. The number of transistors per chip can be elegantly expressed in terms of the minimum feature size of a technology F, the square root of die area D, and the packing efficiency PE of a chip defined as the number of transistors per minimum feature area or $N = F^{-2} \circ D^2 \circ$ PE, which is plotted in Figure 5 versus Y. Figure 4 illustrates a pessimistic (???F), a realistic (???G), and an optimistic (???H) scenario which project a 10^9 transistor chip by 2000 and more than 10^{11} transistors per chip before 2020, according to the realistic scenario.

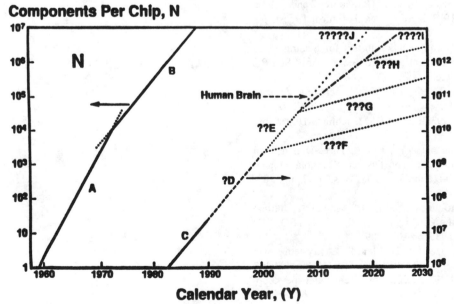

Components Per Chip, N

Calendar Year, (Y)

FIGURE 5. Number of transistors per chip N vs calendar year Y.

Conclusion

The chip performance index or CPI = N/Pt_d has grown by about twelve decades since 1960 and is realistically projected to grow by about another six decades before 2020 [1]. This magnitude of productivity improvement is unprecedented. Because Si MOSFETs can be expected to scale in an essentially predictable manner from the present state-of-the-art design rule of 0.5 microns to 0.05 microns and perhaps even smaller, the feasibility of Si technology continuing as the most powerful driver of the information revolution is not in doubt. The paramount issue which remains unresolved is the capability to engage manufacturing technology and metrology that can continue to drive downward the cost per function of GSI.

References

[1]   J. D. Meindl, "The Evolution of Solid-State Circuits: 1958 - 1992 - 20??," IEEE ISSCC Commemorative Supplement, pp. 23-26.

[2]   J. D. Meindl, et al., "Prospects for Gigascale Integration Beyond 2003," 1993 IEEE ISSCC Digest of Technical Papers, pp. 124-125.

[3]   D. Halliday, R. Reswick, and J. Walker, "Fundamentals of Physics," J. Wiley and Sons, 4th Edition, 1993, p. 623.

[4]   R. Laudauer, "Dissipation and Noise Immunity in Computation and Communication, Nature, vol. 335, October 27, 1988, pp. 779-784.

[5]   M. Ono, et al., "Sub-50 nm Gate Length N-MOSFETs with 10 nm P Source and Drain Junctions," 1993 IEEE IEDM Technical Digest, pp. 119-122.

[6]   B. Agrawal, et al., "Opportunities for Scaling FETs for Gigascale Integration," Proceedings of ESSDERC '93, pp. 919-926.

[7]   V. De, et al., "A Physical Model Including Velocity Overshoot for Drain Current of Sub-150 nm MOSFETs, 1993 IEEE IEDM Technical Digest, pp. 717-720.

[8]   H. W. Wong and Y. Taur, "Three-Dimensional Atomistic Simulation of Discrete Random Dopant Effects in Sub-0.1 micron MOSFETs," ibid., pp. 705-708.

[9]   H. B. Bakoglu and J. D. Meindl, "Optimal Interconnection Circuits for VLSI," IEEE TED, ED-32-(5), May 1985, pp. 903-909.

[10]   H. B. Bakoglu and J. D. Meindl, "A System-Level Circuit Model for Multi- and Single-Chip CPUs," 1987 IEEE ISSCC Digest of Technical Papers, pp. 308-309.

18

Circuits, Transistors, Processing, and Materials for 0.1-Micron CMOS Technology

Y.G. Wey*, J. Weis, K.F. Lee*, D. Monroe*, F.H. Baumann, E.H. Westerwick, and R.H. Yan

AT&T Bell Laboratories, Holmdel, NJ 07733
AT&T Bell Laboratories, Murray Hill, NJ 07974

Building a complete CMOS technology from material, processing equipment, transistor, technology integration, design tools, to product needs continuous investment of resource and development effort of up to 10 - 15 years, as illustrated in Figure 1. Although it is not necessary for all the integrated circuit manufacturers to establish the whole spectrum of capability, they must acquire the whole set through equipment purchasing, technology licencing, and/or joint development with other companies. Since each one of these elements requires substantial focus of resource, the technology development has traditionally been sequential as shown in Figure 2 (a). Processing groups transfer know-how in materials and processing to technology integration groups which develop complete technologies including transistors and interconnects for circuit design groups. This model has worked quite well in the past. However strong competitions in today's semiconductor market force IC manufacturers to bring more value to their customers in order to stay in business. Therefore, the trend is to provide more engineering integration between these different levels from materials to circuits, as shown in Figure 2 (b) and (c). For example, in the 0.5μm and 0.35μm CMOS generations, fabless companies have begun to participate in their technology vendors' technology development, and technology development branches in vertically integrated companies are more and more guided by their product families. We believe this trend will continue in the 0.25μm generation towards the end of this century and into the 0.1μm regime of the 21st century.

The lessons learned from 0.5μm and 0.35μm generations are the basis for the 0.1μm generation, including optimizing gate dielectric thickness between reliability, performance and power consumption, the decision between single n^+-poly gates and dual n^+/p^+-poly gates, the trade-offs between high threshold voltage for less standby current and low threshold voltage for higher drive current and their related circuit issues, and the decision on the number of metal levels between higher

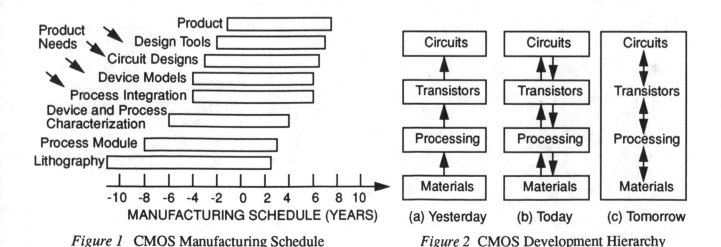

Figure 1 CMOS Manufacturing Schedule

Figure 2 CMOS Development Hierarchy

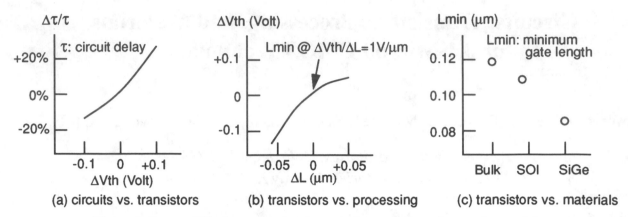

Figure 3 Interactions between circuits, transistors, processing, and materials

performance/better signal integrity and higher processing yield/lower cost. There are also issues that will become more important as the technology dimension approaches 0.1μm. For example, the reduction in power supply voltage makes circuits more sensitive to threshold voltage variation while the increase in threshold voltage variation resulting from linewidth variation [1] and processing variation [2] is making the situation even worse. The reduced supply voltage together with the reduced gate dimension also make the 0.1μm transistor design more challenging than ever. It might be necessary to greatly reduce the processing thermal budget which could then cause increased processing defects and degraded controllability. Tighter trade-offs between higher speeds and lower standby and active power might also promote the need for different material systems such as Silicon-On-Insulator (SOI), for its ideal subthreshold characteristics and reduced capacitive parasitics, and SiGe, for its extra dopant confinement capability and therefore better transistor scaling

behavior[3]. Figure 3 shows schematically the relationship between circuits, transistors, processing, and materials. In addition to optimizing the circuit perform (1) for the average case, understanding its statistical characteristics with respect to the manufacturing environment is also critical.

It might be very costly to fully establish the experimental design space from materials to circuits. Here computer-aided design tools could provide enormous leverage. Due to the increased complexity, our characterization capability in resolving details tends to be the strongest at the material level and gradually weakens as the level moves up to the circuit level. Since circuits are what customers feel directly, the value of characterization is actually the highest at the circuit level and decreases towards the material level, as illustrated in Figure 4. TCAD (Technology CAD) tools, which capture our knowledge at all levels, are ideal for presenting the lower level measurement information in the format

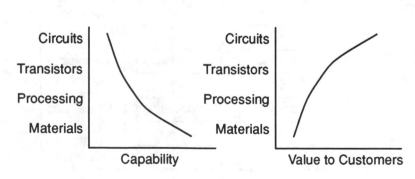

Figure 4 Capability and value of characterization

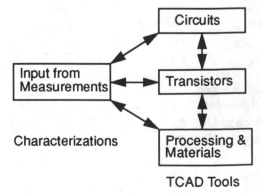

Figure 5 Characterizations and TCAD tools

of higher level meaning (Figure 5.) Today, the technologists and circuit designers engaged in forward-looking CMOS technology and circuit studies have to rely on TCAD to design experiments to optimize the probability of getting the first lot right, to manage the inflated development cost due to increased processing complexity. Circuit designers are beginning to use TCAD to understand the various trade-offs between circuit styles and transistor characteristics, making the boundary between circuit design and technology design gradually disappear.

Improvements on our characterization capability should be made by improving our ability in both mechanical and electrical measurement, and in interpreting the measurement results in terms of circuit performance and its variation, yield, and manufacturing cost. These could then form the building blocks for engineering integration of materials, processing, transistors, and circuits, maximizing added value to the customers of today's and tomorrow's IC manufacturers.

Acknowledgment The authors would like to thank the following people for useful discussions and continuous encouragement: Bryan Ackland, Joze Bevk, Jim Boddie, Bill Brinkman, T-Y Chiu, Bill Cochran, Len Feldman, Rich Howard, Young Kim, Horng-Dar Lin, Colin McAndrew, Aon Mujtaba, Abbas Ourmazd, Mark Pinto, and Bob Swartz.

[1] Y.G. Wey, et al., *"Defining 0.1μm Features Using Deep Ultra-Violet Lithography for CMOS Applications: Processing Technique and Linewidth Control"* (unpublished)

[2] T. Mizuno, et al., *"Experimental Study of Threshold Voltage Fluctuation due to Statistical Variation of Channel Dopant Number in MOSFETs"*, IEEE Trans. Electron Devices, **41**(11), 2216, 1994.

[3] J. Bevk, et al., *"Suppression of Oxidation Enhanced Diffusion in Silicon via Germanium Incorporation"* (unpublished)

METROLOGY REQUIREMENTS FOR BEYOND 0.35-μM GEOMETRIES

Critical Metrology and Analytical Technology Based on the Process and Materials Requirements of the 1994 National Technology Roadmap for Semiconductors

Alain C. Diebold

SEMATECH
2706 Montopolis Drive
Austin, TX 78741

The 1994 National Technology Roadmap for Semiconductors (NTRS) has placed a new emphasis on metrology. The NTRS is assembled by leading technologists from manufacturers of silicon Integrated Circuits (ICs) and suppliers of manufacturing equipment. The NTRS provides a fifteen year horizon for viewing consensus requirements for the manufacturing of leading edge ICs. A Metrology Roadmap has been developed to supplement the NTRS. The main function of both documents is to focus national resources such as government laboratories and academic institutions on critical manufacturing and development issues. The scope of the Metrology Roadmap is to summarize the off-line, in-line, and in-situ physical analysis requirements. The key message of the Metrology Roadmap is that metrology activities must experience a paradigm shift that supports the industry wide movement toward in-situ process control. A discussion of the Metrology Roadmap is presented in this paper.

INTRODUCTION

Over the past several years, communication and transfer between the research and manufacturing communities of the IC industry has significantly increased. In particular, metrology, which is a critical enabler for all stages of process and tool development as well as rapid yield learning in pilot and established IC fab lines, has been hampered by inadequate communication in the past. In 1993, the very successful Manufacturing Science and Technology Topical Conference at the National Symposium of the American Vacuum Society brought leaders in research and development and in manufacturing together. (1–4) The key challenges to off-line and in-line analysis/metrology were reviewed at the topical conference, and the proceedings provide extensive reference material that supplements this chapter. (3) The Materials Research Society has also presented an excellent series of topical symposia, published proceedings, and review articles in the MRS Bulletin which focus on process development issues. (5) Langer and Bullis have also led a SEMI Workshop on process control measurements. (6) The International Workshop on Semiconductor Materials Characterization: Present Status and Future Needs and its proceedings should help close the communications gap for the metrology community.

In an effort to communicate the metrology development needs for the IC industry, a Metrology Roadmap has been developed by a group of representatives of the member companies of SEMATECH, NIST, Sandia NL and suppliers of metrology tools. (7, 8) (In this paper, "metrology tool" refers to the measurement equipment.) The process-based measurement requirements were taken from the needs specified by the technology working groups that wrote the 1994 National Technology roadmap for Semiconductors (NTRS). (7) The Metrology Roadmap is a supplemental document to the NTRS. (8) The following statement, written by Greed and Seiler during the roadmapping activities of the Materials and Bulk Processes Technical Working Group, introduces the theme of the Metrology Roadmap. (8)

"Metrology – A Paradigm Shift

The timely achievement of evolving requirements of the NTRS requires a paradigm shift in the role of metrology from off-line sampling to on line control. The most important enabler for the shift is the realization by management executives that metrology tools must transition to the same level of robustness and hence development support as is accorded to process equipment. A key to establishing this new paradigm is combining the use of in-situ and in-line metrology with off-line capabilities for advanced process control and rapid yield learning."

One way of viewing metrology and analysis is to analyze its place in the economics of manufacturing and development. Clearly, materials characterization is a key enabler for materials process and process tool research and development. This value needs to be recognized. There is a larger issue for Fab Metrology. Each process step fabricates part of the IC and therefore adds value to the partially processed wafer. Measurement of physical and electrical parameters for the purpose of manufacturing process control and product quality control is typically viewed as not adding value to the wafer. The financial impact of savings from minimization of yield loss due to poorly controlled processes and from fully fabricating wafers that (should have been scrapped after a failed process step) is often considered proprietary information. Often, manufacturing problems are not discovered until electrical testing of the completed IC. Due to the manufacturing cycle time, it is likely that there will be a considerable loss due to subsequent wafer lots with identical problems. Often, fab cost/resource models do not account for the value of loss reduction. All of these factors have hindered the establishment of a new metrology paradigm which would add value by minimizing wafer scrap.

Another aspect of the financial impact of metrology and analysis is its contribution to yield learning between product introduction and maturity. Present technology generations can increase profitability through yield learning. Rapid yield learning in pilot line and research activities can increase profitability of future technology generations by decreasing the cycle time for product development and by improving the robustness of processes and tools. Off-line, in-line, and in-situ metrology can reduce the cycle time for yield learning and thereby help establish the new metrology paradigm. Cycle time reduction is facilitated by networking of analysis tools and use of data management and analysis software.

Future fabrication facilities may base metrology tool and method selection on both physical measurement needs and cost of ownership. In conjunction with the paradigm shift moving metrology to more in-line and in-situ operation are the requirement for robust tools and methods and the increased awareness of cost of ownership and its impact on cost/resource for the entire manufacturing facility. Existing cost of ownership models are primarily aimed at in-line process tools and fab facilities. **A new emphasis on metrology requires a new methodology for comprehending the true value of development, pilot line and production fab metrology.** Improving the contribution of metrology to rapid yield learning requires factory wide data analysis and management systems. These systems must be comprehended in cost/resource models.

The lack of statistical methodologies hampers process control metrology. The impact of infrequent events on yield is well known, and the metrology tools must be accompanied by statistically sound sampling, testing, and correlation with physical parameters, Statistical methods associated with detection limits, non-normal data, and distributional data must be developed. Triplett has pointed to gate dielectric quality testing as an excellent example of the challenge facing statistical methods. (9)

In-line metrology measurements maintain high yield when they used for *statistical process control*. The process is under control when the physical value of an appropriate property remains between an upper and lower process specification limit. The metrology tool must have a repeatability value that results in a precision to tolerance ratio , $P/T = 6\sigma /(UL-LL) \leq 30\%$. Here, UL is the upper process specification limit, LL is the lower process specification limit, and σ is the measurement standard deviation. σ includes the contributions of short term repeatability and long term reproducibility. σ is calculated from repeated measurement of a known reference sample. The ISO and ASTM standards use σ to refer to measurement uncertainty which is a scientifically based error analysis. SEMATECH has typically used $P/T \leq 30\%$. (10)

Each IC manufacturer employs a different strategy when implementing fab metrology. (3) A detailed review of this topic can be found in reference (3), and here we briefly discuss some essentials. Some ICs are manufactured in high volumes, while others are low volume specialty products. One approach to metrology in high volume manufacture is that a high yield, easily manufactured process flow is developed using pilot line metrology. Bartelink has championed and reviewed an electrical test structure based version of this approach called *"Statistical Metrology"*, which is discussed in the Process Integration, Devices and Structures section of this paper. This strategy requires little if any metrology when manufacturing a well developed high volume process flow. Another strategy employs frequent process monitoring during low volume manufacture. This is especially useful during "flexible manufacturing" such as that developed by Texas Instruments' Microelectronics Manufacturing Science and Technology (MMST) program. Due to the lower yield frequently observed when a process is transferred from pilot line to volume manufacture, many companies employ a strategy that embraces both approaches. *It is also important to note that test wafers and process monitoring are progressively removed as a volume manufacturing line matures. The industry trend is also toward non-destructive, in-line and in-situ product wafer monitoring to reduce the cost of test wafers.* This trend may be emphasized when 300 mm wafer IC fabricators are built. Both approaches imply the need for metrology to act as a means toward rapid yield learning and process control. This author is an advocate of whole wafer sample stages for analysis tools that are used in conjunction with networked analysis systems.

Networked Analysis Systems

Materials analysis methods are an enabling element of materials, process, and tool development. As the process moves from development to pilot line to volume manufacture, yield must increase. Again, the need to improve the cycle time for yield learning will drive developments in analytical technology. Whole wafer materials analysis tools (such as scanning electron microscopes equipped with energy dispersive spectroscopy (SEM/EDS), Auger, secondary ion mass spectrometers, etc.) and data management systems will become part of a networked analytical system. Future pilot line and FAB data management systems will network in-line inspection tools such as critical dimension SEM (CD-SEM), optical defect detection tools (ODD), new optical defect classification tools, transistor gate dielectric thickness measurement, electrical test, whole wafer defect review tools (SEM/EDS, Auger, ToF-SIMS, etc.), and total reflection x-ray fluorescence. Electrical device and circuit characteristics are the final measure of IC yield, and thus they will drive yield learning. Databases and model based yield learning software are critical parts of rapid yield learning. The Network system is discussed in the Factory Integration section.

Metrology Roadmap

The Metrology Roadmap summarized off-line, in-line, and in-situ physical analysis requirements associated with the NTRS document. In this paper, we employ the same general format used in the Metrology Roadmap. The NTRS projects these needs for the next 15 years according first volume shipment of a generation of DRAM technology. First shipment of DRAMs having 0.25μm design rule features will be in 1998. The Metrology Roadmap contains measurement requirements taken from the NTRS, indicates the challenge

to existing technology, and projects consensus potential solutions when appropriate. A Needs and Potential Solutions Roadmap is presented with each section of this paper. The reader may expect this paper to be divided into measurement sections such as particles, thin films, surface contamination, etc. Instead, the organization of the both this paper and the Metrology Roadmap follows that of the NTRS. Although this results in paper sections that have overlapping metrology requirements, it facilitates understanding process based metrology needs. Hopefully this promotes communication between metrology research and development and the IC manufacturing community. Future Metrology Roadmaps should include sections on Electrical Test and Packaging Metrology requirements.

Our discussion of metrology and analytical technology needs is divided into the following sections: (1) Sensors and Methodology for In-situ Process Control; (2) Process Integration, Devices, and Structures; (3) Materials and Bulk Processes; (4) Interconnects; (5) Lithography; and (6) Factory Integration. While sensors and in-situ process control is not a separate section of the NTRS, its importance in the new metrology paradigm requires this emphasis here. This paper will end with a summary and conclusions. Hosch reviews Sensors and Process Control in this Volume. (11)

SENSORS AND METHODOLOGY FOR IN-SITU PROCESS CONTROL

In-Situ sensors shorten the learning time feedback loop, reduce fabrication errors, and improve knowledge of tool maintenance requirements. In this section, general requirements for process control sensors including the associated methodology are presented. As a basis for understanding this area, process control theory is briefly described. Recently, Rubloff, Scace, and Hosch developed a systematic picture of sensors and control methodology. (8,12) They partitioned process control sensors into the categories of equipment state, process state, wafer state. Process control methodology was divided into equipment/process model/ design; fault detection/ classification; and adaptive process control. The roadmap for control sensors and methodology further describes the evolution toward real time control. Early development of pilot line and factory level data management software is a key enabler for successful implementation of control sensors. At each stage of deployment, process control implementation should be strategically utilized to improve the timeliness of yield learning. Industry experience has shown that there is a decrease in use of in-line metrology especially test wafers as a process line matures. Therefore, it is likely that many sensors used in tool and process development and pilot lines will not be used in mature wafer fabs.

Aspects of Control Theory

Recently, Butler reviewed the application of process control in the semiconductor industry. (13) Butler describes the key concepts as process response, the time scale, control structure, and multivariable controller type. Process gain is the change in output with change in input. Process control is optimally done when the process output responds monatonically to the process input variable being controlled. Control is more easily done when the output/input does not change sign or respond too strongly. (13) This is illustrated in Figure 1. The time scale of control is either process run to process run (run to run) or real time. The control structure is either regulatory or supervisory. The multivariable controller type is either a system of single input—single output (SISO) controllers or a multiple input—multiple output (MIMO) controller. A system of SISO controllers does not recognize the correlation between the variables, which results in oscillations in the controlled response. Butler uses a pressure controller as an example of a real time regulatory controller. As process control moves toward real time intervention, it is important to prioritize possible applications. Here, we describe adaptive process control as evolving from run to run supervision to real time regulation. Real time regulation implies the ability to measure a variable that is related to a process disturbance. In some cases, process disturbances can not be measured using existing technology, and in-situ sensors may not be appropriate. Using present technology, off-line and in-line measurements must be combined with in-situ sensors for complete process control.

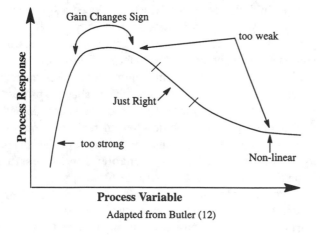

Adapted from Butler (12)

Figure 1. Process Control Fundamentals

Priorities (8, 11, 12)

* Run to run control and process state sensors are the highest priority for in-situ sensors.
* Migrate to real time control except for high priority items such as RTP temperature control needs.
* Migrate wafer state to in-situ from in-line in evolutionary way.

All process control is model based. There are few fundamental or empirical models for the physical/chemical reactions of present generation (and future) processes and tools. Existing models have shown that fundamental work in process and tool modelling will drive process control strategies. Process models allow selection of relevant physical parameters for control of microfeature and wafer level process such as via filling. (14) Tool models allow visualization of particle and process fluid flow patterns and can be used to improve tool design, reduce particle deposition, and pin point effective sensor location. (15) Control software for local tool control must evolve to integrate with control and database at the factory level. Contamination control sensors and process control will follow similar paths.

Current Technology Status

The first fabrication line use of sensors appears to be particle sensors. Models and standards required for process control are under development. Existing sensors can already be applied to many of the near term requirements. Cost of ownership models which suggest how to apply sensors and in/off-line analysis must be developed to guide sensor and process control model development and application.

Needs and Potential Solutions Roadmap

This section has discussed a systematic vision for In-Situ Sensor based Process control. The Needs and Potential Solutions Roadmap developed by Rubloff, Scace, and Hosch (12) provides a picture for their vision. It is important to note that the Materials and Bulk Processes TWG has indicated that when possible, in-line or in-situ non-destructive analysis of product wafers is preferable. A strategy for real time process methodology is required for cost effective implementation. Equipment state sensors monitor mechanical and electrical status of the equipment. Equipment state information will move from local tool to global control as connection to the CIM framework are developed. This implies that the local tool decision information will be transmitted to CIM framework to allow control of wafer moves, etc. Process state sensors monitor chemical/physical parameters, temperature, and spatial distribution, and process models allowing control will be developed and improved. Wafer state sensors will monitor product parameters and uniformity. Process control methodology will require significant development if real time control is to be achieved. Development of cost of ownership models is another critical need. The Needs and Potentials Solutions for Sensors and Methodology for In-situ Process Control is shown in Figure 2.

Gap: Although real time process control is frequently mentioned in the NTRS, the technical community would profit from a roadmap with prioritized applications and process integration requirements. As Butler discusses, the need for advanced process control should be established prior to development. (13)

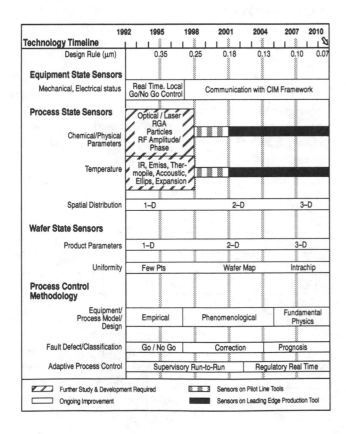

Figure 2. Overview of In Situ Process Sensors/Control [8]

Timeline for In-situ process control based on most advanced design rule used in IC manufacture. This roadmap is further described by Hosch in this volume. Some specific sensor requirements are mentioned in the Interconnects section. Progress toward real time process control is expected to be evolutionary.

PROCESS INTEGRATION, DEVICES AND STRUCTURES

The Process Integration, Devices and Structures roadmap covers overall technology characteristics, a discussion of the current status of memory and high performance logic technologies, technology computer aided design (T-CAD), and future needs for memory and logic. (7) This section contains a discussion of key issues effecting the development of a strategy for metrology such as short flow methodologies, known good die, yield analysis and defect allocation strategy, reliability characterization and models for devices and structures. Process integration issues faced by future materials systems such as those used to fabricate copper based metallizations are covered in the Interconnects section.

Both physical and electrical metrology tools are challenged by the requirements found in the Process Integration, Devices and Structures roadmap. (7) Physical metrology is driven by the need to provide electrical characteristics. In-line electrical test tools must be improved if they are to evaluate wafers according to future specification for known good die. Known good die refers to a die that has been shown to have the same quality standards as it would have in the fully packaged form. A complete strategy for metrology comprehends short flow methodology. (3,4) Short flow methodology refers to fabrication of test structures using a selected set of process steps. For example, a short loop involving only the first two levels of metallization would be used to test for uniform fabrication of contacts between levels of metallization. Standardized test structures are routinely used for short loop tool and process development. Test structures are discussed in Process Integration, Devices and Structures and Interconnects sections of the NTRS. (7)

The term *"Statistical Metrology"* is used to describe a fundamental method of short flow equipment characterization that can isolate the statistically significant variation of process tool output. (4) Due to the infrequent nature of tool induced failures, *"Statistical Metrology"* is based on electrical testing. The key message of *"Statistical Metrology"* is that a very robust, high yield full flow process can be developed before a high volume process is transferred from pilot line to production FAB. (4) In this paper, we discuss the need for statistical methods for physical metrology tools (see introduction) for all stages of development and for manufacturing. In both cases, the intent is to reduce the learning cycle time and allow high yield manufacturing.

Comments and Priorities

Only one physical metrology need is given a priority by the NTRS, and it is mentioned as the fourth priority in the table on T-CAD Top Priority Needs. This is 3D dopant profiling. It should be mentioned that existing technology such as focused ion beam tools (FIB) will continue to provide critical support for process integration and circuit development. (3) These systems are capable of locating specific coordinates provided by optical defect detection tools, and they allow development engineers to perform some circuit rewiring using wafer navigation and CAD circuit layout. Existing FIB tools may now be equipped with electron beam columns and x-ray detectors for rapid analysis of buried particles and defects. These FIB systems are referred to as dual column FIBs. In addition to whole wafer ToF SIMS, Auger, SEM/EDS based defect review tool, and NFOM, FIBs and dual column FIBs are critical components in a networked analysis system for rapid yield learning. Voltage contrast microscopy, which is typically done on single chips, is frequently used for locating the failed section of a circuit for further physical analysis.

Metrology Needs:

Several physical characterization needs were discussed in the unpublished TCAD characterization roadmap. These include two/three dimensional defect and dopant profiles, imaging for very small features, and quick, inexpensive imaging methods for measurement of features across die and wafer. The consensus version of the potential solutions for two/three dimension dopant and defect profiling is presented in Figure 3. In this volume, Griffith and Feenstra review scanning probe microscopy. (16)

Several reviews of two dimensional dopant profiling serve as resources for detailed information on this topic. (17, 18) One example of the need for calibration of process simulators in one dimension was discussed in reference 3. (3) The 1D predictions of two different T-CAD programs show different depths for charge carrier channel region of 0.25 μm design rule PMOS transistor. At the time of writing, some disagreement exists over the status of simulation in one dimension. It is likely that the methods suggested for two dimensional profile characterization will be useful for 1D calibration of defect profiles and co-implant effects discussed by Tasch and Law. (19,20) In their recent review of 2-dimensional dopant profiling, Subrahmanyan and Duane (18) modeled the effect of changes in substrate and LDD dopant concentration and effective channel length on the electrical properties of 0.25 μm design rule transistors. (18) These effects are summarized in Table 1 using the results of reference (18). This work sets requirements for the accuracy of process simulation and physical analysis of transistor dopant profiles to 10% concentration sensitivity down to $2 \times 10^{17}/cm^3$, and sub-10 nm spatial resolution. (18) The data in Table1 was calculated using an idealize transistor with no LDD structure. This data indicates that a 10% change in channel carrier concentration alters the model transistor leakage current (I(off)) by a factor of 2. (18) A transistor with a LDD structure was used to model the effect on changes in LDD carrier concentration. This set of idealized transistors indicated that a sensitivity of at least $2 \times 10^{17}/cm^3$ is required. Subrahmanyan and Duane determined the requirements for spatial resolution of transistor characterization using realistic modelling of a 0.25 μm effective gate length p-channel MOSFET. When the 330 nm junction depth was changed to 290 nm, a 200x increase in leakage current was observed. The resulting value (> 1×10^{-9} A) of the leakage current was unacceptable.

Table 1

Idealized 0.25 μm n-channel MOSFET having 6.5 nm gate oxide, 50 nm spacer, 100 nm deep source/drain with peak surface carrier concentration of $3 \times 10^{20}/cm^3$ as modelled by Subrahmanyan and Duane. (18)

Channel Dopant Conc.	$3.0 \times 10^{17}/cm^3$	$2.7 \times 10^{17}/cm^3$	$1.5 \times 10^{17}/cm^3$
L_{eff} (μm)	0.210	0.208	0.202
V_t (V)	0.662	0.628	0.464
I(off) (A)	1.25×10^{-13}	2.82×10^{-13}	3.56×10^{-11}
I(lin) (A)	1.12×10^{-4}	1.15×10^{-4}	1.29×10^{-4}
I(sat) (A)	5.44×10^{-4}	5.62×10^{-4}	6.5×10^{-4}

L_{eff} (μm) is the effective gate length, V_t is the threshold voltage, C_{gd} is the gate capacitance , I(off) is the current between the source and drain when the gate voltage is below V_t, I(lin) is the linear drain current, I(sat) is the saturation current.

Advances in scanning probe microscopy methods are reported in this volume. (21) Neubauer, et al, present the first 2-D dopant profiles obtained by scanning capacitance microscopy and Vandervorst, et al, report on Atomic Force Microscopy of etched cross-sections and nano-spreading resistance probe dopant profiling. (21)

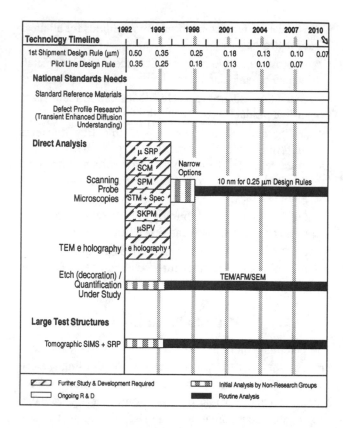

Figure 3. 2-D Dopant Profiling Roadmap [8]

MATERIALS AND BULK PROCESSES METROLOGY

The 1994 Materials and Bulk Processes TWG proactively recognized the need for specifying metrology and contamination control. Materials and Bulk Process (MBP) section of the NTRS is divided into Starting Materials, Wafer Surface Preparation, Doping Technologies, and Thermal/Thin Film Processing, Metrology and Contamination Free Manufacturing (CFM). Several "cross cutting" Metrology needs such as product wafer level particle and metallic contamination analysis are discussed in the MBP section of the NTRS. As mentioned in the Introduction, other sections of the NTRS such as Interconnects also utilize this type of metrology.

Current Technology Status

Starting materials metrology tools measure crystal orientation, wafer diameter, wafer thickness and flatness, wafer bow and warp, oxygen levels, resistivity, surface metallic levels, minority carrier lifetime, and microroughness. Silicon surface metallic contamination and microroughness

must be controlled prior to gate oxide growth. Two reviews discuss process requirements and characterization methods. (22, 23) Recent work indicates the need for microroughness metrology only during development of wafer cleaning processes. (23) The lack of standard methods and reference materials for analysis of metallic and organic contamination is impeding cost effective process development for surface preparation and starting materials.

Off-line metrology for MBP during manufacturing includes gate dielectric reliability testing and dopant profiling (by secondary ion mass spectrometry (SIMS) and spreading resistance probe (SRP)). In-line metrology includes implant dose, gate dielectric thickness control (by ellipsometry), and defect and particle detection and review (defect review refers to compositional analysis) on unpatterned test wafers (laser scanners) and patterned product wafers (automated optical laser scanners, image analysis, and manual defect/particle classification by optical microscope, some in-line and off-line defect review by whole wafer SEM/EDX equipped with coordinate location software and sample stage). (3) Replacement of manual microscopy by automated, commercial defect/particle classification is under development. Forensic science practitioners have long used optical microscopy for identification of particles and fragments, and McCrone has provided training opportunities in this methodology to the semiconductor industry. Total reflection x-ray fluorescence (TXRF ot TRXRF) using test wafers is becoming an accepted method of in-line surface metallic control, and TXRF is routinely used during process development. (3) Vapor phase decomposition (VPD) sample collection followed by atomic adsorption spectroscopy (AAS) or inductively coupled mass spectrometry (ICP–MS) is also used to control surface contamination. Although AFM measurement of surface microroughness will require some improvements in probe tip technology (smaller, reproducible tip radius) and analysis software, wafer cleaning process development has utilized existing commercial systems. (3, 22, 23) Helms and Higashi review surface preparation processes in this volume. (24) In-situ control of contamination may be considered to have entered the the initial stages of manufacturing control for particles and moisture.

The key to *rapid yield learning* during pilot line and mature manufacturing is the use of a networked analysis system. This system is discussed in the Factory Integration section of this chapter. Numerous off-line analysis tools (scanning Auger microprobe, transmission electron microscopy, Rutherford backscattering spectrometry, SIMS,

FIB, etc.) are used to control manufacturing process excursions and during tool and process development and pilot line manufacturing. (3)

Comments and Priorities

Each section of the MBP roadmap listed priorities for unique metrology needs that are unique to that section. Standard electrical test procedures for gate dielectrics is the highest priority need for Materials and Bulk Processes. Triplett has reviewed the effect of extrinsic defect density on gate dielectric oxide quality and the scaling of oxide thickness. (23) These test procedures will foster both process and gate dielectric tool development. Priorities that appear in the Metrology roadmap but are not discussed the remainder of the MBP section are listed below. (8)

Surface Preparation
      Real time, in-situ sensors for chemical, contaminants, and dissolved gasses.

Doping Technologies
      Charge monitor wafers
      Junction Leakage test wafers
      SOI Characterization

Thermal/Thin Films
      Temperature Measurement for Rapid Thermal Processing

Contamination Free Manufacturing
      Particle Reference Materials

As mentioned in the Process Integration, Devices and Structures section, existing equipment will continue to provide critical support. Again, dual column FIB systems will allow more rapid analysis of product and short flow wafers.

Needs Roadmap

When possible, in-line or in-situ non-destructive analysis of product wafers is preferable. A strategy for real time process methodology is required for cost effective implementation. The IC community has determined that 300 mm is the next wafer diameter, and this will require tools capable of handling 300 mm wafers by 1996. The following major gaps in in-line and off-line metrology capabilities were listed in the Metrology Roadmap. (8)

* Data management systems that are an integral part of process control, defect detection and sensors, and data reduction methods.

* Standard reference materials and methods for all areas of metrology.

* Improved metrology for particle identification, metallic and organic contamination.

* Extending existing particle identification methods to in-line product wafer capability.

* In-line and in-situ metrology of thin film thickness uniformity and composition for gate dielectric (requirement is for control of voltage threshold distributions) and other layers must be extended to 4 nm layers.

* Off-line analytical technology also requires improved cycle time, sensitivity and resolution.

* Off-line user facilities with unique capabilities allowing calibration of other methods and characterization not available using in-house tools.

Potential Solutions

The potential solutions discussion for Materials and Bulk Processes is divided into sub-sections on Starting Materials, Thin Film Metrology Needs, Contamination Free Manufacturing, Particle composition Analysis, Metallic and Organic Contamination Analysis, and Implant Calibration & 1D Dopant Profiling. Near field optical microscopy (NFOM) systems capable of spectroscopic analyses have the potential of extending in-line defect identification and composition analysis to technology generations having 0.1 micron design rules and beyond. Therefore, NFOM should be given special attention for research and development efforts. Other techniques are listed in the solutions roadmap. User facilities that allow a broad based access to unique capabilities include the synchrotron x-ray centers and heavy ion backscattering for trace analysis of surface metallic impurities, and accelerator mass spectrometry for analysis of metallic impurities in the region of polished, epi, and SOI (silicon on insulator) wafers used in the fabrication of IC devices. Off-line analytical equipment is needed to improve analysis cycle time and meet the challenge imposed by decreasing device size by smaller spot ion guns and addressable sample stages.

Starting Materials. The time lines for the priority needs for starting materials metrology are listed in the detailed potential solutions Figure 4a. Development of large diameter wafer metrology tools is a critical need. Light scattering from surface defects such as microroughness limit applicability of existing optical particle/defect detection equipment to development and qualification of future starting materials, large wafers, silicon on insulator wafers, and epi-wafers. Huff and Goodall, Kimerling, and Rozgonyi provide reviews of Starting Materials Technology in this volume. (25, 26, 27)

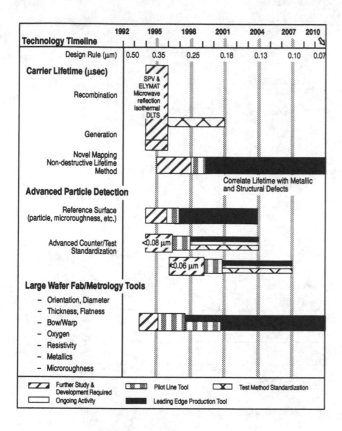

Figure 4a. Starting Materials [8]

Thin Film Metrology Needs. Precise measurements of thin film parameters for future gate dielectrics less than ten atomic layers thick will be required for process control. Control of an area averaged sample to less than one atomic layer will require improvements in tools, methods, and reference materials for precision and accuracy. Thickness uniformity for gate dielectrics is required for keeping tight distributions of threshold voltage. Potential metrology solutions for control of both gate dielectric and thin epitaxial layer processes are illustrated in Figure 4b. Low Z X-ray Florescence Spectroscopy (XRF) provides an alternative to ellipsometric measurement. (28)

Major issues.

*   The understanding and modelling of the materials science of the films, interfaces, and their effect on film parameters.

*   The development of adequate tools and methodology by equipment manufacturers.

*   The push for in-situ controls.

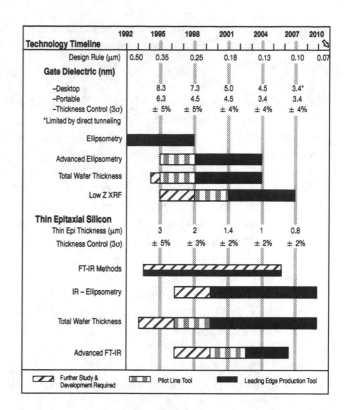

Figure 4b. Thin Film Metrology [8]

The gate dielectric and thin epitaxial Si thickness and uniformity requirements from the MBP roadmap of the NTRS are listed. Low Z XRF refers to low energy x-ray fluorescence measurement of oxide thickness. (21) continuous advancement in data analysis has pushed epi analysis capabilities of exist FT-IR hardware.

Contamination Free Manufacturing. In-Line optical defect detection (ODD) equipment is a key enabler of contamination free manufacturing, and roadmapped in the Contamination Free Manufacturing subsection of the Materials Bulk Processes section of the NTRS. ODD tools locate particles and "visual" defects on unpatterned and patterned wafers. The term "visual" refers to defects that can be observed under high magnification. Starting materials development is dependent on ODD tools for unpatterned wafers, and this need is illustrated in the potential metrology solutions for Starting Materials, Figure 4a. The CFM section of the Materials and Bulk Processes roadmap in the NTRS indicates the possible migration to SEM based (or other technologies such as Near Field Optical Microscopy) tools in order to map the particle/defect sizes that may adversely effect future technology generations. (7) The wafer maps from ODD tools after key process steps are stored and compared to wafer maps of electrical defects. The electrically active defects are partitioned into the process step and location where the particle or defect was first observed. In future fabs, in-situ sensors, in-line metrology tools, and optical defect detection equipment will all communicate with a central manufacturing control system. These data management and fab control systems are discussed in the Factory Integration section.

Particle Composition Analysis. In-line particle and defect characterization on whole product wafers is presently done by manual inspection with optical microscopes and in some instances by energy dispersive spectroscopy (EDS) in scanning electron microscopes (SEM). Whole wafer SEM/EDS systems with automated stages that locate particles/defects using wafer maps from optical defect detection tools are often referred to as Defect Review Tools (DRT). DRTs have been used as in-line and off-line tools. In FAB manual inspection must become automated, and the defect classification algorithms will require information from in/off-line systems such as SEM/EDS. The size limitations of SEM/EDS composition analysis varies according to the composition of the particles and the background layer (Si substrate, oxide layer, aluminum metal lines, etc.). The limit of 0.5 μm particles/defects listed on the potential solutions roadmap is an estimate for patterned wafers. The process based requirements by technology generation specify a need for analysis of 0.08 μm particles/defects for ICs having 0.25 μm design rules. The potential solutions for Particle Composition analysis are found in Figure 4c. Here it is noted that the consensus roadmap lists Auger, ToF-SIMS, and NFOM as potential near term technology for whole, product wafer particle/defect characterization. Two approaches to Auger analysis are under investigation. SEM/EDS DRT systems may be capable of analyzing small particles if residual gas content of the vacuum can be improved. UHV Auger is also under consideration. (3) Consensus information indicated that more than one method would be employed by several IC manufacturers. Anthony presents some recent ToF-SIMS analysis of sub 0.5μm Al and Al_2O_3 particles on silicon wafers fabricated for SEMATECH by NIST in this volume. (29) The capability and commercial viability of synchrotron based X-ray Absorption Near Edge Spectroscopy (XANES) should be explored for sub 0.1 μm particles. New x-ray detector technology may allow high energy resolution analysis of x-ray fluorescence excited by low energy electron beams. The new detector is based on a liquid helium cooled thermistor, and has been shown to have high energy resolution (18 ev), high quantum efficiency (95% at 6 KeV), and a broad energy range (0.5 to 7 KeV). (30) This technology should be developed for high count rate applications as a replacement for EDS in SEM/EDS, TXRF, and XRF.

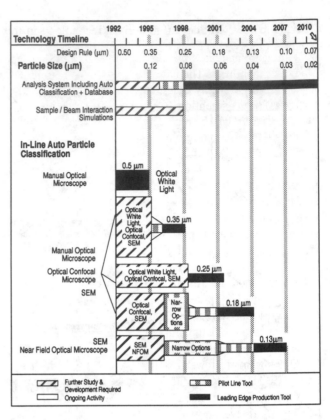

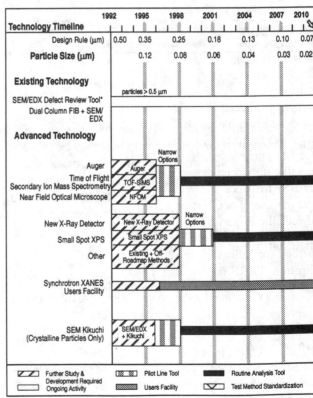

Figure 4c. 1 & 2 – Product Wafer Particle Composition Analysis [8]

Both figures (4c1 & 2) list the NTRS specifications for the minimum size of yield reducing particles

Metallic and Organic Contamination Analysis. The process driven requirements of decreasing levels of surface contaminants increases present and future gaps in metals analysis capabilities. The nature of metallic contamination from implanters may often differ from less energetic sources. Special attention to ultra-clean sample handling and preparation will be required especially for alkali metal analysis. Potential metrology solutions for process control of metallic and organic contamination are listed in Figure 4d. Three areas require special attention for both starting materials and surface preparation. The lack of standard methods and reference materials for analysis of metallic and organic contamination and carrier lifetime is an impediment to effective commerce in starting materials and process development for surface preparation and starting materials. Several analysis methods are routinely used, and consensus standards for cross-calibrating methods are in development. Non-destructive mapping of light element and transition metal (at detection limits required by future specifications) contamination has been specified as a major requirement. The use of Vapor Phase Decomposition (VPD) sampling followed by High Mass Resolution Inductively Coupled Plasma Mass Spectrometry (ICP-MS) provides destructive characterization of alkali and transition metals at levels lower than 1×10^9 at/cm^2 (200 mm and 300 mm). (3, 31) VPD is considered destructive because of the microroughness that results from VPD. (31) Great care should be exercised when VPD-TXRF is used due to uneven metal concentrations at the VPD spot resulting form droplet evaporation. VPD on 300 mm wafers will be challenging due to evaporation of the collecting droplet during sampling. Special mention should be given to the need for users facilities having analysis equipment with unique capabilities for ultra trace characterization. These include the use of synchrotron x-ray center for particle and surface metallic impurities, heavy ion backscattering spectrometry (HIBS) for trace analysis of surface metallic impurities, and the use of accelerator mass spectrometry for analysis of metallic impurities in the region of SOI wafers used in the fabrication of IC devices. HIBS provides a convenient method for calibration of reference materials. (3, 32) Synchrotron TXRF has the capability of selecting radiation that allows analysis of light element contamination on silicon surfaces. The need for a commercial synchrotron based analysis capability using existing beam lines should be evaluated. Off-line analytical technology also requires attention. Time of Flight Secondary Ion Mass Spectrometry (TOF-SIMS) may provide destructive mapping of surface contamination. Smaller spot ion guns and addressable sample stages for surface analysis equipment are two examples. The application of TXRF, VPD-ICP-MS, HIBS, ToF-SIMS, and post-ionization of sputtered neutral atoms (by resonant and non-resonant laser excitation) to

analysis of surface metallic contamination was recently reviewed. (3, 32)

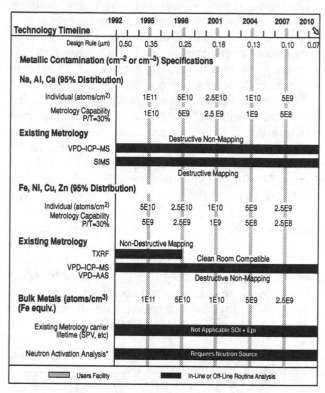

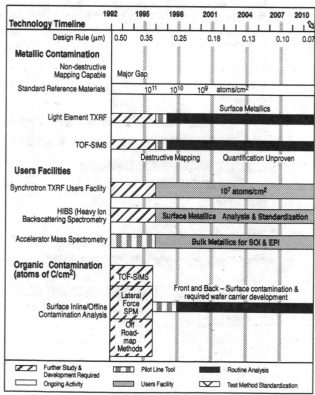

Figure 4d. 1 & 2 – Metallic and Organic Contamination Analysis [8]

Implant Calibration and 1D Dopant Profiling. Larson has lead a team roadmapping in-line implant process control. In-line, non-destructive implant process control methods are challenged by existing manufacturing needs. Consensus requirements indicate that repeatability should be given priority over accuracy. Implant reference materials are primarily needed for process transfer and for industry reference. Four point probe resistivity measurement will continue to provide critical characterization. Methods that monitor the amount of implant induced lattice damage provide non-destructive characterization and low dose sensitivity, monitor stability and sensitivity are critical needs in this regime. Junction depth control is typically monitored off-line by SIMS and Spreading Resistance Profile (SRP). Existing SIMS tools that utilize low energy probe ion beam have demonstrated very shallow junction measurement that meets roadmap requirements beyond 0.25 μm design rules. SRP technology that meets future requirements has been demonstrated but is not yet typical. Potential metrology solutions for 1D dopant profiling are listed in Figure 4e. Resonant and non-resonant ionization of sputtered neutrals will increase sensitivity and reduce matrix effects when compared to SIMS. (3)

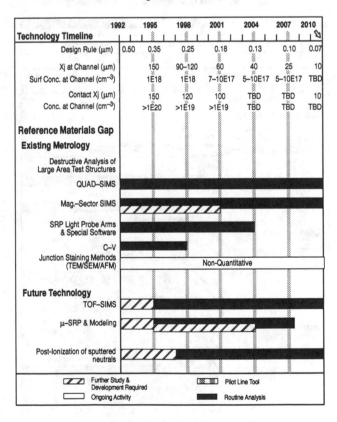

Figure 4e. 1D Dopant Profiling [8]

All types of SIMS instruments are capable of analyzing shallow junctions. The development activity shown on the magnetic sector SIMS indicates that improved magnetic sector instruments are being made commercially available.

LITHOGRAPHY

Lithography steps are considered the most critical process steps. Lithography process control is done in-line during IC manufacture. Critical Dimension measurement refers to the verification of the width of critical features such as the polysilicon gate and metal interconnect lines. Another measurement is the verification of the overlay alignment of subsequent lithography steps. Lithography metrology is presently done on product and test wafers (wafer level) and on lithographic masks (mask level). A Lithography Metrology group lead by Hershey has developed a roadmap that was used in this section. (7) This information is contained in the 1994 NTRS.

Typical Off-line and In-line Metrology

* Chemical purity of photoresist is routinely monitored by supplier.

* Mask manufacturer provides extensive characterization information with each mask.

* In-line, the photoresist exposure, CD, and overlay are carefully checked for each mask change. CD and overlay are monitored using a sampling plan that matches requirements of the process step (implant masks vs metal + via steps) and the IC product (Memory vs logic).

* Lithography process steps include: implant masks, poly Si gate, ᵛmetal and via, and interconnect memory capacitor processes.

Current Technology Status

Wafer level critical dimensions are typically measured using CD-SEMs, and this is expected to continue to at least 0.25 μm design rule technology. The extension from 0.5 μm metrology to smaller dimensions will be done by averaging a larger number of CD values instead of greatly improving the accuracy and precision of each measurement. Present generation Atomic Force Microscopy tools have very poor throughput (wafers per hour) characteristics when compared to CD-SEMs. In addition, new technology is not readily accepted into a production environment. The measurement of the registration of a subsequent lithographic step with the initial pattern is called overlay metrology. This is typically checked by printing a smaller box pattern over an existing box pattern. (33). The error in centering the box is an indication of the placement error for the entire pattern. Overlay metrology is done by brightfield optical overlay measurement using SEMI Standard "box-in-box" targets. Again, modifications such as phase contrast imaging are expected to extend optical overlay metrology to 0.13 μm design rule technology. Metrology is a critical part of mask manufacture. The term "output Metrology" refers to measurement of CD and accuracy of feature placement on the mask. Each mask is carefully inspected prior to shipment to an IC manufacturer. Current CD and registration metrology systems for mask manufacturing are capable for process control through 0.25 μm design rules.

Comments and Priorities

The NTRS calls for improved standardization and improved reproducibility for CD and overlay metrology. Again, the key is reduced cycle time for process setup and control. Typically, several lithography steps are done using the same tool. Process induced variation of measurements require additional time for calibration at each lithography step. Hershey calls for development of measurement techniques that are independent of feature density (eg., closely spaced metal lines) and process history.

Overlay control will become both more difficult and critical as design rules shrink. Present overlay metrology is sensitive to key processing steps such as planarization. Hershey has identified the key issues as: target asymmetries associated with resist coat over topography, radial asymmetry of deposited metal and dielectric films, and design of measurement structures that allow process control when overlaying after planarization by chemical mechanical polishing. (7)

The NTRS calls for in-situ sensors for process control after 0.25 mm design rules. Resist coat thickness and uniformity, concentration of photoresist components, uniformity of the development, and temperature uniformity pre/post baking. The goal is to provide real time process control that results in 100% lithographic yield.

Clearly, Lithography metrology is a critical success factor in IC manufacture. Although this discussion is condensed, the emphasis placed in the NTRS is an indication of the high priority that should be given to development in this area.

In-situ Lithography metrology is in the research and development phase.

INTERCONNECTS

Interconnect technology is often considered the part of an IC that contains the competitive edge of the manufacturer. The interconnects section of the NTRS covers the dielectric, metal film formation, and etch processes which are sometimes referred to as the back end of the line fabrication steps. Development activities include planarization processes such as chemical-mechanical polishing and advanced materials systems such as copper interconnects and high dielectric constant polymers. The biggest issue for etch and deposition technologies is fabrication of low aspect ratio features for 0.18 μm design rule ICs. Areas of overlapping metrology activity include particle detection and identification, surface metallic and organic contamination analysis, and critical dimension measurement.

Current Technology Status

Process development is done using electrical test structures, and process control is mostly off-and in- line. New interconnect tools and processes are evaluated for stress migration and electromigration, plasma etch damage of gate oxides, and failure inducing particles/defects. Work is underway to standardize electrical test structures which allows collection of data for development of predictive interconnect failure models. SEMATECH's SPIDER test structure has become an accepted standard method for evaluation of gate oxide damage. Non-destructive, surface charge analysis is being evaluated for in-line etch damage control. (3)

Comments and Priorities

The priority needs for Interconnect technology is summarized in the Crosscut Technology section of the Interconnects roadmap. They are listed by the priority listed in the NTRS. (7)

1) Critical Dimension (CD), profile, and/or edge roughness for poly silicon, high aspect ratio (dielectric) vias, and metal lines.

2) Film Thickness on patterned wafers of barrier layers (eg.,TiN), conductive films such as $TiSi_2$ contacts, new inter-metal dielectrics.

3) Ex-situ and in-situ evaluation of cleanliness of high aspect ratio contact/via

4) Sensors for in-situ process control.

5) Surface Planarity (flatness and topology) within lithographic field to insure depth of focus at lithography.

6) Post etch residue control to prevent corrosive precursors left on metal surfaces and contamination induced delamination.

7) Film Morphology

Metrology Needs

Measurement of properties of product wafers must be developed, and process and tool models should be used to guide application of in-situ sensors. These needs include critical feature size, film thickness, contamination, surface planarity, and critical process conditions. Metrology will find its widest use as a means of rapid yield learning, and only selected measurements will be used in routine manufacture. Existing modelling of particle formation and deposition inside a plasma has the potential of guiding both chamber design and particle sensor deployment. End point control for chemical mechanical polishing is a short term, high priority issue. The detailed solutions chart shows known metrology activity. The in-situ light scattering approach to CD measurement may find its initial use in the development of Interconnects tools and processes.

There is a near term need for reference materials for existing and emerging x-ray thickness measurements. A new x-ray thin film measurement tool shows promise for in-line process control for SiO_2 and may be extended to TiN and other thin films. (28)

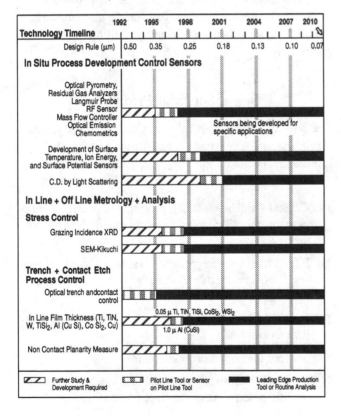

Figure 5. Interconnects [8]

FACTORY INTEGRATION

The Factory Integration section discusses the implementation of all other process and design sections into a cost effective factory. This section covers the manufacturing processes of the wafer fab, assembly, packaging, and test. Of key importance to metrology development are the "fab level" discussions of process control, material handling, environmental control (especially wafer environment), and information systems for the entire fab. The data management and process control issues are examples of information systems needs that will evolve at the "fab level". These systems are a key to use of metrology for rapid yield learning. (3) Factory Integration systems must comprehend off-line analysis needs. This is described in Figure 6. New software that models correlations between device electrical behavior and physical metrology ("Model based yield learning") must become part of pilot line data management. (34–36) Presently, networked analysis systems reduce the time for location and characterization of electrically active defects and experienced analysts determine fault types. (3) Many samples must be physically characterized by FIB/SEM/EDS and then sorted into electrical defect type. Electrical defect types have electrical test signatures. Walker, et al, have developed "model based yield learning" software that will also determine fault types prior to physical characterization (by FIB, SEM/EDS, Auger, TEM, etc.) This further reduces the time required for sampling and analyzing the origin of fault types (see Figure 7). (34–36) Anderson also discusses software system requirements for failure analysis in this volume. (37)

Comments and Priorities

The metrology needs can be found in the section on wafer environment control, and facilities technology. Minienvironments systems must be certified for cleanliness levels, and organic contamination is a well documented concern. Fluid delivery systems may utilize sensors to monitor excursions in contamination levels resulting from fluid transport. A fluid purity requirements roadmap is being developed during 1994, and it should provide point of use fluid purity requirements by technology generation.

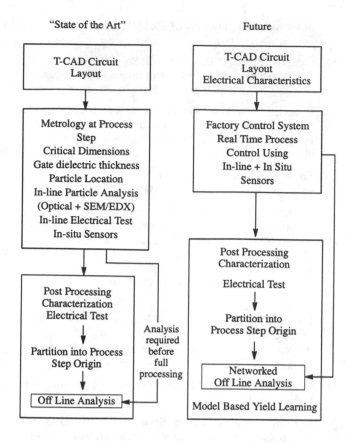

Figure 6. Networking Metrology for Pilot Line + FAB Rapid Yield Improvement [3]

The idealized "state of the art" networked analysis system is a composite of capabilties found in commercial and "in-house" systems. The networking reduces the contribution of defect and fault analysis to the cycle time for yield learning. Data management is a key issue due to the large data sets obtained in pilot line and volume manufacture. The T-CAD circuit layout is used to provide information about multilevel structure locations and to navigate FIBS to specific circuit features.

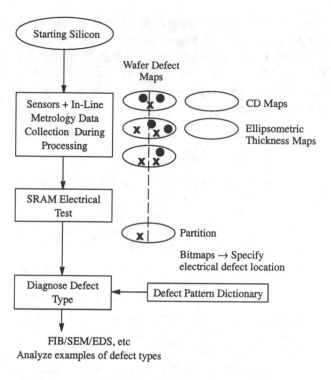

Figure 7. Model Based Yield Learning on SRAM Local ("Spot") Defects

Part of a Future Networked Analysis system that uses "Model Based Yield Learning" software is shown. Partioning refers to sorting through the wafer defect maps to determine which process step caused the electrically active defect. Presorting electrical defects into defect type reduces time for and number of samples required for physical characterization of each defect type.

Roadmapping Process and Acknowledgements

The Metrology Roadmap is a consensus document which was compiled using the inputs of several groups including: the Analytical Laboratory Managers Working Group, the Metrology group of Silicon Council and the Metals Task Force of the Silicon Council, suppliers of analytical equipment. The Materials and Bulk Processes Technical Working Group (MBP-TWG) added a Metrology discussion to the MBP section of the National Technology Roadmap for Semiconductors, and the inputs of the entire MBP-TWG were very important to the development of this roadmap. At the MBP-TWG meetings, Jim Greed provided insights from the SEMI community, and Dave Seiler provided inputs from NIST. The Materials and Bulk Processes Metrology Workshop held May 5 and 6, 1994 further developed key sections of this roadmap. The in-situ process control section was developed by Gary Rubloff, Jimmy Hosch, and Bob Scace at the MBP Metrology Workshop. The Lithography Metrology section in this document is based on the Lithography Metrology Workshop by Robert Hershey.

References

1) Liehr, M. and. Rubloff, G.W., "Concepts in competitive microelectronics manufacturing," J. Vac. Sci. Technol. **B12**, 2727-2748 (1994).

2) R.W. Schmitt, "Do manufacturing technologies need federal policies?," J. Vac. Sci. Technol. **B12**, 2721-2726 (1994).

3) Diebold, A.C., "Materials and failure analysis methods and systems used in the development and manufacture of silicon integrated circuits," J. Vac. Sci. Technol. **B12**, 2768-2778 (1994).

4) Bartelink, D.J., "Statistical metrology: At the root of manufacturing control," J. Vac. Sci. Technol. **B12**, 2785-2794 (1994).

5) Li, J., Blewer, R., Mayer, J.W., "Copper-Based Metallization for ULSI", Mater. Res. Soc. Bull. **XVIII**, 18 (1993) and Li, J., Seidel, T.E., and Mayer, J., "Copper Metallization in Industry," MRS Bulletin, Volume **XIX**, 15-18 (1994).

6) Langer, P.H. and Bullis, W.M., "Proceedings of the International Workshop on Process Control Measurements for Advanced I.C. Manufacture," Semiconductor Equipment and materials International, 805 East Middlefield Rd., Mountain View, CA, 94043.

7) *National Technology Roadmap for Semiconductors*, Semiconductor Industry Association, 4300 Stevens Creek Boulevard, Suite 271, San Jose, CA, November 1994.

8) Diebold, *A.C., Metrology Roadmap*, SEMATECH Technology Transfer Document #94102578A–TR, SEMATECH 2706 Montopolis Drive, Austin, TX 78741.

9) Triplett, B.B., private communication.

10) Horrell, K., *Introduction to Measurement Capacity Analysis*, SEMATECH Technology Transfer Document #91090709A-ENG (1991), Eastman, S.A., *Evaluating Automated Wafer Measurement Instruments*, SEMATECH Technology Transfer Document #94112638A–XRF, SEMATECH 2706 Montopolis Drive, Austin, TX 78741.

11) Hosch, J., "In-Situ Sensors for Monitoring and Control of Process and Product Parameters," this volume.

12) Rubloff, G.W., Scace, R., and Hosch, J., "In-Situ Process Control," *Materials and Bulk Processes Metrology Workshop Meeting Minutes*, A.C. Diebold and D. Simmons, SEMATECH Technology Transfer Document #94062400A (1994), SEMATECH 2706 Montopolis Drive, Austin, TX 78741.

13) Butler, S.W., "Process Control in Semiconductor Manufacturing," 41st National Symposium of the American Vacuum Society, Denver, CO, Oct. 24-28, 1994 to appear in J. Vac. Sci. Technol.

14) Cale, T.S., Chaara, M.B., Hasper, A., "Estimating Local Deposition Conditions and Kinetic Parameters Using Film Properties," Mat. Res. Soc. Symp. Proc. **Vol. 260**, 393–398 (1992).

15) Choi, S., Ventzek, P.L.G., Hoekstra, R.J., Kushner, M.J., "Spatial distributions of dust particles in plasmas generated by capacitively coupled radio frequency discharges," Plasma Sources Sci. Technol. 3, 418–425 (1994).

16) Feenstra, R.M. and Griffith, J.E., "Semiconductor Characterization with Scanning Probe Microscopies," this volume.

17) Diebold, A.C., Kump, M., Kopanski, J., and Seiler, D.J., "Characterization of 2-Dimensional Dopant Profiles: Status and Review," Diagnostic Techniques for Semiconductor Materials and Devices 1994, Eds. D.K. Schroder, J.L. Benton, and P. Rai-Choudhary, The Electrochemical Society Proceedings Volume **94–33**, 78–97, (1994).

18) Subrahmanian, R., Duane, M., "Issues in Two-Dimensional Dopant Profiling", Diagnostic Techniques for Semiconductor Materials and Devices 1994, Eds. D.K. Schroder, J.L. Benton, and P. Rai-Choudhary, The Electrochemical Society Proceedings Volume **94–33**, 78–97, (1994).

19) Tasch, A.F., SEMATECH SIMS Workshop, Set. 14, 1993 (unpublished).

20) Law, M.E., "Challenges for Accurate 3-Dimensional Process Simulation", (Simulation of Semiconductor Devices and Processes), Eds Selberherr, S., Stippel, H., Strasser, E., Vienna, Springer Verlag, 1993, pp. 1-8.

21) Neubauer, G., Erickson, A., Williams, C.C., Kopanski, J.J., Rodgers, M., and Adderton, D., "2D-Scanning Capacitance Microscopy Measurements of Cross-Sectioned ULSI Devices," this volume and Vandervorst, W., DeWolf, P., Clarysse, T., Trenkler, T., Hellemans, L., Snauwaert, J., "Carrier Profile Determination in Device Structures Using AFM-Based Methods," this volume.

22) Diebold, A.C., and Doris, B., "A Survey of Non-Destructive Surface Characterization Methods Used to Insure Reliable Gate Oxide Fabrication for Silicon I.C. Devices," Surf. Interface Anal. **20**, 127–139 (1993).

23) Triplett, B.B., "The limitation of Extrinsic Defect Density on Thin Gate Oxide Scaling in VLSI Devices", Proceedings of the Seventh International Symposium on Silicon Materials Science and Technology 94-10, 333-345 (1994).

24) Helms, R. and Higashi, G., "Si Surface Preparation and Wafer Cleaning: The Critical role, Status, and Future Needs in Characterization," this volume.

25) Huff, H. and Goodall, R.K., "A Silicon Materials: Critical Concepts for Optimal I.C. Performance in the Gigabit Era," this volume.

26) Kimerling, L.C. "Microdefect analysis Si: Theoretical Foundations," this volume.

27) Rozgonyi, G., "Defect Engineering Diagnostic Options for SOI, MeV Implants, and 300 mm Wafers," this volume.

28) Kirby, R.E., Wherry, D., Madden, M., "Oxygen detection in silicon dioxide layers by Low Energy X-ray Fluorescence Spectroscopy", J. Vac. Sci. Techol. **A22**, 2687-2693 (1994).

29) Anthony, J.M., "Ion Beam characterization of Semiconductors: Critical Diagnostics for Research, Development, and Manufacturing," this volume.

30) LeGross, M., Silver, E., Madden, N., Beeman, J., Goulding, F., Landis, D., and Haller, E., "Microcalorimeter for Broad Band High Resolution X-Ray Analysis," Nuclear Instruments and Methods **A345**, 492 (1994).

31) Gupta, P., Pourmotamed, Z., Tan, S.H., and McDonald, R. "Ultra-Sensitive Methods for Microcontamination Analysis in Semiconductor Process Liquids and on Surfaces," this volume.

32) Diebold, A.C., Maillot, P., Gordon, M., Boylis, J., Chacon, J., Witowski, R., Arlinghaus, H.F., Knapp, J.A., and Doyle, B.L., "Evaluation of Surface Analysis Methods for Characterization of Trace Metal Surface contaminants Found in Silicon Integrated Circuit Manufacturing," J. Vac. Sci. Technol. **A10**, 2945–2952 (1992).

33) Stairkov, A., Coleman, D.L., Larson, P., Lopata, A.D., Muth, W., "Accuracy of overlay measurements: tool and mark asymmetry effect," Optical Engineering **31**, 1298–1310 (1992).

34) Walker, D.M.H., "Rapid Yield Learning", 41st National Symposium of the American Vacuum Society, Denver, CO, Oct. 24-28, 1994.

35) Gaitonde, D., and Walker, D.M.H., "DEFAM: A Tool for Hierarchical Mapping of Defect to Fault", TECHON '93, 277-279 (1993).

36) Gaitonde, D., and Walker, D.M.H., "Test Quality and Yield Analysis Using DEFAM Defect to Fault Mapper," Proceedings of the IEEE International Conference on CAD 1993, 78-83 (1993).

37) Anderson, R.E. Soden, J.M., Henderson, C.L. "Failure Analysis: Status and Future Trends," this volume.

Cost Benefit of Process Metrology

Don Rose

Intel Corporation
2200 Mission College Blvd., Santa Clara, CA 95051

The semiconductor industry has grown from ground zero in 1960 to being a business in excess of $100B in 1994. The compounded annual growth rate over this period is about 17%. Capital investment has always been a large portion of the revenues of this industry. Estimates for capital investment for 1994 center around $25B. Investment rates as high as this are signs that high growth rates are forecasted for the next several years.

The large size of the capital investment required to be on the leading edge of semiconductor business requires that decisions undergo economic analysis that is of equal importance to the technical analysis.

This paper will show that the investment in metrology and analytical capability is growing faster than the rest of the capital costs for the industry, which are themselves escalating at unsustainable rates. The functional areas in which this large capital investment is made are identified, with defect metrology holding the most opportunity.

CAPITAL INVESTMENT SCENARIO

In Figure 1, the cost of building a new, leading edge semiconductor factory is depicted covering the period from 1970 to 1994. Note that this is a logarithmic scale. The current cost of a modern wafer fabrication facility is well over $1B and about 300 times the cost of such a plant in 1970.

Another way of looking at the capital investment of these factories is to normalize the cost to a unit of capacity. A de-facto standard unit of capacity in our industry is "wafer start per unit of time" or more specifically wafer starts per week (WSPW).

Figure 2 shows the capital cost of a large wafer fabrication facility per unit of WSPW (normalized to the 1.0 um generation) for the last four generations of technology. The corresponding DRAM technology is also noted. This increase exceeds five times over three generations, or about 60% per generation. The normal time period between introduction of successive generations is about 3 years.

Escalating Cost of Semiconductor Wafer Fabrication Plants

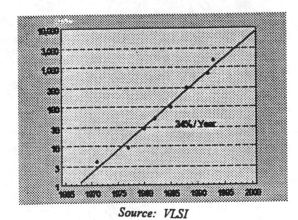

Source: *VLSI*

Figure 1

Total Factory Capital Cost $/WSPW

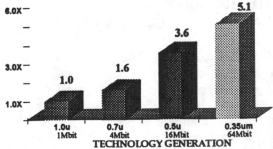

Figure 2

The first two technologies used 150 mm wafers, while the latter two use 200 mm wafers, it is also useful to measure unit capacity in terms of semiconductor chips. In Figure 3, two simultaneous trends are depicted that keep the number of chips on a wafer approximately constant; These trends are larger die sizes and the switch to larger wafer sizes. The DRAM is commonly used as a benchmark in determining the number of units (die) on a wafer. Note the sawtooth pattern depicts those units remaining between 100 and 300 die per wafer.

Die Per Wafer Trend - Wafer Size

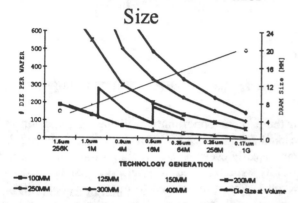

Figure 3

We can conclude from this data that if the capital cost content of a DRAM is increasing with each successive generation of technology, other leading edge products fit the same economic model. This is a trend that must be broken if the industry will continue to grow. The market requires an ever-decreasing cost-per-function. Capital escalation makes that requirement very difficult. For this reason, all capital cost associated with wafer fabrication will come under tighter scrutiny; especially those that are increasing at a higher rate than the others.

METROLOGY AND ANALYTICAL COST

In Figure 4, the capital associated with metrology and analytical capability are charted over the last four generations in terms of costs per WSPW normalized to the 1.0 um generation of technology. Note that the highest cost has increased 6.4 times over the same period; the total capital costs of a wafer fabrication facility has increased by 5.1 times. This increase is becoming a significant part of the total factory cost.

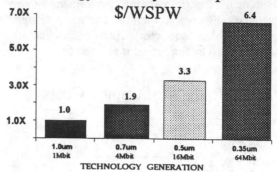

Figure 4

In Figure 5 the cost of metrology and analytical capability is represented as a portion of the total facility costs. To install a metrology capability into a wafer fabrication facility the capital cost can be broken down into three sources of capital spending: tool cost, cost of space, and cost of automation.

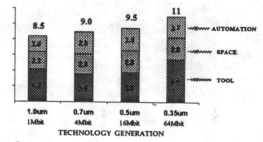

Figure 5

The tool cost comprises the procurement price of the metrology equipment and the cost of installing that tool. The space cost is that of the total facility, including all of the support systems, prorated to the portion of space allocated to metrology and analytical equipment. The automation costs cover the investment required to store and move test wafers as well as the portion of the computer system used for collecting, storing, and analyzing the data collected by the metrology and analytical equipment. This simple cost model does not capture some costs: specifically that portion of process tool investment that is dedicated to measurement, the portion of the cleanroom space allocated to process tool measurements, storage and movement of test wafers, and the computation capability. Estimates of how much the total metrology and analytical costs are understated because of these missing items ranges from 10-30% at 1.0 um to 40-80% at the 0.35 um generation.

It is fairly straight forward to look at the measurement functions being performed by this investment. Figures 6, 7 and 8 provide a "Pareto" analysis of the tool cost, space cost and the automation cost by functional measurement.

Pareto by % of Capital Costs

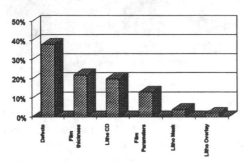

Figure 6

Pareto by % of Facility Costs

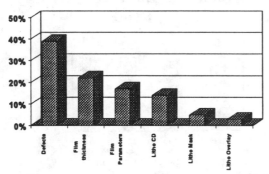

Figure 7

Pareto by % of Automation Costs

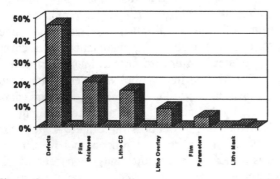

Figure 8

Figure 9 presents the summation "Pareto" by functional measurement. The position of each function within the Pareto is surprisingly similar. Defect measurement is nearly half of all of our

metrology capital costs and is double that of film thickness measurements.

Pareto by % of Total Costs

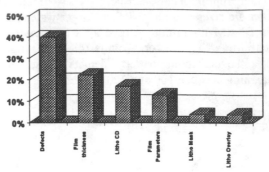

Figure 9

DEFECT MEASUREMENT

If there is one opportunity available for reduction in capital spending, defect measurement clearly stands out. Nearly every process tool in a wafer fabrication facility is monitored for defect or contamination levels. The data is used to qualify or dequalify a process tool for production, initiate or predict preventive maintenance events, and to support yield improvement activities within the factory. The measurements are taken both on test wafers and product wafers, using a variety of tools over the process flow. Efforts to date to use on-line defect monitoring, e.g., scattering of a laser beam in a vacuum chamber, have had a mixed success in decreasing our investment in defect metrology. Admittedly, the preventive maintenance of some vacuum tools has been optimized by using such real time monitors, but the success is not generally applicable. This lack of success is driven by poor correlation between scattered light signals and contamination on production wafers with most process tools.

There may be other opportunities, all of them smaller, to decrease the capital cost associated with measurements of thickness and critical features with lithographic steps. The technology in these two areas changes at a more rapid rate than that required for defect inspection. Such change makes it more difficult to envision a large opportunity for significant improvement.

CONCLUSION

One proposal that needs more than simple consideration is to eliminate the need for inspection. If process tools were capable of delivering and maintaining the technical capability by design rather than by nearly continuous inspection, we might see a reversal in this escalation of metrology capital cost. Defect and contamination control is a candidate for such design practices. Modeling may be one approach. Such modeling might lead to the elimination or significant reduction of the need for sophisticated and expensive measurement.

ACKNOWLEDGMENT

This paper could not have been completed without the support of Bruce Auches, Genaro Mempin, and John Morrissey.

REFERENCES

1. VLSI
2. Data in figures 2 through 9 are from internal Intel Corp databases.

Flexible Manufacturing/Cluster Tools for Semiconductor Manufacturing

Zachary J. Lemnios

Advanced Research Projects Agency
Microelectronics Technology Office
3701 N. Fairfax Drive
Arlington, VA 22203-1714

Critical elements of conventional semiconductor fabrication technologies are expected to reach their limits in the next several generations of device technology. Extending the present technologies will be very difficult in some areas. In technologies for the 0.25 μm generation and below, thin films, such as gate oxides, near the tunneling limit will be required. Sub-0.25 μm generation processes will require very tight control of multiple parameters, further reducing process latitude. Coupled with the stringent processing requirements is the increasing capital costs and risks to maintain leading-edge manufacturing capabilities. A new approach toward semiconductor fabrication is very attractive for managing the fabrication challenges and reducing overall risks. The approach is built on a foundation of rapid, accurate simulation tools coupled to the design framework and to a new class of flexible manufacturing tools.

INTRODUCTION

Trends over the past decade indicate that advancements in high-performance integrated circuits have been driven by continuous improvements in both design and manufacturing. Industry projections indicate that these trends are expected to continue in a similar manner for the foreseeable future. With each new technology generation, product drivers have placed a premium on greater functionality-per-unit area, lower cost and higher performance. Maximizing productivity will be one of the keys to meeting these requirements.

As device and process technology continue to advance [1], requirements on equipment and methodologies are becoming ever more stringent, greatly limiting the process latitude. These requirements have driven the capital costs and complexity of new fabrication facilities. New technology and business approaches are needed to address these issues. One approach to meeting these new requirements is to extend the flexibility of conventional manufacturing technologies with the goal of enabling rapid development, process-centering and deployment of leading-edge semiconductor processes which have scalability in volume over several technology generations. This flexible manufacturing approach will couple the conventional design environment with a new programmable set of process tools for increased productivity.

Device technology and manufacturing approaches for the sub-0.25 μm technology generation must address and solve an emerging class of very difficult challenges. These include the patterning of feature sizes (0.25 μm) near the expected exposure wavelength (248 nm), the growth of uniform, high quality gate oxide films near the tunneling limit (30 Å), the formation of low resistance, shallow junctions (<50 nm), and the tight control of all process parameters over very large area wafers (300 mm diameter). The development of new source materials for novel dielectrics and Chemical Vapor Deposition (CVD) metallization will be a key enabler for many of these physical processes. A new class of sophisticated Technology Computer Aided Design (T-CAD) tools that couple the design and manufacturing environments to maximize yields and minimize cycle times will be necessary to meet these challenges. New flexible manufacturing tools that have embedded intelligence and real time control strategies and schedulers will also likely be required.

The Advanced Research Projects Agency (ARPA) has been working with industry, academia and national laboratories to solve the critical technology issues that will eliminate major technology obstructions to sub-0.25 µm fabrication. This paper discusses the technology approaches that ARPA is pursuing in order to develop advanced semiconductor manufacturing capabilities. These approaches include development of a flexible, agile tool set and integrated processing.

PROCESS INNOVATIONS

The costs of new fabrication facilities have increased tremendously during the past decade (figure 1). Most new facilities are "megafabs," and are optimized for high volume production of a small number of products using a limited set of process flows. Typically, the process development time for a leading-edge technology can take as long as three to five years. The first fabrication of a leading edge technology will occur in a pilot line, where yield can be improved prior to volume manufacturing. Product line yields are almost always very low at the beginning of a manufacturing cycle, and are gradually improved through experimentation and incremental adjustments to the process over time. The potential value of the product is so large and the risks are so great that usually the most conservative approaches are followed for yield improvement, severely limiting process innovation. Those procedures lead to long development times and are also very costly. ARPA is interested in developing capabilities to develop and improve processes through modeling and simulation.

Extrapolations of the manufacturing complexities that may limit productivity in the next decade include, 350 million interconnected transistors, 0.1 µm features, 700 cm^2 wafers, up to 7 layers of metal, perhaps 500 process steps, and just dollars per square centimeter in allowable manufacturing costs. The present challenges in developing highly uniform and controllable

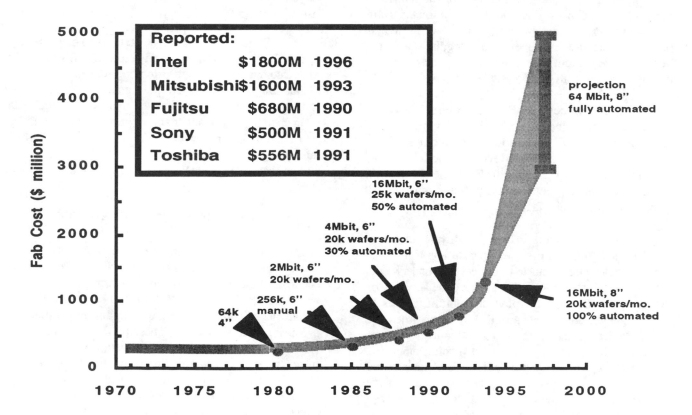

Figure 1 : Cost of New Fabrication Facilities

bulk processes that can produce these transistors consistently are tremendous. Coupled to this are the increasing challenges on the back-end, depositing conformal layers to cover dense topologies, filling high aspect ratio features, and depositing and etching new metallizations. In meeting these challenges to date, processes have become much more complex and difficult to manage. Now, fundamental physical limits are adding to the already daunting challenges.

The length of the anticipated exposure wavelength (248 nm) and minimum size feature for the 0.25 μm generation are about equal. An exposure source of 193 nm is the likely candidate for 0.18 μm technology. Lithography has almost always been an area with tremendous technology challenges, and now will have the additional complications caused by diffraction and scattering. There are challenges in developing new photoresists that are sensitive to those radiation wavelengths. Also, the required gate dielectric thickness for next generation technologies is approaching the SiO_2 tunneling limit, about 3 nm. Some means of maintaining near total isolation between the gate and channel are required. There are also tremendous challenges in other front-end areas, such as shallow junction fabrication and close control of the threshold voltage. In the back end, product requirements are driving the interconnect system to 7 layers of metal and interlevel dielectric - a challenge to lithography systems, deposition and etch, material adherence, as well as packaging and overall component design.

These process challenges require innovative solutions which include novel equipment approaches to new processes and integration technologies. The opportunity now exists to leverage modeling and simulation tools for integration into the fabrication process in order to better manage the complexity of manufacturing in future factories.

ARPA is supporting the development of sophisticated codes and frameworks for applications to semiconductor manufacturing. Projects are underway to mature T-CAD technology and improve the accuracy of the compact models and physical simplifications to provide greater predictive capabilities. Approaches that couple existing point tools for advanced process modeling and virtual lot splits are also being supported. The ability to run large numbers of computer-based process experiments and use computer tools to analyze and correlate parameters is a potentially very powerful tool for process and product developers. This approach will decrease process development cycle times by

removing the need for physical experiments in every case. Once determined, unit processes could be stored in database libraries and reused in other flows. In this approach, transferring a process would be as trivial as transferring a computer file.

These advanced simulation codes may be used to design new process chambers and reactors. Recently reactor suppliers have begun to look seriously at modeling and simulation for new chamber designs. Quite sophisticated codes developed for computational fluid dynamics are being adapted for use by equipment designers. The reception so far has been quite positive and as these codes and interfaces are further refined, equipment development cycles and capabilities should begin improving dramatically.

This computer-based equipment modeling will enable users to generate object models of equipment performance. New processes, flows, or operational parameter spaces could be rapidly defined without exercising the equipment through loop experiments or sacrificing products through lot splits. Customers could evaluate the performance or impact of tools using those models integrated into the virtual factory. Equipment settings could be determined and even controlled from this type of framework.

There are substantial advantages to be gained by closely linking design and fabrication. This is probably best accomplished through extending these T-CAD tools to link design and manufacturing frameworks. Among the challenges here are to manage the real experimental data in an unambiguous manner and to import it automatically into the T-CAD framework. As a first step, however, integrating existing point tools and filling in the gaps will be needed to enable the virtual semiconductor factory.

EQUIPMENT INNOVATIONS

The ARPA vision of the operation of a future semiconductor facility is shown in figure 2. To meet the demands of the process and maintain throughput, a factory with flexible, programmable tools will be integrated with the advanced simulation factory running in parallel and producing a virtual product. These two environments will be in close communication, constantly exchanging information by comparing real and virtual metrology results and feeding back model corrections or process modifications. The fab of the future is likely to support integrated processing technologies, "intelligent" process control methodologies, and flexible work cell configurations. These tools and configurations will be

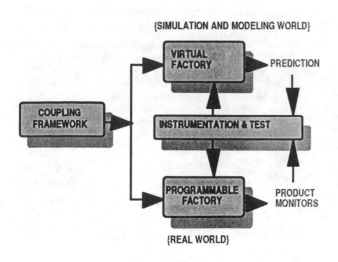

Figure 2 : The Future Semiconductor Facility

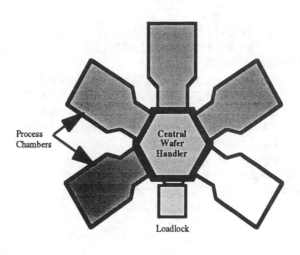

Figure 3 : Cluster-Based Manufacturing Tool

necessary to provide uniform parametric control over large wafer areas and to support reductions in process complexity through innovation. The concept of integrated processing is already being supported by cluster tools, available from a few equipment suppliers.

The core manufacturing tool set will be cluster based as shown in figure 3. This configuration consists of a central wafer handler and a set of process modules, along with a loadlock to accept wafers. Within the cluster tool, a vacuum or low pressure inert ambient is maintained to reduce particulate contamination and unwanted surface reactions. Compared to stand-alone single wafer process tools, cluster tools offer the potential to increase manufacturing throughput via parallel processing since integrated processing with a cluster platform allows substantial reduction in tool set up times. The cluster approach also reduces the number of process cleans. As an example, a typical polycide process that consists of two depositions, poly-silicon followed by WSi, often requires five steps in a traditional process flow. These steps include: poly deposition, diffusion doping with PCl3, HF dip, and WSi deposition usually through CVD. For 100 wafers, this sequence may take 24 to 48 hours. In a cluster tool, the process may be done in two steps since poly deposition and doping can be performed in one chamber. No HF cleans are necessary and WSi is done in vacuum. By comparison, a cluster environment would only require 5 hours to process the same 100 wafers.

Presently, cluster tools are used mainly in metallization processes, where elemental metals tend to be highly reactive. Integrated processing will have the greatest impact when all processes are similar (such as CVD) and require similar processing times within a chamber. The efficiency of the tool and its associated cost-of-ownership begin to degrade quickly if any module's process times are disparate from the others or if extensive gas purging cycles are required due to the process chemistry. Among the processes that are most applicable for cluster tool integration are vapor-phase pre-cleans, metallizations, fabrication of gate stacks with multiple chemistries, silicide, and sidewall spacer processes.

Cluster tool development is not moving forward without issue. At the heart are standards for the interface between the unit chambers and the central handler. Some cluster tool vendors are aggressively pursuing proprietary interfaces while others have opted to promote a standard. MESC is a standard interface defined by SEMI to allow interchange of chambers by various vendors. This approach will allow, in principle, the best of breed modules for any particular technology to be integrated on a single cluster. Although some suppliers are aggressively pursuing development of advanced process modules using MESC, it has not been universally accepted and its longterm acceptance is not yet clear. In addition, the semiconductor industry has not yet totally embraced the standard and the best-of-breed purchasing approach, preferring single vendors for integrated systems. As of early FY1995, no Japanese equipment suppliers were offering MESC interfaces on cluster platforms.

DESIGN-COUPLED MANUFACTURING

As previously described, integrated processing can be supported by a flexible tool set that can also be coupled with a powerful modeling framework and an advanced CIM system to allow substantial improvements in process development. But a factory of the future will not only require a discrete set of stand-alone world class tools, it will also require a technology to integrate these tools seamlessly and efficiently to maximize productivity and decrease development cycle times.

This approach to manufacturing, will tightly integrate a new set of programmable factory tools with a set of embedded manufacturing models and computer aided design tools. The resulting architecture will provide the capability to concurrently design a circuit along with its associated process, significantly improving the yield learning curve for new products and technologies. The ability to make appropriate tradeoffs in performance, reliability, manufacturability, life cycle cost and cycle time will be possible with the availability of a process synthesis framework.

In support of this vision, ARPA and the semiconductor industry have embarked upon a comprehensive program to develop a set of design tools which will provide a "first principle" understanding of semiconductor processes and equipment performance. ARPA and SEMATECH are also working with equipment and materials suppliers to develop a tool set that has these advanced capabilities and extendabilities to begin making use of predictive tools in both process and equipment designs, as described earlier. In addition, sensor technology is beginning to mature which will enable in-situ control to improve the uniformity and control of manufacturing equipment. The coupling of these frameworks and capabilities will enable true process innovations to occur.

The design-coupled manufacturing concept is shown in figure 4. Seamless, efficient information exchange between powerful frameworks and the manufacturing CIM system will enable this new paradigm. Specific IC designs may be compiled from reusable libraries, and the geometric layout information then passed to a device engineering framework that checks design rules against physics-based constraints, simulates performance and functionality, and also provides guidance on manufacturability and reliability. These simulations may then be used to update the libraries and also be imported to technology development

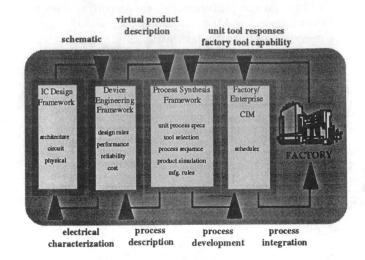

Figure 4 : The Design-Coupled Manufacturing Concept

tools, where processes and recipes are determined. The unit process recipes may then be stored in libraries and reused by compilation into new flows. The CIM system integrates and coordinates all activities associated with the factory and ensures each wafer lot is processed in an optimal manner with absolutely zero mistakes in recipes. As shown in the concept of figure 4, the factory in turn provides a reduced, unambiguous data stream back to the technology, process, and device development frameworks to update models and assist in parameter control.

Coupling advanced simulation with manufacturing tools is one of the key technologies here. A complete capability to accurately and rapidly model the performance, parametric sensitivity, functionality, reliability, and manufacturability of an electronic component is critical. A powerful design environment such as this might support all phases of electronic design and production in a hierarchical fashion, from manufacturing tool design, to unit process, to process integration, to factory modeling, to product performance modeling. In this model, transferring a process will be as simple as transferring a data file. Sharing data between distributed nodes (or factories) in an enterprise is also easily accomplished.

This is a very powerful environment for future manufacturing. It is well suited to meeting the needs of both commercial and military markets and represents a key paradigm shift. Flexible factories will leverage this concept first, although there are still significant payoffs for traditional manufacturing. The virtual prototyping will allow low risk process modifications and rapid yield improvements. Also, this concept will allow optimization of the manufacturing cycle time necessary to produce a high mix of products on a single manufacturing line. Integration of design and manufacture will also enable trade-off assessments without necessarily having to fabricate structures first, ameliorating a lengthy and expensive component in process design. This ability will serve to minimize the risks in construction of new fabrication facilities and will allow much higher equipment utilizations and longer factory lifetimes.

CONCLUSION

As each new generation of semiconductors is developed, the associated device and process technology will increase in complexity. New approaches are necessary to manage the complexities and to decrease manufacturing costs, while maintaining productivity. While modeling and simulation tools are advancing and gaining acceptance, integrated frameworks will increase their utility and power. Advanced sensor technology and integration with control systems is improving, although high costs are still a barrier to general acceptance. Integrated processing and cluster tools may provide a critical enabling technology for future technology generations, allowing repeatable, uniform thin films and fine features over large wafer areas. It also allows elimination of some non-value added steps like pre-cleans, although not all processes are yet integratable. It appears, however, that the approach is attractive for future technology generations and in fact may be the only means to fabricate some structures, such as quantum wells. Finally, coupling design and manufacturing will be necessary for future technology generations to manage process and design complexity while meeting all constraints (rapid integration, high yield, rapid cycle time, etc.) as process latitudes decrease.

REFERENCES

1. The National Technology Roadmap for Semiconductors,
Semiconductor Industry Association (1994).
4300 Stevens Creek Blvd, Suite 271, San Jose, CA.
November, 1994

In-Situ Sensors for Monitoring & Control of Process & Product Parameters

Jimmy W. Hosch

SEMATECH
2706 Montopolis Drive
Austin, Texas 78741-6499

The availability of appropriate sensors will limit the implementation of closed loop process control in the semiconductor industry. A brief explanation is given of how semiconductor process monitoring and control is done today. The methods used for machine control are described.

Fault Detection, Classification & Advanced Process Control are defined. SEMATECH's vision of the industry's evolution from SPC to APC is given with an example of an APC system under development.

The role of the four types of sensors that will be required to accomplish APC is explained with examples of each type sensor. The criteria for determining the need for real time sensing vs. measurements made before or after the process is given. Operational definitions that highlight the similarities and differences between metrology tools, diagnostic sensors and process control sensors is presented with examples of each.

The semiconductor industry appears to be following an evolutionary approach to adding sensors to process equipment. The benefits and risks of this approach are discussed.

Business issues have a great impact on the timely delivery of the sensors needed for APC. A conceptual model for compressing the development time of process control sensors is given. SEMATECH activities to facilitate the development of the sensors needed for process control are described.

CURRENT PROCESS CONTROL PRACTICE

Semiconductor manufacturing is a piece part manufacturing operation that uses batch process techniques in all operations. The batch size varies from one piece in single wafer processing to one hundred or more in furnace operations. This observation is useful when looking to other industries for solutions to process control problems.

Process control in the semiconductor industry is based on statistical process control (SPC) techniques. Simple SPC charts are maintained for the critical outputs of the process. Since semiconductor manufacturing tools, today, are controlled with single-input-single-output (SISO) controllers, the job of the process engineer is to develop a recipe of fixed values for each of the tool's SISO controller inputs that will give optimum SPC performance for the process.

The goal of SPC is to find the causes for process variance and eliminate them. In this endeavor, process engineers are beginning to develop process models such as a Response Surface Models (RSM) by using Design of Experiment (DOE) techniques. Once the recipe of setpoints has been shown to produce an acceptably wide "process window," frequently, the recipe becomes the specification for the process. If the tool drifts over time, resulting in violation of an SPC rule, the processor is taken off line and "repaired" until it will again produce acceptable product with the specified recipe of setpoints.

Today, the ability to control a process is often limited by the precision with which the product feature measurements can be made. SEMATECH has established a standard method for evaluating measurement tools[1] and methods for judging if the measurement tool is adequate for controlling the process using SPC methods.[2] The guideline calls for $0.1 > P/T.$, where P is the 6σ precision of the measurement tool, and T is the tolerance specification of the product. Look, for example, at top-down SEM measurements of resist linewidths. The 6σ precision of a state-of-the-art SEM is +/-12 nm. For 0.35 μm line widths, the tolerance is +/-35 nm.

$$P/T = 12 \text{ nm}/35 \text{ nm} = 0.34$$

Historically, 0.3 is about as large as the P/T ratio can be to achieve profitable yields. This measurement precision issue is of critical importance, independent of the technique for controlling the process. A process cannot be controlled better than the ability to measure the product feature of interest.

Product feature measurements are made from a sampling of the product wafers using off-line stand-alone metrology tools. If the feature cannot be measured on the product wafer, a special "monitor" wafer is typically run through the processor periodically, measured off-line and the results plotted on SPC charts to access the condition of the process. All of these techniques run the risk of unknowingly building defective product between measurement samples. Because measurements for SPC tracking of process performance cost money and are considered to add no value to the product, reduction of measurement frequency is aggressively applied as the process stabilizes with maturity. So the risk of not knowing the current state of the process and building bad product is weighed against the cost of making measurements.

Because these SPC measurements are not automated, the various work routines of the operators produce a wide variation of lag time between production and measurement of the product. This variable lag time must be considered when interpreting archived SPC measurement data. Automation has been applied to the collection of SPC data by the Computer Integrated Manufacturing (CIM) system installed in most wafer fabs. The CIM system is also typically responsible for down loading the specified recipe into the process tool to reduce human error.

With increasing pressure to use a single tool to process a wider range of different products, "set-up" problems have grown in importance. Frequently at change-over to a new product, the first wafer in a new lot is sent through the process and measured to insure that the process settings are optimum before processing the rest of the lot. This technique is used frequently in lithography on immature processes, preventing tool utilization while the send-ahead wafer is being processed and measured.

The bottom line is that many technologists are concluding that SPC techniques will be incapable of maintaining the process tolerances required for the 0.25 μm and smaller devices. The costs of scrapping just one 200 mm diameter wafer increases the industry's interest in continuously knowing the production tool is on target. Systems using multiple sensors with multivariate correlation techniques to monitor the process are more likely to immediately detect malfunctions. Sensor and model based closed loop feedback / feedforward control systems are more likely to be successful in "controlling to target."

FAULT DETECTION, CLASSIFICATION, AND ADVANCED PROCESS CONTROL (FDC & APC)

In this paper, FDC & APC describes three distinct but related functions. Fault detection is concerned with automatically detecting a fault to reduce the number of wafers misprocessed or prevent misprocessing any wafers. Once a fault is detected, the automated classification should identify the cause of the fault to reduce the time to diagnose the problem. In this context, "faults" include **anything** that can cause misprocessing of a wafer - this includes hardware failures as well as process drifts.

FDC will utilize multivariate analysis techniques to analyze the available sensor data from a process tool (sensor fusion). Part of the J-88-E Project is aimed at evaluating real-time SPC,[3] Chemometrics (partial least squares - PLS),[4] and neural networks[5] for fault detection.

[1] Gage Study. SEMATECH Technology Transfer Document 91090709A-ENG (1991), SEMATECH 2706 Montopolis Drive, Austin, TX 78741.

[2] K. Horrell, Introduction to Measurement Capacity Analysis, SEMATECH Technology Transfer Document 91090709A-ENG (1991), SEMATECH 2706 Montopolis Drive, Austin, TX 78741.

[3] Costas Spanos, UC Berkeley.
[4] Harold Anderson, University of New Mexico.

Multiple sensor data will present a different "signature" or "fingerprint" reflecting the condition of the tool and process. The goal of fault detection is to recognize the difference between the sensor signatures for normal vs. abnormal tool/process conditions. The goal of fault classification is to identify the seriousness of the fault and the cause of the fault from the sensor signature.

Several problems must be solved to implement FDC. A significant task will be developing techniques robust to type one and type two errors. Fault detection schemes, today have great difficulty dealing with the wide range of process settings for different products to be run on the same tool. Methods are needed to account for the accumulation of non-lethal faults and the aging effects of the tool between maintenance. Limits must be established beyond which further accumulation of non-lethal faults will trigger a lethal fault shut-down of the process.

Faults must be divided into two classes: "lethal" faults are serious failures that justify immediate termination of operation and repair of the equipment. Under no circumstances should the APC attempt to compensate for a lethal fault. "Non-lethal" faults can be compensated for using the APC. These faults can have either identifiable or unknown causes. Within limits set by the process control engineer, the APC will attempt to use the process model to determine the appropriate corrective action to re-center the process.

APC is a multi-in-multi-out (MIMO) model based closed loop control system that relies on sensor and metrology inputs. The MIMO process model correlates the process inputs to the results produced on the product features of interest. Reverse solving the process model produces the "new" combination of setpoints appropriate for returning the process to the center of its process window. This reverse solving of the process model determines what to adjust and by how much. The final decision to change the process setpoints, however, is tempered by a "control strategy" that automatically applies judgment criteria determined ahead of time by the process control engineer. Many different control strategies have been developed over the years. Standard control strategies can be tailored by the engineer or a new control strategy can be developed for unique circumstances to meet goals of noise rejection vs. responsiveness to drift without over correction.

WHERE FDC & APC "FITS" IN THE FACTORY

FDC & APC will be applied to single tools initially. In addition, the early implementations will likely be executed on "piggy-back" controllers that are attached to the process tool via. a communication link, such as SECS-II, built into the production tool by the manufacturer. The current limitations of controllers on legacy process tools precludes running new applications such as controller code capable of supporting FDC & APC techniques. The alternative to a piggyback control implementation is to rip out the factory installed control system and install a capable controller[6] - an unattractive option in most wafer fabs.

A piggyback controller will fit between the tool and the factory CIM system. Stand-alone metrology tool measurements will likely be used in control schemes for a long time to come. The FDC & APC controller will need access not only to new sensors added to the process tool, but also need access to metrology tools used to measure product features on wafers processed in the production tool being controlled. The Sensor Bus[7] developed by SEMATECH would be an attractive way to provide ready access to this type data.

FDC & APC should be able to advise the CIM system of current process tool capability. This type data would be useful for scheduling maintenance on a needs basis instead of the time based maintenance used today. In addition, the process capability information would be available for process synthesis modeling activities.[8]

---

[5] Don Sofge, NeuroDyne, Inc.

[6] Commercial controllers capable of supporting FDC & APC are available. Four controllers used in SEMATECH projects are 1) ProcessWORKS to be provided in the future by Texas Instruments, contact: Carl Fiorletta (214) 575-3217; 2) Pro-H developed with SEMATECH funding by Honeywell, contact: James Orrock (612) 951-7518, 3) Integrated Systems, Inc., contact: Steve Spain (408)980-1500 X204; 4) Adaptive Tool Platform (ATP) from IBM, contact: Connie Araps (407) 443-9470.
[7] SEMATECH Sensor Bus Project S-102 has proposed a sensor bus to SEMI Standards Generic Sensor Bus Task Force, which is driving standardization proceedings. SEMATECH contact: Dale Blackwell (512) 356-3598.
[8] Get a reference from Harold Hosack or PK

Evolution of Control

Piggy-back controllers connected to legacy tools through the tool's SECS-II communication port will be restricted to run-to-run control. The limitation comes from the fact that today's process tool controllers are not capable of executing a mid-course correction. Much like the intercontinental ballistic missile, one must line them up very carefully. Once launched, only two courses of action are available: press the abort button, or go down-range and see where it lands.

Fortunately, run-to-run control will produce significant improvements for a large majority of processors in the near term. The fact is that these processors are now stable enough to run open loop for extended periods without drifting so far out of control as to break SPC limits. Run-to-run control will add the ability to make adjustments to the machine setpoints based on sensor inputs detecting current performance. Sensor inputs used with process models and control strategies, will enable FDC & APC to add value to the product by keeping the process centered in its process window for longer periods of time.

For unstable processes, such as RTP and some furnace processes, real time control is required simply because of the fast drift rate or complex process trajectory required for the process to work at all. In these cases, controllers capable of running real time model based control strategies are already being installed. Of course, real time control requires real time sensor inputs. Whereas, run-to-run control can be sustained with a longer lag time between processing time and product feature measurement time.

Real time control has other less obvious benefits, however. It is well known to process control engineers that any complex physical system equipped with independent SISO controllers is subject to oscillatory behavior. The SISO controllers can be affected by each other's operation through the hysteresis of the physical system. This oscillatory behavior can be seen in vacuum processors equipped with several mass flow controllers and a pressure controller - each having its own SISO controller. Under certain conditions of gas flow and pressure, the SISO controllers will begin to oscillate, or "fight" each other causing the pressure to oscillate. Real time control can eliminate this kind of behavior because the MIMO control model accounts for the hysteresis induced interactions of physically separated actuators and controls. Thus, moving to real time control will add

benefit when run-to-run "supervisory" control of low level SISO controllers is no longer good enough.

SEMATECH is involved with one full blown FDC & APC project on the Lam Research 9600 Aluminum Metal Etcher. Code named J-88-E, this project is being performed by the SPDC laboratory at Texas Instruments. The project is laid out in five steps. First, install three additional sensors, a particle detector, RF sensors, and a spatially resolved optical emission sensor. Second, show how to collect sensor data from the Lam tool and from the add-on sensors to provide data for modeling. Third, Develop FDC capabilities. Fourth, develop APC capabilities. And fifth, demonstrate over-all performance. Detailed reports will document the methodologies developed. Both neural networks and real-time SPC are being tested for FDC. Both neural network and response surface modeling (RSM) is being compared for process modeling to support APC. This project is completing phase two and is scheduled for completion in 1996.

Type of Sensors Needed for FDC & APC

Four types of sensor measurements are required to perform FDC & APC: equipment state, process state, wafer state, and environmental state.

Equipment State Sensors

Equipment state sensors provide a quantitative indication of the current integrity and performance of the process tool hardware. These sensors are in-situ and usually real time.

In general, equipment sensors are those that indicate the "position" of the actuators under SISO control. The sensors that feedback information to the SISO controller on the status of the controlled parameter are of less value for monitoring and predicting the performance of the equipment. This fact is a direct result of the SISO controller doing its job of maintaining the controlled parameter equal to the setpoint.

The example of a vacuum chamber pressure control as shown in Figure 1 provides a simple example of the relative value of sensors for detecting equipment state. Chamber sensor readings over the same time period are shown in Figure 2 with upper and lower limit alarm guard-bands for each chart. The scenario for the time period displayed by the charts is the mass flow controller goes out of calibration, injecting more gas than expected into the chamber, causing the chamber pressure to rise.

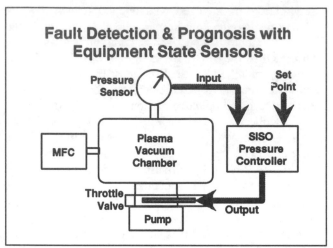

Figure 1.

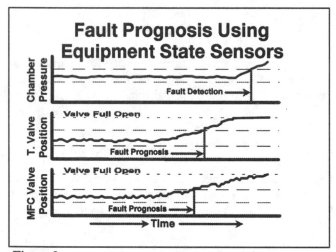

Figure 2.

As can be seen on the pressure chart, the pressure sensor eventually detects the over pressure condition - **detection** of a fault. But the plot of the throttle valve position over time shows evidence of the problem much earlier - **prognosis** of a fault. Only after including the plot of the MFC valve position, can one clearly identify that the cause of the problem is the MFC and not the pump - **classification** of the fault.

The only impractical idea in this example is the assumption that the MFC reads out its valve position which is critical for tracking its performance. Because the MFC manufacturers still do not provide this important piece of information. other sensor correlation's will have to be found to classify the fault.

Examples of other effective equipment state sensors include RF tuning vane position on plasma tools - in preference to the reflected RF power reading; drive motor current on CMP tools - in preference to the RPM reading; shutter time on lithography expose tools - in preference to the integrated expose "dose;" heater current on a bake plate - in preference to the thermocouple temperature.

Other equipment state information can be obtained by in-situ calibrations and tracking the calibration factors over time. For example, periodically performing an in-situ "rate-of-pressure rise" calibration for MFCs and tracking the calibration factors over time provide a check on the operating condition of the MFC. Time drifts of the calibration table values can prognosticate MFC failure.

The selection of equipment state sensors generally meet the current needs. However, robust techniques to convert equipment state sensor data into meaningful FDC information are lacking.

Process State Sensors

Process state sensors provide a quantitative indication of the integrity and performance of the process chemistry that is producing the product feature on the wafer. These sensors are in-situ and real time. The word "indication" is deliberately used to emphasize that quantitative measurements of the chemical species at the surface chemistry site is **not required**. What is required is a sensor signal that is closely **correlated** to the surface chemical environment that is acting on the wafer surface.

A familiar process state sensor is the optical emission spectrometer (OES) used extensively for process endpoint detection for plasma etch tools. These monochromators spectrally isolate a single wavelength of light emitted by an atomic or molecular species that changes concentration during endpoint. Although monitoring a single wavelength is useful, a broader view of the plasma chemistry can be obtained by monitoring emitted light with a spatially resolved OES[9] as shown in Figure 3. This kind of sensor system adds two more dimensions to the sensor capability - spatial discrimination and wavelength resolution between 250 nm and 800 nm. This is the kind of multidimensional sensor observations that will provide the information needed for FDC &

---

[9] Developed by Harold Anderson at the Univ. of New Mexico, in cooperation with Chromex, Inc., contact: John Algio (505) 344-6270.

APC. Multivariate techniques, like PLS, deal with the 3,072 pixels of spectral data from the three locations in the plasma exactly like that many independent sensors. The technique has been shown to correlate to process input parameters as well as etch rate and uniformity.[10]

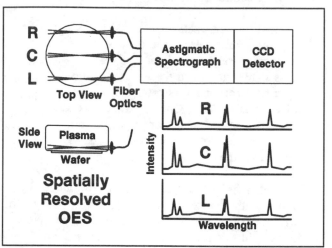

Figure 3.

Another example of an information rich process state sensor is the RF sensor systems available commercially[11] capable of monitoring voltage, current and phase for the first five harmonics in the external RF circuitry of RF plasma tools. These kind of sensors have been correlated to etch endpoints as well as the film stress in PECVD tools. It is not always obvious what empirical correlation that a given sensor will demonstrate.

Wafer State Sensors

Wafer state sensors provide a quantitative indication of what the process does to the wafer. Wafer state is the critical feed forward/feedback information for closed loop control. Without wafer state information there is no information to feed back or feed forward. Again the word "indicator" is used to emphasize that **accurate** measurements of the product feature on the wafer are not needed, since the sensor is **correlated** to the wafer product feature of interest. However, as with all process control sensors, sensitivity, and reproducibility are required and will set the upper limit of achievable process control.

[10] G. G. Barna, B. Wangmanerat, Application of Chemometrics to Plasma Etching Diagnostics, Texas Instruments Technical Journal, vol. 9, no. 6, 1992.
[11] Fourth State Technology, contact: Terry Turner (512) 343-9507.

Where in-situ sensors are lacking and the process is relatively stable, stand-alone metrology tools can be used to acquire the critical wafer state information needed to implement APC. The longer measurement lag time of metrology tools will have adverse effects on the ability of APC to control the process.

Two commercial sensors have recently come on the market that are good examples of wafer state sensors that deserve to be used in FDC & APC projects. The Whole Wafer Interferometer[12] has been demonstrated capable of near real-time measurements of etch rate and uniformity. A small spot scanning ellipsometer[13] will soon be available to measure very thin dielectric films in the load-lock of deposition tools. The point to emphasize is that these sensors provide the first opportunity to implement APC using in-situ wafer state sensors for etch and dielectric deposition on production wafers.

A scatterometer sensor[14] under development for measuring resist and etched linewidths demonstrates a calibration scheme based on physical constants and the ability to accurately model from first principles the physical line width structure of interest. The technique performs a reflectometry measurement from a grating test structure shown in Figure 4. The technique collects about 40 separate measurements over a range of incident angles between 2 and 42 degrees. Plotting the relative reflectivity as a function of incident angle provides a "signature" shown in Figure 5. A first principles model of the grating structure is used to generate a library of signatures for gratings varying only slightly in line width over the expected range of the measurements to be made. By using simple curve fitting techniques, the best fit is determined between the signature of the unknown sample grating and one of the reference signatures.

[12] T. Dalton, H. H. Sawin, Full Wafer Interferometry: In-situ Process Monitoring of Uniformity, End Point, and RIE Lag for Plasma Process Development and Control, SRC Technology Transfer Course Report (May, 1994) Contract Numbers 91-MC-503 and 93-MC-503. Commercial contact: Low Entropy Systems, William Conner (617) 776-5381.
[13] Nova Measuring Instruments Ltd., Sales representative: Dan Kahan (408) 252-7644.
[14] J. R. McNeil, S. S. Naqvi, Univ. of New Mexico SEMATECH Center of Excellence for Metrology and On-Line Analysis for Semiconductor Manufacturing - Scatterometry Contract Annual Review Oct 12, 1994. Commercial contact: Sandia Systems, Scott Wilson (505)-343-8112.

Measurements using this technique compare very favorably with the high resolution SEM measurements shown in Figure 6. The technique is essentially calibrated against the physical constants of the material that makes up the line and space test target, and Maxwell's Equations used to develop the physical model of the interaction of light with the line and space grating.

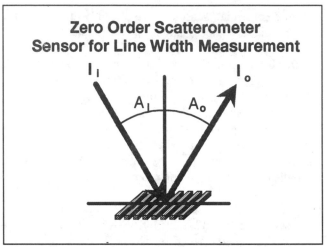

Zero Order Scatterometer Sensor for Line Width Measurement

Figure 4.

Spectral ellipsometry[15] uses a similar approach making a relative measurement and finding a best fit to psi-delta curves generated with a first principles model of the film stack. Self-calibrating sensors are very desirable.

A significant next hurdle for both of these wafer state sensor techniques is to measure product wafer features instead of special test structures.

Virtual Wafer State Sensors

Virtual wafer state sensing is attempted when no acceptable wafer state sensor is available to give a short lag time measure of the wafer state. A MIMO model is constructed to link multidimensional process sensor readings with the wafer state produced by the process. The technique assumes that the responses of the process state sensors in fact correlate to the wafer features of interest. Since these models are often empirically developed, the correlation question can only be answered by performing the experiments and analyzing the data. The SEMATECH J-88-E project is attempting to use the spatially resolved OES and the RF sensors to develop a virtual wafer state sensor to predict etched line width reduction. Prior work has shown that the spatially resolved OES correlates to etch rate and etch uniformity. However, line width reduction is the wafer feature of interest for control of the process.

A virtual wafer state scheme has some obvious drawbacks. First the correlation may not exist or be a very poor correlation. Poor correlation between the process state sensors and the wafer state results in small process state sensor errors producing large errors in the predicted wafer state. One cannot escape the fact that wafer state measurement error limits the ability to control the process. Secondly, this adds one more level of complexity to the control problem.

When and where to measure wafer state is dictated by the stability of the process. Most processes, today, drift slowly over time. Such slow drift can be substantially reduced with control based on measurements made immediately after the process on every wafer. Therefore, in the majority of cases, placing wafer state sensors in the load locks of single wafer processors is the direction the industry should pursue in the near term. The physical access and space constraints in the loadlock is less severe than in the process chamber.

Besides lower measurement lag-time, real time sensors can offer unique capability for process control. For example, the use of a spectral ellipsometer on a two step oxide etch provided the ability to reliably stop a fast anisotropic etch step before reaching endpoint[16]. By then changing to a highly selective slower isotropic etch chemistry, the precise etch characteristics could be quickly achieved.

Environmental Safety & Health Sensors

Environmental, safety and health sensor requirements are increasing with the responsibility of each process for control of its own waste effluent. The sensors for point of origin waste control are just now being defined. It seems reasonable to expect that waste control will be just one more task that FDC & APC will be assigned in the near future. The required sensor detection levels for contaminants will be in the PPM to PPB range.

[15] The spectral ellipsometer developed and used by Texas Instruments in the MMST Program is commercially available from Verity Instruments, Inc., contact: Paul Whelan (214)446-9586.

[16] S. W. Butler, et al., Intelligent Model Based Control System Employing In-situ Ellipsometry, J Vac Sci Tech A, vol. 12, no. 4, July-Aug. 1994.

BUSINESS ISSUES OF PROCESS CONTROL

An Evolutionary Approach to Adding Sensors

An evolutionary approach to adding sensors and implementing FDC & APC techniques is likely. The "go slowly" approach is driven by several issues. First, a full compliment of the needed sensors is not on the shelf today, and will require some time to develop the sensors and build users' confidence. Second, as a group, semiconductor process engineers do not have the process control skill set to immediately utilize sensors with FDC & APC techniques even if the sensors and controllers were installed. It will require time to teach the needed skill set to the people who are daily responsible for process control in production. Third, only now have FDC & APC projects started that will provide hard evidence of the financial gains of implementing FDC & APC. Where results are being produced, the results are not in the public domain. Without some concrete examples of how FDC & APC brings down Cost Of Ownership[17] of both new and legacy tools, decision makers will be reluctant to commit serious money. Fourth, The majority of the installed base of process tools are profitably building 0.5 µm and larger products for their owners. The real demand for a wider process window than SPC can deliver will begin with 0.35 µm and smaller device production. Finally, the architectural issues of implementing FDC & APC in a factory with legacy tools and an existing CIM system are just now beginning to be addressed at SEMATECH[18] and some semiconductor manufacturers.

A real potential for inefficiency exists in taking an evolutionary approach of adding sensors one at a time on a few tools scattered through out the factory to first monitor the process, then do some FDC, and finally attempt APC. Failing, in the beginning, to comprehend the information flow issues and control software requirements to achieve the final goal of fully integrated FDC & APC can cause serious under estimation of the requirements resulting in system re-design and re-build several times along the path to the FDC & APC goal.

---

[17] D. Dance, SEMATECH Cost of Ownership Model, SEMATECH Technology Transfer Document 91020473-B-GEN (1992), SEMATECH, 2706 Montopolis Drive, Austin, TX 78741.

[18] Advanced Equipment Control Framework Project MSDA-001; contact: John Pace.

Process Control Sensor Development

Development of sensors for semiconductor process control is hindered by several conflicting situations in the industry. First, the industry looks at metrology as a non-value-added activity - which is true when those measurements are used only for SPC tracking of the process. This attitude translates into an unwillingness to pay very much for a sensor for every tool. Added to the low sensor cost expectation of the customer is the fact that the total size of the sensor market is relatively small. The result is that it is difficult to recover R&D costs for high tech sensor development. This scenario makes the semiconductor process control market unattractive to the really large high-tech sensor companies. Smaller companies often lack the financial resources to undertake development of a new sensor. Small companies continually face survivability issues, especially in economic recessions.

The most effective solution to these dilemmas will be the successful completion of FDC & APC projects that can truly show substantially lower cost-of-ownership using the SEMATECH CoO Model. Such a success will show that sensor measurements can be value added operations if the information is used for keeping the tool performing with smaller product variation for longer periods of time.

An effective role that NIST could play, is to fund the proposed ATP for Development and Use of Process Control Sensors.[19] Government could obtain large leverage for creating high wage electronic (and other high tech) industry jobs through cost sharing process control sensor R&D costs.

Shorter Sensor Development Time

Compressing sensor development and implementation time can also have a large impact on the implementation of FDC & APC in the semiconductor industry. Today, it takes five years or longer to introduce a new sensor technology into manufacturing after its conception. The idea generator has to convince a semiconductor R&D lab to test a prototype. A wafer fab (potential customer) has to show enough interest that a sensor company will

---

[19] NIST declined to sponsor the ATP for Process Control Sensor Development at its first presentation in Nov. 1994. The proposal will be reviewed for funding in 1Q95.

undertake engineering the sensor for the manufacturing environment. A production tool manufacturer has to agree to install the sensor on a production tool - an operation that usually requires modification to both the tool and the sensor. The industry could shorten the last three steps if they were executed in parallel. SEMATECH does not know how to accomplish this now.

SEMATECH is attempting to improve the communication among the players with the activities of its Sensor Coordinating Council.[20] A NewsGroup[21] has been established to provide a forum for rapid communication between all parties interested in sensors for process control. Currently the Sensor Coordinating Council is performing a survey within SEMATECH to list the sensors needed by the industry. SEMATECH is starting with the latest Texas Instruments sensor needs survey and identifying sensor needs overlooked. The intent is to expand the survey to the SEMATECH member companies.

Conclusion

The implementation of automated fault detection, fault classification and closed loop process control is limited by the availability of the needed sensors. The four kinds of sensors needed must measure equipment state, process state, wafer state and environmental, safety and health parameters. Sensors are most needed to make measurements of wafer features either before or after the processing of the wafer. In some thermal processing cases real-time sensors are needed. The robustness and ease of use required for use in manufacturing, differentiates process control sensors from diagnostic sensors used in fundamental scientific research of the process physics and chemistry. Unlike stand-alone metrology tools required infrequently for high accuracy measurements, the process control sensor is expected to measure every wafer with high sensitivity and reproducibility. Finally, the solution of specific business issues has great potential for accelerating the delivery of the sensors needed to quickly introduce FDC & APC into semiconductor manufacturing.

---

[20]To join, send an e-mail request to Nancy.Paine@SEMATECH.org
[21] To subscribe, send an e-mail request to Robert.Soper@SEMATECH.org. To post an article describing your use of a sensor, need for a sensor, or capabilities of your new sensor, send the article via e-mail to sematech.sensors@pulitizer.eng.sematech.org.

Linewidth Prediction Example

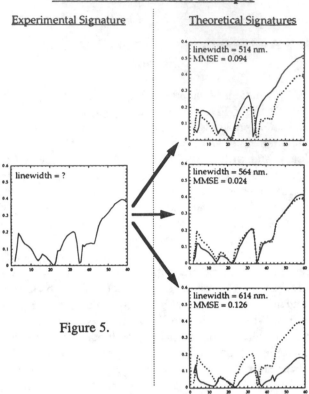

Experimental Signature   Theoretical Signatures

Figure 5.

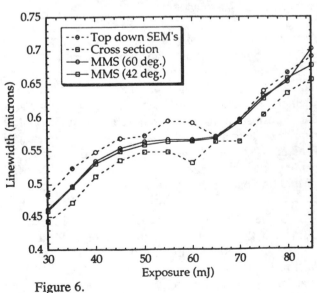

Linewidth vs. Exposure
MMS Prediction Algorithm

Figure 6.

Cost of Ownership Analysis for Metrology Tools

Daren L. Dance

SEMATECH, Inc.

Austin, Texas

Cost of ownership (COO) was developed to address the economic and productive performance of a fabrication tool by estimating the total life-cycle cost of a specific semiconductor process step. But a model for equipment required to support manufacturing such as metrology tools is also needed. Cost of ownership analysis for metrology tools has two parts. First, the cost of operating the metrology tool is estimated. Second, the cost impact of metrology on the processes being measured must be estimated. Comparing the cost impact of metrology on the processes with the metrology COO allows estimating the return on investments in metrology.

COST OF OWNERSHIP: BACKGROUND

Purchase decisions for semiconductor equipment are often based on initial purchase price and installation costs. However, purchase prices do not consider the effect of equipment reliability, calibration and utilization. These factors may have a greater impact on cost of ownership than the costs of purchasing the system. With equipment costs rising every technology generation, semiconductor manufacturers became increasingly sensitive to equipment cost. Thus SEMATECH began developing a COO model for wafer fabrication equipment in 1990. Dean Toombs brought the concept of COO to SEMATECH as an assignee from Intel. Based on his ideas, COO was developed as a sophisticated spread sheet model that could be applied to wafer processing equipment [1].

Use of the SEMATECH COO model has been well accepted by the semiconductor industry. Over 1000 copies were distributed to SEMATECH member and supplier companies. COO standards have been developed with Semiconductor Equipment and Materials International (SEMI) [2] and a commercial COO models is being prepared through a joint development project.

The cost of metrology tools for the semiconductor industry has also been rising. Increased equipment costs are driven by smaller semiconductor device geometries, larger wafer sizes, and the need for increased accuracy and precision. Purchase decisions for metrology systems have been driven more by performance requirements than by purchase and installation costs. However, metrology purchase decisions may not have considered reliability, calibration and utilization. With a few modifications, COO can be applied to metrology systems as well.

COO FOR METROLOGY

The COO model was developed to address the economic and productive performance of a wafer fabrication tool at a specific processing step. COO for metrology equipment is also required since metrology is required in many processes to support and calibrate production tools. Metrology is also required to characterize new and existing products and processes. Thus metrology is integral to manufacturing.

COO analysis for metrology equipment is more complex than for fabrication equipment, requiring a two part analysis. First, the cost of operating the metrology tool is estimated. This step is similar to equipment COO analysis. Second, the cost impact of metrology on the processes being measured must be estimated. The impacts of metrology may include both additional costs and benefits. The two part analysis allows comparing the benefits of metrology with the metrology COO to estimate the return on investment (ROI) in metrology.

STEP 1: COST PER MEASUREMENT

Estimating the cost per measurement is similar to estimating the COO of wafer processing tools or equipment. Equipment COO is defined by SEMI as the "full cost of embedding, operating, and decommissioning in a factory environment a process system needed to accommodate the required volume of product material." [2] For metrology, this definition should include the full cost of embedding, operating, and decommissioning a metrology system in a factory or laboratory environment.

The significant COO inputs for metrology include:

- Equipment cost
- Operating cost
- Yield
- Down time
- Throughput rate
- Value of completed device
- Cost of discarding a good device
- Cost of shipping a bad device

These factors are combined in the COO equation [3]:

(1)

$$COO = \frac{C_F + C_V + C_Y}{TPT \times Y_m \times U}$$

where:

COO = Cost per Measurement
C_F = Fixed Cost
C_V = Variable Cost
C_Y = Cost of Measurement Loss
TPT = Measurement Throughput Rate
Y_m = Measurement Yield
U = Utilization

For metrology, COO may be described in terms of cost per measurement. For a 100% sample, cost per device equals cost per measurement, but for less than 100% samples, the cost per device is some fraction of the cost per measurement.

Fixed costs are incurred once during the life of the system and are associated with the acquisition and installation of equipment. Fixed costs include costs such as equipment purchase, installation, facilities modification, initial training, and calibration costs. Variable costs are incurred on an accrued basis. Variable costs such as material, labor, repair, standards, recalibration, utility and overhead expenses are costs that are incurred during equipment operation. Cost of measurement loss is the value of scrap caused by the measurement methods such as destructive measurements or mismeasurement losses. Process scrap identified by the measurement but caused by previous processing is part of the process tool COO. Thus, yield losses caused by the processing tool must be clearly separated from losses caused by measurement. The sum of costs form the numerator of the COO equation.

The denominator of the COO equation is an estimate of the number of measurements performed during the life of the metrology system. Measurement throughput rate is based on measurement and handling times such as sample preparation, loading and unloading, reporting, and other overhead operation. Measurement throughput rate excludes training, repair, and calibration times since these are included in utilization. Measurement yield may be defined as the ratio of completed measurements providing useful information compared to the total number of measurements performed including repeated measurements:

(2)

$$Y_M = \frac{Total\ Measurements\ -\ Rework}{Total\ Measurements}$$

Utilization is the ratio of actual usage compared to total available time. This shows the impact of non-productive time on cost and normalizes ideal throughput to a realistic estimate. Utilization includes repair and maintenance time, both scheduled and unscheduled; setup and calibration time; and standby time. This should be estimated using SEMI E10 definitions for availability, reliability and maintainability [4]. A simple estimate of utilization is illustrated by the following equation:

(3)

$$U = 1 - \frac{R + C + S}{168}$$

where:

R = Repair and maintenance time
C = Setup and calibration time
S = Standby time

Thus with a few significant details about costs and operation, users can determine the COO in terms of cost per measurement for a metrology system from initial purchase to decommissioning.

STEP 2: IMPACTS ON PROCESS OR PRODUCT

The next step of COO analysis for metrology tools is to estimate the impacts on metrology on the process or product being measured. These impacts may have additional costs and benefits. Some of the impacts of metrology on processes or products include sample sizes, sampling methods, work in progress (WIP) implications, and costs of risk. Dr. Don Phillips, of Texas A&M University, has developed a representation of a generic process step [5]. This representation, which has been modified to include sampling, is shown in Figure 1.

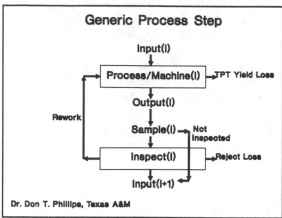

Figure 1: Generic Processing Step

Device input to the *i*th processing step is processed by machine *i*. A throughput yield loss may occur during this processing step. From the output of machine *i*, a sample is selected for inspection step *i*. The three possible results of this inspection are reject, rework or pass. Devices passing inspection and devices not selected for inspection become the input to the *i*th+1 processing step. Devices rejected by inspection are included in the yield lost due to machine *i*. Reworked devices reduce the production capacity of machine *i*. This representation illustrates the relationship between metrology and the processing step.

Since the process and metrology are in series, process throughput depends on metrology methods. Further, since the process requires measurement, there is an impact of measurement on WIP. WIP inventory between process step and inspection is at risk if the process drifts. Several operating methods minimize that risk. Send ahead (or look ahead) samples eliminate WIP but reduce process throughput and utilization. Just-in-time operation minimizes risk with little impact on utilization.

Since product or process yield at subsequent steps depends on the accuracy of metrology, we must consider the costs of discarding a good device and the cost of accepting a bad device. Measurement risk is illustrated in Table I [6]. Minimizing the cost of shipping a bad device is one purpose of metrology. However, if the sampling plan or methods are insufficient, bad devices will be shipped. But if specifications are too restrictive, then good devices may be rejected. Guard banding specifications increases α probability in order to decrease β probability.

Table I
Measurement Risk

True State	Measured Result	Error
Good	Good	None
Bad	Bad	None
Good	Bad	Type I (α)
Bad	Good	Type II (β)

The cost of discarding a good device is estimated by:

$$(4)$$

$$C_\alpha = \alpha \times WIP \times V_p$$

where:
 C_α = Cost of discarding good device
 α = Probability of discarding good device
 V_p = Value of device at metrology

and the cost of shipping a bad device may be estimated by:

$$(5)$$

$$C_\beta = \beta \times WIP \times V_c$$

where:
 C_β = Cost of shipping bad device
 β = Probability of shipping bad device
 V_c = Value of replacement device

The probabilities of discarding a good device and of shipping a bad device are related to the variance of the measurement. These probabilities may be reduced by reducing variance, increasing sample size, or developing more robust processes.

ROI OF METROLOGY

The two part COO for metrology analysis allows comparing the benefits of metrology with the metrologyCOO to estimate the return on investment (ROI) in metrology. Most of the costs of metrology are captured by the basic COO equation. These costs are expressed in terms of cost per measurement. The benefits of metrology are estimated by considering the impact of metrology on a process or product. The knowledge gained by characterizing a process or product lead to the following benefits:

- Reduced cost of shipping bad device
- Reduced cost of rejecting good device
- Improved sample methods
- Improved process throughput
- Reduced impact of mis-measurement on WIP

Return on metrology investment may be described by the following simple equation:

(6)

$$ROI = \frac{Benefit}{Cost}$$

$$ROI = \frac{B_{Product} + B_{Process}}{Metrology\,COO}$$

Knowledge gained through metrology adds value to the process or product through continuous learning and improvement. Thus if characterization information improves the process, then metrology is a value added step.

SUMMARY

Cost of ownership (COO) was developed to address the economic and productive performance of a fabrication tool by estimating the total life-cycle cost of a specific semiconductor process step. But COO for equipment required to support manufacturing such as metrology tools is also needed. With a few modifications, COO can be applied to metrology systems as well. COO analysis for metrology equipment is more complex than for fabrication equipment, requiring a two part analysis. First, the cost of operating the metrology tool is estimated. This step is similar to equipment COO analysis. Second, the cost impact of metrology on the processes being measured must be estimated. The benefits of metrology are estimated by considering the impact of metrology on a process or product. Since characterization information improves the process, then metrology is a value added step.

REFERENCES

1. R. L. LaFrance and S. B. Westrate, "Cost of Ownership: The Suppliers View," *Solid State Technology*, July 1993, pp 33 - 37.

2. "E33: Cost of Ownership for Semiconductor Manufacturing Equipment Metrics," *1995 Book of SEMI Standards*, in publication.

3. D. L. Dance and D. W. Jimenez, "Applications of Cost of Ownership," *Semiconductor International*, Sep. 1994, pp 6-7.

4. "E10: Guideline for Definition and Measurement of Equipment Reliability, Availability, and Maintainability," *1994 Book of SEMI Standards*, Vol 1.

5. D. T. Phillips, "TASK III: Cost Modeling -- An Algorithmic Structure for Lot/Device Cost," in *Texas SEMATECH Center of Excellence Manufacturing Systems Research 1991 - 1992 Research Review*, Presented at SRC Manufacturing Systems SCOE Review, September 2, 1992, Melbourne Florida.

6. D. L. Dance, "Modeling Test Cost of Ownership," *3rd Workshop: Economics of Design, Test, and Manufacturing*, May 16, 1994.

SILICON WAFERS, GATE DIELECTRICS, AND PROCESS SIMULATION

Silicon Materials and Metrology: Critical Concepts for Optimal IC Performance in the Gigabit Era

Howard R. Huff and Randal K. Goodall

SEMATECH
2706 Montopolis Drive
Austin, TX 78741

The technical significance and relevance of several silicon material trends in the *National Technology Roadmap for Semiconductors* (NTRS) will be critically examined in the context of a recent analyses of 600 epitaxial wafers from six 200 mm silicon suppliers. Several directions and opportunities for silicon-based research to ensure compliance with the NTRS will be discussed, including opportunities for: (a) silicon materials, (b) specifications, metrology and standards, and (c) cost-of-ownership (CoO) considerations, includng alternatives to epitaxial structures. Model-based improvements in both silicon materials and metrology instrumentation will be illustrated. In particular, development of a gettering/microcontamination process simulator, applicable for arbitrary thermal budgets, process sequences and device structures, including SOI, is required to more realistically determine the starting wafer requirements.

INTRODUCTION

The 1992 *National Technology Roadmap for Semiconductors* (NTRS) report noted, "semiconductor technology offers the potential for rapid, continued progress in electronics performance and functionality. The advances [in semiconductor technology] are the driving force for the information age" (1). Indeed, electronics has become the world's largest industry with 1993 global revenue of $820B. This is expected to reach $1.5T by the end of the century. Semiconductor materials comprise the smallest component of the electronics-based revenue stream, accounting for about $4B in 1993 and projected to reach about $10B by the year 2000. Although semiconductor materials revenues comprise only about 0.5% of the electronics industry revenue, these materials are the cornerstone upon which the industry is built.

Crystalline silicon and silicon-based materials such as silicon-germanium will continue, through sophisticated device design and processing technologies, to be the materials engine driving the microelectronics revolution to the design rule regime of approximately 70 nm (1)(2). Table 1 summarizes a selection of relevant trends from the 1994 NTRS into the 21st century. Table 2 summarizes a selection of the starting silicon material trends, appropriate for both polished and epitaxial wafers. The requisite improvements in (a) silicon materials, (b) specifications, metrology and standards and (c) cost-of-ownership (CoO) analyses, including alternatives to epitaxial structures, will be discussed. The general trend in the IC industry is to increasingly tighten wafer specifications to match the current sensitivity limits of the metrology tools. The need for the "best wafer possible," however, must be carefully balanced against CoO considerations. Indeed, CoO will become more important as 200 mm wafers proliferate in the industry and 300 mm wafers are introduced at the end of this decade.

The technical significance and relevance of several silicon material trends in the 1994 NTRS will be critically examined relative to a recent assessment of 600 epitaxial wafers, complementing a previous analysis of 2500 polished wafers, from six 200 mm silicon suppliers (3). Representative examples of surface contaminants (particles and metals), surface microroughness, surface topography (wafer flatness) and epitaxial layer properties will be reviewed in the context of the NTRS report. Several background issues for each wafer topic will be summarized, followed by representative data from the epitaxial wafer assessment.

TABLE 1. Technology Trends for ULSI and GSI (1)

Parameter	Units	1995	1998	2001	2004	2007	2010
Resolution (Design Rule)	μm	0.35	0.25	0.18	0.13	0.10	0.07
Bits/Chip							
– DRAM		64M	256M	1G	4G	16G	64G
Chip Area	mm^2						
– DRAM		190	280	420	640	960	1400
– Logic/Microprocessor		250	300	360	430	520	620
Cell Area	μm^2						
– DRAM		1.5	0.6	0.24	0.096	0.038	0.015
Wafer Diameter	mm	200	200	300	300	400	400
Site Flatness (SFQD)	μm	0.23	0.17	0.12	0.08	0.08	0.08
Gate Dielectric (Desktop)*	nm	8.3	7.3	5.0	4.5	3.4**	?→
Junction Depth***	nm	70–150	50–120	30–80	20–60	15–45	10–30
Electrical Defect Density	d/cm^2	0.024	0.016	0.014	0.012	0.01	0.0025
Interconnect Levels							
– DRAM		2	2–3	3	3	3	3
– Logic/Microprocessor		4–5	5	5–6	6	6–7	7–8

*   Gate oxide equivalent thickness
**   Limited by direct tunneling
***   Drain extension/adjacent to channel

TABLE 2. Technology Trends for Silicon Wafers (1)

Attribute	Units	0.35 μm/ '95	0.25 μm/'98	0.18 μm/'01	0.13 μm/'04
Wafer Diameter	mm	200	200	300	300
Equivalent DRAM Generation		64Mb	256Mb	1Gb	4Gb
		Specification at 95% Compliance Level			
Particles per Unit Area [a]	$\#/cm^2$	0.17	0.13	0.075	0.055
– Size	μm	≥0.12	≥0.08	≥0.06	≥0.04
Individual Surface [b] Al, Ca	at/cm^2	1×10^{11}	5×10^{10}	2.5×10^{10}	1×10^{10}
Individual Surface [b] Fe [c], Ni, Cu, Zn, Na	at/cm^2	5×10^{10}	2.5×10^{10}	1×10^{10}	5×10^{9}
Total Bulk Fe [c]	at/cm^3	1×10^{11}	5×10^{10}	2×10^{10}	1×10^{10}
Initial Organics (atoms of C) [d]	at/cm^2	5×10^{14}	3×10^{14}	1×10^{14}	5×10^{13}
Front Surface Microroughness (RMS) [e]	nm	0.2	0.15	0.1	0.1
Site Flatness (SFQD) [f]	μm	0.23	0.17	0.12	0.08
– Site Size [g]	mm x mm	22 x 22	26 x 26	26 x 30	26 x 36
Epi Layer Thickness (± % Uniformity)	μm	3.0 (±5%)	2.0 (±3%)	1.4 (±2%)	1.0 (±2%)
Recombination Lifetime (low-level injection) [h]	μs	150	200	350	500
Generation Lifetime [i]	μs	600	750	1000	1250

(a) Particles decrease by a factor of 3 per generation; particle size = 1/3 of design rule; assume square law cumulative particle distribution

(b) Inital/pre-gate

(c) Numbers are best estimates and cannot be transposed via wafer thickness.

(d) Wafer-carrier/storage ambient critical.

(e) 1 x 1 μm atomic force microscopy (AFM); power spectral density (PSD) required

(f) SFQD = 2/3 of design rule

(g) Partial sites included

(h) P-type polished/etched back wafer, 0(i) <26 ppma

(i) MOSCAP-IC process dependent

Rather than exclusively discuss material trends (3)(4), however, significant improvements in metrology techniques envisioned for 200 mm and 300 mm wafers will also be emphasized. This is necessary because the long-range trends in the NTRS closely match the current sensitivity limits of the metrology tools and, therefore, significant improvements in measurement capability appear to be required. Specifications, metrology and standards have become the essential context in which silicon wafer trends are expressed. Models of the relationships among wafers, processes and measurements are the underpinnings of this expression.

Until recently, the prevailing methodology of measurement used a physical gauge. That is, the object under measurement was constrained by the physical measurement system. In this approach, the metrology models are separated from the information usage environment, so an "unnecessary" data consolidation step (information loss) is required. In physical gauging, there is also an emphasis on the physical phenomena of measurement (the measurement model domain).

Measurement values are more useful, however, when reported within the application context rather than the measurement tool context. In this sense, the measurements become parameters within the application model. Historically, these application models (e.g., process or product models) have been non-existent, intuitive, or not shared with or known by metrology suppliers or users.

A new era of measurement is required in which the object under measurement must not be destroyed, contacted, or altered in any way. The measurement system must conform to the object; data must be obtained by non-contact, in-situ, non-destructive sensors and *transformed* by modeling into *information* about the object. In this style of measurement, there is a strong emphasis on data processing and information extraction. The possibility of an intimate integration of metrology at its point of use might eventually result in the disappearance of production metrology tools.

The enabler of this new era is the digital computer. Digital computers, because of their combined attributes of large volumes of storage data and rapid calculation, are bringing forth this change in metrology because they can support *model management* at a vast computational level. Through the use of computers, models can be applied to a measurement situation quickly, and non-linear, regressive, or adaptive models can be used. Furthermore, extensive historical, retrievable information can be utilized to support statistical and predictive models. Computers allow the integration of sensor monitoring and information extraction which is necessary for the wide spread use of *in-situ* process control measurement. The vast storage capacity of computers eliminates the need for continuously throwing away information merely to keep data volumes manageable.

EPITAXIAL SILICON WAFERS (5)

An analysis of state-of-the-art epitaxial wafers was implemented to assess and elucidate the challenges of the next generation. Six silicon wafer suppliers each generally provided one hundred 200 mm p/p^+ {100} vicinal epitaxial wafers for this analysis. The wafers were randomly sampled from lots targeted for "leading edge" customers (c. late 1993). The general prescription for the data analysis was to calculate the mean and standard deviation for a parameter, P, then plot $<P> \pm \sigma_P$ for all suppliers as a modified bar graph. The statistics of the representative data presented indicate that the wafers received were not artificially prepared by the suppliers for this analysis. Each supplier was given a unique code letter. Supplier Z provided two sets of wafers (ZA and ZB) differing principally in the substrate preparation process. Both sets were included in the analysis.

Localized Light Scatterers

Background (6)(7). Polished and epitaxial silicon surfaces exhibit a plethora of defects which must be controlled to achieve high-yielding IC performance. These defects include particles, haze (minute shallow pits and/or surface microroughness), residual surface chemical residues or structural defects such as the occasional epitaxial stacking fault. Unfortunately, discrimination of these contributions to the localized light scatterer (LLS) specification is generally not possible with today's laser surface scanners. The interference of surface microdefects ("crystal-originated pits", COPs) with the detection of particles ≤ 0.2 μm exacerbates this challenge (7). If these surface microdefects are uncontrolled, their density is expected to far exceed the allowable LLS specification. The measurement (sizing), counting, composition, morphology, removal and prevention of these defects is a state-of-the-art challenge in silicon wafer technology. Laser scanning instruments are widely used to monitor these defects. An inspection technique that has also become widely utilized in the examination of polished wafers is collimated light. The light illuminates the wafer surface, is reflected to a light-receiving screen and transformed to a visual image on a television monitor. By adjusting the contrast and focal depth, a variety of surface anomalies can be qualitatively observed(8).

Data. Light point defects were measured by a Tencor 6200 laser surface scanner using total integrated scattering (TIS) with polystyrene latex (PSL) sphere calibration. The illuminating wavelength, λ, was 488 nm, the edge exclusion was 7 mm and the capture efficiency was 90% and 95% for LLSs ≥ 0.125 μm and ≥ 0.200 μm, respectively. This tool measures all LLSs; no effort is made to discriminate non-particle scattering events such as surface microroughness. The TIS signal is referred to as "particles" in this analysis and represents an upper bound reported as "counts greater or equal to a specific size." Figure 1 indicates most suppliers are capable of averaging less than 50 particles/wafer ≥ 0.125 μm, with a sigma of a few 10's. In addition, all suppliers are capable of producing wafers with fewer than 10–20 particles/wafer (sigma less than 10) with diameter ≥ 0.200 μm. The particle data for epitaxial

wafers manufactured in late 1993 are comparable to the best quality polished wafers examined in 1992 (3).

Measurements in the 0.125 through 0.100 μm regime require attention because of possible interferences near the sensitivity limit of the instrumentation, although LLS measurements as low as 0.100 μm on commercial tools have been reported(9). Individual particles much smaller than 0.100 μm, however, may be difficult to detect by optical scattering because the scattering cross-section is small enough that the intensity can be overwhelmed by interferences such as surface microroughness. The Gbit era will present an even greater challenge in eventually measuring and counting particles as small as 0.02 μm.

Surface Metals

Background (2). Transition metals in silicon exhibit a significant component of interstitial, as well as substitutional, solubility. This results in a mixed diffusion process, with a fast interstitial component. The 3d transition metals, such as Fe and Cr, mainly occupy and diffuse via interstitial sites. They accordingly exhibit high diffusion coefficients. Cobalt, Ni and Cu diffuse sufficiently fast,

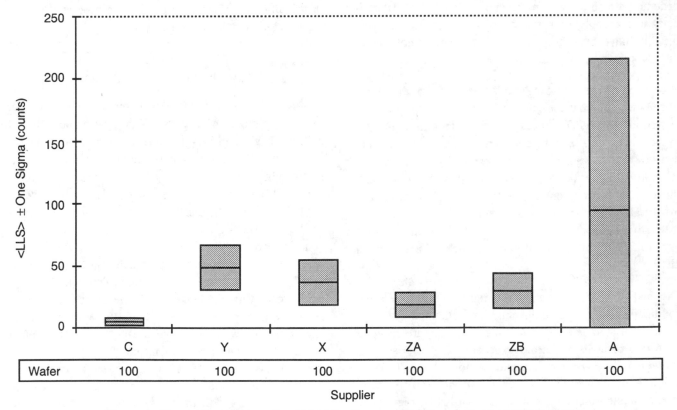

Figure 1. Dependence of Localized Light Scatterer count with supplier (≥ 0.125 μm) (5).

70

even at 300 K, that these elements can transfer from interstitial sites onto dislocation sites as well as form clusters and precipitates. The 4d metals in silicon, such as Ag, exhibit a solubility on substitutional sites comparable to interstitial sites and, therefore, diffuse slower than the 3d transition metals. The 5d metals, such as Au and Pt, exhibit a dominant subsitutional solubility. The 5d metals accordingly take longer until their substitutional solubility is attained throughout the wafer. The U-shaped diffusion profile of Au (higher concentration at the surfaces), however, is clear evidence that there is also a fast diffusing interstitial component. The presence of interstitial sinks limits the kinetics of the transfer of Au onto substitutional sites (kick-out mechanism) during diffusion(10). The temperature dependent solubility and diffusion coefficient, taking into account kinetic considerations during cool-down from high-temperatures, are especially important in determining the state of aggregation of Au at room-temperature. The properties of transition metals have been reviewed in silicon with particular emphasis on their diffusion, solubility and electrical activity (11)(12)(13). Size effects and electronegativity differences between the impurity and host atom are also important considerations and must be taken into account for these analyses.

Although metallic impurities can affect the electrical resistivity of silicon if they are present in sufficiently large enough concentration, they are more important as traps or generation-recombination (g–r) centers which can degrade the carrier lifetime(14). The carrier lifetime is an extremely important example of the structure sensitivity of material properties in semiconductors. Metallic elements exhibit deep energy levels (i.e., far from the band-edges) in silicon with ground states typically 0.4–0.6 eV from the nearest band-edge. In general, metals are considered undesirable in silicon even at levels approaching 1 part per trillion atomic (1 ppta = 5×10^{10} atoms/cm^3). The sensitivity of surface diagnostic techniques varies, but ranges between $\approx 10^8$–10^{10} atoms/cm^2 for a variety of transition metals(15).

Since most transition elements have very small distribution coefficients (10^{-5}) in silicon(16), their direct incorporation from the melt during crystal growth is unlikely if normal precautions are taken. Metallic contamination is much more likely to be introduced onto the wafer surface during wafer preparation or IC fabrication. Contamination by Fe, Ni and Cu, for example, in the range of 10^{11}–10^{13} cm^{-2} during ion implantation, dry etching and plasma CVD have been reported(17)(18). The surface metals diffuse into the wafer bulk during subsequent thermal processing, The surface and bulk descriptions, however, cannot be simply transposed via the wafer thickness due to surface-bulk segregation during thermal processing. In view of this, prescriptions are under development to determine the equivalent bulk concentration from a given surface concentration during subsequent thermal processing(19). An extensive methodology has developed for the removal of surface metals by cleaning with suitable solutions of high purity chemicals(20).

Complications often arise, however, because metals can precipitate out during IC process cool-down, consistent with kinetic considerations, when their solubility is exceeded (21). The metals are often present as precipitates decorating structural defects, resulting in a localized electric field enhancement at: (a) stress-induced locations such as LOCOS(22) structure edges, (b) within or adjacent to p–n junctions (23) or (c) at the Si–SiO$_2$ interface (24). These extended defects modify and exacerbate the g–r characteristics of the elemental metallic impurities, increasing device leakage currents and degrading carrier lifetime and IC performance such as DRAM refresh time. The identification, cataloging and control of point and extended chemical/structural defect complexes and their energy levels/charge states in the silicon band-gap (i.e., degree of electrical activity) continues to be of paramount importance for reducing leakage currents.

Data. Surface contamination by transition metals and other elements was measured by TXRF (25) using a Rigaku 3276. An 8 mm beam diameter, measurement time of 500 sec., incident angle of 0.12° and a tungsten rotating anode set at 30 KV and 300 mA were utilized. Iron, Ni, Cu and Zn were measured with a conservative lower detection limit of 1–2 x 10^{10}/cm^2 and Ca was measured with a conservative lower detection limit of 5 x 10^{10}/cm^2. Figure 2 indicates all suppliers are capable of achieving both a mean and sigma of a few x 10^{10}/cm^2 for Fe, Ni and Cu. Calcium and Cl were consistently detected at about 10^{11}/cm^2 and 10^{12}/cm^2, respectively. The higher value for Ca may be of concern, since Ca has been reported to form a more stable oxide than SiO$_2$ and could degrade gate oxide integrity (26). The dependence of elemental deposition on the silicon surface rather than retention of the elements in solution, with solution pH, has been discussed (27). A more extensive post-epitaxial clean may be effective in removing the Cl.

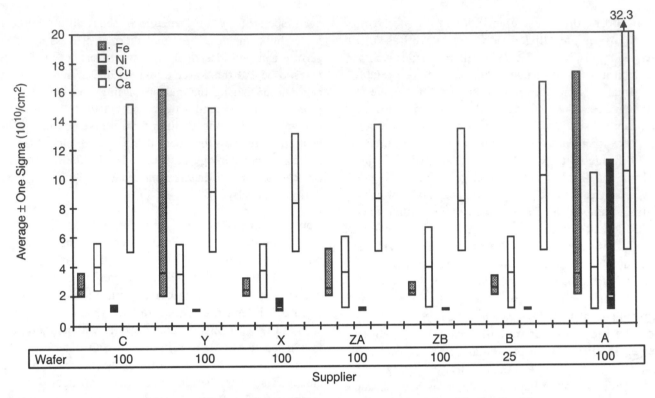

Figure 2. Dependence of surface Fe, Ni, Cu and Ca with supplier (5).

All suppliers achieved low 10^{11} atoms/cm^2 Al and low 10^{12} atoms/cm^2 Na by quasi-static SIMS measurements on a Perkin Elmer 6300 (Quadropole Mass Analyzer based SIMS). A primary oxygen beam (2 keV/10 μA) with a sample bias of ≈ 20 V is rastered over 1500 μm^2. The signal is integrated over a 100 μm radius and the element of interest (i.e., Al, Na) is ratioed to the ^{28}Si signal. This procedure and instrument combination have a detection limit of ≈ 10^{10} atoms/cm^2.

The long-range metal trends in Table 2 are approaching the current sensitivity limits of the metrology tools. The NTRS recommends the metrology tool detection limit should be one-tenth the metallic impurity requirement (1). Conservative detection limit estimates of transition metals by TXRF is ≈ 10^{10} atoms/cm^2. Detection limit estimates of light element such as Al and Na is ≈ 10^{11} atoms/cm^2. The proposed ASTM and ISO standard methods for TXRF and quasi-static SIMS are the first steps in improving the bias and precision of surface metallic contamination metrology (28)(29). The NTRS, furthermore, requires non-destructive, surface metallic contamination wafer mapping metrology tools. Only TXRF, however, provides non-destructive, mapping analysis of wafers. Time-of-flight

SIMS provides wafer mapping capability, but is destructive. VPD collection for ICP-MS and AAS is both destructive and does not provide wafer mapping capability.

SURFACE MICROROUGHNESS

Background (25)(30). Surface microroughness is a characteristic of polished and epitaxial silicon surfaces. It can degrade metal oxide semiconductor (MOS) gate oxide integrity (GOI)(31). Recent advances in Atomic Force Microcopy (AFM) technology have resulted in improvements in both the bias and precision of microroughness measurements. Specifically, TappingMode™ AFM is one method that has overcome the limitation of poor probe tip-sample interactions which are present for certain surfaces using contact mode AFM. In addition, improvements in AFM probe tip geometry and related technologies, which utilize single crystal oriented (111) etched silicon, have improved the control of AFM measurements.

The R_q (RMS) and R_a (absolute average) data presented for a given surface are dependent upon the spatial frequency range of the particular configuration of the measuring instrument used to collect the data (i.e., pixel density, image area). In using an AFM, the width of the image scan,

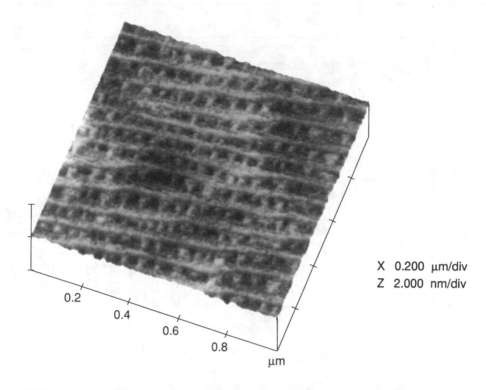

X 0.200 µm/div
Z 2.000 nm/div

Figure 3. 1 µm x 1 µm AFM image showing terrace structures for a 150 mm diameter, 5 µm epitaxial layer (30).

the density of the data (e.g. 512 x 512 pixels) and the size and the shape of the probe tip in principle define the spatial frequency range of the measurement.

AFM measurements were taken on a Digital Instruments Nanoscope III model in the Tapping Mode for a range of surface scans from 1 µm x 1 µm to 20 µm x 20 µm. Figure 3 illustrates terrace structure in a 1 µm x 1 µm image of a 150 mm diameter, 5 µm p-type epitaxial film deposited on a vicinal (100) p^+ substrate. The measurement was made in an air ambient and, therefore, a native oxide was present. The observations are in agreement with UHV experimental observations on the structure of silicon homoepitaxial surfaces which exhibit terrace structures of single atom steps and growth "fingers" in the <110> direction on alternating terraces(32). The observation of a terrace structure is also strong evidence of the conformal nature of the native oxide on the silicon surface. The average periodicity in Figure 3 is approximately 83 nm, which consists of two terraces of unequal width (average width (W) of about 42 nm). Using the relation (33), (34), $\tan \theta = h/W$, where h is the experimentally measured step height (single atomic steps), an angular miscut of about 0.2 degrees is obtained, in rather good agreement with the supplier's data.

The R_q value for a surface, such as a polished or epitaxial silicon wafer, is the square root of the integral of the power spectral density (PSD) over a defined spatial frequency range (35). The PSD analysis is advantageous because it allows comparison of microroughness data taken over different spatial frequency ranges. The PSD methodology also offers a convenient representation of direct space periodicities and the amplitude of the microroughness. Thus, R_q should be reported with the spatial frequency range over which the measurements were taken. Figure 4 is the PSD for a 5 µm x 5 µm scan of the same region of the wafer as in Figure 3, illustrating a sharp peak at approximately 12 μm^{-1} in spatial frequency, corresponding to the periodicity of 83 nm discussed above. Since terrace structures on vicinal (100) silicon surfaces have been well characterized by methods such as LEED, these surfaces may also be used as a standard reference surface for AFM microroughness measurements.

More complicated structures can be observed when the epitaxial film is grown on a substrate miscut in two crystallographic directions as illustrated in Figure 5 for a 1 µm x 1 µm image (R_q of 0.07 nm) from a 200 mm diameter, 5 µm p-type epitaxial film deposited on a vicinal (100) p+ substrate. The "fingers" now point in both <110>

73

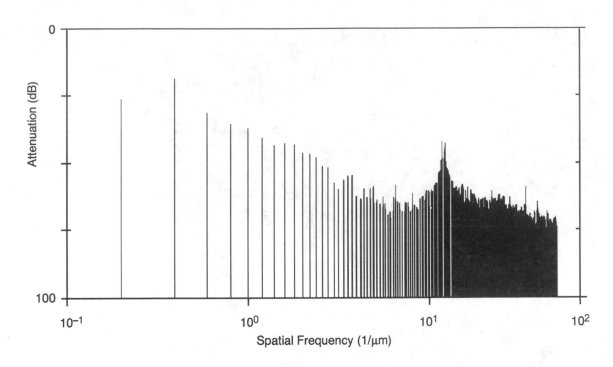

Figure 4. Power Spectral Density (PSD) for a 5 μm x 5 μm scan of the same region as Figure 3 (30).

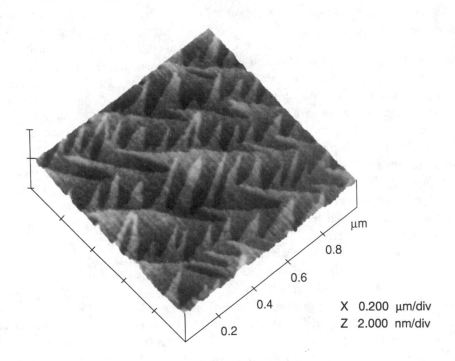

X 0.200 μm/div
Z 2.000 nm/div

Figure 5. 1 μm x 1 μm AFM image showing terrace structures for a 200 mm diameter, 5 μm epitaxial layer, miscut in two crystallographic directions (30).

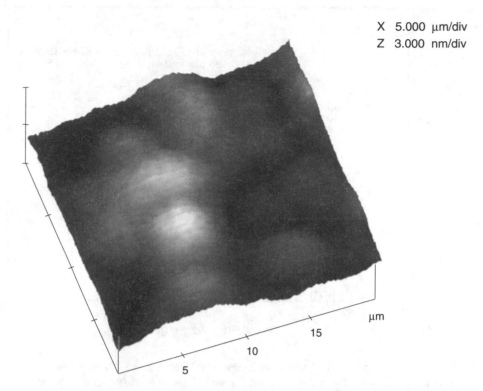

X 5.000 µm/div
Z 3.000 nm/div

Figure 6. 20 µm x 20 µm AFM image of the same region as Figure 3 (30).

directions for this particular wafer. Taking the periodicity of 200 nm (average width (W) of about 100 nm) from Figure 5, and the experimentally observed single step height, an angular miscut of about 0.1 degree is obtained.

An R_q value of 0.06 nm is observed in Figure 3, while an R_q value of 0.5 nm, along with a visually obvious waviness, is observed in a 20 µm x 20 µm scan for the same region of the same wafer (see Figure 6). It is believed that the feature observed in Figure 6 illustrates a polished wafer residual waviness that is replicated in the 5 µm epitaxial film. The difference in R_q values for the two scan areas is dependent on the actual details of each wafer, the spatial frequency range associated with the measurement as well as the detrending mathematics.

Figure 7 illustrates the PSD constructed from a two-dimensional scatterometer Bidirectional Reflectance Distribution Function (BRDF) (35) for a wafer from the same cassette as in Figures 3 and 6. Integrating the PSD for curve 1 from 0.0125 to 1 cycles/µm (surface wavelengths of 1 to 80 µm) an R_q value of 0.46 nm is obtained, which is comparable to the AFM R_q value of 0.50 nm in Figure 6. It should be noted, however, that the AFM R_q value of 0.50 nm is calculated over a 20 µm x 20 µm area. This corresponds to

a range of 0.05 to 12.8 cycles/µm (surface wavelengths of 0.08 to 20 µm) which is different than for the BRDF case. Resolution of the similarity of the R_q values for such dissimilar spatial frequency ranges and related analyses are in progress. The measurement of silicon surface microroughness will be an increasingly important metrology issue for microelectronics technology (36).

Data. The epitaxial study indicated all suppliers were capable of R_q values less than 0.10 nm and 0.15 nm for scan sizes of 1 µm x 1 µm and 20 µm x 20 µm, respectively. These scan sizes correspond to spatial frequency ranges of 1 to 256 and 0.05 to 12.8 cycles/µm, respectively. Both R_q and R_a were computed from detrended topography. Crystal lattice terraces were only observed on the epitaxial wafers which did *not* receive a post-epitaxial wafer clean. The influence of wafer cleaning in obfuscating the terrace structure, methods for its recovery and the impact of terraces on gate oxide integrity (GOI) is an extremely active area of research (37).

Surface Topography

Background. DRAM die size is typically increasing by a factor of 1.46 per generation while the cell area is

75

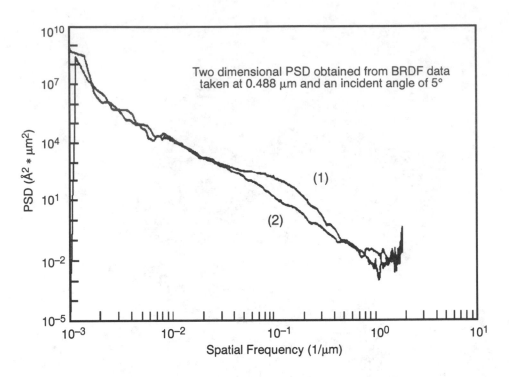

Figure 7. PSDs constructed from scatterometer Bidirectional Reflectance Distribution Function (BRDF) data for two epitaxial wafers. Curve 1 is for a wafer from the same cassette as in Figures 3 and 6 (30).

decreasing, when innovation such as cell shrinkage is taken into account, by 0.33 per generation (38). A 64M CMOS DRAM typically requires 18–20 mask steps while it is anticipated that high density Gbit CMOS DRAMs may require as many as 20–25 mask steps. This increase in bit density is generally achieved by driving IC manufacturing to smaller design rules and geometries, which results in smaller lithographic depth of focus and, accordingly, necessitates improved wafer flatness.

It is unclear, however, how to ensure that the wafer site flatness is compatible with these severe shrinks in IC design rules and microlithography considerations, especially if one takes into account the cumulative thermal processing effects of a ULSI fabrication process. It has been estimated, for example, that fluctuations in wafer site-flatness can contribute as much as 0.5 µm to the random error component of the total focus error of a state-of-the-art, high numerical aperture (NA) microlithographic system (39). Simple analytical approaches for modeling the focus budget using the sum of the squares, a linear methodology or a similar combination, however, are often fraught with assumptions (40). The Rayleigh resolution limit ($R = (\lambda/NA)K_1$) and depth of focus ($DOF = (\lambda/NA^2)K_2$), where K_1 and K_2 are normalizing constants

(41), criteria are also not necessarily reliable guides. One industrial practice suggests that the site total indicator range for a front-surface least squares focal plane (SEMI notation SFQR (42)) is approximately one-third to one-half of the depth-of-focus of the lithographic exposure tool (38). In view of the several, non-satisfactory methodologies available, an empirical procedure that SFQR should equal the IC design rule has been utilized. The NTRS utilizes the site focal plane deviation for a front-surface least squares focal plane (SEMI notation SFQD (42)) as the primary site flatness parameter; extensive measurements indicate that the ratio of SFQD to SFQR is approximately two-thirds (43). This is a characteristic of the current lithography site size convolved with the surface wavelength distributions associated with current polish procedures (44). Further clarification of the mathematical relationship between SFQD with SFQR and its dependence on both the density of data taken and surface wavelength distribution, which describes the surface topography resulting from the polishing process, is required. Clarification of this relationship might also ensure more effective correlation between wafer flatness and stepper performance (45). These studies have been supplemented by examining the wafer surface topography resulting from a specific polishing process by Fourier methods (46).

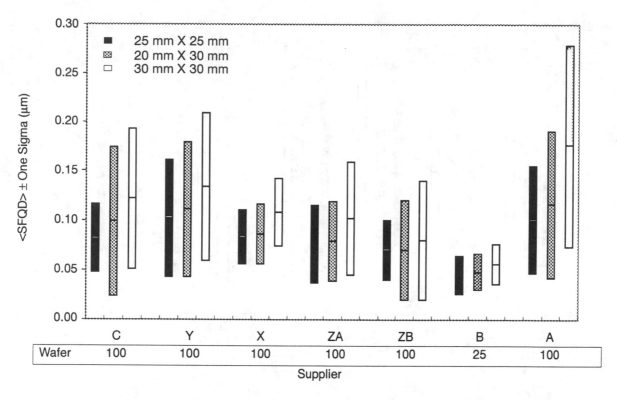

Figure 8. Dependence of SFQD (front-reference site flatness) with site size and supplier (5).

Data. Wafer dimensional information were derived using an ADE 9300. A 4 mm square probe, 3 mm edge exclusion and 8000 data points per wafer were utilized. The global dimensional parameters flatness and bow/warp were computed from the two-sided, capacitive gauging measurement of wafer thickness variations and wafer shape, respectively. Different suppliers excel at different flatness parameters, as previously observed for the polished wafer case (3)(5).

Most suppliers produced epitaxial wafers with a global total indicator range for a back-surface ideal (global) reference plane (SEMI notation GBIR (42)) in the 1–2 μm range with a sigma of less than 1 μm. The global total indicator range for a front-surface global least squares focal plane (SEMI notation GFLR (42)) is typically less than 1.5 μm, with a sigma less than 0.50 μm. The similarity of the front-surface and back-surface referenced flatness indicates that substrate taper is not an issue for these wafers. Double-side polished (DSP) wafers and advanced slicing methodologies such as grind-slice-grind or ductile mode grinding (47) are expected to further reduce the non-parallelism of the back- and front-surfaces (improved GBIR). In conjunction with a polished wafer-edge contour, DSP wafers may also minimize a source of particles.

Wafer warpage is generally 10–25 μm with a sigma less than 10 μm for the wafer back-surface film conditions measured. An intentionally controlled substrate curvature (convex or concave bow) (48) and a minimally warped wafer is beneficial in fabricating device structures with multiple material layers of differing coefficients of thermal expansion.

Site flatness is computed by modeling the focal plane deviations derived from the two-sided, capacitive gauging measurement of wafer thickness variations of the surface within each exposure site. SEMI standard site flatness parameters were measured for site sizes of 25 x 25, 20 x 30, and 30 x 30 mm on a side. These site sizes correspond to 32, 32 and 24 complete, non-partial sites on 200 mm wafers. Figure 8 indicates that all suppliers are capable of average SFQD of 0.10–0.15 μm with a sigma of ≈0.05 μm, consistent with 0.25–0.35 μm design rules and site sizes (1). Several suppliers appear capable of achieving 1G bit flatness requirements. The site flatness degrades with the site longest dimension as well as with the area. Extensive distributional analysis of the site flatness data was also performed to assess the variation of SFQD as a function of radial location on the wafer (5). The relation-

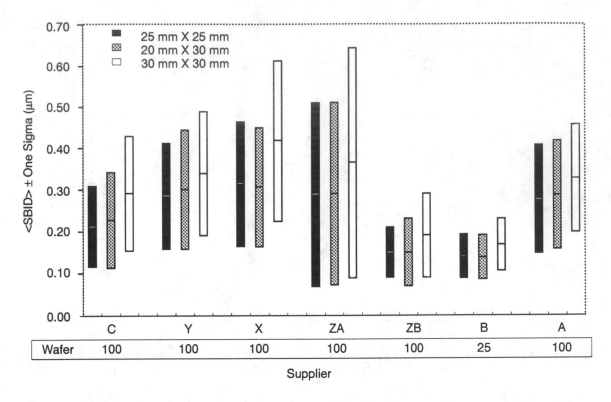

Figure 9. Dependence of SBID (back-reference site flatness) with site size and supplier (5).

ship of site flatness distributions to more general wafer surface models is also under development. Figure 9 illustrates the variation of the site focal plane deviation for a back-surface ideal (global) reference plane (SEMI notation SBID (42)) for the same wafers and site sizes. The larger values of SBID, compared to SFQD, reflects the influence of GBIR, as expected. The smaller SBID data for supplier Z's second set of wafers (ZB) is indicative of an improved slicing methodology.

Epitaxial Layer Properties

Background (49). Epitaxial films will continue to be required for bipolar, CMOS and BiCMOS applications, albeit with thinner films and tighter tolerances (see Table 2). Silicon wafers with epitaxial layers less than 10 and 5 μm thick have been the work horse of the bipolar and CMOS high performance memory and logic industry, respectively, for many years. The NTRS forecasts sub-2 μm epitaxial layers for CMOS applications by the end of the century. The necessity of improved control of the magnitude, tolerance and uniformity of epitaxial wafer characteristics, including controlled epitaxial flat zones (and a sharp transition at the epitaxial-substrate interface)

will be more important than ever before. Contamination Free Manufacturing (CFM) is also critically important in developing defect-free epitaxial films and epitaxial active devices. This includes improvements in the epitaxial chamber and susceptor construction materials as well as control of oxygen and moisture contaminants in process gases (49).

Low-temperature epitaxial deposition (including limited reaction processing (small √Dt), photo-epitaxy, remote plasma CVD and reduced pressure deposition (in single, batch or cluster reactors) offers significant opportunities for improved epitaxial film properties. Selective epitaxy applications may become more pervasive as selectively deposited well configurations, trench isolation, contact fill, elevated source/drain configurations, etc., are incorporated into advanced IC designs.

Data. The thickness of the epitaxial layer was measured by four different techniques: (a) SIMS depth profiles using a Cameca IMS–4F, (b) Spreading Resistance Probe measurements performed at Solecon, Inc., (c) FTIR measurements over the mid-IR range using a Nicolet 800 with a Pike epi analysis accessory and a two-layer model with multiple points per wafer and multiple wafers per supplier, and (d) FTIR measurements over the mid-IR range using a Bomen

78

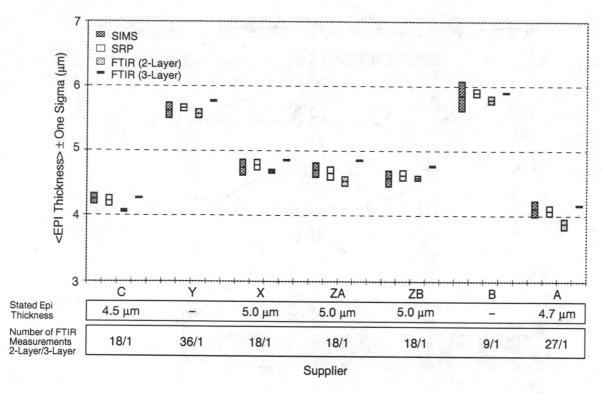

Figure 10. Dependence of epitaxial layer thickness with measurement method.

MB155 FTIR analyzed by a three-layer model for one point per wafer and one wafer per supplier (50). Epitaxial layer thickness variations across the wafer were measured using a Nicolet ECO 8SN with Cepstrum Method software. Interface boron and carbon were also measured by SIMS depth profiles.

The epitaxial layer thickness measurements shown in Figure 10 indicate all the measurement techniques yielded a similar value for the layer thickness, although all techniques yielded a value less than the suppliers' process target values. The good agreement among all the dissimilar measurement techniques and instruments, however, gives credence to the experimental data. Additional measurements are in progress for the 3-layer model.

SPECIFICATIONS, METROLOGY AND STANDARDS

Specifications, metrology and standards have become inextricably intertwined and, belatedly, have assumed an increasing and well deserved importance. In addition to ensuring consistent terminology, reference materials and procedures, standards have also recently become a mechanism to enable productivity cost savings.

The present standards environment for silicon materials is generally divided into two sectors, standards which define and specify values for wafer attributes (SEMI) and standards which specify measurement methodologies (ASTM). Standards for silicon materials are becoming increasingly sophisticated and the role of metrology is of the utmost importance in achieving reduced CoO.

To understand the relationship among specifications, metrology and standards and the materials, processes and products which make use of them, a schematic framework for relating the *models* in these various domains is required. In Figure 11, the right portion of the figure represents "real" reality (the actual factory and the material flowing through it). Most of what is dealt with in specifications, metrology and standards, however, is in a "modeled" reality (the left portion of Figure 11). Models in the areas of materials, processes and products (horizontal layers in this diagram) represent the behavior and intended outcomes for the real factory. Metrology is the connection between these modeled and real behaviors. Metrology models are developed to relate the behavior of real metrology tools to the models of the materials, processes

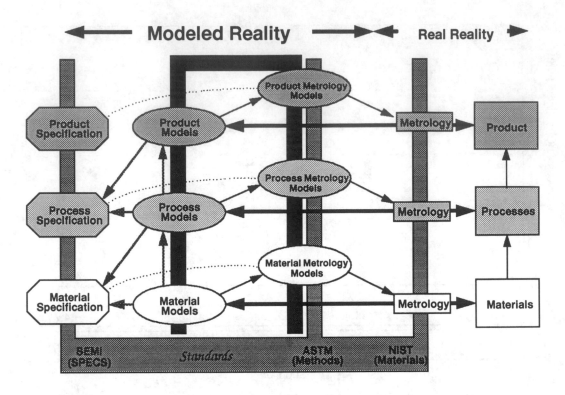

Figure 11. Schematic framework relating modeling domains of materials, processes and products to specifications, metrology models and real metrology tools.

and products being measured. This supports consistent interpretation of results. The models which underlie the various domains and measurements in them come principally from the disciplines of science. Standards provide mutually agreed connectivity between specifications, metrology models and real metrology tools (SEMI specifications, ASTM test methods and NIST reference materials, respectively).

In this section, three important examples of the breakdown of models presently in common use are detailed. Several modeling issues from the interacting domains of the material, process and metrology tools are described for each example. Actions underway to rectify these "ripples in the fabric of understanding" are then described.

Model breakdowns demand improvements of two types which are discussed below. Type I improvements occur when a material specification is tightened beyond existing metrology tool limits, requiring a more sophisticated metrology model and, possibly, technological improvements. Type II improvements occur when increased sophistication in application models requires increased sophistication in metrology models.

Localized Light Scatterers

Background. The detection of particles on wafer surfaces has been accomplished for decades by observation of the wafer under an intense light source using full-wafer illumination and visual inspection and, more recently, automated, focused-beam scanning and detection systems. Because the front surface of the wafer is polished, it reflects light specularly unless an irregularity exists in the surface. An irregularity scatters light away from the specular beam so that a detector (human eye or optical detection system) viewing away from the specular direction will only "see" light at the locations where an irregularity exists. The number of LLSs greater or equal to a specific size remains a critical specification for starting wafers as well as for monitoring process equipment. As defectivity requirements have become more stringent, however, this model of particles no longer suffices.

Two principal forms of model breakdown have been observed. First, the apparently perfect wafer surface, aside from any particles is, in fact, non-ideal. The microscopic roughness of the surface, random topographic ripples which are microns in length and angstroms in height, scatter small amounts of light (relative to the specular beam) in all directions. When attempting to measure

80

particles below a certain size, this surface-scattered light obscures the signal due to the particle-scattered light, inducing a correlation between minimum detectable size and the surface condition. Secondly, the beam size of a particle detector is 10's of microns, so any surface light scattering anomaly smaller than this is counted as a particle, even if it is of no consequence in device performance. Several modeling issues from the interacting domains of the material, process and metrology tools are described below.

A) **Material Domain.** The imperfections at the surface of a silicon wafer have been extensively studied and classified (51). Real particles can exist on the silicon surface or attached to or buried in the native oxide. Particles can be of any shape and constituted of almost any material used in IC fabrication, some more harmful than others. COPs, microroughness and microscopic scratches on the surface resulting from a variety of mechanical contacts are additional surface effects.

B) **Process Domain.** The NTRS assumes that a "true" particle with a diameter of one-third the design rule can be a killer defect. The models of how defects affect process and device yield are complicated and generally statistically based.

C) **Metrology Domain.** The standard model for particle detection captures as much scattered light, TIS, as possible into a single, large signal pulse to enhance the signal-to-noise ratio. By doing this, detailed information derivable from the scattered light is lost, making distinction between various particles and surface anomalies impossible. Information on the true size of the particle is also lost, since particles of disparate sizes may have similar scattering if they have different indexes of refraction.

Opportunities—SEMATECH Particle Counting/Surface Microroughness Task Force(52)(53). Standardized definitions of terminology for laser light scattering, in conjunction with clarification of the role of PSL reference materials and the development of a standard haze reference surface, is playing an important role in assessing surface anomalies determined by advanced laser-light scattering techniques.

The SEMATECH Particle Counting/Surface Microroughness Task Force was created to study the problem of accurately counting and sizing sub-0.10 micron particles on rough surfaces using scattered light. Four initial areas investigated were:

(1) The methodology and results of the deposition of both PSL spheres and "real-world" silicon particles on silicon surfaces and their subsequent individual detection by angle resolved scattering (ARS),

(2) The relationship of the ARS of individual particles to the ARS of large (2 mm) surface areas, both with and without particles,

(3) The possibilities and limitations of the ARS technique for separating particle counts from nuisance counts caused by surface microroughness, and

(4) The development of both procedural and physical standards to support the calibration and operation of laser surface scanners.

As a preliminary, exploratory experiment, PSL spheres or "real-world" silicon particles of size 0.204 µm or 0.305 µm were deposited on 150 mm p-type {100} vicinal silicon wafers. The initial particle count for these wafers was less than 50 for particles greater than 0.125 µm. Three Tencor 6200 laser surface scanners at three different facilities were used to characterize samples of the four material/size combinations. The sizing statistics from each tool were compiled. The results of the study and an analysis of variances for within-run and run-to-run variance components were also determined. Figure 12 summarizes the response of the three laser surface scanners for the prepared surfaces. Since PSL spheres are used for calibration of TIS measurements on a Tencor 6200, PSL spheres are seen to be sized correctly. Silicon particles, however, are systematically undersized at 0.305 µm and oversized at 0.204 µm. This is in qualitative agreement with light scattering models indicating a cross-over in the scattering cross-section of Si and PSL spheres in this size regime (55) (see Figure 13 for an idealized case of TIS in free space). Commercial TIS laser surface scanners can, therefore, erroneously size particles of different materials. Repeatability analysis of the count data also indicated that PSL spheres yield better counting precision than silicon particles. This is believed to be the result of the size and shape uniformity of PSL spheres, allowing for a narrow sampling bin width (53). This benefit of PSL sphere standards must be considered, even as concern is expressed that the PSL material is different from any material normally measured. Similar laser surface scanner measurements of calibrated 0.100 µm PSL and silicon particles are currently in progress.

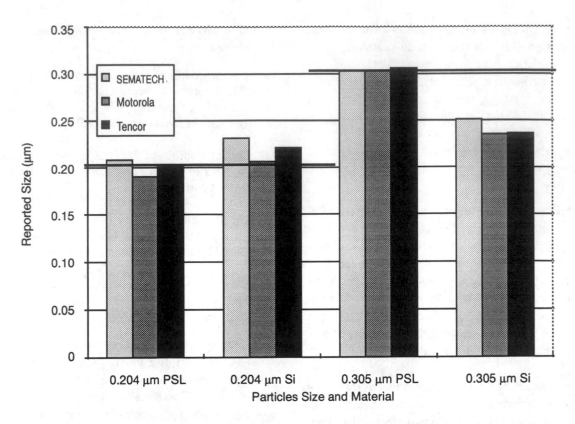

Figure 12. Physical size compared with size reported by three "identical" laser scanners for 0.204 μm and 0.305 μm PSL and Si particles (53)(54).

Light scattering intensity as a function of angle was also collected by ARS for various sizes of individual silicon particles using a 10 μm spot size. SEM measurements were used to verify the particle physical size and shape. A light scattering model was used to interpret the results (54)(55). Similarly prepared samples were also measured by ARS with a 2 mm spot size light scattering instrument, capturing hundreds of particles within the scattering field. The discrimination of 0.305 μm Si particles from surface microroughness is illustrated in the latter case in Figure 14a. Figure 14b illustrates that excellent agreement is obtained between the experimentally measured aggregate particle response for 0.204 μm Si particles and a first principles theoretically synthesized response developed from the individual particle responses (56). This research has recently been extended down to 0.06 μm Si and TiO$_2$ particles (53)(56).

A robust set of haze standards will be required to quantify and verify models for sub-0.10 μm particle detection. An initial step in this agenda was achieved by lithographically etching a random array of mesas, each about 1.7 μm in diameter with a density of about 4 x 10^4/mm^2, onto a silicon wafer. The mechanical shape of the mesas was verified using AFM; the light scattering properties of the mesas were measured by both a commercial TIS tool and ARS using a 2 mm spot size light scattering instrument at two incident angles. Figure 15 illustrates the excellent agreement of the optical properties of the haze standard with the mechanical AFM measurements, both transformed into the PSD descriptor. These results suggest that the fabrication of a lithographically etched surface is a viable initial method for creating a robust haze standard and offers the opportunity to study the discrimination of sub-0.10 micron particles from surface microroughness (53). This is an example of the importance of establishing a well-defined standard to further advanced research in the pursuit of the NTRS goals for particle detection.

To support the NTRS requirements for measurement of 0.08 μm and 0.06 μm particles in 1998 and 2001, respectively, the task force is studying ARS as a replacement for the TIS method used in conventional laser surface scanners. ARS also has the potential of providing additional information necessary to improve particle sizing/counting performance. Development of reference material standards for particle detectors will require both robust haze reference surfaces and "particle" artifacts as well as a

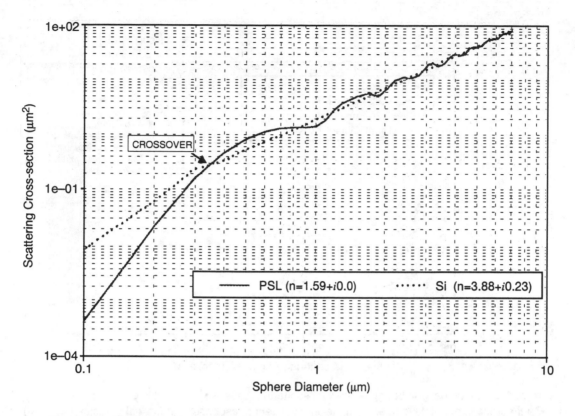

Figure 13. Total Integrated Scattering (free space) for PSL and silicon spheres (52)(53).

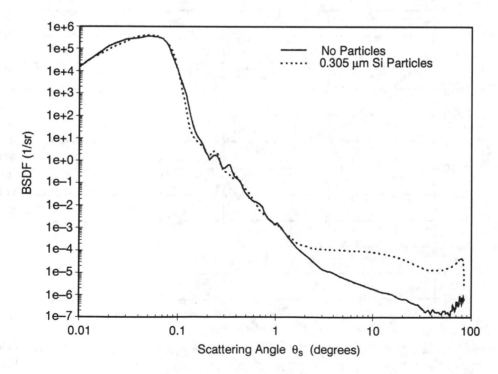

Figure 14. (a) BSDF versus the scattering angle from the specular beam for a wafer with and without 0.305 μm silicon particles

83

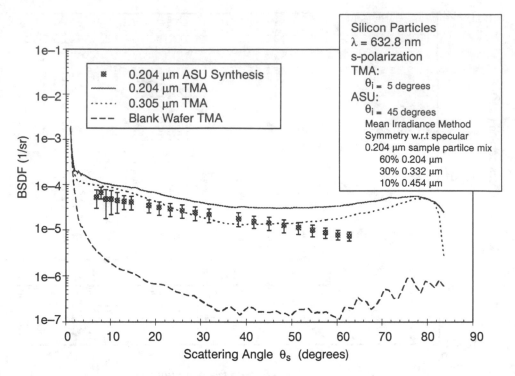

Figure 14. (b) Comparison of aggregate light scattering response from multiple 0.204 μm particles with a theoretical model developed from the light scattering response of individual particles (see text).

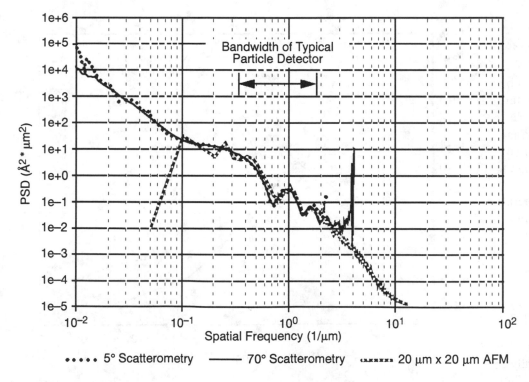

Figure 15. PSDs for etched mesa haze standard constructed from both AFM topographic data (20 μm x 20 μm scan) and scatterometer data (5° and 70° angles of incidence).

reliable deposition methodology for sub-0.10 μm particles of various materials. Developing metrology equipment to ensure production-worthy counters capable of eventually distinguishing 0.02 μm particles from haze and surface topological features is required. Work to standardize terminology, test procedures, and reference materials for particles and microroughness is continuing by the SEMATECH Task Force in conjunction with international standards committees.

Surface Metals

Background (57). Carrier-lifetime affords an extremely sensitive technique to quantify the bulk electrical effects, in parts per trillion, as a consequence of a given coverage of surface metals which are driven into the bulk during thermal processing, consistent with thermodynamic segregation effects.

The carrier recombination lifetime is useful for monitoring the metallic contamination for both *starting wafers and in-process* modification of the wafers due to IC fabrication. The carrier recombination lifetime measures the collective quantitative effects of multiple impurity elements. It is difficult to relate this information to a specific impurity species, however, if more than a single species is present, except for the case of Fe. In this case, the iron can be reversibly cycled between two different electrically active centers: positively charged, deep level interstitial iron centers and shallow, neutral iron-boron pairs(58). This is accomplished by either thermal processing below 250 C or optical excitation at the appropriate wavelength(59). Combining the recombination lifetime measurements (diffusion lengths) for these two cases yields the bulk Fe content(60)(61). A detection limit of approximately 10^9 Fe/cm^3 has been reported by this technique(59).

Carrier generation lifetime is usually the key *product* parameter since uncontrolled leakage currents degrade IC performance such as DRAM refresh time at room temperature. At higher temperatures, such as 80 C, one must also consider the effect of the diffusion current(57)(62).

A) **Material Domain.** The condition of the wafer front- and back-surfaces (surface recombination velocity) is a confounding factor. That is, the surfaces must be passivated such that the carriers diffusing to the surfaces do not readily combine (low surface recombination velocity). The influence of the interface is especially important in wafers with a thin epitaxial layer and/or de-

nuded zone (DZ). In general, the bulk carrier recombination lifetime is much shorter than in the epitaxial layer or DZ and, therefore, dominates the measurement. In this case, the interface acts similarly to a surface with a high recombination velocity(57).

B) **Process Domain.** The surface and bulk descriptions for metallic concentration cannot be simply transposed via the wafer thickness due to surface-bulk segregation during thermal processing. Prescriptions are under development to determine the equivalent bulk concentration from a given surface concentration during thermal processing(19).

C) **Metrology Domain.** The detailed experimental conditions such as the injection level and range and contact versus non-contact sensors complicates the determination of the carrier recombination lifetime. When several unknown impurity centers are present in the material, furthermore, the resulting carrier recombination lifetime contains the collective contributions from all the centers and deconvolution is virtually impossible, except for Fe. Although deep level transient spectroscopy (DLTS) can indicate the presence of deep levels, DLTS is primarily a majority-carrier diagnostic technique and it is difficult to relate this information to the carrier recombination lifetime if more than a single impurity species is present(38).

Whereas the carrier recombination lifetime is a function of the carrier density and capture cross-sections of the impurity center, the generation lifetime additionally depends on the occupancy of the center(57). The relationship between the two lifetimes depends on the particular conditions in the sample being tested(57). In silicon, the generation lifetime is usually much larger than the recombination lifetime(63) although it is difficult to obtain an accurate value for the ratio of the two lifetimes. The reduction of and limiting effects of interface state density on the carrier lifetime also is a state-of-the-art challenge.

Opportunities—SEMATECH Wafer Defect Science and Engineering Task Force (19). The SEMATECH Wafer Defect Science and Engineering Task Force was created to update the 1993 Starting Materials and Surface Preparation portions of the NTRS. Six sub-tasks were identified in developing this update: (a) gettering fundamentals, (b) literature survey, (c) trends, (d) metrology, (e) diagnostics and (f) surface preparation. A detailed summary of the task force's analyses is in progress(19).

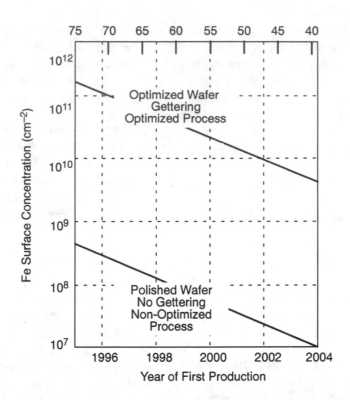

Figure 16. (Upper curve) Increased tolerance to metals due to industrial practices such as utilization of epitaxial wafers, sophisticated process sequences and gettering. (Lower curve) Allowable Fe concentration in the gate oxide region, based on short loop processed wafers (64).

An example of the need for implementing the research recommended in the NTRS is illustrated in Figure 16 (19)(64). The figure illustrates two curves of the sensitivity of gate oxide device structures to Fe-induced defects. The allowable Fe concentration in the gate oxide region, based on short loop processed wafers, is illustrated by the lower curve. Industrial practices such as the utilization of epitaxial wafers, sophisticated process sequences and gettering, however, are expected to increase the tolerance to metals, as indicated by the upper curve. The lowest CoO, to be discussed in more detail below, is obtained somewhere between these two curves. The operating point is identified by the robustness of the IC design to both leakage currents and the specific fabricator process. Optimization of the metallic concentration in the starting material, for a specific wafer type (CZ, epi, SOI, H_2 annealed, etc.), is based on the specific process sequence (extent of gettering) and the allowable metallic concentration after surface preparation.

The importance of developing a gettering/microcontamination process simulator applicable for arbitrary thermal budgets, process sequences and device structures, including SOI, to more realistically determine the starting wafer

requirements is discussed below. The necessity of a non-destructive, in-line technique for mapping carrier lifetime to enable correlation of the lifetime with the metal/structural complexes discussed previously was identified as a critical metrology requirement in the NTRS.

Surface Topography

Background. The dominant method by which flatness is measured today is based on capacitive thickness measurements of the silicon wafer. There are several essential but increasingly questionable model assumptions which make up this information chain.

A) **Material Domain.** The wafer back-surface conforms perfectly to the nominal surface of a clamping vacuum chuck by classical thin-plate bending and/or vertical shearing.

B) **Process Domain.** The stepper has a perfectly flat chuck; for site-to-site leveling, the stepper focus/level system orients the wafer front-surface topography to yield the lowest value of SFQD for each site.

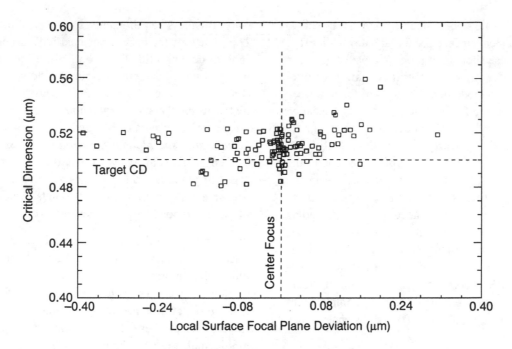

Figure 17. Critical Dimension vs. local surface focal plane deviation.

C) Metrology Domain. A capacitive measurement can be computationally and linearly converted to distance; two distance sensors fixtured in opposition yield thickness. Mathematically removing a least-squares-fit plane from thickness data within each site yields the site focal plane deviation (vis-à-vis the process tool model).

These models have been used for years to great benefit. The increase in ULSI integration level, however, necessitates a measurement technology for monitoring and qualifying wafer manufacturing processes which models the stepper operation more effectively—a type II improvement. The silicon wafer, IC fabrication process, metrology and process models must all be upgraded. Examples of confounding issues include:

(a) The non-flatness of the stepper chuck contributes to increased non-flatness of the wafer.

(b) The inability of the vacuum chuck to completely flatten the back surface of a wafer contributes to unpredicted focal plane deviations. This effect occurs because (i) the small wavelength localized substrate warp ($\lambda \leq 5$ mm, amplitude ≤ 1 μm) is stiff enough to resist the downward force of 1 atmosphere of pressure and/or (ii) the wafer warp is sufficiently high that the vacuum

chuck cannot appropriately seal the entire surface of the wafer, leaving regions unclamped.

(c) The spatial frequency range of the measuring tool must be wide enough to capture features with amplitudes relevant to lithographic flatness. Any surface features with amplitudes exceeding 10 nm are relevant to flatness characterization for photolithography. The sensor size and the sampling density must be engineered to capture all relevant features.

(d) For small edge exclusions, in particular approaching within 1 mm of the wafer edge, the stepper chuck terminates and has no physical contact with the wafer. Accordingly, the measured thickness in this region becomes irrelevant to computation of the focal plane deviation.

While a vacuum chuck-free measurement has many benefits (especially immunity to particulate contamination), spatial frequency range and accuracy requirements will likely require a new synthesis of capacitive and optical techniques to meet the demands of ≤ 0.18 μm lithography.

Even with improved metrology tools, however, it is unclear how to relate the wafer site flatness to lithographic performance in the case of a 20–25 mask gigabit ULSI fabrication process. Figure 17, for example, illustrates that electrical (Prometrix) critical dimension (CD) measurements of 0.5 μm isolated polysilicon lines are constant, to within 5%, over a local surface focal plane deviation range of ≈ 0.5 μm for these short-loop processed wafers (3). The flatness measurements were obtained using a novel software procedure to model the deviation of the processed wafer's surface from the focal plane at the precise location of each CD feature (65). The IC process appears to blur the influence of the starting wafer site flatness quality. Of course, it may also be that non-optimization of the localized lithographic process (i.e., photoresist thickness, exposure, focus off-set, etc.) may be transparent in an electrical CD measurement. Expanded analyses, including SEM measurements, are in progress to clarify this issue. The modeling of wafer surface flatness measurements relative to subsequent lithographic process requirements is a high priority NTRS opportunity. In that regard, it should be kept in mind that steppers also introduce complications due to stepper chuck irregularities (i.e., the stepper chuck signature) (66).

Opportunities—An example of an alternate technology which may be beneficial in expanding the lithographic depth of focus and alleviate the drive to smaller values of SFQD, for example, is phase shift masks. This technology is being extensively assessed to facilitate the approach towards the 64 Mb and 256 Mb DRAMs (67)(68). Phase shift mask technologies may be especially helpful in that they can decouple the K_1 and K_2 normalizing constants in the resolution and depth of focus equations, respectively(69). Advances are also being made in surface imaging, resist and planarization technologies, off-axis illumination and optical lens improvements. The NTRS details the methodology for advanced optical and, eventually, non-optical lithographic techniques.

COST-OF-OWNERSHIP OPPORTUNITIES

Model Based Specifications. The need for the "best wafer possible" must be carefully balanced against CoO considerations. For example, a general trend in the IC industry is to increasingly tighten wafer specifications to match the current sensitivity limits of the metrology tools. From a user benefit point-of-view this is sensible, since nearly all measurements are performed to detect some form of defectivity. This defectivity is relative to a model which defines perfection (*not* the optimum) in every aspect of the production process. Examples include zero particles and metals on the wafer surface and perfectly flat wafers. The product designer is motivated to simply specify problems

away. Business requirements, however, dictate a macroscopic view of the costs associated with driving specifications to detectability limits and often indicates that the *optimal* value of defectivity is not the "lowest possible" but some higher value. *The relevance of this issue to metrology occurs when there is mutual understanding and proper interpretation of material, process and product models in the context of specifications, metrology and standards.* This is often difficult in today's infrastructure because certain stages of manufacturing, such as supplier/customer interfaces, are "information pinch-off points" where substantial information/knowledge becomes lost to down stream users.

The emphasis on distributional specifications (target and process measurement precision) rather than specification limits will continue (70). The processes producing silicon wafers must yield parameter distributions with at least 95% of the product produced having values listed in Table 2. Since many of these distributions are single-sided and non-normal, the 95% compliance level cannot be interpreted as a 2σ variation. A well characterized process, however, is required. The transfer of wafer specifications to a distributional rather than limits approach will afford significant cost-reduction opportunities, if wafer customers are willing to purchase distributional product *and* not also insist on 100% inspection. This will require a major initiative of standards committees during the next decade. Adoption and extension of capability metrics like C_{pk} are rapidly becoming more widely used for both equipment characterization and fab processes, albeit assuming a normal distribution.

Metrology is a value adder when used correctly. That is, the sensitivity, bias, precision and utilization of any measurement must be commensurate with the silicon material and IC processes as well as the circuit technology. In a literal sense, however, measurement never adds value directly to the product. The information derived by the measurement process, however, does add value in the following manner: except for small, non-repeated lots, the *knowledge* that the product/process conforms to specification at a particular stage of manufacturing reduces the (statistically non-zero) probability that "anti-value" (i.e., losses) will occur downstream due to defective or marginal product. The differential "value of information" can be estimated by modifying an attribute of a metrology step (e.g., sampling frequency (including zero), precision, bias, etc.) at some point in the process and noting the long-term yield changes. If yield does not change, the measurement has no value. [This experiment (somewhat hypothetical) can only be performed on processes which are behaving according to a known model. If the measurement is used for *sorting* (i.e., the target cannot be hit at all without measurement), then the measurement must be considered a *process* step, not a measurement step.]

The key information goal for producers and users of silicon materials into the next century is to expand and unify material, process, metrology and product models throughout the entire course of silicon wafer processing. This goal presents the following challenges:

(a) Extensive paradigm shift to model-based metrology and application-oriented interpretation.

(b) Material and measurement standards which include explicit model references and embody computer algorithms, not just simple calculations.

(c) Traceability throughout the production process, including: (i) measurement information management, distribution, and interchange associated with the history of the wafer and (ii) IC fabrication models related to wafer manufacturing process and materials physics models. Examples include gettering vis-à-vis the oxygen content of the CZ crystal and metallics introduced during IC fabrication.

Gettering/Microcontamination Process Simulator Model (1)

Gettering is utilized to remove the metals inadvertently introduced into silicon during IC fabrication. Gettering strategies offer an insurance against unavoidable IC fab altercations. The homogeneous and controlled incorporation of interstitial oxygen (and substitutional carbon to ≤ 0.2 ppma, or essentially non-detectable levels, to prevent spurious oxygen precipitation (71)) is required to facilitate the formation of precipitation dislocation complexes (PDC) and associated bulk micro-defects, for internal gettering (72). PDC formation, in conjunction with optimization of the DZ, continues to be of paramount importance.

Bulk PDC and the attendant DZ formation enhance the room temperature carrier generation lifetime and IC performance. Measurements at IC operating temperatures of approximately 80 C, however, indicate degraded device performance due to g–r current from the PDCs (73)(74) (75). The influence of the PDCs on leakage current, even at room temperature, should be re-examined. Optimization of bulk microdefects and the DZ is a function of the

robustness of the IC design to both leakage currents and the fabricator process.

The importance of "defect engineering" to optimize the formation of bulk PDCs and the DZ for effective gettering, while concurrently ensuring controlled wafer warpage, is a critically important NTRS opportunity. Methodologies emphasizing segregation (i.e., solubility gettering) will become especially important in the sub-1000 C ULSI processing era (13). The high concentration of p-type dopant in the substrate of p/p$^+$ epitaxial structures, for example, is an important example of solubility enhanced gettering (19)(76)(77)(78). A comprehensive program to examine gettering technologies and the formation of a microstructure-free DZ, appropriate for low temperature ULSI processing in the gigabit era, is required. The technology must emphasize low-temperature, element-specific gettering phenomena and the stability of gettered impurity structures during subsequent thermal processing. Quantitative determination of the gettering efficiency with the density, morphology and distribution of microdefect structures—including the role of the nuclei size distribution for condensed matter aggregation phenomena—continues to be one of the most scientifically challenging agendas in the development of high-performance ICs. A concurrent agenda is the development of silicon surface preparation methodologies based on the development of alternate chemistries for the effective removal of both metals *and* particles. Long-range applications of this research will require the development of a gettering/microcontamination process simulator, applicable for arbitrary thermal budgets, process sequences and device structures (i.e., SOI), to more realistically determine the starting wafer requirements.

In that regard, understanding the important role of silicon interstitials and vacancies, managing the mechanism (and mutual interaction) of oxygen precipitation with external gettering—such as polysilicon—and relating the silicon material thermal history to bulk crystal uniformity (79)(80) are issues requiring concurrent consideration. Improved homogeneity of oxygen and control of incipient micro-defect structures such as oxidation induced stacking faults (OISF) in the grown crystal is actively being pursued. Research in advanced crystal growth techniques includes magnetic CZ (MCZ), especially for 300 mm wafers (81), continuous crystal growth and double-layer CZ (82). MCZ has already been used for charge-coupled device (CCD) applications. Continued crystal growth research for both 200 mm and 300 mm is essential. It should be emphasized that the interstitial-vacancy concentrations in the silicon wafer do not necessarily represent their concentrations at the silicon-melt interface during crystal growth (83)(84)(85). Rather, they reflect conditions subsequent to the complete crystal growth process due to the complicated set of microscopic, thermally dependent defect reactions in silicon during crystal growth (83)(84)(85)(86) and during IC fabrication (86).

Interstitial oxygen in the 20–26 ppma range for polished wafers will be required to minimize degradation such as shorting the storage capacitor in trench structures. Reduced interstitial oxygen will also reduce wafer warpage from supersaturated interstitial oxygen precipitation in sub-1000 C ULSI IC processes, while still achieving some degree of relaxation (i.e., internal gettering), assuming "similar" IC thermal cycles. The re-emergence of external gettering, however, such as polysilicon or soft back-surface damage, in conjunction with internal gettering may be warranted with the trend towards lower interstitial oxygen (87), although the benefits of DSP wafers may not then be achieved. The characterization of the wafer back-surface has become more important to the preparation of ULSI quality wafers. For example, the extent of "soft" back-surface damage has often been characterized via a test based on assessing the surface micro-defect density for a given oxidation cycle. Additionally, the determination of the allowable back-surface rms microroughness and appropriate spatial frequency range require new metrology tools and standards for their utilization.

The precipitation of interstitial oxygen during internal gettering varies with the 6–8th power of the initial interstitial oxygen concentration during IC thermal processing (88)(89). This dependence, however, was observed when typical IC processing temperatures were greater than 1000 C and the interstitial oxygen was in the high 20 and low 30 ppma regime. To control the oxygen precipitation to $\pm 10\%$ at 30 ppma, for example, requires the interstitial oxygen to be controlled to approximately ± 0.5 ppma. The measurement technique to achieve ± 0.5 ppma resolution is formidable, especially since two measurements (before and after) are required. Continued improvement in the precision of SIMS measurements and gas fusion analysis is required. This particular exponent dependence, however, may be due to the initial operating position on the S-shaped curve of precipitated oxygen versus initial interstitial oxygen. Since the initial operating position on the transfer curve will decrease to the lower and mid-20s ppma

interstitial oxygen range for low temperature ULSI processing, both the amount of oxygen precipitation and the exponent is expected to decrease.

300 mm Diameter Wafers

Conversion from 200 mm to 300 mm diameter wafers at the end of the decade appears to be necessary to achieve the required economies of scale in the fabrication of large die (90). The introduction of 300 mm wafers, however, will affect virtually every IC process step. Large diameter wafer production and IC process and metrology equipment costs must all be considered to ensure the financially healthy conversion to 300 mm. Lower process costs will be driven by improved CoO analyses and single wafer processes as well as by 300 mm wafer specifications developed through international standards committees. Global coordination and partnerships among wafer, IC metrology and IC equipment suppliers, in conjunction with IC manufacturers, is absolutely imperative. In this regard, international standards will play an increasingly important role in facilitating cost reduction.

The management of thermal-mechanical stresses during ULSI processes for 300 mm (and 200 mm) wafers in vertical furnaces as well as optimization of the synergistic interaction of wafer spacing, insert/withdraw dwell temperature, ramp rates and interstitial oxygen concentration on oxygen precipitation and wafer warpage will continue (91). These studies have become especially important with the advent of rapid thermal processes (92)(93).

Alternative Materials and Processes

Selection of the starting silicon material involves choosing either Czochralski (CZ) polished or epitaxial wafers and generating the material specifications. Logic devices are universally manufactured on higher cost epitaxial wafers, primarily because of latch-up suppression and reduced gate oxide defect density. The higher yield, however, is considered to offset the higher cost of epitaxial material. Memory circuits, however, are commonly manufactured on CZ polished wafers because of lower cost and less stringent demands on IC performance.

High-energy (MeV) ion implanted wafers have been demonstrated as a potential low-cost alternative to epitaxial wafers (94). The technique introduces buried dopant profiles after critical thermal oxidation steps, with gettering benefits, latch-up control and improved transistor performance (94). Defects introduced at MeV energies must be controlled.

Hydrogen annealing of polished or MeV ion implanted wafers may also be required to produce GOI performance equivalent to epitaxial wafers (95). The hydrogen ambient results in a reduced oxygen partial pressure above the polished wafer surface, compared to oxidizing environments, thereby facilitating an enhanced oxygen concentration out-diffusion profile with a concurrent reduction of spurious micro-structures in the DZ. The quality of the near-surface region of the polished wafer accordingly approaches that of an epitaxial wafer. The detailed microscopic reactions involving oxygen and intrinsic point defects in silicon, as well as consideration of the ambient such as hydrogen, argon and nitrogen, is an intense area of research.

Silicon-on-insulator (SOI) obviates the concern over latch-up while offering the potential of low-power applications, fully depleted CMOS devices, fewer process steps, faster device speed, increased device density and perhaps, most importantly, the opportunity of fabricating a 0.18 μm IC utilizing a 0.25 μm equipment tool set. Advances in bonded wafer technology with thickness uniformity better than ±5% will be required in 0.10 μm silicon films. The availability of SOI wafers with flatness and defect characteristics approaching bulk wafers is essential. Understanding the relative benefits of SOI technologies, for example, bonded wafers, bond and etch back SOI (BESOI) and separation-by-implantation-of-oxygen (SIMOX) is required to determine whether SOI will ever become more than a niche technology. Alternative metal gettering schemes (96)(97), cost, IC design/ layout, process tool issues, performance and reliability for SOI are major considerations.

Wafer Cleaning

Contamination-free manufacturing and optimization of the elements of the manufacturing chain for cost-effective minimization and control of contaminants are essential. The contaminants may be organics, metals, anions or particles as well as undesired structural/chemical complexes in the near-surface regions of the silicon wafer. Solutions to this challenge will require improved process equipment, in—situ monitoring and assessment of the efficacy of higher purity materials, chemicals and gases. IC fab process operations must be continuously monitored to ensure that no unwanted sources of particles, contamination or other condensable materials come into contact the silicon surface. Such contamination might superficially appear to be a defect in the starting wafer. Strategic partnerships between IC fabs and material, chemical, gas, equipment and metrology suppliers are crucial to realizing mutual CoO benefits.

Current and future IC designs will require more efficient cleaning procedures to remove both metals *and* particles (19). These procedures have been receiving increased attention as they can modify the silicon surface and impact IC performance. The chemical nature of the wafer surface (i.e., hydrophobic versus hydrophilic) has also become critically important for sub-10 nm gate dielectrics. A hydrophilic surface (or an equivalent chemically passivated surface that is readily dissolved before IC processing) prevents spurious metals and hydrocarbons from adsorbing onto the silicon surface. The benefits of an HF-last clean, however, have been reported (98). The wafer back-surface must be prepared with consideration of the particle and surface metal levels on the front-surface. Wafer edge polishing will reduce a source of particles. Increased understanding of the wafer/carrier interaction is also essential. All of these issues will become more important with the accelerated utilization of reclaim, repolished wafers. In particular, the CoO for the wafer cleaning performed on incoming wafers in the IC facility needs to be addressed, inasmuch as a final clean, often regarded as superior, is also done by the wafer supplier (64).

SUMMARY

Crystalline silicon and silicon-based materials, such as silicon-germanium, will continue to be the most intensively studied material systems in the scientific literature. The opportunities for further silicon-based advancements appear boundless, certainly quite sufficient to support numerous research and development thrusts well into the 21st century. Managing logic, memory and CCD capabilities on the same chip, with selected islands of heterostructures on silicon, offers additional market opportunities (99). The synthesizing of unique material configurations on an atomic scale for novel device/circuit applications and the possibility of efficient light emission from silicon-based materials will also afford stimulating and challenging opportunities in the electronics industry as we approach the 21st century. These advances in device and IC microstructures, moreover, will also be required to support the revolutionary metrology sensors of the future.

The general trend in the IC industry is to continually tighten wafer specifications to match the current sensitivity limits of the metrology tools. Specifications, metrology and standards are now realized to be the essential context in which silicon wafer trends are expressed. The need for the "best wafer possible," however, must be carefully balanced against CoO considerations. Models of the relationships among wafers, processes and measurements provide the tools for identification of the optimal specifications. Although the several epitaxial silicon wafer parameters discussed are consistent with the 0.35–0.25 μm design rule trends noted in the NTRS, the necessity of these ever tightening starting wafer trends is questionable in view of the complex cumulative processing effects of a 20–25 mask, ULSI fabrication process. The development of a gettering/microcontamination process simulator, applicable for arbitrary thermal budgets, process sequences and device structures (i.e., SOI), is an example of an initial step in expanding our ability to more realistically determine the starting wafer requirements.

The control, characterization and, in some cases, utilization of both grown-in and process-induced defects in silicon are critical to ensure the desired IC performance (86). This is an ever-evolving scenario as IC process and device/circuit models offer invaluable insight into silicon material requirements and advances in metrology make measurements more meaningful. Improved control of the magnitude, tolerance and uniformity of silicon material characteristics, coupled with an improved understanding of the inter-relationships among silicon material characteristics, IC fabrication processes, device parameters and IC performance will continue to be the key to superior IC product characteristics, reliability and yield (86).

The achievement of cost-effective selective scaling and component miniaturization is critical as the electronics industry moves toward 21st century technologies. Lower process costs will be driven by equipment standardization, single-wafer processes and larger diameter wafers. Advances in single-wafer processing will be needed to introduce 300 mm wafers. This critical path issue will require economic analysis. Equipment tool sets and equipment qualification programs will require an integrated global vision and plan. In conjunction with IC process/device/circuit innovations for cost-effective production of ICs (100), electronics will offer unsurpassed opportunities for an improved quality of life for the world's citizens.

ACKNOWLEDGMENTS

The assistance of Philippe Maillot and the SEMATECH Analytical Laboratory team in the various measurements is appreciated. Peter Solomon's support in the three-layer epitaxial film thickness measurements is also appreciated. Alain Diebold's clarification of a number of diagnostic and related instrumentation issues is deeply appreciated. The members of the SEMATECH Wafer Defect Science and Engineering Task Force and Particle Counting/Surface Microroughness Task Force are acknowledged for their assistance in the development of a number of issues discussed in this manuscript. Finally, we thank Vishnu Bhat for preparation of the graphics and Kim Upshaw for compiling and formatting the manuscript.

REFERENCES

(1) *National Technology Roadmap for Semiconductors*, Semiconductor Industry Association, 4300 Stevens Creek Boulevard, Suite 271, San Jose, CA., November, 1994

(2) H.R. Huff, "Semiconductors, Elemental – Material Properties," *Encyclopedia of Applied Physics*, VCH Publishers (to be published)

(3) H.R. Huff, "Silicon Materials For The Mega–IC Era," *ULSI Science and Technology/1993*, (edited by G.K. Celler, E. Middlesworth and K. Hoh), 103–132 (1993), The Electrochemical Society, Inc., Pennington, N.J.

(4) W.M. Bullis and H.R. Huff, "Silicon For Microelectronics," *The Encyclopedia of Advanced Materials* (edited by D. Bloor, R.J. Brook, M.C. Flemings and S. Mahajan), pp. 2478–2496, Pergamon Press (1994)

(5) H.R. Huff, R.K. Goodall and V. Bhat, "Competitive Analysis of 200 mm Epitaxial Silicon Wafers," *ECS Extended Abstracts*, **95–1**, 269–270, (1995), The Electrochemical Society, Inc., Pennington, N.J.

(6) E. Morita et. al., "Distinguishing COPs From Real Particles," *Semiocnductor International*, July, 156–162 (1994)

(7) H. Yamagishi et. al., "Evaluation of FPDs and COPs in Silicon Single-Crystals," *Semiconductor Silicon/94,* (edited by H.R. Huff, W. Bergholz and K. Sumino), 124–135 (1994)

(8) K. Kugimiya, et.al., "Characterization of Mirror Polished Silicon Wafers Using "Makyoh," the Magic Mirror Method, *Semiconductor Silicon /1990* (edited by H. R Huff, K.G. Barraclough and J-i Chikawa), 1052–1067 (1990), The Electrochemical Society Inc., Pennington, N.J.

(9) C.T. Larson and S. Arsenault, "Single 0.100 Micron Spheres are Detected by a Laser Inspection System as Verified by Review in a Scanning Electron Microscope," *Microcontamination '92 Conference Proceedings*, 14–25 (1992)

(10) W. Taylor, U. Gosele and T.Y. Tan, "Present Understanding of Point Defect Parameters and Diffusion in Silicon: An Overview," *Process Physics and Modeling in Semiconductor Technology,* (edited by G.R. Srinivasan, K. Taniguchi and C.S. Murthy), 3–19 (1993), The Electrochemical Society, Inc., Pennington, N.J.

(11) E.R. Weber, "Diffusion and Solubility of Transition Metals," *Properties of Silicon,* EMIS Datareviews Series, **4**, p. 409–451 INSPEC, London, (1988)

(12) H. Lemke, "Characterization of Transition Metal-Doped Silicon Crystals Prepared by Float-Zone Technique," *Semiconductor Silicon/1994,* (edited by H.R. Huff, W. Bergholz and K. Sumino), The Electrochemical Society, Inc., Pennington, N.J., 695–710B (1994)

(13) K. Graff, *Metal Impurities in Silicon-Device Fabrication,* Springer-Verlag (1995)

(14) D.K. Schroder, *Semiconductor Material and Device Characterization,* John Wiley & Sons, Inc., New York (1990) (see chapter 7)

(15) A.C. Diebold et. al., "Evaluation of Surface Analysis Methods For Characterization of Trace Metal Surface Contaminants Found in Silicon Integrated Circuit Manufacturing", *J. Vac. Sci. Technol. A,* **10,** 2945–2952 (1992)

(16) H.R. Huff, "Semiconductor Silicon," *Concise Encyclopedia of Semiconducting Materials and Related Technologies"* (edited by S. Mahajan and L.C. Kimerling), 478–492 (1992), Pergamon Press

(17) Sugiura, *Proceedings of the 19th Workshop on ULSI Ultra Clean Technology,* Sponsored by the Ultra Clean Society, 29–30 September, 1992

(18) M. Nagase, M. Kimura, Y. Okui, K. Inino and M. Kanazawa, "Metal Contamination in Ion Implantation Process," *ECS Extended Abstracts,* **94–2,** 742–743 (1994)

(19) *SEMATECH 1994 Wafer Defect Science and Engineering Task Force Report,* (edited by H.R. Huff and C.R. Helms), see chapter by C.R. Helms and H.G. Parks (to be published)

(20) *Handbook of Semiconductor Wafer Cleaning Technology,* (edited by W. Kern), Noyes Publications, Park Ridge, N.J. (1993)

(21) M. Aoki, A. Hara and A. Ohsawa, *Fundamental Properties of Intrinsic Gettering of Iron in a Silicon Wafer,* J. Appl. Phys., **72,** 895–898 (1992)

(22) E. Kooi, "The Invention of LOCOS", *IEEE Case Histories of Achievement in Science and Technology,* **1,** IEEE Press, (1991)

(23) Y. Ichida, T. Yanada, S. Kawado, "Influence of Impurity-Decorated Stacking Faults on The Transient Response of Metal Oxide Semiconductor Capacitors," *Lifetime Factors in Silicon,* ASTM, **STP 712,** 107–118, (1980)

(24) P.S.D. Lin, R.B. Marcus and T.T. Sheng, "Leakage and Breakdown in Thin Oxide Capacitors—Correlation With Decorated Stacking Faults," *J. Electrchem. Soc.,* **130,** 1878–1883 (1983)

(25) A.C. Diebold and B. Doris, "A Survey of Non–destructive Surface Characterization Methods Used to Insure Reliable Gate Oxide Fabrication for Silicon IC Devices," *Surface & Interface Analysis,* **20,** 127–139 (1993)

(26) S.Verhaverbeke et. al., *The Effect of Metallic Impurities on the Dielectric Breakdown of Oxides and Some New Ways of Avoiding Them,* IEDM, 71–74 (1991)

(27) C.R. Helms et. al., "Fundamental Metallic Issues for Ultraclean Wafer Surfaces From Aqueous Solutions," *Proceedings Second International Symposium on Ultra–clean Processing of Si Surfaces,* (edited by M. Heyns) 205–212, Acco, Leuven, Belgium (1994)

(28) ASTM F1.06 Document F 1526–94: Test Method for Measuring Surface Metal Contamination on Silicon Wafers by Total Reflection Xray Fluorescence Spectroscopy, R.S. Hockett, Draft 1, June, 1994

(29) Working Draft "Surface Chemical Analysis—Determination of Contamination Element Contents of Silicon Wafer—Total Reflection Xray Flourescence Spectroscopy (TXRF)," (Document ISO/TC201/WG2 N5, Annex N9 and U.S. Comments N10), Professor Y. Gohshi, Department of Applied Physics, University of Tokyo, Convener, A.C. Diebold, U.S.A.

(30) Y.E. Strausser et.al., "Measurement of Silicon Surface Microroughness by AFM," *ECS Extended Abstracts,* **94–1,** 461–462 (1994)

(31) T. Ohmi et.al., "Dependence of Thin–Oxide Films Quality on Surface Microroughness," *IEEE Trans. Electron. Devices,* **ED–39,** 537–545 (1992)

(32) O.L. Alerhand et. al., "Finite–Temperature Phase Diagram of Vicinal Si(100) Surfaces," *Phys. Rev. Letts.,* **64,** 2406–2409 (1990)

(33) E. Pehlke and J. Tersoff, "Nature of The Step–Height Transition on Vicinal Si (001) Surfaces," *Phys. Rev. Letts.,* **67,** 465–465 (1991)

(34) E. Pehlke and J. Tersoff, "Phase Diagram of Vicinal Si(001) Surfaces," *Phys. Rev. Letts.,* **67,** 1290–1293 (1991)

(35) J.C. Stover, *Optical Scattering,* SPIE (to be published)

(36) W.M. Bullis, "Microroughness of Silicon Wafers," *Semiconductor Silicon/94,"* (edited by H.R. Huff, W.Bergholz and K. Sumino), 1156–1169 (1994), The Electrochemical Society, Inc.

(37) S. Verhaverbeke et. al., in "Semiconductor Silicon/94," (edited by H.R. Huff, W.Bergholz and K. Sumino, 1170–1181 (1994), The Electrochemical Society, Inc.

(38) *Proceedings of The International Workshop on Silicon Materials for Mega–IC Applications*, (edited by H.R. Huff et. al), (1991), Sponsored by SEMATECH, SEMI, NIST and ASTM

(39) C.A. Mack, "Understanding Focus Effects in Submicron Optical Lithography, Part 3: Methods for Depth-of-Focus Improvements," *Optical/Laser Microlithography V*, (edited by J.D. Cuthbert), SPIE, **1674**, 272–284 (1992)

(40) H.R. Huff and H. Weed, "Experimental Assessment of 150 mm P/P$^+$ Epitaxial Silicon Wafer Flatness for Deep Sub-Micron Applications," *Integrated Circuit Metrology, Inspection and Process Control V*, (edited by W.H. Arnold), SPIE, **1464**, 278–293 (1991)

(41) B.J. Lin, "The Optimum Numerical Aperature For Optical Projection Microlithography," *Optical/Laser Microlithography IV*, (edited by V. Pol, SPIE, **1463**, 42–53 (1991)

(42) 1990 SEMI International Standards, SEMI M1–90, *Specifications for Polished Monocrystalline Silicon Wafers*, Figre A1.1 Flatness Decision Tree

(43) H.R. Huff, G.H. Popham and R.W. Potter, "Correlation of 150 mm P/P$^+$ Epitaxial Silicon Wafer Flatness Parameters for Deep Submicron Applications," *J. Electrochem. Soc.*, **140**, 229–241 (1993)

(44) R.K. Goodall, (to be submitted)

(45) H.R. Huff et. al., "Correlation of 150 mm Silicon Wafer Site Flatness With Stepper Performance For Deep Sub-Micron Applications," *IC Metrology, Inspection and Process Control VI*, (edited by M.T. Postek), SPIE **1673**, 357–368 (1992)

(46) L. Denes and H.R. Huff, "A Fourier Analysis of Silicon Wafer Topography," *J. Electrochem. Soc.*, **139**, 558C (1992)

(47) T. Abe et. al., "The Ductile Mode Grinding Technology Applied to Silicon Wafering Process," *Semiconductor Silicon/94*, (edited by H.R. Huff, W. Bergholz and K. Sumino), 207–217 (1994), The Electrochemical Society, Inc.

(48) E. W. Hearn et. al., "Film-Induced Stress Model," *J. Electrochem. Soc.*, **133**, 1749–1751 (1986)

(49) R. Wise et. al., "Silicon Epitaxial Layers: Defects," *The Encyclopedia of Advanced Materials*, (edited by D. Bloor, R.J. Brook, M.C. Flemings and S. Mahajan), 2469–2478 (1994) Pergamon Press

(50) B.W. Fowler et. al., "The Measurement of Sub-Micron Epitaxial Layer Thickness and Free Carrier Concentration by Infrared Reflection Spectroscopy," *Diagnostic Techniques for Semiconductor Materials and Devices/94*," (edited by D.K. Schroder, J.L.Benton and P. Rai-Choudhury), 254–265 (1994)

(51) K. Takami, *International Fine Particles Task Group (SEMI)*, Tokyo, Dec. 1, 1992

(52) H.R. Huff et.al., "Measurement of Silicon Particles by Laser Surface Scanning and Angle-Resolved Light Scattering," *Microcontamination '94 Conference Proceedings*, 121–131 (1994)

(53) H.R. Huff et al., "Measurement of Silicon Particles by Laser Surface Scanning and Angle-Resolved Light Scattering," *Fifth International Symposium on ULSI Science and Technology* (edited by E.M. Middlesworth and H.Z. Massoud), (to be published)

(54) H.C. van de Hulst, *Light Scattering by Small Particles*, Dover Publications, Inc., New York, (1981)

(55) D. Hirleman, private communication

(56) J.C. Stover, "Scatterometry: Principles, Applications, Limitations and Future Prospects," *International Workshop on Semiconductor Characterization: Present Status and Future Needs*, (edited by W.M. Bullis, A.C. Diebold and D. Seiller), (to be published)

(57) W.M. Bullis and H.R. Huff, "Applications of Carrier Lifetime and Diffusion Length Measurements," *Fifth International Symposium on ULSI Science and Technology* (edited by E.M. Middlesworth and H.Z. Massoud), (to be published)

(58) L.C. Kimerling and J.R. Patel, "Silicon Defects: Structures, Chemistry and Electrical Properties," *VLSI Electronics: Microstructure Science*, (edited by N.G. Einspruch and H.R. Huff), **12**, 223–267 (1985)

(59) J. Lagowski et. al., "Iron Detection in the Part Per Quadrillion Range in Silicon Using Surface Photovoltage and Photodissociation of Iron-Boron Pairs," *Appl. Phys. Lett.*, **63**, 3043–3045 (1993)

(60) G. Zoth and W. Bergholz, A Fast, Preparation-Free Method to Detect Iron in Silicon, J. Appl. Phys., **67**, 6764–6771 (1980)

(61) L. Jastrzebski et. al., Monitoring of Heavy Metal Contamination During Chemical Cleaning with Surface Photovoltage in "Proceedings of 2nd Symposium on Cleaning Technology in Semiconductor Device Manufacturing," (edited by J. Ruzyllo and R.E. Novak), 294–313 (1992), The Electrochemical Society, Inc., Pennington, N.J.

(62) D.K. Schroder, J.D. Whitfield and C.J. Varker, "Recombination Lifetime Using the Pulsed MOS Capacitor," *IEEE Trans. Electron Devices*, **ED–31**, 462–467 (1984)

(63) D.K. Schroder, "The Concept of Generation and Recombination Lifetimes in Semiconductors," *IEEE Trans. Electron Devices*, **ED–29**, 1336–1338 (1982)

(64) R. Helms, "Si Surface Preparation and Wafer Cleaning: The Critical Role, Status and Future Needs in Characterization," *International Workshop on Semiconductor Characterization: Present Status and Future Needs,* (edited by W.M. Bullis, D. Seiller and A.C. Diebold), (to be published)

(65) R.K. Goodall and N. Poduje, "Data Point Selection For Site Qualification of Wafers For ULSI Lithography," *IC Metrology, Inspection and Process Control IV,* (edited by W.H. Arnold), SPIE, **1261**, 240–252 (1990)

(66) R.K. Goodall and F. Alvarez, "Characterization of Stepper Chuck Performance," *IC Metrology, Inspection and Process Controll VII,* SPIE **1926**, (edited by M.T. Postek), 236–248 (1993)

(67) H. Shirai et. al., "64M-bit DRAM Production With i-Line Stepper," *Optical/Laser Microlithography IV,* (edited by V. Pol), SPIE, **1463**, 256–274 (1991)

(68) H. Fukuda et.al., "New Approach to Resolution Limit and Advanced Image Formation Techniques in Optical Lithography," *IEEE Trans. Electron Devices,* **ED–38**, 67–75 (1991)

(69) P.K. Vasudev, "Phase Shifting Masks for Application to 0.25 μm Lithography Using Both DUV and I-Line Exposure," presented at the Photopolymer Science and Technology Conference, Chiba University, Tokyo, Japan, June 1993

(70) D. Bruner et. al., "Silicon Substrate Properties and ULSI Yield," *Semiconductor Silicon/90,* (edited by H.R. Huff, K.G. Barraclough and J. Chikawa), 912–923B (1990)

(71) T. Fukuda, "Mechanical Strength of Czochralski Silicon Crystals With Carbon Concentrations From 10^{14} to 10^{16} cm^{-3}," *Appl. Phys. Lett.,* **65**, 1376–1378 (1994)

(72) T.Y. Tan, E.E. Gardner and W.K. Tice, "Intrinsic Gettering by Oxide Precipitate Induced Dislocations in Czochralski Si," *Appl. Phys. Letts.,* **30**, 175–176 (1977)

(73) S.N. Chakravarti et.al., "Oxygen Precipitation Effects on Si n⁺p Junction Leakage Behavior," *Appl. Phys. Lett.,* **40**, 581–583 (1982)

(74) D.K. Schroder et.al., *VLSI Science and Technology/85,* (edited by W.M. Bullis and S. Broydo), 419–428 (1985)

(75) A. Ohsawa et.al., "Effects of Impurities on Microelectronic Devices," *Semiconductor Silicon/90,* (edited by H.R. Huff, K.G. Barracloughn and J. Chikawa), 601–613 (1990)

(76) D. Gilles and H. Ewe, "Gettering Phenomena in Silicon," *Semiconductor Silicon/1994,* (edited by H.R. Huff, W. Bergholz and K. Sumino), 772–783 (1994)

(77) M. Aoki, T. Itakura and N. Sasaki, "Gettering of Iron Impurities in P/P⁺ Epitaxial Silicon Wafers with Heavily Boron-Doped Substrates," *Appl. Phys. Letts.,* **66**, 2709–2711 (1995)

(78) J.L. Benton, et. al., "Enhanced Segregation Gettering of Iron in Silicon by Boron Ion-Implantation," *Fifth International Symposium on ULSI Science and Technology* (edited by E.M. Middlesworth and H. Z. Massoud), (to be published)

(79) S. Miyahara et.al., "Thermal History Analysis of Crystal by the Transient Global Heat Transfer Model for CZ Crystal Growth," *Semiconductor Silicon/90,* (edited by H.R. Huff, K.G. 'Barraclough and J. Chikawa) 94–104 (1990)

(80) S. Miyahara et. al., *Computer Aided Innovation of New Materials,* (edited by M. Doyama, T. Suzuki, J. Kihara and R. Yamamoto), 561–564 (1991)

(81) T. Masui, "Supply of Large Diameter Wafer," *Technical Proceedings SEMI Technology Symposium,* SEMICON/Japan, 345–347 (1994)

(82) S. Kobayashi et. al., "Double Layered CZ (DLCZ) Silicon Crystal Growth," *Semiconductor Silicon/94,* (edited by H.R. Huff, W. Bergholz and K. Sumino), 58–69 (1994), The Electrochemical Society

(83) R.A. Brown, D. Maroudas and T. Sinno, "Modelling Point Defect Dynamics in The Crystal Growth of Silicon," *J. Cryst. Growth,* **137**, 12–25 (1994)

(84) T. Sinno and R. A. Brown, "Point Defect Dynamics in Silicon and the Connection Between Microdefect Formation and Operating Conditions in the Bulk Growth of Silicon," *Semiconductor Silicon/1994* (edited by H.R. Huff, W. Bergholz and K. Sumino), 625–634B (1994)

(85) T. Sinno, Z. Kurt Jiang and R.A. Brown, *Atomistic Simulation of Point Defects in Silicon at High Temperatures* (submitted for publication)

(86) J.E. Lawrence and H.R. Huff, "Silicon Material Properties for VLSI Circuitry," *VLSI Electronics: Microstructure Science,* (edited by N.G. Einspruch), **5**, 51–102 (1982), Academic Press, NY

(87) W.M. Bullis and W.C. O'Mara, "Large–Diameter Silicon Wafer Trends," *Solid State Technology,* April, 59–65 (1993)

(88) H.R. Huff et. al., "Some Observations on Oxygen Precipitation/Gettering in Device Processed Czochralski Silicon," *J. Electrochem., Soc.,* **130**, 1551–1555 (1983)

(89) W. Bergholz, J.L. Hutchison and G.R. Booker, "Oxygen-Related Defects in CZ-Silicon After Annealing at 635°C," *Semiconductor Silicon/86,* (edited by H.R. Huff, T. Abe and B. Kolbesen), 874–888 (1986)

(90) H.R. Huff et.al., "SEMATECH Large Diameter Wafer Perspective," *Forum on Silicon Wafer Diameter: Is There a Change in The Future?*, SEMICON/WEST 1994, p. 50–68

(91) C.O. Lee and P. Tobin, "The Effect of CMOS Processing on Oxygen Precipitation, Wafer Warpage and Flatness," *J. Electrochem. Soc.*, **133**, 2147–2152 (1986)

(92) M. Schrems et. al., "Simulation of Temperature Distributions During Fast Thermal Processing," *Semiconductor Silicon/94,* (edited by H.R. Huff, W. Bergholz and K. Sumino), 1050–1063 (1994)

(93) The 145th Committee for Crystal Processing and Evaluation Technology, 68th Seminar Materials, sponsored by the Japan Society for the Promotion of Science, Dec. 21, 1994, Tokyo, Japan (In Japanese, translated by SEMATECH)

(94) K. Tsukamoto et. al., "High Energy Ion Implantation for ULSI: Well Engineering and Gettering," *Solid State Tech.*, **35** (6), 49–55 (1992)

(95) M. Gardner et. al., "Hydrogen Denudation for Thin Oxide Quality, Device Performance and Potential Epitaxial Elimination," *1994 VLSI Technology Symposium*, 111–112 (1994)

(96) G.A. Rozgonyi et. al., "Low Temperature Impurity Gettering For Giga-Scale Integrated Circuit Technology," *Semiconductor Silicon/94,* (edited by H.R. Huff, W. Bergholz and K. Sumino), 868–883 (1994)

(97) S.V. Koveshnikov and G.A. Rozgonyi, "Iron Diffusivity in Silicon: Impact of Charge State," *Appl. Phys. Lett.*, **66**, 860–862 (1995)

(98) M. Heyns editor, M. Meuris and P. Mertens, coeditors, *Proceedings of the Second International Symposium on Ultra-clean Processing of Silicon Surfaces*, Acco Lauven/Amersfoort, 1994

(99) *Heterostructures on Silicon: One Step Further With Silicon* (edited by Y.I. Nissim and E. Rosencher), NATO ASI Series, 160, Kluwer Academic Publishers (1989)

(100) "2001 Semiconductor Manufacturing," *Break Through Journal Special Issue #71* (1992)

Microdefect Analysis of Silicon: Tools and Strategies

Lionel C. Kimerling, Jurgen Michel, H. M'saad, and Gerd J. Norga
Materials Processing Center
77 Massachusetts Avenue, Room 13-4118
Cambridge, MA 02139

Defect analysis is applied to silicon materials at two levels of detail: 1) simple, rapid measurements for process monitor and control; and 2) complex, design matrices for process diagnostics. In every case, it is desirable to measure the property dictating circuit functionality. Thus, for microelectronics, electrical measurements are preferred for in-line monitors, microstructure analysis and spectroscopy. The physical foundations of this characterization methodology with examples of application in the laboratory and the fabline are presented.

INTRODUCTION

The increasing complexity of integrated circuit fabrication, involving hundreds of unit processes, demands in-line control of processes. Electrical test at the final product stage places significant costs of operations at risk. In fact, the value of yield losses increase with the "sunk costs" of prior wafer processing. In-line sensing of silicon quality is essential to increasing both line and die yield. A monitor and control strategy must involve two key issues: 1) the property or attribute to monitor and 2) the level of information content to track.

The performance of integrated circuits depends on electron transport. Monitors of any other attribute are indirect and require validation of significance for each installation and technology. Similarly, all information is not directly relevant to yield. High quantities of information require both complex data handling and complex decision making. The ideal monitor issues an alarm immediately upon violation of control limits. In summary, process monitors should sense the properties controlling product performance and should emit low information content.

Characterization tools can be classified in terms of hierarchies of information content and relative invasiveness. Table 1 so classifies some common silicon characterization tools. In-line monitors such as light scattering for particle detection and RF-PCD for sensing metal contamination are noninvasive, low information sensors. These Level 1 methods are valuable for statistical process control, but they require supplemental investigations for determination of the root cause of a deviation from specification. Level 2 and Level 3 methods are usually part of materials process development activities. They are employed off-line and, hence, are time consuming. However, they give spatial mapping or spectroscopic information that is vital for design (but irrelevant for control).

This work reviews case studies in each level of the hierarchy: Level 1 (RF-PCD); Level 2 (PL); and Level 3 (DLTS, SEM-EBIC). The purpose of the review is to illustrate that fundamental understanding of both the method and the process are required to transform data into knowledge.

APPLICATIONS

Radio-Frequency Photoconductance Decay is a contactless method of measuring minority carrier lifetime.(1) As shown in Figure 1, a coil broadcasts rf energy into the wafer and detects reflected power. A pulsed light source injects minority carriers which introduce power absorption through induced eddy currents. The time constant of the

TABLE 1.

EXAMPLES OF METHODS HIERARCHY

Level 1: In-Line
- Light scattering
- RF-PCD

Level 2: Monitor Wafer: nondestructive
- FTIR
- M-PCD
- PL
- SPV
- X-ray Topography

Level 3: Monitor Wafer: destructive
- DLTS
- EPR
- Etch
- SEM-EBIC
- TEM

recovery of reflected power is the minority carrier lifetime. Typical rf frequencies are 50-500 MHz. The lower carrier frequencies have the advantage of longer working distances and greater skin depth for sensing heavily doped substrates. Figure 2 gives a comparison skin depths for various carrier frequencies.

Wafer Cleaning: Bulk Contamination

Table 2 shows data for wafer lifetimes following gate oxidation.(2) The data represent average values for two wafer vendors (A, B) and two preoxidation cleaning sequences (SC1 last, SC2 last). The SC2 last clean is clearly preferable; and vendor A material is superior for either clean. These measurements were on 8" diameter wafers, in cleanroom air with no sample preparation following oxidation. The SC1 clean is designed for particle and hydrocarbon removal, and the SC2 clean dissolves transition metals from the wafer surface. The data of Table 2 suggest lifetime reduction by indiffusion of transition metal contaminants during oxidation heat treatment.

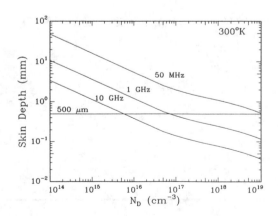

Figure 1. The Radio-Frequency Photoconductance Decay (RFPCD) measurement system.

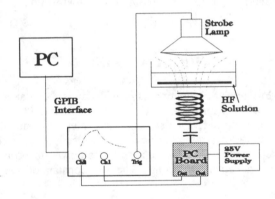

Figure 2. Skin depth as a function of doping concentration in n-type Si for three modulation frequencies in the rf and microwave regions.

TABLE 2.

METALLIC DISSOLUTION BY SC2

Clean		Gate Oxididation		τ_{means}	(μs)
				A	B
HF/SC1/SC2	→	Dry/no Cl_2	→	592	249
HF/SC1	→	Dry/no Cl_2	→	33	3.2

B-doped Si(100), 5.0×10^{15} cm^{-3}

Detection of a contamination problem is sufficient to stop wafers-in-process until the problem is fixed. Identification of the contaminants and their source is appropriate for the higher overhead, Level 2 and Level 3 methods.

Table 3 gives DLTS data for typical transition metal recombination centers in silicon. Figure 3 shows the DLTS spectrum from an SC1 last sample. A small 1 cm^2 piece is cleaved from the wafer, the oxide is stripped in dilute HF, and a Ti Schottky barrier junction is formed by evaporation through a shadow mask. The feature labeled H(0.1) is derived from the (Fe_iB_S) pair. This identification is confirmed by gently heating the sample to 200°C and remeasuring the DLTS spectrum. The pairs are dissociated by the heat treatment and the H(0.1) spectrum is replaced by an equal concentration of the H(0.4) feature which is derived from isolated (Fe_i). Whereas the DLTS spectral features are necessary to identification, the combination of the spectrum and the change in spectra under heat treatment are sufficient for unambiguous identification.(3)

Wafer Cleaning: Surface Contamination

The measured RF-PCD lifetime consists of two components: bulk recombination and surface recombination. Much of the value of the measurement is lost when the signal decay is fit to a single exponential, and a single number is taken as the quality indicator. An additional complication is the deep trap which lengthens carrier lifetime by acting as a temporary bound state (until the carrier is thermally emitted back to the band) in series with recombination transitions. Dislocations, grain boundaries and clusters of point defects exhibit temporary trap behavior. Thus, three channels of minority carrier transitions must be determined, before a PCD experiment can be properly understood: 1) bulk recombination, 2) surface recombination, and 3) thermalization from temporary traps.

TABLE 3.

ELECTRONIC PARAMETERS OF SOME COMMON DEFECTS AND IMPURITIES IN SILICON

Defect	$E_T(eV)$[a]	$T_{1.8\ msec}$ (K)[b]
Ti [−/0]	E(0.09)	51
Ti [0/+]	E(0.28)	116
Ti [+/++]	H(0.28)	198
Cr [0/+]	E(0.22)	108
Cr-B [0/+]	H(0.28)	123
Mn [−/0]	E(0.11)	68
Mn [0/+]	E(0.42)	216
Mn [+/++]	H(0.25)	207
Fe$_i$ [0/+]	H(0.43)	267
(Fe$_i$ • B$_s$) [0/+]	H(0.10)	59
(Ni$_i$ • B$_s$) [0/+]	H(0.14)	88
(Cu$_i$ • B$_s$) [0/+]	H(0.22)	112
Au [−/0]	E(0.53)	288
Au [0/+]	H(0.35)	173
Mo	H(0.28)	191
O donor	E(0.07)	Below freeze out
	E(0.15)	58
Dislocation	E(0.38)	225
	H(0.35)	206
Glide debris	E(0.63-0.68)	288

a E_T: activation energy for electron (E) or hole (H) emission to respective band edge

b $T_{1.8\ msec}$: temperature at which the carrier emission time constant is 1.8 msec.

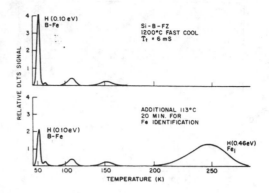

Figure 3. (a) DLTS spectrum of B-doped FZ Si after heat-treatment in an epitaxial reactor at 1200 C for 30 min in H_2 with subsequent fast cool. (b) DLTS spectrum after an additional heat-treatment of 113 C for 20 min in He to produce a decrease in the Fe-B signal, and the associated increase in the Fe$_i$ signal.(3)

exactly equal to the bulk lifetime. When S is infinite, then the measured lifetime for high-quality silicon is the time for a minority carrier to diffuse to the surface. This limiting time can be estimated as $\tau(S=\infty) \approx \dfrac{d^2}{D_n \pi^2} \approx 30\ \mu s$ for a 600 μm thick wafer. Thus, measurement of τ on a wafer with a ground surface cannot reveal useful information concerning the bulk quality unless that quality is very poor ($\tau < 1$ μs). For small S, the limiting contribution of the surface is $\tau_{surf} = \dfrac{d}{2S}$ where d is the wafer thickness.

The spread in measured lifetime values due to changes in surface quality is most pronounced for materials with high bulk lifetimes. Under these conditions one can determine the surface recombination center density N_T with no additional characterization overhead.

$$\frac{1}{\tau_{meas}} = \frac{1}{\tau_{bulk}} + \frac{1}{\tau_{surface}} \qquad (1a)$$

$$= \frac{1}{\tau_{bulk}} + \frac{S}{2d} \qquad (1b)$$

$$S = \sigma\, v_{th}\, N_T \qquad (2)$$

An excellent proof of surface state quantitation is detection of the adsorption isotherm. Figure 5 shows the classic plot of surface coverage θ_s against solution composition for iodine in methanol. The polar CH_3OH molecule

We have found the quality of commercial 8" silicon wafers to be extremely high. Bulk carrier lifetimes of 10-20 ms are typical. At this level of perfection, the surfaces of the wafer must be electrically passivated in order to measure the true bulk lifetime. This passivation is best accomplished by monovalent termination of silicon surface bonds.(4) In this way the sp^3 bulk hybridization is relatively undisturbed at the surface. Near perfect termination (SRV < 1 cm/s) has been achieved by hydrogen (immersion in concentrated HF) and halogens (immersion in an iodine:methanol solution).(5)

The competition between bulk and surface recombination is illustrated in Figure 4. The plot assumes single crystal silicon with no temporary traps. When the surface recombination velocity (S) is zero, the measured lifetime is

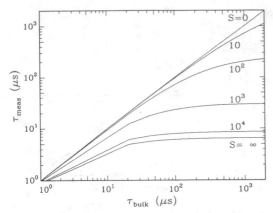

Figure 4. The measured lifetime with RFPCD is a function of both bulk lifetime and surface recombination velocity. At low τ_{bulk}, τ_{meas} is independent of S because the lifetime is limited by bulk recombination. At high τ_{bulk}, τ_{meas} is limited by recombination at the surface.

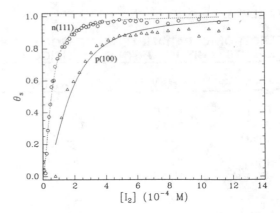

Figure 5. Adsorption isotherms for iodine on Si surfaces.

maintains recombination centers at the silicon surface, yielding $S \approx \infty$. The iodine replaces the methanol by forming a covalent Si-I bond, satisfying the surface bond charge requirement. The coverage is defined as $\theta_s = \frac{N_T(0)-N_T(t)}{N_T(0)}$ where $N_T(0)$ and $N_T(t)$ are determined from S by Equations 1 and 2. The data show an excellent fit to the classic Langmuir isotherm:

$$\frac{\theta_s}{1-\theta_s} = \left(\frac{k_a}{k_d}\right)^{1/n} [I]^{-1/n} \qquad (3)$$

where k_a and k_d and the adsorption and desorption rate constants, and n is the reaction order. The fit of the data give n=0.87 for Si(100) and n=0.98 for Si(111). Values of n=1 denote independent atom-by-atom chemisorption. The scatter for Si(100) is believed to relate primarily to surface roughness. The remarkable fit and consistency of the data confirm quantitation, especially at low coverages.

A related example, critical to silicon fabrication technology is the deposition of transition metals from cleaning solutions. To develop an in-line monitor, one must employ both a sensitive detection method and an understanding of the cleaning process chemistry. To first order, acid solutions should deposit only metals which lie above hydrogen in the electromotive series. The metal must be reduced upon deposition, and the oxidation potential is a crude measure of the free energy change ΔG upon deposition. A negative free energy charge is required for the metal ion to leave solution.

The deposition tendency of a metal from a solution of arbitrary composition can be predicted by evaluation of the driving force for reactions which yield reduced metal atoms or insoluble metal oxides.(6) An accurate and instructive

approach to this problem is the Pourbaix diagram shown in Figure 6. The oxidation potential of the solution E is plotted against the pH. For the industry standard SC-1 and SC-2 solutions, E is determined by the reduction reaction of hydrogen peroxide to water, i.e. the H_2O_2 (oxidant) content. The pH reflects the acid/base balance, i.e. the H^+ content.

Each metal system has a unique phase diagram which determines its stability in solution. A phase boundary demarcates between two regions: 1) where ΔG is positive for precipitation and the metal ion remains in solution; and 2) where ΔG is negative and the deposition reaction proceeds. The power of the diagram is visualization of the interaction between cleaning solution composition and metal contamination. From Figure 6 one can readily conclude that low pH and high E are required to sustain Fe in solution. Since SC-1 does not meet these requirements, Fe purity is a dominant concern. The data of Table 2 confirm this prediction: The SC-1 last clean yields a low lifetime (due to Fe contamination) while the SC-2 last deposits no Fe.

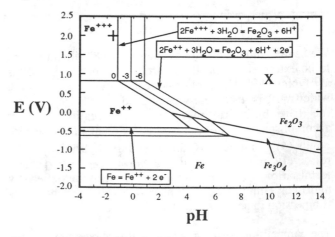

Figure 6. E-pH (Pourbaix) diagram for Fe in aqueous solutions at T=75°C. Phase boundaries for 1, 10^{-3}, 10^{-6} molal concentration in solution are shown (+=SC-2; x =SC-1).

Crystal Growth: Microdefects

The determination of spatial patterns and quantitative distributions of microdefects require scanning probes. The spatial maps are generally used for one of two purposes: 1) determination of the root cause of a yield problem by pattern recognition, and 2) quantitative analysis of heterogeneity in doping. Choice of methods involve evaluation of the measurement overhead against the information content.

SEM-EBIC is the most powerful of the transport property mapping tools.(7) The resolution is given by a convolution of the e-beam spot size and the carrier

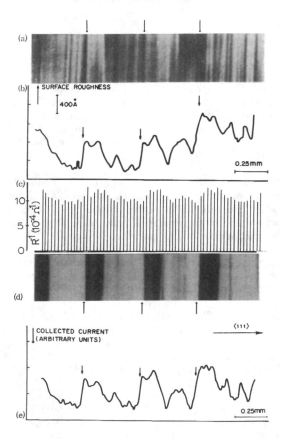

Figure 7. Longitudinal section of a phosphorus-doped swirl-free crystal. Doping level: 2.5 x 10^{15} cm^{-3} (a) Nomarski interference micrograph after preferential etching. (b) Talystep plot of height variations produced by preferential etching and measured along a track perpendicular to the striae. (c) Spreading resistance profile of region shown in (a), measured along a line adjacent to the Talystep trace shown in (b). (d) Charge collection display (inverted contrast) of same areas shown in (a) after repolishing (20 kV, 0 bias). (c) Collected Current measured along same track as Talystep profile of (d).(7)

diffusion length. Temperature can be continuously varied from 4.2–300 K to scan the Fermi level (defect state occupation). Sample preparation is destructive, but simple: application of a Schottky barrier by metal

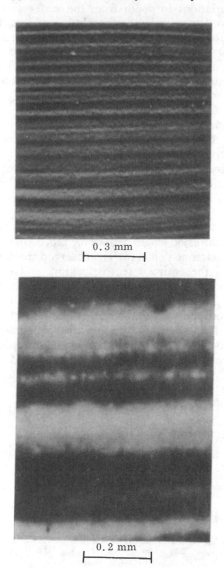

Figure 8. Charge collection micrographs of the center of a silicon crystal containing swirl defects. Longitudinal section. Doping level n = 10^{13} cm^{-3}. (a) shows dopant striations (10 kV, 0 bias); (b) shows swirl defects (30 kV, -2V bias).(7)

evaporation onto the silicon surface. The source of EBIC contrast depends on several factors: 1) injection level (beam voltage and current); 2) carrier transport mechanism, drift (beam range (voltage) < depletion width (bias)) or diffusion (beam range >> depletion width); and 3) lifetime (in secondary electron probe volume) relative to carrier collection times. Figures 7 and 8 show direct applications of EBIC to reveal dopant striations and microdefects. Since spatial distributions are normally of interest in research or root cause studies, the overhead in sample preparation and vacuum pump down is not significant.

Processing: Reactive Ion Etching

Spatial variations in depth from the wafer surface are difficult to determine with a resolution of less than 0.1 μm. Electron beam probe volumes are limited to this length scale. For DLTS depletion widths of less than 0.1 μm, the required heavy doping limits sensitivity and the large required fields exhibit premature breakdown by tunneling or field emission.

Surface defect state measurements combined with anodic stripping can be effective on a 0.02 μm scale. Spectroscopic probing of surface defects can be achieved by photoluminescence in conjunction with anodic stripping. This powerful technique has been applied to RIE damage as shown in Figure 9. The principles of application are as follows. 1) Higher energy UV light is employed to minimize the probe volume. 2) In the defective region, a very short diffusion length defines the probe volume as the photon penetration depth. 3) The observed spectrum is an integral of the entire defective region, and the depth distribution is determined by subtraction at anodic strip intervals. The Figure shows the penetration of interstitial carbon produced by the low energy (200 eV) RIE. The anomalously large (for low energy ions) penetration depth is now understood in terms of trap limited diffusion where C_iC_s is the trap in FZ and C_iO_i is the trap in CZ material.

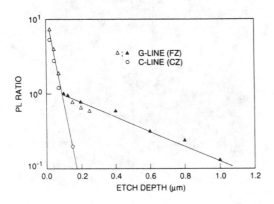

Figure 9. Photoluminescence intensity ratio of silicon after RIE in $CF_4 + 8\% \, O_2$. The circles are a measure of the C-line (C_iO_i) in CZ material, and the triangles measure the intensity of the G-line, (C_iC_s) in FZ Si, both normalized by dividing by the boron, bound exciton, B_{TA} intensity. The open markers are taken on two samples processed simultaneously, and the solid markers are data taken on a third sample.(8)

CONCLUSION

We have reviewed both the methods and methodology for microdefect analysis. Transport methods are used to monitor the variables controlling yield. A hierarchy of methods are appropriate for detection (rapid, sensitive, low

information content) or analysis (spatially resolved, spectroscopic). An effective microdefect monitor and control strategy is applied in-line for detection and off-line for analysis with the goal of global reduction of test overhead.

ACKNOWLEDGMENTS

The authors would like to acknowledge Gregg Higashi and Janet Benton for helpful discussions. This work was supported by the National Renewable Energy Laboratory, Golden, Colorado, under contract No. XD-2-11004-4 and the MIT Leaders for Manufacturing Program.

REFERENCES

1. Boone, T., Higashi, G. S., Benton, J. L., Kistler, R. C., Weber, G. R., Keller, R. C., and Kakris, G., "HF-immersion Inductively Coupled Carrier Lifetime Characterization of Furnace-oxide Growth," *MRS Proceeding of the Conference on Surface Chemical Cleaning and Passivation for Semiconductor Processing*, 359 (1993).

2. With R. Stoner, Intel – Portland Technology Development; Ph.D. Thesis, H. M'saad, *The Role of Surface and Bulk Perfection in the Processing and Performance of Crystalline Silicon*, MIT, Department of Materials Science and Engineering, September 1994.

3. Benton, J. L., Kimerling, L. C., "Capacitance Transient Spectroscopy of Trace Contamination in Silicon," *J. Electrochem. Soc.* **129**, 2098 (1982).

4. Yablonovitch, E., Allara, D. L., Chang, C. C., Gmitter, T., and Bright, T. B., *Phys. Rev. Lett.* **57**, 249 (1986).

5. M'saad, H., Michel, J., Lappe, J. J., and Kimerling, L. C., "Electronic Passivation of Silicon Surfaces by Halogens," *J. Electron. Materials* **23** 487 (1994).

6. Gerd J. Norga and Lionel C. Kimerling, "Metal Removal from Silicon Surfaces in Wet Chemical Systems," *J. Electron. Materials* **24**, 397 (1995).

7. de Kock, A. J. R., Ferris, S. D., Kimerling, L. C., and Leamy, H. J., "SEM Observation of Dopant Striae in Silicon," *J. Appl. Phys.* **48** 301 (1977).

8. Benton, J. L., Michel, J., Kimerling, L. C., Weir, B. E., and Gottscho, R. A., "Carbon Reactions in Reactive Ion Etched Silicon," *J. Electron. Materials* **20** 643 (1991).

Defect Engineering Diagnostic Options for SOI, MeV Implants and 300 mm Wafers

George A. Rozgonyi

Department of Materials Science and Engineering
North Carolina State University
Raleigh, NC 27695-7916

The challenges presented by Giga-Scale Integrated (GSI) circuit processing for cost effective structural, chemical and electrical diagnostics are discussed in the context of: i) substrate defects/gate oxide integrity, ii) the vulnerability of an SOI platform to contamination, and iii) damage control, gettering, and high temperature annealing for MeV implants.

INTRODUCTION

The materials science and defect characterization advances required for each new generation of silicon integrated circuits have been made for over two decades – spanning LSI, VLSI and ULSI – without encountering any insurmountable scientific, technological, or economic barrier. Although this progress will most certainly extend into the Giga Scale Integrated Circuit (GSI) era, the choices involved with long range planning [1] are becoming exceedingly more complex; while the financial risks associated with potential failure in solving basic science problems, choosing an optimum process integration technology, or miscalculating consumer market trends may determine the survival of individual companies or regional economies. In this report we address a limited, but vital, aspect of the forthcoming GSI era regarding defect engineering choices and associated diagnostic options for substrates. Three initially disparate issues will be discussed; namely, 300 mm diameter bulk, silicon-on-insulator (SOI), and mega-volt ion implanted (MeV) wafers.

Although breakthroughs in defect engineering are as unpredictable in the IC world as in other advanced technologies, they are in fact often built on a disciplined process optimization - material/device characterization - basic science insight feedback loop. Recognizing that empirical process optimization alone has had many successes in IC development, it is becoming increasingly apparent that the GSI era will require a more cost effective, less disaster prone progression from R&D to profitable manufacturing. The value of each individual wafer will preclude much of the empiricism used to enhance yields today. Thus, the mechanisms for collection of meaningful diagnostic data relevant to materials and processes will have to be designed into each wafer, much as test chip wafer real estate has traditionally been allocated. The availability of appropriate high throughput, non contacting diagnostic probes is always the most cost-effective approach. However, configuring the actual device, or a larger area "on-board" monitor, as a space charge probe of its local environment can provide defect density and impurity concentration data through the capacitance, leakage current, carrier lifetime, breakdown voltage, etc. Simply stated, the most versatile and sensitive diagnostic tool is often the device itself, which provides complementary data to off-line defect microscopy and impurity spectroscopy.

DIAGNOSTIC PATHWAYS
From Device Failure to Point Defect Engineering

To illustrate how the interplay between on-board monitors and off-line diagnostic tools can provide a feedback to crystal growth and a feed-forward to a defect engineering process enhancement, we examine the source of "B-mode" oxide breakdown failures. B-mode gate oxide integrity (GOI) failures exhibit a 4 to 8 MeV-cm^{-1} breakdown distribution which is readily distinguished from A-mode GOI failures at 1 or 2 MeV-cm^{-1} (surface particles), and C-mode failures at 10 to 12 MeV-cm^{-1} (theoretical breakdown). The relevance of the B-mode failures is that they can be attributed to crystal growth related sources, such as SiO_2 precipitates and the so-called "D-defect", which are intimately (and inversely) related to

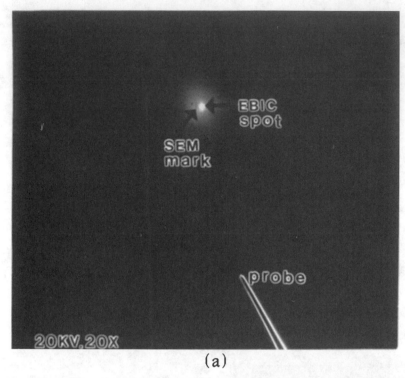

(a)

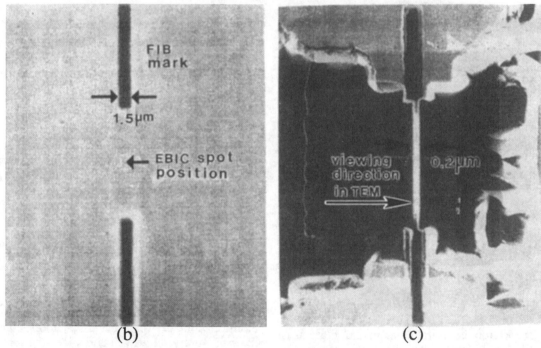

(b)   (c)

Fig.1(a). MOS/EBIC localization of electrically active D-defect, (b) subsequent FIB alignment, and (c) TEM foil configuration followed by FIB etching using Ga source at 300KV for rough and %KV for fine etching.

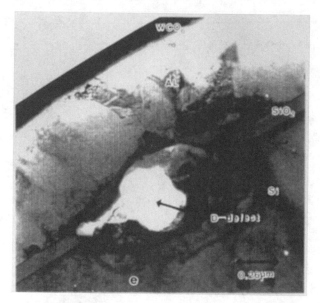

Fig. 2(a) X-TEM image of the D-Defect causing a B-mode type oxide breakdown (4 MV/cm). Note the thin oxide over the top of the cavity area.

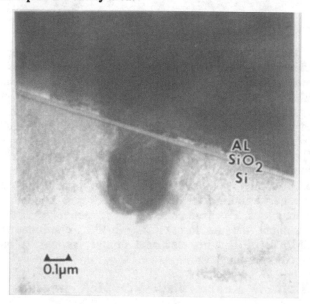

Fig. 2(b) X-TEM image of polyhedral oxygen precipitate incorporated uniformly into the SiO_2 of a MOS capacitor exhibiting BV_{ox} of ~3 MV/cm.

the original ingot pulling rate through the local thermal gradient's impact on vacancies and interstitials. Now, the diagnostic challenge – finding the needle in the haystack – is to isolate individual GOI failure sites when their bulk densities are only 10^3 to 10^7 cm^{-3}.

Park, et al [2] describe such a procedure starting with a full 150 mm wafer array of large area (0.4 cm^2) capacitors which are mapped to isolate individual "failed" units, i.e., those with a leakage current exceeding 40 μA-cm^{-2}. These MOS devices were then examined with a SEM operating in the electron beam induced current (EBIC) mode to locate and precisely mark the B-mode failure site, see Fig. 1(a).

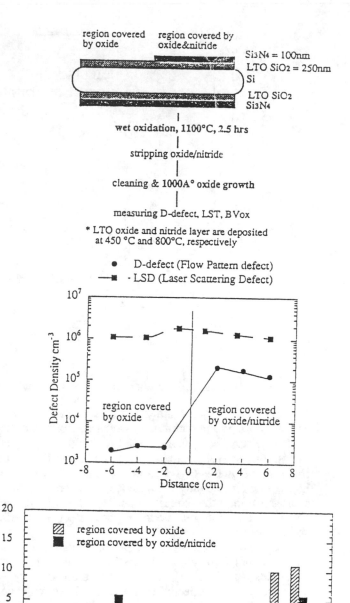

Fig. 3 Effect of interstitial injection via wet oxidation on D-Defects, LSD, and 100 nm capacitor BV_{ox}. The MOS capacitor area was 0.4 cm^2.

Further enhanced marking took place by transferring the sample to a Focused Ion Beam (FIB) system where trenches were etched and the failure site was capped with a FIB deposition of WCO_6, see Fig. 1(b). At this point the sample was cut into a 2x3 mm die, attached to a dummy glass slide, lapped, polished to ~40 μm thickness, mounted on a TEM sample grid and returned to the FIB, where it was thinned (in cross section) to ~200 nm, see Fig. 1(c). Finally, the isolated B-mode failure site was analyzed in the TEM, as shown in Fig. 2(a), revealing a large cavity defect. Local field enhancement at the thinned oxide surrounding the cavity is the probable cause of failure.

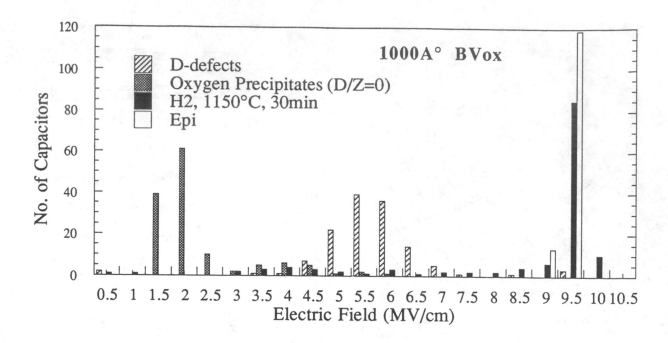

Fig. 4 "B-mode" failure distribution for wafers with various characteristics.

Additional X-TEMs of other samples reveal interfacial defects (bulk SiO2 precipitates) which do not perturb the oxide layer thickness; as shown in Fig. 2(b).

Having arrived at a reasonable understanding of the character of these two defects via the micrographs in Fig. 2, one representing the absence of material (vacancy aggregate), the other excess material (interstitial oxygen complex), it is now possible to propose defect engineering options for their elimination or reduction [4]. For example, wafer annealing which favors injection of interstitials (wet oxidation), vacancies (nitridation), or out-diffusion of oxygen (hydrogen) can be selected to aid in the defect reduction process. Thus, the defect engineering diagnostic pathway from capacitor evaluation to near surface point defect/impurity modification during high temperature annealing can be validated by returning to GOI "B-mode" failure analysis on a partially Si3N4 masked wafer, which was processed as outlined in Fig. 3. The half wafer top surface which had the Si3N4 removed continued to grow more oxide, thereby injecting Si interstitials, during a subsequent 1100°C, 2.5 hr wet oxidation; whereas the Si3N4 masked wafer half remained unchanged. Following stripping of all surface layers and cleaning, a 1000Å oxide was grown. Capacitors with an area of 0.4 cm² were then formed, oxide breakdown field plotted, and then the "D-defect" etch pit density mapped across the wafer, following a preferential Secco etch, see Fig. 3. It is apparent that injection of Si interstitials has effected a two orders of magnitude reduction in the D-defect density and an elimination of the B-mode VB$_{ox}$ failures below 7 MeV/cm.

An alternate defect engineering strategy is to modify the original crystal growth thermal balance by adjusting the

crystal pull rate such that the dominant point defect balance and its subsequent effect on GOI can be monitored.

Fortunately, it is not always necessary to repeat the bad device to off-wafer TEM sequence once the basic defect identifications and differentiations have been established, and then correlated with a device characteristic, process perturbation, or on-line structural/chemical tool. It is expected that preparing for the era of 300 mm wafers will benefit from open and interactive partnerships between wafer suppliers and device makers, with inputs from the National Laboratories/Sematech/SRC and academic communities. For example, can we collectively ask the instrument manufactures if an in-line, combination SEM/EBIC/FIB with electrical contact probes is on anybody's drawing board?

Epitaxy Replacement Via MeV Implant

The complex interplay between extended and point defects, dopants, and impurities in CZ substrates has been kept under control for many devices by using carefully engineered epitaxial layers on gettering substrates. Achieving process simplification and performance advantages without an epitaxial layer, initially projected for high energy (MeV) ion implanted wafers, have recently been aided by controlled ambient, high temperature annealing of CZ substrates prior to MeV implantation, see Ref. 5 and the references therein. This process attempts to create an epitaxial quality denuded zone, while a triple or quadruple implant provides proximity gettering sinks and vertically modulated dopant control. Thus, successful defect engineering at one stage, i.e. improving the denuded zone quality, enables a downstream implantation processing option to be implemented. Once again we use MOS breakdown field distributions to compare the impact of various processes and substrate options. Figure 4 presents data for a reference epitaxial substrate, a sample subjected to 1150°C hydrogen anneal, a wafer containing D-defects (as in Fig. 3), and an internally gettered sample whose denuded zone (DZ) has been removed by etching. We note first that epitaxial material is essentially ideal in having all C-mode failures, while allowing excessive SiO_2 bulk precipitation to reach the surface by eliminating the denuded zone yields totally unacceptable A-mode behavior. The distribution for the sample with D-defects shows a classical B-mode curve centered at 5.5 MeV/cm, typical for samples from a fast crystal pull rate ingot. Finally, we show the beneficial effect of subjecting the D-defect sample to a 30 min, 1100°C hydrogen anneal in shifting the breakdown distribution towards the reference epitaxial or intrinsic C-mode breakdown. The mechanisms underlying this very positive defect engineering enhancement of substrates via hydrogen annealing are not yet fully understood. It is known from SIMS analysis that thermal annealing will definitely enhance the out diffusion of oxygen and is likely to lower the near surface precipitate density. However, it is not clear how the annealing will impact the large D-defect cavities. It is also important to note that the hydrogen annealed sample in Fig. 4 still had a significant number of capacitor failures below 8 MeV/cm.

Summarizing the effect of hydrogen annealing we note the obvious benefits over somewhat "vulnerable" bulk substrates, but the extent to which this alternate wafer/process sequence will impact the use of epi-substrates will depend on such varied materials issues as oxide microroughness following H_2 annealing, stability of gettered product at implantation defect sites, and as always, the electrical impact of near-surface point defects, impurities and their complexes. Complications arising from superimposing an MeV process on a 300 mm wafer will be an additional impetus for integrating device based electrical measurement schemes which can be correlated with process induced defects.

The ability of MeV ion implantation to selectively position near surface buried damage which can then be used to getter impurities from the device region has been demonstrated by several groups [5,6]. We have recently characterized the deep level traps associated with the buried damage using X-TEM and depth dependant DLTS [7]. Figure 5 shows a TEM image of the buried defects resulting from of a 1e15 cm^{-2} Si$^+$ implant which was annealed for 1 hour at 900°C. The buried layer of dislocation loops located at the projected Si$^+$ range correlates with the peak of the depth profile of the deep levels found in the sample. We have also, for the first time, electronically quantified the Cu gettering ability of the MeV induced damage. Cu was diffused at 500°C from the backside of a Si$^+$ implanted wafer previously annealed at 1000°C for 1 hour. DLTS measurements using Au diodes fabricated on the front side of the contaminated samples showed that the device region, i.e. the near-surface region above the dislocated buried layer, a contained reduced Cu related defect concentratiom of 10^{10} cm^{-3}, while

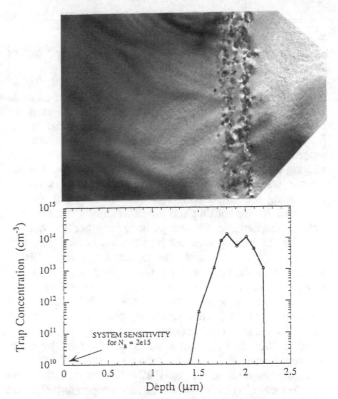

Fig. 5 X-TEM image with depth profile of deep levels found in Si implanted with a 1e15 cm^{-2} dose Si$^+$ at 2.0 MeV and annealed at 900°C for 1 hour [7].

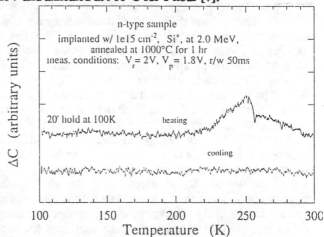

Fig. 6 Comparison of DLTS spectra collected during cooling and heating revealing metastable electronic trap behavior of MeV implantation damage related defects.

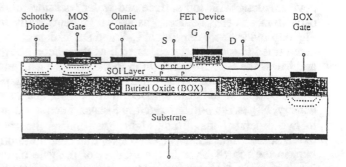

Fig. 7 Schematic illustration of SOI Space Charge Probes for Materials Property Analysis.

elsewhere in the sample these levels were at least 3 orders of magnitude higher.

Possible unforseen issues related to MeV implantation may be illustrated by the example of an MeV damage related instability observed using DLTS. A unique metastable trap behavior has been observed, similar to that previously found in Si/Si(Ge) heteroepitaxial layers [8], where electron traps become electrically active upon cooling to temperatures below 200 K, and are then detected during warming above 230 K. This can be seen by comparing the two DLTS spectra shown in Fig. 6, one collected during cooling and the other during heating. The defects responsible for the peak during heating can be modeled as having two configurations, one of which is electrically active and the other inactive. At low temperatures the electrically active configuration becomes more stable, due to a lower configurational entropy. During warming to temperatures above 260 K the active configuration becomes unstable once again and the electrically inactive defect configuration is preferred. This configurational phenomena has been observed by other workers, see Ref. 5 and the references therein, and is of fundamental interest not only to materials scientists, but also to technologists considering low temperature electron device applications.

Silicon on Insulator

Finally, we discuss the third advanced wafer technology, namely, silicon-on-insulator. The list of existing performance and potential cost savings advantages of an SOI substrate platform is seriously modulated by the number of defect and impurity obstacles to be overcome. We are particularly concerned that the presence of the buried oxide layer also isolates traditional bulk or backsurface gettering sinks from the active device volume, thereby making SOI devices highly vulnerable to inadvertent exposure to sources of contamination. In an SOI structure the most likely sinks for impurities are the actual device or buried oxide (BOX) interfaces and the source/drain implantation damage. Therefore, defect engineering options emphasizing a lateral gettering configuration, e.g., diffusion from gate to source/drain capture sites, are under study at NCSU [9]. Both low temperature processing and the need to measure very low impurity concentrations are particularly important GSI considerations. In our initial work we have configured SOI/FETs as monitor test devices, along with top surface and wafer backside MOS capacitors. These devices, as shown schematically in Fig. 6, provide depth dependent C-V and DLTS data on the gate buried oxides interfaces, as well as the silicon above and below the BOX. As shown in Fig. 8 both the gate and upper BOX interfaces curtain broad spectrum interface traps which peak near the midgap. Note that the interface state density, N_{it}, at the buried oxide/top Si layer interface is up to two orders of magnitude higher than that at the gate oxide/Si layer interface. The deep level trap concentration, N_T, in the top Si layer in the vicinity of the gate oxide interface was found to be below the sensitivity limit,

whereas two deep level traps were detected near the buried oxide interface, as shown in Fig. 8. Figure 8 also shows that both the vendor, A-1 vs. A-2, and the process used, A vs B, yield material with significantly different properties. The relative material properties of unprocessed SOI wafers from the same three vendors were also significantly different, as shown in Fig. 9 which plots the temperature dependence of the effective carrier lifetime obtained from laser/microwave photoconductance decay (LM-PCD) measurements. The lifetime in the A-1 wafer was found to be an order of magnitude lower than in the A-2 and B samples which were essentially the same at room temperature, see Fig. 9. A qualitative explanation for reduced carrier lifetime in the A-1 wafer was traced to the detection of metallic impurities; specifically Fe contamination, which has a specific signature due to Fe-B pair dissociation above 150°C, see Fig. 9. The release of isolated Fe atoms which have a stronger recombination activity, leads to the dip in lifetime [6], which is stronger in A-2 then in B DLTS measurements on these samples confirmed the presence of the FeB trap and revealed additional deep level traps in the substrate near the lower buried oxide interface. The Fe concentration in the A-1 sample was about 3×10^{12} cm^{-3}, which was at least three times higher than in the A-2 sample. However, it appears that other bulk traps, whose concentration is an order of magnitude higher than the iron concentration, are responsible for the strongly reduced carrier lifetime in the A-1 sample. Thus, a review of the data in Fig. 9 provides a direct comparison of samples from Vendors B, A-1, and A-2. Combining this with the "on-board" data obtained on the SOI top layer in the processed device wafers, see Fig. 8, gives a strong diagnostic attack to SOI materials evaluation.

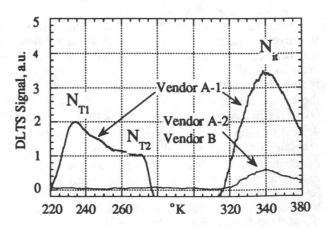

Fig. 8 DLTS probing on three SOI wafers from different vendors.

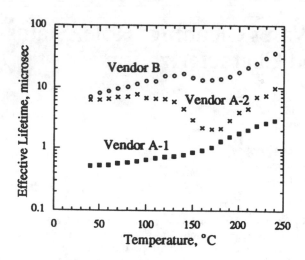

Fig.9. Temperature dependent effective carrier lifetime in three unprocessed SOI wafers from different vendors.

ACKNOWLEDGEMENTS

The author wishes to thank Dr. S. Koveshnikov, Dr. J. G. Park, and Mr. A. Agarwal for their assistance in preparing this paper.

REFERENCES

1. The National Technology Roadmap for Semiconductors SIA, San Jose (1994).
2. J. G. Park, et al, p. 57 in ECS PV. **94-1** and p. 53 in PV. **94-33**, The Electrochemical Society (1994).
3. J. G. Park, " Nature of D-Defects in Czochralski Silicon," Ph.D. Thesis, North Carolina State University (1994).
4. J. O. Borland and R. Koelsch, *Sol. St. Tech*, **12** (1993).
5. H. Wong, N. W. Cheung, et al, *Appl. Phys. Lett.*, **52** 1023 (1988).
6. K. Tsukamoto, T. Kuroi, S. Komori and Y. Akasaka, *Sol. St. Tech.*, **6** (1992).
7. A. Agarwal, et al., to be presented at 1995 Spring MRS Meeting, San Francisco.
8. A. Agarwal, et al, *Mat. Sci. and Engr.* **B24**, 43 (1994).
9. G.A. Rozgonyi, et al, p 868 in ECS Proc. PV **94-10**, The Electrochemical Society (1994)
10. Y. Hayamizy, T. Hamaguchi, S. Ushio, T. Abe, F. Shimura, *J. Appl. Phys.* **69**, 3077 (1991)

Silicon Surface Preparation and Wafer Cleaning: Role, Status, and Needs for Advanced Characterization

C. R. Helms

Department of Electrical Engineering
Stanford University
Stanford, CA 94305

Surface preparation, conditioning, and wafer cleaning are a critical part of any semiconductor manufacturing operation. In front end of line (FEOL) the state of the wafer surface just prior to gate oxidation determines to a large extent the electrical defect density of the grown oxides and their long term reliability. In this paper we will first compare the device requirements and show how they drive the surface prep requirements. Next we will briefly review the state of the art in surface prep and associated metrology, indicating where the key technology gaps are. Finally we will suggest some integrated solutions to the overall problem of surface prep, microcontamination and associated metrology which will lead to the lowest cost of ownership (CoO) solutions.

INTRODUCTION

The development of the technologies and technology enablers for Giga Scale Integration (GSI) will require significant R&D all the way from fundamental atomic level science to factory integration and computer integrated manufacturing. Requirements for the 4 gigabit technology node projected for first production ten years hence include 12" wafers, die > 1" on a side, 0.13 μm design rules, 3.5 - 4.5 nm gate oxides, with 0.01 gate oxide defects/cm^2 (1). All of these requirements pose significant roadblocks for wafer surface preparation, microcontamination in general, and the metrology required to qualify and verify surface prep processes. In the past the application of high level, but empirical, engineering has provided the solutions to microcontamination problems and future generations of advanced GSI devices will benefit from this same level of engineering skill. In the future, however, the level of process integration required for an adequate Cost of Ownership (CoO) and the complexity of each individual step requires a new look at the most effective way to attack microcontamination issues. For example we currently require the wafer suppliers to meet the same metal and particle levels we expect for the Si surface just prior to the gate oxidation, even though we know that numerous unit step processes prior to the gate oxidation will lead to higher levels of both particles and metals. This lack of integration leads to higher wafer costs for little value added. For the 10 year out technology node (we will use these requirements throughout this paper for comparison purposes) the continuation of such strategies will lead to artificially high cost of ownership (CoO) which we will not be able to afford.

In this paper we address a part of the overall problem related to surface preparation. This combined with a coordinated consideration of bulk contamination issues and solutions, such as gettering or integrated cleaning will lead to effective overall lowest CoO solutions (2-11).

We will consider front end of line (FEOL), through source/drain doping where specific critical showstoppers relate to the gate insulator thickness and defect levels required, either for 300 mm wafer sizes or advanced 200 mm SOI designs. By far the most critical factor limiting gate insulator defects is the presence of particles, with metallics a close second. Additional issues related to surface microroughness, organics, and anionics, not dominant currently, are expected be become increasingly important for thinner gate oxides due to more stringent yield requirements.

One of the dilemmas faced by the surface prep community is illustrated in Figure 1 which shows a plot of both Fe solubility in oxidizing solutions such as those containing peroxide (4,11) and surface zeta potential as a function of solution pH (12). High pH conditions are favorable for particle removal but lead to low Fe solubility and precipitation driven adsorption whereas low pH leads to high metal solubility but particle adhesion. The extension of current approaches to reduce particle levels in low pH surface preps and reduce metallic concentrations in high pH surface preps will lead to very high CoO solutions in the 10-year-out time frame.

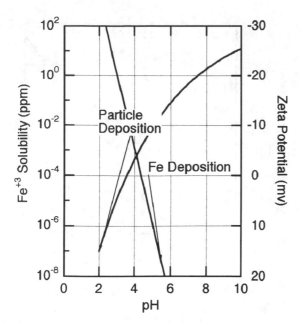

FIGURE 1. Example of the combined effect of metals and particles in aqueous solutions. The solubility of Fe^{+3} is plotted as a function of pH (curve labeled "Fe deposition"); in high pH solutions Fe^{+3} is insoluble leading to precipitation driven adsorption (Fe^{+2} is soluble, but unstable in peroxide containing solutions). The negative zeta potential is also plotted as a function of pH in the curve labeled "particle deposition". For Si surfaces positive zeta potentials lead to particle adhesion for many typical particles (4,11,12).

GENERAL REQUIREMENTS

The requirements for FEOL surface prep are controlled by the requirements on the gate oxide and junction leakage properties. The reader is referred to Reference 1 for more specifics; some of the values are listed below for the 4 gigabit DRAM scheduled for first production in 2004:

Design rule: 130 nm
Wafer size: 300 mm
Gate oxide : 3.5 - 4.5 nm
Voltage: 1.8 - 2.5 V
Junction depth: 20 - 60 nm
Low breakdown gate oxide defects: 0.01 cm^{-2}
QBD: 10^4 cm^{-2}
Organic: 10^{13} cm^{-2}
Surface Al, Ca: 10^{10} cm^{-2}
Surface Fe, Cu, Ni, Na: 5×10^9 cm^{-2}
Bulk Fe: 10^{10} cm^{-3}
Particles: size: 40 nm; density (cm^{-2}):
    0.06 initial; 0.04 added, pre gate clean
Surface roughness (1x1 μm field): 0.1 nm RMS
Recombination lifetime: 500 μs
Generation lifetime: 1.25 ms

The gate oxide defect requirements, which are determined by the acceptable yield for the number of gates specified per die factor determine the maximum acceptable particle levels. The commonly accepted standard for particle size leading to killer gate oxide defects is 1/3rd of the design rule. For the 0.35 μm technology node of today this is just over 0.1 μm; for the 0.13 μm technology node this is only 40 nm No mention is made of the nature or chemical composition of the particles. It is clear however that particles of differing composition will have different effects. The measurement of particles in this size range is a clear metrology show-stopper.

In addition to particles, metals present on the surface prior to gate oxidation can lead to GOI defects such as silicide precipitates or metal induced and decorated stacking faults near the gate oxide Si interface (10,13,14,15). The numbers listed above are related to the expected state of the art in wafer technology (currently p-/p+ epi wafers), gate oxide processes, and gettering strategies. The range of allowable metal concentration is illustrated in Figure 2 where the allowable pre gate oxidation Fe concentration is plotted vs. year of first production for the optimized case (upper line) and for the case of MOSdot capacitors using polished wafers (lower line, ref. 14). For the requirements represented by the lower line, the only convenient metrology methods available are based on wet chemical dissolution, followed by mass spectroscopic analysis (so called (VPD ICP mass spec). The development of other tools is clearly needed for the future.

The importance of the other contaminants (organics and anions) as well as microroughness on gate oxide integrity and reliability is less certain. The value listed for organic contamination is based on the sensitivity of HF-last surfaces to organic contamination and the possible formation of SiC inclusions during the gate oxidation process (16,17). Depending on the details of the gate oxidation, this sensitivity may be enhanced or reduced; more work is clearly needed here to develop better specifications. From a metrology perspective severalf methods are available to measure organics and anions; the more critical need here is to establish a clear relationship between the contaminants and device performance and yield.

It has not been clearly established that random roughness with relatively featureless power spectral densities leads to degradation in either gate oxide integrity or reliability. However mechanisms associated with the formation of excessive roughness can also cause defect singularities leading to gate oxide defects. The roughness specification is therefore an indirect measure of the dominant effect, which is clearly undesirable.

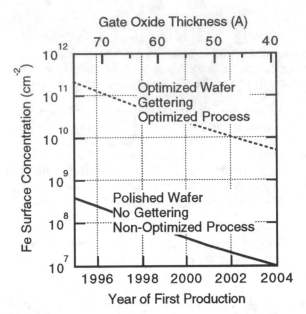

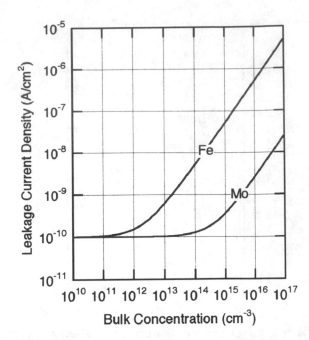

FIGURE 2. The effect of Fe on gate oxide defects is illustrated for two cases as a function of year of first production (bottom axis) and approximate equivalent gate oxide thickness (top axis). The values for the lower curve were taken from an extrapolation of Ref. 14 measured for polished wafers and a MOSdot structure. The values for the upper curve were estimated for p-/p+ epi wafers, with a state of the art CMOS device structure and gate oxidation process (4).

FIGURE 3. Junction leakage currents calculated as a function of bulk concentration for Fe and Mo impurities; the baseline leakage value corresponds to a generation lifetime of 1 ms for a 2×10^{17} cm^{-3} acceptor concentration (18).

Finally, anionic contamination has not been specified for FEOL. It is clear however that such contamination (especially that associated with sulfur residues) is undesirable and should be kept under control. Other anionic contamination, such as chlorides and fluorides may actually be beneficial for some gate oxide properties.

The discussion to this point has related to surface properties since surface prep, of course, is a surface process. However, metals on the surface prior to a thermal step can diffuse into the Si during the thermal process leading to lifetime degradation and junction leakage. This is illustrated in Figure 3 where the junction leakage has been calculated as a function of bulk metal concentration for Fe and Mo, assuming a background value of 10^{-10} A/cm^2 (18). The relationship between the bulk value of the metal concentration and the equivalent surface value (assuming the metal originated from the two wafer surfaces and is equally distributed through the bulk). is twice the surface concentration for one surface divided by the wafer thickness; this simple conversion gives a conversion factor of approximately 30 (surface concentration (cm^{-2}) x 30 ≈ bulk concentration (cm^{-3})). This is always an over estimate due to the effects of gettering, surface segregation, and desorption into the gas phase during the thermal process.

SURFACE PREP TECHNOLOGIES

Discussing the whole of all possible surface prep processes projected is clearly beyond the scope of this paper (see Ref. 12 for a good review through 1993). We will focus our attention on the pre gate oxidation cleaning sequence which has features in common with other FEOL cleans. In addition, since incoming wafer specifications are similar to those required for pre gate cleaning, the final cleaning sequences used at the wafer manufacturers will have similarities to pre gate oxidation cleaning sequences. An additional limitation to the present discussion is that we will not consider vapor or dry cleaning processes, which have significantly different requirements and are not expected be in major use for FEOL in the next few years.

Even with this restriction, many sequences are used in the industry. Most sequences will involve sulfuric acid/hydrogen peroxide/water mixtures (SPM) or equivalent for heavy organic and metal removal, dilute hydrofluoric acid (HF) for native oxide and metal removal, ammonium hydroxide/hydrogen peroxide/water mixtures (APM) for light organic, particle, and metal removal, and hydrochloric acid/hydrogen peroxide/water mixtures (HPM) for metal (especially noble metal) removal. APM is sometimes replaced by choline or similar compounds. HPM is sometimes replaced with aqua regia or dilute hydrochloric acid (without the hydrogen peroxide).

Given the chemistries, various sequences are also used. An SPM/HF/APM/HPM sequence is common but the last 2 steps are often reversed with additional HF steps inserted as well (5). Depending on the application, these chemistries can be applied in a number of tool configurations using immersion, spray, full flow, vapor, & dry processes.

Each clean is used to remove a specific type of contamination but may deposit another, as illustrated in Figure 1. The gravity of the situation is shown in Figure 4 where the steady state concentration of Fe on a Si surface is plotted vs. the amount of Fe present in an SC1 bath (19,20). Initial SC1 bath concentrations range from 100 ppt Fe to about 1 ppb, leading to surface concentrations from 10^{11} to 10^{12} cm^{-2}. As the bath ages during use, additional metals will build up so that it is not uncommon to find > 10 ppb Fe in an SC1 bath. For metals additional examples where ppb levels in solution lead to large surface concentrations are Al and Ca from SC1 and Cu from HF or NH_4F solutions. Given the sensitivity of the gate oxide process to metals (e.g. Figure 2), it is considered difficult to use SC1 last processing without stringent controls which is why SC2 cleaning (or HF) typically follows SC1. These low pH solutions will remove the metals deposited from SC1 but result in higher particle levels. Additional issues relate to sulfur residues from SPM and organics in the rinses or related to drying processes.

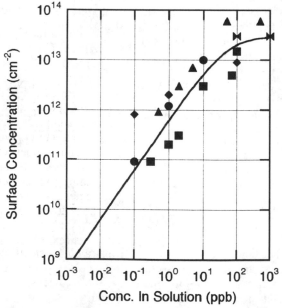

FIGURE 4. Deposition of Fe onto Si surfaces as a function of Fe concentration in SC1 solutions. The points are data complied from the literature (21); the line is a model developed based on precipitation driven adsorption coupled with etching driven desorption (19,20).

Numerous strategies are used to insure effective surface prep and mitigate against the build up of contaminants. By far the best approach is to measure contaminant levels on the wafer surface or at least in the process fluids so that adequate performance can be assured. However in-situ or near real time in line monitors are not universally available for the range and level of impurities of importance.

Off-line monitors are also needed for R&D and process development; if rapid enough off-line monitors can provide enough real time feedback to prevent severe line crashes due to process excursions.

METROLOGY REQUIREMENTS

Given the requirements placed on surface prep and surface properties in general, extensive metrology is needed (21). Needs include off-line and in-line characterization, and in-situ metrology is also desirable as introduced above. Off-line requirements are summarized in Table 1. for the 4 gigabit technology node. The requirements listed relate only to the surface even though bulk analysis for species such as metallics is also clearly necessary (21,22). In addition, electrical evaluation is not considered here.

The table has been structured to list the contaminant, its requirement, and optimistic and pessimistic views of both the state of metrology as well as possible changes in the requirements that may drive the metrology. For the cases of organic and anionic contamination, as well as hydrogen surface termination, the perceived 10^{13} cm^{-2} requirements can be met by combinations of electron and ion spectroscopies (AES, XPS, SIMS, etc.) currently available. In addition, optical methods such as spectroscopic ellipsometry and attenuated total internal reflection FTIR can be utilized. On the pessimistic side, past device generations have been relatively insensitive to these contaminants and it is not completely certain what the future requirements will be. Tighter specifications will lead to more difficult metrology tasks. In one area, pre gate oxide surface termination, it is possible that native oxide terminated surfaces will be acceptable for some time to come. However, this being so, the uniformity of the native oxide thickness will potentially become a major issue, given that it represents over 25% of the total gate oxide thickness for thin gate oxide devices.

Metallics have been a particular problem for two reasons. First the surface levels required to give low gate oxide defect levels and adequate long term reliability are down to 1 ppm of an atomic layer! In addition, this requirement is very sensitive to the actual process and wafer type employed as illustrated in Figure 2. The most convenient technique for measuring most metals is TXRF, which currently is sensitive down to about 10^{10}

TABLE 1. Off-line Metrology Requirements for the 4 Gigabit Technology Node

"Contaminant"	Requirement	Optimistic View	Pessimistic View
Organic	10^{13} cm^{-2}	Electron/Ion spectroscopies in place; rapid optical/IR may be possible	Specification not tight enough
Metallic	5×10^9 cm^{-2}	Extension of TXRF; "wet" chemical	Polished wafer specification may be 10^8 cm^{-2}; may require synchrotron radiation based TXRF
Particles	< 50 40 nm particles per 300 mm wafer	Far vacuum UV light scattering	Not possible
Roughness	0.1 nm RMS for 1x1 micron field	Conventional atomic force microscope	Specification not tight enough over specific wavelength ranges
"Anionic"	Unknown (10^{13} cm^{-2} ?)	Electron/Ion spectroscopies in place; rapid optical/IR may be possible	Specification not tight enough
Surface termination	H preferred; unknown number of unterminated sites allowed (10^{13} cm^{-2} ?)	Ion spectroscopies in place; rapid optical/IR may be possible	Sever specification may be required; if native oxide acceptable, uniformity may be critical

cm^{-2}. Extensions of this technique imagined in the near future can meet the requirement in the table except for specific elements such as Al, Na, and Mg. Various "wet" chemical methods such as VPD ICPMS have demonstrated sensitivities down to 10^8 cm^{-2}. This technique however requires a wet chemical extraction of the impurities over the whole surface and may not provide 100% recovery of the metallics adsorbed there (23). The requirements for polished wafer surfaces are clearly tighter than specified in the table and may be down to the 10^8 cm^{-2} level for the 4 gigabit technology node. Such TXRF analysis requires synchrotron radiation based instrumentation.

By far the most difficult metrology task relates to particulate contamination. Even current requirements for the 0.35 µm technologies relating to particles $\geq$ 0.13 µm in size cannot be met with current wafer scanner technologies. Extension of the conventional technologies will require far UV laser scatterometry which will also require the tools to be operated in vacuum. This is a major show stopper with regard to metrology for future technologies.

Surface roughness can have many effects on device and circuit properties depending on its wavelength. This includes atomic level roughness which can affect gate oxide reliability, all the way up to longer wavelength features which lead to "haze" (6,24,25,26). The major issue here is that the roughness will have spectral components over many orders of magnitude and it is still unclear what the requirements are for specific spectral regions to provide adequate device yield and performance. The specifications listed in Table 1 are a clear over simplification for all the roughness related requirements and current work is underway to develop measurement and data analysis protocols so that robust correlations to device related properties can be performed (27,28). From a

measurement perspective, convenient tools currently exist down to measurement wavelengths of just under 10 nm. Extending this down another order of magnitude in a robust tool is still required, especially for gate oxide defects and reliability.

In-line as well as in-situ metrology is also needed. Some of the off-line capabilities listed above have their place for in-line applications as well. These include particle scanners, ellipsometry, and other optical techniques. Other rapid in-line techniques for organics, metallics, roughness, anions, and surface terminations, would provide enormous if benefits, especially in microscopic versions.

The final requirement relates to in-situ metrology, which for most surface prep implies measurements of the liquid state. Of prime importance are particles and metals. Progress in liquid borne particle monitors and electrophoretic techniques for metal analysis show promise (23).

SHOW-STOPPERS, INTEGRATION, SIMULATION, & TCAD

The above discussion has illustrated some of the future requirements for surface prep and the state of the metrology available to meet the future needs. The optimum technology solution, of course, is to develop processes and tools that have low contaminant levels below the threshold for potential problems (truly contamination free manufacturing). This requires all dry surface prep and/or ppt grade chemicals which would lead to CoO solutions higher than what is projected if integrated microcontamination strategies are adopted, including advanced metrology.

114

There is still no question that performance is a key metric by which surface prep solutions must be measured. Performance relates to such issues as defects mentioned above as well as other device factors such as uniformity in threshold voltages and long term reliability. Other CoO issues include throughput and cost of chemicals, DI water, and recycling and disposal of chemical waste. Key show-stoppers here include the supposition that rinsing is not easily scaleable to larger wafer sizes and that no real time monitors (in-situ or not) are available to evaluate chemical bath impurity levels to determine changeout frequency; even reprocessed chemical strategies need significantly improved real time monitors.

Another key metric very appropriate to surface prep as well as all of microcontamination is cycle time from first silicon to steady state yield. Microcontamination issues have always been the rate limiting step in the climb up the yield curve and solutions that expedite this climb will provide a key competitive edge to those who implement them.

In some ways these challenges may seem insurmountable. However, history tells us that we will find our way, so that Moore's law will continue to be obeyed for future generations to come. If a paradigm shift is required, it will be in the way we integrate our R&D activities so that atomic level research can be applied rapidly to manufacturing. Although this paper has targeted only a small piece of the total GSI technology pie, the message is clear for other aspects of technology development as well.

For surface prep specifically, in addition to the higher level considerations just mentioned, there are some key show-stoppers for the 10 year out technologies. These include the aforementioned need for simultaneous tighter particle and metallic control and surface prep in general for larger wafer sizes. Finally the > 10 year out need for vapor or all dry surface prep for FEOL is a major challenge for future R&D.

These surface prep requirements in turn drive the metrology requirements where particle and metal measurement needs cannot be easily extrapolated form current technologies. This coupled with uncertainties in requirements for other surface contaminants, microroughness, and surface terminations, leads to a major challenge to the metrology community to "keep up" with the whole surface prep field.

Better integration across the whole microcontamination arena gives us a major opportunity for lowering the CoO, including contributions from faster cycle time, higher throughput, higher yields, lower tool cost, and lower chemical and water consumption and cost. This can be illustrated if we look at figure 5. Specifications on wafer surface properties for virgin wafers are often similar to those desired just prior to gate oxidation, leading to higher cost due to the extensive cleaning that is necessary to achieve these requirements. As the wafers pass from the wafer manufacturer to the semiconductor fab, they will go through numerous steps where contamination is removed, deposited, and removed again. In many cases the "memory" of the condition of the wafer surface as delivered is lost in the series of steps performed.

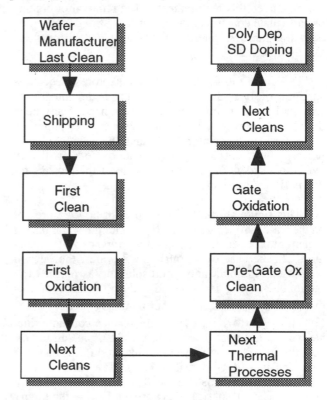

FIGURE 5. Illustration of FEOL steps where contamination will be added and/or removed; in many cases one contaminant will be removed in a step and another contaminant will be deposited in the same step.

Fe is a key example, where incoming wafer requirements may be in the 10^{10} cm^{-2} range. Any subsequent cleaning sequence employing SC1 even with one ppb Fe will deposit 10 times that much on the surface; the 10^{10} cm^{-2} specification on incoming Fe cannot therefore be justified on these grounds. One of the reasons that such a tight specification is used may relate to the bulk contamination issue. Bulk purity is often determined after a thermal cycle to passivate the surface so that a lifetime method can be conveniently applied; Fe initially present on the surface can be driven in during that step, interfering with the bulk measurement. The surface Fe requirement is therefore driven more from metrology considerations rather than performance requirements! For those who choose to perform an initial thermal step before any cleaning step, the low surface Fe level is also a requirement for the same reason.

This example illustrates how a more integrated approach could lead to a lower overall CoO. In the example we could imagine a relaxed Fe specification on incoming wafers (so long as bulk purity could be guaranteed), leading to a lower ultimate wafer cost. This would require the semiconductor fab to perform an initial clean prior to any thermal step that might tend to drive the Fe into the bulk where it would be trapped (the use of more effective gettering strategies as discussed below may alleviate even this requirement). The removal of organic or particulate contamination that might have deposited on the wafer during shipping would be an added advantage of this initial clean.

Designing surface prep and thermal process sequences simultaneously is the next area where better integration will lead to optimum solutions. This has already been discussed in part in reference to Figure 2. Depending on the nature of the wafer itself and the gate oxidation process, the specification on pre gate oxide Fe will vary over three orders of magnitude. One of the reasons for this is shown in Figure 6.

This figure shows two examples of how Fe present on the wafer surface will redistribute into the wafer during a thermal process, depending on the type of wafer being used. In the case of a polished CZ wafer, the Fe will diffuse into the bulk until solid solubility is reached (no internal gettering is assumed). For the case of this example at 750 C, this is approximately 5×10^{11} cm^{-3} (29); if we start with a 10^{11} cm^{-2} surface concentration, we will be left with nearly that much on the surface (or near the interface if this were an oxidation), leading to a large density of GOI defects (13,14). The other example is for a p+ wafer substrate , with a 10 μm p- epi layer present. The p+ substrate "getters the Fe due the higher solubility of Fe (29,30) in the heavily doped material (this is not related to the B-Fe pairing mechanism that occurs at lower temperatures). This leads to lower surface (interface) concentrations and lower GOI defect densities as well as low Fe concentrations in the active device region and therefore longer lifetimes.

These examples show the power of an integrated strategy including microcontamination in the process design arena to provide optimized processes sequences and starting materials specifications for truly design coupled manufacturing (31), which brings us to the final consideration for this paper: the need for microcontamination simulation tools and the inclusion of microcontamination into technology computer aided design, TCAD, methodologies. The beginnings of this have been illustrated here with reference to Figs. 4 and 6. Figure 4 illustrates the development and application of a model to simulate metal (Fe) deposition from cleaning solutions.

The diffusion/segregation results of Figure 6 were obtained using a modified point defect/impurity diffusion simulator designed for compound semiconductors (32).

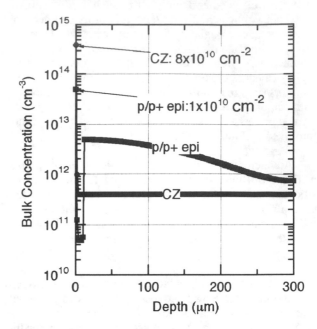

FIGURE 6. Result of a simulation comparing 10^{11} cm^{-2} Fe in-diffusion into CZ vs. p-/p+ epi (10 μm epi layer) wafers at 750 C for 4 hours. Half of the wafer is shown, assuming that the contamination is present on both surfaces. Values for the diffusion coefficients and solubilities are taken from the literature (29,30); the heat of segregation to the surface was unknown and a value of 1.4 eV was used.

Even though new technology solutions are required for some of the critical surface prep and microcontamination problems, the capture of the existing knowledgebase and application to current problems as well as a test bed for evaluating new solutions through TCAD methodologies is powerful indeed. This is a key area where TCAD approaches can be a major enabler-possibly the only enabler- to the overall solution. It is critical in surface prep and gettering process design to be able to track impurities all the way from birth on the wafer through a complex process sequence. In so doing optimized strategies can be developed leading to an overall best solution.

ACKNOWLEDGEMENTS

This work was supported primarily by Semiconductor Research Corporation Contract SJ-350 (W. Lynch). Support from Intel (B. Triplett), ARPA (Z. Lemnios), SEMATECH (H. Huff), and Santa Clara Plastics (M. Hall), is gratefully acknowledged.

The author would like to acknowledge Gregg Higashi, Baylor Triplett, Howard Huff, John Lowell, Sean O'Brien, Skip Parks, Eicke Weber, and George Rozgonyi for critical discussions.

REFERENCES

1. *The National Technology Roadmap for Semiconductors 1994*, Semiconductor Industry Association, San Jose, 1994.
2. Huff, H., & Goodall, R. K., "Si Materials: Critical Concepts for Optimal IC Performance in the Gigabit Era," this volume.
3. Rozgonyi, G. A., "Defect Engineering Diagnostic Options for SOI, MeV Implants, & 300 mm Wafers," this volume.
4. Helms, C. R., Park, H.-S., Dhanda, S., Gupta, P., & Tran, M., "Fundamental Metallic Issues for Ultraclean Wafer Surfaces from Aqueous Solutions," in *Proceedings of the Second Intl. Symp. on Ultra-clean Processing of Si Surfaces*, Leuven Belgium: Acco, pp. 205-212 (1994).
5. Hattori, T., "Trends in Wafer Cleaning Technology," ibid. , pp. 13-18.
6. Triplett, B. B., "The Limitation of Extrinsic Defect Density on Thin Oxide Scaling in VLSI Devices," in *Semiconductor Silicon/1994*, Electrochem. Soc. **P V 94-10**, pp. 333-345 (1994).
7. Park, J.-G., Kirk, H., Cho, K.-C., Lee, H.-K., Lee, C.-S., & Rozgonyi, G. A., "Structure & Morphology of "D-Defects" in CZ Si, ibid. , pp. 370-378.
8. Rozgonyi, G. A., Koveshinikov, S., & Agarwal, A., "Low Temperature Impurity Gettering for Giga-Scale Integrated Circuit Technology, ibid. , pp. 868-883.
9. Tan, T. Y., Gafiteanu, R., Gosele, & U. M., "Diffusion-Segregation Equation & Simulation of Diffusion-Segregation Phenomena," ibid. , pp. 920-930.
10. Helms, C. R., & Park, H.-S., "Generalized Model of Metal Bonding & Cleaning from Wafer Surfaces," in *Surface Chemical Cleaning & Passivation for Semiconductor Processing*, Materials Res. Soc. **315**, pp. 287-297 (1993).
11. Helms, C. R., Park, H.-S., "Electrochemical Equilibria of Fe in Acid/Base/Peroxide Solutions Related to Si Wafer Cleaning," in *Cleaning Technology in Semiconductor Device Manufacturing*, Electrochem. Soc. **PV 94-7**, pp. 26-33 (1994).
12. Donovan, R. P., Menon, V. B., "Particle Deposition & Adhesion," in *Handbook of Semiconductor Wafer Cleaning Technology*, Parkridge NJ, Noyes, pp. 152-200 (1993).
13. Park, H.-S., Helms, C. R., Ko, D.-H., Tran, M., & Triplett, B. B., "Effect of Surface Fe on Gate Oxide Integrity & its Removal from Si Surfaces," loc. cit. 10., pp. 353-358.
14. Henley, W., B., Jastrzebski, L., & Haddad, N., F., "Monitoring Fe Contamination in Si by Surface Photovoltage & Correlation to Gate Oxide Integrity", ibid., pp. 299-312.
15. Correia, A., Ballutaud, D., & Maurice, J.-L, " Microstructure, "Electrical Properties, & Passivation of Defects at the Si-SiO$_2$ Interface," loc. cit. 6., pp. 358-369.
16. Henderson, R. C., "Carbon Contamination on Si Surfaces," *J. Electrochem. Soc.* . **119**, 772-779 (1972).
17. Kasi, S. R., Liehr, M., Thiry, P. A., Dallaporta, H., & Offenberg, M., "Hydrocarbon Reaction with HF-cleaned Si(100) & Effects on Metal-Oxide-Semiconductor Device Quality," *Appl. Phys. Lett.*. **59**, 108-110 (1991).
18. Parks, H. G., Kosier, S. L., Schrimpf, R. D., Henley, W. B., Jastrzebski, L., "First Order Specification of Liquid Chemical Purity Requirements Based on Contaminant Deposition & Advanced DRAM Architecture," in *Micro 94*, Santa Monica, Canon Communications, pp. 132-142 (1994).
19. Helms, C. R., Dhanda, S., Parks, H. G., "Model of Fe Deposition from SC1 Solutions," to be published.
20. Dhanda, S., Helms, C. R., Gupta, P., Triplett, B. B., & Tran, M., "Fe Deposition from SC1 on Si Wafer Surfaces," Materials Res. Soc., to be published (1995).
21. Diebold, A. C., "Critical Metrology & Analytical Technology Based on the Process & Materials Requirements of the 1994 NTRS,", this volume.
22. Schroder, D. K., "Contactless Semiconductor Measurements," loc. cit. 6., pp. 1071-1082.
23. Gupta, K., Tan, S., H., Pourmotamed, Z., Cristobal, F., Oshiro, N., & McDonald, R., "Novel Methods for Trace Metal Analysis in Process Chemicals, DI Water, & on Si Surfaces," in *Contamination Control & Defect Reduction in Semiconductor Manufacturing III*, Electrochem. Soc. **PV 94-9**, pp. 200-221 (1994).
24. Bullis, W., M., "Microroughness of Si Wafers," loc. cit. 6., pp. 1156-1169.
25. Malik, I., J., Vepa, K., Pirooz, S., Martin, A., C., & Shive, L., W., " Surface Roughness of Si Wafers: Correlating AFM & Haze Measurements," ibid., pp. 1182-1193.
26. Koga, J., Takagi, S.-I., Toriumi, A., "A Comprehensive Study of MOSFET Electron Mobility in Both Weak & Strong Inversion Regimes," in *IEDM Technical Digest*, IEEE, pp. 475-478 (1994).
27. Brigham, R. N., & Bauman, S. M., "Statistical Process Control for AFM Microroughness Measurements," loc. cit. 23., pp. 277-284.
28. Halepete, S., & Helms, C. R., "Analyzing AFM's Using Spectral Methods," Materials Res. Soc., to be published (1995).
29. Weber, E. R., & Gilles, D., "Transition Metals in Si: Fundamentals & Gettering Mechanisms," in *Semiconductor Silicon/1990*, Electrochem. Soc. **P V 90-7**, pp. 585-600 (1990).
30. Gilles, D., Schroeter, W., & Bregholz, W., "Impact of the Electronic Structure on the Solubility & Diffusion of 3d Transition Metals in Si," *Phys. Rev.* **B 41**, 5770-5781 (1990).
31. Lemnios, Z. J., "Flexible Manufacturing/Cluster Tools for Semiconductor Manufacturing," this volume.
32. Melendez, J., & Helms, C. R., "Process Modeling of Point Defect Effects in HgCdTe," *J. Elect. Mats.* **22**, 999-1008 (1993).

Specification Of Liquid Chemical Purity Requirements: the knowns, unknowns, and need for technology development

H.G. Parks

Electrical and Computer Engineering Department
The University of Arizona, Tucson, AZ 85721

The current state of knowledge, limitations, and necessary developments required for the accurate specification of chemical purity are addressed. This requires knowledge of metal deposition from process solutions, effectiveness of cleaning sequences, interfacial segregation and distribution of metals, and accurate modeling of the resulting device degradation. The development of simulation tools for process design and optimization throughout the semiconductor industry abounds; however, current generations do not consider the effects of contamination. The computational framework is available so that process modeling strategies can be effectively applied to the area of microcontamination, surface preparation and gettering in an integrated manner. The application of an expanded fundamental knowledge base to these problems through simulation would undoubtedly lead to solutions that cannot be arrived at by experimental lot splits, no matter how effective the design of experiments can be made. This represents an opportunity to significantly escalate the technology through revolutionized process modeling with enormous potential benefits in process simplification and lower cost of ownership (CoO).

INTRODUCTION

Metal ions are a major source of poor electrical performance in solid state devices (1). Metal impurities create generation-recombination centers in silicon that increase reverse-bias junction leakage (2) and also affect oxide breakdown strength and metal-oxide-semiconductor (MOS) capacitor leakage by dislocation decoration, and stacking fault and silicide formation (1,3). Such degradation of device performance adversely affects the function of ultra large scale integrated (ULSI) circuits. Sources of trace metal contaminants include the process equipment, process materials, and even process chemicals used in wafer cleaning operations. It is important to know what levels of contamination are low enough to be acceptable for a particular application and how effective the wafer cleaning solutions and gettering strategies are. There is a need for data-driven input design to produce viable processes for the deep submicron era at the end of this decade. In fact, the data are needed now. The current update of the National Technology Roadmap for Semiconductors (NTRS) was forced to rely on empirical-extrapolation estimates for metal-concentration levels for 1998 and beyond. Wafer surface prep is a complex issue. Specification of chemical purity requires knowledge of metal deposition from process solutions, effectiveness of cleaning sequences, interfacial segregation and distribution of metals in silicon and insulators and accurate modeling of device degradation due to the resulting metal distributions so that overall technology effects (e.g. DRAM performance and trends) can be ascertained. Most of these issues have been addressed to some degree; however, the studies have not been coherently developed. For example, deposition studies have received considerable experimental attention but theoretical considerations for the mechanisms and processes are sorely lacking. Hence, only the simplest of quantitative theories of the deposition process has even been proposed and not fully substantiated. Estimation of device effects requires knowledge of the efficiencies of cleaning steps and cleaning procedures, as well as the effects of further processing, such as interfacial segregation, surface degradation (microroughness), impurity distribution, internal and external as well as intentional and unintentional gettering, and fundamental electrical degradation mechanisms.

Accurate specification of chemical purity requires development of models and simulation tools for physical and electrical consequences of the behavior of metal contaminants in silicon. Physical models should be based on experimental investigation of the deposition of metals from process solutions, the removal efficiencies of cleaning sequences, and segregation and distribution of impurities during subsequent thermal processing steps. The resulting contaminant distributions could then be incorporated into device models for oxide- and junction-leakage degradation due to the metal contamination through controlled experiments based on test-structure metrology. This paper addresses the current state of knowledge, limitations, and necessary developments for the realization of such a simulation/

modeling tool.

DEPOSITION OF METALS FROM PROCESS SOLUTIONS

Three possible mechanisms through which metal contaminants can deposit from solution onto silicon wafers are:

1. Adsorption of metal species without chemical change
2. Adsorption of metal species with chemical decomposition or precipitation
3. Adsorption of metal species with electrochemical reduction to either metal or to a partially reduced state.

These deposition mechanisms have been the subject of recent work (4-6). Currently there is no general agreement on which mechanism will dominate and it appears likely that there may be a different mechanism for each metal and each chemical solution which in conjunction with other complexities of the deposition process has inhibited the development of a general deposition theory.

Thus, it is still necessary to experimentally quantify the deposition characteristics of metals from process solutions. As a first step to this end an extensive data base has been compiled based on data extracted from the literature and supplemented by experiments at the University of Arizona, representing the state of knowledge of metal contaminant deposition on Si surfaces from chemical solutions (7). The results of this study show that, in general, the concentration of wafer surface contamination increases with the time the wafer is in the solution and the concentration of contaminant in solution, but eventually saturates at high wafer surface contamination levels. However, for typical values of process times and metal concentrations in process solutions, the depositions are essentially linear in time and solution concentration at a given temperature. In the case of HF based process solutions (8), the depositions show an Arrhenius behavior with temperature and verify that only noble metals deposit, in agreement with the electrochemical mechanism 3 above. Data for metal contaminants from other process solutions indicate that at a constant temperature and for low surface contaminant concentrations, as expected in semiconductor processing, the depositions are also linear with time and solution concentration. For example, Ryuta et-al (9) have shown that iron, nickel, copper and zinc deposit from the RCA clean SC1 solution ($NH_4OH/H_2O_2/H_2O$) in this manner although the mechanism of deposition is believed to be different (i.e. mechanism 1 for SC1 solution and Mechanism 3 for HF solutions).

From graphical representation of the deposition data, transmission equations relating areal wafer concentration of contaminant ($\#/cm^2$) to bath concentration of contaminant (ppb) for the depositions can be developed and incorporated in a user-friendly computer data base. Additional data on metal solubilities in cleaning chemicals could also be included. Further development of such a data base mandates continued literature searches for existing deposition data, accompanied by fundamental experimental and theoretical research to understand the mechanistic behavior of single component and synergistic depositions.

A key missing link in the information available in the open literature is related to the cleaning sequences favored in industry and what the effects of multiple cleaning steps, i.e. step efficiency, are. The dominant factor in surface prep over the years as been particles. Particles lead directly to gate oxide defects and have therefore needed to be a priority. For front end of line (FEOL), the dominant method for particle removal has been the SC1 clean (RCA1, APM,..., $NH_4OH:H_2O_2:H_2O$). This basic solution with high pH oxidizes the Si surface producing a native oxide as well as providing a solution with negative zeta potential impeding the deposition of most particles of interest. When operated at elevated temperatures or in conjunction with megasonic excitation, it leads to effective particle removal and a surface passivated with respect to additional particle adsorption. It is used as a "last" clean prior to gate oxidation in a number of semiconductor fabrication facilities.

The SC1 clean has one major drawback. The high pH conditions that lead to low particle adsorption rates and effective particle removal also provide an environment with very low metal solubility, especially in the presence of a strong oxidizing agent such as H_2O_2. The Si surface provides an ideal heterogeneous nucleation site for the precipitation of metals that may be present in the solutions at supersaturated concentrations. Conditions which lead to low particles lead to the potential for high metal levels and vice versa. A key challenge is to develop a strategy which provides for low particle and low metal concentrations at the same time. In addition, interactions with other types of contamination and surface-prep-related effects require integrated solutions to the overall problem. Another example of this is the finding that metallics, both on the surface, or in the cleaning solutions can catalyze the generation of surface and interface microroughness. Other effects with respect to anions and organics must also be considered.

IMPURITY SEGREGATION AND DISTRIBUTION

High metal concentrations that can be present on a wafer surface just prior to insertion into the gate oxidation furnace have led to the development of work-arounds to achieve high yields and performance. Some of these work-arounds relate to gettering strategies, such as the use of p/p$^+$ epi wafers. In this case, the p$^+$ substrate provides a low chemical potential for metals, especially Fe, due to the high solubility of Fe in the p$^+$ material. At 700°C the solubility of Fe in p$^+$ material is about 100 times higher than in lightly doped Si corresponding to a reduction in the chemical potential of about 0.4 eV. Also the details of the wafer structure and oxide process itself can provide a significant margin. One such factor is the use of Cl containing ambients during or just prior to the gate oxide growth. The Cl reacts with the metals, forming volatile metal chlorides which evaporate from the surface. Development of a contamination simulation tool would require determination of the mechanisms responsible for these process insensitivities. In the past most of these strategies have been developed independently of surface prep processes. To understand these effects in detail will require the use of numerical modeling and simulation so that all of the complex interactions can be tracked at the same time during a thermal process. Such modeling has been very successful in understanding the motion of dopants in compound semiconductors as well as the interaction of Si point defects during phosphorus diffusion, however, they have not been applied to contamination processes as yet.

The utility of numerical computational approaches to gettering and other metal related effects is clear. Such algorithms could be incorporated into modern process simulators to track metal concentrations all the way through a device process sequence. Considerable improvements in process flows to minimize metal effects would be forthcoming. The development of these capabilities will require significant R&D work. This is due to the need for better models of metal diffusion, bulk, surface, and interface segregation, and interactions with the Si point defects (vacancies and interstitials generated during processing); accurate values for the parameters associated with these models; and finally the implementation of the models into an appropriate process simulator.

Consideration of surface prep, gettering, the effect of sacrificial oxidation, etc., in an integrated fashion will lead to new optimized solutions which will be necessary for future technology nodes. The ability to perform simulations of metallic effects all the way from starting material through all the high temperature thermal processes will be a major enabler to an integrated approach to implement high yield, low CoO process flows. It requires a knowledge of the chemical potentials of the metals throughout the structure as a function of temperature, ambient, and doping level. In addition the kinetic parameters such as diffusion coefficients and point defect interaction coefficients must be known as well. Even for the important contaminant Fe, such information is not complete.

DEVICE EFFECTS/MODELING

A one-dimensional leakage current model, based on Shockley-Read-Hall deep-trap theory (10,11), has been developed at the University of Arizona to estimate the effects of metal contaminants (12). The model considers multiple contaminants, multiple trap levels, and assumes metallic impurities are homogeneously distributed throughout the bulk silicon. The density of generation-recombination centers is assumed to be equal to the contaminant density, and minority-carrier generation lifetimes are calculated using the trap density and experimental values of capture cross sections. An overall generation lifetime is obtained by reciprocal summation of the individual lifetimes. In the absence of measurable contamination, minority carrier lifetime is limited by the process baseline lifetime due to impurities present in the silicon. The process baseline lifetime is reciprocally added to the contamination lifetime to obtain a total generation lifetime. Because of the reciprocal summation, SRH lifetime dominates at high metallic impurity levels, and the process baseline lifetime dominates when the contamination levels are low. The model calculations were verified using a 1.25 μm CMOS device test structure lot in which the pre-gate buffered oxide etch (BOE) was intentionally contaminated with copper (13). Calculated diode leakage current using the deep-trap leakage model with the capture cross-section for copper extracted from lifetime measurements, shows excellent agreement with the measured leakage values. Recently the model has been extended to explain strain induced leakage currents due to copper contamination outside of the diode depletion region, as reported in the literature (14), by introducing a scattering cross section proportionally dependent on the contamination. Although plausible and providing excellent correlation of calculations with experimental results, this model needs further development through research and verification experiments.

At present there is no general model for contamination related defects for gate oxides. The effects of iron contamination on thin oxide dielectric properties were investigated at the University of South Florida using a modelling and experimental approach (15). For the electrical modelling of thin oxide structures, PISCES (16) computer simulation models were developed which represented conductive defects at the Si-SiO$_2$ interface. The defect region was modeled as a highly doped (1 x 10^{22} cm^{-3}) n-type region interface which penetrated into the oxide region at the lower electrode. The simulations indicate enhancement in field magnitude due to the conductive region is far higher than can be accounted for by oxide thinning alone in general agreement with the work of Honda (17) et. al., where thin oxide breakdown in the presence of iron precipitates was approximately 2.5 times lower than that predicted by oxide thinning considerations.

The experimental effort consisted of fabricating MOS capacitors of different area (MOSDOTs) and I-V testing. Float zone wafers were used to eliminate any unwanted internal gettering uncertainties. Specific wafers were contaminated with iron by immersion in a contaminated H$_2$O$_2$ solution following the final pre-oxidation wafer clean. Wafers were immersed in the solution for 1 minute and then spun dry. Thermal oxides of 10, 13, 16, and 20 nm were grown at 900°C in dry O$_2$ with a post oxidation anneal in N$_2$ for 10 minutes. Aluminum electrodes were formed by thermal evaporation to define the MOSDOT capacitors. Results from the MOSDOT experimental wafers indicate a distinct threshold for detrimental effects of iron contamination on oxide breakdown. As would be expected, the contamination threshold is a function of oxide thickness, with thinner oxide layers being more susceptible to contamination.

A recent joint paper by researchers at the University of Arizona and The University of South Florida demonstrates how junction leakage and oxide defect data can be combined to specify chemical purity specifications in light of DRAM architectural trends (18). However, this paper further indicates distinct knowledge gaps to more adequately address this problem such as: effects of cleaning step efficiencies, distribution and electrical activity of metals in Si due to post deposition processing steps, internal and external gettering, and surface micro-roughness, as discussed above. It also demonstrates the need for the development of an analytical model, analogous to the junction leakage model, or a simulator to accurately describe contamination effects on oxides is required. This oxide degradation model and further development of the junction leakage mode as noted above are planned research activities in this program.

CONCLUSION

Considerations of the requirements for materials, processes, and process tools for near 0.1 μm geometries leads to the conclusion that micro-contamination and processes to remove or mitigate microcontamination such as surface preparation and gettering will be major show stoppers. The drivers leading to 0.1 μm design rules also lead to large die sizes, large wafers, ultrathin gate insulators, and severe constraints on defect levels (related primarily to GOI "shorts" and interconnect opens). These requirements cannot be met with conventional technology or extrapolations of conventional technology, especially with respect to metals and particles.

As discussed in this paper, given the fundamental knowledge concerning mechanisms and thermodynamic and kinetic parameters, simulations of microcontamination deposition, removal, and interactions in the wafer and grown or deposited films is well within reach. Applications of process simulations to other aspects of processing have considerably shortened cycle times (factor of 2 estimates are typical) and led to the rapid implementation of optimized processes for high yield, performance, and overall low CoO.

In the microcontamination area, the ability to track the contamination all the way through the process sequence is critical. In one case, a seemingly harmless contaminant introduced in an early process step, may play a dominant role in degrading device performance if released or activated in a subsequent step. In another example, a large quantity of a seemingly harmful contaminant may be introduced in one step, only to be removed or passivated later. Only through the application of a micro-contamination module in a process simulator can these effects be easily understood, and optimum processes designed. It is clear that additional resources need to be allocated to build the models, database, and implement and integrate them into a process simulation capability. This will lead to a significant return on investment to provide a minimum CoO for future technology nodes.

ACKNOWLEDGEMENT

This work was supported by SEMATECH, administered by the Semiconductor Research Corporation under contracts # 88 & 91 MC-501, by the SEMATECH Silicon Materials Council, and by the University of Arizona Center for Microcontamination Control under contract # 12891.

121

Special recognition to Sandia National Laboratories MDL for the contamination lot processing. Special recognition is also given to interactions with colleagues through the SEMATECH Silicon Materials Council's Metals Task Force during 1994.

REFERENCES

1. Jastrzebski, L. "An Overview of Effects of Heavy Metal Contamination, Wafer Characteristics and Gettering on Device/Circuit Performance," in *Proceedings of the Sixth International Symposium on Silicon Materials Science and Technology: Semiconductor Silicon 1990*, Electrochemical Society, 1990, p. 614.

2. Miyazaki, M., Sano, M., Sumita, S., and Fujino, N., "Influence of Metal Impurities on Leakage Current of Si N^+P Diode," *Jpn. J. Appl. Phys.*, **30**(2B), 295 - 297 (1991).

3. Hattori, T., "Contamination Control: Problems and Prospects," *Solid State Technology*, 51-58, July (1990).

4. Kern, F. W., Jr., Itano, M., Kawanabe, I., Miyashita, M., Rosenberg, R., and Ohmi, T., "Metallic Contamination of Semiconductor Devices from Processing Chemicals the Unrecognized Potential," presented at the 37th Annual IES Meeting, San Diego, CA, May 6-10, 1991.

5. Helms, C. R., and Park, H.-S., "Generalized Model for Metal/Si Interactions Related to Microcontamination in Surface Cleaning & Passivation for Semiconductor Processing," G. Higashi, ed., *MRS Proc.* **315**, 287 (1993).

6. Helms, C. R., and Park, H.-S., "Electrochemical Equilibria of Fe in Acid/Base/Peroxide Solutions Related to Si Wafer Cleaning," in Cleaning Technology in Semiconductor Device Manufacturing, J. Ruzyllo & R. Novak, eds., *Electrochem. Soc. Proc.* **94-7**, 26 (1994).

7. Craigin, R., et-al; "Metal Contaminant Deposition on Silicon Surfaces from Solutions," SRC Technical Report, Contract 90-MC-501, ID #C91581, June 1991.

8. Parks, H. G., Hiskey, J. B., and Yoneshige, K., "Mechanistic Study of the Deposition of Metals from HF Solutions onto Silicon Wafers," in Interface Control, of Electrical, Chemical, and Mechanical Properties, *MRS Proc.* **318**, 245 (1993).

9. Ryuta, J., Yoshimi, T., Kondo, H., Okuda, H., and Shimanuku, Y., "Adsorption and Desorption of Metallic Impurities on Si Surfaces in SC1 Solution," *Jpn. J. Appl. Phys.* **31**, 2338 (1992).

10. Hall, R. N., "Electron-Hole Recombination in Germanium," *Phys. Rev.* **87**, 387 (1952).

11. Shockley, W. and Read, W. T., "Statistics of the Recombination of Holes and Electrons," *Phys. Rev.* **87**, No. 5, 835 (1952).

12. Parks, H. G., Schrimpf, R. D., and Craigin, R., "Mapping Contamination from Process Chemicals to Device Degradation," presented at the Microcontamination '92 Conference, Santa Clara, CA, Oct. 26-30, 1992.

13. Parks, H. G., and Schrimpf, R. D., "Metal Deposition, Metrology, and Device Degradation from Contamination in Semiconductor Processing Fluids," presented at the 40th Annual Technical Meeting of the IES, Chicago, IL, May 1-6, 1994.

14. Miyazaki, M., Sano, M., Sumita, S., and Fujino, N., "Influence of Metal Impurities on Leakage Current of Si N^+P Diode," *J. Jour. Appl. Phys.* **30**(2B) L295 - L297 (1991).

15. Henley, W. B., Jastrzebski, L., and Haddad, N. F., "The Effects of Iron Contamination on Thin Oxide Breakdown - Experimental and Modeling," presented at the Spring MRS Meeting, San Francisco, CA, April 1992.

16. Pinto, M. R., et-al., *PISCES II B* Stanford Elec. Lab, Stanford Univ. Tech. Report. Sept. 1985.

17. Honda, K., Nakanishi, T., Ohsawa, A., and Toyokura, N., *J. Appl. Phys.* **62**, 1960 (1987).

18. Parks, H. G., Kosier, S. L., Schrimpf, R. D., Henley, W. B., and Jastrzebski, L., "First Order Specification Of Liquid Chemical Purity Requirements Based on Contaminant Deposition and Advanced DRAM Architecture," presented at the Microcontamination '94 Conference, Santa Clara, CA, Oct. 4-6, 1994.

Gate Dielectrics for High Performance ULSI:
Practical and Fundamental Limits

L. Manchanda

AT&T Bell Laboratories, Holmdel, NJ 07733

One of the simplest and yet most attention demanding process module in the development and manufacturing of ULSI is the gate dielectric. Si has a native oxide i.e. SiO_2 which has been the primary reason for the existence of Si VLSI. After 30yrs of research on MOS physics/technology, there is much still to learn about this amorphous dielectric. In this paper, we make an attempt to understand the system and ULSI technology requirements for the gate dielectrics. From the fundamental point of view, if resistivity was the ultimate scaling factor, a simple classical calculation would show that SiO_2 can be scaled to ~1nm. However, the direct tunneling may set this limit to much higher value. In this paper, we take a technology view to understand the practical scaling limits of SiO_2 from a thickness of 8nm to 1nm and some benefits of replacing SiO_2 with oxynitrides for ULSI and GSI. Then we will address the issue about alternate gate dielectrics for GSI.

I. System and Technology Considerations:

In the last 32 yrs, the level of silicon integration has evolved from a single MOSFET to billion transistors on a chip (figure 1)[1-2]. This high level of integration can be attributed to two major factors: (1) Advances in lithography have given an ability to print fine dimensions in the rage of 0.25μm to 0.1μm and (2) Silicon has a native oxide, i.e. SiO_2. For gate applications, SiO_2 has been scaled from few hundred nanometers to 8nm in manufacturing and 1.5-8nm in R &D.

related to the thickness of the gate oxide,

$$I_D = q\, C_{ox}\, K(V_G - V_{th})^2 \qquad (1)$$

where K is related to the device parameters.

Reduction of the gate oxide thickness, improves drive current/transconductance and provides faster switching capability to charge external capacitance and therefore enhances circuits speed and reduces the operating voltage and power. Reducing the thickness of the gate oxide, also reduces the threshold voltage and the transistor/device can

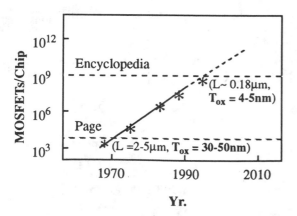

Figure 1: Progress in silicon integration (DRAMs).

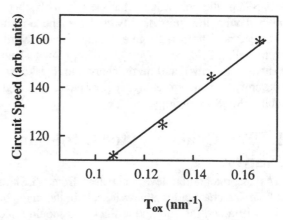

Figure 2: Impact of the gate oxide thickness scaling on a two-input nand circuit (T_{ox} = 6-9 nm).

The scaling of the gate oxide in VLSI or ULSI circuits is driven by performance, density and low power. The current drive of a MOSFET is directly proportional to the gate oxide capacitance and therefore is inversely

be switched with low voltage/power. As shown in figure 2, for small channel length transistors, a few angstrom change in the thickness of the gate oxide has a large impact on the performance of transistors and circuits speed [3]. The scaling of the gate dielectrics have created a revolution in many families of MOS technologies. Thin gate oxide permits high performance and low power logic [4], high density DRAMs [5] and non-volatile memory [6] with low programming voltage. In this paper, we will focus on the gate dielectrics for high performance CMOS ULSI. There is more than 30yrs of research done on this subject. Many thousands research publications, numerous review articles, many conference proceedings and books have been written on the physics, material science and reliability of the Si/SiO_2 interface [7-13]. In this paper we take a technology angle to understand the process, manufacturing and reliability limits on the scaling of SiO_2 from 8nm to 1nm. At the end we will briefly discuss alternate gate dielectrics for GSI.

II. Why SiO_2 has Taken Si Integration from SSI to VLSI:

A simple one step heating of Si in oxygen at high temperatures (600-1200°C) forms a thin insulating film of SiO_2 on Si. The high temperature growth of SiO_2 on Si creates thin films with very low interface stress[13]. The subsequent optimized anneals can quench the non-bridging bonds and reduce the oxide fixed charge and bulk neutral traps to $10^{10}/cm^2$ and the interface state density to $10^9/cm^2$ [14]. The MOS structures with poly gate have the following unique properties:

- High resistivity of SiO_2 (> 10^{16} ohm - cm).
- Large band-gap of SiO_2 (~9eV).
- Large electron affinity of Si (~4.15eV).

For modern day poly/SiO_2/Si structures with SiO_2 thicknesses >8nm, the controlled interface with polysilicon combined with low trap density and the above-mentioned electrical properties give a large Fowler Nordheim tunneling field (>6MV/cm) with very low leakage currents(< $10^{-9}/cm^2$) and high charge to breakdown (>10C/cm^2) and therefore high performance and high reliability MOS technology.

III. ULSI Challenges for Ultra-thin SiO_2:

As the integration level increases from VLSI to ULSI, device channel lengths will scale in the range of 0.25 - 0.1μm and the gate oxide thickness should scale in the range of 7-3nm. In this thickness range SiO_2 may face the following manufacturing, processing and even the fundamental limits:

III.1 Processing and Manufacturing Limits:

For such ultrathin oxides there may be many processing and manufacturing constraints. In this section we will focus on the three major process issues which can affect

the scaling of SiO_2 and the advantages of replacing SiO_2 with oxynitrides.

III.1A Thickness, Impurity Diffusion and V_{th} Control:

Sub-0.25um technologies require large channel doping. Therefore, small changes in the thickness of the gate oxide can have a large impact on the threshold voltage variation. Figure 3 shows the impact of the gate oxide thickness variation on threshold voltage variation of NMOS devices with t_{ox} = 6.5nm.

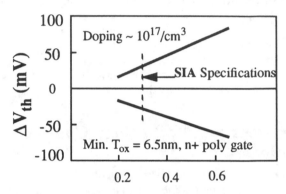

Figure 3: Impact of the gate oxide thickness variation on threshold voltage shift for NMOS.

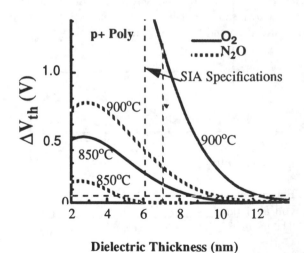

Figure 4: Threshold voltage shift vs the gate dielectric thickness for PMOS devices.

Only a 0.1nm (= 1σ) variation in the thickness of the gate oxide can enhance the threshold voltage by 30mV and reduce the current drive and therefore the circuit performance by 10%. This problem becomes much worse for PMOS devices. Advanced CMOS technologies targeted for low power ASICs and portable applications require low and symmetric threshold voltage for n- and p-channel devices. It is easy to design a surface-channel PMOS device with p+ poly gate. But, SiO_2 has an open network. Boron from the p+ poly gate can easily diffuse into the channel and affect the process margin. Since boron penetration depends on the thickness of the gate dielectric, any changes in the thickness will change the amount of boron that diffuses through the oxide. When small thickness variations (0.1-0.3nm) of the gate oxide are coupled with boron penetration, the threshold voltage variations can become very significant. As shown in Figure 4, using the SIA specifications for 6.5nm gate oxide, depending on the thermal budget of the post-gate processes, 0.1nm (=1σ) variation in gate oxide thickness coupled with boron penetration can increase the spread in threshold voltages to +/- 250mV. This affects the scaling of SiO_2. To solve this problem, SiO_2 may have to be replaced by oxynitrides [15]. It has been shown by M.L Green et al. [16] that small amount of N in the N_2O grown oxides is very effective in retarding the boron penetration.

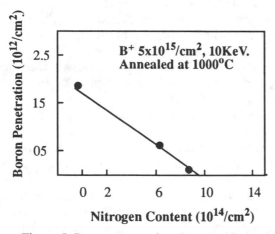

Figure 5: Boron penetration through thin layers of oxynitride grown by RTN_2O [16].

Figure 5 shows the measured boron penetration into silicon as a function of the N content of the N_2O-oxide. The nitrogen was measured by nuclear reaction analysis. Only N content equivalent of 0.1 monolayer of SiN is sufficient to retard diffusion of $\sim 2 \times 10^{12}/cm^2$ boron atoms through the oxynitride layer. Due to this significant retardation of boron from the p+ poly gate, the N_2O grown oxynitrides are very effective in enhancing the process margin of CMOS ULSI. When N_2O grown dielectrics are used the minimum required t_{ox} can be 2-4nm lower than SiO_2. This is a significant enhancement in the scaling of the gate dielectrics. As shown in figure 2, 2-4nm reduction in thickness of the gate dielectrics

can improve the system performance by 25-40%.

III.1B Polysilicon depletion and inversion layer effects:

The polysilicon depletion effect in MOS devices increases the effective gate oxide thickness and therefore requires more aggressive scaling of the gate dielectrics. As described in the last section, the high performance/low power CMOS technology requires dual-poly gates. The simplest way to form the dual-poly gates is to dope the poly layers by ion-implantation either before the gate patterning or during the S/D formation. Due to the reduced thermal budget of deep submicron devices, the available thermal activation may not provide sufficient doping near the poly/SiO_2 interface [17]. The poly-gate doping levels can fall below the degeneracy level. The poly depletion effect and the series inversion layer capacitance from the substrate can increase the effective gate oxide thickness by 10-40% [17-18]. To maintain the circuit performance, one solution of this problem is the gate oxide may have to be scaled 40% more than the gate oxide thickness without the poly depletion effect. This rate of scaling is more aggressive than the scaling for the successive generation of technologies and becomes a manufacturing and reliability issue for sub-0.25μm technologies. The second solution may be a metal-gate technology. The effects of metal gates directly on such ultrathin oxides is also a reliability concern and requires research on mechansims of breakdown of MOS structures.

III.1C Interface Roughness:

As the gate oxide thickness is reduced to < 6-7nm, surface microstructure and the interface roughness may have significant impact on the scaling of SiO_2.

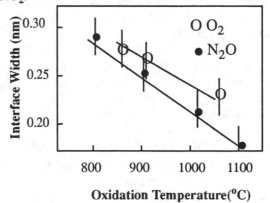

Figure 6: Interface roughness of the Si/SiO_2 and the Si/oxynitride interface as a function of oxidation temperature. The interface roughness measured using synchrotron x-ray source [19].

The interface roughness may reduce the high field-channel mobility and therefore the device performance. The interface roughness may also degrade the reliability of MOS devices. The starting Si roughness before oxidation depends on the quality of the starting substrate, cleaning and the processing before the gate oxide. Evans-Lutterodt, et al. [16, 19] have shown that the oxidation smooths the Si surfaces, but leaves interface roughness in the range of 0.1-0.3nm. As shown in figure 6, the oxide/Si interface becomes smoother with increasing oxidation temperature. The N_2O-oxide inter-

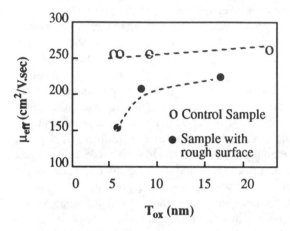

Figure 7: High field mobility as a function of T_{ox}, (E=1MV/cm), surface roughness ~ 0.5-0.7nm [20].

face shows a similar trend. but is smoother. Thus, the Si-O-N layer in some way influences the local structure of the interface.

It has been shown that the large values of the interface roughness can reduce the effective mobility. As shown in figure 7, this impact becomes significant for ultrathin oxides [20]. The large values of the interface roughness can also reduce the oxide-charge to breakdown by a factor of 10 [21]. The state of the art Si-processing can reduce the interface roughness to 0.1-0.2nm (RMS values). However, a quantitaive correlation of these real values of the interface roughness on device properties such as channel mobility, statistical threshold voltage variations and the insulator breakdown has not been established.

III.2 Reliability Limits:

To write few paragraphs on the reliability of SiO_2 is like condensing an encyclopedia into a 1 paragraph write-up! For thick SiO_2, interface charges played a significant role on the threshold voltage control of MOS structures[8-9]. For ultrathin oxides, charges in the range of 10^{10} and $10^{11}/cm^2$ do not cause a large threshold voltage shift. The aggressive gate oxide scaling for

ULSI may be limited by the presence of neutral defects and interface roughness. The oxide defects/ neutral traps charge with time and can reduce the charge to breakdown and cause an early dielectric failure. Depending on the nature of these defects, there may be various failure mechanisms and therefore reliability limits on scaling of SiO_2. In this paper we will limit to three critical reliability limits which have been faced by the gate dielectrics for submicron technologies and expected to continue to be a problem for sub-0.25μm technologies.

III.2A Hot-Carrier Degradation:

The hot-carrier degradation in MOS devices depends directly on the generation rate of hot-carriers in the device and the trapping probability of carriers in the gate oxide. For increasing integration level, MOSFET scaling is being done at the faster rate than the scaling of the system or chip power supply. Therefore, the peak-electric fields in the channel are increasing to values >400KV/cm and it is possible to observe impact-ionization or hot-carrier generation at very low voltages (0.6 to. 0.7V) [22]. However, the aggressive scaling of devices is also leading towards the aggressive scaling of the thickness of the gate oxide. The trapping probability of hot-carriers is directly proportional to t_{ox}^n, where n = 2-3. Decreasing the thickness of the gate oxide very rapidly reduces the bulk hot-carrier trapping. Therefore, for ICs with single level of metal the net-degradation due to hot-carriers becomes less of a concern for devices approaching 0.09μm [23].

The high level of integration is also demanding multi-levels of metal interconnects. Therefore, the total processing steps after the gate oxide are increasing with increase in the metal levels. These processes increase the radiation-induced and charging damage in the oxide creating more neutral traps. There are limits to the thermal budget after the first level of metal and the process temperature are too low to quench the neutral traps. Therefore, for ICs with multilevel metal, the net hot-carrier damage may remain an important problem. The level of this problem can be reduced by low-voltage metal etching.

III.2B Process Induced/Plasma Induced Damage:

Plasma processes are very widely used in the manufacturing of MOS products. If a large area conducting antenna is connected to the gate oxide, during the plasma processing, the charge collected by the antenna can pass through the thin gate oxide. It has been observed that the charge can cause various types of damages in the gate oxide. The observed damages include, early breakdowns, excessive leakages, an increase in interface states and decrease in charge to breakdowns [24]. However, a first-cut on modeling of the oxide thickness dependence of charging damage by plasma processing shows that during the plasma processing MOS devices are subjected to electrical stress from plasma which acts as a current source and therefore the stress is weakly dependent on the thickness of the gate oxide [25]. Therefore, ΔD_{it} becomes smaller as the oxide thickness is decreased. The plasma-induced damage is very much dependent on the configuration of the etching systems and therefore, it is difficult to predict the oxide scaling limits. Also, the impact of plasma damage on statistical-insulator breakdown is unknown.

III.2C F. N. Tunneling and Breakdown:

The practical scaling limit of SiO_2 would depend on the SiO_2 ability to sustain itself as an insulator for the lifetime of the circuit/chip. As the ULSI devices are scaled towards sub-0.25μm, the operating device fields across the gate oxide may enhance from 3.3MV/cm to 5-7MV/cm [2]. The increasing oxide fields may enhance leakage currents, Fowler Nordheim tunneling and cause intrinsic time dependent breakdown. These degradation mechanisms may impose a challenge on the scaling of SiO_2. However, experimentally these values have not been established. As shown in figure 8, the oxide scaling

such as RTO or N-containing dielectrics.

III.3 Fundamental Limits:

Moore's Law is an exponential extrapolation and these usually run into natural barriers at some point. As discussed in the beginning of this paper, SiO_2 and lithography have been two major driving forces for Si integration. Therefore, it is expected that the same forces may be the limiting factors for rapid integration. In this section we briefly address the fundamental issues which may impact the scaling of SiO_2.

III.3A Pinholes or Micropores [28]:

By the year 2010, high yield ULSI would require D_o < 0.01/cm². As discussed in section 2, the reason for such a high level of integration of Si is that it has a gate dielectric which can be fabricated by simple thermal oxidation of Si. The same characteristic of SiO_2 i.e. thermal growth may generate pinhole density larger than 0.1/cm². To estimate pinhole density generated during the growth, we make the following assumptions: Every atom which lands on Si reacts with Si, No diffusion and viscous flow and Poisson statistics are valid. Using these assumptions,

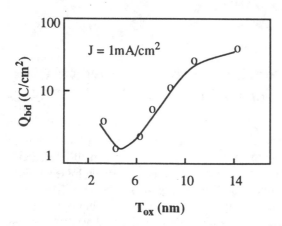

Figure 8: 50% charge to breakdown as a function of the oxide thickness[26].

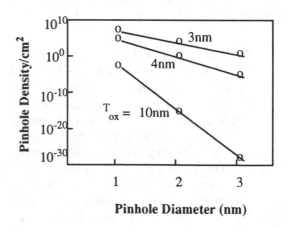

Figure 9: Calculated values of the pinhole density.

may be limited by the statistical beakdown[26]. C. Hu etal [27] have shown that if all defects in gate oxide are modelled as an effective thickness less than the nominal thickness and assuming that the gate hole injection model is valid, for 2.5V technology at 125°C, the minimum gate SiO_2 thickness may be limited to 3.4nm. This 3.4nm value is for today's Si quality and today's gate oxide technology. This limit may be reduced by the better quality of starting Si and new gate oxide processes

figure 9 shows the plot of the calculated pinhole density as a function of pinhole diameter. For a 3nm thick oxide the pinhole density for 3nm diameter pinholes can be as large as 30/cm² (requirement being <0.01/cm²). Due to the assumption that there is no viscous flow, these calculations are over-estimation of the pinhole density. However, the pinhole density as small as 0.03/cm² may impose a serious challenge on the scaling of SiO_2 to thicknesses less than 4nm. The multilayer or stack CVD oxide may be a solution to reduce the pinhole/micropores

[29-30]. This method requires control of two layers of oxides and may not be a viable solution for ultrathin oxides <5-6nm.

III.3B Resistivity and Tunneling:

The ultimate limit of the scaling of SiO_2 would depend on its ability to function as an insulator just after device fabrication and the ultimate practical limit of scaling would depend on the SiO_2 to sustain itself as an insulator during the entire lifetime of the system (3-30 yrs). The resistivity of SiO_2 is $> 10^{16}$ohm-cm. If we assume, a resistivity of 10^{17} ohm-cm, a very crude classical calculation would show that for a 1V operation, the limit of SiO_2 would be 1nm. However, before we reach this limit, for electric fields >1MV/cm, the quantum mechanical direct tunneling between the gate and Si may limit the SiO_2 scaling to 1.8-2.5nm [with WKB approximation, 31]. Until very recently, it was believed that the onset of direct tunneling will set the ultimate limit on the scaling of SiO_2. However, efforts are being made to scale sub-0.05μm devices with 1.5nm gate oxide[23]. The devices are being designed with gate oxide in the tunneling regime. For very low voltages, the direct tunneling may not impose a gate leakage design constraint, but may impose a charge to breakdown reliability constraint. *Therefore, the scaling of SiO_2 may not be limited by the fundamental physics, but by the practical-reliability limits.*

IV. Alternate Dielectrics:

In the last 2 decades, there has been a continues search for an alternate gate dielectric. The question, for gate applications, "can any other dielectric replace SiO_2", has not been answered. As we discussed in section 3, the SiO_2 already has been replaced by slightly modified SiO_2, i.e. oxynitrides. The search continues for a dielectric with dielectric constant much higher than 4 [32-34]. Table 1 shows a potential dielectric material, their dielectric constant and breakdown strength. Let us define the figure of merit of a dielectric = dielectric constant x breakdown strength. The figure of merit for the oxynitride and SiN are larger than the figure of merit of SiO_2. Therefore, it is clear that the next competitor of SiO_2 may be SiN. There has been significant research done on SiN. SiN fabricated with CVD and plasma methods are dominated by traps and therefore these materials could not be used for the gate application. The recent invention of JVD SiN with very low trap density is a looking very promising candidate for replacing SiO_2 for thicknesses less than 3nm [35]. Other high dielectric materials (metal oxides) such as Y_2O_3 and Ta_2O_5 have large electron traps and therefore may be limited to

capacitor applications. At present, there is not enough data base to judge the future of ferroelectric PZT films.

Table 1

Figure of Merit = Dielectric Constant x Breakdown Strength			
Dielectric	Dielectric Constant	Breakdown Strength@	Figure of Merit (A. U.)
SiO_2	3.9	12-15	48-60
SiO_xN_y	~4	15-16	60-64
SiN	7-9	10-11	70-99
Y_2O_3	16-20	4-5	64-100
Ta_2O_5	20-25	3-5	60-125
TiO_2	80-120	0.5	40-60

@ In MV/cm.

V. Conclusions and Characterization Challenges for Ultrathin Gate Dielectrics:

The increasing level of Si integration will force MOS industry to push SiO_2 to its practical limit. In the past SiO_2 scaling has been mainly driven by technology improvements. The scaling at and beyond 5nm would very heavy depend on our ability to characterize and understand ultrathin SiO_2 or the possible alternate gate dielectrics. Therefore, gate oxide scaling for sub 0.25μm technologies faces the following characterization challenges:

Measurements of Thickness Variation:
0.1nm variation in the thickness of the gate oxide has a strong impact on the performance of ULSI circuits. Therefore, it is critical to develop instrumentation for the measurements of absolute thicknesses and statistical thickness variation to an accuracy of 0.1nm over a 8-12" wafer. Device/circuits community needs to specify the correlation lengths for the thickness variation.

Interface Roughness Characterization:

It has been shown that the large values of interface roughness (0.5-1nm) have a strong impact on the performance and reliability of devices with ultrathin oxides. However, the physics of ultrathin-oxides with real values of interface roughness (0.1-0.2nm) has not been understood. To understand the implications of these values of interface roughness, we need to develop tools to measure the roughness with correlation lengths in the range of 10-20nm.

D_o and Pinhole Density Measurements:

High yield ULSI would require defect density $< 0.01/cm^2$. This number translates into less than 4-5 defects and pinholes over a 12" wafer. For the development of defects free gate oxide process, it is important to understand the nature of these defects and the kinetics generated pinholes.

Poly Dielectric Interface:

For oxide thickness less than 6nm, poly depletion effect can reduce the device performance by 30-40%. Therefore, it is very important to understand the poly/SiO_2 interface. The effects of this interface on the reliability of ultra-thin SiO_2 needs to be investigated.

Characterization of Oxynitrides:

The presence of small amount of nitrogen (~0.1 monolayer of SiN) in the oxynitrides described in this paper has a large impact on the process margin of ULSI. Therefore, it is important to accurately measure the nitrogen profile in oxynitrides.

Characterization of Reliability:

The ultimate practical limit of the gate dielectric depends on the reliability of the dielectric. Physics based models are required for an efficient testing of reliability of ultrathin SiO_2.

Collaborators: *F. Baumann, D. Brasen, K. Evans-Lutterodt, M. L. Green, L. C. Feldman, K.S. Krisch, S. Lytle and A. Ourmazd from AT&T Bell Labs. C.Hu from University of California Berkeley. A. Ishitani from NEC Co., Japan and A Toriumi from Toshiba Co., Japan.*

References:

1. C. A Warwick, R. H. Yan, Y. O Kim and A. Ourmazd, AT&T Technical Journal, p 50, Sept. (1993).

2. The National Technology Roadmap for Semiconductors, Semiconductors Industry Association (1994).

3. C. Hu, Proceedings of the IEEE, p682, May (1993).

4. M. Bohr, S. U. Ahmed, L. Brigham, R. Chau, R. Gasser, R. Green, W. Hargrove, E. Lee, R. Natter, S. Thompson, K. Welden and S. Yang, Digest of the International Electron Device Meeting, p 273, (1994).

5. K. Itoh, IEEE J. Solid- St. Circuits, vol. SC-25, pp. 778, June 1990.

6. J. Van Houdt, D. Wellekens, L. Faraone, L. Haspeslagh, L. Deferm, G. Groeseneken and H. E. Maes, IEEE Trans on Electron on Components, Packaging and Manufacturing Technology, V17, p 380, p380 (1194).

7. E. H. Nicollian and J. R. Brews, *MOS Physics and Technology*, John Wiley and Sons(1982).

8. G. Barbottin and A. Vapaille, *Instabilities in Silicon Devices*, V1 and V2, North Holland Press (1986).

9. J. R. Davis, *Instabilities in MOS Devices*, Gordon and Breach Science Publishers (1980).

10. S. K. Pantelides, *The Physics of SiO_2 and its Interfaces*, Pergaman Press (1978).

11. G. Lucovsky, S. K. Pantelides and Frank L. Galeener, *The Physics of SiO_2 and its Interfaces*, Pergamon Press (1980).

12. *Proceedings of the INFOS Conference*, Elsevier Press (1979-1995).

13. C. J. Sofied and A. M. Stoneham, Semiconductor Science and Technology, p215 (1995).

14. G. Declerck, Proc. of the Workshop on Nondestructive characterization of Semiconductors, p1 (1978).

15. K. Krisch, L. Manchanda, F. H. Baumann, M. L. Green, D. Brasen, L. C. Feldman, A. Ourmazd, Technical Digest of IEDM, p325 (1994).

16. M. L. Green, D. Brasen, K. W. Evans-Lutterodt, L. C. Feldman, W. Lennard, H. -T. Tang, L. Manchanda, M. -T. Tang, Applied physics Letters, p 848 (1994).

17. C. Y. Lu, J. M. Sung, H. C. Krisch, S. J. Hillenius, T. E. Smith and L. Manchanda, IEEE Electron Device Letters V10, p192 (1989).

18. R. Rios and N. D. Arora, Technical Digest of the IEDM,

p613 (1994).

19. M. T. Tang, K. W. Evans-Lutterodt, M. L. Green, D. Brasen, K. Krisch. L. Manchanda, G. S. Higashi and T. Boone, Applied Physics Letters, p748 (1994).

20. J. Kaga, S. Takagi and A. Toriumi, Technical Digest of the IEDM, p475 (1994).

21. T Ohmi, M. Miyashita, M. Itano, IEEE Trans. Electron Devices, V39, p537 (1992).

22. L. Manchanda, R. Storz, R. Yan, K. Lee and E. Westerwick, Technical Digest of the IEDM, p 994 (1992).

23. H. S. Momose, M. Ono, T. Yoshitomi, T. Ohguro, S. Nakamura, M. Saito and H. Iwai, Technical Digest of IEDM, p593 (1994).

24. S. Fang, S. Murakawa and J. P. McVittie, IEEE Trans. Electron Devices, p 1849 (1994).

25. H. Shin, K. Noguchi and C. Hu, IEEE Electron Device Letters, V14, p 509 (1993).

26. E. Hasegawa, K. Akimoto, M. Tsukiji, T. Kubota and A. Ishitani, Ext. Abstract of SSDM, p86 (1993).

27. K. F. Schuegraf and C. Hu, IEEE Trans. Electron Devices, V41, p761 (1994).

[28]Readers may be surprised why we have chosen to discuss pin-holes limitations in the section of fundamental limits. The pin-holes we have discussed here are due to the basic kinetics of thermal SiO_2 and not due to the starting process induced defects.

29. P. K. Roy, R. H. Doklan, E. P. Martin, S. F. Shive and A. K. Sinha, Technical Digest of the IEDM, p714 (1988).

30. H. -H. Tseng and P. J. Tobin, J. D. Hayden and K. M. Chang, IEEE Trans. Electron Devices, p 613 (1993).

31. J. Maserjian, Physics and Chemistry of the Si/SiO_2 interfaces, Eds. C. R Helms and B. E. Deal, Plenum Press, p497 (1989).

32. L. Manchanda and M. Gurvitch, IEEE Electron Device Letters, p180 (1988).

33. K. W. Kwon, I. S. Park, D. H. Han, E. S. Kim, S. T. Ahn and M. Y. Lee, Technical Digest of the IEDM, p 835 (1994).

34. S C. Sun and T. F. Chen, Technical Digest of the IEDM, p 333, (1994).

35. D. Wang, T.-P. Ma, J. W. Golz, B. L. Halpern and J. J. Scmitt, IEEE Electron Device Letters, p 482 (1992).

TCAD Status and Future Prospects

Donald L. Scharfetter

SRC/SEMATECH/Intel
Los Alamos National Labs.
MS: B297 P.O.Box 1663
Los Alamos, New Mexico 87545

The TCAD technology advancement pace must accelerate. Tools adequate for modeling and simulating each new generation of technology must be available to process developers before their development task begins. Historically, TCAD capability has lagged leading-edge process development needs by one process generation. Fundamental improvements must be made to TCAD tools if they are to adequately support design and technology developments called for in the National Roadmap on Semiconductor Technology.

CURRENT STATE OF TCAD TECHNOLOGY [1]

Accurate materials, process and device modeling and simulation are critical to timely, economic development of future integrated circuit technology. The time, risk and cost of continuing to rely primarily on experimental verification of process and device characteristics for performance and reliability are incompatible with the pace of world-class technology advancement. TCAD tools that accurately predict device and interconnect properties in anticipated wafer fabrication technologies are indispensable to reducing reliance on extensive experimentation. Materials characteristics and process model accuracy (in both 2 and 3 dimensions) must improve to predict adequately the performance and reliability of device structures with future roadmap technologies. Ability to measure directly 2-D doping profiles is critical to the validation of new models. The interest and availability of the talent in our National Laboratories to work on first-principles process models must be encouraged and directed. Statistical modeling capabilities must be developed that accurately reflect materials and process variations in device and interconnect models for circuit design. Sources of systematic variations as we approach lithographic limits must be identified and characterized. Because of the lack of truly physics-based models, statistical process simulation remains in its infancy. Sufficient direct physical data is not available to calibrate widely used wafer fabrication processes models in regimes required by roadmap technologies. Traditional process engineering has relied on short loop and split-lot experiments to separate the effects of individual control factors. These experiments take time and can yield ambiguous results. Calibrated TCAD process engineering tools can reduce the need for such experiments, shorten the time needed to solve engineering problems and help separate independent control variables. Multilayer interconnect structure, electrical, and reliability tools must be calibrated on a hierarchical basis. Interconnect TCAD tools now available cannot handle the complexity of roadmap interconnection structures.

TCAD tools pose uniquely difficult technical challenges in the area of grid generation. Device simulation requires careful treatment of grids at interface layers, while process simulation has the problem of grid creation with moving boundaries. Furthermore, adaptive grid techniques have yet to be fully developed for TCAD tools. These grid problems place real obstacles to the progress of TCAD. Yet, the grid problems are solvable. Progress will definitely be achieved by an organized, systematic effort.

The current generation of TCAD tools are based on relatively old codes, often developed by a single graduate student. These codes have been adopted by the TCAD vendors and by industry companies and other universities. Extensive modifications have been made. The overhead in maintaining and modifying these old codes is stifling development in the TCAD field. New tools should incorporate the experience gained during the last decade in process and device simulation. They should be modular, so that innovative models or algorithms can be easily added by third-party developers. Finally, these new tools should be widely deployed as the base codes in universities, vendors, and industry, to improve the productivity of the entire TCAD effort.

TCAD tools must provide solutions to materials, process and device problems quickly enough to meet real-time process line engineering needs. This means that tools for day-to-day use must run on workstation platforms that do not rely on super computer resources to speed calculation time. Production engineers must routinely work with calibrated tools and efficient models for effective line control. Direct use of TCAD tools by production engineers will most likely be prohibited by computational expense of these tools. Response surface models, calibrated by TCAD tools, provide a viable alternative.

Simulating Performance

The interface between electronic and technology CAD limits performance prediction. The division between electronic and technology computer-aided design tools (ECAD and TCAD) is based on the idea that circuit designers make use of ECAD and process designers make use of TCAD. The typical communication is that process designers use TCAD tools to provide circuit designers with design rules or transistor model parameters for circuit simulation. This interface is a result of the fact that circuit designers can only modify transistor widths and lengths. The situation is different with interconnect, because circuit designers can drastically modify interconnect performance through layout modifications, and process designers can drastically modify interconnect performance by changing process parameters like dielectric thickness.

In order for circuit designers to manage the worsening problem of interconnect or layout optimization, they need three types of interconnect tools:

1. For "hand" optimization of medium-sized (thousand device) critical cells, full-cell, coupled circuit-electromagnetic analysis will be required.

2. For non critical circuitry, layout synthesis tools will be used, and such tools need computationally efficient delay estimators as well as signal-integrity, ground-bounce-avoidance, and reliability design rules.

3. Finally, full-chip verification tools will be needed to insure signal integrity, low substrate noise, and electromigration tolerance.

It is clear that the present development paths will, over the next five years, provide circuit designers with optimization tools that allow them to zoom-in on entire critical cells in a VLSI layout, automatically generate a very simplified 3-D geometry, extract the circuit-simulator compatible

description, and then compute the circuit wave forms. The same set of tools will also be used by technology designers to optimize process performance and signal integrity. It is also true that these tools will move seamlessly to analyzing associated advanced packaging, because these tools will already be performing the higher frequency analyses presently only required for packages. The main issue that requires additional effort is generating better approximate 3-D geometries and developing approaches that can handle the non-ideal semiconductor substrate.

The tools needed for layout synthesis are in mixed shape. If current trends continue, simplified models for interconnect delay will improve steadily and provide reasonably accurate models even for IC's with multiple metal levels and gigahertz clocks. The signal integrity problem is much harder, because it concerns the coupling between nets rather than just the behavior of an individual net. In addition, signal-integrity-preserving design rules have yet to be established. These rules will be hard to develop because they will have to include sufficient geometry to insure that long-range effects have been screened, and also be easily checked so that they can be used as part of layout synthesis. The difficulty with providing electromigration design rules is in understanding the physical mechanisms and is not an interface issue.

Providing reasonably fast, full-chip signal integrity, substrate noise, and electromigration analyses will require substantial additional efforts. Any successful approaches will certainly require a combination of techniques including: efficiently eliminating non critical nets, using libraries of solved geometries, and developing approaches for hierarchically patching together separate analyses.

CHARACTERIZATION [2]

Although 2-D dopant profiling techniques have progressed in the last 5 years, they are inadequate in providing reliable information for today's technologies. Nonetheless technology development has not slowed, i.e., scaling from 1.0 um to 0.30 um has continued in the past 5 years.

It is clear that the major customer for accurate quantifiable dopant profiling is TCAD. Even here there are two usage's. One mode of TCAD applications is using semi-empirical models to fit 1-D profiles, then supplementing it with device measurements to capture 2-D effects. This must be done with full loop wafers with functional devices, and, often several iterations are required to understand and capture the complex 2-D effects. This encompasses other modeling issues such as carrier transport models in device simulation. This mode of operation is used to support technology development, in the sense that successive split runs capture the knowledge of the previous runs, therefore, although not predictive in the true sense, it helps define process options (e.g. rule them out). This TCAD mode does not directly need 2-D dopant profiling.

The second mode of TCAD as a customer is in physical model development. Here short loop experiments are run (no functional devices to measure) which are intended to sort out the physical models. As an example, studying the interactions of dislocation loops with point defects in a 2-D structure or the nature of the silicon/oxide interface interaction with point defects will need accurate 2-D dopant profiling. This TCAD activity is the true customer of 2-D dopant profiling. This is the ultimate TCAD goal to have true predictive capability.

The current state of the art in 2-D profiling is $>\sim500A$ of lateral spatial resolution plus limits on the sensitivity, i.e., only a limited range from 1e15 to 1e21 can be captured. This may have been useful for 1.0 um technology but is useless for sub 0.25 um. Major problems exist in extending the range. Moreover many problems remain in extracting dopant profiles from 2-D measurements of resistance, etch stains, capacitance maps, STM images, SEM images, etc. On the other hand, device measurements are naturally sensitive. Even in 1-D, while we argue whether SIMS or SRP is good enough to resolve +-20% around 1e17 channel doping range, Vt measurements can be easily used to distinguish this. The message is that for TCAD to assure predictability much improved characterization is required coupled with additional fundamental understanding.

FUTURE PROSPECTS

SEMATECH, SRC and the National Labs

In the fall of 1993 SEMATECH formed a TCAD department, and in January of 1995 the DoE/SRC CRADA launched The Center for Semiconductor Modeling and Simulation (CSMS). The SEMATECH program is naturally focusing on addressing the roadmap needs (discussed above), but with a deliverable schedule generally in 6 months intervals.

The CSMS is attacking basic fundamentals which up until now were not funded in our TCAD university programs. In fact in most cases the deliverables from the CSMS projects will be to the universities. To the extent these programs are successful, one can make the statement that "research in silicon is just beginning", and true TCAD predictability is on the horizon

THE FIVE CSMS THRUSTS[3]

1-Grids and Computational Science

Overview/Introduction. The computational science needed for the next generation of semiconductor process and device models will depend on advanced techniques for girding, and will sometimes require parallel computation and a high-performance computing environment. Visualization, frameworks, and the user environment are significant factors in providing useful simulation tools for the semiconductor engineer. Effective management of the complexity of the computational environment will be essential; methods will need to be developed that permit efficient exploration of the design space so that the most desirable solution may be produced within the required constraints.

With respect to grids, the next generation of simulation codes will require the ability to automatically generate and optimize 3-D moving surface and volume grids as well as local adaptive mesh refinement to resolve time-dependent, feature-scale physical phenomena. We plan to use 3-D moving adaptive surface grids in order to capture the time-dependent curvature, orientation, and surface area of deposited (or etched) features. We will use 3-D moving adaptive volume grids in order to handle the intersection, collision, and growth of surfaces as well as to provide locally orthogonal grid orientations for tracking moving internal discontinuities and fronts. Grid reconnection and optimization will be used to assure the quality of the time-dependent grids. To develop the initial starting grids we will use hybrid element shapes generated by a combination of two tools: The National Grid Project (NGP) mesh generation system coupled to the Los Alamos Mesh Generation System (LAGEN). By using this combination of tools we will be able to automatically generate and maintain 3-D unstructured hybrid grids for arbitrarily complex 3-D geometries.

This grid technology fits into the overall structure of complex simulators as client objects in frameworks that

133

are design-built using an object-oriented analysis & design; this should be coupled to an object-oriented database system that manages object persistence, inter-object communications, and concurrent object execution. The overall design of the Los Alamos simulators will allow active objects to interact and communicate through a client/server relationship. The design will support any of the three major parallel processing programming paradigms: message passing, work sharing, and data parallel.

The goal of the Mesh Generation, Optimization, and Moving Adaptive Meshes project is to develop utilities for mesh generation, adaptive mesh refinement, dynamic mesh reconnection, and moving adaptive mesh algorithms. These are essential for efficient and accurate deposition and etch simulators as well as for other semiconductor process applications. The moving mesh algorithms will also provide a capability for tracking highly distorted surface motion, which often occurs in surface and volume physical processes that involve radiation or chemistry. Software development, language considerations, and parallel distributed computing are also important considerations. A C++ control flow will be used for the grid codes in order to facilitate both software quality assurance and communication interfaces between other code or physics modules.

2-Bulk Process Models

Overview / Introduction Bulk processing is defined by the SIA Roadmap as "IC fabrication steps such as etching, diffusion, and implantation that alter the semiconductor substrate." Like the 1994 TCAD roadmap, in this project we have separated out a distinct thrust area called the Topographic Simulator to handle the important area of etching and plasma deposition, and their effects on feature size and topography. The remaining topics such as diffusion, implantation, and the properties of thin films in direct contact with the silicon (e.g., oxidation, silicidation, nitridation, and polysilicon) are included in the bulk processes thrust area.

More than 15 years of work on TCAD tools that can simulate bulk processes has resulted in a significant capability today. Nonetheless, as the industry continues to move towards smaller (submicron) devices, this area is becoming increasingly critical and today's models are less and less adequate. Thus, the continued advancement of IC

technology is more than ever before predicated on the ability to precisely control device structures, particularly dopant distributions, on an ever finer scale. Moreover, the need for two and three dimensional profiles is becoming equally important, as well as the time dependence of non equilibrium processes during rapid thermal anneals.

In addition, equipment costs and demands for shorter cycle times for new products is driving the requirements for such modeling and simulation to include a true predictive capability. Such a capability can only be acquired through the development of an accurate and deep understanding of the basic physical equations and underlying physics, leading to the development of sophisticated physically based models. This will require the gradual replacement of today's tools by a much more complicated set of codes.

The best hope for the development of this new tool kit is a hierarchical approach. In such an approach fundamental *ab initio* calculations will provide a deep understanding of the phenomena. To achieve work-station-level codes that operate on the length and time scales of wafer-state processes requires scaling up the fundamental codes through a hierarchy of higher-level physically based codes that are more phenomenological, although with parameters calculated from or based on the understanding derived from more fundamental calculations.

The goal of this project is to provide such a hierarchy of understanding and tools that span from a deep and detailed microscopic understanding of the basic processes to high-level tools that can be used in TCAD work stations.

One area that is very important for bulk processes, girding, has been separated out for special treatment (as is also done in the 1994 TCAD roadmap). The manpower and technical requirements for this area are treated in the girding thrust area, discussed above. Girding developments and results will, of course, be explicitly included and used for bulk process tools.

Ion Implantation. The ultimate goal of the ion implantation project is to supply a predictive physically based code capable of calculating ion damage and amorphization as a function of dosage and ion energy, point-defect and extended-defect creation and annihilation, transient enhanced diffusion, and effects on dopant and defect concentrations due to thermal annealing (including RTA). Dopant concentrations need to be accurately determined in both two and three dimensions at each step in the process, including both implant range and lateral straggle. Multiple targets (implanting through other layers) and multiple implants (dopant-dopant interactions) will also need to be addressed. It is important that this project be closely coupled to the dopant and point-defect diffusion project.

Dopant and Point-Defect Diffusion. The goal of the dopant and point-defect project is twofold: (1) to develop new girding and partial differential equation solvers in two and three dimensions that accurately and efficiently solve the coupled diffusion equations needed in high level codes such as SUPREM, and (2) to focus on the microscopic origins of the diffusion process in order to at first guide current dopant diffusion modeling and ultimately to bring about the required predictive capability at the deep submicron length scales required for the next generation of devices. Transient enhanced diffusion, effects of high dopant concentrations, and non-equilibrium aspects and processes during rapid thermal annealing will also need to be addressed. To achieve all of this will require gaining a deeper understanding of the processes which control impurity redistribution on the atomic level and applying powerful new techniques for *ab initio* calculations of the interactions of dopants and defects in silicon to the modeling of integrated circuit fabrication processes. This project will maintain a close connection with the ion implantation project.

3-Topography Models

Reactive Processing. The requirements imposed by continual reductions in feature scales and increases in the complexity of microelectronics devices, make traditional design approaches, based on trial and error experimentation, prohibitively expensive, thus necessitating the use of accurate simulation tools. Considerable advances have been made in the simulation of fluid flow and heat and mass transfer in equipment for reactive processes such as chemical vapor deposition (CVD), but chemical kinetics have not been characterized, save for a few key systems. Although large efforts by several groups have been directed towards obtaining a nearly complete picture of silane kinetics, it is not realistic to assume that experimental rate data will soon become available for the wide variety of chemical systems used throughout the microelectronics industry. Therefore, this program aims to augment ongoing equipment modeling efforts with semi-

empirical and ab initio computational chemistry approaches to obtain kinetic rate parameters.

Computational chemistry has developed rapidly over the past five years, and it is becoming possible to simulate gas-phase molecules, as well as molecule surface interactions, with a variety of techniques based on semi-empirical, molecular orbital, and local density functional (LDF) methods. The techniques have already been used with success to estimate thermodynamic parameters for the Si-Cl-H and Si-F-N-H systems. LDF theory, in particular, has been shown to be a potentially powerful technique for simulating gas-surface interactions. A systematic effort is needed to develop and apply these tools to reactions relevant to semiconductor processing. Currently available computational chemistry techniques could already provide much needed kinetic data, but in order for computational chemistry tools to become accepted within the industrial community, their usefulness must be demonstrated on relevant problems, and the predictions must be evaluated against experiments. In addition, the kinetic information must be compiled into evaluated mechanisms that can be used in equipment and profile evolution simulation strategies.

The goal of these modeling studies is to provide accurate information on kinetic rate processes involving molecular species in the gas phase and on semiconductor surfaces. This information will need to be derived from thermochemical properties and reaction barriers calculated from computational chemistry methods and from solid state physics approaches. A key issue is the development and assessment of methods that can adequately describe gas phase molecules, molecular species adsorbed on semiconductor surfaces, and the reactions that these species can undergo during deposition processes. The modeling of deposition processes described above ultimately must be distilled into topographic simulators through phenomenological models in the form of reduced kinetic schemes for the gas phase and surface chemistry.

Computational chemistry and solid-state density functional theory approaches will be applied to predict deposition of intrinsic and in situ doped polycrystalline silicon from silane and chlorosilanes. Doping sources will be diborane, phosphine, and arsine. This deposition system is chosen because of its industrial importance, known difficulties in controlling uniformity of deposition and doping rates, and the already existing understanding of silane chemistry. The relatively well established gas-phase and surface chemistry of silane will provide a means for evaluating computational techniques, and make it possible to focus on silicon-species and dopant interactions complicating deposition of doped polysilicon.

Etching. The ever decreasing size and increasing aspect ratio of features to be etched in the fabrication of microelectronics is imposing increasing demands on directional plasma etching. As a result, a much broader range of feature aspect ratios must be simultaneously etched with greater critical dimension and sidewall angle control which has driven the industry to develop high density plasma sources for directional etching. The development of these processes is largely a systematic, but empirical experimental program which is costly in terms of both time and money. At present, topographic modeling tools for etching are based on data which is gathered from the analysis of SEM cross sections after wafer processing. As the data input to these programs is confounded by the many physical and chemical effects which affect etching rate and directionality, these programs are largely interpretive and without validated chemical models have limited use in the prediction of the topography for new process development.

The evolution of topography during directional plasma etching is complex because many physical and chemical processes which contribute to the etching and passivation of the feature surfaces. These processes include a) ion-enhanced etching of surfaces, b) spontaneous etching of surfaces by neutrals, c) transport and reactive loss at surfaces within the features of both reactants and products, d) passivation of surfaces within the feature by deposition of plasma species and by redeposition of etch products, e) ion scattering within features, f) charging of features which deflects the ion trajectory, g) the angular dispersion of the impinging ions, etc. As many of these mechanisms give similar qualitative results, simultaneously contribute to the profile evolution, and/or are balanced to off-set each other, the measurement of a profile cannot be deconvoluted to yield the correct kinetics causing the profile evolution.

To develop the topographic design tools needed to facilitate the development of directional etching processes, several developments must occur; 1) the development of equipment and process models which are capable of predicting the fluxes of ionic and neutral species to a surface during processing, 2) the development of topographic simulators which are capable of including all the physical and chemical phenomena which occur at a surface during etching, 3) the determination of the physical and chemical mechanisms and rate coefficients which occur at the surface, and 4) the testing of the these models individually and in combination.

It is the goal of this sub-project to address the measurement and modeling of the surface kinetics in a manner suitable for inclusion in topographical simulators. The results of this project will be transferred to a topographic simulator, which will be under development in the Etch Process Simulator Development project. This collaboration will assure the viability of these surface kinetic formulations in topographic simulators and the transfer of the information created by this task into a form that is of value to the industry. In addition under this task, testing will be performed to verify the correctness and applicability of the kinetics by experimental simulation of the etched features in both beam studies (where the fluxes are well known) and in plasma processes .

The surface mechanisms and kinetics will be measured using molecular beam scattering in which well defined beams can be combined to simulate the flux on the wafer during etching. The relative fluxes of ions and neutrals, the impingement angles, and the ion bombardment energies can be varied independently to probe the surface mechanisms and to observe the etching rates. The products can be analyzed using mass spectroscopy and their sticking probabilities measured using a quartz crystal micro balance.

The independent manipulation of well defined fluxes, the observation of surface state and products, and the deposition of products and beam species on surfaces minimize the convolution of the surface kinetic processes so that the fundamental kinetics can be characterized. The surface state can be assessed using surface analytical tools such as XPS and laser induced temperature programmed decomposition.

The fundamental kinetics of ion-enhanced etching will be studied using computational techniques such as molecular dynamics simulations in combination with transition state calculations . With molecular dynamics calculations, the collision cascade which leads to the production of sputtering products and weakly bound species can be computed. Transition state calculations can then be used to compute the desorption of the weakly bound species which occurs on a time scale which is much greater than can be simulated with molecular dynamics. Using this approach, a fundamental understanding of the dominant kinetic processes can be assessed. Based on this understanding and experimental data, simplified models will be constructed which are suitable for implementation in profile simulators. In addition, this modeling will assist in the experimental beam work by reducing the amount of data acquisition which is necessary for the characterization of a chemical system. The computer facilities and parallel processing expertise at the National Laboratories will greatly enhance the productivity of this effort.

The development of inter atomic potentials for use in these molecular dynamics and transition state computations is critical. Potentials exist for the Si-F system, but do not exist for Si-X systems, where X represents C, O, and all of the halogens (except F). These potentials are necessary to predict and understand the kinetics of oxide etching, selectivity, and sidewall passivation. Strategies for generating potentials for covalently-bonded systems (Si, C, H, halogens) are fairly well defined, whereas a satisfactory potential form that describes ionic-bonding with covalent-bonding (such as that in a oxide/poly-Si interface) does not presently exist. Exploratory studies to solve the latter will commence later in this project.

Etch Process Simulator Development. As feature sizes shrink and more levels of metallization are added, topography simulators are being increasingly used to provide insight into how surfaces are affected during deposition and etch processes. Early topography simulators included very little physics and were instead mostly concerned with generating idealized surfaces. Idealized in this context means 'as drawn' in a schematic device cross section. Over the years, more and more physics has been incorporated into topography simulators; i.e., the transport of reactant and product species and the chemical and physical processes involved in specific processes have been incorporated. The majority of the focus has been on the development of simulators which consider transport in 3-D, but are limited to 2-D structures (long trenches/lines and via/contact-like features). More recently, there has been a drive towards 3-D topography simulators, which deal with processes on general surfaces.

2-D (3-D transport, 2-D surfaces) topography simulation can be considered well in hand, with efforts focusing more on the use of these simulators to develop transport and reaction rate models from experimental information (physical science). Sophisticated 2-D simulators have been developed.. Much of the initial work in this project will focus on further development of this software. However, through the entire course of this work, we expect to investigate the other profilers so that the evolving product

of this development will be a 2-D (and later 3-D) simulator which inherits significant contributions from several sources. We expect this final software to earn a new identifier, with a documented lineage from its seminal ancestors. Further 2-D simulator (code) development will focus on improving efficiency and accuracy; however, we can expect that specialized transport and/or reaction kinetic models will be needed for selected processes. The exception is topography evolution of multiple material substrates, with wide variations in local surface velocities and material boundaries.

3-D topography simulation is not very well developed relative to 2-D simulators. Algorithms have been developed and tested for evolution of single material substrates, but considerable work remains in this area. Evolution of 3-D multiple material substrates, required for a general etch process simulator, can be considered largely unexplored. This is the goal of the this project.

The development of reliable, physically-based topography simulators has a high priority on the technology roadmap, because these simulators can be used for predictions of topography evolution. To be generally useful, topography simulators should handle the evolution of substrates comprised of a number of materials. This is particularly important for etching, but can also be important for depositions in which the reactions and/or transport models depend upon the substrate material; e.g., selective deposition.

In order to use physics-based process simulators, reasonable estimates for the models and model parameter values for transport and reaction kinetic models are needed. Unfortunately, these models have not been developed for most reactive processes. To be widely used, topography simulators should access one or more standardized reaction kinetic and transport databases. There should also be guidelines for the development of reaction and transport models, so that industrial colleagues can develop models for internal' use. Ideally, a single kinetic and transport database could be used for several reactor-scale and feature-scale simulators, with careful guidelines provided so that users can reliably choose the best representations for their application.

For the most part, the determination of reaction kinetic and transport models will continue to be done using 2-D topography simulation and experiments. This is due to both experimental and computational limitations. However, these models will subsequently need to be used in 3-D simulations for simulation of device fabrication steps. There will be additional situations for which 3-D simulations will be needed to interpret experimental results and establish models. Thus, we need to begin to develop methods to validate the predictions of 3-D topography simulators, by comparison with data from 3-D surfaces. As general 2-D and 3-D topography simulators are being developed, versions should be developed for use by TCAD engineers on workstations of reasonable capacity. These latter codes should be commercialized through TCAD vendors.

The etch simulator development effort described in this paper focuses on the development of general 2-D and 3-D etch topography simulators which deal effectively with general, multiple material (MM) substrates and the variety of chemistries which that implies. It is appropriate to focus on etch simulator development, because a 'general' etch process simulator will also deal effectively with deposition and combined etching and deposition. For this reason, we will employ the term 'etch process simulator' in this more inclusive sense.

The algorithms needed for topography evolution during deposition and etch process simulation can be written largely independent of experimental information; i.e., they can be written in terms of general reaction kinetic and transport relationships. This is useful for modularity, and it also provides a useful distinction between the theoretical development of surface representation and moving algorithms and the empirical development of the kinetic and transport models required for accurate simulation of specific processes. Extensions to 3-D, which mandate more complex moving boundary algorithms but retain many of the 2-D physical models, will benefit greatly from this approach.

Link To Equipment. A top priority of this effort is to develop a simulator for a "closed loop plasma reactor" which is a complex piece of equipment used to manufacture multilayer semiconductor devices. LANL, SNL, and LLNL have decided to coordinate their activities in modeling and simulation on plasma processing to develop and implement a simulator of plasma reactors for the semiconductor industry. Such a (two-way) link, from the feature scale to the equipment scale, is the key to making topography simulation more widely useful. If the etch/deposition equipment settings could be used as the inputs for predicting feature-scale profiles, the applicability of topography simulation would rise dramatically.

At present, there are no sophisticated models that link feature-scale physics to reactor-scale models. Although they have proven to be valuable to process engineers, both 2-D and 3-D topography simulators are limited by their need for fluxes from the reactor (source) volume to the local wafer surface. In some situations, it is sufficient to use the predictions of a reactor-scale model as inputs to a feature-scale simulator; however, this is not valid in general. Relating wafer surface conditions to reactor set points requires linking them to reactor-scale codes. Therefore, while the emphasis should be on the development of feature-scale topography simulators and reactor-scale simulators as distinct tools, linkages between them should be built in and developed as the tools mature.

This approach is consistent with the industry goal for equipment vendors to be able to design, build, and deliver next-generation plasma reactors much more quickly and cost-effectively than they can today. To be useful as design tools, simulators must describe generation of the plasma species, transport of the plasma from the source to the wafer, interaction of the plasma with the silicon wafer and reactor walls, and removal of the excess gaseous and particulate material. It is believed that the DOE/DP Labs currently have software and algorithms which can be of immediate use to the equipment and software vendors.

The first step in developing this simulator is implementation of a common "framework" among all the plasma reactor equipment and software vendors. This framework should be common among the participating Labs and industry will serve as the foundation into which all improved physical, chemical, and engineering software models will be inserted as they are developed. This common framework will be substantially better than the best currently in use by any of the vendors and, because of its modular design, it will be flexible enough to allow each vendor to develop its own proprietary concepts.

4-Device Models

The need for a predictive device simulation tool treating nonlocal hot electron effects in silicon, such as impact ionization and emission of electrons and holes into the silicon-dioxide, can be satisfactorily addressed by full band Monte Carlo simulations. Because the inelastic mean free path for these hot electrons/holes is smaller than 0.01 microns, quantum effects other than those related to the band structure, which are included in this method, are of minor important. It is therefore imperative to develop a plan for the next five years to cast this method into a standard form and make it available for use on stand-alone workstations or clusters. To achieve this goal, two major existing shortcomings need to be addressed:

The importance of high energies (up to 4 eV into the bands) opens a vast parameter space that calls for standardization by comparing existing codes and developing a parameter set that optimally explains all existing experiments. A comparative study has been initiated by the National Center for Computational Electronics (NCCE) including 47 groups. However, a standardization effort also requires the development of numerical tools and algorithm that can "convert" convoluted experimental data into distribution functions, which in turn provide information on the parameter space.

Monte Carlo codes are CPU-time intensive. In this respect, great progress has been made by J. Bude, AT&T Bell Labs, to remove the bottleneck in momentum space trajectory generation, using the simplices method and refined griding in k-space with speedups by a factor of 100 or more. Development of adaptive grids embedded between energy isosurfaces of the band structure, and implementation on workstations of standardized Monte Carlo platforms are necessary. One also must investigate the possibility of parallel computing on compatible systems which may be suitable for Monte Carlo applications, for example HP workstation networks connected with an ATM switch or SP-2 by IBM. For distributed parallel network computing, neural net representation of band structures can potentially provide the advantage in speed and memory requirement since the band structure can be replaced by a finite number of neurons which are trained in advance to represent the band structure of a material with a desirable accuracy. In addition, adaptation and development of three-dimensional electrostatic field solvers and advanced variance reduction techniques should be included in the Monte Carlo simulation platform.

Finally, an important aspect of the computational physics involved in these various projects is the integration of all of the codes that are either currently used or in the process of development. For commonalty, ease of use, and to permit work to be performed by teams whose members are geographically disjoint, we will place each code on a new database management system. This system has been designed to be a completely portable library of routines common to all of the host codes. This design has already been tested in the areas of mesh generation, computational electromagnetics, dynamic memory management, and visualization. The common library is object oriented, written in C/C++ and is portable, without change, across all common machine boundaries and operating systems. The physics codes to be developed can likewise be made portable with the use of this advanced database management system since the physics algorithms will not be burdened by any other software such as i/o and individual graphics packages. This flexible modular code structure will enable code flow control, data management, and communication between different algorithms to greatly reduce the burden of code management and modification. This approach will facilitate efficient implementation and evaluation of new physics models and numerical algorithms. Such an environment will allow rapid testing of multiple algorithms developed under our joint efforts to create a standard Monte Carlo full band simulation platform.

We also expect that it will be possible to integrate this effort with earlier work on distributed environments based on NCSA MOSAIC and supported on the Internet/World Wide Web. Interactive capabilities have been added to MOSAIC to enable real time remote collaboration and distributed computation/visualization, all of which can be exploited to improve scientific interaction and technology transfer. In the interactive MOSAIC environment, it is possible, for instance, to control a complete process of computation and visualization, where the control interface, the numerical computation and the visualization can take place on separate platform. The message passing to control the interprocess communication is accomplished by the Data Transfer Mechanism (DTM) protocol developed by NCSA. Large scale simulations running on a super computer may generate enormous data files which could not be efficiently transferred across the network, and DTM provides a mechanism to segment the data access to create the visualization, without physically move the data file to the graphics engine. The distributed interactive MOSAIC system could be configured as a computer equivalent of teleconferencing for remote collaboration, where for instance a computer simulation could be carried on at one site, the visualization process performed at the other site, and the graphical output displayed at both sites.

Feedback mechanisms can be provided, so that one can monitor, from visualized data, the intermediate steps of large scale computations (visualization steering). The user has the possibility to modify interactively input parameters on-the-fly, without having to restart the simulation, and then steer the simulation with the guidance of the visualized data. The integration of the MOSAIC environment to the portable database management system developed at LANL will greatly enhance our ability to effectively develop, maintain, and distribute computer applications.

5-Interconnect Models

In the deep submicron regime, interconnect, contact, and via failure mechanisms are strongly related to specific grain structure features. Therefore, the statistics of failure depend on the statistical characteristics of the metal grain structure. Rare micro structural features that lead to rare, low-reliability metallization elements can control the reliability of the large population of elements (e.g., vias or interconnect segments) in a modern integrated circuit. It has become clear that micro structural "weak-links" generally become less common in deep-submicron metallization. This provides both an opportunity, and a risk. The opportunity is that, given accurate assessment of the true reliability of large populations of metallization features, it should be possible to develop new, aggressive design rules that allow higher current densities in small features, leading, in turn, to the design of integrated circuits with greatly improved performance.

This great potential for performance improvements comes with a risk. If fatal micro structural features or weak links are rare, they are unlikely to appear in the small populations of test elements used in current reliability assessment procedures. It then becomes essential to independently assess the probability of occurrence of these weak links.

This probability is not only a strong function of a specific metallization choice, but also of the process history for interconnects, vias, and contacts. The grain structure in lines evolves during all process steps, not just during deposition. Assessment of the statistical characteristics of grain structures requires modeling which accounts for micro structure development and evolution during metallization and subsequent processing.

Such modeling, coupled with micro structural characterization and lifetime testing, as part of a new reliability assessment methodology, is already needed for accurate assessment of current technology. It will become even more critical as interconnect dimensions continue to decrease. Creation of these new modeling capabilities and methodologies is essential for development of new design rules for high performance integrated circuits.

In the last 10 years, there have been significant advances in the computer simulation of grain structure evolution, including simulation of nucleation and growth during film deposition, and grain growth during post-deposition processing. Two main simulation techniques have emerged: (i) front-tracking (FT) methods, and (ii) Monte-Carlo (MC) methods. Only the FT technique has been used to simulate grain structure evolution specifically in continuous and patterned thin films. Simulations of grain structure evolution that occurs during post-deposition and post-patterning processing allows prediction of the distribution of polygranular cluster lengths. Long polygranular clusters are associated with early failure of contacts and vias, as well as long lines. These clusters are therefore the "units" most likely to fail in large populations. Prediction of the statistical characteristics of these units, in conjunction with conventional reliability testing, will allow more accurate assessment of reliability, and will allow the use of significantly less conservative design rules, especially for narrow lines and small vias and contacts. The goal of this project is to develop reliable design rules to allow higher current densities address the statistical characteristics of the metal grain structure for reliability.

SUMMARY AND CONCLUSION

A cooperative program between Industry, our TCAD University Researchers and the new National Labs CSMS has begun. As this program attacks the industry's considerable lack of fundamental understanding at the atomic level, one can say that:

Research in Silicon is just beginning and true TCAD predictability is on the horizon.

REFERENCES

[1] TCAD working group supplemental report for the National Technology Roadmap for Semiconductors 1994
[2] private communication, Rajiv Mathur
[3] DoE/SRC CRADA, LA94C10163, December, 1994

Computational Physics for Microelectronics--
The Old and the New Challenges

Sokrates T. Pantelides

Department of Physics and Astronomy
Vanderbilt University, Nashville, TN 37235

The basic physics of semiconductors, primarily the behavior of electrons and holes in the vicinity of pn junctions led to the invention of the solid-state transistor. Integrated circuits (chips) entail the fabrication of transistors and other devices (capacitors, rectifying diodes etc.) together with properly insulated interconnecting metal lines. Enormous advances have occurred over the last four decades in the physics that underlies the performance and electrical reliability of semiconductor devices. The physics that underlies processing steps for manufacturing (e.g. etching, oxidation, gettering of unwanted impurities, etc.) is less developed. The physics that underlies the mechanical reliability of integrated circuits, stress- and current-induced mechanical failures is even less developed. These inadequacies are a serious limitation in the ability of engineers to use computer simulation to optimize processing windows and to optimize circuit designs for both performance and reliability.

INTRODUCTION

The solid-state transistor was invented in 1948[1] at a time when semiconductor physics was in its infancy. The key physics was the recognition that electrical conduction in semiconductors was possible by both electrons (n-type) and holes (p-type) and that the level of each type of conductivity can be controlled by impurity concentrations. The behavior of electrons and holes in the region of p-n junctions under externally applied electric fields was the key ingredient of the transistor concept. The first transistor[1] was made out of a block of Ge with point contacts. For the next ten years, the quantitative theory of transistor action was developed while experimental methods for fabricating transistors were explored. Silicon quickly became the semiconductor of choice, largely because of the properties of its native oxide (silicon dioxide) and the fact that it could be produced in highly pure form

The next major breakthrough for microelectronics came in 1958 when the first integrated circuits were fabricated.[2] For the fabrication of integrated circuits, arrays of electronic devices are made simultaneously on the surface of a thin Si wafer by a series of processing steps (oxidation, impurity diffusion, etching) performed through lithographic masks. A schematic of such a device (a bipolar transistor) is shown in Fig. 1. Subsequently, the devices are connected with thin metal lines encased in insu-

lating material (Fig. 2), thus producing entire circuits. Rectangular pieces of silicon wafers containing integrated circuits became known as chips.

This "planar" silicon technology has remained fundamentally the same in the last 35 years. The dimensions of devices and metal lines have been shrinking at a very rapid pace, however, resulting in rapid increases in device and circuit densities with a concomitant rapid increase in performance. A convenient measure of this progress is the number of bits per chip in dynamic random access mem-

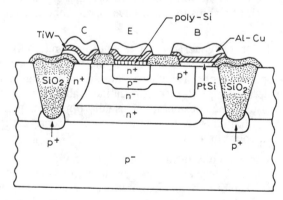

Fig. 1 Schematic cross section of a modern bipolar transistor fabricated at the surface of a Si wafer

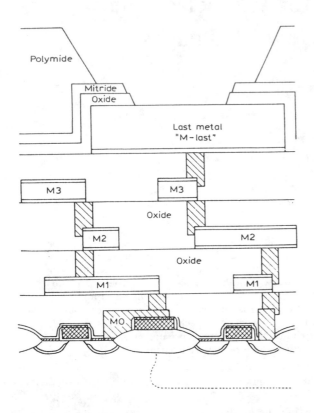

Fig. 2. Schematic of metal lines on a chip. Three levels of metal are shown.

ory chips (DRAMs). Figure 3 illustrates the evolution of this number in the last 20 years and shows projections that the same pace will continue. Currently, chips with 0.5 micron device features and 1 micron metal lines are mass produced. There are millions of devices and hundreds of meters of metal wire on a chip. By the turn of the century there will be billions of devices and kilometers of metal wire on chips that will have an area about a square inch.

For each generation of technology, materials and processes need to be chosen and fine tuned to meet the new specifications. The choices are driven by four major considerations: performance, manufacturability, reliability and cost. These considerations usually impose contradictory requirements so that the final choices represent substantial compromises.

For example, even though GaAs devices are much faster than Si devices, Si remains and will remain unchallenged because of manufacturability constraints. The density of devices one can achieve with Si is more than an order of magnitude higher than can be achieved with GaAs, which means that one needs at least ten GaAs chips for every Si chip. The concomitant signal delays in chip intertconnections eliminate any advantage that faster GaAs devices may have.

Another example is the choice of metal. Aluminum has remained the material of choice even though copper and silver have higher conductivities. Finally, silicon dioxide is used as the insulator in capacitors, where one needs a high dielectric constant, and as the insulator surrounding the metal lines, where one needs a low dielectric constant. Clearly, there are many materials with significantly larger dielectric constants, but they cannot be adopted for manufacturability reasons.

As choices are made for materials and processes, balancing the conflicting needs arising from several considerations, basic science provides valuable guidance. Increasingly, this guidance is in the form of computer simulations. In this paper we will assess the state of the science that underlies issues of performance, manufacturability and reliability in the design and production of microelectronics. We will pay particular attention to the availability of the relevant science for computer simulations.

In the area of performance and electrical reliability, the relevant physics is that of electrons in semiconductors and metals. In the area of manufacturability and mechanical reliability, the relevant physics is that of dynamical rearrangements of atoms as they are manifest at different length scales, from the microscopic atomic scale to the macroscopic continuum. We will examine the state of the corresponding physics and computational implementations in detail in the remainder of this paper. Simply stated, our overall assessment is that the physics of electrons is in excellent shape and well ahead of current needs. The physics of atomic rearrangements is in excellent shape at the microscopic scale, in quite good shape at the macroscopic continuum, but, with a few exceptions, in rather poor shape at intermediate or length scales.

ELECTRICAL PERFORMANCE

Modeling of **circuit designs** to optimize performance has been indispensable in microelectronics. The

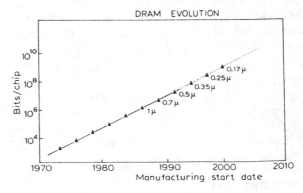

Fig. 3. Evolution of DRAMs.

142

underlying physics and mathematics is well under control. The physics is essentially Maxwell's equations of classical electrodynamics. For circuit modeling, electronic devices such as transistors are represented in terms of conventional "equivalent circuits" which makes classical electrodynamics applicable to entire circuits, in fact the whole chip. Such computer programs have existed for some time and microelectronics companies for years have depended on mainframe computers to design circuits.

The "equivalent circuits" that represent **electronic devices** are defined in terms of a number of quantities that depend on the structure of the devices themselves. This is the interface between circuit designers and device designers. In the early stages of a new generation of technology, circuit designers rely on device modeling to produce the device parameters they need to model circuits. The relevant physics is contained in Boltzmann's transport equation which describes the time evolution of the distribution function f(r,k,t), where r is a position variable, k is momentum and t is time. The equation basically says that the time rate of change of f is equal to a drift term, a diffusion term and a collision term. Once the equation is solved for f for the complex geometry of a particular device, all macroscopic electrical observables can be calculated as integrals of f.

Boltzmann's transport equation, which is highly non-linear, can be solved in a variety of approximations [3,4]. The simplest is the so-called drift-diffusion approximation which amounts to an equation for particle conservation and an equation for momentum conservation. Energy is automatically conserved through the assumption that the electron and hole temperatures are constant and equal to the lattice temperature. This approximation has been the basis for modeling devices for several decades. It is quite adequate except in some hot-electron cases. The next level of approximation, referred to as the hydrodynamic model, corresponds to including one more moment in a moments expansion of Boltzmann's equation. The temperature is no longer constant but is to be determined from an energy conservation equation plus the particle and momentum conservation equations. Finally, it is possible to solve Boltzmann's equation exactly by Monte Carlo techniques. In Fig. 4 we show results obtained by Fischetti and Laux [4] that illustrate the range of application of the drift-diffusion approximation. The figure shows the electron velocity in a field effect transistor of different gate lengths. "Current tool" refers to the drift-diffusion approximation and MC refers to the exact Monte Carlo solution of Boltzmann's equation. Note that the drift-diffusion approximation deviates form the experimental data only when the device length is smaller than 0.2 micros. Such devices will not be manufactured until after the turn of the century (Fig. 3).

Monte Carlo solutions of Boltzmann's equation in real semiconductors have demonstrated a "universality principle" [4]. In very short devices, the carriers acquire substantial velocities (Fig. 4) and hence occupy energy

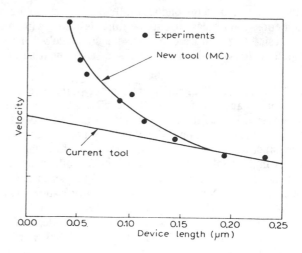

Fig. 4. Electron velocity in an FET as a function of the channel length. "Current tool" indicates results calculated in the drift-diffusion approximation. "New Tool" indicates results from an exact solution of the Boltzmann transport equation by Monte Carlo methods. From Ref. 4.

states away from the valence and conduction band minima. At these higher carrier energies, the band structures of most common semiconductors are quite similar. In fact results for Si and GaAs are virtually indistinguishable. Thus, as devices get smaller, GaAs loses the one advantage over Si, namely the higher mobility for carriers at the conduction-band minimum. It seems that, for microelectronics, GaAs is no longer the material of the future.

In order to solve Boltzmann's equation, one must specify the geometry of the devices and the energy band structure of the semiconductor. Energy band theory, based on quantum mechanics, is a mature discipline. Most calculations today employ the density functional theory developed by Hohenberg and Kohn [5] and by Kohn and Sham [6] in the mid 1960's. According to this theory, the total energy of a many-electron system is a functional of the electron density. The many-electron problem can then be rigorously replaced by an effective single-electron problem. Most calculations are carried out using the local-density approximation for the exchange-correlation part of the energy functional. The details of the energy bands are usually not quite right, unless one goes beyond the local-density approximation. Such calculations are feasible today in a variety of ways [7].

One key quantity is the forbidden energy gap. By alloying the semiconductor, one can customize the band gap, or grade it across the device to achieve desirable electrical behavior. The possibility of alloying used to be a unique advantage of III-V compound semiconductors. In recent years, however, it has been possible to alloy variable amounts of Ge in Si and fabricate devices whose performance far exceeds that of devices made out of pure Si [8]. Theoretical modeling to guide "band-gap engineering" is very much possible today by calculating the energy

143

band structure of semiconductor alloys. The modeling of the deposition process to produce the desirable band gap and desirable grading, are, of course, a completely different matter. We will return to that issue in the next subsection.

The effect of impurities and other point defects on electrons is another important topic. Si has the big advantage over other semiconductors that it has a good set of donor and acceptor impurities that have not raised any issues since their initial understanding by effective-mass theory in the 1950's[9]. Modeling dopant profiles has been done successfully and will be discussed further in the following section. Nevertheless, occasionally issues related to doping arise. For example, a number of years ago, it was found that post-processing anneals at moderate temperatures produced unwanted donors[10]. They were attributed to oxygen clusters and came to be known as oxygen thermal donors. A considerable amount of research has taken place to understand the atomic structure of these defects but the problem remains open to this date. Clever engineers in the meantime figured out how to avoid them. Another "doping problem" in Si was the finding about ten years ago that the conductivity of B-doped p-type Si in MOS devices degraded substantially under certain conditions (electron irradiation or high electron injection levels)[11]. The effect was traced to the release of hydrogen in SiO_2 which made its way into the p-type Si where it paired with a boron impurity, eliminating a hole. The physics of H in pure and doped Si was subsequently studied extensively both experimentally and theoretically. A fairly complete picture of the properties of H and dopant-H pairs has since emerged[12]. All of this provides background knowledge that may prove useful to engineers, but does not have any immediate and direct impact.

Other semiconductors usually have more serious doping problems. For example, GaAs and other III-V compounds and alloys such as AlGaAs have the so-called DX centers that limit n-type doping, cause persistent photoconductivity, and a variety of other interesting phenomena[13]. DX centers have been investigated extensively both experimentally and theoretically for about 15 years and by now are fairly well understood. Whereas the original thinking was that DX centers were pairs of a dopant impurity with something else (the X in the DX name), it has now been established that simple substitutional donors in these materials are actually unstable against a local distortion that renders them deep. Theoretical first-principles calculations played a major role in elucidating the structure and properties of these centers[13].

Wide-gap semiconductors have yet different doping problems that again have been well understood in recent years. For example, II-VI compounds usually can be easily doped only p-type or only n-type but not both. Recently, these difficulties were overcome by successful p-type doping of ZnSe[14]. Detailed theoretical calculations have demonstrated that the fundamental limitation is one of solubility and that N, which is the impurity that led to successful p-type doping, is indeed the impurity with the highest solubility[15]. Blue-green lasers and diodes are currently routinely fabricated with ZnSe pn junctions, but their life times are still short. These degradation issues remain a current challenge. Current research is also focusing on wide-gap III-V compounds such as GaN with promising successes.

Point defects and impurities with energies levels that are deep in the band gap, the so-called deep centers, also play a role in microelectronics. Usually they are undesirable as they shorten carrier life times. In the last 15 years extensive advances in both theory and experiments have elucidated most of their properties. The biggest issue in microelectronics is getting rid of them, e.g. removing transition-metal impurities from the active region where devices are fabricated.

We conclude that both the science and the corresponding computational implementations are in excellent shape for electronic and electrical properties. Modeling electronic transport and other electronic properties for technology development has been and will continue to be valuable. Perhaps the one area were new advances are needed is in implementing statistical variation in modeling device properties.

PROCESSING AND MECHANICAL RELIABILITY

Whereas for performance the issues are electronic properties and electronic transport, for processing (manufacturability) and mechanical reliability the issues are atomic arrangements and rearrangements. In order to address these issues, we need to identify different relevant length scales and time scales. At the one extreme is the microscopic scale where the characteristic distance is the interatomic spacing and the characteristic time is that defined by the atomic vibrational frequencies. At the other extreme is the macroscopic scale, in which the solid can be treated as a continuum obeying the laws of continuum mechanics. There exists, however, an intermediate regime where characteristic distances are longer than those of the atomistic regime but shorter than macroscopic. We can refer to this regime as "mesoscopic." In the case of Si, the mesoscopic regime is that of dopant profiles. In the case of the polycrystalline metal interconnects, the mesoscopic regime is that of grain boundaries, dislocations, and voids. In the case of polycrystals, Fig. 5 illustrates schematically the microscopic, mesoscopic, and macroscopic regimes. In this Section, we will assess the state of the physics and corresponding modeling at these three different regimes.

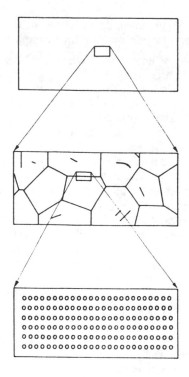

Fig. 5. Schematic of the microscopic, mesoscopic and macroscopic length scales in a polycrystal.

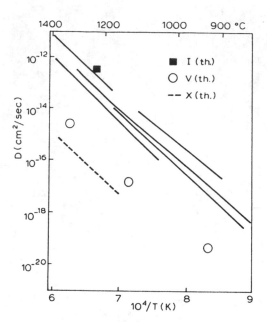

Fig. 6. The coefficient of self-diffusion for three different mechanisms in Si (from Ref. 24)

The microscopic regime

All processing steps and all mechanical deformations ultimately amount to atomic arrangements and rearrangements. Thus, in principle, all processing steps and all mechanical reliability issues can be described by atomic-scale theories.

Density functional theory [5, 6], as we already mentioned, provides a framework to compute the electronic structure of solids. Given any particular atomic arrangement, one can calculate the electronic structure and the net forces exerted on each nucleus by all the other nuclei and all the electrons. If one is interested in the lowest energy configuration, the atomic positions can be changed until the net force on each atom is zero. Alternatively, one can construct a dynamical theory in the manner first introduced by Car and Parrinello [16]: The nuclei (actually the nuclei plus the core electrons which are assumed to be attached to individual nuclei) are treated as classical particles and obey Newton's laws of motion under the forces exerted by the other nuclei (atomic cores) and the forces exerted by all the electrons. The latter can be obtained quantum mechanically in terms of a density-functional formulation. Such calculations are feasible on today's computers if the total number of atoms is restricted to a maximum of a few hundred. By using periodic boundary conditions, one can simulate an infinite solid. One can also study disturbances in infinite solids (defects, surfaces) if the disturbance is restricted to a portion of the repeating "supercell".

A recent example of such first-principles atomic-scale dynamical calculations is the calculation of the self-diffusion coefficient of Si for three different mechanisms, namely self-diffusion mediated by self-interstitials, by vacancies, and by direct exchange[24]. Figure 6 shows the theoretical results and compares them with available experimental data. The results show that self-interstitials yield the largest diffusion constant.

Another approach at microscopic length scales is conventional molecular dynamics[17]. In this approach, all electrons are permanently assigned to individual atoms. The resulting effective atoms are then interacting through classical potential fields. These "interatomic potentials" are usually constructed empirically by requiring that they reproduce total-energy results obtained with full electronic-structure calculations or by experiment. Typically one may construct interatomic potentials by fitting to lattice constants, elastic constants, cohesive energies, defect formation energies, etc. Once the interatomic potentials are known, a system of atoms is viewed as a statistical mechanical ensemble and its dynamical evolution is described by Newton's laws of motions. These calculations are much faster than the full electronic structure calculations and today's computers can handle hundreds of thousands of atoms, stretching it to several hundred million atoms and beyond. If the number of atoms is kept at the level of thousands, the time of simulation can be as much as 10 nanoseconds.

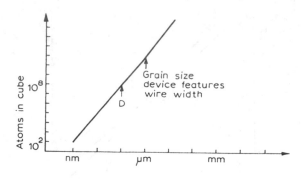

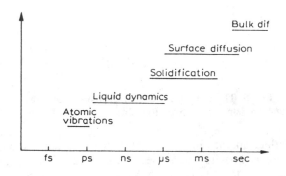

Fig. 7. Top: Number of atoms in a cube whose edge is the abscissa. D indicates the typical dislocation spacing. Bottom: Characteristic times for various processes in condensed matter.

In fig. 7, we plot the number of atoms contained in a cube as a function of the cube edge. We also show the characteristic time constants for various processes. Fig. 7 shows that the relevant length scales in microelectronics (defined by device features, widths of metal interconnects or grain sizes), lie beyond the reach of atomic scale calculations. In today's technologies, the relevant features range from 0.5-1 m and beyond. It will be after the turn of the century that the minimum device features are as small as 0.1 m. Even with the shrinking dimensions, the time scales of processing steps and deformations remain unchanged. Diffusive processes remain slow, as many as 10 orders of magnitude slower than the characteristics times of atomic vibrations.

Atomic-scale calculations and experiments are very valuable in elucidating the microscopic mechanisms that underlie technological applications. Elucidation of the microscopic mechanisms, however, does not immediately imply impact on technology. It all depends whether the microscopic mechanisms can serve as input to theories operating at length scales and time scales that are relevant for technology. In most cases, these links between atomic-scale theories and the relevant length and time

scales of technology have not been developed. We will illustrate this important point with three examples.

First, we give an example where the links between the atomic-scale properties and the mesoscopic regime of longer length and time scales are well developed, namely diffusion of dopant impurities. The theory of diffusion in crystalline solids is in excellent shape. At the atomic level, microscopic mechanisms have been described in detail (e.g. vacancy mechanism, interstitial mechanism, direct exchange, ring mechanisms, etc.) and the expressions for the corresponding diffusion constants are well known [18]. Once the atomistic mechanisms are known, one can write down expressions for diffusion fluxes and the corresponding conservation laws. The latter are known generically as Fick's laws [18. They can be solved for specified geometries and initial conditions. In microelectronics, one of the key issues is the evolution of the dopant impurity profiles under various processing conditions. Computer implementations of diffusion theory to model dopant profiles have existed for several decades. Atomic-sale theory has had an impact on the field by elucidating the atomistic mechanisms. In the early days, modeling of dopant profiles was based on the assumption impurities diffuse by the vacancy mechanisms [19]. There were advocates of the interstitial mechanism as an alternative to vacancies [20, but it really did not matter much because the diffusion equations are virtually identical if only one defect is involved. In the 1980's, atomic-scale calculations [21] and, independently, experimental investigations [22] demonstrated that both vacancies and interstitials play a role. When both defects are included, the diffusion equations are more complex, but they are capable of reproducing the actual data more reliably[23]. It should be remarked, however, that the calculations of diffusion constants for individual mechanisms remains a delicate task. It has mostly been possible to calculate activation energies. For diffusion constants, one needs to track all the entropy terms that go into the pre-exponential. In Fig. 6, we showed the first calculations of the diffusion constants for three different mechanisms of self-diffusion in Si. They were obtained with full Car-Parrinello quantum molecular dynamics [24]. The error bars remain relatively large, i.e., one to two orders of magnitude. For impurities, the calculations would be even more complex. It would require significantly larger computers before calculations can produce diffusion constants that are accurate enough to be used in programs modeling impurity profiles. In the meantime, of course, modeling programs used in technology rely on adjusting parameters to reproduce available experimental data. The impact of atomic-scale calculations is to provide information about the relevant mechanisms.

As a second example, we take the process of etching Si with F ions. The atomistic mechanisms have been investigated by both experiments and theory. The details of how F ions enter the Si lattice and break subsurface bonds has been described by atomic-scale theory [25]. Theory even accounted for the observed Fermi-level dependence of the etch rate. Photoemission experiments [26] have identified the presence of Si atoms with one, two, or three F neighbors. Other experiments [27] measured the evolution rate of SiF_2, SiF_3 and SiF_4 molecules with and without energetic beams of other particles. All of this illustrates that atomic-scale physics is capable of describing in detail the atomistic processes that go on during etching. Technology can clearly benefit from such knowledge, e.g., in choosing the range of energies, the composition of the etching gases, etc. Ultimately, however, technology is interested in controlling the shape of the etched feature. This question is beyond the reach of atomistic calculations in both length and time scales. As in the case of diffusion profiles, one again has to deal with atomic densities and fluxes. In this case, however, the problem is far more complex. One must take into account the fluxes of particles arriving at the surface, surface and subsurface reactions, surface diffusion, surface stresses, and the effect of the subsurface damage. A comprehensive theoretical framework has not been developed. Existing modeling efforts focus mostly on the geometric aspects of particles arriving at the surface and on simple parametric phenomenological models. [28].

As a third example, we take metallic interconnects on chips and in packages. These are polycrystalline metal lines. Polycrystals are not equilibrium arrangements of atoms so that their microstructure (i.e., distribution of grain sizes and orientations, dislocation networks, precipitates, microvoids, etc.) is evolving under the effect of thermal stresses, temperature gradients, current, etc. In the 1960's, it was recognized that current can cause voids and extrusions to grow to the point that open or short circuits would occur, presenting a reliability problem. The problem was controlled by empirical remedies, e.g., by adding small amounts of copper in aluminum lines. [29]. To this date, the role of copper and even its distribution in the grains and grain boundaries are not known. In recent years, as the metal lines are getting narrower, it has been recognized that stress alone can cause similar void problems [30]. There is no doubt that atomic-scale calculations can probe the local properties of individual features of the microstructure, namely grain boundaries, dislocations, voids, etc. The rearrangement of the microstructure, however, which produces voids and extrusions, is a collective property and occurs over length scales and time scales that are beyond the reach of atomic-scale calculations. Again, one needs a comprehensive framework to describe microstructure evolution at the relevant length and time scales, derived from and taking advantage of atomistic theory. We will see later that some advances have recently been made along these lines.

The macroscopic regime

Continuum mechanics, thermodynamics, and electrodynamics are old and venerable disciplines that go back to the eighteenth and nineteenth centuries. They describe the properties of materials without any assumptions about their atomic composition. They rely on general laws that were formulated on the basis of extensive experimental observations. In the case of continuum mechanics, the material's properties are captured in a set of empirical "constitutive equations" that describe the response of the material to external stresses. Materials can be elastic, viscoelastic, plastic, etc. In each case, one has a relationship between the strain or strain rate and the applied stress. Once such a relationship is assumed, mathematics takes over. The theory of continuum deformations goes back to the days of Cauchy and Euler and is mathematically a very rigorous discipline [31]. Today, this theory is used extensively to compute deformations of real materials under external loads. computer programs were pioneered by the aerospace, aircraft and automobile industry. In microelectronics, the same programs are used to compute deformations of structures, primarily at the package level. For a given geometry, given constitutive equations and material parameters, and given loads, the equations can be solved to produce the distribution of stresses and strains. These theories, however, do not describe any of the consequences of the stresses and strains, e.g. crack or void nucleation and growth, delamination, etc. They also do not contain any deformations that are caused by long-range diffusive processes or the interaction between diffusive and displacive deformations.

Continuum theories have been used extensively to study materials at finer length scales. The most successful of these theories is the continuum theory of dislocations [32]. Though the concept of a dislocation requires an atomistic picture of a crystal (i.e., a dislocation is a half plane of atoms in an otherwise perfect crystal of "full" planes of atoms), it can be defined in continuum theory as a line where the displacement field is discontinuous by a vector equal to the Burgers vector. With this definition, one can use standard elasticity theory to calculate the strain field. The result gives rise to infinite energies unless an arbitrary cutoff at both small and large distances is imposed.

There also exist formulations that extend continuum concepts to describe deformations of grains in a polycrystal, crack propagation, and so on. These approaches are sometimes referred to as micromechanics [33]. Complications arise by diffusive processes which somehow must be incorporated into the constitutive equations. There are also hybrid formulations where some diffusive processes are treated explicitly while at the same time other diffusion-mediated deformations such as creep are treated in terms of continuum constitutive relations [34].

The mesoscopic regime

It is clear from the above two subsections that there exists a well-defined intermediate scale, lying between the

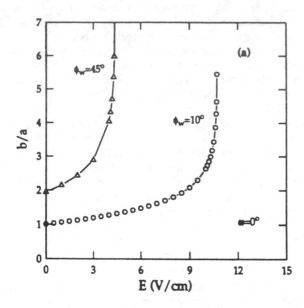

Fig. 8. Aspect ratio of the void as a function of the electric field for certain angle choices.

microscopic, atomic-scale world and the macroscopic world. In crystalline semiconductors, the mesoscopic scales are defined by the dopant profiles. The dynamical evolution of such profiles are describable by diffusion equations. These equations are usually derived phenomenologically, i.e. by defining atomic densities and fluxes and relating them by conservation laws.

In polycrystals, with grain boundaries, dislocations, microvoids, precipitates, etc. the problems are considerably more complex as diffusive processes, dislocation climb and glide, grain-boundary migration and sliding, and elastic deformations are occurring at the same time under the influence of external loads (stresses, electric fields, etc.). Typically one finds in the literature phenomenological models that blend continuum and atomistic concepts and are constructed for particular problems. There is no comprehensive computer-ready theory that can describe the evolution of the so-called microstructure of a polycrystal (grain boundary network, microvoids, dislocations, precipitates, etc.). This is particularly serious for microelectronics, especially the area of mechanical reliability of metal interconnects. The big issues there are electromigration[29] and stress migration[35], namely rearrangements of the microstructure induced by electric fields (currents) and stresses. Typically these rearrangements lead to void growth or extrusions that cause reliability problems.

In the last few years, this author has undertaken the development of a comprehensive mesoscopic theory that is derived entirely from first principles beginning with nuclei and electrons and the laws of quantum mechanics and statistical mechanics. Such an exercise had previously been done only for fluids. In the case of a heterogeneous solid, the resulting dynamical field equations contain the three basic equations of continuum mechanics -- conservation of mass, momentum and energy -- that are normally derived phenomenologically, plus a set of truly mesoscopic equations that track the dynamics of the microstructure and couple with the three basic equations. Diffusion equations in a crystal are a special case of the equations. Similarly, many of the phenomenological models of specific problems are recovered. Fundamentally the equations present a first-principles formulation of plasticity.

The general theory may be found in Ref. 36. A more recent account, containing initial applications may be found in Ref. 37. The two applications described in Ref. 36 are relevant for microelectronics. The first is the evolution of the shape of a void at the interface between a polycrystalline metal and an insulator in the case where a grain boundary emanates from the void. It was found that the void elongates because of surface diffusion (driven by the external electric field and countered by capillarity forces) and the aspect ratio is a strong function of the electric field. Figure 8 shows the aspect ratio as a function of the electric field for two different values of the "wetting angle", namely the angle at which the void meets the interface (this angle is determined by the thermophysical properties of the interface and can thus be modified and controlled). The angle w is the angle that the grain boundary makes with the line normal to the interface. For the results of Fig. 8, $\omega=0$. In both cases, the aspect ratio grows asymptotically at a critical value of the field. Such a phenomenon could be catastrophic for reliability in microelectronics. Figure 9 shows the dependence of the critical field on the angle ω and the wetting angle. It is clear that the critical value hovers around small values that are in the range of fields typical in microelectronics. The critical fields are quite sensitive to both angles, namely to both

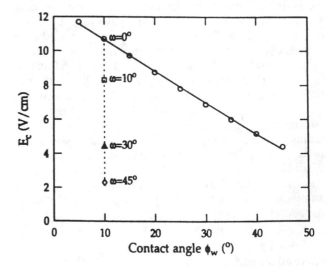

Fig. 9. The critical value of the electric field as function of the wetting or contact angle ϕ_w and the grain-boundary/interface angle ω.

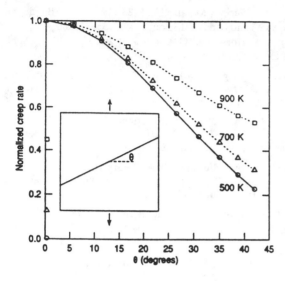

Fig. 10. Creep rate as a function of the grain boundary angle shown in the inset for three different temperatures.

the thermophysical properties of the relevant interfaces and the geometric properties of the microstructure.

The second example, also described in Ref. 37, is stress-induced diffusive creep. The case studied was a bicrystal under uniaxial stress as shown in the inset of Fig. 10. Figure 10 shows the normalized creep rate for three temperatures as a function of the angle θ defined in the inset. Again one sees that the creep rate is a strong function of the geometry of the microstructure.

Ultimately one would like to have a theory that can describe quantitatively the microstructure evolution in a heterogeneous material under external mechanical, thermal and electrical loads. The work of Refs. 36 and 37 present only a beginning.

CONCLUSIONS

We have assessed the state of the science that underlies issues of performance, processing (manufacturability) and reliability in microelectronics. Our net conclusions are that the science for electronic properties and electronic transport is in excellent shape in all relevant length and time scales. The science for atomic arrangements and rearrangements, on the other hand, is in excellent shape only at the microscopic scale where individual atoms are tracked and at the macroscopic continuum scales where all atomic structure and microstructure is swept into empirical relations. The science in the mesoscopic regime, where one needs to deal with microstructure and fabricated structures in the submicron regime, the science needs considerable development.

REFERENCES

1. J. Bardeen and W. H. Brattain, Phys. Rev. 74, 230 (1948); W. Shockley, Bell System Tech. J. 28, 435 (1949).

2. See e.g., J.S. Kilby, IEEE Trans. Electron Devices 23, 648 (1976).

3. See e.g., K. Blotekjaer, IEEE Trans. Electron Devices 17, 38 (1970); M. Rudan and F. Odeh, COMPEL 5, 149 (1986).

4. For a review, see C. Jacoboni and L. Reggiani, Rev. Mod. Phys. 44, 645 (1983); for some recent developments, see M.V. Fischetti and S.E. Laux, Phys.. Rev. B 38, 9721 (1988); IEEE Electron Devices 38, 650 (1991).

5. P. Hohenberg and W. Kohn, Phys. Rev. 136, 864 (1964).

6. W. Kohn and L.J. Sham, Phys. Rev. 140, A1137.

7. X. Zhu and S.G. Louie, Phys. Rev. B 43, 14142 (1991).

8. See e.g., G.L. Patton, J.M.C. Stork, J.H. Comfort, E.F. Crabbe, B.S. Meyerson, D.L. Harame and J.Y. Sun, IEEE Electron Devices Tech. Digest 13 (1990).

9. W. Kohn, Solid State Physics 5, 257 (1957).

10. W. Kaiser, H. L. Frisch, and H. Reiss, Phys. Rev. 112, 1546 (1958).

11. C. T. Sah, J. Y. C. Sun, and J. J. T. Tzou, Appl. Phys. Lett. 43, 204 (1083); J. Appl. Phys. 54, 5864 (1983).

12. C. H. Van de Walle, in *Deep Centers in Semiconductors*, edited by S. T. Pantelides, 2nd Edition, (Gordon and Breach, New York, 1994).

13. P. M. Mooney, in Ref. 12.

14. M. A. Haase, H. Cheng, J. M. DePuydt, and J. E. Potts, J. Appl. Phys. 67, 448 (1990).

15. C. G. Van de Walle, D. B. Laks, G. F. Neumark, and S. T. Pantelides, Phys. Rev. B 47, 9425 (1993).

16. R. Car and M. Parrinello, Phys. Rev. Lett. 55, 2471 (1985).

17. See e.g., M. Parrinello and A. Rahman, J. Appl. Phys. 52, 7182 (1981); M.W.Ribarsky and U. Landman, Phys. Rev. B 38, 9522 (1988); J. Huang, M. Meyer and V. Pontikis, Phys. Rev. Lett. 63, 628 (1989) .

18. For the classical theory of diffusion, see e.g. J.R. Manning, Diffusion Kinetics of Atoms in Crystals (Van Nostrand-Reinhold, Princeton, 1968).

19. See e.g., R.B. Fair, in: Impurity Doping Processes in Silicon, ed. F.Y.Y. Wang (North-Holland, New York, 1981).

20. See e.g., W. Frank, Festkoerperprobleme 21, 221 (1981).

21. R. Car, P.J. Kelly, A. Oshiyama and S.T. Pantelides, Phys. Rev. Lett. 52, 1814 (1984); C.S. Nichols, C.G. Van de Walle and S.T. Pantelides, Phys. Rev. Lett. 62 1049 (1989).

22. See e.g., P. Fahey, G. Barbuscia, M. Moslehi and R.W. Dutton, Appl. Phys. Lett. 46, 784 (1985), and references therein.

23. E. Rorris, R.R. O'Brien, F.F. Morehead, R.F. Lever, J.P. Peng and G.R. Srinivasan, Proceedings of the Second International Symposium on Process Physics and Modeling in Semiconductor Technology, edited by G.R. Srinivasan, J.D. Plummer and S.T. Pantelides (The Electrochemical Society, Pennington, NJ, 1991) p.703.

24. P.E. Bloechl, E. Smargiassi, R.Car, D. Laks, W. Andreoni and S.T. Pantelides, Phys. Rev. Lett. 70, 2435 (1993).

25. C.G. Van de Walle, F.R. McFeely and S.T. Pantelides, Phys. Rev. Lett. 61, 1867 (1988).

26. F.R. McFeely, J.F. Morar and F.J. Himpsel, Surf. Sci. 165, 277 (1986).

27. See e.g., H.F. Winters and D. Haarer, Phys. Rev. B 36, 6613 (1987), and references therein.

28. V. Singh, E.S.G. Shagsfeh and J.P. McVittie, J. Vac. Sci. Technol. B 10, 1091 (1992).

29. See e.g., F. D'Heurle and R. Rosenberg, Physics of Thin Films (Academic, New York, 1973).

30. See e.g., J. Klema, R. Pyle and E. Domangue, Proc. 22nd Reliability Physics Symp. (IEEE, Las Vegas, 1984) p. 1; F.G. Yost, Scripta Metall. 23, 1323 (1989); K. Hinode, N. Owada, T. Nishida and K. Mukai, J. Vac. Sci. Techn. B 5, 518 (2987); M.A. Korhonen, C.A. Paszkiet and C.Y. Li, J. Appl. Phys. 69, 8083 (1991).

31. See e.g., C. Truesdell and R.A. Toupin, in: Encyclopedia of Physics, ed. S. Fluegge (Springer, Berlin, 1960) Vol. III/1, p. 226.

32. A.E.H. Love, Mathematical Theory of Elasticity (Dover, New York, 1944); J.P. Hirth and J. Lothe, Theory of Dislocations (Krieger, Malabar, FL, 1992).

33. A. Needleman, in: Theoretical and Applied Mechanics, eds. P. Germain, M. Piau and D. Caillerie (Elsevier, Amsterdam, 1989) p. 217.

34. A. Needleman and J.R. Rice, Acta Metall. 28, 1315 (1980); T.-J. Chuang, K.I. Kagawa, J.R. Rice and L.B. Sills, Acta Metall. 27, 265 (1979).

35. K. Hinode, N. Owada, T. Nishida, and K. Mukai, J. Vac. Sci. Technol. B 5, 518 (1987).

36. S. T. Pantelides, J. Appl. Phys. 75, 3264 (1994).

37. S. T. Pantelides, D. Maroudas, and D. B. Laks, in Computational Material Modeling, edited by A. K. Noor and A. Needleman, (The American Society of Mechanical Engineers, New York, 1994), p. 1.

INTERCONNECTS AND FAILURE ANALYSIS

Interconnects and Contacts: Issues and Strategies

Robert S. Blewer

Advanced Silicon Projects Department
Contamination Free Manufacturing Research Center
Sandia National Laboratories
Albuquerque, NM 87185-1077
(505) 844-6125

Over the last few device generations, ever greater numbers of metallization levels in devices have increased process complexity and forced interconnect technology to the forefront of IC processing concerns. Currently, more than half of the device processing steps and a majority of yield loss can be traced to interconnect related issues. Chip performance is also increasingly dominated by delays intrinsic to interconnect material limitations. For these reasons, most companies are seeking improved interconnect architectures, deposition processes and materials. The requirements for interconnect technology contained in the National Technology Roadmap for Semiconductors will guide the semiconductor industry in focusing interconnect R&D where it will be most productive and in coordinating environment, safety and health (ES&H) approaches, metrology needs and contamination concerns.

INTRODUCTION

As we move into the next decade, the challenges for successful implementation of high performance, high yielding and reliable interconnect are enormous. Minimum feature size is projected to shrink to 0.07 μm over the next 15 years and the levels of metallization will exceed seven or eight for logic devices. Because of the continual shrink in feature size and the increase in chip dimensions, defect control requirements will increase approximately ten times with every new device generation (1). A premium will be placed on rapid yield learning, development of robust processes, and designed-in reliability for interconnect processing, with potentially huge rewards for those companies that can meet the challenge of defectivity control quickly and successfully. As shown in Figure 1, a swing of hundreds of millions in revenue can occur during the first year of a

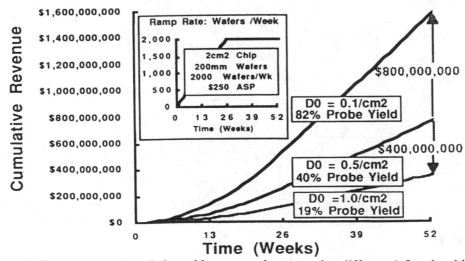

Figure 1: Factory revenue variation with ramp-up time assuming different defect densities (2).

new fab's operation, dependent mainly on the impact of defectivity during start-up (2). This figure illustrates the cumulative revenue to be expected from a new factory that is ramping production during the first year of manufacture for a new technology and/or product family. If an initial overall defect density of 0.5 per square centimeter (40% probe yield) is assumed, the middle curve indicates that an estimated $800M in revenue will be generated in the first four quarters, assuming the factory is producing a two square centimeter chip on eight inch wafers with a capacity of 2000 wafers per week ramped to full volume in the first two quarters. If the defect density is a factor of five lower (0.1 defects per square centimeter), an additional $800M can be obtained during the first year of production. Likewise, if the overall defect density achievable is only 1.0 per square centimeter, the lost revenues, compared to the middle curve, is $400M in the first year of fab operation.

It is clear from this illustration that defect reduction is a major driving force in profitability. The National Technology Roadmap for Semiconductors (NTRS) indicates that die size will increase dramatically while minimum feature size shrinkage will put chip yield at risk because ever smaller particles can cause "killer defects" (1). Already, more than half of the complexity in integrated circuit processing occurs in the back-end-of-the-line (BEOL), and a majority of the yield loss and reliability-based failures as well. Overall reliability requirements will also be difficult to meet, increasing by a factor of 80 by the year 2010. Since interconnection layers are the source of a majority of yield loss and significant reliability degradation, attention must be focused on the BEOL to achieve acceptable interconnect defectivity targets and, in turn, desired revenues and profitability.

	1995 0.35 μm	1998 0.25 μm	2001 0.18 μm	2004 0.13 μm	2007 0.10 μm	2010 0.07 μm
Number of metal levels						
DRAM	2	2 - 3	3	3	3	3
Logic: microprocessor	4 - 5	5	5-6	6	6 - 7	7 - 8
Interconnect density						
DRAM (Metal 1) (meters/cm^2/level)	60	80	109	150	222	300
Logic (Semi-global)	35	50	70	105	125	155
Maximum interconnect length Logic (meters/chip)	380	840	2100	4100	6300	10,000
Cost (without photo) ($/cm^2/level)	0.29	0.23	0.23	0.18	0.18	0.14
Particle size (μm)	≥ 0.12	≥ 0.08	≥ 0.06	≥ 0.04	≥ 0.04	≥ 0.03
Integrated particle density per module (Particles/meter^2)	125	125	125	125	125	125
Reliability - Logic (FITS[*]/meter) X 10^{-3}	16.0	4.7	1.1	0.5	0.4	0.2

[*] *failure units*

	1995 0.35 μm	1998 0.25 μm	2001 0.18 μm	2004 0.13 μm	2007 0.10 μm	2010 0.07 μm
R[†] - (Metal 1) (Ohms/μm) [*]	0.15	0.19	0.29	0.82	1.34	1.34
C[‡] - (Metal 1) (fF/μm)	0.17	0.19	0.21	0.24	0.27	0.27
Crosstalk (V/mm) ~20% of V_{dd}	0.5	0.4	0.4	0.4	0.3	0.3

[*] *R assumes Al - 0.5 weight % Cu Metal 1 with 100 nm refractory thickness, resistance per μm of length*
[†] *resistance*
[‡] *capacitance*

Figure 2. "Ishikawa Diagram" representation of the interconnect requirements in support of the National Technology Roadmap for Semiconductors (1).

NTRS GUIDANCE FOR WIRING

Specific requirements for all facets in future semiconductor manufacturing are listed in the NTRS (1). From the discussion above, advances in every aspect of multilevel interconnect will be required to meet the challenges discussed within the NTRS. In addition to interconnect processing needs, metrology and materials issues in interconnect technology are key to success in the effort to move forward at the pace outlined in the NTRS. The implications of these requirements and a listing of the research and development activities necessary to address these needs will be discussed below.

A summary of interconnect requirements contained within the NTRS is illustrated in Figure 2, in which specific requirements are divided into categories to aid in viewing the full range of needs at a glance. This listing of requirements for BEOL processing is drawn not only from the NTRS section which bears the name of "Interconnect", but also in many other parts of the document, such as those in which "crosscut technologies" (e.g., contamination free manufacturing and metrology) are discussed. The phasing of NTRS solutions to current and future processing requirements are graphically indicated for a 15 year period in Figure 3. Implementation of these technologies and the issues associated with them will also be included in this paper.

TRENDS IN HORIZONTAL METALLIZATION

For horizontal metallization, signal propagation delays caused by narrower (down to 0.07 µm) and longer (up to 10 km) wiring are even now beginning to gate advances in device performance and will force the use of up to seven levels of metallization for advanced design logic devices by the year 2010. Most companies agree that aluminum alloy interconnect will be the material of choice through at least the 0.25 µm logic device generations. Thus much of the current research and development effort will continue to remain focused on improving aluminum alloy interconnects, though many companies across the world are investigating copper for more advanced interconnect applications.. Nonetheless, the high resistivity of aluminum relative to copper is stimulating evaluation of the latter material, in combination with low permittivity dielectrics to improve signal propagation, and reduce power requirements and heat generation.

Ironically, the driving force toward implementing copper as an interconnect material may well be as much related to expected improvements in reliability as to improved electrical performance. Though CVD aluminum (in an alloyed form) is still under investigation as a horizontal interconnect material, copper metallization should scale through more device generations, where shrinking feature size will result in higher current density, and where total wiring length will approach several [~10] kilometers. Reliability data on copper interconnects is still sparse but significant improvements in electromigration resistance have already been demonstrated. Resistance to stress migration effects in copper is less certain, since observation of stress voids in copper lines has been reported.

Problems in patterning copper interconnect using RIE can be largely circumvented by use of the damascene (inlaid metal) process, which has been made possible by advances in chemical mechanical polishing technology for planarization.

Chemical vapor deposited aluminum may be poised to make a comeback despite the difficulty of incorporating alloying agents and problems with surface roughness. However, if significant development efforts for a new interconnection deposition technique are undertaken by IC manufacturers, some believe such efforts will focus on damascene patterned copper, because it has lower resistivity and better reliability characteristics than aluminum. Nonetheless others argue that because of inherent reliability concerns, other approaches to aluminum deposition, such as UHV sputtering or high pressure elevated temperature techniques will prevail.

PLUG TECHNOLOGY

Because of voiding at interfaces between dissimilar materials such as aluminum alloy and tungsten, future interconnect development will emphasize materials which can simultaneously serve as plug fills and horizontal interconnect, an architecture often call "integrated interconnect". Those materials that can be deposited by CVD are especially favorable because a blanket deposition provides both horizontal and vertical metallization in a single processing step. Perhaps the strongest candidate for integrated interconnect is CVD copper. CVD aluminum could be used, but as mentioned above, its ability to scale into deep sub-half micron device generations is questionable. CVD tungsten, the current standard for plugs, is adequate for horizontal local interconnect and for first level metallization in

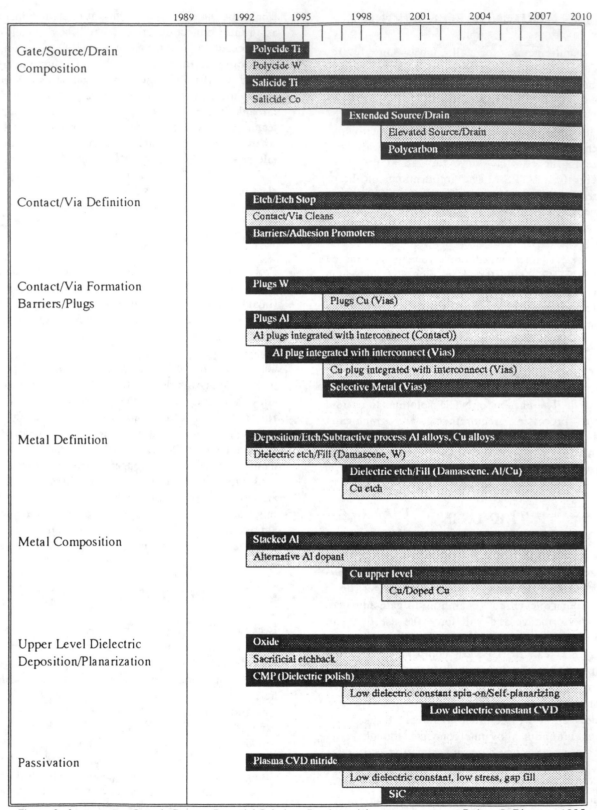

Figure 3. Interconnect Generic Process Potential Solutions Roadmap (1).   Robert S. Blewer - 1995
Invited Paper to be published in the Proceedings of the International Workshop on Semiconductor
Characterization: Present Status and Future Needs, Jan. 30-Feb. 2, 1995, NIST, Gaithersburg, Md.

some current applications, but will be unsuitable as an integrated metallization in future device generations because of its resistivity.

CONTACT ENGINEERING

Much of the challenge in advanced interconnect will be encountered in the design and process development of vertical metallization that is capable of meeting the requirements of the NTRS. Contact resistance is becoming a larger problem since contact area decreases as the inverse square of the scaling factor. In addition, achieving interfacial cleanliness at the base of contacts and vias becomes more difficult in high aspect ratio structures. The relentless move toward shallower junctions is at odds with current metal salacide technology, because design requirements for low contact resistivity require a minimum thickness of titanium at the base of contact holes to achieve reasonable contact

resistance results. However, in the salacide process, silicon consumption is large, and depends on the initial thickness of titanium at the base of the contact as well as on the details of the silicidation/nitridation anneal. Variations in the repeatability of this process and the need for uniformity across the wafer and wafer-to-wafer, in combination with the shallower junction depths which will occur in future device generations will make successful junction engineering an ever larger problem.

The need to clear and clean all contacts before metallization is deposited is a further and increasingly difficult problem. The combination of contact etch and over-etch, post-etch cleaning, salacidation and pre-plug soft etch can consume 80 nm of silicon, as shown in Figure 4 (3). Elevated source/drains, alternate contact metallization for which the metal, rather than the silicon, is the diffusing species (e.g. cobalt silicide), or self-limiting CVD metallization will be required to address this issues inherent in deep-submicron shallow junctions.

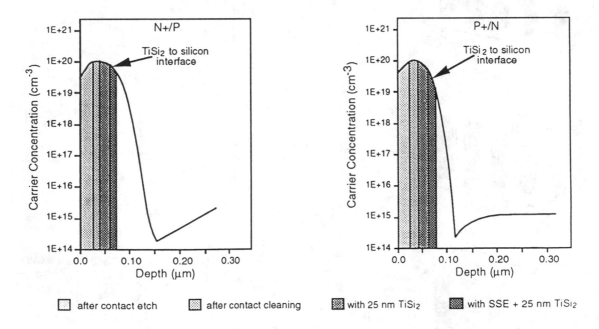

Figure 4. Silicon removal and junction penetration issues for titanium salacide processes (3).

For current and near-future needs, Ti/TiN bilayers will continue to be used as low contact resistance liners for contacts. Because of requirements for good step coverage, adequate bottom coverage of titanium for low resistivity contacts, and continuous sidewall nucleation layers, an alternative to collimated PVD will be of interest. Recently, a successful process for CVD titanium and TiN was published (3). Figure 5, which compares the yield of 10,000 contact chains for several contact sizes, indicates that a PVD Ti + RTN process has increasingly poor yield below 0.5 μm dimensions. In contrast, the CVD Ti + CVD TiN process retains approximately 100% yield even at 0.35 μm dimensions. Figure 6 shows that the leakage in test diodes is superior for parts made using the dual CVD process. Likewise, contact resistance of parts made with the CVD process are lower than the PVD process at 0.35 μm, as shown in Figure 7.

A similar conclusion was reached at Texas Instruments in a process integration study for 0.5 μm parts. Figures 8, 9, and 10 show side-by side comparisons of PVD versus CVD TiN depositions for contact and via applications. Again, in this example, problems are observed for the PVD approach at dimensions of 0.35 μm and below.

Because contact area is decreasing as the square of the shrink factor with each new generation of devices, contact resistance for sub- 0.5 μm and smaller vias can exceed bulk via resistance values. The upper limit for the contact resistivity requirement for 0.25 μm vias is approximately 10^{-9} Ω-cm^2. To achieve such low values in a repeatable manner, new techniques are needed to etch and clean contact and via holes. In-line techniques

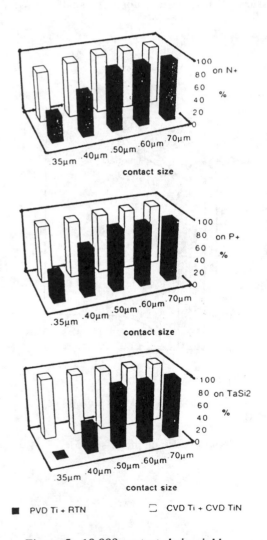

Figure 5. 10,000 contact chain yield versus contact size (3).

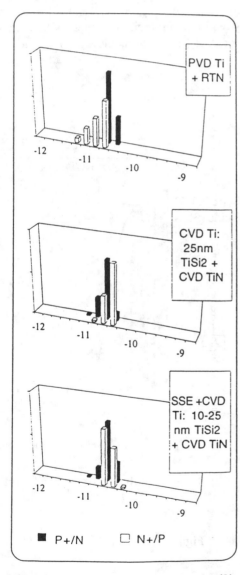

Figure 6. Leakage current at 5 volts (3).

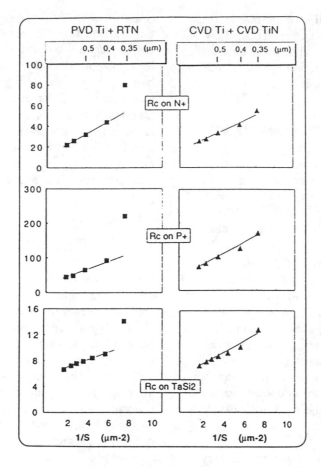

Figure 7. Kelvin contact resistance for CVD Ti/TiN films on n and p doped silicon and on tantalum silicide (3).

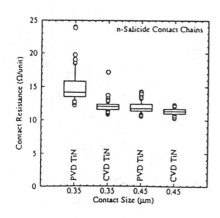

Contact resistances measured from n-salicide contact chains for PVD TiN and CVD TiN contact liners.

Figure 8 (4).

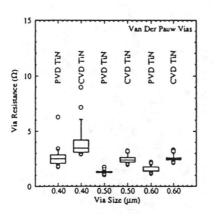

Via resistance distributions for various sizes of Van der Pauw vias.

Figure 9 (4).

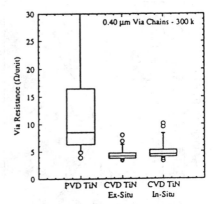

Via resistance distributions for 0.4 μm via chains.

Figure 10 (4).

are also needed to monitor removal of impurities at the base of vias and contacts prior to metal deposition. Low, repeatable contact resistance is a must in multilevel metallization technology where, for future devices, one open contact per billion is unacceptable.

DIFFUSION BARRIER DEVELOPMENT

Use of hot sputtered aluminum for possible near term plug technology and the drive toward realization of copper metallization for the longer term has been hampered by lack of an adequate diffusion barrier technology. If copper is used in via applications, most IC manufacturers believe that 10 μm - 25 μm is the thickness limit for a diffusion barrier in a 0.25 μm via if the composite plug resistivity (copper + diffusion barrier) is to be lower than that of a aluminum alloy plug with no diffusion barrier. Recently, however, a promising ultra-thin barrier technology has been reported which involves the use of refractory binary/ternary (small grain or amorphous) alloys. This development gives hope that effective barriers for use in copper interconnect implementation may well be integrated into some product lines by the 0.18 μm (or even the 0.25 μm) device generation.

The excellent thermal stability of sputtered ternary amorphous diffusion barriers has received widespread attention (4). However, because near-perfect conformality is needed in high aspect ratio via and contact holes, CVD techniques will be needed to meet future applications. Recent research has demonstrated the use of plasma nitrided CVD tungsten amorphous layers for this application (5). As shown in Figures 11-13, a barrier of 20 nm is sufficient to prevent penetration of copper at 600° C for one hour. Moreover, the nitrided tungsten layer, which is the actual barrier, is only 2 μm thick in these experiments, as shown in Figures 14 and 15. Because the tungsten can be deposited conformally on copper or any other metallic nucleation layer, ultra-thin, conformal, self-aligning barriers seem to be within reach.

ADVANCED DIELECTRICS

Despite continuing research to develop lower resistivity metallization, greater gains in interconnect performance could be realized from reduction in the dielectric constant of insulators. In addition to new materials like fluoropolymers, which have been investigated extensively in the U.S., conventional dielectrics like TEOS, treated with fluorine, have shown promise, as indicated in Figure 16 (6). Though moisture absorption above 1% is a concern with respect to hot carrier reliability (Figure 17(6), Si-F/Si-O bonding controlled to adequate levels can reduce the dielectric constant by about 10-15%.

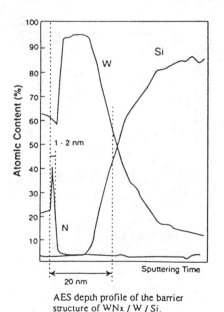

AES depth profile of the barrier structure of WNx / W / Si.

Figure 11 (5).

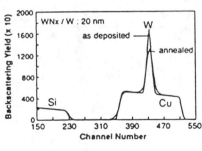

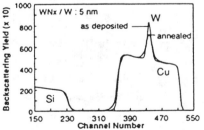

RBS spectra of Cu / WNx / W / Si multilayer before and after annealing at 600 °C for 1 h.

Figure 12 (5).

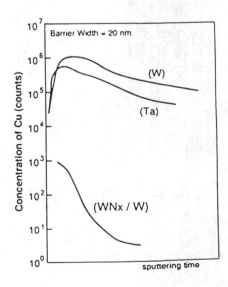

SIMS depth profiles of Cu diffused through Ta, W and WNx / W barriers into Si substrate after annealing at 600 °C for 1 h.

Figure 13 (5).

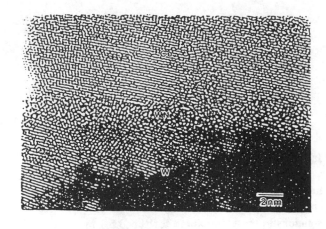

Figure 14. TEM image of Cu/WN/W/Si before annealing (6).

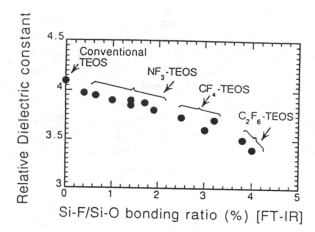

Figure 16. Si-F Bonding versus dielectric contant (5).

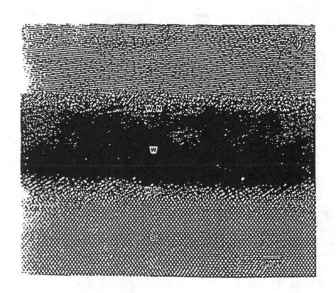

Figure 15. TEM image of Cu/WN/W/Si after annealing (6).

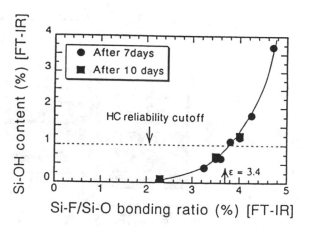

Figure 17. Moisture Absorption as a function of Si-F/Si-O bonding ratio for treated TEOS (5).

NOVEL APPROACHES

Non-conventional approaches, such as optical interconnect technology, are not likely viable on-chip interconnect candidates for silicon devices, at least until the longer term. Exotic approaches such as superconducting interconnection are being given little notice because of critical current effects and because of the necessity to maintain such devices at cryogenic temperatures. Tungsten may continue to be used for some time for applications such as local interconnects and first level metal applications in selected designs. Materials that exhibit corrosion resistance (such as gold) and low resistivity (such as silver) are regarded as having no real prospects for near term implementation in large scale manufacturing because of electromigration, cost or other considerations.

CHALLENGES AND POSSIBLE SOLUTIONS FOR 0.25 μm LOGIC DEVICES

In the areas discussed above, a series of requirements and a discussion of advances that may meet some of these needs has been outlined. Figure 18 illustrates schematically the specific challenges and suggested solutions to these problems for the 0.5 μm logic device generation (2).

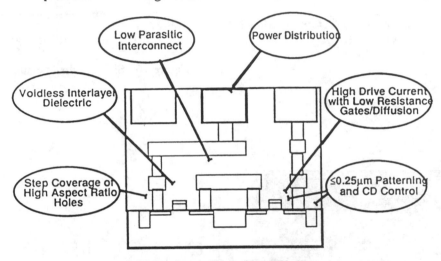

a. 0.25 μm Logic Technology Challenges

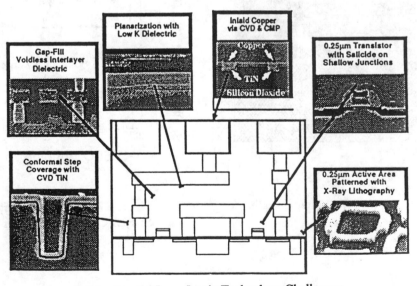

b. Meeting 0.25 μm Logic Technology Challenges

Figure 18. Technology challenges for 0.25 μm devices: challenges and promising solutions (2).

REFERENCES

1. *National Technology Roadmap for Semiconductors*, Semiconductor Industry Assn., 4300 Stevens Creek Blvd., Suite 271, San Jose, CA 95129, publishers, 1994.

2. Parillo, L., "Issues in Developing A Manufacturble Interconnect Technology," *Proc. Advanced Metallization for ULSI Applications,* 1994 (to be published by Mat. Res. Soc., Pittsburgh).

3. Arena, C., et.al. "CVD Ti and CVD TiN: Electrical Performance as Contact Metallurgy," *ibid.*

4. Dixit, G., Jain, M. Chisholm, F, Weaver, T., Havemann, K., Littau, K., Eizenberg, M., et.al., "Chemical Vapor Deposited TiN Films for Sub-0.5 μm Contact and Via Applications", *ibid.*

5. Nakano, T., Ono, H., Ohta, T., Oku, T., and Murakami, M., "Diffusion Barrier Properties of Transition Metals and Their Nitrides for Cu Interconnections", VLSI Multilevel Interconnections Conference (VMIC) Proceedings, June 7-9, 1994, Santa Clara, CA.

6. Anand, M., et. al., *ibid.*

Interconnect Reliability and Stress

Walter L. Brown

AT&T Bell Laboratories Murray Hill, NJ 07974

There are tens to hundreds of meters of submicron aluminum wires in four or more separate levels in modern ULSI chips. Insulated from one another by dielectric, connected to the active silicon devices through windows and between levels through vias, these wires are subject to two major sources of stress that compromise their reliability. The stresses, their consequences in the interconnect structure, the methods used to characterize them and some of the principal issues that are still unsatisfactorily understood are the content of this chapter.

The Stresses

Thermally-induced stress arises from the large difference in the thermal expansion coefficients of Al, Si and SiO_2 and the difference between room temperature and temperatures encountered in processing the multilevel interconnect structure. Figure 1 illustrates the point for a continuous Al film sandwiched between SiO_2 layers on a Si substrate. The upper dielectric layer is typically deposited by a plasma enhanced TEOS process at about 350C. As the structure cools to room temperature the Al contracts much more than the thick Si substrate (or the oxides) and is put in a state of biaxial tension greater than its yield stress. In the continuous film, plastic deformation allows the Al to shrink in thickness as its lateral dimensions are constrained to match those of the Si.

Figure 2 shows the situation for Al wires encased in SiO_2 (as in a real circuit). Now, assuming the Al adheres to the surrounding SiO_2, as the structure cools from the dielectric deposition temperature, the Al is brought into hydrostatic tension. It does not have the "thickness" degree of freedom with which plastic deformation can accommodate tension in the other two directions. If the hydrostatic tension exceeds the cavitation stress of Al it can only be accommodated by formation of voids. With thermal expansion and temperature differences as in Figure 1, and taking account of elastic deformation in the SiO_2 encapsulation, the void volume may be ~1%, an unimportant fraction if the voids are nm's in size. If they coalesce into a few large voids, however, these can produce a complete break in a wire giving stress void failure.[1]

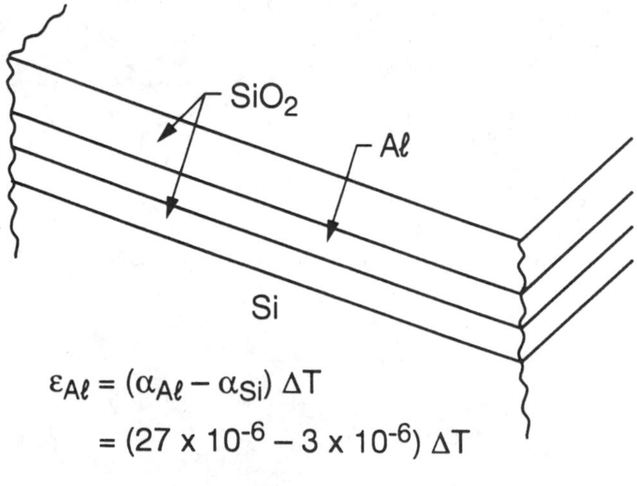

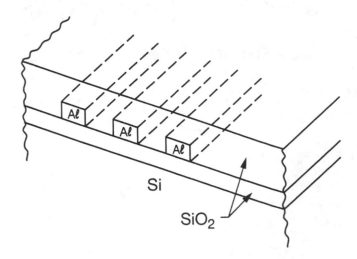

$$\varepsilon_{Al} = (\alpha_{Al} - \alpha_{Si}) \Delta T$$
$$= (27 \times 10^{-6} - 3 \times 10^{-6}) \Delta T$$

FIGURE 1. Schematic of SiO_2/Al/SiO_2 blanket films on a Si substrate. The strain ε in the Al with a temperature change ΔT depends on the difference in the thermal expansion coefficients of Al and Si. For $\Delta T \sim 320$ C, $\varepsilon \sim 0.8\%$. The stress is relieved by plastic deformation which thins the Al film.

FIGURE 2. Schematic of Al lines confined on all sides by SiO_2. The lines are brought into hydrostatic tension with the temperature decrease from SiO_2 processing temperature, leading to void formation.

The second source of stress is electromigration, the movement of atoms due to the small but persistent momentum transfers given to them by electrons.[2] These momentum transfers bias the random motion of vacancy-mediated atomic diffusion to move atoms in the direction of electron flow, as shown in Figure 3. This results in build up of compressive stress at the anode end of a conductor and leaves vacancies at the cathode end to aggregate into voids. At the high current densities required by ULSI circuits, electromigration may produce breaks in lines or failure at contacts or vias as the voids grow with time. It may also produce failure by short circuits due to fracture of the dielectric and extrusion of Al at the anode end. As indicated in the equation in Fig 3, the electromigration drift velocity is the product of the electric field in the wire (ρJ) and the atomic mobility (De/kT) where D is the thermal diffusivity of atoms. In the expression, eZ^* is the effective charge on which the electric field is acting to produce a drift force. The factor Z^* contains the details of the effectiveness of the biased momentum transfers of electrons to atoms. For aluminum $Z^* \sim 10$. Aluminum has a relatively high atomic diffusivity and through D, electromigration is a thermally activated process.

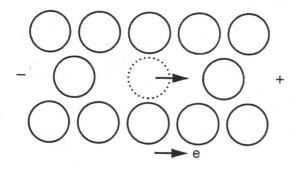

Electromigration Drift Velocity

$$v = DeZ^*\rho J/kT$$

ρ = Resistivity
J = Current Density

FIGURE 3. Momentum transfers from electrons bias the normally random vacancy-mediated diffusion of Al atoms so that they electro-migrate in the direction of electron flow. See text.

Characterizing Thermal Stresses

In blanket films biaxal stress is commonly measured using wafer curvature techniques.[3] For Al films about 1μm in thickness on conventional Si wafers, commercial laser reflection wafer curvature equipment allows routine measurement of radii of curvature as large as a kilometer and determination of stresses in the Al films as small as a few megapascals. A typical stress temperature "loop" measured in this way is shown in Figure 4. Starting at room temperature with a high tensile stress of about 360 megapascals (see Figure 1), the stress decreases with increasing temperature as the aluminum expands more rapidly than the Si substrate. The decrease is linear and has the slope expected for totally elastic changes, dependent on the thermal expansion coefficients of Al and Si. The stress changes sign to compressive at about 230C and saturates (plastically) at a compressive stress of about -100Mpa. Elastic behavior is evident in the early part of the decreasing temperature half of the loop, but plastic deformation starts as early as 350C at a tensile stress of even less than 100 Mpa. The stress temperature behavior at lower temperatures is controlled by temperature dependent and strain-rate dependent dislocation-mediated glide.[4]

Wafer curvature can be used to measure the average stress in arrays of parallel Al lines, either along their length or across their width. Even if they are covered with a conformal or a planarized dielectric, information about their stress state can be deduced from wafer curvature by taking account of the geometry of the lines and the dielectric surrounding them using finite element analysis.[5]

X-ray diffraction provides another way of characterizing the mechanical stress in either blanket films or arrays of lines. In this case the lattice constant of Al is precisely determined and its deviation from that of stress-free material provides a measure of the strain in the metal. By choice of the x-ray diffracting geometry, strain in three orthogonal directions can be measured.[6] Figure 5 illustrates the measurement of two strain components in a line (actually an array of parallel lines) along and perpendicular to its (their) length. The sketches in the figure illustrate, in plan view, the <220> planes in grains of typically highly <111> textured Al. There is no orientational preference for grains in the plane of the film. Fig 5a is for an unstrained line and Fig 5b for a line that has been strained in tension along its length. Fig 5c shows the measured d-spacing of <220> planes as a function of the azimuthal angle of the x-ray scattering-vector for lines strained as in Fig 5b.[7] The strain varies by about 0.6% between <220> planes parallel and perpendicular to the length of the line. In an unstrained case, the d-spacing would be independent of azimuthal angle.

X-ray diffraction measurements have been used to measure the 3D strain in dielectric coated Al lines as a function of temperature.[8] From these the hydrostatic component as well as the shear components of the stress can be deduced.

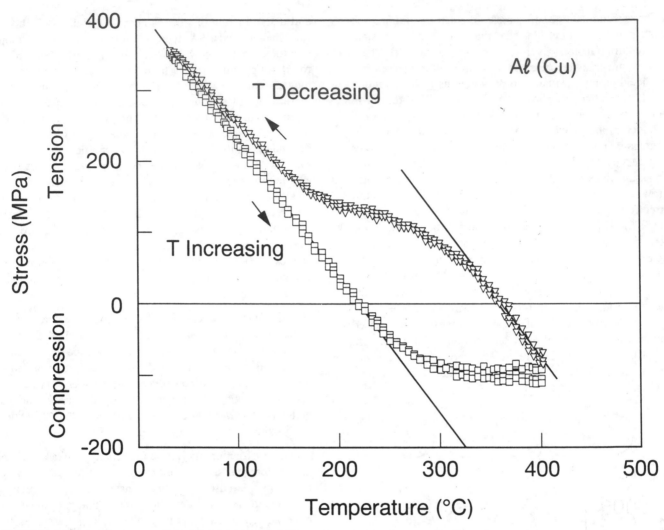

FIGURE 4. Stress-temperature "loop" for a blanket Al film such as in Fig 1, measured by wafer curvature. The arrows indicate the increasing and decreasing temperature portions of the loop.

Characterizing Electromigration

The most common approach to characterizing electromigration is by accelerated testing of conducting structures that simulate those encountered in real circuits. Measurements are typically made of the resistance of lines, contacts or vias as a function of time at current densities and temperatures higher than those encountered in-service in order for changes to be observable in reasonable experimental times.[9] Statistics of failure are obtained by stressing a number of similar structures, with failure defined, for example, as a certain change or percentage change in resistance or as an open circuit. To develop statistics with a small number of test devices, one procedure is to use test structures consisting of a number of lines in parallel, for example fifty lines of equal width connecting two end pads. Operated with constant voltage between the pads, the current density in each line is the same no matter how many are conducting. Measurement of the total current identifies the number of still-conducting lines. Each line break reduces the original current by 2%. Statistics of two sets of fifty-line testers with lines 0.5 and 0.8μm wide are shown in Figure 7.[10] The distribution of failures is approximately log-normal based on the straight lines which represent the data quite well. The mean time to failure is larger for the narrower lines, but the width of the distributions is quite similar. The increase in mean time to failure has been attributed to a larger grain size/line width ratio and hence a more favorable microstructure in the narrower lines.

Experiments of this kind carried out at different current densities and different temperatures are used to determine the activation energy and the functional dependence on current density of the mean time to failure. Activation

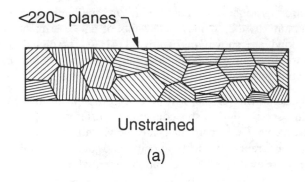

<400> planes

Unstrained

(a)

Strained in Tension

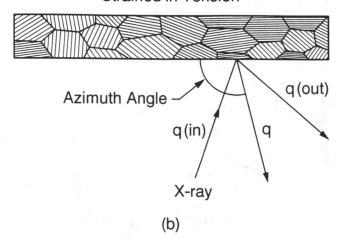

Azimuth Angle

q (out)

q (in) q

X-ray

(b)

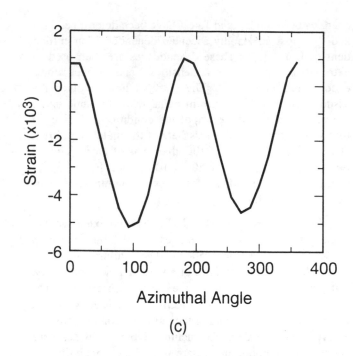

(c)

FIGURE 5. Schematic plan views of a polycrystalline Al line (a) stress free and (b) under tension along the line length. (c) The <220> planar spacing measured for case (b) as a function of the x-ray scattering vector direction with respect to the line length.

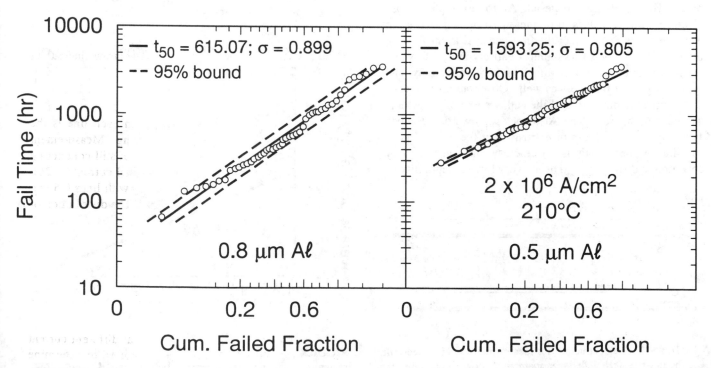

— $t_{50} = 615.07$; $\sigma = 0.899$
-- 95% bound

0.8 µm Al

— $t_{50} = 1593.25$; $\sigma = 0.805$
-- 95% bound

2×10^6 A/cm^2
210°C

0.5 µm Al

FIGURE 6. Statistics of failure (opens) for two 50-parallel-line test structures with 800 um long lines of 0.5 and 0.8 um line width.

energies between 0.5 and 1.4eV have been determined [11] and the most commonly accepted dependence on current density is J^{-2}.[12] These dependences are then used to extrapolate to current densities and temperatures encountered in service. The extrapolation involves the assumption that the failure mechanism is the same under the measured and the extrapolated conditions. It is clear that at least the thermal stress state of the metal is different at a test temperature of 210C than it is at an anticipated operating temperature of 80 to 100C (See Figure 3, for example). The validity of the extrapolation to service conditions is a continuing issue.

Failure analysis by SEM and TEM is used extensively to characterize the nature of the sites where failure has been induced by electromigration. The failure is often associated with the presence of a void at a grain boundary, but intragranular voids are also observed. The understanding of why a void forms where it does and not at another seemingly similar location in a conductor is still elusive. The local microstructure of the metal has been thought to be important. There is also the possibility that the failure is due to some imperfection or impurity at the Al-SiO$_2$ interface. This is a particularly difficult possibility to identify, and ultimately control.

A characterization that lies closer to the definition of electromigration involves measurement of the material drift under a current stress. A typical test structure for such measurements is illustrated in Figure 7.[13] A thin (0.05-0.1μm, for example) TiN conductor joins contact pads. The conducting material, Al for example, whose electromigration drift is to be measured lies on top of the TiN, but does not contact the end pads. Current flowing between the contacts flows almost entirely in the Al where the Al exists because of its much higher conductivity, and typically greater thickness as well. Observations are made of the movement, "drift", of the cathode end of the Al line segment with time. The Al displaced toward the anode typically forms hillocks or extrusions at the anode end of the line segment. TiN is a refractory material with very low atomic diffusivity and hence very low susceptibility to

electromigration. Its purpose is to maintain current continuity as the Al moves. Results of an experiment of this kind are shown in Figure 8.[14] As predicted by electromigration theory, the drift velocity is proportional to the current density.

Measurements of the drift velocity in Al with different grain sizes and also containing different alloying constituents, Cu in particular, are shown in Figure 9. The enormous range in drift velocity and of its activation energy are evident. The activation energy for bulk Al is

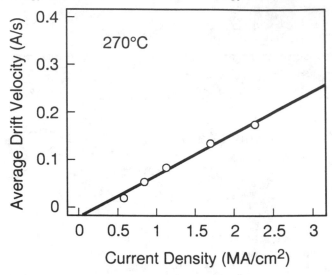

FIGURE 8. Drift velocity vs current density determined from structures such as in Fig 7.

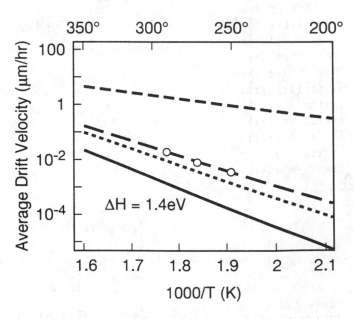

FIGURE 9. Drift velocity in Al vs 1/T determined from structures such as in Fig 7. —— bulk; - - - Al-Cu large grain; – – –Al-Cu small grain; — — Al-Si-Cu.

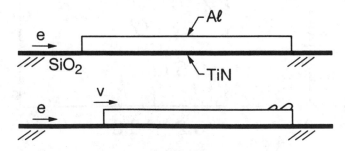

FIGURE 7. A test structure for electro-migration induced drift. Although the TiN underlayer maintains current continuity, the current flows dominantly through the Al where it exists.

168

approximately 1.4eV while that for the small grain Al-Cu is <0.5eV.[14] Such differences influence the anticipated values of drift velocity at a temperature of 100 C by many orders of magnitude.

Recently there have been reports of in-situ microscopic studies of the dynamics of void formation, growth and migration under current stress.[15,16] To maintain the interfacial conditions as close as possible to those of real circuits, the aluminum conductors are covered with dielectric. Their detailed examination through the dielectric requires either high energy backscattering in an SEM or a thin enough layered membrane structure to allow TEM. Both approaches are being used. To speed up the dynamics to a reasonable rate for these observations both elevated temperatures and current densities are required. Particularly in membrane structures, care has to be given to extracting the Joule heat from the lines under test so that the temperature (and temperature gradients) are under control. Great variety in the void behavior is observed. Its interpretation is a current challenge.

Characterizing Microstructure

There is clear evidence that microstructure is one important parameter in the electromigration response of Al. The change in mean-time-to failure with line width (see Figure 6) is one example of it. The diffusivity of Al in grain boundaries is much higher than in the bulk. However, if the grain structure in a conducting line is "bamboo", i.e. the grain boundaries are perpendicular to the line length and hence to the flow of electrons, atomic movement on grain boundaries will not result in transport of material along the line. The narrower the line compared to the grain size the less likely it is that more than one grain will be contained in the line width and provide a longitudinal grain-boundary. Lines patterned from a random array of grains in a blanket film will not have boundaries which are perpendicular to the line, but if the line is filled with single grains, the energetic driving force to reduce grain boundary area will tend to result in a bamboo structure when a line is post-patterned annealed.

Grain size distributions are obtainable using SEM, TEM or focused ion beam (FIB) secondary electron microscopy. TEM is a tedious approach because of the need to prepare thin specimens although it is capable of providing the clearest and most accurate results. SEM generally requires some enhancement of the topographic contrast at grain boundaries, for example by etching, to make boundaries clearly visible. The crystallographic contrast provided by FIB removes the need for pre-processing the samples, but images at more than one incident direction are generally needed to remove possible ambiguity between grains of

nearly identical orientation. Automatic techniques for obtaining grain size distributions from SEM or FIB images often require considerable hand editing to avoid missing boundaries and hence missing small grains in the distribution. Since small grains may have particularly significant roles in void formation, this is an area in which improved image processing techniques would be valuable.

Grain size isn't the only microstructural consideration. The crystallographic orientation of grains and grain boundaries may also be important in determining void formation from either thermal or electromigration stress. This information can also be obtained with TEM, but electron backscatter diffraction, EBSD, [17] offers another approach that does not require preparation of thin specimens. Using an SEM to illuminate a single grain with electrons, an electron diffraction pattern is obtained in reflection geometry. This pattern defines the orientation of the grain. The determination takes only a few seconds so it is feasible to examine a statistically significant number of grains in a test structure. For EBSD to be effective, it is, however, necessary to remove a covering dielectric if one is present. Figure 10 is a stereogram of a collection of 155 aluminum grains.[18] Each cross in the stereogram is the orientation of a single <111> pole of a crystallite. Each crystallite is represented by 4 such crosses. Three sets of four have been joined with lines in the figure. All of the remainder are like the one of those that looks like a 3 armed star. If the Al had perfect <111> texture there would be a cross for each grain exactly at the center of the diagram, and the remaining three <111> poles would lie on a circle at 70° from the center. This is nearly the case for all but two of the 155 grains. The texture is not perfect <111> for 153 grains in the set, but one <111> pole for each of these grains lies close (within ~10°) to the surface normal. Two grains are far from this orientation. None of their <111> axes lie close to the surface normal as is evident by the square pattern they present when joined. It turns out that these are particularly small grains in a distribution whose mean size is about 1μm. An x-ray texture measurement would be unlikely to reveal them. Whether grains of this type play a special role in failure of lines is yet to be determined, but their presence as about 1-2% of the grains in number has been repeatedly observed in films, or in patterned lines before post-patterned annealing.[18]

Characterizing The Role Of Additives

It has been recognized for more than 20 years that adding copper to aluminum greatly increases the electromigration mean-time-to-failure of micron-width Al conductors.[2] Copper has typically been added in concentrations between

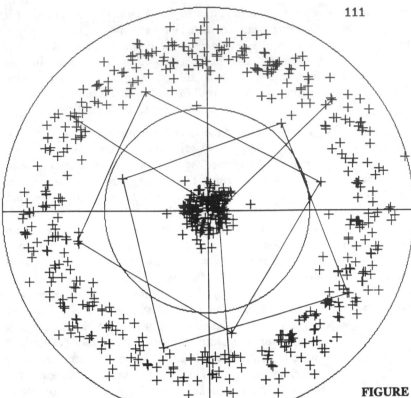

FIGURE 10. Stereogram of 155 Al grains measured with EBSD. The +'s are the orientation of <111> poles, four per grain. Three sets of four are connected with lines. The remaining 152 sets are similar to the three-armed star of one of the marked sets.

0.5 and 2wt%. Such addition has two undesirable effects. Cu increases the resistivity of Al by impurity scattering, about $0.8\mu\Omega$cm for each 1wt%. At high concentrations, Cu also interferes with the reactive ion etching used to pattern Al in the process used by many manufacturers. This latter difficulty is of minor importance below 1wt%.

The solubility of Cu in Al is low. Cu precipitates as $CuAl_2$ and at high Cu concentration precipitates are readily observed and identified by TEM.[19] The precipitates are not preferentially located at Al grain boundaries. At 0.5wt% precipitates are often difficult to observe at all, even though the solubility of Cu is less than 0.1wt% at room temperature. The nucleation and growth of precipitates is, of course, very sensitive to both temperature and concentration. EXAFS measurements are currently being carried out with the objective of measuring the fraction of Cu precipitated at low concentration.[20]

The influence of Cu on electromigration has been attributed to decreasing the diffusivity of Al atoms in grain boundaries. Cu atoms may be present as a monolayer filling the boundaries. In Al with an average grain size of 1μm, less than 0.1wt% Cu is needed to coat all the grain boundaries at a monolayer level. The rest of the Cu, in excess of its solubility, is presumably present as precipitates. In electromigration experiments with Al(Cu), it has been found that Cu atoms electromigrate faster than Al atoms and in fact all the Cu in an Al line may be found at the anode end of the line after extended EM testing.[21] Cu acts as a sacrificial agent that retards (prevents?) Al diffusion on boundaries as long as it is there. The extra Cu provides a continuing source of Cu to replenish the grain boundaries as the Cu atoms electromigrate toward the anode.

For Al line widths larger than the grain size, so that the microstructure is not bamboo, this picture of Cu may be appropriate and it was in this line-width regime that its beneficial effects were established. For submicron lines and bamboo microstructure what does Cu do? How beneficial is it? Such questions are currently under study.

Other alloying additives have been suggested as alternatives to Cu, scandium in particular.[22] The EM benefits of such substitutions are still not clear. It is worth noting that there is great reluctance on the part of IC manufacturers to change from a material that they have learned how to process successfully to one that has to be examined for all its potential negative side effects, however attractive it might be for, say, electromigration.

Modeling Stress Induced Reliability

There is a strong desire to be able to anticipate the reliability of a particular design of multilevel interconnect when it is operated at different current densities and temperatures. The performance of a circuit can be enhanced if limitations on acceptable current densities can be raised above present design rules which were established to avoid failure due to electromigration a number of years ago. These rules are likely to be overly conservative. A better understanding of actual reliability trade-offs is needed. Modeling efforts are being made to address these issues, for example, to follow the resistance consequences of depletion of Al at a contact by electromigration. Modeling that takes account of the actual microstructure of the material in more than an average way is just beginning. Understanding microstructural behavior of alloying constituents well enough to be able to model them is another objective. Of course, the ultimate goal of modeling is to be able to predict the consequences of a design without having to build it, with all the costs and delays that such construction involves. That goal is still rather distant in predicting reliability.

Characterization Wishes

There are two wishes in addition to those expressed in the paragraphs above, that stand out in terms of characterization of materials for reliability in the face of thermal and electrical stress. The first of these is the wish for an indicator of potential reliability problems that could be used at close-to-use temperature and current densities: a precursor of failure. Attempts to use the low frequency noise spectrum for this purpose continue to be pursued.[23] Establishing the connection between changes in the spectrum or in the magnitude of components of it and real "failure" is clearly difficult and requires understanding that does not yet exist.

The second characterization wish is for a technique capable of measuring stress or strain on a submicron scale in structures that are as nearly like real devices as possible. It would be very helpful to be able to determine strain distributions within individual grains and to be able to measure changes as the material is stressed with temperature changes or current. The strain sensitivity of Raman scattering in the underlying Si is being explored as a way of deducing the strain in metal lines nearby.[24] A

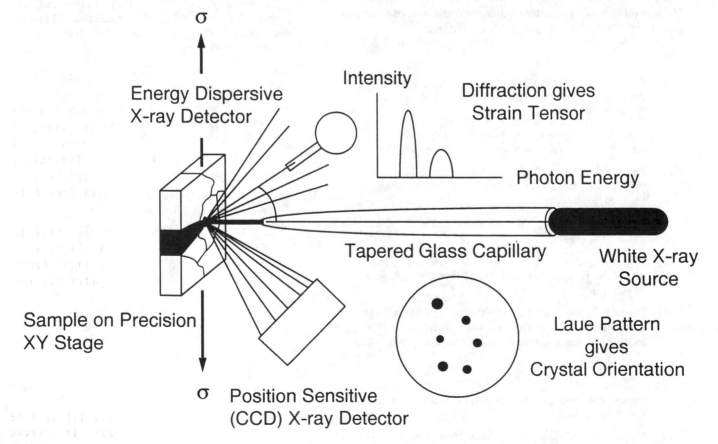

FIGURE 11. Schematic of a submicron white beam x-ray scheme for determining both grain orientation and strain in individual Al grains in structures under test.

micron-sized laser beam is used to excite Raman signals in the Si. The spatial variation in the measured strain is then compared with finite element modeling of the strain in the coupled mechanical system. The measurement is indirect and not ideal in this respect, and also has marginal spatial resolution in its present manifestation.

An attractive alternative which requires access to a major synchrotron facility to be effective is shown schematically in Figure 11.[25] It uses a submicron white beam of x-rays and Laue diffraction from individual Al grains. As illustrated, the beam is formed by condensation through a tapered capillary, but optical schemes using mirrors to form reduced size images of small x-ray source apertures are also possibilities. The symmetry of the diffraction spots will define the crystallite orientation. Careful measurement of the x-ray energy in a spot will provide a measure of the strain. This type of approach is in an early stage of exploration.

Acknowledgements

The author acknowledges stimulating discussions with Dave Barr, Matthew Marcus, Judy Prybyla and Cynthia Volkert on the subject matter of this paper. Several of their as yet unpublished results are also included in it.

[1] Flinn, P.A. , in "Materials Reliability in Microelectronics II", Mat. Res. Soc. Proc. **265**, Pittsburgh, 1992, pp 15-26.

[2] For a review see d'Heurle, F.M. and Ho, P.S. in "Thin Films-Interdiffusion and Reactions" ed. John Poate, K.N. Tu and J.W. Mayer, New York: Wiley, 1978, ch 8.

[3] Flinn, P.A. Gardner, D.S. and Nix, W.D., IEEE Trans. Elec. Dev. **ED-34**, 689 (1987); Volkert, C.A., J. Appl. Phys. **70**, 3527 (1991).

[4] Volkert, C.A., Alofs, C.F., and Liefting, J.R., J. Mater. Res **9**, 1147 (1994).

[5] Witvrouw, A., Proost, J., Deweerdt, B., Roussel, P.L., and Maex, K., Mat. Res. Soc. Symp. Proc. **356** (1995) to be published.

[6] Flinn, P.A. in "Thin Films: Stresses and Mechanical Properties II" ed M.F. Doemer, W.C. Oliver, G.M. Phar and F.R. Brotzen, Mat. Res. Soc. Proc. **188**, Pittsburgh, 1990, pp 3-13..

[7] Marcus, M.A., private communication.

[8] Besser, P.R., Brennan, S., and Bravman, J.C., J. Mater. Res. **9**, 13 (1994).

[9] Lloyd, J.R. and Koch, R.H., Appl. Phys. Lett. **52**, 194 (1988).

[10] Prybyla, J.A., private communication.

[11] English, A.T. and Kinsbron, E., J. Appl. Phys. **54**, 260, 1983.

[12] Shatzkes, M. and Lloyd, J.R., J. Appl. Phys. **59**, 3890 (1986).

[13] Blech, I.A., J. Appl. Phys., **47**, 1203 (1976).

[14] Oates, A.S., J. Appl. Phys. **70**, 5369 (1991).

[15] Marieb, T.N. et al. AIP Conference Proc., **305** "Stress Induced Phenomena in Metallization" ed. P.S. Ho, C.Y. Li, P. Totta, p 1, 1993.

[16] Riege, S.P., Hunt, A.W. and Prybyla, J.A., Mat. Res. Soc. Proc. "Materials Reliability in Microelectronics V" Spring 1995, to be published.

[17] Dingley, D.J., and Randle, V., J. Mater. Sci. **27**, 4545 (1992); Dingley, D.J., Scanning Electron Microscopy, (1984) p 569.

[18] Barr, D.L., Brown, W.L., Marcus, M.A. and Ohring, M., Mat. Res. Soc. Proc. "Materials Reliability in Microelectronics V" Spring 1995, to be published.

[19] d'Heurle, F.M., Ainslie, N.G., Gangulee, A., and Shine, M.C., JVST **9**, 289 (1971).

[20] Marcus, M.A., private communication.

[21] Prybyla, J.A., Barr, D.L. and Fitzgerald, E.A. Mat. Res. Soc. Symp., "Materials Reliability in Microelectronics V", Spring 1995, p 365.

[22] Ogawa, S., Nishimura, H., Tech. Digest of the International Devices Meeting, 227, 1992.

[23] Koch, R.H., Lloyd, J.R., and Cronin, J., Phys. Rev. Lett., 55, 2487 (1985); Alers, G.B., Oates, A.S., and Beverly, N.L., Appl. Phys. Lett., to be published.

[24] Ma, Q. Chiras, S., Clarke, D.R., Suo, Z., J. Appl. Phys., to be published.

[25] Cargill, G.S. and Wang, P.C. in "X-ray Microbeam Techniques and Applications" Workshop Proceedings Spring '94, July 1994.

Back-end Simulation: Etching and Deposition

F. H. Baumann

AT&T Bell Laboratories, Holmdel, NJ 07733

In modern VLSI technology, more than half the processing steps involve the 'back-end' to build the interconnects. The development of a new process module takes more than 3 years, up to 100 experiments, and several million dollars. Reliable etching and deposition simulation tools would expedite this difficult and expensive process. In general, etching and deposition simulators have to develop physically based models for the incoming particle fluxes, for the interaction between the substrate and the deposited particles, and for the evolution of the topography. The complexity of the models used in each case is determined by our knowledge of the physics involved, by the customer's needs, and by the computational efficiency of the algorithms. In this paper, I assess the status of back-end simulation, show examples of the way it can facilitate interconnect development and improve equipment design, and point out some of the future challenges.

I. Introduction

In modern VLSI processing, fabrication of the interconnects requires up to 4 layers of metal, thus starting to dominate the total processing effort. For a typical multi-level technology, more than 50% of the processing steps are used in the back-end to build the interconnects.

In addition, back-end processing continues to present new challenges due to trends in device integration, because lateral dimensions of features (windows, vias, spaces) shrink roughly 40% per generation, but vertical dimensions of films (e. g. glue and barrier layers, contact metal films, and interlevel dielectrics) have to be maintained. This inevitably leads to high aspect-ratio windows and vias, which cannot be filled reliably by conventional processes [1].

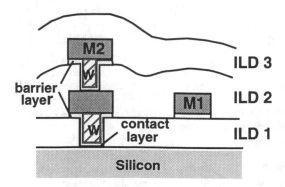

Figure 1. Scheme of typical multilayer interconnect. Areas in need of back-end process simulation are window and via etch (aspect ratio dependent etching), side-wall and bottom coverage of contact and barrier layers, recess of W-plugs after etching, conformality of metal layers (M1,M2) and interlevel dielectrics (ILDs), and the resulting topography after resist etchback or chemo-mechanical planarization.

In isolated instances, a modified process can deliver a cost effective solution (e. g. collimated sputter deposition). Often, however, a completely new process has to be developed (e. g. W-CVD for window and via fill). The development of a new process module takes more than 3 years, up to 100 experiments in the development line, and can cost several million dollars.

The development of back-end simulation tools aims to assist and accelerate the incorporation of a modified or new process by predicting suitable operating conditions. Ideally, back-end simulation should allow the process engineer to predict the outcome of a number of specific processes to optimize the manufacturing of the interconnects, as indicated in fig. 1.

The paramount criteria for the usefulness of a simulator are predictive capability, accuracy, reliability, and ease of use. Predictive capability requires physically-based models [2], and thus the understanding of the physics and chemistry involved in the process under consideration.

In this paper, I first describe the general principle of a deposition or etch simulator (Section II). In Section III, I show, using a specific example, how deposition modeling can be employed to predict the outcome of sputter deposition processes and to improve equipment design. In Section IV, I assess the current status of back-end processing, and in Section V discuss future needs for back-end processing.

II. General Principle of Simulator

In general, deposition and etching simulators have to develop physically-based models for (1) the incoming particle flux, (2) the interaction between substrate and

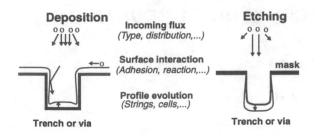

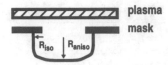

Figure 2. General principle of deposition or etching simulator. Three basic steps have to be simulated: The type and distribution of the incoming particles, their interaction with the substrate, and the evolution of the topography.

Figure 3. For simulation of a complex process like reactive ion etching, the surface chemistry has to be treated phenomenologically. Several reaction steps are grouped into an isotropic chemical reaction component, R_{iso}, and an anisotropic etch-rate enhancement component, R_{aniso}.

particles, and (3) the evolution of the topography, as indicated in fig. 2. In this section, these three aspects are described in some detail. The complexity of the model used in each case is determined by the knowledge of the physics involved, by customer needs, and by the computational efficiency of the algorithms.

II.1. Incoming Particle Flux

For the incoming particle flux, one needs to model the angular and energy distribution for each species.

The angular distribution of the particles reaching a particular spot on the surface can range from completely isotropic (e.g. in the case of LPCVD[3]) to rather directional (e.g. in collimated sputter deposition[4,5]), or can be a mixture of both (e.g. in reactive ion etching[6,7]).

For the relatively simple case of sputter deposition, the flux can be modeled by assuming a certain distribution[8], or by implementing flux distributions obtained from experiments[9]. For collimated sputtering, the modification of the angular distribution of the incoming particles can be modeled using simple geometric considerations [4,10].

For the more complicated case of Reactive Ion Etching (RIE), the flux consists of several different species of neutrals and ions. Here, the particle distribution can be simulated by means of Monte Carlo techniques[6,11], which give the energy and angular distribution of the incident particles.

II.2. Surface Interaction

The surface interaction is the most critical part of process simulation, since it determines the local etch or deposition rate. Incident particles can interact with the surface in a number of different ways. Simulation tools must be able to describe processes such as adhesion[12], chemical reactions, energy deposition[6], surface diffusion[13], re-emission[14] of the particles, and surface

passivation [15] to properly model the process under consideration.

The number of possible interactions can be reduced for certain processes. In sputter deposition, only adhesion and surface diffusion have to be taken into account. Another relatively simple case is Chemical Vapor Deposition (CVD), where, in addition to adhesion and surface diffusion, chemical surface reactions have to be considered.

In RIE, however, an abundance of possible reactions can occur, and in many cases the underlying physics and chemistry of particle-substrate interaction are only partly understood. (For a detailed discussion, see [6,16]). Here, process simulation has to treat the surface chemistry phenomenologically. The total etch rate is divided into a chemical etching (isotropic) component and an etch-rate enhancement (anisotropic) component (see fig. 3). A listing of how these two etch rate components are treated in a simulation tool is given in table 1 (after ref. [6]). Although the exact nature of the etch rate enhancement mechanism is not known, the phenomenological model allows the simulation of such complex processes as reactive ion etching. However, the etch rate components have to be inferred from experiments for substrate and mask

Table 1. Phenomenological etch components for RIE modeling

Isotropic component	Anisotropic component
Chemically active, neutrals	Etch rate enhancing ions
Diffuse in plasma, reach substrate in all different directions	Enhancement mechanism unknown Sputtering? Surface Heating? Lattice defect generation? Ion aided desorption?
Rate limited by slowest reaction	
Etch rate proportional to flux	Etch rate prop. to energy deposited
Etch rate uniform and perpend. to local surface	Etch rate dependent on geometry

Table 2. Implementation of some prominent back-end simulation tools.

Simulator	Developed by	Industrial Partners	Models	Deposition	Etching
SAMPLE	UC Berkeley	IBM	2-D strings	Al sputtering[18] SiO_2 sputtering[18]	SiO_2 sputter etch[18]
SPEEDIE	UC Stanford	LSI, AMD, LAM, Intel	3-D flux 2-D strings 2-D PDE	SiO_2-CVD[19] SiO_2-PECVD[23]	Si RIE[6,14] Si plasma etch[6,14] W etchback[24]
EVOLVE	ASU	SEMATECH Motorola, SRC	3-D flux 2-D strings 2-D PDE	Ti-W sputtering[25] Al sputtering [13] PETEOS CVD[2] W CVD[26]	
ESPRIT	Hitachi	Hitachi	2-D flux 2-D strings		Si RIE[8] SiO_2 RIE[8]
SIMBAD	U. Alberta	Varian	2-D flux 2-D Hard-disk Monte Carlo	Al sputtering[12] Ti-W sputtering[27] W CVD[28] TiN CVD[28]	Si RIE[29] W plasma etch[30]

materials, and for each new set of processing parameters.

A key challenge is a more complete understanding of the individual processes to allow the development of improved, physically-based models to describe the local etch or deposition rate.

II.3. Profile Evolution

The last step in simulating the outcome of an etching or deposition process is the calculation of the topographical changes according to the local etch or deposition rate.

First, this requires algorithms to determine the geometrical shadowing of the flux at each surface element due to the topography of the structure (fig. 4a). For trenches, 2-D calculations are sufficient[14]. Two-dimensional calculations can also be used for highly symmetric 3-D structures, using e. g., the rotational symmetry of a circular via (so called "smart 2-D" [17]). However, for more

complex structures like corners, edges or overhangs, full 3-D simulations are required.

Secondly, each point on the surface has to be moved according to the amount and nature of the deposition or etch rate. Most commonly, 2-D string [18,19] or cell [20,21] algorithms are used to track the time evolution of the surface in small time steps (fig. 4b and table 2). In these approaches, the surface is divided into individual segments (strings) or treated as a chain of points (nodes). The advance of theses strings or nodes for a small time step Δt is calculated according to the local etch or deposition rate. Recently, completely different approaches such as solid modeling [22] or hard disk/hard sphere simulators [4,12] have been proposed and implemented.

As shown in table 2, most back-end simulation tools have been developed by university based research groups. To verify the validity of the simulations, and to couple the simulations to processes and conditions found in manufacturing, close collaboration with industrial partners is essential.

II.4. Extraction of Physical Parameters

The models used to describe the individual steps in etching or deposition require the extraction of the physical parameters governing the process. These parameters (e. g. sticking coefficients, isotropic/anisotropic etching components, re-emission) can be deduced from experiments on specially designed test structures (over-hang structures [19,31]) or from specially designed short-loop experiments[7]. However, in many cases measurements of important process parameters have proved to be a difficult task.

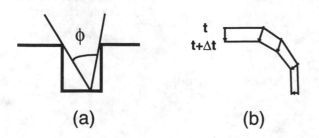

(a)   (b)

Figure 4. (a) Angle ϕ open for incoming particles determines geometrical contribution to local etch or deposition rate. (b) In 2-D string simulators, the local surface evolution is modeled using moving strings or segments for small time steps Δt.

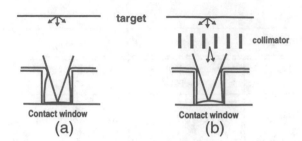

Figure 5. (a) In conventional sputtering, the opening of the window closes significantly during deposition. (b) Using a collimator, the increased directionality of the incoming particle flux reduces the closing and allows higher bottom coverage in the window or via.

III. Examples of Back-end Simulation: 3-D Simulation of Sputter Deposition

Deposition of contact films, glue layers and barrier layers is usually performed by sputtering. However, during the sputter process in small windows and vias, a significant evolution of the topography takes place. This gradually closes the opening of the window or via, which leads to an increase of the aspect ratio during the deposition and thus to poor bottom coverage. A possible solution to this problem lies in the use of collimated sputtering to increase the bottom coverage in contact film deposition (fig. 5).

As examples, I describe two different approaches to modeling sputter deposition of thin films. The first approach shows how a full 3-D Monte Carlo simulation with hard spheres is able to predict accurately and in detail the via side-wall and bottom coverages, and the evolution of the topography on the microscopic scale. For sputtering of contact films, where a minimum thickness is required to achieve reliable contact, this is vital to predict processing windows. In a second approach, 3-D geometric calculations are used to determine macroscopic effects of sputtering, such as shadowing from the collimator.

III.1. Microscopic Model: 3-D Ballistic Monte Carlo Simulation

Principle

Fig. 6 shows a center-cut through a 3-D simulation to illustrate the principle of the simulation program. First, an arbitrary surface is defined by close-packed spheres, each sphere representing a cluster of material (diameter ~ 100-200 Å). The surface is then showered with additional spheres (cosine angular distribution), which are relaxed into stable positions at the surface after a specified amount of surface diffusion. The simulation continues until the desired film thickness is achieved.

Comparison of Simulation with Experiment

Fig. 7(a) shows a High Resolution SEM image of a window after collimated sputter deposition of 1650 Å Ti/TiN (collimation aspect ratio = 1:1). Fig. 7(b) shows the center-slice of the corresponding 3-D ballistic simulation. The simulation closely reproduces the experimental features. Bottom coverage, side-wall coverage and even details in the window corners and on the steps are accurately reproduced. Experimental data (SEM) of the bottom coverage versus window aspect ratio for uncollimated and 1:1 collimated sputtering are plotted in fig. 8. Open symbols and closed symbols represent the measured aspect ratio at the beginning and the end of the deposition, respectively. The window aspect ratio increases,

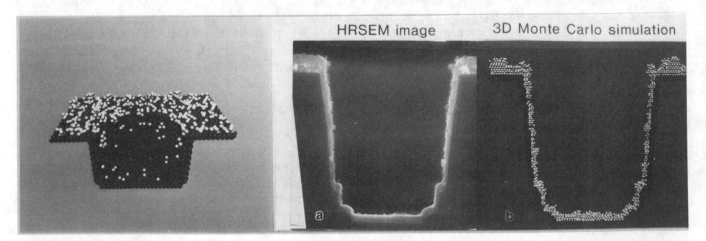

Figure 6. Principle of 3-D Monte Carlo simulation: Deposition of hard spheres (white) on user defined surface (black).

Figure 7. (a) SEM image of 1 μm window after deposition of 1650 Å Ti/TiN. (b) Center slice of corresponding ballistic simulation. Side wall and bottom coverage are predicted within 10% margin.

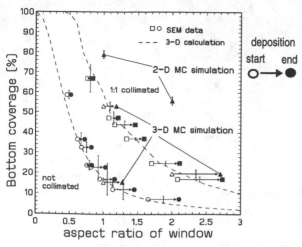

Figure 8. Bottom coverage in the middle part of a window as predicted by the 2-D and 3-D Monte Carlo simulation (triangles) and measured by SEM (circles and squares). Dashed lines show simulations using the 3-D geometric model sketched in fig. 10.

because material is added both on the top and on the side, resulting in a gradual closing of the window. A 2-D hard-disk Monte Carlo simulation, taking into account only shadowing in a plane, does not predict correctly, since bottom coverage values are too high by up to a factor of 3. However, the 3-D hard-sphere simulation accurately predicts the measured bottom coverage. Furthermore, the 3-D simulation is able to predict the evolution of the topography of the window, represented in fig. 9 as an increase in the window aspect ratio. Here, the increase in window aspect ratio after deposition of Ti/TiN is plotted as measured by SEM, for various windows and different

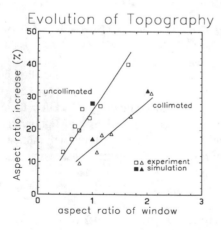

Figure 9. Development of aspect ratio of various windows with ongoing deposition for conventional and collimated sputtering. 3-D MC simulation (solid) predicts evolution of window well in both cases.

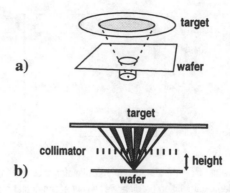

Figure 10. a) Principle of 3-D geometric calculation for sputter deposition. The part of the target which contributes to the flux into the window is determined numerically. b) Shadowing effect of collimator located between wafer and target. The flux distribution depends on the position (height) and the grid size of the collimator

deposition conditions (film thickness = 12.5% of window depth). The reduced closing for collimated sputtering, and the rate of narrowing are well predicted by the Monte Carlo simulation.

III.2. Macroscopic Model:
3-D Geometric Simulator

The Monte Carlo simulation reveals the fine scale features of the deposition process. A simple geometric 3-D model can be used to determine the macroscopic features, such as bottom coverage, collimator shadowing, etc., at a specific point on the wafer surface. The principle is shown in fig. 10a. The flux at any point on the wafer, e.g. in the middle of a window, is determined by calculating the area of the target "seen" from this specific point. The effect of a collimator is sketched in fig. 10b, pointing out that the collimator is opaque for specific ranges of angles. Calculations were performed for a 3-D geometry (as sketched in fig. 10a)), using a square grid collimator with different collimation aspect ratios. In all calculations, the distance between wafer and target was fixed at 7 cm.

Comparison of Calculation with Experiment

A comparison between experiment and this simple geometric model shows good agreement with experimental data for the bottom coverage for conventional and 1:1 collimated sputtering, as indicated by the dashed lines in fig. 8. (For a more detailed discussion see [32].) The geometric model readily yields the transmission of a collimator by calculating the solid angles open between target and wafer (see fig. 10b). Fig. 11 shows the strong decrease of the transmission of a collimator as a function of

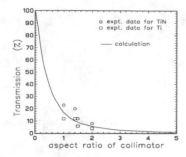

Figure 11. Transmission of square collimator vs collimation ratio. 3-D geometric calculation predicts well within experimental error. Experimental data taken from Ref. [33].

its aspect ratio. For comparison, experimental data obtained for Ti and TiN are also included (experimental data were taken from ref. [33]). Within experimental error, the calculated transmission agrees well with the measured values.

Prediction of Film Uniformity

Using this geometric model, we can now explore the influence of several collimator parameters on the deposited film. First, the uniformity of the film thickness can be determined, both on the wafer surface (top coverage) and at the center of the bottom of a window. The simulation for the bottom coverage in windows with different aspect ratios (AR) is shown in fig. 12. For a fixed collimation ratio of 1:1, the simulation indicates that the bottom coverage varies in different windows from 59% to 49% (window AR = 1) and from 22% to 16% (window AR=2). In the second case, the shadow effects of the collimator can become important, since shadowing produc-

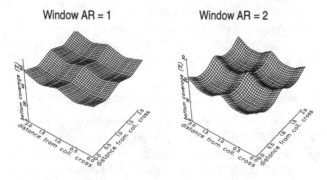

Figure 12. "Shadow" of square 1:1 collimator in the bottom coverage of windows with aspect ratios of 1 and 2. For a window aspect ratio of 2, a significant dependence of the bottom coverage on the position under the collimator (± 15%) can be observed.

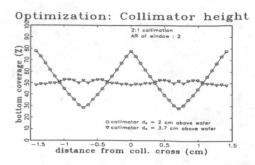

Figure 13. Effect of collimator height on homogeneity of bottom coverage. The optimum height of the collimator grid can be determined depending on the window size in the VLSI process and on the collimator used in the deposition process.

es inhomogeneities of ± 15%. This can affect electrical performance of contacts locally, if the processing window is small.

Optimization of Equipment Design

The geometric model can be used to vary parameters like collimator height, grid size, etc., to optimize the design of the equipment (for details, see [4]). As an example, fig. 13 shows the variation of the bottom coverage in a window (window AR=2) as a function of the position under a 2:1 collimator for different collimator heights. Clearly, an optimum values for the collimator height can be found, resulting in a homogeneous bottom coverage throughout the wafer. This optimum value depends on the aspect ratio of the windows used in the specific VLSI process, as well as on the equipment used to deposit the film. As figure 13 shows, the geometric 3-D model can provide optimum values for equipment design, which can increase the bottom coverage in the windows and thus the throughput of the tool by up to a factor of 2.

In conclusion, the Monte Carlo model of hard-sphere deposition and the 3-D geometric model describe the data surprisingly well. However, sputtering at low temperatures represents a situation where only limited physics is needed to model the outcome of the process. Greater physical understanding is needed to simulate more complex processes, such as high temperature sputtering, where surface diffusion plays a dominant role, or aluminum reflow, where surface tension acts as an additional driving force.

IV. Status of Back-end Simulation

The state of the art in back-end process simulation significantly lags behind modeling of front-end processes. Taking ion implantation and point defect diffusion as examples, we have predictive, accurate, and reliable simulation tools at hand [see e. g. 34,35]. Here, in many cases, simulations are used to predict the outcome of a process before the process is carried out in the development line. The reasons for the success of front-end process modeling are manyfold, and only a few can be mentioned here.

First, front-end processes occur primarily in one material: silicon. Secondly, in most cases modeling remains within the starting simulation boundaries (no topography evolution), which makes moving grids or meshes unnecessary. Third, due to strong interaction with experiments, many crucial parameters are reasonably well known, and lookup tables exist for many implantation and diffusion systems. All the above has been achieved by the research of numerous groups over a time period of more than 20 years.

Compared to the status of front-end process modeling, back-end simulation is much less developed. Although process simulation for etching or deposition has predictive capability in isolated instances (e.g., sputter deposition [4] or CVD [26]), successful modeling is still restricted to rather simple systems. Reasons for the so-far limited success of back-end simulation include the following:

a) Deposition and etch simulators have to treat several coupled problems simultaneously: particle flux, surface interaction, and evolving topography. To treat the above phenomena, plasma, surface and material science have to be involved. In many cases, the underlying physics or chemistry is poorly understood, and the systems contain various unknowns, which are difficult to separate or to measure.

b) Process simulation often relies on phenomenological parameters, e. g., etch rates or sticking coefficients. In principle, these parameters have to be remeasured for every change in process conditions.

c) As indicated by table 2, the community has developed several different approaches to model the same physical process. This can be taken as a sign for a lack of common agreement on the underlying physics of the processes under consideration. (As an example, the validity of a hard disk/ hard sphere approach, where a system of 10^9 atoms is modeled with 10^6 clusters, still has to be proven.) However, the development of different approaches to model the same process can also be seen as an indication of an emerging field, which will find the different approaches to converge as the field matures.

V. Future challenges

In contrast to the front-end, over the last few years entirely new processes have been introduced in the back-end. For the generations below 0.8 µm, CVD window and via fill had to be utilized, and for the generations below 0.6 µm, chemo-mechanical polishing (CMP) is used to keep the topography within limits. For upcoming VLSI technologies, copper metallization and low-k dielectrics are already under investigation [36].

As soon as new processes are introduced, predictive simulations can have a major impact on reducing the time to manufacturing and thus the development costs. The biggest needs to accomplish the task of developing predictive, physically based deposition and etch simulators are:

a) Systematic experiments to measure the most critical parameters. For metal deposition, these are surface diffusion coefficients and sticking coefficients under various processing conditions

b) A better understanding of the physics and chemistry involved in the processes, to reduce the gap between atomistic interactions and empirical models

c) The development of practical models to handle the more complex interactions, e. g., in RIE. To reduce the number of possible unknowns, the dominant parameters controlling the individual process have to be extracted. Furthermore, metrics have to be developed to prove the validity and accuracy of the simulation.

Due to the complexity of many processes which simulation has to deal with in the back-end, a link must be forged between the complicated physics and chemistry taking place at the atomic level and real-life processing in the manufacturing line.

The ultimate goal of every process modeling effort is to transform the simulation from the interpolative curve fitting domain to the extrapolative, predictive regime. Much remains to be done to achieve this.

VI. Conclusions

The center of gravity for process development continues to shift towards the back-end. Already in current VLSI technologies, more than 50% of the processing steps are in the interconnects. In addition, major new processes are introduced in (almost) every new generation, like W-CVD and CMP. While the front-end follows an evolutionary trend, new processes and materials used in the back-end involve more radical changes [37].

Most importantly from the manufacturing point of view, back-end processing is a major yield limiting factor, as most technologies have their highest yield loss in the interconnects.

Increasing costs, rapid change, and impact on yield make back-end process simulation an attractive field for sustained effort.

Acknowledgments

The author acknowledges stimulating discussions with W. L. Braun, C. B. Case, M. Joyce, G. H. Gilmer, R. C. Kistler, W.-Y. Lai, R. Liu, A. Ourmazd, M. R. Pinto, C. S. Rafferty, G. P. Schwartz, and R. K. Smith.

The author gratefully acknowledges helpful discussions with J. P. McVittie (Stanford University) and A. J. Toprac (SEMATECH).

References

1. Kikkawa, T., Kikuta, K., Tsunenari, K., Ohto, K., Aoki, H., Drynan, J. M., Kasai, N., and Kunio, T., *Jpn. J. Appl. Phys,.* Vol. 32, p. 338, 1993

2. Cale, T. S., Raupp, G. B.,and Gandy, T. H., *J. Vac. Sci. Technol.* A 10 (4), p. 1128, 1992

3. Cale, T. S., *J. Vac. Sci. Technol.* B 9 (5), p. 2551, 1991

4. Baumann, F. H., Liu, R., Case, C. B., and Lai, W. Y.-C., *IEDM Techn. Dig.*, p. 861, 1993

5. Yamada, H., Shinmura, T., Yamada, Y., and Ohta, T., *IEDM Techn. Dig.*, p. 553, 1994

6. Ulacia F., J. I. and McVittie, J. P., *J. Appl. Phys.*, Vol. 65, No. 4, p. 1485, 1989

7. Tazawa, S., Matsuo, S., and Saito, K., *IEEE Trans. Sem. Manuf.*, Vol. 5, No. 1, p. 27, 1992

8. Yamamoto, S., Kure, T., Ohgo, M., Matsuzama, T., Tachi, S., and Sunami, H., *IEEE Trans. Comp. Aid. Design,* Vol. CAD-6, No. 3, p. 417, 1987

9. Blech, I. A. and Vander Plas, H. A., *J. Appl. Phys.*, 54 (6), p. 3489, 1983

10. Dew, S. K., Liu, D., Brett, M. J., and Smy, T., *J. Vac. Sci. Technol.* B 11 (4), p. 1281, 1993

11. Dalvie, M., Farouki, R. T., and Hamaguchi, S., *IEEE Trans. Electr. Dev.*, Vol. 39, No.5, 1992

12. Smy, T., Westra, K. L., and Brett, M. J., *IEEE Trans. Electr. Dev.* Vol. 37, No. 3, p. 591, 1990

13. Cale, T. S., Jain, M. K., Taylor, D. S., Duffin, R. L., and Tracy, C. J., *J. Vac. Sci. Technol.* B 11(2), p. 311, 1993

14. Singh, V. K., Shaqfeh, E. S. G., and McVittie, J. P., *J. Vac. Sci. Technol.* B 10(3), p. 1091, 1992, and *J. Vac. Sci. Technol.* B 12(5), p. 2952, 1994

15. Hamaguchi, S. and Dalvie, M., *J. Vac. Sci. Technol.* A 12 (5), p. 2745, 1994

16. Coburn, J. W., *J. Vac. Sci. Technol.* A 12(4), p. 1417, 1994

17. McVittie, J. P., private communication

18. Oldham, W. G., Neureuther, A. R., Sung, C., Reynolds, J. L., and Nandgaonkar, S. N., *IEEE Trans. Electr. Dev.,* Vol. ED-27, No. 8, p. 1455, 1980

19. McVittie, J. P., Rey, J. C., Bariya, A. J., IslamRaja, M. M., Cheng, L. Y., Ravi, S., and Sarawat, K. C., *SPIE*, Vol. 1392, p. 126, 1990

20. Ikegava, M. and Kobayashi, J., *J. Electrochem. Soc.*, Vol. 136, No. 10, p. 2982, 1989

21. Fujinaga, M., Kotani, N., Kunikiyo, T., Oda, H., Shirahata, M., and Akasaka, Y., *IEEE Trans. Electr. Dev.*, Vol. 37, No 10, p. 2183, 1990

22. Tazawa, S., Leon, F. A., Anderson, G. D., Abe, T., Saito, K., Yoshi, A., and Scharfetter, D. L., *IEDM Techn. Dig.*, p.173, 1992

23. Li, J., McVittie, J. P., Ferziger, J., and Saraswat, K.,C., *1994 VMIC Conference*, p. 539, 1994

24. Hsiau, K., Bang, D. S., McVittie, J. P., Dutton, R., Saraswat, K. C., Tripathi, S., Bariya, A., and Kao, D. B., *1994 VMIC Conference*, pp. 545, 1994

25. Rogers, B. R., Tracy, C. J., and Cale, T. S., *J. Vac. Sci. Technol.* B 12 (5), p. 2980, 1994

26. Cale, T. S., Gandy, T. H., and Raupp, G. B., *J. Vac. Sci. Technol.* A 9 (3), p. 524, 1991

27. Liu, D., Dew, S. K., Brett, M. J., Smy, T., and Tsai, W., *J. Appl. Phys.* 75 (12), p. 8114, 1994

28. Dew, S. K., Smy, T., and Brett, M. J., *J. Vac. Sci. Technol.* B 10 (2), p. 618, 1992

29. Tait, R. N., Dew, S. K., Smy, T., and Brett, M. J., *J. Vac. Sci. Technol.* A 12 (4), p. 1085, 1994

30. Tait, R. N., Dew, S. K., Smy, T., and Brett, M. J., *J. Vac. Sci. Technol.* A 10 (4), p. 912, 1992

31. Liu, D., Dew, S. K., Brett, M. J., Janacek, T., Smy, T., and Tsai, W., *J. Appl. Phys.* 74 (2), p. 1339, 1993

32. Baumann, F. H., Liu, R., Case, C. B., and Lai, W. Y.-C., *1993 VMIC Conference*, p. 412, 1993

33. Joshi, R. V. and Brodsky, S., *1992 VMIC Conference*, p. 253, 1992

34. Pinto, M. R., Boulin, D. M., Rafferty, C. S., Smith, R. K., Coughran, W. M., Jr., Kizilyalli, I. C., and Thoma, M. J., *IEDM Techn. Dig.*, p. 923, 1992

35. Pinto, M. R., Rafferty, C. S., Smith, R. K., and Bude, J., *IEDM Techn. Dig.*, p. 701, 1993

36. Roberts, B., Harrus, A., and Jackson, R. L., *Solid State Technology,* Vol. 38, No. 2, p. 69, 1995

37. Pintchovski, F., *IEDM Techn. Dig.*, p. 97, 1994

METROLOGY AND PROCESS CONTROL ISSUES IN CHEMICAL MECHANICAL POLISHING

Rahul Jairath[a] & Lucia Markert[b]

SEMATECH
2706 Montopolis Drive
Austin, Texas 78741-6499

Depth of focus requirements for fabricating integrated circuits with 0.35 µm geometries and smaller are driving the need for global planarization. Chemical mechanical polishing (CMP) is becoming the process of choice not only because of its unique ability to planarize globally, but also because of its ability to reduce electrical defect densities.(1,2) Characterization of CMP requires evaluation of both planarity and surface defect density. The standard array of metrology techniques includes spectrophotometry, profilometry, scanning electron microscopy (SEM), and light scattering. Wafer-to-wafer and run-to-run process variability in CMP is significant, making process control crucial. Several endpoint detection schemes have been proposed, but only one is commercially available. The lack of *in-situ* and in-line metrology has hampered efforts to improve process control. Manufacturing-compatible metrology techniques will be crucial for successfully transforming CMP into a well-controlled well-accepted process.

INTRODUCTION

Chemical mechanical polishing (CMP) has become the planarization technology of choice for sub-0.5 µm devices. Here, polishing of the dielectric layers is used to reduce topography on the substrates over large planarization lengths. Traditional planarization techniques such as spin-on glass (SOG), resist etch back (REB) or deposition-etch-deposition provide only local or intermediate planarity. Patterned resist etch back (PREB) is able to provide intermediate to global planarity, but requires an extra patterning step for each dielectric layer, and is therefore expensive. CMP of dielectric films results in topographical reduction across distances of several millimeters. Local planarization achieved (across 100 µm) enables denser packing of interconnect lines, while global planarization aids subsequent lithography steps.

CMP of metal films has also gained acceptance as a pattern definition technology in the formation of vertical and horizontal conductors. Polishing of tungsten films has become a viable, low cost alternative to the formation of tungsten plugs/studs by etch back techniques. Also, in the absence of a viable copper dry etch process, the formation of vertical and horizontal copper interconnects using dual damascene based architectures is an attractive option.

Table 1 shows the CMP related roadmap considerations as developed by semiconductor technologists.(3) It can be seen that the manufacture of a 256 Mb DRAM circuit in 1998 with a critical dimension of 0.25 µm will require planarization such that the usable depth of focus requirement of 8000 Å over the exposure field is met. For a 1 Gb circuit at the 0.18 µm technology generation to be manufactured in 2001, the usable depth of focus requirement is expected to decrease to 7000 Å. There is also some evidence that CMP of dielectric layers may reduce defect density. On the other hand, CMP of metal films is required mostly due to defect density and process architecture requirements. For example, plasma etchback of tungsten films to form contact or via plugs has traditionally been a defect prone process. As result,

---

[a] On assignment from National Semiconductor Corporation
[b] On assignment from Symbios Logic (formerly AT&T GIS, NCR Microelectronic Products Division)

TABLE 1. CMP Related National Technology Roadmap For Semiconductors Considerations(3)

First Year Of Shipment	1995	1998	2001	2004	2007
Minimum Feature Size (μm)	0.35	0.25	0.18	0.13	0.10
DRAM Memory (Bits/Chip)	64M	256M	1G	4G	16G
Max. Wafer Diameter (mm)	200	200	300	300	400
Max. No. Of Wiring Levels (Logic)	4-5	5	5-6	6	6-7
Electrical Defect Density (defects/ sq m)	500	300	100	40	20
Device Size (mm)	16 X 16	18 X 18	19 X 19	21 X 21	23 X 23
Minimum Field Size (mm)	22 X 22	26 X 26	26 X 30	26 X 36	26 X 44
Usable Depth Of Focus (μm)	1.0	0.8	0.7	TBD	TBD

it could potentially be the yield-limiting step in a given process flow. CMP of tungsten films to form the plugs offers the advantage of higher yield due to lower defect densities.(2) This is consistent with the roadmap requirements for defect density shown in Table 1. Additionally, polishing of other metal films such as aluminum and copper to form horizontal interconnects may be required for advanced metallization schemes. This is consistent with the need for a copper wiring pattern definition technology in the absence of a viable copper etch process.

TECHNOLOGY OF CMP

In a typical polishing process, the wafer is mounted on a rotating carrier which is held down on a rotating polishing pad in the presence of a polishing slurry (see Figure 1). The slurry reduces friction between the wafer and polishing pad and promotes chemically induced erosion of the film. Typically, the slurry used to polish oxide films consists of fine (<0.5 μm) silica particles dispersed in an alkaline medium such as hydroxides of potassium or ammonia. The slurry used in the polishing of metal films is generally acidic in nature, and is based upon alumina or silica particles dispersed in an aqueous solution. An oxidant is added to the slurry in order to provide the chemical action. Two pad types are commonly used in chemical mechanical polishing: cast and sliced polyurethane or urethane-coated polyester felt.

METROLOGY

Characterization of CMP is both time and labor intensive. The most common metrics for evaluating

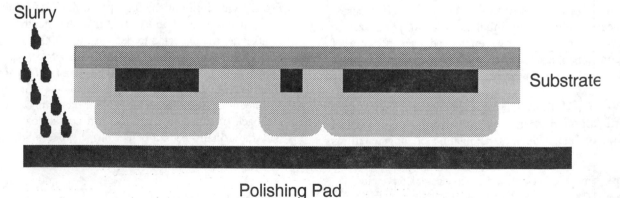

FIGURE 1. Schematic of the CMP process

CMP process performance are defined in Table 2, along with an approximate range of values, standard metrology techniques, and typical modes of application. Of these metrics, only removal rate and defect density are used for both process development and process control. This is because those metrology tools are already automated and easy to use in a manufacturing environment, not because the other metrics are not important. Currently topography measurements are used primarily for process development; an easy-to-use automated tool would definitely be a plus for CMP process control.

Within-wafer and wafer-to-wafer sheet film removal rates are important indicators of process nonuniformity and stability. However, in order to truly assess polishing performance, topographic measurements are essential.

Local planarity, across distances of 1 to 100 µm, is crucial for eliminating metal residues, or stringers, along the bottom edge of steps; it also provides a measure for the likelihood of step coverage problems with subsequent deposited films. Global planarity, across distances of 1 mm or more, is essential for ensuring that the entire lithographic field of view is within the depth of focus.

Dishing, the film thickness difference between the edge and center of a nominally flat area, depends on the width of the area, the mechanical properties of the polishing pad, and the processing parameters, pressure in particular. Oxide erosion describes a particular type of dishing. When polishing arrays in metal CMP, where the metal and oxide removal rates differ by one to two orders of magnitude, there is not only dishing of the metal between the oxide walls, but also dishing of the oxide walls across the entire array, i.e., oxide erosion. Dishing and oxide erosion limit the degree of planarity achievable.

Among the four metrics for topographic variation, global planarization is the most difficult to measure. The lateral distances involved make SEM and AFM impractical, and the vertical differences after polishing are close to the detection limit for profilometry. In addition, true measurements of global planarity must account for effects such as wafer bow and warpage.

Physical defects such as scratches, micropits, and particles can lead to electrical defects in the finished circuit, such as shorts or bridging. The high particle content of slurry, and the chemically and mechanically aggressive nature of polishing itself, drive concerns about physical defect densities before and after CMP. In light of this the improvement in electrical defect densities due to CMP, as mentioned earlier, is truly unexpected and remarkable.

According to the National Technology Roadmap for Semiconductors,(3) the evolution from 0.35 to 0.18 µm geometries will decrease the depth of focus from 1.0 to 0.7 µm, corresponding to planarity requirements

TABLE 2. Common Process Metrics In CMP

Metric	Definition	Values	Techniques and Application
removal rate	film erosion averaged over polish time	500 to 5000 Å/min	spectrophotometry (oxides) resistivity (metals) process development & process control
local planarization	topographic variation across 1 to 100 µm	0.02 to 2.0 µm	profilometry, SEM, AFM process development
global planarization	topographic variation across 1 mm or more	0.02 to 2.0 µm	profilometry process development
dishing	film thickness difference between the edge and center of a nominally planar area	50 to 5000 Å	profilometry, SEM process development
dielectric erosion	dielectric thickness difference between the edge and center of an array after metal CMP	50 to 5000 Å	profilometry, SEM process development
defect density	density of particles, scratches, and micropits larger than the smallest feature size	100 to 10,000 m^{-2}	light scattering process development & process control

TABLE 3. Metrology Techniques For CMP

TECHNIQUE	PRECISION & ACCURACY	DESTRUCTIVE?	SPEED	COST OF OWNERSHIP
spectrophotometry	good	no	rapid	low
resistivity	fair-good	contact	fair	fair
profilometry	poor	contact	slow	high
SEM	good	yes	slow	high
AFM	good	contact	slow	high
light scattering	good	no	fair	fair

changing from 2000 to 1400 Å. At the same time, the electrical defect density limit will decrease from 500 to 100 m^{-2}. With respect to CMP, today's film thickness metrology basically meets the expected needs. However, physical defect size and density detection limits must become better than 0.18 μm and 100 m^{-2}. Topography metrology must advance in both measurement capability and manufacturing compatibility to evaluate planarities of less than 1400 Å and dishing and oxide erosion as low as 50 Å.

Ideally, all metrology techniques would be precise, accurate, non-destructive (preferably non-contact), rapid, and inexpensive. Table 3 summarizes how the standard metrology techniques for CMP compare. The most pressing needs are to find a better approach for measuring planarity and to improve detection limits for submicron physical defects, and at the same time increase measurement speed and reduce cost of ownership. In addition to the polishing performance metrics listed in Table 2, there is also interest in monitoring the surface chemistry for residual slurry, pad material, or cross-contamination.

Non-contact capacitance measurements are often used to assess substrate thickness and flatness after polishing, with a thickness resolution of about 0.5 μm and a lateral resolution of about 1000 μm. To the best of our knowledge, this technique is not widely used to evaluate planarity. A key consideration for integrated circuit applications is the uniformity of total film thickness with respect to the substrate surface. Non-contact capacitance measurements do not measure film thickness with respect to the substrate surface and do not have sufficient topographic resolution to measure post-polish planarities of 1400 to 2000 Å.

One promising technique for CMP metrology is confocal microscopy. Light from a very narrow focal plane is used to form two-dimensional images of submicron features. Sequential imaging from different focal plane locations provides information about three dimensional topography, with vertical resolutions of 50 Å or better. In order for a new metrology technique to become widely accepted in the semiconductor industry, it must provide significant advantages not only in measurement capability, but also in cost of ownership.

PROCESS CONTROL

Currently there is no *in-situ* and little in-line process monitoring equipment commercially available, in part because CMP tools are still relatively new, and also because insertion of an *in-situ* probe or sensor into the mechanically and chemically aggressive CMP environment is difficult. The lack of these technologies results in an increase in cost of ownership of CMP processes.

Efforts to develop a viable endpoint detection scheme have resulted in only one commercially available tool to date. The Luxtron system monitors the current to the motor which rotates the wafer carrier. Exposure of the underlying dielectric film at the end of metal polishing results in an increase in friction. There is a corresponding slight but measurable increase in carrier current.

Another endpoint detection approach, developed at AT&T, is to insert an electrode into the polish platen to measure and monitor the capacitance of the dielectric layer on the wafer.(4) The electrode makes contact to the wafer through the slurry-soaked polishing pad. The electrical properties of slurry are frequency dependent,

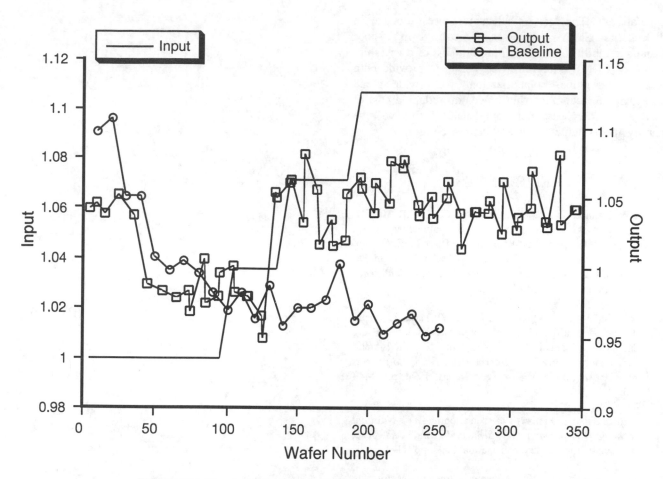

FIGURE 2. Simple SISO Run-To-Run Control Of Oxide CMP Process

and for this purpose low frequencies are recommended. The capacitance corresponds to the remaining dielectric thickness, and depends on oxide type and circuit pattern density, making calibration curves necessary.

Other proposed methods include detecting the presence or absence of a metal layer by current flow(5) or by impedance,(6) and using acoustic wave reflection to monitor oxide thickness.(7)

In the absence of a viable end-point technique for oxide CMP processes, "Run-to-Run" (R2R) techniques have been developed and used to control CMP processes. These techniques combine response surface, statistical process control, and feedback control techniques. There are two major steps: [a] creation of a linear regression model based on off-line experiments and [b] on-line process estimation and process control, during which the model is continuously updated or tuned based on observed process data. Avoiding overcompensation for process variation is an important consideration.

SEMATECH has recently tried the equivalent of a single-input/single-output (SISO) gradual mode control on the CMP process. Figure 2 shows the results from one of these experiments(8) which consisted of running 250 wafers at a standard or "baseline" process without the SISO/GM controller; the tool consumables were then refreshed, and a further 345 wafers were run under SISO/GM control. Although the controller response to initial drift was slow, the process performance did ultimately improve over the baseline.

SUMMARY

In summary, both metrology and process control have become increasingly high contributors to the overall cost of ownership of CMP processes. Rapid developments in manufacturing compatible low cost metrology techniques, especially for measurement of defects, are key to achieving the cost targets for future device generations. The lack of adequate process control in dielectric CMP processes results in poor manufacturability, and therefore high cost of ownership. The use of CMP is driven by its ability to provide both local and global planarization, but wide-spread acceptance in manufacturing hinges on reducing cost of ownership. An essential component of this will be improvements in metrology and process control.

REFERENCES

1. C.W. Kaanta, S.G. Bombardier, W.J. Cote, W.R. Hill, G. Kerszykowski, H.S. Landis, D.J. Poindexter, C.W. Pollard, G.H. Ross, J.G. Ryan, S. Wolff, and J.E. Cronin, "Dual Damascene: A ULSI Wiring Technology," in *8th International VLSI Multilevel Interconnection Conference Proceedings*, 1991, pp. 144–152.

2. C. Yu, S. Poon, Y. Limb, T.-K. Yu, and J. Klein, "Improved Multilevel Metallization Technology Using Chemical Mechanical Polishing of W Plugs and Interconnects," in *11th International VLSI Multilevel Interconnection Conference Proceedings*, 1994, pp. 144–150.

3. Semiconductor Industry Association, *The National Technology Roadmap for Semiconductors*, San Jose, Semiconductor Industry Association, 1994, pp. 11–13, 83.

4. G.L. Miller and E.R. Wagner, "In-Situ Monitoring Technique and Apparatus for Chemical/Mechanical Planarization Endpoint Detection," U.S. Patent 5,081,421, Jan. 14, 1992.

5. C.W. Kaanta and M.A. Leach, "In-Situ Conductivity Monitoring Technique for Chemical/Mechanical Planarization Endpoint Detection," U.S. Patent 4,793,895, Dec. 27, 1988.

6. M.A. Leach, B.J. Machesney, and E.J. Nowak, "Device and Method for Detecting an End Point in Polishing Operation," U.S. Patent 5,213,655, May 25, 1993.

7. C.C. Yu and G.S. Sandhu, "Chemical Mechanical Planarization (CMP) of a Semiconductor Wafer Using Acoustical Waves for In-Situ End Point Detection," U.S. Patent 5,240,552, Aug. 31, 1993.

8. E.D. Castillo and A.M. Hurwitz, "Run to Run Process Control: A Review and Some Extensions," submitted to J. Quality Technology.

Failure Analysis: Status and Future Trends

Richard E. Anderson, Jerry M. Soden, and Christopher L. Henderson

Failure Analysis Department
Electronics Quality/Reliability Center
Sandia National Laboratories
Albuquerque, NM 87185-1081

Failure analysis is a critical element in the integrated circuit manufacturing industry. This paper reviews the changing role of failure analysis and describes major techniques employed in the industry today. Several advanced failure analysis techniques that meet the challenges imposed by advancements in integrated circuit technology are described and their applications are discussed. Future trends in failure analysis needed to keep pace with the continuing advancements in integrated circuit technology are anticipated.

INTRODUCTION

Integrated circuit technology has advanced rapidly as design and manufacturing approaches have become more sophisticated. This paper explores the challenges for IC (integrated circuit) failure analysis in the environment of present and future silicon IC technology trends. Each of these trends has a significant impact on FA (failure analysis).

Some trends necessitate evolutionary changes in FA apparatus, which generally increase the cost of the equipment and the required lab space. Larger wafer sizes require larger wafer stages in probe stations and microscopes and larger vacuum chambers and load locks in SEMs (scanning electron microscopes) and other equipment employing a vacuum. Increased I/O (input/output) pin counts and higher operational frequencies dictate the use of higher performance electrical test equipment and redesign of the electrical interconnection scheme between the tester and the FA apparatus (1).

However, many of these IC technology trends necessitate revolutionary advances in FA technology. For example, the implementation of additional interconnection levels, power distribution planes, or flip chip packaging may completely eliminate the possibility of employing visible light optical FA techniques without destructive deprocessing. New paradigms for FA are needed to overcome these limitations.

FA has always been employed in the semiconductor industry. The process of performing FA involves verifying that there is a failure, characterizing the failure, verifying that the symptoms of the failure are consistent with the initial observation of the failure, localizing the failure,

determining the root cause of the failure, suggesting corrective action, and documenting the results of the analysis (2). FA has traditionally been a "postmortem" activity used only when devices failed final testing or failed in the field. The results were often not fed back to the production line for corrective action. The physical location of the FA laboratory was often remote from the fabrication facility. There was little interaction between the FA staff and the design, process, device physics, reliability, test, and product engineering staffs. As a result, the value added from FA activities was often not significant and many manufacturers invested only at a token level in FA facilities and personnel.

However, as the semiconductor industry matures there has been constantly increasing emphasis on yield, quality, and reliability of the products. The technology and process complexity of today's ICs as well as the economics of IC manufacturing demand that testing to monitor yield, performance, and reliability be performed during the manufacturing process. The results must be immediately fed back for corrective action. In this environment, traditional FA activities are finding an expanded role: manufacturing or "process" FA (3).

Manufacturing FA redefines what, how, and where FA is performed. Manufacturing FA includes identification of yield limiters such as microcontamination, defects, process equipment damage, technology shortcomings, and design errors. These activities are often called yield analysis. Unpatterned test wafers, short loop monitor wafers, test circuits on completely processed wafers, and completed integrated circuits are all candidates for analysis. Because many ICs are now extremely complex, it may not be cost effective to perform FA on them. In that case, the suite of test structures must dependably predict effects that will be manifested in the IC. Manufacturing FA also includes the traditional FA activities, such as field return analysis.

Because rapid feedback of results is essential, manufacturing FA activities are often performed at the wafer level. FA laboratories are increasingly being located adjacent to or on the manufacturing floor to improve turnaround time and communication. Also, FA engineers are teaming with their counterparts in design, process, device physics, reliability, test, and product engineering. Manufacturing FA is a high value added activity, and increased investment in FA equipment and personnel provides a large return.

CHALLENGES FOR FAILURE ANALYSIS

The following trends have developed for the past several generations of IC technology and are expected to continue for at least the next several generations, although the rate of change may slow in some cases.

> larger wafer sizes
> larger die sizes
> smaller feature sizes
> higher integration levels
> additional interconnection levels
> power distribution planes
> completely planarized surfaces
> mixed analog and digital circuitry
> increased operating frequencies
> reduced operating voltages
> increased I/O pin counts

There have also been trends toward higher density packaging technologies for ICs including:

> smaller package profiles
> flip chip
> MCM (multichip modules)

This paper reviews developments in FA techniques that help meet the challenges presented by advancements in IC and packaging technologies. The focus is on defect localization and root cause analysis using techniques based primarily on the electrical behavior of devices and defects. Analyses which involve detailed materials characterization using physical, chemical, and optical techniques are not discussed here since they are reviewed thoroughly in other papers in this workshop. The discussions are restricted to silicon IC technology, but the applicability of the FA techniques extends to other semiconductor technologies.

FAILURE ANALYSIS TECHNIQUES TO MEET THE CHALLENGES

Electrical Testing

Larger IC die sizes and higher integration levels have made FA impossible using purely physical techniques. There is simply too much silicon circuitry to analyze. Even for test circuits and short loop monitors, it is not cost effective to apply only physical analysis techniques. Therefore, FA techniques based on electrical signatures are becoming the preferred approach. Traditional FA relied primarily on electrical testing at the device pins using curve tracer type instruments. More recently, sophisticated digital and analog test equipment similar to that used in IC test laboratories has been employed. The complexity of parametric and digital instrumentation needed for FA is determined primarily by the requirement to stimulate the device sufficiently to observe and characterize the failure condition. Technology trends having a significant impact on test methodology are mixed analog and digital circuitry, increased operating frequencies, lower operating voltages, and increased I/O pin counts.

The new paradigm for FA based on electrical testing is increasing reliance on software fault isolation and testing-based diagnostic approaches. These may be employed by testing only at the IC I/Os or by using a combination of I/O and internal probing techniques. This new paradigm can be based on digital techniques such as logic backtracing, on parametric techniques such as I_{DDQ} (quiescent power supply current) testing, or on a combination of these approaches. Parametric techniques are emerging for CMOS IC diagnosis. These include methods that take advantage of the global circuit aspects of the power supply current. I_{DDQ} (I_{SSQ}) testing measures the current of the V_{DD} (V_{SS}) power supply in the quiescent logic condition (the stable condition between logic state transitions). In addition to increased defect and fault coverage, I_{DDQ} testing enables rapid identification and physical localization of many design, layout, and fabrication problems (i.e., processing problems of a non-defect nature, such as excessive lateral diffusion) (4). Software tools have been written that relate I_{DDQ} test vectors to logic fault and physical defect localization (5,6).

Powerful techniques based on the application of internal IC probing using electron beam, optical beam, and force probe techniques combined with electrical testing at the IC pins are enabling dramatic improvements in rapid defect localization. In many cases the level of integration of the IC is not an important factor in quickly isolating the defect. These techniques will be discussed later in this paper.

Design for FA

As IC complexity evolved, it became economically imperative for circuit designers to consider testability early in the product concept stages. It was found that the cost of testing, including the cost of test escapes, could be reduced only by bringing together the design and test activities. Design for testability is now an established practice, a successful example of concurrent engineering.

Continued advances in complexity and reduction in the IC manufacturing cycle time make design for FA and defect diagnosis as critical as design for testability. Work teams that include design, test, and FA personnel must be formed early to assure that IC defects can be successfully diagnosed quickly and efficiently. Approaches to designing for FA involve diagnostic software and IC design concepts. Software issues include moving away from models, such as the stuck-at fault, that are abstract representations of defects towards defect classes that inherently link to the circuit architecture (7,8). Test pattern generation or selection approaches based on more realistic models provide the capability to map from failing test vectors to physical structures based on netlist and layout information (5,6). Merging defect classes with improved algorithms for fault dictionaries created during simulation strengthens software diagnosis (9). IC design concepts include extension of techniques that improve controllability and observability, such as partitioning, on-chip self test circuitry, and internal test access points.

Benefits from incorporating design for FA occur immediately, resulting in the ability to detect design problems during simulation and test pattern generation, analysis of first production wafer lots, as well as throughout the product life cycle. Designing for FA benefits include enabling real time diagnosis during production testing of wafers. This provides the ability to rank order the occurrence of "killer" defects, by type and location, with minimal or no impact on production throughput. Such information is needed to immediately plan and implement corrective action, which is the essence of manufacturing FA.

Linking to CAD Databases

Larger die, higher integration levels, completely planarized surfaces, additional interconnection levels, and power distribution planes all create navigation difficulties while performing FA. CAD navigation software is available for a number of FA tools (10). This software links layout and schematic information from the CAD database with images and internal probe data from the FA tools. Linkage assists in positioning probes at the point of interest and correlates the defect site to its physical and electrical location. CAD navigation software is typically provided on electron beam probe systems and can be installed on focused ion beam systems and other FA equipment. However, full integration throughout the FA laboratory is uncommon. Further utilization of the CAD database, such as providing cross-sectional diagrams of any point on the IC, will greatly assist the failure analyst. Ultimately, networking of all the FA tools in the laboratory and linking with the design and test databases will be required for effective FA.

Scanning Optical Microscopy

Optical microscopy techniques locate many physical defects and have been a mainstay in FA for years. However, smaller feature sizes are pushing optical microscopy towards its resolution limit, around 0.2 - 0.3 μm using visible light. Other technology trends that have a large impact on optical microscopy include additional interconnection levels, power distribution planes, and flip chip packaging. Obviously, an optical image from the top of an IC reveals few features deep in the structure if there are many interconnection levels and power distribution planes which obscure a large percentage of the IC surface.

The SOM (scanning optical microscope) provides several advantages over standard optical microscopy. The SOM uses a laser as a source of monochromatic light and forms a pixel by pixel image in several imaging modes that provide a wide variety of information about the IC (11). The confocal mode yields image resolution and contrast that are improved over conventional optical microscopy. "Optical sectioning" in the confocal mode permits observation of discrete planes in the IC without blurring by structures above and below the plane of interest. The extended focus mode provides an apparent large depth of focus.

The use of an IR laser extends the usefulness of the SOM by permitting observation through the wafer. Images of the active regions of an IC can be obtained using reflected IR microscopy from the back of the die (usually after appropriate back surface preparation such as metal removal and mechanical polishing). This circumvents problems caused by additional interconnection levels, power distribution planes, and flip chip packaging. Lightly doped silicon is quite transparent for wavelengths beyond 1100 nm, and heavily doped silicon has sufficient transparency to allow IR microscopy for wavelengths in the range from about 1000 - 1400 nm. Reflected IR SOM images typically have higher resolution (about 0.6 μm) than normal IR microscopy images since the light wavelength is shorter and the surface roughness of the back of the IC die has much less effect than in conventional IR microscopy.

Light Emission Microscopy

Light emission microscopy analysis is often the primary tool for localizing many types of common defects, such as gate oxide shorts and degraded *pn* junctions, and for identifying MOS transistors in saturation due to interconnection shorts and open circuit defects (12). This is the first FA technique discussed in this paper that combines electrical stimulation of the IC with other types of "internal probes". Light emitting regions of biased ICs are imaged using CCD (charge coupled device) cameras with "night vision" image intensifiers or slow scan CCD cameras. The technique is fast, permitting localization of light emitting defects on an IC at low magnification, and is nondestructive. It can be performed using static or dynamic IC operating conditions. Because the emitted photons have energies in both the visible and near IR wavelengths, the light emission can be observed from the backside of the die, circumventing optical obscuration caused by additional interconnection levels, power distribution planes, and flip chip packaging. Light emission microscopy is a valuable FA technique for localizing defects on large die with small feature sizes and high integration levels.

Photon generation results from hot carrier production and subsequent energy release (12,13). Transistor saturation, oxide leakage, and junction leakage are common sources of hot carriers. CMOS IC transistors are in saturation briefly during normal operation when their gate bias switches during logic changes. The time that transistors are in saturation can be increased greatly by design errors or defects such as interconnection open and short circuits. Figure 1 shows a low magnification light emission image of a 1.25 µm feature size, two level metal, 32-bit microprocessor with four gate oxide shorts (highlighted by the four white boxes).

Scanning Electron Microscopy

As IC feature sizes decrease and defects of smaller size become more important, imaging techniques with higher resolution than optical microscopy are required. SE (secondary electron) imaging in a SEM (scanning electron microscope) generates a high resolution (around 5 nm), large depth of field image that depicts the surface topography of an IC and locates physical defects that are too small or difficult to image with optical microscopy. Cross-sections through IC structures are usually imaged in the SEM. Sample charging from absorbed electrons reduces spatial resolution on dielectric surfaces and can be eliminated by coating the sample with a thin conductive film or by careful selection of the primary electron beam energy. Electron irradiation at beam energies above a few

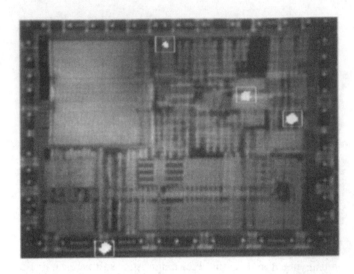

FIGURE 1. Low magnification light emission microscopy image showing four gate oxide shorts (bright areas in the four boxes) in a 32-bit microprocessor.

keV can damage an IC, causing effects such as threshold voltage shifts in MOS transistors.

FESEMs (field emission SEMs) extract electrons from a sharp tip with a high electric field rather than from a heated filament. Because of this, FESEMs have superior spatial resolution at low beam energies, allowing nondestructive imaging of in-process wafers and high resolution imaging of insulating layers without the need for conductive films. The change in secondary electron emission with material is also pronounced at low beam energies, so different layers in a cross-section can be identified in a FESEM without using special wet or dry etches to produce topology differences. The resolution of FESEMs used for IC work is about 2 nm at high beam energies and about 5 nm at low beam energies, as compared to about 100 nm for conventional SEMs at low beam energies.

Although SE imaging in the SEM will not have difficulty resolving features the size of future lateral IC dimensions, the increasing use of planarization techniques negates the effectiveness of SEM imaging. Unless deprocessing approaches are employed, the images of planar surfaces will provide little information. The vertical dimensions which must be imaged in cross-sections have already decreased to near the FESEM resolution limits, and defects that can cause failure may be much smaller. Detailed root cause analysis may require the use of a TEM (transmission electron microscope) or an AFM (atomic force microscope).

Voltage Contrast Techniques

The family of electron beam VC (voltage contrast) techniques have an important role in FA since they provide a nondestructive method for observing internal IC operation and the electrical effects of defects. These techniques encompass the most commonly employed ways to combine electrical stimulation of an IC with internal probing. The resolution of voltage contrast techniques is about the same as the SE image resolution of the electron beam instrument employed.

The basic VC technique creates an image in which the contrast is largely determined by the voltages on IC interconnections (14). By analyzing the variations in brightness, the logic levels of a digital IC can be determined and the voltage on internal test nodes can be measured. VC imaging takes advantage of differing SE emission efficiencies with applied bias on an IC. Basically, conductors at ground potential will be bright in the image and conductors at a positive potential will be dark. In order for the basic VC technique to work, the conductors of interest must be depassivated and only the topmost metallization layer can be imaged. Such images present a qualitative view of the voltages on an IC and are effective only at dc or very low frequencies.

The CCVC (capacitive coupling voltage contrast) technique enables nondestructive imaging of dynamic voltages beneath passivation layers with negligible electron beam effect on the IC (14). CCVC uses the passivation layer as a discharging capacitor to generate a dynamic image of changing subsurface voltages. A low primary beam energy of around 1 keV is used. CCVC imaging for qualitative voltage information is performed at fast electron beam scan rates to increase the time resolution of the dynamic signal. Since the voltages on conductors are imaged on the top of the passivation, lower metallization levels may be analyzed if they are not covered by other metal layers. However, for deeper layers the voltage resolution is lower and the interference from the voltages of nearby conductors is more problematic.

By merging information available from CCVC imaging and I_{DDQ} testing, images can be produced that show the logic state responsible for high or anomalous I_{DDQ} (15). This can be a very effective technique to quickly localize a defect in a complex IC, since a limited number of stimulus vectors need to be applied.

The VC and CCVC techniques are primarily employed for voltage waveform measurements on internal IC conductors. The importance of these techniques for FA can be measured by the prevalence of commercial electron beam test systems (10) in FA laboratories around the world. Quantitative voltage measurements are enabled by employing an energy spectrometer to measure the energy spectrum of the emitted SEs. Voltage waveforms on internal IC conductors are obtained by using a pulsed electron beam and stroboscopically sampling a repetitive waveform. These systems can acquire waveforms with voltage and timing resolutions of about 10 mV and 10 ps, respectively. Advanced electron beam test systems provide integration of CAD databases and permit comparison of control and failing devices through image processing. Software tools are available that reduce the time to locate and analyze logical faults through an automated backtracing approach (16).

All these VC-based FA techniques are impacted by the following IC technology trends: larger die sizes, smaller feature sizes, higher integration levels, additional interconnection levels, power distribution planes, lower operating voltages, and flip chip packaging. Evolutionary improvements in electron beam test equipment will continue, providing better voltage, timing, and spatial resolution as well as more powerful software control. However, there are fundamental limitations imposed by advances in IC technology that must be addressed in order for VC to remain useful.

Obviously, the equipment for VC analysis must provide a method for making electrical connection to the IC under test inside a vacuum chamber. This problem is not trivial for packaged ICs and is quite challenging for high I/O pin count die. Perhaps the most difficult challenge is the increasing number of interconnection levels and the use of the upper metallization level or levels as power planes. Effective application of VC analysis on fully processed ICs without deprocessing will be ineffective. Solutions to this problem include designed-in internal test points on the topmost metallization layer and the use of a focused ion beam system (to be discussed later) to open holes to lower levels or create probe pads at the point of interest. If anisotropic depassivation procedures are used, lower layers may be analyzed as long as they are not covered by higher level conductors. However, the decreasing feature size trend also means that there is more interference in the VC measurements from adjacent conductors. Since manufacturing FA is moving away from deprocessing approaches, it appears that a design for FA philosophy will be necessary in order to maintain the utility of VC.

Charge-Induced Voltage Alteration and Light-Induced Voltage Alteration Imaging

CIVA (Charge-Induced Voltage Alteration) (17) and LIVA (Light-Induced Voltage Alteration) (18) are techniques which provide a fast, simple method for locating open conductor, contact, via, and junction defects, which are increasingly important reliability problems for complex circuits. Open circuit failures traditionally have been very difficult to localize in complex ICs since the failure signature is not obvious. CIVA and LIVA simplify localization of these defects on an entire IC in a single image. Both techniques employ the same electrical approach but use different probes. CIVA employs an electron beam while LIVA uses a photon beam. CIVA analysis may be performed on upper and lower level conductors in multilevel interconnection ICs, and LIVA may be performed from the front or back of the IC die.

A CIVA image is generated by monitoring the voltage shifts in a constant current power supply providing power to an IC as an electron beam is scanned over its surface. When electrons are injected into an electrically floating conductor, the voltage of the conductor becomes more negative. This abrupt change in voltage on the floating conductor generates a shift in the voltage demand of the constant current source supplying bias to the IC. The shifts observed in the power supply voltage, even for open circuits with significant tunneling current, are relatively large and produce images in which the contrast is dominated by the open conductors (17). Figure 2 shows an example of low magnification CIVA imaging of a 1 μm feature size, passivated, two-level metal gate array with an open metal conductor. The CIVA image (bright white lines) of the conductor network that is open circuited is superimposed on a secondary electron image. CIVA analysis may be performed through multiple layers of dielectrics and metals by increasing the electron energy, but care must be exercised to avoid irradiation damage to the IC.

CIVA may be performed nondestructively on passivated ICs using low electron beam energies. This technique is called LECIVA (low energy CIVA) (19). The LECIVA signal is produced by changing bound charges induced at the conductor-dielectric interface which induce voltage pulses on the conductor. Figure 3 shows a low magnification LECIVA image of the conductor network connected to an open level two to level three via on a 0.8 μm feature size, three-level-metal Intel 486 microprocessor. Application of LECIVA is subject to the same limitations that apply in CCVC, since lower metallization levels may be analyzed only if they are not covered by other metal layers.

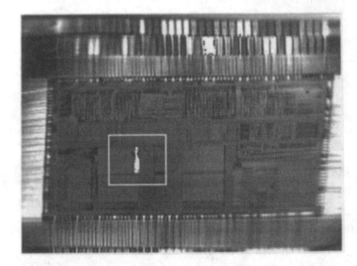

FIGURE 3. Low magnification LECIVA image of conductors connected to an open via (bright area in box) superimposed on a SEM image of an Intel 486 microprocessor.

LIVA is a recently developed technique that uses a SOM to quickly localize either defective *pn* junctions or biased junctions that are electrically connected to defects such as open circuits (18). LIVA can also identify the logic states of transistors with much greater sensitivity than other optical techniques. LIVA analysis is performed from the front of an IC die using a visible laser or from the back of the die using an infrared laser. LIVA images are produced by monitoring the changes in the constant current power supply voltage as the SOM beam scans across the IC. Voltage changes occur when the recombination current increases or decreases the operating voltage of the IC. Figure 4(a) shows a backside IR SOM image of an I/O port region of a 1.25 μm feature size, two level metal microcontroller, and Figure 4(b) shows the backside LIVA logic state image when the port is in the "1" state. The dark contrast areas indicate p-channel

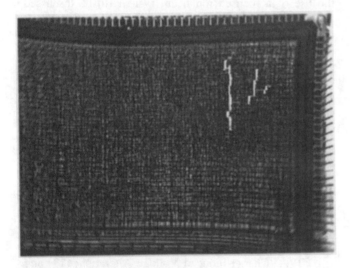

FIGURE 2. Low magnification CIVA image of an open conductor network (bright white) superimposed on a SEM image of a gate array IC.

transistors that are "off". The "fuzzyness" of the image results from the diffusion length of the optically generated electrons and holes. Note that the transistors in the image are completely covered by level two metal, so no front side imaging or logic state analysis is possible.

The capability to rapidly locate defects in complex ICs make CIVA and LIVA extremely effective for FA of technologies with high integration levels, completely planarized surfaces, dense interconnection layers, or flip chip packaging.

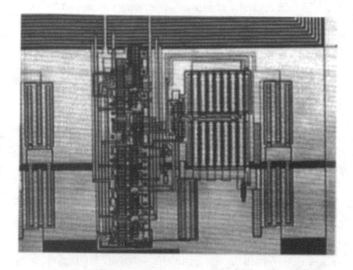

FIGURE 4(a). Backside IR SOM image of an I/O port on a microcontroller.

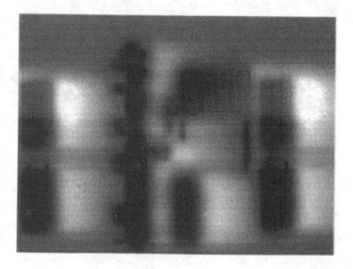

FIGURE 4(b). Backside LIVA logic state image of the I/O port shown in Figure 4(a).

Scanning Probe Microscopy

SPM (scanning probe microscopy) is a rapidly growing field that is just beginning to have an impact in the FA of ICs. Samples are imaged in an SPM by scanning a sharp probe tip in close proximity to the sample surface and detecting the interactions between the tip and the sample. Scanning probe instruments provide an entirely new capability for topographical imaging and analysis of submicron structures, extending spatial resolution well beyond the limits of optical tools.

AFM (atomic force microscopy) provides a topographical map of a sample with atomic resolution (20). Variations of the AFM technique have been used to perform precise potentiometry, measure internal currents on ICs, and map thermal gradients. The AFM may be operated in contact or non-contact modes. In the contact mode AFM, the tip is brought close enough to the surface that there is a repulsive interaction between the atoms in the tip and in the surface. Contact mode AFM has recently been used to image IC cross-sections of polished samples with nanometer resolution, better than that obtainable with optical microscopy or SEM (21).

In non-contact mode AFM, the tip is moved 10 to 100 nm away from the sample surface. At this distance, longer range interactions, such as Van der Waals, electric, or magnetic forces may be used to modulate image contrast. Some spatial resolution is lost in this mode because the tip is at a greater distance from the sample surface. CFM (charge force microscopy) can be used to perform potentiometry within ICs with submillivolt sensitivity, and has also been used to develop an AFM-based voltage contrast technique for measuring voltage waveforms up to 20 GHz (22). MFM (magnetic force microscopy) uses a magnetized tip to detect magnetic field gradients. MFM/CCI (MFM current contrast imaging) can be used to analyze internal IC currents with a sensitivity of ~ 1 mA dc and ~1 µA ac (23). The combination of CFM and MFM in a single instrument may enable simultaneous measurement of internal IC voltages and currents. These techniques will experience technological limitations similar to those discussed for voltage contrast analysis, except that MFM should be capable of detecting currents on conductors located beneath other conductors.

Figure 5(a) shows an optical micrograph of two metal interconnections on a test structure; the white box indicates the area scanned for the MFM/CCI image shown in Figure 5(b). A 10 µA ac square wave is applied to the right interconnection and its phase is shifted halfway through the top-to-bottom scan. The contrast in Figure 5(b), which indicates the magnetic force applied to the MFM tip, reverses when the phase shift occurs. The left interconnection is not electrically driven and is imaged due to Van der Waals forces.

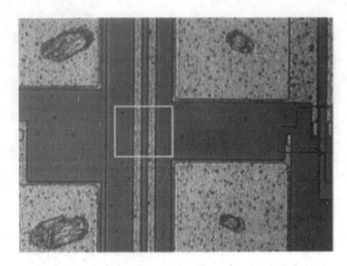

FIGURE 5(a). Optical micrograph of two interconnections on a test structure; the white box is the area scanned for MFM/CCI.

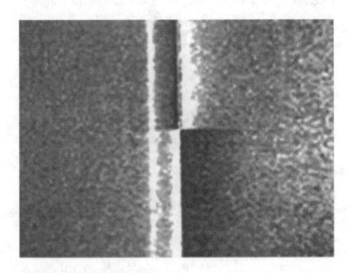

FIGURE 5(b). MFM/CCI image of the two conductors in Figure 5(a) showing contrast from a 10 µA square wave current in the right conductor.

Thermal Imaging Analysis

Thermal imaging techniques are helpful in defect localization when there are no obvious detectable symptoms such as light emission or CIVA/LIVA signals. Thermal imaging techniques rely on detecting a defect on an operating IC through a local temperature increase caused by increased power dissipation. For example, a metal to metal short may behave in this manner. These techniques are nondestructive and are generally unaffected by feature size, integration level, number of interconnection levels, and planarization. However, thermal conductivity causes a somewhat diffuse "hot spot" and the spatial resolution of the imaging system may not be the limiting factor in defect localization.

Infrared thermography provides a temperature map of the IC surface. A system consists of an IR microscope with computer aided emissivity correction. The spatial resolution is relatively poor, ranging from about 4 to 20 µm (depending on the IR detector employed) while temperature resolution is quite good (0.025 to 0.1 °C) (24).

Liquid crystal techniques are commonly used for "hot spot" detection (25). A thin film of liquid crystal material with a phase transition temperature just above the IC ambient temperature is applied on the IC surface and observed with polarized light. A phase transition is observed at the "hot spot". The spatial resolution (several µm) and temperature resolution (about 0.1 °C) of this technique are low, and there is no temperature mapping capability. After thermal analysis, the thin surface film may be easily removed.

FMI (fluorescent microthermographic imaging) uses the temperature dependent fluorescence quantum yield of a rare earth chelate to provide a direct, quantitative conversion of surface temperature into detectable photons (26). A thin film containing the rare earth chelate is applied to the IC surface. The surface temperature of the IC is determined by imaging the intensity of the visible light fluorescence from the film when it is pumped with an ultraviolet light source. The spatial resolution of this technique is diffraction limited to 0.3 µm and the temperature resolution is hardware limited to about 0.01 °C. As with liquid crystals, the thin film is easily removed after analysis.

Focused Ion Beam Techniques

The capabilities of FIB (focused ion beam) systems are rapidly becoming critical for FA of submicron technology ICs. FIB systems use a focused beam of Ga^+ ions for imaging, milling, and deposition of metals and dielectrics (27,28). No other tools provide these capabilities with such precise control. FIB manufacturers offer optional CAD navigation, which provides a convenient operator interface for complex ICs.

Ion beam imaging provides both atomic number and ion channeling contrast (29). Atomic number contrast enables identification of metal, dielectric, and passivation layers in a cross section. Ion-channeling contrast occurs because of the differences in grain orientation with respect to the ion beam and is useful in determining grain size and structure.

FIB system ion milling produces in-situ cross sections with a high degree of spatial control, providing an effective approach for analyzing the root cause of a failure. This provides a great advantage over conventional mechanical cross-sectioning of ICs because the location and depth of the cross section can be accurately specified (within about 0.2 μm) and multiple cross sections can be made on a sample. Sequential cross sections can be made through a feature of interest.

FIB systems may be used to mill through upper layers to expose underlying conductors for mechanical or electron beam probing, emission microscopy, or LIVA analysis. They also can be used to cut metallization and polysilicon lines to isolate interconnections and devices. The location of cuts can be controlled very accurately, and the cuts are cleaner than those made by mechanical or laser techniques.

FIB systems also can be used to precisely deposit conductors (tungsten or platinum), and dielectric (SiO_2) deposition is under development. This capability enables local modifications in the electrical interconnections of the IC, or the deposition of additional contact areas that can be used to probe circuit functions and measure voltages on functioning ICs (30). Design changes can often be made in less than an hour, avoiding the cost and delay of a new mask set and wafer lot. If electrical feedthroughs are provided into the FIB chamber, electrical measurements and voltage contrast imaging may be used for in-situ monitoring of the device modification. Additional FIB system capabilities being developed include in-situ SIMS for precision elemental analysis, gas assisted etching for enhanced material removal and decorative etching, and dual ion/electron beam columns for in-situ high resolution SEM imaging and energy dispersive x-ray analysis.

Failure Analysis Databases and Expert Systems

FA is a multidisciplinary activity. The effective failure analyst must understand circuit design, IC architecture, semiconductor device physics, IC processing, and IC testing. This broad set of skills takes years to acquire, perhaps through rotational assignments in different departments of a company, while constant advances in technology further complicate the process. Analyst training is usually accomplished on the job and through conference attendance, since few universities provide applicable laboratory experience. There are several short courses and seminars that provide useful general information, but the depth of coverage may be insufficient to provide immediate help back at the workplace. FA books also have broad applicability, but rapidly become dated as technology advances.

A new paradigm for training failure analysts is the use of failure analysis databases, hypertext help systems, and expert systems. These software products also add value by preserving the knowledge of experienced failure analysts, which is usually lost through retirement or job change. A large "smart database" has been described that uses all of a company's FA records to provide a prediction of the failure mechanism and a proposed FA course of action (31). Interactive FA expert systems and an associated hypertext help system have been implemented to help train inexperienced analysts as well as guide and assist experienced analysts while performing analyses (32,33). As the information superhighway evolves, new training and information assistance technologies will be available. The World Wide Web (a graphical, hypertext view of the Internet) can be used to provide up-to-date, comprehensive multimedia information and training to analysts at their work locations.

CONCLUSION

FA in today's IC industry is squeezed between the need for very rapid analysis to support manufacturing and the exploding complexity of IC technology. However, the FA community is responding to these challenges. FA laboratories are implementing a number of advanced FA techniques and tools and are addressing the information and training needs of failure analysts. Continued support for research on new FA technologies and for innovative new approaches for supporting failure analysts is a critical requirement for the future.

ACKNOWLEDGMENT

The authors thank Daniel L. Barton, Ann N. Campbell, and Edward I. Cole Jr. of Sandia National Laboratories for their contributions and review. This work was performed at Sandia National Laboratories and supported by the U. S. Department of Energy under contract DE-AC04-94AL8500.

REFERENCES

1. Argyrakis, S. N. ,"A Novel High Speed, High Channel Electrical Interface to Remotely Connect a Device Under Analysis (DUA) to a Tester," *Proc. Int. Symp. for Testing and Failure Analysis*, 1994 , pp. 57-62.

2. Soden, J. M. and Anderson, R. E., "IC Failure Analysis: Techniques and Tools for Quality and Reliability Improvement," *Proc. IEEE* **81**, 703-715 (1993).

3. Burggraaf, P., "Failure Analysis: From 'Postmortem' to 'Preventive'," *Semiconductor Int.* **15**, No. 10, 56-61 (Sep. 1992).

4. Soden, J. M., Hawkins, C. F., Gulati, R. K., and Mao, W., "I_{DDQ} Testing: A Review," *J. Electronic Testing: Theory and Applications* **3**, No. 4, 291-303 (1992).

5. Mao, W. and Gulati, R. K., "QUIETEST: a Methodology for Selecting I_{DDQ} Test Vectors," *J. Electronic Testing: Theory and Applications* **3**, No. 4, 349-357 (1992).

6. Aitken, R. C., "Diagnosis of Leakage Faults with I_{DDQ}", *J. Electronic Testing: Theory and Applications* **3**, No. 4, 367-375 (1992).

7. Soden, J. M. and Hawkins, C. F., "I_{DDQ} Testing and Defect Classes - A Tutorial," to be presented at the *Custom Integrated Circuits Conf.*, 1995.

8. Jee, A. and Ferguson, F. J., "Carafe: A Software Tool for Failure Analysis," *Proc. Int. Symp. for Testing and Failure Analysis*, 1993, pp. 143-149.

9. Ryan, P. G. and Fuchs, W. K., "Addressing the Size Problem in Fault Dictionaries," *Proc. Int. Symp. for Testing and Failure Analysis*, 1993, pp. 129-133.

10. Concina, S. and Richardson, N., "Workstation Driven e-Beam Prober," *Proc. Int. Test Conf.*, 1987, pp. 554-560.

11. Bossmann, B., Baurschmidt, P., Hussey, K., and Black, E., "Failure Analysis Techniques with the Confocal Laser Scanning Microscope," *Proc. Int. Symp. for Testing and Failure Analysis*, 1992, pp. 351-361.

12. Hawkins, C. F., Soden, J. M., Cole Jr., E. I., and Snyder, E. S., "The Use of Light Emission in Failure Analysis of CMOS ICs," *Proc. Int. Symp. for Testing and Failure Analysis*, 1990, pp. 55-67.

13. Shade, G., "Physical Mechanisms for Light Emission Microscopy," *Proc. Int. Symp. for Testing and Failure Analysis*, 1990, pp. 121-128.

14. Cole Jr., E. I., Bagnell Jr., C. R., Davies, B. G., Neacsu, A. M., Oxford, W. V., and Propst, R. H. "Advanced Scanning Electron Microscopy Methods and Applications to Integrated Circuit Failure Analysis," *Scanning Microscopy* **2**, 133-150 (1988).

15. Bottini, R., Calvi, D., Gaviraghi, S., Haardt, A., and Turrini, D., "Failure Analysis of CMOS Devices with Anomalous IDD Currents," *Proc. Int. Symp for Testing and Failure Analysis*, 1991, pp. 381-388.

16. Noble, A., "IDA: a Tool for Computer-Aided Failure Analysis," *Proc. Int. Test Conf.*, 1992, pp. 848-853.

17. Cole Jr., E. I. and Anderson, R. E., "Rapid Localization of IC Open Conductors Using Charge-Induced Voltage Alteration (CIVA)," *Proc. Int. Reliability Physics Symp.*, 1992, pp. 288-298.

18. Cole Jr, E. I., Soden, J. M., Rife, J. L. Barton, D. L., and Henderson, C. L., "Novel Failure Analysis Techniques Using Photon Probing with a Scanning Optical Microscope," *Proc. Int. Reliability Physics Symp.*, 1994, pp. 388-398.

19. Cole Jr., E. I., Soden, J. M., Dodd, B. A., and Henderson, C. L., "Low Electron Beam Energy CIVA Analysis of Passivated ICs," *Proc. Int. Symp. for Testing and Failure Analysis*, 1994, pp. 23-32.

20. Rugar, D. and Hansma, P., "Atomic Force Microscopy," *Physics Today*, 23-30 (Oct. 1990).

21. Neubauer, G., Dass, M. L. A., and Johnson, T. J., "Imaging VLSI Cross Sections by Atomic Force Microscopy," *Proc. Int. Reliability Physics Symp.*, 1992, pp. 299-303.

22. Hou, A. S., Ho, F., and Bloom, D. M., "Picosecond Electrical Sampling Using a Scanning Force Microscope," *Elect. Let.* **28**, pp. 2302-2303 (1992).

23. Campbell, A. N., Cole Jr, E. I., Dodd, B. A., and Anderson, R. E. "Internal Current Probing of Integrated Circuits Using Magnetic Force Microscopy," *Proc. Int. Reliability Physics Symp.*, 1993, pp. 168-177.

24. Burggraaf, P., "IR Imaging: Microscopy and Thermography," *Semiconductor Int.* **9**, 58-65 (July, 1986).

25. Geol, A., and Gray, A. "Liquid Crystal Technique as a Failure Analysis Tool," *Proc. Int. Reliability Physics Symp.*, 1980, pp. 115-120.

26. Barton, D. L., "Fluorescent Microthermographic Imaging," *Proc. Int. Symp. for Testing and Failure Analysis*, 1994, pp. 87-95.

27. Boylan, R., Ward, M., and Tuggle, D., "Failure Analysis of Micron Technology VLSI Using Focused Ion Beams," *Proc. Int. Symp. for Testing and Failure Analysis*, 1989, pp. 249-255.

28. Nikawa, K., "Applications of Focused Ion Beam Technique to Failure Analysis of Very Large Scale Integrations: A Review," *J. Vac. Sci. Technol.* **B9**, 2566-2577 (1991).

29. Olson, T. K., Lee, R. G., and Morgan, J. C., "Contrast Mechanisms in Focused Ion Beam Imaging," *Proc. Int. Symp. for Testing and Failure Analysis*, 1992, pp. 373-382.

30. Van Doorselaer, K., Van den Reeck, M., Van dem Bempt, L., Young, R., and Whitney, J., "How to Prepare Golden Devices Using Lesser Materials," *Proc. Int. Symp. for Testing and Failure Analysis*, 1993, pp. 405-414.

31. Bellay, L. M., Ghate, P. B., and Wagner, L. C., "Computers in Failure Analysis," *Proc. Int. Symp. for Testing and Failure Analysis*, 1990, pp. 89-95.

32. Henderson, C. L. and Soden, J. M., "ICFAX, An Integrated Circuit Failure Analysis Expert System," *Proc. Int. Reliability Physics Symp.*, 1991, pp. 142-151.

33. Henderson, C. L. and Barnard, R. D., "The Advent of Failure Analysis Software Technology," *Proc. Int. Reliability Physics Symp.*, 1994, pp. 325-333.

CMOS IC Layers: Complete Set of Thermal Conductivities

O. Paul, M. von Arx and H. Baltes

Physical Electronics Laboratory, ETH Zurich, HPT H6, CH-8093 Zurich, Tel: +41 1 633 3671, Fax: +41 1 633 10 54

The thermal conductivities κ of all thin films of a commercial CMOS process were determined in the temperature range from 130 to 400 K. The κ values of the silicon oxides are reduced from bulk fused silica by 0 to 16%. The average κ of all silicon oxide layers is 1.26±0.05 W/mK at 300 K. The passivation, a sandwich of PECVD silicon oxide and nitride, has a thermal conductivity of 1.65±0.05 W/mK at 300 K and lies between published data for LPCVD nitride and silicon oxide. The κ values of two polysilicon layers are 23.5±1.2 and 17.2±0.8 W/mK, respectively, at 300 K and are reduced from undoped bulk silicon by factors of 7 and 9 For the lower and upper metal layers we found thermal conductivities of 194±8 and 173±8 W/mK, respectively, at 300 K, to be compared with 238 W/mK of bulk aluminum. The observed discrepancies between thin film and bulk data demonstrate the importance of determining the process-dependent thermal conductivities of CMOS thin films.

INTRODUCTION

As the dimensions of IC structures shrink and dissipated power densities increase, thermal considerations have a growing importance in the development of advanced microelectronic components [1]. Optimal thermal management requires the precise knowledge of the thermal conductivities κ of their component thin films. The same conclusion applies to integrated thermal microtransducers fabricated using commercial CMOS processes [2]. The sensitivity of such microstructures is directly determined by the κ of their constitutive dielectric and conducting materials. Often thermal conductivity values are not available for thin films and values determined with bulk samples have to provide the basis for numerical simulations or optimizations. The use of bulk data is questionable, however, in view of the small thicknesses of IC layers and of the expected process dependence of their microscopic properties. The measurement of thin film thermal conductivities is therefore highly desirable.

Thermal conductivities of materials related to CMOS processes have been reported in Refs. 3 to 8. The κ of SiO_2 has been measured with a dynamical thermal method [3,4]. Mastrangelo et al. have reported the κ of silicon nitride thin films [5]. Thermal conductivities of polysilicon thin films have been reported in Refs. 6 and 7, and of selected CMOS films in Ref. 8.

This paper reports the κ values of all thin films of a commercial CMOS process, namely the 1.2 µm double-poly double-metal CMOS process of Austria Mikro Systeme (AMS). This required the characterization of four dielectric films, two degenerately doped semiconductor layers, and two metal layers. In the order of their appearance during the CMOS process, these films are:

- field oxide (thermal),
- gate polysilicon,
- capacitor polysilicon,
- contact oxide (CVD),
- first metal (metal 1),
- intermetal isolation oxide (CVD),
- second metal (metal 2),
- passivation sandwich (CVD).

EXPERIMENTAL

Test structures

In view of the expected process-dependence of the properties of the thin films, we designed special diagnostic struc-

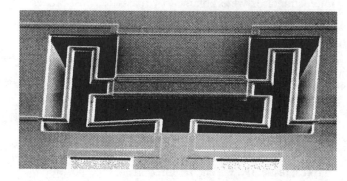

Figure 1: SEM micrograph of one of ten microstructures used to determine the thermal conductivities of CMOS IC thin films. The cantilever width is 200 µm.

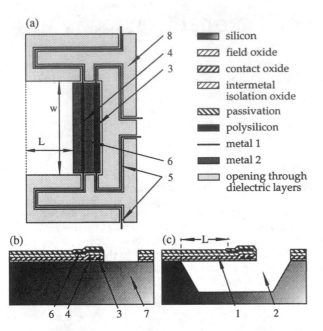

(a)

■	silicon
▨	field oxide
▨	contact oxide
▨	intermetal isolation oxide
▧	passivation
■	polysilicon
—	metal 1
■	metal 2
▨	opening through dielectric layers

Figure 2: Schematic top view (a) and cross-sections (b,c) of the microstructure to determine the thermal conductivities of CMOS materials. (b) Cross-section after CMOS process, (c) after post-processing. 1: cantilever, 2: etched cavity, 3: polysilicon heater, 4: polysilicon temperature monitor, 5: metal connections to pads, 6: metal cover for temperature homogenization, 7: silicon substrate, 8: windows in CMOS dielectrics.

tures compatible with commercial CMOS processes. The structures are similar to those used by Völklein et al. to characterize CMOS polysilicon [6]. They were co-fabricated with IC designs of other research groups, using the complete CMOS process of AMS. This guaranteed that the processing condition of each characterized layer were identical to those of an integrated circuit or a CMOS microsensor. After completion of the CMOS process the microstructures were obtained by mask-less post-processing consisting of silicon micromachining. One of the test structures is shown in Fig. 1. It consists of a 200 μm wide and 150 μm long cantilever suspended over a micromachined cavity. Four thin arms extend from the rim of the cavity to the tip of the cantilever.

The post-processing is made possible by the appropriate layout of the field, contact, via, and pad masks. Cuts through the four dielectrics down to the silicon substrate are defined by the superposition of these mask layers. The layout and resulting cross-section are shown in Figs. 2 (a) and (b), respectively. The required mask superposition has no equivalent in IC design and violates several CMOS design rules. However, these violations do not jeopardize the CMOS process, and integrated circuits operate as expected. The exposed silicon is anisotropically etched

with an aqueous solution of ethylenediamine/pyrocatechol/pyrazine at 95 °C for 4 hours. Such post-processing [9] results in the underetched, suspended structure shown in Figs. 1 and 2 (c). The structure is composed of the sandwich of the CMOS dielectrics. The geometry of the four arms was designed for reduced heat loss and effective stress relaxation.

By appropriate layout design, two resistors made of gate polysilicon were integrated into the cantilever, as shown in Fig. 2. Both are connected to contact pads with integrated lines of the metal 1 in the four arms. The resistor close to the edge of the cantilever has a resistance of 1.0 kΩ at 300 K and is used as a heater. The second is used as a temperature monitor, with a corresponding resistance of 4.2 kΩ. The temperature distribution over the two resistors is homogenized with an integrated rectangular cover made of CMOS metal.

When a power P is dissipated in the heater, the temperature of the cantilever tip is increased by ΔT with respect to the substrate. The heat power P is conducted away along the cantilever and the four arms. P is related to ΔT by $P / \Delta T = G = G_a + G_c$, where G, G_a, and G_c denote the thermal conductances of the entire structure, of the four arms, and of the cantilever, respectively. Under conditions discussed in Ref. 7, G_c is given by

$$G_c = \sum \frac{\kappa_i d_i w}{L} \tag{1}$$

where the summation runs over the component layers of the cantilevers, and κ_i, d_i, w, and L denote their respective thermal conductivitiy, thickness, width, and length, respectively (for w and L see Fig. 2). To determine κ_i of the individual thin films, we systematically varied the composition of the cantilever. Ten different structures were designed, with cross-sections shown in Fig. 3.

The first structure of Fig. 3 (a) consists of the sandwich of the four dielectric CMOS layers and is used as a reference. For the characterization of the dielectric layers, one layer is removed in each test structure in Figs. 3 (b) to (e), by appropriate layout design of the field, contact, via, and pad mask layers, respectively. The missing film reduces the thermal conductance of the cantilever sandwich. The whole sandwich is removed from the structure in Fig. 3 (f).

Adding the gate and capacitor polysilicon layer to the dielectric sandwich allows to measure the κ values of these two semiconductor layers. The resulting cross-sections are shown in Figs. 3 (g) and (h). Metal/poly contacts ensure optimal heat transfer.

The thermal conductivities of the metal layers are measured with the two structures shown in Figs. 3 (i) and (j). They contain a metal 1 or metal 2 layer, respectively, extending from the heater to the substrate. Optimal heat transfer is guaranteed by metal/substrate contacts.

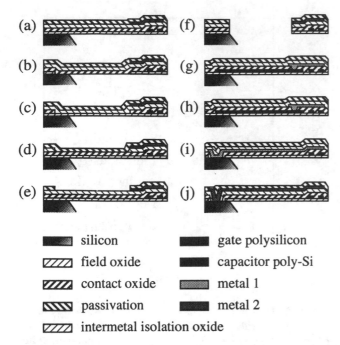

(a)

(b)

(c)

(d)

(e)

(f)

(g)

(h)

(i)

(j)

- ▬ silicon
- ▨ field oxide
- ▨ contact oxide
- ▧ passivation
- ▨ intermetal isolation oxide
- ▬ gate polysilicon
- ▬ capacitor poly-Si
- ▨ metal 1
- ▬ metal 2

Figure 3: Schematic cross-sections of microcantilevers to determine the thermal conductivities of all CMOS thin films. (a): Reference structure, (b) to (j): modified structures for field oxide (b), contact oxide (c), intermetal isolation oxide (d), passivation sandwich(e), all dielectrics (f), gate polysilicon (g), capacitor polysilicon (h), first metal (i), second metal (j).

Measurement procedure

The thermal characterization of the structures was performed with a liquid nitrogen cryostat. Conductive and convective heat losses through air were negligible under the used vacuum conditions ($p \approx 10^{-6}$ mbar). The chips with the test structures were cooled to 90 K and then ramped up to 420 K with a rate of 2 K/min.

The measurements consisted of determining the thermal conductances G_0 and G_m ($m = 1,...,9$) of the reference structure and its nine modifications. The thermal conductances were obtained by measuring the temperature increases ΔT of each structure under a dissipated heat power. The temperature elevation ΔT of each cantilever was determined using the temperature-dependent resistance $R(T)$ of its temperature monitor, via its temperature coefficient of resistance $\alpha(T)$ defined as

$$\alpha(T) = R(T)^{-1}\frac{\partial R}{\partial T}(T) .\tag{2}$$

$R(T)$ was measured at a set of equidistant temperatures T_n of the cryostat (every 10 K), with a probing current of 0.1 mA. With these data, $\alpha(T)$ was calculated as the finite difference

$$\alpha(T_n) = R(T_n)^{-1}\frac{R(T_{n+1})-R(T_{n-1})}{T_{n+1}-T_{n-1}} .\tag{3}$$

As an example $R(T)$ of the temperature monitor on the reference structure and the calculated $\alpha(T)$ are shown in Fig. 4.

0.2 K below and above each probing temperature T_n a heating current was passed through the heating resistor. The heat dissipation increased the cantilever temperature to $T_n - 0.2 + \Delta T$ and $T_n + 0.2 + \Delta T$, respectively. Simultaneously, corresponding resistance values $R_n^{(-0.2)}$ and $R_n^{(0.2)}$ of the temperature monitoring resistor were recorded. The temperature elevations ΔT of the cantilever at the probing temperatures T_n were then calculated as

$$\Delta T(T_n) = \frac{\left(R_n^{(-0.2)}+R_n^{(0.2)}\right)/2 - R(T_n)}{\alpha(T_n)R(T_n)} ,\tag{4}$$

where use was made of the Taylor expansion of R. At 300 K heating currents of the order of 0.3 mA, 0.5 mA, and 1.5 mA were required to heat the structures in Fig. 3 (a) to (f), (g) to (h), and (i) to (j), respectively, up by typically 10 K. Finally the thermal conductances G were calculated as $P/\Delta T$, using the measured heat powers. Carrying out this procedure with the ten test structures, we obtained the ten temperature-dependent thermal conductances $G(T)$.

The temperature-dependent κ of the CMOS layers were obtained by subtraction of the reference structure from the modified structures, using the fact that G_a is identical for all test structures. In the case of the field oxide and the polysilicon and metal layers, by virtue of Eq. (1), κ_m is given by

$$\frac{\kappa_m d_m w}{L} = |G_0 - G_m| .\tag{5}$$

In the case of the CVD dielectric layers, one has

$$(\kappa_m d_m + \kappa_{oe}d_{oe})\frac{w}{L} = |G_0 - G_m| ,\tag{6}$$

where κ_{oe} and d_{oe} denote the thermal conductivity of the subjacent dielectric layer and the overetch depth into it, respectively.

The thicknesses of the polysilicon and the metal layers were determined with a mechanical surface profiler. For the thicknesses of the dielectric layers, nominal values provided by the IC manufacturer were taken. The overetch depths were determined as the difference of the nominal oxide thicknesses and surface profiler results. With these data, the temperature-dependent thermal conductivities were finally obtained using Eqs. (5) and (6).

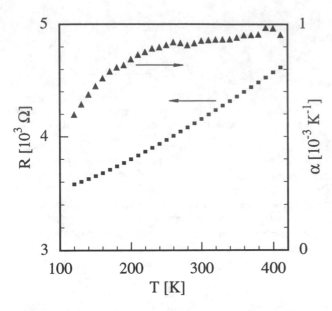

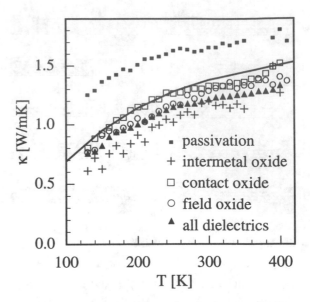

Figure 4: Resistivity R and temperature coefficient α of the temperature monitoring resistor on the reference structure.

Figure 5: Thermal conductivities of CMOS IC dielectric layers vs. temperature. Recommended values [10] for fused silica are shown as solid line. Errors in κ are between 9 and 21%.

RESULTS

Experimental κ values for the dielectric layers are shown in Fig 4. The thermally grown field oxide has a thermal conductivity of 1.28±0.11 W/mK at 300 K. The two CVD oxides, i.e. the contact and intermetal oxides, have thermal conductivities of 1.32±0.18 W/mK and 1.16±0.24 W/mK, respectively, at 300 K. The $\kappa(T)$ data are reduced by 0 to 16% from those of bulk fused silica [10], shown as solid line in Fig. 4. The passivation sandwich consists of silicon oxide and silicon nitride. Its thermal conductivity is 1.65±0.25 W/mK at 300 K and lies between published data for LPCVD nitride [5] and silicon oxide [10]. The structure in Fig. 3 (f) provided the thermal conductivity of the sandwich of all dielectric layers. We obtained a value of 1.22±0.14 W/mK.

As shown in Fig. 6, the gate and capacitor polysilicon layers have κ values of 23.5±1.2 W/mK and 17.2±0.8 W/mK, respectively, at 300 K. At this temperature, they are reduced from pure bulk silicon by factors of 7 and 9. The measured sheet resistances are 21.4 Ω/sq. (n$^+$-doped) and 213 Ω/sq. (p$^+$-doped), respectively, at 300 K.

The two aluminum-based metal layers have thermal conductivities of 194±8 W/mK and 173±8 W/mK, respectively, at 300 K. Temperature-dependent data are shown in Fig. 7. For comparison, the values recommended in Ref. 10 for pure aluminum are also shown. The $\kappa(T)$ curves of both metal layers are significantly reduced from aluminum.

A large part of the error in κ originates in the thickness measurements. Thickness values were measured with errors from 2 to 8%, depending on the surface roughness. A second error contribution comes from the determination of G with an accuracy between 1 and 4%. These errors are amplified in each κ due to the required subtracton of thermal conductances. For this contribution, errors between 5 and 21% were obtained. Calculated errors are indicated in Figs. 5 to 7.

CONCLUSIONS

We have measured the complete set of thermal conductivities of a commercial CMOS process. The measurements were performed with CMOS compatible diagnostic structures fabricated with the complete CMOS process under investigation.

The measured data deviate significantly from available bulk values. The observed discrepancies demonstrate the importance of such thermal measurements.

In view of the CMOS compatibility of the diagnostic structures, variations from wafer to wafer, batch to batch, and process to process are easily studied by systematically integrating them in parallel with the usual CMOS test structures.

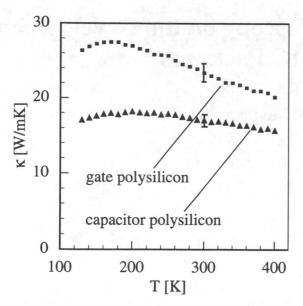

Figure 6: Thermal conductivities of gate and capacitor polysilicon vs. temperature.

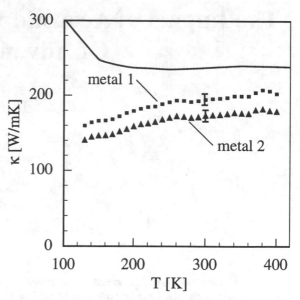

Figure 7: Thermal conductivities of both metal layers vs. temperature. Recommended values [10] for aluminum are shown as solid line.

ACKNOWLEDGEMENTS

This work has been funded by the Swiss Federal Priority Program Power Electronics, Systems, and Information Technology (LESIT) and by the Bundesamt für Bildung und Wissenschaft through grant Nr. 93.0161.

REFERENCES

[1] Sauter, A., Nix, W. D., "Finite element calculations of thermal stresses in passivated and unpassivated lines bondet to substrates", *Mat. Res. Soc. Symp. Proceedings* **188**, MRS, 1990, pp. 15-20.

[2] Baltes, H., "CMOS as sensor technology", *Sensors and Actuators A* **37-38**, 51—56 (1993).

[3] Lee, S. M., Cahill, D. G., Allen, T. H., "Heat transport in sputtered optical coatings", *41st National Symposium of the AVS Conference Abstracts* , Denver, USA, 1994, p. 228.

[4] Cahill, D. G., "Thermal conductivity measurement from 30 to 750 K: the 3ω method", *Rev. Sci. Instrum.* **61** (2), 802—808 (1990).

[5] Mastrangelo, C. H., Tai, Y. C., and Muller, R. S., "Thermophysical properties of low-residual stress, silicon-rich, LPCVD silicon nitride films", *Sensors and Actuators A* **23**, 856—860 (1990).

[6] Völklein, F., and Baltes, H., "A microstructure for measurement of thermal conductivity of polysilicon thin films", *J. Microelectromech. Syst.* **1**, 193—196 (1993).

[7] Paul, O. M., Korvink, J., and Baltes, H., "Determination of the thermal conductivity of CMOS IC polysilicon", *Sensors and Actuators A,* **41—42**, 161—164 (1994).

[8] Paul, O., and Baltes, H., "Thermal conductivity of CMOS materials for the optimization of microsensors", *J. Micromech. Microeng.* **3**, 110-112 (1993).

[9] Parameswaran, M., and Paranjape, M., "Layout design rules for microstructure fabrication using commercially available CMOS technology", *Sensors and Materials* **5** (2), 113-123 (1993).

[10]Touloukian, S., Powell, R. W., Ho, C. Y., and Klemens, P. G., *Thermophysical Properties of Matter* **1—2**, New York: Plenum, 1970.

The Impact of Acoustic Microscopy on the Development Of Advanced IC Packages

Thomas M. Moore

Texas Instruments, Inc., MS-147
13588 N. Central Expressway
Dallas, Texas 75243
moore@resbld.csc.ti.com

During the 1980's, integrated circuit manufacturers converted from conventional through-hole plastic packaging to innovative surface mount technology. Surface mount packages offered technological and economic advantages, but unexpected reliability problems were encountered during board assembly that could not be addressed with traditional x-ray radiographic inspection. During the period of time in which surface mount packaging was being implemented, high frequency scanning acoustic microscopes became commercially available. It was demonstrated that acoustic inspection nondestructively revealed such critical package defects as package cracks and delaminations. These microscopes provided a platform for the development of a sophisticated hybrid acoustic microscope dedicated to integrated circuit package inspection. A successful synergistic relationship has developed between the implementation of surface mount packaging and the development of the integrated circuit package acoustic microscope that has accelerated the resolution of the reliability issues encountered during surface mount board assembly. Acoustic inspection has revealed the critical importance of internal delamination in package reliability and has rapidly been incorporated into the process flow for environmental testing, process optimization and new package development.

INTRODUCTION

By 1985, the major integrated circuit (IC) manufacturers were packaging large die, high pin count products in surface mount technology (SMT) packages. This type of plastic packaging is attractive for several reasons. The practical limits of the conventional dual in-line package (DIP) had been exceeded by the requirements of the new generation of VLSI devices. DIPs are limited to a pin count of only 64, while SMTs can accommodate large die with much higher pin counts. Having leads on all four sides of the package results in a significant reduction in space for SMTs as compared to their through-hole DIP predecessors.

SMT packaging provides advantages to board designers by eliminating the need for through-hole mounting of the leads. This allows for a reduction in lead pitch down to 635μm (.025") and enables higher component density and improved electrical performance. The component density advantage of SMT is made even greater by the fact that SMTs can also be mounted on both sides of the board. The

increases in board functionality and density offered by SMT can produce a 5X reduction in board cost. For these reasons, SMT is projected to become the dominant plastic packaging and assembly technology by the year 2000 (1,2).

Although the SMT package is made of the same materials and uses the same package assembly process as the DIP, the SMT package posed a unique set of reliability issues not encountered in DIP packaging. These issues appeared during board assembly but had not appeared during component level reliability testing. Although compelling economic and technical advantages of SMT packaging were clear, they would become meaningless if board assembly could not be completed reliably. Package cracking during the board soldering operation was the first major failure mode identified in plastic SMTs. X-ray radiography was The conventional means of nondestructive inspection of IC packages, but it proved inadequate because it was insensitive to package cracks. Failure analysis engineers turned to mechanical cross sectioning to find out what was happening inside SMT packages during board assembly. However, cross sectioning is not preferred for routine inspection for several reasons: it is destructive, it provides a limited view of the sample and it may introduce new damage that confuses the results. Failure analysis engineers, thus sought a characterization technique that would nondestructively detect plastic package cracks. The ideal solution would be an inexpensive apparatus, both simple to use and reproducible, that could be widely distributed to all packaging locations within a company.

Scanning acoustic microscopes (SAMs) became commercially available at the time that the IC industry was converting to SMT packaging (mid 1980's). These microscopes were based on the Stanford SAM design and used high frequency sound (1GHz) to provide a spatial resolution that is comparable to that of optical microscopes (3). Acoustic image contrast is based on changes in acoustic impedance, a feature that is advantageous in many applications. The SAM could be modified to operate at the lower frequencies (10-100MHz) needed to penetrate plastic packaging. Experiments with SAMs also demonstrated that defects such as package cracks could be imaged nondestructively. Digital data acquisition systems and software algorithms dedicated to IC package inspection were developed which provided sophisticated data analysis and excellent reproducibility between microscopes. Phase analysis of the reflected acoustic pulse, a technology underutilized in the past, became of primary importance in recognizing and labeling internal delaminations. The result of such innovations was an inexpensive hybrid acoustic microscope, referred to as a C-mode acoustic microscope (C-AM). It combines advanced data analysis with the depth of penetration of traditional nondestructive acoustic inspection (C-scan) (4-6).

Moisture Sensitivity

The advent of SMT assembly introduced a reliability issue that was relatively unimportant in DIP assembly: the SMT board assembly operation exposes the entire body of the package to solder reflow temperatures (up to 80 seconds at 215°C to 260°C). By comparison, only the leads of a DIP are exposed to the high temperature of the solder assembly operation while the body of the DIP is shielded from this temperature by the board (Figure 1). The mold compound used in plastic packaging absorbs moisture from the air during shipping and storage. The temperature ramp-up during the SMT soldering operation is very rapid, often heating the package body from room temperature to 215°C or higher in less than one minute. The rapid expansion of absorbed moisture at internal interfaces during the SMT assembly operation can produce internal delamination and package cracking. The tendency of a plastic packaged IC to suffer damage during board assembly due to absorbed

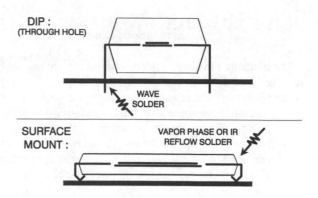

Figure 1. Comparison of DIP and SMT solder assembly operations.

moisture and heat is referred to as its "moisture sensitivity" (7).

The problem of package cracking was identified early in the development of SMT applications and became known as the "popcorn effect" because of the characteristic popping sound the packages made in the vapor phase reflow soldering chamber as they cracked. It was suspected that these cracks could increase the risk of contamination-related failure. Figure 2 shows a model that was developed to explain the back-side package cracking mechanism in thick (3.6mm) surface mount packages (8). Absorbed moisture at the mold compound/die pad interface expands to produce a pressure dome that is visible at the package surface. The mechanical strength of the mold compound is exceeded at the edge of the die pad resulting in internal package cracks. Cracks that intersect the surface will vent the trapped moisture to the ambient and the dome then collapses. The moisture sensitivity of a package, as well as the expected location of package cracks, is strongly dependant on the package type and size, internal geometry and the materials used.

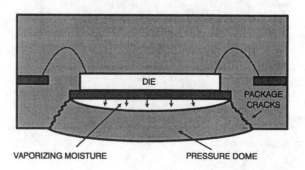

Figure 2. Model for popcorn crack formation in typical large-die SMT package (7).

Temperature Cycling

The moisture sensitivity phenomenon is limited primarily to the soldering operation during board assembly. Assuming the package is not reworked, the package will not experience temperatures high enough to produce moisture-related mechanical damage after assembly. However, the initial damage due to absorbed moisture that occurs during solder reflow can have a dramatic effect on the long-term performance of the package during subsequent temperature cycling. In particular, delamination at the die surface can lead to wire bond degradation, thin film cracking at the die surface and metal conductor displacement (smear) during temperature cycling (9-12).

A typical moisture sensitivity and temperature cycling evaluation is diagramed in Figure 3. This accelerated test regimen is designed to simulate assembly of the product and its application in the field. Several solder reflow cycles are used to simulate assembly and rework of the package. The package experiences the minimum internal stresses at the mold compound cure temperature (typically 150°C to 175°C). The high temperature of the assembly operation is responsible for the initial moisture-related damage. However, the highest interfacial shear stresses are encountered during the low temperature extremes of temperature cycle testing. Because of its relatively higher coefficient of thermal expansion (CTE), the rapidly shrinking mold compound places shear and compressive stresses on the die as the temperature drops (for temperatures below the mold cure temperature).

The packaging of large die in SMTs has created packages with very large internal interfaces. This development has highlighted a fundamental weakness of the molded plastic

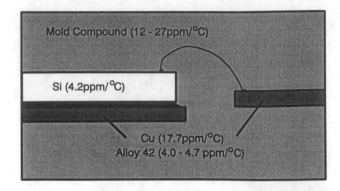

Figure 4. CTEs for the different materials involved in the molded plastic package (7).

package: materials having very different CTEs are adhered together in a system that is expected to experience large temperature variations during assembly and application (Figure 4). The larger mold compound/die interfaces in SMTs develop dramatically higher stresses than those found in conventional DIPs during temperature cycling.

ACOUSTIC INSPECTION OF IC PACKAGES

Surface-breaking cracks in plastic packages are usually difficult to detect optically because of the minimal surface relief they produce. Furthermore, package cracks and delaminations are, for all practical purposes, invisible during x-ray inspection because the extra layer of air in the crack or delamination produces an insignificant difference in x-ray absorption (13). (However, the high spatial resolution and convenience afforded by real-time x-ray radiography make it ideal for other package inspection applications such as wire damage and wire sweep inspection.)

C-AM is an excellent tool for identifying package cracks and delaminations. Unlike x-ray radiation, sound is a matter wave and relies on the transmission of vibrations from one molecule to the next. Package cracks are good reflectors of acoustic waves and are easily detected during nondestructive C-AM inspection. Voids in the mold compound reflect sound and can be detected as well. Acoustic inspection can discriminate between bonded and delaminated interfaces through phase analysis of the acoustic echo pulse. C-AM inspection also gives the packaging engineer depth information, a capability useful in identifying the interface where a defect is located, and in visualizing the shape of a package crack.

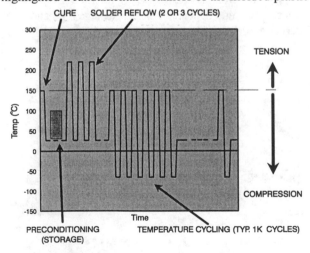

Figure 3. Typical moisture sensitivity and temperature cycling evaluation diagram (7).

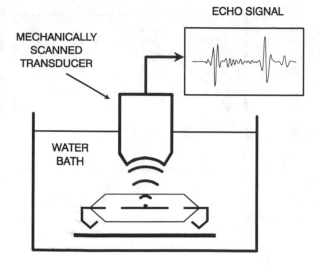

Figure 5. The pulse-echo technique used in C-mode acoustic microscopy.

The C-mode Acoustic Microscope

The C-AM uses a single piezoelectric transducer that is mechanically scanned in a plane over the surface of the package (Figure 5). The transducer focusses acoustic pulses to a tiny spot at the depth of the lead frame or die surface within the package. The same transducer then receives the echo signals back from the package. The echo signal is analyzed, and characteristics of the signal, such as amplitude, phase and interface depth, are used to form images of internal structures and defects. A water bath is used to couple the sound between the transducer and the package.

Typical acoustic echo signals from a plastic packaged IC

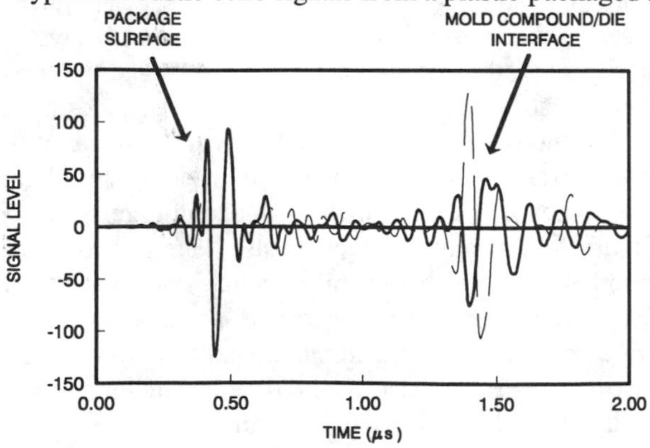

Figure 6. Typical acoustic echo signals from a 68PLCC package. Reflections from a bonded area (solid trace) and a delaminated area (dotted trace) at the mold compound/die interface are shown.

are shown in Figure 6. Echo signals from both a bonded and a delaminated portion of the mold compound/die interface are shown. Several images can be formed from the information contained in these signals. The die and lead frame can be imaged using the intensity of the sub-surface echoes. A time-of-flight image showing the depth of various interfaces can be formed by recording the time delay between the package surface echo and the sub-surface echo at each pixel. An image of voids in the mold compound can be made by detecting echoes that appear between the package surface echo and the lead frame or die echo. Delaminations can be detected by analyzing the phase of the sub-surface echo. As can be seen in Figure 6, the echo from the delaminated portion of the interface has inverse polarity relative to the echo from the bonded portion. The time-of-flight image and delamination detection will be discussed in more detail later in this article.

The center frequency of the broad-band acoustic pulse in C-AM is in the range of 15-100MHz. This range of frequencies is a balance between small spot size and depth of penetration. Most plastic mold compounds, loaded with irregularly shaped filler particles, are strong absorbers of acoustic energy. The acoustic absorption increases dramatically with increasing frequency. The spatial resolution may vary from 50-400µm depending on the center frequency of the acoustic pulses, the type of mold compound and the thickness of the package. Since it is a reflection technique, C-AM typically detects defects that are smaller than the spot size of the acoustic probe.

Delamination Detection

An important advantage of acoustic inspection of IC packages is the ability to discriminate between bonded and delaminated sub-surface interfaces. These interfaces can be distinguished by determining the phase of the returning echo pulse. The explanation of phase inversion in acoustic pulse-echo testing ca be seen by considering the ideal acoustic interface in Figure 7. The amount of incident acoustic energy that is reflected back to the transducer at an internal interface is referred to as the reflectivity of the interface. Based on a simplified plane wave model with normal incidence, the ideal reflectivity is a function of the acoustic impedances (Z_i) of the adjoining materials (14). The acoustic impedance is defined as the ratio of the acoustic pressure to the particle velocity per unit area and can be approximated by the product of the density (ρ_i) and the speed of sound (v_i) in layer i.

The reflected pressure amplitude (P_R) is described in Equation 1 in terms of the incident pressure amplitude (P_I) and the acoustic impedances (Z_i) of the two materials. Layer 1 is on the incident side and layer 2 is on the

205

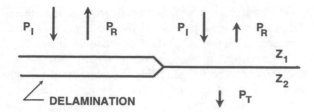

Figure 7. Ideal acoustic interface based on a simplified plane wave model.

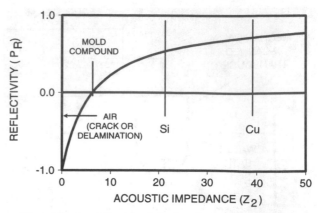

Figure 8. Acoustic reflectivity based on the ideal interface model in Figure 6 (Based on Equation 1 assuming mold compound as the first medium).

transmitted side of the planar interface.

$$P_R/P_I = (Z_2 - Z_1)/(Z_2 + Z_1), \quad \text{where } Z_i = \rho_i v_i \quad (1)$$

For convenience we will assume that $P_I = 1$, and, therefore, P_R indicates the fraction of the incident amplitude that is reflected back to the transducer. Figure 8 is a plot of reflectivity versus the acoustic impedance of the material on the transmitted side of the interface (Z_2). The acoustic impedance of the material on the incident side of the interface (Z_1) is assumed to be that of plastic mold compound. Figure 8 shows that at bonded interfaces between the plastic mold compound and the die or Cu lead frame, the transition is from lower to higher acoustic impedance. P_R is positive at these interfaces and there is no phase inversion.

A delamination or a package crack is assumed to be an interface between mold compound and air, which has a very low acoustic impedance. In this case, $P_R = -1$, 100% of the pulse amplitude is reflected, and the phase of the reflected pulse is inverted relative to the incident pulse. Because of the high reflectivity at a delamination or crack, and because air will not transmit sound well at typical C-AM frequencies, the thickness of the air gap at typical package cracks or delaminations cannot be measured.

The ideal plane wave model presented above is useful for describing the basics of delamination detection. Under the non-ideal conditions of practical inspection, other factors can affect the apparent phase of the echo pulse. These factors include apparent phase shifts due to multi-layer interference effects, frequency-dependent attenuation and spatial resolution limitations. Studies are underway to determine if intermediate phase shifts may provide additional information on the integrity of an interface (15).

Time-of-flight Images

Because the speed of sound in materials is relatively low (<10,000m/sec), the time delay between returning echoes can be measured easily and images with three-dimensional information can be displayed. This unique advantage of a reflection acoustic technique is often used to identify

failure mechanism. The three-dimensional information is typically displayed as either a surface plot or as a cross section through the sample. Examples of these images are shown in the following section.

EXAMPLES OF IC PACKAGE INSPECTION

Figure 9 shows several views of a 132 pin plastic quad flat pack (132PQFP) package that was damaged during a simulated board soldering operation. Figure 9a shows an optical image of the package surface. The surface relief produced by the intersection of the package crack with the package surface is visible under low angle illumination. In the micro-focus x-ray image of the same package (Figure 9b) the crack cannot be seen. However, in the acoustic images in Figures 9c and 9d, the crack is clearly visible. Figure 9c is an image of the intensities of the sub-surface echoes. Those areas that produced echoes with inverted polarity are labeled in black. The package "vented" through cracks along the lead frame on three sides of the die. On the fourth side, the crack leaves the plane of the die and extends to the surface. The topography of this crack is seen in the time-of-flight image in Figure 9d.

Figure 9c indicates that there is also delamination at the corners of the die surface. This is an important observation because reliability studies involving acoustic inspection have demonstrated that delamination at the die surface is the primary cause for electrical failure during temperature cycling (9,10,12). This delamination is often created during the board soldering operation and then propagates across the die surface during subsequent temperature cycling. The delamination typically initiates at the die corners where the shear stress produced by the CTE mismatch is the greatest (11,16). Once the area surrounding a ball bond has delaminated, the shear

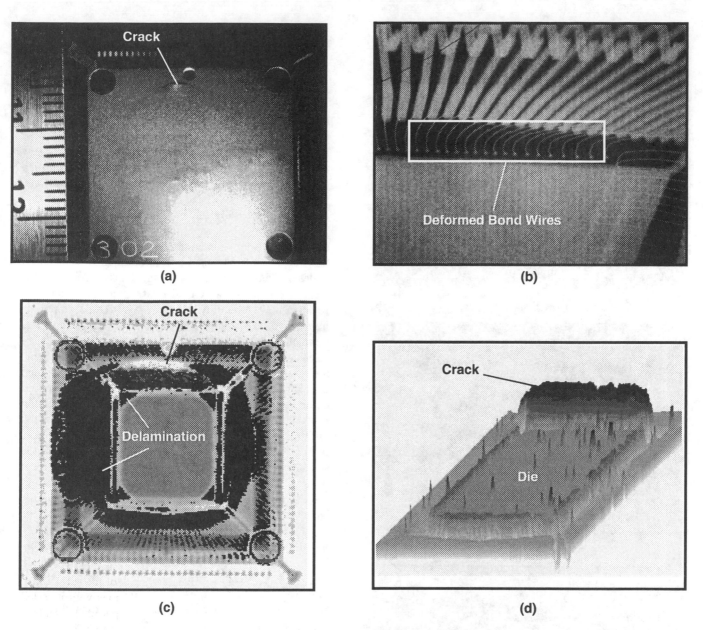

Figure 9. Top side acoustic inspection of a 132PQFP package cracked during simulated board assembly: a. optical, b. x-ray, c. acoustic image showing delamination, d. acoustic time-of-flight surface plot (7).

displacement produced by temperature cycling can separate the ball bond from the bond pad. Simple electrical testing is often unsuccessful at detecting wire bond degradation because the compressive stresses produced by the package at room temperature tend to rejoin the severed wire ball and bond pad connection. For this reason, acoustic detection of die surface delamination is considered a better indicator of wire bond degradation than simple electrical testing, even though acoustic inspection typically cannot resolve the wire bonds themselves.

Figure 10a shows the acoustic intensity image of the

bottom side of another SMT device, a 68 pin plastic leaded chip carrier (68PLCC). This package has developed popcorn cracks from the board soldering operation. Reflections from these cracks form the circular feature that surrounds the die pad. The horizontal line in Figure 10a indicates the position of the acoustic time-of-flight image shown in Figure 10c. This "B-scan" image shows the echo intensity versus depth along the indicated line and is similar to a mechanical cross section through the device. The B-scan image positively identifies the cracks and shows their trajectories. The B-scan image was

207

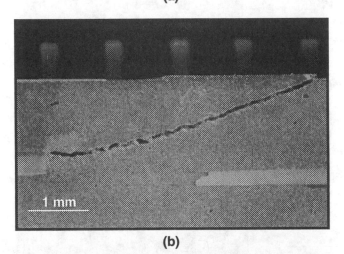

(a)

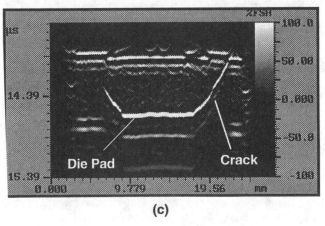

(b)

(c)

Figure 10. Bottom side acoustic inspection of a 68PLCC package that has "popcorn" cracks after simulated board assembly: a. intensity image, b. mechanical cross section, c. B-scan time-of-flight image (7).

subsequently confirmed by mechanical cross sectioning.

SUMMARY

The conversion in the IC plastic packaging industry from through-hole DIP packages to the more economical and technically superior surface mount packages presented package designers and failure analysis engineers with new reliability issues. The new SMT packages are often characterized by large die. The reliability of the package was now directly related to the integrity of internal interfaces, especially the mold compound/die interface. The possibility of damage produced by the expansion of absorbed moisture at internal interfaces during the rigorous SMT board assembly operation further increased the need to detect delamination and package cracks. Conventional inspection methods were inadequate in providing answers for the new reliability issues. X-ray radiography is insensitive to package cracks and delaminations, and mechanical cross sectioning is difficult and the results are often uncertain. The appearance of commercially available high frequency Scanning Acoustic Microscopes provided the platform for the development of an acoustic microscope dedicated to IC package inspection. Acoustic inspection proved to be sensitive to package cracks, enabled discrimination between bonded and delaminated interfaces through phase analysis of the echo pulse, and provided sufficient spatial resolution. As a result, the development of the IC package acoustic microscope has been closely linked to the development of SMT packaging. Nondestructive C-AM inspection has demonstrated that the most critical moisture-related reliability issue during SMT board assembly is delamination at the mold compound/die interface rather than package cracking, as originally suspected. C-AM inspection has been rapidly incorporated into manufacturing process flows, reliability testing and new package development efforts in IC manufacturing centers around the world.

IC manufacturers are working to improve the rsistance of SMT packages to damage during board assembly. The moisture sensitivity issue with many SMT processes has led many IC manufacturers to store and ship components in dry bags. Time limits have been established regulating the duration of exposure of these components to the atmosphere before the board assembly operation. C-AM inspection has become a critical tool in the understanding of the development and propagation of damage during the board solder operation and temperature cycling. C-AM has also become a key component in the standard proceedures for characterizing and handling of moisture sensitive packages (17).

208

REFERENCES

1. *Semiconductor Group Package Outlines Reference Guide* (Publication SSYU001A), Dallas, TX: Texas Instruments, Inc., 1995, ch. 1, 1.17-1.25.

2. *Packaging Databook*, Mt. Prospect, Il: Intel, 1995, ch. 7, 7.1-7.5.

3. C.F. Quate, Physics Today, Aug. 1985, 34-42 (1985).

4. T.M. Moore, "Identification of Package Defects in Plastic-Packaged Surface Mount IC's by Scanning Acoustic Microscopy", *Proceedings International Symposium for Testing and Failure Analysis*, 1989, 61-67.

5. T.M. Moore, R. McKenna, S.J. Kelsall, "The Application of Scanning Acoustic Microscopy to Control Moisture/Thermal-Induced Package Defects", *Proceedings Int. Symp. Testing and Failure Analysis*, 1990, 251-258.

6. T.M. Moore, R. McKenna, S. Kelsall, J. Surface Mount Technology, (3) 4, 31-38, 1990.

7. T.M. Moore, S.J. Kelsall, R.G. McKenna, *Characterization of Electronic Packaging Materials*, T.M. Moore and R.G. McKenna (eds.) New York: Butterworth/Manning, 1992, ch. 4, 79-96.

8. *Packaging Databook*, Mt. Prospect, Il: Intel, 1995, ch. 8, 8.1-8.7.

9. T.M. Moore, R. McKenna, S.J. Kelsall, "Correlation of Surface Mount Plastic Package Reliability Testing to Nondestructive Inspection by Scanning Acoustic Microscopy", *Proceedings Int. Reliability Physics Symposium*, 1991, 160-166.

10. K. Van Doorselaer and K. De Zeeuw, *IEEE Trans. Comp. Hybrids Man. Tech.*, CHMT-13, **4,** 879-882, (1990).

11. T.M. Moore, S.J. Kelsall and R.G. McKenna, *J. Surface Mount Tech.*,. 973-980 (1991).

12. T.M. Moore, S.J. Kelsall, "The Impact of Delamination on Stress-Induced and Contamination-Related Failure in Surface Mount ICs", *Proceedings Int. Reliability Physics Symposium*, 169-176 (1992).

13. A. van der Wijk, K. van Doorselaer, "Nondestructive Failure Analysis of ICs Using Scanning Acoustic Tomography (SCAT) and High Resolution X-ray Microscopy (HRXM)", *Proceedings Int. Symp. for Testing and Failure Anal.*, 69-74 (1989).

14. L.A. Kinsler, *Fundamentals of Acoustics*, New York: John Wiley & Sons, 1982, ch. 6, 121-140.

15. T.M. Moore, "Reliable Delamination Detection by Polarity Analysis of Reflected Acoustic Pulses", *Proceedings Int. Symp. Testing and Failure Analysis*, 49-54 (1991).

16. D. Edwards, et al., *IEEE Trans. Comp. Hybrids Man. Tech.*, CHMT-12, **4,** 618-627, (1987).

17. *Procedures for Characterizing and Handling of Moisture/Reflow Sensitive ICs* (IPC-SM-786A), Lincolnwood, Il: The Institute for Interconnecting and Packaging Electronic Circuits, 1995.

CRITICAL ANALYTICAL METHODS

ELECTRICAL METHODS

Electrical Characterization of Materials and Devices

Dieter K. Schroder

Department of Electrical Engineering
Arizona State University
Tempe AZ 85287-5706

Semiconductor measurements can be classified into electrical, optical, electron beam, ion beam, X-ray, and probe techniques. All of them are covered in this Workshop. I address electrical measurements here. In particular I will concentrate on recent advances in traditional techniques and point out limitations that have frequently been overlooked in the past, but will also point out some recent techniques that promise to add to the repertoire of electrical measurement techniques. The limitations become important as junction depths and oxide thicknesses decrease, for example. I also discuss the effect of metallic contamination on future devices and the characterization of such defects when defect densities in the 10^9 cm^{-2} and 10^9 cm^{-3} are demanded by the end of the century. In spite of increasing importance of methods that can measure very small dimensions directly, *e.g.*, TEM and probe methods, electrical characterization techniques will continue to remain important since semiconductor devices ultimately have to perform electrical functions.

INTRODUCTION

The range of measurements for a typical semiconductor device is illustrated by the generic MOSFET in Fig. 1. On it are indicated those parameters most important for device operation that are commonly measured electrically. These parameters span the range from device dimensions, *e.g.*,

effective channel length, to density of metallic contamination. Junction depths and junction doping profiles are increasingly difficult to measure as junction depths approach 50 nm by the end of the century according to projections by the SIA Roadmap as illustrated in Fig. 2 (1).

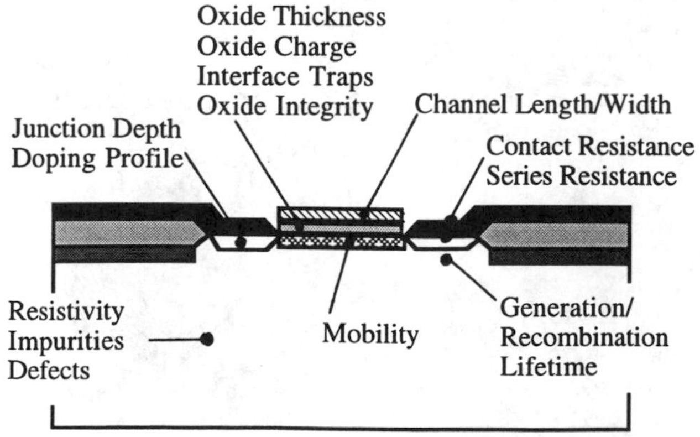

Figure 1. A MOSFET and key parameters that need to be determined.

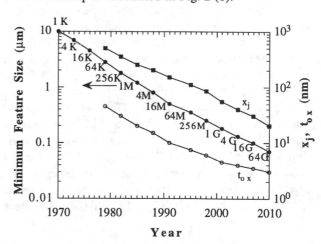

Figure 2. Minimum feature size, junction depth, x_j, and gate oxide thickness, t_{ox}, predictions according to the SIA National Technology Roadmap for Semiconductors.

Contamination issues are increasingly important as device dimensions decrease and chip dimensions increase. As supply voltages are reduced, consistent with lower critical dimensions and reduced power consumption, threshold voltage variations must be more tightly controlled, calling for low oxide charge densities. Metallic contaminants are of concern chiefly because they degrade oxide integrity. With reduced oxide thicknesses, the levels of detrimental contaminants enter the regime where only electrical techniques are sufficiently sensitive to detect them.

Many of the material and device parameters called for in the SIA Roadmap for the year 2001 that can be, already have been characterized electrically. For example, a critical parameter is the 50 nm junction depth. Such shallow junctions have already been profiled by SIMS, spreading resistance profiling, and surface resistance analyzer. In fact, even shallower junctions have been characterized. However, the different techniques frequently do not agree with one another (*e.g.*, see Fig. 6). What then is the true junction depth? Similar statements can be made about other parameters. Then there are examples where we do not have reliable experimental data at all, such as *lateral* junction depth and doping profiles. Yet it is the vertical *and* the lateral profiles that are critical for accurate modeling.

Measurement equipment and sample configuration (*e.g.*, bulk versus surface recombination during lifetime measurements) are sometimes forced to operate at the very limits and accuracy or repeatability may be in question. So, while the goals for electrical measurements for the end of the century have, for the most part, been met, it has frequently been done by highly trained specialists and not through routine measurements. In some cases we need to expand the measurement capability of the equipment, in others it must become more routine or easier to use. Development of contactless techniques is highly desirable.

As device dimensions shrink, submicron measurements become more critical. This points to increasing use of TEM and probe microscopies, all of which are discussed during this workshop by others. TEM becomes especially powerful when combined with focused ion beam cross sectioning. Recent reviews of material and device characterization are given by Shaffner (2) and Diebold (3). Ultimately, however, a semiconductor device performs an electrical function and its ultimate performance is determined by electrical measurements such as I-V, I-t, V-t, C-V, and others.

RESISTIVITY

Wafer resistivity, traditionally measured by the *four-point probe* technique, is also measured by non-contacting *Eddy current* methods (4) and other techniques to be discussed. Among the latter is the recently developed applied current tomography method. In the Eddy current technique the wafer constitutes a lossy element when brought near a tuned circuit and its sheet resistance is determined. An independent thickness measurement yields the resistivity. Besides resistivity measurements, this method is frequently used to determine the thickness of metallic layers on a semiconductor wafer. Since the sheet resistance of a metal deposited on a semiconductor is determined chiefly by the metal with its sheet resistance much less than that of the semiconductor, the metal thickness t_m is determined from the relation

$$t_m = \rho/\rho_s \qquad (1)$$

where ρ_s is the sheet resistance and ρ the metal resistivity. By measuring ρ independently, one can determine t_m.

Applied Current Tomography

Recently, *applied current tomography* has been proposed and implemented to determine wafer resistivity (5). Contact to the wafer is made at the wafer periphery with a fixed number of contacts – typically 16 or 32. A constant current, injected at one pair of contacts, generates a potential distribution pattern that is dependent on the sample geometry and resistivity. Data acquisition time is reduced since there are no moving probes. With the probes confined to the wafer periphery, the usable area of the wafer is neither contacted nor contaminated.

The current source is then moved to another pair of electrodes, the potentials are determined and the measurement is repeated until all contacts have been used. The resistivity distribution is obtained through data transformation. There are a number of different ways of doing this. In one particular mode, the potential difference between two

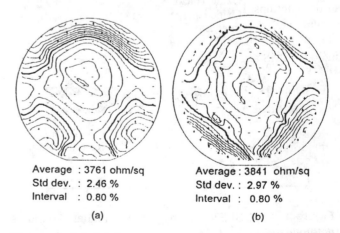

Average : 3761 ohm/sq	Average : 3841 ohm/sq
Std dev. : 2.46 %	Std dev. : 2.97 %
Interval : 0.80 %	Interval : 0.80 %
(a)	(b)

Figure 3. (a) Applied current tomography and (b) four-point probe sheet resistance maps of an n-type Si epitaxial layer with $\rho=10$ Ω-cm.

equipotential lines on the surface is back projected to the resistivity between these two equipotential lines (6). An example of an applied current tomography generated plot is shown in Fig. 3, where the sheet resistance map of an n-type Si epitaxial layer is compared with a four-point probe map.

JUNCTION PROFILING

With continued scaling of Si devices to smaller critical dimensions, junctions become shallower to reduce punchthrough current and gate oxides become thinner to increase drain current and reduce threshold voltage variations. Knowledge of both vertical and lateral junction depth and profile is important for device modeling. For example, recent work shows the leakage drain current at $V_G=0$ to increase from 10^{-11} A to 2×10^{-8} A, when the junction depth of a 0.25 μm channel length MOSFET increases from 0.26 μm to 0.33 μm (7).

Measurement techniques for junction and threshold voltage control implant profiling include *spreading resistance profiling* (SRP) (8), *differential Hall effect* measurements (9), *C-V* (10), *threshold voltage-substrate bias* (11), *microwave surface resistance analyzer* (SRA) (12), and *SIMS* (13). The sheet resistance can be determined by *modulated photoreflectance* (14), *optical dosimetry* (monitoring optical changes of implants into resist-coated glass wafers) (15), *four-point probe* [4], *SRA,* and *applied current tomography.*

Equally important is a knowledge of the lateral doping profile. None of the conventional vertical profiling methods lend themselves easily to lateral profile measurements. Profiling methods that have been most successful thus far are based on the etch rate dependence of semiconductors with varying doping densities as occurs in the lateral region of pn junctions. Etching the lateral junction portion causes varying etch depths which in turn are measured, for example, by TEM (16), STM (17), or AFM (18). Other methods include SIMS and scanning probe methods (19).

SRP, SIMS, C-V

SRP and SIMS are the most common methods for vertical junction profiling. Spreading resistance, through very shallow angle lapping and good surface preparation, is able to profile junctions shallower than 100 nm. Forward modeling has extended the SRP technique to higher accuracy for shallow junctions (20). In this approach, Poisson's equation and a multilayer algorithm are used to generate a calculated SRP profile that is compared to the measured profile. The input doping components are adjusted

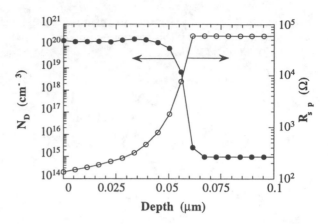

Figure 4. High resolution doping and spreading resistance profiles. Data courtesy of S. Weinzierl, Solid State Measurements, Inc.

until an acceptable match between calculation and measurement is achieved.

A recent high resolution SRP plot in Fig. 4 shows both raw SRP data and the calculated doping profile. This junction, fabricated by the gas immersion laser doping (GILD) technique, produces very shallow and very abrupt profiles that are well suited for characterization by SRP. In this case no forward modeling was necessary to extract the doping profile from the SRP data.

SIMS is routinely used for dopant profiling. Of all the dopant types, it is most sensitive for B in Si and has been shown to be sensitive to densities as low as 5×10^{12} cm^{-3} (21). As most of the sputtered atoms are neutral, SIMS becomes even more sensitive when combined with post-ionization of sputtered neutral atoms (22).

Conventional C-V profiling in which the capacitance is measured as a function of applied bias voltage, is not suitable for profiling heavily doped regions, such as sources and drains, but is very useful for determining the profile of threshold voltage control implants. The same comments apply to the threshold voltage (V_T)-substrate bias (V_B) method in which V_T is measured as a function of V_B. The doping profile is extracted from these data. In the electrochemical C-V technique, the capacitance is measured at a constant voltage with depth information obtained by etching the sample (23).

Surface Resistance Analyzer

A well-known microwave technique has recently been applied to semiconductor doping profiling. In this contactless method, known as the *surface resistance analyzer*, a resonant cavity is formed between a sample and a spherical

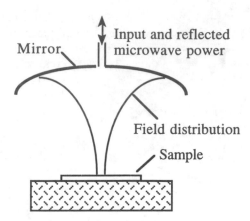

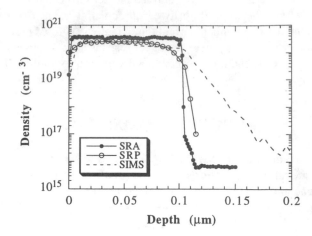

Figure 5. SRA confocal resonator. For f_o=94 GHz, the mirror radius is 6.1 cm and the coupling hole radius is 0.4 mm.

Figure 6. Doping profiles obtained by SRP, SRA, and SIMS. Data courtesy of E. Ishida, Sematech.

mirror, as illustrated in Fig. 5 (12). With the structure in resonance, the standing electromagnetic wave is strongly affected by the dielectric and conductive properties of the sample. In particular, the resonant frequency f_o and the quality factor Q, depend on the geometry of the cavity and the resistivity and the dielectric properties of the sample. Resistive samples reduce Q while dielectric samples change f_o. The lateral resolution is a few mm for f_o=94 GHz.

SRA measures surface resistance, defined by (24)

$$ R_s = \frac{\pi f_o \mu_o r}{2Q} - R_{sm} \qquad (2) $$

where μ_o is the free space permeability ($4\pi \times 10^{-7}$ H/m), r the mirror radius of curvature, and R_{sm} the mirror surface resistance. From R_s the resistivity is given by

$$ \rho = \frac{R_s^2}{\pi f_o \mu_o} \qquad (3) $$

For metals the dominant change is largely in Q while for semiconductors both Q and f_o are changed. Frequency measurements made at microwave frequencies of typically 94 GHz have a precision of 1 ppm resulting in resistivity and epi-layer thickness repeatability of 1% or better. The skin depth varies with frequency. Doping profiles are determined by using the frequency dependence of the conductivity.

A Comparison of Techniques

The results of a comparison of junction doping profiling with SRP, SRA, and SIMS are shown in Fig. 6. The

n$^+$ phosphorus-doped region was formed by gas immersion laser doping (25), in which, a spatially homogeneous pulsed laser drives adsorbed gas phase dopants into the Si substrate during a laser-induced melt regrowth step.

Since the regrowth occurs epitaxially, dopants are incorporated substitutionally. Furthermore, the total heating cycle lasts only about 1 μs, making the doping profile very abrupt. The SRP and SRA results in Fig. 6 agree quite well for this shallow junction, with the SIMS curve giving a slightly deeper profile. The SIMS tail is likely caused by cascade mixing and knock-on of dopant atoms by the sputtering beam.

MOS MEASUREMENTS

MOS measurements to determine oxide charge, interface trap density, oxide thickness, doping density, capacitor storage time, and generation lifetime have been in use for the past thirty years. The interpretation of C-V data has been carried out by conventional theory for most of this time. As oxides become thinner and oxide charge densities are reduced, effects that have in the past been neglected, can become significant. Among these are quantization of carriers in the inversion layer, the necessity of using Fermi-Dirac statistics in the analysis, and the finite doping density of poly-Si gates (26). These effects lead to modifications in the interpretation of experimental MOS C-V data as illustrated in Fig. 7. Neglecting these effects can lead to incorrect oxide thickness extraction from C-V data and erroneous interface trap densities, for example.

The inclusion of Fermi-Dirac, rather than Maxwell-Boltzmann, statistics predicts a more precise oxide thickness from MOS capacitor measurements in accumulation or

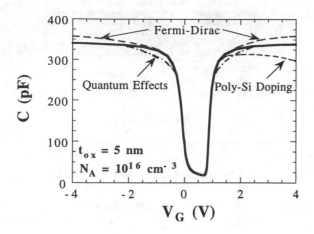

Figure 7. MOS capacitor C-V_G curve showing effects that are often neglected.

inversion. The difference in modeled oxide capacitance between these two models increases with decreasing oxide thickness amounting to about 4% for 42 Å oxides (26). For high electric fields, the electrons in the two-dimensional electron gas are spatially delocalized, with the electron centroid located away from the SiO_2-Si interface. Furthermore, the electron energy is quantized. Failure to consider this effects results in reduced C_{ox} when comparing data with theory. Both of these effects can influence the interface trap densities extracted from C-V curves.

A third effect usually neglected is the finite doping density of the poly-Si gates. In the past, with sufficiently thick oxides, the capacitance due to the narrow space-charge region in the partially depleted gate was sufficiently high to be negligible. However, as gate oxides become thinner, the gate capacitance can no longer be neglected and the experimental C-V data are lower than predicted. This effect impacts those parameters extracted from the low-frequency C-V curve, such as interface trap density. Poly-Si gate depletion also affects time dependent dielectric breakdown measurements, because n^+ poly-Si gates deplete with positive gate voltages. The band bending in the poly-Si saturates at approximately 1.2 V. Hence, 1.2 V of the applied voltage is dropped across the gate, rendering t_{BD} measurements unreliable. For a 68 Å oxide it was found that the extrapolated t_{BD} at 5V was overestimated by a factor of 1200 by neglecting the voltage drop across the oxide (27).

Mobile Ions

Mobile ions are still a concern in the semiconductor industry even today because they cause threshold voltage instabilities. Threshold voltage variations should be less

than 40 mV by the end of the century (1). If we attribute threshold voltage variations due to mobile ions as being 10% of the allowed 40 mV, this requires ionic contamination densities, N_m, to be below 10^{10} cm^{-2}. Analog circuits require even lower levels. Bias-temperature stress (BTS) measurements have traditionally been used to determine mobile ions in oxides (28). The reproducibility of such measurements becomes questionable as mobile charge densities approach 10^9 cm^{-2}. For example, the flatband voltage shift in a 10 nm thick oxide due to the drift of a 10^9 cm^{-2} mobile ion density is 0.5 mV. This is difficult to measure. Changing the gate area does not help the situation since we are looking for voltage shifts, not capacitance.

Triangular voltage sweep (TVS) is based on measuring the charge flowing through the oxide at an elevated temperature in response to an applied time-varying voltage. The charge flow is detected either as a current (29) or as a charge (30). For a mobile ion charge density of 10^9 cm^{-2}, the resulting current is $I=3.4\times10^{-11}$ A for a sweep rate of 0.01 V/s and gate area of 0.01 cm^2. The charge in a charge sensing measurement is $Q=1.6$ pC. Both of these are within today's measurement capability.

In the simultaneous high-frequency (hf), low-frequency (lf) or quasistatic C-V measurement, both hf and lf curves are determined simultaneously. An example is shown in Fig. 8. It clearly shows the ability to detect mobile ion densities in the low 10^9 cm^{-2} range. TVS has the additional advantage that it can also detect mobile ions in interlayer dielectrics where BTS C-V measurements no longer work. While mobile charges in such dielectrics do not affect the threshold voltage directly, it was shown recently that mobile ions tend to plate preferentially on metal line edges from where they can migrate into gate and field oxides (31). Once that happens, the threshold voltage becomes unstable.

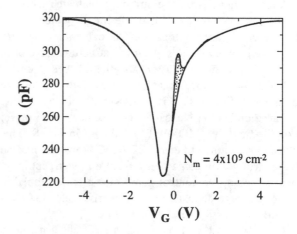

Figure 8. Mobile ion charge density determination by the simultaneous h-f, l-f C-V technique. t_{ox}=100 nm, A=0.01 cm^2. Data courtesy of L. Stauffer, Keithley Instruments.

Coupled with measurement of charges and traps in oxides are oxide integrity measurements. Typically these include gate current measurements to determine time-to-breakdown and charge-to-breakdown. Recent advances in low-noise instrumentation allow gate current measurements to be extended into the femto amp regime - currents that have been masked by noise in the past.

WAFER CHARGING

Charging of wafers during processing is encountered during ion implantation, reactive ion etching, plasma-enhanced CVD, pre-metal sputter clean, photoresist stripping, and plasma etching (32). When charge is deposited on an oxide during processing, a voltage develops across the oxide and the resulting current flow through the oxide leads to weakening of the oxide. It can even lead to oxide breakdown. Although wafers are annealed during subsequent processing thereby eliminating many of the charging problems, once the oxide is weakened by current flow through the oxide, it does not recover its full pre-charge strength and is permanently weakened (33).

Wafer charging problems are generally not encountered in medium current (tens of μA beam current) implanters at doses less than 10^{15} cm^{-2}. However, high-dose implantation for high throughput, coupled with reduced oxide thicknesses, have led to wafer charging problems. Hence, many high-current implanters are operated with reduced beam current to avoid yield loss due to wafer charging. Unfortunately, this also gives lower throughput. Wafer charging also occurs during other process operations in which there is the possibility of charge being deposited on a wafer.

Wafer charging is most commonly evaluated with test chips similar to those used for oxide integrity studies, including variable-area antenna MOS capacitors. *In-situ* probes have been used and special test structures have been developed. Among the latter are the charge monitor CHARM-2 (34) and the Sematech process induced damage effect revealer (SPIDER) (35). CHARM-2 contains varying area charge collection electrodes connected to the control gate of floating gate MOSFETs as shown in Fig. 9. This structure is very much like an EEPROM. However, instead of the control gate voltage being from a power supply, in this test structure the control gate voltage is developed in response to charge deposited on the charge collection electrode by the process causing the charging.

Charge collected on the collection electrode generates a voltage between the control gate and the substrate, which, in turn, induces a voltage on the floating gate. For sufficiently high floating gate voltage, charge is injected from the substrate onto the floating gate or vice versa. The floating gate assumes a potential which changes the threshold

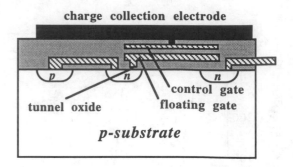

Figure 9. Charge monitor using the EEPROM principle.

voltage of the structure, with the threshold voltage measured between the control gate and the substrate. Current densities from 40 μA/cm^2 to 100 mA/cm^2 can be detected. These measurements can be converted to surface potentials and charge fluxes. The SPIDER test structure contains various MOS capacitors and transistors as well as gated and ungated diodes.

METALLIC DEFECTS

The defect density allowed in and on devices is continually decreasing as device dimensions shrink. This is true for metallics as it is for particles. Metallic contamination affects device operation in various ways. Alkali metals are of concern because they cause MOSFET threshold voltage shifts, Al and Zn are important because they affect the oxidation rates of Si, and heavy metals, of which Fe, Cr, Cu appear to be particularly important, cause junction leakage currents and oxide integrity degradation. A major requirement for front end processing is low metal densities on the wafer surface before thermal processing, because such metals can lead to silicide precipitation, stacking fault nucleation and decoration. When incorporated into the growing oxide or at the oxide/Si interface, metals lead to poor oxide reliability. Iron is believed to be the most important heavy metal impurity in Si today. Projections of allowed Fe densities to the year 2007, based on projections in the SIA Roadmap, are illustrated in Fig. 10.

The importance of Fe impurities on oxide integrity is illustrated by the relationship between oxide breakdown electric field $\mathcal{E}_{BD}$ and Fe density N_{Fe} (36)

$$\mathcal{E}_{BD} = 11.677+(6.51x10^{-12}-1.63x10^{-8}/t_{ox}^3)N_{Fe} \quad (4)$$

where $\mathcal{E}_{BD}$ is in MV/cm, N_{Fe} in cm^{-3}, and t_{ox} is the oxide thickness in nm. A plot of Eq. (4) is shown in Fig. 11. It clearly illustrates the sensitivity of oxide breakdown field to Fe contamination. For example, an increase in N_{Fe} from

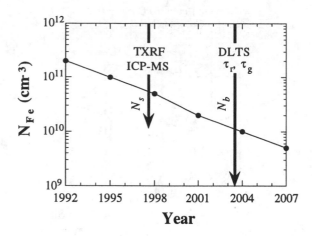

Figure 10. Projected allowable Fe densities in Si; also shown are several characterization techniques for metallic contamination. N_s=surface density, N_b=bulk density.

5×10^9 cm^{-2} to 5×10^{11} cm^{-2}, reduces the breakdown field from 11.6 MV/cm to 6.8 MV/cm for a 10 nm thick oxide.

How can we detect the low densities demanded of future IC generations? If the impurities are on the surface, the most sensitive techniques are *total reflection X-ray fluorescence* (TXRF) and *inductively coupled plasma - mass spectroscopy* (ICP-MS). The efficacy of wafer cleans is usually determined this way. The sensitivity of TXRF and ICP-MS is enhanced by vapor phase decomposition in which the impurities on the entire sample surface are collected in a small area using HF (37). The measurement sensitivity is enhanced approximately by the ratio of the wafer area to the collecting area allowing surface concentrations on the order of 10^9 cm^{-2} to be determined (37).

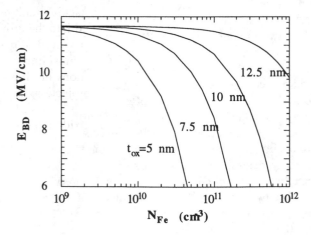

Figure 11. Oxide breakdown electric field as a function of Fe contamination. N_{Fe} is determined from surface photovoltage measurements.

Once the metallic impurities are driven into the wafer, only electrical techniques are sufficiently sensitive to detect them. Deep-level impurities are chiefly determined by two methods: *deep-level transient spectroscopy* (DLTS) and *recombination* or *generation lifetime* methods. DLTS is a direct method to determine density N_T, energy level E_T, and capture cross section $\sigma_{n,p}$ of a given impurity. While it yields spectroscopic data, it is more complex than lifetime measurements. However, it is a well established, commercial characterization tool with a sensitivity related to the substrate doping density by $N_T \approx 10^{-4}$-$10^{-5}N_A$ cm^{-3}. Its chief strength lies in the spectroscopic nature of the measurement.

Lifetime Characterization

Recombination (τ_r) and generation lifetime (τ_g) measurements are easier to carry out than DLTS, but generally they are an indirect measure of impurities, because both recombination and generation are influenced by *all* electrically active defects in a device or material. Hence, a lifetime measurement is a measure of the effect of all impurities and structural defects on the lifetime. However, lifetime measurements can be element-specific for several elements. Of the various metals, Fe and Cr in p-type Si lend themselves to direct determination by lifetime measurements. Both Fe and Cr form complexes with p-type dopants in Si, *i.e.*, B, Al, and In. Fe forms a Fe-B complex at room temperature with an energy level E_{Fe-B}-$E_v = 0.1$ eV.

When a p-Si sample containing Fe is heated to about 200°C or exposed to white light, the Fe-B complex dissociates into interstitial Fe (E_{Fe}-$E_v = 0.4$ eV) and substitutional B. The recombination properties of interstitial Fe differ from those of Fe-B pairs and this difference is the basis for Fe determination by recombination lifetime or minority carrier diffusion length measurements. By measuring the minority carrier diffusion length before, L_{Fe-B}, and after, L_{Fe}, dissociation, the Fe density is determined from the relation (38)

$$N_{Fe} = 1.06 \times 10^{16}(1/L_{Fe}^2 - 1/L_{Fe-B}^2) \qquad (5)$$

Lifetime measurements are very sensitive. If we take the recombination lifetime expression to be

$$\tau_r = \frac{1}{\sigma_n v_{th} N_T} \qquad (6)$$

and use $\sigma_n = 10^{-15}$ cm^2, $v_{th} = 10^7$ cm/s, and $N_T = 10^9$ cm^{-3}, then $\tau_r = 0.1$ s, which is certainly easy to measure. The dif-

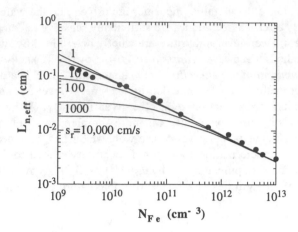

Figure 12. Minority carrier diffusion length of Fe-contaminated Si versus Fe density as a function of surface recombination velocity. Data from refs. 38, 39, and 40.

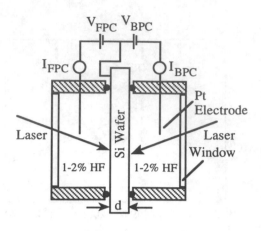

Figure 13. Schematic cross section of the ELYMAT electrolytical double cell.

ficulty with long lifetimes is not the measurement but its interpretation due to surface effects. As metallic impurity densities decrease and τ_r and τ_g increase, surface recombination and surface generation become more important and can dominate the lifetime measurement. What is measured is the effective lifetime τ_{eff}, given by (41)

$$\tau_{eff} = \frac{\tau_r}{1 + \tau_r/\tau_s} \qquad (7)$$

where $\tau_s = t/2s_r$ is the surface lifetime with t the sample thickness and s_r the surface recombination velocity. Unless care is taken in the measurement interpretation, it is the interface trap-dominated surface that is characterized and not the metallic impurity contaminated bulk, as illustrated in Fig. 12. This figure illustrates that lifetime or diffusion length measurement techniques have the required sensitivity for the metallic contamination the SIA National Technology Roadmap for Semiconductors calls for.

The surface recombination velocity can be reduced by passivating the Si surface. For example, an oxidized and well-annealed Si wafer has significantly lower s_r than a bare surface. It turns out that even lower s_r can be achieved by immersing the sample in HF (42) or dilute iodine/methanol solutions (43) during the lifetime measurement. One can also use the surface-dominated τ_{eff} to determine the effect of surface treatments by using Si with very high τ_B so that τ_s dominates the measurement (44).

Minority carrier diffusion lengths are most readily mapped with the *electrolytical metal tracer* (ELYMAT) shown schematically in Fig. 13 (45). The wafer is immersed in an electrolyte (normally, but not necessarily, 1%-2% HF). The electrolyte serves the dual function of photo-current collection as well as wafer surface passivation. The front and back induced photocurrents are measured in response to laser beam excitation of the sample. Using laser excitation on both sides of the sample, coupled with two different wavelengths and voltage bias, allows the surface recombination velocity and the minority carrier diffusion length as well as depth-dependent diffusion length to be determined. Scanning the laser produces diffusion length maps.

CONTACTLESS MEASUREMENTS

Contactless characterization techniques are gaining in importance for obvious reasons as contamination levels need to be continually reduced. Optical methods are, of course, largely non-contacting as are many of the beam techniques. But there are also several electrical methods that exist today or are being developed that fall into this category. Among these are the microwave surface resistance analyzer for doping profiling, discussed earlier, and several lifetime measurement methods. Contactless recombination lifetime techniques are the radio frequency-photoconductance decay (43) and the microwave reflection-photoconductance decay (46) methods. The minority carrier diffusion length is commonly determined by surface photovoltage (47). A promising technique for oxide quality and semiconductor impurity measurements, is corona charging in which the metallic or poly-Si contact in MOS or Schottky devices is replaced by ions (48). There are a number of other methods not discussed here. These include capacitive wafer thickness and flatness measurements (49). From the wafer flatness, the wafer stress can be determined. The surface charge analyzer, while not truly contactless, uses temporary contacts to determine oxide charge, interface trap density, and doping density (50).

Corona Charging

Corona charging uses ions deposited in a selected area on a semiconductor sample to deduce a variety of material or device parameters. For example, charging an oxidized Si wafer by ions drives the MOS capacitor, where now the "M" is a deposited charge layer, into deep depletion. As the deep-depleted device relaxes to its equilibrium state, the capacitance changes as a result of thermal electron-hole pair generation. With charge Q constant, once it is deposited, the transient voltage ΔV is determined from the relationship

$$\Delta V = Q/\Delta C \qquad (8)$$

The voltage is measured with a vibrating Kelvin probe by positioning the sample first under the corona gun and then under the Kelvin probe as illustrated in Fig. 14. By measuring the voltage as a function of time, the device leakage current or storage time can be determined. Depositing negative and positive charges adjacent to each other during the corona discharge makes it possible to create devices with guard rings to inhibit perimeter leakage currents. In this way, one can produce ideal guard rings, with zero spacing between the guard ring and the device to be measured.

Combining optical excitation with corona charging and contactless voltage measurements, a number of semiconductor device parameters can be determined that are related to defect densities in the material. For example, shining weak light on a pn junction diode creates a slight forward bias. If the forward voltage is kept below kT/q, then measuring the resulting voltage decay, after cessation of the light pulse, by a Kelvin probe, one can determine the junction leakage current contactless.

Combining corona charging with strong optical excitation allows the oxide thickness and surface potential of an MOS device to be determined as a function of "gate" voltage. Oxide charge, interface trap density, and doping density can be extracted from such data. Furthermore, the oxide integrity can also be determined by incrementally increasing the deposited charge. Combining Kelvin probing with wavelength variable light, gives the minority carrier diffusion length by the surface photovoltage technique using either the constant surface photovoltage or the constant photon flux density version. Other variations allow the mobility and sheet resistance to be measured (51). Corona charging can be considered an *in-line* characterization tool allowing wafers to be measured and reinserted into a product line (51). In this capacity it fulfills the goal of a contactless, *in-line* characterization tool.

A recent contactless characterization tool uses a metal-air-oxide-semiconductor device (52). An electrode is held a fraction of a micron above the oxidized wafer and the capacitance is measured. The data are analyzed with conventional

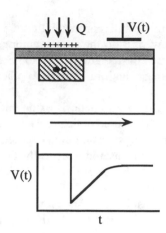

Figure 14. Schematic of corona charging, Kelvin probe voltage measurements.

MOS theory. By varying the air gap between the electrode and the sample, one can distinguish between charge at the SiO_2/Si interface, charge in the oxide, and charge at the surface of the oxide. The technique has been used recently to determine wafer charging due to plasma and ion implantation (53).

THE FUTURE

What is the future of electrical characterization? Much like process equipment, measurement equipment is being adapted to the demands of future ICs. Let me take junction profiling as an example for further discussion, because such profiling is a critical issue. Junction depths, projected to be 20-30 nm by 2010, already exist today and have been characterized by SRP, SRA, differential Hall effect, and SIMS. These measurements are so far very specialized and not routine. Sample preparation becomes more critical, but the equipment exists for these measurements. A major challenge here is the *lateral* characterization of junction profiles that has not been solved yet. But even for vertical profiling, issues need to be addressed, as evident from a recent workshop devoted to this topic (54).

Fig. 6 shows a disagreement in junction depth for the three methods used there. Why is that? Do we measure different properties? Certainly that is one consideration. SRP and SRA measure electrically active carriers, while SIMS detects ions after they have been removed from the sample by sputtering. Were these ions electrically active when they were in the sample? What do we know about the fundamentals of the metal/semiconductor probe contact in SRP measurements? Do these differ from deep junctions? Is SRA the contactless profiling method we have sought for a long time and do we understand it thoroughly? What are the precision, accuracy and reproducibility of each of the techniques and how well do the results among the techniques correlate? These considerations need to be addressed as we push each of these, and, of course, all other techniques as well, to the limit.

Another critical parameter is oxide thickness which is most commonly measured optically. However, by considering Fermi-Dirac statistics and quantization, accurate oxide thicknesses can be measured electrically. Oxide charges and interface traps can be determined by existing techniques even for densities as low as 10^9 cm^{-2}. Triangular voltage sweep has the sensitivity for such low densities even for interlayer dielectrics. In most MOS devices, mobile charges are most frequently found in the interlayer dielectrics, not so much in the gate oxide.

Channel length and width are often measured by SEM, but electrical test structures exist for the narrow lines of future devices. The specific contact resistance must decrease as contacts become smaller. For example a 1 μmx1 μm contact with a specific contact resistance of $\rho_c=10^{-6}$ $\Omega\cdot$cm^2 has a contact resistance $R_c=\rho_c/A=100$ Ω. When that contact size is reduced to 0.2 μmx0.2 μm, R_c increases to 2500 Ω if ρ_c remains constant. Obviously, ρ_c must be reduced to maintain reasonable contact resistance. Then it becomes necessary to use four-terminal or Kelvin techniques to measure ρ_c. This technique is well established. Mobility measurements are routinely made electrically and no changes are necessary.

One of the severest challenges is the characterization of defects. Particulates are most commonly measured by light scattering methods, where wavelength and angle-resolved measurements look promising. Metals, when on the surface, are measured by X-ray or chemical techniques. The lower practical limit is on the order of 10^{10} cm^{-2} today. Allowed densities will be reduced at least one to two orders of magnitude in the future, straining surface-sensitive techniques, but improvements will no doubt be made to meet that challenge.

Metals in a wafer at the required low densities can only be detected electrically. Already, densities as low as 10^9 cm^{-3} have been detected and I see no difficulty in continuing that trend. Detecting specific elements is more of a challenge. DLTS can do this, but it is not routinely used. Hence metal detection remains the domain of lifetime measurements. Only Fe and Cr can be determined directly so far utilizing the pairing/dissociation with boron in p-Si. It remains to be seen whether other elements can be identified by lifetime methods. Measuring lifetimes and defects *locally* is possible through generation lifetime and leakage current measurements, where the generation volume is confined to the space-charge region. This will be increasingly important to characterize gettering effectiveness. Gettering implementation will become more difficult as process temperatures are reduced and technologies such as silicon-on-insulator are introduced. New gettering techniques need to be developed and these have to be characterized electrically since the density of defects is too low for chemical and beam technique detection.

SUMMARY

In this review I have highlighted a number of electrical characterization techniques. I have emphasized those methods that have been added recently to the repertoire of characterization techniques as well as those well-established techniques where recent advances have extended the measurement capabilities. Parameter extraction, not discussed in this review, is continually improving to be able to model and predict device performance. New approaches and extensions of existing methods continue to be developed for parameter extraction, which is largely a fitting of parameters to a particular model with the required complexity set by the particular model. Research is underway to use first-order principles modeling wherever possible.

In general, the measurement community is well poised for the challenges of the coming decade. This challenge, however, calls for continued improvements in measurement sensitivity and precision. Many of the demands of the National Technology Roadmap for Semiconductors for the year 2001 with respect to characterization have already been met. But there are exceptions. For example, the Roadmap calls for junction depths of 50-60 nm. SIMS, SRP, and SRA profile measurements down to around 30 nm junction depth have already been made. However, these measurements have been made with careful sample preparation and "well tuned" equipment, not routine, operator controlled equipment. Bulk metal densities of 2×10^{10} cm^{-2} are called for. Recombination lifetime and minority carrier diffusion length measurements consistent with Fe densities of 2×10^9 cm^{-3} have been made. Lifetime measurements are relatively easy to make. It is the interpretation that is difficult due to surface and bulk recombination, with surface recombination usually not well defined.

Most of the existing measurement techniques are *off-line* sampling methods. A few measure *in-line*, but rarely are *in-situ*, process control measurements made. With a few exceptions, there is a general lack of *contactless, in-line* characterization tools. Development of these tools is the greatest challenge for the measurement community. One of the few techniques meeting these criteria is ellipsometry, that can and has been used as a process-control tool. The ultimate, of course, is not to make any measurements at all during processing. Unfortunately this requires perfect processing at all stages.

REFERENCES

1. SIA, *The National Technology Roadmap for Semiconductors*, Semiconductor Industry Association, 1994.
2. T.J. Shaffner, "Beam Characterization Techniques for ULSI," in *Diagnostic Techniques for Semiconductor Materials and Devices 1994* (D.K. Schroder, J.L. Benton, and P. Rai-Choudhury, eds.), Electrochem. Soc., Pennington, NJ, 1994, 295-312.
3. A.C. Diebold, *J. Vac. Sci. Technol.* **B12**, 2768-2778 (1994).
4. D.K. Schroder, *Semiconductor Material and Device Characterization*, New York, Wiley-Interscience, 1990, Ch. 1.
5. F. Djamdji, I.C. Mayes, S.R. Blight, I.L. Freeston, A.C. Gorvin, and R.C. Tozer, "Resistivity Mapping of Wafers Using Applied Current Tomography," in *Diagnostic Techniques for Semiconductor Materials and Devices 1994* (D.K. Schroder, J.L. Benton, and P. Rai-Choudhury, eds.), Electrochem. Soc., Pennington, NJ, 1994, 108-119.
6. D.C. Barber and A.D. Seagar, *Clin. Phys. Physiol. Meas.* **8**, Suppl. A, 47 (1987).
7. R. Subrahmanyan and M. Duane, "Issues in Two-Dimensional Dopant Profiling," in *Diagnostic Techniques for Semiconductor Materials and Devices 1994* (D.K. Schroder, J.L. Benton, and P. Rai-Choudhury, eds.), Electrochem. Soc., Pennington, NJ, 1994, 65-77.
8. R.G. Mazur and D.H. Dickey, *J. Electrochem. Soc.* **113**, 255-259 (1966).
9. R. Galloni and A. Sardo, *Rev. Sci. Instrum.* **54**, 369-373 (1983).
10. D.K. Schroder, *Semiconductor Material and Device Characterization*, New York, Wiley-Interscience, 1990, Ch. 2.
11. D.W. Feldbaumer and D.K. Schroder, *IEEE Trans. Electron Dev.* **38**, 135-140 (1991).
12. J.S. Martens, V.M. Hietala, D.S. Ginley, T.E. Zipperian, and G.K.G. Hohenwarter, *Appl. Phys. Lett.* **58**, 2543-2545 (1991).
13. C.G. Pantano, "Secondary Ion Mass Spectroscopy," in *Metals Handbook* (R.E. Whan, coord.), Am. Soc. Metals, Metals Park, OH, 1986, 610-627.
14. W.L. Smith, A. Rosencwaig, and D.L. Willenborg, *Appl. Phys. Lett.* **47**, 584-586 (1985).
15. J. Golin, N. Schnell, J. Glaze, and R. Ozarski, *Nucl. Instrum. Meth.* **B21**, 542-549 (1987).
16. D.M. Maher and B. Zhang, *J. Vac. Sci. Technol.* **B12**, 347-352 (1994).
17. N. Takigami and M. Tanimoto, *Appl. Phys. Lett.* **58**, 2288-2290 (1991).
18. V. Raineri, V. Privitera, and W. Vandervorst, *Appl. Phys. Lett.* **64**, 354-356 (1994).
19. A.C. Diebold, M. Kump, J.J. Kopanski, and D.G. Seiler, "Characterization of Two-Dimensional Dopant Profiles: Status and Review," in *Diagnostic Techniques for Semiconductor Materials and Devices 1994* (D.K. Schroder, J.L. Benton, and P. Rai-Choudhury, eds.), Electrochem. Soc., Pennington, NJ, 1994, 78-96.
20. H.L. Berkowitz, D.M. Burnell, R.J. Hillard, R.G. Mazur, and P. Rai-Choudhury, *Solid State Electron.* **33**, 773-781 (1990).
21. P.K. Chu, R.J. Bleiler, and J.M. Metz, *J. Electrochem. Soc.* **141**, 3453-3456 (1994).
22. S.W. Downey and A.B. Emerson, *Surf. Interface Anal.* **20**, 53-59 (1993).
23. P. Blood, *Semicond. Sci. Technol.* **1**, 7-27 (1986).
24. J.S. Martens, S.M. Garrison, and S.A. Sachtjen, *Solid State Technol.* **38**, 51-54 (1994).
25. K.H. Weiner, P.G. Carey, A.M. McCarthy, and T.W. Sigmon, *IEEE Electron Dev. Lett.* **13**, 369-371 (1992).
26. K.S. Krisch, J. Bude, and L. Manchanda, "Non-Idealities in the Characterization of Thin SiO_2 Films Using Capacitance-Voltage Techniques," in *Diagnostic Techniques for Semiconductor Materials and Devices 1994* (D.K. Schroder, J.L. Benton, and P. Rai-Choudhury, eds.), Electrochem. Soc., Pennington, NJ, 1994, 12-23.
27. S.J. Wang, I.C. Chen, and H.L. Tigelaar, *IEEE Electron Dev. Lett.* **12**, 617-619 (1991).
28. D.K. Schroder, *Semiconductor Material and Device Characterization*, Wiley-Interscience, New York, 1990, Ch. 6.
29. M.W. Hillen and J.F. Verwey, "Mobile Ions in SiO_2 Layers on Si," in *Instabilities in Silicon Devices: Silicon Passivation and Related Instabilities* (G. Barbottin and A. Vapaille, eds.), Elsevier Science Publ., North Holland, 1986, 403–439.
30. T.J. Mego, *Rev. Sci. Instrum.* **57**, 2798-2805 (1986).
31. M.A. Chonko, R. Khamankar, T. Tiwald, T. Allen, and B. Vasquez, *J. Electrochem. Soc.* **141**, 1862–1866 (1994).
32. T.C. Smith, "Electrical Characterization of Ion Implant Charging," in *Diagnostic Techniques for Semiconductor Materials and Devices 1994* (D.K. Schroder, J.L. Benton, and P. Rai-Choudhury, eds.), Electrochem. Soc., Pennington, NJ, 1994, 1-11.
33. H. Shin and C. Hu, *IEEE Trans. Semic. Manufact.* **6**, 96-102 (1993).
34. W. Lukaszek, W. Dixon, E. Quek, W. Weisenberger, and S. Ho, *Nucl. Instrum. Meth.* **B74**, 301-305 (1993).
35. P.K. Aum, "Oxide Wearout Testing," UC Berkeley Short Course, 1993.
36. W.B. Henley, L. Jastrzebski, and N.F. Haddad, "Monitoring Iron Contamination in Silicon by Surface Photovoltage and Correlation to Gate Oxide Integrity," in *Surf. Chem. Clean. for Semic. Proc.* (G. Higashi, E. Irene, and T. Ohmi, eds.), Mat. Res. Soc. Symp. Proc. **315**, 1993, 299-304.
37. Y. Mizokami, T. Ajioka, and N. Terada, *IEEE Trans. Semic. Manufact.* **7**, 447-453 (1994).
38. G. Zoth and W. Bergholz, *J. Appl. Phys.* **67**, 6764–6771 (1990).
39. G. Borionetti, M. Porrini, P. Gernazani, R. Orizio, and R. Falster, "Influence of Polysilicon and Crucible Purity on the Minority Carrier Recombination Lifetime of Czochralski Si Crystals," in *Semiconductor Silicon/94*

(H. Huff, W. Bergholz, and K. Sumino, eds.), Electrochem. Soc, 1994, 104-110.

40. A.L.P. Rotondaro, T.Q. Hurd, P.W. Mertens, H.F. Schmidt, M.M. Heyns, E. Simoen, J. Vanhellemont, G. Vegh, C. Claeys, and D. Gräf, "Limitations of MCLT as a Parameter for Evaluating Fe in Si", presented at Electrochem. Soc. Meet., Miami Beach, FL, 1994.

41. D.K. Schroder, *Semiconductor Material and Device Characterization*, Wiley-Interscience, New York, 1990, Ch. 8.

42. E. Yablanovitch, D.L. Allara, C.C. Chang, T. Gmitter, and T.B. Bright, *Phys. Rev. Lett.* **57**, 249–252 (1986).

43. H. M'saad, J. Michel, J.J. Lappe, and L.C. Kimerling, *J. Electron. Mat.* **23**, 487-491 (1994).

44. H. M'saad, J. Michel, A. Reddy, and L.C. Kimerling, *J. Electrochem. Soc.*, to be published.

45. J. Carstensen, W. Lippik, and H. Föll, "Mapping of Defect Related Silicon Bulk and Surface Properties With the ELYMAT Technique," in *Semiconductor Silicon/94* (H. Huff, W. Bergholz, and K. Sumino, eds.), Electrochem. Soc, 1994, 1105-1116.

46. M. Kunst and G. Beck, *J. Appl. Phys.* **60**, 3558-3566 (1986).

47. A.M. Goodman L.A. Goodman, and H.F. Gossenberger, *RCA Rev.* **44**, 326-341 (1961).

48. M.S. Fung and R.L. Verkuil, "Process Learning by Nondestructive Lifetime Testing," in *Semiconductor Silicon/90* (H. Huff, K.G. Barraclough, and J.I. Chikawa, eds.), Electrochem. Soc, 1990, 924-950.

49. J.L. Kawski and J. Flood, *IEEE/SEMI Adv. Man. Conf.*, 106-110 (1993).

50. V. Murali, A.T. Wu, A.K. Chatterjee, and D.B. Fraser, *IEEE Trans. Semic. Manufact.* **5**, 214-222 (1992).

51. M.S. Fung and R. Verkuil, IBM Corp., private communic.

52. T. Sakai, M. Kohno, S. Hirae, I. Nakatani, and T. Kusuda, *Japan. J. Appl. Phys.* **32**, 4005-4011 (1993).

53. S. Hirae, M. Kohno, H. Okada, H. Matsubara, I. Nakatani, T. Kusuda, and T. Sakai *Japan. J. Appl. Phys.* **33**, 1823-1830 (1994).

54. *Third International Workshop on the Measurement and Characterization of Ultra-Shallow Doping Profiles in Semiconductors*, Research Triangle Park, NC, 1995.

Lifetime Measurements for Routine QC/QA of SOI Wafers

J. L. Freeouf and N. Braslau

Interface Studies, Inc.
27 East Mountain
Katonah, NY 10536

Silicon-on-Insulator (SOI) technology is a crucial aspect of space hardening microelectronic components, and is likely to become a mainstay of the computer industry. We report the development of a system to provide rapid, non-invasive, non-contact characterization of the electrical quality of the surface silicon layer that is the active layer of SOI devices. The same technique provides useful information about surface damage or contamination for bulk Si wafers. This system has been used to determine lifetimes as short as 15 nanoseconds (for a SOI wafer) and as long as 350 μseconds (for a bulk wafer). The reproducibility is about 10% for these measurements.

INTRODUCTION

Silicon-on-Insulator (SOI) technology is a crucial aspect of space hardening microelectronic components. The same technology appears poised to become a mainstay of the computer industry[1]. This technology is based upon the difficult process of fabricating thin silicon films on an insulating layer. There are many ways of achieving this film, all of which can lead to low quality silicon if not all steps go properly. This thin silicon layer is the active layer of the device, but it is difficult to obtain electrical characterization of this layer in a rapid and non-invasive fashion. We have developed a technique for Quality Control and Quality Assessment (QC/QA) by providing rapid, non-invasive, non-contact characterization of the electrical quality of the surface silicon layer that is the active layer of these SOI devices. The same technique provides useful information on the surface properties of bulk silicon wafers. This technique is to measure the apparent lifetime of carriers excited by pulsed optical excitation.

EXPERIMENTAL

A schematic diagram of the apparatus is shown in Figure 1. The specific approach used is to impinge a pulse of photons upon the sample which excites carriers within the silicon. The wafer sits on an RF resonator, whose "Q" factor is altered by the change in resistance of the sample due to these excess free carriers. For photons of energy greater than ≈ 3.5 eV, all photons will be absorbed in the topmost ≈ 1000Å, so all carriers will be generated in this layer. If the insulating layer is impervious to those carriers, they are all confined to remain in that layer, so the decay time of the excess conductance due to those excess carriers will tell us the effective lifetime of those carriers. This effective lifetime depends upon the

electrical quality of the silicon in which the carriers reside; the observed value will also depend upon the quality of the interfaces (the interface recombination velocity) since the silicon layer is very thin. By using a pulsed laser source we obtain larger signal and can measure shorter lifetimes than were possible in previous attempts to perform this measurement on SOI samples[2]. The present apparatus uses a pulsed nitrogen laser, operating at 337 nm, with a pulse width of order 10 nsec. The spot size limits the spatial resolution; we currently operate at sizes ranging from $\approx 1 \times 1$ mm to $\approx 3 \times 8$ mm. We move the sample manually to obtain spatial mapping.

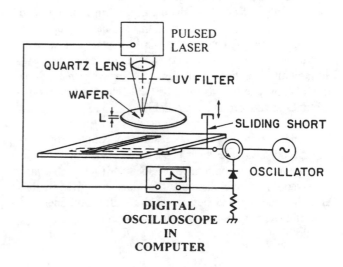

FIGURE 1. Schematic illustration of experimental layout.

Research is sponsored by SDIO/IST and managed by ONR.

DISCUSSION

The technique used provides a signal that decays in an exponential fashion. The next task is to use this signal to provide useful information about the sample being measured. The first step is simply to obtain a characteristic decay time, the lifetime of the excited carriers. We have developed three distinct approaches to this task, both for redundant checking and in search of a robust implementation for automated applications. In all these implementations we have assumed that the signal is a **single** exponential; a disagreement between model results then provides an indication that this assumption is incorrect. The simplest approach, and one we used in a previous publication, is to perform an iterative least-squares-fit of an exponential to the measured signal. This approach works, but it requires manual interaction to obtain an appropriate starting point for the parameters, and it is iterative, so the time to obtain a solution is not fixed. Both aspects of this approach are drawbacks for an automated tool. We have therefore sought other techniques, and compared them with the results of this fit. The first technique implemented is the use of modulating functions, as reported by Ransom et al[3]. We have programmed this technique with two modulating functions that emphasize different portions of the data; as discussed in Ransom et al, agreement between the two resulting lifetimes suggests that our single exponential assumption is acceptable. The final approach to determining lifetime is to find an "instantaneous lifetime." We first approached this by calculating the ratio of the second derivative to the first derivative; for noise-free data, this provides the lifetime, but our data did not provide sufficient signal-to-noise. Our current technique is to use the slope of the logarithm of the data; since we perform a background subtraction in the data acquisition, this is a natural approach. The modulating function approach has so far appeared to provide the most robust solution, but in most cases the least-squares-fit definition of lifetime agrees with that of modulating functions to roughly 10%. The "instantaneous" lifetimes are noisier and less reliable.

The relationship between this result and sample properties depends upon variables such as the thickness (L) of the wafer and the surface recombination velocity S (i.e., the surface passivation). If we assume that the wafer is sufficiently thin and that the top and bottom of the wafer have identical recombination velocity, then the measured lifetime[4] τ_{meas} should be

$$1/\tau_{meas} = 1/\tau_{bulk} + 2S/L \qquad (1)$$

If we assume that the bulk lifetime τ_{bulk} is infinite, then $\tau_{meas} = L/2S$. In Table I we show possible results for these assumptions.

Table 1.

Surface Effect on Lifetimes for Films

Thickness (micron)	Recomb. Velocity (cm/sec)	τ_{meas} (sec)
100	1	5×10^{-3}
	100	5×10^{-5}
	10,000	5×10^{-7}
1	1	5×10^{-5}
	100	5×10^{-7}
	10,000	5×10^{-9}
0.1	1	5×10^{-6}
	100	5×10^{-8}
	10,000	5×10^{-10}

Complications added to this analysis for SOI films include the probability that the top and bottom interfaces will have different values of a recombination velocity. However, we can treat the results of this table as providing a lower limit for τ_{bulk}, given knowledge of film thickness and surface treatment. For bulk wafers, even if the two surfaces are treated identically, we cannot assume uniform distribution of carriers -- especially for our apparatus, where the carriers are all generated in the first few hundred Å. We shall demonstrate surface sensitivity on bulk silicon at a later point in this paper.

RESULTS

Before analyzing signals from very thin films, it is appropriate to demonstrate that the signal from bulk silicon is at least reasonable. In Figure 2 we show the results of a measurement on bulk silicon that has been given a passivating dip. The same figure shows the quality of the exponential fit to this sample. Note that the modulating functions gave a lifetime of $125\mu sec$ while the iterative fit gave a value of $133\mu sec$. The agreement between these two results (as well as between the two modulating functions used) strongly suggests that a single exponential treatment is justified for this sample.

Bulk Silicon

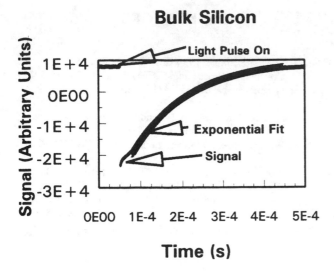

FIGURE 2. Data and exponential fit for passivated bulk silicon wafer. Modulating functions gave a lifetime of 125μsec whereas the least-squares-fit gave a value of 133μsec.

In Figure 3 we show results for the same sample, before and after wiping it off with a Kimwipe® and acetone. Note that the measured lifetime has decreased by a factor of 3! The bulk characteristics of this wafer have not changed. What has changed is the recombination velocity; this effect is enhanced by the extreme surface sensitivity of the carriers generated in our apparatus.

Bulk Si

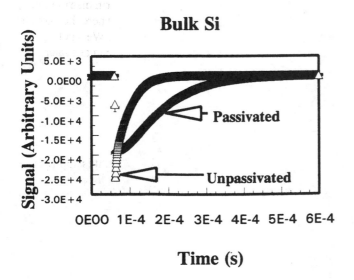

FIGURE 3. Data for bulk silicon wafer before and after its surface passivation is removed. Note strong effect of increase in surface recombination velocity.

SIMOX (Retains Oxide from Thinning)

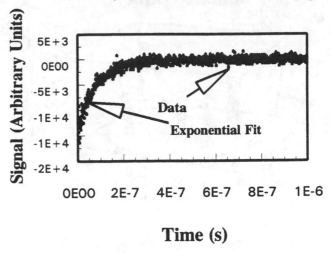

FIGURE 4. Data and exponential fit for a thin (≈2,000Å) SIMOX sample with the oxide from the thinning process still present.

In figure 4 we show results from a thin SIMOX sample, and in figure 5 we show similar results from a thin BSOI sample. In both cases the sample had a passivating oxide on the surface and the film was ≤2,000Å thick. Without some such passivation we have still been able to obtain a signal for samples thicker than 1μ, but thinner samples have not offered much success unless some passivation is performed. The results in Table 1 explain the reasons for this limitation. An alternative passivation technique is an HF dip[4], which we used in our previous lifetime study.[2]

Oxidized BSOI

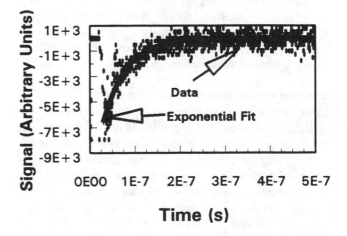

FIGURE 5. Data and exponential fit for a thin (≈2,000Å) BSOI sample with a thin passivating oxide.

Bulk Si Wafer

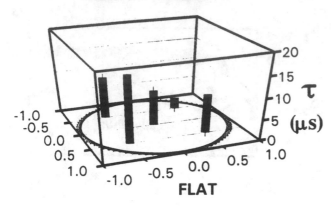

FIGURE 6. Distribution of lifetime results across a 2" diameter bulk wafer.

Many of the useful conclusions to be drawn from such a technique will be strengthened if the spatial variation of these results can be obtained. In figure 6 we show the results of one such measurement. This bulk silicon sample was apparently processed with a source of contamination either on the wafer or in the furnace: It was clearly degraded in the upper right hand corner. Such process analyses are expected to prove useful in SOI process development as well.

SUMMARY

We have demonstrated that we do indeed obtain a signal from both bulk wafers and thin ($\approx 2,000\text{Å}$) films, although the latter usually require passivation for reasonable results. These thin film lifetime results were typically shorter than 100 nanoseconds, indicating that the measurements were dominated by interface recombination at the imperfect interfaces involved. Our automated lifetime calculation has demonstrated reproducibility of order 10% for different measurements of the same sample.

Our oscilloscope uses a sampling mode of operation for measurement of sub-microsecond lifetimes. This becomes a rather time-consuming operation (over 8 minutes per measurement). We have therefore altered our oscilloscope to permit faster capture of the short decay spectra that we have observed in SOI samples so far (current acquisition is less than 3 minutes). We have also designed and tested a new resonator built on a transparent substrate. Since the sample need not sit on the resonator, it can be placed on an automated X-Y table below the resonator. This means that a wafer handling system can be used in conjunction with this system, thus permitting full automation of the tool.

REFERENCES

1. See, for example, 1994 IEEE International SOI Conference Proceedings (IEEE Catalog #94CH35722) Piscataway, NJ 08855-1331, IEEE 1994.

2. Freeouf, J. L., Braslau, N., and Wittmer, M., , "Lifetime Measurements on Silicon-on-Insulator Wafers," *Applied Physics Letters* **63**, 189 - 190 (1993).

3. Ransom, C. M., Chappell, T. E., Freeouf, J. L., and Kirchner, P. D., "Modulating Functions Waveform Analysis of Multi-exponential Transients for Deep- Level Transient Spectroscopy," *Materials Research Society Symposium Proceedings* **69**, 337-342 (1986).

4. Yablonovitch, E., Allara, D. L., Chang, C. C., Gmitter, T., and Bright, T. B., "Unusually Low Surface-Recombination Velocity on Silicon and Germanium Surfaces," *Physical Review Letters* **57**, 249-252 (1986).

Application of Surface Photovoltage and Contact Potential Difference for in-line Monitoring of IC Processes.

K. Nauka

Hewlett-Packard Company, Palo Alto, CA 94304

Surface Photovoltage and Kelvin probe based Contact Potential Difference techniques have been employed for in-line, real-time monitoring of the manufacturing operations of a 0.35 µm IC (Integrated Circuit) CMOS process. Applications included monitoring of metallic impurities in front-end operations, measurement of dielectric charges, and detection of a dielectric damage induced by plasma processing. Ability of the employed techniques to conduct non-contact, non-destructive measurements allowed to test product wafers, and return wafers free of misprocessing back to the manufacturing line. Quick response, low cost, clean room compatibility, and versality makes these techniques very attractive for monitoring of IC manufacturing processes. Their extensions for future generation of IC processes are also discussed.

Introduction

A continuing increase of the complexity of an integrated circuit combined with the rapidly raising manufacturing costs requires implementation of the in-line monitoring techniques detecting in real-time misprocessed wafers so that immediate corrective action can be undertaken. Such techniques must be fully clean room compatible, fast, reliable, and integrable with the in-fab information system. They should also minimize the monitoring added cost by not requiring any additional wafer processing, and by returning tested wafers to the manufacturing line when deemed free of misprocessing. This paper presents examples of the application of a commercial monitoring tool (1) integrating Surface Photovoltage (SPV) and Contact Potential Difference (CPD) measurements to monitor unit operations in a 0.35 µm IC CMOS manufacturing process. Extension to future IC processes is discussed.

Experimental

The SPV signal V_{SPV} is obtained by illuminating one side of a Si wafer with photons having energies higher than the Si bandgap, while the other side remains in darkness. Nonequlibrium carriers generated near the illuminated surface introduce a charge imbalance resulting from a partial collapse of the surface potential barrier and causing a potential difference V_{SPV} between the illuminated front and the dark back surfaces. At low light intensities V_{SPV} is directly related (2) to the minority carrier diffusion length and the surface recombination rate. Since metallic impurities frequently act as efficient recombination centers decreasing the diffusion length, SPV can be employed for qualitative detection of metallic contaminants introduced during the IC processing. In the case of Fe, the most frequent metallic contaminant, further advantage is taken of the well understood (3) transformation between Fe interstitials acting as very efficient recombination sites and much less efficient Fe-B complexes. By suitable means, SPV can quantify the amount of Fe impurities (4). At high light intensities surface barrier disappers and V_{SPV} becomes saturated. Thus, the SPV value obtained under saturation conditions provides information about the magnitude of a surface barrier. The surface barrier value can be translated into a surface charge or, in the case of an oxidized Si wafer, into the mean value of oxide charges (5).

CPD measurement employs a vibrating Kelvin probe to measure the difference (V_{cpd}) between the work function of the reference metal probe and the work function of a measured wafer (6). In the case of an oxidized Si wafer, the last quantity is determined by the oxide surface charge. Thus, when the work function of a Kelvin probe is known, CPD can provide information about the electrical charges stored on the oxide surface.

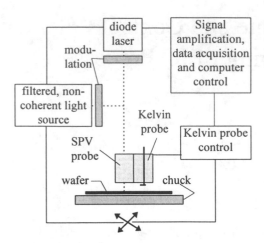

Figure 1. Experimental set-up for the Surface Photovoltage and Contact Potential Difference measurements.

Both, SPV and CPD were integrated (1) on a common platform (Figure 1) facilitating robot wafer handling and rapid, computer controlled measurements in the clean room environment. Both measurements were contactless, and they did not required any surface preparation. They were accomplished by using ac signal, allowing for capacitive coupling between the wafer's chuck, the wafer, and the SPV and CPD probes. The SPV signal was obtained using the filtered, non-coherent light source for the minority carrier diffusion length measurements, and a diode laser was employed for the surface barrier measurements.

Results and Discussion

This section presents examples of an application of SPV and CPD for three different types of the IC manufacturing operations: detection of the metallic impurities in front-end processes, monitoring dielectric charges in a manner complementing the Capacitance-Voltage (C-V) measurements, and sensing dielectric damage introduced by plasma processing.

1. Monitoring of Metallic Impurities

A large number of succesful SPV applications for the monitoring of metallic impurities in IC processes have been reported. They include: furnace monitoring (7), detection of metallic impurities in incoming chemicals (7), control of resist ashing operations (8), and implementation of the SPV - based Statistical Process Control (SPC) employed to front-end operations (9). Figure 2 demonstrates the application of SPV to optimize a pre-epi clean. It shows the effect of a reversal of the cleaning sequence (brush scrub followed by Piranha clean vs. Piranha clean followed by brush scrub) on

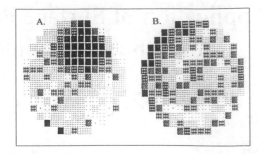

Figure 2. Wafers maps of a diffusion length: light - large diffusion length, dark - small diffusion length. A. scrub followed by Piranha clean, B. Piranha clean followed by scrub.

deposition of an epitaxial layer (at 1100° C in the presence of HCl). The "Piranha last" clean gave a larger mean value, but also caused a larger variation of the diffusion length, as demonstrated by appearance of a low diffusion length region (dark spot) in Figure 2.A. A detailed investigation showed that sulphuric acid used in the Piranha clean contained some metallic impurities (mostly Fe) which were left in "spots" when the Piranha clean was used as a last step (subsequent DI H_2O rinse did not remove the contaminants). Reversal of the cleaning sequence caused the impurities deposited by the Piranha clean to be dispersed by the following brush scrub giving a lower mean diffusion length and avoiding the formation of low diffusion length spots.

As a part of present studies SPV has also been employed for routine evaluation of the Fe impurities introduced by high temperature furnace processes (10). Since SPV is a "diffusion length" technique, information about Fe is collected mostly within the bulk of a Si wafer rather than in the region adjacent to the wafer's surface where IC devices are fabricated. Application of the SPV as a Fe monitoring tool required to establish a correlation between the SPV results and the concentrations of Fe impurities vicinal to the wafer's surface. Figure 3 compares the bulk Fe data obtained by SPV with the Fe concentrations in the IC device region measured by Deep Level Transient Spectroscopy (DLTS). Discrepancy between the SPV and DLTS results observed at low Fe concentrations was due to a complex process involving out-diffusion of Fe atoms from the bulk towards the surface and in-diffusion of Fe atoms from the contaminated furnace into the Si wafer, mediated by the Si oxidation. It was specific for a given process, and it changed when the type of furnace, furnace operating condictions, and wafer grade were changed. At high Fe concentrations one of these processes prevailed resulting in an agreement between the DLTS and SPV data. Understanding of discrepancy observed at low Fe impurity

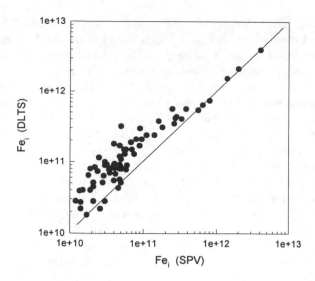

Figure 3. Correlation between the Fe_i results obtained by SPV and DLTS

concentrations allowed to employ SPV as a reliable Fe monitoring tool.

2. Monitoring of Dielectrics

C-V is a routine way of evaluating quality of dielectrics employed in IC devices. Since it requires fabrication of MOS capacitors suited for C-V analysis and time consuming testing, it is inherently slow and costly. SPV offers a fast and contactless alternative to the C-V monitoring of IC dielectrics. However, results obtained by both techniques are not necessarily equivalent. C-V is a spectroscopic technique that can distinguish between the different types of charges in dielectrics, while SPV can provide information about the mean dielectric charge (mean

charge (Q_{mean}) equals sum of fixed (Q_f), mobile (Q_m), interfacial (Q_i), and trapped (Q_t) charges). Additionally, due to difference in work functions (C-V requires conductive gate, SPV is performed on bare surface of a dielectric) the corresponding data obtained by both techniques are shifted by a constant value. Figure 4 compares C-V and SPV results obtained for 100 nm thick, thermally grown field oxides (11). The C-V measurements were conducted using Al electrodes, and the SPV data were obtained in the vicinity of measured C-V gates. Since Q_f and Q_m are dominant dielectric charges, it is assumed that $Q_f + Q_m$ represents the mean dielectric charge value. Results shown in Figure 4 demonstrate that despite of the aforementioned limitation SPV can be succesfully employed to monitor field oxidation. It shows that an increase in some of the oxide charge components leads to an increase of the Q_{mean}, thus allowing for SPV detection of faulty oxides with a high degree of accuracy. Furnace grown oxides are commonly qualified in terms of the amounts of fixed and mobile charges. The horizontal line in Figure 4 separates wafers that passed an arbitrarily selected C-V spec of $Q_f < 1 * 10^{11}$ cm^{-2} and $Q_m < 4 * 10^{10}$ cm^{-2} from those that failed the test. Its intersection with the SPV: Q_{mean} axis yields the corresponding SPV spec.

SPV has been also employed to monitor charges in thin gate oxides and in deposited dielectrics. Because of its virtue of contactless measurement this technique is particularly well suited for analysis of thin dielectric films, where physical contact, or electrical stress during C-V probing, or processing required to fabricate conductive C-V gates can easily damage the oxide or perturb the distribution of oxide charges. Figure 5 demonstrates an excellent agreement between the C-V and SPV results obtained for the 10 nm thick thermally grown gate oxides. Since, in this experiment

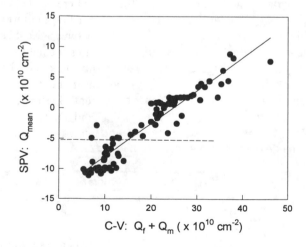

Figure 4. Correlation between SPV: Q_{mean} and C-V: $Q_f + Q_m$ for the 100 nm thermal field oxide. C-V results were obtained using Al gates.

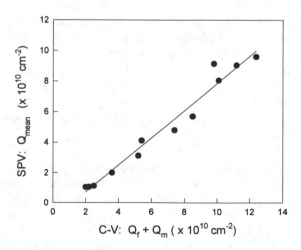

Figure 5. Correlation between SPV: Q_{mean} and C-V: $Q_f + Q_m$ for the 10 nm thermal gate oxide. C-V results were obtained using polysilicon gates.

Table 1. Comparison of the SPV: Q_{mean} and C-V: $Q_f + Q_m$ for deposited dielectrics. Densification was conducted for selected dielectrics at 800°C in N_2 ambient.

Deposited Dielectric	C-V: $Q_f + Q_m$ (q/cm^2)	SPV: Q_{mean} (q/cm^2)
PSG ([P] = 8%)	$3.60 * 10^{11}$	$3.04 * 10^{10}$
PSG densified	$3.00 * 10^{11}$	$2.79 * 10^{10}$
TEOS oxide	$2.70 * 10^{11}$	$1.59 * 10^{10}$
TEOS oxide densified	$3.20 * 10^{11}$	$2.90 * 10^{10}$
PECVD nitride	$3.40 * 10^{11}$	$3.00 * 10^{10}$
LPCVD nitride	$2.40 * 10^{11}$	$1.20 * 10^{10}$

polysilicon was used instead of Al to fabricate conductive gates, the shift between the C-V and SPV data is different than in Figure 4. Similarly to thermally grown oxides, a very good agreement between the C-V and SPV results (Table 1) was observed for dielectrics fabricated by the Chemical Vapor Deposition (CVD). Comparison of the results obtained for the films subjected to the post-deposition densification annealing demonstrates that SPV can reliably predict changes in dielectric charges caused by prolonged densification annealing.

3. Monitoring of Plasma Damage in Dielectrics.

Since plasma processing has become a mainstream part of IC maufacturing, its impact on dielectrics is one of the major factors determining the quality of IC devices. Plasma induced dielectric damage is caused by the lack of local balance between the positive and negative charges deposited on the surface of a dielectric (12). Local charge build-up can lead to a very high voltage stress across the dielectric and subsequent dielectric damage. This effect is particularly severe in the case of thin gate dielectrics, where, because of the small thickness of a dielectric layer, damage can occur at relatively low dielectric surface charges. Although indirect evaluations of the plasma deposited dielectric charges in devices are available, there has been not a fast, clean room compatible, technique that would facilitate their direct imaging (13,14). Present work demonstrated for the first time application of the CPD for imaging surface dielectric charges deposited by plasma processing.

Two plasma processes were investigated: Ar plasma clean preceding metal deposition, and O_2 plasma post-etch ashing. They were conducted using state-of-the-art commercial equipment: metal deposition system in the first case, and dielectric plasma etcher in the second case. Two types thermal oxides were used: thick, 100 nm oxides were used for dielectric charge mapping with the CPD, and 8 nm gate oxides were used to fabricate antenna test devices (their fabrication was accomplished using a 0.35 μm IC CMOS

process). Devices were tested for the oxide leakage at 3.3 V and the transistor threshold shift before and after Fowler-Nordheim stress. Since very similar results were obtained for both tests, the following discussion is limited to the leakage measurements only. Additionally, C-V measurements were conducted on damaged oxides using Al gates. Charges deposited on an oxide's surface by plasma processing were mapped with the CPD. Simultaneously, in a manner previously described, SPV was used to map charges within the bulk of an oxide. Aformentioned 100 nm thick oxides were used for the CPD and SPV measurements in order to allow for deposition a large quantities of surface charges before dielectric damage occured. A V_{cpd} map of a wafer subjected to the Ar plasma clean is presented in Figure 6.A. It shows a large charge in the wafer's center and a smaller charge near the edge. The corresponding V_s map (Figure 6.B) is similar, except for the ring of high V_s approximately half way from the wafer's edge. C-V measurements of the flatband voltage shift showed an increase in the fixed and trapped charges in the ring area (Figure 6.B - point A) as compared to the area outside the ring (same Figure - point B). Since similar features has been previously reported for thermal oxides grown under faulty gas flow conditions, it is proposed that the ring pattern is created by defects that exist as neutral centers in as-grown films and become charged during the plasma processing. When plasma processed wafers were rinsed with DI water and dried, the V_{cpd} values decreased 4 - 10 times demonstrating that plasma deposited charges resided on the oxide's surface, and that they can be observed with the CPD. The distribution of charges measured by CPD correlated to device degradation. 3-dimensional projection of device leakage (Figure 6.C) shows that high dielectric leakage occurs within the areas where V_{cpd} is large. A quantitative agreement between V_{cpd} and device leakage is demonstrated in Table 2, which compares the $\Delta V_{cpd} = V_{cpd}^{max} - V_{cpd}^{min}$ with the average value of leakage current for different Ar plasma conditions. The average leakage value was obtained from the 1000 point mapping of a 6'' wafer with the leakage of antenna devices measured at 3.3 V bias. V_{cpd}^{max} and V_{cpd}^{min} represent, respectively, maximum and minimum values of the V_{cpd} signal for a

Table 2. Correlation between the V_{cpd} and leakage current of antenna devices after pre-metal Ar plasma cleaning

V_{cpd}^{max} (V)	V_{cpd}^{min} (V)	ΔV_{cpd} (V)	$\log(I_{leakage})$ (A/cm^2)
11.91	-8.38	20.29	-8.599
12.40	0.70	11.70	-10.648
11.00	-2.30	13.30	-10.855
4.22	0.01	4.21	-11.454
5.60	1.10	4.50	-11.615
8.70	2.20	6.50	-11.597
8.80	3.70	5.10	-11.612

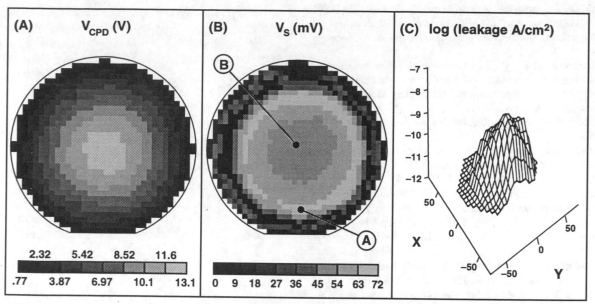

Figure 6. (A) V_{cpd} map. (B) V_s map ($V_s = V_{SPV}$ when SPV signal is saturated); C-V measurements were taken at point A ($Q_f + Q_m = 3.5 * 10^{12}$ cm^{-2}) and B ($Q_f + Q_m = 4.2 * 10^{11}$ cm^{-2}). (C) antenna capacitor leakage map; vertical axis represents log ($I_{leakage}$).

given wafer. Since plasma nonuniformity determines dielectric damage induced by a plasma treatment and CPD measured net charge on a dielectric surface is equal V_{cpd} * constant (15), it is assumed that ΔV_{cpd} represents a degree of plasma induced damage in a dielectric. Comparison between the V_{cpd} and V_s maps, and device results for wafers that had undergone O_2 plasma post-etch ashing yielded similar results, except for the absence of the ring caused by bulk oxide traps. This is probably due to the fact that deposited charges were lower here than in the case of the Ar plasma.

reaching the range that is unobtainable by a routine C-V monitoring. However, the largest gain is expected in the area of plasma damage monitoring. Single wafer processing and sub - 0.25 μm device will require sophisticated plasma etching and increased number of plasma cleans. Additionally, shrinking thickness of gate oxides will make devices more sensitive to the plasma damage. All these factors indicate that CPD monitoring of dielectrics subjected to plasma processing could play an increasingly important role. It is also expected that other applications of the SPV and CPD for IC process monitoring will emerge.

Summary

Three applications of the SPV and CPD for monitoring of the IC manufacturing processes have been shown. Because of their non-contact, non-destructive nature and clean room compatibility, they offer a low cost opportunity for fast and reliable detection of faulty IC manufacturing processes. Since the Semiconductor Technology Roadmap indicates raising importance of in-line monitoring techniques for the further development in IC industry (16) and need to increase their detection capabilities, extendability of the described applications for future generations of IC processes offers an additional advantage. Extension of the SPV to monitor Fe concentrations as low as $6 * 10^9$ atoms / cm^{-3} has already been demonstrated (17). It is also expected that improvements in the SPV probe design will lower detection limits of dielectric charges below 10^{10} cm^{-2},

Acknowledgments

The author is grateful to J.Lagowski and L.Jastrzebski (Semiconductor Diagnostics, Inc.) for their help and support in development of applications for SPV and CPD. Collaboration of J.B.Kruger, W.W.Dixon, W.E.Greene, B.W.Langley (Hewlett-Packard Company) in monitoring of the plasma induced damage in dielectrics is also acknowledged.

References:

1. Contamination Monitoring System, Model CMS-IIIA/R, Semiconductor Diagnostics, Inc., Tampa, FL.
2. Lagowski, J., Edelman, P., Dexter, M., Henley, W., *Semicond. Sci. Technol.* **7**, A185 - A192 (1992).

3. Zoth, G., Bergholz, W., *J.Appl.Phys.* **67**, 6764 - 6771 (1990).

4. Zoth, G., in *Tech.Proc. SEMICON Europa '90*, Zurich, Switzerland, March 1990, p.24; also Bergholz, W., *Siemens Forsch.* **16**, 241 (1987).

5. Edelman, P., Lagowski, J., Lastrzebski, L., "Surface Charge Imaging in Semiconductor Wafers by Surface Photovoltage", presented at the Materials Research Society Meeting, San Francisco, CA, April, 1992.

6. Brattain, W.H., Bardeen, J., *Bell System Tech. J.* **32**, 1 - 41 (1953)

7. Jastrzebski, L., Milic, O., Dexter, M., Lagowski, J., DeBusk, D., Nauka, K., Witowski, R., Gordon, M., Persson, E., *J. Electrochem. Soc.* **140**, 1152 - 1159 (1993).

8. Hoff, A.M., Persson, E.J., "Surface Photovoltage Measurement of Contamination Introduced by Resist Ashing", *Electrochem.Soc.Ext.Abstr. Vol.93-2*, 1993, p.494.

9. Lowell, J., Wenner, V., Thomas, J., Lastrzebski, L., Lagowski, J., DeBusk, D., Edelman, P., Nauka, K., "In-line, Real-time non-destructive monitoring of Fe Contamination for Statistical Process Control by Surface Photovoltage", presented at the Internat. Electron. Manufact. Symposium, Santa Clara, CA, October, 1993.

10. Nauka, K., Gomez, D.A., *J.Electrochem.Soc.* **142**, L98 (1995).

11. Nauka, K., Chang, R.R., Cheney, E.A., Jastrzebski, L., "In-line monitoring of the oxide charges with Surface Photovoltage", *Electrochem.Soc.Ext.Abstr. Vol. 93-1*, 1993, p.1116-7.

12. Fang, S., McVittie, J.P., *IEEE Electron Dev. Lett.* **13**, 288 - 290 (1992).

13. McVittie, J.P., "Murakawa, S., "A New Probe for Direct Measurement of Surface Charging", *Electrochem.Soc.Ext. Abstr. Vol.94-1*, 1994, p.562.

14. Lukasek, W., Quek, E., Dixon, W., "CHARM2: Towards an Industry Standart Wafer Surface Charge Monitor", in *Proceeding of the IEEE/SEMI Advanced Semiconductor Manufacturing Conference*, 1992, p.148.

15. Edelman, P., Hoff, A.M., Jastrzebski, L., Lagowski, J., "New approach to measuring oxide charge and mobile ion concentration", presented at the Internat. Soc. for Optical Engineering (SPIE) Microelectronics Manufacturing '94 Conference, Austin, TX, October, 1994.

16. The National Technology Roadmap, Semiconductor Industry Association, San Jose, CA, (1994).

17. Jastrzebski, L., Lagowski, J., Lowell, J., DeBusk, D., Persson, E., Nauka, K., "Application of Surface Photovoltage to monitor metallic contamination with sub-ppt sensitivity in the IC processing environment", presented at the Institute of Environmental Sciences Meeting, Chicago, IL, May, 1994.

236

Improvement of Isothermal Capacitance Transient Spectroscopy for Deep Level Measurement Including Interface Trap

H. Yoshida, H. Niu, T. Matsuda, and S. Kishino

Department of Electronics, Faculty of Engineering, Himeji Institute of Technology, Shosha, Himeji 671-22, Japan

Two kinds of improvements of isothermal capacitance transient spectroscopy (ICTS) technique applied to metal-oxide-semiconductor (MOS) devices are described. One is clear identification of various signals obtained by ICTS measurements. It is determined from the dependence of ICTS signals on reverse bias voltage. As a result, we have found that three kinds of ICTS signals originated in bulk traps, interface traps, and occurence of inversion show absolutely different behavior by adding the reverse bias voltage. The other is a proposal of a simple technique for the simultaneous measurement of both the majority and minority carrier traps. It is only necessary for the measurements of minority carrier traps in the proposed technique to reverse the injection pulse polarity. The validity of the improvements have been demonstrated with the use of gold-diffused MOS diodes.

1. INTRODUCTION

The deep levels in semiconductor are very harmfull to the device characteristics and so the study on the deep levels has been widely conducted. Deep level measurements including interface traps (interface states) are very important for small-sized metal-oxide-semiconductor (MOS) devices. This is because interface traps could be enhanced by the hot carriers which are easily induced in small-sized MOS devices.

Recently, the transient capacitance techniques have been widely used for the study of the deep levels because of the accuracy and the sensitivity of the measurements. As is well known, isothermal capacitace transient spectroscopy (ICTS) (1) is one of the transient capacitance techniques, the principle of which is the same as the deep level transient spectroscopy (DLTS) (2). ICTS techniques has the advantage that the measurements are possible at constant temperature, being different from DLTS techniques. However, various kinds of unknown signals appear when the ICTS technique is applied to Si MOS devices. Therefore, there are a few application of the ICTS techniques to MOS devices (3) in comparison with Schottky diodes or p-n junctions. Also, the ICTS measurements using MOS diodes are restricted to the emission transient spectroscopy of the majority carrier traps. In order to easily apply the ICTS techniques to the MOS devices, we must solve the above demerits of ICTS techniques.

In this paper, two kinds of improvements of ICTS technique are reported under the condition where ICTS technique is applied to a Si MOS diode. One is clear identification of various signals obtained by ICTS measurements. The other is a proposal of a simple technique for the simultaneous measurement of both the majority and minority carrier traps.

2. CLEAR IDENTIFICATION OF ICTS SIGNALS

Figure 1 shows a representative overall ICTS spectrum of a gold-diffused n-Si MOS diode. All the signals A, B, and C in Fig. 1 are not necessarily those

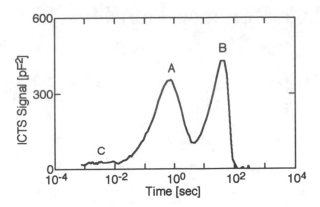

FIGURE 1. Representative ICTS spectrum of gold-diffused MOS diode at 230 K. Injection pulse voltage V_b and reverse bias voltage V_a applied to the diode are 1 V and -5 V, respectively.

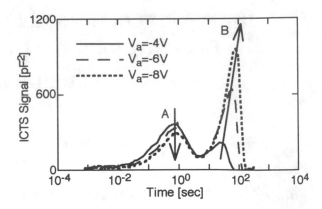

FIGURE 2. Dependence of ICTS signals A and B on reverse bias voltage at 230 K (experimental results).

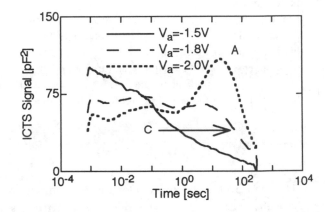

FIGURE 3. Dependence of ICTS signals C on reverse bias voltage at 200 K (experimental results).

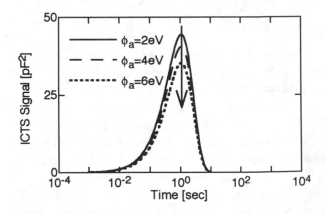

FIGURE 4. Dependence of ICTS signal due to bulk traps on reverse bias voltage (simulation results). ϕ_a is surface potential corresponding to the reverse bias voltage V_a.

from deep levels. In order to identify these signals, we examined the dependence of these signals on reverse bias voltage. The examination was carried out experimentally with the help of a numerical simulation method. As a result, we found that three kinds of ICTS signals showed absolutely different behavior by adding the reverse bias voltage.

First, we show the dependence of the ICTS signals A, B, and C on the reverse bias voltage V_a. Figure 2 shows the dependence of the ICTS signals A and B on the reverse bias voltage V_a, where the injection pulse voltage V_b is 1 V for all the measurements. The height of the signal A decreases with increasing the reverse bias voltage and the lateral position (emission time) does not shift as seen in Fig. 2. On the other hand, the height of the signal B sharply increases and the lateral position slightly shifts to the longer emission time with the increase of the reverse bias voltage.

Figure 3 shows the dependence of the ICTS signal C on the reverse bias voltage. The lateral position of the signal C sensitively shifts to the longer emission time with increasing the reverse bias voltage as shown in Fig. 3. The behavior of these signals indicates that origins of the signals are different among the signals A, B, and C.

In order to identify these signals, a numerical simulation for the ICTS signals of MOS diodes was performed in which the presence of both the bulk and the interface traps is assumed. Figure 4 shows simulation results of the dependence of the ICTS signal due to the bulk traps on the reverse bias voltage, where ϕ_a is the surface potential corresponding to the reverse bias voltage. Since the carrier emission rate is dependent on the energy level of the traps, the ICTS signal due to the bulk traps with a discrete energy level is observed as a peak and the peak position (emission time) does not shift regardless of the reverse bias voltage value as seen in Fig. 4. Also, the height of the peak decreases with increasing the reverse bias voltage. This is because the capacitance transition decreases with increasing the reverse bias voltage. The dependence of the signal A in Fig. 2 on the reverse bias voltage is same as that of the bulk traps. As a result of Arrhenius plot, it is found that the signal A is of the gold acceptor-like traps at E_c-0.55eV.

Figure 5 shows simulation results of the dependence of the ICTS signal due to the interface traps on the reverse bias voltage. Since carriers are emitted from the traps located above the Fermi level, the ICTS signal due to the interface traps with continuous energy distribution should be observed as a plateau as shown in Fig. 5, being different from the bulk traps with a discrete energy level. It is to be noted that the Fermi level at the semiconductor surface changes with the reverse bias voltage. Therefore, the lateral position of the ICTS signal due to the interface traps shifts to the longer emission time with increasing the reverse bias voltage. The signal C resulted in being originated in the interface traps.

As for the signal B, it was found from the reverse bias dependence that the signal B was neither the signal due to the bulk traps nor that due to the interface traps. In the MOS diodes, the capacitance transition pertaining to the formation of the inversion layer is indispensably present. In the deep depletion mode corresponding to the initial state in the ICTS measurements, the width of the space-charge

region widens with increasing the reverse bias voltage. On the other hand, in the inversion mode corresponding to the steady state in the ICTS measurement, the space-charge region is constant regardless of the reverse bias voltage. Therefore, the larger the reverse bias voltage is, the larger the capacitance transition pertaining to the formation of the inversion layer is. As a results, the height of the ICTS signal pertaining to the formation of the inversion layer increases with the increase of the reverse bias voltage. Also, the lateral position of the signal slightly shifts to the longer emission time with the increase of the reverse bias voltage. This is because the time for the formation of the inversion layer is lengthened with the increase of the reverse bias voltage. It is evident from the above consideration that the signal B is originated in the capacitance transition pertaining to the formation of the inversion layer.

From the above mentioned investigations of the bias dependence, it was clearly determined that the ICTS signals A, B, and C are originated in bulk traps with a discrete energy level, the formation of the inversion layer, and interface traps with continuous energy distribution, respectively.

3. SIMULTANUOUS MEASUREMENT OF MAJORITY AND MINORITY CARRIER TRAPS

As is well known, in the conventional ICTS technique, the injection of majority carriers into the deep

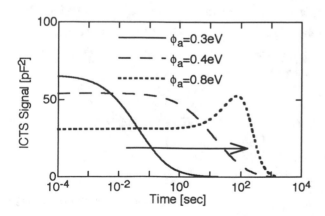

FIGURE 5. Dependence of ICTS signal due to interface traps on reverse bias voltage (simulation results).

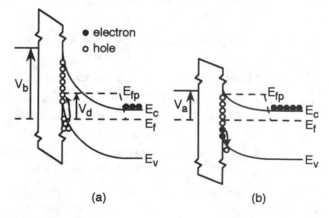

FIGURE 6. Energy-band diagrams corresponding to (a) minority carrier injection and (b) minority carrier emission processes for the gate-pulse technique. E_{fp} and E_f are Fermi level for holes and electrons, respectively.

239

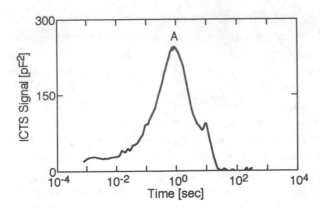

FIGURE. 7   ICTS signal due to the electron (majority carrier) traps at 230 K.

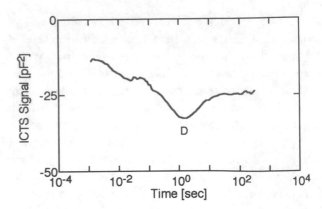

FIGURE. 8   ICTS signal due to the hole (minority carrier) traps at 140 K.

levels is performed by applying the pulse voltage to the gate so that the diode is in the accumulation mode. In order to measure the minority carrier traps, minority carriers must be injected into the traps before the ICTS measurements. In MOS structures, the injection of minority carriers could be carried out by forming the inversion layer. In the proposed gate-pulse technique, n-Si MOS diode encircled with a p^+-diffused layer was used as a source of carriers to rapidly form the inversion layer.

Figure 6 shows procedures of the gate-pulse technique. The reverse bias voltage V_d is kept being applied to the p^+-diffused layer during the ICTS measurements. The injection of minority carriers is achieved by adding the reverse-biased pulse voltage to the gate as shown in Fig. 6(a). The measurement system of the gate-pulse technique is same as that of the conventional ICTS technique except for applying the reverse bias voltage V_d to the p^+-diffused layer. It is only necessary for the measurements of minority carrier traps to reverse the injection pulse polarity in the gate-pulse technique.

In order to demonstrate the validity of the gate-pulse technique, a gold-diffused MOS diode with the p^+-diffused layer was used. Figures 7 and 8 show the ICTS signals due to the majority and minority carrier traps, respectively. The peak A in Fig. 7 is of the gold acceptor-like traps (the majority carrier traps) at E_c-0.55 eV. A downward peak D in Fig. 8 is of the gold donor-like traps (the minority carrier traps) at E_v+0.35 eV. It should be noted that the space-charge region decreases with the majority-carrier emission and increases with the minority-carrier emission.

Therefore, the polarity of the ICTS signal in the case of the majority-carrier emission is opposite to that in the case of the minority-carrier emission. It is clear from the above results that both the majority and minority carrier traps are easily measured with the use of the gate-pulse technique.

4. CONCLUSIONS

Two improvements of ICTS technique applied to MOS devices were carried out. The origines of the ICTS signals were clearly determined from the detailed investigations of the bias dependence. Also, the simultaneous measurement of both the majority and minority carrier traps was achieved with the use of the gate-pulse technique.

The improved ICTS technique is also useful for the measurement of interface traps, which are sometimes enhanced by heavy metal impurities. The enhancement is dependent on the elements of the impurity (4). Sensitivity of the present technique is ~10^{n-3} cm^{-3} (10^n: carrier density of wafer) for the bulk traps and ~10^9 cm^{-2} eV^{-1} for the interface traps, respectively.

1. Okushi, H., and Tokumaru, Y., *Jpn. J. Appl. Phys.* **28**, pp.L335-L338 (1980).
2. Lang, D., V., *J. Appl. Phys.* **45**, pp.3023-3032 (1974).
3. Yoshida, H., Niu, H., and Kishino, S., *J. Appl. Phys.* **73**, pp.4457-4461 (1993).
4. Kishino, S., Iwamoto, S., Yoshida, H., Niu, H. and Matsuda, T., *J. Appl. Phys.* 1 March (1995) to be published.

CONTAMINATION FREE MANUFACTURING

Optimization of Wafer Surface Particle Position Map
Prior to Viewing with an Electron Microscope

Man-Ping Cai, Yuri Uritsky

Microcontamination Dept., Applied Materials, Inc.
3050 Bowers Ave, Santa Clara, CA 95054

Patrick D. Kinney

Particle Technology Laboratory
Department of Mechanical Engineering, University of Minnesota
111 Church St. SE, Mpls., MN 55455

The combining of laser surface particle detectors (LSPD) with scanning electron microscopes (SEM) allows contamination particles to be analyzed on the surface of a semiconductor wafer. The LSPD provides a map of particle positions on the wafer surface, and this map is used to find the particles at high magnification under the SEM. In this paper we report on several techniques for optimizing the accuracy of particle position maps. Optimizations are accomplished using coordinate system transformations to correct particle position maps for systematic errors. These transformations are improved using corrections based on reference particles. Higher accuracy is achieved by averaging multiple particle position maps. With these techniques, particle positions are routinely predicted with positioning errors of less than 40 μm.

INTRODUCTION

In developing future semiconductor processes and improving existing ones, it is critically important to identify and eliminate the sources of particle contamination. One approach for accomplishing this task is to analyze the particles as they lie on the wafer surface. A number of different laser surface particle detectors (LSPD) exist that measure the number, location, and size of particles on the wafer surface, but this information alone is seldom useful for identifying the source of the contaminants. A scanning electron microscope (SEM) equipped with an x-ray spectrometer (EDS) works well for measuring the morphology and chemical composition of particles, but it is nearly impossible to find particles under the SEM on a relatively clean wafer surface. A number of years ago Applied Materials developed a technique that uses the LSPD to locate particles on the wafer surface, and the SEM/EDS system to analyze particles. The combined system is known as the Particle Analysis System (PAS), and similar versions of the technique are now used throughout the semiconductor industry.

The PAS has proven to be extremely useful in identifying and solving contamination problems (1). However, as chip technology advances and device geometries shrink, the size of a "killer" particle shrinks, and it is necessary to analyze smaller particles. The challenge in analyzing small particles with the PAS is in finding the particles with the SEM. The limiting factor is the positioning accuracy of the LSPD, which may be insufficient to provide the minimum SEM magnification needed to see the particles. A minimum of 1500X magnification is usually required on the SEM in order to see a 0.16 μm particle (the current industry standard). For a typical CRT screen, this magnification translates to a field of view of only 70 x 70 μm, indicating that the particle's position must be known with an error of less than 35 to 40 μm. As the error increases, analysis time is wasted searching for particles. An impractical amount of time is spent searching for particles if errors rise above 100 μm.

The PAS's ability to analyze small particles has been significantly enhanced by optimizing the accuracy of particle position maps. In the remainder of the paper, we provide a basic description of the PAS, a description of the methods used to reduce positioning errors, and experimental data that demonstrates the improvements which have been obtained so far.

BACKGROUND

PAS Description

The current PAS at Applied Materials includes a Tencor 6200 LSPD, and a Jeol 848 SEM equipped with a Kevex EDS. A PC is used for data transfer and manipulation. These instruments are unmodified except that a special specimen holder was developed to handle eight inch wafers in a preferable orientation. The LSPD detects particles using a light scattering technique. A robot arm is used to load a wafer from a wafer cassette into the LSPD chamber. A laser is raster-scanned over the wafer surface while the wafer is moved orthogonally to the scan direction. When the laser intersects a particle, light is scattered by the particle onto a detector. The magnitude of the light scattering signal provides information about the particle size. The measurement of a particle's position is more complicated; in short, by knowing the position of the laser as a function of time, particle positions can be determined by timing the light scattering events. For each wafer scan, the LSPD produces a data file that contains the size and x-y coordinates of each particle detected on the wafer surface.

The concept of the PAS is to use the LSPD to locate the particles on the wafer surface and the SEM/EDS to analyze the particles. The procedure by which this is accomplished is conceptually simple. A wafer is scanned with the LSPD to obtain a map of particle positions. The wafer is then removed from the LSPD and manually loaded into the SEM. The coordinate system used by the SEM's x-y stage is inconsistent with the coordinate system used by the LSPD; the wafer's orientation in the SEM coordinate system differs from its orientation in the LSPD coordinate system. The inconsistency is eliminated by transforming the particle map from the LSPD coordinate system to the SEM coordinate system. With several optimization steps, the transformed particle map is used to find particles under the SEM.

Reduction of Targeting Error

One of the main objectives in optimizing PAS's performance has been to improve the accuracy of particle position maps. The difference between particle positions predicted by the LSPD and the particle positions observed on the x-y stage is referred to as the *targeting error*.

Sources of Targeting Error

Since the SEM's x-y stage is accurate to within several μm's, nearly all of the targeting error is caused by uncertainties in the LSPD particle map. According to its manufacturer, the LSPD measures particle positions with a resolution that exists as a rectangular region of 10 μm X 26 μm, but these positions are referenced to a less-accurately determined coordinate system. The coordinate system is aligned to the wafer in a specific orientation with respect to the wafer's center and notch positions. These positions are determined using a lower resolution (26 μm X 120 μm) measurement of the wafer's edge geometry. The alignment is (ideally) insensitive to the wafer's orientation during scanning. From the uncertainties produced by resolution limits, we expect the LSPD to provide a relatively accurate particle position map, but one that is referenced to a significantly mis-aligned coordinate system. Thus, a key step in reducing targeting error is to eliminate coordinate system misalignment. The second approach for reducing targeting error involves averaging multiple particle position maps to reduce the influence of random uncertainties. These techniques are described below.

Elimination of Coordinate System Misalignment

The first step in reducing targeting error involves fine-tuning the coordinate transformation required to transfer the particle map from the LSPD frame to the SEM frame. The technique has been reported elsewhere (2, 3), but its significance in the present context justifies its presentation here. The coordinate transformation is conducted in a multiple step process. A *first-order* transformation is accomplished by measuring the orientation of the wafer in the SEM coordinate system, and transforming the LSPD map data to reflect the same wafer orientation. At this point the targeting errors are large, apparently due to errors in alignment of the LSPD coordinate system to the wafer. This alignment error can be eliminated once several particles (termed *reference particles*) are located with the SEM. By comparing the coordinates of two particles in the SEM and LSPD frames, a *second-order* coordinate transformation is made. The second-order transformation effectively eliminates the influence of LSPD coordinate system alignment uncertainties. Further improvements in accuracy are obtained by using more than two reference particles, and averaging the coordinate transformation parameters. The data presented here were obtained using three reference particles.

Targeting Error Reduction by Map Averaging

Reduced targeting errors are observed when two or more LSPD maps of the same set of particles are averaged. If two independent measurements of the same position are averaged, the random uncertainty associated with the average value should be less than the random uncertainty associated with the individual measurements. The main source of random error appears to be the limited resolution of the LSPD position measurement. However, if a wafer is scanned repeatedly without disturbing its position inside the LSPD chamber, the position measurements are extremely repeatable (4), and averaging does not provide any benefit. If the wafer is removed from the chamber between scans, significant variability in position measurements is observed. Part of the variability is due to coordinate system misalignment, but when this factor is eliminated with a second-order transformation, variability is still observed in the particle positions. The map averaging technique takes advantage of this apparently random variability.

There are varying degrees of complexity in the method of averaging (5), but here we present the simplest case. Two scans of a wafer are first transformed to the SEM coordinate system using a first-order transformation. Next, the two scans are combined by averaging particle positions that are common to both scans. The averaged data file is used with the SEM to find reference particles, which are then used to perform a second-order transformation on the averaged file.

EXPERIMENTAL METHOD

Targeting errors were measured for two different bare silicon wafers. The first wafer had been exposed to a semiconductor process, and was contaminated with real-world particles. The second wafer was contaminated with 0.9 µm PSL spheres using an aerosol deposition process. Measurements of the first and second wafer were performed before and after a calibration of the LSPD, respectively. Each wafer was analyzed a single time under the SEM to determine the actual particle positions. Each wafer was repeatedly scanned to obtain multiple LSPD maps of the same set of particles. Between scans, the wafer was removed from the LSPD and rotated so that the notch was in a different position for each scan. The first wafer was scanned five times using notch orientations from the 8:15 o'clock position to the 5:30 position, with the 12:00 o'clock position corresponding to the notch facing up in the wafer cassette used to load the LSPD. The second wafer was scanned six times with notch orientations covering the entire range of rotation. Analysis was performed for a set of 26 particles on the first wafer, and 28 particles on the second wafer. Targeting errors were calculated for each particle in the set by determining the difference between the actual (SEM) and predicted (LSPD) positions. Each LSPD map resulted in a unique set of targeting errors. Targeting errors were also calculated for averages of two and three LSPD maps. The average targeting error and standard deviation in targeting error were calculated for each LSPD map.

RESULTS AND DISCUSSION

Tables 1 and 2 show the targeting errors for the first and second wafers, respectively. Targeting error are shown for single LSPD maps (scans), and for averages of two and three scans. The columns marked 'scan' and 'average scans' indicate the wafer orientation(s) for the LSPD scan(s) used for the analysis. The '1st order' and '2nd order' columns refer to the coordinate transformation method (described above). The major trends of the data indicate that a second-order transformation significantly reduces the targeting error as compared to only a first-order transformation. It is also clear that targeting errors

TABLE 1. Average targeting errors for single and averaged scans computed for the first wafer. Targeting errors are shown as a function of wafer scan-orientation. Each table entry represents an average result for 26 particles.

Scan	Single Scan Targeting Error (µm) 1st Order	2nd Order	Average Scans	Two Scan Average Targeting Error (µm) 1st Order	2nd Order	Average Scans	Three Scan Average Targeting Error (µm) 1st Order	2nd Order
8:15	125 ± 32	65 ±42	8:15/8:00	83 ± 37	48 ±26	8:15/8:00/7:00	105 ±30	50 ±24
8:00	147 ± 76	58 ±21	8:15/7:00	111 ± 28	54 ±25	8:15/8:00/6:30	124 ±34	44 ±22
7:00	164 ± 28	74 ±35	8:15/6:30	148 ± 32	46 ±23	8:15/8:00/5:30	108 ±33	43 ±15
6:30	229 ± 42	64 ±34	8:15/5:30	173 ± 57	51 ±20	8:00/7:00/6:30	167 ±28	58 ±28
5:30	273 ±111	71 ±36	8:00/7:00	148 ± 38	62 ±27	7:00/6:30/5:30	132 ±50	53 ±27
Ave.	187	66	8:00/6:30	171 ± 37	54 ±27	Ave.	169 ±51	50 ±26
			8:00/5:30	126 ± 63	47 ±24		134	49
			7:00/6:30	194 ± 33	66 ±32			
			7:00/5:30	175 ±79	61 ±34			
			6:30/5:30	209 ± 79	58 ±31			
			Ave.	154	54			

TABLE 2. Targeting errors as in Table 1 but for the second wafer. These data were measured immediately following LSPD calibration. Each table entry represents an average result for 28 particles.

Single Scan Targeting Error (µm)			Two Scan Average Targeting Error (µm)			Three Scan Average Targeting Error (µm)		
Scan	1st Order	2nd Order	Average Scans	1st Order	2nd Order	Average Scans	1st Order	2nd Order
12:00	76 ±29	50 ±17	12/10	43 ±12	30 ±14	12/10/8	63 ±30	30 ±15
10:00	88 ±47	54 ±24	12/8	82 ±26	38 ±20	12/10/6	26 ±13	27 ±14
8:00	146 ±89	68 ±31	12/6	44 ±20	40 ±19	12/10/4	54 ±31	30 ±18
6:00	75 ±40	66 ±29	12/4	74 ±20	30 ±20	12/10/2	65 ±21	29 ±19
4:00	144 ±51	60 ±37	12/2	107 ±23	40 ±27	12/8/6	56 ±20	32 ±20
2:00	157 ±27	63 ±51	10/8	100 ±55	46 ±20	12/8/4	37 ±21	34 ±18
Ave.	114	60	10/6	54 ±26	39 ±20	12/8/2	94 ±22	41 ±29
			10/4	88 ±56	49 ±26	12/6/4	44 ±20	28 ±14
			10/2	78 ±30	40 ±27	12/6/2	60 ±16	37 ±24
			8/6	86 ±39	45 ±27	12/4/2	72 ±23	32 ±23
			8/4	71 ±25	53 ±24	10/8/6	70 ±35	34 ±21
			8/2	123 ±42	61 ±38	10/8/4	74 ±26	44 ±22
			6/4	75 ±27	41 ±22	10/8/2	85 ±45	45 ±26
			6/2	62 ±22	51 ±34	10/6/4	61 ±32	37 ±18
			4/2	87 ±33	49 ±32	10/6/2	38 ±23	36 ±24
			Ave.	78	43	10/4/2	35 ±32	40 ±25
						8/6/4	48 ±20	38 ±21
						8/6/2	74 ±32	45 ±35
						8/4/2	64 ±35	51 ±28
						6/4/2	46 ±16	39 ±28
						Ave.	60	36

are lower for the two- and three-scan averages. The targeting errors obtainable using map averaging and second-order transforms allow 0.16 µm particles to be immediately located under the SEM without searching. A number of more subtle characteristics are also observed. The targeting error after a first-order transformation varies with the wafer's scan-orientation. This variation is about a third as great after a second-order transformation, which eliminates LSPD coordinate system misalignment. Thissuggests that the variation of first-order targeting error with wafer orientation reflects the degree of scan-to-scan variability in alignment of the LSPD's coordinate system to the wafer. Since this variability is not observed unless the wafer is removed from the LSPD chamber between scans, the cause of the variability appears to involve the method of detecting the wafer's edge to align the LSPD's coordinate system. The magnitude of the variability is about the same as the resolution of edge measurements.

Both the first-order and second-order targeting errors drop significantly when averaged scans are used. Immediately following calibration (Table 2), a first-order transformation of a three-scan average provides targeting errors on the same level as a second-order transformation of a single scan. Improved first-order targeting error is important because it expedites the process of finding the first few reference particles. With the higher first-order targeting errors of a single scan, the process of finding reference particles can be extremely difficult, especially when large, easily detectable particles are not present.

Further analysis has been done to study the source(s) of variability in the targeting error for second-order transform data. Figures 1 and 2 demonstrate two factors that appear to influence targeting errors. Figure 1 shows a plot of the residual targeting error for a second-order transformation of a single LSPD map. The X and Y components of targeting error vary systematically with the Y and X position on the wafer, respectively. The trends of these data suggests that there is a relative rotation between the SEM and LSPD coordinate systems, but to correct the ΔY values would require rotation in the opposite direction of that required to correct the ΔX data. A simple coordinate transformation will not eliminate the systematic error shown in the figure. The likely cause for this problem is a slight deviation from orthogonality in the LSPD coordinate system. The deviation of the coordinate axis from orthogonality can be estimated from Figure 1, which indicates a deviation of about 0.1°. By analyzing similar data for other scan orientations, the deviation was found to depend on the orientation of the wafer during scanning, indicating that it is not caused by the SEM's x-y stage. Another factor that appears to influence targeting error is illustrated in Figure 2. The targeting error shows discontinuities with jumps of about the same amount as the resolution by which the wafer edge is measured. These discontinuities might be reduced if the edge was measured with higher resolution.

CONCLUSION

Analysis of contamination particles on the surface of a semiconductor wafer is ideally done using an electron microscope. In this paper we described two methods for improving the accuracy of particle position maps produced by a laser surface particle detector (LSPD).

These maps are essential for finding particles at high magnification under an electron microscope. The two methods involve fine tuning the transformation of the LSPD coordinate system to the SEM coordinate system, and averaging multiple LSPD particle maps. Fine tuning the coordinate system transformation using reference particles provides a two to three times reduction in targeting error. LSPD map averaging provides additional reduction in targeting error depending on the number of maps that are averaged. Using both techniques, particles can be located under the SEM with targeting errors of under 40 μm.

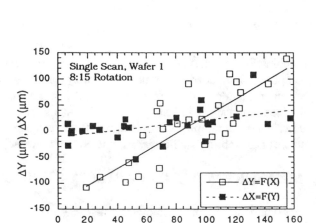

FIGURE 1. Error between predicted and actual particle positions in the y-direction (ΔY) and x-direction (ΔX) as a function of the x-coordinate and y-coordinate, respectively. Rotation of LSPD coordinate system to reduce ΔY magnitude will increase ΔX magnitude. The data indicates a 0.1° deviation of the coordinate system from orthogonality.

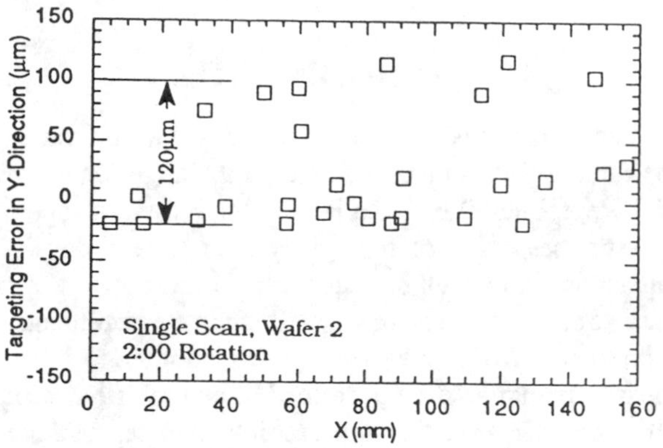

FIGURE 2. Y-direction component of targeting error as a function of the x-coordinate. The error appears to oscillate by an amount equal to the resolution of wafer edge measurements.

REFERENCES

1. Uritsky, Y., Rana, V. Ghanayem, S., and Wu, S., *Microcontamination*, 12(5), 25-29 (May, 1994).
2. Miller, S.J., Baker, E.J., and Ahn, K-H, *Microcontamination*, 7(10), 35-39, (Oct. 1989).
3. Uritsky, Y., Lee, H.Q., "Particle Analysis on 200 mm Notched Unpatterned Wafers", in *Proceedings of the Symposium on Contamination Control and Defect Reduction in Semiconductor Manufacturing III*, the Electrochemical Society, 1994, 154-163.
4. Cooper, D.W., Haller, K.L., and Batchelder, J.S., "Repeated Mapping of Environmental Particles on Surfaces to Evaluate Location Precision and Detection Efficiency", in *1992 Proceedings - Institute of Environmental Sciences,* 1, 258-262.
5. Uritsky, Y., Lee, H.Q., Kinney, P.D., and Ahn, K-H, U.S. Patent # 5,267,017, *"Method of Particle Analysis on a Mirror Wafer"* (Nov. 30, 1993).

Evaluating Automated Wafer Measurement Instruments

Steven A. Eastman

SEMATECH, Austin, TX 78741-6499

This paper focuses on effective metrology methodology for automated wafer measurement instruments. As wafer measurement instruments become increasingly automated, the methods of evaluating them have changed. Instrument operators are no longer the major contributors to measurement variation. Instead, variation is more commonly due to multiple systematic and random sources related to automation. The emerging trend is toward simpler data collection plans. Unfortunately, such plans often do not collect the information required to accurately identify and quantify the sources of variation. This paper demonstrates a method that is based on a series of experiments, each of which uses a nested data collection scheme. The plan consists of repeated measurements within a wafer loading cycle, repeated wafer loading cycles per day, and repeated data collection over many days. At SEMATECH, we have efficiently achieved the goals of metrology studies for automated wafer measurement instruments using this experimentation model.

INTRODUCTION

A standard method of evaluating a measurement instrument uses a procedure called a Gauge Repeatability and Reproducibility (R&R) Study. Several operators measure several parts repeatedly over several days. The data shows how well operators repeat their own results on the same part, and how well the measurements are reproduced from one operator to another. The total measurement variation is defined to be precision, which is the combined effect of repeatability and reproducibility. In the semiconductor industry, many wafer measurement instruments are highly automated. A cassette of wafers is loaded on the instrument, and then the wafers are automatically loaded, measured, and unloaded. The automation features reduce the operator influence on measurement variation, but the sources of variation are no longer simple to understand and simple to measure. For example, reproducibility of particle measurements is affected by factors such as focusing optics, wafer orientation, environmental influences such as electrical fluctuations, temperature changes, and contamination levels in the area of the instrument. There is a need to study the automated measurement instrument using a different method than that used in classical R&R studies.

NESTED DESIGNS WITH VARIANCE COMPONENTS ANALYSIS AND CONTROL CHARTS

There are many references to the use of variance components and control charts for the study of measurement instruments (1,2,3). However, the combination of these two techniques, using a nested sampling plan, appears to not be widely practiced in evaluating automated wafer measurement instruments in the semiconductor industry. The following sections describe useful templates for planning effective capability studies and stability studies for automated instruments.

MEASUREMENT CAPABILITY STUDIES

Measurement capability studies are used: 1) to estimate repeatability, reproducibility, and precision over a short time span, 2) to assess linearity of accuracy and precision over several wafer types or thicknesses, and 3) to decide whether improvements need to be made before proceeding with a stability study. Environmental factors such as humidity, temperature, and contamination are not expected to influence measurements taken in capability studies, but could be influential in stability studies where the sampling is carried out over a longer period of time. Depending on the goals of the evaluation, it may not be necessary to quantify the specific components of reproducibility. Contributors to reproducibility variation can be individually evaluated later, if needed, using more detailed methods as described by John(4), and by Montgomery and Runger (5).

First Capability Experiment Plan

The goal of the first capability experiment is to determine initial rough estimates of repeatability and reproducibility. One or two wafers are used, representing the most difficult wafer types. The sampling plan could consist of 2 to 5 load/unload cycles per wafer, with 6 to

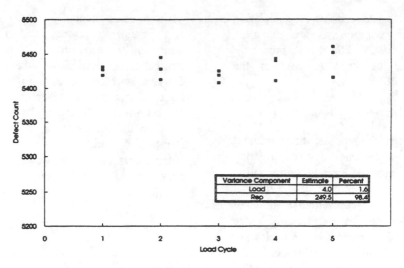

Figure 1   Data from a First Capability Experiment for Particle Counts

15 repeated measurements per cycle. These are balanced to achieve a total of about 30 measurements per wafer, over 1 to 3 days. Figure 1 illustrates the nested sampling structure in a first capability experiment.

Second Capability Experiment Plan

If repeatability and accuracy are acceptable, a second capability experiment is performed. The goal here is more broad: to understand the repeatability and reproducibility of many different types of wafers that might be measured using the instrument. It is also important to determine if repeatability and reproducibility are linear across the range of measurements. The relative magnitudes of repeatability and reproducibility found in the first capability experiment are used to determine the balance of sample sizes in the second capability experiment. The sampling plan for a second experiment would consist of 5 to 20 load/unload cycles, with 3 to 10 repeated measurements per cycle. The balance is based on results of the first experiment. Aim for a total of about 50 measurements per wafer. Examples are: 20 cycles of 3 repeats, 10 cycles of 5 repeats, 5 cycles of 10 repeats. In the example in Figure 1, the major source of variation is within-load repeatability, not between-load reproducibility, so it would make sense to plan 5 cycles of 10 repeats each. This experiment should allow the reproducibility component to include variation due to

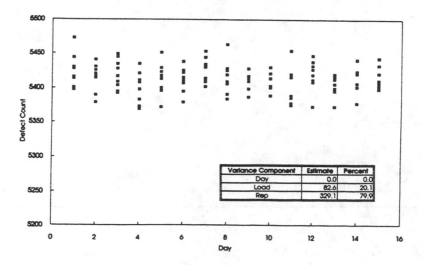

Figure 2   Data from a 15-Day Stability Study for Particle Counts

short term day-to-day effects. Algorithms for calculating estimates of the variance components are given in Montgomery(6) and Beyer(7). The SAS[1] procedure PROC NESTED (8) can be used to perform the calculations.

The Stability Study

The stability study will determine whether the instrument can maintain its capability over time. Factors affecting the instrument's stability might be: ambient temperature, humidity, contamination, as well as drift of the internal electronics of the instrument. A representative subset of wafers can be used in the stability evaluation. There should be a minimum of 15 working days, 3 cycles per day per wafer, and 3 repeated measurements per cycle. Previous experiments may provide information that suggests adjusting these sample sizes. This study provides enough data to adequately estimate reproducibility and repeatability each day. Figure 2 shows wafer surface defect data for 15 days. Since the between-load effect is relatively small compared to within-load repeatability, this graph only shows data grouped by day. The relative contribution of day-to-day variation is so small that its estimated percentage contribution to total variation is zero.

Data Analysis for Stability Studies

The data analysis steps for the stability study are: 1) plotting the raw data, 2) calculating the means and variance components, 3) creating control charts for means and variance components, and 4) judging the stability and reviewing the capability. Of particular interest are the trend charts of daily means, daily standard deviations, and daily variance components. Figures 3, 4, and 5 show these trend charts for a different study on film thickness measurements. In these charts, we can see an upward trend in the means, plus an out-of-control point in the standard deviations. We can see from the variance components chart that variation in overall precision is primarily affected by changes in reproducibility, not repeatability.

SUMMARY

The nested sampling plan provides enough information to estimate the variance components (repeatability, reproducibility, precision), even when there are systematic problems with the instrument or sample material. In addition, the method provides information about accuracy and linearity. The stability study provides a way to measure the consistency of the variance components over time. Another benefit of the method is that it provides more data with which to perform detective work when results are suspicious. Also, it can reduce the total amount of work by allocating the right amount of resources for achieving each of the separate goals of a measurement instrument evaluation.

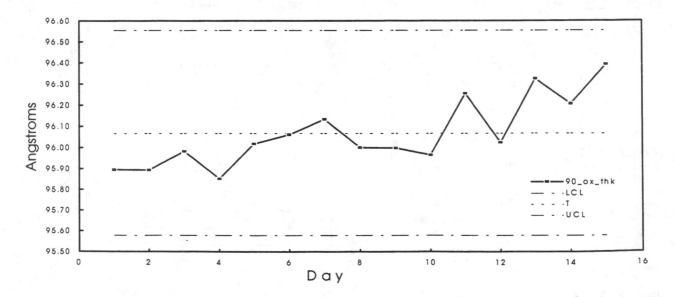

Figure 3 Trend Chart of Daily Mean of Film Thickness Data

---

[1] SAS is a registered trademark of SAS Institute, Cary, NC.

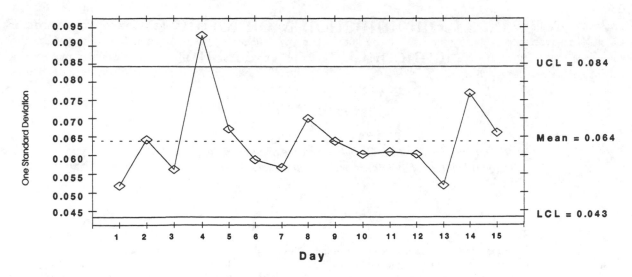

Figure 4 Trend Chart of Daily Standard Deviations of Film Thickness Data

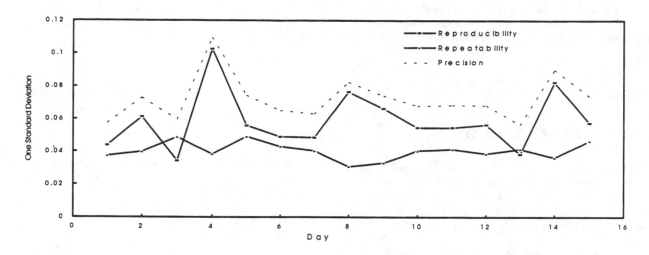

Figure 5 Trend Chart of Variance Components from Film Thickness Data

REFERENCES:

1. Wheeler, D. J., and Lyday, R. W., *Evaluating the Measurement Process*, SPC Press Inc., 2nd Ed, 1989, pp. 35-47.

2. Wheeler, D. J, "Problems with Gauge R&R Studies", 46th Annual Quality Congress, ASQC, 1992, pp. 4-5.

3. Automotive Industry Action Group (AIAG), *Measurement Systems Analysis Reference Manual*, ASQC, 1990, pp. 38-46, 65-70,

4. John, P. M., *Alternative Models for Gauge Studies*, SEMATECH technical transfer document 93081755A-TR, 1993, pp. 41-44.

5. Montgomery, D. C., and Runger, G. C., *Gauge Capability and Designed Experiments, Parts I and II*, Arizona State University, Publication Numbers 92-11, 92-12.

6. Montgomery, D. C., *Design of Experiments*, Wiley, 3rd Edition, 1991, pp. 439-450.

7. Beyer, W. H., *CRC Handbook of Tables for Probability and Statistics*, 2nd ed., CRC Press, Inc., 1988, pp. 28,31,451-454.

8. SAS Institute, *SAS/STAT User's Guide, Volume 2, GLM-VARCOMP, (The NESTED Procedure)*, 1990, ch 28.

Gas Contamination Monitoring for Semiconductor Processing

Charles R. Tilford

Thermophysics Division, Chemical Science and Technology Laboratory
National Institute of Standards and Technology, Gaithersburg, MD 20899

Many semiconductor processes are limited by low-level impurities introduced with process gases or generated within the process chamber. Both impurities and reaction products are often monitored using Residual Gas Analyzers (RGAs) or Partial Pressure Analyzers (PPAs). Unfortunately, the sensitivity of these instruments can be affected by a number of instrument and vacuum-environment variables, so that their performance can differ significantly from the expected. The magnitude of the resultant errors varies greatly for different instruments, and can reach orders of magnitude in extreme cases. The Vacuum Group of the National Institute of Standards and Technology is addressing these problems with a three-part program: It is testing RGAs to determine the magnitude of unwanted effects and how best to minimize them; it is developing in situ techniques to allow the calibration of RGAs without removing them from process chambers; and it is developing new types of vacuum standards for reactive gases for which there are presently no standards.

INTRODUCTION

The quality of vacuum-processed devices can be significantly affected by low levels of molecular contaminants. Classic examples are the changes in the resistivity and mechanical strength of sputtered aluminum films caused by part-per-million (ppm) levels of water, nitrogen, or oxygen in the argon sputtering environment. Low-level contaminants can enter the processing chamber as impurities in the process gas, or can be generated directly in the process chamber by leaks, outgassing, or chemical reactions.

Residual Gas Analyzers (RGAs) or Partial Pressure Analyzers (PPAs), most commonly mass spectrometers of the quadrupole type, indicate the partial pressures, as a function of molecular weight, of all the gases in a vacuum environment. Although originally developed for the qualitative diagnosis of residual gases in ultra high vacuum systems, the growing need for better vacuum process control has encouraged the application of these instruments to the quantitative monitoring of process gases, reaction products, and impurities. However, the performance of these instruments varies significantly, depending on design and conditions of use. In the best cases, after proper calibration, they can be accurate to within a few percent; in other cases they can have orders-of-magnitude errors. The unpredictable performance of some instruments has caused misleading results, frustrated users and probably discouraged the wider use of a potentially extremely valuable instrument.

The National Institute of Standards and Technology (NIST) Vacuum Group has a three-part program to address these problems. The first part is an ongoing effort to evaluate the performance of RGAs. The purpose is to inform users of the magnitude of possible problems so that they can interpret their results accordingly, and to determine optimum operating procedures and possible design improvements for these instruments.

Initial RGA test results indicate that instrument performance is so dependent on instrument and application variables that calibration near the time of use and under the conditions of use will be required. Prompted by these results, the second part of the NIST program is directed towards the development and testing of in situ calibration systems for RGAs. These calibration systems will be designed to be integrated directly into process chambers.

Many semiconductor processing gases are not compatible with existing vacuum standards and may adversely affect the performance of RGAs. Quantitative measurements with these gases will require new measurement techniques. The third part of the NIST program is the development of quantitative, species-specific, optical techniques that can be used as reference standards for reactive gases. These same techniques may also lead to quantitative non-intrusive sensors. Promising results have already been obtained using multi-photon, resonance-enhanced laser ionization, and efforts are underway to exploit a new optical absorption technique.

RGA PERFORMANCE

Several studies of RGA performance have been carried out in the last few years (1-6). The studies at NIST (5,6) have involved the repeated calibrations of more than a dozen different RGAs with partial pressures of different gases between 10^{-7} and 10^{-1} Pa (1 torr = 133.3 Pa). The performance of these instruments, especially mass resolution and sensitivity as a function of mass-to-charge ratio, will depend on a number of factors, including ion energy, electron emission current, and both the dc and rf potentials in the quadrupole filter. In most instruments these factors can be varied by adjusting different instrument operating parameters. Unfortunately, most manufacturers provide little or no information on the overall performance effects of varying these parameters. Therefore, in the NIST study the calibrations of some instruments were repeated with different combinations of instrument operating parameters. In some cases the calibrations were performed with pure gases, in others with combinations of two or three gases, each of which could be independently controlled and measured. The calibrations for inert gases were referenced to NIST primary vacuum standards (7), either directly or with calibrated transfer gages. Water calibrations used a new NIST primary water vapor standard (6, 8). Brief examples of the NIST results are presented below. Further details and results obtained by other workers can be obtained from the references.

Sensitivity as a Function of Pressure

As a first approximation, the sensitivity of a hot-cathode ionization-type instrument is independent of pressure, i.e., the ion current is linear with pressure. However, with the more complicated RGAs the linearity of the detected ion current, after it passes through the mass filter, can be strongly influenced by instrument design, operating parameters (5), and even history of use (2). The most influential operating parameter is the ion-extraction voltage (labeled "ion energy" in many instruments), the potential difference between the center of the quadrupole filter and the ion-source anode. The observed nonlinearities can be broadly categorized as "low-pressure", typically occurring at pressures of 10^{-4} Pa and below during operation with high ion-extraction voltage settings, and "high-pressure", typically occurring at pressures of 10^{-3} Pa and above during operation with low ion-extraction voltages.

The range of observed performance can be appreciated from Fig. 1. The data in this figure, discussed in detail in (5), were selected from 27 different data sets obtained for each instrument with different ion-source operating parameters: emission current, electron-accelerating voltage, and ion-extraction voltage. To emphasize the changes in linearity, the data from each set have been normalized to a value of 1 at 10^{-4} Pa. The data presented were chosen to illustrate extremes of behavior: best linearity, maximum low-pressure nonlinearity, and maximum high-pressure nonlinearity. For PPA-D the performance illustrated by the line with no symbols (sensitivity constant to within a few percent up to 10^{-2} Pa) was obtained for a wide range of operating parameters, the high-pressure nonlinearity was observed only under extreme settings of the operating parameters, and virtually no low-pressure nonlinearity was observed. For PPA-A significant nonlinearities were observed for all combinations of operating parameters. The line with no symbols was the best observed performance, and large

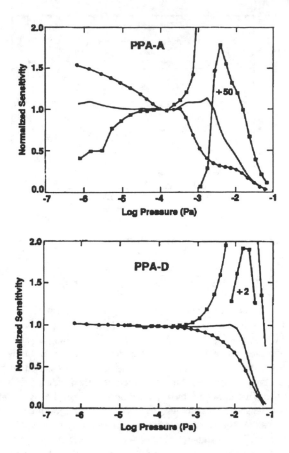

Figure 1. Argon sensitivities (normalized to 1 at 10^{-4} Pa) for two different RGAs operated with different combinations of ion source parameters. The data selected represent the best linearity and worse linearities for each instrument.

low- and high-pressure nonlinearities were observed for a wide range of operating parameters. In particular, when operated at low ion-extraction voltages the sensitivity of this instrument changed by more than two orders of magnitude as a function of pressure. For both types A and D, less-extensive calibrations of additional instruments of the same designs gave similar results.

The most probable explanation for these effects focuses on the effects of space charge in the ionizers. If this explanation is correct, then the sensitivity for any gas will depend on the total pressure. This problem is discussed in the next section.

Background Gas Effects

It is implicitly assumed that an RGA measures the partial pressure of one gas independent of the pressures of other gases. This assumption has been tested using calibration systems in which two or three gases can be simultaneously and independently controlled and measured (2, 5). The assumption is not always valid.

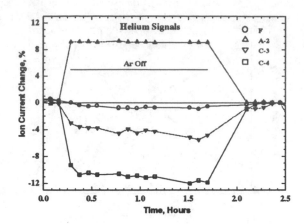

Figure 2. Change in the helium signals for four different RGAs when the helium pressure (1.2×10^{-4} Pa) is held constant and an argon background pressure is changed from 8.6×10^{-5} Pa to "zero" and then restored.

Figure 2 illustrates the results of such a test with two inert gases. The helium responses of several RGAs were simultaneously monitored as a function of the pressure of a second gas - argon. A helium pressure of 1.2×10^{-4} Pa was maintained constant to within 1 % throughout the experiment. The experiment started with an argon pressure of 8.6×10^{-5} Pa, which was then reduced to "zero" for the time indicated by the horizontal line on the figure, after which the argon pressure was restored. The changes in the helium responses of four instruments are shown; three are quadrupoles and instrument F is a magnetic-sector. As can be seen, the responses of all four instruments were stable with time, and that of instrument F was little affected by the change in the argon pressure. However, the helium sensitivities of the three quadrupoles variously increased or decreased by up to 12 % when the argon was "turned off". Not included in the figure were data for two other quadrupoles tested at the same time; their response was similar to F, i.e., they showed virtually no change in their helium sensitivity.

The results of Fig. 2 illustrate the variety of responses observed for different instruments. In general, the responses depend not only on the particular instrument, but also on the combination of gases and on the total pressure. Typically, argon sensitivities are much less affected by changes in a helium pressure. On the other hand, instruments that are unaffected by changes in argon or helium background pressures can be significantly affected by a change in water pressure. Generally, with higher background pressures the sensitivity changes are larger; in some cases we have seen order-of-magnitude changes for background pressure changes in the 10^{-3} to 10^{-1} Pa range. This is of particular concern since for many semiconductor-process-monitoring applications the total pressures are in this range, and small changes in the background pressure can cause large changes in the indicated partial pressures of low-pressure gas species.

Other Effects

The sensitivity of an RGA for different gas species will depend not only on the ionization potential of the species, but also on the resolution settings of the instrument and the ion optics

between the ionizer and the mass filter. Because of this, the ratios of the sensitivities for different gases can differ significantly from the relative gas sensitivities published for ionization gages - in some cases the differences can be as large as an order of magnitude. The actual sensitivities for a particular RGA should be determined by calibrating with the gases of interest.

For higher sensitivities, many RGAs amplify the ion signal with secondary electron multipliers (SEMs). SEMs are susceptible to gain changes, particularly upon exposure to reactive gases. If SEMs are used they should be periodically recalibrated by comparison with a Faraday-cup detector.

Although the root causes are not well understood, exposure to reactive gases, including water, oxygen, and hydrogen, can cause RGA sensitivities for all gases to change, even during operation with Faraday-cup detectors. In our experience these changes can be positive or negative, and as large as 20%. When the reactive gas is removed we typically find that the sensitivities will asymptotically return back to original values with time constants of many hours.

IN SITU CALIBRATION SYSTEMS

Laboratory calibrations can be used to detect anomalous RGA behaviour and assist in the adjustment of the instrument to minimize undesirable characteristics. However, many RGA characteristics tend to be time and use dependent, and may depend on the particular combination of gases in a process chamber, so quantitative monitoring of semiconductor processes will probably require that RGAs be periodically calibrated with a combination of gases approximating the process conditions. In a manufacturing environment it is desirable (essential) that this be done without removing the RGA from the process chamber. Reference 9 briefly discusses in situ RGA calibration procedures using the common Bayard-Alpert ionization gage as a reference standard. However, this is not an optimal solution since it is difficult to perform calibrations with mixtures of gases with this technique.

We believe that a better and more usable solution is to adapt the techniques used for primary vacuum standards. The most-common technique is to generate standard pressures by passing a known flow of the calibration gas through a calculated conductance (known pumping speed). When implemented for the best accuracy (7), this technique, particularly the flow generation, is expensive and complicated. However, in recent years we have found that it is possible to generate known flows, with reduced but adequate accuracies, using simple flow elements and pressure gages (10). If the gas flow is introduced into a process chamber at the inlet of the vacuum pump, it is possible with inert gases to generate stable and uniform pressures throughout the process chamber. If the pumping speed is determined using a calibrated vacuum gage, the pressure in the calibration chamber can be known with accuracies approaching 5%. If additional flow generation systems are added and the system is operated in molecular flow conditions, it is possible to simultaneously generate known partial pressures of different gases.

We are currently working to implement this approach in a manner that can be automated and operated by technical personnel, and we plan to test it in a processing environment.

STANDARDS FOR REACTIVE GASES

Primary vacuum standards are pressure or density generators that depend on conservation of a flow or quantity of the calibration gas. Therefore, their performance can be seriously degraded when operated with surface-active or reactive gases, particularly at lower pressures. While we have successfully applied the orifice-flow (7) technique to a primary standard for water vapor (8), the response time of this system is of the order of hours, and it is clear that response time will be unmanageable for more reactive gases, e.g. oxygen. As alternatives to conventional techniques, we have been developing quantitative species-specific optical measurement techniques that can be calibrated at higher pressures, and then extrapolated to establish reference standards at much lower pressures or densities.

A variety of techniques have been used as qualitative optical diagnostics, often to monitor chemical reactions or for low-level species detection. The NIST Vacuum Group has initiated work with two of these techniques to enhance detection levels and determine their reproducibility and independence from background or gas interference effects.

Initial results obtained with the first technique, Resonance Enhanced Multi Photon Ionization (REMPI) are described in Refs. 11 and 12. With careful control and monitoring of laser power and beam shape, it was possible to achieve detection levels of 10^{-10} Pa for CO and a two-sigma reproducibility of +/- 15%.

Since the REMPI photoionization process depends on a virtual energy level, that in turn depends on specific molecular characteristics, REMPI should be species specific, i.e., it should be unaffected by the presence of other gases. For CO this has been tested, most notably with N_2, which has the same nominal mass as CO, and cannot be distinguished from CO by most RGAs. The results of such a test are shown in Fig. 3. In this experiment a constant pressure of CO, 2×10^{-6} Pa, was maintained and the REMPI signal monitored while a N_2 background was changed over seven decades. As can be seen, the CO signal is unaffected until relatively high N_2 pressures, where we believe the signal decreases due to molecular scattering in a time-of-flight detector. Work is underway to quantify the REMPI technique for water and other reactive gases.

Optical absorption has long been used for gas diagnostics, both in a vacuum environment and in atmospheric ambients. The sensitivities of conventional absorption techniques are often limited by the difficulties in measuring small intensity differences between a probe beam and a reference beam. This difficulty can be ameliorated by a new approach that measures the exponential decay of a light pulse stored in a low-loss optical cavity - a so-called ring-down cavity (12). The decay time constant is determined by the optical losses in the cavity, which will be dominated by losses in the optical media (gas) if high-reflectivity mirrors are used. The time constant measurement is effectively immune to detector drifts, gain changes, and absorption outside the cavity. Experiments are underway to evaluate this technique with oxygen and water.

ACKNOWLEDGEMENTS

Much of the work described here was carried out by my colleagues; Albert Filippelli, Laszlo Lieszkovszky, J. Pat Looney and Stuart Tison. Continuation of the work on RGAs is supported by the National Semiconductor Metrology Program.

REFERENCES

1. Mao, F.M., Yang, J.M., Austin, W.E. and Leck, J.H., Vacuum **37** 335-338 (1987).

2. Austin, W.E., Fu Ming Mao, Jin Man Yang, and Leck, J.H., J. Vac. Sci. Technol. A **5**, 2631-36 (1987).

3. Reid, R.J. and James, A.P., Vacuum **37**, 339-342 (1987).

4. Austin, W.E., Leck, J.H. and Batey, J.H., J. Vac. Sci. Technol. A **10**, 3563-67 (1992).

5. Lieszkovszky, L., Filippelli, A.R. and Tilford, C.R., J. Vac. Sci. Technol. A **8**. 3838-54 (1990).

6. Tilford, C.R., in James B. Breckinridge and Alexander J. Marker III (eds.), SPIE Vol. 1761, <u>Damage to Space Optics and Properties and Characteristics of Optical Glass</u>, Vol. 1761, SPIE, Bellingham WA, 1992, p. 119-129.

7. Tilford, C.R., Dittmann, S. and McCulloh, K.E., J. Vac. Sci. Technol. A **6**, 2853-59 (1988).

8. Tison, S.A. and Tilford, C.R., in Benjamin A. Moore and Joseph A. Carpenter, Jr. (eds.), <u>RL/NIST Workshop on Moisture Measurement and Control for Microelectronics</u>, NIST Internal Report 5241, NIST, U.S.A., 1993, p. 19-29.

9. Tilford, C.R., Surf. Coat. Technol. **68/69**, 708-712 (1994).

10. Tison, S.A., Vacuum **44**, 1171-75 (1993).

11. Looney, J.P., Harrington, J.E., Smyth, K.C., O'Brian, T.R., and Lucatorto, T.B., J. Vac. Sci. Technol. A **11**, 3111-20 (1993).

12. Looney, J.P., J. Vac. Soc. Jpn. **37**, 703-710 (1994).

13. Romanini, D.. and Lehmann, K.K., J. Chem. Phys. **99**, 6287-6301 (1993).

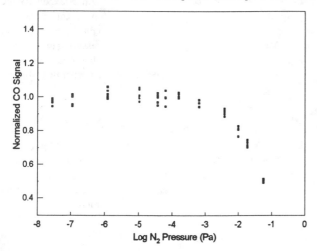

Figure 3. REMPI signal for a constant CO pressure of 2×10^{-6} Pa as a function of nitrogen background pressure.

Liquid-borne Sub-Micron Particle Detection through Acoustic Coaxing

Sameer I. Madanshetty
Aerospace and Mechanical Engineering
Boston University
110 Cummington St., Boston, MA 02215

A novel approach to the detection of liquid-borne sub-micron particles (extendible to ultra-clean liquids) is described. The key concept is to coax the sub-micron particles to soft cavitate and then detect the ensuing transient bubble activity acoustically rather than the particle itself (which has only a weak scattering signature). The method, therefore, relies on facilitating acoustic microcavitation through acoustic coaxing. Acoustic microcavitation is brought about by low megahertz acoustic fields giving rise to micron size bubbles that live a few microseconds. Liquid-borne microparticles do not, ordinarily, cause any cavitation when exposed to strong sound fields (of 1 MHz). If, however, a very weak, high frequency auxiliary acoustic field (e.g. 30 MHz) is added to this sound field, cavitation by the microparticles is readily facilitated. We term this technique of facilitating cavitation, "acoustic coaxing". Results of preliminary experiments indicate that even smooth spherical micro-particles can be coaxed to cause cavitation. An explanation of the "Acoustic Coaxing Effect" is offered. The physics seems not to be limited by the smallness of microparticles (up to 50 nm). This novel method based on the acoustic coaxing of microcavitation promises to be a good basis for an on-line, real time monitor of liquid-borne submicronic particulate presence. This method is not limited to small sensing volumes and, unlike optical methods, it has an intrinsic, location specific, signal enhancement at the source particle.

INTRODUCTION

Presence of particulate contaminants severely limits the manufacturing of semiconductors. With line widths being reduced by 70% every three years, the smallest tolerated size of particulates (a third of the line width) keeps becoming proportionately finer. The 1995 road-mapped specifications for a DRAM chip (minimum feature size 0.35μm) require the control of all particulates larger than 0.12 μm (10). Traditional methods are finding it difficult to cope with the increasingly stringent demands on water purity. Further, on-line, real-time detection of such sub-micron liquid borne particles in the clean water systems used for processing silicon wafers remains a formidable task requiring new methods. Addressing this challenge, we present a novel method of particle detection based on the newly found acoustic interaction—"acoustic coaxing".

The outline of topics discussed is as follows: 1. The limitation of conventional (pulse-echo backscattering off of particles) acoustic detection. 2. Solution principle of the new method via cavitation: acoustic microcavitation using "acoustic coaxing". 3. A brief explanation of the physics of "acoustic coaxing". 4. Experiments to support this detection method. 5. Conclusion.

THE LIMITATION OF CONVENTIONAL ACOUSTIC DETECTION

For very small scattering objects ($a << \lambda$, where a is the radius of the particle, and λ is the wavelength of the incident plane acoustic wave) with a strong density-compressibility contrast with respect to the host medium, following Kino (3) one may assume quasi static pressure, p, field variations in the neighborhood of the object. Then the ratio of the scattered intensity I_s, to the incident intensity I_i, at a radial distance r from the center of the scattering particle, is given by:

$$\frac{I_s(\theta)}{I_i} = \frac{k^4 a^6}{9r^2} \left[\frac{3\left(1 - \frac{\rho'_{mo}}{\rho_{mo}}\right)}{1 + \frac{2\rho'_{mo}}{\rho_{mo}}} \cos\theta + \left(1 - \frac{\kappa}{\kappa'}\right) \right]^2 . \quad (1)$$

Here ρ'_{mo}, and κ' are the density and compressibility factors, respectively, inside the scattering object, while the corresponding parameters without the ' superscript refer to the host medium. For forward scattering $\theta = 0$, while for backward scattering $\theta = \pi$, θ being the angle between the scattering direction and the incident wave. Since intensity involves the square of scattered pressure, this equation implies that scattered pressure varies as the square of the frequency and as the cube of the scattering sphere's radius (its volume).

To appreciate the limitations of the conventional acoustic detection of microparticulates—pulse-echo backscattering off of particles—one needs to compare the scattering from the particles, and air bubbles. Because of the strong density and compressibility contrast for air bubbles in water, the backscattering from an air bubble in water is approximately 80 dB greater than that from a similar-sized, spherical sand particle in water. Detecting bubbles, instead of the particles themselves, involves a natural signal enhancement by a factor of 10,000.

SOLUTION PRINCIPLE OF THE NEW METHOD VIA CAVITATION

The above example highlights the stark contrast between the ease of detectibility (using acoustic scattering methods) of bubbles as opposed to the particles themselves. With this insight one is prompted to consider liquid-borne sub-micron particle detection differently. The problem becomes: How can a particle become a bubble, i.e., how might one associate a bubble with a microparticle? At once, one realizes that cavitation may hold the key.

Cavitation involves the formation of cavities, bubbles, within the bulk of a liquid in response to a local reduction in pressure, e.g., the alternating pressures of a sound wave. The site of bubble formation is invariably an incompletely wetted crevice-like feature on a liquid-borne particle or the container wall. Such adventitious particulate motes, conditions with the potential to nucleate a cavitation event, are quite infrequent. Cavitation is a rare phenomenon. The key design question is thus: How can one coax the liquid-borne sub-micron particles to give rise to cavitation? Recent research by Madanshetty identifies "acoustic coaxing" as the needed break-through (4—7).

For detailed reviews in acoustic cavitation the reader is referred to Flynn (2), Neppiras (8), Apfel (1), and Prosperetti (9). Typically one thinks of cavitation in terms of a pre-existing gaseous presence either as a bubble or a stabilized pocket of gas trapped in a suitable container crevice. Cavitation can occur if a sound wave imposes sufficiently strong tensile fields. The oscillating pressure that a sound wave is implies restrictions on the time duration of the available tensile field (within a period of oscillation). At high frequencies it becomes increasingly difficult to bring about cavitation because insufficient time is available for the bubble growth, which precedes the cavitational implosion. Consequently the cavitation thresholds of bubbles become greater at higher frequencies,

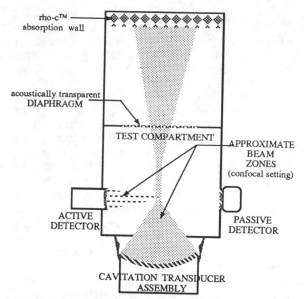

FIGURE 1. SCHEMATIC OF THE TEST CELL
The test cell is divided into two compartments so that the rho-c™ impedance matching rubber wall immersed in water provides a non reflecting boundary while remaining isolated from the cavitation chamber. An acoustically transparent stainless shim stock forms a water tight seal between the two compartments. To detect cavitation two kinds of acoustic detectors are used. The first one is an unfocused, untuned 1 MHz receiver transducer which serves as a passive detector. The other one is a focused 30 MHz transducer which is used in pulse-echo mode and is called the active detector. Cavitation itself is brought about by a focused 0.75MHz PZT-8 crystal driven in pulse mode (tone bursts), typically 10μs long pulses at 1kHz PRF. The active detector is arranged confocally with respect to the cavitation transducer. Both the interrogating pulse and the cavitation pulse arrive simultaneously at the common focus, which is the region of cavitation.

and for practical purposes, it becomes very difficult to bring about cavitation (in water) at frequencies in excess of 10 MHz [cf. ...in excess of 3 MHz, (8)]. Therefore, it appears paradoxical that a high frequency field may actually help lower the thresholds and facilitate cavitation. Madanshetty (7) presents the first report of this lowering of thresholds in the presence of high frequency fields.

Acoustic coaxing effects were found while researching acoustic microcavitation in water primarily at 0.75 MHz frequency and 1% duty cycle. See figure 1 for schematic of the test cell and a description of the apparatus. With the test chamber filled with clean water (distilled, de-ionized and extensively filtered) no cavitation was observed, even when the cavitation transducer was driven to give its peak output in excess of 22 bar peak negative.

When the test chamber with polystyrene latex microparticles suspended in a clean water host was irradiated with short (10 μs long) acoustic pulses from the focused cavitation transducer, the cavitation thresholds indicated by the passive detector were around 15 bar peak negative. When the 30 MHz focused active transducer was switched on to operate in the pulse-echo mode, it measured thresholds of 5 bar peak negative. This observation in itself, may amply speak for the superior sensitivity of the active detector. However, when the thresholds were measured with the passive transducer, while the active transducer field was left switched on, the passive thresholds dropped to around 7 bar peak negative. This difference in the measured passive thresholds, 15 bar peak negative with the active detector off, and 7 bar peak negative with the active detector on, suggests that the cavitation environment was being influenced by the detection field of the active detector. In the experimental setup, the active detector focal pressure amplitudes were invariably less than 0.5 bar peak negative. Even with the crudest superposition, this cannot account for the lowering of passive detector thresholds from 15 bar peak negative to 7 bar peak negative. In fact, it becomes increasingly difficult to initiate cavitation at increased acoustic frequencies and it was found that the active detector did not cause cavitation on its own.

The polystyrene latex microparticles of 0.984 μm mean diameter, added to the host water were monodispersed, spherical, and smooth. Extensive observation of the particles under the scanning electron microscope did not reveal any significant surface flaws or crevices down to about 500 Å, the resolution limit of the device used. These observations of the smoothness of the latex microspheres in the context of the crevice model of cavitation raise the questions: *How can smooth spherical particles help initiate cavitation? How does the acoustic field of the active detector facilitate the process?*

A BRIEF EXPLANATION OF "ACOUSTIC COAXING"

In this section we present a conceptual model to answer these questions. A detailed theory is presented in Madanshetty (4) (see figure 2—the scenario). A reasonable starting premise is that the energetics of the solid-liquid interface must preclude wetting at some scale of fineness. At some scale of fineness, roughness must persist and result in unwet sites on the particle surface. Although the surfaces of the polystyrene latex particles are smooth down to 500 Å, surface gas pockets may exist at the nano scale. If a single nano-gas-dot of diameter 50 nm (500 Å), were to cavitate alone in the tensile environment of the cavitation transducer, the estimated threshold would be around 60 bar peak negative. The observed lower thresholds in the presence of the active detector field suggest that the nano-gas-dots aggregate to form sufficiently larger gas patches.

In water, the wavelengths at 30 MHz and 0.75 MHz are 50 μm and 2 mm, respectively, while the largest particles we have used are less than 1 μm in diameter. In these circumstances, a particle feels an essentially uniform

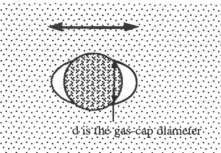

d is the gas-cap diameter

4. For cavitation to occur, the negative pressure in the tensile environment should overcome the surface tension force.

$$\sigma \pi d = p(\pi d^2/4)$$

FIGURE 2. ACOUSTIC COAXING EXPLAINED

pressure over its surface. Then, in water, the active detector operating at 30 MHz at a modest pressure level setting of 0.5 bar peak negative can give rise to a particle (fluid parcel) acceleration of about 6.47×10^6 m/s^2, or equivalently 6.5×10^5 g's. At any given point a particle denser than water will inertially lag the acceleration, while a less dense particle will move in the direction of acceleration.

In the present case we are considering polystyrene spheres whose density is 1.05 g/cc, which are reasonably density matched with water. Air density, on the other hand is about 1.2×10^{-3} g/cc. Coupled with high acceleration fields, this density contrast, a factor of 830, will enhance the kinetic buoyancy and urge the nano-gas-dots towards the fore and aft regions on the sphere where they agglomerate and form gas caps. One can visualize these gas caps as lens-like regions located at the extremities of an oscillating (to and fro) spherical particle. A cavitation event may be expected to occur when the force due to surface tension on the perimeter of the gas cap is overcome by the tensile forces effective on the gas region.

In the case of the polystyrene spheres of mean diameter 0.984 μm, the measured passive detector threshold is 7 bar peak negative (6) in the presence of the active detector field. The surface tension (0.073 N/m) force acting on the perimeter of a circle of diameter 0.984 μm can be overcome by a tensile force due to a pressure of just 3 bar peak negative acting on the corresponding projected area. Here we have assumed a gas patch of the same diameter as the particle. In reality, the gas patch would be smaller than the particle dimensions. This would, as observed, invariably give rise to thresholds higher than the lower limit corresponding to a gas patch as large as the particle diameter, or a bubble exactly encapsulating the particle. Recall that for a single nano-gas-dot of size 50 nm, to cavitate alone in the tensile environment of the cavitation transducer the estimated threshold would be around 60 bar peak negative.

The above mechanism of gas collection due to kinetic buoyancy effects helps us explain how cavitation is prompted by smooth particles and why one finds reduced thresholds in the presence of the active detector field—"acoustic coaxing". We are not yet in a position to predict the exact threshold values for a given particle/host context. More detailed experiments will help. For practical applications one will also need to separate the two tasks—(active) detection, and acoustic coaxing. In what follows, the effect of the active detector field is referred to as "acoustic coaxing".

EXPERIMENTS TO SUPPORT THE METHOD OF DETECTION

In the following we mention only the passive detection of cavitation and understand the effect of active detector as an incidental, unoptimised coaxing field. Two experiments illustrate the viability of acoustic coaxing for the detection of liquid-borne sub-micron particles:

Acoustic Coaxing Effect on the Passive Thresholds for Various Particles

Figure 3 shows how the passive thresholds vary with the strength of the active detector field, the acoustic coaxing field. Note that in these experiments the coaxing field is still very weak, and is on for a very short time (1% duty factor). (Zero on the abscissa corresponds to zero active detector field; active detector switched off.)

Essentially monodispersed, polystyrene particles of three different sizes, having mean diameters of 0.245 µm, 0.481 µm and 0.984 µm were tested. Particle number density was maintained the same at 1.9×10^8 particles/cm^3. The dissolved air saturation in the test cell was held around 70%. Experiments with increased number density of particles in the test cell (10^5 particles/cm^3 to 10^9 particles/cm^3) failed to produce any saturation effects; the passive thresholds in the presence of the active field remained lower than those in the absence of the active field.

When the active detector is switched off, it does not participate in the cavitation process; the thresholds are solely due to the main cavitation field. The observed passive thresholds are around 15 bar peak negative with the smaller particles giving a slightly greater value (figure 3). It should be noted that if the main cavitation transducer were to act alone, in order to give rise to the acceleration levels as high as those due to the active detector in normal setting, the main cavitation transducer

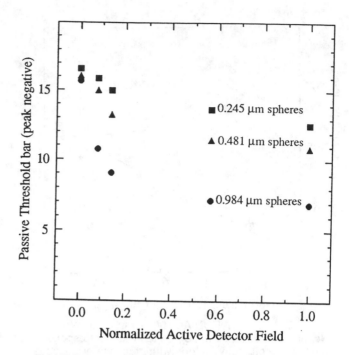

FIGURE 3. ACOUSTIC COAXING REDUCES CAVITATION THRESHOLDS

ought to be operating at the pressure level 15 bar peak negative. At similar acceleration levels it is reasonable to expect that the particles will be processed for gas cap formation similarly.

Note that in figure 3, the effect of particle size becomes more pronounced at the stronger active detector fields. This suggests that a proper choice of the coaxing field may enable one to say something about the size of the particles, beyond their mere detection.

Effect of "acoustic coaxing" on cavitation activity post-threshold conditions

When the acoustic coaxing field is present, liquid-borne microparticles exposed to acoustic fields stronger than the threshold strengths should cause cavitation quite readily. As explained above the coaxing field reduces the cavitation threshold and processes the particles uniformly to cause cavitation. In order to study the effect of acoustic coaxing on the cavitation activity post-threshold conditions, one will have to hold fixed the main cavitation transducer field at a higher level above the threshold value and monitor the cavitation activity by counting (for a fixed duration) the number of cavitation pulses during which cavitation is present.

Count rate experiments indicate that cavitation activity is low and sporadic in the absence of the "acoustic coaxing" field. The mere presence of the coaxing field increases the cavitation activity significantly. In the case of 0.75 µm silica particles at a number density of 3.6×10^8 particles/cm^3 the mean count rate per minute jumped

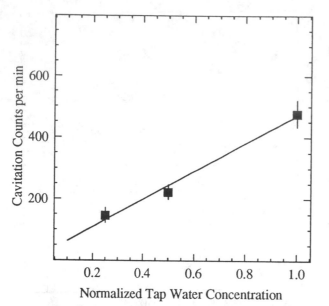

FIGURE 4. CAVITATION ACTIVITY FOR TAP WATER
Normalized concentration of the tap water corresponds to the test cell fully filled with tap water. Other concentrations of tap water imply that a fraction of the test cell was drained off and made up for volume by adding ultra-pure reference water (cavitation free host). The dissolved air saturation in the test cell was maintained at 85%.

from 50 without the active field to 12000 with the normal active field, when the over threshold was 4 bar. Most importantly, though, this activity seems to be directly proportional to the particle number density. Experiments with various particles—0.984 μm, 0.481 μm, and 0.245 μm polystyrene particles, and 0.75 μm silica spheres—and tap water (see figure 4) confirm this finding.

CONCLUSION

We have attempted to conceptualize a mechanism by which any particles, even smooth, spherical particles may nucleate cavitation events. Liquid-borne microparticles will ordinarily cause little cavitation at threshold conditions when exposed to strong sound field (the main cavitation field). But, if in addition to the sound field there exists a weak, high frequency auxiliary field, such as the coaxing field, cavitation by the microparticles is readily facilitated, and cavitation thresholds are significantly reduced. The mechanism of "acoustic coaxing" conceptualized here explains how the observed thresholds might come about and accounts for the observed dependence of post-threshold cavitation activity on the particle number density present in the detection region. When perfected, measurements of cavitation activity should lead to the counting of liquid-borne sub-micron particles. They will thus enable one to continuously monitor process purity. While we have attempted to establish the reasonableness of acoustic coaxing of microcavitation, future research will be directed to precisely quantifying the coaxing effect. The physics

of acoustic coaxing as envisaged here is not intrinsically limited by the smallness of the particles (at least up to 50 nm) to be detected as long as one is willing to use acoustic fields of commensurate strength—increasing the exposure to coaxing field, using greater duty factors, will reduce the cavitation thresholds needed for detection.

ACKNOWLEDGMENTS

This work was supported by the NSF grant CTS-9110920, and a grant from Community Technology Fund made available by the Trustees of Boston University.

References

1. Apfel, R. E., "Acoustic Cavitation," *Methods of Experimental Physics*, New York, Academic Press, 1981, **19**, pp 355—411.
2. Flynn, H.G., "Physics of Acoustic Cavitation in Liquids", *Physical Acoustics*, New York: Academic Press, 1964, **1 B**, pp. 57—172.
3. Kino, G. S., *Acoustic Waves: Devices, Imaging and Analog Signal Processing*, New Jersey: Prentice Hall, 1987, pp. 303—308.
4. Madanshetty, S. I., "A Conceptual Model for Acoustic Microcavitation," *J. Acoust. Soc. Am.* (Submitted), 1994.
5. Madanshetty, S. I., Roy, R. A., Apfel R. E., "Acoustic Microcavitation: Its Active and Passive acoustic detection," *J. Acoust. Soc. Am.* **90** (3), 1515—1527 (1991).
6. Madanshetty, S. I., Apfel R. E., "Acoustic Microcavitation: enhancement and applications," *J. Acoust. Soc. Am.* **90** (3), 1508—1514, (1991).
7. Madanshetty, S. I., *Acoustic Microcavitation*, Doctoral Dissertation, Yale University, 1989.
8. Neppiras, E. A., "Acoustic Cavitation", *Phys. Rep.* **61**, 160—251, (1980).
9. Prosperetti, A., "Physics of Acoustic Cavitation", *Frontiers in Physical Acoustics*, **XCIII**, 145—188, (1986).
10. Semiconductor Industry Association, *The National Technology Roadmap for Semiconductors,* 1994.

X-RAY METHODS

X-ray Scattering for Semiconductor Characterisation

B K Tanner

Department of Physics, University of Durham,
South Road, Durham DH1 3LE, U.K.

This paper presents a review of the status and potential of high resolution X-ray diffraction, reciprocal space mapping, X-ray topography and grazing incidence X-ray scattering techniques for semiconductor analysis. The application of these non-destructive techniques to compound semiconductor and silicon characterisation is described. Together with photoluminescence, high resolution X-ray diffraction is presently a key characterisation tool in compound semiconductor production lines and potential applications in silicon production are examined.

INTRODUCTION

In the early 1970s, there was a widespread belief amongst my fellow graduate students that electron diffraction would provide all the answers to semiconductor structural problems and that X-ray diffraction ended its development in the 1930s. Due to the large wavelength and narrow intrinsic width of the Bragg reflection, instruments were difficult to align and remained the province of a few, highly skilled and patient, specialists. Yet in the 1980s X-ray scattering was rejuvenated, principally through the introduction of low cost computer-control systems. With computer control and improvements in linear bearing technology, it has proved possible to transfer some of this instrumentation to the production line itself. X-ray diffraction has found a key niche in the total quality assurance process. It is rapid, it is non-destructive and, for compound semiconductors, the sample can be returned to the production process after analysis. While the silicon industry has yet to adopt this strategy, for reasons we will see later, the cost sensitivity of the compound semiconductor materials has led to X-ray diffraction and photoluminescence becoming standard characterisation techniques on the most advanced III-V production lines. The key strength of X-ray diffraction techniques is the very high strain sensitivity achievable when appropriate X-ray optics are employed. However, the intrinsically weak scattering also means that the depth of penetration cannot

simultaneously be small. Thus surface sensitive, grazing incidence diffraction techniques have a relatively low strain sensitivity compared with bulk methods. Figure 1 shows the wide applicability of X-ray diffraction in a comparison of techniques sensitive to near surface-strains.

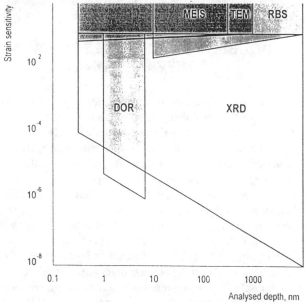

Fig 1 Minimum detection limits versus depth resolution for a number of techniques. XRD: X-ray diffraction, DOR: differential optical reflectivity, RBS: Rutherford back scattering, MEIS medium energy ion scattering, TEM: transmission electron microscopy. [Courtesy D. K. Bowen (1)]

SINGLE AXIS DIFFRACTION

Slit collimated X-ray diffraction techniques are principally used with single crystal semiconductors only for substrate orientation. The Laue technique, although used industrially to check the orientation of superalloy turbine blades, does not have the accuracy required for current wafer orientation specification. With considerable effort, a precision of 0.1^o is possible (2), just within the specification for compound semiconductor substrates. The method developed by Bond (3) provides an order of magnitude higher precision. A pinhole-collimated beam of X-rays is shone onto the crystal, carried on a V block and barrel, mounted on a two circle goniometer. The Bragg reflection is initially found for the planes parallel to the face to be cut and then the crystal rotated about its axis in the V block. If the axis is not perpendicular to the Bragg planes, the intensity will fluctuate sinusoidally. The crystal alignment with respect to the barrel or V blocks is adjusted until the intensity remains constant. Slit collimation provides precision to better than one arc minute, but care must be taken if this is to be transferred to the cutting machine. Although silicon has long been cut by such precision techniques they do not yet seem to be universally adopted by compound semiconductor vendors. As it is becoming clear that, for some important devices, the quality of epitaxy depends critically on the orientation of the substrate (4) this needs to be addressed.

DOUBLE AXIS (HIGH RESOLUTION) DIFFRACTION

Use of one or more beam-conditioning crystals results in Bragg peak widths being reduced to a few arc seconds, and comparable with the intrinsic, theoretical, reflecting range of a perfect crystal. This has proved particularly crucial in determination of the composition of Al in AlGaAs on GaAs (5,6) and n or p doped epitaxial silicon where the mismatch is so low that the individual Bragg peaks of substrate and epilayer are not resolved in a slit collimated instrument. The key to the high strain sensitivity, which may be between 10^{-4} and 10^{-8}, depending on the X-ray wavelength and reflection chosen, lies in the use of a beam conditioner to limit the angular divergence of the X-rays at the sample. This is a minimum when the beam conditioner and specimen are of the same material, and the same reflection is used (Fig 2). (Then the rocking curve width is the convolution of the two plane wave reflecting curves.) Such a simple two crystal arrangement, in which the beam conditioners are factory-set, is used in the Bede QC2a diffractometer (7) designed specifically for use in the production environment and incorporating a minimum of adjustments.

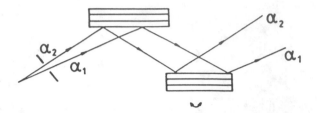

Fig. 2 The double crystal arrangement for high resolution XRD

Use of a monochromator in addition to a beam conditioner provides a compromise which enables one or two X-ray optical conditions to be used for most purposes. This removes the need to match specimen and beam conditioner each time and a wide range of specimens and reflections can be used without major adjustment to the instrument. Although pre-aligned beam conditioners which can be changed in a few minutes provide an excellent solution to the production line control problem (7), in a research environment, the flexibility that a monochromator brings is most important. The most widely used design is due to duMond (8) in which one pair of Bragg reflections is used to limit the angular divergence and a second pair to limit the wavelength dispersion. By use of pairs of reflections, the exit beam is made to be collinear with the incident beam. Bartels (9) exploited channel cut Ge crystals which could be used with either the 022 or 044 reflections for different sensitivities. Fig. 3 shows a design which exploits the change in reflecting range as the angle of incidence to the surface is reduced by off-cutting the crystal surface from the 022 Bragg planes of a monolithic Si crystal.

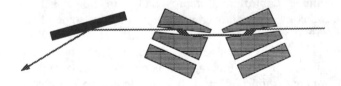

Fig 3 (a) Double axis diffraction with beam-conditioning monochromator. Asymmetric reflection, high intensity setting.

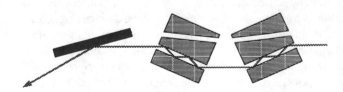

Fig 3(b) High resolution beam conditioning monochromator using multiple Bragg reflections in the channel-cut crystals.

By simultaneous translation of the beam conditioner and monochromator, and a few arc second rotation of the monochromator to allow for refractive index effects, high intensity or high resolution settings can be selected (10).

One major adjustment in high resolution diffraction is aligning the Bragg planes accurately normal to the plane of incidence. This means that the Bragg planes of specimen and reference crystal can then be brought exactly into parallelism. Failure to do this results in broadening of the rocking curve. Fig 4 shows the variation of the rocking curve as a function of tilt and we see that the function is very sharp. Rapid algorithms have now been developed (11) which permit this adjustment to be made automatically in a few minutes.

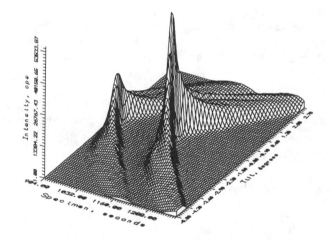

Fig 4 Experimental measurement of the diffracted intensity for a single AlGaAs layer on GaAs as a function of specimen rotation (rocking curve) angle and tilt out of the dispersion plane (11).

High resolution diffraction is particularly important in characterisation of epitaxial layers of compound semiconductors where it can provide data on composition, layer thickness, perfection, relaxation and uniformity. Two recent reviews provide an excellent means of access to the literature (12,13). The diffraction experiment provides a measure of the strain in the epilayer relative to the substrate and for ternary systems it is necessary to assume that Vegard's Law applies both to lattice parameter variation and the elastic constants in order to convert this to composition. For the AlGaAs on GaAs system there is still disagreement as to the correct parameters to use (6).

Dynamical diffraction of X-rays from perfect crystals is now well understood and the formulation of Takagi and Taupin has been particularly powerful for the study of III-V epitaxial structures. Development of fully dressed,

menu-driven PC code (14) has permitted widespread use of simulations for materials analysis. By calculation of the scattering from a model structure and matching with experiment, quantitative determination of the epitaxial layer parameters can be obtained. The quality of current III-V epilayers is such that excellent agreement between simulation and experiment is achieved (Fig. 5).

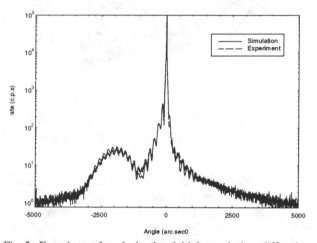

Fig 5. Experimental and simulated high resolution diffraction rocking curves. The structure used in the Bede RADS program was GaAs (substrate)/ 103nm $Al_{0.23}Ga_{0.77}As$/ 16nm $In_{0.12}Ga_{0.88}As$/ 34nm $Al_{0.23}Ga_{0.77}As$/ 70 nm GaAs. A Lorentzian diffuse scatter component and a background of 0.7 c.p.s. were included in the simulation.

Superimposed on the sharp peak (close to zero) and the broad InGaAs epilayer peak are well defined interference fringes. They correspond to a period determined by the total epilayer stack thickness. A significant development, achieved over the past decade, has been to use these fringes to bring the sensitivity of the high resolution diffraction method to a state where the thickness of strained layers can be measured to sub-monolayer precision. Such sensitivity arises when, as in Fig 4, a very thin highly strained epitaxial layer is sandwiched between two thicker layers of equal lattice parameter. The structure then forms an X-ray interferometer, and exactly in analogy with the optical case, a standing wavefield is set up between the top and bottom layers with a period of the Bragg plane spacing. Increase of thickness of the sandwiched layer by this amount leads to an oscillation of the intensity of the diffracted beam from these two layers. When the sandwiched layers is strained with respect to these layers, addition of a monolayer does not exactly correspond to a 2π phase shift and the fringes seen in Fig 4 move with respect to the sharp substrate peak (15). The sensitivity is inversely proportional to the mismatch between the strained layer and the top and bottom layers. For 20nm $In_{0.2}Ga_{0.8}As$ on GaAs and capped with 100nm

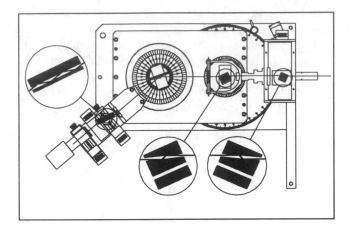

Fig 6. Schematic diagram of the triple axis technique in which an analyser is added to limit the angular range of X-rays accepted by the detector.

It is easy to show (12) that when the analyser is kept fixed and the specimen only rotated, the rocking curve is only sensitive to tilts and not dilations. Thus broadening of the rocking curve arises only from tilts and in imperfect materials such as II-VI compounds the contribution from these may be distinguished. Fig 7 shows a specimen-only scan of an epitaxial layer of (HgMn)Te grown on GaAs. In this system, the tilts dominate the double axis rocking curve broadening and, as evident from the narrow widths of the peaks in "θ-2θ" scans, there is little strain within the misoriented sub-grains (20). The tilts show a Gaussian distribution, the width of which decreases characteristically as the layer thickness increases.

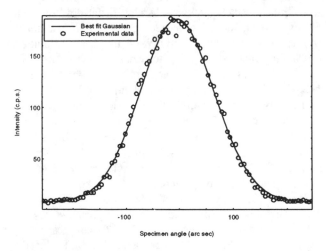

Fig. 7. Triple axis, specimen-only scan of a thick (Hg,Mn)Te epilayer on GaAs.

The intensity recorded in Fig 7 is extremely low and arises because the angular range of rays accepted by the analyser was too small. For the problem under investigation, the resolution was too high. In the study of relaxation in such highly mismatched systems as here, or for example InGaAs on GaAs, slit collimation of the detector may provide appropriate resolution and restores the intensity. Two slits are essential, as they act to limit the angular acceptance of the detector. The approach is similar to use of an area detector to perform quasi-triple axis analysis of imperfect materials at low resolution.

The realisation that incorporation of an analyser permits the intensity scattered around the reciprocal lattice point to be mapped as a function of position in reciprocal space has resulted in a shift of perspective in crystal characterising scientists. Whereas, most crystal growers had confined their attention to real space studies of inhomogeneities, the use of reciprocal space for the presentation of data has moved from being the preserve purely of physicists to that of the materials scientist. Transformation from angular motions to reciprocal space coordinates is simple. The θ-2θ scan, where the detector moves at twice the rate of the specimen, corresponds to movement in a direction q_z parallel to the reciprocal lattice vector. The specimen-only scan corresponds to movement in a direction q_y at right angles to this line. For detector angular movement $\delta\varphi$ and specimen angular movement $\delta\psi$, the values of q_z and q_y (all referenced to the reciprocal lattice point) are

$$q_z = \delta\varphi \; \cos\theta_B / \lambda$$

$$q_y = (2\,\delta\psi - \delta\varphi)\,\sin\theta_B / \lambda$$

where θ_B is the Bragg angle and λ is the X-ray wavelength. From these formulae it is easy to see that no change in φ, i.e. $\delta\varphi = 0$, means no change in q_z. Similarly, stepping ψ by one unit and ϕ by two units results in no change in q_y. Thus a combination of motions of the detector and specimen can be designed to map out the volume in reciprocal space around the reciprocal lattice point from which diffraction is occurring. (Note that the length of the slit normal to the dispersion, i.e. diffraction, plane results in an integration of the scattering in the q_x direction.) Fig. 8 shows a typical reciprocal space map of the scattering around the 004 reciprocal lattice point from a (001) oriented, polished, GaAs wafer. The vertical streak of scatter is the "surface streak" from the "truncation rod" which arises due to the truncation of the crystal at the surface. On a logarithmic scale it dominates the scattering from a perfect crystal. The diagonal streaks arise due to the fact that beam conditioner and analyser diffract over a range of angles. These streaks decrease dramatically when multiple reflection grooved crystals are used, but can never be totally suppressed. Fig 9 shows the equivalent plot for a lapped GaAs wafer. The circular feature around the

266

GaAs, it is straightforward to determine the layer thickness to monolayer precision (16).

Sub-monolayer precision is also achieved in studies of superlattices. Here, not only can the composition, the superlattice period and well-to-barrier ratio be determined, but subtle information about the interface shape can be obtained. Matyi and co-workers (17) have shown outstandingly good agreement between experimental and simulated structures in Si/GaAs superlattices. In order to fit the data a monolayer of alloy at the interface was necessarily included. Similar sensitivity has been found by Müller (18) in MOVPE-grown InGaAs/InP superlattices where a monolayer of InAs was identified at the InP/InGaAs interface and a 20Å linearly graded layer at the InGaAs/InP interface.

The need to control relaxation in strained layer optoelectronic device structures is critical and high resolution diffraction provides a powerful method for such monitoring. Key to this is the use of at least one asymmetric reflection in which the Bragg planes are not parallel to the specimen surface, thereby measuring a component of strain parallel to the interface. A surprisingly large number of practitioners still use only symmetric reflections, which intrinsically cannot distinguish between relaxation and variation in composition of ternary or quaternary systems. Assuming no tilts between epilayer and substrate, the symmetric plus one asymmetric reflection is sufficient, and provided that the layer is thick, an analytical expression can be found which yields directly the relaxation from the epilayer and substrate peak splitting. The next few years will see much more extensive use of this form of analysis.

Thus, together with photoluminescence, high resolution diffraction is a key non-destructive assessment tool in compound semiconductor technology. Realisation that for high volume MOVPE production lines, these two measurements may be sufficient for quality assurance has led to the recent commercial development of a single instrument which combines PL and X-ray mapping over an area up to 4 inches square.

In silicon technology, high resolution X-ray diffraction has been less extensively used. For Si-Ge on Si epitaxy, there is an extensive literature and the resolution requirements and the information available are similar to that for III-V systems. Fewster (13) notes that significant deviation from Vegard's Law is found in this system. High resolution diffraction is sensitive to very thin layers of highly strained material and thus can be used to characterise delta doping of silicon (19). It has also been applied to the study of doped epitaxial layers of Si on Si. However, for typical device-relevant doping levels, the epilayer peaks is very close to that of the substrate. Use of multiple reflections in the beam-conditioner enables the tails of the Bragg peaks to be reduced and the epilayer peaks resolved. For the study of oxygen levels in Si, high order Bragg peaks and short wavelength radiation are necessary. The 0 0 12 reflection with MoKα radiation provides a means of measuring strains to parts in 10^7, but control of temperature to parts in 10^2 of a degree is necessary.

TRIPLE AXIS DIFFRACTION (RECIPROCAL SPACE MAPPING)

Use of an analyser crystal (or collimating slits) after the specimen and before the detector provides a means of determining the direction of the scattered X-rays. This technique, which has grown dramatically in popularity in the past few years, provides both a way of separating coherent and diffuse scatter, and also separating the effects of dilations and tilts in the broadening of the rocking curve. The three dimensional maps of the scattering in reciprocal space also provide means of identifying layer-substrate misorientation and the specific origin of diffuse scatter. It is shown that the technique is applicable to a wide range of semiconductors and that, with correct selection of the X-ray optics, useful data from both highly perfect materials such as silicon and relatively imperfect materials such as CdTe can be obtained.

Fig. 6 shows a schematic diagram of the triple axis method. As the third axis is concentric with the specimen axis, the analyser stage is conveniently carried on the detector axis and thus features as an accessory on commercial systems. As with double axis diffraction, it is essential both to have a means of rotating the analyser in the dispersion plane and tilting to bring the Bragg planes normal to the dispersion plane. The former precision must be one arc second or better, as for the specimen axis. Although this type of measurement is not new, problems of alignment confined use to a few specialised laboratories. However, the recent availability of well engineered commercial instruments has now put the technique into many crystal characterisation laboratories. By use of precision slides on both detector and channel-cut analyser (fig. 6), the switch between double and triple axis settings takes only a few seconds. Once aligned by a very straightforward procedure, the analyser does not need significant readjustment. Although the intensity is substantially reduced by the addition of the extra crystal, tens of thousands of counts per second can be recorded from low order Bragg peaks off high perfection crystals with a conventional generator.

reciprocal lattice point arises from scatter associated with the strain around the defects present. Its extent provides a quantitative measure of the polishing quality which, as our own study of III-V substrates has shown, varies from vendor to vendor. Use of highly asymmetric reflections from the specimen reduces the depth to which X-rays penetrate and thus enhances surface sensitivity.

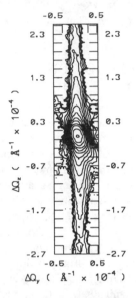

Fig. 8 Reciprocal space map around the 004 lattice point from a (001) oriented polished GaAs wafer. [Courtesy I Pape]

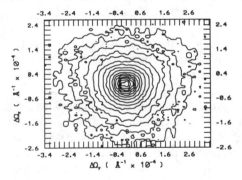

Fig 9 Reciprocal space map of a lapped GaAs wafer.

Reciprocal space maps provide a means of measuring the in-plane roughness of superlattices from the broadening of the satellite peaks in the q_y direction. Off-cut of the wafer from the Bragg planes results in the surface streak being slewed away from the q_z direction (Fig 10). This is a structure similar to that of Fig. 5 and note that the interference fringes now appear as islands of scatter. The extensive diffuse scatter close to the substrate peak arises from imperfection of the GaAs cap layer. The origin of the small (reproducible) island of scatter to the far right is not presently understood.

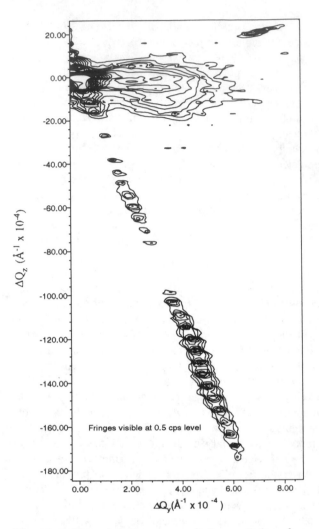

Fig 10 Reciprocal lattice map of the scattering from a GaAs/AlGaAs/InGaAs/GaAs structure, similar to that of fig. 5. [Courtesy of Dr. N Loxley, Bede Scientific.]

Quite spectacular reciprocal space maps arise from quantum wire structures where the in-plane periodicity results in satellites in both q_z and q_z directions. An example is shown in Fig 11.

While still in its infancy, through comparison between work presented at the 1st and 2nd European Conferences on High Resolution X-ray Diffraction and Topography in 1992 and 1994, it is clear that reciprocal space mapping is currently a major growth area whose impact is now being strongly felt in the compound semiconductor area. felt in the silicon industry. There will certainly have major impact in the silicon industry for polishing quality control but this is an area where few secrets are given away!

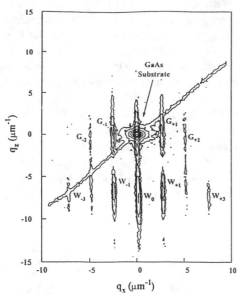

Fig 11. Reciprocal space map of the scattering from a quantum wire array [Courtesy of Prof. L. Tapfer (21)]

LATTICE PARAMETER COMPARATORS

Using advanced X-ray optics and high stability goniometry, techniques have been developed for rapidly comparing the lattice parameters of separate, bulk, crystals. The seminal review of these techniques by Hart (22) has not received the attention it deserves. Apart from the Bond method, which relies on accurate knowledge of the $CuK\alpha_1$ wavelength, all are comparative methods. Methods of achieving this at a resolution of a part in 10^6 can be devised using standard commercial diffractometers. The Bragg angles for the specimen and a known, standard, reference crystal are measured and from this the lattice parameter of the specimen is immediately deduced. Although the best standard is silicon, secondary standards may be appropriate for some applications. Crucial to all the methods is removal of the zero error introduced when moving from specimen to reference standard and alignment of the crystals to bring the diffraction vectors in the dispersion plane. These have recently been described in both double and triple axis geometries and applied to the measurement of Zn concentration in ZnCdTe (23). For the important measurement of carbon in silicon, a precision approaching parts in 10^9 is required. Hausermann and Hart (24) have described a monolithic device which uses simultaneous double and triple reflections from reference and specimen to reach this precision. Tight control of the temperature is necessary. It is unfortunate that these elegant and quite rapid methods have yet to be widely adopted in the silicon industry.

X-RAY TOPOGRAPHY

X-ray topography is the X-ray equivalent of transmission electron microscopy and use of an spatially resolving area detector enables topographic information to be obtained from any diffraction experiment. In high resolution settings, a spatial resolution down to 1μm, large field of view (now up to 200mm) and transmission through typical wafer thicknesses makes it highly appropriate for the study of semiconducting materials. Insertion of photographic film in the diffracted beam after the specimen or after the analyser (13) provides an easy means of performing moderate resolution topography with commercial instrumentation. In the former double axis case, the presence of misfit dislocations which are responsible for relaxation in strained layer systems can be easily detected. Although individual dislocations may not be resolved, the bunching of misfit dislocations results in a cross-hatch pattern of strain which can be imaged with quite fast X-ray film or an X-ray sensitive TV imaging system. For quality assurance such low resolution is perfectly adequate.

High-resolution *in-situ* X-ray topography studies of the relaxation process in MBE-grown InGaAs on GaAs are underway at the SRS at Daresbury. Whitehouse *et al* (25), using a standard MBE chamber equipped with Be windows, have shown that the asymmetry of the relaxation process in GaAs is an artefact of the substrate dislocation distribution. Initial dislocations are nucleated at threading dislocations according to the mechanism proposed by Matthews and Blakeslee and all are the fast α type. A knee is found in the relaxation above the critical thickness leading to a region of metastability prior to multiplication of the misfit dislocations at higher strain.

Ultra-high resolution "plane wave" X-ray topography provides a means of measuring minute strains in highly perfect silicon. So far these measurements have been the preserve of the Japanese workers at the Photon Factory synchrotron radiation source at Tsukuba. After a symmetric, broad band-pass symmetric 111 reflection to remove the heat load, the synchrotron radiation is diffracted successively off two asymmetrically cut crystals to deliver a beam of less than 0.01 arc second divergence at the specimen. In the transmission (Laue) geometry, rapid oscillations are observed in the rocking curve from the specimen with a sub-arc second period. These provide a very sensitive means of measuring strains such as those around defects in float zone (FZ) silicon. Recently Kimura *et al* have measured the stain fields around D type defects in FZ silicon crystals (26). They showed that the 220 lattice spacing shrank by parts in 10^6 in the D region in which these vacancy-type defects were present. It is clear that this type of experiment provides new insights into the properties of defects with small strain fields in both FZ and Czochralski silicon (27). Such experiments are highly appropriate to the 3rd generation of synchrotron radiation sources such as the ESRF at Grenoble, the APS at Argonne or SPring 8 in Japan. Such intense X-ray beams, will have an important impact on the use of ultra-high strain sensitivity topography within the silicon community.

Section topography was the earliest of the high resolution topography techniques and as it uses a ribbon collimated beam, provides information on the energy flow through the crystal (Fig 10). Although it has a reputation for being slow, this arises principally due to a desire for high spatial resolution. Then, fine grained and therefore slow photographic emulsions have been used. For quality control much faster films or TV imagers can be employed.

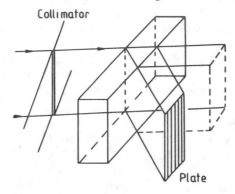

Fig. 12 (a) The X-ray section topography technique.

Section topography has the major advantage that simulation of images can now be performed quite rapidly on workstations and fast PCs, thereby providing a means of determining the microscopic strain fields from the scattered intensity (Fig 13)..

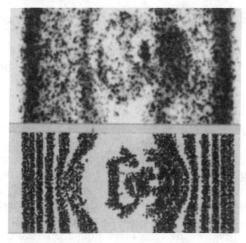

Fig 13. Experimental and simulated section topography images of an isolated precipitate formed in MCZ silicon after annealing for 1 hour at 650°C.

A potentially powerful application within the silicon industry is in the non-destructive monitoring of the thickness of denuded zones in intrinsically gettered silicon (28). As shown in Fig 14 (a), no direct image forms from the defect free regions and thus the margins of the section topograph CD and EF provide a direct measure of the extent of the denuded zone. Although there is a dynamical image which appears in the region EF, simulations show that the region CD remains completely free of defect contrast of the denuded zone is clear (29). The method may be used at low resolution for quality control. There is no need to image at high resolution when, as seen from the simulations, individual images are fundamentally not resolved. Fast, low resolution film or better, an image plate or a direct viewing TV detector is highly preferable.

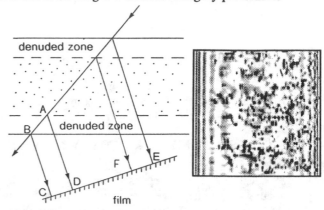

Fig 14. (a) Schematic diagram, showing the formation of defect images from denuded zones in silicon (b) Simulation of contrast

In the compound semiconductor area a number of North American industrial laboratories are using X-ray topography very successfully for substrate characterisation. Noteworthy is Bassignana (30) at Bell Northern, Ottawa, who is using double axis topography with asymmetric reflections to expand the beam from 3 inch LEC GaAs wafers and perform routine goods-inward inspection.

Grazing incidence specular reflectivity

At very low angles of incidence X-rays are totally externally reflected from the surface of a material. Beyond the critical angle, the intensity falls characteristically and by fitting to computer simulations, the surface electron density and the r.m.s. roughness can be measured. The sensitivity of roughness measurement by the method is about 1Å, although this is averaged over a macroscopic area. Presence of a thin film, (which need not be single crystal nor yet crystalline,) results in interference fringes which provide a measure of the layer thickness to an accuracy of better than a monolayer (Fig 15)

270

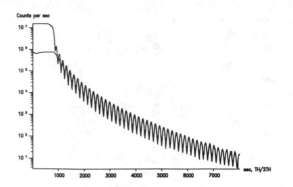

Fig 15 Experimental and simulated reflectivity curves of a 978A (±1A) Polymethyl methacrylate film spun on a silicon substrate at the Durham Polymer IRC. Surface roughness of 4.5A assumed on both top and bottom surfaces. 100m radius of curvature and 25 sec instrument function. PMMA density assumed 1.45 g/cc

A powerful method of data display is to multiply the specularly scattered intensity by the fourth power of the scattering angle. If the surface is perfectly smooth, beyond about twice the critical angle, the curve is parallel to the abscissa. A negative slope provides an accurate method of fitting roughness (Fig. 16).

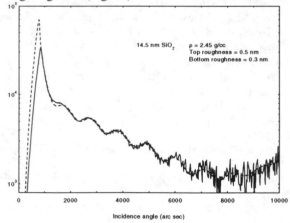

Fig 16. Specular reflectivity from a single oxide layer on silicon (solid curve). Inset parameters used to simulate the dashed curve using software (31) based on the theory of Parratt (32).

This shows experimental and simulated data from a 14.5 nm silicon oxide film on silicon. As well as providing an accurate measure of the oxide film thickness, the interference fringes can be used to determine the critical angle θ_c. Although the period remains unchanged, change in θ_c results in change of the absolute angular position of the maxima and minima.

Reflectometry can be performed on the recent generation of commercial high resolution diffractometers. However,

for good results off small samples, a small beam width is needed as the "footprint" of the beam at low angles is large. Without the right system configuration, intensity can be disastrously low for all except test samples!

GRAZING INCIDENCE DIFFUSE SCATTERING

From the specular scatter, it is not possible to distinguish between surface roughness and interdiffusion. This can be achieved by measurement of the diffuse scatter. and by fitting to simulations based on a fractal description of the interface (33), roughness σ, correlation length ξ and fractal parameter h can be determined. Exactly in analogy with the triple axis diffraction described earlier, it is necessary to limit the angular range of scattered X-rays reaching the detector. For reflectometry this may be done with slits and equivalent transverse and longitudinal scans in reciprocal space may be performed by scanning respectively specimen only or specimen and detector coupled in a 1:2 ratio. Fig. 17 shows a specimen-only scan of a polished GaAs wafer, with a theoretical fit using the fractal description.

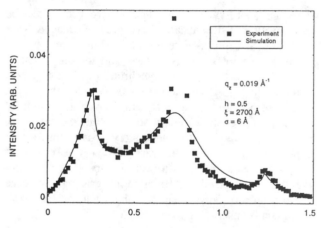

Fig. 17 Grazing incidence diffuse scatter from a polished GaAs wafer, rotated with fixed detector. Inset parameters used for the simulation . [Courtesy I. Pape and M. Wormington.]

In multilayer systems, the uncorrelated roughness and that correlated through the multilayer can be separated. Full reciprocal space maps of the low angle scatter provide a powerful means of analysis yet to be used widely.

Measurement of the fluorescence X-ray or photoelectron yield for incidence angle below the critical angle provides a means of detecting minute quantities of impurity in the very near surface region. This is now a mature technique but use of a variable incidence angle X-ray probe is not so widespread. As the angle of incidence is increased beyond the critical angle, the X-ray wavefield penetrates into the sample in a characteristic, calculable fashion (34). The resulting variation in fluorescence yield enables the impurity distribution to be examined as a function of depth. There are presently diverging opinions as to the scope of this technique and the next few years will certainly see definitive model experiments.

GRAZING INCIDENCE (TWO DIMENSIONAL) SURFACE DIFFRACTION

Surface diffraction occurs when the incident beam is at an angle comparable to the critical angle, but the diffraction vector lies in the surface. Scanning the detector in the specimen plane provides data on the in-plane strains in the very near surface region. Most work has been undertaken at synchrotron radiation sources, with many experiments set up to study growth *in-situ*. Very few attempts have been made to perform such experiments in the laboratory.

THE NEXT GENERATION OF SOURCES

The merits of synchrotron sources have been discussed extensively and they clearly have great advantages over conventional sources in terms of brightness and power. However, complex infrastructure and off-site location make X-ray scattering with synchrotron radiation almost unthinkable in a production context. On the other hand, developments are taking place rapidly in the fields of multilayer optics and capillary optics. While the latter may not be so useful for laboratory based diffraction experiments, graded multilayers can provide orders of magnitude gain in intensity in a parallel beam (35). Parallel beam powder diffraction at grazing incidence angle from thin polycrystalline films such as aluminium on silicon is undergoing a revolution.

REFERENCES

1. Bowen, D.K. and Tanner, B.K., *High Resolution X-ray Diffraction and Topography*, London, Taylor & Francis, 1995
2. Bowen, D.K., in Tanner, B.K. and Bowen, D.K., *Characterization of Crystal Growth Defects by X-ray Methods*, New York, Plenum, 1980, pp. 333-348
3. Bond, W.L., *Crystal Technology*, New York, John Wiley, 1976
4. Nakamura, M., Katsura, S., Makino, N., Ikeda, E., Suga, K. and Hirano, R., *J. Crystal Growth* **129** 456-64 (1993)
5. Bartels, W. and Nijman, W., *J. Crystal Growth* **44** 518-525 (1978)
6. Herres, N., *J. Phys. D :Appl. Phys.* (1995) in press
7. Loxley, N., Bowen, D.K., and Tanner, B.K., Mater. Res. Soc. Symp. Proc. **208** 119-124 (1991)
8. DuMond, J.W., *Phys. Rev.* **52** 872-883 (1937)
9. Bartels, W., J. Vacuum Sci. Tech. **B1** 338 (1983)
10. Loxley, N., Tanner, B.K. and Bowen, D.K., *J. Appl. Cryst.* (1995) in press
11. Loxley, N., Cockerton, S., Tanner, B.K., Cooke, M.L., Gray, T. and Bowen, D.K., *Mat. Res. Soc Symp. Proc.* **324** 451-456 (1993)
12. Tanner, B.K. and Bowen, D.K., *J. Crystal Growth* **126** 1-18 (1993)
13. Fewster, P.F., *Semicond. Sci. Tech.* **8** 1915-34 (1994)
14. Bowen, D.K., Loxley, N., Tanner, B.K., Cooke, L. and Capano, M., *Mat. Res. Soc. Symp. Proc.* **208** 113-8 (1991)
15. Holloway, H., *J. Appl. Phys* **67** 6229-6236 (1990)
16. Tanner, B.K., *J.Phys.D: Appl. Phys.* **26** A151-5 (1994)
17. Gillespie, H.J., Wade, J.K., Crook, G.E. and Matyi, R.J., *J. Appl. Phys.* **73** 95-102 (1993)
18. Müller, R., Univ. of Munich, private communication
19. Powell, A.R., Kubiak, R.A.A., Whall, T.E. and Bowen, D.K., *J. Phys. D: Appl. Phys* **23** 1745-7 (1990)
20. Hallam, T.D., Halder, S.K., Hudson, J.M., Li, C.R., Funaki, M., Lewis, J.E., Brinkman, A.W. and Tanner, B.K. *J. Phys. D: Appl, Phys* **26** A161-6 (1993)
21. Tapfer, L., Sciacovelli, P, and De Caro L., *J. Phys. D: Appl. Phys* (1995) in press
22. Hart, M., *J. Crystal Growth* **55** 409-427 (1981)
23. Bowen, D.K, Tanner, B.K., Hudson, J. M., Pape, I, Loxley, N and Tobin, S., *Adv. X-ray Anal.* **37** 129-133 (1995)
24. Hausermann, D. and Hart, M., *J. Appl. Cryst.* **23** 63-9 (1990)
25. Whitehouse, C.R., Barnett, S.J., Usher, B.F., Cullis, A. G., Keir, A.M., Johnson,.A.D., Clark, G.F., Tanner, B.K., Spirkl, W., Lunn, B., Hagston, W.E. and Hogg, J.C.H, *Inst. Phys. Conf. Ser.* **134** 563-8 (1993); *J. Phys. D (Appl. Phys)* (1995) in press
26. Kimura, S., Ishikawa, T. and Matsui, J, *Phil. Mag. A* **69** 1179-1187 (1994)
27. Kimura, S., Ono, H., Ikarashi, T. and Ishikawa, T., *Jap. J. Appl. Phys.* **32** L1074-7 (1993)
28. Tuomi, T., Tilli, M., and Anttila, J., *J. Appl. Phys.* **57** 1384-6 (1985)
29. Holland, A.J. and Tanner, B.K. *J. Phys. D (Appl. Phys.)* (1995) in press
30. Bassignana, I. and Macquistan, D.A., *7th Int. Conf. on III-V Semi-Insulating Materials, Ixtapa, Mexico,* (1992)
31. Wormington, M., Bowen, D.K. and Tanner, B.K., *Mat. Res. Soc. Symp. Proc.* **238** 119-124 (1992)
32 Parratt, L.G., *Phys. Rev.* **95** 359-369 (1954)
33.Sinha, S.B., Sirotta, E.B., Garoff, S and Stanley, H.B, *Phys. Rev. B* **38** 2297-2311 (1988)
34. DeBoer, D.K.G., *Phys. Rev B* **44** 498-511 (1991)
35. Goebel, H., *Adv. X-ray Anal.* **38** (1995) in press

Silicon Wafer Trace Impurity Analysis Using Synchrotron Radiation

S. S. Laderman, A. Fischer-Colbrie, A. Shimazaki,* K. Miyazaki,*
S. Brennan,** N. Takaura,** P. Pianetta,** and J. B. Kortright***

Hewlett-Packard Laboratories, Hewlett-Packard Company,
3500 Deer Creek Road, Palo Alto, CA 94304, USA
*Integrated Circuit Advanced Process Engineering Department, Toshiba
Corporation, 1, Komukai Toshiba-cho, Saiwai-ku, Kawasaki 210, Japan
**Stanford Synchrotron Radiation Laboratory, Stanford Linear Accelerator Center,
P.O. Box 4349, Stanford, CA 94309, USA
***Center for X-Ray Optics, Lawrence Berkeley Laboratory, University of California,
Berkeley, CA 94720, USA

We explored total reflection x-ray fluorescence spectroscopy (TXRF) with a synchrotron source as a means of achieving high sensitivities to surface metals on silicon wafers. Most recently, we demonstrated a sensitivity of 3×10^8 atoms/cm^2 for third-row transition metal elements.[1] The configuration appears to be capable of high sensitivities for a wide range of elements. Even higher sensitivities are being pursued by exploring detector systems capable of higher counting rates and concomitant modifications of the chamber geometry. Routine use of the capability would require additional equipment modifications, a stable configuration, industry-wide calibration, and adequate user access.

INTRODUCTION

Due to the sensitivity of VLSI circuit yields to low levels of metallic contamination, total reflection x-ray fluorescence (TXRF) analysis based on rotating anode sources is routinely used throughout the semiconductor industry. Conventional TXRF equipment is typically reported to have a sensitivity to iron or nickel near 5×10^9 atoms/cm.^{2} This high sensitivity, efficient sample handling procedures, mapping and depth profiling capabilities, and the ease of reliably quantifying the data have all contributed to the success of the method. However, current advanced wafer surface preparation methods reproducibly create silicon surfaces free of transition metal impuritie to a level below the detection limit of widely available measurement methods, including conventional TXRF. Without more sensitive measurement methods, it is difficult to maintain, evaluate, and improve the processing technology. The motivation for the work reported here was this need to develop improved wafer surface analysis methods.

Prior synchrotron radiation TXRF studies have demonstrated detection limits near 10^{10} atoms/cm^2 using single crystal monochromators.[2,3,4] Here, we review how more than an order of magnitude enhancement in sensitivity over conventional equipment and prior synchrotron radiation studies was achieved with a configuration based on synthetic multilayer monochromators. This configuration provides high sensitivities for a much wider range of elements than is possible when using a rotating anode source. We also describe some steps to be taken to achieve further improvements in sensitivity and usefulness.

SAMPLE PREPARATION

Two types of samples were studied. The first type was calibration standards used to determine sensitivities and to calibrate intensity levels. These standards were made at the Toshiba Corporation using well-developed dip contamination methods based on acidic solutions of transition-metal salts. The lateral uniformity and fine contaminant dispersion in these samples were verified with conventional TXRF mapping and synchrotron radiation TXRF angle scans. The concentrations were determined with calibrated conventional TXRF and wet chemical methods. The excellent lateral uniformity made it possible to compare conventional and synchrotron radiation data directly without geometric correction. The fine dispersion of the contaminants insured that the samples were directly relevant to the practical cases of highest interest. The second type of sample was nominally clean samples from wafer vendors, Toshiba,and the Hewlett-Packard Company.

TXRF CONFIGURATIONS

The conventional TXRF data reported here were acquired at the Toshiba Corporation and at the Hewlett-Packard Company using commercially available equipment.

The synchrotron radiation TXRF data were collected at the Stanford Synchrotron Radiation Laboratory using a 15 period, 1.4 Tesla permanent magnet wiggler source. The x-ray beam was focused horizontally and vertically using a toroidal mirror. The horizontal divergence was set to a value comparable to the vertical divergence found in conventional equipment. The beam was monochromated by reflection from two sequential, identical synthetic multilayers in parallel geometry. The primary x-ray energy emerging from the multilayer assembly was easily tuned. Here, primary energies between 10 keV and 11 keV were explored. At a storage ring energy of 3 GeV and a storage ring current of 100 mA, approximately 10^{13} photon/sec entered the experimental chamber in a beam close to 2 mm high and 2 mm wide. Only the central, brightest portion of this beam contributed to the measured signals. Even so, the single element Si(Li) detector with analog electronics was run near its count rate limit under typical experimental conditions. In the case of the synchrotron configuration used here, the surface area sampled was about 0.1 cm^2. In contrast, the conventional equipment sampled about 1.0 cm^2. Further details concerning the monochromator and source characteristics are published elsewhere.[5] Further details concerning the experimental geometry, detector windows, and collimator parts are also reported elsewhere.[6] A schematic of the stainless steel sample chamber has been published.[7]

The incident angle was calibrated for each sample by curve fitting the Si fluorescence signal as a function of incident angle below, close to, and above the critical angle for total reflection to theory. Such incident angle scans also demonstrated that the contaminants on the standard samples were not significantly particle-like but had a very narrow distribution in height, consistent with atomically disperse contamination. In general, impurity fluorescence data were acquired at an angle a few hundredths of a degree below the critical angle. This is also typical of many conventional rotating anode

experimental configurations, including the one used to collect the data shown in Figure 1.

EXPERIMENTAL DATA

Figure 1 shows TXRF data obtained from a wafer having 1×10^{11} atoms/cm^2 of Fe, Ni, and Zn added as intentional contaminants to the surface. The top portion of the figure shows data collected using a rotating anode source. The bottom portion shows data from the same wafer obtained using a synchrotron source. The data in the bottom portion appear smoother because the signal to noise ratio is higher when the synchrotron source is used. The factor of 200 difference in the ordinate scale reflects this large difference in useful intensity. Using a 3σ rule and assuming the background level is given by the signal levels next to the fluorescence peak, the detection limit for Ni derived from the conventional spectrum is 5×10^9 atoms/cm^2. For the case of the synchrotron, the detection limit so derived is 3×10^8 atoms/cm^2. This improvement is commensurate with the increased signal level. It should be noted that this increased signal to noise is achieved along with about an order of magnitude improvement in spatial resolution.

Another difference between the conventional and synchrotron radiation data shown in Figure 1 is the lower background near the Zn peak in the case of the synchrotron radiation source. This is a simple example of a general capability made possible by the broadband nature of the synchrotron source and the tunability of the multilayer monochromator. The conventional source has high intensity only near its anode fluorescence lines. The line used in the spectrum shown here is so close to the Zn fluorescence line as to contribute a high background. The synchrotron source, however, has high intensity over a broad continuum of energies. It is straightforward and convenient to tune the incident x-rays close to an energy which gives high Zn signal rates yet low background rates near the Zn peak. In principal, this broadband nature of the synchrotron source and the easily tunable monochromator permit high sensitivity study of many different impurity elements. Comparison of the absorption cross-section and fluorescence yields for elements from Na through Rn to those for Ni support this hypothesis.[6,8]

A third difference between the spectra shown in Figure 1 is the appearance of the background near 2.5 keV. The trend in the synchrotron radiation case is due to absorption in the Teflon filter. The smooth background throughout the middle region of the spectrum is due to photoelectron bremsstrahlung.[9] The bremsstrahlung

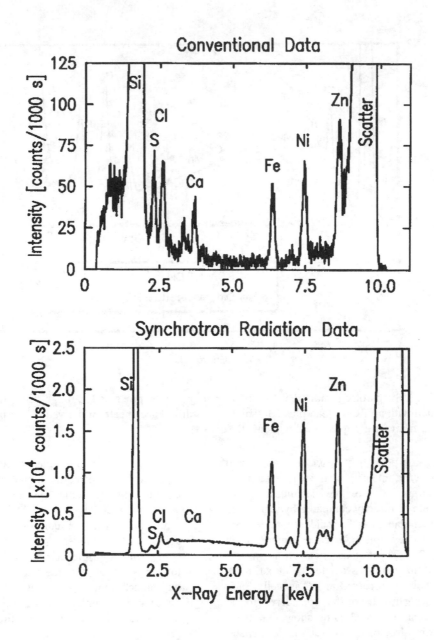

Figure 1. Conventional (top) and synchrotron radiation (bottom) TXRF spectra for a sample with 10^{11}

x-rays are strongly absorbed at lower energies by the 25 micron thick Teflon filter. This determination of the origin of the background shows us how to engineer further improvements in signal to noise in the future.

Repeated study of nominally clean wafers from different sources revealed that peaks always appeared in the vicinity of Fe, Ni, Cu and Zn. By successively placing thick low Z filters at strategic positions in the apparatus, it was determined that the detector assembly itself was the source of the peaks.[1] The detector is being redesigned for this application in order to remove these parasitic fluorescence peaks. In the meantime, the analysis is improved by quantitatively accounting for these background peaks, as shown in Figure 2 and described next.

For concentrations of impurity atoms as low as those studied here, the x-rays entering the detector assembly are dominated by x-rays scattered by the silicon wafer.

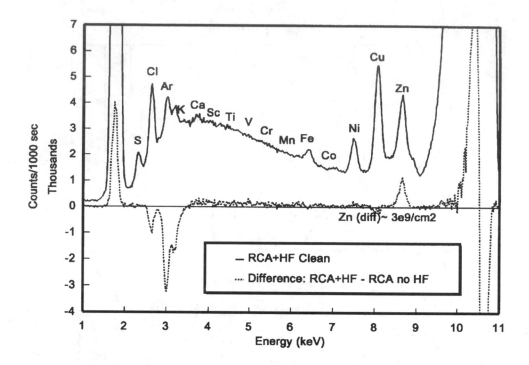

Figure 2. Synchrotron radiation spectrum from a wafer processed through wet chemical baths associated with an RCA clean plus HF (solid) and a difference spectrum derived by comparison to the spectrum from a wafer processed without HF (dotted).

Therefore, the parasitic peaks just described were found to be proportional to the intensity of the x-rays scattered from the wafer. In such a case, the cleanliness of two different wafers could be directly compared by scaling the data sets to the scatter peak near the incident x-ray energy and subtracting. An example of such a comparison is shown in Figure 3. Those data come from wafers processed through the baths associated with an RCA clean.[10] One wafer was processed in HF as well. Such single wafer studies should not be considered to be statistically significant. Figure 3 does show that these particular wafers were free of metals such as V, Cr, and Mn to a very high precision. The difference data show a difference in Zn of about 3×10^9 atoms/cm^2. Using a similar methodology, we also observed differences in the cleanliness of various wafers received from substrate vendors.[8]

FUTURE PROSPECTS

The data described here show that synchrotron radiation sources make possible higher sensitivities to impurities on Si wafer surfaces than obtained with current conventional TXRF equipment. This improvement offers significant practical advantages to the semiconductor industry. With multiple detector arrays and faster detectors, the configuration could be improved to achieve even higher sensitivities. For example, multiple detectors would immediately provide higher average sensitivities. Since the synchrotron configuration used here samples a significantly smaller area than the conventional equipment, multiple detector arrays would still sample total areas about the same size as conventional equipment with much higher sensitivities. As another example, detectors with higher count rate capability would permit more efficient use of the total flux available in the total synchrotron beam. With sufficiently fast detector systems, vertical configurations may yield higher overall sensitivities than horizontal arrangements due to a sufficiently larger increase in the overall count rate than in the relative background rate. Testing such possiblities and improvements are part of currently funded efforts in this field.

A limited number of practical applications of synchrotron radiation TXRF have been and will continue to be performed. However, widespread and routine practical use of synchrotron radiation TXRF requires further improvements in the sample handling, stable configurations, industry-wide calibration, and greater access to synchrotron radiation facilities. Synchrotron radiation TXRF has now been shown to bring TXRF analysis into a new regime of sensitivity. Automatic and clean wafer handling equipment and protocols capable of working in this new regime of sensitivity have not yet been demonstrated. Yet such equipment is essential to widespread practical use of the method. While such developments appear to be straightforward, undertaking them involves significant risk. As yet, it is not clear that such developments will be funded.

In summary, synchrotron radiation is capable, in principal, of bringing TXRF analysis into a new and useful regime having important applications in the semiconductor industry. Practical developments are underway which may achieve even greater sensitivities. Additional equipment development activities are also required to overcome the technical barriers to widespread practical implementation.

ACKNOWLEDGMENTS

This research was performed, in part, at the Stanford Synchrotron Radiation Laboratory (SSRL) which is operated by the Department of Energy, Office of Basic Energy Sciences. Partial financial support for the synchrotron radiation TXRF sample chamber was provided by the Intel Corporation. The solid state detector system used at SSRL was provided by the Fisons Corporation. We acknowledge useful discussions with D. Wherry and experimental support from W. Tompkins and R. Smith.

REFERENCES

1. S. S. Laderman, Bull. Am. Phys. Soc. **39** 514 (1994).

2. A. Iida, Advances in X-Ray Analysis, edited by C.S. Barett (Plenum, New York, 1992), vol. 35, p. 795.

3. M. Madden et al., Proc. Mater. Res. Soc. **307** 125 (1993).

4. S. S. Laderman and P. Pianetta, Proceedings of the Workshop of Applications of Synchrotron Radiation to Trace Impurity Analysis for Advanced Silicon Processing, Stanford Linear Accelerator Center Technical Report (1992).

5. P. Pianetta et al., Rev. Sci. Instrum. **66** (2) (1995).

6. S.S. Laderman et al., to be published in the proceedings of the 5th Workshop on Total Reflection X-Ray Fluorescence Spectroscopy and Related Spectroscopical Methods, held in Tsukuba, October 1994.

7. S. Brennan et al., Nucl. Instrum. Methods **A347** 417 (1994).

8. A. Fischer-Colbrie et al., Proceedings of the Second International Symposium on Ultra-clean Processing of Silicon Surfaces, edited by M. Heyns (Acco, Leuven/Amersfoort, 1994), page 57.

9. N. Takaura et al., to be published in the proceedings of the 5th Workshop on Total Reflection X-Ray Fluorescence Spectroscopy and Related Spectroscopical Methods, held in Tsukuba, October 1994.

10. W. Kern and D. Puotinen, RCA Rev. 31 187 (1970).

Characterization of Defects in Silicon Carbide Single Crystals by Synchrotron X-ray Topography

Shaoping Wang, Michael Dudley, Wei Huang
Department of Materials Science and Engineering,
State University of New York at Stony Brook,
Stony Brook, NY 11794-2275

Calvin H. Carter, Jr., Valeri F. Tsvetkov
Cree Research, Inc.
2810 Meridian Parkway,
Durham, NC 27713

C. Fazi
U.S. Army Research Laboratory,
2800 Powder Mill Road,
Adelphi, MD 20783

Synchrotron white beam X-ray topography (SWBXT) and its versatility in characterizing defect structures in SiC single crystals are discussed, emphasizing applications in the semiconductor industry as an efficient quality control tool in the research & development of bulk crystal growth as well as epitaxial thin film growth techniques. Examples of defect characterization in 6H-SiC single crystals grown by the physical vapor transport (PVT) technique are discussed.

1. INTRODUCTION

X-ray diffraction topography is a powerful, non-destructive tool for characterizing defect structures in large semiconductor crystals [1]. X-ray topographs recorded using X-rays from conventional generators, such as Lang topographs, can have exposure times from several minutes at best to many hours due to low X-ray intensities. With the advent of dedicated synchrotron X-ray sources in the mid-seventies, however, the X-ray topography technique has been significantly enhanced by the fact that the intensities from synchrotrons are several orders of magnitude higher than those from conventional generators, reducing exposure times to just a few seconds. Among the various synchrotron topography techniques, synchrotron white beam X-ray topography (SWBXT) [2] is particularly suitable for both routine screening of defects in large single crystal wafers and detailed examination of defects in localized regions of crystals. In addition, due to the non-destructive nature of this technique and the absence of a necessity for extensive sample preparation, defect characterization can be readily carried out on crystals at various stages throughout the device fabrication procedure, i.e., from as-grown crystals to final devices.

SiC has drawn a great attention in recent years as a wide bandgap, high temperature semiconductor that has many promising device applications. There have been increasing efforts in developing various devices made from SiC single crystals, mainly due to advances in growth by the sublimation physical vapor transport (PVT) technique [3] so that 6H (as well as 4H) SiC wafers of over one inch in diameter with sufficient crystalline perfection are commercially available. Nevertheless, SiC single crystals still contain significant crystallographic defects such as hollow tubes [4] latterly referred to as "micropipes" and dislocations. It has been shown that these defects can strongly affect the performance of various devices and, in some cases, they have actually hindered the application of SiC to some particular types of semiconductor device.

The research reported here is part of a joint effort to

study defect structures in commercial SiC single crystals with SWBXT. The goal of this work is to provide structural defect information to crystal growers, through a feed-back loop which is created in order to aid in the improvement of bulk crystal quality and hence lead to better yields and improved device performance.

2. EXPERIMENTAL

X-ray topography using synchrotron white radiation, synchrotron white beam X-ray topography (SWBXT), makes defect characterization of large single crystals a very efficient, routine process. Because of the wide spectrum of the synchrotron source (a usable wavelength range 0.3 Å—1.6 Å for the synchrotron facility utilized in this research), there are no precision operations necessary for orienting the crystals and usually several useful topographs, or images, can be recorded simultaneously on a single sheet of X-ray film (8×10 inch Kodak SR-5 with a linear resolution around 2-3 μm). The spatial resolution of X-ray topographs is usually about a few microns, which translates into a capability of resolving dislocation densities up to around 10^5 cm^{-2}, covering the common dislocation density levels of most compound semiconductor single crystals.

In SWBXT, topographs can be recorded in two different diffraction geometries: transmission geometry and reflection geometry. Figure 1 shows a schematic diagram of the transmission geometry for SWBXT, where each of the diffraction "spots" is a topograph. An

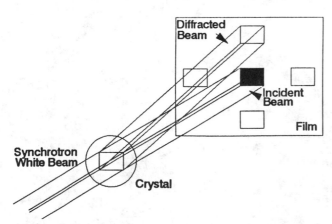

FIGURE 1. Schematic diagram showing transmission geometry for recording synchrotron white beam X-ray topographs.

example of synchrotron white beam transmission topograph, i.e. diffraction pattern, is shown in figure 2, which was recorded from a (0001) 6H-SiC single crystal.

Contrast on each diffraction topograph reveals defects in the crystal.

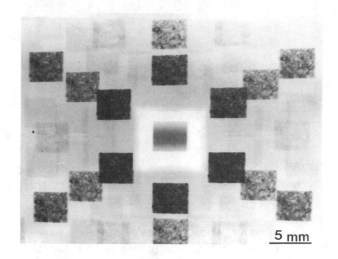

FIGURE 2. An example of synchrotron white beam topographs recorded in Laue transmission geometry from a (0001) 6H-SiC wafer.

Reflection topographs, on the other hand, reveal defects that are located close to the surface of crystal wafers. In particular, synchrotron white beam grazing incidence Bragg-Laue reflection geometry [2] is particularly useful in the study of defects in thin epitaxial films.

Although an area-filling beam is usually used in SWBXT to enable the recording of synchrotron topographs from areas up to 5 mm × 50 mm, industrial applications can demand fast turn-around and whole-wafer coverage from crystal areas which can be much larger than this. Much larger coverage can be achieved by scanning the sample and the recording film simultaneously. Therefore, a prototype scanning mechanism was built for recording synchrotron topographs, which is schematically shown in the drawing in figure 3, which is similar to a Lang topography set-up. Utilizing the fact that the horizontal synchrotron X-ray beam size is about 50 mm (while the vertical source size is only about 5 mm), the scanning direction is chosen to be vertical. The first set of slits, S1, is used to obtain a narrow horizontal line source and the second set of slits, S2, is used to let only one desired diffraction spot pass through and register on the X-ray film placed behind it. Using such a scanning mechanism, SiC single crystal wafers up to 2 inches in diameter have been imaged in just a single few minute exposure. This scanning mechanism can also be used for recording synchrotron reflection topographs.

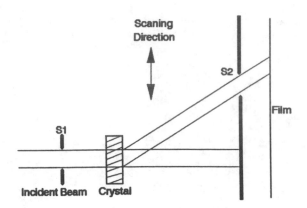

FIGURE 3. Schematic diagram showing the scanning mechanism for recording synchrotron white beam topographs.

A variety of 6H-SiC single crystal wafers were cut from PVT boules which were grown at Cree Research, Inc. and subsequently polished on both sides to about 250 μm thick so that they could be examined by SWBXT in transmission geometry. Most of the 6H-SiC wafers are of about one inch in diameter or larger and they are transparent with color ranging from colorless to dark green or dark blue. In order to disclose the three dimensional distributions of defects in crystal boules, the crystals under investigation were basically cut in two different geometries: one perpendicular to [0001] axis (i.e. (0001) wafer) and the other parallel to the [0001] axis (the longitudinal cut).

Synchrotron white beam X-ray topography experiments were carried out at the Stony Brook Synchrotron Topography Facility, Beamline, X-19C, at the National Synchrotron Light Source (NSLS), Brookhaven National Laboratory. The transmission geometry, or Laue geometry, was used for imaging bulk defect structures, while the reflection geometry, or Bragg geometry, particularly the surface sensitive grazing Bragg-Laue geometry, was employed to exclusively image defects located near the crystal surface regions.

In addition, synchrotron white beam topographs recorded in *back-reflection* geometry were found to be useful for imaging line defects running approximately parallel to the [0001] axis in (0001) PVT 6H-SiC wafers. Such back-reflection topographs were also compared with the corresponding etch-pit maps recorded from wafers after molten KOH etching.

3. RESULTS AND DISCUSSIONS

An example of a scanned synchrotron topograph recorded from a 6H-SiC wafer of one inch in diameter is shown in figure 4(a). A montage of topographs recorded with the standard SWBXT technique from the same wafer is shown in figure 4(b) for comparison. Clearly, the scanning provides a uniform topographic image over the whole wafer, while the montage image has significant variations of intensity between adjacent topographs mainly due to the non-uniform distribution of the incident synchrotron beam in the vertical direction. In fact, the scanning in the vertical direction evens out the incident beam intensity in this direction. By using the scanning mechanism, the efficiency of SWBXT technique is estimated to be increased by a factor of at least 50 in terms of time savings, not to mention savings on photographic materials.

FIGURE 4. (a) Scanned synchrotron white beam topograph (**g**=10$\overline{1}$1, λ=0.86 Å) and (b) montage of pictures of the same topograph recorded from a (0001) 6H-SiC wafer.

Apart from achieving whole wafer coverage in a single exposure, there are several other advantages to utilize such a scanning mechanism to record synchrotron topographs. First, the background due to scattered X-rays can be dramatically reduced, simply because only a very small area of the X-ray film that is limited by the second set of slits is exposed to the X-rays. Secondly, since only a narrow line source (typically about 1mm in width) is used in the scanning technique, the specimen—film distance can be reduced significantly, provided no image overlap occurs, which also improves spatial resolution.

Transmission synchrotron topographs reveal intricate dislocation structures in the PVT 6H-SiC (0001) wafers. A typical example of an enlargement of such a topograph is shown in figure 5. There are two different kinds of contrast feature observable on this topograph: the dark line contrast forming a network parallel to the basal plane (i.e. the image plane) and the dark "dot-like" features, e.g. indicated by **M**, that appear to be the nodes

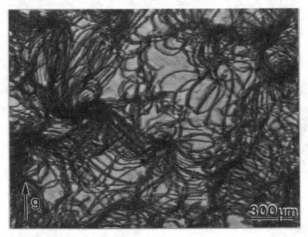

FIGURE 5. Enlargement of synchrotron white beam topograph recorded in transmission geometry ($g=10\bar{1}1$, $\lambda=0.86$Å) from a (0001) 6H-SiC wafer.

of the network features. The network line features are associated with dislocations lying on the basal plane, referred to as basal plane dislocations. Burgers vector analysis shows that most of these basal plane dislocations are perfect dislocations [5]. The "dot-like" features are "so called" "micropipes". In fact, "micropipes" are hollow tubes of a fraction of a micrometer to a few micrometers in diameter and they are approximately parallel to the [0001] axis in PVT SiC single crystals. "Micropipes" are believed to be most detrimental to device performance. Our studies with SWBXT [5,6,7] have proved that "micropipes" are indeed hollow core screw dislocations with large Burgers vectors usually ranging from three to seven times of the lattice parameter c (15.17Å for 6H).

"Micropipes" can be imaged more clearly in longitudinal cut wafers. Figure 6 shows an example of transmission topographs recorded from such wafers, where the broad, bi-modal images indicated by **M** are associated with "micropipes". The bi-modal images are characteristic of screw dislocations and the image widths indicate the magnitude of the Burgers vector [6]. It should be pointed out that there are also some narrow, bi-

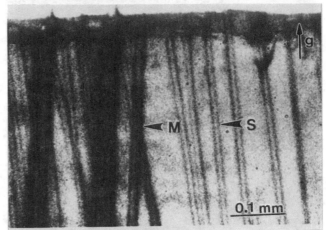

FIGURE 6. Enlargements of synchrotron white beam topographs recorded in transmission geometry ($g=0006$, $\lambda=0.89$Å) from a longitudinal-cut 6H-SiC wafer, showing micropipes (**M**) and screw dislocations (**S**).

modal images of approximately constant image width (e.g. indicated by **S**) that were determined to be associated with screw dislocations with Burgers vectors equal to one times the lattice parameter c, the basic lattice translation in the basal plane normal direction of the hexagonal SiC structure.

X-ray topographs recorded in the reflection geometry are suitable for studying defects within a small depth, for example a few microns beneath the sample surface. In particular, the highly surface sensitive grazing Bragg-Laue geometry has been used extensively for imaging defects in 6H-SiC homoepitaxial thin films [8]. Reflection topographs were also employed for imaging defects in standard (0001) SiC wafers which are polished only on one side. In fact, it was shown that reflection topographs recorded in back-reflection geometry with synchrotron white radiation are particularly suitable for imaging "micropipes" and screw dislocations that are approximately parallel to the [0001] axis [6,8]. An example of such back-reflection topographs recorded from a standard 6H-SiC wafer is shown in figure 7(a), while the corresponding etch-pit map recorded from the same region of the wafer after KOH etching is shown in figure 7(b). The large circular contrast features (e.g. **M**) on the topograph are associated with "micropipes", while

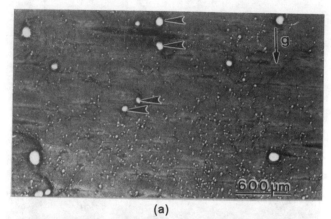

(a)

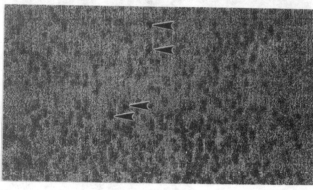

(b)

FIGURE 7. (a) Synchrotron topograph recorded in back-reflection geometry (principal g=00024, λ=1.25Å) (b) the corresponding etch-pit map recorded from the same region of a (0001) 6H-SiC wafer.

the small circular features (e.g. S) of approximately the same diameter are associated with screw dislocations with Burgers vectors equal to 1*c*. Although "micropipes" and screw dislocations were also revealed by KOH etching, as can be seen on the etch-pit map, synchrotron back-reflection topographs are superior because they also reveal the magnitude of Burgers vectors of "micropipes" and screw dislocations, because the diameters of the circular contrast features on the topograph are proportional to the strength of the screw dislocations [6].

4. CONCLUSION

The applicability and versatility of SWBXT for defect characterization in SiC single crystals are clearly demonstrated. A prototype of a scanning mechanism developed exclusively for recording synchrotron white beam topographs was tested and found to offer much larger wafer coverage without compromising spatial resolution. Examples of defect characterization using synchrotron topography for PVT 6H-SiC crystal wafers

were discussed. It is also found that back-reflection topographs are superior to etch-pit maps in revealing the strength of screw dislocations and "micropipes".

ACKNOWLEDGMENTS

This research is supported in part by the National Institute of Standards and Technology (NIST) through the Advanced Technology Program via Cree Research, Inc. and by U.S. Army Research Office (grant numbers DAAH04-94-G-0091 and DAAL04-94-G-0121, contract monitor Dr. John Prater). The scanning stage was designed and constructed in conjunction with Grumman Corporate Research Center.

REFERENCES

1.   M. Dudley, "Characterization by X-ray Topography", in *Encyclopedia of Advanced Materials,* **Vol. 4**, D. Bloor, R. J. Brook, M. C. Flemings and S. Mahjan (Eds), Pergamon, 2950-2956, (1994), *Commissioned Review Article.*

2.   M. Dudley, "X-ray Topography", in "Applications of Synchrotron Radiation Techniques to Materials Science", D. L. Perry, R. Stockbauer, N. Shinn, K. D'Amico and L. Terminello (Eds.) *Mat. Res. Soc. Symp. Proc.,* **307**, 213-224, (1993).

3.   Yu. M. Tairov and V. F. Tsvetkov, *J. Cryst. Growth,* **52**, 146 (1981).

4.   G. Ziegler, P. Lanig, D. Theis and C. Weyrich, *IEEE Trans. Electron Devices,* **ED-30**, 277(1983).

5.   S. Wang, M. Dudley, C. H. Carter, Jr., D. Asbury and C. Fazi, "Characterization of Defect Structures in SiC Single Crystals Using Synchrotron X-ray Topography", in "Application of Synchrotron Radiation Techniques to Materials Science", D. L. Perry, R. Stockbauer, N. Shinn, K. D'Amico and L. Terminello (Eds.) *Mat. Res. Soc. Symp. Proc.,* **307**, 249-254 (1993).

6.   M. Dudley, S. Wang, W. Huang, C. H. Carter, Jr., V.F. Tsvetkov and C. Fazi, *J. Phys. D: Appl. Phys.,* (accepted) (1995)

7.   S. Wang, M. Dudley, C. H. Carter, Jr., V.F. Tsvetkov and C. Fazi, "Synchrotron White Beam Topography Studies of Screw Dislocations in 6H-SiC Single Crystals", in "Application of Synchrotron Radiation Techniques to Materials Science", D. L. Perry, R. Stockbauer, N. Shinn, K. D'Amico and L. Terminello (Eds.) *Mat. Res. Soc. Symp. Proc.,* **375**, (1994) (to be published).

8.   S. Wang, M. Dudley, C. H. Carter, Jr., and H. S. Hong, "X-ray Topographic Studies of Defects in PVT 6H-SiC Substrates and Epitaxial 6H-SiC thin films", in "Diamond, Silicon Carbide and Nitride Wide-Bandgap Semiconductors", C. H. Carter, Jr., G. Gildenblat, S. Nakamura and T. J. Nemanich (Eds.), *Mat. Res. Soc. Symp. Proc.,* **339**, 735-740 (1994)

Depth-Dependent Non-Destructive Analysis of the Oxide Surface of InP(100) Using Grazing Incidence X-Ray Photoemission

Terrence Jach

National Institute of Standards and Technology
Gaithersburg, MD 20899

Stephen Thurgate

School of MPS
Murdoch University,
Murdoch, WA 6150 Australia

Terenceo LaCuesta

University of the Philippines at Los Baños,
College, Laguna, Philippines 4031

By looking at the x-ray photoemission spectra from collimated x-rays which excite the surface in total reflection, we are able to obtain elemental concentrations as a function of depth which is controlled by the penetration of the x-rays into the surface. The dependence of chemically shifted peaks as a function of incidence angle provides data about the chemical composition as well. We have applied this technique using a laboratory Mg Kα source with adjustable incidence angle to study the chemical makeup of an oxidized (100) surface of indium phosphide. We are able to obtain information about the chemical compounds present, their thickness, density, and stoichiometry. This information is obtained by correlated fits to the photoemission peaks from different chemical states over a range of angles of incidence near the critical angle.

INTRODUCTION

The current trend in the production of metal-insulator-semiconductor devices is toward smaller device areas and thinner insulating layers. There is a demand for better analytical techniques to determine the concentration of constituents and impurities in the layers as well as the thickness of the layers. One of the methods now used for the analysis of thin layers is angle-resolved x-ray photoemission (ARXPS) (1), in which the sampling depth is varied by adjusting the takeoff angle to the electron spectrometer. There is an alternative method of controlling the sampling depth which takes advantage of additional information present in the x-ray optical constants of the sample and its overlayers. The total reflection of x-rays from a smooth surface at grazing angles allows some control over the penetration depth of the radiation (2). We describe here the application of x-ray photoemission performed near or at total reflecion for the purpose of obtaining thicknesses and chemical constitutents of one or more layers on a semiconductor surface. The range of layer thickness for which these measurements are applicable is 0.5-5nm, a range which encompasses thin oxide layers, surface passivation layers, and spike doping of surfaces.

Grazing Incidence X-Ray Photoemission Spectroscopy (GIXPS) depends on the fact that at x-ray energies, the real part of the index of refraction is slightly less than 1. As a result, there is a small angle of incidence below which the x-rays undergo total external reflection (3). The penetration of the resulting evanescent wave decreases as the angle of incidence is diminished. The sampling depth is determined by the penetration of the x-rays and the escape of photoelectrons. The range of sampling depth is therefore quite short, roughly 1-2 nm (4). However, photoelectrons can be obtained from depths greater than one sampling length, hence the method is useful over a wider range of layer thicknesses. The principal advantage of the method is evident in the case of one or more layers on the surface with different optical constants. The highly nonlinear behavior of the x-ray fields in these layers provides intensity shifts of the various components that are further constraints in the analysis of the layers. In all cases, the variation of the sampling depth by means of the incident x-rays rather than extreme electron trajectories minimizes the effect of surface layers with a high degree of heterogeneity.

We report here the results of the method applied to the surface layer resulting from the etching of an InP (100) surface with an acid etch. It had been previously reported by Wilmsen and Kee that thick oxides of InP were separated into layers of different components (5). This work is part of a comprehensive study of the oxide layers which form on InP as a result of different etches and oxidation techniques.

EXPERIMENTAL DESCRIPTION

The apparatus which is used to obtain photoemission spectra is shown in Fig. 1. It has been described in detail elsewhere (4,6). The angle of incidence is determined by a slit in front of the x-ray source and the position of the sample. Because of the small solid angle of the sample intercepted by the beam at a grazing angle, the angular resolution is 0.1°.

We use a Mg Kα laboratory x-ray source on a movable track to adjust the angle of incidence. The sample position is held fixed in front of the electron spectrometer, a double-pass cylindrical mirror analyzer, to avoid any artifacts due to changes in the electron collection efficiency.

The measurements reported here were made on a sample consisting of a small cleaved block of InP which had a (100) mechanically-polished surface. The samples showed sharp x-ray reflections with little scatter when observed with the channel plate. The surface was prepared by etching with a 0.1% Br in methanol solution, followed by an etch consisting of 1 M HNO_3 for 2.0

minutes. The sample was rinsed in distilled water, dried in a stream of dry nitrogen and immediately inserted into the photoemission chamber. Spectra were taken at six angles of incidence between 9 and 55 mrad for each of the following lines: In $3d_{5/2}$, O 1s, P 2p, and C 1s. Data were obtained at a base pressure of 1×10^{-9} torr, a pressure at which the oxide characteristics and surface contamination are not observed to change noticeably over a period of weeks

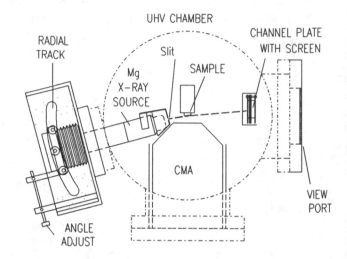

FIGURE 1. Experimental apparatus.

Each of the above spectra were taken in sequence until 480 s of counting time per data point was accumulated, with 0.25 eV spacing. Our instrumental resolution for the photoemission lines is 1.8 eV. We fit the photoemission lines to Gaussian-Lorentzian lineshapes. In some cases, the chemical shifts of the different species are smaller than the instrumental resolution. There is currently a wide variation in the binding energies reported for the photoemission lines listed above for InP and its oxidation compounds (7-10). We used the binding energies assigned by Thurgate and Erickson (9), which are slightly modified from the values of Hollinger et al. (8).

DATA ANALYSIS AND RESULTS

The data analysis is carried out in two steps. In the first step, the photoemission peaks are fitted to the spectra at each angle of incidence. The photoemission peaks were fitted for the compounds reported by Hollinger et al. (8) to be stable constituents of the surface of InP. These were In_2O_3, $InPO_4$, $In(OH)_3$, H_2O, and C. Since there must be a complete correlation between the photoemission intensities of different elements in the

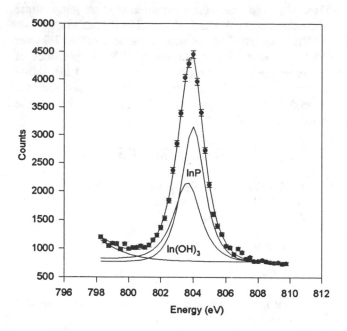

FIGURE 2. Decomposition of the Indium $3d_{5/2}$ line with an incidence angle of 55 mrad. The In component of $InPO_4$ is so small that it has not been included. The peak labelled "$In(OH)_3$" also includes the contribution from In_2O_3, which is at nearly the same binding energy.

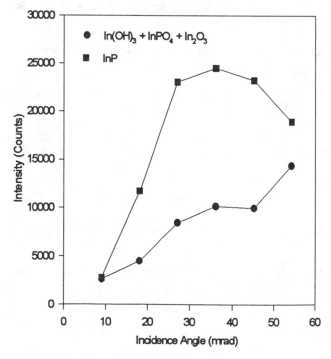

FIGURE 3. Intensity of In XPS peaks as a function of incidence angle.

same chemical compound, fits which reflect this criterion are considerably more constrained than they might

otherwise be, and our fitting routine takes advantage of this fact. Figure 2 shows a sample spectrum for the In $3d_{5/2}$ line taken at an incidence angle of 55 mrad. The spectrum shows both the photoemission peak and the fits of the different chemically shifted constituents. Fig. 3 shows a summary plot of all the In fits as a function of the angle of incidence.

In the second step, the resulting intensities as a function of incidence angle are fitted by a multiple-layer model which assumes that the substrate is covered with individual layers of each chemical species. This model incorporates layer thicknesses, densities, roughness, and the order of the layers. Since each layer has its own optical constants, it is necessary to calculate the x-ray field transmitted and reflected by each layer boundary (3) as well as the field inside the layers (2).

The photoemission intensities of elements in and on this InP sample are shown in Fig. 4 along with the results of our fits. The photoemission peaks from the substrate and from the overlayer displayed in Figs. 3 and 4 show local maxima near 31 mrad, the critical angle for total reflection from InP. This is because the x-ray field at this angle is reinforced by the coherent addition of the incident wave field and the reflected wave field. The method of calculating the fields in the layers has been previously described (11). The photoemission intensities are calculated by integrating with depth the intensity which occurs within each layer.

In our original analyses, layer thicknesses were fitted individually to each element (11). We have improved upon that routine by eliminating many of the original approximations and by performing the fits simultaneously to the intensity versus incidence angle curves associated with elements of a single compound.

In fitting this data we immediately noticed that the intensity of the photoemission peak corresponding to $InPO_4$ was extremely weak. Secondly, models which included multiple layers gave fit results inferior to those favoring a single overlayer. We assumed an inelastic mean free path of 2.0 nm in the overlayer for an electron with a kinetic energy of 1250 eV(12).

In this particular case, the fitting procedure strongly indicated that the real part of the optical constants for the overlayer were indeed the same as for the substrate. This was consistent with a very thin overlayer. If we used the optical constants for In_2O_3 or $InPO_4$, the intensity versus angle curves became unreasonably peaked. On the other hand, we found it necessary to use values for the imaginary part of the optical constants that were 25% greater than the bulk in the overlayer. This could be due to the overlayer being rougher than the substrate, an effect that our model is not yet able to account for. Both

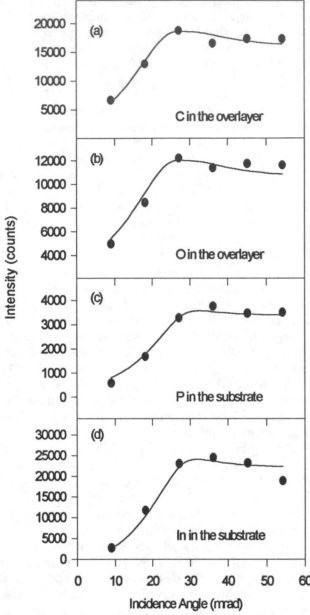

FIGURE 4. Model fits to the intensity versus angle curves for overlayer (a,b) and substrate (c,d).

of these details is consistent with a thin and irregular overlayer.

In conclusion, we deduce that the overlayer resulting from the etching process described above is a heterogeneous mixture of In_2O_3, $In(OH)_3$, and C, with a very small quantity of $InPO_4$. The oxide had not separated into distinct component layers of $InPO_4$ and In_2O_3 as reported for thicker oxides (5). The overlayer had a thickness of 1.8 nm, a composition of 83% C, 13 % O and 4% In, and a density equal to the density of the substrate. The overlayer characteristics were similar to those observed on surfaces prepared at room temperature and examined by ARXPS, reported by Zemek et al. (13).

The potential advantages of the use of GIXPS over ARXPS are the ability to distinguish the presence of layers by their different optical constants and the ability to vary the sampling depth in a manner that does not depend on the homogeneity of long escape paths of the exiting electrons.

REFERENCES

1. P. H. Citrin, G. K. Wertheim, and Y. Baer, *Phys. Rev. Lett.* **41**, 1425 (1978); M. Pijolat and G. Hollinger, *Surf. Sci.* **105**, 114 (1981).

2. B. L. Henke, *Phys. Rev.* A **6**, 94 (1972).

3. L. G. Parratt, *Phys. Rev.* **95**, 359 (1954).

4. T. Jach, M. J. Chester, and S. M. Thurgate, *Rev. Sci. Instrum.* **65**, 339 (1994).

5. C. W. Wilmsen and R. W. Kee, *J. Vac. Technol.* **14**, 953 (1977).

6. M. J. Chester, T. Jach, and S. Thurgate, *J. Vac. Sci. Technol.* B **11**, 1609 (1993).

7. D. T. Clark, T. Fok, G. G. Roberts, and R. W. Sykes, *Thin Solid Films* **70**, 261 (1980).

8. G. Hollinger, E. Bergignat, J. Joseph, and Y. Robach, *J. Vac. Sci. Technol.* A **3**, 2082 (1985).

9. S. M. Thurgate and N. E. Erickson, *J. Vac. Sci. Technol.* A **8**, 3672 (1990).

10. J. C. Woicik, T. Kendelewicz, K. Miyano, R. Cao, P. Pianetta, I. Lindau, and W. E. Spicer, *Physica Scripta* **41**, 1034 (1990).

11. M. J. Chester and T. Jach, *Phys. Rev.* B **48**, 17262 (1993).

12. S. Tanuma, C. J. Powell, and D. R. Penn, *Surf. Interface Anal.* **17**, 927 (1991).

13. J. Zemek, O.A. Baschenko, and M.A. Tyzkhov, *Thin Solid Films,* **224**, 141 (1993).

Calibration of Binding-Energy Scales of X-Ray Photoelectron Spectrometers

C. J. Powell

Surface and Microanalysis Science Division,
National Institute of Standards and Technology, Gaithersburg, MD 20899

X-ray photoelectron spectroscopy (XPS) is in common use in the semiconductor industry for measuring the composition of the outermost atomic layers of device materials. A strong advantage of XPS over other techniques for surface analysis is that useful information on the chemical state of detected elements can be obtained from small "chemical shifts" of the characteristic peaks in a photoelectron spectrum. Reliable determination of chemical state, however, depends on adequate calibration of the instrumental binding-energy (BE) scale.

We report the development and validation of an improved procedure for calibrating the BE scales of XPS instruments with Cu, Ag, and Au foils. A draft calibration procedure was prepared for this purpose for consideration by ASTM Committee E-42 on Surface Analysis. An interlaboratory intercomparison was then organized to evaluate the draft procedure and different methods of peak location. Briefly, the calibration was performed at two points on the BE scale (the Au $4f_{7/2}$ and Cu $2p_{3/2}$ lines), and checks were made of the assumption of BE-scale linearity from measurements on other lines. It was found that small systematic differences in peak position (up to approximately 0.05 eV) could occur if peaks were located with assumed backgrounds of non-zero slope or if multi-peak fits were made to Cu L_3VV and Ag M_4VV Auger spectra. The two parameters in the linear calibration equation could vary with instrumental parameters. Check measurements made with Auger lines on XPS instruments equipped with x-ray monochromators showed an average offset of 0.13 eV from the expected positions; this offset is due to different average x-ray energies from monochromatized and non-monochromatized sources.

The calibration procedure has also been used to provide retroactive calibrations in three of four large sets of elemental BE data that had significant inconsistencies. The mean BE values for each element appear to be a useful reference set for the determination of chemical shifts in XPS.

INTRODUCTION

X-ray photoelectron spectroscopy (XPS) is one of several techniques that are used extensively in semiconductor and other application areas for determining the composition of the outermost several atomic layers of a material. A substantial advantage of XPS in these applications is that it can often give information on the chemical states of the detected elements. This information is deduced from small shifts, called chemical shifts, of observed peaks in a photoelectron spectrum from the positions found for the pure elements or for reference compounds. In favorable cases, the chemical shifts can be 1 to 10 eV but in other cases the chemical shifts may only be a few tenths of an eV. For these latter cases in particular, reliable determination of chemical state necessitates calibration of the instrumental binding-energy (BE) scale to better than 0.1 eV. Interlaboratory comparisons (1, 2) have shown, however, that BE measurements are frequently not measured and reported with this level of accuracy.

Inadequacies in BE-scale calibration have also led to the publication of inconsistent BE data. For example, four sets of BE data (3-6) that each contain BE measurements for many elements made on a single instrument show differences of up to 0.33 eV among the values for the BE of the Cu $2p_{3/2}$ line which is often utilized for instrument calibration. As a result, there are greater differences in BE values for a given element than might be the case if all four instruments had been calibrated in a consistent way.

A major impediment to adequate BE-scale calibration has been the lack of a satisfactory calibration procedure with which an instrument could be routinely calibrated. To rectify this deficiency, a draft "Standard Practice for the Calibration of the Electron Binding-Energy Scales of X-Ray Photoelectron Spectrometers" has recently been prepared for consideration by Committee E-42 on Surface Analysis of the American Society for Testing and Materials (ASTM). A synopsis of this calibration procedure is given in the following section. Before formal balloting of the draft standard was initiated, it was decided to conduct an interlaboratory comparison to evaluate the convenience and reliability of this standard practice. It was also desired to evaluate the adequacy of different methods of peak location in performing the calibration. The third section of this report contains a summary of the work to evaluate peak-location procedures, while information on the interlaboratory comparison is presented in the fourth section; further details are presented elsewhere (7).

A review has been made (8) of the four sets of elemental BE data (3-6). The draft standard has been used to perform retroactive calibrations of the BE scales of three of these instruments with the result that BE data from all four instruments were placed on a common scale. The fifth section of this report contains a summary of a comparison of these BE values and a table of mean elemental binding energies (8).

CALIBRATION PROCEDURE

It is assumed initially that the BE scale of each instrument is linear; this assumption is believed reasonable for many instruments (2). The measured BE on an instrument E_{meas} can then be related to a reference value E_{ref} by the following equation:

$$E_{meas} = a + b E_{ref} \qquad (1)$$

where the parameters a and b are constants for the particular instrument. These parameters were determined from BE measurements made for the Au $4f_{7/2}$ and Cu $2p_{3/2}$ peaks and corresponding reference values reported by the UK National Physical Laboratory (NPL) (9). The NPL reference values have reported measurement uncertainties of 0.02 eV (one standard deviation), but these reference values are regarded here as just two suitable numbers (without error) that are needed to establish an internally consistent BE scale for XPS. After values for the

parameters a and b have been determined for an instrument (and a given set of operating conditions), it is then simple to calculate corrected BE values E_{corr} from other BE measurements using $E_{corr} = (E_{meas} - a)/b$. The assumption of BE-scale linearity can then be checked from BE measurements at a few intermediate energies using photoelectron and Auger-electron peaks for Cu and Ag and the corresponding NPL reference values (9).

PEAK-LOCATION PROCEDURES

The locations of peaks in the NPL data (9) were determined from least-squares fits of a quadratic function to experimental data points comprising a specified fraction of each peak (e.g., the top 10% of the peak). The position of the peak maximum was found from parameters of the quadratic equation (when the first derivative equals zero).

It was suggested to participants in the interlaboratory comparison that peak-synthesis software available on many instruments could also be used for peak location. It was thought that different functions (e.g., Lorentzian, Gaussian, or mixed Gaussian-Lorentzian functions) could be used for fits to the data points comprising a peak. The instructions here were deliberately non-specific so that participants

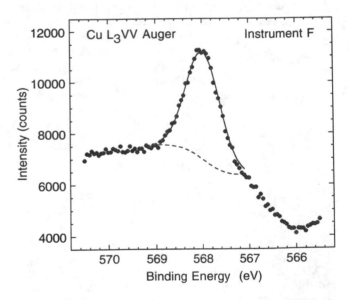

FIGURE 1. Experimental Cu L_3VV Auger spectrum measured with instrument F in the interlaboratory comparison (7). The solid circles show the experimental data, the dashed line indicates an assumed Shirley background, and the solid line shows the overall fit based on the use of a mixed Gaussian-Lorentzian function to represent the measured Auger-electron lineshape.

288

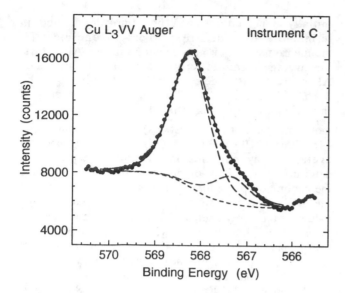

FIGURE 2. Experimental Cu L_3VV Auger spectrum measured with instrument C in the interlaboratory comparison (7); see caption to Fig. 1 and text.

could apply the procedures in common use for this purpose in their laboratories.

Figure 1 shows an example of the use of instrumental software to locate the position of the maximum in the Cu L_3VV Auger spectrum (7). The dashed line in Fig. 1 is a so-called Shirley background that was selected by the operator; this particular function is a simple and convenient means for making an approximate correction for inelastic scattering but its implementation is subject to some arbitrariness. A mixed Gaussian-Lorentzian function was then fitted to the difference intensity (corresponding to the measured intensity less the Shirley background), and the solid line indicates the overall fit to the measured spectrum in the vicinity of the maximum. The fitting function had its peak maximum at 567.973 eV on the BE scale while the peak position found by the quadratic-fit method to data points comprising the top 10% of the peak was 568.007 eV, a difference of 0.034 eV. This difference in peak positions is due to the finite slope of the Shirley background in the vicinity of the maximum. The difference is small compared to other sources of uncertainty in a BE calibration (see below), but it is not neglible. This source of systematic error can be eliminated by avoidance of Shirley or other backgrounds of non-zero slope during peak location with instrumental software. Such backgrounds should thus not be used during calibration of the energy scale.

Figure 2 shows a Cu L_3VV Auger spectrum measured on another instrument. The operator chose to fit this spectrum with a Shirley background (short-dashed line) and *two* Gaussian-Lorentzian peaks (long-

dashed lines) to represent the known asymmetry of the main peak; the solid line is again the overall fit to the data. The position of the maximum is found here to be 568.231 eV using the quadratic-fit method, while the position of the larger Gaussian-Lorentzian peak is found to be 568.234 eV. The difference of these two values is indeed small, but is the result of an almost complete cancellation of two systematic errors in this particular example. One source of error arises from the slope of the Shirley background (as in Fig. 1) and the other is associated with the use of two overlapping Gaussian-Lorentzian functions to represent a complex lineshape. The latter source of error can be avoided by using only a single function to fit a measured peak in the vicinity of its maximum. It has been estimated that systematic offsets in peak location of up to about 0.05 eV can occur if the two sources of systematic error discussed here are not corrected (7).

INTERLABORATORY COMPARISON

The interlaboratory comparison involved measurements made on eleven instruments (7). Each participant was asked to make repeated measurements of the BE for the Cu $2p_{3/2}$ photoelectron peak. The average value of the repeatability standard deviation was found to be 0.007 eV, with the repeated measurements usually made after the specimen was translated in one or more directions parallel to the surface. One participant pointed out that a better measure of the standard deviation would be obtained following repeated mounting of the specimen foil, introduction into the vacuum system, cleaning, and alignment. The repeatability standard deviation found from this procedure was 0.027 eV compared to an average value of 0.018 eV with the shorter procedure on this instrument.

Figure 3 is a plot of $\Delta E_n = E_{corr\ n} - E_{ref\ n}$ versus BE where $E_{corr\ n}$ and $E_{ref\ n}$ are a corrected BE value from the calibration and a NPL reference value, respectively, for particular photoelectron or Auger-electron peaks used to evaluate the adequacy of the calibration; the Ag $3d_{5/2}$ photoelectron peak and the Cu L_3VV and Ag M_4VV Auger peaks were used for this purpose. The Cu and Ag Auger-electron peaks each have characteristic kinetic energies but appear at different positions on the BE scale depending on whether x rays from Al or Mg anodes are used for the XPS measurement.

The solid circles in Fig. 3 show average ΔE_n values based on data from instruments without an x-ray monochromator and the solid squares show average ΔE_n values from instruments with a monochromator. The error bars denote one standard deviation.

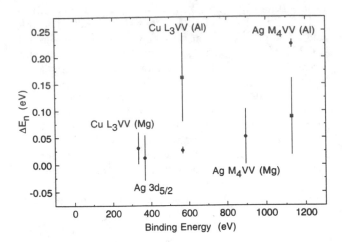

FIGURE 3. Plot of average values of ΔE_n versus BE for the indicated photoelectron and Auger-electron lines for Mg and Al characteristic x rays. The error bars are unrealistic for the two Auger points corresponding to Al x rays and no monochromator; each point is the average of only two ΔE_n values (7).

In assessing the values of ΔE_n and the corresponding error bars in Fig. 3, it should be noted that there could be systematic offsets in peak location of up to about 0.05 eV in measurements on some instruments (as discussed in Section 3). It was also estimated that the confidence limits (at the 95% level) in values of $E_{corr\ n}$ and thus in ΔE_n would be about 0.09 eV (7).

The average value of ΔE_n for the Ag $3d_{5/2}$ line in Fig. 3 is 0.013 eV with a standard deviation of 0.041 eV. These two values are considered sufficiently small to conclude that the instrumental BE scales are, on the average, close to linear for this portion of the scale.

The average values of ΔE_n for the Cu L_3VV (Mg), Cu L_3VV (Al), and Ag M_4VV (Mg) lines as measured with Mg x rays or with non-monochromated Al x rays are considered acceptably small in comparison with the estimated confidence limits and possible peak location errors. For the Ag M_4VV (Al) line, however, there are only two values of ΔE_n (both obtained on the same instrument), and these appear to be abnormally large. It is therefore believed that the BE scale of this instrument is slightly nonlinear for binding energies greater than the high-BE calibration point (the Cu $2p_{3/2}$ line).

The average values of ΔE_n for the Cu L_3VV and Ag M_4VV Auger lines excited by monochromatized Al x rays are 0.162 ± 0.082 eV and 0.089 ± 0.071 eV, respectively. These ΔE_n values are much larger than

initially expected but can be simply explained in terms of the different average energies for monochromated and non-monochromated Al x rays. The average (centroid) energy for non-monochromated Al x rays has been recently determined to be 1486.568 ± 0.010 eV (10). The corresponding energy of the $K\alpha_1$ component (usually selected by the monochromator on XPS systems) has been found to be 1486.706 ± 0.010 eV (10). This slight difference in average x-ray energies is of no consequence in determining the BEs of photoelectron lines because the calibration procedure ensures that the zero-point on the BE scale corresponds to the particular average value for the x-ray energy. That is, the kinetic energy of a given photoelectron will vary with the x-ray energy, but the BE remains the same. The position of Auger-electron lines on the BE scale, however, will vary with the particular average x-ray energy.

The average Al $K\alpha_1$ energy is thus 0.138 eV larger than the Al $K\alpha_{12}$ centroid energy, and it is this difference that accounts for the generally larger ΔE_n values found for the Cu L_3VV and Ag M_4VV Auger lines on instruments equipped with x-ray monochromators. The standard deviations found for these ΔE_n values are larger than for the other lines. In addition to the peak location uncertainties discussed previously, it is possible that the x-ray monochromator on a given instrument may be detuned slightly (e.g., due to mechanical misalignments) so that the actual average x-ray energy may differ from the average $K\alpha_1$ energy. A measurement of ΔE_n could thus be useful in determining the average energy of x rays emerging from the monochromator.

In addition to minor improvements suggested by participants, the laboratory intercomparison was useful in indicating how the draft standard should be modified in two important areas. First, peaks need to be located by procedures that avoid the two systematic errors discussed in the previous section. Second, the positions of the Cu L_3VV and Ag M_4VV Auger lines cannot be used to check the adequacy of BE-scale calibration on XPS instruments with x-ray monochromators.

MEAN ELEMENTAL BINDING ENERGIES

Four sets of elemental BE data (3-6) were recently examined to check the degree of consistency of the elemental BE values (8). One of the XPS instruments in this group had been calibrated using the NPL reference BE values (9) but the calibration history of the other three instruments was less well established. The BE scales of these three instruments were

TABLE 1. Mean binding energy values for the indicated elements and photoelectron lines (8). Recommended values from the UK National Physical Laboratory are shown for Cu, Ag, and Au. The values in parentheses for B, S and Se are less reliable (see text).

Element	Line	Mean BE (eV)
Be	1s	111.85
B	1s	(188.63)
C	1s	284.44
Mg	$2p_{3/2}$	49.79
Al	$2p_{3/2}$	72.87
Si	$2p_{3/2}$	99.34
P	$2p_{3/2}$	130.01
S	$2p_{3/2}$	(163.82)
Ar (implant)	$2p_{3/2}$	241.82
Ca	$2p_{3/2}$	346.61
Sc	$2p_{3/2}$	398.53
Ti	$2p_{3/2}$	453.98
V	$2p_{3/2}$	512.20
Cr	$2p_{3/2}$	574.35
Mn	$2p_{3/2}$	638.89
Fe	$2p_{3/2}$	706.86
Co	$2p_{3/2}$	778.35
Ni	$2p_{3/2}$	852.73
Cu	$2p_{3/2}$	932.67
Zn	$2p_{3/2}$	1021.81
Ge	$3d_{5/2}$	29.39
As	$3d_{5/2}$	41.65
Se	$3d_{5/2}$	(55.13)
Kr (implant)	$3d_{5/2}$	86.97
Y	$3d_{5/2}$	155.90
Zr	$3d_{5/2}$	178.78
Nb	$3d_{5/2}$	202.34
Mo	$3d_{5/2}$	227.95
Ru	$3d_{5/2}$	280.08
Rh	$3d_{5/2}$	307.20
Pd	$3d_{5/2}$	335.12
Ag	$3d_{5/2}$	368.26
Cd	$3d_{5/2}$	405.12

TABLE 1 (continued)

In	$3d_{5/2}$	443.88
Sn	$3d_{5/2}$	484.98
Sb	$3d_{5/2}$	528.25
Te	$3d_{5/2}$	573.03
Xe (implant)	$3d_{5/2}$	669.62
Pr	$3d_{5/2}$	931.89
Nd	$3d_{5/2}$	980.90
Eu	4d	128.19
Gd	4d	140.37
Tb	4d	146.01
Dy	4d	152.35
Ho	4d	159.59
Er	4d	167.28
Tm	4d	175.39
Yb	4d	182.40
Lu	$4f_{7/2}$	7.19
Hf	$4f_{7/2}$	14.31
Ta	$4f_{7/2}$	21.83
W	$4f_{7/2}$	31.37
Re	$4f_{7/2}$	40.34
Os	$4f_{7/2}$	50.69
Ir	$4f_{7/2}$	60.84
Pt	$4f_{7/2}$	71.12
Au	$4f_{7/2}$	83.98
Hg	$4f_{7/2}$	99.85
Tl	$4f_{7/2}$	117.73
Pb	$4f_{7/2}$	136.85
Bi	$4f_{7/2}$	156.96

retroactively calibrated using the procedure described in the second section. Since the BE data from the four instruments were now on a consistent scale, it was possible to calculate mean elemental BE values (listed in Table 1) for those elements and core levels for which there were two or more measurements. Individual differences from the mean values were then calculated to examine possible differences among the BE values from each source and possible trends in the difference values versus BE for each instrument.

Figure 4 is an example of the trends in the difference of the BE values reported by one data source (5) from the corresponding mean elemental BE values (8). It should be emphasized that this plot shows trends and outliers that are not only associated

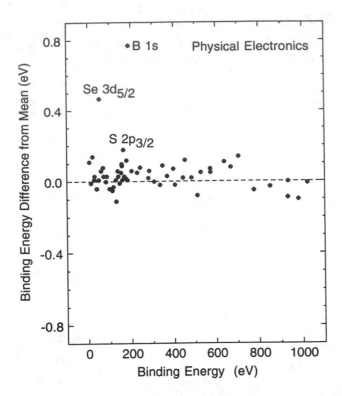

FIGURE 4. Plot of differences in the BE values (solid circles) reported by Physical Electronics (5) from the corresponding mean BE values (Table 1) versus BE (8).

with BE data from this source but also with the mean energies determined from the data for the other instruments. Figure 4 nevertheless shows that most differences are very close to zero (indicated by the horizontal dashed line) and that there does not appear to be any significant nonlinearity in the BE scale of this instrument.

There are three outliers indicated in Fig. 4. These outliers are associated with photoelectron lines of B, S, and Se which are difficult materials to handle. That is, the outliers are probably due to differences in surface chemical state rather than to instrumental mistakes. The mean BE values for these three elements are shown in parentheses in Table 1 to indicate that they have much higher uncertainties than the other values. Two other outliers were found in the other data sets that were believed due to surface impurities or possible mistakes; in each of these cases, the BE values from three other measurements for the particular elements (V and Ta) were in close agreement, and were used to determine the mean BE values for these elements shown in Table 1.

It was implicitly assumed in the comparison of data that the four data sets were of equal precision and accuracy. This assumption is unlikely to be correct

since differences in specimen purity, differences in measurement conditions, and different peak-location algorithms will influence the accuracy and precision of any BE measurement. Nevertheless, there was a high degree of consistency in the BE data from the four sources after the recalibration. The mean and median of the BE differences for each data set were less than half the corresponding standard deviations, and it is therefore believed that the remaining systematic effects are acceptably small. The recalibration was thus successful in reducing the systematic BE differences in the four data sets by nearly an order of magnitude.

Table 1 shows mean elemental BE values (8) for 61 elements. With the exclusion of data for B, S, and Se, the combined standard deviation for the remaining 58 values is 0.061 eV. It is believed that the BE data in Table 1 should be useful for the calculation of chemical shifts in XPS.

REFERENCES

1. Powell, C. J., Erickson, N. E., and Madey, T. E., J. Electron Spectrosc. **17**, 361-403 (1979).
2. Anthony, M. T., and Seah, M. P., Surf. Interface Anal. **6**, 107-115 (1984).
3. Svensson, S., Martensson, N., Basilier, E., Malmqvist, P. A., Gelius, U., and Siegbahn, K., J. Electron Spectrosc. **9**, 51-65 (1976); Berndtsson, A., Nyholm, R., Martensson, N., Nilsson, R., and Hedman, J., Phys. Stat. Sol. (b) **93**, K103-K105 (1979); Martensson, N., Berndtsson, A., and Nyholm, R., J. Electron Spectrosc. **19**, 299-301 (1980); Nyholm, R., and Martensson, N., J. Phys. C: Sol. State Phys. **13**, L279-L284 (1980); Nyholm, R., Berndtsson, A., and Martensson, N., J. Phys. C: Sol. State Phys. **13**, L1091-L1096 (1980); Lebugle, A., Axelsson, U., Nyholm, R., and Martensson, N., Phys. Scripta **23**, 825-827 (1981).
4. Ikeo, N., Iijima, Y., Niimura, N., Sigematsu, M., Tazawa, T., Matsumoto, S., Kojima, K., and Nagasawa, Y., *Handbook of X-Ray Photoelectron Spectroscopy*, Tokyo: JEOL, 1991.
5. Moulder, J. F., Stickle, W. F., Sobol, P. E., and Bomben, K. D., *Handbook of X-Ray Photoelectron Spectroscopy*, Eden Prairie: Perkin-Elmer, 1992.
6. Crist, B. V. (private communication).
7. Powell, C. J., Surf. Interface Anal. **23**, 121-132 (1995).
8. Powell, C. J. Appl. Surf. Science (in press).
9. Seah, M. P., Surf. Interface Anal. **14**, 488 (1989).
10. Schweppe, J., Deslattes, R. D., Mooney, T., and Powell, C. J., J. Electron Spectrosc. **67**, 463-478 (1994).

SCANNING PROBE MICROSCOPIES

Semiconductor Characterization with Scanning Probe Microscopies

R. M. Feenstra
IBM T. J. Watson Research Center, Yorktown Heights, NY 10598
J. E. Griffith
AT&T Bell Laboratories, 6F225, Murray Hill, NJ 07974

Scanning probe microscopes are valuable tools for semiconductor surface characterization. The scanning tunneling microscope can be used to study surface reconstruction, epitaxial growth, and heterostructures in cross section. The scanning capacitance microscope can be used to measure dopant profiles. Topography measured with the atomic force microscope reveals information about processes such as strained layer growth, and it is a valuable tool for dimensional metrology.

1. INTRODUCTION

From its inception in 1982, the scanning tunneling microscope (STM) has proven to be a valuable tool for semiconductor surface characterization. Early studies concentrated on atomic scale properties of surface reconstructions in ultra-high-vacuum (UHV) (1), and such studies have continued to be important in understanding various aspects of epitaxial growth (2). More recently, the method of cross-sectional STM has been developed, in which semiconductor heterostructures are viewed on a cleavage face, thus providing unique results for dopant profiling, alloy segregation, and structural properties of buried interfaces (3-11). Examples of such studies are presented in Section 2 of this paper, including results from GaAs doping structures, low-temperature-grown GaAs containing arsenic-related point defects, and InAs/GaSb superlattices.

The STM has spawned many other scanning probe microscopies, which measure different physical quantities and/or operate in other environments. For example, the scanning capacitance microscope (SCM) has enabled quantitative dopant profiling in silicon devices, by utilizing the dependence of capacitance on doping concentration in a metal/semiconductor junction (12-14). Results from such work are discussed in Section 3.

Perhaps the most widely used scanning probe instrument is the atomic force microscope (AFM), which provides direct surface topographic images, as well as information on tip/sample forces (e.g. magnetic, friction, etc.). The topographic information is useful in understanding semiconductor growth and processing in a wide range of applications. An example of such work in given in Section 4, where we discuss effects of plastic and elastic strain relaxation in growth of strained SiGe films.

The topographic information provided by AFM is also useful for determining the size of tiny fabricated structures. The relative insensitivity of the AFM to sample characteristics other than topography is actually an advantage because it eliminates sources of error that affect dimensional measurements with other microscopes. We describe recent advances in proximity detection, probe fabrication, and data analysis.

2. CROSS-SECTIONAL SCANNING TUNNELING MICROSCOPY

The method of cross-sectional STM has been utilized in the past several years for studying a variety of epitaxial semiconductor structures (3-11). Such structures generally consist of a number of layers, with thickness ranging from Å to µms, grown by molecular-beam epitaxy (MBE) on a single crystal substrate. In the method, a cross-section of the structure is prepared by cleaving, and STM studies are performed on this cleavage face. For layered systems (i.e. superlattices), one gains information on the interfaces between layers, as well as the properties of the individual layers themselves.

Most studies to date have been performed on III-V semiconductors (GaAs and related materials), although some work has appeared on the group IV materials (Si and Ge) (9-11). For GaAs, the cleavage face is a (110)

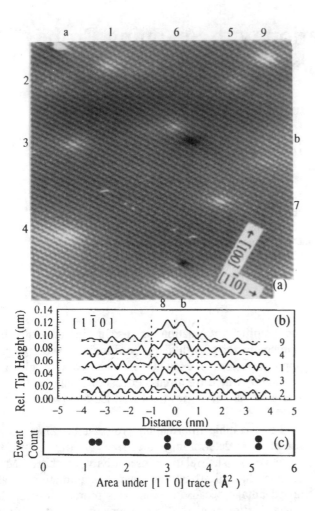

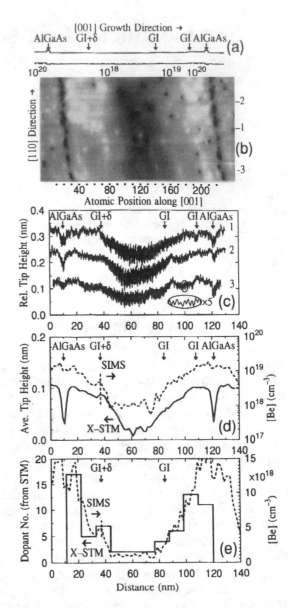

Figure 1(a) STM image of a (110)-cleaved, 1×10^{19} cm^{-3}, Be-doped GaAs surface. Image size is 31x29 nm^2, acquired at a sample voltage of —2.1 V and current 0.1 nA. The relative tip height is given by a grey scale, from 0 to 0.2 nm. Nine hillocks (dopants) are identified using numbers at the closest point on the perimeter. (b) Tip height trace along the [1$\bar{1}$0] direction of a selection of the hillocks identified in (a). (c) Scatter plot of area under the [1$\bar{1}$0] tip height traces (integrated intensity) of all nine hillocks —note the uniform distribution. (From (4).)

plane, and for cleavage in UHV this face has the convenient property that the dangling bond energies do not lie within the band gap. Thus, spectroscopic studies of this surface reveal bulk-like properties for the band gap, band offsets, and other features. Alternatively, for Si, cleavage can be accomplished on either the (110) or (111) faces, but in both cases the spectroscopy of as-cleaved surfaces is dominated by dangling bond surface states. Thus, some chemical treatment is required to passivate these states, and perfect passivation may be difficult to attain.

Figure 2. STM image of GaAs and AlGaAs layers with variable dopant concentration and various depth profiles. (a) Calculated bulk-band structure for the intended heterostructure. In registry below, (b) shows the topographic image, 100x50 nm^2 in size, of the layers taken with sample voltage of —2.1 V and current 0.1 nA. The relative tip height is given by a grey scale, from 0 to 0.2 nm. (c) Several topographic lines scans across the image in (b), along the [1$\bar{1}$0] direction (AlGaAs layers vertically aligned); atomic corrugation shown in inset. (d) Line scan (solid) corresponding to the averaged line scan over the top half of (b), smoothed slightly to take out atomic corrugation, and compared to the Be concentration measured by SIMS (dotted, right axis). (e) Dopant histogram determined by counting the white hillocks (dopants) in a series of rectangles across (b) (solid, left axis) and compared to the Be concentration measured by SIMS (dotted, right axis). (From (5)).

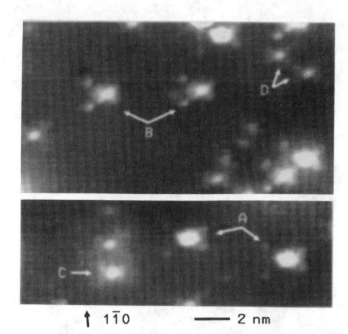

↑ 1̄10   —— 2 nm

Figure 3. STM images of the (110) cleaved surface of unannealed low-temperature-grown GaAs, acquired with 0.1 nA tunnel current and at a sample voltage of —2.0 V. Various point defects can be seen, and they are classified as types A, B, C, and D as indicated. The defects arise from arsenic atoms located on gallium sites in the GaAs lattice. (From (6)).

Cross-sectional STM offers several methods for dopant profiling. The most direct is to simply count the number of dopant atoms appearing in (or near) the surface atomic plane. An example is given in Fig. 1, where we show an STM image obtained from a 1 x 10^{19} cm^{-3} Be doped GaAs sample (4). Individual dopant atoms can be clearly resolved, with their contrast in the image determined simply by considering them as charged scattering centers for the carriers in the semiconductor. Thus, for the case in Fig. 1, the negatively charged dopant atoms are attractive to the holes in the valence band, and the dopant atoms appear as protrusions. From the number and magnitude of the observed protrusions, it is concluded that dopant atoms in the top several atomic planes are being imaged. Using this method of dopant atom counting, profiles for dopant concentrations in the range (1-15) x 10^{18} cm^{-3} have been determined, as shown in Fig. 2 (5). Panel (e) of that figure show a comparison of dopant concentrations deduced by direct counting, with those obtained by secondary ion mass spectrometry (SIMS) measurements; reasonable agreement between the two measurements is obtained. Fig. 2(d) shows an alternative measure of doping concentration by STM —

examining the tip height at constant current (proportional to the logarithm of the current at constant tip height). In this case, band bending induced in the semiconductor by the presence of the probe tip causes a decrease in the tunnel current. The tip height traces seen in Fig. 2(d) are in remarkable agreement with the SIMS profiles for dopant concentration. This same physical mechanism of tip induced band bending is the basis for capacitive determination of dopant concentration, discussed in Section 3.

In addition to dopant atoms, defects of any sort can, in principle, be imaged with the STM. The main requirement is the preparation of ideal, atomically flat lattice planes, on which the defects will be readily identifiable. A concentration of defects greater than about 10^{16} cm^{-3} is also required. As example of such work is shown in Fig. 3, which is an STM image acquired from GaAs which was grown at relatively low temperature (near 200° C), thereby containing an excess arsenic concentration of about 1 at.% (6). This material has found application for buffer layers in field-effect transistors (FETs), and for fast photodetectors (15). For material with no additional annealing, as in Fig. 3, the excess arsenic forms point defects, mainly arsenic antisites (an arsenic atom on a site normally occupied by gallium). Numerous defects are visible in Fig. 3, and they are classified into several types as labeled in the figure. These various types of images are interpreted as all arising from the same type of defect, the arsenic antisite, with the core of the defect being located in differing planes relative to the (110) cleavage plane (6). Two distinct satellite features can be seen on the left-hand side of the type B image, and these satellites can be faintly seen around the type A images as well. The satellite features arise from long-range tails of the defect wave-function. The defects have electronic states located in the middle of the GaAs band gap, and it is the wave-functions of these states which form the image of Fig. 3 (6).

Roughness of interfaces in semiconductor superlattices has significant consequences in device operation. Electron and hole mobility can be adversely affected, and carrier lifetime may be influenced by defects at interfaces. Intermixing at interfaces, as well as alloy clustering, will both contribute to roughness. In addition, atomic steps present on the surface during growth (e.g. due to islanding) will make a major contribution to the roughness. These effects have been studied by STM for AlGaAs/GaAs interfaces by several groups (16,17). A particularly simple case for determining interface roughness with cross-sectional STM is for superlattice made from binary materials (i.e. avoiding complication of alloy fluctuations and clustering, as in AlGaAs/GaAs).There, differing band energies in the layers lead to strong contrast between layers in STM images, and interface positions can be easily defined simply as the midpoint in the transition between observed layers. An example of this type of imaging is given in Fig. 4, where we show data from an InAs/GaSb superlattice, a material which has application for infrared photodetectors (18). For filled state images such as Fig. 4(a), the GaSb layers are bright and InAs layers dark. Faint fringes, with spacing of 0.6 nm, arise from the atomic planes in the superlattice. The interface between InAs and GaSb layers can be more clearly defined by expanding the grey-scale of the image and taking a derivative in the horizontal direction, as shown in Fig. 4(b). Interface roughness, with step heights of 0.3—0.6 nm, is clearly visible there. The interfaces of GaSb grown on InAs are seen to have relatively long flat sections, interrupted by steps separated by about 5 nm. Alternatively, interfaces of InAs grown on GaSb are considerably rougher, with average step spacing of 1—2 nm. Quantitative spectra of interface roughness have been obtained by Fourier transforming interface profiles of the type shown in Fig. 4(b) (7,8). Another feature of the InAs/GaSb growth apparent in Fig. 4(a) are the atomic-size bright dots seen in the InAs layers. These features are interpreted as Sb atoms, substitutional on As sites in the InAs. Growth studies indicate that a layer of Sb rides up on top of the InAs layer during MBE growth, and some of this Sb is incorporated in the InAs (7,8). Measurements of this Sb incorporation by cross-sectional STM provide greater precision than that available with SIMS.

Another aspect of STM measurements which we have not yet commented on is the ability to measure tunneling spectra, i.e. tunnel current vs. voltage, thus providing a measure of state-density for states located a few eV on either side of the Fermi-level. For the III-V semiconductors, the spectra provide a direct measurement of band gap, Fermi-level position within the gap, possible defect states within the gap, and also higher lying band features such as the L-valley in the conduction band (19). Furthermore, for superlattice structures, the quantum confined subbands in the system can be detected in the spectra (8). For silicon, the spectra are generally dominated by surface state features; passivation of the surface states produces some sensitivity to the bulk energy bands (9-11), although perfect passivation is difficult to obtain. As discussed above, some of the applied voltage between probe tip and sample may be dropped in the semiconductor, thus producing band bending in the semiconductor. This tip-induced

band bending provides a sensitivity to doping concentration in the semiconductor, as illustrated in Fig. 2 above, and as seen for capacitive measurements, discussed in the following section.

3. SCANNING CAPACITANCE MICROSCOPE

In contrast to the measurement of direct tunneling between a metal probe tip and semiconductor sample, one may also measure the capacitance of this junction. This

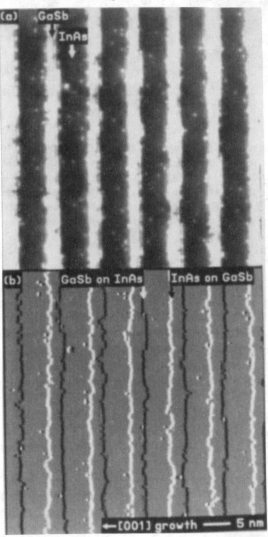

Figure 4. STM image of InAs/GaSb superlattice, grown by MBE at 380° C. Growth direction is from right to left, as indicated. Image was acquired with sample voltage of —2.0 V. (a) Constant-current topograph, with grey-scale range of 2.0 Å. Sb incorporation in the InAs gives rise to the bright spots in the InAs layers. (b) Derivative image, computed from (a). (From (7)).

technique, known as scanning capacitance microscopy (SCM), has been developed by Williams and co-workers (12-14). An atomic force microscope is used to position a nanometer scale tip at a semiconductor surface, and local capacitance is measured as a function of sample bias. A high frequency bridge is used to measure the capacitance, with sensitivity of about $3 \times 10^{-22} F / \sqrt{Hz}$. The semiconductor surface is oxidized, to provide a constant (low) density of interface states. Each capacitance-voltage (C-V) curve provides information on the carrier density in the semiconductor, in the same way as for standard metal-insulator-semiconductor devices.

Inversion of the C-V curves to obtain carrier density generally requires detailed knowledge of parameters such as tip radius-of-curvature and oxide thickness, which must be incorporated into a 3-dimensional model for the electrostatics. However, a new scheme has been developed in which the measured capacitance is maintained at a constant value while the AC voltage applied to the tip is varied (14). In that case, the depletion layer thickness in the semiconductor is relatively constant, thus permitting the use of a simplified 1-dimensional band bending model to obtain the carrier density. Results are shown in Fig. 5, for a lateral doping structure prepared by ion implantation. Simulation of the

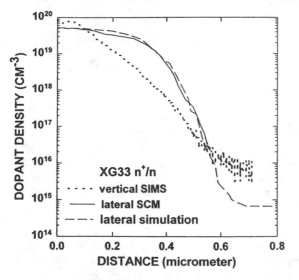

Figure 5. Comparison of an SCM measured lateral doping profile with a Suprem IV generated profile near the edge of an implantation region. The vertical SIMS profile is shown for reference. (From (14)).

lateral doping profile by SupremIV is shown by the dashed line, and the SCM result is shown by the solid line. Inversion of the SCM data requires one known point in the dopant density as input, and this was taken to be the highest concentration point (at zero lateral distance in Fig. 5). With this calibration, the SCM displays impressive agreement with the lateral simulation, over the doping range 10^{17} to 10^{20} cm^{-3}. The eventual intended application of SCM is for 2-dimensional profiling of structures (i.e. depth and lateral information, obtained by cross-sectioning the device), thus providing information which is not attainable by other methods.

In addition to cross-sectional STM and SCM, other scanned probe based methods have been developed for measuring dopant concentrations. Shafai et al. have demonstrated the use of an AFM with conducting tip to probe the resistance between a metal tip and semiconductor surface (20). This method can be viewed as an extension of standard spreading resistance measurements (21), but with smaller probe tips and better controlled force between tip and sample. As with spreading resistance measurements however, the properties of the tip/sample contact must be reproducibly established before quantitative results are possible.

4. STUDIES OF STRAINED LAYER GROWTH BY ATOMIC FORCE MICROSCOPY

Use of the AFM has become ubiquitous in many areas of thin film growth. Surface morphology often provides a useful measure of the underlying physical processes occurring during growth. In this section, we provide a few examples of the application of AFM to the characterization of semiconductor thin film growth, specifically, growth of strained and strain relaxed SiGe layers on Si substrates. This system has application for both bipolar and FET devices (22), the latter of which requires thick SiGe layers with graded Ge content such that the strain at the top of the layer is completely relaxed. This strain relaxation comes about from the formation of misfit dislocations in the layers, and particular arrangements of dislocations known as "pile-ups" form, with the resulting density of dislocations threading up to the surface being minimized (23).

For SiGe, AFM and STM have been applied to the study of 3D growth (24,25), dislocation formation in buffer layers (26,27), and quantification of surface roughness by Fourier analysis (28). An example in shown in Fig. 6, which displays results from a $Si_{0.84}Ge_{0.16}$ film grown by ultra-high-vacuum chemical vapor deposition (UHV-CVD) (22) at 560° C, with a thickness of 875 nm. A large-scale AFM image is shown in Fig. 6(a). The distinctive morphology seen there, consisting of surface undulations on the μm-scale extending along [110] and [1$\bar{1}$0] directions, is known as a "cross-hatch" pattern. This type of pattern is often observed in the morphology of strained or strain-relaxed films. It has been found to

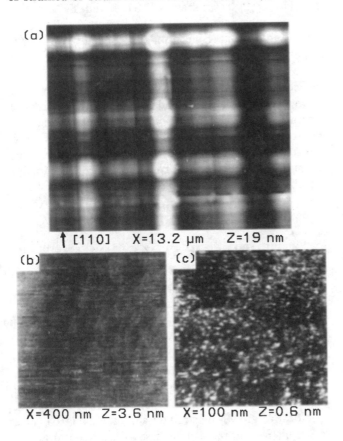

↑ [110]   X=13.2 μm   Z=19 nm

(b)   (c)

X=400 nm Z=3.6 nm   X=100 nm Z=0.6 nm

Figure 6. (a) AFM image of 875 nm thick $Si_{0.84}Ge_{0.16}$ film grown at 560° C, displaying large scale cross-hatch morphology. Expanded views of same sample are shown in (b) and (c), which are STM images displaying atomic-scale roughness on the surface. The lateral (X) and vertical (Z) full-scale ranges are listed below each image. (From (28)).

form in this system by the creation of surface step bunches due to the intersection of dislocation glide planes with the surface (27). In addition to the cross-hatch morphology, the film of Fig. 6 also has roughness on the 100 Å length scale, arising from monoatomic steps and

other atomic scale features on the surface. This type of roughness is illustrated in Figs. 6(b) and (c), which are STM images obtained in UHV. In the 400 nm wide image of Fig. 6(b), monoatomic steps can be faintly seen extending roughly vertically up the image. A further expanded view of the surface morphology is shown in Fig. 6(c), in which individual monoatomic steps, and other disordered atomic features, are clearly resolved.

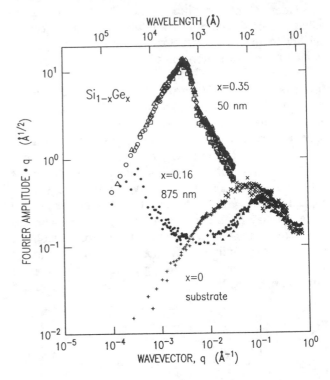

Figure 7. Fourier transforms of SiGe surface morphology, for a Si substrate (+ and x-marks), 16% Ge film (solid symbols) and 35% Ge film (open symbols). The Fourier amplitude multiplied by the wavevector q, is plotted vs. q. The wavelength $\lambda = 2\pi / q$ is shown on the upper axis. For a given sample, different symbols refer to images of different sizes. (From (28)).

New methods for analyzing surface morphology have been developed in recent years (29,30). The methods fall generally into two classes, fractal based analysis and Fourier analysis. In principle, equivalent information is obtained with either method, and we illustrate the Fourier analysis here, as shown in Fig. 7. Spectra are shown there for the 16% Ge film of Fig. 6, and also for a Si substrate and for a 50 nm thick 35% Ge film. In each case the spectra include results from many images acquired at different length scales, and results from the various images are plotted using different symbols. The Fourier amplitude multiplied by the wavevector q is plotted in Fig. 7, with the multiplication by q done to remove the 1/q dependence of the amplitudes which typically occurs between roughness components of different lateral length scales. In this plot, individual roughness components appear as spectral peaks. For the 16% Ge film, two spectral peaks appear in Fig. 7, one at a wavelength of 2 μm arising from the cross-hatch pattern, and another at a wavelength of about 50 Å arising from atomic-scale roughness. The Si substrate spectrum reveals a single spectral peak centered at about 100 Å, with amplitude slightly larger than that for the 16% Ge film. In contrast, the spectrum for the 35% Ge film reveals a large peak centered at about 2000 Å. This film displayed a rough, 3D morphology, arising from higher strain in the film due to its greater Ge content (25). In each case above, the AFM provides useful information on the growth and relaxation of the films. This information is often complementary to that obtained by scanning or transmission electron microscopy, but the simplicity of the AFM measurements and the ability to quantify roughness provide advantages over the traditional techniques.

5. DIMENSIONAL METROLOGY

So far, in this review, we have discussed characterization of surfaces that deviate very little from a plane. The fabrication of an integrated circuit also produces topography that can be truly rugged. Deep trenches and holes with vertical sidewalls abound. The size, shape and placement of these features must be tightly controlled for the completed integrated circuit to function properly. Size is, in fact, the defining characteristic of each new generation in semiconductor lithography. As the size decreases, the difficulty of actually knowing the size to within the required tolerances increases.

In recent years the dimensional metrology capabilities of scanning probe have improved significantly. The improvements include advances in surface proximity detection, stylus fabrication, and image analysis. This work has been motivated by a rapidly growing need to perform dimensional measurements with nearly nanometer uncertainty. Probe microscopes are unique in their ability to provide this level of performance simultaneously in all three dimensions.

Though scanning force microscopes have several advantages over other metrology tools, reliability of a dimensional measurement is not automatically obtained (31). Many aspects of the microscope's behavior must be tightly controlled. In the remainder of this article we discuss those aspects of a probe microscope's design that

affect dimensional measurement. First, the proximity sensor must allow the probe to move across a surface in close proximity to it without suffering damage. Second, position sensors are required to assure that the probe position is known within the required tolerances. Third, the probe tip must have a stable, well controlled shape that is appropriate for the feature scanned. Fourth, the system must be calibrated. Finally, the data must be correctly analyzed. Many workers have contributed to achieving these requirements, but we have space to present that of just a few.

Sensors

Proximity Sensors

The large number of surface proximity detectors invented in recent years attests to the importance, and the difficulty, of surface detection in scanning probe microscopy (32). We present here two systems that illustrate most of the characteristics of these sensors. The first is a relatively large capacitance-based sensor that measures contact forces in one dimension. The second employs a resonating microcantilever to measure non-contact forces in two dimensions.

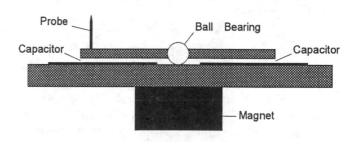

Figure 8. Diagram of a magnetically constrained rocking beam force sensor. (From (35)).

The capacitance-based force sensor, shown in Fig. 8, is a beam held balanced on a weak pivot by electrostatic force balance. The two sides of the rocking beam form capacitors with the base. This rocking beam force sensor (33), also known as the interfacial force microscope (34), is an inherently unstable mechanical system stabilized by a servo loop. By using force balance rather than a weak spring, this method of force sensing decouples the sensitivity from the stiffness. This servoed sensor does

301

not suffer from the instabilities, such as "jump to contact", that can affect weak cantilevers.

The pivot, developed by Miller and Griffith (35), is a pair of steel ball bearings, one on each side of the beam, held to the substrate with a small magnet.. The ball bearings are rugged, so a beam can be used almost indefinitely. The magnetic constraint suppresses all degrees of freedom except the rocking motion, so the sensor is quiet. This system has been used to measure surface roughness with RMS amplitude less than 0.1 nm.

Figure 9. Scan of a contact hole taken with the rocking beam force sensor.

The capacitors serve two purposes: they both detect and control the position of the beam. The detector is a 500 kHz bridge circuit, which compares the relative values of the two sides. The bridge circuit controls the DC voltage, $V \pm \Delta V$ across the two capacitors to keep the beam balanced. The capacitors set the size of the balance beam, since the electronics are designed for a capacitance of a few picofarads. The size of the beam is 5 mm × 10 mm. Mounting probe tips, or even samples, on a structure of this size is easy. The force F impressed on the tip can be directly calculated from the dimensions of the sensor and the difference ΔV in the voltage on the two sides. For a typical sensor , $\Delta V = 1$ mV implies $F = 2 \times 10^{-8}$ N.

The rocking-beam sensor is sensitive only to the vertical component of the force between the probe tip and the sample. Lateral forces that occur, for instance, between a cylindrical probe tip and a vertical side wall are not detected. To protect the probe from damage, it must not be moved laterally toward a vertical side wall. The sensor is used with a scan algorithm developed by Lee and Harrison (36), and digital electronics developed by

Surface/Interface, Inc. Lateral motions occur only when the probe is retracted a fixed distance from the surface. A scan of a contact hole with this scan algorithm is shown in Fig. 9.

The rocking beam force microscope has several advantages. The most important of these arises from its large size. It can accept almost any probe tip. The sensor consequently does not constrain either the probe material or its mode of fabrication. The tip can be several millimeters long so the sensor does not have to be extremely close to the surface. The sensor can be mounted on the scanner piezos, so the fixed sample can be large. The sensor performs well in vacuum. It can operate simultaneously as a tunneling microscope. Magnetic constraint allows easy tip exchange, and the pivot is exceptionally rugged. The sensor also allows the tip to be accurately aligned perpendicular to the sample, which is important when measuring vertical sidewalls.

Most atomic force microscopes employ a microcantilever as the force sensor (37,38) The flexing of this cantilever is usually monitored optically (39), though there exists a design based on piezoresistive sensing (40). The size of the cantilever is small to achieve a high resonant frequency.

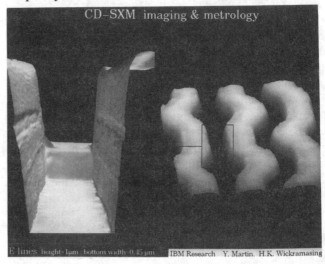

Figure 10. Image of a 16-Mbit DRAM taken with the boot tip and 2D force sensor. Sidewalls are ≈1µm high, angles are 83°, and the space at the bottom is 0.45µm wide. (Reprinted with permission, Y. Martin and H. K. Wickramasinghe. From (41)).

Martin and Wickramasinghe have developed a sophisticated two-dimensional force sensor based on a microcantilever (41). The system calculates, in real time, the surface normal and moves the probe accordingly. The data set collected is stored as a three-dimensional mesh that maintains a relatively constant areal density of

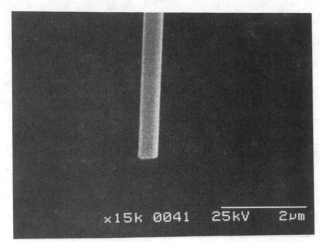

Figure 11. Cylindrical probe tip formed by a wet-chemical etch of an optical fiber. Diameters less than 100 nm have been achieved. (From (51)).

data points irrespective of the sample slope. It is especially good for measuring the angle of a sidewall or its roughness. This system is supplied with a flared probe tip called a boot tip, which allows improved access to vertical sidewalls. This probe will be shown in the section on probe tips. An example of a scan made with this probe microscope is shown in Fig. 10.

The force sensor developed by Martin and Wickramasinghe is a noncontact sensor that exploits resonance enhancement in two dimensions. A microcantilever with a spring constant of ≈ 10 N/m is vibrated both vertically and laterally with an amplitude of ≈ 1 nm. The Q of the system is about 300. The vibration is detected optically. The force sensitivity of the sensor is 3×10^{-12} N, which allows it to detect the presence of the surface at a distance of a few nanometers.

When scanning nearly vertical topography with a tip having nearly vertical side walls, the scan speed is necessarily slower than when scanning a nearly flat surface. Since protection of the probe tip is of paramount importance, the system should not be driven faster than its ability to respond. This is a source of complaint from those used to faster microscopes. In many instances the probe microscope is, however, providing information unavailable from any other tool, so the choice is between slow and never. In addition, when comparing the speed of a probe microscope against cross-sectional SEM, the sample preparation time for the cross sectioning should be taken into consideration along with the fact that the sample has been irretrievably altered.

Position Sensors

The piezoceramic actuators that generate the probe motion exhibit hysteresis and creep. Though it is possible to compensate for the nonlinear behavior of the actuator by driving it with an algorithm having canceling nonlinearities, a better method is to monitor the probe motion with a reliable sensor. It is often convenient to servo the lateral motions to the lateral sensors. This is not necessary in the vertical direction: the output of the vertical monitor simply becomes the data for the topograph. Barrett and Quate developed an optical

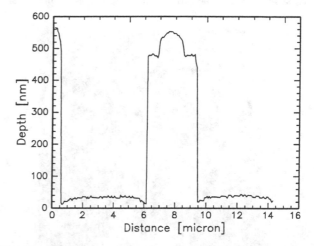

Figure 12. Force microscope scan of a phase-shifting mask with the rocking beam sensor. The vertical scale was stretched to show the 20 nm dips at the sidewalls. (From (53)).

position monitor for use with a tube scanner (42). Griffith, et al. used capacitors to monitor the position of the scanner tube in all three dimensions (43).

A serious source of error is the tilting of the scan head as the probe tip moves. Unless special precautions are taken, the position sensors will not respond to this tilting. This is an example of Abbé offset error (31,44). In systems built around piezoceramic tubes, the head tilting induces an $\approx 1\%$ nonlinearity into the monitor response. In tube scanners, the magnitude of the tilting is not a very stable function of the lateral motion of the tube, which makes calibration of the tube unreliable. To minimize this error the sensors are designed to be as close as possible to the probe. One can, alternatively, add extra sensors to measure the tilt. Because of this problem with tube scanners, a design being developed at the National Institute for Standards and Technology employs a flexure stage to produce more reliable motion (45,46). A design goal of this tool is to reduce the Abbé offset as much as possible. Ideally, the position of the probe apex should be directly measured; a new tool developed by

Marchman achieves this ideal with an optical sensor (47).

Probe Tips

There exists no universal probe shape that is appropriate for all surfaces. Two probe types are most common: cones and cylinders. Cones are used in measurements of surfaces with relatively gentle topography. A cone with a small radius of curvature at the apex can perform surface roughness measurements that cover regions of wavelength-amplitude space unavailable to other tools (48-50).

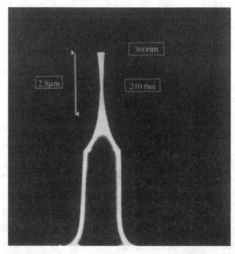

Figure 13. SEM image of the boot-shaped tip for profiling sidewalls. It was fabricated by T. Bayer, J. Greschner and H. Weiss at IBM Sindelfingen. (Reprinted with permission, Y. Martin and H. K. Wickramasinghe. From (41)).

Cylinders are used when steep sidewalls are to be measured. A cylindrical probe can be formed by chemical etching of an optical fiber with buffered hydrofluoric acid (51, 52). This procedure produces a nearly perfect cylinder that can have a diameter less than 0.1 μm. Since the elastic modulus of glass is typically 70 GPa, the probes are stable against flexing as long as the slender section is not too long. During scanning these probes have proved to be very durable. An SEM image of one of these probes is shown in Fig. 11. This technique also allows flared quartz probes to be made.

A scan taken with the quartz cylindrical probe is shown in Fig. 12 (53). In this series of scans, the walls were shown to be within 2° of vertical. The most interesting feature in the scan is, however, the 20 nm dip at the base of each wall. This feature was accessible because of the cylinder's ability to hug the sidewall all of the way to its foot. The dip arises from an enhancement of the reactive ion etch that formed the wall. Because of the subtle slope

at the dip, it is very difficult to detect with SEM even in cross section. This is an example of the importance of the high vertical resolution of probe microscopes.

A cylindrical probe that widens, or flares, at the apex provides the most complete rendition of side wall shape. An example of such a probe tip is shown in Fig. 13, it was developed by T. Bayer, J. Greschner and H. Weiss at IBM Sindelfingen (41). This probe is fabricated on a silicon microcantilever. The tolerances in the fabrication of this probe are very tight. With all cylindrical or flared probes, the accuracy of a width measurement depends on knowledge of the probe width. The width measurement can be no better than the uncertainty in the probe diameter.

Though a conical probe tip may not be able to access all parts of a feature, its interaction with an edge has some advantages over cylindrical probes (54). At the upper corners of a feature with rectangular cross section, size measurement becomes essentially equivalent to pitch measurement. The uncertainty in the position of the upper edge becomes comparable to the uncertainty in the radius of curvature of the probe apex, which can be very small. If the sidewalls are known to be nearly vertical, then the positions of the upper edges gives a good estimate for the size of the feature.

To produce sharp conical tips one can employ a focussed ion beam technique in which a Ga beam is rastered in an annular pattern across the apex of an etched metal shank (55). This method routinely produces tips having a radius of curvature at the apex of 5 nm and widening to no more than 0.5 μm at a distance of 4 μm from the apex. Occasionally the FIB sputtering generates a tip with nearly cylindrical shape.

The probe tip can affect the uncertainty of a measurement in several ways. The radius of curvature of a conical probe must be determined to know the region of wavelength-amplitude space that can been reached. If the width of a cylindrical probe is uncertain, then there is a corresponding uncertainty in the width of a measured object. The durability of the probe tip is especially important. If the probe is changing during a measurement, it affects both the accuracy and precision of the measurement. Finally, the stability of the probe against flexing is important in determining the precision of a measurement. If a slender probe is too long, it will flex under the lateral forces that it encounters. Susceptibility to flexing sets a fundamental limit how deep and narrow a measured trench may be.

Calibration and Data Analysis

The probe-sample interaction is a source of error for all measuring microscopes, but the interaction of a stylus with a sample offers several advantages, especially when high accuracy is needed. The most important advantage arises from the relative insensitivity of a force microscope to sample characteristics such as index of refraction, conductivity and composition. Optical and electron beam tools are sensitive to these characteristics (56, 57), so errors can arise if a reference artifact has a composition different from the sample to be measured. For instance, calibrating an SEM CD measurement with a metal line may give inaccurate results if the line to be measured consists of photoresist.

A comparison between probe microscope measurements of photoresist lines and scanning electron microscope (SEM) measurements of the same lines was made by Marchman, et al. (58). The probe microscope measurements, made with a commercial atomic force microscope (59), agree closely with the cross-sectional SEM measurements, while the plan view measurements deviate from them by up to 0.1 μm, depending on the magnification and how the SEM signal is interpreted.

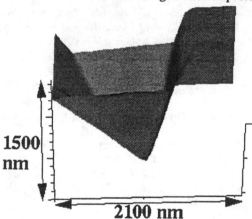

Figure 14. Reconstruction of a probe tip using Vizprobe™. (Reproduced with permission, G. S. Pingali and the University of Michigan.)

In the calibration of a probe microscope there are two fundamental problems: measuring the behavior of the position sensors and finding the shape of the probe tip. Calibrating the position sensors is the easier of the two chores; we discuss it first.

Lateral position sensor calibration is equivalent to a pitch measurement, and an excellent reference material is a grating, a two-dimensional one if possible. Periodic reference materials are available commercially, or they can be reliably fabricated through, for instance, holographic techniques (60). Some samples are self-calibrating in that they contain periodic structures with a well known period.

Vertical calibration is more troublesome because most step height standards contain only one step. This makes it impossible to generate a calibration curve. In addition, the step height sometimes differs by orders of magnitude from the height to be measured. Surface roughness measurements often involve height differences of less than a nanometer, though they are sometimes calibrated with step heights of 100 nm or more. The vertical gain of a piezoscanner may be substantially different in the two height ranges.

When a size or surface roughness measurement is performed with a probe microscope, it is essential to know the size and shape of the probe tip. A common practice is to measure the probe with an SEM, but we have found that calibration of the SEM image is difficult. In addition, SEM scans of insulating probes suffer from charging that obscures the true probe shape. It is often possible to extract quite a lot of information about the shape of a probe from a probe microscope scan (61).

Some sample shapes reveal more about the probe than others (62, 63). Those shapes specially designed to reveal the probe shape are called probe tip characterizers. The characterizer shape depends on the part of the probe that is to be measured. To measure the radius of curvature of the apex, a sample with small spheres of known size might be used. If the sides of the probe are to be measured, then a tall structure, preferably with reentrant sidewalls should be used.

Because of the strongly nonlinear behavior of the probe-sample interaction, some powerful analytical tools have been developed to untangle the probe shape from the image (61, 64-66). These tools can also be used to extract the shape of the probe. An example of such an extraction is shown in Fig. 14 (66). If the probe shape is known, then parts of the scan representing the shape of the line can be extracted. The algorithms used to correct for probe shape are also capable of revealing those areas that are not reconstructable. It is the responsibility of the microscope user to choose a probe shape that minimizes the unreconstructable regions.

There are many sources of error in surface measurement in addition to uncertainty in probe shape. The spacing in the data points and noise in the position sensor affect the precision of the measurement. Another source of imprecision is flexing of the probe tip (31, 62). Lateral forces will bend a probe tip that is too slender. This bending is usually erratic, so the effect appears as extra noise in the image: a sidewall will, for instance, appear to have an unreproducible jagged edge. This effect is a major limiting factor in trying to image deep, narrow features.

Probe microscopes excel as metrology tools because they achieve both high performance and simplicity. The probe-sample interaction in a stylus profilometer is much less complicated that that of optical and electron beam microscopes, even with the nonlinearities that are intrinsic to a probe microscope's image formation. Especially important is the atomic force microscope's insensitivity to the type of material scanned. In a surface profile measurement, it is advantageous to employ a tool that responds principally to the sample's topography but not to the sample's composition.

ACKNOWLEDGMENTS

J. E. Griffith thanks H. M. Marchman, G. L. Miller and L. C. Hopkins of AT&T Bell Laboratories for their ongoing assistance with the metrology work.

REFERENCES

1. J. A. Stroscio and W. J. Kaiser, *Scanning Tunneling Microscopy* (Academic Press, Boston, 1993), chap. 5,6.
2. For review, see M. G. Lagally, *Jap. J. Appl. Phys.* **32**, 1493 (1993).
3. For review, see R. M. Feenstra, *Semicond. Sci. Technol.* **9**, 2157 (1994).
4. M. B. Johnson, O. Albrektsen, R. M. Feenstra, and H. W. M. Salemink, *Appl. Phys. Lett.* **63**, 2923 (1993); **64**, 1454 (1994).
5. M. B. Johnson, H. P. Meier, and H. W. M. Salemink, *Appl. Phys. Lett.* **63**, 3636 (1993).
6. R. M. Feenstra, J. M. Woodall, and G. D. Pettit, *Phys. Rev. Lett.* **71**, 1176 (1993); *Mat. Sci. Forum* **143-147**, 1311 (1994).
7. R. M. Feenstra, D. A. Collins, and T. C. McGill, *Superlattices and Microstructures* **15**, 215 (1994).
8. R. M. Feenstra, D. A. Collins, D. Z.-Y. Ting, M. W. Wang, and T. C. McGill, *Phys. Rev. Lett.* **72**, 2749 (1994); *J. Vac. Sci. Technol. B* **12**, 2592 (1994).
9. M. B. Johnson and J.-M. Halbout, *J. Vac. Sci. Technol. B* **10**, 508 (1992).
10. E. T. Yu, M. B. Johnson, and J.-M. Halbout, *Appl. Phys. Lett.* **61**, 201 (1992).
11. E. T. Yu, J.-M. Halbout, A. R. Powell, and S. S. Iyer, *Appl. Phys. Lett.* **61**, 3166 (1992).
12. C. C. Williams, J. Slinkman, W. P. Hough, and H. K. Wickramasinghe, *Appl. Phys. Lett.* **55**, 1662 (1989).
13. Y. Huang and C. C. Williams, *J. Vac. Sci. Technol. B* **12**, 369 (1994).
14. Y. Huang, C. C. Williams, and J. Slinkman, *Appl. Phys. Lett.* **66**, 344 (1995).
15. See e.g. M. R. Melloch, N. Otsuka, K. Mahalingam, C. L. Chang, J. M. Woodall, G. D. Pettit, P. D. Kirchner, F. Cardone, A. C. Warren, and D. D. Nolte, *J. Appl. Phys.* **72**, 3509 (1992).
16. H. W. M. Salemink, O. Albrektsen, and P. Koenraad, *Phys. Rev. B* **45**, 6946 (1992); M. B. Johnson, U. Maier, H.-P. Meier, and H. W. M. Salemink, *Appl. Phys. Lett.* **63**, 1273 (1993).
17. S. Gwo, K.-J. Chao, C. K. Shih, K. Sadra, and B. G. Streetman, *Phys. Rev. Lett.* **71**, 1883 (1993).
18. D. L. Smith and C. Mailhiot, *J. Appl Phys.* **62**, 2545 (1987); C. Mailhiot and D. L. Smith, *J. Vac. Sci. Technol. A* **7**, 445 (1989).
19. R. M. Feenstra, *Phys. Rev. B* **50**, 4561 (1994).
20. C. Shafai, D. J. Thompson, M. Simard-Normandin, G. Mattiussi, and P. J. Scanlon, *Appl. Phys. Lett.* **64**, 342 (1994).
21. R. J. Hillard, R. G. Mazur, H. L. Berkowitz, and P. Rai-Choudhury, *Solid State Technology* **32**, 119 (1989).
22. B. S. Meyerson, *Proc. IEEE* **80**, 1592 (1992).
23. F. K. LeGoues, B. S. Meyerson, J. F. Morar, and P. D. Kirchner, *J. Appl. Phys.* **71**, 4230 (1992).
24. A. J. Pidduck, D. J. Robbins, A. G. Cullis, W. Y. Leong, and A. M. Pitt, *Thin Solid Films* **222**, 78 (1992).
25. M. A. Lutz, R. M. Feenstra, P. M. Mooney, J. Tersoff, and J. O. Chu, *Surf. Sci. Lett.* **316**, L1075 (1994).
26. J. W. P. Hsu, E. A. Fitzgerald, Y. H. Xie, P. J. Silverman, and M. J. Cardillo, *Appl. Phys. Lett.* **61**, 1293 (1992).
27. M. A. Lutz, R. M. Feenstra, F. K. LeGoues, P. M. Mooney, and J. O. Chu, *Appl. Phys. Lett.* **66**, 724 (1995).
28. R. M. Feenstra, M. A. Lutz, M. Copel, *Mat. Res. Soc. Symp. Proc. "Evolution of Thin-Film and Surface Structure and Morphology"*, 1994, to be published.
29. E. A. Eklund, E. J. Snyder, and R. S. Williams, *Surf. Sci.* **285**, 157 (1993).
30. L. Spanos and E. A. Irene, *J. Vac. Sci. Technol. A* **12**, 2646 (1994).
31. Griffith, J. E. and Grigg, D. A. , *J. Appl. Phys.* **74**, R83-R109 (1993).
32. Sarid, D. *Scanning Force Microscopy, 2nd*, New York, Oxford University Press, 1994.
33. Miller, G. L., Griffith, J. E., Wagner, E. R. and Grigg, D. A., *Rev. Sci. Instrum.* **62**, 705-709 (1991); Griffith, J. E., Marchman, H. M., Miller, G. L., and Hopkins, L. C., *J. Vac. Sci. Technol.* (in press).
34. Joyce, S. A. and Houston, J. E., *Rev. Sci. Instrum.* **62**, 710-715 (1991).
35. Griffith, J. E. and Miller, G. L., U. S. Patent No. 5,307,693, (1994).
36. Lee, D. and Harrison, R. G., U.K. Patent No. 2,009,409 (1979).
37. Albrecht, T. R., Akamine, S., Carver, T. E. and Quate, C. F., *J. Vac. Sci. Technol. A* **8**, 3386-3396 (1990).
38. Wolter, O., Bayer, Th. and Greschner, J., *J. Vac. Sci. Technol. B* **9**, 1353-1357 (1991).
39. Meyer, G. and Amer, N. M., *Appl. Phys. Lett.* **53**, 1045 (1988).
40. Tortonese, M., Barrett, R. C., and Quate, C. F., *Appl. Phys. Lett.* **62**, 834-836 (1993).
41. Martin, Y. and Wickramasinghe, H. K. *Appl. Phys. Lett.* **64**, 2498-2500 (1994).
42. Barrett, R. C. and Quate, C. F., *Rev. Sci. Instrum.* **62**, 1393-1399 (1991).

43. Griffith, J. E., Miller, G. L., Green, C. A., Grigg, D. A. and Russell, P. E., *J. Vac. Sci. Technol. B* **8**, 2023-2027 (1990).

44. Teague, E. C., *The Technology of Proximal Probe Lithography*, Vol. IS10, C. K. Marrian, ed., Bellingham, WA, SPIE Institute for Advanced Technologies, 1993, 322-363.

45. Schneir, J., McWaid, T., Vorburger, T. V., *Proc. SPIE* **2196**, 166 (1994).

46. Schneir, J., McWaid, T., Alexander, J. and Wifley, B. P., *J. Vac. Sci. Technol. B* (to be published).

47 Marchman, H. M., Griffith, J. E. and Trautman, J. A. (to be published).

48. Stedman, M., *J. of Microscopy* **152**, 611-618 (1988).

49. Stedman, M. and Lindsey, K., *Proc. SPIE* **1009**, 56-61 (1988).

50. Westra, K. L., Mitchell, A. W., and Thomson, D. J., *J. Appl. Phys.* **74**, 3608-3610 (1993).

51. Marchman, H. M., Griffith, J. E., and Filas, R. W., *Rev. Sci. Instrum.* **65**, 2538-2541 (1994).

52. Marchman, H. M., U. S. Patent (1994).

53. Griffith, J. E., Marchman, H. M., Hopkins, L. C., Pierrat, C., and Vaidya, S., *Proc. SPIE* **2087**, 107-118 (1993).

54. Griffith, J. E., Marchman, H. M., Miller, G. L., Hopkins, L. C., Vasile, M. J., and Schwalm, S. A., *J. Vac. Sci. Technol. B* **11**, 2473-2476 (1993).

55. Vasile, M. J., Grigg, D. A, Griffith, J. E., Fitzgerald, E. A., and Russell, P. E., *J. Vac. Sci. Technol. B* **9**, 3569-3572 (1991); Hopkins, L. C., Griffith, J. E., and Harriott, L. R., *J. Vac. Sci. Technol.* (in press).

56. Nyysonen, D. and Larrabee, R. D., *J. Res. Nat. Bur. Stand.* **92**, 187-204 (1987).

57. Postek, M. T. and Joy, D. C., *J. Res. Nat. Bur. Stand.* **92**, 205-228 (1987).

58. Marchman, H. M., Griffith, J. E., Guo, J. Z. Y., Frakoviak, J., and Celler, G. K., *J. Vac. Sci. Technol. B* (to be published).

59. Digital Instruments, Inc. Nanoscope III.

60. Anderson, E. H., Boegli, V., Schattenburg, M. L., Kern, D., and Smith, H. I., *J. Vac. Sci. Technol. B* **9**, 3606-3611 (1991).

61. Villarrubia, J. S., *Surf. Sci.* **321**, 287-300 (1994).

62. Griffith, J. E., Grigg, D. A., Vasile, M. J., Russell, P. E., and Fitzgerald, E. A., *J. Vac. Sci. Technol. B* **9**, 3586-3589 (1991).

63. Grigg, D. A., Russell, P. E., Griffith, J. E., Vasile, M. J., and Fitzgerald, E. A., *Ultramicroscopy.* **42-44**, 1616-1620 (1992).

64. Keller, D. J., *Surf. Sci.* **253**, 353-364 (1991).

65. Keller, D. J. and Franke, F. S., *Surf. Sci.* **294**, 409-419 (1993).

66. Pingali, G. and Jain, R. *Proc. SPIE* **1823**, 151-162 (1992).

Scanning Capacitance Microscopy Measurements and Modeling for Dopant Profiling of Silicon

Joseph J. Kopanski, Jay F. Marchiando, and Jeremiah R. Lowney

Semiconductor Electronics Division
National Institute of Standards and Technology
Gaithersburg, MD 20899-0001 USA

A scanning capacitance microscope (SCM) for the two-dimensional profiling of dopants in silicon has been implemented by interfacing a commercial atomic force microscope (AFM) with a high sensitivity capacitance sensor. The AFM is used to control the position of a conducting tip and measure surface topography, while other electronics are used to control and record a simultaneous measurement of the capacitance-voltage (C-V) response between the tip and a semiconductor. The operation of the SCM has been quantified by acquiring differential capacitance images of a variety of lateral pn-junctions and uniformly doped silicon samples. The shape of the C-V response between the AFM tip and oxidized silicon has been measured. Also, a computer code has been written to solve Poisson's equation for the SCM geometry in three dimensions by using the method of collocation of Gaussian points. The purpose of the SCM modeling is to extract the dopant profile from the SCM measurement.

INTRODUCTION

Initial implementations of a scanning capacitance microscope (SCM) were analogous to the scanning tunneling microscope (STM) in that tip height was controlled by maintaining a constant capacitance (1). More recent SCMs have been based on an atomic force microscope (AFM) with a conducting tip and a parallel and essentially independent capacitance measurement (2). We have implemented a robust and sensitive SCM along these lines using a commercial AFM and other off-the-shelf electronics for processing the capacitance signal.

SCM is a leading candidate for a "next generation" technique for determining the dopant profile in silicon. Determination of the two-dimensional dopant profile in processed silicon with a 20-nm spatial resolution and 10% accuracy is identified in the National Technology Roadmap for Semiconductors as a critical measurement need for the development of "next generation" integrated circuits (3). A number of techniques have been suggested as potential "next generation" dopant profilers (4), including tomographic, secondary, ion mass spectroscopy (SIMS), electron holography, and various scanning-probe microscopies. Scanning capacitance

microscopy (SCM) has an advantage over other scanning probe techniques for junction profiling in that its probe is much less surface sensitive and does not depend on establishing a current-carrying metal-semiconductor contact. Dopant information would be extracted from capacitance changes resulting from electric-field-induced changes of the volume depleted of carriers within the semiconductor. Other potential applications of SCM include mapping and manipulation of the charge distributions on insulators (5), and mapping of the native and process-induced lateral variations in the doping and defects of bulk and layered semiconductor materials.

The emphasis of our program is to develop the measurement techniques, capacitance sensor calibrations, and theoretical interpretation that are essential for SCM to become a practical method of junction profiling by the semiconductor industry. This paper reports details of an implementation of an SCM based on a Digital Instruments Inc. Multimode AFM (6). Test samples, which include both lateral pn-junctions and uniformly-doped silicon with a patterned oxide, have been designed and fabricated for establishing the repeatability and accuracy of the SCM technique. Operation of the SCM is demonstrated through differential SCM images of these test samples. The effect of a dc-bias offset on differential-capacitance images is discussed. Measurement of the shape of the capacitance-voltage response between the conducting AFM tip and oxidized silicon has also been demonstrated.

Dopant profiles extracted from SCM measurements can be no more accurate than the model used to interpret the data. We have given equal effort to the development of both the SCM instrumentation and appropriate models to interpret SCM measurements. Scanning capacitance microscopy measures the capacitance between a nominally hemispherical SCM probe and the planar semiconductor surface, and involves the charge distribution within the semiconductor as well as on the surfaces. Thus, the conventional one-dimensional, parallel-plate model for the C-V profiling of semiconductors is not appropriate, and a numerical solution of a nonlinear Poisson's equation in three dimensions is necessary. Details of a computer code that has been written to solve Poisson's equation for the SCM geometry in three dimensions by using the method of collocation of Gaussian points are discussed.

EXPERIMENTAL

Figure 1 shows a schematic drawing of the SCM. The SCM uses as its basis a Digital Instruments Inc., model Dimension 5000 scanning probe microscope (SPM), which is usually operated as a contact-mode AFM. Like other SCMs, the capacitance between the tip and sample is conveniently measured by using a capacitance sensor from an RCA Video Disc player (7-8), which is electrically connected to the tip. The capacitance measurement is made independent of and simultaneous with the AFM measurement of topography. The RCA sensor measures capacitance at around 1 GHz; this high frequency allows small variations in capacitance to be quickly resolved. Applied to SCM, this ultrasensitive

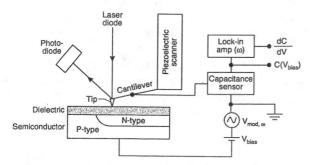

FIGURE 1. Block diagram of the scanning capacitance microscope.

capacitance sensor can detect *relative* variations in capacitance in the range of 10^{-18} F (attofarads) around a total input capacitance of about 0.1 pF (7-8).

The input of the capacitance sensor is connected to a conducting AFM cantilever by using silver paint and a thin ($\approx$0.005-cm diameter) enamel-coated wire. The AFM head has been modified so that all material in contact with the cantilever is insulating ceramic or plastic. The conducting tips were made by coating commercially available cantilevered AFM tips with metal. Silicon nitride cantilevers coated with approximately 20 nm of Cr or Ti have proven to have useful lifetimes during probe scanning. Commercially available Cr-on-Fe-coated, highly-doped silicon cantilevers, such as those used for magnetic force microscopy (MFM), have also been successful. A degree of electrical isolation of the SCM from the environment is achieved by enclosing the entire microscope in a grounded, acoustic-isolation hood.

The SCM has been operated in three modes. A direct-capacitance image-mode simply monitors the capacitance sensor output on one of the SPM control station's auxiliary channels during an AFM scan. In order to resolve capacitance changes from the inherent sensor noise, it is necessary to amplify and bandpass-filter the sensor output. Metal lines on a photolithographic mask (insulating glass substrate) have been detected in this manner with the AFM tip both in contact and in a "lift-mode" where the tip is uniformly lifted above the surface topography based on a previous scan in contact. Metal lines could still be detected with the tip lifted up to 80 nm above the surface.

A differential-capacitance mode has proven useful for imaging semiconductor junctions (9). Figure 1 shows the SCM configured for this mode of operation. The AFM tip, separated by an oxide layer from a semiconductor, is used to form a metal-oxide-semiconductor (MOS) capacitor. A modulation voltage at a frequency ω, $V_{mod, \omega}$, is applied between the sample and the AFM tip, with the casing of the capacitance sensor grounded. V_{mod} is typically 1 V_{p-p} at 20 to 80 kHz plus a dc bias voltage, V_{bias}. Since the MOS capacitance is voltage dependent, the ac part of the sensor output signal is proportional to the differential capacitance, and the magnitude of this signal can be measured with a lock-in amplifier. In a third mode, the shape of the capacitance-voltage (C-V) response between the tip and sample is measured. A digital oscilloscope is used to record and average the sensor output at a number of bias-voltage points. The bias voltage is applied as triangular wave at ≈ 1 kHz with the magnitude such that the MOS capacitor is rapidly swept between accumulation and depletion. Many signal waveforms need to be averaged to reduce the noise in the capacitance sensor output.

RESULTS

A series of test structures was fabricated with which to characterize the performance of the SCM. Lateral pn-junction test structures were formed by ion implantation of 50-keV boron to a dose of 10^{15} cm^{-2} through 10 nm of silicon dioxide into n-type silicon. The test structure pattern consists of 5-μm-wide lines which were subject to ion implantation alternating with 5-μm-wide spaces which were masked from implantation. Additional structures were fabricated from a series of wafers having bulk dopings ranging from 7 x 10^{15} to 2 x 10^{20} cm^{-3}. The bulk-doped wafers were processed to produce a pattern consisting of 5-μm-wide lines of 20-nm oxide thickness alternating with 5-μm-wide spaces of 10-nm oxide thickness. These wafers were initially characterized by their four-point-probe resistivity and "macroscopic" C-V curves measured by using a mercury contact 732 μm in diameter.

Figure 2 shows simultaneously acquired AFM and SCM images of the lateral pn-junction test structure. The AFM image is to the left and shows only the small 1 nm step in the oxide resulting from physical swelling due to the boron implant. The SCM image is to the right and shows higher differential capacitance for the lightly doped n-type substrate regions than for the more heavily doped p-type implanted regions. The SCM image was acquired by using a modulation signal of 1 V_{p-p} at 30 kHz and a dc bias voltage of -1.0 V. The lock-in sensitivity was 50 μV with a time constant of 3 ms.

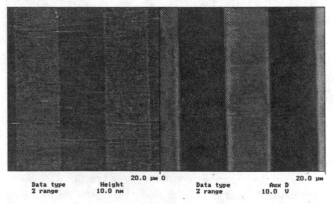

FIGURE 2. AFM image (left) of lateral pn-junction test structure and simultaneously acquired SCM image (right).

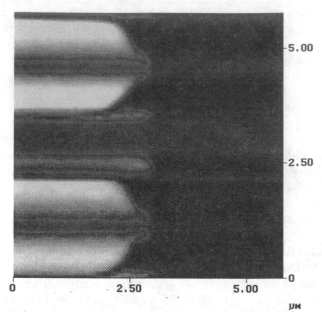

FIGURE 3. SCM image in the vicinity of a lateral junction showing the effect of a bias-offset voltage.

Although the surface of the test structure is almost flat, the underlying junctions can be sharply resolved by the SCM. While STMs have also delineated pn-junctions in both silicon and GaAs, STMs require a surface that is both unpinned and with no or very thin oxide. Significantly, the SCM has delineated the junction through a 10-nm-thick silicon-dioxide layer. This relaxed requirement on surface conditions gives SCM an experimental advantage as a junction profiling technique.

A bias offset voltage will change or even invert the contrast of the SCM image of a junction. Figure 3 shows an SCM image, near a test-structure junction edge, with the addition of a bias voltage that is slowly varied linearly between -3 and +3 V over the time it takes to acquire a complete image. In Fig. 3, the bias voltage is 0 V at the top of the image, decreases linearly to -3 V (substrate relative to the tip), reverses and increases linearly to +3 V at the middle of the image, and then reverses again to repeat the pattern and reach 0 V near the bottom of the image.

The bias voltage induces additional contrast in the SCM image because it determines where on the C-V curve the differential capacitance is measured. For a typical MOS C-V curve, a 1 V_{p-p} modulation signal would cause larger differential capacitance when the device is biased in depletion than when it is biased in accumulation. Notice also that the width of the interfacial region between the n- and p-type (right side) region is a function of the bias voltage. The applied bias voltage modulates the width of the built-in depletion region at the junction. That the

310

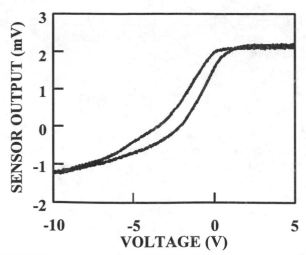

FIGURE 4. Sensor output (capacitance) versus bias voltage measured with SCM tip stationary over p-type silicon with $N_a \approx 10^{15}$ cm^{-3} and a 20-nm-thick oxide layer.

sample-tip bias can modulate the built-in depletion region indicates that it may be possible to use SCM to extract dopant information close to pn-junctions.

The contrast induced in SCM images by the bias voltage is a consequence of the MOS-capacitor's voltage-dependent capacitance. The shape of the complete C-V curve between the AFM tip and oxidized silicon can be measured when the scanning of the AFM tip is stopped. Figure 4 shows a plot of the ac part of the capacitance-sensor output versus bias voltage for a silicon wafer with a dopant concentration of 10^{15} cm^{-3} and a 20-nm oxide layer. The hysteresis in sensor output between bias-voltage sweep directions may be due to oxide traps or to the limited bandwidth of the digital oscilloscope used to measure the curve. While the shape of the C-V curve can be measured, the sensor output is not calibrated in units of farads. In Fig. 4, the oxide capacitance is estimated to be of the order of 10^{-18} F from the known oxide thickness and tip shape. More detailed C-V measurements on the complete range of test samples will be conducted by using a digital oscilloscope with wider bandwidth and greater averaging capability.

MODELING

Within the integrated circuit industry, planar silicon MOS capacitors are well-known test structures used to measure dopant profiles and oxide interface properties (10). A simple one-dimensional model of the MOS capacitor extracts the dopant profile, $n(x)$, as a function of carrier depletion width, x, from the variation

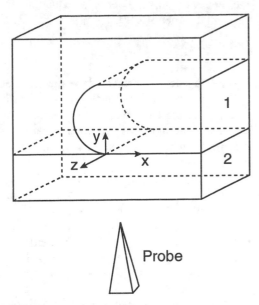

FIGURE 5. Geometry used to model pn-junctions samples and the SCM measurement. Region 1 is the silicon with a distribution of ion implanted dopants represented by the curved region; region 2 includes the surface oxide and ambient.

of capacitance with voltage (10):

$$n(x) = \frac{C^3}{q\epsilon_{Si} A^2 \frac{dC}{dv}} , \qquad (1)$$

where q is the electron charge, ϵ_{Si} is the dielectric constant of silicon, A is the capacitor area, and the depletion depth is related to the capacitance via $C = \epsilon_{Si}A/x$. The situation is rather more complex for SCM since the capacitance is measured between the nominally hemispherical probe tip of the SCM and a planar surface with a lateral dopant gradient. The electrical field distribution is inherently three-dimensional, and calculation of the capacitance requires a numerical solution of a nonlinear Poisson's equation in three dimensions by using iterative methods.

A computer code has been written to solve an elliptic system of coupled nonlinear partial differential equations on a rectangular-shaped three-dimensional domain by using the method of collocation of Gaussian points (11). The SCM and the pn-junction under examination are modeled according to Fig. 5. A presumed two-dimensional dopant profile is found from a Monte Carlo calculation by using the known ion implantation parameters. The electron and hole distributions, n and p, in the doped semiconductor region are determined by using the code to solve Poisson's equation for the electric-potential function (Ψ):

$$\nabla \cdot (\epsilon_r \nabla \psi) = -(q/\epsilon_o)(N_d - N_a + p - n). \quad (2)$$

Here ϵ_r is the dielectric constant of the medium, and N_d and N_a are the donor and acceptor concentrations, respectively. The electric potential in the insulator and ambient regions is found from Laplace's equation, which is of the same form as Eq. 2, except that the right-hand-side of the equation is zero, and by matching the boundary conditions along the interface. When applied to calculate the potential distributions in the semiconductor and the insulator, the software algorithms have proven to converge. Capacitance can then be calculated from the change in the potential in the semiconductor due to an incremental change in applied bias on the probe.

The inverse problem, calculation of the dopant profile from the capacitance, is intractable. Utilization of the model output to determine the dopant profile will require interpolation between known solutions. Solutions, based initially on the known process parameters used to create the pn-junctions, can be adjusted to bracket the measurements with ever tighter precision. There will also be a need for experimental determination of calibration factors with which to relate the measured capacitance data with the model output.

CONCLUSIONS

A scanning capacitance microscope was constructed by using a commercial atomic force microscope as a platform. The operation of the SCM was confirmed by differential capacitance imaging of lateral pn-junctions and oxide steps over bulk-doped silicon. C-V curves between the AFM tip and oxidized silicon have also been measured. Preliminary results show that the SCM is sensitive enough to measure the changes in capacitance resulting from small changes in bias voltage or oxide thickness even for silicon substrates doped as highly as 10^{20} cm^{-3}. This sensitivity appears to be adequate for SCM to function as a dopant profiling technique. A computer code has been written that solves Poisson's equation for the SCM geometry in three dimensions.

A series of systematic measurements of C-V curves and SCM images of the test structures is in progress. Likewise, the modeling effort aims to calculate C-V curves as a function of SCM tip position in the vicinity of a junction. The ultimate goal is to combine SCM measurements and modeling to determine a high-accuracy dopant profile.

ACKNOWLEDGMENTS

This work was supported in part by the National Semiconductor Metrology Program at the National Institute of Standards and Technology; not subject to copyright. The assistance of Dennis Adderton, Digital Instruments, Inc., Santa Barbara, Calif., in interfacing the capacitance sensor to the AFM is acknowledged.

REFERENCES

1. Williams, C. C., Hough, W. P., and Rishton, S. A., *Appl. Phys. Lett.* **55**, 203-205 (1989).

2. Huang Y., and Williams C. C., *J. Vac. Sci. Technol. B* **12**, 369-372 (1994).

3. The National Technology Roadmap for Semiconductors, Semiconductor Industry Association, 4300 Stevens Creek Boulevard, Suite 271, San Jose, CA 95129.

4. Diebold, A. C., Kump, M., Kopanski, J. J., and D. G. Seiler, "Characterization of two-dimensional dopant profiles: status and review," in *The Electrochemical Society Proceedings* **94-33**, 1994, pp. 78-97.

5. Barrett, R. C., and Quate, C. F., *Ultramicroscopy* **42-44**, 262-267 (1992).

6. Certain commercial equipment, instruments, or materials are identified in this paper in order to adequately specify the experimental procedure. Such identification does not imply recommendation or endorsement by NIST, nor does it imply that the materials or equipment used are necessarily the best available for the purpose.

7. Clemens, J. K., *RCA Rev.* **39**, 33-59 (1978).

8. Palmer, R. C., Denlinger, E. J., and Kawamoto, H., *RCA Rev.* **43**, 194-211 (1982).

9. Williams, C. C., Slinkman, J., Hough, W. P., and Wickramasinghe, *Appl. Phys. Lett.* **55**, 1662-64 (1989).

10. Nicollian, E. H., and Brews, J. R., *MOS Physics and Technology*, New York: John Wiley & Sons, 1982.

11. J. F. Marchiando, "On using collocation in three dimensions and solving a model semiconductor problem," submitted to *SIAM J. Scientific Computing* (1995).

Progress Toward Accurate Metrology using Atomic Force Microscopy

T. McWaid[1], J. Schneir[1], J.S. Villarrubia[1], R. Dixson[1], and V.W. Tsai[2]

[1]National Institute of Standards and Technology, Gaithersburg, MD 20899.
[2]Dept. of Matls. Engr., University of Maryland, College Park, MD 20742-4111.

Accurate metrology using atomic force microscopy (AFM) requires accurate control of the tip position, an estimate of the tip geometry and an understanding of the tip-surface interaction forces. We describe recent progress at NIST towards accurate AFM metrology. We begin with a brief introduction to the potential applications of accurate AFM metrology within the semiconductor industry. Next, we describe the technological infrastructure required before industry personnel can successfully use AFM for repeatable, reproducible, and accurate measurements. NIST's programmatic response to the identified needs is then outlined.

We give particular attention to the development of a calibrated atomic force microscope (C-AFM), since the development of this instrument is essential to the successful completion of most of our specific projects. Measurements of a prototype pitch/height artifact are presented. Next we summarize our efforts toward the development of a viable method of characterizing probe tip geometry. Finally, we present a linewidth measurement that has been bracketed through the use of mathematical morphology for tip geometry estimation.

INTRODUCTION

Scanned probe microscopy (SPM) denotes the family of measurement techniques which use microscopic mechanical probes to obtain very highly resolved three-dimensional images of surface features. Atomic force microscopy (AFM) is the only type of SPM that is being used as a metrology tool in the semiconductor industry. There is a strong interest in the potential of scanning capacitance microscopy (SCM) for lateral dopant profiling (1); however, a suitable SCM is not yet commercially available.

AFM has been used to a limited extent in quality assurance, reliability, and production metrology applications. AFM applications include: critical dimension (CD) measurement; microroughness measurement of bare and patterned wafers; grain size measurement of films; inspection after planarization; defect review; measurement of trench and via depth; measurement of edge profile/roughness; metrology of phase shift masks; and metrology of x-ray masks. An excellent review discusses the limitations and issues relevant to the use of AFM for metrology (2).

We will first describe the technological infrastructure required for the successful use of scanned probe microscopy in industry. This infrastructure includes pitch, height and width calibration artifacts, the development of standard practices for various measurements, and tip metrology techniques. NIST's programmatic response to these needs is then summarized.

We give particular attention to the development of a calibrated atomic force microscope (C-AFM); the development of this instrument is essential to the successful completion of most of our specific projects. The C-AFM will be used to aid the development of, and to calibrate, AFM calibration artifacts.

INDUSTRIAL NEEDS

X,Y,Z Calibration Standard

Suitable X,Y,Z calibration standards are essential to the use of AFM in metrological applications, and several AFM calibration standards are commercially available. One type of artifact contains pitch values ranging from 1.8 to 20 μm, and heights ranging from 100 to 18 nm (3,4). The pitch of this standard is calibrated optically; the height is calibrated using a stylus profiler. Although these standards have proven to be valuable to the AFM community, additional X,Y,Z calibration standards are required by the semiconductor industry.

Ideally, this family of standards should be expanded to include artifacts with pitch values down to 0.1 μm and step heights below 10 nm. It is essential that the standards are calibrated pitch calibration becomes increasingly difficult. Moreover, even where the optical and stylus techniques can be used, methods divergence between these calibration techniques and AFM could introduce uncertainty into the calibration of a force microscope. The impact of methods divergence on the calibration of AFM standards is discussed in a recent publication (5).

Tip Standards

A scanned probe microscope image is a "convolution" (more correctly, a dilation) of the surface with the probe tip (6). Tips can wear or be damaged as they are used. Industry needs a standard that, when imaged in an AFM, can be used to determine the shape of the tip. If the tip shape is known exactly, it can be "deconvolved" (eroded) from the original image to obtain some or all of the actual specimen geometry. Of course, the actual specimen geometry can only be determined in those areas where the tip interacted with the specimen surface.

Whereas, in general, only the very tip of a probe interacts with a surface during a surface roughness measurement, imaging high aspect ratio features on patterned wafers or masks results in the interaction of a larger portion of the probe with the sample. Thus, several different types of tip calibration standards are probably needed.

AFM Linewidth Standard

There is great interest in developing an AFM system that can accurately measure CDs. This would allow non-destructive cross sectioning. Perhaps the greatest immediate benefit of accurate AFM CD measurement would be the cross sectional measurement of photoresist.

Although AFMs do not yet have sufficient measurement throughput to replace SEMs for CD measurement, they could be very valuable in calibrating SEMs on photoresist product. If an AFM can be used to make accurate CD measurements on product wafers these same wafers could probably be used to calibrate a SEM for CD measurement. Many current problems in SEM metrology could be circumvented by calibrating SEMs on real product.

Surface Microroughness Standard

The semiconductor community makes use of the AFM for surface microroughness measurements. Surface microroughness can affect device performance and laser particle detection on bare wafers. It is important that surface microroughness measurements are precise and reproducible.

A recent industry sponsored surface roughness round-robin using silicon wafers found variations in AFM measured rms roughness of 700% (7)! The opinion of most vendors, industry members, and NIST is that tool to tool reproducibility of AFM microroughness measurements can be improved by taking the following steps: (1) develop a standard method for calculating AFM microroughness; (2) improve available X,Y,Z calibration standards; (3) develop a tip calibration standard. The first item is best addressed in an American Society for Testing and Materials (ASTM), American Society of Mechanical Engineers (ASME) or Semiconductor Equipment and Materials International (SEMI) standards committee.

A NIST certified microroughness standard would be useful to the semiconductor industry. Even after an instrument has been calibrated using an X,Y,Z calibration artifact, and the probe tip geometry has been determined using a tip calibration standard, the roughness standard would still provide a useful final check that the AFM measurement is under control.

NIST PROGRAMS

Important work is being performed throughout NIST on problems of importance to the semiconductor industry. We will only discuss the work being performed in the Surface and Microform Metrology Group of the Precision Engineering Division that is relevant to the dimensional metrology issues described above. The successful resolution of these issues requires the development of a metrology AFM. The calibrated atomic force microscope will therefore be discussed in some detail. We present measurements of a grating fabricated using laser-focused atomic deposition, a prototype improved X,Y,Z calibration artifact, nominally 5 nm diameter gold particles on mica, and the width of a line on a holographic grating. The linewidth measurement accounts for the influence of tip geometry through the use of algorithms based on grayscale morphology.

Calibrated Atomic Force Microscope

Several AFMs with integrated metrology have been constructed. Two distinct approaches have been taken to designing a metrology AFM: (1) record the X-Y tip position when each Z data point is taken (8,9); (2) use a feedback loop in conjunction with a position sensor to provide closed loop control of the X-Y tip position (10). We have implemented the second approach using a digital signal processor (DSP) based control system and heterodyne laser interferometers as the X and Y displacement transducers.

The various components and subsystems that comprise the C-AFM have been described in the literature (5,11).

Commercially-available components and subsystems have been used to minimize the development time and cost, and to maximize the upgradeability of the C-AFM. The precision and accuracy of commercial metrology equipment have improved considerably over the last five years. We expect the requirements of ultraprecision metrology to continue to drive this trend, thereby influencing the evolution of the C-AFM.

We have conducted several experiments to characterize the performance of the C-AFM system in its current configuration. Several of these preliminary measurements will now be presented. A detailed quantitative uncertainty analysis has not yet been performed on the C-AFM. The uncertainties given below have been estimated from our knowledge of the measurement system. Type A uncertainties are those uncertainties evaluated by statistical methods; Type B uncertainties are those which are evaluated by other means. In all cases the expanded uncertainties include a coverage factor of two (12).

We have measured the pitch of a specimen fabricated using laser-focused atomic deposition (10). The nominal average pitch of this specimen is 212.78 nm (10). Based on cursory analysis of a profile from a 4 μm image, the nominal height of the deposited features is 6 nm ± 2 nm, and the fundamental peak in the profile power spectral density (PSD) function is located at a spatial wavelength of 214 nm ± 1.5 nm. The peak in the PSD is related to the average pitch of the specimen. The agreement between the measured and nominal pitch values is very encouraging.

Measurement of X,Y,Z Calibration Artifact

The first AFM calibration artifact calibrated with the C-AFM will be a combined two-dimensional pitch and height standard (i.e., an X,Y,Z calibration artifact). An image of a prototype VLSI Standards pitch and height surface topography artifact is shown in Figure 1. The nominal pitch and height of this specimen are 1.8 μm and 18 nm, respectively.

We measured the height of this artifact to be 19 nm with an expanded uncertainty of 2 nm. This height value is the average of three sets of six measurements. Each set of measurements was made at a different location on the specimen. The expanded uncertainty for the height measurement includes a Type A expanded uncertainty of 0.1 nm, a Type B expanded uncertainty of 2 nm.

Preliminary pitch measurements have been performed on the prototype pitch artifact in order to gauge our present pitch calibration capabilities. The pitch was measured three times at each of four different locations using 1.9 μm profiles. The pitch values were calculated using an unsophisticated edge detection criterion. As shown in Table 1 of the raw data, we found significant variations in the pitch measurements.

Table 1

Location	Measured Pitch [μm]		
	First Measurement	Second Measurement	Third Measurement
1	1.807	1.805	1.809
2	1.805	1.804	1.802
3	1.802	1.797	1.792
4	1.787	1.784	1.790

We estimate that the expanded uncertainty surrounding these measurements is 10 nm. This estimation is based on the assumption that the Type A expanded uncertainty is 4 nm, and the Type B expanded uncertainty is 3 nm. Our goal is to calibrate sub-micrometer pitch artifacts with an expanded uncertainty, including a coverage factor of two, of three nanometers. The attainment of this goal requires the development of extremely uniform artifacts (very little edge roughness, form error, etc.), further refinement of the C-AFM, and the development of appropriate edge location criteria and sampling procedures.

Probe Characterization

Probe geometry affects feature width and surface microroughness measurements. Although no consensus has been reached on the best type of tip characterization standard for a given application, many candidate standards are being considered. In parallel, researchers are developing the algorithms required to determine the tip shape from an image of an appropriate artifact.

Colloidal gold particles might prove to be valuable artifacts for characterizing AFM probe tips. An excellent paper by Vesenka, et al. discusses the use of colloidal gold particles for assessing the compressibility of biomolecules (13). The colloidal gold particles are incompressible, monodisperse and spherical. The spheres are available from Ted Pella, Inc. in quoted diameters of 5.7, 14.3 or 28 nm (4,14).

We have used the C-AFM to image both the 5.7 nm and the 28 nm diameter gold spheres deposited on mica. Thus far, we have only been able to image the spheres using non-contact and intermittent contact ("tapping") imaging modes; the tip pushed the particles aside when we attempted to image the specimens in contact mode.

Figure 2 shows an image of the nominally 5.7 nm diameter spheres that was obtained in non-contact mode. Although, the heights of the imaged spheres are approximately 6 nm, the widths vary between 40 nm and 80 nm. The measured widths of the imaged spheres are related to the shape of our AFM probe tip. We currently have no traceable measure of the size of the gold spheres either prior to or subsequent to being deposited on the mica surface. Since we are relying on the manufacturer's size data, we are hesitant to deduce the size of our probe tip from these measurements. We are working on algorithms that might enable one to calculate the tip geometry from such images; however, this work is still in its infancy.

Linewidth Measurement

We have used grayscale morphology algorithms (6) in an effort to bracket the width of a line on a nominally 200 nm pitch holographic grating. An image of this grating is presented as Figure 3. Since a line profile obtained from a C-AFM image is a dilation of the actual line profile and the probe tip, the width obtained from a C-AFM image is an upper bound on the actual feature width. Grayscale morphology can be used to first obtain an upper bound of a probe's geometry, and, then, erode the probe geometry from an image. The resulting reconstructed surface provides a lower bound on a feature width over those regions of the line that actually interacted with the probe.

Figure 4 presents two average line profiles. Each of these profiles represents the average of approximately 200 individual traces across the line. One average profile was calculated using raw C-AFM data. The second profile was calculated after eroding the probe geometry from the C-AFM data set. Based on this analysis we estimate that the average linewidth is 79 nm ± 8 nm along the height indicated in the figure. Since the sidewalls are not vertical, the linewidth is a function of height above the substrate. As before, the reported expanded uncertainty is based on estimations of the Type A (1 nm) and Type B (4 nm) expanded uncertainties, and includes a coverage factor of two. This analysis is based on the assumption that the tip contacted the line at the relevant height. In general, one can not obtain a lower bound on the line width over its entire height if the line sidewall angle is greater than the included half-angle of the probe tip.

SUMMARY

We began with a brief overview of scanned probe microscopy. We then presented a summary of the semiconductor industry's metrology requirements for AFM. We have described the impediments to successful use of AFM based on these requirements. Successful resolution of these technical issues will benefit not only the semiconductor industry, but all industries attempting to use SPM in metrological applications. NIST's programmatic response to these technical roadblocks was then outlined. We gave particular attention to development of a calibrated atomic force microscope, since the development of this instrument is essential to successful completion of almost all of our specific projects.

ACKNOWLEDGMENTS

This work was supported in part by SEMATECH and the National Semiconductor Metrology Program.

REFERENCES

1. Huang, Y. and. Williams, C.C., *J. Vac. Sci. Technol. B* **12**, 1(1994).
2. Griffith, J.E. and Grigg, D.A., *J. Appl. Phys.* **74**, R83 (1993).
3. VLSI Standards, Incorporated, 3087 North First Street, San Jose, CA 95134-2006.
4. Certain commercial equipment is identified in this report in order to describe the experimental procedure adequately. Such identification does not imply recommendation or endorsement by NIST, nor does it imply that the equipment identified is necessarily the best available for the purpose.
5. Schneir, J., McWaid, T.H. and Vorburger, T.V., "An Instrument for Calibrating Atomic Force Microscope Standards," SPIE Vol. 2196, 166 (1994).
6. Villarrubia, J.S., *Surface Science* **321**, 287 (1994).
7. Raheem, R. and Adem, E., "Evaluation of Wafer Micro-Roughness with Sub Å Resolution on the AFM. Results of an AMD Sponsored Round Robin," presented at the NIST Conference on Industrial Appl. of SPM, March 24-25 (1994).
8. Nyyssonen, D., "Recent Developments in 2D AFM Metrology," SPIE 1926, 324 (1993).
9. Barrett, R.C. and Quate, C.F., *Rev. Sci. Instrum.* **62**(6), 1393 (1991).
10. Griffith, J.E., Marchman, H.M., Miller, G.L., Hopkins, L.C., Vasile, M.J. and Schwalm, S.A. , *J. Vac. Sci. and Technol. B.* **11**, 2473 (1993).
11. McWaid, T.H., Schneir, J. and Vorburger, T.V. "A Calibrated Atomic Force Microscope," in Proceedings of 3rd Intl. Conf. Ultraprec. in Manuf. Eng., 1994, pp. 342-345.
12. Taylor, B.N. and Kuyatt, C.E., *Guidlenes for Evaluating and Expressing the Uncertainty of NIST Measurement Results*, NIST Tech. Note 1297, 1994.
13. McClelland, J.J., Scholten, R.E., Palm, E.C. and Celotta, R.J., *Science* **262**, 877(1993).
14. Ted Pella, Inc., P.O. Box 492477, Redding, CA 96049-2477.
15. Vesenka, J., Srinivas, M., Giberson, R., Marsh, T. And Henderson, E., *Biophysical Journal* **65**, 992-997 (1993).

316

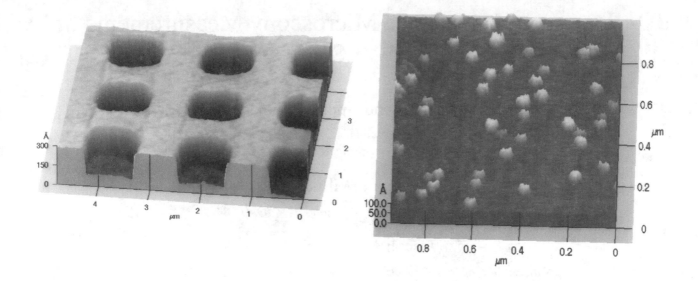

Figure 1. Image of prototype 1.8 μm pitch calibration artifact.

Figure 2. Image of 5 nm Au spheres on mica.

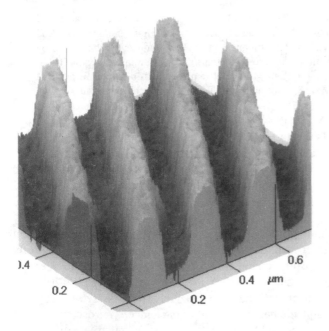

Figure 3. Holographic grating. The tip used to image this grating was analyzed using grayscale morphology. This gave an upper and lower bound on the linewidth.

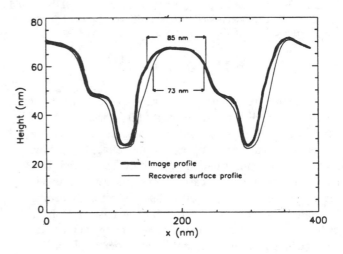

Figure 4. Average line profiles from raw data (outer profile) and from recovered surface (inner profile).

2D- Scanning Capacitance Microscopy Measurements of Cross-Sectioned VLSI Teststructures

Gabi Neubauer and Andrew Erickson*

Intel Corporation, Materials Technology Dept., Santa Clara, CA 95052
*) University of Utah, Dept. f. Electr. Engineering, Salt Lake City, UT 84105

Clayton C. Williams

University of Utah, Dept. of Physics, Salt Lake City, UT 84105

Joseph J. Kopanski

National Inst. of Standards and Technology, Gaithersburg, MD 20899

Mark Rodgers and Dennis Adderton

Digital Instruments, Santa Barbara, CA 93103

We have developed a setup which uses scanning capacitance microscopy (SCM) to obtain electrical data of cross-sectioned samples while simultaneously acquiring conventional topographical AFM data. The results presented here include 2D SCM maps of cross-sections of blanket implanted, annealed Si wafers as well as teststructures on Si. We found the technique to be sensitive over several orders of magnitude of carrier density concentrations <E15 to E20 atoms/cm^3, with a lateral resolution of 20-150 nm, depending on probe tip and dopant level. We find excellent agreement of total implant depth obtained from SCM signals of cross-sectioned samples with conventional Secondary Ion Mass Spectrometry (SIMS) profiles of the same sample.

INTRODUCTION

Over the past few years, an increasing need has been identified for direct two-dimensional measurement of carrier densities in semiconductor devices to provide a means for the calibration of process modeling. Up to now, quantitative 2D- information could only be inferred from one-dimensional dopant/carrier level measurements via Secondary Ion Mass Spectroscopy (SIMS), Spreading Resistance Profilometry (SRP), and Capacitance-Voltage (C-V) measurements. A variety of different approaches have been taken to close this gap experimentally [1]. Many of these techniques, such as chemical etching and staining methods [1,2] or SIMS based tomography techniques [3,4], can only indirectly measure carrier densities or depend on the reproducibility of fairly lengthy sample preparation steps. Application of electrical measurement schemes, which are well suitable to directly yield the desired carrier distributions, appear more promising. This area has greatly benefited from the development of scanning probe technology, which offers access to the submicron scale in two dimensions.

Derivatives of both principal techniques, i.e., Scanning Tunneling (STM) and Atomic Force Microscopy (AFM), have been applied [6-22]. Whereas STM and workfunction-based concepts prove difficult on VLSI structures due to their largely non-conductive nature, an AFM setup with a conductive tip/cantilever assembly allows for simultaneous input of topographical and electrical data. Research in this area has mostly concentrated on developing setups feasible to measure capacitance [14-17], spreading resistance [18-20], and contact potential using a Kelvin probe setup [21]. We implemented Scanning Capacitance Microscopy (SCM) into an AFM setup to measure relative capacitance changes when scanning cross-sectioned implanted Si-wafers and -teststructures. Earlier experiments had shown the AFM to be well capable of imaging VLSI cross sections [22]. Contrast in AFM images of polished VLSI cross sections is due to the topography which develops from the material dependent removal rates during polishing. In our new setup this topographical information is acquired simultaneously to the electrical data and is used to spatially correlate the data.

EXPERIMENTAL SETUP

Our experiments are conducted on a Digital Instruments Nanoscope III Large-Stage AFM, operated in air. Topography information is obtained with the probe tip scanning under constant repulsive force via a feedback network. Tip and part of the cantilever substrate are coated with 300-500 Å Cr, followed by ~100 Å SiO_2 to ensure insulating properties.

Capacitance measurements are accomplished by an RCA capacitance sensor circuitry which was originally designed for the RCA Video disk player /23/. Its operation is based on a 915 MHz oscillator driving a resonance circuit which is tuned in part by the external capacitance to be measured. As the resonant frequency is moved off the oscillator frequency by the external capacitance, the amplitude of the oscillation is decreased. The peak oscillation of the resonant circuit is detected and, rectified into a DC voltage, forms the sensor output. An AC bias (typically 5-10V, 10kHz) is superimposed on a DC sample bias (2-3V), while the tip is at DC ground. The AC bias is chosen large enough to alternately deplete and accumulate the surface region. The modulated relative surface capacitance changes ΔC_{rel} under the probe tip are registered via lock-in technique, simultaneously with all topographical data, while the probe tip is raster-scanned across the surface. This setup measures ΔC_{rel} across the entire C-V curve and does not allow us to extract any information which wouyld be accessible in full C-V curves. However, it is less dependent on shifts in the flat band voltage due to oxide or surface charges.

Sample preparation was optimized during our earlier experiments /22/ and is similar to standard metallographic Secondary Electron Microscopy cross section preparation: A protective glass cover slide is glued to the top of the sample; subsequent polishing steps produce a smooth cross section consisting of a glass/glue/Si-substrate stack. The glue layer is routinely less than 1μm in thickness. During these preparation steps, a native oxide is formed on the exposed cross-section surface. No attempts have been made at this point to further passivate the sample. Better sample passivation schemes will be pursued in future experiments through HF treatment or growth of low temperature oxides.

Although our setup is capable of detecting junctions /24/, we have decided here to concentrate on samples with no junctions to allow for a more systematic approach. Results presented include measurements from cross-sections of a blanket p-type test-wafer, implanted with 1E14 atoms/cm^2 Boron at 100 keV, and a special teststructure, manufactured at NIST. The teststructure consists of 5mm wide stripes of lightly bulk doped p-type Si (~1E16 atoms/cm^3 B), covered with ~16 nm of oxide, alternating with highly doped p-type Si (dose 1E15 atoms/cm^2 B, 50keV), covered with 10nm of oxide. Cross-sections were prepared by cutting through the line pattern and taking measurements at the exposed implanted area.

RESULTS

Figure 1 compares SCM measurements (a) obtained when scanning across a cross-section of a blanket Si-wafer implant (1E14 atoms/cm^2 B at 100keV into a p-type substrate of 3E14 atoms/cm^3) to a SIMS depth profile (b) from the same sample. The glue/Si interface coincides with the vertical axis. Its location was identified from the simultaneously acquired topography image (not shown).

Comparison of the data shows an excellent agreement for the total implant depth between the two techniques: The background of the SIMS measurement is at about 3E14 atoms/cm^3, which corresponds to the bulk dopant concentration of the p-type substrate. We can see from the SCM plot that it reaches its plateau at the same depth (top cursor at 1 μm) as the SIMS signal levels off. A further qualitative comparison shows that the rise in the SCM graph correlates to the maximum of the implant profile at 0.35 μm. The arrows in the SCM plot indicate the width of the signal rise which can be used to estimate lateral resolution (see Discussion).

True 2D- capability of this technique is demonstrated when applied to a cross-section of the NIST teststructure. Figure 2 shows the result of a 1mm x 1mm scan covering part of the implant region at the edge of an oxide line. The sample surface is at the left side. The thin oxide layers of 8nm (in the opening) and 16nm (oxide mask), are not resolved in the topography image (a) due to the limited point-to-point resolution. However, the glue/Si interface can clearly be identified and is used to spatially correlate the SCM data in Figure 2b (displayed in a contour plot) and the SIMS data in figure 2c, which were obtained from a large (1mm^2) implanted area on the same sample. The origin of the SIMS profile is placed at the location of the interface. The contour lines in fig. 2b represent equi-ΔC_{rel} lines and have been labeled with the respective chemical dopant concentrations obtained from fig. 2c. In this way, a 2D SCM map is obtained outlining carrier/dopant concentrations over several orders of magnitude.

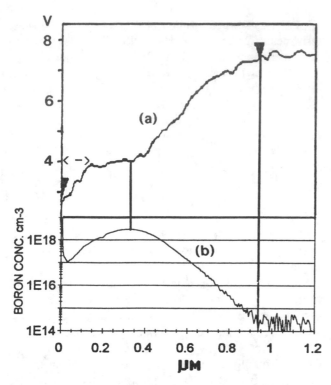

Figure 1. Comparison of a single SCM line scan (a), and a SIMS depth profile (b), obtained on the same sample. The origin of the SCM profile is at the sample/glue interfae. The top cursor indicates the total implant depth (1mm). The implant maximum coincides with the rise in the SCM data (0.35 mm). The arrows indicate the width of the signal rise (see Discussion).

DISCUSSION

Figure 2 represents our key finding. Differences can be expected between concentrations of total chemical dopant as measured by SIMS and electrically active carriers which are probed by the SCM. Nonetheless, we believe that a comparison is justified within the resolution of these micrographs and on the base of the results of figure 1, which serves as proof of concept through the excellent qualitative agreement between SIMS and SCM results. The measurements are relatively easy to perform and have a remarkable stability: we have successfully scanned one such cross-sections for over an hour without a change in signal, only limited by the quality of the tip coating. This makes this technique ideal for obtaining reliable measurements under ambient conditions.

Overall noise levels can be estimated from SCM single line sections, such as in figure 1, to be less than 10%. Refinement of the electrical setup should further improve this value. The sensitivity of the SCM technique covers

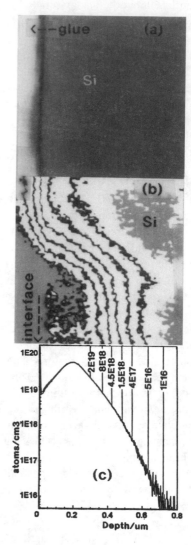

Figure 2. Simultaneously acquired topographical (a) and 2D SCM data (b) of the cross- sectioned NIST teststructure, scanned over an area of 1μm x 1μm; signal ranges are 40nm and 5V, respectively. The arrows indicate the location of the glue/Si interface. The dark areas represent equi-ΔC_{rel} lines and are labeled with the respective chemical dopant concentrations from the SIMS profile in (c).

several orders of magnitude of carrier concentrations, as demonstrated in figure 1. Our measurements on the top surface of a highly bulk doped sample (9.5E19 atoms/cm³) with an oxide pattern have also shown the technique to be sensitive at very high dopant levels.

An estimate for lateral resolution can be obtained from single SCM lines, such as in figure 1. The width of the SCM signal rise (arrows) depends on the probe tip and on the dopant concentration. The corresponding values for figure 2 were 40nm in the high dopant region between the oxide lines and 150nm for the low dopant region below the oxide mask. Measurements on other samples have shown values as low as 20nm. We interpret this as the "electrical tip diameter" because it is the combined effect of actual geometric tip diameter and carrier concentration: Due to the difference in carrier concentration the applied AC bias results in a much larger depletion depth and interaction volume for the lower concentration. This results in a lateral "spread" of the signal and consequently a loss in lateral resolution. This effect is addressed in an improved setup using a feedback system which allows to maintain a constant depletion depth by varying the AC bias amplitude /15/.

CONCLUSION

Our work represents an important milestone on the way to direct 2D- measurements of local carrier densities. The current setup demonstrates that the SCM technique can resolve carrier concentrations on cross-sections in two dimensions with a single raster scan, using a standard sample preparation technique. Two-dimensional quantitative data are easily obtained by correlation to SIMS profile data, acquired on the same sample. Spatial correlation to structure geometry is obtained through simultaneously acquired AFM topography data. Our future experimentation will concentrate on the development of a better defined sample passivation, application of the AC bias feedback system, and modeling of the tip/sample geometry to obtain a direct correlation between SCM signal output and absolute concentration.

ACKNOWLEDGEMENTS

The authors wish to express their thanks to their respective Departments for having made possible this cooperation between the four sites of Intel, NIST, University of Utah and Digital Instruments. Prof. L. Sadwick, University of Utah, has contributed through many fruitful discussions in interpreting C-V curves. A.E. would like to express thanks for the opportunity of summer internships with Intel Corporation which made this work possible. Special thanks go to Regina Campbell, Intel, for her patience, assistance and "magic tips" to allow the authors to successfully prepare a variety of cross-sectioned samples, as well as to Jerry Hunter, Intel, and Clyve Jones, Philips Materials Analysis Laboratories, for measuring the SIMS depth profiles.

REFERENCES

1. R. Subrahmanyan, J. Vac. Sci. Technol. B **10** (1), 358 (1992).
2. H. Cerva, Proc. 1st Intern. Workshop on the Measurem. and Charact. of Ultra-shallow Doping Profiles, Eds. C. M. Osburn, G.E. McGuire, Vol. 2, 286 (1991).
3. S. H. Goodwin-Johansson, Y. Kim, M. Ray, H.Z. Massoud, J. Vac. Sci. Technol. B **10**, 369 (1992).
4. R. Chapman, M. Kellam, S. Goodwin- Johansson, J. Russ, G. McGuire J. Vac. Sci. Technol. B **10**, 502, (1992).
5. T. Tagikami, M. Tanimoto, Appl. Phys. Lett. **58**, 2288 (1991).
6. S. Kordic, E.J. vanLoenen, D. Dijkkamp, A.J. Hoeven, H.K. Moraal, J. Vac. Sci. Technol. B **10**, 496 (1992).
7. J.V. LaBrasca, R.C. Chapman, G.E. McGuire, R.J. Nemanich, J. Vac. Sci. Technol. B **9**, 752, (1991).
8. M.B. Johnson, J.M. Halbout, Proc. 1st Intern. Workshop on the Measurem. and Charact. of Ultra- shallow Doping Profiles, Eds. C. M. Osburn, G.E. McGuire, Vol. 2, 286 (1991).
9. R.M. Feenstra, J. A. Stroscio, J. Vac. Sci. Technol. B **5**, 923 (1987).
10. W.F. Tseng, J.A. Dagata, R.M. Silver, J. Fu, J.R. Lowney, J. Vac. Sci. Technol. B **12**, 373 (1994).
11. R.M. Silver, J.A. Dagata, W. Tseng, 41st Nat. Symp. AVS (1994), J. Vac. Sci. Technol. B, to be published.
12. C. K. Shih, Appl. Phys. Lett. **64**, 493 (1994)
13. O. Albrektsen, M.B. Johnson, R. M. Feenstra, H. Salemink, Appl. Phys. Lett. **63**, 2923 (1993).
14. Y. Huang, C.C. Williams, J. Vac. Sci. Technol. B **12**, 369 (1994).
15. Y. Huang, C.C. Williams, J. Slinkman, J. Vac. Sci. Technol. B, to be published.
16. J.J. Kopanski, J.F. Marchiando, J.R. Lowney, D.G. Seiler, NIST, Sematech, ASTM, E42.14, and AVS Workshop: Industrial Applications of Scanned Probe Microscopy, eds. J. A. Dagata, A.C. Diebold, C.K. Shih, R.J. Colton (1994).
17. Y. Huang, C.C. Williams, Proc. 2nd Intern. Workshop on the Measurem. and Charact. of Ultra-shallow Doping Profiles, Eds. R. Subrahmanyan, C.M. Osburn, P. Rai-Choudhury, Vol. 2, 286 (1993).
18. P. DeWolf, T. Clarysse, W. Vandervorst, J. Snauwert, L. Hellemans, M. D'Olieslaeger, D. Quahagens, 41st Nat. Symp. of the AVS, Denver, CO (1994), J. Vac. Sci. Technol. B, to be published.
19. T. Clarysse, W. Vandervorst, J. Vac. Sci. Technol. B **12**, 290 (1994).
20. C. Shafai, D.J. Thomson, M. Simard- Normandin, G. Mattiussi, P.J. Scanion, Appl. Phys. Lett. **64**, 342 (1994).
21. M. Nonnenmacher, M. P. O'Boyle, H. K. Wickramasinghe, Ultramicroscopy **42-44**, 268 (1992).
22. G. Neubauer, M.L.A. Dass, T.J. Johnson, Mat. Res. Soc. Symp. Proc. **265**, 283 (1992).
23. J. R. Matey, J. Blanc, J. Appl. Phys. **47**, 1437 (1985).
24. G. Neubauer, A. Erickson, NIST, Sematech, ASTM, E42.14, and AVS Workshop: Industrial Applications of Scanned Probe Microscopy, eds. J. A. Dagata, A.C. Diebold, C.K. Shih, R.J. Colton (1994).

Carrier profile determination in device structures using AFM-based methods.

W. Vandervorst, P. De Wolf, T. Clarysse, T. Trenkler

Imec, Kapeldreef 75, B-3001 Leuven, Belgium

L. Hellemans and J. Snauwaert

KULeuven, Celestijnenlaan 200, B-3001 Leuven, Belgium

V. Raineri

CNR, Stradale Primosole 50, 95121 Catania, Italy

Direct observation of carrier distributions in devices requires nanmometer resolution in the vertical and lateral direction as well as high sensitivity. At the same time it is also important to im age the surrounding layers such that the registration between dopants and mask edges can be determined. A first method appraopirate for this purpose is selective etching combined with AFM imaging. The selective etch induces a concentration dependent topography which can be imaged by AFM. Quantification is based on calbiration curves allowing a conversion from etch depth to concentration. Examples taken from CMOS-processes illustrate the visual power of this technique which is however limited to concentration higher than 10^{18} at/cm^3. A second method, Nano-SRP, is based on a conductive AFM-tip which is used to measure the local spreading resistance under the tip. Being a resistance measurement this method has a very high sensitiivity and good dynamic range and provides easy quantification of the results. The method can be applied directly to the cross section of a device and, if combined, with standard AFM-imaging, provides information on the dopant mask edge registration as well. Examples are presented on profiles taken from B-implants in p- and n-type doped substrates which are in nice agreement with standard 1D-SIMS and SRP profiles. These profiles illustrate the capabilioty to probe two-dimensional carrier distributions with Nano-SRP.

Introduction

Direct observation of 1D/2D-carrier distributions in devices is the only way to understand detailed device operation. The latter does impose however extremely stringent requirements on the measurement tools used since nanometer resolution in the vertical as well as the lateral direction is required. Moreover high sensitivity and dynamic range are required as the carrier concentration can range from 10^{14} at/cm^3 to 10^{21} at/cm^3. As small concentration variations influence device performance significantly (for instance the surface concentration has a direct impact on the threshold voltage), the measurement technique must be very sensitive, preferentially having a response linearly proportional with the concentration variation.[1] Finally but equally important is the need to localise the dopants/carriers with respect to other device characteristics such as mask edges, gate oxide, spacers,..etc.which should be imaged at the same time as well. [2]

Whereas scanning probes provide adequate spatial resolution, the research has concentrated around finding the appropriate signal which is at the same time characteristic for the carrier concentration and does exhibit the required properties of sensitivity and dynamic range. Additional problems which need to be addressed are the repeatability, the quantification properties, the correlation with topography images and the ease of selecting the device of interest.

In this paper we present two approaches for carrier profile determination using AFM. The first one is based on selective etching combined with AFM imaging whereas the second relies on a miniaturised version of the spreading resistance probe.

CHEMICAL DELINEATION AND AFM-IMAGING

In this first method the signal corresponding to the carrier concentration originates from the observation that the etch rate of Si in certain etchants can be mediated by the carrier concentration in the sample. [3-5] The latter implies that when these etchants are applied to the cross section of a device, a topography will appear due to differences in etch rate. The basic concept of the chemical delineation methods is then to etch the cross section of a device with these etchants and to analyse the resulting topography with high resolution techniques such as TEM, SEM or AFM [3-5]. In particular the use of the AFM is attractive since it provides a quantitative measurement of the topography. Given appropriate control over the etching itself, its reproducibility and the calibration of the etch rate, a quantitative carrier profile might result from the AFM-image.

P-type doping

The most common etchant reported for p-type material so

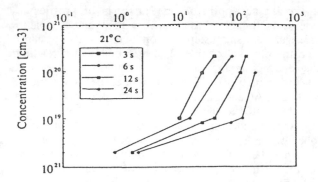

Figure1: Calibration curve of carrier concentration as a function of etched silicon for different etch times at 27°C. (For etching at 21°C see ref 5)

far is a mixture of HF (40%) - HNO_3 (65%) - H_2O - CH_3COOH (96%) in a ratio 1:3:8 [4-7]. Prior to the etching (3-1s, under UV-illumination) a cross section of the device of interest is made by polishing the sample with different polishing compounds and as final step using colloidal silica. It is important that a nearly atomically flat surface is obtained since scratches, dust and organic residues on the surface can create artifacts during the etching step.

Other sample preparation procedures to be investigated are cleaving or Focused Ion Beam milling. The latter is most attractive if one particular (defective?) device needs to be analysed since positioning of the cross section on a complex circuit becomes a crucial problem. Detailed studies are still required to determine the influence of the ion beam damage on the delineation procedure.

In order to correlate the etched depth with carrier concentration, a series of uniformly doped substrates was etched in certain regions whereas the step height between the etched and unetched regions was measured afterwards with the AFM. The calibration curves ([5] and fig 1) resulting from these experiments point out that for concentrations below 10^{18} at/cm^3 the etch differences become extremely small (less than 1 nm for a 3 s etch) implying a poor, or even completely absent sensitivity. This limit in sensitivity has been a common observation in most studies except in ref. 6 where they manage to obtain a contour line corresponding to the junction position as well. It is also important to note that different etch times or temperatures lead to different calibration curves with different sensitivity for certain concentration regions. (fig. 1) The latter is important to realise since in structures with widely varying and very steep carrier profiles, the measurement of the etched topography might become limited by the dimensions of the AFM-tip shape. Hence the selection of the etch conditions will depend on the sample under investigation. Present investigations centre around increasing the repeatability by increasing the (short) etch time either by lowering the etch temperature or by further diluting the etchant solution whereby further improvements regarding the sensitivity are necessary as well.

Despite its limitations, the selective etching can provide interesting information because it generates a visual impression of the carrier concentrations. In fig.2 an image is shown for a 20 keV 2 10^{15} at/cm^2 B-implant. The implant region was defined by a mask consisting of a thin oxide and a polysilicon (200 nm) layer. The image in fig.2 quite clearly shows the ability to directly determine the registration between the carriers in the substrate and the mask edge. Note that because the polysilicon layer was doped it is etched as well thereby enhancing the definition of the carrier profile/mask edge overlap.

Due to its high spatial resolution the technique is ideally suited for device analysis. An important benefit is indeed that the other layers (oxides, metals,...) are becoming apparent as well. An example of that can be seen in fig.3 which shows an MOS transistor. The device is completely processed showing the details of the gate and spacing structure, the metal layer contacting the polysilicon gate and the carrier profile from the source drain regions. Again by combining this image with the calibration curve of fig.1 a quantitative analysis of the carrier distributions can be made.

N-type doping

For n-type material a different etchant based on $HF:HNO_3:H_2O$ in a ratio 1:100::25 must be used where the addition of H_2O slows down the reaction making the timing more manageable and hence the whole procedure more reproducible. In this case the etching does not require any UV-light. Under those conditions the sensitivity seems to be poorer (limited to 5 10^{18} at/cm^3) as compared to p-type etching.

Surprisingly enough the presence of UV-light affects the selectivity of the solution such that with UV-light p-doped layers are preferentially etched. If we apply this solution to a n$^+$pn-structure we find that now the p-region is strongly etched whereas the n$^+$-doped region etches a lot slower. In this way again a topography related to the n-type doping is induced but now in the inverse way i.e. the higher the n-type doping the lower the etching rate. Hence now the n-type doped regions will appear as hills instead of as valleys in the AFM- topograph. The image obtained using this inverse etch shows a delineation of the n-doped region down to the 2 10^{17} at/cm^3 level. It is clear that the reverse etching is a more sensitive way of delineating the n-doped regions than the normal etching and hence the preferred method. The only case where it is not applicable is for n$^+$n structures since then there is no p-doped region to etch.

Using a correct sample preparation a reproducibility of better than 40% can be obtained if all the experimental conditions (surface quality, light, temperature,...) are controlled very carefully. Under those conditions the technique is ideally suited to image the carrier distributions in device structures and to visualise at the same time the surrounding material. The main limitation is the limited sensitivity of around 5 10^{17}- 10^{18} at/cm^3 for p- as well as n-type material. Since the carrier distribution is etched and not the dopant distribution, one must realise that all limitations due to carrier diffusion and Debye length smearing do apply. Further work is required to improve the etching characteristics in terms of reproducibility, sensitivity and quantification properties.

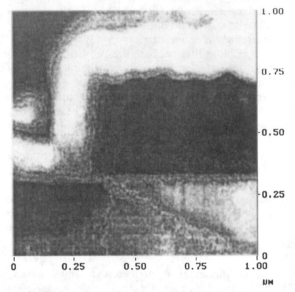

Figure 2: AFM image of chemically delineated B-implant. The B-doped poly layer was used as an implant mask. Darker regions correspond to higher concentrations.

Figure 3: AFM-image of chemically delineated MOS-structure showing source/drain implant regions as well as gate oxide, spacers and metal connections.

NANO-SRP

In order to obtain better sensitivity and quantification properties, an additional method is under development using the measurement of the spreading resistance between one or two tips as a measure for the local resistivity. If these tips are small enough such that they can be placed directly on the cross-section of a device,

they might be applicable for analysis of profiles in devices. As this tip needs to be very small and capable to move with very small steps, a conductive AFM-tip seems to be the obvious choice. The concept of a spreading resistance measurement is quite important since it implies that the sensitivity of a resistance measurement is maintained. An additional benefit is that the resistance can easily be related to the local resistivity. In practice there are now two concepts possible i.e. a one probe or a dual probe system. The dual probe system is the direct replica of the standard SRP and looks the most obvious but it does require a large engineering effort to construct a dual head AFM with tips which are very close together (ideally again only a few nm). A more relaxed concept is to use only one tip and to use a large backcontact on the other side of the cross section. [8] This one tip system is easier to design but does require that the structure under investigation must be long enough such that it is uniform over the thickness of the cross section (~100 μm).

The problem of the conductive tip is not so simple as it may seem because one of the important requirements is that it should make a good electrical contact with the sample and its contact resistance should be sufficiently low such that the measured resistance is dominated by the local resistivity in the sample and not by the contact resistance. Initial attempts with a metal tip with the required dimensions (nm-size contact) have shown that a rather high load (~ 100-200 μN) is required before good electrical contact can be obtained.[9] Under these high pressures the fine metals tips are not sufficiently rigid and they deform rather quickly. A more stable operation has been found through the use of B-doped diamond tips coated with W which have been mounted on a specially made cantilever system. The whole cantilever is then fitted to a standard AFM-system (Nanoscope II or III).[8] The advantage of the diamond tip is that it is extremely hard and does withstand the necessary pressures.

Also with these tips the measured resistance is strongly dependent on the applied pressure and stable operation can only be achieved when a minimum load is applied. For instance on Si as well as on Pt-substrates it was found that only a low resistance can be measured when the force is larger than 50-70 μN. When measuring now on Si-substrates homogenously doped to different levels, the I-V curves change drastically with substrate resistivity. For the high concentrations the I-V curves are very much Ohmic like whereas for lower concentrations they are of the Schottky Barrier type. The I-V curves are similar for p- and n-type material with of course the expected change in polarities. This behaviour can be rationalised within the standard concept of a probe-semiconductor contact whereby a plastically deformed region and an elastically stressed part play a role in the contact behaviour. [9]

From these I-V curves it is possible to determine the resistance around zero bias (5 mV as also done in standard SRP). When these resistance values are plotted as a function of resistivity of the substrate, we obtain calibration curves which look very similar to a standard SRP calibration curve i.e. showing a nice monotonic relation between measured resistance and sample resistivity.(fig 4) At the moment a saturation around 7-9000 Ω for the lowest resistivities is found which indicates that the measurements in this interval are limited by a series resistance probably from the diamond which is apparently still not sufficiently doped.

AFM-images of the probe imprint show a contact size of roughly 50 nm. At present the main obstacle in improving this further arises, from the necessity to apply a minimum pressure in order to get a good electrical contact. The latter might be due to the pressure required to break through the native oxide present on the sample surface. Modelling of the conventional SRP contact indicates that the pressure under the tip is extremely important and induces changes in material properties (such as bandgap narrowing) which aid in getting good electrical contact. Modifications to the tip geometry might lead to other stress distribution and perhaps smaller contact radii and hence higher spatial resolution. The problem of the native oxide might be circumvented by working under a controlled atmosphere or in ultrahigh vacuum. Evidently further work is required to improve the characteristics of these diamond tips (shape,

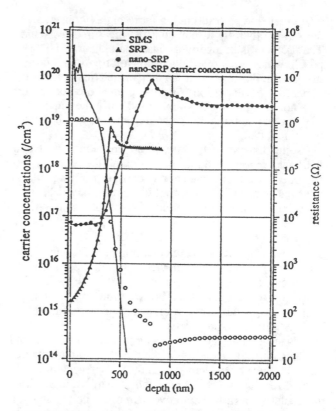

Figure 5: Resistance variation mesured with Nano-SRP and SRP on a B-implant into n-type substrate. Also shown is the Nano-SRP concentration profile calculated using the calibration curve of fig 4 and the SIMS profile.

the gradual transition of n-type in the substrate to p-type near the surface. Again the series resistance of the tip limits the resistance values in the highly doped region as well.

The resistance behaviour is completely identical to the resistance variation obtained by a standard SRP analysis (which is however performed on the bevel surface) (fig 5) except that the absence of a bevel in the Nano-SRP case, leads to a totally different behaviour with respect to carrier spilling. The standard SRP profile gives an "apparent" junction position of 470 nm, whereas the Nano-SRP indicates a junction position at 800 nm. The SIMS profile places the metallurgical junction at a depth of 570 nm. These differences are in complete agreement with our understanding of carrier spilling. The bevelling process in standard SRP leads to a displacement of the electrical junction towards shallower depths whereas the Nano-SRP determines the electrical junction position in the vertical direction which for this structure is larger than the metallurgical junction in agreement with a simple solution of the Poisson equation.

In principle it is possible to convert these resistance values to resistivities using the calibration curve reported in fig.4. However a word of caution is required here in terms of interpretation of data taken on a cross section. Indeed when the tip is contact with a lowly doped part of the sample it is quite possible that the current will follow

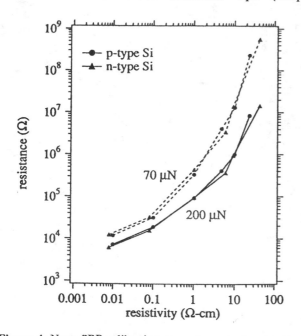

Figure 4: Nano-SRP calibration curve (measured resistance vs substrate resistivity) measured with two tip loads.

conductivity,..) and the experimental procedure.
Nevertheless using the present tips it is already possible to make a scan across the cross section a device and to demonstrate the capabilities. In fig 5 we show the resistance variation measured on the cross section of a B-implanted n-type substrate. The data show the expected low resistance in the near surface region where the largest amount of boron is located and a rise towards a high value in the depletion region decreasing again towards the substrate value. The stepsize in these measurements is 50 nm. In addition an analysis of the I-V curves does show

a less resistive path through a more highly doped part of the sample thereby influencing the measured resistance. The problem is in a certain sense analogue to a 1-D SRP profile which does require a deconvolution step, but now the deconvolution must be performed using a three-dimensional calculation. At present there are not enough data nor calculations available to identify the importance of this problem but a dedicated deconvolution scheme has been identified.

Despite the presently unresolved problems, the Nano-SRP has great potential in achieving all the requirements for one- and two-dimensional carrier profiling and is believed to be the ultimate tool for ULSI-analysis because of its combination of sensitivity and spatial resolution.

CONCLUSIONS

At present two methods are emerging as appropriate tools for carrier profiling in devices. From these two the chemical delineation combined with AFM-imaging provides the most visual information on the carrier distribution at the expense of sensitivity, reproducibility and quantification properties.

The Nano-SRP is far more sensitive and quantitative and its capabilities to measure on the cross section of a device have already been demonstrated. The present limits in spatial resolution (30-50 nm) and dynamic range can be overcome by a further improvement of the conductive tip. Data interpretation seems straightforward using a calibration curve although some additional 3-D verifications are still required to evaluate possible deviations in complex structures. A major point which still needs to be addressed for both methods is the generation of the cross section through that one particular defective device in a sea of structures. Computer aided selection and Focused Ion Beam milling are probably the most elegant routes in achieving the latter objective.

REFERENCES

1.*R.Subrahmanyan and M.Duane* : Proceedings Electroch. Soc. Fall meeting, 1994
2.*A.C.Diebold, M.Kump, J.J.Kopanski and D.G.Seiler* : Proceedings Electroch. Soc. Fall meeting, 1994
3.*D.M.Maher and B.Zhang* : Journ. Vac. Sci. Techn. B12, 347 (1994)
4.*L.Gong, A.Barthel, J.Lorenz and H.Ryssel* : Proc. ESSDERC-89, pp198 (1989)
5.*V.Raineri, V.Privitera, W.Vandervorst, L.Hellemans and J.Snauwaert* : Appl. Phys. Lett. 64 (3), 354 (1994)
6.*L.Gong, S.Petersen, L.Frey and H.Ryssel* : Nucl. Instr. Meth. Phys. Res. B (1995, in press)
7.*C. Spinella, V.Raineri, M.Saggio, V.Privitera and S.U. Campisano* : Nucl. Instr. Meth. Phys. Res.B (1995, in press)
8.*P.De Wolf, T.Clarysse, W.Vandervorst, J.Snauwaert and L.Hellemans* : presented at AVS-meeting, Denver, Oct 1994.
9.*P.De Wolf, T.Clarysse, W.Vandervorst, J.Snauwaert and L.Hellemans* : Appl. Phys. Lett. (submitted)

NUCLEAR METHODS

Neutron Activation for Semiconductor Materials Characterization

S. C. McGuire and T. Z. Hossain

Ward Laboratory, Cornell University, Ithaca, NY 14853-7701

Albert J. Filo, Craig C. Swanson

Analytical Technology Division, Eastman Kodak Co., Rochester, New York 14650-2155

and

James P. Lavine

Microelectronics Technology Division, Eastman Kodak Co., Rochester, New York 14650-2008

For the past three decades the analysis of trace impurities in silicon materials by nuclear techniques such as neutron activation analysis (NAA) has been successfully used by the semiconductor industry and others. The presence of transition metal contamination in silicon wafers, for example, can have a detrimental effect on the electrical characteristics and yield of integrated circuits. The contamination present in the device may come from the starting material (silicon wafer) or can be introduced during one of the many processing steps employed in the manufacture of the integrated circuits. It is therefore necessary to be able to identify the transition metal contaminants present in the material, their concentration, and relative position within the matrix. This may be accomplished by using NAA in conjunction with autoradiography to obtain two dimensional information on the location of the impurities present in the silicon wafer. Sandwiching the silicon wafer between two phosphor imaging screens provides a third dimension in the location of the impurities by comparing the image intensity between the two screens. To date, many different processing steps have been examined in this collaboration, including oxidation, ion implantation, and plasma deposition. Large samples have been used to obtain ultra low detection limits for a number of elements such as Au, Co, Cr, Cu, Fe, K, Mn, Na, Th, U and Zn. Further, because of its ultra low sensitivity and high reliability NAA is an excellent candidate as a reference method that can certify and provide quality control (Q/C) for other recently developed in-plant techniques for trace element analysis. One such method is total reflection x-ray fluorescence (TXRF) which can be standardized against NAA. In this respect another nuclear method, neutron depth profiling (NDP) already provides standard reference materials for calibrating depth profiling measurements by such techniques as secondary ion mass spectroscopy (SIMS). NDP has been demonstrated to provide non-destructive and quantitative boron depth profiles with a resolution of ~180 Å. The recent development of cold neutron facilities may extend the capabilities to other light elements such as N and O. Furthermore, the depth resolution may be enhanced to less that 100 Å by the use of time of flight techniques that are being developed at Cornell.

INTRODUCTION

It is the purpose of this paper to describe recent developments in the use of nuclear methods for semiconductor characterization and to relate those methods to the semiconductor industry's needs for the next decade. There are many examples of the effectiveness of using neutron activation as a probe of semiconductor materials in the recent literature (1-3). We have included in this discussion those nuclear methods that are based on the use of thermal and cold neutrons. Thus, we have focused on well established conventional as well as evolutionary applications of neutron activation analysis (NAA). Additionally, neutron depth profiling (NDP) of light elements such as boron involving time-of-flight (TOF) may also yield improvements in spatial resolution required as device dimensions continue to shrink. Boron provides a particularly important example for discussion given its widespread usage as a dopant in microelectronics technology. Our emphasis is on employing nuclear methods as a means of providing primary standards for other techniques used for product monitoring during processing.

The utility and importance of the methods described in this paper are underscored by the requirement for contamination-free manufacturing (CFM) during the next decade. In order to remain technologically competitive into the next century it is projected for example that tolerable contamination levels for transition metals will be lowered to less than 2.5×10^9 atoms/cm^2 (4). Similar tolerance levels are being projected for alkali metals. In order to meet these requirements additional research effort leading to improvements in impurity detection sensitivity and reliability are required.

As an illustration of a recent outcome from our application of NAA, we have achieved significant reductions in contamination levels by preparing wafer samples in the class 100 clean rooms of Cornell's National Nanofabrication Facility (NNF) Further, identification of preferred scribing tools has been seen to play an important role as well. Our primary objective is to take full advantage of the sensitivity and reliability offered by nuclear techniques generate standards against which other, less proven methods, can be measured.

DETECTION LIMITS FOR INDIVIDUAL METALS

By using large sample sizes (1-10 g) and long irradiation times (~30 h) in thermal neutron fluxes of 1-2 x 10^{12} n/cm^2·s we have been able to obtain detection limits of about 10^{10}-10^{11} atoms/cm^2 for transition metals and mobile ions required to be minimized for achieving contamination-free silicon wafer processes. Key to realizing these results are the precise steps employed for preparing the samples for irradiation. Of importance here is the use of clean room areas and physically handling specimens with instruments that do not introduce contamination. In this regard, we have found, for example, that the choice of tweezers for handling samples can be critical. Although sample preparation for NAA is minimal, it is important to select correct handling tools such as diamond scribe, and tweezers. Experiments conducted at Cornell University, and at Eastman Kodak Company have revealed that the diamond scribe and green teflon coated tweezers could be a serious source of contamination.

NAA of Si wafers sectioned with diamond scribes showed the presence of Cu, Ag, Au, and Fe. While wafers manipulated with green teflon coated tweezers were found to have Fe and Cr on the surface. Such contamination by green tweezers has also been noted by other workers. The diamond scribes, near the tip were seen to have a dark ring (scum), and when analyzed with microspot (scum) EDX showed presence of Cu, Ag, Au, and Fe.

It is therefore recommended that these two implements be avoided for high purity sample handling, and perhaps not used at all in the clean rooms. At Cornell, high purity suprasil quartz tools have been used as scribes for wafers, and white plastic tweezers were used for handling samples. Using this combination of tools it has been possible to to avoid contamination during sample preparation. Figures 1 and 2 represent two spectra obtained from the same lot but prepared differently as discussed above. The energy window shown is from approximately 150 to 2000 keV.

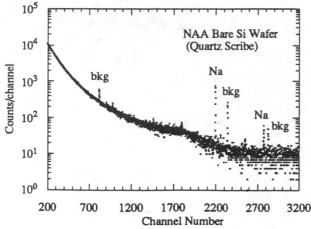

FIGURE 1. Gamma spectrum of virgin silicon wafer section cut using a quartz scribe.

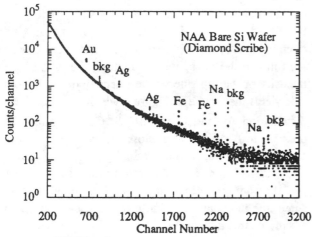

FIGURE 2. Gamma spectrum of virgin silicon wafer section cut using a diamond scribe.

NAA WITH AUTORADIOGRAPHY

As part of an effort to improve the quality of integrated circuit products, we have found it useful to combine standard NAA with autoradiography to characterize transition metal contamination in silicon wafers. This class of elements is well known to have a detrimental effect on the electrical properties and product yield at concentrations of ~ 10^{10} atoms/cm^2.

EXPERIMENTAL

The silicon wafers analyzed were 4-inch diameter n-type Czochralski wafers with a (100) orientation and a resistivity of 30 Ω-cm. The Silicon wafers were irradiated at the University of Missouri Research Reactor for 150 hours at a thermal neutron flux of 1.5×10^{12} n/cm^2/sec. During the exposures the samples were housed in a high purity graphite container (manufactured by Poco Graphite of Decatur, Texas) that was coated with pyrolitic graphite. The addition of the pyrolitic graphite layer greatly reduced the removable surface contamination that would have been transferred to the samples from the graphite container. The samples placed in the graphite containers were separated from each other by a clean silicon spacer that was in the shape of an annular ring. The samples were allowed to decay for 48 - 72 hours prior to data acquisition. The concentrations of the contaminants were determined using the parametric method of NAA using the fundamental nuclear parameters (5,6). The cross sections, resonance integrals, and isotopic abundances were from (7), and the γ-ray intensities and half-lives from the tables by Erdtmann (8).

The radiographs were made using storage phosphor plates (Eastman Kodak Model SO230) and were developed on a Molecular Dynamics Model 425E Phosphor Imager using the ImageQuant Software. The resolution for each pixel in the image measures 176 μm. Each analyzed silicon wafer was sandwiched between two storage phosphor plates in order to simultaneously obtain an image of both sides of the wafer.

AUTORADIOGRAPHY

The gamma rays released as a result of the decay of a radionuclide comprise only one component of the radiations typically produced. In addition, there are also X rays, β particles, and electrons emitted during the decay process. It is a combination of all the radiations emitted by the decay process, but primarily the β particles, that will be used to perform the autoradiography.

Well established as a method capable of ultra low level determination of impurities, NAA has been used to characterize the contamination introduced during device processing by determining the total concentration of the contaminants. This is very important information in determining if a process is being carried out properly and not acting as a source of contamination which would result in reduced device performance and yield.

While this information is necessary, it does not give all of the information that is needed to understand what is happening with the process. NAA, as with most analytical techniques, assumes that the contamination present is uniformly distributed and this is reflected in the units that the results are reported (atoms/cm^2 or atoms/cm^3). In most cases this is not true. Although NAA will report the presence of the metal, it does not indicate how it is distributed on the material or if it has penetrated into the matrix. As an example, we have examined patterns produced by dipping a wafer in an etchant and allowing the etchant to simply drain off while the wafer is in a carrier. The images that we have obtained clearly show a two dimensional distribution of the residual contamination. Sandwiching the irradiated silicon wafer between two storage phosphors, allows us to obtain a radiographic image of contamination on both sides of the wafer simultaneously. We note that it is the combination of all of the radiation emitted by the decay process, but primarily the β-particles, that are used to perform the autoradiography (9). The technique offers several important features: first, we can visually see where the contamination is, ie. is it from the sample or is it from our spacer. Second, it shows us if the contamination is on the polished side or the back side of the wafer, and third, by comparing the position and intensity of the image we can infer if the contamination is surface contamination or if it is bulk contamination. Thus, we have achieved the ability simultaneously acquire a γ-ray spectrum, to determine what elements are present, and acquire the two radiographic images which allows us to obtain a three dimensional picture of the contamination. Typical surface contaminations that we have observed on the backside of wafers include Cr and W.

In addition to the radiographic image obtained from our samples, we may also obtain a line trace through the image. This can give us information regarding the intensity of the image obtained as well as the width or size of the image. Figure 3 is an example of a line trace through the image of a Cu foil that had been irradiated. The measure with of the image at the FWHM was calculated to be 0.256" compared to the actual 0.253" punched sample (this also shows no distortion in the image).

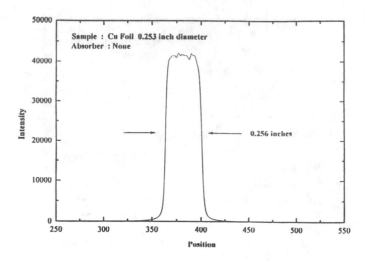

FIGURE 3. Line trace from an automated laser scan of the sample showing its width.

In another set of experiments, we have simulated the observation of the depth distribution of contaminants. This was done by sandwiching an irradiated Cu foil between two phosphors and interposing varying thicknesses of Al absorber between the foil and the phosphors. Plotting the ratio of the intensity of the top side vs the intensity of the bottom side yields the results in Figure 4 which represents the distribution for the decay of ^{64}Cu. It should be possible to obtain similar plots for each of the more common contaminants present during device fabrication. Thus, the combination of the γ-ray spectra by NAA and the radiographic images we should be able to determine the contamination present in the wafer and where (in three dimensions).

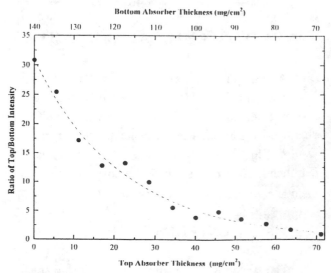

FIGURE 4. Simulation of the measurement of contamination intensity with depth.

REFERENCE METHODS

In addition to being a high sensitivity analytical tool, nuclear methods can play a very important role in developing standards. The ability for the nuclear methods to provide high reproducibility, and accuracy stems from their inherent signal-to-noise ratio (radioactivity vs. no radioactivity), and the non-contact, non-destructive nature of the analysis. As has already been demonstrated, NDP was crucial in providing a standard reference material (SRM) for depth profiling of boron by such technique as SIMS (10). Standards for trace analysis by total reflection x-ray fluorescence (TXRF), for example, can be developed using NAA as the reference technique. Both methods have ultra low detection limits, and the NAA values are already well established.

FUTURE WORK

In looking towards the future we point out that sources of cold neutrons (11) provide an especially attractive advantage in that improved reaction rates are attainable and, when used with guides tubes, can yield enhanced signal-to-background ratio in a particular experiment. In the form of a beam they are ideally suited for prompt neutron activation analysis (PGNAA) measurements of low level concentrations of light elements such as H, Li, C, N, and O.

Many applications of neutron depth profiling have appeared in the literature in recent years (12), however, results on the use of time-of-flight (TOF) as a means of determining depth distribution has been lacking. It can be anticipated that the spatial resolution of conventional or standard boron depth measurements can be extended from ~ 180 Å to less than 100 Å through the use TOF methods. A promising approach can be illustrated through the well known reaction $^{10}B(n,\alpha)^{7}Li + E_\gamma(478 \text{ keV})$. It makes use of the detection of the 478-keV gamma ray in delayed coincidence with its source, the ^{7}Li fragment. An important underlying feature of this idea is that the 478-keV gamma ray is emitted shortly after the excited fragment is born, a condition satisfied by the 73 fs half life of the excited nucleus. It is also recognized that the 478-keV γ-ray must be observed with good efficiency in order to achieved practical coincidence rates. Further, it is necessary that both detectors provide fast output signals.

A possibility is using BaF_2 scintillators for detection of the gamma ray and a thin plastic scintillator for the ^{7}Li fragment. Both will be coupled to photomultiplier tubes the timing signals for which are subsequently processed by commercially available electronics and computer hardware. System timing resolution can be adjusted through choice of the flight path length. Our initial studies are targeted at the examination of boron distributions in silicon at maximum depths of approximately 0.5 µm.

Another scheme that has been suggested involves using the secondary electrons ejected from the sample surface as the charged fragment leaves to produce a start signal for the delayed coincidence measurement (13).

ACKNOWLEDGMENTS

The authors express their appreciation to David D. Clark for his continued interest and support of this work, to H. C. Aderhold and P. Craven for assistance with reactor operations, and to Y. Shacham and R. Soave for help with the use of the NNF clean room facilities. One of us, (SCM), expresses gratitude to the Cornell University College of Engineering and the General Electric Foundation for partial financial support.

REFERENCES

1. T. Z. Hossain, *Encyclopedia of Materials Characterization*, C. R. Brundle, C. A. Evans, S. Wilson (eds.), Butterworth-Heinamann, Boston (1992) p. 671.

2. S. C. McGuire, K. Wong and J. Silcox, *Trans. Am. Nucl. Soc.* **66**, 137-138 (1992).

3. S. C. McGuire, T. Z. Hossain, C. Golkowski, N. D. Kerness and J. D. Sulcer, *J. Radioanal. and Nucl. Chem.***192** (1), 65-72 (1995).

4. *The National Technology Roadmap for Semiconductors*, Published by SIA, (1994), p. 116.

5. P. F. Schmidt, D. J. McMillan, Anal. Chem. **48**(13), 1962-1969 (1976).

6. P. F. Schmidt, J. E. Riley,Jr., D. J. McMillan, *Anal. Chem.* **51** (2) 189-196 (1979).

7. S. F. Mughabghab, M. Divadeenam, N. E. Holden (eds), *Neutron Cross Sections*, Academic Press, New York (1981).

8. G. Erdtmann, W. Soyka, *The Gamma-Rays of the Radionuclides*, vol 7, Verlag Chemie, New York (1979). G. Erdtmann, *Neutron Activation Tables*, vol 6, Verlag Chemie, New York (1976).

9. E. W. Haas, R. Hofmann, *Solid-Sate Electronics* **30**(3), 329-337 (1987).

10. P. M. Zeitzoff et al., *J. Electrochem. Soc.* **137** (12) p. 3917 (1990).

11. B. N. Taylor (ed.), *J. Res. of NIST* **98** (1) (1993).

12. See, for example, G. P. Lamaze, R. G. Downing, J. K. Langland, S. T. Hwang, *J. Radioanal. and Nucl. Chem.* **160**, 315-325 (1992); J. P. Lavine, T. Z. Hossain, D. L. Losee, R. G. Downing, T. R. Carducci, M. Mejra and M. J. Cumbo, *Proceedings of the Sixth International Symposium on Silicon Materials Science and Technology:Semiconductor Silicon*, Edited by H. R. Huff, K. G. Barraclough and J. I. Chikawa, The Electrochemical Society, Pennington, NJ (1990) pp. 1041-1051.

13. Private communication from R. G. Downing, Nuclear Analytical Methods Group, NIST,(1/95).

International Intercomparison for Trace Elements in Silicon Semiconductor Wafers by Neutron Activation Analysis

D. A. Becker, R.M. Lindstrom
Nuclear Methods Group, Analytical Chemistry Division
National Institute of Standards and Technology
Gaithersburg, MD 20899

T. Z. Hossain
Cornell University - Ward Laboratory
Ithaca, NY 14853

Members of ASTM Task Group E10.05.12 (on Nuclear Methods of Chemical Analysis) conducted an interlaboratory comparison on the determination of trace elements in high purity silicon semiconductor wafers. The purpose of the intercomparison was to investigate methods for establishing accuracy in such analyses. Nine laboratories from six countries participated by analyzing one wafer each from two different boules of high purity silicon by neutron activation analysis (NAA), mostly by instrumental NAA. Four of the impurity elements most often reported were Au, Na, As, and Br. Data for these elements are reported along with the techniques used, including cleaning and/or etching techniques. Comparison is difficult due to different sample pretreatments and also widely differing sensitivities (due to differing irradiation times, neutron fluence levels, and counting sensitivities). Au appears to be primarily found as a surface contamination, and is reported down to 3×10^{-5} μg/kg $(2 \times 10^8$ atoms/cm$^3)$ for etched bulk silicon. Arsenic had the best overall agreement, at 0.034 ± 0.019 μg/kg (wafer 1, n=5).

INTRODUCTION

Members of ASTM Task Group E10.05.12 (Nuclear Methods of Chemical Analysis) and a number of other scientists in the nuclear analytical field expressed interest in an interlaboratory comparison of the analysis of high purity silicon and silica. Despite the great economic importance of the silicon device industry, and the probability that large numbers of compositional analyses are performed on these materials for characterization by industry, there have been almost no interlaboratory evaluations of the quality of these analyses. The only recent comparison prior to the start of this effort involved four samples analyzed sequentially in four laboratories in Hungary, the USSR, and the GDR [1]. The results agreed within a factor of two for As, Cu, and W, and worse for Au. A workshop organized by the International Atomic Energy Agency at NBS in 1987 called attention to the lack of standardization in this field, and suggested that systematic interlaboratory comparisons and compositional reference materials would be of value in assessing the ability of analysts to obtain demonstrably accurate data [2].

At the time, one of the authors (T.Z.H.) was working in industry (Eastman Kodak Co.) and was unable to verify the accuracy of his instrumental neutron activation analysis (INAA) results on a set of silicon semiconductor wafers. He requested the assistance of other members of the Task Group in looking at the overall problem of demonstrating accuracy in these types of analyses. As a beginning toward evaluating this problem, the Task Group arranged to distribute four samples of silicon and silica to anyone who indicated an interest in participating in the intercomparison. The four samples were as follows: 1. a silicon wafer of "moderate purity" - 10 cm diameter x 525 μm thick; 2. a silicon wafer of "high purity" - similar size; 3. a 50 mg chip of high purity silicon; and 4. a 0.5 gram sample of high purity silicon dioxide powder. The wafers were donated by Eastman Kodak Co., and the other samples by another party. Twenty five wafers from each of two different boules of high purity silicon were provided, supplied in a sealed container. Only results for analysis of the wafers will be considered in this paper.

The wafers were received from the manufacturer in a 25-wafer cassette, sealed and inside a sealed polyethylene bag. Working in a Class 10 clean air hood at NIST, each wafer was transferred with high purity nitric acid cleaned Teflon FEP forceps from the manufacturer cassette to a new polyfluoroethylene (PFE) single-wafer shipper. Wafers were numbered serially from the cassette. Neither the wafers nor the shippers were cleaned at NIST. Each packaged sample was then heat sealed in a polyethylene bag while still in the Class 10 clean hood.

A total of 19 sets of samples (one each of the two types) were sent out to requesting laboratories. Participants were asked to determine the trace element composition of each of the samples according to their best standard procedures and to report the results. In order to intercompare the data, details of the methods employed were requested as well. Reflecting the makeup of the Task Group, all of the participants used nuclear activation analysis for analyzing the wafers although other analytical techniques were also welcome.

Nine sets of results were returned, from six different countries: Canada, Finland, Germany, India, The Netherlands (2), and the U.S. (3). These represented government laboratories (3), universities (2), semiconductor industry laboratories (2) and analytical laboratories (2). Results were reported by one or more laboratories for a total of 15 elements (Au, Na, As, Br, Ag, Co, Cr, Sb, Zn, Mn, Cu, Fe, K, O, and P) , and included data on both surface and bulk impurities. A variety of post irradiation techniques were used including acid cleaning without etching, and heavy etching of the wafers after irradiation for determination of both surface contamination concentrations in the etch solutions and the bulk silicon impurities.

EXPERIMENTAL

All participants used INAA for the analyses; one laboratory also used charged particle activation analysis for oxygen determination, and radiochemical NAA for phosphorus. Since there were numerous variations in the specific procedures used, this paper will describe below in some detail the NIST INAA procedure, with comments about other procedures included.

As described above, one each of wafers #1 and #2 were received in their single PFE holder (wafer numbers were 1-5 and 2-5). Each wafer was broken into pieces by a sharp tap of a hammer on the top of the wafer holder prior to opening the holder. Individual pieces were placed into an acid washed "2/5 dram" polyethylene snap-cap vial, using new, acid washed Teflon FEP tweezers. Where pieces were still too large, they were placed inside folded Whatman #42 filter paper and broken again with finger pressure using gloved hands. Two vials were filled from each wafer. The same tweezers was used for loading all samples, with acid washing between wafers. All breaking and loading operations were carried out inside of a Class 100 clean hood.

Standards used included two Standard Reference Materials, SRM 1633a (Coal Fly Ash) and SRM 1571 (Orchard Leaves), plus a known amount of Au pipetted unto a filter paper. The fly ash was sealed in a polyethylene bag and snap-cap vial and irradiated at the same time as the wafers. The other two standards were irradiated separately. Both irradiation containers had iron

foil fluence monitors placed at fixed locations to verify the neutron exposure of samples and standards.

The silicon wafer samples were irradiated for 16 hours in the RT-4 pneumatic tube facility at a neutron fluence rate of 1.3×10^{13} n•cm^{-2}s^{-1}. The NIST reactor was at the 15 MW power level during this time. After irradiation, each vial containing a sample (multiple pieces) was removed from the irradiation container, the silicon wafer pieces washed vigorously for 5 minutes in 4:1 H_2O: HNO_3 (both high purity reagents); then rinsed in high purity H_2O three times for 5 minutes each, and finally rinsed in reagent grade ethyl alcohol, air dried, and placed in a disposable polystyrene Petri dish for counting of induced radioactivation products. Sample weights were determined by weighing after cleaning.

It should be noted that a number of participants used no cleaning at all after irradiation, while others used a planar etch (HF:HNO_3:$HC_2H_3O_2$) and/or a cyanide etch, with or without Au and Cu holdback carriers. Two laboratories counted the entire wafers. Details of the irradiation and clean/etch procedures used by each participant are found in Table 1.

For the NIST wafer samples, counting was initiated almost immediately after the end of irradiation and transfer. Most participants counted the irradiated samples several times at varying decay times. The HPGe detector used at NIST was 35 % efficient and had a resolution of 1.66 keV for the 1332.5 keV line of ^{60}Co. Samples and standards were counted on the same detector at different times, some counts for as long as 260,000 s (3.0 d). All spectra were analyzed using a commercial peak search and integration program. Actual concentration calculations were made on a spreadsheet program written by one of the authors (DAB), and data were corrected for neutron fluence variations, live time of count, decay during counting, pulse pileup, and background interferences if present.

RESULTS AND DISCUSSION

Initial results presented here are for four elements only: Au, Na, As, and Br. These four elements were reported by all nine participants, either as a concentration or as a "less than" value, and were four of the most found elements. These data are plotted in Figures 1-4. "Less than" values are indicated by a downward pointing arrow connected to the data point. It should be noted that the units of concentration plotted are μg/kg, on a log scale. Each of the four elements will be discussed individually.

Gold was the element found at the lowest concentration level (Figure 1). Five participants reported Au in the total sample, and three of those after cleaning or etching. Au appears to be found primarily as a surface

TABLE 1. Neutron Activation Analysis Analytical Procedure Information

Lab. No.	Sample Size (g)	Pre-clean?	Neutron Fluence	Post-irradiation Clean?	Post-irradiation Etch?	Elements Found Bulk Only	Elements Found Total Sample
1	0.8	No	8E17 [b]	Yes	No	---	Au,As,Na,Br
2	0.5	No	NR [c]	---	Yes	Au,As,Cr,Mn Sb,Zn,O,P	---
3	2.7	No	4E18	Yes	No	---	As,Cu,K,Na Ag,Fe,Sb,Zn
4	8.7	No	1E16	No	No	---	Br,Na,Zn
5	0.2	No	7E19 - 5E16	No	Yes	Na	---
6	0.6	No	1E18	No	No	---	Au,Ag,Na,Zn
7	8.7	No	8E18	No/Yes	No	Au,Br,Ag,Sb Zn	Au,Ag,As,Br,K Co,Cr,Cu,Na Fe,La,Sb,Sc,Zn
8	5.4	No	3E16	No	No	---	Au,As,Br,Na K,Sb,W,La Cr,Sm
9	NR	No	5E17	No	Yes	Au,Na,As	Au,As,Na,Br K,Cr,Zn,Cu Mn,Ni

[a] Results labeled "Bulk" are after etching or very vigorous cleaning in a mixed acid bath with holdback carriers present in the bath. Results labeled "Total Sample" are either the sample after irradiation prior to etching or the sum of the bulk and etch solution results.

[b] Units are: $n \cdot cm^{-2}$

[c] NR = not reported

contamination at 0.01 µg/kg or above, with bulk silicon Au concentration levels several orders of magnitude lower. The lowest reported level was 3×10^{-5} µg/kg (2×10^8 atoms/cm^3) for bulk silicon after etching. Note that for Laboratory 1, two sets of data for each wafer are included. The two filled circles are separate samples from wafer #1, the two open circles are separate samples from wafer #2. This data suggests significant inhomogeneity in the surface contamination, using samples sizes of 0.7 to 1.1 g each.

Sodium was found in these wafers by all the participants (Figure 2). This is due to the indicator isotope for sodium, ^{24}Na, also being produced by a fast neutron reaction (n,αp) on the silicon matrix [3]. This is supported by the results shown, in which Laboratory 1 (which has a very low fast neutron component in their reactor fluence) also obtained the lowest value for Na. Most reactors will provide an apparent Na concentration of 0.6 to 2 µg/kg from the fast neutrons.

Arsenic was also found by most participants (Figure 3). This element had the best overall agreement for total concentration, at 0.034 ± 0.019 µg/kg (wafer 1, n=5; this is equal to 6×10^{11} atoms/cm^3). Two additional laboratories reported "less than" values close to this number.

Bromine was found by 5 laboratories in the total samples, but none after etching. Agreement between these labs is reasonably good (± 100%) at about 0.1 µg/kg (2×10^{12} atoms/cm^3) (Figure 4).

Finally, Figure 5 shows data for 9 elements from one participant (Laboratory 9) who severely etched the wafers (including use of holdback carriers) and then counted the etched bulk material as well as the etchant. They found substantially higher levels of impurities in the etchant than in the bulk material for the three elements seen in the bulk

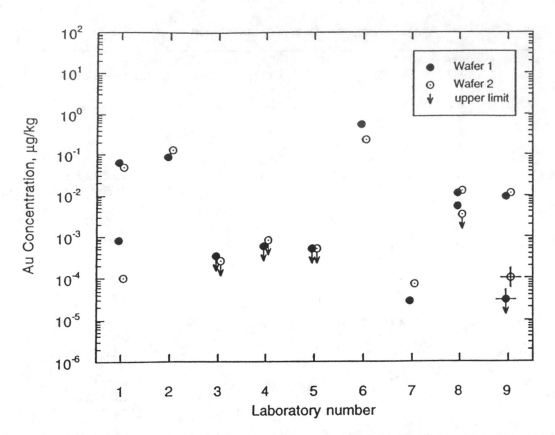

Figure 1. Gold in Silicon Wafers (filled circles are wafer #1; open circles are wafer #2; data points with superimposed + are for the bulk silicon after heavy etching)

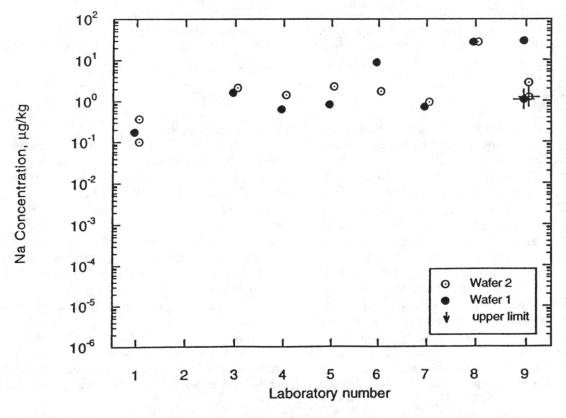

Figure 2. Apparent Sodium in Silicon Wafers (see text) (filled circles are wafer #1; open circles are wafer #2; data points with superimposed + are for the bulk silicon after heavy etching)

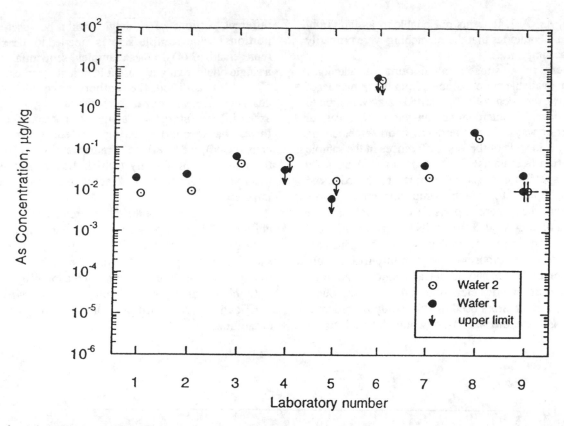

Figure 3. Arsenic in Silicon Wafers (filled circles are wafer #1; open circles are wafer #2; data points with superimposed + are for the bulk silicon after heavy etching)

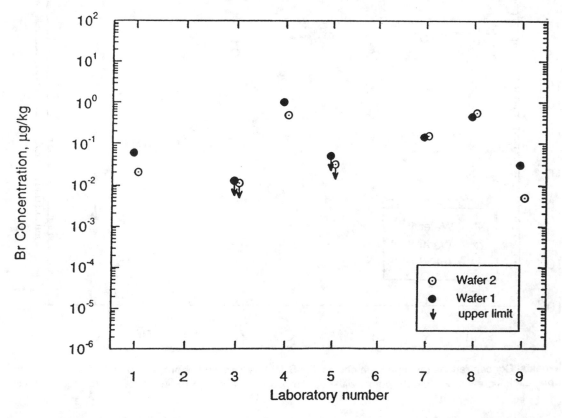

Figure 4. Bromine in Silicon Wafers (filled circles are wafer #1; open circles are wafer #2)

silicon (Na, As, Au). It seems reasonable to assume from this data that laboratories that did no etching were primarily seeing surface impurities.

There is a substantial amount of additional information available to be extracted from the remainder of the data. However, even with the limited data evaluation to date, we believe this intercomparison was very successful in a number of ways. First, there was good agreement on some elements in spite of the vast differences in the sample post-irradiation treatments. Second, it has identified a number of specific questions that need to be addressed among participants to reach some consensus before possible future such intercomparisons can actually answer the question of how well different INAA laboratories can agree when analyzing the "same" sample. These include: (1) which are most important, surface impurities, bulk impurities, or both? (2) should surface contamination be determined before or after cleaning, and how vigorous a cleaning? (3) sample sizes used in this study ranged from 0.2 g to 8.7 g the entire wafer) - - - would the analysis of

different portions of a single wafer be useful? which portions? what sample size is needed to be reasonably representative? (4) do these samples, sequential slices from a single boule, really represent identical samples?

In addition, the authors have identified some analytical needs in our laboratory which would materially assist in any future such analyses. These include a much lower background counting system (currently being constructed), and a series of irradiations in the NIST reactor thermal column facility (which has a negligible fast neutron component), to look at the interference-free sodium impurity.

In conclusion, the data presented here confirms the ability of INAA to provide multielement, high sensitivity ultra trace analyses for impurities in high purity silicon samples. The capability for varied post-irradiation sample treatments is also important. The capability for RNAA with chemical separations and/or INAA with half-life verification are additional useful quality assurance capabilities.

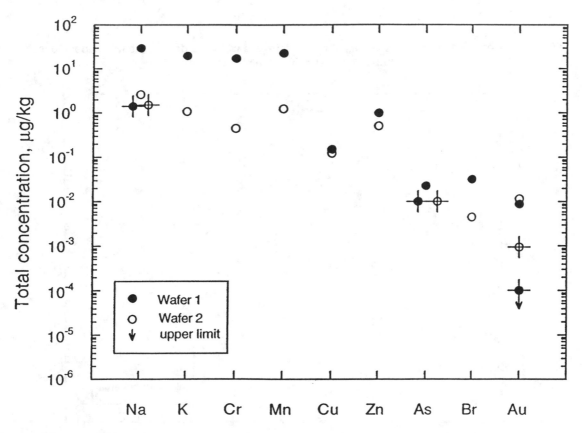

Figure 5. Elemental Concentrations for Nine Elements in the Silicon Wafers by Laboratory #9 ((filled circles are wafer #1; open circles are wafer #2; data points with superimposed + are for the bulk silicon after heavy etching)

ACKNOWLEDGMENTS

The authors would like to acknowledge the participation and cooperation of the following organizations and individuals for the analytical portion of this intercomparison:

Bhabha Atomic Research Center - Bombay, India (S. Gangadharan)

Becquerel Laboratories, Inc. - Mississauga, Ontario, Canada (C. Stuart)

DSM Research - Geleen, The Netherlands (D. Bossus)

Phillips Research Laboratories - Eindhoven, The Netherlands (J.Vrakking)

Texas Instruments, Inc. - Dallas, Texas, USA (J. Keenan)

Univ. of Missouri/Columbia - Columbia, Missouri, USA (M. Glascock)

University of Ulm - Ulm, Germany (V. Krivan)

VTT Reactor Laboratory - Espoo, Finland (R. Rosenberg)

NIST - Gaithersburg, Maryland, USA (D. Becker)

Without their efforts and cooperation this study would not have been possible. Please note that laboratory numbers used in this paper do not correspond in any way to the order of the above listing.

REFERENCES

1. M.N. Schulepnikov, et al., "Interlaboratory Comparison of INAA for Semiconductor Silicon", *J. Radioanal. Nucl. Chem.* **122** (1988) 261-264.

2. R.M. Lindstrom and R.J. Rosenberg (Eds.) "Analytical Chemistry in Semiconductor Manufacturing: Techniques, Role of Nuclear Methods and Need for Quality Control", (IAEA-TECDOC-512), International Atomic Energy Agency, Vienna, 1989.

3. M.L. Verheijke, "Instrumental Neutron Activation Analysis Developed for Silicon Integrated Circuit Technology", Ph.D. Thesis, Tech. Univ. Eindhoven (1992) p.117

Applications of Cold Neutron Prompt Gamma Activation Analysis to Characterization of Semiconductors

Rick L. Paul and Richard M. Lindstrom

Analytical Chemical Division, National Institute of Standards and Technology, Gaithersburg, MD 20899

Cold neutron prompt gamma-ray activation analysis (CNPGAA) has proven useful for the analysis of both major and trace elements in semiconductor materials. The analysis is nondestructive, measures the entire sample, and the results are independent of the chemical form of the element being measured. We have used CNPGAA to determine stoichiometry of samples of mercury cadmium telluride and to measure trace levels of hydrogen in samples of quartz and germanium. Future developments will serve to improve the measurement capability of the technique and to create new applications for semiconductor characterization.

INTRODUCTION

Prompt gamma-ray activation analysis (PGAA) is a versatile multielement technique which has proven useful for analysis of a wide variety of materials. The basics of the technique, described in detail elsewhere (1), are as follows. When a sample is bombarded by a beam of neutrons, nuclei of many elements in the sample capture neutrons and are transformed to an isotope of higher mass number. Upon de-excitation, these nuclei emit prompt gamma-rays, which may be measured using a high resolution gamma-ray detector. Qualitative analysis is accomplished by identification of gamma-ray energies, while comparison of gamma-ray intensities with those emitted by standards yields quantitative analysis. The analysis is nondestructive and, since both the neutron and gamma radiation are penetrating, the entire volume of sample irradiated by the neutron beam is analyzed. Because the analytical signal results from a nuclear and not a chemical reaction, the results are independent of the chemical form of the element being measured. Although most elements may be detected using PGAA, the technique is particularly well-suited for the nondestructive analysis of light elements (H, B, C, N, Si, S, Cl) and elements with high neutron capture cross-sections (Cd, Sm, Gd) which are difficult to measure using other techniques.

Because of poor detection limits, PGAA was once considered a subsidiary technique to traditional neutron activation analysis (NAA). However, due to recent improvements, such as the development of high efficiency, high resolution germanium detectors, PGAA has become an important analytical tool in its own right. Two recent innovations include the use of guide tubes to efficiently extract neutrons from the reactor and the use of low energy "cold" neutrons to increase capture rate. Implementation of these two features produces an instrument with substantially lower background and better detection limits. Such an instrument has been constructed at the National Institute of Standards and Technology.

The NIST cold neutron prompt gamma activation analysis (CNPGAA) instrument has been used to characterize a variety of materials of industrial importance (2- 8). Major element analyses of fullerene derivatives, solid acid catalysts, and hydrofluorocarbons have provided information on the stoichiometry of these compounds. A number of elements (H, B, Cl, Cd, Sm, Gd) have been determined at microgram levels. The nondestructive measurement of trace hydrogen has been of particular importance, since hydrogen is a common impurity which can alter the properties of semiconductors, metals, and other materials (3, 4, 8).

In this paper we discuss the application of CNPGAA to the characterization of semiconductor materials. We discuss the applicability of the technique for determination of major elements and for measurement of trace amounts of hydrogen in a variety of semiconductor materials. We also discuss the impact that new developments, such as the implementation of a neutron focusing lens, will have on future measurements.

APPARATUS

The instrument, located in the Cold Neutron Research Facility (CNRF) at NIST has been described in detail elsewhere (2, 9). Neutrons from the core of the NIST research reactor, slowed by passage through a D_2O ice moderator at 30 K, travel through a guide tube to the CNPGAA workstation in the CNRF. The neutron beam emerges from the guide, and is collimated to 20 mm diameter before striking the sample. Prompt gamma-rays are measured by a high resolution, boral and lead-shielded germanium detector positioned perpendicular to the neutron beam. The Ge detector is surrounded by a single crystal bismuth germanate (BGO) Compton shielding detector.

Gamma-ray signals from the Ge and BGO detectors are processed using appropriate electronics; singles and Compton suppressed spectra up to 10 MeV are displayed on a VAXstation 3100[1]. Data acquisition and display are performed using Canberra Nuclear Data acquisition and display software.

APPLICATIONS

Major Element Analyses

PGAA is suitable for determination of major elements in semiconductor devices. The technique is capable of measuring Si, Ge, Ga, As, In, P, Hg, Cd, Te, and other elements that make up these devices. A future application of PGAA may involve use of the technique for on-line determination of major elements in compound semiconductors. One such application is discussed later in this paper (see **Portable Neutron Sources**).

Measurement of Trace Hydrogen

The presence of trace amounts of hydrogen in semiconductors and related materials can produce changes in electrical and optical properties (10). Although hydrogen detection limit is largely dependent upon the sample matrix, the NIST CNPGAA instrument is capable of detecting less than 10 μg of hydrogen in some materials. We have previously reported the measurement of hydrogen in crystals of hydrothermal quartz and of high purity germanium (11). The results of some of these analyses are discussed below.

Five crystals of hydrothermally grown quartz (weighing from one to four grams) were analyzed for hydrogen by CNPGAA. The crystals had been grown and analyzed by infrared spectroscopy by researchers at the U. S. Army Research Laboratory in Fort Monmouth, New Jersey. Table I gives H concentrations determined by CNPGAA, along with H concentrations estimated from the IR data. Upper limits for CNPGAA analyses are expressed as $\leq 2\sigma$, based on sample counting statistics. Uncertainties were evaluated using guidelines given by Taylor and Kuyatt [12]. Expanded experimental uncertainties were evaluated by combining Type A

uncertainties (those evaluated by statistical means) in quadrature and multiplying by a coverage factor of 2 to give a 95% confidence level. Type A uncertainties originate from sample counting statistics and background subtraction (>25%), standard counting statistics (<1%), sample positioning and neutron fluence variation (~1%), and sample weighing (<1%). Type B uncertainties (those not determinable by statistical means) were negligible. Hydrogen concentrations determined for the quartz samples do not differ within statistical uncertainties from each other or (for 4 of 5 samples) from H concentrations determined by IR data.

In a second study, five samples of semiconductor grade germanium, believed to contain as much as 1 atom of hydrogen per atom of germanium, were also analyzed by CNPGAA. The samples consisted of three ~ 1-gram pieces of germanium which had been electrolytically etched prior to analysis, and two pieces of unetched germanium of the same approximate size. Since germanium is believed to absorb hydrogen readily upon etching, higher hydrogen concentrations were expected in the etched samples. Hydrogen concentrations determined in the samples ranged from 50 ± 40 to 80 ± 55 mg H/kg Ge, or 0.4 to 0.6 atomic % of H . Uncertainties were evaluated as discussed above, except for the addition of a type B peak fitting uncertainty, added to account for uncertainties in peak integration due to the complex baseline of the germanium spectrum (11). No significant differences were observed between etched and unetched germanium samples.

It should be noted that hydrogen concentrations in many of the materials analyzed were near or below the detection limit for CNPGAA. Nevertheless the data have allowed us to determine upper limits for the hydrogen content of these materials. Future development of the instrument will result in lower background and better hydrogen detection limits.

Table I. H concentrations in samples of hydrothermal quartz estimated from both CNPGAA and IR data.

Sample	mg/kg H (CNPGAA)	mg/kg H (IR)
TD4	6 ± 12	5.4
TD7C	≤ 5	16
TD7D	10 ± 12	12
A3LZ	≤ 5	4.9
B4LZ	≤ 6	4.6

---

[1]Certain commercial equipment, instruments, or materials are identified in this paper in order to specify the experimental procedures in adequate detail. Such identification does not imply recommendation by the National Institute of Standards and Technology, nor does it imply that the materials or equipment identified are necessarily the best available for the purpose.

Analysis Of Other Trace Elements

A variety of other elements occur as dopants and impurities in semiconductor materials. Although concentrations of many impurities are too low to be currently measured by CNPGAA the measurement capability is expected to improve with future development of the instrument. Spectral interference free limits of detection for selected elements in quartz are given in Table 2. Poorer limits of detection are expected for these elements in other semiconductor materials.

Table 2. Spectral interference free limits of detection for selected elements by CNPGAA, determined using formulae given by Currie (13). Determined for 1 gram quartz sample counted for 24 hours.

Range, μg	Elements
0.01 - 0.05	B, Gd
0.05 - 0.5	Cd, Sm
0.5 - 5	Hg, Eu
5 - 50	H, Cl, Sc, Ti, Mn, Co, Ag, In, Nd
50 - 100	Na, K, V, Cr, Cu, As, Se, Br, Mo, Te, Au
100 - 500	Al, S, Ca, Fe, Ni, Zn, Ga, Ge, Sr, Y, I
500 - 1000	Sb, Ba, La
1000 - 5000	N, F, Mg, P, Rb, Zr, Sn, Pb
> 5000	C

NEW DEVELOPMENTS

A number of recent innovations will serve to improve the measurement capability of PGAA as well as to create new applications for characterization of semiconductors. Some examples are given below.

Neutron Focusing Lens

A neutron lens, which uses capillary optics to focus the neutron beam onto a spot size of 0.5 mm, has been tested at the CNPGAA work station at NIST. Use of the lens has resulted in an 80-fold increase in neutron current density (relative to the unfocused beam) and a 60-fold increase in gamma-ray signal from submillimeter particles of Gd and Cd (14). A focused neutron beam may be useful for analysis of submillimeter samples, and for compositional mapping of semiconductor materials, in order to look for regions of high concentrations of neutron absorbing elements (such as boron).

Combining PGAA with Other Nuclear Techniques

More complete characterization of semiconductor materials may be obtained by supplementing PGAA data with analyses by other nuclear techniques. These include NAA (which yields better detection limits than PGAA for many elements), small angle neutron scattering, neutron reflectometry, and neutron depth profiling (NDP). A combination of PGAA and NDP has already been used in an attempt to characterize a phosphosilicate glass containing an adsorbed layer of ^{17}O labeled H_2O. In this sample, H was measured by PGAA and ^{17}O by NDP. Future plans include using the two techniques to measure Ti (by PGAA) and N (by NDP) in a titanium nitride thin film.

A unique application of nuclear techniques to the analysis of thin films has been obtained by combining measurement of prompt gamma-rays and neutron reflectivity (15, 16). A polymer film with an embedded Gd thin layer was exposed to a grazing beam of monochromatic, cold neutrons, creating a resonance-enhanced neutron standing wave within the film. Simultaneous measurements of neutron reflectivity and prompt gamma emission from the buried Gd layer enabled determination of a depth profile for the sample.

Portable Neutron Sources

A major limitation to the use of PGAA to characterize samples is that a source of neutrons is needed for the analysis. For this reason PGAA measurements are performed primarily at nuclear reactor facilities. Although PGAA may be performed using neutrons from portable isotopic sources, e. g. ^{252}Cf, sensitivities from instruments using these sources are generally poorer than those which use reactor neutrons, primarily because the neutron flux is lower. Better sensitivities for non-reactor based PGAA may eventually be achieved by using of an isotopic source in conjunction with a cold neutron moderator (17). The use of a PGAA instrument of this type has been proposed for such industrial applications as on-line sorting of aluminum alloys for recycling purposes (5). Such an instrument might also be useful for on-line quality control of semiconductor devices.

A unique possible use of PGAA in manufacturing is the determination of the Hg/Cd ratio in (Hg,Cd)Te. Because both Hg and Cd have large capture cross sections and high gamma-ray yields, the sensitivity for these elements is extraordinarily high. Calculations and model experiments indicate that it should be possible to measure accurately the composition of a HgCdTe ingot as it is being pulled, even with the modest neutron flux available with a ^{252}Cf neutron source. If the composition of the melt from which the crystal is pulled is monitored, then all three components can be measured simultaneously. For a melt that is 17 wt % Hg and 0.7 wt % Cd, the peaks of Cd and Te are in the ratio of 4:1.

CONCLUSIONS AND FUTURE PLANS

Cold neutron prompt gamma-ray activation analysis has proven useful for the analysis of major, minor, and trace elements in semiconductor materials. Future innovations, such as the implementation of a neutron focusing lens and alternate neutron sources may increase the applicability of the technique to semiconductor characterization.

Future improvements in the CNPGAA spectrometer at NIST will serve to improve the measurement capability of that instrument. The replacement of the D$_2$O cold source with a liquid hydrogen source will result in a 5-10 fold increase in neutron capture rate, with a corresponding increase in gamma-ray signal. The implementation of improved neutron and gamma-ray shielding will lower background for hydrogen and other elements, resulting in improved detection limits.

ACKNOWLEDGMENTS

We gratefully acknowledge the cooperation of the NIST Reactor staff during these experiments. We also thank Sam Treviño, R. Aaron Murray, Richard Deslattes and Hayden Wadley for providing samples for analysis.

REFERENCES

1. Lindstrom, R. M., *J. Res. Nat. Inst. Stand. Techn.* **98**, 127-133 (1993).

2. Paul, R. L., Lindstrom, R. M., and Vincent, D. H., *J. Radioanal. Nucl. Chem.* **180**, 263-269 (1994).

3. Lindstrom, R. M., Paul, R. L., Vincent, D. H., and Greenberg, R. R., *J. Radioanal. Nucl. Chem.* **180**, 271-275 (1994).

4. Neumann, D. A., Copley, J. R. D., Reznik, D., Kamitakahara, W. A., Rush, J. J., Paul, R. L., and Lindstrom, R. M., *J. Phys. Chem. Solids* **54**, 1699-1712 (1993).

5. Gardner, R. P., Dobbs, C. L., and Paul, R. L., *Trans. Amer. Nucl. Soc.* **71**, 165-166 (1994).

6. Crawford, M. K., Corbin, D. R., and VerNooy, P. D., *Trans. Amer. Nucl. Soc.* **71**, 168-169, (1994).

7. Krug, F. A., Schober, T., Paul, R. L., and Springer, T., submitted to *Solid State Ionics*.

8. Paul, R. L., Privett, H. M. III, Lindstrom, R. M., Richards, W. J., and Greenberg, R. R., submitted to *Metallurgical Transactions*.

9. Lindstrom, R. M., Zeisler, R., Vincent, D. H., Greenberg,, R. R., Stone, C. A., Mackey, E. A., Anderson, D. L., and Clark, D. D., *J. Radioanal. Nucl. Chem.* **63**, 121-126 (1993).

10. Myers, S. M., Baskes, M. I., Birnbaum, H. K., Corbett, J. W., DeLeo, G. G., Estreicher, S. K., Haller, E. E., Jena, P., Johnson, N. M., Kirchheim, R., Pearton, S. J., and Stavola, M. J., *Rev. Mod. Phys.* **64**, 559-617 (1992).

11. Paul, R. L., and Lindstrom, R. M., *Diagnostic Techniques for Semiconductor Processing*; MRS Symposium Proceedings Volume **324**, ed. O. J. Glembocki, S. W. Pang, F. H. Pollak, G. M. Crean, and G. Larrabee, Materials Research Society, Pittsburgh, PA, 1994, pp. 403-408.

12. Taylor, B. N., and Kuyatt, C. E., "Guidelines for Evaluating and Expressing the Uncertainty of NIST Measurement Results", NIST Technical Note 1297, National Institute of Standards and Technology, U. S. Government Printing Office: Washington, DC, 1993.

13. Currie, L. A., *Anal. Chem.* **40**, 586-593 (1968).

14. Chen, H., Sharov, V. A., Mildner, D. F. R., Downing, R. G., Paul, R. L., Lindstrom, R. M., Zeissler, C. J., and Xiao, Q. -F., *Nucl. Instrum. Meth. B* **95**, 107-114 (1995).

15. Zhang, H., Gallagher, P. D., Satija, S. K., Lindstrom, R. M., Paul, R. L., Russell, T. P., Lambooy, P., and Kramer, E. J., *Phys. Rev. Lett.* **72**, 3044-3047 (1994).

16. Satija, S. K., Zhang, H., Gallagher, P. D., Lindstrom, R. M., Paul, R. L., Russell, T. P., Lambooy, P., and Kramer, E. J., "Resonance Enhanced Neutron Standing Waves in Thin Films", presented at the Materials Research Society Fall Meeting, Boston, MA, November 28 - December 2, 1994.

17. Clarke, D. D., and Hossain, T., *Trans. Amer. Nucl. Soc.* **68**, 141 (1993).

Nondestructive Characterization of Semiconductor Materials Using Neutron Depth Profiling

R. G. Downing and G. P. Lamaze

Nuclear Methods Group
National Institute of Standards and Technology
Gaithersburg, MD 20899

Neutron depth profiling is an isotope-specific, nondestructive technique for the determination of several nuclides of importance to the semiconductor industry. Concentrations of boron, lithium, and nitrogen (and in special cases beryllium, sodium, and oxygen) are determined as a function of depth. Profiles are generated in real time, analyzing the first few micrometers of depth in materials such as silicon, silicon dioxide, telluride compounds, and diamond films; profiles across multilayer insulating device structures are possible. An introduction to the technique is given, with examples from the analysis of semiconductor materials.

INTRODUCTION

The development of the neutron depth profiling (NDP) technique was originally motivated by the need to better determine light element distributions in electronic materials. In 1972, Ziegler and co–workers (1,2) introduced NDP by determining the range and shape of boron implantation distributions in silicon wafers. Biersack and co–workers (3,4) at the Institut Laue-Langevin in Grenoble subsequently improved upon the technique and continue to apply it for material analysis. The most frequent use for NDP remains the study of semiconductor related materials but it is also used to ensure the calibration and in a complementary role for the quality control of other analytical instrument (5-8).

Today the U.S. has NDP facilities at the University of Michigan (9), Texas A&M University (10), University of Texas at Austin (11), and North Carolina State University (12). Furthermore, since 1982 NIST has operated a NDP facility that uses thermal neutrons, (13) and since 1990 a second NDP instrument (14) that uses cold neutrons and is part of the national users facility at the Cold Neutron Research Facility (CNRF). This paper briefly describes the technique and some uses of NDP in the semiconductor industry.

FUNDAMENTALS OF THE TECHNIQUE

Neutron depth profiling is an isotope-specific, nondestructive technique for the measurement of concentration-versus-depth distributions in the near-surface region of solids. The first few micrometers of nearly any material can be probed nondestructively. Lithium, beryllium, boron, nitrogen, sodium, and a few other elements have an isotope that, upon capturing a neutron, undergoes an exoergic charged particle reaction. These reactions produce either a proton or an alpha particle, depending upon the isotope, and a recoiling nucleus. Each particle emitted initially has a characteristic energy defined by the kinematics of the reaction which serves to identify the element.

To obtain a neutron depth profile, a collimated beam of low energy neutrons ($<10^{-2}$ eV) is used to illuminate uniformly a sample volume. The illuminated area may be as large as several cm^2 or less than a mm^2. While most of the neutrons continue through the sample volume, some nuclides capture neutrons in proportion to their capture cross section and produce the isotropic emission of monoenergetic charged particles. The particles travel in nearly straight paths and lose energy primarily through numerous interactions with the electrons of the matrix. The difference between the well-known initial energy of the particle and its residual energy upon

TABLE 1. A Compilation of Nuclear Reactions, Data, and Detection Limits for Neutron Depth Profiling.

Reaction	% Abundance or (atoms/mCi)*	Energy of Emitted Particles (keV)		Useful Range in Silicon (μm)	Cross Section (barns)	Detection Limit† (atoms/cm²)
$^3He(n,p)^3H$	0.00014	572	191	4.7	5333	1.5×10^{12}
$^6Li(n,\alpha)^3H$	7.5	2055	2727	40	940	9.0×10^{12}
$^{10}B(n,\alpha)^7Li$	19.9	1472	840	3.8	3837	2.1×10^{12}
$^{14}N(n,p)^{14}C$	99.6	584	42	5.0	1.83	4.5×10^{15}
$^{17}O(n,\alpha)^{14}C$	0.038	1413	404	3.6	0.24	3.5×10^{16}
$^{33}S(n,\alpha)^{30}Si$	0.75	3081	411	11	0.19	6.0×10^{16}
$^{35}Cl(n,\alpha)^{35}S$	75.8	598	17	5.2	0.49	1.7×10^{16}
$^{40}K^*(n,p)^{40}Ar$	0.012	2231	56	54	4.4	1.9×10^{15}
$^7Be^*(n,p)^7Li$	(2.5×10^{14})	1438	207	26	48000	1.7×10^{11}
$^{22}Na^*(n,p)^{22}Ne$	(4.4×10^{15})	2247	103	55	31000	2.3×10^{11}
$^{59}Ni^*(n,\alpha)^{56}Fe$	(1.3×10^{20})	4757	340	21	12.3	7.0×10^{14}

†Detection limit based on 0.1 counts per second, 0.1% detector solid angle, and a neutron intensity of 6×10^9 s^{-1}
*Radioactive species

emerging from the surface of the sample is nearly proportional to the depth of origin for the particles (i.e., the site of the parent atom). The target chamber is kept under vacuum so that no additional energy is lost by the particle as it travels between the sample surface and the detector. Because the low-energy neutron carries very little momentum, the reaction center-of-mass is entirely coincident with the site of the parent atom.

The depth corresponding to the determined energy loss for the emitted particle is determined by evaluating the characteristic stopping power of the material, as compiled by Ziegler (15) and others (16) or, for compounds, by estimating the stopping power using Bragg's law (17) for addition of the stopping powers for the individual elemental constituents. Only the concentration of the major elements in the material is used to establish the depth scale through the relationship of stopping power and the volume density of the sample. These calculations are discussed in detail elsewhere (18-20).

Damage to the sample is usually negligible as is the temperature rise (<1°C) during an NDP analysis. However, minor lattice damage to the sample can occur during extended analysis in crystalline materials as the passage of charged particles displaces matrix atoms.

DETECTION LIMITS

The detection limit achieved using the NDP method is directly proportional to the incident neutron fluence and to the cross section of the nuclide of interest. The number of charged particle counts collected by the analyzer into a data channel, of energy width dE, is directly proportional to the concentration of target atoms located within that corresponding depth interval of the sample. Upon calibrating the facility for a given nuclide reaction, concentrations can be measured for that isotope (or other similar reactions) in subsequent samples, independent of the matrix, the concentration level, or sample depth, provided that the induced particles can escape the sample surface and be detected. Table 1 lists several properties for target atoms and the nominal detection limits obtained using the CNDP facility at NIST. As can be seen for the case of boron in silicon, concentrations down to the level of 10^{12} atoms/cm² can be readily determined.

The time required for an analysis is a function of the element, its concentration, and the desired precision. A boron implant of 1×10^{15} atoms/cm² typically takes a few hours to obtain a total uncertainty of the order of 2.5% (1σ) and in more optimum cases at the level of 2%. Since background interference is almost negligible, a sample can be counted for a longer period of time to obtain additional definition in the profile shape. A computer-based data acquisition system displays the depth profile in real time, to enable the analyst to ascertain when sufficient statistics have been obtained.

DEPTH RESOLUTION

Deviation of the measured elemental profile from the actual distribution is significantly limited by the energy resolution of the detector and associated electronics. Because the reactions are nuclear, neither the chemical or electrical state of the target material has an effect on the measured profile in the NDP analysis. However there are several factors that contribute to the depth resolution including: i) energy straggling of the charged particles, ii) small-angle scattering of the particles within the sample and iii) the finite acceptance angle of the detector giving a spread in path lengths for particles from the same depth. Contributions to the degradation of resolution are discussed by Biersack et al. (18) and others (20,26).

Each material has a characteristic stopping power and, therefore, the resolution and the profile depth will vary accordingly. Deconvolution algorithms used to unfold the system response function from collected energy spectra (1,18,21-26) have contributed substantially to the improvement of depth resolution. Nonetheless, the more a priori knowledge about the sample, the better modeling of the spectrum can approach the true profile distribution.

THE NIST COLD NDP INSTRUMENT

The thermal equivalent neutron fluence rate at the target position was 1.2×10^9 cm^{-2} s^{-1}. However, a liquid hydrogen moderator is currently being installed at the NIST reactor. When the reactor resumes operation the NDP instrument will be equipped with a curved neutron guide and remain in the reactor experimental hall. It is anticipated the neutron reaction rate will increase at the target position by a factor of 5 or greater.

The NDP instrument is a high vacuum stainless steel chamber. It is evacuated with a 180 l/s magnetic bearing turbo-molecular pump that enables rapid pump down during sample changes. The chamber itself is a 610 mm diameter cylinder with numerous flanges making it possible to add new features to the chamber as needed.

The chamber is being modified to accept samples up to 300 mm in diameter. A rotary base selects the angle of the sample with respect to the beam. Additional motor control allows any location on the surface of the sample to be analyzed. The detectors are transmission-type surface barrier detectors in a ring mount. A detector can be placed by motor control at any angle and additional detectors can be mounted at 10 degrees intervals. All the positioning devices are controlled by a microcomputer which permits unattended sample scans. A separate multi-user minicomputer can simultaneously process data from both the thermal and the cold NDP facilities.

A number of measures are taken to ensure quality control during each analysis. A pancake fission chamber mounted on the entrance port of the NDP chamber provides a run-to-run neutron monitor. Furthermore a reference pulse is introduced into the system to monitor the stability of the electronics. Elemental concentrations are determined by taking a ratio between the signal of the sample and a rigorously evaluated boron concentration standard.

APPLICATIONS

Determinations by NDP induce negligible damage to most materials. Sample surfaces are neither sputtered, as observed with SIMS, nor is the sample matrix greatly altered. Consequently, the same sample may be subjected to a sequence of processing conditions and examined by NDP at each stage. Alternatively the sample may at any point be submitted for analysis by another analytical method to obtain complementary data.

As an example, recently NDP was used to certify the concentration and confirm the profile of boron in silicon for the NIST Standard Reference Material (SRM 2137). The SRM is for the calibration of secondary ion mass spectroscopy (SIMS) measurements. Details of this work will be described elsewhere by Simons et al. (27).

For a thin dielectric, such as a borosilicate glass on a silicon wafer, a single NDP spectrum is able to reveal the film thickness, the boron distribution profile, the total amount of boron present and the weight percent of boron at any given depth. In addition, NDP has been used to follow the effect of reflowing the glass film on the original boron profile, including boron loss and diffusion into the substrate (28) during subsequent wafer annealing. The latter determination can be used to infer the amount of trapped reaction products and the volume of trapped voids occluded in the glass film.

Figure 1 is a plot of NDP analyses for a series of boro-phosphosilicate (BPSG) films, each of about 0.8 micrometer thickness deposited on Si wafers. The series

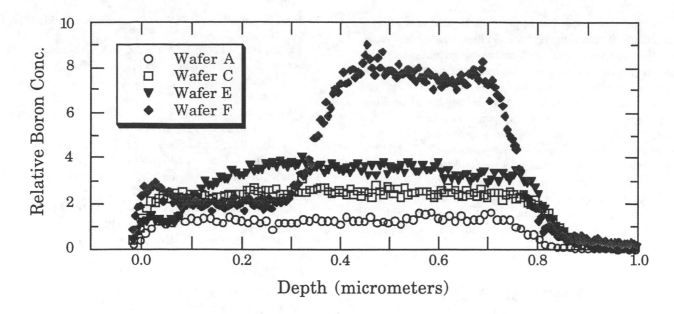

Figure 1. An NDP analysis of a series of borophosphosilicate films on silicon wafers. Boron depletion occurs at the near surface of the those wafers with the highest boron concentration.

of films are of increasing boron concentration until the solubility of the boron in the glass is exceeded. The two profiles obtained from the films of highest B concentration show evidence of near-surface boron loss. This and similar studies (29) have shown that the phosphorus concentration has a direct influence on the chemical stability of BPSG film.

Other applications of NDP include studies of the implantation profiles of ^{10}B in mercury cadmium telluride, an infrared detector material (8,25), or photoresists to determine implant range values necessary for device fabrication. Using the reaction ^{17}O(n,α)^{14}C, the diffusion of water enriched with ^{17}O can be followed through dielectric layers by the use of NDP. Neutron depth profiling is particularly well suited for measurements that require traversing interfacial boundaries of material with different ionization or electrical characteristics. Kvitek et al. (22) and others (30) have studied profiles of boron implanted and diffused across the interfacial region of Si/SiO₂ (31). Likewise, Downing et al. (32) have shown that the native oxide (1-1.5nm) that appears on Si surfaces is contaminated with boron at a level of 10^{12} atoms/cm^{2}.

Nitrogen profiling by NDP is becoming an important activity in the analysis of oxynitrides, TiN, BN and Si₃N₄ compounds. Independently, or in combination with other analytical methods, NDP is able to determine elemental stoichiometry. This has been of particular importance in the study of BN and other diamond-like films as discussed by Lamaze et al. in these proceedings (33).

NEW CAPABILITIES

Improvements to detection limits for NDP are being achieved by the development of more intense neutron beams and changes in basic instrumental design.

The development of two different coincidence techniques for NDP (37-39) substantially reduces low energy background interferences. The depth resolution is somewhat improved because the dependency upon solid angle of acceptance to the detector is reduced. Moreover the collection efficiency can be increased by a factor of 10 or more by bringing the detector closer to the sample. The major disadvantage is that only select sample conditions permit the use of these two methods.

Both neutron intensity and gain in spatial resolution are greatly improved using a neutron focusing device recently developed at NIST (40). Long wavelength neutrons are guided to an area less than a mm in diameter providing a locally high neutron fluence rate. The development of 2- and 3-dimensional neutron depth profiling is an obvious next step.

SUMMARY

NDP provides an isotope-specific, nondestructive technique for the measurement of concentration-versus-depth distributions in the near-surface region of solids. The simplicity of the method and a few of its applications

349

have been described. Major points to be made for NDP as an analytical technique include: i) it is nondestructive; ii) isotopic concentrations are determined quantitatively; iii) profiling measurements can be performed in essentially all solid materials with depth resolution and depth of analysis being material dependent; iv) NDP is capable of profiling across interfacial boundaries; and v) there are few interferences. The profiles are generated in real-time, analyzing depths of up to tens-of-micrometers. NDP is applicable to diverse areas of semiconductor research, as presented here and in the references. With the CNDP facility the ability to obtain profiles for chlorine is now possible. Neutron focusing optics will add improved detection limits and applicability to the technique of neutron depth profiling.

ACKNOWLEDGMENTS

The authors gratefully acknowledge the numerous colleagues, collaborators, and reactor personnel both within and external to NIST, that have influenced the successful development of the NDP program at NIST.

REFERENCES

1. Ziegler, J.F., Cole, G.W. and Baglin, J.E.E., *J. Appl. Phys.* **43**, 3809-3815 (1972).

2. Ziegler, J.F., et al., *J. Appl. Phys.* **21**, 16-17 (1972).

3. Biersack, J.P. and Fink, D., *Nucl. Instr. and Meth.* **108**, 397-399 (1973).

4. Fink, D., Wang, L., Biersack, J.P., and Jahnel, F., *Rad. Eff. and Defects in Solids* **15**, 93-112 (1990).

5. Ehrstein, J.R., et al., "Comparison of Depth Profiling B-10 in Silicon Using Spreading Resistance Profiling, Secondary Ion Mass Spectrometry, and Neutron Depth Profiling" in *Proceedings of Third Symposium on Semiconductor Processing* pp. 409-425 (1984).

6. Fink, D., *Radia. Eff.* **106**, 231-264 (1988).

7. Cox, J.N., Hsu, R., McGregor, P.J. and Downing, R.G., "NDP and FTIR Studies of Borophosphosilicate CVD Thin-Film Glasses" *Proceedings of the American Nuclear Society* **55** pp. 207-209 (American Nuclear Society, Los Angeles, CA, 1987).

8. Jamieson, D.N., et al., "Study of Boron Implantation in CdTe" pp. 299-304 (Materials Research Society, Pittsburgh, PA, 1988).

9. Myers, D.J., Halsey, W.G., King, J.S. and Vincent, D.H., *Radia. Eff.* **51**, 251-252 (1980).

10. Khalil, N.S., Design, Installation and Implementation of a Neutron Depth Profiling Facility at the Texas A&M Nuclear Science Center (Texas A&M University, 1989).

11. Ünlü, K. and Wehring, B.W., *Nucl. Instr. Meth.* **353**, 402-406 (1994).

12. Parikh, N.R., Chu, W.K., Wehring, B.W. and Miller, G.D., "Boron-10 Distribution in Silicon, $TiSi_2$, and SiO_2 Using Neutron Depth Profiling" pp. 211-212 (American Nuclear Society, Los Angeles, CA, 1987).

13. Downing, R.G., Fleming, R.F., Langland, J.K. and Vincent, D.H., *Nucl. Instr. Meth.* **218**, 47-51 (1983).

14. Downing, R.G., Lamaze, G.P., and Langland, J.K., *J. Res. Natl. Inst. Stand. Technol.* **98**, 109-126 (1993).

15. Ziegler, J.F., "*The Stopping and Ranges of Ions in Matter*" Pergamon Press Inc., New York, 1977.

16. Janni, J.F., Atom. *Nucl. Data Tabl.* **27**, 147-529 (1982).

17. Thwaites, D.I., *Radia. Res.* **95**, 495-518 (1983).

18. Biersack, J.P., Fink, D., Henkelmann, R. and Müller, K., *Nucl. Instr. Meth.* **149**, 93-97 (1978).

19. Fink, D., Biersack, J.P. and Liebl, H. in *Ion Implantation: Equipment and Techniques* (eds. Ryssel, H. and Glawischnig, H.) 318-326 (Springer-Verlag, Berlin, 1983).

20. Maki, J.T., Fleming, R.F. and Vincent, D.H., *Nucl. Instr. Meth.* **B17**, 147-155 (1985).

21. Bogáncs, J. et al., *Joint Instit. for Nucl. Res. Ann. Rpt.* **1**, 59-64 (1979).

22. Kvitek, J., Hnatowicz, V. and Kotas, P., *Radiochem. Radioanal. Lett.* **24**, 205-213 (1976).

23. Ryssel, H. et al., *IEEE Trans. on Elect. Dev.* **ED27**, 1484-1492 (1980).

24. Nagy, A.Z. et al., *Physica Status Solidi* (a) **61**, 689-692 (1980).

25. Cervená, J., et al. *Tesla Elect.* **14**, 16-20 (1981).

26. Coakley, K.J., Downing, R.G., and Lamaze, G.P., "Modeling Detector Response for Neutron Depth Profiling" (to be published NIM 1995).

27. Simons, D.S., Chi, P.H., Downing, R.G., and Lamaze, G.P., to be published

28. Downing, R.G., Maki, J.T. and Fleming, R.F., in *Microelectronics Processing: Inorganic Materials Characterization* (ed. Casper, L.A.) 163-180 (ACS, Washington, D.C., 1986).

29. Zeitzoff, P.M., Hossain, T.Z., Boisvert, D. M., and Downing, R.G., *J. Electrochem. Soc.* **137**, 3917-3922 (1990).

30. Nagy, A.Z. et al., *J. Radioanal. Chem.* **38**, 19-27 (1977).

31. Vandervorst, W., Shepherd, F.R., and Downing, R.G., *J. Vac. Sci. and Techn.* **A3**, 1318-1321 (1985).

32. Downing, R.G., et al., *J. Appl. Phys.* **67**, 3652-3654 (1990).

33. Lamaze, G.P. and Downing, R.G., These Proceedings.

37. Parikh, N.R. et al., *Nucl. Instr. Meth.* **B45**, 70-74 (1990).

38. Fink, D. et al., *Nucl. Instr. Meth.* **B15**, 740-743 (1986).

39. Welsh Jr. , J.F., Schweikert, E. A., Lamaze, G. P., and Downing, R.G. (submitted NIMB 1995)

40. Xiao Q.F., et al., *Rev. Sci. Instrum.* **65**, 3399-3402 (1994).

Boron Analysis in Synthetic Diamond Films Using Cold Neutron Depth Profiling

G.P. Lamaze and R.G. Downing

Nuclear Methods Group
National Institute of Standards and Technology
Gaithersburg, MD 20899

We have been engaged in an effort to measure the concentrations and depth profiles of boron-doped synthetic diamond and diamond-like films. These materials are of interest because of their potential as semiconducting materials. Diamond samples were obtained from the Pennsylvania State University, the University of Konstanz and the National Institute of Standards and Technology. Some of the samples were homoepitaxially grown on natural diamonds and others were grown on silicon substrates. The boron contents were analyzed at the Cold Neutron Depth Profiling Instrument of the Cold Neutron Research Facility at the NIST research reactor. Through the use of neutron depth profiling (NDP), both the concentration and the distribution of certain elements in the surface layers of the diamond can be determined. The most easily analyzed elements by this technique are boron, lithium, and nitrogen, all elements that have been suggested as dopants for diamond semiconductors. The NDP technique is non-destructive, allowing distribution measurements before and after any modification processes, such as annealing. Results of the measurements are presented as well as some details of the analysis technique.

INTRODUCTION

Because of their physical properties, much interest has been shown in diamond and diamond-like materials to construct high-current semiconductor devices. Among the most important of these physical properties are the highest known thermal conductivity (20 W/cm/K), wide energy gap (5.5 eV) and high breakdown fields (10^7 V/cm) (1). Natural type II diamond crystals are known to be semiconductors with boron as the dominant acceptor (2,3) with an activation energy of approximately 0.3 eV. Recent efforts have concentrated on introducing the boron during the synthesis of thin diamond and diamond-like films. Fujimori et al (4) have shown that boron doping can be accomplished during gas phase growth by adding B_2H_6 to the gas mixture. Geis et al (5) report the fabrication of Schottky diodes on high-temperature high-pressure process synthetic diamonds that operate in air up to 580 °C and in inert atmospheres up to 700 °C. Gildenblat et al (6) report testing Schottky diodes on chemical vapor deposited (CVD) diamonds at temperatures up to 580 °C and field-effect transistors at temperatures above 300 °C.

The technique of neutron depth profiling (NDP) allows one to measure both the concentration and distribution of boron in CVD diamonds. Depth profiles of up to one micrometer are measured with little difficulty. The technique is non-destructive, which allows sample modification and remeasurement or comparison with other profiling techniques. This paper will review NDP measurements made at NIST on thin diamond films.

NDP MEASUREMENTS

The depth profile measurements were made at the Cold Neutron Depth Profiling (CNDP) Instrument at the National Institute of Standards and Technology (NIST) research reactor. The CNDP instrument is part of the Cold Neutron Research Facility (CNRF) which consists of a cold neutron source, a neutron guide hall with seven guides and the associated instruments. Both the CNDP instrument and the CNRF have been described in detail elsewhere (7,8). "Cold" neutrons improve the detection limit of the NDP method, the sensitivity of which is directly proportional to the cross section of the reaction of interest. In the low energy region, these cross sections are inversely proportional to the square root of the neutron energy. That is, the lower the neutron energy, the greater the reaction rate. By using "cold" neutrons one can approximately double the reaction probability.

More details of the NDP technique are given in an accompanying article in these proceedings (9). The technique was originally applied in 1972 by Ziegler et al (10). The method involves positive Q-value nuclear reactions of neutrons with low-Z elements that yield charged particle decay products. Because the neutron undergoes few interactions as it penetrates these relatively thin targets, the neutron fluence rate is essentially the same at all depths. Since the neutron energies are in the meV range and the Q-values of the reactions are in the MeV range, the momentum and kinetic energy of the neutron can be ignored. To date, we have examined diamond layers for both boron and nitrogen concentration. For the boron reaction, $^{10}B(n,\alpha)^{7}Li$, there are two outgoing alphas with energies of 1.472 (93% branch) and 1.776 MeV (7%), and two corresponding recoil ^{7}Li nuclei with energies of 0.840 and 1.014 MeV. For the nitrogen reaction, $^{14}N(n,p)^{14}C$, there is a 584 keV proton and a 42 keV ^{14}C recoil. Particles escaping from the surface of the sample are detected by a silicon surface barrier detector. The charge deposited in the detector is directly proportional to the energy of the incoming particle.

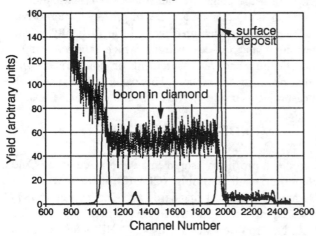

FIGURE 1. Spectra of a boron surface deposit and a boron doped CVD diamond.

Figure 1 gives typical energy spectra for samples used in this experiment. The line indicated as a surface deposit is the spectrum for a thin boron sample evaporated onto a silicon wafer. This well characterized sample provides both a concentration standard and an energy calibration. The other line is for a CVD diamond uniformly doped with boron. The reactions occur at and below the surface. The charged particles lose energy in exiting the material and the energy loss is directly related to the depth of the originating boron atom. This depth can be

determined by using the stopping power of the material, as compiled by Ziegler (11) or Janni (12). Figure 2 gives the calculated relationship between the residual energy of the charged particle and the originating depth of the reaction, in this case for boron in diamond. Using this relationship, the energy spectra can be converted to a plot of concentration vs depth.

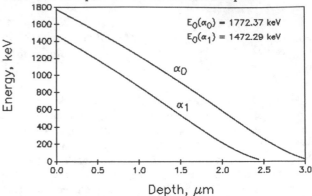

FIGURE 2. Residual energy vs originating depth for boron in diamond.

The example shown in figure 3 is of three different concentrations of boron in CVD thin diamond films, supplied by Ed Farabaugh of the NIST Ceramics Division. In this plot, the energy scale has been converted to a depth scale. Comparisons over two orders of magnitude are easily made and one of the samples is seen to vary by about a factor of two near the surface.

Since NDP is non-destructive, a sample can be modified and then remeasured. The next sample was prepared to test the effect of annealing on the stability of the incorporated boron. Sample 496 is a diamond multilayer on a silicon substrate with a doped layer between two undoped layers. The sample was prepared at the Inter-departmental Materials Research Laboratory at the Pennsylvania State University (13). The profile was first measured as prepared, after which the sample was annealed in vacuum to a temperature of 900 °C for five hours. The profile was again measured and the sample annealed again in vacuum to a temperature of 1100 °C for five more hours. A final depth profile was then measured. Figure 4 shows the results of the three depth profile measurements. The profiles of the as prepared and the 900 °C annealed sample are indistinguishable. After the 1100 °C anneal, a slight shift (~30 nm) of the boron toward the sample surface was observed. The Raman spectra of this sample did not show any changes as the film was annealed. The diamond peak at 1332 cm^{-1} remained dominant with no discernible onset of peaks at 1580 cm^{-1} and 5890 cm^{-1} (GR-1). It was concluded that the diamond film was not graphitized

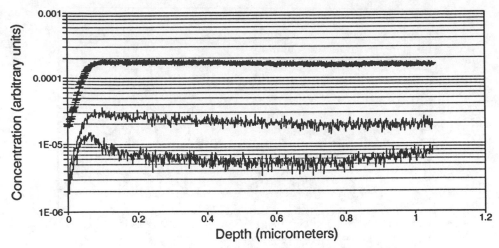

FIGURE 3. Boron distribution in three different CVD diamond films.

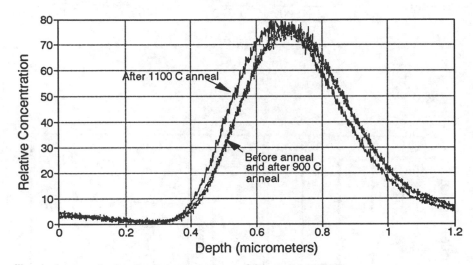

FIGURE 4. Boron profiles before and after heat treatments at 900 and 1100 °C.

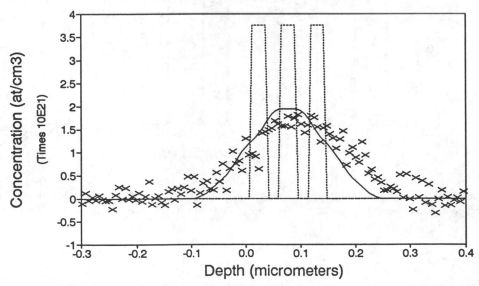

FIGURE 5. Nitrogen profile of ion implanted DLC film. The dotted line represents the expected nitrogen distribution, the solid line the expected distribution after smearing with the resolution function, and the points the measured distribution.

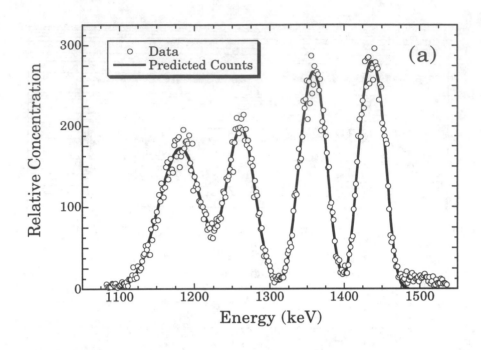

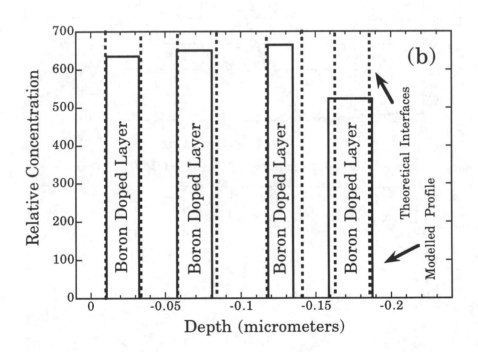

FIGURE 6. a) Measured boron distribution in a DLC multilayer together with a fit to this distribution based on a modified Bohr model for straggling. **b)** The extracted boron distribution based on the data in part a) and the modified Bohr model.

354

with annealing at these temperatures in a vacuum nor was there appreciable diffusion of boron away from the vacancy interstitial pairs that commonly yield a GR-1 band in diamond. Further tests are being performed to analyze the effects of this shift in boron concentration.

An example of nitrogen measurements in diamond-like carbon (DLC) is shown in figure 5. This sample was produced at the University of Konstanz (14) and is a DLC multilayer with alternate layers doped with nitrogen. Because the energy loss per unit depth of the proton measured in the nitrogen analysis is small, the depth resolution for nitrogen is poor, about 140 nm FWHM. However, we were able to measure the concentration and confirm the distribution within the limits of the instrumental resolution. A time-of-flight NDP system is under development (15) that will greatly improve the depth resolution for near surface nitrogen. This system measures the flight time of the ^{14}C recoil nucleus giving much higher energy resolution.

Another example from work on diamond-like carbon is shown in figure 6. Here a DLC multilayer was produced on a silicon substrate at the University of Konstanz (16). Boron doping occurred in alternate layers. To improve the depth resolution, the outgoing alpha particle was measured at 70° to the normal. This increases the path length for each depth (by a factor of $1/\cos(70°)=2.9$) giving greater energy loss per unit depth. The four boron containing layers were resolved and the amount of boron in each layer was determined. The increased path length does however lead to increased straggling, making the original distribution less well defined. By modeling the detector response function, the boron distribution, the stopping power of the material and the particle straggling, we can improve our ability to determine the original boron distribution from the alpha particle spectrum. Kevin Coakley of the Applied Mathematics Division at NIST has developed a program to unfold boron distributions from the alpha particle spectra (17). Figure 6a shows the as measured spectrum plotted as a function of the outgoing particle energy together with a fit using a modified Bohr model for the straggling term; figure 6b shows the boron distribution obtained from the fit along with the positions of the layers as given by the manufacturer. This good agreement gives us increased confidence in our ability to unfold complicated distributions.

SUMMARY

Several examples have been given of the use of neutron depth profiling with cold neutrons to analyze the distribution of boron in synthetic thin diamond films. The technique is nondestructive and even after many hours of irradiation, no detectable residual radiation is present in the diamond films. For samples of sufficient boron content, off angle counting can be used to obtain 20 nm depth resolution. Concentrations of boron in diamond of 10^{18} at/cm^3 are measured routinely. Measurements of nitrogen in diamond-like carbon have also been made. Measurements of lithium in diamond would be easily performed, unfortunately no one has yet succeeded in incorporating lithium in thin diamond films during the growth phase.

ACKNOWLEDGEMENTS

The authors would like to acknowledge the invaluable contributions of our coworkers on these various measurements: Ed Farabaugh at NIST; Larry Pilione, Andre Badzian and Theresa Badzian at PSU; and Hans Hofsass and Carsten Ronning at the Univ. of Konstanz.

REFERENCES

1. Grot,S.A., Gildenblat,G.Sh., Hatfield,C.W., Wronski,C.R., Badzian,A.R., Badzian,T. and Messier,R.: *IEEE Elec. Device Lett.*, 11(1990)100.
2. Custers, J.F.H.: *Physica (UTR)*, 18 (1952) 489.
3. Chrenko, R.M.: *Phys.Rev. B*, 7 (1973) 4560.
4. Fujimori,N., Imai,T., and Doi,A.: *Vacuum*, 36(1986) 99.
5. Geis,M.W., Rathman,D.D., Ehrlich,D.J., Murphy,R.A., and Lindley,W.T.: *IEEE Elec. Device Lett.*, 8(1987)341.
6. Gildenblat,G.Sh., Grot,S.A., and Badzian,A.: *Proc. of the IEEE*, 79(1991)657-668.
7. Rowe, J.M., *NIST J. of Res.* 98(1993)1.
8. Downing,R.G. and Lamaze,G.P., Langland,J.K., and Hwang,S-T.: *NIST J. of Res.* 98(1993)109.
9. Downing,R.G. and Lamaze,G.P., these proceedings
10. Ziegler,J.F., Cole,G.W., and Baglin,J.E.E: *J. Appl. Phys.* 43 (1972) 3809.
11. Ziegler,J.F.: "The Stopping and Ranges of Ions in Matter" Pergamon Press Inc.: New York, 1977.
12. Janni,J.F.: *Atomic Data and Nuclear Data Tables*, 27(1982)147.
13. Lamaze,G.P., Badzian,A., Badzian,T., Pilione,L., and Downing,R.G.: *Advances in New Diamond Science and Technology*, accepted for publication.
14. Ronning, C., Griesmeier, U., Gross, M., Hofsass, H.C., Downing, R.G., and Lamaze, G.P.: *Diamond and Related Materials*, submitted
15. Welsh,J.F., Schweikert,E.A., Lamaze,G.P., and Downing, R.G.: to be submitted to N.I.M.
16. Hofsass,H.C., Biegel,J., Ronning,C., Downing,R.G., and Lamaze,G.P.: *Proc. of the Materials Research Society*, Boston (1993) in press.
17. Coakley, K.J., Downing, R.G., Lamaze, G., to be submitted to N.I.M.

ELECTRON AND ION BEAM TECHNIQUES (INCLUDING **SIMS**)

Transmission Electron Microscopy in VLSI: Strengths and Limitations

Young O. Kim

AT&T Bell Labs, Holmdel, NJ 07733

The Transmission Electron Microscope (TEM) has been a standard tool for the characterization of semiconductor materials and devices. Especially as device dimensions decrease into the deep submicron regime, electron microscopic diagnostics of devices and processes become increasingly important. The role of microscopy varies from assisting module development, which requires rapid turn-around and high throughput, to complete device diagnostics, which requires examination of precisely determined regions, often at atomic resolution. Sometimes chemical information from very small areas is needed to solve unexpected puzzles. Moreover quantitative studies are necessary to establish a correlation between the physical and the electrical characteristics of the same device. In this paper, I will discuss the current capabilities and future potential of electron microscopy.

INTRODUCTION

Very large scale integration (VLSI), which packs millions of transistors on to a chip, has been achieved by the continued reduction of the minimum device size and increased complexity. The manufacturability of this advanced technology requires high reliability and yield. Total process yield (Y) depends on the yield of each process step (y_n), and can be roughly estimated by the product of each process step yield ($Y = y_1 y_2 y_3 \cdots y_n$). Since the total number of the process steps is more than 1000 in a typical CMOS process, each process step needs to be controlled tightly to reach a reasonable yield. As a rough estimate, to obtain 80% total yield for 1000 process steps, the individual process yield should be better than 0.9997. Most of all, this needs to be achieved through the understanding of each process step and the related effects.

Microscopy is an essential tool in this task. This can range from assisting module development and examining complete devices, to fundamental quantitative studies which can give direct correlation between the electrical properties and the physical characteristics of the same device. The equipment we can use ranges from optical microscopes to scanning electron microscopes (SEM) and TEM with advanced sample preparation equipment. Especially, in VLSI process development, high resolution TEM is often required to solve the important problems.

We have developed and applied techniques ranging from TEM sample preparation methods to new analytical techniques which make it possible to measure dopant distributions, roughness at buried interfaces, and the composition change across interfaces quantitatively. This paper will discuss how microscopy can be used through examples derived from current frontier technologies, and introduce some of new techniques which might become useful in near future.

MODULE DEVELOPMENT

In module development, many samples need to be examined quickly to determine the optimum process parameters in each process step. SEM has been widely used for module development, because sample preparation and the operation of the microscope are both relatively simple and fast. As an example, consider the poly-buffered LOCOS (PBLOCOS) process in the deep submicron regime. This process is used to isolate devices from each other, and the purpose is to make the smallest possible isolation with low leakage. As shown in a schematic diagram (Fig. 1), a nitride layer is deposited on the poly-Si and oxide layers, patterned, field oxide grown, and the nitride removed. To optimize this process, 70 different areas in 14 sets of sample had to be examined. This can be done in about two days by SEM. One of the optimized isolations is shown in Fig. 2.

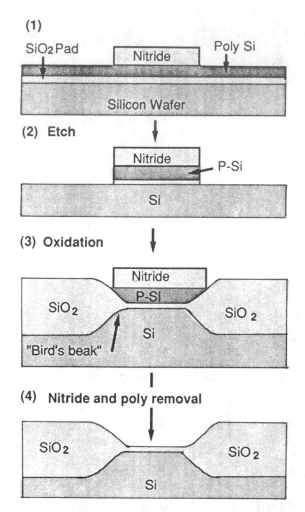

Fig. 1. A schematic diagram showing the poly-buffered LOCOS process.

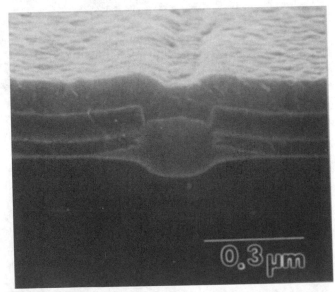

Fig. 2. SEM micrograph of a PBLOCOS with leakages as low as 1 pA/μ.

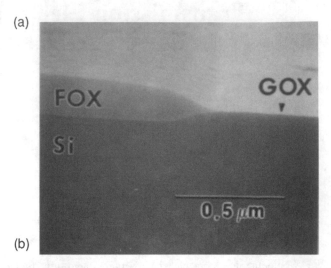

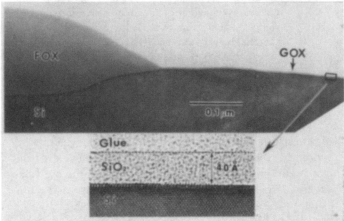

Fig. 3. (a) SEM and (b) low magnification TEM and HRTEM images of a PBLOCOS isolation.

SEM is useful in giving rapid turn-around, but to obtain very fine scale information (especially in the deep submicron regime), TEM examination is necessary. Fig. 3 shows SEM, low magnification TEM, and high resolution TEM (HRTEM) images of a PBLOCOS isolation. The bird's beak in the SEM looks short and generally fine, but low magnification TEM and HRTEM images reveal substantial interface roughness on micron and 100 Å length scales. This kind of roughness is highly undesirable.

COMPLETE DEVICES

After each lot is completed and the electrical characteristics are measured, physical examination is done with TEM testers. A TEM tester typically includes MOSFETs, capacitors, windows, etc, with various lateral dimensions (lengths), but all stretched in one direction to

200 to 500 μm width, to make TEM sample preparation easier. However the results obtained from TEM testers can sometimes be misleading, because the outcome of a process can depend on the size of region being processed. Therefore, it is important to design TEM testers considering this kind of proximity effect, or investigate the possibility of conducting electrical measurements on TEM testers. Real devices can also be examined, although sample preparation is rather time consuming. Fig. 4 shows a lattice image of an actual 0.1 μm MOSFET. The channel is only about 200 atoms long and about 10 μm wide. The image reveals unexpected bird's beak under the gate, and gate oxide thickness variations, which can affect the device characteristics. Excellent electrical characteristics have been measured from experimental 0.13 μm N and P-MOSFETs produced at AT&T Bell Labs [1-3]. The TEM cross-sections of these MOSFETs have shown the expected gate length and excellent gate definition.

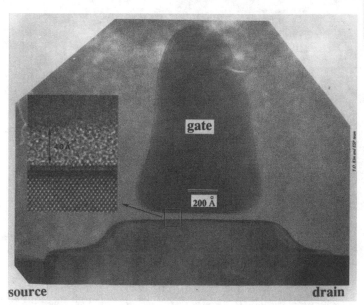

Fig. 4. High resolution TEM of a deep-submicron n-channel MOS transistor. The inset shows the lattice image of the 40 Å thick gate oxide, sandwiched between the poly-Si gate and the underlying Si.

TEM is useful not only for revealing the physical structure, but also for quantitative studies. Quantitative correlation of the physical and electrical properties of the same devices can elucidate the fundamental mechanisms for device operation, and help determine the margin of critical processes. As an example, in advanced bipolar devices consisting of poly-Si emitters on the top of single crystalline Si emitters, an interfacial oxide layer (~10-20 Å) is naturally formed between poly-Si emitter and single crystalline emitter during the process. Charge transport through this interfacial oxide layer determines gain and emitter resistance, directly affecting device speed.

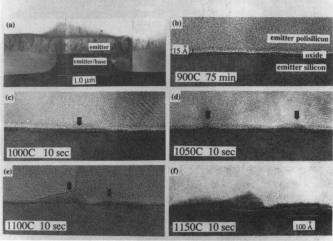

Fig. 5. (a) Low magnification and (b-f) high resolution lattice images of the polysilicon/silicon interface before and after RTA at four different temperatures.

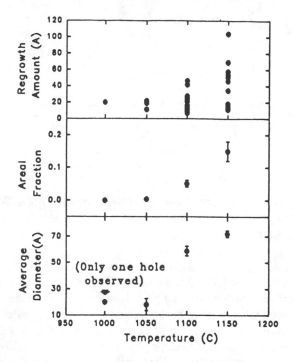

Fig. 6. The areal fraction of pinholes, the radius of pinholes, and the amount of epitaxial growth as a function of annealing temperature.

361

Therefore, it is important to optimize the physical properties of the interfacial oxide between poly and single crystalline Si. We found the interfacial oxide is broken and Si regrows through the pinholes during annealing. Fig. 5 are low magnification TEM and high resolution lattice images, showing the interfacial oxide layer between poly and single crystalline Si before and after rapid thermal annealing (RTA) at different temperatures. Before RTA, the interface was covered by a uniform ~15 Å thick oxide. By annealing for 10 seconds at each temperature, the interfacial oxide is broken and Si regrows through the pinholes. The number of pinholes and amount of epitaxial growth increases as the annealing temperature increases.

and the amount of epitaxial growth as a function of annealing temperature. Fig. 7 shows the correlation between the physical (areal fraction of pinholes) and electrical (emitter resistance and base current) characteristics of the same devices [4]. The results show that the emitter resistance is lowered substantially when only 2% of the interface is covered by the pinholes [4-6]. Any further formation of interfacial pinholes simply increases the base current without lowering the emitter resistance significantly.

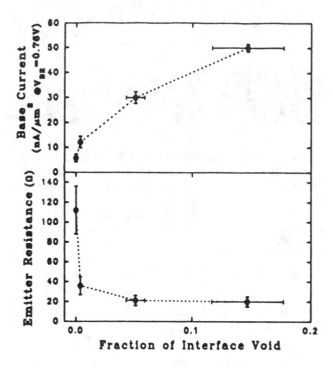

Fig. 7. Correlation between the electrical properties of bipolar transistor and the structure of the poly-Si/Si interface.

To find out how extensively the interface oxide needs to be broken to reduce the emitter resistance as much as possible, without increasing the base current significantly, the total areal fraction of the interfacial pinholes needed to be measured at each temperature. This was obtained by measuring the distribution of the distances between pinholes from high resolution lattice images and analyzing the distribution with the assumption that the pinholes are formed randomly following Poisson statistics [4]. Fig. 6 shows the areal fraction of pinholes, the radius of pinholes,

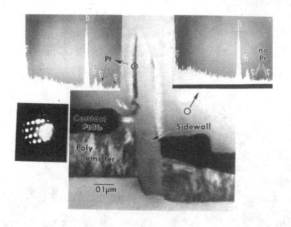

Fig. 8. TEM, X-ray, and microdiffraction analysis of PtSi residue on the sidewall of a bipolar transistor.

CHEMICAL ANALYSIS

Sometimes chemical information from very small regions is required to solve critical problems. There are several chemical analysis methods, but to obtain chemical information from very small areas, energy-dispersive x-ray spectroscopy (EDXS), electron energy-loss spectroscopy (EELS) or microdiffraction can be used. Here, I discuss an example which diagnosed the origin of a high leakage problem in bipolar transistors using EDXS and microdiffraction. To find out the cause of high leakage between emitter and base in bipolar transistors, a cross-sectioned TEM sample was examined, expecting some suspicious defects in the Si. However, metal residues on the sidewall were found and turned out to be responsible for the leakage (Fig. 8). Focusing the electron probe diameter to 200 Å, X-ray analysis revealed the presence of Pt. With microdiffraction patterns obtained from the same areas and the X-ray spectra, the particle was identified to be PtSi. The fact that the phase was PtSi, and not Pt, identified what part of the process was responsible for the Pt contamination.

In the above case, the particle size was about 200 to 300 Å. However to obtain chemical information from smaller regions, parallel EELS or energy filtered TEM can be used. Especially, energy filtered TEM allows the spectroscopic imaging. It includes a prism/mirror/prism type spectrometer (so-called omega filter) after the objective lens in the TEM column; the magnetic field of the prism disperses the electron beam, depending on the energy. By choosing the particular energy loss of the inelastically scattered electrons, spectroscopic images can be obtained. This method is especially useful for chemically mapping light elements like Al, O, N, and Si, which dominate the backend process. It also has a relatively high spatial resolution (~10 Å) and chemical sensitivity (~100 ppm).

FAILURE-MODE ANALYSIS

Failure-mode analysis is one of the essential stages in VLSI technology development. Even when a few devices go wrong out of a million, the performance of the chip can be substantially affected. FMA consists of two parts: One is to identify the physical location of the bad bits. This is usually done by bitmapping methods which give the electrical address of the bad bits. These addresses are then translated into a physical identification of a particular bad bit on the chip. The other part is to identify the defects causing the cells to fail. A coarse method is stripping back layer by layer using chemical solutions or other standard etching methods like RIE, and comparing each layer of good and bad bits by optical microscopy or SEM in plan-view. This requires a detailed understanding of the layout, etching procedures, and etching rates. And through trial and error, by looking at many bad cells, real defects can be identified with some confidence. A better way is cross-sectioning bad and good bits, and comparing these by SEM or TEM. Cross-sectioning of one specific device out of one million devices, although possible, is time consuming by conventional sample preparation methods. However, using focused ion beam (FIB) machines, we can increase the success rate and decrease the amount of time needed to prepare a sample. Fig. 9 is a schematic diagram showing TEM sample preparation using an FIB machine. Fig. 10 are high and low magnification TEM micrographs of a sample prepared by FIB. Most low magnification micrographs show fine features, but high resolution pictures show some material mixing (smearing) by ion beams. Since the FIB has the capability of secondary ion imaging or secondary electron imaging,

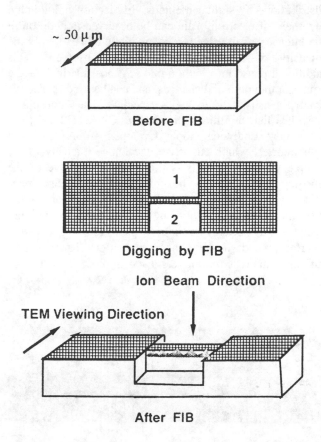

Before FIB

Digging by FIB

Ion Beam Direction

TEM Viewing Direction

After FIB

Fig. 9. A schematic diagram showing TEM sample preparation using FIB.

direct device cross-sections can also be used to reveal large particles (>~500 Å), voids, metal disconnects, etc in a relatively short time.

NEW DIAGNOSTIC METHODS

Here, I will discuss two new methods, currently under development or initial application. One is electron holography for electric field mapping (dopant distribution) and measuring small density differences. The other is quantitative high resolution TEM (QUANTITEM) for quantitative measurement of composition and roughness [7]. There have been many attempts to measure the junction position. Especially for shallow junction (less than 50 nm) the lateral and vertical extent need to be measured accurately. So far, junction delineation by selective chemical etching or anodic oxidation has been widely used, but the methods remain difficult and essentially qualitative. Electron holography has been used to measure the magnetic field distributions [8]. It should be possible to measure electric fields, and hence dopant distributions by similar methods. In electron holography, the phase of electron beam can be observed in addition to the intensity of the electron beam, while in TEM, only the intensity of electron beam can be observed. This additional phase information makes it possible to measure small sample potential changes, and can be used to observe small density differences especially in amorphous materials like multilayer dielectrics. QUANTITEM [7,9] is a new analytical tool developed in AT&T Bell Laboratories, which can relate the image intensity to the sample potential directly without any knowledge of the imaging condition. It can be applied to crystalline samples, which contain no extended defects. With near atomic resolution and sensitivity, it can be used to measure roughness at buried interfaces in plan-view (e.g. Si/SiO_2 interfaces) and measure the composition change across interfaces in cross-section (e.g. $GeSi/Si$ interface) [9].

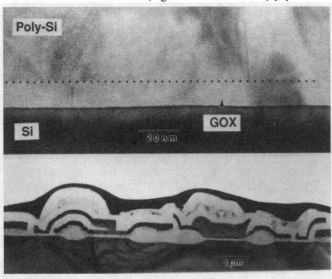

Fig. 10. High and low magnification TEM micrographs of a sample prepared by FIB, showing poly-Si smearing into gate oxide.

SUMMARY

In conclusion, as the device sizes decrease, the combination of SEM and TEM diagnostics becomes increasingly important for process development and failure-mode analysis. Especially with new sophisticated equipment like FIB for TEM sample preparation, FMA has become less time consuming and more reliable. Examination of the same device physically and electrically can give direct correlation and help solve important problems. The understanding and control of solid state processes through electron microscopic studies can make more accurate process and device simulation possible, and may reduce time delay and expense resulting from technology development by on trial and error.

ACKNOWLEDGMENTS

The author acknowledges the collaboration of team members involved in deep submicron CMOS effort and high performance bipolar CMOS effort in AT&T Bell labs. Especial thanks go to J. Rentschler for technical help and A. Ourmazd for stimulating discussions.

REFERENCES

1. Yan, R.H., Lee, K.F., Jeon, D.Y., Kim, Y.O., Park, B.G., Pinto, M.R., Rafferty, C.S., Tennant, D.M., Westerwick, E.H., Chin, G.M., Morris, M.D., Early, K., Mulgrew, P., Mansfield, W.M., Watts, R.K., Voshchenkov, A.M., Boker, J., Swartz, R.G., and Ourmazd, A., "89-GHz f_T Room-Temperature Silicon MOSFET's", IEEE Electron Device Letters, Vol. **13**, No. 5, pp. 256-258, May 1992.

2. Lee, K.F., Yan, R.H., Jeon, D.Y., Kim, Y.O., Tennant, D.M., Westerwick, E.H., Early, K., Chin, G.M., Morris, M.D., Johnson, R.W., Liu, T.M., Kistler, R.C., Voshchenkov, A.M., Swartz, R.G., and Ourmazd, A., "0.1μm p-channel MOSFETs with 51 GHz f_T", IEDM 1992 Technical Digest, pp. 1012-1014 (1992)

3. Lee, K.F., Yan, R.H., Jeon, D.Y., Chin, G.M., Kim, Y.O., Tennant, D.M., Razavi, B., Lin, H.D., Wey, Y.G., Westerwick, E.H., Morris, M.D., Johnson, R.W., Liu, T.M., Tarsia, M., Cerullo, M., Swartz, R.G., and Ourmazd, A., "Room Temperature 0.1μm CMOS Technology with 11.8 ps Gate Delay", IEDM 1993 Technical Digest, PP. 131 (1993)

4. Kim,Y., Liu, T.M., Lee, K.F., Jeon, D.Y., Rentschler, J.A., and Ourmazd, A., Appl. Phys. Lett., Vol. "Microstructure of the emitter polycrystalline silicon/silicon interface in bipolar transistors after rapid thermal annealing", **60**, No. 4, pp. 437-438. (1991)

5. Liu, T.M., Kim, Y.O., Lee, K.F., Jeon, D.Y., and Ourmazd, A., "The Control of Polysilicon/Silicon Interface Processed by Rapid Thermal Anneal", Proc. of The 1991 Bipolar Circuits and Technology, pp. 263-269 (1991)

6. Sung, J.J., Liu, T.M., Kim, Y.O., Chiu, T.Y., "Analytical Modeling of Oxide Break-up Effect on Base Current in N^+ Polysilicon Emitter Bipolar Devices", IEEE Transactions on Electron Devices, Vol. **39**, No. 12, (1992)

7. Schwander, P., Kisielowski, C., Baumann, F.H., Seibt, M., Kim, Y., and Ourmazd, A., "Mapping Projected Potential, Interfacial Roughness, and Composition in General Crystalline Solids by Quantitative Transmission Electron

Microscopy", *Phys. Rev. Lett.* Vol. 71, No. 25, pp. 4150 (1993)

8. Tanomora, A., "Electron holography of magnetic materials and observation of flux-line dynamics", *Ultramicroscopy* 47, pp. 419 (1992).

9. Kisielowski, C., Schwander, P., Baumann, F.H., Seibt, M., Kim, Y., and Ourmazd, A., "An Approach to Quantitative High Resolution Transmission Electron Microscopy of Crystalline Materials", submitted in *Ultramicroscopy* (1994).

Ion Beam Characterization of Semiconductors

J. Mark Anthony

Texas Instruments, P.O. Box 655936, MS 147, Dallas, Tx. 75265

The metrology and characterization of materials plays an increasingly important role in the development and manufacturing of semiconductor devices. Although electrical performance is the ultimate test of manufacturing validity, control of individual processes on a step by step basis often requires direct physical understanding of the materials in question. In order to provide physical insight on an increasingly smaller scale, a variety of analytical methods are required. These methods usually involve some type of particle solid interaction utilizing energetic probes such as electrons, photons or ions. This paper provides insight into some of the ion beam based characterization methods currently utilized in the electronics industry.

INTRODUCTION

Ion beam based analytical methods are routinely used to provide a variety of types of information about semiconductor materials and processes. Impurity distributions (in depth and laterally), film thicknesses and stoichiometries, surface and bulk constituents and crystalline integrity can all be investigated using ion beams. Not surprisingly, this broad range of applications requires a broad variety of ion beam methods utilizing various aspects of the ion solid interaction process.

As an ion beam penetrates a solid target, there are two major energy loss mechanisms at work. For low energies (keV range) the energy loss per unit path length dE/dx is dominated by many high energy transfer collisions with the target atoms (nuclear stopping), which leads to damage in the sample material and extensive sputtering of the sample surface. As the energy is increased, the nuclear stopping component decreases and energy loss is dominated by electronic energy transfer to the outer electrons of the target atoms. The maximum in the nuclear and electronic stopping depends on the particular ion-target combination of interest. For He, Ar and Xe ion beams penetrating a Si target, the peak in the nuclear stopping occurs at ~ 1, 30 and 300 keV[1], respectively, while the peak in the electronic stopping occurs at ~ 0.4, 30 and 400 MeV[2]. The nuclear stopping contribution is very low at energies/mass > 100 keV/amu. Since the sputtering yield is directly related to the nuclear stopping, sputtering is a small effect at these energies.

The details of individual ion-solid interactions and ion generation processes in the nuclear stopping regime are difficult to predict quantitatively from first principles, and thus ion beam analytical methods in this energy region (such as SIMS) usually require standards for quantitation. At higher energies most ion beam methods utilize interactions between a probe ion and an individual target atom and the cross sections for this process are much easier to describe quantitatively. Thus most methods in this region can be considered quantitative without standards.

Ion beam techniques can thus be rather broadly divided into sputter based methods and scattering/excitation based methods, and this paper will attempt to provide an introduction to typical applications for a variety of methods in each area.

SPUTTER BASED METHODS

Possibly the most common ion beam method in use today for semiconductor analysis is Secondary Ion Mass Spectrometry (SIMS). The basic principles are outlined in Figure 1. A primary ion beam (typically O_2, Cs, Ga or Ar) is rastered across the sample of interest, and both neutrals and positive and negative ions are ejected by the sputtering process. (Electrons and light emission also accompany the sputtering process.) The establishment of an electric field near the sample allows collection of these secondary ions, which can then be passed to a mass spectrometer for mass analysis.

The ion yield during the sputtering process varies dramatically depending on the element of interest, the sample material and the sputtering beam. Cs and O_2 primary ions are utilized to enhance the formation of negative and positive ions, respectively, during the sputtering process. Ion yields (ratio of sputtered ions to sputtered atoms) of 0.01% - 1% can usually be achieved for most elements[3,4], although higher (and lower) values will occur. Ga and Ar ion beams provide very little chemical enhancement of the ion formation process, so yields are orders of magnitude lower.

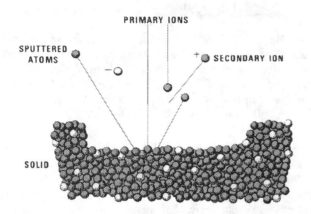

FIGURE 1. Schematic of the SIMS process.

In order to generate depth information, the primary ion beam can be rastered over a square region of the sample and ions collected from the central region of the sputter crater. Rasters of a few hundred microns on a side are typically employed, and the ion collection area may range from ~1-10^4 μm^2.

Figures 2-4 provide examples of SIMS applications of current interest for semiconductors. Figure 2 is a SIMS depth profile showing evidence for Al contamination on a Si wafer after low energy B implantation. Analysis of wafers which have been

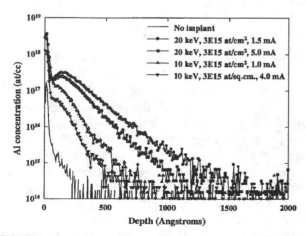

FIGURE 2. Al contamination on a Si wafer after exposure to a 3 x 10^{15} at/cm^2 B implant.

implanted at varying energies and currents provides information about the origin of the contamination, which in this case was due to exposed Al metal in the ion source region. Figure 3 shows the presence of metallic contamination after Sb implantation into a Si wafer. Although the implant was fairly energetic (70 keV), the shape of the contaminant profiles suggests a low energy origin, such as the beam stop or wafer handling equipment. A variety of implant contamination mechanisms relevant to manufacturing have been identified[5].

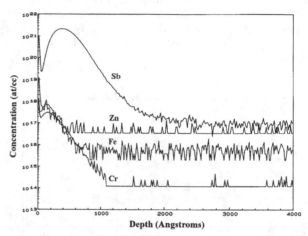

FIGURE 3. Metallic contamination on a Si wafer after exposure to a 70 keV Sb implant.

Figure 4 shows an example of the use of SIMS with Cs based cluster ions to provide compositional information. Although in general SIMS ion yields are dominated by surface chemistry and change dramatically with changes in composition, it is possible to look at ions formed "above" the sample, which in principle should be much less susceptible to matrix effects on ion yield. Ions generated in this fashion (for example, CsO$^+$ positive ions rather than O$^-$ negative ions during sputtering of an oxide) may depend more on sputter rates than surface chemistry, allowing the generation of sensitivity factors for quantification of the layer composition. This has been used quite successfully[6,7] to profile the composition of thin (~10 nm) gate dielectrics with sensitivities not available to other methods. Figure 4 shows the N and O distributions[8] (determined by monitoring CsO$^+$ and CsN$^+$ ions generated during Cs sputtering) of 3 films grown on bare Si in dry O$_2$ followed by N$_2$O exposure at 800, 900 and 1000 °C. Nitrogen incorporation in SiO$_2$ films has been shown to reduce interface state densities[9]. In addition, the nitrided regions of oxide films may act as a B diffusion barrier for p+ poly gates[10] (to prevent shifts in transistor threshold voltage). The buildup of N at the

SiO₂/Si interface can easily be measured using the Cs cluster beam detection.

One limitation of dynamic SIMS instruments in use today is the sequential nature of ion collection in the mass spectrometer. This can be addressed in the Time-Of-Flight SIMS (TOF-SIMS) instrument, which pulses the primary beam and then uses fast electronics to detect in parallel essentially all secondary ions (of a chosen polarity) generated by the primary ion pulse. Mass resolution in this system is governed by flight time, and therefore high mass resolution can be generated without the loss of transmission sometimes found with dynamic SIMS. The TOF-SIMS compromise in this case is the decrease in current and possible increase in spot size of the primary ion beam. Due to the pulsed nature of the TOF-SIMS system, it is ideally designed for examining

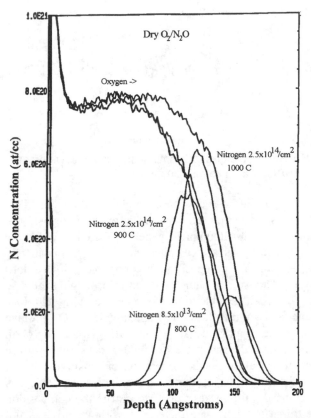

FIGURE 4. Cs based cluster analysis of thin oxynitride layers on Si. Nitrogen pileup at the SiO₂/Si interface can be detected. (Data courtesy of Evans East).

surface structure of the top few monolayers of a sample. In addition, very high mass ions can be collected simply by extending the collection time.

Figure 5 gives an example of the "rich" TOF-SIMS spectra generated from Si wafers which have been coated with photoresist and then cleaned using various wet processes[11]. The data are plotted on a semilog scale,

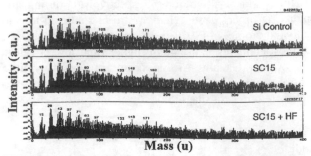

FIGURE 5. TOF-SIMS spectra of Si wafers which have been coated with photoresist, stripped and then cleaned, as well as a Si control wafer. Cleaned wafers were exposed either to SC1 (NH₄OH:H₂O₂:H₂O) for 5 minutes (SC15) or to SC1 for 5 minutes followed by an HF rinse. (Data courtesy of M. Douglas, Texas Instruments.)

and there is clearly a multitude of peaks in the spectra. Figure 6 shows an expansion of a region of the spectrum containing the C₇H₇ ion[11], which is used as a marker for residual photoresist. Differences in the cleaning effectiveness can now be seen. In many cases the complexity of the spectra may require more complicated methods to deconvolute the information.

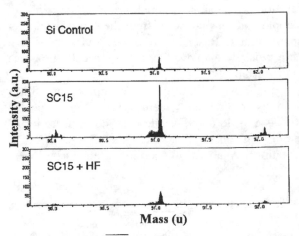

FIGURE 6. Expansion of the spectra shown in figure 5.

The parallel detection of ions also makes TOF-SIMS ideal for analyses of unknown contaminants with limited sample material, such as particles generated during manufacturing. Figure 7 shows different ion images of a Si surface containing a 0.3 µm Al particle. Figures 7a and 7b show high mass resolution scans of the Al ions (mass 26.98154 amu) and a nearby C₂H₃ signal (mass 27.02347 amu). The data show clear evidence for the Al particle, which is well separated from the interfering signal. A major advantage of TOF-SIMS over dynamic SIMS in the identification of particles is the parallel detection capability. Analysis of the material requires sputter erosion, and thus the TOF-SIMS has greatly enhanced "material efficiency" in the analysis of unknown materials.

Although TOF-SIMS utilizes parallel detection to increase the efficiency of the ion collection process, the ion generation is still limited by the secondary ion yields, which are always smaller than the neutral atom yield. This suggests that some type of post-ionization method (ionization separate from the sputtering process) might provide enhanced detection of the sample material. In addition, ionization above the sample in the gas phase should be independent of the surface chemistry and

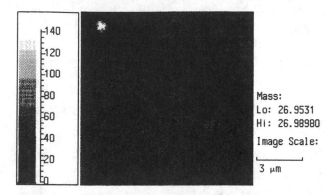

Comments: 0.3um Al, 12um field, Al+

Comments: 0.3um Al, 12um field, C2H3+

FIGURE 7. TOF-SIMS scans of a 0.3 μm Al particle. The Al ion signal is well separated from the C_2H_3 interference, which shows no evidence for a particle signature. (Data courtesy of P. Lindley, Charles Evans and Associates.)

therefore free of the bulk of the matrix effects which are so common in SIMS and TOF-SIMS methods. One of the most promising of the post-ionization methods uses pulsed lasers to ionize the sputtered particles above the sample surface. Other methods of post-ionization have also been demonstrated[13,14].

There are several strategies for laser based ionization methods, but to a large extent they can be divided into resonant and non-resonant methods. The

non-resonant methods[15,16] utilize high photon flux (from high power density lasers) to stimulate ionization. This can be done using either a single short wavelength photon (with photon energy > than the ionization potential), or through a multi-photon ionization process utilizing lower energy photons and excitation of a virtual state[17] in the atom of interest. Since these virtual states have very short lifetimes, high photon flux (laser power density) is required to generate ionization by absorption of a second photon before decay of the state. For sufficient laser powers essentially all atoms will be ionized, but multiply charged ions and molecular interferences will also be present.

Although resonant methods only ionize one element at a time, they offer the possibility of mass selection through careful tuning of the photon wavelength[18,19]. Figure 8 shows a schematic of the resonant process. A pulsed primary beam generates a

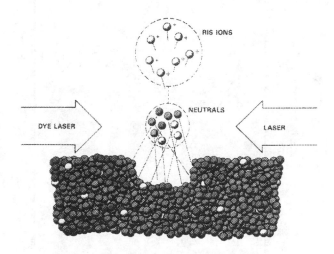

FIGURE 8. Schematic of the resonant ionization spectroscopy (RIS) process for selective ionization.

cloud of sputtered particles above the sample. A tunable laser is tuned to the wavelength necessary to excite the atom of interest from its ground state to a known excited state below the ionization potential. Absorption of a photon from a second laser can then provide enough energy to result in ionization of the atom (as long as the sum of the two photon energies is larger than the ionization potential of the atom). Since the first excitation process is a resonant excitation (longer lifetime), the laser power required is orders of magnitude lower than the non-resonant process. As is suggested in the figure, the uniqueness of the atomic states allows selective ionization of the atom of interest.

Figure 9 shows an example of the application of the resonant ionization method to a familiar SIMS type problem, the detection of a Cu in Si implant profile. A

1 μA O_2^+ beam was used for DC sputtering, and then the beam was utilized in pulsed mode for resonant ionization based data acquisition. The profile acquisition time of ~ 1-2 hours is defined primarily by the pulse repetition rate of the laser and the counting statistics required at low concentration levels. In general the resonant method provides sensitivities similar to SIMS, without the mass interference problems. In addition the signal is not susceptible to the matrix effects which plague the SIMS based methods. This has been successfully exploited in the study of impurity buildup at the Si/SiO_2 interface in silicon on insulator samples[21], which is very problematic with SIMS.

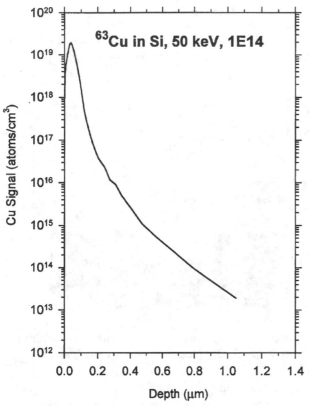

FIGURE 9. Resonance ionization spectrum for a Cu implant in Si. (Data courtesy of H. Arlinghaus, Atom Sciences.)

Although the resonant methods allow complete removal of mass interferences, a different wavelength is required for each element, which limits the survey capabilities of the technique. An alternative method which allows true molecule free mass spectrometry of all masses is accelerator mass spectrometry (AMS)[22]. This method utilizes conventional SIMS hardware for generating negative secondary ions which are injected into a small tandem accelerator. These ions are accelerated to the terminal of the accelerator where they undergo collisions in a gas stripper cell, which removes several electrons. Molecular ions with charge states >+2 are unstable[23] and will dissociate during acceleration away from the accelerator terminal. Magnetic and electrostatic analysis after acceleration will allow detection of atomic ions only, since the molecular fragments will have different kinematics than the atomic ions of interest. The technique thereby provides truly atomic mass spectra.

Figures 10 and 11 show the power of this method[24]. Figure 10 is a mass scan of Si before injection into the accelerator. This is similar to a negative ion SIMS scan on a conventional instrument. All ion signals have been directly ratioed to the $^{28}Si^-$ signal. Figure 11 shows the data for ions of charge state +3 after acceleration and analysis. Once again the ion signals

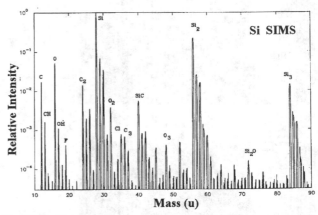

FIGURE 10. Accelerator mass spectrometry mass scan of Si before injection into the accelerator. This data is similar to a conventional negative ion SIMS scan. (Figure courtesy of Journal of Vacuum Science and Technology.)

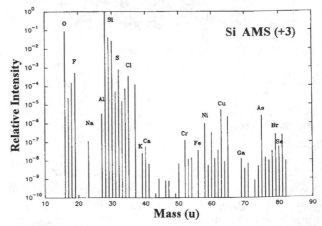

FIGURE 11. Accelerator mass spectrometry mass scan of Si after acceleration and magnetic and electrostatic analysis. All molecular interferences seen in figure 10 have been removed. (Figure courtesy of Journal of Vacuum Science and Technology.)

Table 1. Comparison of the sputter based methods.

	Ion Detected	Quantitative	Matrix Effects	Molecular Rejection	Detection Scheme	1 ppb acq. time (s)*
SIMS	+, –	standards	yes	Small slits	serial	2
TOF-SIMS	+, –	standards	yes	Timing	parallel	2×10^4
non-resonant	+	semi	no	Timing	parallel	2×10^3
resonant	+	yes	no	Resonant ionization	single element	2×10^3
AMS	–	standards	yes	Coulomb explosion	serial	10

* Assuming 10^{-6} A primary beam, sputter yield of 1, 1 kHz primary beam pulse rate (TOF-SIMS), 10x primary ion beam bunching, 100 Hz laser pulse rate, 10^{-8} s ion pulse, 10^{-3} ion yield, 10% collection of laser generated ions, 15% AMS transmission.

have been ratioed to the $^{28}\text{Si}^{+3}$ signal. The scan shows dramatically improved sensitivity and complete removal of interfering molecular signals. This method has demonstrated bulk detection limits for transition metals in Si in the 10^{13} at/cc range[24], and has also been utilized for depth profiling in some cases[25,26].

Table 1 attempts to compare the details of these sputter based methods. In general, SIMS and AMS offer the best absolute sensitivity, since the pulsed methods are limited by the duty cycle of either the primary beam or the lasers. From a material efficiency perspective (ions detected/atoms sputtered), the laser based methods are preferable since they allow direct access to the sputtered neutrals. General removal of molecular interferences is perhaps most straightforward with the AMS method. The table also attempts to estimate the acquisition time necessary for 10 counts from a 1 part-per-billion impurity (assuming no mass interferences). Longer primary ion pulses will decrease the acquisition time for the resonant technique, but will adversely affect mass resolution for the TOF-SIMS and non-resonant methods.

SCATTERING/EXCITATION METHODS

As the energy of the primary ion beam is increased, hard collisions will still occur, but with reduced probability. Most of the energy transfer will be via electronic stopping through excitation of the target electrons. The decreasing probability of interaction has an advantage, however, since it simplifies the analysis of particles scattered from the sample. At high energies, the scattering probability[27] is proportional to $[Z_1 Z_2/E]^2$, where Z_1 and Z_2 are the atomic numbers of the incident and target atoms, respectively, and E is the energy of the incident beam. If a given detection angle is chosen, the kinematics of the scattering event are well defined, and therefore the energy transfer from the collision can be calculated directly[27].

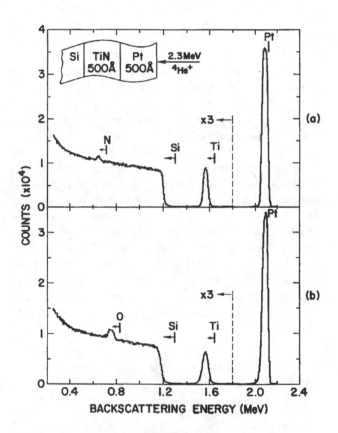

FIGURE 12. RBS spectra of Pt/TiN/Si structure before (a) and after (b) annealing in oxygen. Oxidation of the TiN is easily detected using RBS. (Figure courtesy of Journal of Materials Research.)

For large scattering angles, the incident beam can be scattered back out of the target and detected. Since the kinematics of the scattering process and the energy loss of the incident beam are well understood, the energy of the scattered ion gives information about the mass of the scattering center as well as its depth within the sample. Energy analysis of the backscattered particles therefore provides depth distributions and stoichiometries to be determined with high accuracy without the need for standards. This process is routinely utilized in the Rutherford Backscattering (RBS) technique.

Figure 12 shows an example of the use of RBS analysis to study the evolution of materials after annealing[28]. As DRAM densities increase, the conventional dielectric (primarily SiO_2) will not be able to provide the needed capacitance at reduced geometries. Alternative materials which have been proposed (such as barium strontium titanate) may require deposition at high temperatures in an oxygen ambient, which places severe restrictions on the electrode and barrier materials used in the capacitor. Pt and TiN have been proposed as possible electrode and barrier materials, respectively, for (Ba,Sr)TiO$_3$ capacitors. Figure 12 examines the behavior of a Pt/TiN structure before (a) and after (b) annealing at 650 °C for 30 minutes in oxygen. The RBS spectra clearly show the oxidation of the TiN layer, and the increase in width of the Ti signal reflects the increase in roughness caused by the oxidation (which would lead to leakage in the capacitor).

In some cases the large signal due to scattering of the primary beam from the substrate can mask features of interest, and it would be advantageous to reduce this substrate effect without changing the information from other layers. This can be implemented on crystalline samples by carefully orienting the primary ion beam to "channel" it along the crystalline axis. Scattering from amorphous or poly-crystalline regions will not be affected, but the scattering from the crystal will be greatly reduced[29]. This effect is demonstrated in figures 13 and 14[30]. Figure 13 shows the random and channeled

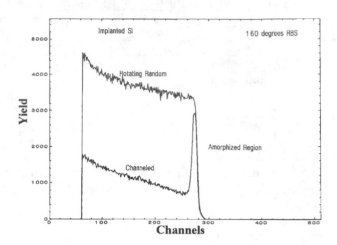

FIGURE 14. Random and channeled RBS spectra from an implanted Si wafer. (Data courtesy of J. Kirchhoff, Charles Evans and Associates.)

spectra from a "virgin" Si wafer, while figure 14 shows spectra from a similar wafer which has been ion implanted. The implant creates an amorphous region near the sample surface. The signal from the underlying crystalline substrate is dramatically reduced by channeling, while the signal from the top amorphous layer is unaffected (no channeling is possible in the amorphous layer). Thus the signal to noise from the amorphous region is dramatically enhanced by the sample alignment. Conversely, the method can also provide information about the crystalline quality of surface layers.

The form of the scattering probability discussed above provides insight into the sensitivity of the backscattering methods. Decreasing the beam energy and increasing the beam atomic number both act to increase the scattering cross section (and therefore the sensitivity of the method), but in conventional RBS detectors (Si surface barrier detectors) the resolution of the method suffers greatly due to poor energy resolution in the detector. The introduction of high speed data acquisition electronics, however, has recently allowed the implementation of new detection schemes based on time-of-flight methods which provide high energy resolution under these conditions. This allows the system sensitivity to be dramatically increased without sacrificing depth and mass resolution capabilities. Figure 15 shows an example of a TOF design for backscattering currently being optimized for detection of trace quantities of surface impurities on semiconductor wafers (Heavy Ion Backscattering - HIBS[31]). Electrons generated by passage of the backscattered ion beam through a thin foil are collected in an "electron" microchannel plate, and

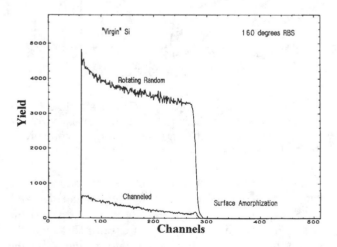

FIGURE 13. Random and channeled RBS spectra from a clean Si wafer. (Data courtesy of J. Kirchhoff, Charles Evans and Associates.)

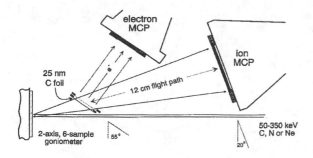

FIGURE 15. Schematic of a time-of-flight detector used for high sensitivity RBS at low energies. (Figure courtesy of J. Knapp, Sandia National Laboratory.)

the ions are collected in an "ion" microchannel plate. The difference in time between the two signals is related to the ion velocity and thereby the scattered energy. Figure 16 shows spectra collected using this scheme from Si wafers which had been intentionally contaminated with Ni[31] at surface doses ranging from 1.1×10^{12} at/cm^2 to 1.4×10^{11} at/cm^2. In addition to the Ni signal, various contaminants can be seen at low levels on the sample. A second generation version of this system is under development, and detection limits of $\sim 5 \times 10^9$ at/cm^2 for

Fe and 1×10^8 at/cm^2 for Au are expected[32] for surface contamination on Si wafers. In addition to the high system sensitivity, these levels can be quantified without standards due to the knowledge of the scattering cross sections involved. (At low energies the bare nuclei are "screened" somewhat by the presence of the atomic electrons, but the effect can be calculated[33].)

In addition to high sensitivity, a slight modification of the detection system[34] can provide excellent resolution as well. Figure 17 shows analysis of four titanium nitride on SiO$_2$ samples using this strategy[35]. The thicknesses varied from 43 - 47 nm, and differences of 2 nm in the film thickness can easily be detected. Figure 18 shows the oxygen spectra from a 15 nm SiO$_2$ sample[35], and depth resolution of 1 nm is expected after optimization. (Even higher resolution can be achieved by electrostatic analysis of the ion beam after

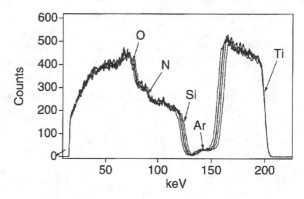

FIGURE 17. Low energy backscattering spectra using a high resolution time-of-flight detection geometry. Four TiN/SiO$_2$ structures were measured, with TiN thicknesses varying from 43 - 47 nm. The 2 nm thickness differences are easily determined. (Data courtesy of R. Weller, Vanderbilt University.)

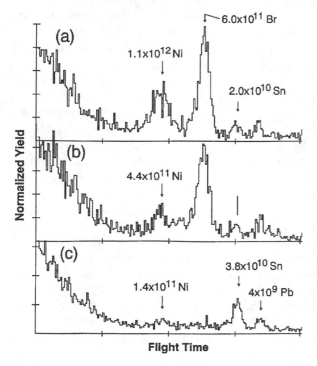

FIGURE 16. HIBS spectra from Si wafers contaminated with Ni at 3 different surface concentrations. Unexpected heavy elements were also detected. (Data courtesy of J. Knapp, Sandia National Laboratory.)

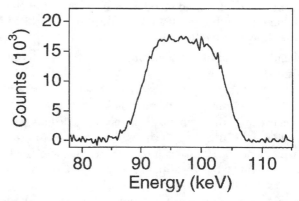

FIGURE 18. Low energy backscattering spectra for the oxygen distribution in a 15 nm SiO$_2$ film on Si (after background subtraction). (Data courtesy of R. Weller, Vanderbilt University.)

373

scattering, but this dramatically increases data collection time due to the serial nature of the acquisition versus the parallel detection with TOF.)

Although backscattering is more well known as a technique, it is possible to tilt a sample of interest and scatter the sample atoms in the forward direction. This geometry is illustrated in Figure 19[36]. This method is best suited for detection of low mass elements, since the maximum scattering angle (for scattering the target atoms) decreases with mass. Detection can be performed either with a surface barrier detector, or with TOF hardware discussed above. Figure 20 shows the results of the analysis of the H, C and O contamination on Si, before and after treatment with UV/ozone and HF immersion[35] using an Ar beam to forward scatter the light elements. The control wafer exhibits a native oxide as well as slight H and C contamination. UV/ozone treatment reduces the hydrocarbon signal but increases the oxygen thickness, while the subsequent HF dip removes the native oxide but deposits a large surface H peak as well as some residual C (presumably due to impure chemicals). Detection limits for this geometry are expected to be ~ 10^{13} at/cm^2 for light elements on Si. Once again, this method can be quantified without

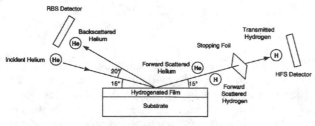

FIGURE 19. Schematic of the forward scattering geometry. Light elements in the sample are scattered into a detector at forward angles to the incident beam. A foil is sometimes used to stop forward scattered particles from the incident beam. (Figure courtesy of J. Kirchhoff, Charles Evans and Associates.)

standards due to the understanding of the scattering interaction.

As the energy of the probe beam is increased into the MeV range, depth distributions for these light elements can be determined. This method, Elastic Recoil Detection (ERD)[37], has been used extensively in the measurement of H distributions. The glancing angle geometry limits the maximum analysis depth and also degrades depth resolution somewhat due to energy straggling during passage through the sample. Both these issues can be addressed through the use of Nuclear Reaction Analysis (NRA)[L38,39], which utilizes resonance reactions that occur between the probe beam and H in the

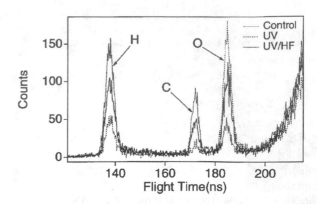

FIGURE 20. Forward scattered spectra from bare Si wafers exposed to various surface treatments. Quantitative detection limits of 10^{13} at/cm^2 for H, C and O on Si surfaces are expected with this method. (Data courtesy of R. Weller, Vanderbilt University.)

sample. Commonly used reactions include $^1H(^{15}N,\alpha\gamma)^{12}C$ (6.385 MeV resonance energy) and $^1H(^{19}F,\alpha\gamma)^{16}O$ (6.418 MeV resonance energy). Since these are resonance reactions, the reaction only occurs for a well defined energy range of the incident beam. This offers the opportunity of increasing the energy of the probe beam above resonance and then allowing the incident beam to lose energy during passage through the sample. At some depth the beam will reach the resonance energy and will react with any hydrogen present at that depth. Varying the energy of the beam thus provides the ability to quantitatively depth profile the hydrogen in the sample for depths up to ~ 1 um and sensitivities <0.1%. This is an excellent method for determining depth distributions, especially at interfaces[40,41] where techniques such as SIMS experience ion yield variations. Unfortunately this technique requires slightly higher energies than are available with many small accelerators, which has limited the availability to some extent.

In addition to the methods discussed here, the broad range of energies and ion beams available allows a variety of other methods, which, although perhaps not commonplace, offer unique information. Backscattering with heavy ions at higher energies increases the mass resolution of the method, activation analysis using charged particles allows quantitative bulk information on many elements including gaseous species like C and O, the use of resonance reactions will greatly enhance the sensitivity to elements such as oxygen, scattering of ions from surfaces at very low energies provides unique structural information, and the use of field ionization can provide atomic resolution of materials of interest. These and other techniques are well described in a variety of publications[42,43,44]

TOOLS

In addition to the analytical methods themselves, there are many computational tools relevant to ion beam issues. TRIM[45] is a Monte Carlo code for determining the effects of ion penetration in amorphous materials. Multi-element targets and multiple films can be constructed, and the final distribution of the original ions as well as redistribution of target atoms can be simulated. Figure 21 shows a TRIM simulation of the redistribution of 10^{12} at/cm^2 of surface boron on Si after implantation of 3.5 x 10^{15} at/cm^2 As at 150 keV. The As knock-on produces substantial amounts of B even at large depths. Since B is known to be present even on "clean" Si wafers[46,47], this redistribution could be important in some cases. Other codes available include MARLOWE[48], for calculating the penetration of ions in crystalline materials, and SIMION[49], for design and simulation of ion optical elements. These and other programs are often quite useful for gaining insight into either instrument operation or the ion-solid interaction process itself.

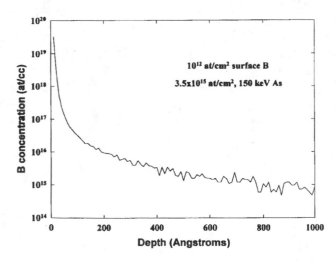

FIGURE 21. TRIM simulation of the knock-on effect of a 150 keV As implant on a Si wafer coated with B surface contamination. Measurable amounts of B will exist even at 100 nm.

SUMMARY

The field of ion beam analysis is quite extensive, with techniques ranging from eV to MeV, and applications ranging from starting materials to finished devices. Relatively mature methods such as SIMS and RBS are routinely in extensive use in the electronics industry, and new methods are on the horizon to extend the capabilities of these methods. Developments in hardware including improvements in the brightness and lifetimes of ion sources, new accelerator and high voltage designs and the use of high speed electronics to enable novel detection schemes are also pushing the capabilities of the ion beam methods. This will be necessary to allow ion beam based analytical methods to continue to serve a crucial function in the development of semiconductor processes and materials.

ACKNOWLEDGMENTS

The author would like to gratefully acknowledge helpful data and/or discussions with H. Arlinghaus (Atom Sciences), S. Downey (AT&T), J. Kirchhoff and P. Lindley (Charles Evans and Assoc.), D. Dahl (Idaho Nat. Eng. Lab), C. Magee (Evans East), A. Grill (IBM), J. Knapp and B. Doyle (Sandia Nat. Lab), A. Diebold (Sematech), T. Aton, R. Beavers, G. Brown, M. Douglas, T. Gray, J. Julien, J. Keenan, T. Moore and T. Shaffner (Texas Instruments), F. McDaniel (Univ. of North Texas) and R. Weller (Vanderbilt).

REFERENCES

1. Behrisch, R., editor, *Sputtering by Particle Bombardment 1*, New York: Springer-Verlag, 1981.
2. Ziegler, J., *Handbook of Stopping Cross Sections for Energetic Ions in All Elements*, New York: Pergamon Press, 1980.
3. Wilson, R.G., Stevie, F. and Magee, C.W., *Secondary Ion Mass Spectrometry*, New York: John Wiley, 1989.
4. Benninghoven, A. , Rudenauer, F.G. and Werner, H.W., *Secondary Ion Mass Spectrometry*, New York: John Wiley, 1987.
5. Stevie, F.A., Wilson, R.G., Simons, D.S., Current, M.I., Zalm, P.C., *J. Vac. Sci. Technol.* **B12** (4), 2263 (1994).
6. Okada, Y., Tobin, P.J., Hegde, R.I., Liao, J. and Rushbrook, P., *Appl. Phys. Lett.* **61** (26) 3163 (1992).
7. Novak, S.W., Frost, M.R. and Magee, C.W., these proceedings.
8. Analyses provided by Evans East analytical services.
9. Hwang, H., Ting., W., Kwong, D-L. and Lee., J., *IEEE Elec. Dev. Lett.* **12** (9) 495 (1991).
10. Lo, G.Q. and Kwong, D-L., *IEEE Elec. Dev. Lett.* **12** (4) 175 (1991).
11. Data courtesy of Monte Douglas, Texas Instruments.
12. Data courtesy of P. Lindley, Charles Evans and Associates.
13. Williams, P. and Streit, L.A., *Nuc. Instrum. and Meth. in Phys. Res.* **B15**, 159 (1986).
14. Maclaren, S.W., Loxton, C.M., Sammann, E. and Kiely, C.J., *J. Vac. Sci. Technol.* **A7** (1), 17 (1989).
15. Welkie, D.G., Daiser, S. and Becker, C.H., *Vacuum* **41** (7-9), 1665 (1990).

16. Becker, C.H., Mackay, S.G. and Welkie, D.G., *J. Vac. Sci. Technol.* **B10**, 380 (1992).

17. Chin., S.L. and Lambropoulos, P., *Multiphoton Ionization of Atoms*, New York: Academic Press, 1984.

18. Lucatorto, T.B. and Parks, J.E., editors, *Resonance Ionization Spectroscopy 1988*, Philadelphia, Pa.: Institute of Physics, 1988.

19. Pellin, M.J., Young, C.E., Calaway, W.F. and Gruen, D.M., *Nucl. Instrum. Meth. Phys. Res.* **B13**, 653 (1986).

20. Data courtesy of H. Arlinghaus, Atom Sciences.

21. Private communication with H. Arlinghaus, Atom Sciences.

22. Yiou, F. and Raisbeck, G.M., editors, *Nucl. Instrum. and Meth. in Phys. Res.* **B52** (3,4) 1990.

23. Purser, K.H., Litherland, A.E. and Rucklidge, J.C., *Surf. Interface Anal.* **1**, 17 (1979).

24. Anthony, J.M., Kirchoff, J.F., Marble, D.K., Renfrow, S.N., Kim, Y.D., Matteson, S. and McDaniel, F.D., *J. Vac. Sci. Technol.* **A12** (4), 1547 (1994).

25. McDaniel, F.D., Anthony, J.M., Renfrow, S.N., Kim, Y.D., Datar, S.A. and Matteson, S., *Proc. of the 13'th Inter. Conf. on App. of Accel. in Res. and Ind.*, Nov. 7-10, 1994, to be published in Nucl. Instrum. and Meth. in Phys. Res., 1995.

26. Gove, H.E., Kubik, P.W., Sharma, P., Datar, S., Fehn, U., Hossain, T.Z., Koffer, J., Lavine, J.P., Lee, S-T. and Elmore, D., *Nucl. Instrum. and Meth. in Phys. Res.* **B52**, 502 (1990).

27. Chu, W-K., Mayer, J.W. and Nicolet, M., *Backscattering Spectrometry*, Orlando: Academic Press, 1978.

28. Grill, A., Kane, W., Viggiano, J., Brady, M., Laibowitz, R., *J. Mater. Res.*, **7** (12), 3260 (1992).

29. Mayer, J.W. and Rimini, E., *Ion Beam Handbook for Materials Analysis*, New York: Academic Press, (1977).

30. Data courtesy of Joe Kirchhoff, Charles Evans and Associates.

31. Knapp, J.A., Banks, J.C., Doyle, B.L., Sandia National Laboratories internal report SAND94-0391 (1994).

32. Private communication from J. Knapp, Sandia National Laboratory.

33. Mendenhall, M.H. and Weller, R.A., *Nucl. Instrum. Meth. Phys. Res.* **B58**, 11 (1991).

34. Mendenhall, M.A. and Weller, R.A., *Nucl. Instrum. Meth. Phys. Res.* **B47**, 193 (1990).

35. Data courtesy of R. Weller, Vanderbilt University.

36. Figure courtesy of Joe Kirchhoff, Charles Evans and Associates.

37. Tesmer, J.R., Maggiore, C.J., Nastasi, M., Barbour, J.C. and Mayer, J.W., *Proceedings: High Energy and Heavy Ion Beams in Materials Analysis,* Pittsburgh: Materials Research Society, (1990).

38. Barnes, C.A., Overley, J.C., Switkowski, Z.E. and Tombrello, T.A., *Appl. Phys. Lett.* **31**(3), 239 (1977).

39. Lanford, W.A., Trautvetter, H.P., Ziegler, J.F. and Keller, J., *Appl. Phys. Lett.* **28**(9), 566 (1976).

40. Briere, M.A. and Braunig, D., *IEEE Trans. Nuc. Sci.* **37**(6), 1658 (1990).

41. Marwick, A.D. and Young, D.R., *J. Appl. Phys.* **63** (7), 2291 (1988).

42. Duke, C.B. editor, *Surf. Sci.* 299/300 (1994).

43. Duggan, J.L. and Morgan, I.L., *Nucl. Instrum. Meth. Phys. Res.* **B79** (1993).

44. Anderson, H.H. and Rehn, L.E., *Nucl. Instrum. Meth. Phys. Res.* **B85**(1-4), 1994.

45. Ziegler, J.F., IBM Research Center, Yorktown Heights, N.Y., TRIM computer code.

46. Downing, R.G., Lavine, J.P., Hossain, T.Z., Russell, J.B., Zenner, G.P., *J. Appl. Phys.* **67** (6), 1990.

47. Stevie, F.A., Martin, E.P., Kahora, P.M., Cargo, J.T., Nanda, A.K., Harrus, A.S., Muller, A.J., Krautter, H.W., *J. Vac. Sci. Technol.* **A9**(5), 1991.

48. Robinson, M.T. and Torrens, I.M., *Phys. Rev.* **B9**, 5008 (1974).

49. Dahl, D.A., Delmore, J.E., Appelhans, A.D., *Rev. Sci. Instrum.*, **61**(1), 1990.

Trace Nanoanalysis by Analytical Electron Microscopy/ Electron Energy Loss Spectrometry: Achieving High Spatial Resolution and High Sensitivity Analysis

Dale E. Newbury

Surface and Microanalysis Science Division
National Institute of Standards and Technology
Gaithersburg, MD 20899

Analytical electron microscopy with parallel-detection electron energy loss spectrometry (AEM/PEELS) can detect uniformly distributed trace constituents while simultaneously achieving nanometer scale lateral resolution. PEELS spectra are collected in the second difference mode with the count rate matched to detector characteristics through control of the incident beam current. AEM/PEELS measurements performed on NIST Standard Reference Materials have demonstrated limits of detection as low as 10 parts per million with simultaneous achievement of lateral resolution of 10 nm.

INTRODUCTION

The development of semiconductor devices with progressively finer features, penetrating ever further into the sub-micrometer size range, has raised new challenges to chemical characterization methods. These challenges are especially severe when low concentration levels, so-called "trace" constituents, are to be measured at high spatial resolution. Figure 1 shows a plot of the achievable limit of detection versus spatial resolution for a selection of analytical techniques. (In considering Figure 1, the boundaries of the field of operation for a technique should not be considered sharp as drawn since the actual limit of detection for a particular element depends on additional factors such as the physical characteristics and the matrix in which it is dispersed.) When features in the 100 nm range are to be measured, Figure 1 shows that the choice of available techniques is extremely limited when dilute constituents are to be measured.

ANALYTICAL ELECTRON MICROSCOPY

When lateral spatial resolution is required, analytical electron microscopy (AEM) is a highly promising technique (1). AEM is based upon focusing a fine probe (~1-20 nm) of high energy (100-400 keV) electrons incident on a thin foil (20 to 200 nm thick) to provide spatial selectivity. Inelastic scattering of the beam electrons with core shell electrons creates vacancies with subsequent transitions leading to the emission of electrons (Auger) and photons (x-rays) of characteristic energy. Three analytical spectrometries are possible in the AEM: (1) Electron energy loss spectrometry (EELS) where the energy losses suffered

directly by the beam electrons are measured; (2) Auger electron spectrometry (AES), which is much more commonly performed as a first surface characterization method with a low energy probe (2 - 20 keV); and (3) X-ray spectrometry, by far the most common chemical analysis technique implemented on the AEM and almost always performed with the energy dispersive x-ray detector (EDS).

The limit of detection of an analytical technique in terms of the minimum mass fraction (MMF) has been described by Wittry with the following equation (2):

$$MMF = [1/(P/B)] \, (1/\sigma \, \epsilon \, \beta \, t)^{1/2} \quad [1]$$

where P/B is the peak-to-spectral background, σ is the cross section, ϵ is the efficiency of signal collection and measurement, β is the brightness of the primary exciting source, and t is the measurement time. Note that the peak-to-background ratio, which is determined by the physics of particle interaction and is a measure of the spectral quality, appears as a linear term and therefore has a strong influence. Trace-sensitive techniques must have a high P/B. Comparing the AEM EELS and EDS situations, the cross section is the same for both processes, and the primary source brightness is the same. EELS would seem to have a considerable edge with regard to detection efficiency because: [1] The EDS case suffers the partitioning of de-excitation between the Auger and x-ray branches, with the fluorescence (x-ray) yield typically being between 0.01 and 0.3 for most x-ray energies. [2] X-rays are emitted isotropically into 4π steradians, and the typical AEM EDS detector has a solid angle of 0.2 steradians, resulting in a geometric efficiency for EDS of less than 2%. Moreover, attenuation of the x-ray signal due to spectrometer window absorption becomes significant below 2 keV, further reducing the x-ray efficiency. However, the actual situation

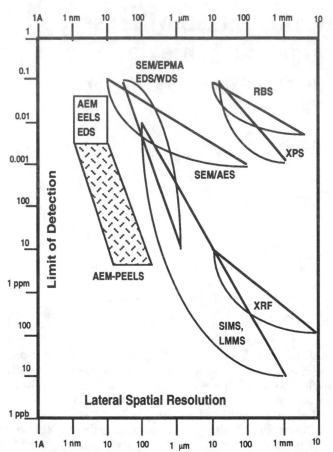

Figure 1. Limit of detection as a function of lateral spatial resolution for secondary ion mass spectrometry (SIMS), laser microprobe mass spectrometry (LMMS), x-ray fluorescence (XRF), x-ray photoelectron spectrometry (XPS), Rutherford backscattering spectrometry (RBS), scanning electron microscopy (SEM) or electron probe microanalysis (EPMA) with Auger electron spectrometry (AES), with energy dispersive x-ray spectrometry (EDS), and with wavelength-dispersive x-ray spectrometry (WDS), analytical electron microscopy (AEM) with energy dispersive x-ray spectrometry (EDS) and electron energy loss spectrometry (EELS), and AEM with parallel-detection electron energy loss spectrometry (PEELS) in the trace nanoanalysis mode (hatched), the subject of this report.

of the EDS/EELS limits of detection found in practice is illustrated in Figure 2, which compares a portion of the EDS and EELS spectra of NIST SRM 2063a (Thin Glass Film, 76 nm thick). The peak-to-background is found to be much higher for the EDS case than for EELS. In EELS, there are other inelastic scattering events that contribute to the background, and the core edge information is typically spread over a 100 eV energy range. From equation (1), EDS detection limits are generally in the range 0.001 to 0.01, while EELS is generally regarded to have limits of detection in the percent range because of the poorer peak-to-background (2).

IMPROVING EELS DETECTION LIMITS

Recent developments in instrumentation and spectral processing have led to remarkable improvements in the fractional sensitivity which can be achieved with EELS (3). A great improvement in measurement efficiency has been realized with the introduction of the parallel detection spectrometer (PEELS), in which an energy band ($\Delta E = 500$ to 2000 eV) can be simultaneously measured by using a magnetic prism to disperse the spectrum across an array of 1024 parallel photodiode detectors (4).

A second advance has been made in processing the PEELS spectrum, as shown in Figure 3. The raw EELS spectrum is characterized by abrupt changes in intensity ("edges") which correspond to the core level ionization energies superimposed on the continuum of energy losses arising from other inelastic processes. The intensity increase that occurs at an ionization edge extends above the edge to energy losses of 100 eV or more. In conventional EELS analysis, the intensity associated with an edge is extracted by fitting a background to the pre-edge continuum, following an E^{-r} model, and then extrapolating under the edge. An example of this procedure is shown in Figure 3(a) for NIST glass K1012. A minor level of barium (Ba $M_{4,5}$, atom fraction $C_a=0.024$) can be recognized in the raw spectrum, and after background correction and vertical scale expansion, the edges for trace Ce ($M_{4,5}$, $C_a=0.0028$) become visible. While this form of background modeling and subtraction is useful, the low-intensity edge features of trace constituents are eventually overwhelmed by inevitable channel-to-channel detector gain variations, a particularly egregious example of which is noted in Figure 3(a). A powerful alternative to this modeling approach is collecting PEELS spectra in the "difference" mode (5). The difference mode involves shifting the spectrum on the detector array by an amount, ΔE, during accumulation. To minimize sensitivity to beam current variations and/or thickness variations if the beam position wanders during collection, the original and shifted spectra are collected sequentially with a duty cycle of 1 to 10 seconds. For "second difference" processing, the original and $\pm\Delta E$ spectra are collected and combined on a channel-by-channel basis as

$$S_2(E) = 2*S_0(E) - S_{\Delta+}(E) - S_{\Delta-}(E) \qquad [2]$$

An example of a second difference spectrum is shown in Figure 3(b). Two features are noted: (1) Second difference processing acts as a frequency filter that suppresses low frequency spectral components, especially the slowly varying background and the extended tail above the edge, while enhancing high frequency features such as the "white lines", which are narrow resonance structures found at some core edges. (2) The spectral artifacts created by channel gain variations are quantitatively suppressed except, of course, for the variance, as shown in Figure 3(b). Figure 3(b)

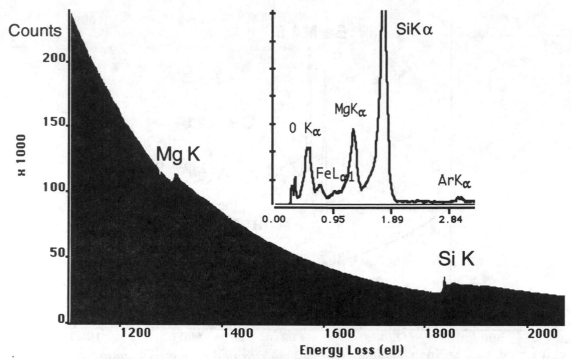

Figure 2. SRM 2063 a (Thin Film Glass) a portion of the spectrum as measured by EELS and by EDS (inset, scale counts *vs*. keV); atomic fractions: O = 0.609; Mg = 0.074; and Si = 0.204.

demonstrates the recovery of an additional element, Cs (Cs $M_{4,5}$, C_a = 0.0035), in the second difference spectrum as compared to the direct spectrum with background modeling.

There are two keys to successful detection of trace constituents (3). The first requirement is the presence of the sharp white line resonance features in the spectrum. In terms of equation [1], improved detection sensitivity is achieved because the second difference processing permits extraction from the spectral artifacts of a characteristic spectral feature with high P/B. Not all elements have white lines, and their presence can be dependent on chemical state. For example, copper in the metallic state does not have white lines, while copper oxides have pronounced white lines at the Cu $L_{2,3}$ edge. Second, the ultimate limit of detection is inevitably imposed by variations controlled by the counting statistics. As is always the case in statistically-limited situations, the only way to improve the limit of detection is to accumulate more counts. The strategy followed is to increase the incident beam current to provide the maximum possible count rate that can be processed by the individual detector elements, about 30 kHz, in the spectral region of interest. (Note that because of the E^{-r} behavior of the spectrum, the spectrum at lower energy loss must be shifted off the detector array to avoid saturation.) Thus, incident beam current is the parameter to control. It is desirable to provide incident beam currents in the range 1 - 50 nA. Trace PEELS measurements can be performed with the high brightness field emission gun, which provides probe sizes in the 1 - 3 nm range, but the lower brightness,

high current LaB_6 source is actually better suited to controlling and maximizing the PEELS detection efficiency.

PEELS AT PARTS PER MILLION LEVELS

Experiments performed on NIST SRM 610 (Trace Elements in Glass) have demonstrated that second difference PEELS measurements can detect uniformly diluted trace constituents present below 100 parts per million (ppm), with possible limits of detection below 10 ppm (3). A particle specimen volume with an estimated maximum thickness of 100 nm and with a cross-sectional area of 10 x 10 nm was selected by scanning a 3 nm, 10 nA, 100 keV beam. This volume contains approximately 1 million total atoms. Figure 4 shows a composite spectrum covering the energy loss range from 0.5-1 keV with low intensity peaks corresponding to several of the trace elements present in SRM 610. Specific information on detection is given in Table 1. The volume sampled contains only about 75-500 atoms for most of the trace constituents. Measurements on different particles revealed consistent detection of the constituents noted in Table 1, but the intensities were found to vary in a manner consistent with the expected statistical distribution of atoms in a homogeneous dilute solution. This achievement of simultaneous trace and nanometer-scale spatially-resolved analysis has been designated "trace nanoanalysis," (3) and its position as AEM/PEELS in this mode relative to AEM/EDS and AEM/conventional EELS as well as the other techniques is indicated in Figure 1 as the hatched area.

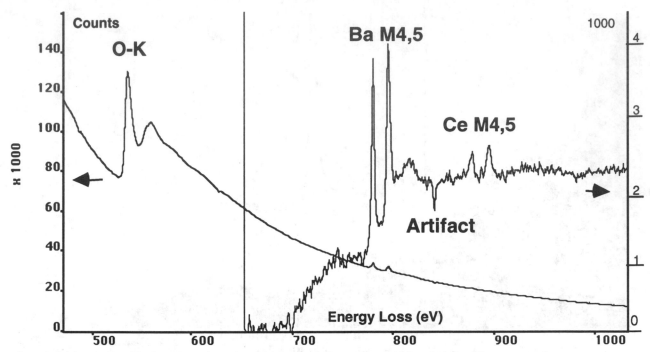

Figure 3(a). PEELS of NIST glass K-1012, showing raw and background corrected spectra.

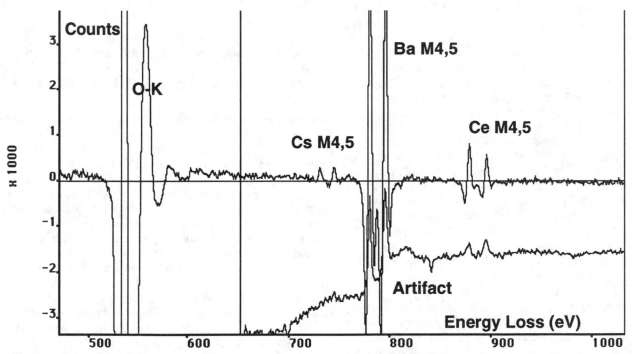

Figure 3(b) Second difference PEELS spectrum superimposed on background corrected spectrum.

AEM/PEELS SENSITIVITY LIMITATIONS

Three principal limitations exist in the application of AEM/PEELS trace nanoanalysis to a particular problem: [1] The element to be measured must have a white line structure at the ionization edge. White lines are somewhat dependent on the chemical environment, so that the matrix can have a strong influence on the behavior observed for a particular element. No database currently exists for white lines in semiconductor matrices. However, it is interesting to note that transition metals such as iron and copper do produce strong white lines in silicate systems. Thus, it may be possible to study the distribution and gettering of these transition metals by oxide precipitates in silicon.

380

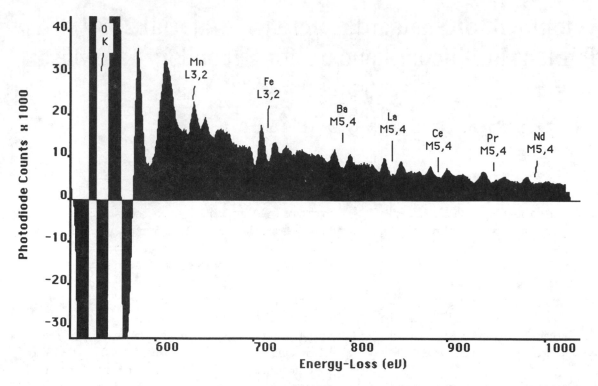

Figure 4 PEELS spectrum of NIST SRM 610 (Trace Elements in Glass) demonstrating trace level detection.

[2] The specimen must be capable of being prepared as a thin section, preferably less than 100 nm thick, and ideally less than 50 nm thick to reduce multiple scattering which lowers the signal received in the PEELS spectrometer aperture.

[3] The sample must be capable of withstanding significant electron irradiation. Typically, each trace constituent atom is ionized tens to hundreds of times to produce a spectrum such as that shown in Figure 4. The specimen must not be significantly damaged during this irradiation. To reduce radiation damage as well as specimen contamination, AEM/PEELS measurements are typically made with a low temperature stage (< 100 K) .

REFERENCES

1. Joy, D. C., Romig, A.D., Jr., and Goldstein, J. I., *Principles of Analytical Electron Microscopy,* New York, Plenum Press, 1986.
2. Wittry, D. B., *Electron Microscopy Analysis, Proc. 9th Intl. Conf. X-ray Optics& Microanalysis,* **3** , eds. P. Brederoo, V. E. Cosslett, EUREM, Leiden, 14-21,1980.
3. Leapman, R. D. and Newbury, D. E., *Analytical Chem.*, **65**, 2409-2414, 1993.
4. Krivanek, O. L., Ahn, C. C., Keeney, R. B., *Ultramicroscopy*, **22**, 103-116, 1987.
5. Shuman, H. and Kruit, P., *Rev. Sci. Instrum.* **56**, 231, 1985.

Table 1
Elements Detected in SRM 610
(concentrations in parts per million, atomic fraction)

Element	Z	Conc.	Edge
F	9	559	K
Mg	12	437	K
Sc	21	236	$L_{2,3}$
Ti	22	194	$L_{2,3}$
Mn	25	187	$L_{2,3}$
Fe	26	174	$L_{2,3}$
Co	27	140	$L_{2,3}$
Ba	56	77	$M_{4,5}$
La	57	76	$M_{4,5}$
Ce	58	76	$M_{4,5}$
Pr	59	75	$M_{4,5}$
Nd	60	74	$M_{4,5}$
Sm	62	71	$M_{4,5}$
Eu	63	70	$M_{4,5}$
Gd	64	68	$M_{4,5}$
Tb	65	67	$M_{4,5}$
Dy	66	65	$M_{4,5}$
Ho	67	64	$M_{4,5}$
Er	68	63	$M_{4,5}$
Tm	69	63	$M_{4,5}$
Tb	70	61	$M_{4,5}$

The Development of Standard Reference Material 2137 - A Boron Implant in Silicon Standard for Secondary Ion Mass Spectrometry

D. S. Simons, P. H. Chi, R. G. Downing and G. P. Lamaze

Chemical Science and Technology Laboratory, National Institute of Standards and Technology, Gaithersburg, MD 20899

Ion-implants are recognized as valuable reference materials for secondary ion mass spectrometry (SIMS), because the integral of the concentration as a function of depth is equal to the retained ion dose or fluence. If the ion dose and sputter rate for a SIMS depth profile are known, then the scaling factor between ion count and atomic concentration, known as the relative sensitivity factor (RSF), can be calculated. This RSF can be used to determine the concentration of the same element in an unknown sample of the same host material.

The ion dose is controlled during implantation by charge integration in the ion implanter. However, for a *certified* reference material, the dose should be independently measured in the target by a reference method based on well-understood physical principles. In the semiconductor industry, implants are most often checked by Rutherford backscattering spectrometry (RBS) if the dose is sufficiently high and the dopant has a higher atomic weight than the matrix. However, a recent study showed that discrepancies in RBS determinations among different laboratories can be as high as 20 % (1).

We chose to developed a Standard Reference Material for SIMS based on ion implantation of boron in silicon. Boron is the most common dopant in silicon semiconductor technology, and the dose of the isotope ^{10}B can be measured by a nuclear reaction method known as Neutron Depth Profiling. NDP satisfies the criteria for a reference method because it relies on a nuclear reaction, which is not affected by the chemical matrix, and the reaction cross section and geometric parameters can be calibrated with a sputter-deposited metallic ^{10}B film of known mass per unit area as determined by gravimetry.

The starting material for the standards was a batch of 24 commercial n-type silicon (100) single crystal wafers of 76-mm diameter, polished on one side. A grid pattern of 5-mm squares was made on the unpolished side of the wafer by a photolithographic method. Each 5-mm square contains row, column, and wafer numbers that provide unique locational identification. The polished side of the wafer was implanted with ^{28}Si ions to render the surface sufficiently disordered that it would appear to be amorphous to the ^{10}B ion beam. The purpose was to eliminate ion channeling of the boron that would otherwise produce a more complicated profile shape, which could vary across the surface of a wafer. The ^{28}Si ions were implanted at three different energies to create lattice disorder over the entire range of the boron implant. The ^{10}B was implanted at a nominal energy of 50 keV and a nominal dose of 1×10^{15} atoms/cm². The wafers were then cut into 1 cm x 1 cm squares with a wafer saw. Twenty-one squares, each at least 7 mm from the outer edge, were obtained from each wafer.

Extensive NDP measurements were made on three wafers to determine the certified boron ion dose. Fifty-one dose measurements were made on 38 individual samples, with 13 replicates. The final certified dose of ^{10}B is $1.018\pm0.035\times10^{15}$ atoms/cm², the stated uncertainty being a 95 % coverage, 95 % confidence tolerance interval that takes into account the spatial variation of the implantation process.

SIMS measurements were made to check the profile shape and to examine the contribution of $^{30}Si^{3+}$ to the signal at m/z 10. An overlay of eight superimposed profiles from two areas on each of two different wafers, with two replicates of each, shows excellent homogeneity, a smooth shape over four decades of concentration, and a low background level.

(1) P. Roitman *et al.*, in A. Benninghoven *et al.*, Eds., *Secondary Ion Mass Spectrometry SIMS VII*, Chichester: John Wiley & Sons, 1990, pp. 115-118.

Quantitative 3-Dimensional Image Depth Profiling by Secondary Ion Mass Spectrometry

Greg Gillen

Surface and Microanalysis Science Division
Chemical Science and Technology Laboratory
National Institute of Standards and Technology
Gaithersburg, MD 20899

A procedure is described for using secondary ion mass spectrometry image depth profiling on an ion microscope to generate quantitative 3-dimensional distribution images of gallium-and boron-patterned ion implants in silicon. The gallium and boron implants were profiled with the SIMS instrument operating as an ion microscope and as an ion microprobe, respectively. The relative merits of these different approaches for 3D imaging are evaluated.

INTRODUCTION

Secondary Ion Mass Spectrometry (SIMS) is the most widely used analytical tool for quantitative depth profiling analysis of dopants/impurities in electronic materials. As an elemental depth profiling technique, SIMS offers typical detection limits of less than one ppm and a depth resolution approaching 5 - 10 nm. One of the limitations of conventional SIMS depth profiling is that information on the lateral distribution of the dopant is lost. As electronic device dimensions continue to decrease, it is becoming necessary to characterize not only the depth distribution of dopant species but also their lateral distribution. This can be achieved simultaneously using image depth profiling SIMS. In this mode of operation, a series of elemental SIMS images are acquired from the same region of the sample as a function of time. Because the sample surface is continually eroded during analysis, each subsequent image corresponds to a "slice" from a slightly greater depth below the original sample surface. After the analysis is completed, this "stack" of images can be processed to produce a 3-dimensional image of the dopant/impurity in the near-surface region of the sample. Although the potential advantages of 3D SIMS imaging have been realized for many years (1) (see ref 1 for a recent review of the technique), relatively few publications have been presented on its use, and the technique is not commonly applied in most SIMS laboratories This lack of progress has been due, in large part, to the requirement for expensive high speed computer graphic workstations and sophisticated 3D rendering software to process and display the large amounts of data that are obtained in a 3D SIMS analysis. However, advances in image processing software, inexpensive and larger storage capacity disk drives and faster personal computers have recently made it feasible to process 3D data sets on the "desktop" without resorting to specialized image processing workstations. Coupled with advances in quantitative imaging detectors for SIMS instrumentation, quantitative 3D SIMS imaging is now becoming a viable tool for practical semiconductor analysis. This paper describes the instrumentation and experimental procedures used in our laboratory to acquire, process and display 3D SIMS images. Examples of quantitative 3D imaging of patterned gallium and boron ion implants in silicon will be presented. The relative merits of two different instrumental approaches for 3D imaging will be briefly evaluated. Finally, some of the factors currently limiting the extension of this technique to the characterization of more complex semiconductor devices will be discussed.

EXPERIMENTAL

All 3D SIMS profiling experiments were conducted on a magnetic sector Cameca IMS 3F or 4F* SIMS instrument with detection of positive secondary ions using one of two different imaging modes. When the instrument was used as an ion microscope, the analyzed area was bombarded by a rastered, large diameter (> 20 μm), O_2^+ primary ion beam with a current of several hundred nanoamps at an impact energy of 8 keV. The stigmatic secondary ion optics of the instrument were used to project a mass-resolved secondary ion image of the circular analyzed area (typically 150 or 400 μm in diameter) onto an imaging detector. For qualitative imaging this detector is typically a channel plate electron multiplier followed by a fluorescent screen; a video or CCD camera is then used to digitize the image projected on the fluorescent screen. For quantitative imaging experiments, a double channel plate electron multiplier followed by a position-sensitive, pulse-counting resistive anode encoder detector (RAE) (2) was used. Because this detector is pulse-counting, quantitative SIMS imaging is greatly simplified as compared to the use of video cameras. The spatial resolution of the images obtained in the microscope mode is ~ 1 μm. The resolution is increased by reducing the size of secondary-beam-defining apertures, which reduces aberrations in the secondary ion image. However, increased spatial resolution is obtained at the expense of reduced secondary ion transmission. The SIMS instrument may also be operated as an ion microprobe. In this mode, a finely focused O_2^+ primary ion beam with a diameter of 0.2-1 μm, primary current of < 10 pA, and an impact energy of 10.5 keV was rastered across the sample surface while the computer stored the secondary ion intensity, as measured on an electron multiplier detector, as a function of the beam position. In this mode, image resolution depends only on the primary beam diameter so that the secondary-beam-defining apertures used to reduce aberrations may be wide open resulting in maximum secondary ion transmission.

After completion of an image depth profile, individual images are transferred from the acquisition computer to a Macintosh IIfx computer for off-line processing and display. Using image processing software written at NIST (3), images are checked for registration, normalized to a matrix ion species, and scaled for display. After processing, the "stacks" can be displayed in a variety of ways using commercially available image processing software. The examples shown in this paper were displayed with VoxBlast (VayTek, Fairfield, IL), a 3D volume-rendering software package using a ray-tracing approach.

Samples used in this study include two types of patterned ion implants into silicon. The boron sample consisted of a series of rectangular B implants (^{11}B, 50.0 keV, 2×10^{15} ions/cm^2) with a width of 6 μm and a length of 0.5 or 1 mm. This sample was prepared using implantation through a mask. The second sample was a test structure prepared by overlaying four rectangular gallium implants (^{69}Ga, 25 keV, $\sim 1 \times 10^{16}$ ions/cm^2) with a 45° rotation between implants. Each implant had a width of 30 μm, a length of 330 μm and was prepared without a mask using a gallium focused ion beam (FIB).

QUANTIFICATION OF 3D SIMS IMAGES

To quantify the 3D data sets, a standard SIMS depth profile was first run on a homogeneous implant standard (in silicon) of the impurity to be analyzed. For ion microscope analysis, the standards were run using the RAE detector. For ion microprobe analysis, the electron multiplier was used as a detector. From this depth profile a relative sensitivity factor (RSF) is calculated using the dose of the implant, integrated counts under the implant peak (after background subtraction) and depth of analysis (determined by stylus profilometry) (4). For analysis of the patterned ion implants, the relative sensitivity factor allows for the conversion of the measured intensity ratio I_i/I_m (impurity /matrix) in each pixel of the 3D volume to atomic concentration using the following relationship:

$$\textit{Impurity density (atoms/cm}^3) = I_i/I_m \times RSF$$

ION MICROSCOPE IMAGING

The ion microscope coupled with a quantitative imaging detector is the most generally useful configuration for quantitative image depth profiling. Because the image resolution depends only on the secondary optics, large diameter, high current primary ion beams compatible with the high erosion rates necessary for depth profiling can be used. For the same reason, lower primary beam energies can be used that decrease the primary ion penetration depth and thereby increase the depth resolution of the analysis. Finally, because the secondary ion signal from each pixel in the illuminated area is acquired and digitized in parallel, image acquisition is rapid which is also useful for depth profiling. A limitation with quantitative microscope image depth profiling is that the RAE detector has a significant dead-time due to the characteristics of the pulse-processing electronics. The dead-time limits input count rates to $\sim 10^4$ counts/s (5). However, this limitation has recently been overcome by a newer "fast RAE" that allows processing of input count rates up to 10^5 counts/s. Also, slow-scan, scientific grade CCD cameras are now available that offer similar performance to the RAE (in terms of linearity with input signal) but are able to process input count rates that are several orders of magnitude higher.

As an example of microscope-based image depth profiling, Figure 1 shows a quantified 3D SIMS image of the patterned gallium ion implant. The image consists of a stack of 91 ^{69}Ga ion images acquired for 60 s each on the RAE detector. The instrument was operated with a circular field-of-view of 400 μm in diameter. The displayed depth is 370 Å. The figure shows 3 different views of the implant. The first shows the implant as viewed looking down on the surface (0°). The volume was then rotated 90° clockwise into the plane of the paper to give the side view. Finally, the volume was rotated a full 180° to show the sample from the underside. The image was quantified using a relative sensitivity factor determined from a homogeneous gallium implant in silicon.

ION MICROPROBE IMAGING

When small areas are to be analyzed, the microprobe imaging approach is preferred. Because no beam-defining apertures are used, secondary ion transmission at spatial resolutions of one micrometer or less are over two orders of magnitude greater than for operation in the microscope mode(6). Unfortunately, the highly focused primary beams needed for microprobe imaging have very low primary ion currents resulting in very low erosion rates. This makes microprobe depth profiling only practical for rastered areas of at most a few tens of micrometers on a side. Figure 2 shows a series of 3D SIMS images of the patterned boron ion implants in silicon. The instrument was operated with an imaged area of 30 μm x 30 μm. Each pixel in the volume was quantified using an RSF determined from a homogeneous boron implant standard in silicon. The figure shows 6 views of the implant beginning with the surface view (0⁰) and then rotating the volume in subsequent steps of 60⁰ clockwise into the plane of the paper.

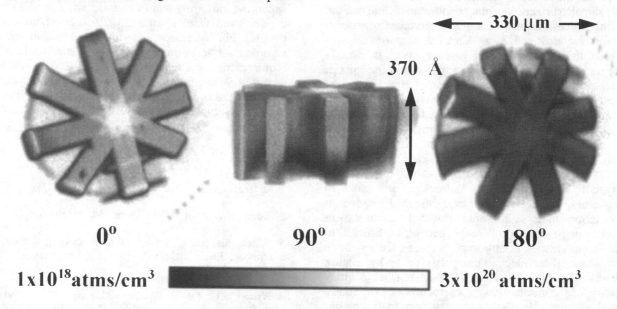

Figure 1. 3D SIMS image of patterned gallium ion implant in silicon.

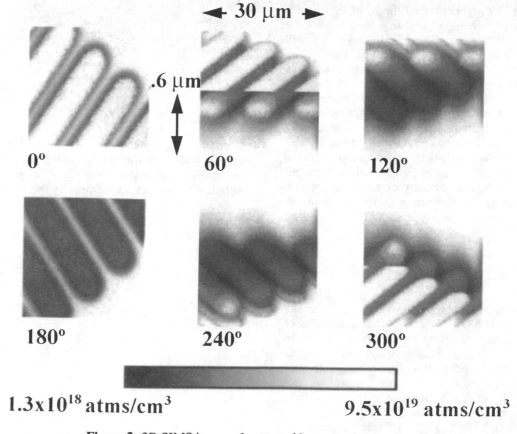

Figure 2. 3D SIMS image of patterned boron ion implant in silicon.

ARTIFACTS IN 3D SIMS

The most significant limitation with using ion beam sputtering as the basis for a 3D visualization technique is that it must implicitly be assumed that the original sample surface was planar and smooth and that the sputtering process removes sample material at a uniform rate across the analyzed area. These assumptions allow a unique depth to be assigned to each voxel in the 3D volume. These assumptions are valid for ion implants in well behaved matrices such as silicon. However, most samples will not exhibit such ideal behavior. For example, the typical microelectronic device may have a non-planar surface (i.e., interconnects, bonding pads) and is most likely composed of a complicated multilayer structure containing several types of materials, each of which may sputter at a different rate under ion bombardment. Also, polycrystalline metals, such as aluminum, are prone to development of surface topography during analysis due to local variations in crystallographic orientation with respect to the incoming ion beam. While it may prove difficult to correct for non-planar sample surfaces (samples may have to be planarized by mechanical polishing or by coating before analysis), there are a number of approaches that may be useful for minimization or correction of spatially resolved sputter rate variations on originally planar samples. We are currently investigating the use of the atomic force microscope (AFM) to image SIMS craters after 3D analysis. By registration of the AFM image with the 3D SIMS image, we hope to be able to determine the local depth scale for each voxel in the 3D image. We will also attempt to use rotation of the sample during analysis to reduce beam-induced topography and spatial variations in sputter rate resulting from crystallographic effects. This experiment will require sophisticated image processing to correct for the effect of the rotation on the acquired images. Finally, we have conducted preliminary experiments using ion implantation of a tracer element into the analytical sample prior to analysis. Assuming crystallographic effects are taken into consideration during implantation, the tracer implant will have a peak concentration that occurs at a uniform depth throughout the sample. The peak of the implant then serves as an internal depth calibration for each voxel in the 3D volume.

CONCLUSIONS

With improvements in the speed and storage capacity of microcomputers, the commercial availability of 3D volume-rendering software, and improvements in imaging detectors, it is now feasible to consider 3D SIMS as a tool for practical depth profiling analysis of samples where both in-depth and lateral information are needed. Even when spatially resolved information is not required, as in the case of homogeneous ion implants, image depth profiling SIMS may be beneficial. For example, once an image depth profile is stored in the computer, the images may be resampled to generate selected area depth profiles that would allow the analyst to improve the dynamic range of the depth profile by removing sample artifacts and crater wall effects. An image depth profile might also serve as a useful diagnostic tool to determine if a sample was sputtering uniformly and to identify artifacts that might not be observed from examination of the standard depth profile. Finally, for depth profiling of small features on complex surfaces, image depth profiling would relax the requirement for exact alignment with the region to be analyzed. A larger area would be analyzed by image depth profiling and subsequent selected area profiling could then be used to localize the region of interest.

REFERENCES

1. Rudenauer, F.G., *Analytica Chimica Acta* **297**, 197-230 (1994).

2. Odom, R.W. et. al., *Anal. Chem.* **54,** 2 (1982).

3. MacLispix, public domain image processing software for the Macintosh, David Bright, Chemical Science and Technology Laboratory, NIST.

4. Wilson, R.G., Stevie, F.A., and Magee, C.W., "Secondary Ion Mass Spectrometry: A Practical Handbook for Depth Profiling and Bulk Impurity Analysis. ", John Wiley and Sons, New York, 1989, pp. 3.1-1-3.2-11.

5 Gillen, G., and Myklebust, R.L., "Secondary Ion Mass Spectrometry, SIMS VIII"", John Wiley and Sons, Chichester, 1992, pp. 509-512.

6. Zeininger, H., and v. Criegern, R., "Secondary Ion Mass Spectrometry, SIMS VII", John Wiley and Sons, Chichester, 1990, pp. 419-424.

New Applications of Quadrupole-based Secondary Ion Mass Spectrometry (SIMS) in Microelectronics

Steven W. Novak, Michael R. Frost and Charles W. Magee

Evans East, 666 Plainsboro Road, Suite 1236, Plainsboro, NJ 08536

We have developed strategies for using quadrupole-based SIMS to measure shallow-junction dopants, ultra-thin gate dielectrics such as nitrided oxides, and surface metal contaminants introduced by cleaning processes or ion implantation. All of these measurements utilize low energy ion bombardment (1-3keV) at a 60° incidence angle which maximizes depth resolution yet maintains sufficiently low detection limits. Measurements for dopants and surface contaminants are typically done by measuring positive or negative atomic ions, whereas measurements of thin oxynitrides are accomplished by monitoring positive Cs-cluster ions.

INTRODUCTION

The push for greater complexity in integrated circuits is driving the need for ever smaller device dimensions, both in area and in depth. These changes in turn place more stringent requirements on the analytical techniques needed to characterize the new processes. In this paper, we will discuss methods for using SIMS in three areas of current intense interest in semiconductor manufacturing: measuring shallow junction dopant profiles, elemental distributions in ultra-thin gate dielectrics, and contaminants on Si surfaces. We have developed methods to optimize measurements of these materials and these methods should find general utility in any measurement of elemental distributions at very shallow depths (≤ 10 nm).

INSTRUMENTATION

All of the analytical work reported here was acquired using an unmodified PHI 6600 quadrupole SIMS instrument equipped with both a duoplasmatron source and a Cs ion source. We will focus primarily on improved results obtained using different operating conditions and analytical strategies, rather than on instrumental modifications.

MEASURING SHALLOW JUNCTION DOPANT PROFILES

As device geometries shrink, it will be necessary to shrink junction depths proportionately. This means that SIMS, the technique of choice for measuring dopant profiles in semiconductors, will be increasingly required to accurately determine dopant profiles within the top 10's of nanometers of a Si wafer, as well as measure the effects of subsequent thermal processing. We have had considerable success in measuring dopant implantation profiles within

this region by using low energy ion bombardment at high incidence angles. A quadrupole-based instrument is well suited for such an application because present designs do not couple the incident bombardment energy with the secondary ion extraction potential, as is commonly done in magnetic sector instruments. This means that the incident bombardment energy can be reduced to as low as 1 keV for both Cs and O bombardment with either positive or negative secondary ion detection. In addition, the angle of incidence of the primary beam can be varied at a given impact energy. Therefore, all of the relevant beam parameters can be varied independently to best optimize depth resolution and sensitivity.

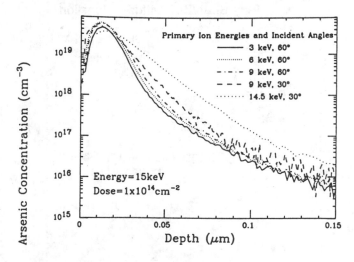

FIGURE 1. Measurements of a 15 keV As implant in Si at different Cs bombardment energies.

The need for bombarding with low energy Cs is clearly shown when profiling As in Si (Figure 1) (1). It is evident from this figure that it is necessary to reduce the beam impact energy to at least 3 keV to minimize ion beam-induced atomic mixing of the As ions and therefore determine the As channeling distributions accurately. We have also used 3 keV Cs bombardment to measure B profiles (Figure 2). However, a comparison of low energy O_2^+ bombardment with low energy Cs bombardment shows that even 3 keV Cs causes enhanced atomic mixing of the B profile. Note that even at these low impact

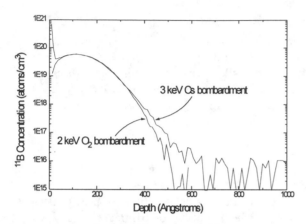

FIGURE 2. Measurements of a 3 keV ^{11}B implant in Si.

energies, the detection limits for B and As are near 1×10^{15} atoms/cm^3, well below the limits needed to measure the depth to a shallow bipolar junction that may have a doping level of closer to 1×10^{19} atoms/cm^3 (Figure 3). Note that

the depth scale in this figure begins at a depth of 0.20 micrometers, and the depth of the original crystalline Si surface is at 0.22 micrometers as shown by the peak in the O profile. Figure 3 shows that we can measure the depth to a p/n junction that is 27 nm beneath the original Si surface. This is shallower than the requirements anticipated by SEMATECH for manufacturing well into the next century (2).

MEASUREMENT OF THIN DIELECTRIC LAYERS

Another area of active investigation is the formation of high strength dielectric layers for MOS gates. As device dimensions shrink, the gate dielectric thickness must also be reduced. However a thinner dielectric layer has a lower breakdown voltage so efforts have concentrated on methods to increase the dielectric strength of conventional thin SiO$_2$ films. In addition, this layer must have improved barrier characteristics to prevent diffusion of dopants or contaminants through the oxide into the overlying gate. One method that has met with particular success is nitridation of oxide layers. This technique could be shown to improve dielectric and barrier characteristics, but the distribution of N within the oxide layers, and hence the mechanism of nitridation, were unknown. By using 1 keV Cs bombardment and monitoring positive Cs-cluster ions (3), we have been able to obtain accurate N profiles from oxide layers thinner than 10 nm (Figure 4). The use

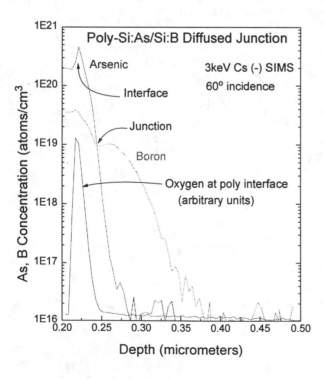

FIGURE 3. Depth profile of a shallow p/n junction in Si.

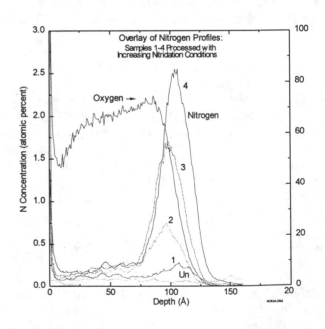

FIGURE 4. SIMS measurements of thin nitrided oxides processed with different nitridation times.

of Cs-cluster ions to reduce matrix effects has been well documented and is discussed in a number of papers in the most recent International SIMS conference (4). Our measurements clearly show that the N-rich layer typically lies at the interface between the oxide layer and the Si substrate, rather than within the Si substrate as has been portrayed in previous publications (5).

These measurements are accurate enough that subtle effects of differing nitridation conditions can be readily measured (Figures 4 and 5). Figure 5 shows that even for oxide layers thinner than 5 nm the nitrogen peak is clearly resolved at the substrate interface. This implies that the depth resolution of these measurements must be less than about 2 nm.

has also become clear in recent years that the ion implantation process can introduce significant amounts of impurities to a wafer surface due to scattering from internal surfaces of the implanter. Measuring surface contamination on Si wafers has been an active area of interest for many years in both the SIMS community and the microelectronics community (6,7). Here we focus on measuring a selected group of elements, Na, K and Al, that have particularly high ion yields. Measurements we have performed show that secondary ions of these elements have nearly identical ion yields from SiO_2 and Si. We can therefore use ion implanted standards in Si to calibrate profiles of these elements and the profiles will be accurate up through the surface native oxide. Their high ion yields also eliminate the requirement to use an oxygen bleed, because the yields are not changed significantly by using the bleed-in. A quadrupole-based instrument is particularly suited to this type of measurement because the time required to switch between masses is very low and allows acquisition of many data points within the shallow region at the wafer surface.

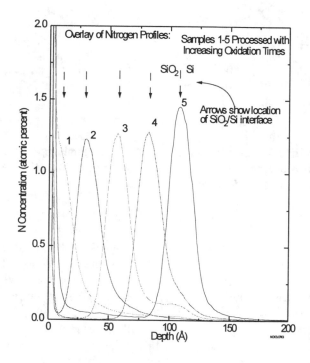

FIGURE 5. SIMS measurements of thin nitrided oxide films having different oxide thicknesses.

Measurements of surface contamination on Si wafers

The cleanliness of the Si wafer surface has always been important in controlling wafer yields. As device dimensions are reduced, cleanliness requirements become increasingly stringent in order to eliminate incorporation of impurities in thin gate oxides and better control oxide thicknesses, as well as a number of other reasons (2). It

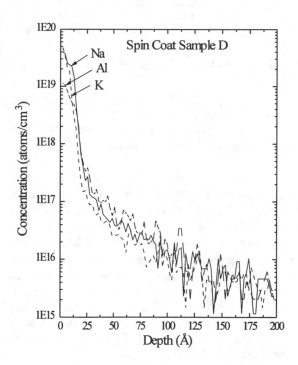

FIGURE 6. Measurement of three elements on the surface of an intentionally contaminated Si wafer.

The analytical technique simply consists of acquiring a shallow depth profile and integrating the signals for each element from the surface (Figure 6). Measurements of intentionally contaminated spin-coated wafers, calibrated using ion implanted Si standards, are compared to VPD-AAS measurements of the same wafers in Table 1. Note

TABLE 1. Average areal densities (atoms/cm^2), standard deviation and percent standard deviation for spin coated Si samples analyzed by SIMS using ion implanted Si standards.

Sample	n	Na	Al	K
A	6	$1.2\pm.06\times10^{11}$ (5.5%)	$3.9\pm.15\times10^{10}$ (4.4%)	$6.6\pm.43\times10^{10}$ (7.1%)
B	7	$3.6\pm.20\times10^{11}$ (6.1%)	$9.2\pm.58\times10^{10}$ (6.8%)	$1.9\pm.43\times10^{11}$ (12%)
C	6	$1.2\pm.09\times10^{12}$ (8.4%)	$2.7\pm.19\times10^{11}$ (7.6%)	$7.3\pm.59\times10^{11}$ (5.9%)
D	6	$3.8\pm.35\times10^{12}$ (10%)	$9.7\pm.06\times10^{11}$ (7.3%)	$2.4\pm.19\times10^{12}$ (8.7%)
D by VPD		3.4×10^{12}	1.2×10^{12}	2.9×10^{12}
E	6	$2.8\pm.24\times10^{10}$ (9.6%)	$1.2\pm.18\times10^{10}$ (17%)	$7.1\pm1.0\times10^{9}$ (16%)

the excellent agreement between the absolute areal densities measured by both techniques, as well as the high precision of the individual measurements.

We have routinely used this technique to examine the amount of impurities introduced to a wafer surface during ion implantation. Figure 7 shows a series of measurements for various contaminants across the surface of an ion implanted wafer. It was necessary to cleave the wafer into small pieces to perform these measurements. It is clear that greater amounts of contamination were introduced at the wafer edges due to scattering from the platen used to hold the wafer. SIMS is the only technique that can provide this type of quantitative information because of its combination of high sensitivity and high spatial resolution, especially near the wafer edge.

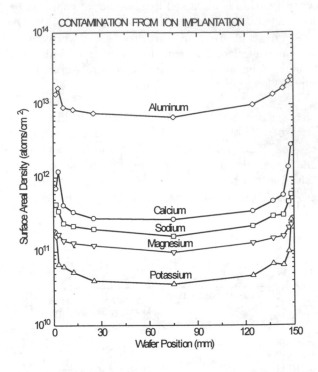

FIGURE 7. Measurements of surface contamination at a series of locations across an ion implanted Si wafer. The center of the wafer is at 75 mm.

CONCLUSIONS

Secondary ion mass spectrometry continues to be the technique of choice for quantitatively measuring dopant and impurity distributions in semiconductor processes. We have demonstrated its utility in three areas of active current interest. By using the Cs-cluster ion method we have extended the range of accurate SIMS measurements to the atomic percentage level. With continued improvements, such as microfocused Ga ion guns capable of sputtering and imaging sub-micron areas, SIMS will evolve along with the demands of semiconductor processing, providing essential information to semiconductor device engineers for years to come.

REFERENCES

1. S.-H. Yang, S. J. Morris, D. L. Lin, A. F. Tasch, R. B. Simonton, D. Kamenitsa, C. W. Magee and G. Lux, *J. Elect. Mat.* **23**, 801-808 (1994).

2. SEMATECH Materials and Bulk Processes Metrology Workshop, May 5-6, 1994, SEMATECH Technology Transfer #94062400A-MIN.

3. C. W. Magee, W. L. Harrington, and E. Botnick, Int. *J. Mass Spect. Ion Proc.* **103**, 45 (1990).

4. Proceedings of the Ninth International Conference on Secondary Ion Mass Spectrometry (SIMS IX), 1994, pp. 375-426.

5. I. Bannerjee and D. Kouzminov, "Anomalous diffusion of nitrogen in SiO_2 under ion bombardment," in *Proceedings of the Second International Workshop on the Measurement and Characterization of Ultra-shallow Doping Profiles in Semiconductors*, 1993, pp. 224-227.

6. G. J. Slusser and L. MacDowell, *J. Vac. Sci. Tech.* **A5,** 1649-1651 (1987).

7. R. S. Hockett and J. C. Norberg, "The practical use of polyencapsulation/SIMS for quantitative surface analysis of silicon substrates," in *Secondary Ion Mass Spectrometry, SIMS VII*, 1990, pp. 491-494.

The Depth Measurement of Craters Produced by Secondary Ion Mass Spectrometry - Results of a Stylus Profilometry Round-robin Study

David S. Simons

Chemical Science and Technology Laboratory, National Institute of Standards and Technology, Gaithersburg, MD 20899

The depth scale of sputter depth profiles obtained by secondary ion mass spectrometry (SIMS) is generally obtained by measuring the depth of the final crater with a stylus profilometer, and assuming a constant erosion rate during the profile. The uncertainty in the depth scale is thus directly related to the uncertainty in the crater depth measurement. This uncertainty has often been quoted in the range of 5 % to 10 % of the crater depth, but our experience suggested that the error could be considerably less. Therefore, a round-robin interlaboratory comparison was organized to assess the repeatability and overall uncertainty of stylus profilometry measurements.

The specimen for the round-robin was a section of a silicon wafer with three separate craters produced by a modern SIMS instrument, with nominal depths of 0.1 μm, 0.5 μm, and 2.0 μm, and nominal areas of 250 μm x 250 μm. Twenty-two SIMS labs in North America agreed to participate by making depth measurements of the three craters with in-house stylus profilometers that had been calibrated with step-height standards. Six separate depth measurements were requested for each of the craters. At NIST, the craters were remeasured several times throughout the study as a consistency check.

The results of the round-robin can be viewed in terms of intra-lab repeatability and inter-lab reproducibility. The median standard deviation for an individual crater measurement by a single lab was 4.0 nm, with a range from 0.5 nm to 54 nm. The high value was anomalous and indicated that the specific profilometer that produced it was in need of repair. The lowest standard deviations for a laboratory were not always associated with the same crater. These observations indicate that the component of uncertainty due to random error is not proportional to crater depth for this range of crater depths in silicon, but is roughly constant for a specific profilometer. The median relative standard deviation is only 0.2 % of the depth for the 2 μm crater, but is 4 % of the 0.1 μm crater.

The inter-laboratory comparison checks calibration as well as repeatability factors. For the 2 μm crater, the relative standard deviation among laboratories was 1.3 %. This is quite impressive considering that many of the step-height standards had systematic errors stated to be only "less than 5 %". At NIST, a commercial step-height standard of 1 μm was compared with a NIST primary standard of similar step height. The disagreement was only 0.6 %, which was within the stated uncertainty of 1 % for this particular commercial standard.

The inter-laboratory relative standard deviations for the other craters were 2.4 % for the 0.5 μm crater and 4.7 % for the 0.1 μm crater. We conclude that the calibration error dominates the overall uncertainty of the depth measurement for the 2 μm crater, whereas the random error component is significant for the 0.1 μm crater. These results have demonstrated conclusively that the uncertainty of SIMS crater depth measurements by stylus profilometry is *not* a simple percentage of the crater depth, and that it can be as small as 1 % of the depth for craters deeper than 1 μm.

OPTICAL AND INFRARED TECHNIQUES

Applications of Near-Field Scanning Optical Microscopy

Harald F. Hess and Eric Betzig

AT&T Bell Laboratories
600 Mountain Ave., Murray Hill, NJ 07974

A variety of applications of near-field scanning optical microscopy, NSOM, are reviewed. Its ability to optically resolve features smaller than the diffraction limit implies a potential for wide ranging impact both for local modification and high resolution characterization. In finding the proper industrial niches for NSOM, it is equally important to take into account its limitations of slowness of its serial process, fragility of the probe, and short depth of focus.

Near-field scanning optical microscopy (NSOM) has recently proven itself through a number of applications to be a powerful new tool in the laboratory. NSOM exploits the optical interaction arising from a sharp probe in close proximity to a sample in order to image surfaces on a scale well beyond the classical diffraction limit. Typically this probe consists of an optical fiber that terminates by tapering into a fine truncated cone. The sides are made optically opaque with an aluminum coating, however, the tip has a small uncoated aperture through which the light can pass. Its diameter commonly ranges from $\lambda/2$ to $\lambda/40$. This locally confines the radiation field to a spot smaller than that achievable with far-field diffraction limited optics. The aperture is freely positioned or raster scanned across the surface allowing highly local-

ized exposures of the sample or generation of high resolution images. As with conventional far-field optics, nanotechnological applications fall into these two broad classes: Surface modification/lithography and characterization/metrology. Each will be considered in turn in their application to different systems. Several topics from a more comprehensive review[1] are described here along with a sampling of more recent highlights.

The potential advantages of NSOM as a modification tool include its adaptability to existing optical resists, its ability to image simultaneously with the exposure process and the possibility of closed loop control of this exposure by monitoring the resultant latent image. Fig. 1 left (from [1]) shows lines with ~100 nm linewidths and spaces that have been written by NSOM in a

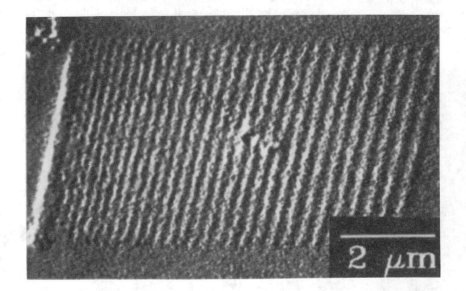

Fig. 1 (left) NSOM written patterns on photoresist, (right) latent images in PMMA

50 nm thick film of conventional photoresist. In Fig. 1 right (also from[1]) a latent image of electron beam exposed polymethylmethacrylate is recorded by NSOM. The contrast here is solely from refractive index-induced changes in the coupling of light into the aperture. The primary disadvantages of NSOM are its limited depth of focus and its limited speed, arising from both the mechanical scanning aspects and the relatively low photon flux involved. Therefore, barring the development of massively parallel arrays of efficient near-field probes, the potential uses of NSOM for modification will be in the laboratory R&D rather than production setting.

One possible exception is in the area of high density near-field optical data storage[2], which in many aspects represents a marriage of magnetic and optical storage technologies. Fig 2 (from [2]) shows how NSOM has been used to both image and record domains in thin-film magneto-optical material. Data densities of ~45 Gbits/in^2 have been achieved, well in excess of current technologies. If this is to become a viable technology a crucial issue of read/write speed must be addressed. Recent development of a high flux near-field fiber laser probe [3] further highlights the potential of this particular application.

In the area of nanostructure characterization, the applications are more broadly based, because NSOM takes advantage of the numerous and powerful contrast mechanisms of conventional optical microscopy. Refractive index differences and polarization contrast have highlighted features of the two examples just given photoresist and magnetic materials. Reflectivity, absorption, and spectroscopy are only the beginning of a growing list of contrast methods. Use of the NSOM probe as a sub-

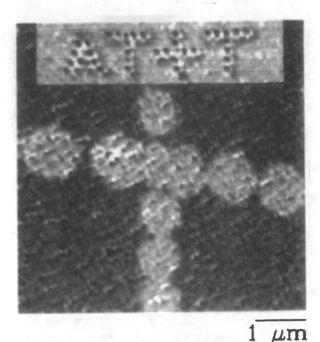

1 μm

Fig. 2 Magnetic domains recorded by near-field optics (top), compared with bits recorded by conventional means.

wavelength optical detector has permitted the characterization of mode profiles of semiconductor lasers[4], and near-field photoconductivity measurements on such lasers have resulted in maps of the p-n junctions therein. Recent demonstration of single molecule detection with NSOM[5] and in particular the ability to determine the orientations of the individual molecules, as illustrated in Fig. 3 (from [5]), may be of some aid in the growing field

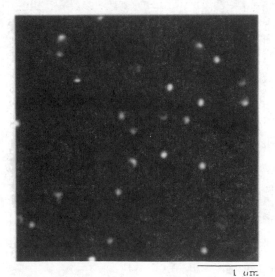 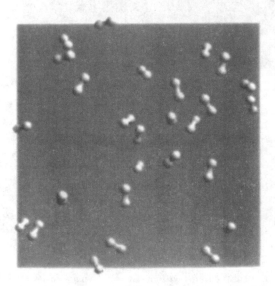

1 μm

Fig. 3 Near-field optical image of individual dye molecules (left) with orientations inferred from the observed patterns (right).

of nanofabrication via molecular self-assembly. The NSOM probe might also offer interesting opportunities for metrology, where position monitoring can be accomplished via the optical signal[6], possibly using the optical path as a leg of an interferometer.

Finally, low temperature near-field scanning optical microscopy/spectroscopy has proven useful in characterization of quantum confined semiconductor structures. In a study of cleaved-edge overgrown quantum wire laser structures, the NSOM spatially resolved the three regions of different luminescent spectra and imaged how local exciton diffusion influences their spatial extent[7]. This has recently been extended to revealing the individual optically active quantum constituents of quantum wells[8]. Specifically, sharp (<0.07 meV), spectrally distinct emission lines of a GaAs/AlGaAs

quantum well can be imaged at a specific spatial location, or as a spectral evolution image as the probe is scanned along a line across the surface, or as a real space image at a specific luminescence wavelengths, see Fig. 4 (from [8]). Temperature, magnetic field, and line width measurements establish that these luminescence centers arise from excitons localized at interface fluctuations. For sufficiently narrow wells, virtually all emission originates from such centers. Local spectral shifts can allow effective well thickness variations to be imaged. This is possibly the most immediately useful development in the quest for a new monitors of optimal crystal growth conditions. Quantities such as diffusion, lateral confinement, and lifetime of the excitation can now be measured at the quantum level. Near-field microscopy/spectroscopy provides a means to access energies and homogeneous line widths

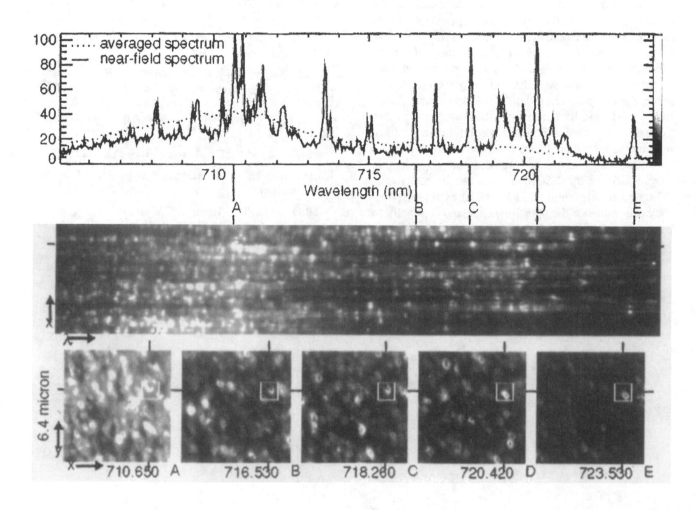

Fig. 4 (Top) Near-field, and spatially averaged photoluminescence spectra from a 23 Å single quantum well. (Middle) Spatial evolution of the near-field spectrum with linear position, presented as the image $L(\lambda,x)|_y$. Bottom) Real space near-field images $L(x,y)|_\lambda$ at the five wavelengths labeled above. Luminescence is found to originate only from certain spatially and spectrally discrete sites, which appear as rings.

for the individual eigenstates of these centers, and thus allows the luminescent components to be identified and characterized with the extraordinary detail previously limited to the realm of atomic physics. Such techniques may soon be extended to map quantum well thicknesses more accurately using a NSOM transmission absorption setup. Likewise when dopant spacings are larger than the NSOM resolution then each donor or acceptor might soon be individually spatially mapped.

Despite the impressive potential of this new technology, it is important not to forget its fundamental limitations. The serial nature of the scanning process means that scanning speed is a severe constraint, commonly from photon flux or mechanical considerations. As an extreme example, to map out all of the eigenstates of a wafer using near-field spectroscopy would take a few millennia. A short depth of focus and the requirement that the probe be much closer to the sample than the required resolution, makes it useful only for exposed surface or near-surface systems. Only flat surfaces are suited to maintaining uniform resolution. Furthermore the near-field probes are extremely delicate. It's very easy to damage them by abrasion with the surface or excessive laser power, thereby degrading their resolution. Many "near-field" experiments have been published using such probes where the images are no higher in resolution than one that could be more easily obtained from a standard diffraction limited optical microscope. As with all microscopies interpretation of the NSOM images can be easily misleading if careful scrutiny is lacking. For example, topography can influence the optical images, and especially so if supplemental tip height control techniques (such as shear force) are utilized. Sharp topographical signals can then give misleading sharp optical signals. Even if the data is well understood, there remains the tremendous

challenge of using it constructively and cost-effectively. Consider that all of the energy states of a device such as a laser can be individually catalogued. (This might be true only for narrow quantum wells (<10 nm thickness).) Then how, if at all, can this highly, possibly overly detailed information be utilized to improve device performance than knowledge of a simpler far-field photoluminescence measurement, where only the overall distribution is known? Knowing and finding the answer to such questions in addition to an honest appreciation of NSOM's limitations will determine the niche that NSOM finds amongst competing technologies.

References.

1. E. Betzig and J. K. Trautman, *Science* **257**, 189 (1992).

2. E. Betzig, J.K. Trautman, R. Wolfe, E. M. Gyorgy, P. L. Finn, M. H. Kryder and C.-H. Chang, *Appl. Phys. Lett.* **61**, 142 (1992).

3. E. Betzig, S.G. Grubb, R.J. Chichester, D.J. DiGiovanni, and J.S. Weiner, *Appl. Phys Lett.* **63**, 3550 (1993).

4. S.K. Buratto, J.W.P. Hsu, J.K. Trautman, E. Betzig, R.B. Bylsma, C.C. Bahr, and M.J. Cardillo, ., *J. Appl. Phys.* **72**, 7720 (1994), S.K. Buratto, J.W.P. Hsu, E. Betzig, J.K. Trautman, R.B. Bylsma, C.C. Bahr, and M.J. Cardillo, *Appl. Phys. Lett.* **65** 2654 (1994).

5. E. Betzig and R.J. Chichester, *Science* **262**, 1422 (1993).

6. H.M.Marchman, J.E.Griffith, J.K.Trautman, *J of Vac Science and Tech*(1995)

7. R.D.Grober, T. D. Harris, J. K. Trautman, E. Betzig, W. Wegscheider, L. Pfeiffer, and K. West, *Appl. Phys. Lett.*, **64**, 1421 (1994).

8. H. F. Hess, E. Betzig, T.D.Harris, L.N. Pfeiffer, and K. W. West, *Science* **264**, 1740 (1994).

Scatterometry: Principles, Applications, Limitations, and Future Prospects

John C. Stover

TMA Technologies Inc.
601 Haggerty Lane
P.O. Box 3118
Bozeman MT, 59715

Scatterometry is now finding expanded use as a metrology tool in the semiconductor industry. In addition to being used as a means to map and estimate particle size on wafers through laser scanning it can also be used to do wafer surface roughness characterization. Roughness is becoming important on the polished side to reduce haze (which prevents small particle detection) and on the backside where it hides particles and effects the precision with which rapid thermal processing is achieved. In addition it may play a role in monitoring roughness of chemical mechanical planarization and in monitoring the quality with which advanced lithography techniques produce ever smaller lines. The paper includes data from a number of different applications.

INTRODUCTION

Light scatter has been used to map and size particulates on silicon wafers for many years, but, in contrast to the optical industry, only recently has it been seriously considered as a source of surface feature characterization in the semiconductor industry. In fact it is only recently that micro-roughness has been considered a serious problem in the semiconductor industry [1-7]. However, roughness on the polished surface of the wafer creates scatter that introduces noise into laser scanner particle detectors. There has been speculation that roughness adversely impacts gate oxide breakdown voltage, and roughness has come under scrutiny on bare wafer backsides because it impacts RTP (through emissivity changes). CMP is another process where surface roughness may play a role in device quality and/or yield. Post process backside grind is another area that has been identified as possibly benefiting from roughness monitoring. The diffraction patterns created by reflecting a laser beam off a patterned wafer change with line width, height and sharpness. Thus, there are a number of wafer and device related processes, in addition to particle mapping, where surface features can be characterized, specified and monitored through the use of scatter metrology.

This paper reviews basic scatter measurement principles and the work that has been done for several varied applications where silicon surface characterization is needed. Because scatter can come from many sources besides surface roughness (particles, subsurface defects, bulk effects present in transparent wafer coatings, etc.), there can be confusion in interpreting measurement results. This is not always a disadvantage as it means that scatter can be used to monitor some of these other properties as well as roughness. However, because some scatter from roughness is almost always present, acting as a noise source (often called haze) it has to be dealt with in many measurement situations. In order to evaluate these issues, it is necessary to have a basic understanding of how scatter is measured, when it can (and cannot) be used to calculate roughness statistics, and how to tell the difference.

REVIEW OF SCATTER MEASUREMENTS

Detailed information on how scatter measurements are taken and interpreted is already available in the literature [8] and various tutorials. It is not the intent of this paper to provide a tutorial on that subject; instead, after a very brief review concentrating on current measurement capabilities, the emphasis of this paper is to review relatively new scatter measurements that are currently being taken in the industry and to suggest measurements that could be exploited in the near future. The two common types of scatter measurement and their relationship to roughness and discreet defects are briefly reviewed here.

TIS Measurements

The easier of the two scatter measurements described here is called Total Integrated Scatter, or TIS. It consists of gathering (or integrating) light scattered by a reflective surface (out of the reflected specular beam) and measuring this signal with one detector. Lenses, mirrors and diffuse integrating spheres have been used to gather the scatter signal in different TIS configurations. Commercial particle scanners are examples of integrating scatterometers, although they do not calculate the TIS ratio defined here. The TIS is evaluated as the ratio of the scattered power (P_s) to the power in the reflected specular beam (P_0). The equation is given in Figure 1. For smooth (mirror like) surfaces the TIS is small (in the range of 10^{-2} to 10^{-6} for visible light), but for rough surfaces, where lots of light is scattered and the specular reflection almost disappears, it can be a very large number. Notice that the ratio is relatively independent of surface color because both the scattered light and the specular light will change roughly in proportion if the surface reflectance changes. Also notice that the ratio will change with the collection efficiency of the scattered light.

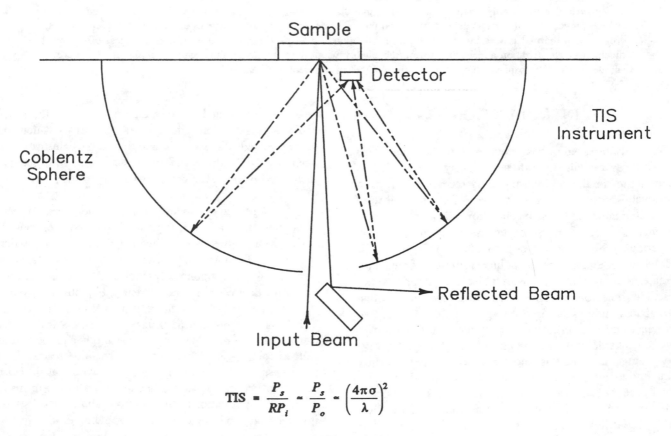

$$TIS = \frac{P_s}{RP_i} \approx \frac{P_s}{P_o} \approx \left(\frac{4\pi\sigma}{\lambda}\right)^2$$

FIGURE 1. Definition of the TIS and its relationship to the RMS roughness, σ.

In particular, systems that gather more of the intense scatter close to the specular beam will have a larger TIS ratio. The meaning of this dependence on integration angles will become more apparent when angular resolved scatter measurements are discussed.

One method of measuring TIS is shown in Figure 1. The measurement of TIS is described in an ASTM standard. The equation in Figure 1 relates the measured TIS to root mean square roughness, σ (which is often referred to as the "RMS"), in terms of the light wavelength, λ, and the incident angle, θ_i. This relationship depends on the surface being an optically smooth source of topographic scatter. Optical smoothness can be defined as the RMS being much smaller than a wavelength. The term "topographic scatter" simply means that the measured scatter must come from surface roughness and not other sources, such as particulates, etc. It would not be reasonable to try to calculate roughness from a scatter signal dominated by non-topographic sources; however, it can be a challenge to know when roughness is dominating the scatter signal.

Notice that because the TIS depends on the collection angles employed in the measurement, the calculated roughness also depends on these angles. This effect, which may be described in terms of spatial frequency bandwidth limits, is present in all roughness measurement systems. Profilometers also have these types of limits imposed on them. That is, a profilometer will not be able to measure surface waves longer than the length of the measured profile (imposing a low spatial frequency limit), or deep undulations spaced more closely than the stylus tip radius (imposing a high spatial frequency limit). When roughness values are compared, it is critical that these issues be taken into account (ie: compare apples to apples, etc.). One advantage of the angle resolved scatter measurements, to be described next, is that they are easily converted to a form that facilitates comparison of these bandwidth effects.

Angle Resolved Scatter Measurements

As the name describes, ARS measurements are simply scatter measured as a function of angle. Measurements can be made anywhere in the hemisphere in front of the reflector; although, they are often taken just in the incident plane. They are quantified in terms of the bidirectional reflectance distribution function (BRDF), which is defined in Figure 2. It is really nothing more than the scattered power density ($dP_s/d\Omega_s$) normalized by the incident power (P_i), with the cosine of the polar scattering angle (θ_s) included to adjust for the reduction in apparent scatter source width at high scatter angles. The BRDF is usually presented against degrees from specular on a log-log plot. If the BRDF signal is integrated over angle and divided by the surface reflectance the TIS may be calculated. If the BRDF is integrated over the near specular region, the specular reflectance may be found. The measurement of BRDF is written up in an ASTM standard and is accepted throughout the scatter measurement community as the way to quantify measured scatter.

Also shown in Figure 2, is the diffraction theory relationship between the BRDF and the surface power spectral density function, or PSD. The PSD may be thought of as the surface roughness power per unit roughness frequency. Long surface waves scatter (diffract) close to the specular beam and high frequency surface waves scatter farther from the specular beam. The intensity of the scattered spot from one sinusoidal surface wave grows as the amplitude of the surface wave grows. As can be seen from the defining equation in Figure 2, the PSD and the BRDF are proportional except for the dependence on θ_s, which is the polar scatter angle. The value Q in this equation is the polarization coefficient which is approximately equal to the specular reflection when evaluated in that direction. The PSD is often presented against spatial frequency on a log-log plot.

If the PSD is integrated over a band of spatial frequencies, the result is the mean square roughness (ie: the square of the RMS). Thus the bandwidth limits of this calculation are readily apparent when evaluating the RMS from the PSD (and the TIS from the BRDF). Conversion of the BRDF to the PSD depends on the surface being an optically smooth source of topographic scatter.

The accuracy of the BRDF to PSD conversion has been the subject of many experiments over the last twenty years. The result is conclusive. The relationship given in Figure 2 provides an excellent way to calculate surface statistics from optically smooth sources of topographic scatter. There are optically smooth sources of non-topographic scatter and this can cause some confusion. For example gold and copper scatter topographically at visible wavelengths, but not in the IR. Aluminum can not be trusted to scatter topographically at any wavelength, but there are exceptions. Fortunately, clean silicon has been shown to be an excellent source of topographic scatter [9]. The next section describes how topographic scatter is identified. The technique is important to some of the semiconductor measurements proposed in this paper.

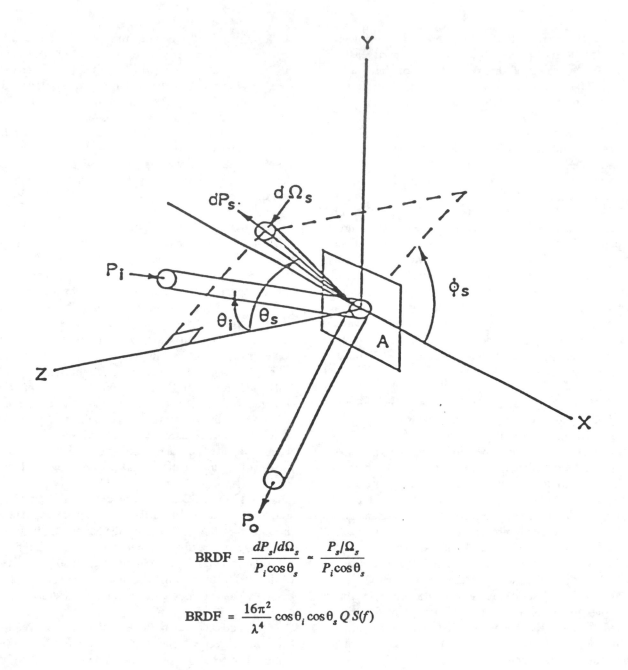

$$\text{BRDF} = \frac{dP_s/d\Omega_s}{P_i \cos\theta_s} \approx \frac{P_s/\Omega_s}{P_i \cos\theta_s}$$

$$\text{BRDF} = \frac{16\pi^2}{\lambda^4} \cos\theta_i \cos\theta_s \, Q \, S(f)$$

FIGURE 2. Definition of BRDF and its relationship to the surface PSD, S(f).

Wavelength Scaling, Topographic Scatter and Silicon

Examination of the relationship between the BRDF and the PSD reveals a dependence on wavelength to the fourth power. Thus for a given sample (that scatters only from surface topography) the BRDF can be expected to change if the wavelength is changed. The converse is more interesting. If the BRDF is evaluated at several different wavelengths, then the same PSD will be found from each data set if the BRDF is due just to topographic scatter. This has to be true because the surface has the same roughness statistics (PSD) regardless of what color laser is chosen for the scatter measurement. This property, which is referred to as "wavelength scaling" in the literature, can be used to verify that a material is scattering predominantly from topography and, thus, that its PSD can be evaluated from BRDF measurements.

Because of the growing recognition that monitoring roughness is important in several semiconductor manufacturing processes, SEMI, SEMATECH and several industry groups have worked not only on roughness specifications (where the PSD has been chosen as a preferred route), but on the issue of whether or not scatter can be used to characterize the roughness of smooth, clean silicon surfaces. This work has resulted in confirming that clean silicon wafers are in fact excellent sources of topographic scatter over the wavelength range 325 nm to 850 nm [9]. Clean, polished silicon is such an excellent source of topographic scatter that a sharp silicon step (of known height and thus known PSD) has been used as a BRDF standard [10].

Figure 3 shows the PSD's calculated from BRDF measurements taken at three wavelengths in the near IR to near UV on a bare silicon wafer. The agreement in the calculated PSD's is excellent. This close agreement has been observed many times for wafers manufactured by different processes and different manufacturers. Wavelength scaling for silicon sputtered on fused silica has been previously reported [11]. Hundreds of wafers have now been measured for a variety of industry, ASTM and SEMI studies. As long as clean, uncoated wafers are used the answer seems to be that silicon exhibits qualities that make it an excellent candidate for roughness characterization by light scatter measurement. In addition, most of these wafers appear to be isotropic over the bandwidth shown in Figure 3. That is, they exhibit the same surface statistics (PSD) regardless of wafer orientation.

The PSD's of wafers made by different processes do not always have the same shape. Many wafers have a nearly straight line form with slope of -1 to -3 on a log-log plot, which implies a fractal, or f^n behavior [12]. Other wafers have more complicated PSD shapes that require a three parameter (ABC) fit [13, 14]. In some cases the shape of the PSD can be obtained from the summation of more than one of these generic forms. The various power law forms have been found in the PSD's of many surfaces (both industrial and natural). When the PSD of a manufactured surface can be modeled by the summation of two or more power law expressions, the implication is that more than one separable mechanism is at work. For example, would the use of two different grit sizes in the polishing process show up in this manner? There is still a lot to be learned about the relationship of the PSD shape to the manufacturing process.

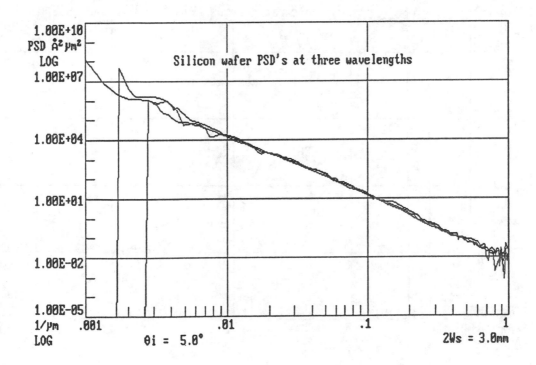

FIGURE 3. The same PSD is found from BRDF data taken at three wavelengths (835 nm, 633 nm and 488 nm) from this silicon wafer. This result is common for smooth, clean, uncoated wafers.

403

An example of a silicon PSD changing dramatically with process parameters is the epitaxial wafer of Figure 4. The addition of this layer to the substrate produces changes that increase the high frequency roughness and decrease the low frequency roughness. These characteristics emphasize the need to pay attention to bandwidth (or angle) limits when comparisons are made. For example, if the rms roughness found via profilometer (which would typically start at about $f = 0.01 \mu m^{-1}$) were used to investigate the effects of epitaxial growth, the conclusion would be that the surface gets smoother during the epitaxial process because the roughness measurement (integral of the PSD) is dominated by the low frequency contributions. Conversely, a haze measurement which often starts at five to ten degrees from specular (or about $f = 0.2 \mu m^{-1}$), would indicate just the opposite. This general behavior has been observed in epitaxial wafers from more than one source.

There are some exceptions to the isotropic nature of polished silicon surfaces. Wafers cut at a small angle from the crystalline plane have been shown, via atomic force microscopy, to exhibit a periodic stepped surface [15]. For many wafers, this periodicity is at a high enough frequency (generally above 5 μm^{-1}) that the effect is not seen in the spatial bandwidth available to many scatter measurements, which has a theoretical maximum less than $2/\lambda \; \mu m^{-1}$ and a practical maximum often closer to $1/\lambda \; \mu m^{-1}$. The presence of these diffraction peaks has been reported for the case of low frequency steps.

The usual topographic nature of silicon wafer scatter over the near IR to UV range can be lost when the wafer is coated. Figure 5 shows a failure to wavelength scale on a wafer with a nitride film. As will be seen similar results have been obtained for SiO_2 films. Whether other films exhibit similar properties and to what extent these measurements can be exploited to characterize films remains to be seen.

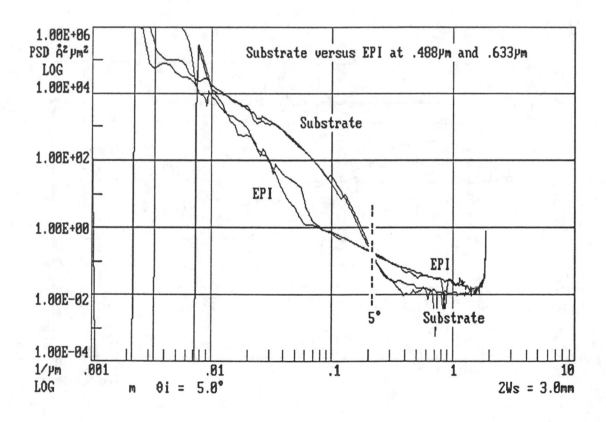

FIGURE 4. Epitaxial silicon also "wavelength scales" (633 nm and 488 nm), but the resulting PSD differs from the underlying silicon substrate.

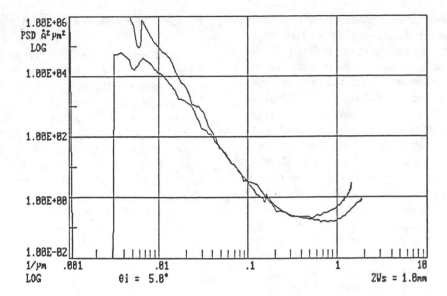

FIGURE 5. Deposition of a nitride coating on this wafer results in the loss of wavelength scaling for data taken at 633 nm and 488 nm. This is interpreted as an indicator of non-topographic scatter from the coated wafer.

Particle Scatter

The BRDF is an excellent way to quantify scatter and it has been shown that it can be related to surface roughness. Illuminated spot size on the surface is not an issue as long as the spot is larger than the lateral extent of the roughness features of interest. Thus a particle scanner illuminating a clean wafer will get the same roughness induced haze (ie: BRDF noise) signal regardless of spot size. The same is not true of the scatter signal from a small particle illuminated by a beam of light larger in diameter than the particle. Most BRDF measurements are taken with illuminated spot sizes on the order of a few millimeters. Most particle scanners increase their particle detection sensitivity by decreasing the illuminated spot diameter to about 100 μm, which increases the incident power density by two or three orders. If a constant power beam is focused (increasing power density) the particle will scatter more light and a larger particle induced BRDF would be reported without any increase in particle size. Obviously a means of quantifying particle scatter is needed that allows particle diameter to be estimated. This is the differential scattering cross-section which is defined in Figure 6. The incident beam of P_i watts is assumed to have a uniform cross-section and illuminates area A on the surface. The incident intensity is then P_i/A. The differential cross-section has units of area/steradian and is easily related to the BRDF by use of the multiplier $A\cos\theta_s$.

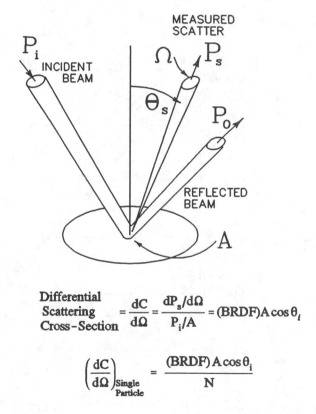

$$\begin{array}{c}\text{Differential}\\\text{Scattering}\\\text{Cross-Section}\end{array} = \frac{dC}{d\Omega} = \frac{dP_s/d\Omega}{P_i/A} = (\text{BRDF})A\cos\theta_i$$

$$\left(\frac{dC}{d\Omega}\right)_{\substack{\text{Single}\\\text{Particle}}} = \frac{(\text{BRDF})A\cos\theta_i}{N}$$

FIGURE 6. The single particle differential scattering cross-section can be defined in terms of the BRDF measured from N particles in illuminated area A.

405

Multi-particle scatter measurements have been reported by members of the SEMATECH Task Force studying the problem of small particle scatter [7]. By choosing multi-particle samples two experimental problems are quickly overcome. First, the particle scatter is multiplied by N, the number of illuminated particles, so that particle signal will not be lost in the roughness haze, even for very small particles. Secondly, the uncertainty introduced by a non-uniform beam cross-section (near Gaussian for laser sources) is not an issue if many particles are uniformly distributed over the illuminated area. Figure 7 shows the measured BRDF from three multi-particle samples (with different particle densities) compared to the background roughness induced noise from the virgin (clean) wafer. The individual particle differential scattering cross-section for very small particles can be calculated from these measurements using the equations of Figure 6 (for known particle density N/A) and subtracting off the measured surface scatter. The effect of surface roughness induced (scatter) noise on a particle scanner can be studied with this technique, and these measurements are expected to become an important step in designing the next generation of particle scanners.

BACKSIDE SCATTER AND ROUGHNESS

Roughness on the wafer backside is an issue for at least two reasons. First, it is a hiding place for (and thus a source of) small particles that may eventually contaminate wafer front sides during device fabrication. Second, it can be a source of problems for rapid thermal processing (RTP), where temperature uniformity as small as one degree per thousand need to be achieved. This latter difficulty is caused because variations in roughness result in variations in wafer emissivity. This makes accurately, and/or uniformly, heating the wafer to high temperatures in a few seconds very difficult because different wafers (and different wafer areas) radiate heat at different rates during RTP. One solution may be to produce backsides with less variation in roughness and thus a more consistent emissivity. One way to achieve this would be to simply produce much smoother backsides, which solves the backside particle problem as well. Even polished backsides have been considered. Whatever the solution(s), it is clear that a specification for (and thus measurement of) backside roughness is probably desirable.

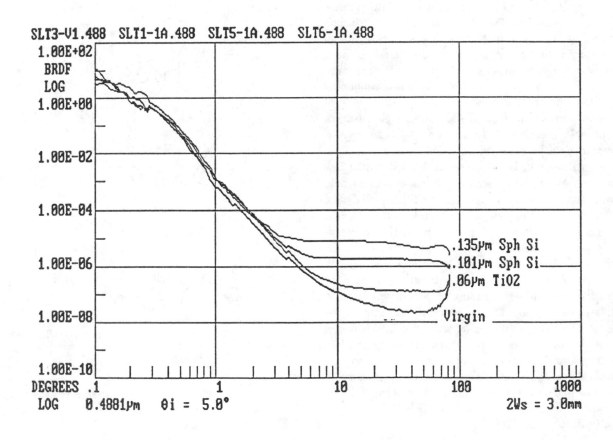

FIGURE 7. BRDF from several multiple particle samples compared to the BRDF from the background wafer roughness.

TIS measurement appears to be a reasonable approach to this problem and several types of wafers have been measured as shown in Figures 8-12. All of the maps show variation across the wafer in RMS roughness as found from the measured TIS. Measurements were made in the manner illustrated in Figure 1 using a 0.633 μm laser. Scatter was collected over a range of 0.9 to 80 degrees from specular, which corresponds to a spatial frequency band of 0.025 to 1.5 μm^{-1}.

Figure 8 shows data taken on a very rough (conventional) backside that has the typical matt, grey finish we have been accustomed to for many years. Because the RMS values exceed 2000 Angstroms, it does not meet the smooth surface requirement, and in fact this surface does not wavelength scale in the visible (data not shown). Even so, Figure 8 demonstrates that there appear to be larger variations in roughness across the wafer surface than in the following measurements of smoother surfaces.

Figure 9 is of a smoother backside. At just over a 1000 Angstroms RMS, this wafer comes pretty close to meeting the smooth surface requirement. Figures 10, 11 and 12 show roughness maps of even smoother backsides that were respectively produced by advanced etching, machining and a post device fab backside grind. This data is not presented in order to allow comparison of the various manufacturing techniques (that work still has to be done), but rather to demonstrate the power of scatter measurement to perform the metrology necessary to help solve this development problem.

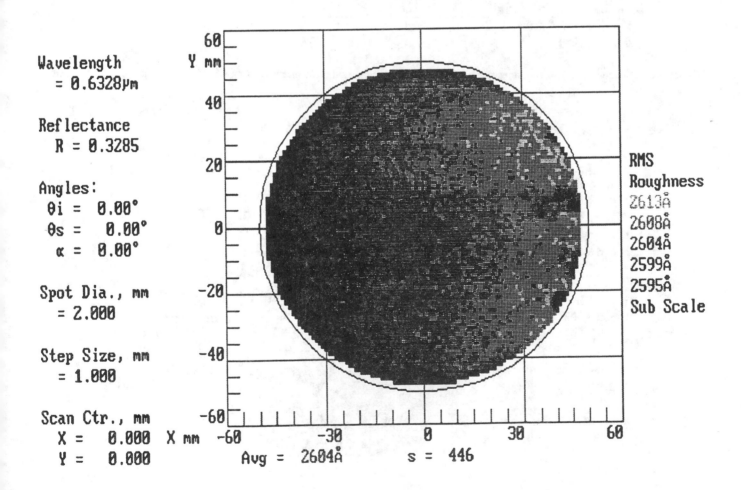

FIGURE 8. A scatter generated roughness map from an etched wafer backside that violates the smooth surface requirement.

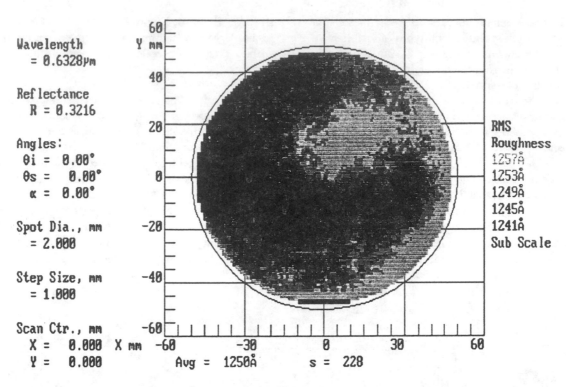

FIGURE 9. A scatter generated roughness map of an etched wafer backside that marginally meets the smooth surface requirement.

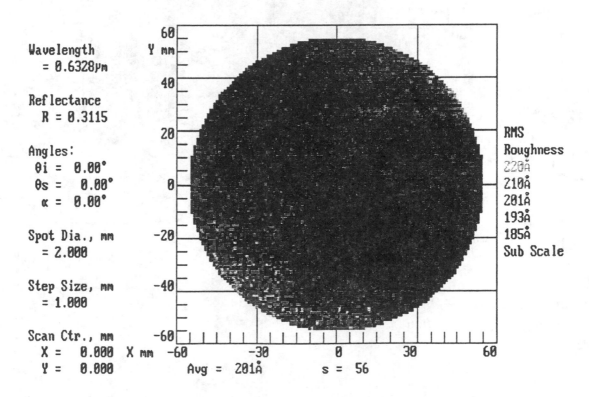

FIGURE 10. A scatter generated roughness map of a smooth etched wafer backside.

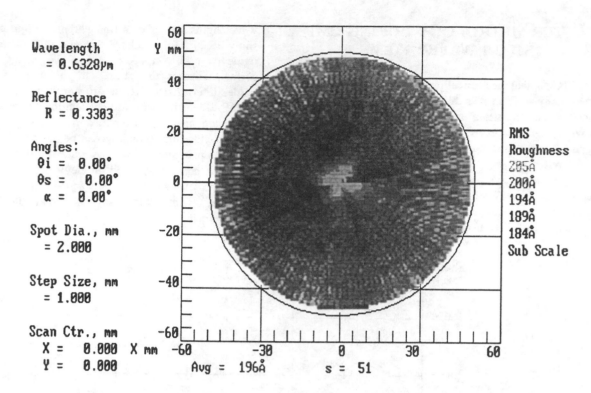

FIGURE 11. A scatter generated roughness map of a smooth machined wafer backside.

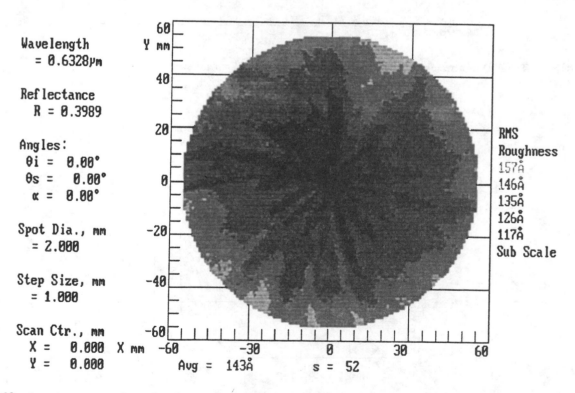

FIGURE 12. A scatter generated roughness map of a smooth ground wafer backside.

409

SCATTER METROLOGY DURING CMP AND ON WAFER FILMS

Chemical mechanical planarization (CMP) is now used as many as five times during fabrication of advanced devices. The resulting surface roughness can effect following fab processes. The surfaces are in the roughness range that can be easily measured by light scatter, but it is still uncertain whether the various materials in question scatter topographically. An obvious exception to the smooth, clean, front surface reflector requirement (to obtain roughness statistics from scatter measurement) is silicon oxide measured at visible wavelengths where it is transparent. Figures 13 and 14 show BRDF plots made of thin and thick layers of silicon oxide over silicon taken at two wavelengths. Neither surface wavelength scales. In the case of the thicker oxide, interference effects are apparent in the measured BRDF. Measurements at wavelengths where silicon oxide is opaque have not been taken. Thus whether or not silicon oxide film roughness can be measured via scatter is still an open question. The same holds true for other materials commonly used in device fabrication. Also unresolved, is whether or not additional information (for example, film thickness) is available from the measurements.

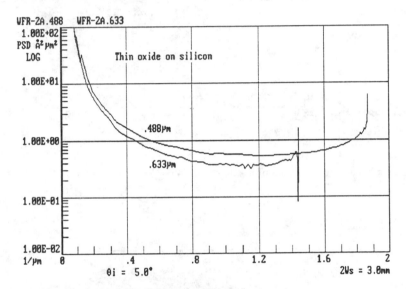

FIGURE 13. The PSD of a wafer with a thin oxide film fails to wavelength scale.

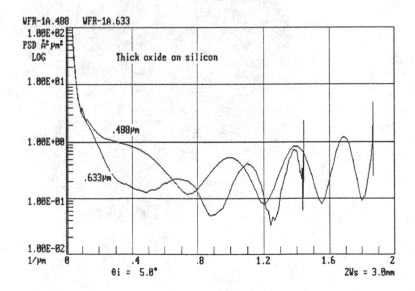

FIGURE 14. The PSD of a wafer with a thick oxide film fails to wavelength scale and exhibits interference effects.

410

LINE WIDTH -
LITHOGRAPHY CHARACTERIZATION

Diffraction (or scatter) from well defined surface shapes, such as a line with square corners, can be calculated with a high degree of precision. In fact, scatter from a square step of known height can be used as a means of producing a known BRDF value [10]. This means that scatter measurement offers potential as a means to monitor various lithography processes. The basic idea is illustrated in Figure 15, which compares the ideal $sinc^2$ diffraction shape that would be obtained from a perfect square cornered, vertical walled line, to the diffraction pattern that was measured from an actual line. The zero locations are determined by line width. The diffraction peak broadens as the line narrows. The amplitude and the diffraction peak depends on illuminated spot diameter and

line height. The deviations between actual line shape and ideal line shape cause the intensity variations between the ideal and actual diffraction patterns.

Work of this type is underway at the University of New Mexico [16], where they inspect the quality of a periodic structure created by several identical parallel lines (which form a diffraction grating). The scatter pattern now becomes a series of discrete diffraction peaks, whose amplitudes fall on an amplified version of the $sinc^2$ function associated with one line. This amounts to sampling the function that characterizes the line shape. Developing this research into a usable inspection tool will require measurement of multiple samples of varying exposure and focus as well as line width and height. Verification of actual line shape will have to be performed by profilometry.

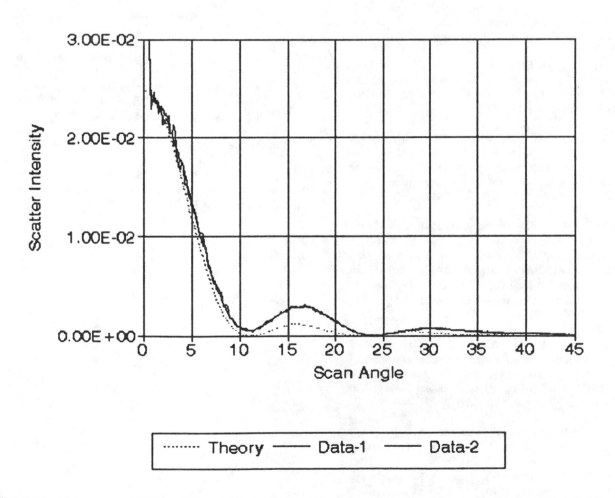

FIGURE 15. Light scattered (diffracted) from a single line should vary in intensity as a sinc squared function (the dotted line), but the actual data (solid lines) takes on a measurably different shape indicating that the line does not have the ideal shape.

SUMMARY

Scatter metrology is powerful non-contact inspection technique that is finding new applications in the semiconductor industry. In particular, it has been proven to be useful for particle detection, estimation of particle sizing and for roughness measurement of smooth silicon surfaces. The measurement of backside roughness, which is becoming more of an issue, is an area where scatter measurement is likely to provide almost immediate benefits. In addition, it is likely to become useful as a measurement of CMP roughness (for at least some film materials), provides the potential for characterizing other film parameters, and is under investigation as a source of monitoring for high resolution lithography.

REFERENCES

1. Abe, T., E.F. Steigmeier, W. Hagleitner, A.J.Pidduck. "Microroughness Measurements on Polished Silicon Wafers." Jpn. J. Appl. Phys. Vol. 31 pp.721-728 Part 1, No. 3, March (1992).

2. Denes, L., H. Huff. "A Fourier Analysis of Silicon Wafer Topography." Abstract 823 RNP, 182nd Meeting of The Electrochemical Society, Toronto, Ontario, Oct. 11-16, 1992.

3. Bawolek, E.J., J.B. Mohr, E.D. Hirleman, A. Majumdar. "Light scatter from polysilicon and aluminum surfaces and comparison with surface-roughness statistics by atomic force microscopy." Applied Optics Vol 32, No. 19 P3377, July (1993).

4. Bullis, W.M.. "Microroughness of silicon wafers."

5. Diebold, A.C., B. Doris. "A Survey of Non-Destructive Surface Characterization Methods Used to Insure Reliable Gate Oxide Fabrication for Silicon IC Devices;" Surface and Interface Analysis V20 p127-139 (1993).

6. Strausser, Y.E., B. Doris, A.C. Diebold, H.R. Huff. "Measurement of silicon surface microroughness by AFM." Abstract 284, Extended Abstracts of the 185th Meeting of the Electrochemical Society, Vol. 94-1 San Francisco, CA, May 22-27, 1994.

7. Huff, H.R., R.K. Goodall, E. Williams, K-S Woo, B. Y. H. Liu, G. Starr, T. Warner, D. Hirleman, K. Gildersleeve, W. M. Bullis, B. W. Scheer, J. C. Stover; "Measurement of Silicon Particles by Laser Surface Scanning and Angle-Resolved Light Scattering"; To be presented at Electrochemical Society Meeting, Reno Nevada, May (1995).

8. Stover, J.C.. Optical Scattering: Measurement and Analysis; Second Edition, SPIE (1995).

9. Stover, J.C., M.L. Bernt, E.C. Church, P.Z.Takacs; "Measurement and analysis of scatter from silicon wafers"; Proc. SPIE. Vol 2260-21, July (1994).

10. Takacs, P.Z., M.X.-O. Li, K. Furenlid, E.L. Church. "Step-height standard for surface-profiler calibration." Proc. SPIE 1995-33, P235 July (1993).

11. Stover, J.C., M.L. Bernt, "Wavelength scaling investigation of several materials", Proc. SPIE Vol 1995-26, July (1993).

12. Church, E.L.. "Fractal Surface Finish." Appl. Opt. V27, N8, April (1988).

13. Church, E.L., P.Z. Takacs, and T.A. Leonard. "The prediction of BRDF's from surface profile measurements." Proc. SPIE, 1165-10, (1989).

14. Church, E.L., P.Z. Takacs. "Optimal estimation of finish parameters." Proc. SPIE 1530-09 p71, July (1991).

15. Izunome, K., Y. Saito, H. Kubota. "Periodic Step and Terrace Formation on Si(100) Surface during Si Epitaxial Growth by Atmospheric Chemical Vapor Deposition." Jpn. J. Appl. Phys. Vol 31 pp.L1277-L1279 Part 2, No. 9A, September (1992).

16. McNeil, J.R., S.S. Naqvi, S.M. Gaspar, K.C.Hickman,, K.P.Bishop, L.M.Milner, R.U.Krukar, G.A. Petersen; "Scatterometry applied to microelectronics processing; Solid State Technology, p29, March (1993).

Spectroradiometers for Deep-Ultraviolet Lithography

J. M. Bridges, J. E. Hardis, C. L. Cromer, and J. R. Roberts

Physics Laboratory, NIST, Gaithersburg, MD 20899-0001

Spectroradiometer systems were designed and built for measuring ultraviolet exposure in an SVGL Micrascan II. An additional spectroradiometer maintains traceability to NIST as part of a mobile calibration unit. The UV measurement probes contain diffusers made of sintered aluminum oxide, which was found to have good transmission and Lambertian character. The systems were calibrated against the NIST argon mini-arc. An overall expanded uncertainty of 7% (2σ) is estimated for spectral irradiance measurements from 230–330 nm.

INTRODUCTION

The trend in the semiconductor industry towards smaller feature sizes increases the need for accurate control of deep ultraviolet (DUV) dose in lithographic tools. An incorrect exposure leads to the creation of poor circuit features, thereby reducing manufacturing yield.

In order to assist the semiconductor industry in ultraviolet lithography, NIST is working to improve the accuracy and reliability of ultraviolet metrology. In addition to advances in fundamental standards and measurement techniques, NIST is also developing methods to transfer radiometric measurements to semiconductor lithographic processes with the best possible accuracy.

As an example of this effort, two spectroradiometers recently have been designed, constructed, and calibrated for measuring ultraviolet exposure in a Micrascan II, manufactured by SVG Lithography Systems, Inc., Wilton, CT [1]. A mobile calibration unit (MCU) was also designed and constructed for periodic recalibrations. These instruments were designed to meet a need for which there were no suitable commercial instruments available.

SYSTEMS DESCRIPTION

Figure 1 is a block diagram of the spectroradiometer systems. A UV radiation collector, or probe, is designed to be mountable on the Micrascan II wafer stage. The ultraviolet radiation is transmitted from the receiver to a spectrograph by a fiber optic cable. The spectrally dispersed output of the spectrograph is imaged onto a CCD camera, called an Optical Multichannel Analyzer (OMA), at its focal plane. The OMA detects the radiation and sends data to the computer for processing and storage. The system as a whole was calibrated at NIST for spectral irradiance responsivity.

UV Radiation Receiver

Figure 2 is a block diagram of the ultraviolet probe. It includes an aperture, diffuser, and connector for the fiber optic cable. The receiver was designed to detect incident radiation up to 30° from the normal to the aperture.

For a Lambertian detector, the response to incident radiation varies as cos θ, where θ is the angle from the normal. This ideal response allows accurate measurement of irradiance from a UV beam of any sized numerical aperture. The spectroradiometer can therefore be calibrated using a source with a collimated incident beam and later used with other optical arrangements. By using a good diffuser between the aperture and fiber optic, the response can be made close to Lambertian.

Earlier, prototype instruments contained diffusers made of PTFE (teflon). After some use, their transmission greatly decreased. A recent study confirms that

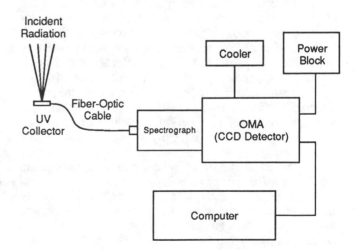

Figure 1: Block Diagram of Spectroradiometers

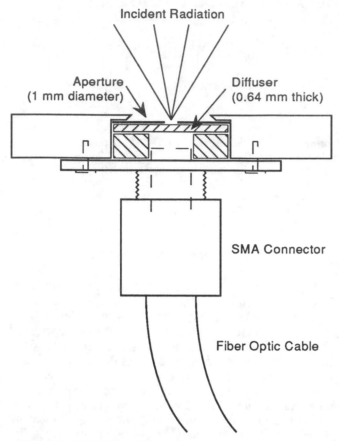

Figure 2: UV Radiation Probe

materials containing PTFE undergo changes in reflectance (and presumably transmittance) after being irradiated with intense UV radiation [2]. Therefore we tested other materials for suitability as diffusers. Requirements for the material included adequate transmission over the spectral range of 220–500 nm, a Lambertian character for the probe response, and improved resistance to aging.

Of the materials tested, sintered aluminum oxide was most satisfactory, and it is now used in the spectroradiometers. A probe containing an aluminum oxide diffuser was exposed continuously for one hour at the wafer plane of a Micrascan II. No change in its transmission was observed over this period, which is much longer than the exposure time (typically a few seconds) needed to make an irradiance measurement.

A diffuser thickness of ≈0.64 mm was selected to balance transmission and good Lambertian angular behavior. Transmission was measured by alternately including and removing the diffuser from the probe, and then ratioing the signals from these two conditions. It varied from 0.0015 at 320 nm to 0.00011 at 220 nm, the primary spectral range of interest. The response of the probe as a function of angle was determined by irradiating the probe and measuring the signals as the probe was rotated. Figure 3 shows the relative signal as a function of angle, measured at 254 nm. At 30° from the

normal, the response deviates from a cosine function by only 3.5%.

Fiber Optics

Each spectroradiometer is equipped with a fiber optic cable 2.7 m in length, containing 40 fibers of 50 μm core diameter. The fibers are of step index, fused silica. They are contained in a flexible, stainless steel sheath. The cables can be bent to a 2 cm radius without affecting the transmission. The entrance end of each cable is composed of a circular array of fibers, about 0.5 mm in diameter, held in an SMA connector. At the exit end of the cable, the fibers are arranged in a linear array, held in a stainless steel ferrule. This ferrule is clamped onto a spectrograph such that the linear end of the fiber array serves as an entrance slit.

Spectrographs

Two different types of spectrographs were required for this project. On one spectroradiometer and on the MCU, the spectrographs were required to span 230–330 nm. On another spectroradiometer, the spectrograph was required to span 220–500 nm. For the 230–330 nm instruments, we obtained double-grating spectrographs custom-made by American Holographic (AH). For the 220–500 nm spectrograph, we used an off-the-shelf unit, the Jobin-Yvon (JY) model CP-140-1604 spectrograph, with a 405 lines/mm holographic grating blazed at 250 nm. We developed the necessary hardware in-house to incorporate the spectrograph into a spectroradiometer.

The AH units were originally chosen for this project because of their double-grating designs. The concern was that a single-grating spectrograph would not be able to reduce stray light on the CCD sufficiently. This would be a problem for calibration, because the UV radiation of interest from argon and xenon arc sources comprises only a small fraction of their output. Experimentally, we find little difference in stray-light immunity between the

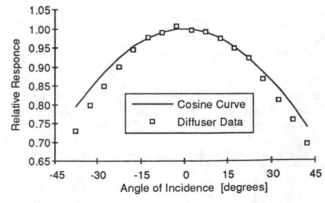

Figure 3: Aluminum Oxide Diffuser Performance

spectrograph designs, and both were adequate. This was determined by putting a microscope slide (which acts as a UV-blocking filter) between an argon-arc and the UV probe to remove the signal while leaving the stray-light background.

The throughput of the JY design is very much higher than the AH designs, due both to its fewer number of internal reflections and its higher numerical aperture. On the AH spectroradiometers, the column of optical fibers at the end of the fiber coupler serves as an entrance slit. The slitwidth is determined both by the fiber size (50 µm core) and the mechanical problem of cementing the fibers into a straight stack (the tolerance appears to be ≈30 µm). On the JY spectroradiometer, the additional throughput allowed us to add a diffuser and a 25 µm physical entrance slit. This significantly improved the spectral resolution of the instrument. Also, the slit height was reduced on the JY spectroradiometer. This further improved the spectral resolution, because fewer pixels on the CCD contribute to a spectral line. (This effect is due to aberrations in the spectrographs that are beyond our control.) As a result, the FWHM linewidth of the JY spectroradiometer is ≈1.3 nm (≈4 channels), compared with ≈1.1 nm (≈10 channels) on the AH spectroradiometers.

OMA (CCD Camera)

The EG&G Princeton Applied Research (PAR) OMA-IV system, model 1530-PUV-1024S, contains a 1024 (horizontal) by 256 (vertical) CCD detector that is mounted in a hermetically sealed enclosure. The CCD contains 19 µm square pixels, and is coated with a fluorescent material to make it sensitive to UV radiation. The signals from all 256 pixels in each column are summed.

The CCD sits on a Peltier cooler to reduce dark current and noise. The CCD is driven, and its output is amplified and digitized, by electronics integral to the detector head. The detector head is driven by a custom controller card that fits into an IBM-compatible PC. The controller card and detector head communicate through fiber optics to isolate the detector head from electrical noise generated by the computer.

SPECTRORADIOMETER CALIBRATION

Wavelength calibration utilized spectral lines from a mercury-pen lamp. The centers of spectral lines were located by fitting parabolas to their five top points. (The difference between this and the centroids of the lines was considered for the error estimate.) Wavelength as a quadratic function of channel number was determined by fitting for each instrument.

The calibration of the absolute spectral response of the spectroradiometers is based on standard sources as well as a standard detector. For highest accuracy, a

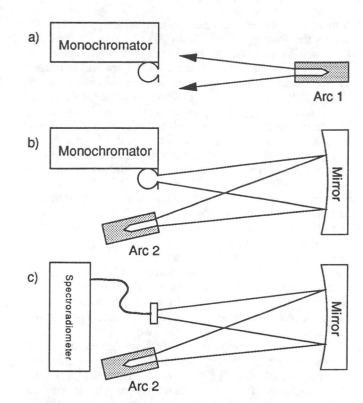

Figure 4: Irradiance Calibration Sequence

calibration should be done with an irradiance comparable to that which the instrument will be measuring, namely the irradiance at the wafer plane of the Micrascan. Since the standard irradiance sources at NIST, the argon mini-arc [3] and the tungsten lamp [4], are much weaker, a transfer source providing greater irradiance was used. This source is the focussed beam from a higher-powered, 7 kW argon arc, which is approximately uniform within the 1 mm diameter aperture of the spectroradiometer probe.

The irradiance of the focussed beam is itself first calibrated as a function of wavelength by comparing it to the irradiance from two standard sources, an argon mini-arc and a tungsten lamp. These two sources cover ranges which partially overlap; the tungsten lamp can be used as low as 250 nm, but its irradiance decreases rapidly in that region. The range of the argon mini-arc extends from below 220 nm to 500 nm, though coverage gaps exist near the argon spectral lines. The measurements were made using a monochromator equipped with a photomultiplier detector and an integrating sphere behind an entrance aperture of 1 mm diameter. This provided Lambertian responsivity.

The calibration procedure is illustrated in Figure 4. Figure 4a shows the argon mini-arc (Arc 1), with a known spectral irradiance, being used to calibrate the monochromator system. The tungsten lamp was also used in this manner. In Figure 4b, this source is replaced by the transfer source, the higher-powered arc (Arc 2) with a focusing mirror. The spectral irradiance of the transfer source is thus determined. Finally, Figure 4c shows the focussed beam from Arc 2 being used as the reference for the calibration of the spectroradiometer.

This procedure, based upon the mini-arc and tungsten lamps, uses standard irradiance sources. Since the range of signal levels covered in the procedure is rather large, we checked it against another scale, one based upon standard detectors [5]. A calibrated photodiode fitted with an aperture was used to measure the irradiance of the focused beam, though a filter was added to limit the spectral bandpass. The filter's peak transmission was ≈300 nm, with a width ≈10 nm (FWHM). The spectral transmittance of the filter was determined with an arrangement similar to that shown in Figure 4b, where the attenuation of the monochromator signal, as a function of wavelength, was noted after the filter was inserted into the beam. This allowed a photodiode current to be predicted based upon the detector's known spectral responsivity and aperture area, the filter's spectral transmittance, and the source's spectral irradiance. The actual photocurrent agreed within 2%, demonstrating excellent agreement between the two scales.

Both the source-based and detector-based measurements described above have uncertainties which we estimate as follows. For the source-based measurement, these include the scale uncertainty of the irradiance standard source (2.5%), the uncertainty in the transfer to the higher-powered source (2.0%), and the variation among measurements made by the spectroradiometer (1.5%). Combining these contributions in quadrature, and using a coverage factor of $k=2$, gives 7% (2σ) for the expanded uncertainty in the calibration of a spectroradiometer. The detector-based check had standard uncertainties in the detector responsivity (2.3%), the aperture size (1.5%), the filter transmission (1.4%), and the variations among signal measurements (1.0%). Combining these individual components in quadrature and using a coverage factor of $k=2$ gives 6.5% for the expanded (2σ) uncertainty in the prediction of the photocurrent.

MOBILE CALIBRATION UNIT (MCU)

The spectroradiometers were calibrated prior to delivery to SVGL. After being irradiated with strong UV radiation, however, some change in response is expected to occur eventually. The Mobile Calibration Unit (MCU) was developed to periodically recalibrate a spectroradiometer where it is used, either on the factory floor or at another location.

The MCU consists of a small, low-power DUV source and an additional calibrated spectroradiometer for measuring an irradiance from the source. Since the MCU spectroradiometer would be used relatively infrequently and thus be exposed to less UV radiation, it should retain its calibration longer than the instrument used in production.

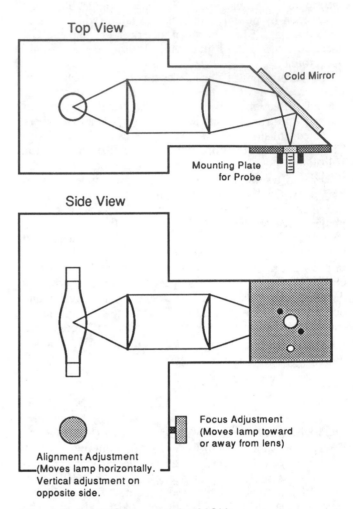

Figure 5: DUV Source of the MCU

In order to provide a recalibration over the spectral range of 230–330 nm, the DUV source should provide continuum radiation over this range with an irradiance of the magnitude found at the wafer plane of the Micrascan II. For this purpose, a 75 W xenon lamp was chosen. The lamp is mounted in a housing containing adjustments for the lamp position, shown in Figure 5. Two fused-silica lenses focus the light to a small spot. The spectroradiometer probes are held by a locknut on a mounting plate. Pins on the plate allow the probes to be precisely positioned repeatedly.

To perform a recalibration, the DUV source is used to irradiate in turn the MCU spectroradiometer and the spectroradiometer being recalibrated. The MCU spectrophotometer is used to determine the spectral irradiance of the lamp, which is then used as the reference for recalibrating the production spectrophotometer. Periodically, the MCU spectroradiometer would itself be recalibrated; this could be done at NIST.

The recalibration procedure was accurate to within 6% (varying with wavelength), as compared to direct calibration against the argon mini-arc. This results in an expanded (2σ) uncertainty in the response of a recalibrated spectrometer of 14%. We believe that the limiting factor was the residual geometric sensitivity of the probes and the non-uniformities of the UV sources.

CONCLUSION

The spectroradiometers described above, together with the MCU, have been delivered to SVG Lithography Systems. Their utility for quality control will be evaluated, in comparison with other techniques.

For the future, work continues on developing and characterizing high-intensity UV irradiance sources, investigating their stability, uniformity, and spectral distribution, comparing UV source- and detector-based radiometric scales, and developing new methods of delivering accurate and precise calibrations from NIST to the semiconductor and other industries using the methods that the customer sees as most useful.

ACKNOWLEDGEMENT

This work was supported by SEMATECH, though contract #3200310100P. Additional support was provided by the NIST Atomic Physics Division and the NIST Radiometric Physics Division, in which this work was done.

REFERENCES

1. Certain commercial equipment, instruments, materials, or software are identified in this report to adequately specify the experimental procedure. Such identification does not imply recommendation or endorsement by the National Institute of Standards and Technology, nor does it imply that the items identified are necessarily the best available for the purpose.

2. Gibbs, D. R., Duncan, F. J., Lambe, R. P., and Goodman, T. M., "Aging of Materials Under Intense UV Radiation," to be published in *Proceedings of the 5th International Conf. on Radiometry*, Berlin, Sept. 1994.

3. Klose, J. Z., Bridges, J. M., and Ott, W. R., J. Res. Natl. Bur. Stand. **93** (1), 21–39 (1988).

4. Walker, J. H., Saunders, R. D., Jackson, J. K., and Mc-Sparron, D. A., "Spectral Irradiance Calibrations," Natl. Bur. Stand. (U. S.), Spec. Publ. 250-20, Sept. 1987.

5. Zalewski, E. F., "The NBS Photodetector Spectral Response Calibration Transfer Program," Natl. Bur. Stand. (U. S.), Spec. Publ. 250-17, March 1988.

Infrared Microspectroscopy of Semiconductors at the Diffraction-limit

G.L. Carr, D. DiMarzio, M.B. Lee and D.J. Larson, Jr.

Research and Development Center
Grumman Aerospace & Electronics
Bethpage, NY 11714-3582

Infrared microspectroscopy, using synchrotron radiation as a source, is demonstrated as a tool for exploring semiconductor materials and devices. The synchrotron provides greater brightness than the standard infrared source, enabling an order of magnitude improvement in practical spatial resolution, i.e. approaching the diffraction limit. Since conventional and microscopic infrared spectroscopy are already used for probing semiconductor materials, we expect that the ability to explore sample regions only a few microns in size will be valuable for analyzing materials at dimensions approaching those of many microelectronic devices. Some examples for inclusions in Si and CdZnTe are presented.

INTRODUCTION

Infrared (IR) spectroscopy is a standard probe for semiconductor materials due to its ability to sense electronic properties as well as detect various impurities. In manufacturing, IR spectroscopy can be used to identify contaminants that may appear during device processing. The spectral region between 600 cm^{-1} and 3000 cm^{-1} is often referred to as the "molecular fingerprint region" due to the large number of molecules (both organic and inorganic) that can be differentiated and identified from extensive catalogs of library spectra.

As the dimensions of individual component structures become smaller, device operation becomes increasingly sensitive to microscopic defects and contaminants. IR microspectroscopy has developed into a valuable tool for characterizing such defects, and high performance microspectrometers are now commercially available. In principle, these instruments are capable of analyzing small specimens (approaching the diffraction limit), however the spectrometer's infrared source (typically a ~1200K blackbody) does not possess adequate brightness to achieve acceptable signal-to-noise ratios (S/N) when probing regions much smaller than about 30 μm. In contrast, the infrared produced as synchrotron radiation is 2 to 3 orders of magnitude brighter than a blackbody source[1]. To exploit this, an Infrared Microspectroscopy Facility (IMF) has been established at the National Synchrotron Light Source (NSLS) of Brookhaven National Laboratory. A primary component of the IMF measurement program involves the characterization of various semiconductor materials and devices.

Semiconductor Infrared Properties

A number of important semiconductor properties can be sensed by IR spectroscopy. At long wavelengths ($\lambda \sim 20$ μm), absorption due to free carriers and phonons occurs, in some cases allowing for quantities such as the carrier density to be determined. The onset of absorption at the band edge is useful for confirming the stoichiometry of alloys such as $Hg_{1-x}Cd_xTe$. Electronic transitions involving dopants or impurities can also be sensed. Even neutral impurities can sometimes be detected through their "local vibrational mode" (LVM) absorption. An important example of the latter is oxygen in silicon[2], which has well-defined absorption modes near $\lambda \cong 9$ μm and $\lambda \cong 8$ μm, depending on whether the oxygen is located at an interstitial site or in the form of an oxide, respectively. Another technically important semiconductor is GaAs, for which a variety of local vibrational modes, corresponding to important dopants and impurities, have been catalogued[3].

Synchrotron Infrared Microspectroscopy

The brightness limitation of the conventional source in IR microspectroscopy is alleviated by the use of infrared synchrotron radiation (IR-SR) as the source. We have installed a commercial IR microspectrometer (a Spectra-Tech "IRμs"™) at beamline U2B of the NSLS[4] and

Synchrotron source

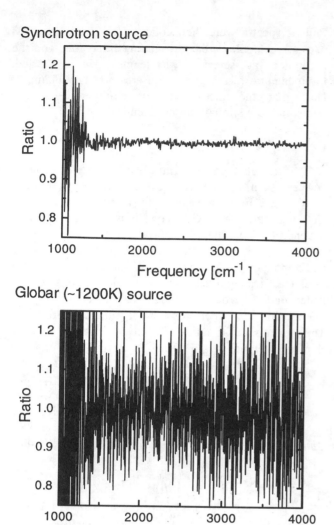

Globar (~1200K) source

Figure 1. Noise comparison of the synchrotron and globar infrared sources for a 3 μm aperture.

tested its performance[5]. The high source brightness enables good S/N even when probing regions smaller than 10 μm in size. Figure 1 shows a comparison of the S/N for the IR-SR and globar sources when the infrared sampling area is limited by a 3 micron diameter aperture. The data represents the ratio of two sequential transmission measurements, and in the ideal case (no noise or drift) would be a perfectly smooth line at unity. Note that the synchrotron source is capable of spectroscopy for frequencies down to nearly 1400 cm^{-1}, which is a wavelength ($\lambda \approx 7$ μm) more than twice the size of the aperture. This demonstrates the synchrotron source's potential for exploiting near-field techniques that could extend the spatial resolution beyond the diffraction limit.

MEASUREMENT & ANALYSIS

The IMF has been used to map oxygen variations (which includes differentiating interstitial atoms from oxide precipitates[2]) in silicon, and for probing the regions around micron-sized Te precipitates in $Cd_{1-x}Zn_xTe$ wafers. The IRμs™ includes a scanning stage to allow for automatic mapping of spectra over a pre-defined area.

Silicon

Figure 2 shows the IR transmission of two silicon wafers. The $\lambda = 9$μm (1110 cm^{-1}) absorption due to interstitial oxygen is present to some degree in most wafers[6]. One

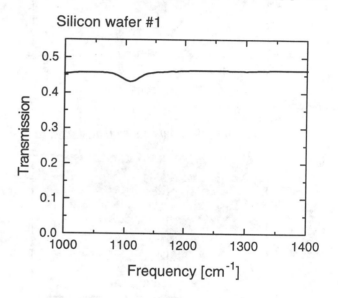

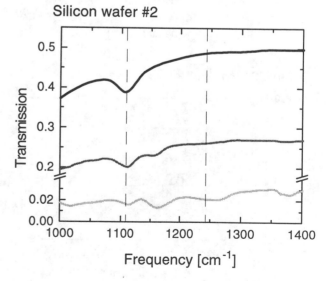

Figure 2. IR transmission for two silicon wafers. The lower panel shows spectra near a major defect. Vertical dashed lines mark two of the oxygen mode frequencies.

particular piece of silicon had a major sub-surface defect (not visually detectable), and a region around the defect was spectroscopically mapped. Transmission spectra for three locations at, and near, the defect are shown in the lower panel of Figure 2.

Overall absorbance

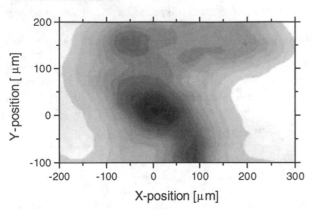

$\lambda=9\,\mu m$ *(1110 cm^{-1}) absorbance - <u>interstitial oxygen</u>*

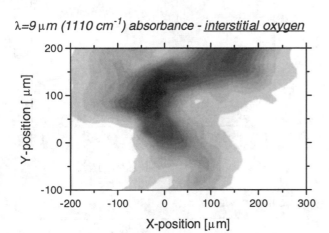

$\lambda=8\,\mu m$ *(1240 cm^{-1}) absorbance - <u>oxide</u>*

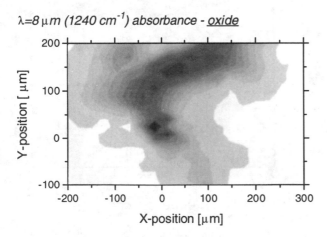

Figure 3. Silicon wafer absorbance maps at three different wavelengths (chosen to differentiate the type of oxygen) for the region surrounding a sub-surface defect. Darker shading indicates higher absorption.

Full IR spectra were then collected at 25 μm intervals over a rectangular area surrounding the defect, and the absorption at selected wavelengths was extracted. Contour maps of the absorption are shown in Figure 3. Note that the distribution of oxide and oxygen interstitials are similar, but not identical.

$Cd_{1-x}Zn_xTe$ (CZT)

CZT is a substrate material for II-VI semiconductor devices, such as the $Hg_{1-x}Cd_xTe$ (MCT) photodiode arrays used in long-wavelength IR imaging systems. Growing high quality CZT crystals is difficult, and it is not entirely clear what defects are specifically responsible for poor array performance. Tellurium precipitates are one commonly encountered defect, but their affect on the local electrical and chemical environment is not well understood. We have begun to investigate these ~ 5 micron sized inclusions by IR microspectroscopy. The transmission (relative to a clear section of CZT) is shown in Figure 4. Diffraction of light around the

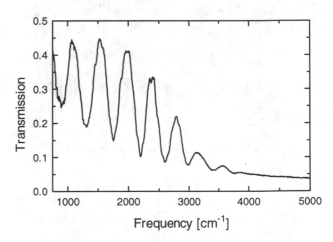

Figure 4. The transmission through a single Te precipitate buried near the surface of a CdZnTe wafer crystal. The IRµs apertures were set to sample a 5 μm square area, approximately matching the size of the precipitate.

precipitate is partly responsible for the increasing transmission at long wavelengths. Also visible are broad interference fringes for wavelengths greater than 3.5 μm, which corresponds to the bandgap for bulk Te. The fringe spacing is consistent with a diameter of 4 μm and a refractive index between 2.5 and 3.0. Contour maps of the absorption at two wavelengths (spanning $\lambda_g = hc/E_g$ of Te, for material contrast) are shown in Figure 5. These maps were extracted from approximately 180 complete spectra, acquired at 10 μm intervals over a

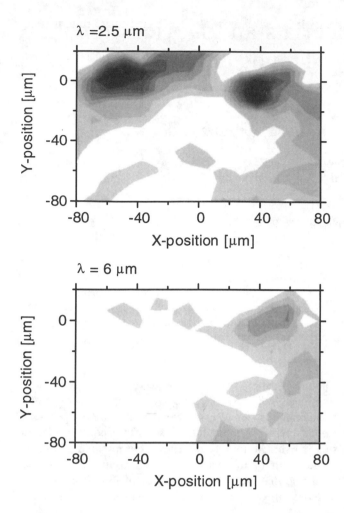

λ =2.5 μm

λ = 6 μm

Figure 5. Absorbance maps of a CdZnTe wafer in the vicinity of two Te precipitates. *Upper panel:* for λ=2μm. *Lower panel:* for λ=6μm. Darker regions represent higher absorbance. The same absorbance gray-scale was used for both panels.

rectangular region of a CZT wafer section. Other regions of less (but non-zero) absorption are also visible.

SUMMARY

The synchrotron radiation source extends the spatial resolution of infrared microspectroscopy to a size scale approaching a few microns. The IMF will continue to be developed and utilized for studies of semiconductor materials and devices. Much of our effort will be directed toward compound and alloy semiconductor systems, including materials and devices for infrared focal plane arrays. Of particular interest will be correlating localized defects with specific device failure mechanisms.

Other features of the infrared synchrotron source, such as the high degree of polarization and the sub-nanosecond pulse duration, are maintained by the IR microspectrometer, though they have not been exploited for the measurements presented here. The strong polarization should prove valuable for grazing-incidence reflection studies of the absorption by very thin layers, such as the gate oxides in silicon MOSFETs. The pulses are useful for time-resolved measurements, such as determining the intrinsic speed of a detector material, or for pump-probe spectroscopy[7].

Finally, synchrotron IR microspectroscopy allows entry into *near-field* techniques that could extend the spatial resolution beyond the diffraction-limit. The work here demonstrates that a 3 μm spatial resolution can be achieved, and improvements to eliminate aberrations from the present beamline mirror system could enable 1 μm spatial sensitivity to be realized. To reach another order of magnitude beyond that will almost certainly require a much brighter source of tunable infrared photons, such as the IR free electron laser.

ACKNOWLEDGEMENTS

Research at Grumman Aerospace was supported by IR&D funds and by NASA through contract NAS8-38147. DOE supports the NSLS through contract DE-AC02-76CH00016. We are grateful to Gwyn Williams (NSLS) and John Reffner (Spectra-Tech) for their assistance, and A. Berghmans (Grumman Aerospace) for the CdZnTe material preparation.

REFERENCES

1. Williams, G.P., *Nucl. Instr. Meth.* **195**, 383 (1982).

2. Ono, H., Ikarashi, T. Kimura, S., and Tanikawa, A., *J. Appl. Phys.* **76**, 621 (1994).

3. Sangster., M.J.L., Newman, R.C., Gledhill, G.A., Upadhayay, S.B., *Semicond. Sci. Technol.* **7**, 1295 (1992).

4. Carr, G.L., Williams, G.P., and Reffner, J.A., *Rev. Sci. Instr.* **66**, 1490 (1995).

5. Gladden, W.K., Baghdadi, A., ASTM STP 960, D.C. Gupta and P.H. Langer, Eds. (1986).

6. Carr, G.L., Hanfland, M., and Williams, G.P., *Rev. Sci. Instr.* **66**, 1643 (1995).

7. Carr, G.L., Reichman, J., DiMarzio, D., Lee, M.B., Ederer, D.L., Miyano, K.E., Mueller. D.R., Vasilakis, A., and O'Brien, W.L., *Semicond. Sci. Technol.* **8**, 922 (1993).

Optical Characterization of Materials and Devices for the Semiconductor Industry: Trends and Needs

S. Perkowitz

Department of Physics, Emory University, Atlanta, GA 30322 - 2430

D. G. Seiler

National Institute of Standards and Technology, Gaithersburg, MD 20899

W. M. Bullis

Materials & Metrology, Sunnyvale, CA 94087 - 4015

Contactless, nondestructive optical methods are used to characterize materials, processes, and devices in the semiconductor industry. In response to industrial needs, the Semiconductor Electronics Division of the National Institute of Standards and Technology conducted a survey of the needs for and use of optical characterization methods within the semiconductor industry. Data from forty-two firms were analyzed to show the impact of the methods, what they measure, their range and precision, and their cost. A significant finding of the study is the need expressed by many industrial users for improved standards and test methods for optical characterization.

BACKGROUND OF SURVEY

Optical methods are contactless, nondestructive means to characterize semiconductor materials and devices. [1,2] Measurements can be made in situ, and different regions of a sample can be probed. A recent workshop on manufacture of integrated circuits notes that in situ or real-time metrology is needed for the next generation of devices. [3] It also calls for the development of new optical techniques, such as means to analyze and image an entire wafer surface to determine etch rate and properties.

In response to such industrial needs, the Semiconductor Electronics Division of the National Institute of Standards and Technology (NIST) conducted a survey of the use of optical methods in the semiconductor industry. Questionnaires were designed to determine and analyze the use of nine methods: ellipsometry, infrared spectroscopy, interferometry, optical microscopy (visible and infrared), photoluminescence, photoreflectance, Raman scattering, reflectometry, and scatterometry. These were chosen because of their known or assumed value in characterizing semiconductors for electronic and photonic devices. Combined, these methods measure most of the properties that are important for such devices.

Seventeen responses were returned by materials manufacturers, 13 from device manufacturers, and 12 from equipment suppliers, primarily makers of characterization equipment rather than processing equipment. Although questionnaires were sent to scientists and engineers in all of the different stages in the research-to-product cycle, the survey overwhelmingly represents the research and development part of the cycle. This unfortunately biases the results so that the extent of the use of these techniques in materials and device production environments is not fully indicated.

This paper presents only a portion of the data gathered. The data reported here represent a mixture of responses from the silicon and compound semiconductor fields. These areas differ significantly in terms of the materials parameters that must be controlled as well as in the mix between R&D and production activity. A more complete report of the survey results will be published as a NIST Special Publication. [4]

USE AND VALUE OF THE METHODS

Table 1 shows the response to a question regarding the industrial use and value of each technique. Respondents were asked to rank each method as "critical," that is, essential for the end application; "useful," meaning that the method gives valuable but not essential information; or "not used." In the table, the methods are listed in order of use, measured by the percentage of respondents who consider the method either critical or useful. The "value rating" is determined by weighting the number of "critical" responses by 2, the number of "useful" responses by 1, and the number of "not used" responses by 0, and dividing the weighted sum by the total number of responses. The last column is the ranking of the impact of the techniques based upon a combination of their value ratings and their percent use.

The most widely used method is optical microscopy and the least used is Raman scattering. Although scatterometry falls well down on the list, it should be noted that this technique is the basis of scanning surface inspection systems which are universally used by silicon materials and device companies for automated particle counting.

In addition to the nine methods listed, photovoltaic and photoconductive methods also were rated as critical or useful by several respondents. Taken together, these methods appear to be less widely used than Raman spectroscopy.

PROPERTIES CHARACTERIZED

Table 2 shows the materials properties reported to be measured optically by the respondents to the survey. These properties were selected as those essential to characterize, design, and fabricate electronic and photonic devices. The table combines responses from both the silicon and compound semiconductor fields and from all sectors of the research-to-product cycle.

Bold-face entries represent a significant intersection of methods and properties. The last row gives the number of listed materials properties measured by each technique. Ellipsometry, infrared spectroscopy, photoluminescence, and Raman scattering are reported to have been used for the measurement of 10 to 12 of the 13 listed parameters.

MEASURED MATERIALS AND DEVICES

Respondents to the questionnaire reported that they examine materials by optical methods about three times as often as they do devices. The materials are measured most often in film form, with bulk form a distant second. Table 3 indicates that of the materials represented, silicon is measured optically most often, closely followed by dielectrics and dielectric films, and Group III-V semiconductors. Group II-VI semiconductors and metals are cited much less frequently by the respondents to the survey. The citations for measurements on compound semiconductors also include ternary and quaternary compounds for photonic devices.

Table 4 lists the electronic and photonic devices that were reported to be examined by optical characterization methods for both electronic and photonic devices. As might be expected from relative market sizes, photonic devices are characterized much less frequently than electronic devices.

RANGE AND PRECISION

Table 5 summarizes the range and precision for some of the measured quantities. The responses indicate that the respondents typically work with 100-mm diameter wafers or larger. Whatever the absolute values of precision and the other parameters, it is essential that they serve the needs of the industry. On average, 70% of the respondents indicated that both the precision and spatial resolution of the methods cited were adequate for their current applications, but there were wide variations in the responses for individual parameters.

In response to the question, "Are standards or test methods needed for the measurement?" 81% answered "yes," but only 27% answered "yes" to the question, "Are such standards or test methods available?" This suggests that there is a strong need for optical measurement standards; this result is an important finding of the survey.

COMPARISON TO OTHER METHODS

The nondestructive nature and high speed of optical techniques are the most important reasons given for preferring these methods over other techniques. Secondary reasons are the reduced preparation effort, and the ability to examine different portions of a sample.

Lack of technical support is the main reason that potentially useful optical methods are not employed. Cost and lack of commercial equipment are the next most important considerations. The importance of the availability of commercial equipment is indicated by the fact that three out of four applications reported by survey respondents utilized commercially available equipment.

COST ISSUES

Table 6 summarizes reported costs for optical characterization. Equipment costs vary greatly. This may reflect the presence of multiple measurement stations, in-house *vs.* commercial costs, and the degree of automation. Operational costs also vary widely, probably for similar reasons, and because different organizations may allocate such costs differently. Some of the diversity in the staff numbers may also reflect how staff time is allocated for internal budgeting, as well as the number of stations. Despite the wide variation in both equipment and operating costs reported by the respondents to the survey, cost is perceived as a major inhibiting factor to the use of optical characterization methods by many in the industry.

COMMENTS FROM RESPONDENTS

In addition to responding to fixed questions, respondents were asked for their comments on any other issues that they considered important but that were not covered by the questions in the survey form. Comments were received from a small fraction of the total number of respondents.

Several respondents proposed spectroscopic ellipsometry and photoluminescence as worthy of further development. Nearly two-thirds of these responses asked that NIST establish reference materials for optical characterization. Examples where such reference materials might be most useful include film thickness standards for SiO_2/Si and multilayer films 2 nm to 10 nm thick, standards for defects and surface microroughness in silicon, and standards for spectroscopic ellipsometry. Requests were also made for standardized measurement procedures and analytical software, a catalog of photoluminescent spectral signatures from impurities, and workshops.

Some respondents were concerned about the lack of fundamental work in the physics of characterization. Many firms are said to be too small to carry out such research, and university work in this area is limited. The database for optical characterization activities was seen as lacking, for instance in energy gap *vs.* alloy content for AlGaAs and InGaAs. Some respondents look to NIST to establish such data. Two respondents consider in situ measurement and optically-guided control of sample growth as important future possibilities not treated in the survey. One felt that the benefits and drawbacks of optical characterization have yet to be clearly laid out to the industry.

SUMMARY

Optical methods are used to characterize a wide variety of material systems. The forty-two responses from different types of firms show that these techniques are heavily used in industrial research and development activities. Of the specific techniques identified, microscopy has had the greatest impact and Raman scattering, the least. The nondestructive nature of optical methods is a major reason both for their present use and for their future importance for in situ characterization. Lack of technical support is the main reason given for not employing optical techniques.

The respondents' largest single application of optical techniques is to silicon, but optical characterization of dielectrics and dielectric films and Group III-V semiconductors is nearly as widespread. A diverse set of devices is examined by these techniques. Among the respondents to the survey, electronic devices are examined by optical characterization methods twice as often as photonic ones.

The range of values that can be measured for many properties is broad, and precision ranges from moderate to excellent. Precision, sensitivity, and spatial resolution are thought adequate for two out of every three current applications, but deficiencies were noted in some specific areas. Ellipsometry and infrared spectroscopy are reported to be the most expensive methods; photoluminescence and photoreflectance, the least expensive. Many users carry out optical characterization with a relatively modest investment, but the reported range of costs involved was very large.

The need for reference materials and standard test methods for optical techniques is an important finding of the survey. A number of respondents would find workshops on these techniques to be useful.

REFERENCES

1. Perkowitz, S., Seiler, D. G., and Duncan, W. M., Optical Characterization in Microelectronics Manufacturing, *J. Res. Natl. Inst. Stand. Technol.* **99** (5), 605−639 (1994).

2. Perkowitz, S., *Optical Characterization of Semiconductors: Infrared, Raman, and Photoluminescence Spectroscopy* (Academic Press, London, 1993).

3. *Proceedings of the International Workshop in Process Control Measurements for Advanced IC Manufacturing*, November 9−10, 1992, Austin, Texas, P. H. Langer and W. M. Bullis, editors (Semiconductor Equipment and Materials International, Mountain View, CA, 1993) pp. 5, 13.

4. NIST Special Publication (to be published).

Table 1. Use and Value of Optical Methods.

Optical Method	Critical	Useful	Not Used	Percent Use	Value Rating	Final Rank
Optical microscopy	23	13	1	97	1.6	1
Ellipsometry	19	11	7	81	1.3	2
Infrared spectroscopy	20	8	7	80	1.4	2
Photoluminescence	10	9	14	58	0.9	3
Interferometry	9	7	14	53	0.8	4
Reflectometry	10	7	15	53	0.8	4
Scatterometry	8	9	15	53	0.8	4
Photoreflectance	4	11	17	47	0.6	5
Raman scattering	2	10	21	36	0.4	6
Total	105	85	111			
Other Methods[*]						
Photoconductive, photovoltaic	5	4				
Miscellaneous[†]	17	12				

[*] Added by respondents.

[†] All other methods added. Each appears only once as either "critical" or "useful."

NOTE — The entries in the table represent a composite of usage on both silicon and compound semiconductor materials and devices.

Table 2. Properties Measured by Optical Methods, Listed Alphabetically by Technique.

Parameter	Optical Method[*]								
	ELL	IR	IN	MIC	PL	PR	RAM	REF	SC
Alloy composition	6	6			12	6	5	3	
Carrier density	3	1			3	3	2		
Carrier mobility		1					2		
Carrier lifetime	1	1			3				
Crystal orientation	1	2					2		
Crystallinity	2	2		5	4	3	6	2	
Defect density	1		2	23	3	1	1	1	8
Defect type	1	1	1	21	3		2		1
Energy band gap	4	5	1		14	5		1	
Film thickness	29	7	4	1	1			10	
Impurity density	2	8		1	4	2	2		2
Impurity type		9		1	9		2	1	1
Resistivity		2					1		
Total	50	45	8	52	56	20	25	18	12
Parameters Measured	10	12	4	6	10	6	10	6	4

[*] ELL, ellipsometry; IR, infrared spectroscopy; IN, interferometry; MIC, optical microscopy; PL, photoluminescence; PR, photoreflectance; RAM, Raman scattering; REF, reflectometry; SC, scatterometry.

NOTE — The entries in the table represent a composite of usage on both silicon and compound semiconductor materials and devices.

Table 3. Materials Characterized by Optical Techniques.

Material	Number of Citations
Silicon	65
Dielectrics, dielectric films	60
Group III-V semiconductors	59
Group II-VI semiconductors	36
Metals	15
Other*	14
Total	249

* Includes composites, contaminants, fluids, glass, high-T_c superconductors, photoresist, and other semiconductors.

Table 4. Applications of Optical Characterization Techniques to Electronic and Photonic Devices.

Electronic Devices	Number of Citations	Photonic Devices	Number of Citations
Unspecified	15	Sources	
Microprocessors, memory	11	Quantum-well lasers	10
Silicon linear and digital ICs	10	Electroluminescent devices	4
HEMT, p-HEMT	9	Light emitting diodes	1
CMOS, MOS capacitor	5	Detectors, Solar Cells	
MESFET	4	Infrared, x-ray, unspecified	9
		Solar cells	1
Other		Other	
Magnetoresistors	2	Optical modulators	4
		Flat panel displays	1
Total	56	Total	30

Table 5. Range and Precision of Measurements, and Maximum Sample Area Examined.

Parameter	Number of Citations	Range	Mean Precision	Maximum Area, cm^2
Alloy composition	11	0% to 100%	0.02	81
Carrier density	3	10^{17} cm^{-3} to 10^{19} cm^{-3}	50%	
Carrier lifetime	7	1 ps to 10 s	10%	Wafer
Defect density	16	10^1 cm^{-2} to 10^{11} cm^{-2}		314
Energy gap	9	0 to 2 eV	0.03 eV	81
Film thickness	24	2 nm to 50 μm	2%	625

Table 6. Source and Cost of Optical Characterization Equipment, and Required Staffing Commitment.

Optical Method	Source of Equipment		Cost of Equipment, $1,000		Staff, FTE
	Commercial	In-house	Capital	Annual Operating	
Ellipsometry	23	4	97 (9 – 375)	4 (0 – 20)	0.5 (0 – 1.5)
Infrared spectroscopy	24	1	94 (20 – 300)	15 (0.5 – 85)	0.6 (0.1 – 2)
Interferometry	10	3	82 (2 – 300)	4 (0 – 16)	0.6 (0.1 – 1)
Optical microscopy	31	2	70 (3 – 1,000)	7 (0 – 80)	0.9 (0 – 10)
Photoluminescence	7	9	52 (15 – 150)	7 (0 – 20)	0.5 (0.2 – 1)
Photoreflectance	5	5	43 (10 – 150)	3 (0 – 10)	0.5 (0.1 – 1)
Raman scattering	4	5	96 (30 – 150)	8 (0 – 20)	0.5 (0.2 – 1)
Reflectometry	10	2	75 (10 – 320)	2 (0 – 10)	0.6 (0.1 – 2)
Scatterometry	9	4	117 (25 – 350)	5 (0 – 35)	0.7 (0.1 – 2)
Other	15	12	87 (8 – 250)	11 (0 – 80)	0.8 (0.1 – 2)
Total	138	47			
Percent	75	25			
Mean			81 ± 22*	7	0.6

* Standard deviation.

NOTE — "Commercial" and "In-house" mean, respectively, commercially available equipment, and equipment constructed or assembled in-house. "Staff" is the number of full-time-equivalent (FTE) employees needed to operate the optical installation. The last three columns show the means of the responses, with lowest and highest values reported placed in parentheses. As can be seen, the typical range was extremely large.

Characterization of SiGe/Ge Heterostructures and Graded Layers Using Variable Angle Spectroscopic Ellipsometry

A. R. Heyd* and S. A. Alterovitz

NASA Lewis Research Center, 21000 Brookpark Road, MS 54-5, Cleveland, Ohio 44135

E. T. Croke

Hughes Research Laboratories, Malibu, California 90265

K. L. Wang and C. H. Lee

University of California, Los Angeles, California 99024

Variable angle spectroscopic ellipsometry (VASE) has been used to characterize Si_xGe_{1-x}/Ge superlattices (SLs) grown on Ge substrates and thick Si_xGe_{1-x}/Ge heterostructures grown on Si substrates. Our VASE analysis yielded the thicknesses and alloy compositions of all layers within the optical penetration depth of the surface. In addition, strain effects were observed in the VASE results for layers under both compressive and tensile strain. Results for the SL structures were found to be in close agreement with high resolution x-ray diffraction measurements made on the same samples.

The VASE analysis has been upgraded to characterize linearly graded Si_xGe_{1-x} buffer layers. The algorithm has been used to determine the total thickness of the buffer layer along with the start and end alloy composition by breaking the total thickness into many (typically > 20) equal layers. Our ellipsometric results for 1 μm buffer layers graded in the ranges $0.7 \leq x \leq 1.0$ and $0.5 \leq x \leq 1.0$ are presented, and compare favorably with the nominal values.

INTRODUCTION

Characterization of semiconductor superlattice (SL) structures is typically done by x-ray diffraction (XRD) and transmission electron microscopy (TEM). Both XRD and TEM can be used to quantify SL ordering and periodicity. In addition, XRD can give the average value of the composition. Both techniques have limitations; TEM is destructive, XRD requires relatively thick samples (~1000 Å), and neither technique can be used to quantify individual interfaces or quantum wells. Recently, variable angle spectroscopic ellipsometry (VASE) has been shown to be a powerful, non-destructive technique for the post-deposition characterization of Si_xGe_{1-x}/Si SLs and other multilayer heterostructures [1,2]. In these studies VASE has been used to determine layer thicknesses, alloy composition, oxide thickness, number of superlattice periods, and sample homogeneity. In this current work we will concentrate on Ge rich Si_xGe_{1-x}/Ge SLs and heterostructures.

Graded composition Si_xGe_{1-x} layers are used in the base of Si_xGe_{1-x}/Si heterojunction bipolar transistors to increase device speed. Graded composition Si_xGe_{1-x} layers (graded to achieve $x \approx 0.3$) are also used as buffers to relieve strain in the growth of n-type Si_xGe_{1-x}/Si modulation doped field effect transistors (MODFET) structures. In previous VASE analysis of MODFETs, only the high energy portion of the ellipsometric spectra was used. Thus, a graded composition layer analysis was avoided, since the graded layer was buried below the optical penetration depth of the probing light at these energies. In this work a simple algorithm for the calculation of the Fresnel reflection coefficients of a linearly graded Si_xGe_{1-x} layer has been developed. This allowed a graded Si_xGe_{1-x} layer to be characterized in terms of its thickness, and Si content at the substrate and ambient surfaces.

BACKGROUND

To determine the properties of multilayer structures the measured ellipsometric angles, $\{\tan\Psi(\lambda), \cos\Delta(\lambda)\}$, must be compared to results calculated from well defined models. Linear regression analysis (LRA) is used to minimize the unbiased estimator, σ:

$$\sigma^2 = \frac{1}{n-m-1}\sum_{i=1}^{n}\left\{(\tan\Psi_i^e - \tan\Psi_i^c)^2 + (\cos\Delta_i^e - \cos\Delta_i^c)^2\right\} \quad (1)$$

---

*This work was performed while the author held a National Research Council-NASA Research Associateship.

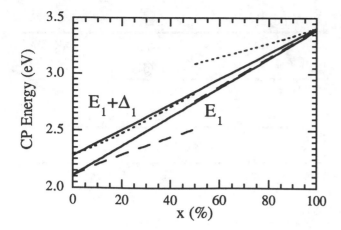

Figure 1: E_1 (long dash) and $E_1 + \Delta_1$ (short dash) critical point energies for relaxed and strained Si_xGe_{1-x} layers. The relaxed curves (Ref. 6) are shown as solid lines. The strain induced splitting (dashed lines) was calculated using Eqs. 2 and 3. For $x > 0.5$ the substrate is Si and the strain is compressive, and for $x < 0.5$ the substrate is Ge and the strain is tensile.

Table I: Comparison of target sample structure with that determined by HRXRD and VASE. The period is the sum of the Ge and SiGe layer thicknesses, and x(avg) is the average silicon content in one period. The parameters of the VASE analysis are the oxide, Ge and SiGe layer thicknesses, and the silicon content, x, of the SiGe layer.

Sample	Source	Period (Å)	x(avg) (%)	d(Ge) (Å)	d(SiGe) (Å)	x (%)
HA57	target	178	6.2[a]	128	50	22.0
HA57	HRXRD	201.1	8.0	141.8[b]	59.3[b]	26.6[b]
HA57	VASE[c]	202.8	8.4[a]	127.5 ±3.9	75.3 ±3.5	22.7 ±1.6
HA58	target	201	7.8[a]	142	59	26.6
HA58	HRXRD	202.1	8.3	141.8[b]	60.3[b]	27.5[b]
HA58	VASE[d]	204.5	8.4[a]	126.9 ±4.7	77.6 ±4.4	22.1 ±2.0

[a] Calculated from d(Ge), d(SiGe), and x.
[b] Estimated from the period, x(avg), and shutter opening/closing times.
[c] $\sigma = 0.0137$, d(oxide) $= 17.9$ Å ± 0.4 Å.
[d] $\sigma = 0.0167$, d(oxide) $= 25.5$ Å ± 0.6 Å.

where n is the number of observed data points and m is the number of free parameters used in the model. The superscripts e and c refer to the experimentally observed data and the corresponding results calculated by the model, respectively. In order to construct a model, the dielectric function of all the constituent materials must be known. The dielectric function of composite layers can be determined using effective medium approximations (EMAs) [3]. The energy shift algorithm [4] is used to interpolate between Si_xGe_{1-x} dielectric functions published at discrete alloy compositions [5]. The algorithm requires a knowledge of the functional dependence of the critical point (CP) energies with the alloy composition, usually the $E_o(x)$, $E_1(x)$, and $E_2(x)$ CP energies. To improve the analysis of Ge rich Si_xGe_{1-x} materials the energy shift algorithm has been modified to include the effects of the $E_1 + \Delta_1$ CP energy [6].

Fig. 1 shows the E_1 and $E_1 + \Delta_1$ CP energies for strained and relaxed Si_xGe_{1-x} layers. The in-plane strain, ε, in the strained Si_xGe_{1-x} layer results from the lattice mismatch between the layer and the substrate. The shifts in the E_1 and $E_1 + \Delta_1$ CP energies for a biaxial (001) strain are given by [7]:

$$\Delta E_1 = \frac{+\Delta_1}{2} + E_H - \frac{1}{2}\left(\Delta_1^2 + 4E_S^2\right)^{1/2} \quad (2)$$

$$\Delta(E_1 + \Delta_1) = \frac{-\Delta_1}{2} + E_H + \frac{1}{2}\left(\Delta_1^2 + 4E_S^2\right)^{1/2} \quad (3)$$

where the hydrostatic shift, E_H, and uniaxial shear, E_S, are given by:

$$E_H = 2\mathcal{E}_1(1 - C_{12}/C_{11})\varepsilon \quad (4)$$

$$E_S = (2/3)^{1/2} D_3^3 (1 + 2C_{12}/C_{11})\varepsilon \quad . \quad (5)$$

C_{ij} are the elastic stiffness constants, $\mathcal{E}_1$ is the hydrostatic deformation potential, and D_3^3 is an intraband deformation potential for the Λ_3 valence band for a [001] uniaxial strain. The energy shift algorithm assumes the functional dependence of the relaxed E_1 and $E_1 + \Delta_1$ CP energies shown in Fig. 1. Therefore, when modeling strained Si_xGe_{1-x} layers the strain induced shifts in the E_1 and $E_1 + \Delta_1$ will cause the value of the Si content, x, to be over- or under-estimated for a compressive or tensile strain, respectively.

SiGe/Ge SUPERLATTICE STRUCTURES ON Ge SUBSTRATES

Two fifty period Si_xGe_{1-x}/Ge SL structures, HA57 and HA58, were grown by Si molecular beam epitaxy (MBE) on Ge (100) substrates at Hughes Research Laboratories. The samples were first characterized by high resolution x-ray diffraction (HRXRD) [8], the results are shown in Table I along with the target structures. Each sample was then measured by VASE at several angles of incidence (69°, 73°, and 77°) which were chosen to increase the sensitivity of the ellipsometric angles to the sample structure parameters [9]. The measured {$\tan\Psi$, $\cos\Delta$} spectra for sample HA57 are shown in Fig. 2 along with the spectra generated from the best fit model. Due to the strong absorption in the low wavelength region, discrepancies between the measured and calculated spectra in this region are a result of devia-

429

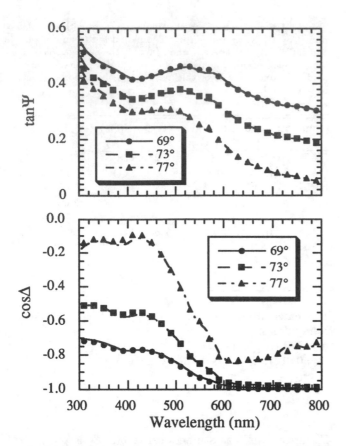

Figure 2: Comparison of measured VASE data (symbols) with that determined from the best fit model (lines) for the Si_xGe_{1-x}/Ge SL sample HA57. Sample HA58 shows similar fitting. The structure parameters found from the best fit model are shown in Table I.

Nominal		VASE	
	Oxide	Si_xGe_{1-x} (49%)	36.5 Å
		GeO (51% ± 11%)	± 6.5 Å
200 Å	$Si_{0.3}Ge_{0.7}$	Si_xGe_{1-x}	206.1 Å
		x = 29.9% ± 0.8%	± 10.1 Å
40 Å	Ge	Ge	34.1 Å
			± 10.7 Å
1 μm	Stepped SiGe Buffer	Si_yGe_{1-y} Substrate y = 25.1% ± 5.2%	
	Si Substrate	σ = 0.0284	

Figure 3: Structure of sample CL141 showing the target structure parameters and those determined from the VASE analysis.

SiGe/Ge HETEROSTRUCTURES ON Si SUBSTRATES

Two Si_xGe_{1-x}/Ge heterostructures, CL141 and CL171, were grown on Si substrates at the University of California at Los Angeles. The nominal structures for sample CL141 and CL171 are shown in Fig. 3 and Fig. 4, respectively. The Si content of the stepped buffer was changed from 100% to 20% or 30% in approximate steps of 25%. Therefore, it would require six parameters to completely characterize the buffer layer. Characterization of the buffer layer is further complicated by the fact that it is buried deep in the structure. As a result sample CL141 was modeled over the range 300 nm to 550 nm to ensure that the top step of the buffer would act as the substrate. Over this range the maximum penetration depth is ~1000 Å and ~200 Å in $Si_{0.3}Ge_{0.7}$ and Ge, respectively. The oxide has been modeled as a mixture of Si_xGe_{1-x} (x same as in underlying layer) and GeO_2 using the Bruggeman EMA; this mixture simulates a surface roughness. The surface of these samples is expected to be rough due to the large lattice mismatch between the Ge layer and the underlying Si substrate (~4%). The VASE results, shown in Fig. 3, agree with in the 90% confidence limits with the nominal structure.

The second sample, CL171, was modeled over the spectral range 300 nm to 760 nm. Over this range the Si_xGe_{1-x} layer will act as the substrate due to the increased thicknesses of both the Ge and SiGe layers The measured VASE data is shown in Fig. 5 along with spectra generated from the best fit model; the resulting structure parameters are shown in Fig. 4. Modeling the

tions in the model that occur near the surface. Similarly, discrepancies in the fitting in the long wavelength region are most likely due to problems in the model which occur deeper in the structure. In this case the model assumes that each period of the SL is identical. Some slight oscillations can be seen in the long wavelength region of the $\cos\Delta$ spectra which are not reproduced by the model. This indicates that the periods of the sample are not identical and that the thickness and/or x(avg) of each period may be fluctuating about some average value.

The agreement between the HRXRD and VASE results is quite good, especially when comparing the period and x(avg). There is, however, a substantial difference between the individual layer thicknesses and x. The Si_xGe_{1-x} layer is under a tensile strain as a result of being sandwiched between the thicker Ge layers. This tensile strain accounts for the lower x value obtained by VASE. The differences in the layer thicknesses may be a result of interfacial layers which are not currently accounted for in the model.

Nominal		VASE	
	Oxide	Ge (38%)/ GeO$_2$ (62% ± 9%)	36.0 Å ± 2.6 Å
1000 Å	Ge	Si$_y$Ge$_{1-y}$ y = 5.8% ± 0.3%	1263 Å ± 11 Å
1200 Å	Si$_{0.2}$Ge$_{0.8}$	Si$_x$Ge$_{1-x}$ Substrate x = 16.5% ± 1.5%	
1 μm	Stepped SiGe Buffer	σ = 0.0269	
	Si Substrate		

Figure 4: Structure of sample CL171 showing the target structure parameters and those determined from the VASE analysis.

Table II: Results of VASE analysis of graded Si$_x$Ge$_{1-x}$ layers on Si substrates. The nominal thickness of all samples is 1 μm. The samples were linearly graded from 100% Si at the substrate to x_n(nom) at the surface. The parameters of the VASE analysis are the oxide thickness, buffer layer thickness (D), and the Si content at the surface (x_n). The Si content at the substrate was held constant at 100%. A value of $n = 30$ was used to model all three samples.

Sample	σ	d(oxide) (Å)	D (μm)	x_n (%)	x_n(nom) (%)
HA82	0.0522	44.8 ± 0.9	1.04 ± 0.04	76.7 ± 0.6	70
HA83	0.0524	59.3 ± 1.1	1.02 ± 0.03	66.7 ± 0.6	50
HA17	0.0202	43.0 ± 0.4	0.97 ± 0.01	58.7 ± 0.3	50

Ge layer as Si$_x$Ge$_{1-x}$ resulted in a ~25% decrease in σ. Since the Si content of this layer should in fact be 0%, the value of 5.8% can be attributed to a compressive strain which results in an over-estimate of the Si content.

SiGe GRADED LAYERS

Three continuously, linearly graded Si$_x$Ge$_{1-x}$ layers were grown by Si MBE on Si (100) substrates at Hughes Research Laboratories; the nominal structures are described in Table II. Graded layers are simulated in the VASE model by breaking the layer in to $n+1$ sub layers. The thickness of the i-th sub layer is:

$$d_i = \begin{cases} D/2n & i = 0, n \\ D/n & 1 \leq i < n \end{cases} \quad (6)$$

where D is the total layer thickness. For a linearly graded layer the Si content of the i-th layer is given by:

$$x_i = \frac{(n-i)x_0 + ix_n}{n} \quad (7)$$

where x_0 and x_n are the Si content at the substrate and surface, respectively. As $n \rightarrow \infty$ this model will more closely approximate a continuously graded layer. However, values of $n > 20$ were found to yield nearly identical results in the VASE analysis for the graded samples used in this study.

The graded samples were measured by VASE at three angles of incidence: 70°, 75°, and 77°. Fig. 6 shows the experimental data for sample HA17 along with the ellipsometric angles generated from the best fit model. The fitting is nearly perfect in the cos Δ spectrum and in the long wavelength region of the tan Ψ spectrum. The poor fitting in the low wavelength region of the tan Ψ spectrum is most likely due to poor surface

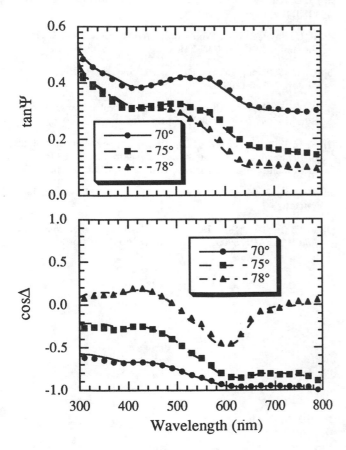

Figure 5: Comparison of measured VASE data (symbols) with that generated from the best fit model (lines) for the thick Si$_x$Ge$_{1-x}$/Ge heterostructure sample CL171. The best fit model is shown in Fig. 4 along with the resulting values of the structure parameters.

431

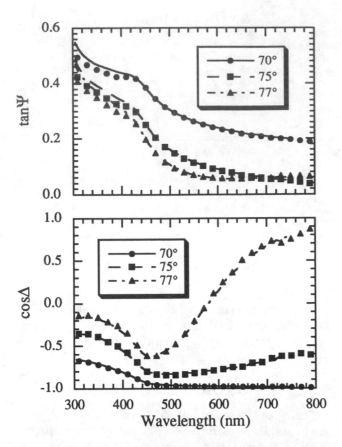

Figure 6: Comparison of measured $\{\tan\Psi, \cos\Delta\}$ spectra (symbols) for the graded Si_xGe_{1-x} sample HA17 with that determined from the best fit model (lines). The structure parameters determined from the model are shown in Table II.

quality. Because the graded layers are uncapped, the surface region of the graded layers are expected to be strained which can cause the surface to become rough. Table II summarizes the results of the VASE analysis. It can be seen from this table that the thickness values are very close to the nominal values. The values for the Si content at the surface (x_n), however, are consistently much higher than the nominal values. These discrepancies can be caused by the compressive strain in the surface region of the layer which would result in an overestimate of x. The growth log notes that the surface morphology of sample HA83 is poor when compared with sample HA82. This may be the cause of the larger discrepancy in x_n for sample HA83.

CONCLUSIONS

VASE has been used to characterize Si_xGe_{1-x}/Ge SL structures grown on Ge substrates. The values of the period and x(avg) determined by VASE show excellent agreement with those same values determined by HRXRD. Due to strain effects or interfacial layers not accounted for in the VASE models, the individual layer

thicknesses and x determined by the two methods do not agree as closely. Thick Si_xGe_{1-x}/Ge heterostructures grown on Si substrates have also been characterized by VASE. Results show close agreement with the nominal structure. In addition VASE analysis provides a qualitative assessment of the surface roughness and strain in the various layers of the structures. Finally, VASE has been used to characterize continuously, linearly graded Si_xGe_{1-x} layers in terms of the layer thickness and Si content at the surfaces. Results agree extremely well with the target structure, especially when strain in the surface region of these uncaped graded layers is taken in to account.

REFERENCES

[1] R. M. Sieg, S. A. Alterovitz, E. T. Croke, and M. J. Harrell, Appl. Phys. Lett. **62**, 1626 (1993).

[2] R. M. Sieg, S. A. Alterovitz, E. T. Croke, M. J. Harrell, M. Tanner, K. L. Wang, R. A. Mena, and P. G. Young, J. Appl. Phys. **74**, 586 (1993).

[3] D. E. Aspnes, Proc. Soc. Photo-Opt. Instrum. Eng. **276**, 188 (1981).

[4] P. G. Snyder, J. A. Woollam, S. A. Alterovitz, and B. Johs, J. Appl. Phys. **68**, 5925 (1990).

[5] G. E. Jellison, Jr., T. E. Haynes, and H. H. Burke, Opt. Mater. **2**, 105 (1993).

[6] A. R. Heyd, S. A. Alterovitz, and E. T. Croke, Mat. Res. Soc. Symp. Proc. **358**, (1995), in press.

[7] F. H. Pollak, in *Strained-layer superlattices: Physics*, Vol. 32 of *Semiconductors and semimetals*, edited by T. P. Pearsall (Academic Press, San Diego, CA, 1990), Chap. 2, pp. 17–53.

[8] V. S. Speriosu and T. Vreeland, Jr., J. Appl. Phys. **56**, 1591 (1984).

[9] P. G. Snyder, M. C. Rost, G. H. Bu-Abbud, J. A. Woollam, and S. A. Alterovitz, J. Appl. Phys. **60**, 3293 (1986).

432

Ellipsometric Characterization of Thin Oxides on Silicon

William A. McGahan and John A. Woollam

J. A. Woollam Co., Inc.
650 J. Street
Lincoln, NE 68508

In this work, we first use a Kramers-Kronig consistent parametric model for semiconductor optical constants to extend the optical constant spectra for silicon measured by Jellison for use over the spectral range 250-1700 nm. Second, we use these results to compare the effects of employing different silicon substrate optical constants and various optical models for the silicon dioxide layer on the characterization of silicon dioxide films ranging from 2 nm (native oxides) to several microns in thickness. The purpose of these comparisons is to compare the systematic errors obtained from the analysis of ellipsometric data from oxide films on silicon. Systematic errors due to errors in the silicon substrate optical constants and to inaccurate parameterizations of the silicon dioxide index of refraction are both studied as a function of the oxide film thickness. Results are presented for both thermally grown and deposited oxide films ranging from 10 to 2500 nm in thickness.

INTRODUCTION

The characterization of thin oxide films deposited or thermally grown on crystalline silicon substrates is a problem of great importance to the semiconductor industry. Ellipsometry is a standard technique for the characterization of oxide films on silicon[1], with both single wavelength and spectroscopic techniques employed at either a single angle of incidence or multiple angles of incidence.

In this work, we first use a Kramers-Kronig consistent parametric model for semiconductor optical constants to extend the optical constant spectra for silicon measured by Jellison for use over the spectral range 250-1700 nm. Second, we use these results to compare the effects of employing different silicon substrate optical constants and various optical models for the silicon dioxide layer on the characterization of silicon dioxide films ranging from 2 nm (native oxides) to several microns in thickness.

EXPERIMENTAL TECHNIQUE

Ellipsometric psi and delta data were measured from silicon dioxide films ranging from native oxides to 2500 nm thick on silicon wafer substrates. Each sample was measured at three angles of incidence (70°, 75°, and 80°) over the spectral range 250-1000 nm using a silicon photodiode detector. The films thicker than 500 nm were also measured from 1000-1700 by 10 nm steps using an InGaAs photodiode detector. The ellipsometric measurements were performed with a J. A. Woollam Co. rotating analyzer VASE® ellipsometer. Measurements were averaged for 40-50 cycles of the analyzer, and the fluctuations of the data over the averaged cycles in conjunction with the uncertainties of the system calibration parameters were used to estimate the standard deviations of each data point, used for weighting of these data points during regression analysis of the data. Typical standard deviations on the experimental data were 0.01° - 0.05° for psi and 0.01°-0.2° for delta.

ANALYSIS OF ELLIPSOMETRIC DATA

In an ellipsometric experiment, the measured quantities describe the polarization state of the light beam entering the detector. Useful physical information about the sample under study is obtained by constructing an optical model for the sample and varying parameters in this model in order to best-fit the experimental data. It is not only necessary that the optical model fit the experimental data well, but it must also be established that the set of best-fit variable parameters is unique.

The measured ellipsometric parameters psi (Ψ) and delta (Δ) are defined in terms of the Fresnel reflection coefficients of the sample under study as follows[2]:

$$\tan \Psi \exp(i\Delta) = \frac{R_p}{R_s}, \qquad (1)$$

where R_p (R_s) equals the ratio of the p-polarized (s-polarized) component of the reflected light beam to the p-polarized (s-polarized) component of the incident light beam. Maxwell's equations are employed along with the matching conditions for the electric and magnetic fields of the light beam at each interface in order to calculate psi and delta as a function of wavelength and angle of incidence. The optical model required for this calculation must specify the thickness of each layer in the model (with the substrate in this work assumed to be semi-infinitely thick) and the index of refraction (n) and extinction coefficient (k) of each layer and the substrate.

Thicknesses in the optical model and optical constants of the layer(s) and/or substrate (or parameters in parametric dispersion models for the optical constants) may then be varied in order to best-fit the ellipsometric data. The J. A. Woollam Co. WVASE® software was employed in this work to perform this data analysis. This software uses the Levenberg-Marquardt non-linear multivariate regression algorithm to obtain the best-fit set of variable parameters for a given model.

The uniqueness of the best-fit solution is established through several tests. First, the sensitivity correlation matrix is calculated for the best-fit set of variables, with any off-diagonal element whose magnitude approaches unity indicating a very strong parameter correlation and a probable non-unique solution. Second, the 90% confidence limits on the variable parameters are calculated, with excessively large confidence limits indicating strong parameter correlations and probable non-uniqueness of the solution. This evaluation is based on statistical analysis of the fit based on the assumption that all errors in the data are random and normally distributed. The final and most conclusive test is to repeat the analysis with differing initial values of the variable parameters. If the fit always converges to the same best-fit result the best-fit set of variable parameters is unique.

EXTENSION OF SILICON OPTICAL CONSTANTS

The silicon optical constant spectra obtained by Jellison[3] are generally considered to be the most accurate in the literature; however, these constants were measured over the range 234 - 830 nm only. We often wish to fit ellipsometric data beyond this range, so in order to obtain accurate silicon optical constants we fit a Kramers-Kronig consistent parametric model developed by Herzinger and Johs[4] to the Jellison silicon optical constant spectra, and

then used this model to predict the silicon optical constants from 830-1700 nm. Most fits in this work are performed from 250-1000 nm; however, for the thicker films it is helpful to acquire data to 1700 nm, as the interference oscillations in the experimental data become less densely spaced in the infrared.

The silicon optical constants from the parametric fit are shown in figure 1, with Jellison's constants for comparison.

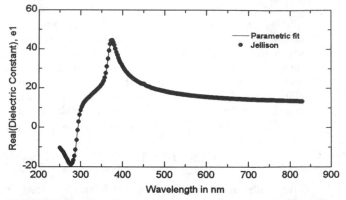

FIGURE 1. Fit of parametric optical constant model to the published silicon constants from Jellison. Real part of the dielectric function is shown as a function of wavelength.

The parametric function which was determined from these fits was then used to describe the silicon substrates in the rest of this work. This optical constants predicted by this model for silicon in the range 840-1700 nm are shown in figure 3.

The real part of the dielectric function shown in figure 3 yields index values which are very accurately represented by a Cauchy function (see equation 2 below) with A = 3.4212, B = 0.13359, and C = 0.018309.

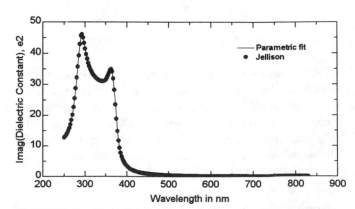

FIGURE 2. Parametric fit to Jellison's silicon optical constants. Imaginary part of dielectric function is shown.

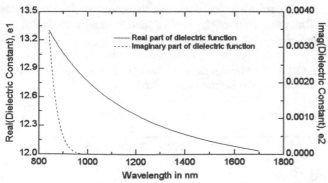

FIGURE 3. Silicon optical dielectric function from 840-1700 nm, from parametric model fit to measured spectra from Jellison.

SYSTEMATIC ERRORS IN THE ANALYSIS OF OXIDE FILMS

We now use the extended silicon optical constant spectra to examine some systematic errors obtained when characterizing oxide films on silicon. We examine errors due to inaccuracy of the substrate optical constants as well as errors due to inaccurate models for the oxide index of refraction.

We first examine the effect of the choice of substrate optical constants. We fit the experimental data from each sample using the extended Jellison constants for the substrate, and repeated the fit with the silicon constants of Aspnes[5] for comparison. Literature silicon dioxide optical constants from Palik[6] were employed to model the native and 10 nm oxide films, while a three term Cauchy dispersion relation was used to model the index of refraction of the thicker films:

$$n(\lambda) = A + B/\lambda^2 + C/\lambda^4, \qquad (2)$$

where the wavelength λ is given in microns, and A, B, and C are fit parameters. Table 1 shows the results for the determination of the oxide thickness.

TABLE 1. Comparison of silicon dioxide thickness vs. choice of substrate optical constants.

Nominal (nm)	Aspnes	Jellison
native	1.63 ± 0.04	1.86 ± 0.02
10	10.66 ± 0.02	10.94 ± 0.02
35	33.85 ± 0.02	33.46 ± 0.03
50	52.57 ± 0.02	52.41 ± 0.03
75	75.19 ± 0.03	75.06 ± 0.02
100	102.08 ± 0.05	101.77 ± 0.05
500	531.9 ± 0.3	530.5 ± 0.4
1500	1477.7 ± 0.4	1480.0 ± 0.8
2500	2493.5 ± 0.5	2493.4 ± 0.5

As expected, the sensitivity of the best-fit film thickness to the substrate optical constants increases with decreasing film thickness. The results for the very thick film are very insensitive to the substrate optical constants.

We also examine errors induced by inaccurate choices for the silicon dioxide film index of refraction spectra. Three different choices for the silicon dioxide index are considered: 1. Index values from the Palik, 2. Three term Cauchy dispersion relation (eq. 2). 3. Six term inverse power series. To demonstrate the value of the three term Cauchy and six term parametric models for the silicon dioxide index, we fit these models to the silicon dioxide index spectrum given in Palik. Figure 4 shows the best-fit obtainable from the three term Cauchy relation, along with Palik's index spectrum for silicon dioxide for comparison.

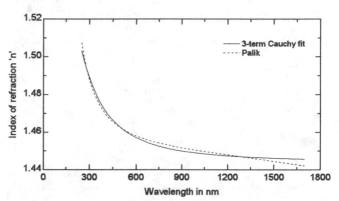

FIGURE 4. Best-fit of three term Cauchy model to Palik's silicon dioxide index spectrum.

The three term Cauchy model does not have sufficient flexibility to reproduce the literature silicon dioxide index spectrum, particularly in the infrared where the Palik spectrum is roughly linear in the wavelength. We found that a simple inverse power series extended to sixth order fits the Palik spectra very well:

$$n(\lambda) = A + B/\lambda + C/\lambda^2 + D/\lambda^3 + E/\lambda^4 + F/\lambda^5 + G/\lambda^6, \qquad (3)$$

where again the wavelength is specified in microns and A-G are fit parameters. The best fit obtainable to Palik's silicon dioxide index spectrum from this model is shown in figure 5.

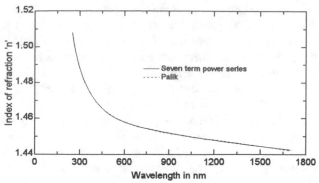

FIGURE 5. Seven term inverse power series fit to Palik silicon dioxide index of refraction spectrum.

The seven term power series fits the Palik silicon dioxide spectrum very well. The coefficients of the power series determined from the above fit were A = 1.413, B = 0.080, C = -0.068, D = 0.031, E = -0.006, and F = 0.0005.

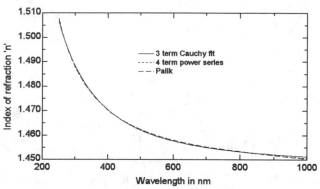

FIGURE 6. Fit to Palik's silicon dioxide index of refraction from 250-1000 nm, using 4 term power series and three term Cauchy model.

In many cases data are only fit to 1000 nm, and in this case a less complicated power series is required. The following power series yields the fit to the Palik index spectrum shown in figure 6.

$$n(\lambda) = A + B/\lambda + C/\lambda^2 + D/\lambda^4. \qquad (4)$$

The fit from the three term Cauchy model is close, but the four term fit is indistinguishable from the Palik spectra. The coeffiecients of the power series (equation 4) determined from this fit were A = 1.442, B = 0.0076, C = 0.0010, and D = 0.00007.

We then use the extended Jellison silicon optical constants for the substrate and fit the oxide film data with these various models to study the error in the results due to the choice of model. For the native oxide and 10 nm films, the index of refraction is very strongly correlated to the thickness of the film, and it is not possible obtain a

unique solution for both the thickness and index of the film, regardless of the optical model employed for the film index.

TABLE 2. Comparison of silicon dioxide thicknesses vs. choice of model for the silicon dioxide index.

Nominal	Palik	3-term Cauchy	4- or 7-term Power series
35	34.53 ± 0.02	33.46 ± 0.03	33.85 ± 0.02
50	52.85 ± 0.02	52.41 ± 0.03	52.57 ± 0.02
75	75.74 ± 0.04	75.06 ± 0.02	75.19 ± 0.03
100	102.29 ± 0.02	101.77 ± 0.05	102.09 ± 0.05
500	538.45 ± 0.05	530.5 ± 0.4	531.7 ± 0.3
1500	1502.6 ± 0.1	1480.0 ± 0.8	1479.1 ± 0.3
2500	2545.7 ± 0.1	2538.3 ± 0.5	2538.6 ± 0.6

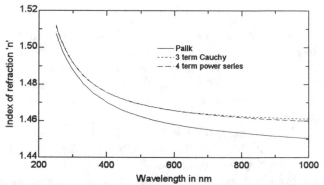

FIGURE 7. Index spectra obtained for 50 nm film from 3 term Cauchy dispersion relation and 4 term power series, with Palik spectrum for comparison.

We applied these models to films ranging from 30-2500 nm in thickness. The results are given in table 2. For films in which data were acquired to 1000 nm wavelength the 4-term power series was employed, while the full 7-term series was used when fitting data for the very thick films out to 1700 nm.

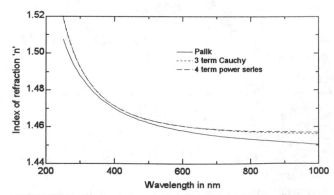

FIGURE 8. Index spectra obtained for 100 nm film from 3 term Cauchy dispersion relation and 4 term power series, with Palik spectrum for comparison.

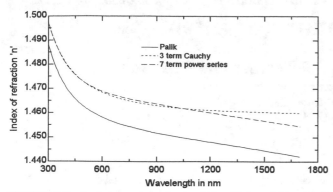

FIGURE 9. Index spectra obtained for 500 nm film from 3 term Cauchy dispersion relation and 7 term power series, with Palik spectrum for comparison.

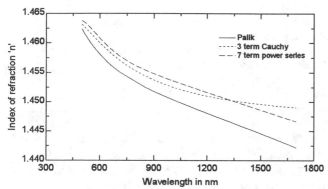

FIGURE 10. Index spectra obtained for 2500 nm film from 3 term Cauchy dispersion relation and 7 term power series, with Palik spectrum for comparison.

Figures 7 - 10 show the index spectra obtained for the 50, 100, 500, and 2500 nm films from the 3-term Cauchy and power series fits, along with the Palik index spectrum for comparison.

DISCUSSION

Several conclusions may be drawn from this work. First, we used a parametric semiconductor model to extend the published silicon optical constants from Jellison out to 1700 nm in the infra-red.

Second, we found as expected that systematic errors in the determination of the film thickness due to the choice of the substrate optical constants increase with decreasing film thickness. As a result the use of accurate substrate optical constants becomes increasingly important as the films under study become very thin, as is the case for gate oxides, for example.

Third, we found that systematic errors in film thickness determination associated with the choice of

silicon dioxide optical constants, or with the choice of parameterization of the film index increase with increasing film thickness. This is due to the fact that the sensitivity of the measurement to the index of refraction of the film is roughly proportional to the distance which the beam propagates in the film. For very thin films, the thickness measurement is less sensitive to the index of the oxide film.

Finally, we show that the errors in the measurement of the film thickness due to the choice of parameterization of the film index of refraction are fairly small for all thicknesses studied. We note however, that as the films become thicker it becomes increasingly more important to use some sort of parameterization rather than literature constants, as the high sensitivity of the measurement for thicker films to the index of the film can lead to significant errors if the literature constants do not match those of the film exactly. Also, if the film index of refraction spectrum is desired accurately, the more general four or seven term power series should be employed to parameterize the film index, particularly when fitting data beyond 1000 nm wavelength where the film dispersion is nearly linear with wavelength.

[1] Woollam, J. A., and Snyder, P. G., *Materials Science and Engineering* **B5**, 279 (1990).

[2] Azzam, R. M. A., and Bashara, N. M., *Ellipsometry and Polarized Light*, New York, North-Holland Publishing, 1983.

[3] Jellison, G. E., Jr., Lowndes, D. H., and Wood, R. F., *SPIE Proceedings* **710**, 24, (1986).

[4] Currently preparing for publication.

[5] Aspnes, D. E., Theeten, J. B., and Hottier, F., *Phys. Rev. B.,* **20**, 3292, (1979).

[6] Palik, E. A., ed., *Handbook of Optical Constants of Solids, Vol. I*, Orlando, FL, Academic Press, 1985.

High-Accuracy Principal-Angle Scanning Spectroscopic Ellipsometry of Semiconductor Interfaces

N.V. Nguyen, D. Chandler-Horowitz, J.G. Pellegrino, and P.M. Amirtharaj

Semiconductor Electronics Division
National Institute of Standards and Technology
Gaithersburg, Maryland 20899

A high-performance spectroscopic ellipsometer has been custom built, in house at NIST, based on the commonly used rotating-analyzer configuration. The data accuracy was highly enhanced by using the principal-angle scanning technique. This technique requires an accurate setting of the angle of incidence which is accomplished by an interferometer and high-precision goniometers for the sample stage and the polarizer. For each wavelength, the principal angle of incidence was automatically searched to obtain a 90ts phase shift of the polarized light upon reflection, i.e., $\Delta = 90°$, and the polarizer azimuth was set to Ψ. At this condition, the ac component of detected intensity is near a null. With zone averaging, systematic errors such as the detector nonlinearity, and the analyzer and polarizer calibration constants, are minimized.

To illustrate the use and capability of this system, we use the results of a recent study of the optical properties of SiO_2/Si system and, in particular, the transitional region, defined as an interlayer between the thermally grown SiO_2 film and the Si substrate. In this study, the complex dielectric function and the thickness of the interlayer were determined. From the dielectric function, the existence of both strain and microroughness at this region was inferred, and the strain was seen to induce a redshift of 0.042 eV of the critical point E_1.

INTRODUCTION

The rapid rate in the reduction of feature size in microelectronic circuits and the concomitant shrinkage in the thickness of constituent films lead to demands for greater accuracy and precision in thin-film measurement. Spectroscopic ellipsometry (SE) is a monolayer-sensitive, optical-characterization technique that has been widely used to study semiconductor materials and thin films. In this article, we show the basic setup of our custom-built, high-accuracy, principal-angle scanning spectroscopic ellipsometer and demonstrate its application to a technologically important class of materials, namely, thin SiO_2 films on Si. The most common SE configuration is a rotating analyzer type without a compensator. The accuracy and precision of this type is greatly enhanced by employing principal angle, polarizer tracking, and zone averaging in the measurement. This technique requires the principal angle of incidence to be set accurately. In our system, an interferometer and goniometers of high precision were used to set accurately the angle of incidence.

For the illustration, we use the results of our recent study of the optical properties of the SiO_2/Si interface.[1] By combining our high-accuracy, principal-angle scanning technique with a unique data analysis scheme on a set six of SiO_2/Si samples, we were able to obtain the complex dielectric function of an extremely thin layer in the interface region, buried under a comparatively thick overlayer oxide. The existence of such thin layers is generally difficult to detect. We briefly discuss the physical characteristic of the interface inferred from its ellipsometrically determined dielectric function.

PRINCIPAL ANGLE SCANNING SPECTROSCOPIC ELLIPSOMETRY

The schematic diagram of the ellipsometer is shown in Fig. 1. Note that the sample holder and the detector arm are mounted, with a precision of 0.001°, on two separate, high-precision goniometers, which enables the angle of incidence to be set with the same precision. In addition to the broad-band Xenon light source used for spectroscopic measurements, a 632.8-nm HeNe laser beam, which is expanded and carefully collimated, is used for high-accuracy, angle-of-incidence alignment. An ellipsometric measurement can also be performed at this particular wavelength.

The procedure for aligning a sample is as follows: The mounted sample is first positioned so that it is normal to the HeNe laser beam (see Fig. 1). Interference fringes resulting from the interference of the reflected beam with the incident beam are usually observed from a beam splitter placed in the beam path. Adjusting the substrate azimuths to null out the fringes brings the incidence beam to a precise normal angle, thus resulting in the 0° angle of incidence. The non-normal angle of incidence is then set by rotating the high-precision goniometers of the sample and the detector arm. The repeatability of the angle of incidence was found to be better than 0.005°.

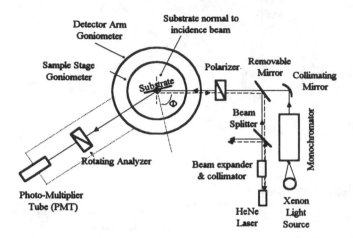

FIG. 1 Schematic diagram of NIST's principal-angle spectroscopic ellipsometer.

The ellipsometric parameters Δ and Ψ are defined by the ratios of the Fresnel complex amplitude reflection coefficients R_p and R_s for the parallel, p, and perpendicular, s, components of polarized light, respectively:

$$\frac{R_p}{R_s} = \tan\Psi \exp(i\Delta). \qquad (1)$$

The intensity detected by the detector, i.e., the photomultiplier tube in our system, varies as the analyzer azimuth angle, A, as follows:[2]

$$I = \frac{I_0(|R_s|^2 \sin^2 P + |R_p|^2 \cos^2 P)(1 + \alpha \cos 2A + \beta \sin 2A)}{2} \qquad (2)$$

where

$$\alpha = \frac{\tan^2 \Psi \cot^2 P - 1}{\tan^2 \Psi \cot^2 P + 1}, \qquad (3)$$

$$\beta = \frac{4 \tan \Psi \cot P \cos \Delta}{\tan^2 \Psi \cot^2 P + 1}. \qquad (4)$$

and P is the polarizer azimuth angle. The two parameters α and β are obtained from an analysis of the intensity and phase of the signal recorded at the photomultiplier tube (PMT).

The principal-angle measurement technique is based on a null method. In this case, the rotating analyzer ellipsometer uses an ac null method in which the rotating analyzer seeks circularly polarized light and hence the null condition. Using the intensity relation (eq. 2), the null signal is obtained when both α and β are zero. This corresponds to the conditions:

$$\Delta = \pm 90°$$
$$\Psi = \pm P.$$

The advantage in using the rotating-analyzer, ac-null method over the conventional null method is that the sensitivity is greater for the former.[2] The measurement procedure to seek a null intensity signal is first to set the angle of incidence at approximately the principal angle and then to adjust the polarizer angle until a null signal is obtained; i.e., P = Ψ. The detailed analysis of this method, published elsewhere,[3] concluded that the linearity of the detector is not critical in the measurement and the optimum accuracy is not wavelength dependent and thus is well suited for spectroscopic studies.

DETERMINATION OF THE OPTICAL PROPERTIES OF SIO₂/SI INTERFACE

To demonstrate the technique described above, we use a recent study[1] of the optical properties of an extremely thin layer, in the interface region, buried under a comparatively thick overlayer. The material system selected here is very important for IC devices, namely, the gate-silicon-dioxide film on silicon. In the following, we show that from the optical constants or the dielectric function determined by SE, the interface region between SiO_2 and Si, is under the effect of strain and is microrough.

Experimental Procedure

A set of SiO_2 films of nominal thicknesses 10, 14, 25, 50, 100 and 150 nm were thermally grown on (100) silicon wafers in an dry O_2 ambient at a temperature of 900 °C. All the measurements, both with the HeNe laser and the Xenon source, were performed at the principal angle (Φ) for each wavelength. The typical measured Δ data for the 10-nm-thick oxide film are close to 90°, as shown in Fig. 2, which also depicts the principal angles (Φ) used in the measurement for this sample. Furthermore, to minimize the systematic errors and to further improve precision and accuracy, a two-zone averaging method was employed for each wavelength.[4]

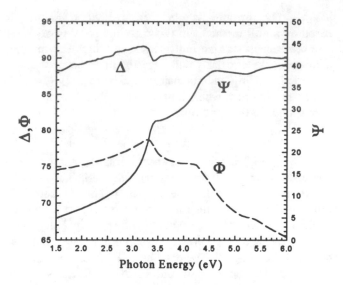

FIG. 2 A representative ellipsometric (Δ and Ψ) data of a 10-nm-thick SiO2 film on Si substrate measured by using principal-angle (Φ) scanning.

Data Analysis and Results

Numerous characterization techniques have been used to study thermally grown SiO_2 films on Si. Of critical concern are the physical properties of the interface between the oxide films and the Si substrate. However, most optical measurements such as ellipsometry usually ignore this region when determining the thickness of the oxide films. As the oxide films get thinner (<10 nm) because of the size reduction of IC devices, this interfacial region becomes a larger portion of the film and must be taken into account for an accurate film-thickness measurement. A four-phase isotropic model was therefore used to analyze the data from these samples. This model consists of (a) air ambient, (b) the SiO_2 layer, (c) the interface region (i.e., herein called the interlayer), and (d) the Si substrate. The parameters used in this model (the thickness and the dielectric function of the interlayer, and the index of refraction of the oxide film), except for the oxide film thickness, are assumed to be identical for all the samples. The detailed description of the experiment and the data analysis can be found in our previous publication.[1]

Even though a few optical studies have established the existence of the interlayer,[5] our accurate ellipsometric data can further confirm it by the comparison of the measured with the calculated values at 632.8 nm for the two-layer model described above and a single-layer model, i.e., without the interlayer. Using the standard nonlinear regression analysis, we show in Fig. 3 the deviations of the measured data from the calculated ones for the two models.

It is obvious that the deviations for the two-layer model are significantly less than those of the single-layer model and are also within the experimental accuracy (0.05°). Thus, Fig. 3 clearly shows the existence of the interlayer.

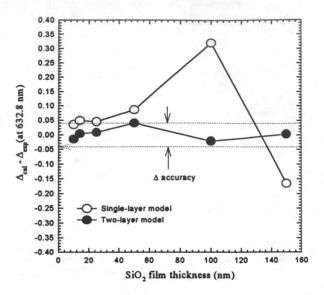

FIG. 3 Deviations from the measured Δ_{exp} of the calculated Δ_{cal} at 632.8-nm HeNe wavelength for the two-layer model (filled circles) and the single-layer model (open circles) for all six samples of different thicknesses.

The next step is to determine the complex dielectric function (ϵ_{int}) of the interlayer. The ellipsometric data taken with the 632.8-nm HeNe laser were the most accurate and were therefore used to provide the initial values for the model parameters. Because the refractive index (n_{int}) of the interlayer at 632.8 nm is not well established, a range of values extending from 1.460 of the oxide to 3.875 of the silicon substrate were investigated. For each assumed value, a set of film thicknesses and one interlayer thickness for all six samples were calculated. The calculated interlayer thickness and the different oxide film thicknesses were then used as known parameters in the least-squares fitting of the spectroscopic data to deduce the refractive index of the oxide film and the dielectric function of the interlayer as a function of photon energy. Fig. 4 shows the refractive index of the oxide (n_{oxide}) as obtained for different 632.8-nm index (n_{int}) values of the interlayer. Since the refractive index of the oxide is known to be featureless and to vary smoothly over the experimental range, it is obvious that only the n_{int} curve corresponding to the selection of 3.65 for the interlayer 632.8-nm index correctly represents the refractive index of the oxide. At this value of n_{int}, the interlayer thickness was found to be 2.2 nm. Accordingly, the calculated real ($\epsilon_{1,int}$) and imaginary ($\varepsilon_{2,int}$) parts of ε_{int} are shown by the dotted curves in Fig. 5.

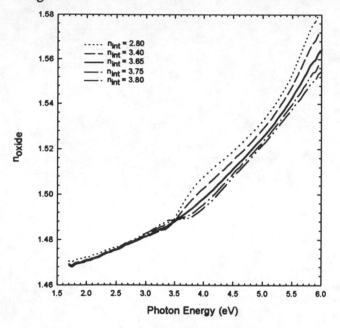

Fig. 4 The refractive index of the SiO$_2$ film obtained from the least-squares fitting of SE data from all the samples.

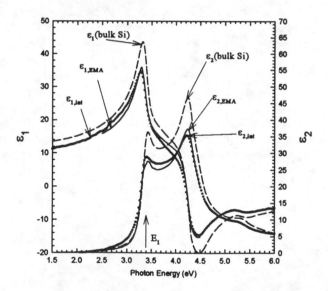

FIG. 5 The dielectric function of the interlayer (filled) and bulk Si (dashed curves). Solid lines are the effective dielectric function of the interlayer obtained from Bruggemann's effective medium approximation (EMA).

A brief discussion on the physical properties of the interlayer is now presented. Compared with the dielectric function of the bulk silicon, ε_{int} obviously contains all the

features of the bulk silicon (dashed curves in Fig. 5). However, there are two distinct features that characterize the interface properties: the lower magnitude of ε_{int} and the redshift of the interband transition E_1. The former is obvious, but the latter requires further analysis. To accurately quantify the amount of the redshift, we determine the energetic position of E_1 of the interlayer from the least-squares fit of the second derivative of ε_{int} to the usual analytic line-shape expression:[6]

$$\frac{d^2 \varepsilon_{int}}{dE^2} = \frac{A \exp(i\theta)}{(E - E_1 + i\Gamma)^2} \qquad (6)$$

where A, θ, and Γ are the amplitude, phase angle, and broadening parameters, respectively, of the critical point E_1. E_1 was found to be 3.314 eV compared to 3.356 eV for the bulk silicon (see Fig. 6). Thus the redshift is 0.042 eV. It has been established that uniaxial stress causes the splitting and shift of the E_1 transition in bulk silicon.[7] For the SiO$_2$/Si system in our study, the splitting of E_1 is not experimentally observed. Therefore, the redshift is due to the interlayer being strongly influenced by the hydrostatic component of the strain.[8]

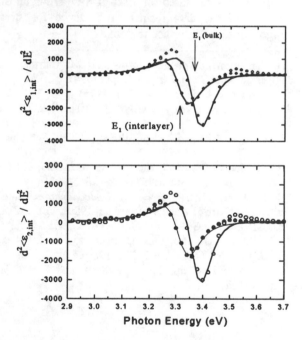

FIG. 6 The second derivatives of the interlayer dielectric function (filled circles) and the bulk silicon substrate (open circles). The solid cures are the least-squares fitting to the analytic line shape (see text).

The fact that the magnitudes of ε_{int} are lower than those of ε_{Si} indicates the presence of the micro-roughness at the transitional region. Bruggemann's effective medium

441

approximation (EMA) was employed to characterize it as a physical mixture of the strained silicon and the oxide. We have calculated the dielectric function of the strained silicon,[9] and this was used in fitting the effective dielectric function (ε_{EMA}) to the interlayer dielectric function determined above (dotted curves in Fig. 5). As a result of this fitting, it was determined that the interlayer effectively contains 12% SiO_2 and 88% strained Si.

SUMMARY

In this report, the technique of the principal-angle scanning spectroscopic ellipsometry was described and demonstrated in the application to the study of SiO_2/Si interfaces. The precision and accuracy of the ellipsometry measurement is greatly enhanced by employing the principal angle of incidence at each wavelength and tracking the polarizer azimuth to obtain a null ac signal measured at the detector. Such a method minimizes the effect of the nonlinearity of the detector. Furthermore, by zone averaging, the systematic errors are also minimized to yield higher accuracy data. This technique produces very accurate data measured on a set of SiO_2 films on Si. The analysis of the data results in evidence for strain and roughness at the interface region.

[1] N. V. Nguyen, D. Chandler-Horowitz, P. M. Amirtharaj, and J. G. Pellegrino, *Appl. Phys. Lett.* **64**, 2688 (1994).

[2] D. E. Aspnes, Appl. Opt. **14**, 1131 (1975).

[3] D. Chandler-Horowitz and G. A. Candela, *Appl. Opt.* **21**, 2972 (1982).

[4] J. M. M. de Nijs and A. van Silfhout, *J. Opt. Soc. Am.* **A5**, 733 (1988).

[5] E. Taft and L. Cordes, J. Electrochem. Soc. 126, 131 (1979); D. E. Aspnes and A. A. Studna, *Phys. Rev.* **B 27**, 985 (1983).

[6] D. E. Aspnes, in *Handbook on Semiconductors*, edited by M. Balkanski (North-Holland, Amsterdam, 1980), Vol. 2, Chapter 4A.

[7] F. H. Pollak and G. W. Rubloff, *Phys. Rev. Lett.* **29**, 789 (1972).

[8] J. T. Fitch, C. H. Bjorkman, G. Lucovsky, F. H. Pollak, and X. Yin, *J. Vac. Sci. Technol.* B 7, 775 (1989).

[9] D. Chandler-Horowitz, N. V. Nguyen, and P. M. Amirtharaj (unpublished).

CHEMICAL METHODS

Application of ICP-MS
for Relating Metal Contamination on Wafers
To Metal Sources and Levels

Marjorie K. Balazs and János Fucskó

Balazs Analytical Laboratory
252 Humboldt Court
Sunnyvale, California 94089

ICP-MS (Inductively Coupled Plasma - Mass Spectrometry) is an excellent tool for quantitative assessment (measurement) of metallic contamination in the production of semiconductor devices. It is suitable for determining metal concentrations on bare silicon wafers, as well as in many dielectric, metallic and metal oxide layers on a wafer. It is also suitable for analysis of liquids of all types. As a consequence, ICP-MS provides useful information on both the source and the levels of metallic contamination at virtually every processing step (see Table 1).

TABLE 1. Processes Where Metallics are Routinely Measured Using ICP-MS.

- Wafer cleaning, e.g., wet, dry, spray
- Oxidation, e.g., thermal, CVD, wet (steam)
- Other CVD Processes
- Photolithography
- Etching, e.g., wet chemical, plasma
- Ion Implantation
- Wafer Handling, e.g., metal tools
- Wafer Transportation, e.g., boxes, cassettes, conveyors
- Reactor Cleanliness
- Atmospheric Contamination
- Process Development

TECHNIQUE

The advantages of using ICP-MS procedures (rather than surface analysis procedures), come from the ability to standardize, accurately quantitate the analyses, do matrix matching, and provide highly accurate and sensitive measurements (down to 10^7 atoms/cm^2) over a wide range of concentrations (six orders of magnitude). All of this is possible in a single measurement run of up to 72 elements. These advantages come from the inherent versatility of mass spectroscopy in general, coupled with a variety of sample preparation and front-end sample introduction techniques.

While techniques such as gettering can be used to manage metallic contamination on the wafer, as device geometrics go to 0.25 μm and 0.18 μm with very thin gate oxides (<50 Å), it becomes increasingly important that these sources of metal contamination are understood so that steps can be taken to eliminate them. ICP-MS techniques are a key tool in this process of contaminant elimination.

This laboratory, being proficient in the analysis of metal at very low level (0.1-10 ppt) in all liquid materials used in processing and most of the thin films produced, embarked on a program to relate sources of metallic contamination to contaminants found on wafer surfaces. While sampling and measuring the liquids involved in processing is relatively simple and straight-forward, procedures for measuring the various films, such as oxides, doped oxides, nitrides, metals, and organics needed to be developed.

Vapor Phase Decomposition (VPD) is a process that has been used to remove metals from bare wafer surfaces and to concentrate them in a small spot on the wafer or a small drop that is removed from the wafer for analysis by tools such as GFAAS, ICP-OES, and ICP-MS using ETV, FIA, MDX, FIA-USN or DIN nebulizers. There are, however, serious problems if VPD is used alone. They include:

- loss of volatile fluorides (B, As, etc.) see Table 2.
- precipitation of insoluble fluorides (La, etc.)

- inability to remove metals above silicon on the electromotive series (Cu, Au, Pt, etc.)
- slowness for thick dielectric films

TABLE 2. Volatile Metal Fluorides.

	MP°C		BP°C
BF_3	-129		-101
AsF_3	-6		51
GeF_4	-15	s@	-37
GeF_2	110	d@	160
PF_3	-151		-101
PF_5	-94		-85
SiF_4	-90	s@	-96
CrF_3	1100	s@	1100
CrF_4	-28		400

s=sublimes
d=decomposes

Other sample removal methods were needed. Two techniques were developed which are referred to as DSE (drop scan etching) and FDC (film desolation and concentration). In VPD, a small amount of pure water is used to pick up the soluble metal fluoride salts for subsequent analyses by ICP-MS. In DSE, the same technique is used, except that the small quantity of liquid referred to as a drop is specifically designed to dissolve metals (e.g., Cu) that do not form soluble fluoride salts. Frequently, nitric acid is used instead of pure water as the appropriate liquid.

For thick film oxides, nitrides or metals, VPD is too slow or even impractical. Consequently, the entire film and all metals are first dissolved in a relatively large quantity of a specific solution. This solution is then chemically processed and concentrated so that specific metals can be measured by ICP-MS.

APPLICATIONS

With sample collecting and processing procedures in place, it is now possible to measure a series of events to determine how they affect residual metallic contamination on wafers, especially for critical steps such as gate oxides. Since the gate oxide is so critical, cleaning has received considerable attention with the result of many papers being published on cleaning solutions, their specific chemistry, and steps.

An example of applying these techniques is shown in Table 3, which describes the correlation between chemical purity and the concentration of metals on wafers dipped into these chemicals. The results illustrate the accumulative effects of metals like aluminum and iron whose insoluble metal oxides precipitate onto a wafer surface. Although SC2 should remove most of these metallic contaminants, any peroxide solution that didn't contain HCl or HF would not.

Table 4 shows another set of measurements comparing the cleaning efficiency of various solutions in removing metallic contamination. Wafers were purposely contaminated in a SC1 bath heavily spiked with 650 ppb of each metal shown. Subsequently, they were cleaned with three different cleaning solutions. The results illustrate that the SC2 and HF peroxide can remove most of the contaminating metals. However, as previously discussed, insoluble metal oxides such as Al, Fe, Sn, and Zn remain on the wafer in considerable quantities. Although hot dilute $HF/H_2O_2/H_2O$ does a significantly better job, concentrations of the 10^{10} atoms/cm² persist.

TABLE 3. Trace Metals in Two H_2O_2 Solutions and on Wafers Dipped into Them

Element	H_2O_2 [M] ppb		Wafer 10^{10} atoms/cm²	
	A	B	A	B
Al	22	2.6	580	92
Cr	3.3	0.1	<1	<1
Fe	6.7	<2.0	82	<5
Ni	2.6	<0.1	1.2	0.9
Na	10	2.3	12	14

Determining metals in thick oxides such as initial oxide, fill oxide, spacers, intermetal dielectrics, or doped planarization or passivation films has become a routine activity at the lab. Table 5 depicts a study of thick film oxide and TEOS samples produced via different techniques and reactors. This Table is interesting in that it illustrates that reactors can contribute significant quantities of metallic contamination which will get dispersed in the etching bath. Also, it appears that TEOS can vary by lot or manufacturer.

Table 6 was compiled by evaluating numerous wafers for metallic contamination that had been through photoresist etching, ion implantation, and various wafer handling procedures. These data are presented as ranges of what has been found to date utilizing data from several reactors and processes at different IC manufacturing sites. Of these, ion implantation studies are particularly interesting since what is measured as metallic contamination caught in a thick oxide test wafer is also what was implanted in the processed wafers. In this case, cleaning is of no help in removing these metals and thus, cleaning up the implanter becomes a real necessity.

When measuring metals in films, both recovery and sensitivity are important. Table 7, the Periodic Chart, contains recovery data by element. A recovery study is done by deliberately contaminating a clean wafer with known quantities of metals or metal salts and using a specific procedure to see what quantity was actually recovered. The percentages given are a measure of the amount of metal recovered versus the amount applied.

As is seen, recoveries are very good for most elements. Some exceptions are Cu, W, and Ag, which are above Si on the electromotive series. They require special treatment to remove them from the silicon wafer. However, Pt, Au, Hg and Bi cannot be recovered even when strong chemical treatment is used.

Different setups for both front-end injection systems and types of ICP-MS units (quadrapole or HR/Magnetic Sector) can result in very sensitive measurements down to 1×10^8 atoms/cm^2 for most elements, and 10^7 or 10^6 for some. Table 8 illustrates the differences in obtainable sensitivity for tin when it is measured as 1 of 30, 6, or 2 elements

from a single 6-inch wafer. Oftentimes, a rough scan for 30 elements will be used for detecting unsuspected metallic contamination, whereas, 6-critical elements are most frequently measured to follow specific processes of cleaning. Only 1-3 elements are measured when a specific study is being done where the greatest sensitivity is desired.

SUMMARY

In summary, because of the ability to put wafer films into solution for very low level, accurate, and NIST traceable tests, it has now become possible to relate metallic contamination on wafers with sources of metallic contamination and to measure the efficiency of wafer cleaning processes. It is now possible to evaluate both liquid and wafer samples from any series of events to determine their metallic contamination. This capability has been accomplished by using different chemistry for metal removal and sample concentration, and by using multiple sample injection systems to enhance ICP-MS performance. Nearly all metals (Au, Pt, Pd, Hg, Te, Ru, and Bi are exceptions), can be analyzed accurately via these techniques.

TABLE 4. Cleaning Efficiency Study — Contamination Measured on Wafer Surface (x 10^{10} atoms/cm^2)

	Starting Wafer	75°C SC2 Clean	25°C Dilute HF:H$_2$O$_2$:H$_2$O Clean	72°C Dilute HF:H$_2$O$_2$:H$_2$O Clean
Al	16,000	18.5	19.5	9
Fe	134	5.1	1.8	2.1
Sn	315	29.4	1.84	1.2
Pb	7.9	0.1	0.1	0.01
V	0.9	0.02	0.02	0.03
Zn	12.1	6.5	3.75	0.6
Ca	5	8.1	7	2

447

TABLE 5. Typical Concentration of Metal Contaminants in Dielectric Oxide Films on Silicon Wafers from Different Sources

Source	Concentration: ppm of oxide					
	A	**B**	**C**	**D**	**E (TEOS)**	**F (TEOS)**
Oxide Thickness Å	3000	3000	3000	6000	2000	8000
Aluminum	2.9	1.2	16	18	0.2	7.3
Sodium					<0.4	0.93
Chromium	1.2	0.53	6.9	0.63	<0.02	
Iron	4.6	4.4	12	3.0	<0.2	0.76
Nickel	2.3	0.68	4.5	0.56	<0.2	-
Copper	2.4	1.8	0.31	-	-	-
Manganese	-	-	-	0.11	-	-
Cobalt	-	-	-	0.01	-	-
Magnesium	-	-	-	-	-	0.66
Zinc	-	-	-	-	-	0.18
Zirconium	-	-	-	-	-	0.02

TABLE 6. Typical Concentrations of Trace Metals Found on Silicon Wafers after Different Processing Steps. Surface Concentrations (x 10^{10} atoms/cm^2)

Process	Na	Al	Ti	Cr	Fe	Ni	Zn	W
Photoresist Etching								
Low		10		1	10	1		
High	500	800		50	200	50		
Ion Implantation								
Low		50	10	10	10	10		2
High		10,000	1,000	5,000	10,000	5,000	100	500
Wafer Handling								
Low		10		1	5	1	5	
High		500		4,000	8,000	8,000	700	

TABLE 7. Recovery of Elements from Bare Silicon Wafers

1	2	3	4	5	6	7	8	9	10	11	12	13	14	15	16	17	18
H																	He
Li 98%	Be 95%											B +>90%	C	N	O	F	Ne
Na 104%	Mg 98%											Al 98%	*Si	P	S	Cl	Ar
*K 100%	Ca 98%	Sc 98%	Ti 95%	V 98%	Cr 98%	Mn 102%	Fe 110%	Co 100%	Ni 98%	Cu 85%	Zn 104%	Ga 100%	Ge 94%	As +<90%	Se	Br	Kr
Rb 100%	Sr 98%	Y 101%	Zr 95%	Nb 82%	Mo 87%	Tc -	Ru 20-80%	Rh 15-80%	Pd 85%	Ag 84%	Cd 96%	In 99%	Sn 105%	Sb 95%	Te 20-80%	I	Xe
Cs 96%	Ba 95%	*La 40-94%	Hf 100%	Ta -	W 78%	Re -	Os -	Ir 78%	Pt 10-50%	Au 1-10%	Hg 25-70%	Tl 95%	Pb 101%	Bi 10-48%	Po	At	Rn
Fr	Ra	Ac **															

*
58	59	60	61	62	63	64	65	66	67	68	69	70	71
Ce 99%	Pr 96%	Nd 96%	Pm	Sm 97%	Eu 97%	Gd 100%	Tb 100%	Dy 100%	Ho	Er 100%	Tm 99%	Yb 99%	Lu 99%

**
90	91	92	93	94	95	96	97	98	99	100	101	102	103
Th 100%	Pa	U 100%	Np	Pu	Am	Cm	Bk	Cf	Es	Fm	Md	No	Lr

+Without VPD, with solution strip technique

TABLE 8. Sensitivity of Tin in VPD Scans of Varying Sizes.

	When Part of a 30 Element Scan	When Part of a 6 Element Scan	When Part of a 2 Element Scan
Sensitivity (10^{10} atoms/cm^2) On 6" Wafer	0.25	0.1 to 0.2	0.01

Ultra Sensitive Methods for Microcontamination Analysis in Semiconductor Process Liquids and on Surfaces

P. Gupta[1], Z.Pourmotamed[1], S.H. Tan[2], R.McDonald[1]

[1]*Intel Corporation, Santa Clara, CA 95052*

[2]*ChemTrace Corp., Hayward, CA*

Several analytical methods have been established for the detection and measurement of microcontamination on surfaces and in process liquids. However, it is becoming increasingly difficult for the current methods of microcontamination analysis to meet today's needs for sensitivity and throughput. To meet these newer challenges, several improved methods for trace metal analysis in chemicals and surfaces are being introduced. For example, the use of a magnetic sector ICP-MS as an alternative to the more commonly used quadrupole ICP-MS for the measurement of trace metals in corrosive chemicals offers both higher sensitivity and higher productivity. Furthermore, this technique allows the analysis of phosphorus, sulfur, chlorine at the ppb detection limits in several chemicals and DI water. Previously these species had to be measured in their anionic form by a separate technique known as ion chromatography (IC). Newer methods of surface analysis, such as VPD -GFAA (Vapor Phase Dissolution followed by Graphite Furnace Atomic Absorption analysis are also discussed. Because of the suitability of measuring very small volume of liquids, GFAA has been the preferred choice for the measurement of VPD solutions. Although ICP-MS is a more sensitive and faster technique than GFAA, a typical ICP-MS measurement requires 10-20 ml volume of solution. New methods of sample introduction into the ICP-MS system have been developed that allow as little as 0.25 ml to be analyzed by ICP-MS. The coupling of VPD techniques to magnetic sector ICP-MS now provides as low as 10^7 atoms/cm^2 sensitivity for many metals. In addition, analysis of non-metallic elements on the surface such as phosphorus can be achieved with this High Resolution ICP-MS - VPD combination.

1. 0 TRACE METAL ANALYSIS IN LIQUIDS

Introduction

Several techniques are typically utilized for the trace analysis of metals in chemicals. These techniques include graphite furnace atomic absorption spectroscopy (GFAAS), inductively coupled plasma-atomic emission spectrometry (ICP-AES) and Inductively coupled plasma-mass spectrometry (ICP-MS). Recently, it is clear that ICP-MS has evolved as the preferred choice for trace metal analysis because of its high sensitivity and rapid analysis time for chemicals.

In an ICP-MS analysis, solutions are neubulized into an atmospheric argon plasma. Dissolved solids in the solutions are vaporized, dissociated and ionized and then extracted into a mass spectrometric system. (1-4). The type of mass spectrometers utilized for semiconductor chemicals analysis are typically quadrupole based instruments (5-7)

One of the problems of using an argon ICP as an ion generation source for mass spectrometry is that spectral interference's can occur between the analytes of interest and background molecular peaks, oxide and doubly ionized ions (8-10). Examples of major interference's are listed in Table 1. The analysis of typical acids used in semiconductor manufacturing such as HF, NH_4OH, HNO_3, HCl, H_2SO_4 can give rise to further interference's due to the formation of molecular species as a result of interactions of argon and oxygen with fluorine, nitrogen chlorine and sulfur (9). Although, the quadrupole mass analyzer is utilized at a resolution sufficient to resolve the mass number of isotopes across the mass range, this resolution is not sufficient to separate the molecular interference's that nominally at the same mass as the analyte metal in question. As a result, these molecular interferences can severely limit the ultimate detection limit that can be obtained.

Several methods have been utilized to reduce or remove some of these molecular interferences. These methods include the adjustment of plasma conditions to minimize the formation of polyatomic species (11) and the use of alternative gases to replace the usual argon gas (12) and using different methods of sample introduction such as electrothermal vaporization (ETV) (13,14).

However, the most common method of reducing the molecular interferences is to remove the matrix from the chemical by evaporating the sample. In this method, the sample is gently heated in either a platinum dish or a Teflon container to a small volume or to dryness. A small volume of nitric acid is added to dissolve the residue and then DI water is added to bring the sample to a final volume. Although this technique allows the removal of matrix induced molecular interferences, the major background molecular interfernces due to Ar (such as ArO interferences with ^{56}Fe, ^{39}ArH interferences with ^{3k}K, and ^{14}N$_2$ interferences with ^{28}Si) still remain. There are also practical issues with this method. Sample throughput is limited by the fact that, each chemical sample needs approximately 3-4 hrs for sample preparation. Large clean areas are required for sample evaporation in order to minimize cross- contamination problems. Furthermore, the analysis of elements such as boron and arsenic is limited by the fact these elements are volatile and are lost during the sample evaporation steps.

An alternative approach for solving the molecular intereference problem is to couple the ICP to a high resolution double focusing magnetic sector analyzer. This ap-

Table 1: *Common molecular intereferences in ICP-MS*

Element of interest	Interfering species
^{28}Si	N$_2$, CO
^{56}Fe	ArO
^{44}Ca	Ar, CO$_2$
^{39}K	ArH

proach has been discussed previously (15). In this previous study, the magnetic sector mass analyzer allowed the separation and resolution of polyatomic ions from analyte ions at the same nominal mass. Lower background is also obtained when this instrument is operated in the low resolution mode compared to a quadrupole based systems (15). This enhancement of signal to noise has allowed the determination of sub ppt levels of actanoids and lathanoid in DI water (16,17).

We have used the same approach for analyzing semiconductor process chemicals (18). In this presentation, the analysis of a 50:1 HF solution and DI water using an ICP coupled to a magnetic sector analyzer is described. This presentation will demonstrate the use of different resolutions as a method of resolving between the element of interest and the interfering molecular species. As a result, the direct analysis of metals in chemicals and DI water to very low detection limits is now possible. Moreover, the analysis of selected anions such as phosphorus, sulfur, chlorine and bromine in chemicals and DI water by ICP-MS for is demonstrated for the first time.

Resolution

Figures 1,2,3 and 4 show the spectra of ^{56}Fe, ^{44}Ca, ^{28}Si and ^{39}K in DI water that were obtained using both low resolution (600R) as well as the high resolution mode at 4000R. ^{39}K was measured at a higher resolution of 6000R. As discussed in the previous section, the analysis of Si, Ca and Fe is limited by the presence of molecular interference that are present at the same nominal mass as the elements of interest. Figure 1a is a spectra of ^{56}Fe at a resolution of 600. This is similar to the spectra obtained from most commercial quadrupole ICP-MS instruments. Figure 1b shows the spectra of the same solution measured at a resolution of 4000. Figure 1b shows that the Fe peak is resolved from the ArO molecular peak.

The measurement of Ca at low resolution is similarly hindered by a molecular interference at ^{44}CO$_2$. Figure 2a shows the measurement of a 10 ppb ^{44}Ca solution at low resolution. The corresponding spectra at a resolution of 4000 is shown in 2b. Figure 3a shows the spectra of a 10 ppb ^{28}Si at low resolution. Figure 3b shows the separation between Si and molecular species such as N$_2$/CO.

The N$_2$ and CO peaks would have been separated if the resolution was increased further. However, as shown in a previous study (18), the ion transmission drops dramatically with increasing resolution. As a result, once a resolution has been reached where the peak of interest has been separated from the molecular interference, increasing the resolution further, only deteriorates detection limits.

Figure 4a shows the low resolution spectrum of a 5 ppb ^{39}K solution. This study shows that the separation of K from ^{38}ArH$^+$ requires a resolution of 6000R. At this resolution, the ion transmission is only about 5 % as compared to ion transmission of about 10 % at 4000R. However, this improved separation of the analyte and the molecular interference translates to vastly improved detection limits as will be seen in the next section.

Detection Limits: Metals

This improved separation between the analyte of interest and molecular interference translates to vastly improved detection limits for these elements as shown in Table II for ultrapure water It can be seen that trace metal detection limits in the ppt and sub-ppt levels are achieved with high resolution ICP-MS. It is noted that the detection limit of many metals was limited by background impurities resulting either from the sample introduction system or the sample blank and but not limited by the signal sensitivity

451

Table II *Detection Limits of Non metallic elements in Ultrapure Deionized Water and 2% HF by Magnetic Sector ICP-MS.*

Element	Detection Limit (ppt)
Aluminum	2.7
Antimony	0.45
Arsenic	3.2
Barium	0.27
Beryllium	6.6
Bismuth	0.21
Boron	40
Cadmium	0.45
Calcium	120
Chromium	20
Cobalt	2.9
Copper	1.4
Gallium	3.3
Germanium	0.03
Gold	6.3
Iron	16
Lead	0.42
Lithium	6.7
Magnesium	3.1
Manganese	8.1
Molybdenum	0.6
Nickel	4.8
Potassium	115
Silicon	2200
Silver	1.2
Sodium	6.9
Strontium	0.3
Tin	1.5
Titanium	4
Vanadium	2.3
Zinc	8.3
Zirconium	3.1

and instrument noise. This means that detection limits

Table II: *Detection Limits of Trace Metals in Ultrapure Deionized Water by Magnetic Sector ICP-MS.*

Element	DI water (ppb)	2% HF (ppb)
Phosphorus	0.24	0.66
Sulfur	6.7	57
Chlorine	3.1	20
Bromine	1	-

for these metals; Al, B, Cr, Au, Ag, Li, Si, Ti and Zn, can be further reduced if the sample introduction and sample blank are cleaner. It is thought that the background signals of Al, B, Si and Li originated from the glass torch and nebulizing system and other elements such as Au and Ag may have came from the Pt sampling cone. It follows that at these extreme low detection limits, impurities that are inherent conventional sample introduction system are limiting analysis. Therefore, more care selection and qualification of sample introduction systems are required in order for the ultimate detection limits to be achieved.

Detection Limits: Non Metals

We also investigated the possibility of analyzing non-metal species such as phosphorus, chlorine, sulfur and bromine. Typically, these species exist in solution as the anionic forms, such as phosphates, chlorides, sulfates, bromides. Traditionally, these anionic species are measured by ion chromatography and more recently, by capillary electrophoretic techniques at the ppb levels (19). Speciation of the anionic form is also possible by chromatographic/electrophoretic techniques. However, one problem about these techniques is the inability to analyze anions in matrices other than relatively pure DI water. Strong acids and bases have to be diluted by at least 10^3 folds with DI water before analysis. Consequently, detection limits are often in the ppm levels.

Table III shows the detection limit of these elements in both DI water and a 2% HF solution. It can be seen that the detection limits in DI water are surprising good, especially for phosphorus. In HF, a very good detection limit was obtained for P. However, the detection limits or S and Cl were higher by an order of magnitude. It is not clear whether the reason for this degradation of detection limit was due to the high blanks or to poor recovery of these two elements in HF. Further studies are underway to determine the baseline blank levels.

2.0 TRACE METALS ANALYSIS ON SURFACES

Introduction

The first part of this paper discussed new techniques for trace metal analysis in liquids/chemicals. In the second part of the paper, we will discuss trace metal analysis on silicon surfaces. Surface metallic contamination has been correlated to device degradation and the lowering of process yields (20-23). For example, at the 1×10^{11} - 1×10^{13} atoms/cm^2 level, degradation of oxide voltage tolerance as well as crystal defects have been observed (23). Minority carrier lifetime was found to be reduced by contamination that was as low as 1×10^{10} atoms/cm^2 of contaminants (21)

Traditional methods for surface analysis using ion and/or electron beam techniques have been utilized extensively for surface analysis in the semiconductor industry. Techniques such as Auger Electron Spectroscopy (AUGER) (24), Secondary Ion Mass Spectrometry (SIMS) (25) have been well characterized. However, as device dimensions continue to decrease, the need for measuring ultra low levels of contamination accurately on surfaces has emerged. It is well known that electron and ion based spectroscopy are limited to detection limits of 1×10^{11} - 1×10^{12} atoms/cm^2. Furthermore, quantification, although possible, is not straight forward with these techniques. For example, quantitative auger requires knowledge of factors that correct for backscattering effects (26, 27)). As a result of these limitations, there has been a lot of activities in the semiconductor field in developing alternative techniques for the measurement of trace metals at levels at levels below the detection limits of SIMS and AUGER.

Vapor Phase Dissolution

In VPD techniques (28), a bare wafer or an oxidized surface is exposed to HF vapor. After the native or thermal oxide has been removed, a droplet (0.5 ml) of water is rolled around the wafer surface to dissolve the metals. In an attempt to improve the collection efficiency, various types of acidic solutions have been used instead of pure DI water (28, 29). The droplet is then collected and analyzed by Graphite Furnace Atomic Absorption methods (21, 28). In order to improve detection limits, quadrupole based ICP-MS have been utilized for the analysis of the droplet (14, 29). Figure 5 shows the detection limits of VPD in

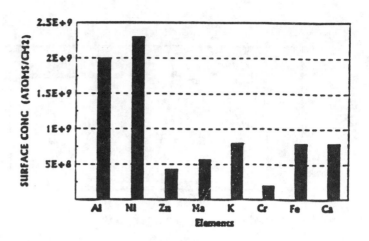

Figure 5: *VPD Detection Limits using GFAAS techniques*

units of atoms/cm^2. In this experiment, the VPD residue is measured using a Graphite Furnace Atomic Absorption Spectroscopy (GFAAS). A Varian A400 GFAAS model was utilized for the experiment. Multiple injection methods were utilized to increase the sensitivity. Figure 5 shows that the detection limits obtained using GFAA techniques vary from 5×10^8 atoms/cm^2 to 2×10^9 atoms/cm^2. Despite the high sensitivity obtained, one disadvantage of using a GFAAS is the long time of analysis. With the use of multiple injection techniques, the analysis for one element can take approximately an hour. Thus the analysis of a suite of elements can take several hours. Moreover, relatively poor detection limits is obtained for the refractory elements such as nickel and titanium.

Although there are more sensitive techniques than GFAA, the reason behind the popularity of VPD-GFAA combination is the ability of the GFAA to analyze very small volumes. ICP-MS would be an alternative choice for the analysis of VPD droplets: unfortunately regular ICP-MS sample introduction techniques (Meinhard Neubulization) require at least 5-20 ml of solution. This is the amount of volume that is required to achieve a steady state ICP-MS signal using the normal methods of sample introduction. However, normal VPD droplets are 0.25 -0.5 ml. Thus to provide sufficient volume, for a VPD-ICP-MS analysis, the solution would have to diluted 5X-40X. Dilution of the solution causes a loss in sensitivity and adds another processing step that may lead to cross contamination/errors.

In order to develop higher sensitivity VPD analysis, we have been developing alternative ICP-MS sample introduction procedures (30). Our objective was to develop a sample introduction system that would produce a steady state signal from a very small volume of solution (i.e. less than 1 ml).

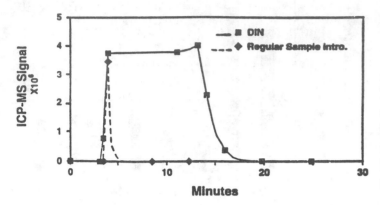

Figure 6: *ICP-MS signal obtained from two different types of sample introduction systems. The same sample volume (0.2 ml) was used for both systems. Over 10 minutes of steady state signal was obtained from the DIN system. In comparison, less than 30 second of ICP-MS signal was obtained using the regular sample introduction system (Meinhard Neubulizer)*

Small Volume ICP-MS analysis

Recently, we experimented with a new ICP-MS sample introduction known as direction-injection nebulization (DIN) system manufactured and developed by CE-TAC Corporation. The innovative DIN system allows a very small volume of sample (10 - 20 *uL*) to be introduced directly into the ICP torch without the use of the conventional nebulization chamber. This is achieved by replacing the Meinhard nebulizer with a very narrow capillary inserted directly into the injector tube of an ICP torch. By this arrangement, essentially all of the sample solution is injected directly into the ICP.

The result is a steady state signal is obtained from as low as 0.25 mL for approximately 10 minutes. For comparison, Figure 6 shows the difference between the ion signal obtained from a DIN system and a conventional Meinhard nebulizer. For an identical volume of 0.25 mL of 10 ppb the ion signal lasted for about 10 minutes versus the signal obtained from a regular sample introduction system (standard Meinhard nebulizer) which lasted for only for a fraction of a minute. A 10 minute steady state signal obtained from this new sample introduction system allows as many as 33 elements to be measured from the same VPD residue. No dilution is required for the VPD residue with this new procedure. The high sensitivity of the ICP-MS is thus preserved, as well as a reduction of sample preparation steps.

3.0 NONMETALLIC ANALYSIS ON SILICON SURFACES

Traditionally VPD has always been limited to trace metals analysis on silicon surfaces. The capabilities of High Resolution ICP-MS have allowed us to explore other elements such as Phosphorus and other non-cations. Section 1.0 describes the ICP-MS analysis of some of these elements in DI and 50:1 HF. We have used similar ICP-MS same procedure for analyzing phosphorus on silicon surfaces .

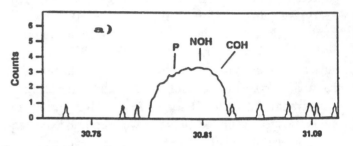

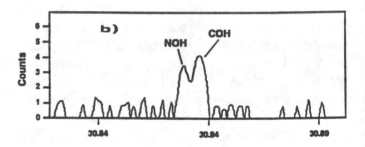

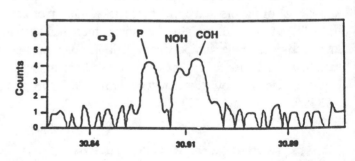

Figure 7 a) 50 ppb Phosphorus: Low Resolution. No separation between Phosphorus and interferences. **b)** Blank DI Water: High Resolution. **c)** 50 ppb Phosphorus: High Resolution.

Analysis of phosphorus, for example, is difficult using quadrupole based instruments because of mass interference of molecular species such as NOH. Figure 7 shows the difficult in analyzing phosphorus using quadrupole instruments. Figure 7a shows a mass spectrum of mass 31 using low resolution. This is the type of spectra that would be obtained from a regular quadrupole instrument. Figure 7b and 7c shows the mass spectra of the same mass region at higher resolution. 7b is a blank solution and Figure 7c is the result of a VPD residue from a 10^{11} atoms/cm^2 Phosphorus contaminated surface. We estimate the detection limit of surface phosphorus using this procedure to be in the 10^{10} atoms/cm^2 range.

References

1. D.J. Douglas and J.B. French, Anal. Chem. **53**, 37 (1981).

2. A.L. Gray, Analyst **100**, 289 (1975).

3. A.L. Gray, Anal. Chem. **47**, 600 (1975).

4. R.S. Houk, V.A. Fassel, G.D. Flesch, H.J. Svec, .L. Grey and C.E. Taylor, Anal. Chem. **52** 2283 (1980).

5. P.J. Paulsen, E. S. Beary, D.S. Bushee and J. R. Moody, Anal. Chem. **60** (10), 971 (1986).

6. S. H. Tan, T. Chu and M. K. Balazs, Chemical Proceedings of Semiconductor Pure Water and Chemicals Conference, (1992).

7. S.H. Tan and M.K. Balazs, Semiconductor International, (1988).

8. A.L. Grey, Spectrochemica Acta, **41B**, 151 (1986).

9. S. H. Tan and G. Horlick, Applied Spectrosco py, **40**, 445 (1986).

10. M.A. Vaughn and G. Horlick, Applied Spectroscopy, **40**, 434 (1986).

11. S. Jiang, S. Houk, and M.A. Stevens, Anal. Chem. **60** 1217 (1988).

12. E.H. Evans and L. Ebdon J. Anal. At. Spectrosc **4**, 299 (1989).

13. W. Shen, J. Caruso, F. Fricke and R.D. Satzger, J. Anal.At. Spec, **5**, 451 (1990).

14. K. Ruth, P. Schmidt, J. Coria and E. Moori, Proceedings of Electrochemical Society, Abstract No. **298**, 448 (1993).

15. N. Bradshaw. E. F.H. Hall and N. E. Sanderson J. Anal. At. Spec. **4** 801 (1989).

16. S. Yamaski and A. Tsumura, Anal. Sci. **7** 1135 (1991).

17. C. Kim, S. Morita, S. Yamasaki, A. Tsumura, Y. Takaku, Y. Igarashi, M. Yamamoto, J. Anai. At. Spec. **6** 205 (1991).

18. P. Gupta, S. H Tan, Z. Pourmotamed, F.Cristobal, N.Oshiro and B. McDonald, Novel Methods for Trace Metal Analysis in Process Chemicals & DI water and on Silicon Surface, Electrochemical Society Proceedings of " Contamination Control and Defect Reduction in Semiconductor Manufacturing III" San Fransisco, 200 (1994).

19. P. Gupta and S.H. Tan, UltraPure Water, Volume 11 (4) 54 (1994).

20 P.F. Schmidt and C. W. Pearce, J. Electrochem. Soc., **128** 630 (1981).

21. T. Shimono, M. Tsuji, M. Morita and Y. Muramatu, Microcontamination 91 Proceedings, Canon Communications Inc., Santa Monica, CA 554 (1991).

22. K. Hiramoto, M. Sano, S. Sadamitsu, N. Fujino, Jap. J. Appl. Phys., **28** 2109 (1991).

23. W. Bergholz, G. Zoth, G. Gelsdorf and B. Kolbensen, Elec Chem Soc. Proceedings (1991)

24 . J.C Tracy in "Electron Emission Spectroscopy"., ed. W. Dekeyser, (D. Reidel Publishing), 295 (1973).

25. C. A. Evans and R.J. Blattner, Ann. Rev. Mater. Sci., 181 (1978).

26. A. Jablonski, Surf. and Int. Sci., **1** (4) 122 (1979).

27. J. W. Coburn, J. Vac Sci. Technol. **13** (5) 1037 (1976).

28. A. Shimazaki, H. Hiratsuka, Y. Matsushita and S. Yoshii, Extended Abstracts of the 16th Conference on Solid State Devices and Materials, Kobe, Jap. Soc. Appl. Phys., Tokyo, 281 (1984).

29. J. Fucsko, S.H.Tan, M. K. Balaz, Electrochem. Soc., **140**, 1105 (1993).

30. P. Gupta, S.H. Tan and Z Pourmotamed (in progress)

IN SITU, REAL TIME DIAGNOSIS, ANALYSIS, AND CONTROL

In-Situ Process Control for Semiconductor Technology: A Contrast Between Research and Factory Perspectives

Or

Observations of a Researcher Who Has Seen the Light... At Dawn... Driving the Interstate to an Out-of-State Factory

John C. Bean

AT&T Bell Laboratories, Murray Hill NJ 07974

In the redirected 90's, many researchers have been cast into the brave new world of the factory. This happened to me, and it happened to the AT&T Bell Labs department I manage. In this paper I draw on that experience and suggest ways in which the researcher can increase his effectiveness in this new environment. I adopt a critical and irreverent factory viewpoint in the hope of better preparing researchers for the adventure they face. I focus on the likely disconnects in areas ranging from attitude to methodology, drawing from our experiences in an optoelectronics factory.

PROBLEM SELECTION

The bulk of this paper focuses on methodology. However, it is critical that we first address the subjects of problem selection and definition - critical because it is at this point that many researchers have already lost the battle to become effective factory contributors. Let's examine the classic research value system. Researchers are taught to pursue problems that offer the possibility of comprehensive and enduring understanding. The global nature of these attributes mean that research and researchers are naturally evaluated by communities extending well beyond a single institution or employer. This process presents a dilemma: how does the individual researcher achieve acknowledgment in a vast, essentially worldwide, community? The proven solution is to specialize and thereby define a community of more manageable dimensions. This process is further expedited by association with a specific problem or methodology, ideally of one's own invention.

However simplified, accept the above description and consider the behavior it stimulates. At the top of the value list is the desirability of knowledge. This knowledge should be complete and all-encompassing, it should be unambiguous, it should be fundamental, and it should be timeless. Researchers are the generators of this knowledge and they are evaluated almost exclusively on the basis of their expertise in a specific chosen field. This evaluation is, in turn, critically dependent on peer recognition. Recognition can be enhanced by tackling what are acknowledged to be the most difficult problems. Recognition can be enhanced by novelty, by invention, invention of analytical approaches, techniques and instrumentation. Further, a portfolio of small inventions is generally less recognizable than a single large invention with which the researcher will be irrevocably associated. This leads naturally to the researcher's role as champion and advocate of said tool or technique.

Contrast the above points with the needs of the factory engineer. A factory process is, of course, based on knowledge. However, that knowledge is largely acquired at an earlier development stage. That foundation should be largely complete by the time a process moves into a factory, at which point the emphasis turns to maintenance of a production status quo, that is, reproduction of a given product again and again with little or no deviation. Problems do, of course, arise but the emphasis will then be on expeditious solution, not on further extension of a knowledge base. This leads to one of the first disconnects I repeatedly observe in discussions between researchers and factory engineers. The engineer looks for a parameter that will define a successful process, or a tool to measure that parameter. The researcher dogmatically insists that a shift in that parameter can be produced by a range of phenomena and that a better parameter must be found. The discussion continues at cross purposes until frustration drives one of the parties to leave. The problem is that while the researcher may believe he is focusing on the engineer's need for a solution he is implicitly, and perhaps unconsciously, working another agenda: the need for unambiguous understanding. At the moment, however, the engineer does not need to resolve that ambiguity, he simply needs a handle, an identifier, a fingerprint of a successful process. In time he, or his colleagues, will attempt to resolve much of that ambiguity either as a part of improving the process or setting up the next product process. That, however, will occur at a future date in what is often a separate off-line investigation.

The intolerance of ambiguity is closely related to another chronic disconnect between researcher and factory engineer: the need for completeness. Research training dwells on those problems, such as the hydrogen atom, for which complete, closed, analytical solutions are possible. Many researchers seek to emulate such examples by choosing problems for which similar depth of understanding appears likely. This leads to the cliché of the scientist treating only very small systems, atoms or molecules, or extremely large systems, e.g. universes. Factory problems, of course, fall at neither extreme and complete models are not possible. To deal with this, modern factories are making increased use of statistics to define a history of common patterns. Statistics can thus be used to fill in the gaps in understanding. Statistical histories can be used to identify the most significant process parameters. Often, statistical histories can be used to define those areas that can be safely ignored, either because parameters are irrelevant or because intrinsic variation is small. Unfortunately, as researchers many of us have been taught that statistics are a poor man's substitute for complete knowledge and we will stubbornly resist not only using statistics but even learning the language (which for a Ph.D. is a relatively straight-forward process). However, in the absence of such supplemental information, the researcher is often left with either an intractable issue or compelled to define a problem domain that is so small as to be irrelevant in either return or time scale.

This leads to a discussion of specialization, invention and advocacy. Most researchers would characterize themselves as problem solvers. However, to the factory, researchers may appear more as solution providers. The distinction is important. In the factory, economic survival dictates that precedence be given to solution of a specific problem. The most appropriate solution is the solution that is fastest and cheapest. It does not matter if the solution is borrowed from someone or somewhere else. In fact this may the best way to ensure that it will be cheap and fast. An obsession with novelty, cleverness, invention and other research icons may in fact be counterproductive or even dangerous in this context. *Further, it is the problem that must take precedence, not the solution.* Again, these statements may seem obvious but I have repeatedly observed researchers ignore these principles. Our roles as expert, as inventor and advocate too often compel us to obsessively focus on our pet solution. Researchers repeatedly tune out aspects of a problem that do not fit their preconceived range of expertise and will in fact often try to turn discussion to new, more appropriate, problems! By surrendering our role as experts we may in fact undercut opportunities for recognition in an external community. It is ironic, however, that this surrender may allow us to use a much wider range of our expertise. Modern laboratory practices mean that the most successful researchers are in fact experts, of a sort, in a wide range of fields: chemistry, electrical engineering, computer programming, mechanical engineering, et cetera. World class experts, no, but experts and jacks-of-all-trades nevertheless. A focus on strictly pragmatic problem solving can allow the researcher unconstrained use of these skills and can, in fact, be surprisingly liberating and fun.

METHODOLOGY

It is time to get down to a more specific list of suggestions on how a researcher can increase factory impact, to define what I would call a path of least resistance. This will not be the only possible path but it is one relevant to our optoelectronic factory experience and one that may nevertheless offer insights to researchers in other production environments.

At the outset, I will lump possible activities into two categories: off-line and on-line. Off-line would include most of the essential work generally described as process or product characterization. These activities use many of the tools and methodologies with which the researcher is already familiar and comfortable. Because of that familiarity and comfort, and given the general thrust of this volume, I would like to focus on the other more foreign area of on-line investigation or what is generally known as process control. Process control has become increasingly important as virtually all semiconductor technologies move to the use of ever larger full wafers. Too many otherwise powerful off-line tools interfere with such processing or are fundamentally incompatible because of their sample destructive nature. Their use is thus delayed until processing is completed or restricted to secondary process monitoring samples. In either case the production process engineer is left with gaping blindspots or blind periods in which valuable product may be lost. Hence the compulsion to identify on-line supplements.

What should the researcher propose in such an on-line tool? First, preference should always be given to the tool that is **NON-INVASIVE**. Factory engineers are well-schooled experts on the myriad ways in which a process can go wrong. They will fiercely resist any proposal, however promising, if it may upset an established, successful process. The simplest way to protect against such a perturbation is by holding the actual processing chamber inviolate. This means that there is a built-in bias against tools that employ ion beams, electron beams, and to a lesser extent X-ray beams. Ion and electron beams have two shortcomings: they require vacuum and they entail introduction of foreign bodies in the process chamber, specifically guns and detectors. An X-ray apparatus can sidestep this problem by penetrating the wall of a process system. Unfortunately it as readily penetrates an operator and the bulky radiation shielding may require redesign of the processing system, the ultimate invasion. These considerations leave the door open for another alternative: optics. Because man is an optical animal, optical viewport materials are well characterized and they are most likely already in use on the processing system. Addition of optical viewports thus entails minimal risk and existing ports may, in fact, provide necessary access for what will then be a totally non-invasive optical tool.

Beyond being non-invasive, an ideal process control tool should be **NON-PERTURBING**. What do I mean by this? Consider the following. In many process systems, samples are only very loosely positioned. This may help to accommodate thermal cycling or it may be implicit in the mechanics of transferring samples through load-locked systems such as cluster tools, MOCVD or MBE systems. In addition, samples may well rest on opaque stages, e.g. heating, cooling and or electrical biasing assemblies. Finally, in many epitaxy and coating systems, samples rotate to promote deposition uniformity.

Combine this with loose positioning and the result is wobble. This wobble can be quite large and I have measured values as large as $\pm 3°$. In principle, all of these shortcomings can be engineered out of a process system and the promise of enhanced process control tools may drive this process. However, for the moment, these shortcomings exist and a process control tool that requires their elimination will be hugely perturbing. For instance, an otherwise non-invasive optical tool such as an ellipsometer may have to be bypassed because it implicitly requires sample alignment to a fraction of a degree, a condition that simply does not exist in many current pieces of equipment. An ideal optical tool should instead accommodate existing equipment shortcomings. This can be accomplished though scanning, phase locking or other synchronization methods. An even more robust tool can be based on what a colleague christened "the lucky photon principle," that is, the design of a fixed optical illumination and collection system with acceptance angles so wide that regardless of the sample dance, at least the occasional photon makes it through the optical path.

As a third attribute, the ideal process control should be truly **ON-LINE rather than IN-SITU**. The distinction here parallels the previous paragraph and may be best explained by a specific illustration. In the last ten years the technique of RHEED oscillations[1] has received intense attention in the research community from which it was proposed as process control tool capable of unambiguously measuring growth of single atomic layers. In this context, the specifics of the acronym and technique are irrelevant. Suffice it to say that it uses an electron beam to detect the roughening of a crystalline surface as a new atomic plane is nucleated and is indeed in-situ and capable of detecting a single such plane. It is, however, off-line in one critical sense: it requires growth on a precisely oriented crystal surface. Most research uses such crystals. Most production does not. Discovering this fact, many research advocates suggested that one could nevertheless grow on one such crystal each day and thereby calibrate the system for later off-orientation growth. Again, this is possible but in effect it takes a very expensive growth apparatus off-line and forces it to serve the role of a comparatively inexpensive thickness measurement tool. This perverts the whole idea of effective capital utilization. Other popular optical techniques, such as Reflection Difference Spectroscopy[2],

[1] J.H. Neave, B.A. Joyce, P.J. Dobson and N. Norton, *Appl. Phys.* **A31**, 1 (1983)

[2] D.E. Aspnes, J.P. Harbison, A.A. Studna and L.T. Florez, *Phys. Rev. Lett.* **59**, 1687 (1987)

may suffer from the same need to operate a production tool in a non-production mode.

As a final generalized attribute, the ideal process control tool should **FOCUS NARROWLY ON CRITICAL PROCESS PARAMETERS.** This focus is dictated by the fact that a broader analysis almost always entails additional expense and complexity, in both instrumentation and interpretation. Again, an illustration is useful. In a transparent material, the optical path is the product of the layer thickness and the layer refractive index. While a large variety of relatively simple optical tools can measure optical path, very few can unambiguously determine the two underlying parameters. Spectral ellipsometry is one such technique. Fueled by the researcher's compulsion for completeness and resolution of ambiguity, ellipsometry has become hugely popular accounting for at least half of the 150 odd papers I browsed in preparing this article. However, if one is only trying to control a process, the ambiguous measure of simple optical path may be more than adequate! While it is possible that there may be compensating errors in thickness and refractive index, it is extremely unlikely, especially when one considers that for semiconductors, refractive index is limited to a rather small range of values and thickness is not. Production is probabilistic process. If the ambiguity latent in a simple measure of optical path means that one will miss a small subset of possible process deviations, so be it, especially, if the tool for measuring path (or another comparably general parameter) is an order of magnitude less expensive or accommodating of process constraints than the full-blown ambiguity-resolving instrument. As I will describe in the final case study, many such opportunities exist. To paraphrase another factory colleague "shoot for the fat rabbits."

IMPLEMENTATION

I would now like to offer a laundry list of specific suggestions on both hardware and software. I will build on the example of an on-line optically based process control tool. This entails a loss of generality, but I again believe at least the flavor of the comments will have broader relevance.

To start, I strongly suggest the use of **OFF-THE-SHELF EQUIPMENT.** From the factory perspective, this has many advantages. Among the less obvious, but most important, is the fact that someone else is responsible for maintaining the expertise implicit in the design and manufacture of such an item. To put it another way, a

$50,000 instrument becomes radically less attractive if its guaranteed operation requires one to pay the $200,000 plus annual loaded salary of a staff Ph.D. The factory may be willing to pay your salary to build the instrument but it will be strongly averse to an open ended service contract.

One should similarly aim for **MODULAR COMPONENTS,** which is another way of saying that off-the-shelf items should mate directly with off-the-shelf items. Same argument as above. In the context of my hypothetical optical process control tool these recommendations lead me to highlight the tremendous opportunities now offered by optical fibers complete with shielding and coax-like end connectors. These connectorized fibers mate with a large and ever-growing array of standard optical components. In a single issue of an optical industry trade-rag, I found ads for light sources (broadband, LED, laser, both CW and modulated), detectors (tunable, miniature), modulators, stabilizers, power meters, rotators, switches, splitters, combiners, isolators, circulators and if that were not enough, there were offers to design and fabricate custom but fully packaged fiber compatible items. One can also purchase fiber compatible subsystems. A dramatic example is provided by the modular spectrometers now being offered by several vendors. These units have no moving parts (not even an on-off switch), they fit in the palm of the hand, and, by borrowing CCD detection elements from copying machine technology, they are available with a price tags under $2000.

To be fair, the classical optical table still has a role in the research lab where it offers a wider range of options than will perhaps ever be available in connectorized modules. That does not, however, reopen the door to its use in the factory when well-engineered alternatives, such as that illustrated in figure 1, exist. Unlike the research-derived optical bench with its singular emphasis on versatility, the system of figure 1 places equal emphasis on ultimately locking in a design in a robust manner that will *not* be easily perturbed and will *not* require repeated expert alignment.

Discussion of alignment brings me to the next suggestion: use **NON-VARIABLE COMPONENTS.** This may seem like an almost trivial suggestion, but it is one at the heart of the research vs. production conflict. A researcher designs for the unexpected and thus incorporates in as much flexibility as possible. On the other hand, the factory engineer knows that an adjustment provides one way to optimize a process but a near infinite number of ways to louse it up. If your job is to maintain a production status quo, you want to design out variability. Again, let me provide an optics illustration. In the lab, one modulates light intensity with a neutral density wheel. However, when the proper intensity is determined and a move to the factory is contemplated, it is time to throw out the wheel. One option is to select a connectorized fiber with a core diameter chosen to pass the optimum light intensity. Yes, this will entail purchase of a selection of fibers and some swapping to find the one that is exactly right. However, when that process is complete, your job is done and the need for your expertise is removed. Anyone can note the part number of that fiber, order it and replace it.

Fig. 1   Rail based modular optical bench system of Spindler-Hoyer Inc, Milford Massachusetts. Among other applications these units have been employed in optical systems built to withstand the vibration and impact loads of launch aboard the U.S. Space Shuttle (by permission of Spindler-Hoyer).

Beyond these hardware suggestions, if you are designing an instrument for the factory, also tailor the data presentation to the factory. As the instrument should focus narrowly on the most relevant process parameters, so the data should **HIGHLIGHT PROCESS DEVIATIONS**. Use ratios, offsets or differences *from the production norm*. Absolute numerical values are in a

sense irrelevant; it is the deviation that is critical in an established production process. The absolute values were determined by other instruments at other stages. A simple idea, again, but I have seen the implementation of this idea crystallize the solution of a production problem that had been festering in a factory for years.

Beyond highlighting deviations, **PRESENT DATA FOR OPERATORS NOT SCIENTISTS**. I am not downplaying the competence of operators, I am simply acknowledging that our favored forms of presentation build on our lust to maximize information, even at the expense of clarity or relevance. The presentation of on-line control information should remove such extraneous information.

As data presentation will now almost certainly occur on a personal computer screen, **USE BROADLY ACCEPTED SOFTWARE PLATFORMS**. There is no longer an excuse for compelling a operator to face a DOS prompt or the manual for a unique single purpose software package. Virtually all factory denizens are familiar with graphical interfaces as implemented in the increasingly indistinguishable Mac or Windows format. Use of such a format, guarantees that most operators will already be familiar and comfortable with the mundane file and hardcopy functions that make up much of every program. Further, the prevalence of these standards means that most instrument vendors are being compelled to offer libraries (in Window's parlance: DLL's) that take care of the detailed hardware interface code. While this is still a somewhat painful work-in-progress, the existence of such libraries will soon make the programmer's job far simpler, and make it possible for the instument originator, who knows the equipment best, to drive that process.

This leads to a final software suggestion, **USE A GRAPHICAL SOFTWARE INTERFACE**. The above packages are intrinsically image based and build on the fact that the human brain is a superb image processor. Mathematical curves may incorporate precision but they have nowhere near the impact or immediacy of graphics. This is especially true for those less comfortable with mathematics, a group that includes virtually all non-scientists except mathematicians. Rather than illustrating this single point, I now offer a case study that illustrates a variety of the above suggestions.

CASE STUDY: CONTROL OF EPITAXIAL GROWTH BY INTERFEROMETRIC REFLECTOMETRY

This study is a prime example of borrowing and building upon another's ideas. In this case I am the borrower, and the borrowees are Kevin Killeen and Bill Breiland of the Sandia National Laboratories. The problem is that of determining when one has correctly grown the desired complex layer sequence of semiconductor materials. The proposed solution is based on interferometric reflectometry.[3] Use is made of penetrating wavelengths such that the reflected signal will be a complex sum of interfering reflections from interfaces throughout the structure. This is one instance in which complexity is a plus because it contributes to the uniqueness of the signal which, when presented properly, makes it easier to detect minute process deviations.

A conventional reflectometry spectrum is presented in figure 2. It is a good example of how not to present data in the factory. The spectrum has plenty of wiggles and without a computer only an idiot-savant would stand a chance of relating such wiggles back to product characteristics. In fact, even the savant would fail because these wiggles are a measure of optical paths, which, as discussed above, do not map unambiguously back to physical characteristics. Now, imagine that one proposes using reflectometry as the real-time monitor of a growth or etching process. Figure 3 presents a subset of the data produced: a superposition of spectra, one for each layer of a complex device (in this case a vertical

cavity surface emitting laser or a resonant cavity photodiode). Things have gone from bad to worse. At this point, I believe Killeen and Brieland made a breakthrough by going graphical. Rather than superimposing such curves in a marginally more intelligible manner, they made each curve into a single scan line of a CRT image. Peaks were converted to white pixels, valleys to black, intermediate points to appropriate gray values. Consecutive curves were thus translated into consecutive scan lines to produce an image such as that in figure 4. What does this convoluted "fingerprint" mean? *It means that if you reproduce it you have almost certainly grown the same structure I grew.* No more, no less. It does not give you fundamental information on the structure. It does not eliminate all ambiguity and in fact the very careful design of an alternate structure could reproduce this image. However, by random process variations, such a false reproduction would be exceedingly unlikely and the fingerprint thus serves as a very effective process monitor.

I now invoke the rule on highlighting process deviations. In figure 4, I superimpose the fingerprint of a growing structure on the background fingerprint of an accepted standard. It is relatively easy to see that the fingerprint of the growing structure (below the horizontal white demarcation line) is offset slightly to the right of the background reference. Because I am measuring optical path, the degree of offset is a measure of the drift in physical dimension: right corresponds to longer or larger. In fact, this information is not necessary. It is only important that the deviation of the fingerprint scale in some smooth way with the error in the process. A statistical history and correlation with product yields would then establish the limits of acceptable alignment.

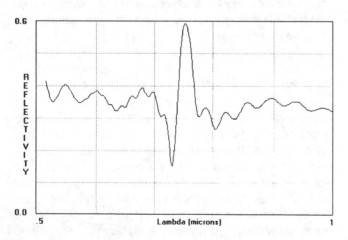

Fig. 2   Conventional spectrum of visible light reflected back from complex semiconductor layer structure.

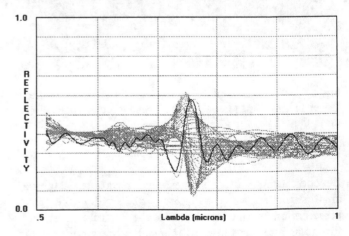

Fig. 3   Superposition of spectra such as those in fig. 3 illustrating nearly incomprehensible build-up of information obtained during growth of structure.

---

[3] K.P. Killeen and W.G. Breiland, *J. Elec. Materials* **23**, 179 (1994)

464

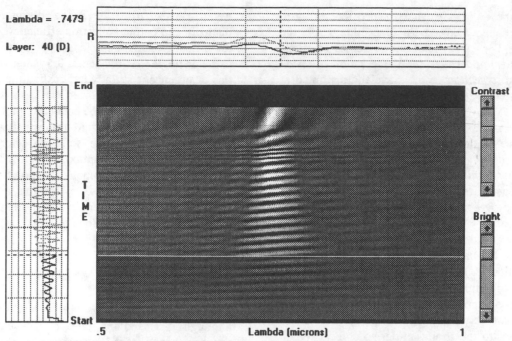

Lambda = .7479

R

Layer: 40 (D)

End

Contrast

TIME

Bright

Start

.5   Lambda (microns)   1

Fig. 4   Translation of conventional spectra into time-dependent gray-scale "fingerprint" after the work of Kevin Killeen and Bill Breiland of the Sandia National Laboratories. Lower section of fingerprint taken from actual growing sample. This is superimposed on reference background fingerprint of known "good" sample. The slight left/right mis-alignment is indicative of process deviation. This deviation highlighted in fingerprint cross-sections in top and side boxes.

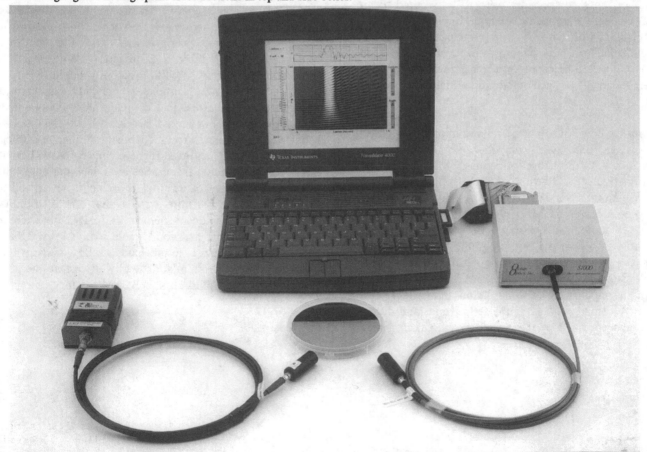

Fig. 5   Completely modular, off-the-shelf, sample tolerant, implementation of interferometric reflectometer.   Appended to a Molecular Beam Epitaxy growth system, this simplified unit produced the data of the above figures.

In this superimposed image, I have added the ability to take cross sections along particular planes of the fingerprints. The top box shows cross sections immediately above and below the demarcation; the side box shows a cross section at the wavelength printed at upper left or indicated by the dotted line in the top box. These are called up by simply clicking on the fingerprint at a point of interest. The particulars are unimportant. The significance is that, in a graphical programming environment (here Microsoft Visual Basic and Windows), a very wide range of graphical options is available and easily implemented. The right format is the one that proves most successful at highlighting process deviations.

Now let's turn to the hardware. In presenting this paper I used an ad entitled "Reflectance Measurements Made Simple." For copyright reasons I'll simply describe it here. It pictures a lamp housing with a beam directed out at a chopper assembly through a beam-splitter to the sample and back to the beam splitter to a focal length matcher appended to a mechanical spectrometer appended to a detector cabled to a digital lock-in amplifier to a PC. Their disconnected nature suggests that all of these components were to be secured to an optical bench. This may be a simple implementation by research standards, but it is entirely inconsistent with the factory principles I have attempted to illustrate. It is too complex; it is too susceptible to misalignment; and it is entirely incapable of accommodating a real-life wobbling wafer.

I did not see how Killeen and Breiland implemented their spectrometer and was compelled to improvise in accordance with the above guidelines. My apparatus is shown in figure 5. Starting from the top, it avoids the use of exotic workstations or languages, custom microprocessors or ROMs. It is driven by an off-the-shelf DOS laptop with a single PCMCIA input/output card running a Visual Basic program, under Windows. The spectrometer, at right, is the simple no-adjustment module described above (with a price less than one half that of the laptop). The illumination source is a light bulb in the small module at the left. Input and output light paths are fixed by the coiled optical fibers, which were selected to achieve the optimum illumination intensity and output collection efficiency. At the ends of these fibers, fixed simple lenses spread the light path out to a 3.5° angle. At the actual 0.6 meter sample distance, this divergence is slightly larger than the sample wobble, ensuring that there will always be a "lucky photon" making it back through the collection path. The software automatically smoothes and re-normalizes the data for each run. Further, the software offers the operator complete default setup options which include automatic start of data collection upon a signal from the process crystal growth apparatus. With the exception of the software, every single item in this system comes straight out of one of several catalogs.

SUMMARY

The instrument above is still a work-in-progress and although it is moving onto our development lines I cannot yet say that it is successfully monitoring product. However, I can say that is has been accepted, indeed embraced, for trial use in a way that simply would not have occurred three years ago. The difference lies in many of the points above. To start with, for this and a number of other instruments, we took the time first to learn the culture, the priorities and the statistical language of our factory customers. We used that knowledge not only to build credibility but to focus on their most important problems, not just our favorite solutions. We learned that complete and unambiguous information was not always necessary and, liberated by that, we were able to focus on much simpler and direct tools. Instead of just inventing, we endeavored to borrow as much as possible, from other scientists and engineers and from catalogs of existing equipment. We tried to deliver non-invasive little systems that offered minimal risk, expense or inconvenience. Finally, we tried to construct tools that had no invisible strings leading back to us, to our expertise, to our loaded salary, systems that could be operated and maintained by anyone, in the hope that such value would ultimately create a much stronger bond.

Real-Time Spectral Ellipsometry Applied to Semiconductor Thin Film Diagnostics

W. M. Duncan, M. J. Bevan, S. A. Henck

Corporate Research and Development
Texas Instruments, Incorporated
Dallas, Texas 75265

Spectral ellipsometry is an ideal sensor methodology for *in situ* real-time monitoring of wafer state properties (*e.g.*, temperature, composition, film thickness and surface properties). We have developed a rapid scan spectral ellipsometer capable of precisely measuring thicknesses and compositions of multilayer structures *in situ* and in real-time (i.e., the time frame of process changes or about one second or less). The spectral ellipsometer applied in this work utilizes phase modulation, multichannel detection and digital signal processing techniques. This approach provides acquisition of 46 spectral points in less than 0.5 second. Numerical algorithms are used for reducing measured spectral ellipsometric data to wafer state properties, i.e., thicknesses and compositions, in real-time based on a "standard model" approach. Successful applications of spectral ellipsometry to *in situ* real-time monitoring and control include several vacuum chambers of a flexible process flow for silicon CMOS circuits (rapid thermal oxidation, plasma etch, and in vacuum metrology), and molecular beam epitaxy of compound semiconductors. *In situ* spectral ellipsometry has been used for analyzing samples at process temperatures as high as 1100 °C. Atomic layer thickness sensitivity and part-per-thousand composition sensitivity have been demonstrated .

INTRODUCTION

Sensors capable of *in situ*, real-time measurements are essential components for flexible manufacturing environments. Ideal sensors are non-contacting and non-invasive. Spectral ellipsometry (SE) possesses these attributes and exhibits high sensitivity to layer thicknesses and complex index of refraction (i.e., n - ik). In addition, index of refraction can be related to composition through the effective medium approximation and to temperature if the temperature dependence of the dielectric function is known. Spectral ellipsometry allows over determination of variables in multilayer stacks, hence, structure models and layer parameters can be verified using statistical methods. Although the principles of ellipsometry were established more than a century ago (1) the technique was impractical for real-time process control until the development of both fast acquisition techniques and fast, inexpensive computers for quickly performing the necessary calculations for reducing experimental data to physical parameters (i.e., film thicknesses and dielectric function, hence, composition and temperature).

Ellipsometry is based on the polarization transformation that occurs when a beam of polarized light is reflected from or transmitted through a medium (2). The transformation is determined by the optical properties of the material which are related to the fundamental material constants (i.e., dielectric function) and consists of two parts: an amplitude change and a phase change. These changes are different for incident radiation with its electric vector oscillating in the plane of incidence (p-state polarization) compared to radiation with its electric vector oscillating perpendicular to the plane of incidence (s-state polarization). An ellipsometer measures the change in polarization state resulting from reflection. The polarization change is typically described in terms of two angles, Δ and Ψ. The first of these, the differential phase retardation, Δ, is simply the phase change on reflection of the two orthogonal polarization states, $\Delta = \delta_p - \delta_s$. The differential amplitude ratio, $\tan\Psi$, is related to the complex Fresnel coefficients, r_p and r_s by $\tan\Psi = |r_p|/|r_s|$. The parameters Ψ and Δ are usually combined into a single complex quantity ρ which is defined as,

$$\rho = r_p / r_s = \tan\Psi e^{-i\Delta} \qquad (1)$$

The complex reflection ratio is used for subsequent calculation of physical parameters (see below).

The spectral ellipsometer used in the current work combines phase modulation, multichannel detection and digital signal analysis techniques, providing an instrument which is high speed, low cost and robust (i.e., no moving parts). Although phase modulated ellipsometry was first described by Jasperson and Schnatterly (3) in 1969, it has been the advent of digital signal analysis techniques and multichannel detection that have allowed this approach to progress significantly (4-8).

RAPID SCAN ELLIPSOMETRY

The phase modulated spectral ellipsometer used in this work has been described previously (6-8). Figure 1 shows diagrammatically the ellipsometer attached to a vacuum chamber used for rapid thermal oxidation of silicon. For in process sensing, an ellipsometer requires optical access through two opposing ports on the chamber. Ellipsometer ports are normally fixed such that an angle between 68° and 75° is formed with respect to the surface normal of the sample. The ports must be terminated with strain free windows.

The spectral ellipsometers utilized here employ a polarizer-modulator-sample-analyzer (PMSA) arrangement. Light from a broad band Xe arc lamp is delivered via fiber optics to polarization components mounted on the chamber. Polarization modulation is accomplished by passing linear polarized light through a photoelastic modulator. The modulator is a quartz bar driven by a 50 kHz piezoelectric transducer. After traversing the vacuum chamber and being reflected from the sample under study, the light passes through an analyzer. The light is then taken off the reactor and delivered via fiber to a grating monochromator and photodiode array detector. In this instrument, the solid state multichannel detector eliminates the need for scanning the monochromator and allows information from different spectral regions to be multiplexed into common signal processing hardware. The detected intensity at each photodiode of the array has the form:

$$I(t) = I\{I_o + I_\omega \sin[\delta(t)] + I_{2\omega}\cos[\delta(t)]\} \qquad (2)$$

where $\delta(t)$ is the phase shift induced by the modulator. For the current arrangement where the maximum modulator retardation is set at 137.8° for each wavelength, the ellipsometric parameters Ψ and Δ can be derived by frequency analysis of the time dependent intensity signals, $I(t)$ and the following equations (4-6):

$$I_o = 1 \qquad (3)$$
$$I_\omega = I_o \sin2\Psi\sin\Delta \qquad (4)$$
$$I_{2\omega} = I_o \sin2\Psi\cos\Delta \qquad (5)$$

where ω is the fundamental modulation frequency of the modulator. The time domain waveforms are digitized at 1 MHz with 12 bit resolution. Time domain waveforms are transformed to the frequency domain using a standard fast Fourier transform algorithm. Waveform digitization, waveform frequency analysis, experiment control and all post-acquisition computations are performed using an industry standard architecture (ISA) buss computer with 80486 processor. The ellipsometer may run either in a stand alone mode where data is displayed and stored on the ellipsometer computer or may communicate with a process host computer using the Semiconductor Equipment Communication Standard (SECS) protocol. In the case of real-time process control, the ellipsometer is operationally a sensor slave device to the process host computer. Film stack models in process control environments are down loaded from the host computer to the ellipsometer, and thickness, composition, and fitting parameters are passed back to the host over the SECS link. Other information such as spectral region, fitting criteria, measured Ψ and Δ values, etc., pass between the host and the ellipsometer over this link if desired.

In order to extract composition from measured spectra, we must solve the so called "inverse problem" where the thicknesses and/or dielectric functions are found which match measured Ψ and Δ values. We use a "standard model" approach for calculating spectra for comparison to measured data. A program flow similar to one published by Hu, et al. (9) and shown in Figure 2 is employed for minimizing the error between calculated and measured spectra. The "trial" values of ρ are calculated from the Fresnel reflection equations for a "model" structure using trial or known values for the independent variables, layer thicknesses, d, angle-of-incidence, q, complex index of refraction, n-ik, and wavelength, λ. This calculation is carried out while iterating on unknown independent parameters (d, n, k or θ) until a desired deviation between the measured and calculated spectra is obtained. For a single layer on substrate, the thickness of the layer can be found in about 0.5 seconds of 80486-50 CPU time for 46 wavelength values using current algorithms. Computation times increase approximately linearly with the number of discrete wavelength values and increase approximately as the factorial of the number of thicknesses determined.

468

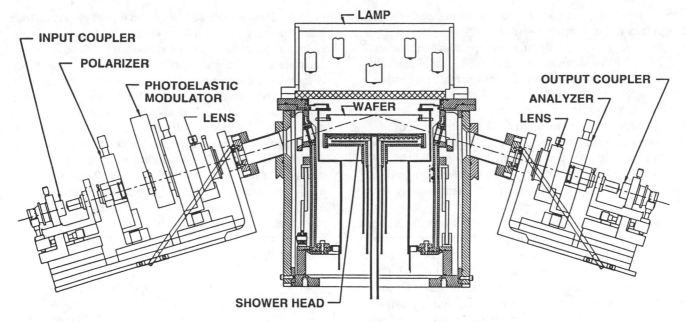

FIGURE 1. Spectral ellipsometer attached to rapid thermal oxidation chamber. From Ref. 8.

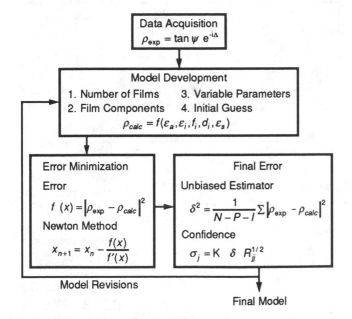
FIGURE 2. Program flow for minimizing error between calculated and measured spectra. From Ref. 8.

In cases where mixtures of semiconductor or dielectric materials are studied, the layers are treated as heterogeneous or composite mixtures of two or more materials. In such cases the dielectric response is calculated from a model of the effective medium. Aspnes (10) has discussed the general and specific forms of effective medium theory for calculation of macroscopic dielectric response. We employ the effective medium approximation (EMA) of Bruggeman (11) where a random mixture is assumed. In this approximation the effective medium dielectric function ε is given by:

$$0 = \sum_{i=1}^{n} f_i \frac{\varepsilon_i - \varepsilon}{\varepsilon_i - 2\varepsilon} \quad \text{and} \quad \sum_{i=1}^{n} f_i = 1 \qquad (6)$$

where the summation is over the number of distinct constituent media in the mixture, f_i is the volume fraction of the ith component, ε_i is the dielectric function of the ith component and ε is the dielectric function of the effective medium. The quantities ε_i must be known for the constituent media. Mixtures are usually formed from constituent media that are similar in composition to the unknown. We have used either published dielectric function libraries such as those found in reference (12) or those generated by measurement of pilot structures.

RESULTS AND DISCUSSION

Silicon Processing

There have been a number of investigations using *in situ* single wavelength ellipsometry for diagnostics during processing of silicon (13-18). However for multilayer structures, an adequate number of dependent variables (i.e., spectral points) must be obtained in order to allow determination of multiple independent variables. Hence, spectral ellipsometry has been applied in a number of studies to the analysis of complex multilayer structures (6-8, 19-27).

Our first example of *in situ* process diagnostics and control using SE will be the study of rapid thermal oxidation (RTO) of silicon. In thermal processing applications such as RTO, knowledge of both the processing temperature and the dielectric function values at that temperature are required for direct inversion of the ellipsometric equations (see Fig. 2). In the case of RTO where processing temperatures in the range of 1100 °C are employed, temperature dependent dielectric function values for SiO_2 and Si existed in the literature only for 6328 Å (28). Although spectral values of the dielectric function could have been derived from pilot samples for the temperature range of interest, we chose simply to use the spectral ellipsometer in a single wavelength mode when the samples were at elevated temperature (>600 °C). Because this was a single layer on substrate problem, adequate information could be obtained from a single wavelength. The full spectral mode was employed when the wafer was first introduced into the reactor at room temperature and the full spectrum inverted for accurate determination of the angle of incidence. Calculating the angle of incidence from measured spectra using well established room temperature dielectric function libraries for SiO_2 and Si yielded a calculated angle-of-incidence confidence typically less than ±0.03° at nominally 75°. Temperatures were provided to the ellipsometer from a multiple wavelength pyrometer. Shown in Figure 3 are real time *in situ* thickness measurements for rapid thermal oxidation at 1100 °C determined from the 6328 Å measurement. With this reactor, because the inner quartz chamber was rotated during heating, inner and outer chamber windows were coincident for only about 75 ms. Although this coincidence and, hence, line of sight through the reactor occurred twice per rotation, rapid data acquisition was synchronized only to one of these occurrences of coincidence.

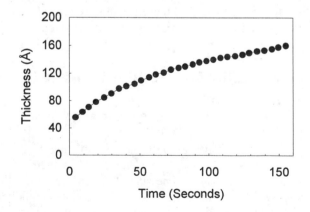

FIGURE 3. Real time thickness measurements at 1100 °C.

Shown in Table I are standard deviations of four RTO process lots using timed and ellipsometric endpointing. In the case of a single temperature endpoint, adequate results could be obtained either with timed or ellipsometric endpointing. However when more complex thermal environments were employed such as multiple temperature setpoints, ellipsometry was essential to obtain good film thickness control. Using ellipsometric endpointing, thicknesses could be controlled to within ±1.5 Å for 70 Å SiO_2 gate oxide films independent of thermal profile or ambient effects. Thicknesses of SiO_2 on Si determined *in situ* at growth temperatures were found to agree within ±1.5 Å at 70 Å when checked *ex situ*. Based on simulations, the thickness error resulting from a ±10 °C temperature error was estimated to be less than ±0.3 Å.

TABLE 1. Comparison of timed and ellipsometer endpointing.

Temperature Setpoint	Endpoint	Thickness Standard Deviation
Constant	Timed	3 Å
Constant	Ellipsometer	1.5
Variable	Timed	10.0
Variable	Ellipsometer	1.5

The second *in situ* application of spectral ellipsometry to silicon processing is that of plasma etch. In the case of plasma etch, sample temperatures remained near room temperature avoiding temperature complications to the analysis. However, in this case, both single and multilayer stacks require monitoring, hence, adequate spectral information is needed for solution of up to six independent parameters. As an example of monitoring film thicknesses during plasma etch, we consider the case of two thin layers on a silicon substrate. This stack structure prior to etch includes nominally a 2800 Å polycrystalline silicon layer on a 1000 Å SiO_2 layer on silicon substrate. Presented in Figure 4 is the result of monitoring the polycrystalline Si film thickness versus etch time. When the wafer is first introduced into the reactor, all the layer thicknesses and compositions (polycrystalline Si, crystalline Si and void fraction) in the structure are treated as free variables to obtain the best fit to measured spectra. We refer to this as a "full stack minimization". In the case of a room temperature process such as plasma etching, it can be assumed that only the surface layer is changing with time. Hence, following the full stack minimization, the ellipsometer operates in a "real time" mode where, during the early part of the etch, only the thickness value of the top layer is minimized, thus allowing more rapid fitting of the spectra. However, as the top layer becomes thin, better information on the buried layers may be obtainable. Hence, while operating in the "real time" mode, the algorithm starts minimizing

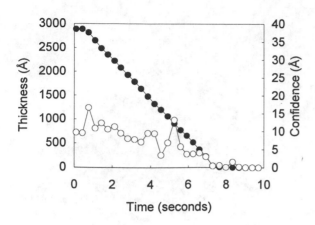

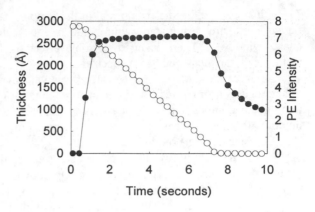

FIGURE 4. Real-time *in situ* polycrystalline Si thickness (solid) and thickness confidence (open) measurements during plasma etching.

Figure 5. Simultaneous ellipsometric thickness (open) and plasma emission intensity (solid) during etch of a 2800 Å polysilicon film.

the thickness values of the top two layers in the stack when the thickness of the top most layer reaches a prescribed thickness, typically set around 500 Å. The ellipsometer software can be set to automatically change the model from two layers to a single layer when passing through the poly/SiO_2 interface. The calculated confidences are controlled by quality of fit parameter, δ, and correlation coefficient (see Fig. 2). For buried layers, particularly when below high index films such as polycrystalline silicon, calculated confidences are degraded by limited layer information content to the total spectrum.

Shown in Figure 5 is simultaneous ellipsometer and plasma emission data taken during the etch of the 2800 Å polycrystalline silicon on 1000 Å SiO_2 on silicon stack. The wafer state ellipsometric data was taken at the center of the wafer. The plasma emission signal is the silicon emission intensity at a photodiode looking into the plasma with a filter at 4070 Å. Whereas plasma emission signatures are used in a number of processes for endpointing, it is apparent from Figure 5 that it is difficult to detect the point where removal of the polycrystalline silicon layer from the SiO_2 is complete based on the plasma emission signal. In fact the plasma emission signal begins dropping well before the polycrystalline silicon layer has cleared and persists after the polycrystalline silicon layer has cleared. With spectral ellipsometry it is possible to definitively detect the interface between the layers. It should be noted that with wafer state monitors such as ellipsometry, interfaces between layers can be anticipated and processes stopped at any desired film thickness.

Compound Semiconductors

In situ ellipsometry has been used by several groups to study epitaxial growth and processing of III-V compounds (29-34). *In situ* real time monitoring during III-V growth is the subject of two articles in these proceedings (35, 36). We will direct the discussion here to II-VI materials as they are the compound system currently emphasized in the authors' laboratory.

Ellipsometry has been used in several previous studies for monitoring epitaxial growth of II-VI compounds. It is the goal of the current work to study the surface properties of these materials under ultra high vacuum conditions and to monitor and control film compositions. Spectral ellipsometry was first used by Demay et al. (37) for *in situ* MBE studies of II-VI films. These authors, however, did not try to monitor the composition of $Hg_{1-x}Cd_xTe$ during growth but rather studied the interdiffusion of HgTe and CdTe. Hartley et al. (38) have used single wavelength ellipsometry to monitor $Hg_{1-x}Cd_xTe$ composition during MBE growth. In the single wavelength work of Hartley, et al., a minimal data approach was used to demonstrate a sensitivity to composition, x, of about ±0.003. Johs, et al., (39) have also used spectral ellipsometry to monitor CdTe growth on GaAs.

In the II-VI work reported here, a DCA MBE chamber was custom-built to accommodate a spectral ellipsometer and other *in situ* sensors. The chamber is fitted with 4 standard Knudsen cells and a 9" Hg diffusion pump to maintain a vacuum of better than 1×10^{-9} mb. The ellipsometer is attached to two opposing conflat flange ports on the MBE chamber. Light propagating through the ellipsometer ports forms an angle of about

471

73° with respect to the surface normal of the sample. The flange ports are terminated with strain free bakeable quartz windows. There is no discernible effect of the windows on the measured Ψ - Δ spectrum.

In case of $Hg_{1-x}Cd_xTe$, we have treated the layers as heterogeneous mixtures (Bruggeman approximation) in finding the composition. We have used the dielectric function libraries published by Arwin and Aspnes (40) for $Hg_{1-x}Cd_xTe$ and for the oxides.

In this study we have been interested in the nature of the surfaces prior to and during growth of $Hg_{1-x}Cd_xTe$. The surface condition of the <112> $Cd_{1-x}Zn_xTe$ substrate (x=0.05) prior to growth derives from a combination of *ex situ* chemical preparation and *in situ* thermal cycling. We can monitor changes in the surface properties of samples during the pregrowth heat up and growth cycle simply by measuring spectral Ψ and Δ values with no additional processing of the data required. For example we show in Figures 6 and 7, Ψ spectra for two $Cd_{1-x}Zn_xTe$ substrates from the same boule, where one sample was chemomechanically polished in 1.5% Br in methanol (Br/MeOH) only and the second was oxidized using deep UV illumination following the Br/MeOH treatment. Both the Ψ and Δ data show similar trends comparing the two surfaces. The oxide free surface exhibits a very smooth Ψ and Δ behavior with temperature and no discontinuities are observed. By contrast, the oxidized surface exhibits higher Ψ and lower Δ values than the unoxidized surface, and shows a greater change in Ψ and Δ with temperature than the unoxidized surface. A discontinuity in the Ψ and Δ behavior is also observed at about 300 °C with the oxidized surface. This discontinuity is possibly due to partial desorption or chemical change in the surface oxide. Even above this discontinuity temperature, the oxidized surface does not return to the Ψ and Δ values of the unoxidized surface.

The surfaces of LPE grown <111> $Hg_{1-x}Cd_xTe$ have been studied in a similar way by monitoring Ψ and Δ behavior during a thermal cycle. The $Hg_{1-x}Cd_xTe$ surfaces were prepared with Br/MeOH and immediately introduced into the MBE chamber. In this case presented in Figure 8, the surface is observed to change at about 177 °C as measured with a pyrometer. This change is likely due to desorption of Te from the Te rich surface formed by the Be/MeOH treatment.

In order to determine composition of $Hg_{1-x}Cd_xTe$ during growth, it is necessary to determine the dielectric constants of the constituent materials in the structure at the growth temperature. Presented in Figure 9 and 10 for $Cd_{1-x}Zn_xTe$ where x = 0.05, are n and k versus temperature. n and k are determined from measured Ψ and Δ values by inverting the Fresnel equations. The noise in the data below about 7000 Å is a result of low

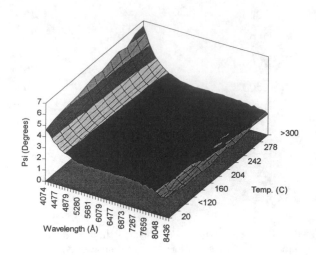

FIGURE 6. Ψ contours versus temperature for chemomechanically polished CdZnTe.

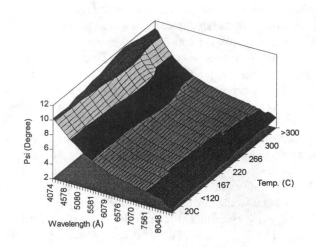

FIGURE 7. Ψ contours versus temperature for oxidized CdZnTe.

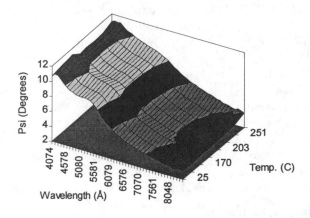

FIGURE 8. Ψ contours versus temperature for chemomechanically polished $Hg_{1-x}Cd_xTe$.

472

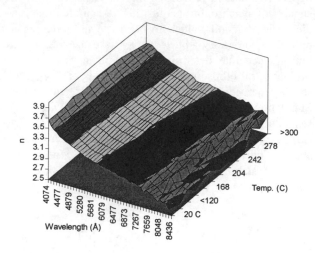

FIGURE 9. Temperature dependent dielectric constant contour plots for CdZnTe.

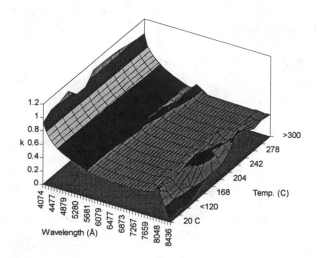

FIGURE 10. Temperature dependent extinction coefficient contour plots for CdZnTe.

reflectivity of the $Cd_{1-x}Zn_xTe$ in this region and a lower optical efficiency of the system due to the grating response.

At this writing we have not begun growth studies in the MBE system. Hence we have undertaken to estimate the sensitivity of spectral ellipsometry to composition from *ex situ* measurements of previously grown $Hg_{1-x}Cd_xTe$ epitaxial samples. Shown in Figure 11 is a representative spectra measured *ex situ* together with the best fit calculated spectra. This fit used a model assuming a 20 Å native oxide and 30 μm thick $Hg_{1-x}Cd_xTe$ film on CdTe substrate. Based on this fit using the Bruggeman EMA, we find a x value of 0.201 for this sample. Compositional determination from EDX

analysis on this same sample yields an x of 0.219. Also based on the quality of fit shown in Figure of 11, we can calculate a confidence for the composition. The calculation as defined in Figure 2, yields a confidence for x of ±0.0016 for this sample.

Shown in Figure 12 is the comparison of compositions determined by spectral ellipsometry to those determined from FTIR cutoff measurements. Because we used literature libraries for the dielectric functions of $Hg_{1-x}Cd_xTe$ in determining composition from SE spectra, it is not surprising that there is some deviation between the measurement methods. Figure 12 is shown with confidences of ±0.002 for the SE data and ±0.004 for the FTIR composition measurement.

We have also used simulations to estimate the sensitivity of *in situ* spectral ellipsometry to compositional variations. To this end, we first calculate the spectra for two compositions around a center composition. We then determine the differential Ψ and Δ parameters, $\delta\Psi$ and $\delta\Delta$, respectively, between two compositions x_1 and x_2 of $Hg_{1-x}Cd_xTe$ as:

$$\delta\Psi = \Psi_2 \text{ (composition } x_2) - \Psi_1 \text{ (composition } x_1) \quad (7)$$
$$\delta\Delta = \Delta_2 \text{ (composition } x_2) - \Delta_1 \text{ (composition } x_1) \quad (8)$$

Shown in Figure 13 are the differential Ψ and Δ values, for a compositional difference of 0.263 and 0.272. The maximum sensitivity to composition x corresponds where either $\delta\Psi$ or $\delta\Delta$ are the largest. While $\delta\Psi$ shows very little variation over the entire spectral range, $\delta\Delta$ exhibits a change as large as -1.1° for a compositional change of 0.263 to 0.272 in the spectral region of 6500 to 7000 Å. Under typical conditions where the noise in Δ is ±0.1°, then it is possible to achieve a sensitivity to composition x of ±0.0008 at a single wavelength operation in this spectral region (6500 to 7000 Å). Since sensitivity scales with the square root of the number of measurements (wavelengths), it should be possible by measuring

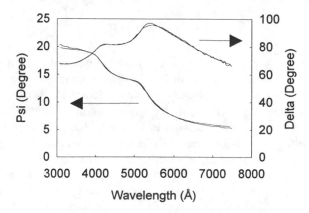

FIGURE 11. Measured and calculated spectra from best fit parameters for $Hg_{.799}Cd_{.201}Te$.

473

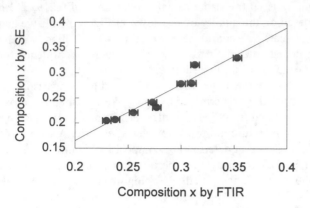

FIGURE 12. Comparison of compositions determined by spectral ellipsometry and FTIR cutoffs. The solid line is the linear least squares fit of the experimental data.

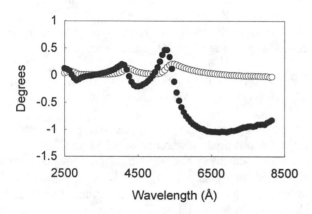

FIGURE 13. Change in Δ (solid) and Ψ (open) to compositional change from 0.263 to 0.272.

multiple wavelengths in the maximum sensitivity region to achieve even better sensitivities to composition than ± 0.0008.

CONCLUSIONS

Phase modulated spectral ellipsometry has successfully been applied to real-time diagnostics in a number of processing environments Examples discussed include rapid thermal oxidation, plasma etching of multilayer stacks relevant to silicon CMOS, and surface and compositional analysis of compound semiconductors. Using fast acquisition techniques and fast algorithms and processors for reducing measured data, materials and layer parameters for multilayer dielectric stacks can be obtained in real time. In order to be useful in realistic processing environments, the spectral ellipsometer must be capable of compensating for temperature effects on the dielectric function, capable of high speed acquisition, and capable of synchronous acquisition. Accurate models and dielectric function values for constituent materials and optical penetration of buried layers are the major factor limiting accurate determination of physical parameters. Multispectral ellipsometric measurements on $Hg_{1-x}Cd_xTe$ show a sensitivity to x between ± 0.002 and ± 0.001. Simulations indicate an ultimate multi-wavelength sensitivity of better than ± 0.0008. Tellurium stabilized $Cd_{1-x}Zn_xTe$ surfaces are free of any apparent surface changes during heating under UHV conditions whereas tellurium stabilized $Hg_{1-x}Cd_xTe$ surfaces show changes occurring at about 177 °C.

ACKNOWLEDGMENTS

We are grateful to Doug Mahlum, Chyi Sheng, and Larry Taylor for their assistance in this work. This work has been supported by Texas Instruments, Inc., the Air Force Wright Laboratory and the ARPA Microelectronic Technology Office under contract F33615-88-C-5448, and by ARPA/NAVAIR Contract No. N00019-93-C-0151.

REFERENCES

1. P. Drude, Ann. d. Phys. u. Chem. N.F. **36**, 532, 865 (1889).
2. R.M.A. Azzam and N.M. Bashara, *Ellipsometry and Polarized Light, North-Holland, Amsterdam* (1977).
3. S.N. Jasperson and S.E. Schnatterly, *Rev. Sci. Instrum.* **40**, 761 (1969).
4. B. Drevillon, J. Perrin, R. Marbot, A. Violet and J. L. Dalby, *Rev. Sci. Instrum.* **53**, 969 (1982).
5. O. Acher, E. Bigan and B. Drevillon, *Rev. Sci. Instrum.* **60**, 65 (1989).
6. W.M. Duncan and S.A. Henck, *Appl. Surface Sci.* **63**, 9 (1993),
7. S.A. Henck, W.M. Duncan, L.M. Lowenstein, and S. Watts-Butler, *J. Vac. Sci. Technol.* **A11**, 1179 (1993).
8. W.M. Duncan, S.A. Henck, J.W. Kuehne, L.M. Loewenstein and S. Maung, *J. Vac. Sci. Technol.* **B12**, 2779 (1994).
9. Y.Z. Hu, M. Li, K. Conrad, J.W. Andrews, E.A. Irene, M. Denker, M. Ray, and G. McGuire, *J. Vac. Sci. Technol.* **B 10**, 1111 (1992).
10. D.E. Aspnes, *Am. J. Phys.* **50**, 704 (1982).
11. D.A.G. Bruggeman, *Ann. Phys. (Leipzig)* **24**, 636 (1935).
12. E.D. Palik, *Handbook of Optical Constants of Solids*, Academic Press, New York, 1985.
13. D.J. Thomas, P. Southworth, M.C. Flowers, and R. Greef, *J. Vac. Sci. Technol.* **B8**, 1044 (1990).
14. M.A. Hopper, R.A. Clarke, and L. Young, *Surf. Sci.* **56**, 472 (1976).
15. S.A. Henck, *J. Vac. Sci. Technol.* **A10**, 934 (1992).

16. H. Kuroki, H. Shinno, K.G. Nakamura, M. Kitajima, T. Kawabe, *J. Appl. Phys.* **71**, 5278 (1992).

17. G.S. Oehrlein, I. Reimanis, and Y.H. Lee, *Thin Solid Films* **143**, 269 (1986).

18. R. K Sampson and H.Z. Massoud, *J. Electrochem. Soc.* **140**, 2673 (1993).

19. R.T. Carline, C. Pickering, D.J. Robbins, W.Y. Leong, A.D. Pitt, and A.G. Cullis, *Appl. Phys. Lett.* **69**, 1114 (1994).

20. H. Yao, J.A. Woollam, S.A. Alterovitz, *Appl. Phys. Lett.* **62**, 3324 (1993).

21. P. Raynaud, J.P. Booth and C. Pomot, *Physica B* **170**, 497 (1991).

22. M. Stchakovsky, B. Drevillon, and P. Roca I Cabarrocas, *J. Appl. Phys.* **70**, 2132 (1991).

23. Y.Z. Hu, J.W. Andrews, M. Li and E.A. Irene, *Nuclear Instruments and Methods* **B59**, 76 (1991).

24. R. Ossikovski, H. Shiroi, and B. Drevillon, *Appl. Phys. Lett.* **64**, 1815 (1994).

25. Ilsin An, Y.M. Li, C.R. Wronski, H.V. Nguyen and R.W. Collins, *Appl. Phys. Lett.* **59**, 2543 (1991).

26. Ilsin An, Y. M. Li, H.V. Nguyen, and R.W. Collins, *Rev. Sci. Instrum.* **63**, 3842 (1992).

27. M. Fang and B. Drevillon, J. Appl. Phys. **70**, 4894 (1991).

28. Y.J. Van der Meulen and N.C. Hein, *J. Opt. Soc. Am.* **64**, 804 (1974)

29. D.E. Aspnes, W.E. Quinn, M.C. Tamargo, M.A.A. Pudensi, S.A. Schwarz, M.J.S.P. Brasil, R.E. Nahory, and S. Gregory, *Appl. Phys. Lett.* **60**, 1244 (1992).

30. R. Droopad, C.H. Kuo, S. Anand, K.Y. Choi, and G.N. Maracas, *J.Vac. Sci. Technol.* **B12,** 1211 (1994)

31. G.N. Maracas, J.L Edwards, D.S. Gerber, R. Droopad, *Appl. Surf. Sci.* **63**, 1 (1993)

32. G.N. Maracas, C.H. Kuo, S. Anand, and R. Droopad, *J. Appl. Phys.* **77**, 1701 (1995).

33. B. Johs, J.L. Edwards, K.T. Shiralagi, R. Droopad, K.Y. Choi, G.N. Maracas, D. Meyer, G.T. Cooney, and J.A. Woollam, *Mater. Res. Soc. Symp. Proc.* **222**, 75 (1991).

34. F.G. Celii, W.M. Duncan, Y.-C. Kao, *J. Electronic Materials*, Accepted for publication.

35. G. N. Maracas, "Real-Time Analysis and Control of Epitaxial Growth," this volume.

36. F.G. Celii, Y.-C. Kao, T.S. Moise, A. Katz, T.D. Harton, and M. Woolsey, "Real-Time Monitoring and Control of Resonant Tunneling Diode Growth Using Spectroscopic Ellipsometry," this volume.

37. Y. Demay, D. Arnoult, J.P. Gailliard, and P. Medina, *J. Vac. Sci. Tech.* **A5**, 3140 (1987).

38. R.H. Hartley, M.A. Folkard, D. Carr, P.J. Orders, D. Rees, I.K. Varga, V. Kumar, G. Shen, T.A. Steele, H. Buskes, and J. B. Lee, *J. Crystal Growth* **117**, 166 (1992).

39. B. Johs, D. Doerr, S. Pittal, I.B. Bhat and S. Dakshinamurthy, *Thin Solid Films* **233**, 293 (1993).

40. H. Arwin and D.E. Aspnes, *J. Vac. Sci. Technol.* **A2**, 1316 (1984).

Real Time Analysis and Control of Epitaxial Growth

George N. Maracas
Motorola Phoenix Corporate Research Lab
2100 E. Elliot Road EL508
Tempe, AZ 85284

C.H. Kuo
Arizona State University
EE Department/ CSSER
Tempe, AZ 85287-5706

This paper will present aspects of real-time monitoring and control of epitaxial III-V semiconductor heterostructures by ellispometry. Although the emphasis here will be on monitoring and control of molecular beam epitaxy (MBE), the techniques presented here are applicable to chemical vapor deposition (CVD) processes as well. This treatment will emphasize spectroscopic ellipsometry and will briefly address other characterization techniques that are mainly used today and have a strong potential for robust growth control.

INTRODUCTION

Epitaxial growth techniques such as molecular beam epitaxy (MBE) and organometallic chemical vapor deposition (OMCVD) have been used to achieve heterostructure devices having complicated multilayer epitaxial structures. The strength of these epitaxial techniques is that alloy composition, thickness and dopant concentrations can be achieved on thickness scales of monolayers. Application of these techniques to quantum devices has resulted in the realization of modulation doped field effect transistors(MODFET), heterojunction bipolar transistors (HBTs), resonant tunneling devices, ridge lasers and vertical cavity surface emitting lasers (VCSEL).

Growth of such structures relies on extensive calibration of either solid or gaseous source fluxes and substrate temperature. These parameters are controlled by individual temperature or mass flow controllers. With this source calibration (growth rate, doping and alloy composition), the desired thickness, doping and alloy composition profiles are obtained by a "dead reckoning" or open loop control approach. This works reasonably well if the system state is constant (stable sources and substrate temperature) and there is a predictable model (first principles or empirical) linking system state to the state of the final epitaxial structure. Systematic deviations from nominal values and model inaccuracies result in epitaxial structures that have electrical and/or optical properties that deviate from the design values. This can adversely affect device yield.

A second approach to achieving nominal design values is to monitor the state of the epitaxial layers (i.e. the final state of the system) during growth and determine whether the targeted values are being grown. Deviations from the design parameters can then be compensated by feedback to the appropriate source controller for example.

Such closed-loop feedback control can increase the sample-to-sample reproducibility over open loop systems.

Several sensors are presently available for use as growth diagnostic tools. Most measure one parameter of the growing film. For example, absorption edge spectroscopy can measure the substrate temperature by monitoring the semiconductor band edge temperature dependence, reflectance spectroscopy can measure the thickness of multilayer structures such as in a Bragg reflector by optical interference, reflectance mass spectrometry can measure surface alloy composition, reflectance difference/anisotropy can measure the chemical nature of the surface as well as monolayer thicknesses by polarization changes, optical absorption can be used to measure incident flux densities, reflection high energy electron diffraction can be used to monitor surface structure and monolayer thickness.

Optical techniques require knowledge of the materials' index of refraction or pseudodielectric functions. A reference database for the As based III-V semiconductors will be discussed and its application to temperature, thickness and composition monitoring and control demonstrated. These temperature and alloy composition dependent reference functions are now generally available and have been used to grow quantum structures and microcavity optical resonators under ellipsometry control.

Epitaxial layer formation is governed by surface kinetics which are not easily controlled by open loop control systems. Standard, in-situ MBE growth sensors have been mainly limited to RHEED (reflection high energy electron diffraction) and pressure gauges. Spectroscopic ellipsometry (SE) is useful as an in-situ sensor because it provides information on the material's dielectric properties which can be used to determine critical material growth parameters. This technique has the potential to measure all of the aforementioned parameters in real-time and thus provide the capability to control them during growth. The intent here is to present techniques which can be useful to the crystal grower for process calibration, monitoring and device structure control. Other research groups working on in-situ monitoring and control techniques are represented in this book so a comprehensive review of the field will not be presented.

ELLIPSOMETRY BASICS

Ellipsometry is a well-developed material diagnostic technique that measures the polarization state of light reflected from a surface [1]. The change in the polarization state of the reflected light is expressed in the form of a complex reflection coefficient ratio which is defined in terms of the reflection coefficients for light polarized parallel (R_p) and perpendicular (R_s) to the plane of incidence as:

$$\rho = \frac{R_p}{R_s} = \frac{|R_p|}{|R_s|}\, e^{i(\delta_p - \delta_s)} = \tan(\Psi)\, e^{i\Delta}$$

where Ψ and Δ are the "ellipsometric parameters" and $\tan\Psi$ and Δ are the amplitude change and phase difference between the p and s components of the electric field respectively. The measured angles Ψ and Δ, are functions of the wavelength and angle of incidence and are sensitive to the structure (layer thickness, compositions, etc.) and the pseudodielectric functions (often referred to as optical constants) of the sample. The dielectric function is expressed as $\mathcal{E} = \langle\mathcal{E}_1\rangle + i\langle\mathcal{E}_2\rangle$ (or equivalently the index of refraction $\langle n'\rangle = \langle n\rangle + i\langle k\rangle$). Information on layer thickness and alloy compositions of individual layers in a multilayer structure can be extracted from the analysis of Ψ and Δ taken as a function of wavelength (spectroscopic ellipsometry). Fitting for thickness and composition requires an accurate knowledge of the dielectric functions at the sample temperature for all layers. Ellipsometric data can also be taken using discrete wavelengths (as opposed to spectroscopically) to monitor changes in material properties during the growth of a layer. In this case any changes observed in Ψ and Δ will be a function of variations in the optical constants and thickness. Analyzing ellipsometric data requires the parameters of a model (i.e. layer thicknesses and indices of refraction) to be varied until the mean squared error (MSE) between the generated data from the model and the experimental data is minimized.

The SE experimental measurements are expressed as $\Psi(h\nu_i, \Phi_j)$ and $\Delta(h\nu_i, \Phi_j)$ where $h\nu_i$ is the photon energy and Φ_j is the external angle of incidence. Measurements in an MBE system [3-6] are most often performed at a single, fixed angle because the optical port geometry is fixed. Varying the angle of incidence provides additional information on a particular structure enabling a more accurate structure model to be achieved. This technique, called variable angle spectroscopic ellipsometry (VASE), is usually performed outside the growth chamber on a goniometer stage.

The measured ellipsometric parameters, Ψ and Δ, are sensitive to the structure (i.e., layer thicknesses, alloy compositions, surface microstructure, etc.) and the optical constants of the sample. A model for the MBE structure is first constructed with initial estimates of layer thickness and alloy composition. Fitting is achieved by minimizing the mean square error (MSE) between the measured data and the generated ellipsometric fit typically using a Levenberg-Marquardt minimization algorithm. The general form of the mean square error is

$$MSE = \frac{1}{N}\sum_{i=1}^{N}\left\{\left(\Psi_i^{calc} - \Psi_i^{meas}\right)^2 + \left(\Delta_i^{calc} - \Delta_i^{meas}\right)\right\}$$

where the sum is taken over all measured wavelengths. Reference [2] describes the minimization scheme used in the following experiments.

MBE SYSTEM CONSIDERATIONS

The hardware necessary to successfully implement an in-situ ellipsometer onto an MBE system is schematically depicted in figure 1. Design criteria for III-V semiconductor MBE systems are discussed in reference [3]. Some other considerations are necessary for the II-VI materials [9-13]. The growth chamber must have optical ports whose axes coincide with the center of the wafer at an angle close to the Brewster angle for the particular material being analyzed. In the case of most III-V semiconductors this angle is approximately 75°. The optical ports must have strain-free glass so no birefringence is introduced into the optical path of the ellipsometer. It is useful to have either shutters or externally pumpable gate valves to protect the window from coating and to enable their cleaning.

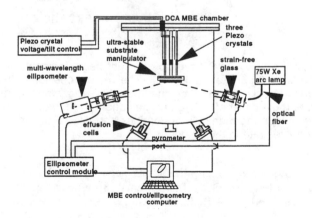

Figure 1. Implementation of a multi-wavelength ellipsometer onto an MBE system.

477

Two major requirements for implementing SE control on a molecular beam epitaxy (MBE) system include a) the ability to record and analyze spectroscopic data on time scales faster than the growth is occurring (~1 monolayers (ML)/sec) and b) having a substrate manipulator that maintains the angle of incidence (typically 75°) under rotation to less than 0.1 degrees [7,8]. The latter is required because it obviates the need to use angle of incidence as a data fitting parameter, thus reducing analysis time and fitting errors.

Stable Substrate Manipulator

In an MBE system, the substrate is rotated at approximately one revolution per second. This is essential to obtain films with high thickness and alloy composition uniformity across the substrate. In polarization sensitive measurements such as SE, small deviations in the sample position and tilt affect the confidence of the extracted layer thickness and composition. To obtain a low level of noise in the ellipsometric data from a rotating epitaxial layer, the plane of incidence (which is normal to the substrate surface) and angle of incidence must be kept constant within 0.1 degree. To achieve this, the stability of the substrate position and of the substrate normal with respect to the incoming light beam must be maintained. This is shown schematically in figure 2.

plane of incidence

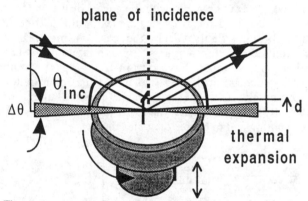

Figure 2.   Schematic of a light beam incident onto an MBE substrate manipulator. The vertical position of the substrate is affected by the thermal expansion of the manipulator shaft. Angular deviations are caused by wobble in the manipulator bearings and tilted mounting of the substrate in the holder.

A commercial manipulator has been developed and is available from DCA Instruments [8] whose "wobble" under rotation is less than 0.1° and whose absolute sample position is unaffected by thermal expansion.

MULTIPLE WAVELENGTH ELLIPSOMETER

MBE process monitoring and control by SE require ellipsometric spectra to be recorded and analyzed approximately two times per second (for growth rates of

approximately 1ML/sec). This has been achieved using a "multi wavelength ellipsometer" designed for in-situ, real time monitoring applications in conjunction with the J. Woollam Co. The rotating analyzer instrument is capable of recording 44 wavelengths in the spectral range of 415 nm < λ < 750 nm at a maximum rate of 25 spectra per second. SE data fitting is performed in real-time and parameters transferred to the MBE control computer via serial link for process feedback. For the following thickness control experiments, the SE computer was used to directly control the MBE shutters via an RS232 line.

PSEUDODIELECTRIC FUNCTIONS

The variables in an ellipsometric model are the thicknesses and indices of refraction (pseudodielectric function) of the layers being analyzed. If the pseudodielectric functions are known a-priori, then the individual material parameters (such as alloy composition and temperature) of a multilayer structure can be determined. Room temperature dielectric functions for Si, Ge, GaP, GaAs, GaSb, InP, InAs, InSb and SiGe can be found in the literature [14-18]. The dielectric functions are temperature dependent so knowledge of these is necessary for real-time growth monitoring applications. These functions have been measured for GaAs from 30°C to 650°C which spans the MBE material growth temperature range [19]. Figure [3] shows the imaginary part of the dielectric function ϵ_2 for GaAs versus temperature [19].

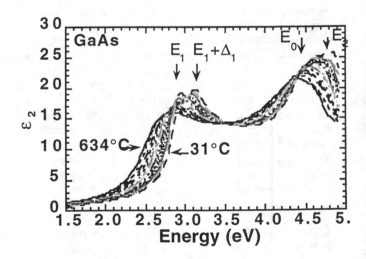

Figure 3. a) Imaginary part (ϵ_2) of the GaAs pseudodielectric function, at temperatures between 31°C and 634°C.

The dielectric function of a crystal can be expressed as a superposition of N Lorentz oscillators. To model GaAs interband transitions, the dielectric function can be represented as [20]

$$\varepsilon = 1 + \sum_1^N A_k \frac{1}{\left(E_k^2 - E^2\right) - iE\Gamma_k}$$

where A_k is the amplitude, E_k is the energy, and Γ_k is the damping coefficient for each oscillator. A seven oscillator model [21] (N=7) was used to fit the GaAs data. The complex refractive index <n'> is then obtained directly from the complex dielectric function relation <n'>^{2} = ε. In figure [3], the two peaks at 2.92eV and 3.14 eV correspond to the E_1 and $E_1+\Delta_1$ interband transitions at room temperature. The peaks at higher energies correspond to the E_0' and E_2 interband transitions. The expected decrease of the interband transitions energies with increasing temperature is evident. The use of seven oscillators is sufficient to obtain good agreement with experiment. Although this model is not an exact representation of the dielectric function, it is sufficient for determining the interband transition energies E_1 and $E_1+\Delta_1$ and their temperature behavior. The temperature dependence of interband transition energies E_1 and $E_1+\Delta_1$ are seen to follow Varshni's relation [22]

$$E(T) = E(0) - \frac{\alpha T^2}{T + \beta}$$

where α and β are empirical constants. Reference [19] compares these and tabulates the calculated critical point energies.

Similar measurements have been made for AlAs and AlGaAs [23] versus composition. Figure [4] shows the pseudodielectric functions of AlAs at selected temperatures between 28°C and 626°C.

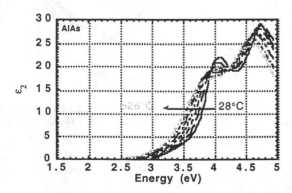

Figure 4.   Imaginary part (ε_2) of the AlAs pseudodielectric function at different temperatures from 28°C to 626°C. For clarity, data for all temperatures measured is not shown.

Figure [5] shows the imaginary part of the dielectric function versus composition for AlGaAs at a typical growth temperature of 630°C. The temperature range for this database is the same as for AlAs.

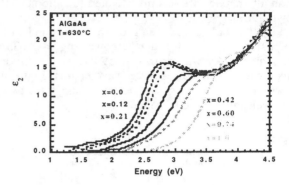

Figure 5.   a) Imaginary part (ε_2) of Al$_x$Ga$_{1-x}$As pseudodielectric function versus composition at a temperature of 630°C.

Because the epitaxial layers were both grown and characterized in an MBE chamber, uncertainties in the analysis from an unknown composition surface layer are not present (i.e. a nearly "ideal" surface allows the use of a two-phase model (vacuum/GaAs) to analyze the data). Applications of this database are in epitaxial layer process monitoring and control and also in modeling of devices operating at elevated temperatures.

MBE PROCESS CALIBRATION USING SE

As mentioned previously, *a priori* knowledge of the dielectric functions versus wavelength, temperature and alloy composition enables the measurement of these critical material parameters using spectroscopic ellipsometry. If the growth of an epitaxial layer is tracked in real-time, then thickness, growth rate and V/III ratio are readily measured. This section demonstrates how SE can be used to quickly calibrate the MBE process in a single layer growth at growth temperature and without necessitating subsequent measurements (e.g. photoluminescence) outside the MBE system.

Growth Rate and Alloy Composition Calibration

Presently the most common in-situ technique available to the MBE crystal grower for growth rate [24] and flux ratio measurements is reflection high energy electron diffraction (RHEED). Substrate rotation during RHEED is difficult and thus subject to inaccuracies in the presence of flux non-uniformities when the substrate is not rotated. Also, because RHEED intensity oscillations typically decay after approximately 20 monolayers, the growth rate measured is overestimated for thick layers because of effusion cell flux transients (cooling after shutter opening). For high growth temperatures, RHEED oscillation measurements are facilitated by using exactly oriented substrates to ensure layer-by-layer growth [25]. As an alternative, SE was investigated to calibrate MBE growth [26].

To calibrate the growth rates and alloy composition by SE, a structure consisting of AlGaAs, GaAs and AlAs was grown at 590°C and monitored dynamically using SE at three wavelengths. Figure [6] shows the ellipsometric parameters Ψ and Δ for two of the wavelengths (4650 and 5500Å).

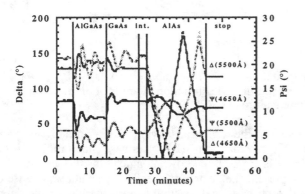

Figure 6.   Dynamic scan of the ellipsometric parameters Ψ and Δ for 4650Å and 5500Å during the growth of AlGaAs, GaAs and AlAs at 590°C. The growth sequence is described in the text. The dotted line is the resulting fit used to extract the growth rates and Al mole fraction.

Data was acquired at a rate of 4 secs for each pair of data, i.e. a total of 12 secs for one set of data at three wavelengths. The test structure consisted of GaAs which began at 2 mins, after which the Al shutter was opened at t = 5 mins. The AlGaAs was then grown for a period of 10 mins before the Al shutter was closed (t = 15 mins). GaAs continued to grow for 10 mins after which the Ga shutter was closed (t = 25 mins) and growth interrupted under an As flux for a period of 2 mins. The Al shutter was then opened at t = 27 mins and AlAs grown for 18 mins. For the first 2 mins and the final 5 mins represented in figure [6] there was no growth (growth interruptions). Dynamic fitting for the growth rates, carried out simultaneously on the data for the 3 wavelengths, are shown as the dotted curves in the figure. The fitting is first started by using the baseline data for GaAs growth during the period 2-5 mins to obtain the optical constants for the "substrate". Here "substrate" refers to the starting material, not necessarily a clean GaAs substrate but, as in this case, a substrate on which other layers were grown. The optical constants for this "substrate" take into account the past history of all the layers grown. These optical constants are then fixed and subsequent layers fitted for the

optical constants and growth rates. Since for each material the optical constants will not vary with time (assuming the substrate temperature remains constant), a growth rate can be extracted for each layer.

The growth rates (R_g) provide all the information required

to derive the composition (x) of the AlGaAs layer from
$x = R_g(\text{AlAs})/R_g(\text{GaAs})$
$R_g(\text{AlGaAs}) = R_g(\text{AlAs}) + R_g(\text{GaAs})$

The second equation shows that the AlGaAs growth rate can be extracted simply from the AlAs and GaAs growth rates and measurement of the AlGaAs can be used as a check. The individual growth rates were measured by SE to be 3.057 Å/sec, 2.239Å/sec and 0.818 Å/sec for AlGaAs, GaAs and AlAs respectively. This gives an Al mole fraction of x=0.268. Note that the calculated growth rate is $R_g(\text{AlGaAs})$ = 3.057Å which agrees with that measured on the AlGaAs layer.

The oscillations in figure [6] are due to optical interference during the change in thickness of the epitaxial layer which has a different index of refraction than the "substrate". Second, the thickness of the layer represented by the period of the oscillations is a function of the optical constants of the growing layer, wavelength and the angle of incidence. Finally, as growth proceeds the amplitude of the oscillation decays (more rapidly for the shorter wavelengths) and disappears altogether for optically thick layers. A graphical presentation that is easier to see is the plot of Δ vs Ψ for the entire growth run in figure [7].

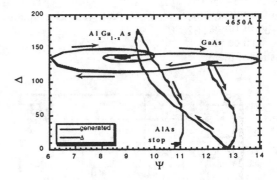

Figure 7.   Experimental and simulated plots of Δ vs Ψ at l= 4650Å showing the trajectory of the ellipsometric parameters during the growth of the AlGaAs, GaAs and AlAs structure at 590°C. The trajectories spiral toward the optically thick values of Δ and Ψ. In the AlAs case, the trajectory is plotted such that the value of Δ is always positive.

There are three spirals which center on three points (Δ,Ψ) on the graph. These points correspond to the optically thick values of the index of refraction toward which the thin epitaxial layers evolve. The arrows denote the trajectory direction with increasing time during the run. Values of Δ <0 are not allowed in the analysis and so are "reflected" back into the Δ>0 plane.

Implicit in this discussion is that SE can be used to monitor the thickness and alloy composition throughout the entire layer growth in contrast to RHEED oscillations which decay after a few seconds and thus and provide growth rate information only for the first few epitaxial layers. Because RHEED only measures the growth rate after a shutter is opened, any long time constant effusion cell cooling effects are not observed until the layer thickness is measured after the growth. These flux transients tend to overestimate the average growth rate in thick layers. One disadvantage with SE is that growth rates can only be established during heteroepitaxial growth and thus a sacrificial calibration wafer is needed. Additional advantages of SE are that it can easily be performed under substrate rotation and under any growth mode (i.e. two or three dimensional growth). An interesting observation in [26] was that the measured growth rate under substrate rotation was 7% higher than without substrate rotation showing how flux nonuniformities can give erroneous results. Also, after growth is completed, the thicknesses and alloy compositions of the various layers can be measured without removing the layer from the system.

V/III Ratio Calibration

An additional application of the growth rate measurement is to determine the V/III flux ratio of a growing film. This procedure [27] is analogous to group V induced oscillations measurement of V/III ratio by RHEED. A few monolayers of group III atoms is deposited onto a surface without a group V overpressure. The ratio of times taken to convert the group III surface to a III-V surface after the group V shutter is opened provides a measure of the V/III ratio.

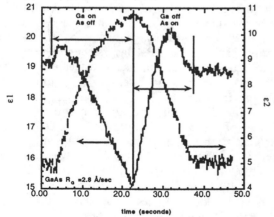

Figure 8.   Real and imaginary parts of the dielectric function (ε_1 and ε_2) are shown versus time. A Ga layer (20ML) is fist deposited and the time taken to be consumed by an As flux is measured. The time to deposit

Ga divided by the time required to consume it gives the V/III ratio.

Figure [8] shows a trace of ε_1 and ε_2 (real and imaginary parts of the dielectric function respectively) versus time. At t=0 ε_1 and ε_2 have the values of the bulk GaAs optical constants. The GaAs growth rate was measured by SE to be 2.83Å/sec. At t=3 seconds the As shutter was closed for 20 seconds which resulted in twenty monolayers (ML) of Ga being grown (1ML = 2.83Å). At t=22 seconds in the figure, the Ga shutter was closed and As shutter opened which initiated the incorporation of As into the Ga film thus growing GaAs. After 15 seconds under an As4 flux, ε_1 and ε_2 recovered to the GaAs bulk values indicating that all the surface Ga (20ML) was consumed by 20ML of As. Since 20ML of As was incorporated in 15 seconds, the V/III ratio is 20/15 = 1.33. It should be noted that the optical constants of the Ga layer need not be known for this experiment because only the times required for deposition and incorporation were required to calculate the V/III ratio. Knowledge of the V/III ratio is important in growing high quality ternary materials especially ones containing aluminum.

GROWTH PROCESS CONTROL

The ability to measure relevant growth parameters on a time scale shorter than that at which the growth occurs leads to the possibility of obtaining closed-loop feedback control. The general scheme is to fit the ellipsometric data for the desired control variable during the growth and feed that variable to a controller. Temperature, thickness and composition have been successfully controlled by SE. Aspnes [28, 29] has successfully controlled alloy composition in a metalorganic MBE (MOMBE) quantum well by growing a nonrectangular profile. Temperature control was achieved using only SE as the control sensor [30]. Thickness control has led to the growth of quantum wells and Fabry Perot cavities [31, 32] consisting of distributed Bragg reflector (DBR) structures.

Temperature Control

Temperature measurement by SE is achieved by using the temperature dependent pseudodielectric function database discussed previously. The ellipsometry spectrum is fitted for temperature and that value used to feed back to the temperature controller.

The growth temperature in many MBE systems is not accurately measured by the system thermocouple because it is not in physical contact with the substrate. Errors on the order of 100-200°C are common [30], necessitating the use of optical pyrometry or absorption edge spectroscopy. SE has been used to measure and control substrate temperature using the temperature dependent optical constant database. To demonstrate this, a thermocouple was embedded in the surface of a SI GaAs substrate while in the MBE system. Temperatures obtained by the surface thermocouple, system thermocouple, optical pyrometer and ellipsometer were measured simultaneously during a thermal ramping schedule controlled by a Eurotherm 820 controller. Figure [9] shows temperature control of a GaAs substrate by SE.

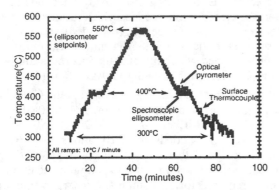

Figure 9.   Plot of substrate temperature versus time during a programmed temperature cycle. The substrate temperature was measured with a surface thermocouple, optical pyrometer, and spectroscopic ellipsometer. The control temperature sensor was the SE which was used for feedback to the substrate temperature controller.

The system thermocouple temperature was consistently ~100°C higher than the others and is not shown in the figure for clarity. The controlling temperature sensor was the SE which measured temperature at three times per second. Its temperature was converted to an analog signal by a digital to analog converter and fed directly into the Eurotherm for control. Setpoints of 300°C, 400°C and 550°C were used with a ramp rate of 10°C per minute. The figure shows good agreement among the three sensors. The optical pyrometer is limited to temperatures above 400°C. Oscillations in the temperature at 400°C and below are a result of the Eurotherm temperature controller's PID settings being optimized only for high temperatures. It is observed that the SE temperature is lower than the others. This is because of radiative heat loss from the surface and also a cooling effect from the desorbing As. The data has a temperature error of approximately 2°C.

Quantum Well Control

Thickness control in chemical beam epitaxy was first achieved by Aspnes [28] using the virtual substrate approximation. This method constrains the underlying epitaxial layer parameters into a virtual "substrate" and fits for properties of only a thin changing layer at the surface. A measure of the growth rate is directly obtained in this way. Tracking of thickness and composition (and other parameters) by SE is performed by monitoring spectra and fitting the data to a structure model consisting of target layer thicknesses and compositions approximately two times per second [32]. The shutter closing time is calculated based upon the target thickness, present thickness and present growth rate. This extrapolation technique reduces the frequency at which the structure model needs to be updated.

Control of quantum well thickness was attempted on a multiple quantum well structure consisting of five 100Å GaAs wells and five 75Å wells separated by 500Å of AlGaAs. The structure was grown on semi-insulating GaAs, all wells had 200Å $Al_{0.3}Ga_{0.7}As$ barriers and the entire structure was capped with 50Å of GaAs. One sample was grown with closed-loop SE feedback control of the shutters while a second sample was grown using growth rates measured by SE and timed shutter sequences (no feedback control). 77K photoluminescence (PL) of the two test structures showed clear excitonic peaks corresponding to the n=1 electron to heavy hole transitions in the GaAs wells. A barrier composition of x=0.3 that was used and also independently measured by PL. The "timed" (dead-reckoning approach based on an initial growth rate) quantum well thicknesses were calculated to be 72Å and 91Å while the "SE control" thicknesses were 75Å and 94Å for the nominally 75Å and 100Å wells respectively. This demonstrates that SE closed-loop feedback can control multiple quantum well thickness to an accuracy of a few percent.

Fabry Perot Cavity Control

Another example of thickness control by SE is the growth of a Fabry-Perot (FP) cavity used for vertical cavity surface emitting lasers (VCSELs) and electro-optic modulators [32]. Such structures consist of a cavity having an optical thickness corresponding to a particular optical mode placed in between two dielectric mirrors. This particular structure was used to test SE control of thick epitaxial layers because the position of the FP mode is a very sensitive measure of the cavity thickness. Distributed Bragg reflectors (DBRs) consist of alternating λ/4 high and low index of refraction layers producing mirrors that have a high reflectance wavelength band. High reflectance indicates high thickness and composition uniformity among the DBR layers. Such structures are used for fabricating vertical cavity surface emitting lasers (VCSELs) and resonant cavity FP electro-optic modulators for example.

The FP cavity design for this experiment consisted of ten period bottom and top DBRs having AlAs/GaAs thicknesses of 829.9Å and 695.8Å respectively. The GaAs 1λ FP cavity was designed to have a mode at 975nm so the thickness of the GaAs was nominally 2783.3Å. The entire structure was grown on an n^+ GaAs substrate. Figure [10] shows the result of growing the FP cavity under SE control.

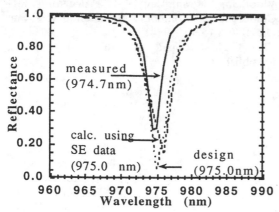

Figure 10. Normal incidence reflectance of a 975.0nm (design) Fabry Perot cavity consisting of 10 period AlAs/GaAs DBRs and 1λ cavity. The wavelength calculated from SE measured thicknesses and compositions was 975.0nm. Superimposed on the calculation is the measured curve showing a FP mode at 974.7nm.

In this experiment all the optical constants for AlAs and GaAs at 600°C used were from the pseudodielectric function database previously mentioned. The angle of incidence was measured before the run and the substrate was rotated for layer uniformity. In figure [10], the calculated normal incidence reflectance for the 975 nm design is shown along with that calculated using the layer thicknesses and compositions measured by SE during the growth run. The two agree. The FP mode measured by normal incidence reflectance was observed at 974.7nm which is 0.3 nm from the design value. This corresponds to a thickness error of 0.3% or 8.6Å in the 2783Å cavity from the design value.

Recent work [33] using closed-loop feedback control of the structure just described showed that a high degree of reproducibility is possible using this method. Five identical Fabry Perot cavities were grown over the period of several weeks under SE control. Four FP cavities were grown using the same effusion cell temperatures, one had the effusion cell temperatures randomly changed by 10°C during the run and a final one was grown with high doping. The FP modes of all five agreed to within 3nm producing an error among runs of approximately 1%.

FUTURE DIRECTIONS AND NEEDS

It is clear that spectroscopic ellipsometry is a potentially powerful, noninvasive sensor for monitoring and controlling epitaxial growth. It is suggested that some areas still need to be more fully developed.

On the hardware side, it would be nice to have an ellipsometer capable of analyzing approximately 100 wavelengths several times per second. The wavelength range should span more than one critical point energy and preferably more than two. This is necessary in order to obtain sufficient data to extract temperature, composition and thickness in real time on complicated multilayer structures with a high degree of confidence. An obvious need is more computing power to reduce the large quantity of data in real time.

In order for new, intelligent sensors to be incorporated into a working laboratory, the control software for the process needs to have the capability of taking inputs from these sensors. Software "hooks" should be written into the control code by equipment manufacturers to enable such integration. This will require some standardization of sensor/controller interface software.

Finally, basic material science work in developing high accuracy databases of dielectric functions versus wavelength, temperature, composition, strain, etc. and effects due to thin layers and surface effects is critical.

ACKNOWLEDGMENTS

Thanks go to D. Aspnes, W. Duncan, F. Celii, T. Zettler and C. Pickering for useful discussions. This work was supported by ARPA/ULTRA under contract No. N00014-92-J-1931. Pseudodielectric functions (on floppy disk) are available upon request from the authors.

REFERENCES

[1] Azzam, R.M.A. and Bashara, N.M. *Ellipsometry and Polarized Light*, North Holland Publishing Co. 1977

[2] Herzinger, C.M., Snyder, P.G., Johs, B. and Woollam, J.A. submitted to J. Appl. Phys. June 1994.

[3] Maracas, G.N., Edwards, J.L., Shiralagi, K. Choi, K.Y., Droopad, R., Johs, B. and Woollam, J.A., J. Vac. Sci. Tech. A 10(4) (1992)

[4] Aspnes, D.E., Thin Solid Films, 89 (1982) 249

[5] Aspnes, D.E., Quinn, W. E., Gregory, S., Appl. Phys. Lett. 56 (1990) 2569

[6] Johs, B., Edwards, J.L., Shiralagi, K.T., Droopad, R., Choi, K.Y., Maracas, G.N., Meyer, D., Cooney, G. and Woollam, J.A. Materials Research Society, Spring 1991 Meeting, Anaheim, CA

[7] Maracas, G.N., Edwards, J. L., Shiralagi, K., Choi, K. Y., Droopad, R., Johs, B. and Woollam, J. A., J. Vac. Sci. Tech. A 10(4), 1832-1839 (1992)

[8] Kuo, C.H., Anand, S., Droopad, R., Mathine, D.L., Maracas, G.N., Johs, B., He, P., Woollam, J.A., and Levola, T., Int'l Conference on Compound Semiconductors, San Diego, CA Sept (1994)

[9] Duncan, W.M., Westphal, G. H., Bevin, M.J. and Shih, H-D, Proceedings of IRIS Specialty Groups on Infrared Materials and Infrared Detectors Boulder, CO August 1994

[10] Hartley, R.H., Folkard, M.A., Carr, D., Orders, P.J., Rees, D., Varga, I.K., Kumar, V., Shen, G., Steele, T.A., Buskes, H. and Lee, J.B., J. Cryst. Growth 1(17), p. 166 (1992)

[11] Hartley, R.H., et. al. B(10), 1410 (1992)

[12] Johs, B., Doerr, D., Pittal, S., Bhat, I.B., Dakshinamurthy, S., Thin Solid Films, 2 (33), 293 (1993)

[13] Demay, Y., Arnult, D., Gaillard, J. P., Medina, P., JVST A(5), 3140 (1987)

[14] Palik, E.D., editor Handbook of Optical Constants of Solids I & II, Academic Press 1991

[15] Aspnes, D.E., and Studna, A.A., Phys. Rev. B, 27, 985 (1983)

[16] Pickering, C. et al., Appl. Phys. Lett., 60, 2412 (1992)

[17] Huumlicek, J. et al., J. Appl. Phys. 65, 2827 (1989)

[18] Jellison, G.E., Haynes, T.E. and Burke, H.H., Optical Materials, 2, 105 (1993)

[19] Maracas, G.N., Kuo, C.H., Anand, S. and Droopad, R., J. Appl. Phys. 77 (44)1995

[20] Ziman, J. M., in Principles of the Theory of Solids (Cambridge, U. K., 1972).

[21] Yao, H., Snyder, P. L., and Woollam, J. A., J. Appl. Phys. 70, 3261 (1991).

[22] Lautenschlager, P., Garriga, M., Logothetidis, S. and Cardona, M., Phys. Rev. B 35, 9174 (1987).

[23] Kuo, C.H., Anand, S., Fathollahnejad, H., Ramamurti, R., Droopad, R. and Maracas, G.N., Proceedings NA MBE Conference, Urbana-Champagne (1994), JVSTB

[24] Neave, J. H., Joyce, B. A., Dobson, P. J., Norton, N., Appl. Phys. A31 (1983) 1.

[25] Joyce, B. A., Dobson, P. J., Neave, J. H., Zhang, J., Surf. Sci. 174 (1986) 1.

[26] Droopad, R., Kuo, C. H., Nithianandan, S., Choi, K. Y., and Maracas, G. N., JVST B, 12(2), 1211-1213 (1994)

[27] Maracas, G. N., Kuo, C.H., Anand, S. and Droopad, R. "Ellipsometry for III-V Epitaxial Growth Diagnostics,"J. Vac Sci. Technol. A 13(3), May/Jun 1995

[28] Aspnes, D.E., Quinn, W. E., Gregory, S., Appl. Phys. Lett. 56 (1990) 2569

[29] Aspnes, D.E., Quinn, W. E., Gregory, S., Appl. Phys. Lett. 57 (1990) 2707

[30] Maracas, G.N., Edwards, J. L., Shiralagi, K., Choi, K. Y., Droopad, R., Johs, B., and Woollam, J. A., J. Vac. Sci. Technol. A 10, 1832 (1992).

[31] Maracas, G.N., Edwards, J.L., Gerber, D.S., and Droopad, R., Appl. Surf. Sci. 63, 1 (1993)

[32] Kuo, C.H., Anand, S., Droopad, R., Mathine, D.L., Maracas, G.N., Johs, B., He, P., Woollam, J.A., and Levola, T., Proceedings Int'l Conference on Compound Semiconductors, San Diego, CA (Sept. 1994)

[33] Kuo, C.H., Droopad, R. and Maracas, G.N., unpublished

Cost Effectiveness in Real-Time, In-Situ Analysis for Deposition Equipment

Norman E. Schumaker

EMCORE Corporation
35 Elizabeth Avenue
Somerset, NJ 08873

In-situ analytical tools operating in real-time have potential for improving the deposition technology for compound semiconductors. Careful consideration of the system requirements leads to a series of constraints that should be met for such techniques to be cost effective. For optical techniques, as the tools of choice, serviceability requirements and the size of the analyzed area demand that the deposition system be well characterized. The throughput demand for production systems requires that analysis time be matched to growth time constraints for cost effective use. Application of in-situ techniques may find appropriate cost effective use in set up, recalibration and system validation in addition to real time growth control.

INTRODUCTION

In-situ analysis techniques for use in deposition equipment are being developed to provide real- time information about thickness, composition, doping and surface morphology. These tools have been generally developed under laboratory conditions and often are focused on high vacuum systems. The applications of these techniques and tools to production deposition equipment require careful consideration of many factors. Mechanical, electrical, geometrical, as well as, software constraints must be analyzed to develop a tool that will meet the needs of production oriented equipment. Ease of use, time constraints, cost of facilities, productivity and serviceability all contribute to the overall cost of ownership. Since these techniques have particular interest and potential value for the compound semiconductor materials, it is worthwhile to focus on a few of the generic problems that must be addressed as these devices evolve. However, in the production arena, the costs associated with the application of these techniques will be the major consideration in their ultimate utility.

AREAS of APPLICATION

Compound semiconductor materials have been applied to a variety of devices for an increasing number of products. For the purposes of these discussion, the device structures that represent the general classes of devices of interest will be: HEMT and/or MESFET, Laser and LED. These three devices span the gambit of thin high precision requirements of the HEMT to the thicker and perhaps less stringent requirements of the LED. Since these types of products are intended for volume production, they represent potential candidates for the new analytical techniques. In the production environment the productivity of the deposition equipment is directly related to the growth rate of the material required for the device. For high precision, low growth rates are preferred. In a device like a HEMT, the high precision is required throughout the structure. In the case of the LED, high precision is dictated only in the vicinity of the abrupt heterojunctions. Consequently, a high growth rate can be used in the majority of the device. However, as we will observe later this places severe constraints on the analytical techniques to be used. Table 1 summarizes the basic structural requirements for these devices.

Table 1 - Device Structural Requirements

Device Type	Thickness (nm)	Number of Layers (No.)	Interface Abruptness (al)*	Wafer Size (mm)
HEMT	500	3	1	150
Laser	3,000	5	1-3	50
LED	10,000	4	1-3	50 - 150

* (al = monolayer)

SYSTEM CONSTRAINTS

Since most deposition systems operate at substantial pressures relative to MBE techniques, optical analytical tools represent the most promising adjunct to production deposition systems. These techniques have;
- little impact on the deposition process,
- are pressure independent
- measurements can be made external to the system geometry.

Examples of such techniques are:
- light scattering for morphology control,
- ellipsometry for thickness measurements
- absorption or reflection spectroscopy for determination of composition and doping
- standard pyrometery can be used for temperature measurements.

All of these systems require careful consideration of the mechanical design relative to the deposition system. Figure 1 shows the geometrical limitations relative to changes in the wafer to deposition head distance. In this Figure, the "chamber separation" represents the distance from the Chamber Top to the wafer surface. As the separation between these two surfaces decreases, the angle "9" decreases so that some analytical techniques can no longer be used. In addition the use of optical techniques presupposes careful alignment and calibration. These requirements will impose additional service issues such as testing and calibration and software compatibility. Concurrently, response time issues and certain size limitations may become critical in the actual utility of a technique. Effective analytical techniques will clearly demand that system configurations and service requirements not impose additional time or service constraints in the manufacturing environment..

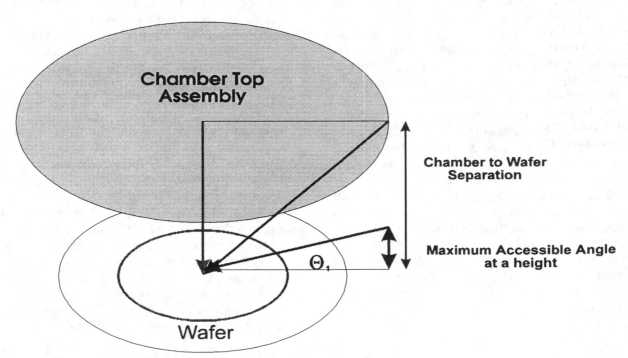

Figure 1 - Geometric Considerations in Chamber Design

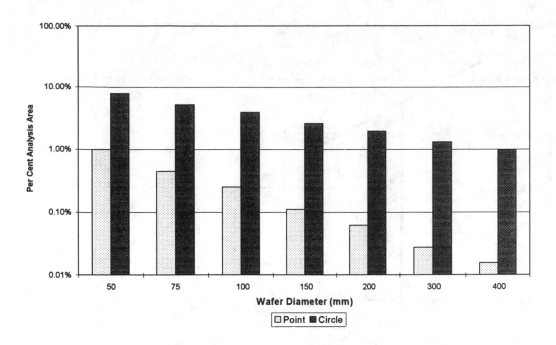

Figure 2 - Analysis Area on Wafer Relative to Wafer Size

ANALYSIS IMPACT

Again referring to Figure 1, the area of the wafer effectively analyzed by an optical probe is dependent upon the geometrical limitations associated with the technique. For stationary wafers or for rotating wafers with the analysis point at the center of rotation, the relative area analyzed on a compound semiconductor wafer becomes vanishingly small as the wafer diameter increases. Figure 2 shows this quite clearly. Figure 2 also shows that for rotating wafers with the analysis region off-center and positioned at one half the wafer radius, a clear advantage exists relative to the single point measurement on a stationary wafer. However, both of these analysis scenarios show a substantial limitation associated with point wise analytical tools; i.e., the probed area is small. Thus any in-situ analytical technique will require that the deposition system be well characterized, exhibit reproducible behavior and possess excellent uniformity across the wafer. Without these requirements, the use of the in-situ analysis will be an inadequate guide for the materials engineers responsible for production line. However, this represents a conundrum. A point wise analysis requires that the system be well characterized and uniform which are the characteristics of a system that does NOT require monitoring.

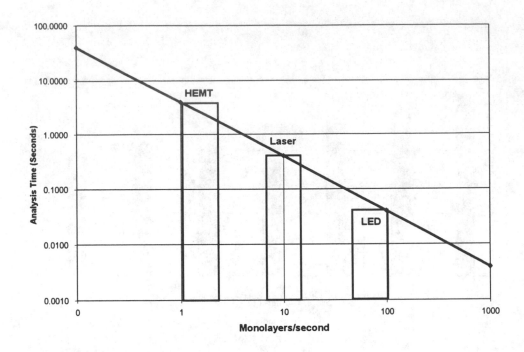

Figure 3 - Analysis Time Relative to Growth Rate

CONTROL CONSIDERATIONS

The device structures characterized in Table 1 have individual materials requirements as were discussed. The cycle time is of paramount concern in the operation of a deposition system. The total cycle time must incorporate all aspects of a deposition step, so the production manager must consider the time necessary to implement an in-situ analysis technique and its impact on the cycle time of his process line. Figure 3 shows the relative time available for the real-time, in-situ process control for each of the generic device structures. In Figure 3 we have used the growth of a monolayer of material as a measure of the growth rate; i.e., the growth rate is monlayers per second. For each of the generic devices the growth rate varies from a slow 1 atomic layer per second for the HEMT device to almost 100 atomic layers per second for LED's.

The time allowed for analysis under these three scenarios must accommodate:

- data collection
- interpretation
- analysis, and subsequent
- process modifications.

Clearly, the time available during the growth of the HEMT type devices is sufficient to allow this sequence of events. It should be clear that the growth of LED structures at the substantially higher growth rate will impose severe time limitation for the same analytical techniques. Under these conditions, perhaps the growth in the vicinity of the heterojunctions would be sufficient to ensure high quality LED devices. The laser structure, with more layers and interfaces will be more complex challenge for in-situ monitors. Fortunately, lasers are rather small devices and do not typically require the volume production of the LED market.

From these simple considerations, we can reach the conclusion that analytical techniques will find application in those deposition processes where the time constraints are compatible with the time constraints of the growth conditions for a device structure. For the more rapid growth conditions, analysis may be used more for calibration purposes.

ECONOMIC CONSIDERATIONS

Of critical concern to the production manager is the potential economic impact associated with the use of any new analytical control system. These issues are highlighted in the Cost of Ownership models which are increasingly used in the semiconductor industry. CoO models attempt to take all the critical cost factors associated with the operation of any piece of equipment into consideration. For the purposes of this discussion, we will highlight three(3) principal factors in the cost of ownership models.

These factors are:
- equipment,
- process and
- facility.

The equipment factors of most interest are
- price,
- reliability/uptime and
- serviceability.

The process factors are:
- impact on the process,
- ease of use,
- quality of information generated, and ultimately
- yield improvement.

Facility factors are:
- size and
- any additional service constraints or
- spatial constraints on the equipment.

In all CoO models time plays a critical role. All system operations are evaluated in a self-consistent manner to achieve a cost per wafer processed. Clearly, the throughput of wafers per unit time is the controlling factor. In addition, the total cost of all aspects of operation from chemicals and gases, power and facility charges, to manpower and purchase price are included. Annual service charges and spare parts and calibration costs are all calculated for a comprehensive cost. In all cases, the availability of a system to perform useful work is critical. Any event, maintenance, repair, accidental downtimes, etc. which decrease the available time for use is to be avoided.

The substantial impact of time in the application of the in-situ analytical techniques is shown in Figure 4. In this Figure, we have made the assumption of a 10% increase in the base price of the deposition equipment by the addition of an analytical tool. We have assumed a 10% increase in the time associated with general maintenance of the complete system This graph shows the required yield improvement that must be achieved for a reference growth rate of 8 monolayers per second. In other words, the reference point is a 3 micron layer grown at an assumed growth rate of 8 monolayers per second.

If there is no change in the growth rate, a 7% process yield improvement will compensate for the 10% increase in price and 10% increase in maintenance. For the situation where the original process requires a growth rate of 12 monolayers per second, the associated yield improvement must be 20%. In a process where the growth rate must be changed from 25 monolayers per second to 8 monolayers per second, the in-situ analytical technique would have to contribute a 41% yield improvement.

Analysis time and its interplay with the total process yield is critically important. The adaptation of an analytical tool to the deposition process must pay close attention to any impact on the total time associated with the deposition process. If the analytical tool cannot accommodate the growth rate of the existing process, other considerations for the economic justification must be examined.

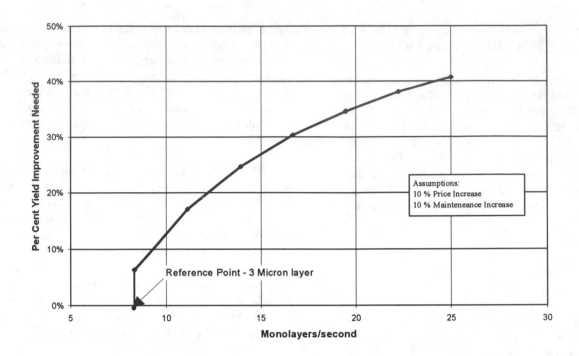

Figure 4 - Yield Improvement Required to Justify Use of In-situ Techniques

CONCLUSIONS

The promise of real time in-situ process characterization must confront the realities of the production environment. The in-situ analysis tool can represent a substantial equipment benefit to the user if such equipment can be demonstrated to be cost effective. To be cost effective requires that the price of the add-on system be below that projected by a cost of ownership model. Cost of ownership models take into account service issues, up-time, and impact on throughput and improved yields. Such considerations will necessarily be customized for each process situation. The limitations associated with a thin HEMT structure grown at a slow rate are substantially different than those associated with a relatively thick LED structure grown at a high growth rate. It is also clear that single point in-situ analytical techniques will demand greater sophistication in the development of the deposition equipment. Increased uniformity of deposition and consistent system behavior will be a prerequisite for effective utilization of these new techniques.

In addition to the traditional view of the role of in-situ analysis in deposition technology, it is clear that such techniques can also be effectively used to:

- Validate the daily status and condition of the deposition equipment.

- Reduce the cost of set-up times associated with the growth of compound semiconductor structures.

- Recalibrate and re-establish the deposition process conditions after maintenance periods.

These applications can be used without consideration of the time limitations identified with the deposition process. Ultimately, these may prove to be the most critical applications of these exciting new techniques.

ACKNOWLEDGMENTS

W. J. Kroll, R. A. Stall and numerous other EMCORE staff contributed valuable and insightful discussion relative to these problems.

Metrology, Sensors, & Process Control Application to Reactive Ion Etch

Arnon Max Hurwitz (ed.)
SEMATECH
2706 Montopolis Drive
Austin, TX 78741–6499

James R. Moyne (ed.)
University of Michigan
Ann Arbor, Michigan

Researchers at the University of Michigan's Department of Engineering and Computer Science [1,2,3] have recently addressed the problem of a lack of adequate control in an important step of semiconductor processing—that of Reactive Ion Etching (RIE). Notable advances were made by these researchers in the control of an Applied Materials 8300 Hexode RIE device by reconfiguring the control system to include more and better sensors, more capable actuators, and a multivariable feedback control scheme.

A re-think of the etch "factory" implicit in the etcher, as well as innovations in sensors and control structure have enabled promising directions for VLSI fabrication to emerge. This paper details some of these advances, focussing on the metrologic issues that were encountered and resolved. These issues, it is believed, are typical of feedback control in a general sense, and are reported here as illustrative of the type of problems to be overcome when considering the feasibility of automated control.

PROCESS DESCRIPTION

A reactive ion etcher is a low pressure, low power plasma system. The plasma is generated by the application of RF power across two electrodes surrounded by gas; in this case CF_4 was the main reactant. This generates a chemically active mixture of electrons, ion, and free radicals. Due to electrons being more mobile than ions, a DC self bias voltage develops across the electrodes. This self-bias voltage accelerates ions towards the surface of the wafer. The free radicals diffuse to the surface of the wafer where they react chemically with the exposed material. One can conceptualize RIE as consisting of two distinct mechanisms, namely the chemical etching caused by radicals, and the physical etching caused by ion bombardment.

Figure 1 sketches the total RIE system; inputs are: RF power, throttle valve position (to regulate the exhaust of

gasses from the chamber) and a gas flow regulator for the input system, while desired outputs are anisotropy, selectivity, and wall angle. These outputs relate to the "goodness" of the etch.

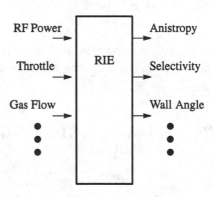

Figure 1. Total RIE system

INPUT ACTUATOR ISSUES

Throttle Valve

It was found that the original throttle valve had several shortcomings from the point of view of dynamic control. These included a large leakage conductance when fully closed, an operating regime near saturation, hysteresis in the motion of the valve, and the lack of a sensor to determine the actual valve position.

The original valve was replaced with an MKS Type 652A Throttle Valve and a Type 653 Throttle Valve Controller. This new valve was sized to be smaller, thus moving its operating region away from saturation. In addition, the valve has a low leakage conductance when fully closed, a good response time, and a measurement of actual valve position. The throttle valve controller allows for the specification of a pressure setpoint to be regulated by an internal PID loop, or the direct control of throttle position.

Gas Flow

The standard gas flow regulator was replaced by a MKS Type 2259C Mass Flow Controller to expand the range through which gas flows can be controlled. All gasses are mixed in a manifold before entering the chamber.

RF Power

The RF power actuator includes a 2000W, 13.56 MHz generation unit and matching network. The tuner in the matching network was left on during all of the experiments.

OUTPUT SENSOR / METROLOGY ISSUES

Etch

Anisotropy, selectivity, and wall angle are the important etch parameters. In this work, however, etch rate was focussed on as a measure of performance for the various control schemes because the first three parameters are not available as real-time measurements. The focus on etch rate as a measure of performance does not mean to suggest that keeping etch rate constant would be the most important use of real-time feedback control in RIE. Indeed, anisotropy will certainly be more important. Etch rate is, however, straightforward to measure *in situ* in near real time. Moreover, etch rate is a sensitive function of the conditions of the plasma which also affect selectivity and wall angle. Thus attenuating the effects of significant disturbance on etch rate is indicative of the potential benefits of feedback control on more important wafer parameters.

The wafers used consisted of a stack of 5600 Å of poly on oxide on silicon. Each etch experiment lasted about 15 minutes. Thick films and long etches were necessary because with a single wafer reflectometer, there is over a minute between peaks and valleys in the reflectometer data. That is, the information on etch depth is available at a "low data rate."

Etch rate is measured using the interference pattern of a HeNe laser reflected at near normal incidence from the wafer. The laser light is modulated to 1 kHz by a mechanical chopper before entering the chamber and the reflected intensity is detected using a Thorlab PDA150 silicon photodiode detector. The signal is demodulated using a Stanford Research SR150 Lock-In Amplifier with a filter time constant of 1 s. The intensity of the detected signal varies as the strength of the interference between light reflected off the front and back of the polysilicon layer changes as material is etched. The time between a peak and a valley in the interference pattern corresponds to the etching of 417 Å of polysilicon. The corresponding etch rate was assigned to the midpoint between the peaks and valleys.

Fluorine

Fluorine concentration is estimated via optical emissions spectroscopy (OES) using actinometry, with argon (5%) as the calibration species: Optical emission from the plasma is modulated to 1 kHz using a mechanical chopper and is collected by two fused silica optical fiber bundles. The 703.7 nm and 750.4 wavelengths in the Fluorine and Argon spectra, respectively, are selected using two Oriel Multispec 125 mm monochromators with 1200 lines/nm holographic gratings with blaze lengths of 600 nm. The light is then converted into electrical signals using Oriel photomultiplier tubes and demodulated with a Stanford Research SR850 DSP Lock-in Amplifiers with a low pass filter time constant of 30 ms.

Formerly, the monochromators used were not sufficiently discriminating between argon and fluorine spectral lines to enable an actinometric system to be applied. Consequently the researchers were forced to use only the very noisy fluorine intensity measurement. One consequence of this metrology noise is that the entries in the estimated transfer function matrix become suspect. The fluorine intensity measurement has been cleaned up by using a higher quality monochromator and a thermoelectrically cooled photomultiplier tube (see Figure 2).

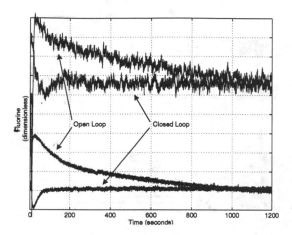

Figure 2. Original, noisy measurement of fluorine (top two curves) and improved measurement (lower two curves). Also shows achievement of closed loop control of [F].

The fundamental idea behind actinometry is that for approximately chosen spectral lines, $[F]/[Ar] = K_o * I_F/I_{Ar}$, where I_F and I_{Ar} are the intensities of the particular fluorine and argon spectral lines, and K_o is the actinometry constant. In this work, the absolute concentration of fluorine [F] in the plasma is estimated from the intensity of the emission lines by $[F] = P * I_F/I_{Ar}$, since the absolute concentration of argon [Ar] is <u>roughly</u> proportional to pressure P. This is still an uncalibrated measurement because it does not include the actinometry constant K_o. A more important issue is that $P * I_F/I_{Ar}$ is not exactly proportional to [F] because it does not account for the dilution of Ar in the plasma due to the dissociation of CF_4 into CF_x and F_y, nor does it take temperature into account. Though the measurement of fluorine now (Figure 2) is much less noisy than formerly, work is proceeding on better dynamic real-time estimates of F from actinometric data.

Therefore, though the measurement of fluorine now is much less noisy than formerly, work is continuing on deriving better dynamic real-time estimates of F from actinometry data.

DC Bias

The DC bias voltage (V_{bias}) is measured through an inductive tap into the powered electrode.

Pressure

Pressure in chamber monitored by an MKS Type 127A Baratron Capacitance Manometer, which is sensitive to pressures between 1 and 100 mTorr; 20mTorr is a typical operating point for the RIE.

PROCESS CONTROL

With existing sensor technology, it is very difficult to measure the key wafer etch parameters, selectivity, anisotropy, etc., in real-time during the etch process. Therefore, for real-time feedback control, an indirect strategy is necessary. Consideration of the etching process leads to a conceptual decomposition of the RIE system into two functional blocks (Figure 3): the Plasma Factory (PF) and the Etch Factory (EF). These sequential "factories" separate the generation of the important chemical and physical species from the action of etching the surface of the wafer.

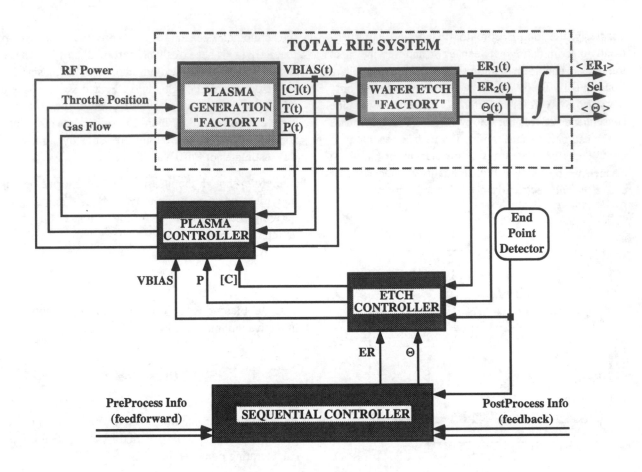

Figure 3. Conceptual decomposition of an RIE system and control structure.

The inputs to the PF are applied power, throttle position, and gas flows; its outputs are V_{bias} which represents the mechanical energy of the impinging ions; [C] (or [F], as in our application) represents concentrations of the various chemical species; T represents thermal energy, and P pressure. The EF is driven by these PF outputs, and its output represent quantities crucial to etch performance. While this decomposition is based on sound physical principles, it is not completely accurate as there will be a certain (small?) amount of feedback coupling from the wafer surface reactions to the plasma.

The above decomposition leads to our control structure. The key idea is to regulate the EF by regulating the key outputs of the PF. In addition, a high-order sequential controller, also shown in Figure 3, may be wrapped around any lower-order realtime controllers [4].

Input signals for control purposes were:

1) throttle (%open)

2) gas flow (sccm) mixed 95 at $\%CF_4$ and 5 at %Ar, and

3) applied RF power (watts).

The output signals were: V_{bias} (volts), 2. P (pressure, mTorr) and 3. [F] (arbitrary units). The transfer matrix resulting has an effective rank of two, not three, the reason being that input flow rate and throttle position are primarily affecting pressure, and then pressure in turn affects the plasma parameters V_{bias} and [F]. Thus these two actuators are affecting the plasma through a single process variable, pressure, and thus are not independent. From a controls point of view, this means that only two of the model's outputs can be independently controlled since we have only two independent actuators: power, and either flow or throttle. Further analysis on the reactor revealed that throttle was the most effective actuator, and so flow was simply held fixed at 30 sccm.

If we wish to make use of our conceptual decomposition of the RIE system (Figure 3) we will specify etch recipes in terms of any two of V_{bias}, P, and [F]; we do not specify the recipe on all three because of our discussion of the previous paragraph. V_{bias} is a good indicator of the physical portion of the etch process, so it remains to choose between P and [F]. It is widely accepted that in a CF_4 plasma, [F] is a good indicator for the chemical portion of the etch process. However, measuring [F] on line and in real time is more expensive than just measuring P. We therefore compare three control schemes: 1. the traditional method of controlling pressure P; 2. control by feedback regulation of P and V_{bias}; 3. control by feedback regulation of V_{bias} and [F].

RESULTS

For each of the three control schemes of the last paragraph, a standard etch recipe [3] was repeated up to the 10 minute point, at which time a 1 sccm O_2 disturbance was introduced to simulate a mass flow controller leak. The results are summarized in Figure 4. The pressure control, and the pressure/V_{bias} control demonstrate the impact on etch rate of chemical noise, such as outgassing, as well as the O, 'step' disturbance.

It is clear that the strategy of controlling V_{bias} and [F] significantly outperforms the other two strategies. It is worth noting that controlling V_{bias} and P does not lead to much better control than the (traditional) method of controlling P. This is because, much like pressure-only control, chemical noise parameters such as chamber wall outgassing, oxygen leaks, etc., which have not been parametrized will impact the relationship between P/V_{bias} and [F]. Consequently, since [F] is a good indicator of chemical etch, a P/V_{bias} controller would be subject to the same chemical noise as a pressure-only controller.

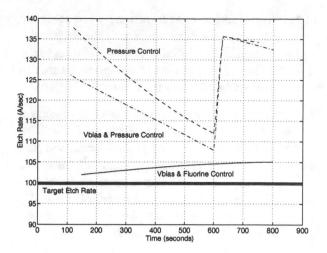

Figure 4. Figure 4: Etch rate comparison of feedback on P only, on V_{bias} and P, and on V_{bias} and [F].

Transfer Function Matrix for V_{bias} and [F]

The V_{bias}/[F] controller transfer functionmatrix used in the above was derived by performing simple step-increment experiments on Power (10% step) and on Throttle Position (16% step). The transfer function matrix thus derived [5] was:

$$\begin{bmatrix} V_{bias} \\ [F] \end{bmatrix} = \begin{bmatrix} 2.72e^{-0.5s}/(s+0.17) & 0.444/(s+1.25) \\ -0.43e^{-0.5s}/(s+0.21) & 0.12/(s+4.93) \end{bmatrix} \begin{bmatrix} \text{Throttle} \\ \text{Power} \end{bmatrix}$$

CONCLUSIONS

Researchers at The University of Michigan have achieved "good" control over etch rate in a reactive ion etch process. In order to achieve this level of control, innovations in sensors, actuators and control structures were incorporated into the system. Specifically, input throttle valve and gas flow actuators were replaced with components that have a more appropriate operating region and range respectively. Less noisy measurements of relative fluorine concentrations were achieved through sensory equipment upgrades. The 2X2 multivariate controller developed regulates Vbias and relative fluorine concentration in the plasma through actuation of applied RF power and throttle (% open).

The entire etch rate feedback controller development process is believed to be representative of feedback controller development in the general sense, and thus the exercise demonstrates the feasibility of automated control. A logical next step in etch process control then is the development of a more robust controller that also controls other etch "goodness" parameters such as selectivity and wall angle.

ACKNOWLEDGEMENTS

The author would like to thank Professor Mike Elta, director of the DTM center at the University of Michigan, Ann Arbor, for permission to use its material in the preparation of this paper. Thanks are also made to the authors of the documents cited in the references below, namely: M.Elta, H.Etemad, J.P.Fournier, J.S.Freudenberg, M.D.Giles, J.W.Grizzle, P.T.Kambamba, P.P.Khargonekar, S.Lafortune, S.M.Meerkov, B.A.Rashap, D.Teneketzis, F.L.Terry, Jr., and T.Vincent.

REFERENCES

[1] Elta, M.E., et al, "Applications of Control to Semiconductor Manufacturing: Reactive Ion Etching," *Proceedings 1993 American Control Conference.*

[2] Rashap, B.A., et al, "Real-time Control of Reactive Ion Etching: Identification and Disturbance Rejection," *Proceedings of the 32nd Conference on Decision and Control*, December 1993; San Antonio, TX.

[3] Elta, M.E., et al. "Real-time Feedback Control of Reactive Ion Etching," *Electronic Manufacturing & Control systems*: Related Publications, Vol. 1. 1993. The University of Michigan, Center in Automated Semiconductor Manufacturing, Solid-State Electronics Laboratory, University of Michigan, Ann Arbor, MI.

[4] Moyne, J.M., et al, "A run-to-run framework for VLSI manufacturing," *Electronic Manufacturing & Control systems*: Related Publications, Vol. 1. 1993. The University of Michigan, Center in Automated Semiconductor Manufacturing, Solid-State Electronics Laboratory, University of Michigan, Ann Arbor, MI.

[5]Rashop, B.A., et al "Control of Semiconductor Manufacturing Equipment: Real-Time Feedback Control of a Reactive Ion Etcher." *(to be published in) IEEE Transactions on Semiconductor Manufacturing* (?Aug 1995).

Improved Gas Flow Measurements
For Next-Generation Processes

Stuart A. Tison

Thermophysics Division, Chemical Science and Technology Laboratory,
National Institute of Standards and Technology, Gaithersburg, MD 20899

Many semiconductor processes require stable and known flows of gas be delivered to the processing chamber. Gas types and flow rates are process dependent, but it is clear that next-generation processes will require flow measurements that are two to three decades lower than current requirements. Specifically, the Semi-Sematech-sponsored Mass Flow Controller Working Group recently identified the need for flow measurements with an accuracy of 1% or better to be extended to cover the range 7×10^{-8} to 7×10^{-5} mol /s[†] (0.1 to 100 sccm, standard cubic centimeters per minute). At least two problems must be overcome if this goal is to be attained. Achieving 1% accuracy in the process chamber will require reference standards with an accuracy of 0.2% or better, but adequate reference standards do not exist over this range. Further, the thermal mass flow controllers (TMFCs) used to measure and control the process gases are generally calibrated with nitrogen and "corrected" for other gases, but the correction factors are not well understood and of questionable reliability. The National Institute of Standards and Technology (NIST) is addressing both of these problems by developing new flow standards and by investigating the performance of TMFC's. This paper will present data on the performance of five TMFC's, from different manufacturers, with full scale ranges of 1.5×10^{-6} mol/s to 3.7×10^{-6} mol/s (2 to 5 sccm).

INTRODUCTION

The measurement and control of gas flow are critical in many manufacturing processes. In particular, semiconductor manufacturers rely upon mass flow measurements for gas admission into semiconductor processing tools or reaction vessels. The thermal mass flow meter (TMFM) is most prevalently used in the semiconductor industry. This meter senses the flow by measuring the thermal transfer between a heated tube wall and the gas stream. The TMFM's operate over a wide range of flow, 0.04 mol/s to 7.4×10^{-8} mol/s (5×10^{4} to 0.1 sccm), and are suitable for use with most gases, including corrosives routinely used in the semiconductor industry. Flow measurement from 0.04 mol/s to 7.4×10^{-5} mol/s (5×10^{4} to 100 sccm) has been routine for a number of years in the semiconductor industry, and the performance of thermal mass flow meters in this range has been investigated[1]. The use of TMFM's in the range of 7.4×10^{-6} mol/s to 7.4×10^{-8} mol/s (10 to 0.1 sccm) is becoming more prevalent, but their performance in this range is not well documented.

Due to the fact that TMFM's are often used with multiple gases or highly toxic gases, it is a common practice to calibrate the instrument with one gas, such as nitrogen, and employ "generic" correction factors to estimate the flow with other gases. Unfortunately, these correction factors are instrument specific and may vary by as much as 10% between instruments of different designs. Because the values of the correction factors may vary from 0.2 to 1.5, uncertainties in the correction factors can add significant uncertainties to measurements that rely upon them. Additionally, it has been suggested[2] that the correction factors may be a function of flow and not constant at all.

This paper addresses two issues of major concern to most users of TMFM's and to the manufacturers themselves. The first issue is how accurately are TMFM's calibrated by manufacturers with the reference gas and the second is the range of variability between the recommended and actual correction factors for other gases. To accomplish this task, instruments from five manufacturers were chosen with full scale ranges between 1.5×10^{-6} mol/s and 3.7×10^{-6} mol/s (2 and 5 sccm). This flow range was selected because of its increasing importance to the semiconductor industry and the lack of knowledge of the performance characteristics of TMFM's in this range.

DESCRIPTION OF TMFM's

The TMFM senses flow by measuring the heat

[†]1 standard cubic centimeter per minute at 0° C = 7.41×10^{-7} mol/s

transferred from a heated tube to the gas flowing inside the tube. While designs between manufacturers vary, there are two measurement techniques that are commonly employed. The first is to provide a constant input power to a section of tubing and measure the temperature of the tube on both sides of the heated section. The flowing gas skews the temperature such that the downstream temperature is larger than the upstream value. This measured difference is linearly dependent upon mass flow to first order. The second technique heats the tube by maintaining a constant temperature independent of flow. The amount of power required to maintain the constant tube temperature is then proportional to the mass flow in the tube. Typically very small stainless steel tubing is used with inside diameter varying from 0.25 mm to 1 mm and wall thickness minimized to lessen axial thermal losses in the tubing. The tube is wrapped with a number of heater windings which have a high resistance and a high temperature coefficient of resistance. This allows the heater to become a temperature sensor as well as a heat source. TMFM's are designed so that the flow is laminar with maximum flows through the tube less than 7.4×10^{-6} mol/s (10 sccm). Larger flow TMFM's are constructed by splitting the flow with a channel which bypasses the sensor. The particular attributes of the TMFM's used in this study, including manufacturers specifications, are given in the appendix. It is NIST policy to identify instruments only by their generic specifications.

APPARATUS AND MEASUREMENT TECHNIQUE

The TMFM's were calibrated by direct comparison with the NIST piston flowmeter. This flowmeter generates and measures flow by advancing a piston of known volume into a vessel at a rate such that the pressure in the vessel remains constant while the gas escapes through an attached leak valve into a vacuum system. This flowmeter generates and measures flow over a range of 1×10^{-11} to 5×10^{-6} mol/s (1×10^{-5} to 7 sccm) and is described in detail elsewhere[2].

The comparison of the flows between the TMFM's and the NIST piston flowmeter was effected in the following manner using spinning rotor gages(SRG's). The pressure readings of two SRG's in the vacuum chamber downstream of the piston flowmeter are recorded with no flow. The flow from the piston flowmeter is directed into the vacuum chamber and evacuated through a 1 cm orifice, which has a stable conductance or "throughput". The equilibrium pressure above the orifice is measured and recorded by two SRG's. The flow is changed and this process is repeated over the flow range of interest of 7.4×10^{-8} mol/s to 4×10^{-6} mol/s (1 to 5 sccm) for nitrogen. A correlation between the known flow from the piston flowmeter and the pressure measured by the SRG's is then determined from this data. For nitrogen

the measured pressures ranged from 0.02 to 1 Pa. To first order the relationship between flow and the observed pressure is linear, but due to small deviations from molecular flow through the orifice and possible SRG nonlinearities, a second order polynomial was used. This process was repeated for argon and sulfur hexafluoride. The calculated total flow uncertainty using this technique is ±0.35%, representing two standard deviations.

The TMFM's, which are labeled A through E to preserve the manufacturers anonymity, were mounted in a temperature controlled enclosure (±0.5 °C) with TMFM's A-D mounted in parallel, and TMFM E mounted in series with TMFM's A-D. The volume between TMFM E and the others was minimized by using 0.4 cm inside diameter tubing with small lengths. Downstream of TMFM E was a 0.1 MPa full scale pressure gage and a variable conductance valve that isolates the TMFM's from the vacuum system. Upstream of the TMFM's was a gas handling system, a 1 μm filter, and a 0.15 MPa full scale pressure gage.

To calibrate the TMFM's, the upstream pressure of TMFM's (A-D) was set to 0.1 MPa and the readings monitored with their regulating valves closed until the signal equilibrated. The "zero" readings were then recorded. The setting of TMFM A was then changed via a remote analog set point with a programmable power supply. The flow was directed into the vacuum chamber and evacuated through the 1 cm orifice. Upon equilibrium, the outputs of TMFM A and E were recorded, along with the observed pressure above the 1 cm orifice (measured with two SRG's). This process was repeated at increments of 10% of the instruments full scale to 100% full scale and at 10% decrements down to 0. The actual flow was computed using the observed pressure readings of the SRG's and the previously described relationship between these values and the flow. This process was repeated with TMFM's B, C, and D.

RESULTS

The TMFM's were first calibrated with nitrogen. The results of this calibration are shown in FIG. 1 for TMFM's A, C, and E. TMFM's A, C, and E were within their manufacturers stated uncertainties of ±1% full scale. The results for TMFM A indicate that the manufacturer's calibration at the full scale value was in very good agreement with the NIST value, but the deviations increased for lower flow rates to a maximum of 0.6% of full scale (1.2% of reading at 50% of full scale). TMFM C gives results that are systematically low by 0.4% from the NIST measured values. TMFM E has no significant systematic trends within the uncertainties of these measurements. TMFM B, not shown, exhibited systematic differences ranging from -10% at 70% of full scale to -17% at 10% of full scale. TMFM D, not shown, exhibited a systematic offset that varied from 8.5%

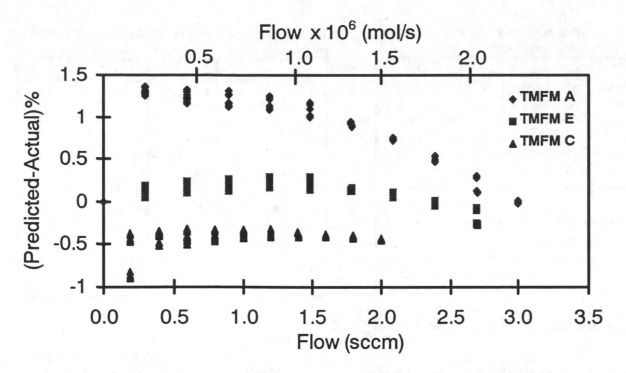

FIG.1 Deviations of TMFM's from the NIST calibration with nitrogen as a percent of reading.

to 6.8% higher than the measured NIST values.

than 1%.

The results of the calibration with argon are given in Table 1. The correction factor between the nitrogen and argon flow is given at 25%, 50%, 75%, and 100% of full scale (F.S.). The correction factor is defined as the indicated argon flow divided by the equivalent nitrogen flow. In general, the correction factor is found to be a function of flow and to first order to increase linearly with increasing flow. The value of the change varies from 0.2% for TMFM E to 1.7% for TMFM B with the other TMFM's having intermediate values. The last row of Table 1 contains the computed correction factors of the TMFM's averaged over all meters. From this data it can be surmised that the correction factor increases with increasing flow with an average change of 0.7% over the given range. The manufacturers (Man.) reported values, column 7, differed considerably from the measured values. The maximum deviations between the manufacturers recommended values and the observed values ranged from 1.6% to 3.6%. These deviations are not surprising, as most manufacturers stipulate that the uncertainties of the correction factors are on the order of a few percent. What is surprising is the agreement of the observed correction factors between different manufacturers. The average correction factor for the TMFM's (averaged over all flows), given in column 6, ranged from 1.426 to 1.418 with an average of 1.423, which represents less than 0.6% maximum variation. Use of the grand average value (average of all TMFM's) of 1.423 results in deviations from the observed values of less

The results of the calibrations with sulfur hexafluoride are given in Table 2. The average correction factor as given in the last row varies little with flow. Individually the TMFM's show variations with flow from no change to a maximum change of -1.8%. The average correction factor for individual TMFM's, given in column 6, ranged from 0.268 to 0.281, which represents a maximum variation of

Table 1. Correction Factors (CF) for TMFM's using argon gas relative to a nitrogen calibration. Manufacturer (Man.)

TMFM	CF 25% F.S.	CF 50% F.S.	CF 75% F.S.	CF F.S.	Ave.	Man. value
A	1.423	1.423	1.426	1.432	1.426	1.443
B	1.413	1.423	1.430	1.437	1.426	1.400
C	1.427	1.422	1.425	1.430	1.426	1.396
D	1.417	1.417	1.423	1.425	1.421	1.398
E	1.415	1.417	1.418	1.420	1.418	1.370
Average	1.419	1.420	1.424	1.429	1.423	1.401

Table 2. Correction Factors (CF) for TMFM using sulfur hexafluoride gas relative to a nitrogen calibration.

TMFM	CF 25% F.S.	CF 50% F.S.	CF 75%F.S.	CF F.S.	Av.	Man. value
A	0.277	0.276	0.275	0.274	0.275	0.275
B	0.267	0.267	0.268	0.270	0.268	0.270
C	0.273	0.273	0.272	0.272	0.272	0.275
D	0.261	0.258	0.257	0.256	0.258	0.260
E	0.284	0.282	0.280	0.279	0.281	0.284
Average	0.272	0.271	0.270	0.270	0.271	0.273

Table 3. Change in nitrogen calibration during a 2 month period expressed as a percent of reading (Rd.).

TMFM	Deviation % Rd. at 25% F.S.	Deviation % Rd. at 50% F.S.	Deviation % Rd. at 75%, F.S.	Deviation % Rd. at 100% F.S.
A	0.2	0.2	0.3	0.3
B	0.4	0.4	0.4	0.3
C	0.5	0.2	0.1	0.0
D	0.5	0.4	0.2	0.2
E	-0.4	-0.2	0.0	0.2

9%. While correction factors of TMFM's varied widely, the manufacturers' recommended values are in good agreement with the measured average correction factors listed in column 6. The maximum deviations between the manufacturers' recommended values and the observed values ranged from 1.8% to -0.7%. Use of a grand average correction factor (an average of all TMFM correction factors) in place of the measured value for the instrument, while not introducing considerable error for the case of argon would lead to significant errors for sulfur hexafluoride.

In addition to quantifying the uncertainties in the TMFM's, it is desirable to know the stability of the instruments over time. All of the TMFM's were recalibrated with nitrogen two months after the first calibration to determine their stability. Between tests the power was maintained to the TMFM's and they remained in the same physical location without exposure to corrosive gases. The results of the recalibration expressed as a deviation from the first calibration at 25%, 50%, 75%, and 100% of full scale are given in percent of reading in Table 3. From the table it can be seen that TMFM's A and B experienced constant calibration shifts whereas TMFM C, D, and E show flow dependent changes. These results with the previous results shown in FIG. 1, indicate that TMFM's A, C, and E remained within their manufacturers' uncertainties of ±1% full scale, whereas TMFM's B and D remain outside their prescribed manufacturers' uncertainties.

The correction factors for argon and sulfur hexafluoride were redetermined after the second nitrogen calibration. A maximum period of 10 days elapsed between the second nitrogen calibration and the argon and sulfur hexafluoride calibrations. There was no significant change, less than 0.2%, in the correction factors for four of the TMFM's. One of the

TMFM's, D, shifted by 2%. This meter's calibration shifted for all gases between the time of the second nitrogen and argon calibration. This shift did not change the value of the argon or sulfur hexafluoride correction factors.

CONCLUSIONS

Five TMFM's with full scale ranges of 1.5×10^{-6} mol/s to 3.7×10^{-6} mol/s (2 to 5 sccm) were investigated to determine their uncertainties with nitrogen gas and their correction factors with other gases. It was found that three of the five TMFM's were within the manufacturers' stated uncertainty of ±1% of full scale. Two TMFM's were well beyond their stated uncertainty, one by as much as 17%. The measured correction factors for argon ranged from 1.437 to 1.415, and generally deviated from the manufacturers' recommended values by 1 to 3.5%. The measured correction factors for sulfur hexafluoride ranged from 0.284 to 0.256, and were in good agreement with the manufacturers' recommend values (±1%). Four of the five tested TMFM's were found to be repeatable to within ±0.5% over a two month period. One TMFM, TMFM D, experienced a 2% shift after two months of operation.

ACKNOWLEDGMENTS

The contributions of Ms. Christian Alavanja in writing software and in the setup of the experiment are gratefully acknowledged. This work was funded in part by the National Semiconductor Metrology Program (NSMP).

APPENDIX

TMFM A

This TMFM has a full scale of 2.2 x 10^{-6} mol/s (3 sccm) for nitrogen and is equipped with a solenoid type control valve The TMFM senses the mass flow using a constant power technique which incorporates one upstream temperature sensor and one downstream temperature sensor with a heated section between the sensors. The instrument has a bypass section which is completely blocked so that the total flow is directed through the sensor. The TMFM has manual adjustments for zero, linearity, and span, and was calibrated by the manufacturer with nitrogen with the following specifications:

Accuracy: ±1% full scale
Repeatability: 0.25% of reading
Zero temperature sensitivity: .075% of full scale per °C.
Span temperature sensitivity: 1.0% over 10 to 50 °C.

TMFM B

This TMFM has a full scale of 3.7 x 10^{-6} mol/s (5 sccm) for nitrogen and is equipped with a solenoid type control valve. The TMFM senses the mass flow using a constant power technique which incorporates one upstream temperature sensor and one downstream temperature sensor with an intermediate heated section. The instrument has a bypass section which is completely blocked around which the flow is directed through the sensor. The TMFM has manual adjustments for zero, linearity, and span, and was calibrated by the manufacturer with nitrogen with the following specifications:

Accuracy: ±1% full scale
Repeatability: 0.2% of reading
Zero temperature sensitivity: not reported
Span temperature sensitivity: 0.1% of full scale per °C.

TMFM C

This TMFM has a full scale of 1.5 x 10^{-6} mol/s (2 sccm) for nitrogen and is equipped with a piezoelectric type control valve. The TMFM senses the mass flow using a constant temperature technique which incorporates two heaters which are also used as temperature sensors. The sensor consists of a straight section of tubing with no bypass. The TMFM has manual as well as an automatic zeroing capability, and has no user adjustments for span or full scale. The instrument was calibrated by the manufacturer with nitrogen with the following specifications.

Accuracy: ±1% of full scale
Repeatability: 0.2% of full scale

Zero temperature sensitivity: not reported
Span temperature sensitivity: not reported

TMFM D

This TMFM has a full scale of 2.2 x 10^{-6} mol/s (3 sccm) for nitrogen and is equipped with a solenoid type control valve. The TMFM senses the mass flow using a constant power technique which incorporates two heaters which are also used as temperature sensors. The instrument has a bypass section which is completely blocked so that the total flow is directed through the sensor. The TMFM has manual adjustments for zero and span, and was calibrated by the manufacturer with nitrogen with the following specifications:

Accuracy:± 1% of full scale
Repeatability: 0.2%
Zero temperature sensitivity: not reported
Span temperature sensitivity: not reported

TMFM E

This TMFM has a full scale of 1.5 x 10^{-6} mol/s (2 sccm) for nitrogen and is a meter only. The TMFM senses the mass flow using a constant power technique which incorporates three heaters, two of which are also used as temperature sensors. The sensor consists of a straight section of tubing with no bypass. The TMFM has manual zero, span, and full scale adjustments and was calibrated by the manufacturer with nitrogen with the following specifications.

Accuracy: ±1% of full scale
Repeatability: 0.2% of full scale
Zero temperature sensitivity: 0.5% of full scale per °C
Span temperature sensitivity: 0.1% of reading per °C

REFERENCES

[1] J. Riddle and J. Hardy, SEMATECH Report #9402373A-XFR (1994)

[2] L.D. Hinkle and C.F. Mariano, J. Vac. Sci. Technol., A9, 2043 (1991)

[3] K.E. McCulloch, C.R. Tilford, C.D. Ehrlich, and F.G. Long, J. Vac. Sci. Technol., A5, 276 (1987)

Application of the Raman Microprobe to Identification of Organic Contaminants and to In-situ Measure of Stresses

Fran Adar and Howard Schaffer

Instruments S.A., Inc.
3880 Park Ave., Edison N.J. 08820

The Raman Microprobe is a tool that provides unique information to the analyst in a semiconductor fabrication support laboratory. Using a conventional optical microscope it can
identify organic contaminants that appear during manufacture,
as well as characterize the silicon itself.
Spatial resolution is determined by the physical diffraction limit - 1μm. Sample preparation is exactly the same as for optical microscopy; samples do not have to be thinned, cast into films, coated with conducting films, etc. and they are examined under ambient conditions.

In contrast to the vacuum microprobes that cannot determine more about organic materials than that they contain carbon, a Raman microprobe provides a detailed fingerprint that has been used to identify chemical species, crystallographic phase, and sometimes the mode of contamination.

The crystallographic phase and mode of contamination that can be inferred from the Raman spectra enable one to do more detailed detective work than what would be provided by simple chemical identification.

The sensitivity of the Raman spectrum of crystalline silicon to strain was originally documented in 1970 and applied to thin films in the early 1980's. Since then it has become clear that the potential for a Raman microprobe to measure strain in silicon devices with 1μm spatial resolution can aid in device engineering. Both the geometry of the device structures and the mismatch of thermal properties of dielectric films in contact with the silicon can produce stresses which will effect the electrical conductivity of the silicon. As devices become smaller, these effects on materials properties become more important making an in-situ stress measurement more useful.

INTRODUCTION

Description of the Effect

The Raman effect is a light scattering phenomenon in which a monochromatic beam of photons is focussed onto a sample and the light scattered by the sample is collected and analyzed. A small amount of light in the scattered beam appears at wavelengths other than that of the laser. It is this light occurring at shifted wavelengths that has the information used for analysis of molecular and condensed phase materials. The light scattering event is energy conserving, and consequently some of the energy in the initial photon is transferred to a transition that can be used for analytical purposes. A large majority of Raman spectroscopy examines vibrational transitions in molecules and crystals. However, many phenomena in solid state physics are amenable to study by monitoring other excitations such as plasmons, single particle excitations, magnons, and spin-flip excitations.

Survey of Instrumentation

Lasers used for Raman spectroscopy generally occur in the visible part of the spectrum. Early in the 1970's it was recognized that lasers with wavelengths of the order or $0.5\mu m$ could be focussed with high n.a. objectives to diffraction-limited spots of the order of $1\mu m$ (1-2). The design of Raman microprobes followed the recognition of this fact. The instrument built at the (then) NBS was based on an ellipsoidal reflector for collecting the light (2). The system designed at the University of Lille was based on an epi-illuminator in a metallographic microscope, and as such took advantage of readily available components.

The original systems were based on the optical instruments of the time - large double monochromators and single channel detectors. During the 1980's efficient, low-noise multichannel detectors were developed that initiated the re-design of Raman instruments. Triple spectrographs were developed with focal lengths that provided a good compromise between resolution and coverage on the multichannel detectors. The purpose of the first two stages of the system was to suppress the elastically scattered light that is typically 5 orders of magnitude larger than the signal of interest. Early in the 1990's holographic notch filters were introduced. These filters suppress the elastically scattered light enabling one to build a Raman spectrograph with a large choice of monochromators. The only loss of capability would be in the very low frequency-shift regime because the best cut-on of the holographic filters is 50-150 cm^{-1}.

Organic Fingerprinting

The Raman spectrum of a complex organic molecule exhibits 3N-6 lines, where N is the number of atoms in the molecule. Simplification of spectral identification can often be achieved by referring to catalogued frequencies of functional groups in the molecule (3). For example, a carbon double or triple bond is usually strong and readily identifiable in a Raman spectrum. CH, OH and NH groups tend to occur in isolation from other features. Based on these types of arguments, unknown contaminating species can be identified by recording their Raman spectra.

In addition, it is important to recognize that the samples need very little pre-treatment before analysis. The s mples are examined on a conventional optical microscope stage, under ambient conditions, whereas the elemental microprobes require conductive coatings before loading the samples into a vacuum environment. In contrast to the vacuum microprobes where the infomation derived will be limited to the presence of carbon, the Raman microprobe has the potential to fingerprint the exact molecular species, even its crystallographic phase.

One can argue that the same molecular information is available from infrared spectroscopy. However, an infrared microprobe will have ten times worse spatial resolution, and may require thinning the sample which can alter its state. In addition, Raman tends to be sensitive to molecular groups to which the infrared can be insensitive (eg., carbon double and triple bonds).

Sensitivity

In general the sensitivity of the technique is related to the spatial resolution. Usually if one can see a particle in the optical microscope, then there is a good chance that it is possible to acquire a spectrum which will provide identification. In order for a Raman microprobe to be effective on contaminants smaller than $1\mu m$, it is necessary to increase the sensitivity.

An extremel;y effective technique for enhancing sensitivity to trace organic contaminants is called Surface Enhanced Raman Spectroscopy (SERS). The application of SERS to trace organic analysis has been reviewed (4, for example). In this technique, carefully prepared activated surfaces with silver, gold, or sometimes copper particles are coated with trace amounts of organic material. Because of an enhancing interaction between the photon electromagnetic field and the metal particles (typically 200 to 500 nm in dimension), the Raman signal from material in intimate contact with the particles is enhanced by as much as 6 orders of magnitude. However, the requirement for a metal colloid substrate is not relevant to contamination studies of silicon wafers.

SERS can be applied to contaminated wafer surfaces by turning the SERS conditions upside down when overcoating the contaminated regions of the wafers with metal colloids. Because the colloids are semi-transparent, some of the laser can penetrate to the colloid-contaminant interface, and some of the Raman light can be collected. This possibility has been demonstrated on a silicon surface on which sub-micron particles of diamond has been deposited by CVD conditions (5).

In-situ Stress Analysis

Anastassakis, et. al. (6) showed that the frequency of the Raman phonon band in silicon responds to the application of stress; strains of the order of 10^{-2} to 10^{-3} can be detected. Zorabedian, et. al. (7) demonstrated the relevance of this measurement to the analysis of a laser-annealed laterally-seeded epitaxial film of polysilicon on silica islands. It has also been demonstrated that stresses due to geometrical features of devices can be measured (8); because the electrical mobility of carriers is effected by stress, the ability to measure stress in-situ can aid in the engineering design of the devices.

In summary it is clear than a Raman microprobe can provide unique capabilities in the arsenal of analytical tools available to the integrated circuit industry.

CONTAMINANT ANALYSIS - EXAMPLES

There are many successful examples of contaminant analyses that have been preformed over the years. Any organic product that can come in contact with a silicon wafer during the manufacture of integrated circuits, or that can come into its environment, is capable of producing contamination problems. Several examples have been chosen that illustrate not only that identification can be made, but the identification adds some unexpected information about the mode of contamination.

Figure 1 show the Raman microprobe spectrum of a hazy film on a wafer undergoing IC patterning that was identified as cellulose in crystallographic form I (9). This material occurred as a film, and not as fibers. But cellulose is normally difficult to dissolve, and only rarely precipitates in crystallographic phase I. However, Atalla was able to show that phase I can be precipitated if the sample is hot when precipitated (9). Therefore, the Raman microprobe not only identified the contaminant, but provided some information about the conditions under which the contamination occurred.

A 5" wafer with completed IC patterning had been submitted for analysis of gelatinous material that was found depositing on (non-passivated) aluminum bonding pads. The spectrum showed OH and C=C functional groups as well as bands of usual aliphatic organic material. This sample had been stored in zip-lock bags for shipment to another site. It was inferred that the contaminant was a long-chain, unsaturated alcohol that was added to the plastic in the bag for one of two reasons - either to protect the wafer from static electricity, or to assure slippage of the extruded plastic film used to make the bags. Apparently the contaminant was depositing only on the unpassivated aluminum because of the attraction of the image charge of the polar alcohol group.

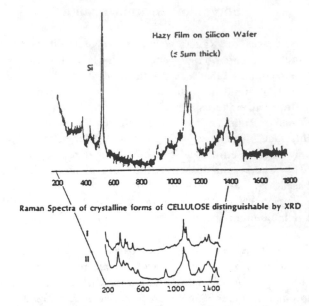

FIGURE 1. Raman microprobe spectrum of a hazy film contaminating a silicon wafer, compared to reference spectra of cellulose I and II. This figure is reproduced from Reference 10.

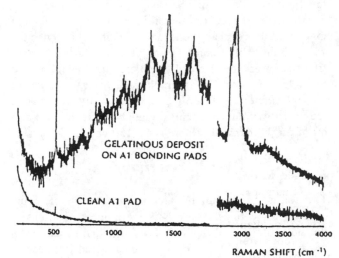

FIGURE 2. Raman microprobe spectrum of gelatinous deposit on non-passivated aluminum bonding pads.

A unpatterned sample wafer that had only been polished and then etched in a plasma exhibited a contaminating film that was identified as a "teflon-like" material. In this case it was concluded that the deposition occurred because of poorly regulated conditions in the plasma.

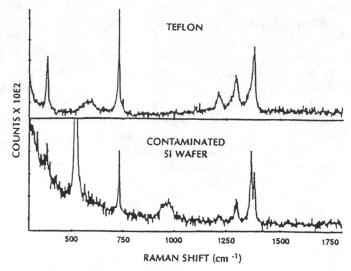

FIGURE 3. Raman microprobe spectrum of film deposited on silicon wafer during plasma etch following mechanical polish. This figure is reproduced from Reference 10.

The results of Raman microprobe analysis of the following two examples are described without figures because of a lack of space. The fingerprint spectra allowed identification of molecular species. The interest here is in describing the source of contamination.

A particle of polyethylene terephthalate (PET) or polybutylene terephthalate (PBT) was identified on a wafer from a facility where wafers are carried in baskets that had been guaranteed not to degrade.

A silicone contaminant appeared on a wafer that was presumed to originate from a degrading rubber gasket. The Raman spectrum specifically identified the silicone polymer (poly[tetramethyl]siloxane) which contains carbon, oxygen, silicon, and hydrogen. Elemental microprobes would have identified the presence of carbon, oxygen and silicon, but not the molecular arrangement holding them together. In fact, carbon is often present in some of the elemental microprobes as a systematic impurity, and it would not be surprising to find silicon because the substrate itself was silicon. Thus, only a molecular microprobe could provide clear identification of the contaminating species.

An IC that had failed during use was submitted for analysis. Inspection revealed the presence of a break in a wire connect. It was suspected that corrosive material had entered the package because of the presence of silica (ie., silica gel) found at the base of the broken wire. This information had come from IR analysis. Raman analysis could not confirm the presence of silica. In fact, Raman identified the presence of microcrystalline silicon. The

inference was that the crystallite size of the silicon added to aluminum wires to increase ductility might have been too large; because of the lower conductivity of silicon, the wires would have overheated leading to electromigration and melting and obvious failure.

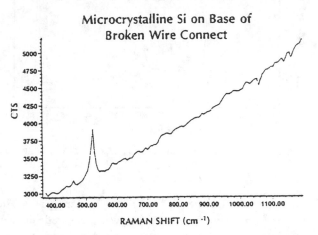

FIGURE 4. Raman microprobe spectrum of material deposited at based of broken wire connect.

The complementarity of the Raman and IR techniques is interesting in this application. Raman is much more sensitive to crystalline Si than IR, whereas IR is much more sensitive to silica than Raman. The information derived from both measurements is "correct" but the inferential conclusion as to the nature of the failure is quite different.

IN-SITU STRESS MEASUREMENTS

The phonon frequency in crystalline silicon will shift with applied stress(4). This shift can be calibrated and used as an in-situ measure of stress in integrated circuits. The importance of this ability is related to the dependence of electrical mobility on crystalline stress. As circuit dimensions become smaller, the effects on crystalline stress become more important and engineering design must be engaged to control these stresses. The stresses can be due to materials mismatch between silicon and overlaying films, or to the geometry of the devices themselves.

Raman frequency shifts smaller than 0.1 cm^{-1} are of interest. In order to measure such small shifts, an atomic lamp is mounted between the microprobe and the entrance slit in order to monitor the mechanical drift of the spectrograph. Figure 5 shows the Raman microprobe spectrum of silicon with Ne lines recorded simultaneously to monitor instrumental drift.

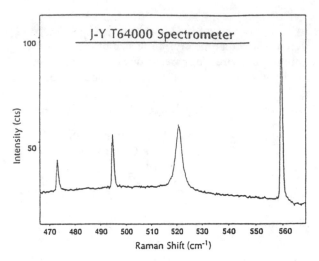

FIGURE 5. Raman microprobe spectrum of silicon wafer with simultaneous illumination with neon lamp.

An integrated circuit was submitted for analysis. The Raman frequency of the Si phonon was measured between aluminum lines, and then compared to the phonon measured far from any aluminum lines. The figure shows the variation. Points 1 to 8 were recorded between lines; point 9 was recorded far away. Point 7 was recorded much farther from a line than points 1-6 or 8. It is quite clear that there is measureable compressive stress close to aluminum lines.

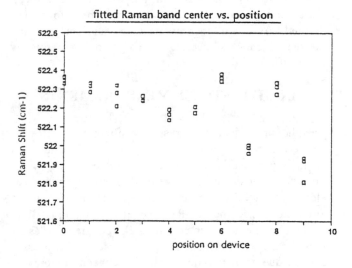

FIGURE 6. Raman microprobe spectrum of narrow regions of silicon between aluminum lines (points 0 to 5), in a wide silicon line between aluminum pads (points 6 to 8), and far from the aluminmum (point 9).

CONCLUSIONS

It is clear that a Raman microprobe has an important role to play in the development and manufacture of integrated circuits. The ability to measure stresses in-situ can be used to engineer the devices. That is, one can determine where stresses are in the devices and include that information in the engineering parameters. This can effect both the geometrical design and the choice of materials. The ability to identify organic contaminants provides an important capability to increase manufacturing yields.

REFERENCES

1. Delhaye, M. and Dhamelincourt, P., *J. Raman Spectrosc.*, **3**, 33-43, 1975

2. Rosasco, G.J., Etz, E.S. and Cassatt, W.A., *Appl. Spectrosc.*, **29**, 396-404, 1975

3. Lin-Vien, D., Colthup, N.B., Fateley, W.G., and Grasselli, J.G., *The Handbook of Infrared and Raman Characteristic Frequencies of Organic Molecules*, New York: Academic Press, 1991

4. Vo-Dinh, T., "Surface-Enhanced Raman Spectroscopy," in *Chemical Analysis of Polycylclic Aromatic Compounds*, ed T. Vo-Dinh, 1989, pp. 451-485

5. Knight, D.S., Weimer, R., Pilione, L. and White, W.B., *Appl. Phys. Lett.*, **56**, 1320-1322, 1990

6. Anastassakis, E., Pinczuk, A., Burstein, E., Pollak, F. and Cardona, M., *Solid State Commun.*, **8**, 133-137, 1970

7. Zorabedian, P. and Adar, F., *Appl. Phys. Lett.*, **43**, 504-506, 1983

8. Brueck. S.R.J., Tsaur, B-Y, Fan. J.C.C., Murphy, D.V., Deutsch, T.F. and Silversmith, D.J., *Appl. Phys. Lett.* **40**, 895-898, 1982

9. Atalla, R.H., *Appl. Polym. Symp.*, **28**, 659-669, 1976

10. Adar, F., "Application of the Raman Microprobe to Analytical Problems of Microelectronics," in *ACS Symposium Series 295 Microelectronics Processing: Inorganic Materials Characterization (ed. L. Casper)*, 1986, American Chemical Society, pp.231-239

Real-Time Monitoring and Control of Resonant-Tunneling Diode Growth Using Spectroscopic Ellipsometry

F. G. Celii, Y.-C. Kao, T. S. Moise, A. J. Katz, T. B. Harton and M. Woolsey

Corporate Research & Development/Technology, Texas Instruments, Inc.
M/S 147, P. O. Box 655936, Dallas, TX 75265

B. Johs

J. A. Woollam Co., 650 J St., Suite 39, Lincoln, NE 68508

We employed spectroscopic ellipsometry (SE) for *in situ* growth monitoring of pseudomorphic AlAs/In$_{0.53}$Ga$_{0.47}$As/InAs resonant-tunneling diodes (RTDs). Vertically-integrated 4-RTD stacks were prepared using molecular beam epitaxy (MBE), and contained designed variations in eight growth variables. Electrical characteristics were obtained from the processed wafers and compared with the sensor data. Post-growth analysis of the SE data using growth-temperature optical constants yielded thicknesses of the RTD active layers. For each of the 28 individual double-barrier RTDs prepared, both AlAs barrier thicknesses and an effective thickness value for the quantum well were determined. A linear model provided a suitable mapping of structural parameters into the room-temperature I-V characteristics. Determination of InGaAs composition using SE at growth temperature was also demonstrated. The layer thicknesses (~20 Å) and lattice strain (> 3% lattice mismatch) in this materials system make pseudomorphic RTDs a challenging case for real-time monitoring and control.

INTRODUCTION

Sensor-based *in situ* monitoring presents numerous advantages for characterization of complex semiconductor structures. Primary is the ability to monitor and analyze each layer in the structure, at the growth front and without interference from intervening capping layers. An additional consideration is the ability to record growth conditions (e.g., substrate temperature history) which are not typically recoverable by post-growth analysis of the sample. These advantages must be weighed against the inherent limitations of characterization at elevated temperature and under the typical constraints (geometrical and environmental) of the growth chamber.

In situ monitoring also forms the basis for real-time growth control. Assuming the sensor has sufficient sensitivity, accuracy and speed to monitor the quantity of interest on a reasonable timescale, and assuming the growth system can quickly respond to this input, real-time closed-loop growth control may be an achievable goal. In fact, closed-loop control may not be necessary (depending on growth reproducibility), but this may not be obvious until the groundwork of *in situ* monitoring is performed.

The other benefits of *in situ* monitoring, such as process or reactor improvement through mechanistic insight, may eventually outweigh the benefits of real-time control. In any case, there is sufficient motivation to investigate the application of *in situ* sensors to growth methods like molecular beam epitaxy (MBE).

Quantum devices represent a class of structures with a potential need for closed-loop growth control because of their sensitivity to layer thickness.[1,2] Resonant-tunneling diodes (RTDs), for example, exhibit a 50% change in peak current density with a monolayer variation in barrier thickness.[3] MBE is capable of monolayer thickness precision, but effusion cell flux transients,[4] long-term flux drift and instabilities can degrade the reproducibility of thin layers. With layer thicknesses on the order of 20 Å (for pseudomorphic AlAs/InGaAs/InAs RTDs), these devices present a formidable challenge for application of real-time monitoring and growth control. Additionally, one ultimately requires control over *device characteristics* (e.g., RTD peak current or peak voltage) rather than *structural parameters*. We are evaluating *in situ* sensors, such as spectroscopic ellipsometry (SE),[5,6,7,8,9] as a means for adaptive or closed-loop MBE growth control for improved device reproducibility.[10]

We report here the *in situ* monitoring of RTD growth using SE. The MBE-grown layers contained planned variations in structure and growth parameters in order to determine the sensitivity of the sensor data (and device characteristics) to these variables. We used previously-determined growth-temperature optical constants[11] and post-growth analysis of the SE data to extract structural parameters. We also explored various models to map the structural data into the room-temperature RTD electrical characteristics.[12] Finally, we summarize our progress towards sensor-based growth control of RTDs.

EXPERIMENTAL

The MBE-SE system used in this work has been described previously.[11,12] Briefly, we used a solid-source MBE system (VG-V90) equipped with conventional single-zone cells and valved As_4 source. The manipulator has low wobble during rotation due, in part, to the track delivery system which eliminates the need for manipulator motion during wafer transfer. The chamber is outfitted with various ports to allow sensor access.

The spectroscopic ellipsometer is a phase-modulated system developed at TI.[13] The output of a 150 W Xe lamp is transferred through a fiber, then collimated, linearly polarized and passed through a photoelastic modulator (PEM). Mild focusing brings the beam through strain-free quartz windows and onto the growth surface at a nominal angle of 75° from normal, resulting in an elliptical probed area of approximately 10 x 25 mm. The analyzer arm consists of a second fixed polarizer and fiber coupler, delivering the specular beam into a 0.32 m monochromator. A 46-element Si diode array, calibrated at the center of its range from 407.4 to 853.2 nm, is used for detection. To minimize problems associated with wobble from rotating substrates, SE measurements were made with synchronous waveform acquisition by triggering at the same position of the wafer rotation (20 rpm). The Ψ and Δ values at each wavelength were extracted from the DC, fundamental and first harmonic intensities (I_0, I_ω and $I_{2\omega}$) of the 50 kHz PEM modulation using fast waveform digitization and Fourier transform analysis, as previously described.[13]

Analysis of the SE (Ψ, Δ) data was made with the WVASE analysis software[14] by constructing appropriate structural models based on our growth-temperature (450 °C) optical constants,[11] and fitting the calculated to the observed values using standard error minimization techniques. For real-time applications, WVASE was integrated with the SE acquisition module and interfaced to both the MBE system hardware and other sensor inputs. A Windows-based graphical user interface (GUI) provides a convenient method for growth recipe building

and other operator input, as well as system status monitoring.

In addition to the SE system, the MBE system is equipped with laser light scattering (LLS)[15] and reflection mass spectrometry (REMS).[4] LLS employs a 5 mW HeNe laser and diffuse light collection through two 65° ports with periscope attachments,[16] and was used in these experiments to verify that lattice coherence was maintained throughout the growth of the 4-RTD stacks. The REMS system consists of an unapertured, line-of-sight quadrupole mass at 70° to the surface normal. REMS was employed here to verify shutter operation and effusion cell flux levels during growth, but can also be used to measure InGaAs composition in real-time.[17] Temperature measurement is performed using single-color optical pyrometry at 940 nm. The pyrometer views the wafer through an effusion cell port at ~35°, using reflection off of a silicon wafer to reduce As window deposits.

Fig. 1(a) shows the structure grown for this study, a vertically-integrated stack of four consecutively-grown RTDs (VIRTD). The substrates were epi-ready 2" InP

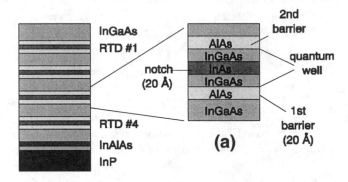

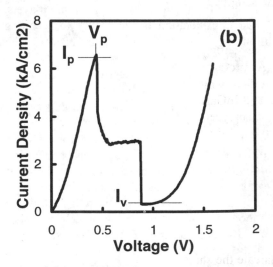

Figure 1. (a) Vertically-integrated 4-RTD stack. (b) Room temperature I-V curve of a single RTD.

(Sumitomo, Fe-doped (100), used as received), with the native oxide desorbed by heating (2 min at 550 °C) under an As_4 flux. Lattice-matched $In_{0.53}Ga_{0.47}As$ and $In_{0.52}Al_{0.48}As$ n[++]-doped (5×10^{18} cm[-3] Si) buffer layers were first deposited, followed by alternating RTD active layers and spacer layers. The standard active structure of a single RTD contains two AlAs barriers (~20 Å) sandwiching a quantum well region consisting of lattice-matched InGaAs and an InAs notch layer (~20 Å), all undoped. The 2000 Å spacer layers between RTDs contained two 500 Å n[+]-layers (1×10^{18} cm[-3] Si) with a central 1000 Å n[++]-layer. We prepared 28 RTDs (in seven 4-RTD samples) with variations, designed using the E-chip software,[18] to the nominal structure in eight parameters, including the AlAs barrier, InGaAs well and InAs well thicknesses.[12] Target thicknesses were: 6, 7 or 8 monolayers (ML) for AlAs, 5, 6 or 7 ML for the InAs notch, and 4, 5 or 6 ML for the two InGaAs quantum well layers. The VIRTDs were processed and I-V curves obtained for each RTD, as described elsewhere.[3] A sample I-V curve, including identification of the peak current density and peak voltage of the resonant feature, is shown in Fig. 1(b). The symmetry of the two resonant peaks observed for each RTD (one for forward bias, one for reverse) was also measured.

RESULTS AND DISCUSSION

Initial SE measurements were made on the oxide-desorbed InP substrate for each sample to determine the exact angle of incidence of the SE beam. The angle varies slightly, due to substrate wobble, and the average value over the SE spectral measurement time was used. This value is typically reproducible within 0.03° for a given sample holder. Once determined, the angle was fixed at this value for analysis of subsequently grown layers.

The buffer and spacer layers of lattice-matched InGaAs produced thin-film interference fringes in the Ψ and Δ data which could be fit to extract average growth rates and InGaAs composition (Fig. 2(a)). The data can also be analyzed using a point-by-point (single spectrum) pseudo-real-time fit to composition and thickness, which shows good agreement with the *ex situ* x-ray value (Fig. 2b). Use of the virtual interface model[19] to analyze various growth regions showed negligible variation in growth rates, indicating the lack of shutter-induced flux transients in our MBE system.

Sensor data from the *in situ* growth monitoring of an RTD active region are shown in Fig. 3. The REMS data demarcate the shutter opened and closed times, including growth interruptions. Most SE wavelengths show good

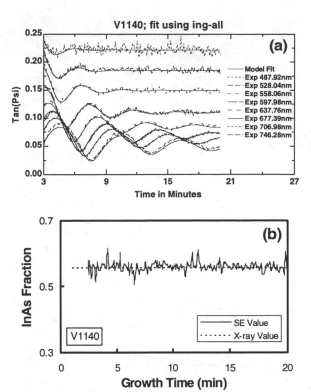

Figure 2. (a) Observed and fitted SE data for InGaAs growth on InP. (b) SE-determined InGaAs composition, from point-by-point fit.

sensitivity to AlAs barrier and InGaAs spacer layer deposition, as seen in the Ψ data (only 4 of 23 recorded wavelengths shown) plotted vs. growth time in Fig. 3(b), while the well region causes smaller changes over a limited spectral range. To extract AlAs layer thicknesses, the Ψ and Δ values at 17 wavelengths (467.8-765.9 nm) were fit using strained AlAs optical constants.[11] The thicknesses of the quantum well layers (InAs notch and 2 InGaAs spacers) could not be determined from the present data, primarily because of low S/N and the absence of growth interruptions between these layers. Treating the 3-layer well region as a single layer, we fit the Δ values over the 7 wavelengths from 488 to 608 nm to effective optical constants which had n and k values intermediate between those of InAs and InGaAs. The effective thickness values which were derived were used for qualitative comparison with the nominal well thicknesses.

Data from post-growth analysis of the 28 RTD sample set are collected in Figs. 4 and 5. Overall, the SE-determined thicknesses followed the trend of nominal values, with 1σ errors for AlAs of ~1Å. The increasing error bars are due to degraded S/N toward the end of the series from arsenic window coating. The second barrier

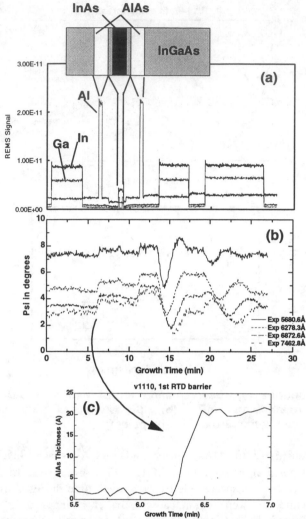

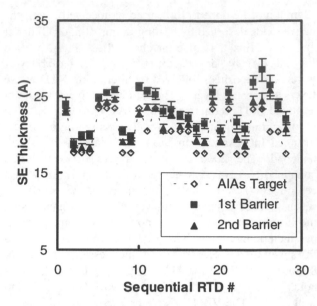

Figure 4. AlAs barrier thickness from SE.

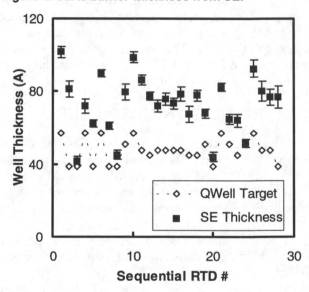

Figure 5. Effective well thickness from SE.

Figure 3. *In situ* **sensor data from RTD active layer growth: (a) REMS, (b) SE. (c) Conversion of SE data to thickness for one AlAs barrier.**

of each RTD pair is consistently thinner than the first barrier, which agrees with the current asymmetry (Ip^-/Ip^+, where Ip^- and Ip^+ are the peak currents under reverse- or forward-bias, respectively), plotted vs. barrier thickness asymmetry (t_1/t_2) in Figure 6. Since Ip^- describes electron flow from the surface toward the substrate, having the current vs. thickness asymmetry data fall in the upper right quadrant is consistent with initial tunneling through a thinner top barrier, as confirmed by the SE measurement. Other effects (e.g., interface roughness), which are not directly measured, may also play a role in determining the peak current asymmetry.

We examined the SE-determined structural parameters for correlation with RTD device properties. Mappings from both the nominal and SE-determined features to the I-V characteristics (peak current, peak voltage, valley current) were examined using a variety of modeling techniques.[12] Because of the sample set size (28) and number of variables (8), a linear model performed the best at predicting device parameters of samples not in the initial 28-sample (training) set. None of the models showed predictive capability for the current or voltage asymmetry. A search for non-linear dependences by analysis of a second designed experiment, with variations in only the most significant parameters (i.e., barrier and quantum well thicknesses), is currently in progress. Signal-to-noise improvements

510

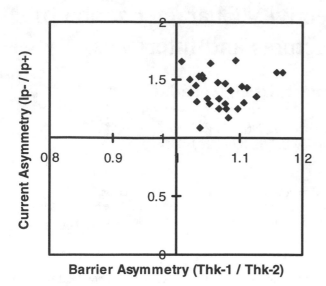

Figure 6. RTD current asymmetry plot.

and the inclusion of growth interruptions should also enable accurate quantum well layer thicknesses to be determined.

We are currently developing real-time, closed-loop RTD growth control by actuating shutter closure based on SE-determined layer thicknesses. Our present capability is depicted by the pseudo-real-time AlAs thickness analysis shown in Fig. 3(c). Significant issues include: the precision obtainable from each single-spectrum measurement; the sampling rate achievable for SE measurement and model analysis; the robustness of the control algorithm in choosing the shutter closed time, and; the timing accuracy between software and hardware components. Success will ultimately be measured by demonstrating an improvement in RTD reproducibility over dead-reckoning growth by using sensor-based growth.

CONCLUSIONS

We have demonstrated the use of *in situ* SE monitoring for MBE growth of RTDs. The SE-determined AlAs barrier thicknesses, extracted using post-growth analysis and growth-temperature optical constants, were in reasonable agreement with the nominal values. A thickness asymmetry in the AlAs double-barrier structures indicated a correlation with the observed RTD peak current asymmetry. Effective thicknesses for the InAs/InGaAs quantum wells followed the trend of nominal thickness values, but the absolute values were not physically reasonable because of the approximate analysis method used in this case. Variations in RTD properties, namely peak current, peak voltage and valley current, were successfully predicted

from a linear model which included only the AlAs, InAs and InGaAs layer thicknesses. Work is in progress to extend this model to include non-linear terms, and to evaluate the advantages of real-time, closed-loop growth control on RTD reproducibility.

ACKNOWLEDGMENTS

We gratefully acknowledge support of SE hardware and software by Walter Duncan, and partial support from ARPA under the Consortium for Advanced Materials, Synthesis and Processing (MDA972-93-H-0005).

REFERENCES

1. A. C. Seabaugh, J. H Luscombe and J. N. Randall, *Future Electron. Dev. J.*, **3**, Suppl. 1 (1993) 9.
2. T. P. E. Broekaert, W. Lee and C. G. Fonstad, *Appl. Phys. Lett.*, **53** (1988) 1545.
3. T. S. Moise, Y.-C. Kao, A. J. Katz, T. P. E. Broekaert and F. G. Celii, *J. Appl. Phys.*, submitted for publication.
4. F. G. Celii, Y.-C. Kao, E. A. Beam, III, W. M. Duncan and T. S. Moise, *J. Vac. Sci. Technol. B*, **11** (1993) 1018.
5. D. E. Aspnes, W. E. Quinn, M. C. Tamargo, M. A. A. Pudensi, S. A. Schwarz, M. J. S. P. Brasil, R. E. Nahory and S. Gregory, *Appl. Phys. Lett.*, **60** (1992) 1244.
6. S. A. Henck, W. M. Duncan, L. M. Lowenstein and S. W. Butler, *J. Vac. Sci. Technol. A*, **11** (1993) 1179.
7. J. L. Edwards, G. N. Maracas, K. T. Shiralagi, K. Y. Choi and R. Droopad, *J. Cryst. Growth*, **120** (1992) 78-83.
8. B. Johs, J. L. Edwards, K. T. Shiralagi, R. Droopad, K. Y. Choi, G. N. Maracas, D. Meyer, G. T. Cooney and J. A. Woollam, *Mat. Res. Soc. Symp. Proc.*, **222** (1991) 75-80.
9. C. Pickering, *Thin Solid Films*, **206** (1991) 275-282.
10. G. G. Barna, L. M. Loewenstein, S. A. Henck, P. Chapados, K. J. Brankner, R. J. Gale, P. K. Mozumder, S. W. Butler and J. A. Stefani, *Sol. State Technol.*, **37** (1994) 47.
11. F. G. Celii, Y.-C. Kao, W. M. Duncan and B. Johs, *Proceedings of the 21st International Symposium on Compound Semiconductors,* September, 1994 (IOP, Bristol, 1995), p. 35.
12. F. G. Celii, Y.-C. Kao, A. J. Katz and T. S. Moise, *J. Vac. Sci. Technol. A*, **13** (1995), in press.
13. W. M. Duncan and S. A. Henck, *Appl. Surf. Sci.*, **63** (1993) 9.
14. The WVASE analysis software is a product of the J. A. Woollam Co., Lincoln, NE.
15. F. G. Celii, Y.-C. Kao, H.-Y. Liu, L. A. Files-Sesler and E. A. Beam, III, *J. Vac. Sci. Technol. B*, **11** (1993) 1014.
16. A. J. SpringThorpe and A. Majeed, *J. Vac. Sci. Technol. B*, **8** (1990) 266.
17. F. G. Celii, Y.-C. Kao and H.-Y. Liu, unpublished.
18. B. Wheeler, *Experiments in a Chip* (ECHIP, Hockessin, DE, 1989).
19. D. E. Aspnes, *Appl. Phys. Lett.*, **62** (1993) 343.

Energy Dispersive X-Ray Reflectivity Characterization of Semiconductor Heterostructures and Interfaces

E. Chason[a], T.M. Mayer[a], Z. Matutinovic Krstelj[b] and J. C. Sturm[b]

a) Sandia National Laboratories, Albuquerque, NM 87185-0350
b) Dept. of Electrical Engineering, Princeton U., Princeton, N.J. 08544

Energy dispersive X-ray reflectivity is a versatile tool for analyzing thin film structures. Layer thickness, interface roughness and composition can be determined with a single non-destructive measurment. Use of energy dispersive detection enables spectra to be acquired in less than 500 s with a rotating anode X-ray generator, making the study of kinetics possible.

Multiple advanced device structures incorporate thin semiconductor heterolayers to obtain enhanced performance, e.g., heterojunction bipolar transistors (HBT's), enhanced mobility FET's and resonant tunneling diodes. These structures require precise control of the layer thickness and interface morphology for optimal performance. We have developed a technique using energy dispersive X-ray reflectivity (XRR) to characterize these structures with high depth sensitivity. The parameters that can be determined from this measurement are layer thickness, composition and surface/interface roughness. Layer thickness can be determined in the range of 3 - 200 nm with ± 5% resolution and surface and interface roughness can be determined in the range of 0.1 - 3 nm with ± 20% resolution.

We have applied the XRR technique to characterize CVD-grown SiGe/Si hetero-structures [1], enabling us to correlate the layer thickness and interface roughness with the growth conditions. We have also performed real-time measurements during low-energy ion sputtering of semiconductors [2,3] and SiO_2 [4] to determine the kinetics of surface evolution. These measurements confirmed the presence of a rapid roughening instability and enabled us to quantitatively determine the value of ion-enhanced transport parameters (diffusivity, viscosity) important for sputter-morphology evolution.

PRINCIPLES OF X-RAY REFLECTIVITY

The origin of XRR is treated in a number of publications, both in an optical multilayer [5] and a diffraction formalism [6], so only a brief treatment will be presented here. X-ray reflectivity is defined as the ratio of the reflected intensity to the incident intensity and is measured as a function of the scattering vector,

$$k = 4\pi/hc\ E\ \sin\theta \qquad (1)$$

where E is the energy of the X-ray and 2θ is the scattering angle. In contrast with X-ray *diffraction*, X-ray reflectivity is performed at small values of k where the reflectivity can be interpreted using the Fresnel equations. The layer is treated as a continuous medium with an index of refraction, n, that depends on the electron density, ρ_{el} [7]. The index of refraction for X-rays in matter is less than 1 so that for sufficiently small incident angles total external reflection occurs.

Above the critical value for total external reflection (k_c), the reflectivity from an ideal interface is given by the Fresnel reflectivity $(R_F(k))$ with the asymptotic form $(k/k_c)^{-4}$. For an imperfect interface, the reflectivity is given approximately by [6]

$$R(k) = R_F(k)|\ F(d\rho_{el}/dz)|^2 \qquad (2)$$

where $F(d\rho_{el}/dz)$ is the Fourier transform of the electron density gradient in the direction normal to the surface of the film. It is important to note that this equation refers to the specular reflectivity, i.e., where the angle of incidence equals the angle of reflection. Under these conditions, the scattering vector is oriented normal to the surface so that the reflectivity is only sensitive to variations in the electron density normal to the film

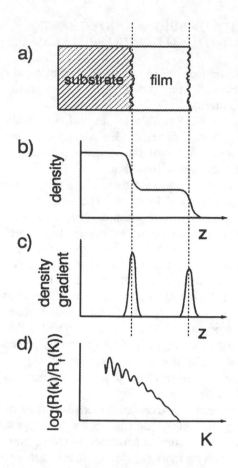

Figure 1. Relationship between thin film structure and X-ray reflectivity as expressed in eq. (2). a) Film structure consisting of homogeneous layer on substrate. b) Electron density normal to surface correspondng to structure in (a). c) Gradient in electron density showing peaks at interfaces (indicated by dotted lines). d) Reflectivity derived from Fourier transform of electron density gradient normalized by Fresnel reflectivity.

surface and does not probe the interface structure in the plane of the film.

The relationship expressed in eq. (2) between the thin film and the reflectivity is shown schematically in figure 1. The structure consists of a single uniform layer on a substrate (fig. 1a). The electron density in the direction normal to the film surface (fig. 1b) is constant where the film composition is uniform and changes at the interfaces between the substrate, the film and vacuum. The gradient of the electron density ($d\rho_{el}/dz$) has peaks at these interfaces as shown in fig. 1c. If the interface is smooth, then the peak in $d\rho_{el}/dz$ will be narrow, while if the interface is rough or diffuse, the peak will be broader.

The normalized reflectivity ($R(k)/R_F(k)$) from this structure (fig. 1d) is given by the Fourier transform of the density gradient. The oscillations shown in the reflectivity spectra come from interference between scattering from the surface and the buried interface. The period of the oscillations is inversely proportional to the layer thickness. The decay in the reflected intensity is determined by the roughness of the interfaces. For rough interfaces, the reflectivity decreases faster with increasing k than for sharp interfaces. The roughness is generally taken to have a Debye-Waller form ($\exp(-k^2\sigma^2)$) where σ is the interface roughness.

EXPERIMENTAL TECHNIQUE

The details of the experimental apparatus are shown in figure 2. The X-ray source is a rotating anode generator operated at 40 kV, 100 mA. In the energy dispersive technique used in this work, the broad range of X-ray energies produced by Bremsstrahlung radiation from a Mo anode impinge simultaneously on the sample at a fixed angle. A solid-state Ge detector is used to measure the reflectivity at each energy and the energy spectrum is converted to wavevector using eq. 1. In order to obtain the reflectivity, the measured spectra is normalized by the incident energy spectrum.

The X-ray beam is incident on the sample at a grazing incidence angle between 0.4 and 1.0 deg. The angular divergence of the ingoing and outgoing beam is typically on the order of 0.01 deg as defined by slits. The angular resolution is chosen to match the energy resolution of the Ge detector, approximately 1-2 % in the energy range of 8 - 35 KeV used. The total area of the sample illuminated by the X-ray beam is approximately 0.5 x 0.5 cm^2; the lateral coherence length of the X-rays is on the order of 1 μm.

There are several advantages to energy dispersive detection over conventional angle scanning for in situ measurements. The fixed angle of incidence means that the sample does not have to be moved during the measurement. The fixed angle also means that the footprint of the beam on the sample is constant so no corrections need to be made for low k values. By using

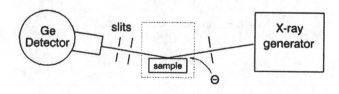

Figure 2. Schematic of apparatus for energy dispersive X-ray reflectivity measurements.

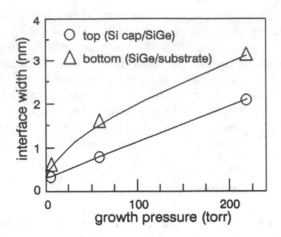

Figure 4. Dependence of roughness on growth pressure for top (Si cap/SiGe) and bottom (SiGe/Si substrate) interfaces.The spectra were fit to a two-layer model; the calculated intensity from the multilayer model is shown as the dashed line in the figures. The intensity of the oscillations in the spectra decreases for the samples grown at higher pressures.

Analysis of these spectra indicate that the decrease in the oscillation intensity corresponds to increased roughness at the interfaces. The roughness at the SiGe/substrate interface and the SiGe/Si cap interface is shown in figure 4 as a function of growth pressure. The increase in surface roughness with increasing growth pressure was determined to be caused by the presence of graded layers at the interfaces created by a transient in the switching time of the gases in the growth reactor. This discovery was used to determine growth conditions to produce more abrupt interfaces.

KINETICS OF SURFACE ROUGHENING AND SMOOTHING DURING SPUTTERING

We have used the in situ measurement capabilities of XRR to study the evolution of surface roughness during sputtering of SiO_2 [4] surfaces. These measurements were performed in real time while the surface was being sputtered. Be windows allowed the X-rays to enter and exit the sputtering chamber with minimal attenuation of the beam.

We initially bombard the surface with heavy Xe ions with an energy of 1000 eV. The surface roughness is shown in figure 5 to rise approximately linearly with ion fluence. These kinetics are much more rapid than the $t^{1/2}$ behavior that is expected if the sputter beam is randomly removing atoms from the surface. The Xe-roughened surface is then bombarded with light (H or He) ions. The change in roughness with time is shown in fig. 6 and indicates that the surface roughness decreases exponentially with ion fluence with a rate that increases with ion energy. XRR spectra corresponding to increasing He fluence are shown in fig. 7; the decrease in the surface roughness is evident by the decreasing slope of the spectra.

These kinetic studies confirmed the presence of a roughening instability during heavy ion sputtering caused by the curvature dependence of the sputter yield that leads to much more rapid roughening than a simple stochastic removal process. The smoothing by light ions was due to ion-enhanced viscous flow of the oxide. Although in principle similar measurments could be obtained by sequences of AFM measurements, these would be very difficult to obtain. The real time capability of the XRR provided critical kinetic data for

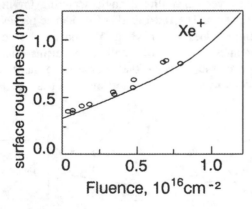

Figure 5. Kinetics of SiO_2 surface roughening induced by low energy Xe sputtering. Line represents fit to model of sputter-induced roughening instability.

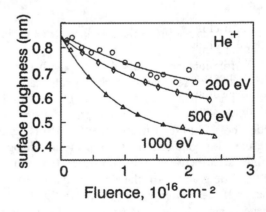

Figure 6. Decrease in surface roughness during He sputtering of SiO_2. Surfaces were initially roughened by Xe sputtering. Lines show fit to exponential decay due to ion-enhanced viscous flow.

the Bremsstrahlung radiation, spectra can be obtained from a laboratory-based X-ray generator in less than 500 s so that kinetics can be measured. The disadvantages to energy dispersive detection are reduced resolution and possible interference by X-ray fluorescence from the sample.

XRR ANALYSIS AND SENSITIVITY

For analysis of the reflectivity spectra, an optical multilayer model that takes into account multiple scattering is used. Optimum values of parameters corresponding to layer density, thickness, surface roughness and buried interface roughness are obtained using a non-linear least-squares fitting routine. It is generally impractical to model a structure containing more than two layers unless some of the parameters can be determined by alternative means.

The sensitivity of the least squares fit to changes in the parameters is used to obtain a value for the error associated with each parameter. The layer thickness can be determined with a resolution of approximately ± 5% in the range of 3 - 200 nm. The surface and interface roughness can be determined in the range of 0.1 - 3 nm with a resolution of approximately ± 20 % of the roughness. For smooth surfaces, this implies a roughness sensitivity of ± 0.02 nm. The determination of electron density is strongly coupled to the degree of roughness, but typical sensitivity is on the order of ± 30%.

XRR CAPABILITIES AND COMPARISON WITH OTHER TECHNIQUES

It is useful to compare XRR with other thin film analytical techniques to determine its advantages and disadvantages. The greatest strength of XRR is the ability to measure thickness, density and roughness in a single measurement. Because the X-rays are highly penetrating, buried layer interfaces can be probed as well as surface roughness. The technique is non-destructive and requires no special sample preparation. The glancing incidence geometry does not interfere with deposition and other processing techniques. Ambient gas processing environments are not a problem since the technique does not require a vacuum like electron diffraction does. The measurement averages over a large area instead of a small fraction of the sample like cross-sectional transmission electron microscopy (XTEM) or scanning probe microscopies (AFM, STM). The depth sensitivity is very high compared to other probes like

Rutherford backscattering (RBS). The technique can be employed in situ and using energy dispersive detection, spectra can be obtained in less than 500 s making studies of kinetics possible. In comparison with ellipsometry, the optical properties of the layers depend only on the total electron density and can be more easily modeled.

The primary drawbacks to the technique are that parameters can not be obtained directly and the measured spectra need to be fit to an optical multilayer thoery. This makes the uniqueness of the parameter set obtained difficult to determine, especially for more complicated structures. In comparison with imaging techniques such as XTEM and STM, XRR provides no information about the in-plane structure of the layers.

CVD GROWTH OF SiGe HETEROSTRUCTURES

We have used XRR to characterize heterostructures of SiGe grown by CVD. Further details of the growth apparatus and experimental conditions can be found in ref 1. The structures consisted of a Si substrate, a SiGe layer and a Si cap. In figure 3, we show the reflectivity spectra from three samples grown at different pressures (the zeroes of the curves have been displaced for clarity).

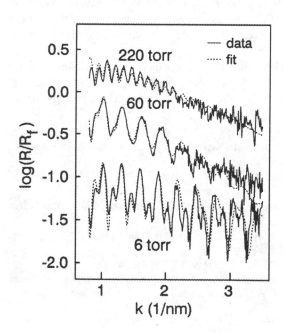

Figure 3. XRR spectra from CVD-grown hetero-structures consisting of Si cap/SiGe/Si substrate. Dashed lines represent fit to optical multilayer model. Growth pressure indicated on figure.

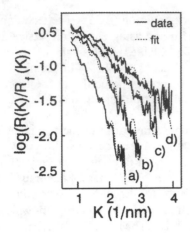

Figure 7. XRR spectra from SiO$_2$ surfaces smoothened by low energy He sputtering. Increasing ion fluence from a) to d). Dashed lines represent fit to optical multilayer model.

development of models of surface roughening and smoothing.

In summary, XRR is a useful in situ probe of semiconductor thin films. It provides information about layer thickness, composition and interface roughness with sensitivity in the nanoscale regime that is becoming increasingly important technologically. Simple non-destructive sample preparation, relatively rapid data acquisition and compatability with deposition geometries make it suitable for in situ kinetic studies of morphology evolution. For complete characterization of the structure, XRR needs to be combined with other in-plane probes.

The authors gratefully acknowledge useful discussions with Bruce Kellerman. This work was supported by the U.S. Department of Energy under contract DE-ACO4-94AL85000.

1. Z. Matutinovic-Krstelj, J.C. Sturm and E. Chason, J. Elect. Mater., submitted.

2. E. Chason, T.M. Mayer, B.K. Kellerman, D.N. McIlroy and B.K. Kellerman, Phys. Rev. Lett. **72**, 3040 (1994).

3. E. Chason, T.M. Mayer, A. Payne and D. Wu, Appl. Phys. Lett 60, 2353 (1992).

4. T.M. Mayer, E. Chason and A.J. Howard, J. Appl. Phys. **76**, 1633 (1994).

5. L.G. Paratt, Phys. Rev. **95**, 359 (1954)

6. J. Als-Nielsen in *Structure and Dynamics of Surfaces*, edited by W. Schommers and P. von Blanckenhagen (Springer, Berlin, 1986), p. 181.

7. J.H. Underwood and T.W. Barbee, Appl. Opt. **20**, 3027 (1981).

Real Time Spectroellipsometry Characterization of the Fabrication of Amorphous Silicon Solar Cells

R. W. Collins, Yiwei Lu, Sangbo Kim, and C.R. Wronski

Materials Research Laboratory and the Department of Physics
The Pennsylvania State University, University Park, PA 16802

Real time spectroellipsometry (RTSE) has been applied to obtain a wealth of information on the microstructure and electronic structure of hydrogenated amorphous silicon-based solar cells in the superstrate configuration (glass/SnO_2:F/p-i-n/Cr). The ellipsometer is based on the rotating polarizer principle and uses a linear photodiode array detector to collect 100-point spectra in the ellipsometric parameters (ψ, Δ) from 1.5 to 4.2 eV. In monitoring the growth of the thin (~200 Å) doped layers of the device, acquisition and repetition times for full spectra are 0.16 and 1 s, respectively. In monitoring the development of the much thicker (5000 Å) intrinsic (i) layers, slower acquisition and repetition times are employed, 3.2 and 15 s, respectively. With 0.16 and 3.2 s acquisition times, the precision of the instrument in ψ is ~0.01° and 0.003°, respectively, for a Cr surface measured at 2.5 eV. In the studies of solar cell fabrication, the ellipsometric spectra are analyzed after deposition and provide the surface roughness and bulk layer thicknesses as well as interface roughness and contaminant layer thicknesses. The dielectric functions of the separate bulk layers are also determined so that the optical gaps can be estimated. We demonstrate the accuracy of the analysis through comparisons of roughness layer thicknesses deduced by RTSE and by atomic force microscopy. The future directions of this work involve characterizing graded layers that are often built into the solar cell p-i interface region.

INTRODUCTION

For complex semiconductor device structures, the properties of very thin layers may differ substantially from those of thick layers. The differences arise from the evolution of structure with thickness, or from size effects on the electronic properties. Also, the layers of a structure may interact during deposition or in subsequent processing. Such effects may not be easily discovered in a post-process analysis of the completed device. With real time measurements of device fabrication, however, small differences or changes in layer properties can be detected and traced to specific processing steps.

Hydrogenated amorphous silicon (a-Si:H) solar cell prepared by plasma-enhanced chemical vapor deposition (PECVD) represent a classic system in which such problems occur (1). First, the ~200 Å doped layers used in these cells are often different than the ~5000 Å layers normally used for electrical evaluation (2). In addition, interactions between the transparent conducting oxide (TCO) substrate layer (usually SnO_2:F) and the hydrides in the plasma used to deposit the overlying layer can modify the properties of the TCO (3,4).

Effects such as these yield device performance characteristics that depend critically on processing procedure details. Thus, the performance of the a-Si:H solar cells is often irreproducible and cannot be predicted from the information obtained on the thicker counterparts of the layers that form the device. In order to expand current knowledge of the relationships between the preparation and properties of devices and to identify sources of irreproducibility, monolayer-sensitive real time probes are needed. Here we briefly review the characteristics and applications of real time spectroscopic ellipsometry (RTSE) developed for this purpose, emphasizing its ability to characterize very thin and graded layers in amorphous semiconductor device structures with precision and accuracy.

EXPERIMENTAL DETAILS

Figure 1 shows a schematic of the real time spectroscopic ellipsometer, interfaced to a single thin film deposition chamber for rf PECVD of a-Si:H solar cells (5). The ellipsometer employs a rotating polarizer

(12.5 Hz) for incident polarization state modulation. A fixed analyzer, spectrograph, and photodiode array are used for polarization state detection at 128 wavelengths in parallel. The typical useful spectral range is 1.5-4.2 eV. At a rotation frequency of 12.5 Hz, the minimum acquisition time for full spectra in (ψ, Δ) is 40 ms. In monitoring the thin (~200 Å) p-layer as well as the p/i interface of the SnO$_2$:F/p-i-n/Cr solar cell structure, data acquisition and repetition times are 0.16 and 1 s, respectively. In monitoring the development of the thicker i-layer after formation of the p/i interface, slower acquisition and repetition times are employed, 3.2 and 15 s. With 0.16 and 3.2 s acquisition times, corresponding to averages of 4 and 80 optical cycles, the precision in the ellipsometric parameter ψ is ~0.01° and 0.003°, respectively, for a Cr surface measured at 2.5 eV.

The a-Si:H solar cells were co-deposited on three types of substrates: textured and specular SnO$_2$:F-coated glass for device measurements, and a specular SnO$_2$:F-coated Si wafer for RTSE at enhanced sensitivity. The substrate temperature was fixed at 240°C. The p- and i-layers were prepared with an rf power flux of 0.07 W/cm^2. The p-layer was a-Si$_{0.91}$C$_{0.09}$:H doped with B using gas flow ratios of [SiH$_4$]:[CH$_4$]:[B$_2$H$_6$]:[H$_2$] =6:4:0.01:0.99 (in sccm). With this mixture, formation of metallic Sn via the reduction of the SnO$_2$:F substrate surface by the hydrides is below detection limits (< 1 Å). A 55 min Ar flush was performed after p-layer and before a-Si:H i-layer deposition. For the latter, pure SiH$_4$ was used. In order to complete the device, 200 Å of P-doped microcrystalline Si (μc-Si:H) was deposited on top of the i-layer using a power of 0.5 W/cm^2 and a gas flow ratio of [SiH$_4$] : [PH$_3$] : [H$_2$] = 0.99 : 0.01 :1 00 (in sccm).

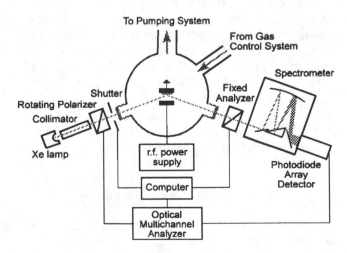

Fig. 1 Schematic of the single-chamber PECVD system and real time spectroscopic ellipsometer for real time monitoring of a-Si:H solar cell fabrication.

Finally, a 1000 Å Cr top contact was evaporated onto the n-layer. The deposition of the n-layer and the Cr top contact were not monitored by RTSE.

RESULTS AND DISCUSSION

a-Si:H Solar Cell Analysis

Figure 2 shows the structural evolution deduced from RTSE data for p-layer growth by PECVD, for p/i interface layer formation during Ar flushing, and for the initial stage of i-layer growth by PECVD. Also shown in Fig. 2 are the dielectric functions $\varepsilon=\varepsilon_1+i\varepsilon_2$ (insets) and the optical gap determinations for the three layers. The gap is obtained by plotting $(\varepsilon_2)^{1/2}$ versus energy, and extrapolating the linear behavior to $\varepsilon_2=0$ (6).

In the RTSE data analysis for a-SiC:H p-layer growth on SnO$_2$:F, we use a three-layer model consisting of a surface roughness layer, a bulk layer, and an interface roughness layer that arises from the surface roughness on the SnO$_2$:F. Thus, we determine (i) the dielectric function of the bulk a-SiC:H p-layer, (ii) the thickness d_b of the bulk p-layer, (iii) the thickness d_s of the surface roughness on the p-layer, (iv) the thickness d_i of the roughness on the SnO$_2$:F, which fills in with p-type a-SiC:H to become interface roughness, and (v) the volume fraction f_i of p-type a-SiC:H filling the SnO$_2$:F roughness layer. Assumptions in the analysis are that the surface roughness layer of thickness d_s has a composition of a-SiC:H:B/void given by 0.5/0.5, and that the interface roughness layer of thickness d_i has a composition of SnO$_2$:F/a-SiC:H:B/void given by 0.5/f$_i$/(0.5−f$_i$). The analyses of p/i interface layer formation and i-layer growth are similar, the only difference being the nature of the substrate onto which the deposition occurs (and thus the thickness, d_i). For the p/i interface layer, the thickness on top of the p-layer is so small, that no distinct bulk density layer can develop.

Figure 2 shows that in the first 55 s of p-layer growth, the modulations in the 98 Å roughness layer on the SnO$_2$:F surface fill with p-type a-SiC:H ($f_i \rightarrow 0.5$). After about 55 s, a distinct bulk layer forms and increases linearly at a rate of 1.5 Å/s. A smoothening effect is observed whereby the roughness thickness on the final p-layer surface is less than that on the SnO$_2$:F. Upon termination of the plasma, d_s=67 Å, d_b=111 Å, and d_i=98 Å. This leads to a final "mass" thickness of $d_{mass} = 0.5d_s + d_b + f_id_i = 194$ Å, which is close to the intended value of 200 Å. During Ar flushing, a contaminant layer develops on the p-layer surface (6). The contaminant material, a Si-B alloy, is somewhat arbitarily partitioned as either falling into the modulations of the 67 Å p-layer roughness (which increases f_i from 0 to 0.035), or forming a distinct roughness layer of thickness d_S (which increases to 3 Å). This partitioning is not critical; however, the important parameter is the final mass thickness, given in this case by $d_{mass}= f_i d_i + 0.5 d_S = 3.8$ Å. Figure 2 shows that in the first 45 s of i-layer growth, the i-layer fills the 67 Å modulations of the p-layer, trapping the contaminant material at the modulated p/i interface. The bulk i-layer growth rate is 1.0 Å/s, and the roughness on the i-layer decreases with time. After 5000 Å of i-layer growth, the roughness thickness has dropped to 21 Å (not shown).

The optical gaps in Fig. 2 provide useful insights, as well as essential input, for modeling solar cell performance. It should be recalled, however, that the gaps in Fig. 2 are characteristic of a sample temperature of 240°C, and are smaller than the room temperature Tauc

gaps by ~0.20 eV. Figure 2 shows that by incorporating C, a p-layer gap 0.08 eV wider than the i-layer has been achieved. However, the p-i interface contaminant layer exhibits an absorption onset that is ~0.27 eV lower than the i-layer. Because of the thinness of this layer, we cannot interpret this onset as a gap in the usual sense, but rather an indication of a high density of electronic states below the band edges of the p- and i- layers.

The growth and flushing procedures applied here are designed to simulate those used in multichamber deposition of solar cells. We conclude that solar cell efficiency improvements are possible through changes in the procedures that either passivate the contaminant layer or avoid its formation. The solar cell co-deposited onto textured SnO_2:F along with the structure of Fig. 2 exhibited an open circuit voltage (V_{oc}) of 0.74 V, a

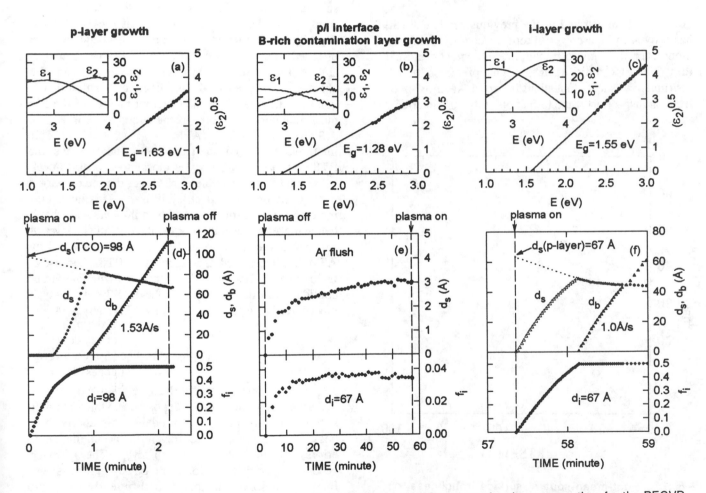

Fig. 2   Optical data at 240°C (top panels) along with thicknesses and void volume fractions versus time for the PECVD p-layer (left), the p/i interface contaminant layer (center), and the PECVD i-layer (right) that make up a SnO_2:F/p-i-n/Cr a-Si:H solar cell structure. The optical gap is determined by an extrapolation of the linear trend in $(\varepsilon_2)^{1/2}$ to zero ordinate, where $\varepsilon=\varepsilon_1+i\varepsilon_2$ is the dielectric function. The thicknesses d_S, d_b, and d_i, are for the surface roughness, bulk, and interface roughness layers, respectively. In addition, f_i is the volume fraction of the film that forms within the modulations of the roughness layer on the substrate material. The substrate roughness is completely filled in when f_i reaches 0.5, and once this occurs, a distinct bulk layer forms. In the bulk film growth regime (i.e., when $d_b> 0$), f_i is fixed at 0.5.

519

short-circuit current (J_{sc}) of 13.7 mA/cm^2, and a fill-factor (FF) of 0.65 for an initial efficiency of 6.6%. As a reference point, similarly-prepared cells fabricated in a commercial multichamber system exhibited V_{oc}=0.85 V, J_{sc}=12.3 mA/cm^2 and FF=0.67 for an initial efficiency of 7.0%. Interface passivation treatments using atomic H have been designed based on RTSE studies. These treatments have led to improvements in V_{oc} and J_{sc} of solar cells prepared by multichamber deposition, such that initial efficiencies of >8% have been achieved. Although the improvements are modest, this work suggests that insights into the monolayer growth processes can lead to improvements in devices.

A Correlation with Atomic Force Microscopy

As demonstrated in the previous section, RTSE has proven to be a powerful tool for characterizing the monolayer scale microstructural development of a-Si:H thin film electronic devices. The precision of the instrument in such applications is ~±0.1 Å, as assessed from an inspection of the formation of the p/i interface

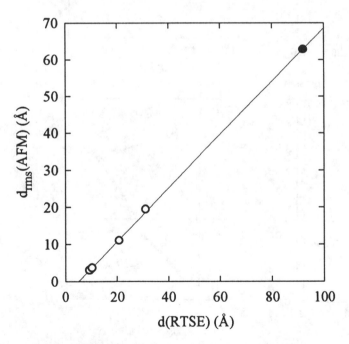

Fig. 3 Root-mean-square surface roughness on a-SiC:H samples measured ex situ by atomic force microscopy (AFM), plotted as a function of the surface roughness layer thickness on the final film measured in real time by spectroellipsometry. The line is the best fit to the data. For the solid point and open points, respectively, a SnO$_2$:F substrate and Si wafer substrates were used.

layer in Fig. 2. The accuracy of the thicknesses is another question, however. Generally, the accuracy is believed to be limited by the ability of the optical models to simulate reality. The uncertainties center on the validity of the effective medium theory (EMT) used to determine the roughness layer thickness and void densities, as well as the meaning of the roughness thickness. Previous studies have demonstrated that the Bruggeman EMT is best for fitting spectroscopic ellipsometry data measured over a wide spectral range, for a-Si samples of different roughness thicknesses (8). Further studies comparing the results of independent optical and structural analyses are needed, however.

Here, we have correlated RTSE and atomic force microscopy (AFM) results from the same set of a-SiC:H samples. The RTSE measurements were performed throughout the deposition of films on Si and SnO$_2$:F substrates. In Fig. 3, the roughness thickness obtained from RTSE data collected in situ at the end of the deposition is compared with the root-mean-square value obtained from an image analysis of AFM performed ex situ on the same sample after removing it from the chamber. The points follow a linear correlation that is quite remarkable given the different natures of the two measurements. The form of the straight line in Fig. 3, $d(RTSE) = 1.4d_{rms}(AFM) + 4$ Å, has a simple interpretation. First, the slope of 1.4 indicates that the RTSE-deduced roughness is closer to a peak-to-peak value than to an rms value. Second, the intercept of 4 Å suggests one of two possibilities. This layer may represent: (i) true roughness with an in-plane scale that is below the AFM resolution or (ii) the continuous layer of SiH$_n$ bonds at the film surface that is not detected by AFM, but interpreted as roughness by RTSE because of its lower Si-Si bond packing density. In summary, our confidence in the accuracy of the RTSE data analysis procedures is reinforced as a result of this study.

Future Directions: Graded Layers

In the analysis of Fig. 2, we followed the full history of the p- and i-layer depositions, which required seven layers in the optical model. This approach is possible if care is taken to prevent the accumulation of errors with the addition of each layer. The approach becomes more difficult for continuously graded layers. In order to solve this problem differently, we have adapted a minimal-data approach to RTSE (9). This approach allows us to obtain the evolution of the optical properties of the top 15 Å of the bulk film and the surface roughness on top of this layer, without any knowledge of the underlying sample structure. Our first test of this approach was to form structures having graded void volume fractions. This was achieved by ramping the H_2-dilution in an a-SiC:H alloy film, keeping all other deposition parameters constant. The hydrogen dilution ratio is characterized by the parameter $R=[H_2]/\{[CH_4]+[SiH_4]\}$. As shown in Fig. 4 (top), we started with R=2 and increased R linearly with time to the optimum value of R=20, returned to R=2, and then back to R=20, all at the same ramping rate. In earlier work, we found that the H_2-dilution ratio strongly affects the void volume fraction in the a-SiC:H film, but has little effect on the C and H contents (4). Because of this, the void fraction is expected to be the primary variable parameter in characterizing the optical property gradients.

The minimal-data approach when used at a single photon energy provides the composition (9), however, the additional data resulting from the spectroscopic capability yields the instantaneous deposition rate and the surface roughness thickness. The details of this analysis will be provided elsewhere. The top part of Fig. 4 includes the instantaneous deposition rate which varies

from ~1.3 Å/s when R=2 to 0.6 Å/s when R=20. By integrating the instantaneous deposition rate, the accumulated thickness can be determined versus time. In the lower part of Fig. 4, the gradient in void volume fraction and the surface roughness layer thickness is plotted continuously versus accumulated bulk layer thickness. The change in roughness is relatively small throughout the preparation process, however, the relative void fraction shows large gradients that track the variation in R. The results here are unique in that they represent the first successful analysis of amorphous semiconductor structures having continuously varying properties with thickness. The procedures we have developed here will be readily adaptable to samples with continuously varying alloy compositions. Once this procedure has been perfected, we plan to apply it to characterize graded alloy layers in solar cells and study the effects of the optical gap profile on cell efficiencies.

Acknowledgements

This research was supported by the AT&T Foundation, the National Science Foundation (DMR-8957159 and DMR-9217169), the National Renewable Energy Laboratory (XAN-4-13318-03), and the New Energy and Industrial Technology Development Organization (NEDO) of Japan. The authors would like to thank Dr. Y. E. Strausser for AFM measurements.

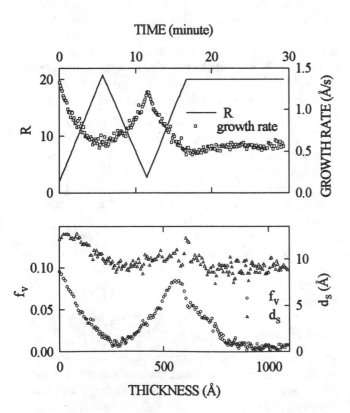

Fig. 4   Results of RTSE analysis of an a-SiC:H film having a graded void density, prepared by ramping the H_2-dilution ratio R between R=2 and R=20 (solid line, top). Also shown at the top is the instantaneous deposition rate. At the bottom, the void volume fraction vs. distance from the substrate film is plotted, along with the surface roughness thickness versus accumulated film thickness.

References

1.   Luft, W., and Tsuo, Y., *Hydrogenated Amorphous Silicon Deposition Processes*, New York: Marcel Dekker, 1993.
2.   Collins, R.W., Clark, A.H., Huang, C.-Y., and Guha, S., *J. Appl. Phys.* **57**, 4566-4571 (1985).
3.   Kumar, S., and Drevillon, B., *J. Appl. Phys.* **65**, 3023-3034 (1989)
4.   Lu, Y., An, I., Gunes, M., Wakagi, M., Wronski, C.R., and Collins, R.W., *Mater. Res. Soc. Symp. Proc.* **297**, 31-36 (1993).
5.   An, I., Li, Y.M., Nguyen, H.V., and Collins, R.W., *Rev. Sci. Instrum.* **63**, 3842-3848 (1992).
6.   Cody, G.D., in *Semiconductors and Semimetals, Vol. 21B*, edited by J.I. Pankove, Orlando: Academic, 1984, ch. 2 pp. 11-82.
7.   Collins, R.W., *Appl. Phys. Lett.* **53**, 1086-1088 (1988).
8.   Aspnes, D.E., Theeten, J.B., and Hottier, F., *Phys. Rev. B* **20**, 3292-3302 (1979).
9.   Aspnes, D.E., *J. Opt. Soc. Am. A* **10**, 974-983 (1993).

Performance capabilities of reflectometers and ellipsometers for compositional analysis during $Al_xGa_{1-x}As$ epitaxy

W. Gilmore III,[a] D. E. Aspnes, and C. B. Lee[a]

Department of Physics, North Carolina State University, Raleigh, NC 27695-8202
[a]Department of Electrical Engineering, North Carolina A&T State University, Greensboro, NC 27411

We perform model calculations to establish performance capabilities for determining compositions of the prototypical semiconductor alloys $Al_xGa_{1-x}As$ from kinetic reflectance, complex reflectance, or ellipsometric data as needed for process control. A virtual-interface (V-I) approach is used. Additional calculations are performed for R where the substrate properties are assumed to be known. Measurement of phase as well as amplitude improves relative sensitivities by over an order of magnitude.

INTRODUCTION

The composition of the outermost subsurface region of depositing material is encoded in its dielectric response, ε_o, which must be determined for sample-driven, closed-loop feedback control of epitaxy. While the determination of the average dielectric response of optically thick films from kinetic reflectometric data is standard practice in optical coatings technology,[1] a stable algorithm for determining ε_o for the very thin films of interest in semiconductor technology has been developed only recently.[2] In this virtual-interface (V-I) approach one makes use of the value and thickness derivative of either $\tilde{r}_p$ or $\tilde{r}_s$, the complex reflectances for light polarized parallel or perpendicular to the plane of incidence, respectively. These are used to determine the values of ε_o and a virtual complex reflectance $\tilde{r}_v^p$ or $\tilde{r}_v^s$, after which the virtual reflectance is discarded.

Although complex reflectometry can be performed at normal incidence by interferometric techniques, the mechanical stability required is beyond present technical capabilities. Consequently, kinetic data are presently obtained by measuring either the power reflectance $R = |\tilde{r}|^2$ by reflectometry or the complex reflectance ratio $\rho = \tilde{r}_p/\tilde{r}_s$ by ellipsometry. Since ρ involves both $\tilde{r}_v^s$ and $\tilde{r}_v^p$, the V-I approach for ρ must either make use of the exact equations[3] or a simplifying approximation. In control situations the latter is the only viable alternative because in the ideal case the film thickness of interest is vanishingly small, of the order of the last several deposited

monolayers, and insufficient information is available to use the exact equations. This difficulty was resolved by the linearized virtual-substrate approximation (LVSA), where $\tilde{r}_v^p$ and $\tilde{r}_v^s$ are assumed to result from a common virtual substrate with dielectric function ε_v.[2] Using the LVSA fully automatic sample-driven growth of complex semiconductor structures has been realized.[4]

Although ellipsometry is well developed and viable algorithms for calculating ε_o exist, many growth chambers do not have ports that provide the oblique-incidence optical access to the sample that is needed to use this technique. Consequently, the first significant applications of real-time optical diagnostics in semiconductor technology are likely to involve near-normal-incidence reflectometry. Applications to relatively thick films have already been reported.[5] As a result, it is important to establish the capabilities of reflectometry for determining ε_o from a V-I analysis of kinetic R data. This information would allow the relative merits of reflectometry and ellipsometry to be compared and would establish design criteria for specific applications.

In this work we examine these issues by using model calculations to evaluate the performance capabilities of reflectometry, complex reflectometry, and ellipsometry for $Al_xGa_{1-x}As$ ($0 \leq x \leq 0.30$) epitaxy on GaAs. This system is important in semiconductor technology, its dielectric properties under growth conditions are known,[6] control data are available,[7] and results for other semiconductor materials should be qualitatively similar. We find that reflectometry and ellipsometry are comparable for optically thick films, mainly because the poorer sensitivity of the reflectance approach for determining ε_0 is compensated by the higher precision to which R data presently can be obtained. Although sensitivity can be improved somewhat for R by assuming that the underlying sample structure is known (Fresnel theory), for film thicknesses below several hundred Å ellipsometry must be used.

THEORY

In the V-I approach the contribution of the underlying layers to the optical response is not evaluated explicitly but is parameterized on a wavelength-by-wavelength basis as a virtual complex reflectance $\tilde{r}_v^p$ or $\tilde{r}_v^s$, which is the ratio of the reflected to transmitted waves of either s- or p-polarized light at a distance d below the surface. In this formulation the normal-incidence complex reflectance $\tilde{r}$ is given by

$$\tilde{r} = (\tilde{r}_{oa} + Z\tilde{r}_v)/(1 + Z\tilde{r}_v\tilde{r}_{oa}), \qquad (1a)$$

where

$$\tilde{r}_{oa} = (n_a - n_o)/(n_a + n_o), \qquad (1b)$$
$$Z = \exp(2 i k_o d), \qquad (1c)$$

where $ck_o/\omega = n_o = (\varepsilon_o)^{1/2}$, and d is the distance between the virtual interface and the surface. If data $\tilde{r}_j$ are obtained at equal intervals during deposition and the deposition rate is known, then ε_o can be obtained in principle from as few as two data points $\tilde{r}_j$ and $\tilde{r}_{j+1}$ as[2]

$$\varepsilon_o = \frac{(1 - \tilde{r})^2 - i \lambda n_a \tilde{r}'/\pi}{(1 + \tilde{r})^2} \varepsilon_a, \qquad (2)$$

where $\tilde{r}' = \partial\tilde{r}/\partial d \approx (\tilde{r}_{j+1} - \tilde{r}_j)/\Delta d$, where Δd is the thickness increment between $\tilde{r}_j$ and $\tilde{r}_{j+1}$. A solution is possible because $\tilde{r}$ and $\tilde{r}'$ provide 4 constraints, which is sufficient to determine ε_o and $\tilde{r}_v$. An analogous equation can be given for ρ in the LVSA.[2]

If only two points are used, the uncertainty $\delta\varepsilon_o$ in ε_o can be calculated directly from Eq. (2) if the relative uncertainty $|\delta\tilde{r}/\tilde{r}|$ of each data point $\tilde{r}_j$ is known. In practice, calculation of the uncertainty is more complicated because multiple data points are generally needed for acceptable precision. The problem is then overdetermined. If only a short segment of the $\tilde{r}$ or $\langle\varepsilon\rangle$ trajectory is involved, an approximate analytic solution can be obtained by least-squares fitting a straight line and substituting the results directly in Eq. (2), in which case

$$|\delta\varepsilon_o| = \frac{2\sqrt{3}\,\lambda\,n_a\,|\tilde{r}|}{\pi\Delta\delta\,|1+\tilde{r}|^2\,\sqrt{N(N^2-1)}}\,\frac{|\delta\tilde{r}|}{|\tilde{r}|}; \qquad (3a)$$

or

$$|\delta\varepsilon_o| = \frac{\sqrt{3}\,\lambda\,|\langle\varepsilon\rangle|}{2\pi\,|\langle n\rangle|\,\Delta d\,\sqrt{N(N^2-1)}}\,\frac{|\delta\langle\varepsilon\rangle|}{|\langle\varepsilon\rangle|}; \qquad (3b)$$

for complex reflectometry and ellipsometry, respectively. Here $\langle\varepsilon\rangle = \langle n\rangle^2$ is the ellipsometrically measured pseudodielectric function, λ is the wavelength of light, $n_a = 1$ is the refractive index of the ambient, Δd is the thickness increment between successive points, N is the number of points being fit, and $|\delta\langle\varepsilon\rangle/\langle\varepsilon\rangle|$ is the relative uncertainty of each data point $\langle\varepsilon\rangle_j$. In Eq. (3a) we have assumed that the real and imaginary parts of $\delta\tilde{r}/\tilde{r}$ are uncorrelated and have the same amplitude (equal to $|\delta\tilde{r}/\tilde{r}|/\sqrt{2}$), so that the uncertainty of any projection, $|(\delta\tilde{r}/\tilde{r})_p|$, of $\delta\tilde{r}/\tilde{r}$ will also equal $|\delta\tilde{r}/\tilde{r}|/\sqrt{2}$. In Eq. (3b) we assume in addition that $|\varepsilon_s| \approx |\varepsilon_o| \gg \varepsilon_a$. We express the results in relative values because they are a more usual representation of precision than $|\delta\tilde{r}|$ and $|\delta\varepsilon|$ alone. Typical relative precisions (not to be confused with accuracies) for reflectometers and ellipsometers are of the order of ± 0.0001 and ± 0.001, respectively. Equations (3) should also contain a Student t-distribution multiplier expressing the confidence limit,[8] for example 1.645 for 90% certainty, but we omit this term in accordance with usual practice. No equivalent analytic approximation exists for R since value and three derivatives are required to obtain the necessary four constraints.

The most accurate procedure, and the only one presently possible for R, is to least-squares fit the exact V-I expressions directly to the data. In this case $\delta\varepsilon_o$ is given in terms of $|\delta\tilde{r}/\tilde{r}|$, $|\delta\langle\varepsilon\rangle/\langle\varepsilon\rangle|$, or $|\delta R/R|$ by multiplying the corresponding relative uncertainty of the data by $|\tilde{r}|$, $|\varepsilon_o|$, or R and the square root of the corresponding diagonal elements of the variance-covariance matrix. The theory is discussed elsewhere.[9] Because this approach is numerical the results are necessarily application-specific. However, the matrix structure shows that the coefficients relating the real and imaginary parts of $\delta\varepsilon_o$ to $|\delta\tilde{r}/\tilde{r}|$ are the same. This is also true for $|\delta\langle\varepsilon\rangle/\langle\varepsilon\rangle|$, but not for $|\delta R/R|$.

RESULTS

As a representative case we consider deposition of an $Al_{0.3}Ga_{0.7}As$ layer on a GaAs substrate, using angles of incidence of 0° for R and $\tilde{r}$ and 67.08° for $\langle\varepsilon\rangle$. The values of ε for these materials at 2.6 eV and 600 °C are 21.1 + i9.4 and 20.5 + i14.5, respectively,[6] which for bare GaAs yields $\tilde{r} = -0.676 - i0.085$. Using the empirical relation[6] $\varepsilon_2 \approx \varepsilon_{2,GaAs} - 16.9x - 0.33x^2$, which is valid for $x \leq 0.3$, we have $|\delta x/x| = |\delta\varepsilon_{o2}|/16.9$, where $\delta\varepsilon_{o2}$ is the imaginary projection of $\delta\varepsilon_o$. To determine the variance-covariance matrix we generated trajectories of $\tilde{r}$, $\langle\varepsilon\rangle$, and R with a point spacing Δd of 1 Å, then fit N successive points of these trajectories to the exact V-I expression (R and $\tilde{r}$) or the LVSA approximation ($\langle\varepsilon\rangle$) to determine the uncertainty δx per unit relative experimental uncertainty in $\tilde{r}$, $\langle\varepsilon\rangle$, and R. We considered three cases: (1) an extending fit for increasing d with d = 0 corresponding to the bare GaAs substrate, where the virtual and physical interfaces remain in coincidence at d = 0; (2) an extending fit for R as in (1) but with the substrate optical properties assumed to be known (a two-parameter fit); and (3) a sliding fit where d remains fixed at 10 Å below the surface for $\tilde{r}$ and $\langle\varepsilon\rangle$ and 300 Å for R.

The results for the extending and sliding fits are shown in Figs. 1 and 2, respectively. In the limit of small d the results of Fig. 1 should reduce to those obtained from Eqs. (3). For N = 11 (d = 10 Å) we find that $\delta x/|\delta\tilde{r}/\tilde{r}|$ and $\delta x/|\delta\varepsilon/\varepsilon|$ are 50.71 and 10.68 for the exact calculation and 51.82 and 10.70 for the linear approximation. Thus the use of Eqs. (3) for very thin films is justified. As discussed in ref. 2, for the data of ref. 4 (N = 11, Δd = 1.5 Å, and λ = 4767 Å) we find $\delta x \approx 0.005$, in reasonable agreement with the experimental estimate of 0.003. The reason that experiment appears better than theory here is due to the neglect of correlations that occur between the real and imaginary parts of $\delta\langle\varepsilon\rangle/\langle\varepsilon\rangle$ in this application.

In Fig. 1 two results are shown for R: that of the full 4-parameter fit, which is labeled R(4), and that for a 2-parameter fit, which is labeled R(2) and is obtained by supposing that the substrate dielectric function is known. The latter will be discussed below. The results for R(4) begin at d = 70 Å because our single-precision matrix-inversion routine would not converge for the nearly singular matrices encountered for thinner overlayers. Although the results could have been extended by double precision, there would be no point in doing this because δx for R(4) is diverging rapidly at this point so that in practice V-I analysis could not be performed for thinner layers under these conditions even if the reflectometer were extremely good ($|\delta R/R| = 10^{-5}$). Corresponding limitations on computational precision also required us to take d = 300 Å to obtain a reasonable range for R = R(4) in Fig. 2, as opposed to 10 Å for $\tilde{r}$ and $\langle\varepsilon\rangle$. To make a rough correspondence meaningful between the scalar R and the complex quantities $\tilde{r}$ and $\langle\varepsilon\rangle$ we assumed that the trend established above 300 Å for the sliding fit could be extrapolated to smaller d, specifically to 70 Å, then used the separation between R(4) and $\tilde{r}$ in Fig. 1 to establish the relative separation in Fig. 2. In fact, if R calculations could have been performed for d = 10 Å the separation

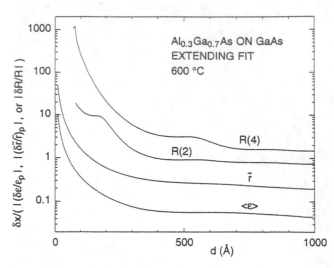

Figure 1. Relative uncertainties in composition x for an $Al_{0.30}Ga_{0.70}As$ layer being deposited on a GaAs substrate in terms of the single-point relative uncertainties $|\delta R/R|$, $|(\delta\tilde{r}/\tilde{r})_p|$, and $|(\delta\varepsilon/\varepsilon)_p|$ for reflectometry, complex reflectometry, and ellipsometry, respectively. The calculation is done for points spaced 1 Å apart starting from an overlayer thickness of 0. For these calculations the virtual and physical interfaces remain in coincidence at d = 0. The curve R(2) was calculated assuming that the substrate optical functions are known (2-parameter fit.)

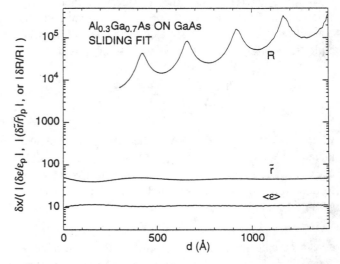

Figure 2. As Fig. 1, except the virtual interface remains 10 Å below the surface for $\tilde{r}$ and $\langle\varepsilon\rangle$ and 300 Å for R. To provide a more representative comparison the results for R have been positioned by extrapolating them to 70 Å then using the separation between R(4) and $\tilde{r}$ in Fig. 1 to locate the relative position.

would be larger.

DISCUSSION

The results show that for thicknesses above about 1/2 an interference cycle (here about 300 Å) an ellipsometer is better by at least 1 1/2 orders of magnitude than a reflectometer of the same relative precision. However, the relative precision of a good reflectometer is usually about an order of magnitude better than that of a good ellipsometer. Thus in practical applications to uniform films at least a few hundred Å thick the capabilities of reflectometry and ellipsometry should be comparable. Not surprisingly, the precision to which δx can be determined improves with increasing thickness, although the improvement for R(4) exhibits sensitivity plateaus.

However, for a sliding fit more appropriate to control applications, Fig. 2 shows that the situation is different. These results show that for a fixed number of points the sensitivity of $\tilde{r}$ and $<\varepsilon>$ is essentially independent of thickness, while that of R deteriorates rapidly with a periodic variation. This is not surprising, because both real and imaginary parts of ε_0 can be determined for a thick film from a single (complex) datum $\tilde{r}$ or $<\varepsilon>$, whereas no such separation is possible for a single (real) datum R. The results show that for practical purposes there exists no usable range of thicknesses for R for which V-I—based compositional control would be possible.

The constraint of not using previously established data to help determine x may seem unnecessarily restrictive, but for continuous closed-loop feedback control of composition it is necessary for stability. As discussed previously,[2] for complex data such as $\tilde{r}$ or $<\varepsilon>$ no accuracy is lost with the V-I approach relative to conventional Fresnel analysis. However, for R the use of previously determined parameters is expected to improve the situation, since a knowledge of previous parameters reduces the number of necessary additional constraints to 2, although the stability issue remains. We have investigated the improvement obtained by reducing the number of determined parameters to 2, i.e., to ε_0 alone, assuming that the substrate parameters are known. Clearly, the only meaningful case here is the extending fit. The results are shown in Fig. 1 as the curve labeled R(2). It is seen that the relative uncertainty in x is reduced if only two parameters need be determined, but the improvement is not sufficient to overcome the disadvantage incurred by not measuring the phase.

For some applications, e.g., distributed Bragg reflectors, where the need to maintain continuous control over composition is less important than maintaining total phase shift, a combined Fresnel—V-I approach may be more appropriate. These considerations will be treated in subsequent work.

SUMMARY

We provide a framework for calculating the capability of a reflectometer or an ellipsometer for determining the composition x of the outermost layer of an epitaxial semiconductor during deposition or etching in terms of the signal-to-noise capabilities of the instrument. Our calculations for the prototypical semiconductor $Al_{0.30}Ga_{0.70}As$ show that with present capabilities reflectometers and ellipsometers are equally effective in determining x for films of thicknesses greater than about 300 Å. However, in compositional control, where x must be assessed within a few Å, an ellipsometer is required.

ACKNOWLEDGMENT

It is a pleasure to acknowledge support of this work by the Office of Naval Research under Contract N-00014-93-1-0255. One of us (WG) also received partial support from the Engineering Research Center for Advanced Electronic Materials Processing, which is funded by the National Science Foundation.

REFERENCES

1. P. Gadenne and G. Vuye, "*In situ* determination of the optical and electrical properties of thin films during their deposition," J. Phys. E. **10**, 733-6 (1977); A. Englisch and J. Ebert, "Refractive index measurement during the deposition of dielectric coatings," SPIE Proc. **401**, 17—24 (1983); E. Pelletier, "Monitoring of optical thin films during deposition," SPIE Proc. **401**, 74—82 (1983); R. Herrmann and A. Zöller, "Automated control of optical layer fabrication processes," SPIE Proc. **401**, 83—91 (1983); C. J. van der Laan, "Optical monitoring of nonquarterwave stacks," Appl. Opt. **25**, 753—60 (1986); T. Skettrup, "Optical monitoring of nonquarterwave stacks," Optical Engineering **26**, 1175—81 (1987); B. T. Sullivan and J. A. Dobrowolski, "Deposition error compensation for optical multilayer coatings. I. Theoretical description," Appl. Opt. **31**, 3821—35 (1992); "Deposition error compensation for optical multilayer coatings. II. Experimental results—sputtering system," **32**, 2351—60 (1993).

2. D. E. Aspnes, "Minimal-data approaches for determining outer-layer dielectric responses of films from kinetic reflectometric and ellipsometric measurements," J. Opt. Soc. Am. **A10**, 974—83 (1993).

3. F. K. Urban III and M. F. Tabet, "Virtual-interface method for *in situ* ellipsometry of films grown on unknown substrates," J. Vac. Sci. Technol. **A11**, 976—980 (1993).

4. D. E. Aspnes, W. E. Quinn, M. C. Tamargo, M. A. A. Pudensi, S. A. Schwarz, M. J. S. P. Brasil, R. E. Nahory, and S. Gregory, "Growth of $Al_xGa_{1-x}As$ parabolic quantum wells by real-time feedback control of composition," Appl. Phys. Lett. **60**, 1244—6 (1992).

5. L. E. Tarof, C. J. Miner, and C. Blaauw, "InGaAsP compositional determination from reflectance spectroscopy of InGaAsP/InP heterostructures," J. Appl. Phys. **63**, 2939—44 (1990); "Epitaxial Layer Thickness Measurements by Reflectance Spectroscopy," J. Electronic Mater. **18**, 361—7 (1990); H. Sankur, W. Southwell, and R. Hall, "In Situ Optical Monitoring of OMVPE Deposition of AlGaAs by Laser Reflectance," J. Electronic Mater. **20**, 1099—1104 (1991); K. Bacher, B. Pezeshki, S. M. Lord, and J. S. Harris, Jr., "Molecular beam epitaxy growth of vertical cavity optical devices with *in situ* corrections," Appl. Phys. Lett. **61**, 1387—9 (1992); D. I. Babic, T. E. Reynolds, E. L. Hu, and J. E. Bowers, "*In situ* characterization of sputtered thin films using a normal incidence laser reflectometer," J. Vac. Sci. Technol. **A10**, 939—44 (1992); Y. Raffle, R. Kuszelewicz, R. Azoulay, G. Le Roux, J. C. Michel, L. Dugrand, and E. Toussaere, "*In situ* metalorganic vapor phase epitaxy control of GaAs/AlAs Bragg reflectors by laser reflectometry at 514 nm," Appl. Phys. Lett. **63**, 3479—81 (1993); W. G. Breland and K. P. Killeen, "Spectral reflectance as an in situ monitor for MOCVD," Mat. Res. Soc. Symp. Proc. **324**, 99—104 (1994); K. P. Killeen and W. G. Breland, "In Situ Spectral Reflectance Monitoring of III-V Epitaxy," J. Electronic Mater. **23**, 179—83 (1994).

6. D. E. Aspnes, W. E. Quinn, and S. Gregory, "Application of ellipsometry to crystal growth by organometallic molecular beam epitaxy," Appl. Phys. Lett. **56**, 2569-71 (1990).

7. D. E. Aspnes, W. E. Quinn, and S. Gregory, "Optical control of growth of $Al_xGa_{1-x}As$ by organometallic molecular beam epitaxy," Appl. Phys. Lett. **57**, 2707-9 (1990).

8. E. S. Keeping, *Introduction to Statistical Inference* (Van Nostrand, Princeton, 1962), Ch. 12.

9. D. H. Loescher, R. J. Detry, and M. J. Clauser, "Least-Squares Analysis of the Film-Substrate Problem in Ellipsometry," J. Opt. Soc. Am. **61**, 1230—5 (1971); J. Humlicek, "Sensitivity extrema in multiple-angle ellipsometry," J. Opt. Soc. Am. **2**, 713—22 (1985).

In situ Pyrometric Interferometry Monitoring and Control of III-V Layered Structures During MBE Growth: Modeling and Implementation

D.L. Sato, X. Liu, Y. Li, A.R. Stubberud, and H.P. Lee
Department of Electrical and Computer Engineering
University of California, Irvine
Irvine, CA 92717

J.M. Kuo
AT&T Bell Laboratory
Murray Hill, NJ 07974

A new theoretical pyrometric interferometry model for layered AlAs/GaAs structures during epitaxial growth is presented. Implementation of an *in situ* monitoring and control system is discussed. Based on this control scheme, highly reproducible PI signals for AlAs/GaAs Distributed Bragg Reflector (DBR) structures have been grown by molecular beam epitaxy without the need for precise determination and control of the material growth rates.

INTRODUCTION

Reproducible control of layer thickness and composition is crucial to the manufacturability of many optoelectronic devices including Vertical Cavity Surface Emitting Lasers (VCSELs). The causes for thickness deviation include: a lack of precise knowledge of the growth rate, fluctuation of the growth rates (both short and long term), and a dependence of growth rate on the growth conditions. Until recently, thickness control during molecular beam epitaxy (MBE) growth was achieved by careful calibration of the growth rate using a combination of *in situ* techniques such as Reflection High Electron Energy Diffraction (RHEED) prior to each device growth and post-growth analysis. While RHEED oscillation measurements are capable of monolayer resolution, it is unsuitable for continuous monitoring during the entire growth. It is highly desirable to develop non-invasive optical techniques capable of continuous monitoring of the composition and thickness of the epilayer. Compared to *in situ* beam flux monitoring such as atomic beam absorption[1], these measurements depend on the thickness and composition of the *deposited* material, rather than the incident beam fluxes. Consequently, these techniques are insensitive to the details of the growth incorporation mechanisms. Since most epitaxial growth is relatively slow compared to the sampling and processing time of the monitored data, it is possible to combine *in situ* monitoring and control in a closed-loop system. Among *in situ* optical monitoring techniques, Pyrometric Interferometry (PI) has received wide attention. It has been used for growth rate and substrate temperature measurements[2,3], and has also achieved considerable success in improving the reproducibility of the DBR mirrors for VCSELs emitting at both 780 and 980 nm wavelengths[4]. Compared with other related optical monitoring methods such as reflectometry[5] and ellipsometry[6], PI can be readily installed in most existing MBE systems without retrofitting, and is relatively insensitive to optical alignment and substrate wobbling during rotation. This paper, presents a new physical model for the PI process and discusses the implementation and experimental results of a closed-loop control system based on PI monitoring during MBE growth.

EXPERIMENTS AND MODELING OF PYROMETRIC INTERFEROMETRY

In essence, PI records the continuous time-evolution of spontaneous thermal radiation, measured through a narrow-band optical filter from a heated sample during epitaxial growth. In the presence of heterostructures, the spontaneous radiation undergoes multiple reflections at the layer interfaces before being detected. Continuous increase in the layer thickness, and hence the optical path length, causes a quasi-sinusoidal modulation of the signal intensity. The PI process is characterized by three parameters: the layers thicknesses, the layer refractive indices ($n_c = n - j\kappa$), and the spontaneous emission rates (R) of all emitting elements in the material. The first two terms govern the propagation characteristics of the emitted radiation, whereas the last term determines the intensity of the radiation source.

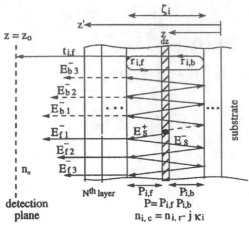

$$E_{f\ total}^- = E_{f1}^- + E_{f2}^- + E_{f3}^- + ...$$
$$E_{b\ total}^- = E_{b1}^- + E_{b2}^- + E_{b3}^- + ...$$

Figure 1. Schematic of the layered structure showing the pyrometric radiation due to a differential radiating element in the ith layer.

Theoretical Model

To calculate the PI signal, consider an infinitesimal radiating source inside the ith layer of a N layer sample as shown in Fig.1. The emitted radiation from this source can be decomposed into plane waves of angular frequency ω, propagating in the forward and backward directions, denoted by $E^+ = E_{i,s}(\omega,z)\exp(-jkz)$ and $E^- = E_{i,s}(\omega,z)\exp(jkz)$ respectively; where $k = (\omega/c)n_c$ is the medium wave vector, and $E_{i,s}(\omega,z)$ is the electric field phasor for photons emitted in either directions. Treating the ith layer as a Fabry-Perot cavity, we obtain:

$$E_f^-(\omega,z,z_o) = \frac{t_{i,f}\,P_{i,f}\,E_{i,s}(\omega,z)}{1 - r_{i,f}\,r_{i,b}\,P^2} \quad (1a)$$

$$E_b^-(\omega,z,z_o) = \frac{t_{i,f}\,P_{i,f}\,E_{i,s}(\omega,z)\,r_{i,b}\,P_{i,b}^2}{1 - r_{i,f}\,r_{i,b}P^2}, \quad (1b)$$

where $t_{i,f}$ is the effective transmission coefficient from the front surface of the ith layer to z_o (the detection plane); $r_{i,f}$ and $r_{i,b}$ are the reflection coefficients of the cavity; $P_{i,f}$ and $P_{i,b}$ are the phase factors from the source to the front and back cavity interfaces; and $P = (P_{i,f})(P_{i,b})$ is the phase factor of the ith layer. If the refractive indices and layer thicknesses are known, the values of $t_{i,f}$, $r_{i,f}$ and $r_{i,b}$ can be calculated using a dynamic matrices method[7]. The total detected E fields inside the ith layer are obtained from the sum of Eqs.(1a) and (1b), integrated over the layer thickness (ζ_i):

$$E_{i,t}(\omega,z_o) = \int_{\zeta i} E_{i,s}(\omega,z,z_o)\,G_i(\omega,z,z_o)\,dz, \quad (2)$$

where $G_i(\omega,z,z_o)$ is the transfer function from the radiating source to z_o. The total radiation intensity, $I_{E,i}(\omega,z_o)$, given by $(1/2)\,|E_{i,t}(\omega,z_o)|^2$ (setting $\mu_o = 1$) can be expressed as:

$$I_E(\omega,z_o,z') = \sum_i I_{E,i}(\omega,z_o)$$
$$= \sum_i \int_{\zeta i} |G_i(\omega,z,z_o)|^2\,R_i(\omega,z,T(z,z'))\,dz, \quad (3)$$

where z' is the total thickness of the sample, $R_i(\omega,z,T(z,z'))$ is the spontaneous emission rate at z, and $T(z,z')$ is the temperature distribution in the sample which in general is a function of z'. Taking into account the optical filter line width, $S_f(\omega)$, the measured PI signal (dropping the z_o term) is given as:

$$I_{E,exp}(z') = P_o \int_{\omega} I_E(\omega,z')\,S_f(\omega)\,d\omega + P_1(z'), \quad (4)$$

where P_o is a proportionality constant related to the measurement sensitivity, and $P_1(z')$ is the background signal due to the scattered radiation from the high temperature Al, Ga, and dopant Knudsen cells.

By considering the optical absorption inside the same infinitesimal element due to an incident monochromatic wave of unity field intensity ($E_o = 1$), it can be shown that:

$$\Delta A_i(\omega,z) = \frac{-\alpha_i(\omega,z)}{R_i(\omega,z,T(z))}\frac{n_{i,r}}{n_o}\Delta I_{E,i}(\omega,z), \quad (5)$$

where $\alpha_i(\omega,z) = 2k_o\kappa_i$ is the absorption coefficient at z. Since the ratio $R_i(\omega,z,T(z))/\alpha_i(\omega,z) = f(\omega,T(z))2\pi/h\omega = (2n^2\omega^2/c^2 2\pi)/(\exp(h\omega/2\pi kT)-1)^8$ is a function of temperature only, independent of materials. I_E at any given thickness z' can be expressed in terms of the absorption and Planck distribution function $f(\omega,T(z,z'))$ as:

$$I_E(\omega,z') = \sum_i \int_{\zeta i} f(\omega,T(z,z'))\,\Delta A_i(\omega,z)\,dz, \quad (6)$$

Equations (3) and (6) establish the general expression for thickness (time) evolution of $I_E(\omega)$ for a layered sample with arbitrary spatial distributions of spontaneous emission rates and refractive indices. Since the effective absorption (emission) depth $(1/\alpha)$ in GaAs at 600°C is only ≈ 0.77 μm^9 for $\lambda_m = 1048$ nm, the temperature distribution can be neglected and Eq.(6) becomes:

$$I_E(\omega,z') = f(\omega,T(z'))\,A(\omega,z')$$
$$= f(\omega,T(z'))\,\varepsilon(\omega,z'), \quad (7)$$

Equation (7) is a re-statement of Kirchhoff's Radiation Law for multiple layer structures. Previous modeling[4] of the PI signal which assumed a constant temperature ($T(z') = T_o$) throughout the growth can be viewed as a special case of Eq. (7). To ease computation, the effect of temperature variation can be approximated by treating the AlAs layers as non-reciprocal absorbers (one that does not emit) with an effective imaginary index $\kappa_{AlAs} = \kappa_e$. The total $I_E(\omega,z')$ is then calculated as:

$$I_E(\omega,z') \approx f(\omega,T_{o,N})\sum_i^N A_{GaAs,i}(\omega,z')$$
$$= \beta_N(\omega)\sum_i^N A_{GaAs,i}(\omega,z'), \quad (8)$$

by assuming that the sample temperature changes only from one layer to the next ($T_{o,N}$). The modeling of the background signal $P_1(z')$ requires a detailed consideration of diffused scattering from the AlAs/GaAs layered structure[10] and will not be discussed here. For simplicity, two constants, $P_{1,GaAs}$ and $P_{1,AlAs}$, were used for the growth of GaAs and AlAs respectively.

Experimental Procedure and Results

The experimental setup for the PI measurement is shown in Fig. 2. An extended periscope made of a 45° Si mirror inside the MBE chamber is used to avoid As coating of the viewport during the extended period of observation. A Si p-i-n detector/preamplifier, followed by a lock-in amplifier, is used for signal detection in lieu of a conventional pyrometer. An optical filter centered at 1048 nm with a bandwidth of 60 nm is used. Data from the lock-in amplifier is fed into a computer which controls the effusion cell shutters. A stable substrate temperature (≈600°C) is maintained during the growth. All samples (a quarter of a 2" GaAs wafer) are indium-mounted on a 2" molybdenum (Mo) block, and PI data are taken when the thermocouple reading, detector signal, and the heater power have all reached steady states after a GaAs buffer layer has been deposited. The temperature of the Mo block is controlled by a thermocouple at the back side of the block, which is not in physical contact with the block. In all the experiments, the PI data are sampled at a rate of 2 Hz with a lock-in amplifier time constant of 1 second.

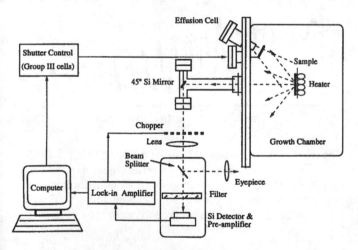

Figure 2. Schematic of the PI measurement setup.

Figure 3 shows the representative PI signals of GaAs/AlAs structures of different layer thicknesses grown on GaAs substrates. The abrupt discontinuity of the PI signal when the layers are switched between GaAs and AlAs is caused by a change in the background non-specular scattering from the Group III effusion cells. The calculated PI signals from Eq.(7), using a constant temperature ($T(z') = T_o$) and a thickness-dependent temperature ($T(z')$) are shown in Fig. 3(b) and (c)

respectively. The calculated PI signal using the effective-index model (Eq.(8)) is shown in Fig. 3(d) for comparison.

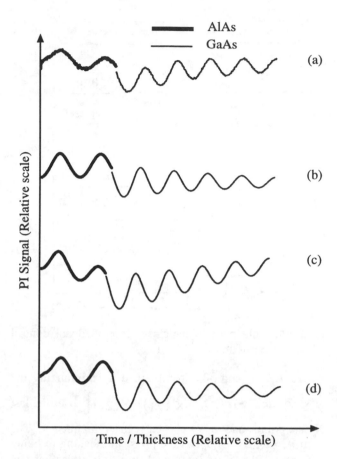

Figure 3. (a) Experimental PI data (vs. time) for a GaAs/AlAs structure grown on a GaAs substrate, (b) and (c) Calculated PI signals (vs. thickness) using Eq.(7) for constant temperature and thickness-dependent temperature profiles respectively, (d) Calculated PI signal using an effective-index model (Eq.(8)).

IMPLEMENTATION OF A REAL-TIME MONITORING AND CONTROL SYSTEM

The effective-index model is used for implementing a closed-loop reference model control system as shown in Fig. 4. The simulated reference PI signal versus thickness, $\mathbf{X}(z_n)$, is calculated using Eq. (8) ($\beta_N = 1$ for all GaAs layers) with a thickness resolution of 0.1 nm. This is comparable to experimental thickness resolution obtained with a sampling time of 0.5s for growth rates between 0.2 to 0.3 nm/s. The $\mathbf{X}(z_n)$ is then filtered through a low-pass filter to give $\mathbf{X'}(z_n)$. Since the PI signal for the DBR structure is quasi-sinusoidal, it is modeled as:

$$Y(z_n) = A_n + D_n \cos(\theta_n)$$
$$= A_n + D_n \cos(\omega n \Delta T + \delta\theta_n), \qquad (9)$$

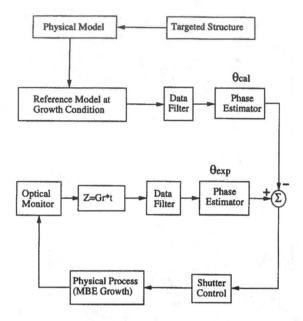

Figure 4. Schematic of the closed-loop feedback control system.

where A_n, D_n and $\delta\theta_n$ are obtained by minimizing the least-square (LS) of $\sum_{i=n-N}^{n} \left[X'(z_i) - Y(z_i) \right]^2$ over the last N data points, and ω is a chosen "carrier" frequency. The term $\theta_n = \omega n \Delta T + \delta\theta_n$ has the physical meaning of being the total "phase" of the signal at z_n. Using this data processing scheme, the simulated phase list $\theta_{s,j}(z_j)$ corresponds to the recorded switching point at the end of the jth layer. The experimental PI data are processed in a similar way by first converting $S(t_n)$ into $S(z_n)$ using $z_n = G_r * t_n$, where G_r is the layer growth rate. Switching of the shutters takes place when $\theta_e(z_n)$ is matched to $\theta_{s,j}(z_j)$. The optimization and trade-off of the parameters (ω and N) used for the LS estimation is discussed elsewhere[11].

To evaluate the performance of this technique, a test structure consisting of a 15 pair AlAs/GaAs DBR mirror, targeted at the peak wavelength of 1040 nm, was grown on top of a pair of GaAs and AlAs calibration layers with a thickness of 280 nm and 240 nm respectively (Fig. 5). The calibration layers were used to estimate the AlAs and GaAs growth rates (from the PI signal waveform itself) for the $z_n = G_r * t_n$ transformation. Fig. 5 (i) and (ii) shows the simulated and experimental PI data obtained with the closed-loop control for three samples grown at different growth rates. Fig. 6 plots normalized layer time, $\tau_j = t_j / t_{average}$, for each sample. If the growth rates are constant for each layer, total optical phase for each DBR pair is $\approx n_{(GaAs)} * t_{j(GaAs)} + n_{(AlAs)} * t_{j(AlAs)}$. The normalized optical

phase, ϕ_j, for a larger DBR pair is also plotted in Fig. 6. A self-compensating mechanism is noted: while the growth time for GaAs and AlAs layers varies with layer number, the total optical phase of each DBR pair remains nearly constant throughout the growth. The results show that a highly reproducible PI signal can be obtained without a precise knowledge of the growth rate. This is significant in view of the fact that until recently, careful growth rates were needed prior to each device growth.

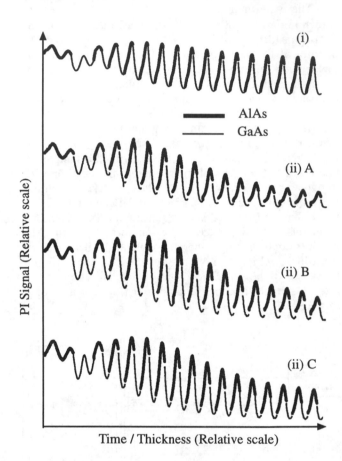

Figure 5. (i) Simulated PI signal for AlAs/GaAs DBR mirrors targeted at a peak reflectance of 1040 nm, (ii) Experimentally monitored PI data for samples A, B, and C grown with closed-loop feedback control but at different growth rates.

	Sample A [nm/s]	Sample B [nm/s]	Sample C [nm/s]
GaAs Growth Rate	0.3141	0.3114	0.2946
AlAs Growth Rate	0.2678	0.2501	0.2744

Table I.   GaAs and AlAs growth rates for samples A, B, and C shown in Figure 5 and 6.

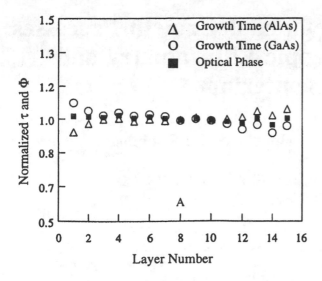

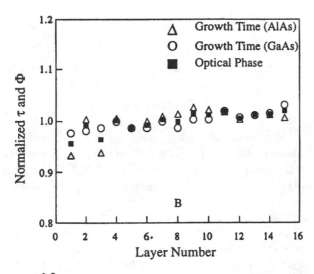

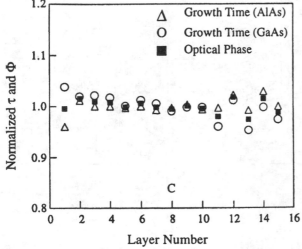

Figure 6. Normalized layer time (τ_j) and total optical phase (ϕ_j) for each DBR pair for samples A, B, and C. Notice the change in scale in each figure. The GaAs and AlAs growth rates for each sample are summarized in Table I.

SUMMARY

In summary, a new method is described for calculating the pyrometric emission of layered structures. Using an approximate form of this model, a closed-loop monitoring and control system was implemented. Experimental results for a 15 pair AlAs/GaAs DBR structure showed that excellent reproducibility of the PI signal can be achieved at varying growth rates. The technique is expected to improve the growth reproducibility of sophisticated layer structures including VCSELs.

Acknowledgment: The authors want to thank Peter Chan for assistance in computer programming. This project is supported in part by CI Systems through the University of California MICRO program and the Irvine Research Fellowship.

REFERENCES

1. Chalmers S.A., and Killeen K.P., *Applied Physics Letters*, vol. **63**, (no.23): 3131-33 (1993).
2. Springthorpe A.J., Humphreys T.P., Majeed A., and Moore, W.T., *Applied Physics Letters*, vol. **55**, (no.20) : 2138-40, (1989).
3. Bobel F.G., Moller H., Wowchak A., Hertl B., Van Hove J., Chow L.A., and Chow P.P., *Journal of Vacuum Science & Technology B* (Microelectronics and Nanometer Structures), vol. **12**, (no.2):1207-10, (1994).
4. Houng, Y.M., Tan M.R.T., Liang B.W., Wang S.Y., and Mars D.E. , *Journal of Vacuum Science & Technology B* (Microelectronics andNanometer Structures), vol.**12**, (no.2):1221-4, (1994).
5. Armstrong J.V., Farrell T., Boyd A., and Beanland R., *Applied Physics Letters*, vol.**61**, (no.23):2770-2, (1992).
6. Aspnes D.E., *Applied Physics Letters*, vol. **62**, (no.4): 343-5, (1993).
7. Yeh P., *Optical Waves in Layered Media*, New York: Wiley-Interscience, 1988.
8. Verdeyen J.T., *Laser Electronics*, Second Edition, Eaglewood Cliffs, New Jersey: Prentice Hall, 1989, chap. 11, pp 381.
9. Maracas G.N., Kuo C.H., Anand S., and Droopad R., *Journal of Applied Physics*, vol. **77**, (no.4):1701-4, (1995).
10. Liu X., Ranalli E., Sato D.L., and Lee H.P., *Journal of Vacuum Science & Technology B* **13**, 742 (1995).
11. Liu X., Ranalli E., Sato D.L., and Lee H.P., submitted to *IEEE Photonic Technol. Letter*, (1995).

In-situ Monitoring of Heteroepitaxial Growth Processes Using Real-time Spectroscopic Ellipsometry and Laser Light Scattering

C Pickering, R T Carline, D A O Hope and D J Robbins

Defence Research Agency, St Andrews Road, Malvern, Worcs WR14 3PS, UK

Laser light scattering and real-time spectroscopic ellipsometry have been used as *in situ* monitors for pre-cleaning and layer growth of $Si_{1-x}Ge_x$ and Si epitaxial layers. The techniques have been used to optimize conditions for growth of smooth, high quality epilayers and the potential indicated for real-time control of morphology, composition and thickness.

INTRODUCTION

Novel electronic devices require growth of complex multilayer structures to a tightly controlled specification. Real-time process control is required to optimize the precision, performance and yield of such devices. The combination of laser light scattering (LLS) and real-time spectroscopic ellipsometry (RTSE) provides the basis of a powerful, *in situ* monitoring system with potential for use in process control, with applicability to a wide range of microelectronics processes, eg surface cleaning, epitaxy, oxidation, implantation, annealing, etc (1)(2). The techniques are complementary (2)(3) since LLS is mainly sensitive to surface non-idealities while RTSE provides information averaged over the optical penetration depth (~ 100Å to several μm depending on wavelength, λ). Thus, RTSE can be used to measure growth rate and composition in real time, while LLS indicates layer surface topography. Deviations from non-ideality can be detected with monolayer sensitivity, owing to the measurement of phase in RTSE and the large diffraction effects which can occur in LLS for surface spatial frequencies which match the scattering vector. Both techniques are sensitive to surface roughness, but on complementary lateral length scales, LLS being mainly sensitive to macroscopic features $\geq \lambda$ while RTSE is sensitive to microscopic features $\ll \lambda$. This paper discusses the implementation of the techniques in a real growth environment, ie low-pressure vapour-phase epitaxy (LPVPE) of $Si/Si_{1-x}Ge_x$ heterojunction bipolar transistor (HBT) structures at 600-850°C. The techniques have been used to optimize pre-cleaning and growth conditions for production of high quality Si epilayers (3)

and smooth, commensurate, strained $Si_{1-x}Ge_x$ layers (3)(4).

EXPERIMENTAL

The epitaxial layers were grown on Si (001) substrates by LPVPE in a UHV-background stainless steel reactor described previously (5). GeH_4 and SiH_4 are used as source gases, together with H_2, at total pressures of ~ 15Pa. Alloy layers with $0.06 < x < 0.25$ were deposited at 610°C or 750°C after growth of a Si buffer layer at 850°C. RTSE and LLS measurements were made using the equipment shown in Fig. 1. A SOPRA RTSE system is used with the light from the Xe

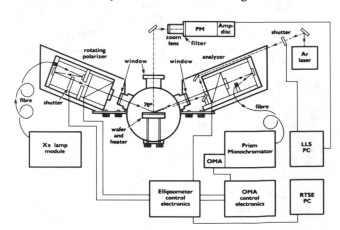

FIGURE 1. Schematic diagram of LLS/RTSE diagnostics interfaced to growth reactor.

lamp, incident at 70°, being dispersed and detected after reflection from the sample by a 0.5m prism monochromator/photodiode array. Typically tan ψ (amplitude ratio), cos Δ (phase difference) spectra can be obtained at 100 energies over the range 1.6-4.7eV with a precision and resolution at 3eV of ~ 5 x 10^{-4} and ~ 0.02eV, respectively, with a 2sec integration time. Optical access to the growth chamber is by low-strain quartz windows and corrections for residual window effects were made to the ψ, Δ data before conversion to pseudo-dielectric function, $\langle \tilde{\varepsilon} \rangle = \langle \varepsilon_1 \rangle + i \langle \varepsilon_2 \rangle$, data. LLS measurements were made simultaneously (3) using the 488nm line from an Ar$^+$ laser, the scattered intensity, I, being monitored normal to the wafer with a photon-counting photomultiplier, sampled once every 2sec.

RESULTS AND DISCUSSION

Figure 2 shows typical results from both LLS (3) and RTSE of the oxide desorption process prior to epitaxy. The peak in I arises due to inhomogeneous oxide removal by pit formation. As the pits expand, they pass through a size/distribution condition which matches the scattering geometry of the detection system. In our arrangement, large diffraction effects occur for lateral surface features ~ 0.52μm. The completion of the oxide removal process is also observed as a change in the $\langle \tilde{n} \rangle$ ($= \langle \tilde{\varepsilon} \rangle^{1/2}$) spectra during the ramp-up in temperature at ~ 840-870°C. The peak in the $\langle n \rangle$ spectrum shifts from ~ 3.35eV at room temperature to ~ 3eV at 890°C and the rise in $\langle n \rangle$ shown in Fig. 2 due to removal of the oxide overlayer is superimposed on the temperature-induced change.

Extensive studies of Si epitaxy have shown that LLS is a sensitive monitor of surface contamination (6). Thus, C impurities can give rise to thermal etching and non-ideal epitaxy which are easily detected as an increase in scattered light. The LLS traces of Fig. 2 were routinely obtained after optimizing pre-cleaning and growth conditions and indicate a stable bare Si surface after oxide desorption with no perturbation of the initial layer growth (3). These surfaces have also been shown to be atomically smooth by atomic force microscopy (AFM) and ex-situ SE (2)(7).

$Si_{1-x}Ge_x$ has a tendency to grow on Si with an undulating surface to minimize the elastic strain energy due to the lattice mismatch (8)(9). The *in situ* diagnostics have been used to identify conditions for growth of smooth, fully strained alloy layers as required for the

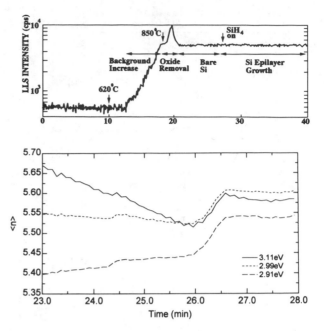

FIGURE 2. (upper) LLS trace obtained during pre-bake and deposition of Si epilayer.

(lower) $\langle n \rangle$ traces obtained during temperature ramp from 680-890°C. $\langle n \rangle$ rise begins at ~ 840°C.

HBT base region. As well as causing large increases in scatter (3)(8), the surface features cause a decrease in $\langle n \rangle$, due to the roughness acting as a density-deficient overlayer. Thus, roughness features with lateral sizes well below and of the order of the wavelength of the light must both be present. This is confirmed by AFM studies (9) which have been used to interpret the features in the LLS curve for x = 0.25 shown in Fig. 3. The roughness is isotropic at (b) but at (c) the formation of μm-sized patches of regular undulations aligned along orthogonal <100> directions is seen. Thus at (c) there is a significant component of roughness on the 0.5μm lateral length scale to which our system is most sensitive, giving rise to the peak in I, as well as at the undulation period (< λ/2), causing the decrease in $\langle n \rangle$. The rate of development of the surface roughness increases as the temperature or Ge content increases. Figure 3 also shows that at 610°C layers with x = 0.20 can be grown with smooth surfaces to thicknesses ~ 1000Å while for x = 0.25 roughening begins after ~ 300Å, before the onset of dislocations due to plastic strain relaxation occurring near X' (4) and (e) (9).

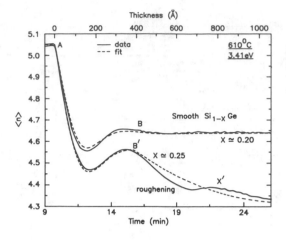

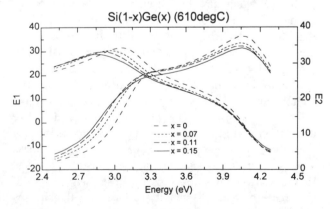

FIGURE 3. (upper) <n> at 3.41eV obtained during growth of Si$_{1-x}$Ge$_x$ layers with x = 0.20 and 0.25.

(lower) LLS trace obtained during x = 0.25 growth. Arrow indicates source gases turned on.

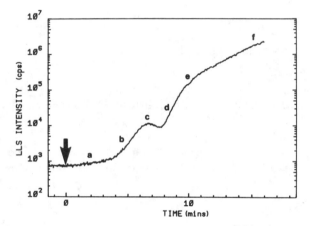

FIGURE 4. $\tilde{\varepsilon}$ spectra determined from Si$_{1-x}$Ge$_x$ layers at the growth temperature of 610°C.

The use of RTSE to monitor composition requires a database of $\tilde{\varepsilon}$ spectra at the growth temperature for a range of compositions. We have determined spectra for x = 0.07, 0.11, 0.15 at 610°C as shown in Fig. 4. These data were obtained from thick (500-800Å) fully strained epilayers by mathematically removing the effect of the substrate, optimizing thicknesses determined *ex situ* to produce smooth spectra. The critical point (CP) energy dependencies, required for interpolation of spectra for arbitrary compositions (10), were determined by lineshape fitting to second-differential spectra, following the procedure used for room-temperature reference spectra (11)(12). Polynomials were obtained for the variation of the E_1 and $E_1 + \Delta_1$ CPs in the 3eV region, the E_2 CP at 4.1eV being kept constant.

Figure 5 shows real-time <ε_1> spectra obtained during growth of a ~ 200Å Si$_{0.85}$Ge$_{0.15}$ layer using 2sec integration times. The peak can be seen to move from the Si substrate position (3.05eV) towards the bulk Si$_{0.85}$Ge$_{0.15}$ position (2.85eV) by following a trajectory first to higher <ε_1> values than those of either material and then to energies below 2.85eV. This behaviour is consistent with growth of a uniform alloy layer, although the effect of an anomalous initial growth region will be discussed later. The effect of the increasing optical penetration depth at lower energies can be seen, with <ε_1> approaching the bulk value for energies $\geq$ 3.3eV (optically thick). Below this energy, the spectra are still a convolution of layer and substrate spectra.

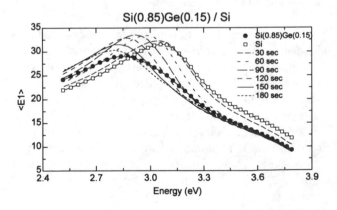

FIGURE 5. <ε_1> spectra obtained in real time (2sec integration time) during deposition of Si$_{0.85}$Ge$_{0.15}$ layer on Si at 610°C. Points indicate bulk spectra.

Similar real-time spectra, from growth of a $Si_{0.89}Ge_{0.11}$ layer, have been analyzed to determine composition as a function of growth time using the pseudo-substrate approximation (13). This uses a virtual interface kept a fixed distance below the surface, with previously-grown material represented, together with the substrate, as a pseudo-substrate. We have used this method, in conjunction with the interpolation model for $\tilde{\epsilon}$ discussed above (10-12), to calculate x at every third data point, the results being shown in Fig. 6. Thus, in this measurement each value of x corresponds to the composition of the last-grown ~ 9Å of material, with the growth rate of 1.2Å/sec assumed constant. Further work is in progress to determine growth rate and composition independently. It can be seen that ~ 35sec is required to reach a constant composition. Measurements of interrupted alloy growth (14), to be discussed in detail elsewhere, have shown that this finite interface region is due to the complex surface chemistry of the Si-Ge-H system. Although these results were calculated post-growth, they indicate the potential for real-time control.

Figure 7 shows the raw tan ψ, cos Δ data at energies in the range 2.5-3eV as a function of time. The data have also been analyzed to determine the optical constants independently at each energy, ie without assuming a spectral dependence. This method requires

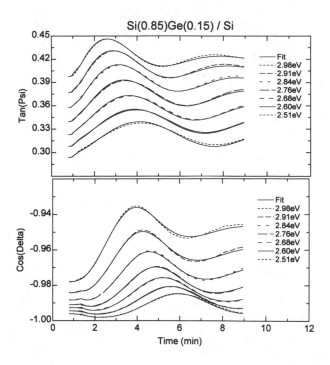

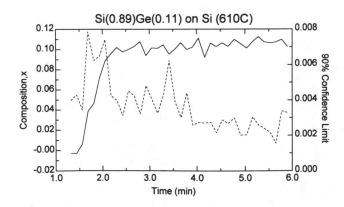

FIGURE 6. Ge content, x, determined from real-time data obtained during deposition of $Si_{0.89}Ge_{0.11}$ layer on Si at 610°C. Data points separated by 7.5sec represent ~ 9Å of material.

FIGURE 7. Tan ψ, cos Δ data obtained during growth of $Si_{0.85}Ge_{0.15}$ layer (as Fig. 5) with fitted data using a two-layer model (see text).

blocks of data in which $\tilde{\epsilon}$ is assumed constant as a function of time and in addition allows determination of the growth rate (14). Fits were attempted to the full data set using a model consisting of a single layer on the Si substrate, growing at a constant growth rate. If the optical constants of the layer were obtained from the data at long growth times, it was not possible to fit the early data. A two-layer model was required to produce the fits shown, with a 23Å interface layer (growth rate 0.86Å/sec) followed by the bulk layer (rate 1.2Å/sec). This interface corresponds to the initial growth region seen in Fig. 6, and the spectral dependence of the independently determined $\tilde{\epsilon}$ values is consistent with a region with reduced Ge content. Beyond this, the data are very well explained by growth of a uniform layer with constant growth rate.

535

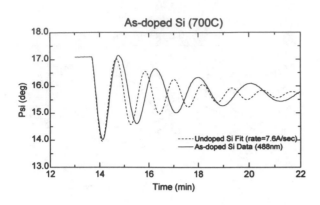

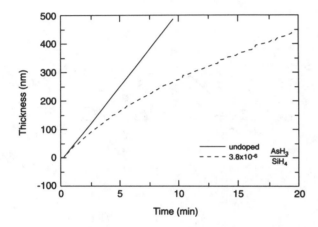

FIGURE 8. (upper) ψ at 488nm obtained during growth of undoped and As-doped Si on $Si_{0.8}Ge_{0.2}$ at 700°C.

(lower) Thickness vs. time determined from the above data.

The growth rates of B-doped alloy and undoped Si layers have also been found to be constant. However, interference fringes observed during the growth of As-doped Si (as required for the HBT emitter layer) spread out with time, compared to the evenly spaced fringes seen for undoped Si. This, as shown in Fig. 8, indicates that the growth rate of As-doped Si decreases with time, starting at the same initial rate but decreasing by a factor of 3 after about 20min. This is due to As surface segregation passivating the growing surface and illustrates the importance of *in situ* diagnostics for monitoring/control of device layer thicknesses.

CONCLUSIONS

LLS and RTSE have proved to be extremely useful monitoring techniques for optimization of growth conditions for heteroepitaxial growth. The results obtained indicate the potential of the techniques for use in real-time control of layer morphology, composition and thickness.

REFERENCES

1. Pickering, C., *"In Situ Optical Studies of Epitaxial Growth"* in *Handbook of Crystal Growth* (ed. D. T. J. Hurle), Amsterdam: Elsevier, 1994, vol. **3**, ch. 19, pp. 817–878.

2. Pickering, C., *"Ellipsometry and Light Scattering Characterization of Semiconductor Surfaces"* in *Electromagnetic Waves : Recent Developments in Research* (ed. P. Halevi), vol. **2** (to be published) (1995).

3. Pickering, C., *Thin Solid Films*, **206**, 275–282 (1991).

4. Pickering, C., Carline, R.T., Robbins D. J., Leong W.Y., Gray, D.E., and Greef, R., *Thin Solid Films*, **233**, 126–130 (1993).

5. Robbins, D.J., Glasper, J.L., Cullis, A.G., and Leong, W.Y., *J. Appl. Phys.*, **69**, 3729–3732 (1991).

6. Robbins, D.J., Pidduck, A.J., Pickering, C., Young, I.M., and Glasper, J.L., *SPIE*, **1012**, 25–34 (1989).

7. Nayar, V., Pickering, C., Pidduck, A.J., Carline, R. T., Leong, W.Y., and Robbins, D.J., *Thin Solid Films*, **233**, 40–45 (1993).

8. Robbins, D.J., Cullis, A.G., and Pidduck, A.J., *J. Vac. Sci. Technol. B*, **9**, 2048–2053 (1991).

9. Pidduck, A.J. Robbins, D.J., Cullis, A.G., Leong, W.Y., and Pitt, A.D., *Thin Solid Films*, **222**, 78–84 (1993).

10. Snyder, P.G., Woollam, J.A., Alterovitz, S.A., and Johs, B., *J. Appl. Phys.*, **68**, 5925–5926 (1990).

11. Carline, R.T., Pickering, C., Robbins, D.J., Leong, W.Y., Pitt, A.D., and Cullis, A.G., *Appl. Phys. Lett.*, **64**, 1114–1116 (1994).

12. Pickering, C., and Carline, R.T., *J. Appl. Phys.*, **75**, 4642–4647 (1994).

13. Aspnes, D.E., *J. Opt. Soc. Am. A*, **10**, 974–983 (1993).

14. Pickering, C., Hope, D.A.O., Leong, W.Y, Robbins, D. J., and Greef, R., *Mat. Res. Soc. Symp. Proc.*, **324**, 53–58 (1994).

Real Time Diagnostics of Semiconductor Surface Modifications by Reflectance Anisotropy Spectroscopy

J.-T. Zettler, W. Richter, K. Ploska, M. Zorn, J. Rumberg,
C. Meyne, and M. Pristovsek

Institut für Festkörperphysik der Technischen Universität Berlin, Germany
Hardenbergstr. 36, D-10623 Berlin, Germany

In most cases the surfaces of the zincblende and diamond type semiconductors with respect to their optical response appear to be anisotropic in contrast to their bulk behaviour. This anisotropy is related to surface reconstructions exhibiting ordered arrays of dimers and thus is leading to an anisotropic optical polarisability. The latter can conveniently be measured in the spectral range from 1.5 to 5.5 eV by eliminating the bulk response through differential measurements. A large number of surface modification processes exhibit significant anisotropic surface signals. In this work an optical technique taking advantage of this surface anisotropy is described: Reflectance Anisotropy Spectroscopy (RAS) (also refered to as Reflectance Difference Spectroscopy (RDS)). We summarize the physical background of RAS and give an overview about applications were RAS has been proven to be an optical in-situ technique sensitive to semiconductor surface modifications on a sub-monolayer scale.

1 INTRODUCTION

While ellipsometry is a well established and widely used optical technique for the in-line and in-situ characterization of thin oxide and semiconductor layers in device production lines [1, 2, 3, 4] its derivative Reflectance Anisotropy Spectroscopy (RAS; also refered to as Reflectance Difference Spectroscopy (RDS)) has yet to develop itself from a surface sensitive tool, developed and utilized in surface science research laboratories, to an accepted in-situ technique in production environments. The potential of RAS as a sophisticated in-process sensor covers a wide range of semiconductor materials (Si [5], Si/Ge [6], GaAs [7], ZnSe [8], etc.) and growth processes (Molecular Beam Epitaxy (MBE) [9], Metal-Organic Chemical Vapour Deposition (MOCVD) [9], Metal-Organic MBE (MOMBE) [10, 11]) While the other optical in-situ techniques like photometry and ellipsometry [4] entered the field of in-situ semiconductor growth monitoring after a long standing tradition in ex-situ characterization, the present situation of RAS is different: due to the physical origin of the reflectance anisotropy this method is predominantly applicable in-situ. Therefore this work is aimed at summarizing the physical background and at giving an overview about applications were RAS already has been shown to be an optical in-situ technique sensitive to semiconductor surface modifications on a sub-monolayer scale. In Sec. 2 we outline the experimental details of the RAS technique covering both the optical setup and the adaptation to the growth process. The present status of the available database of RAS spectra and the procedure of gaining reference data are disscussed in Sec. 3. Sec. 4 contains a set of examples where RAS is already used as a sensitive surface monitor, and Sec. 5 gives an outlook at the future potential of RAS.

2 EXPERIMENTAL

2.1 The Optical Setup

RAS is based on a normal incidence ellipsometric setup. One way of introducing the experimental details of RAS is starting from the well known standard ellipsometer [1, 2, 3, 4] where the ellipsometric angles Ψ and Δ are measured at an angle of incidence in the 60° to 80° range. Ellipsometry takes advantage of the fact that the semiconductor surface has different reflectivities for light polarized perpendicular and parallel to the plane of incidence. The amplitude ratio and phase difference of both polarization components is measured in order to determine parameters such as film thickness and alloy composition. If the angle of incidence is shifted from oblique incidence

a)

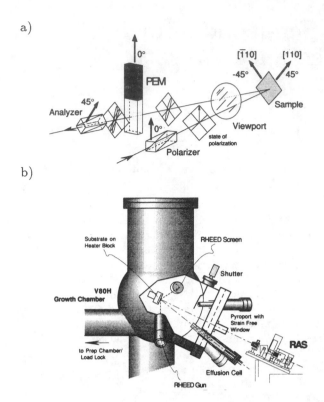

b)

Figure 1: The RAS optical setup (a) and an RAS system attached to the pyroport of an MBE growth chamber (b)

to normal incidence, the plane of incidence is not defined any more. The measured intensity ratio and phase difference approach 1.0 and 0.0, respectively, thus yielding always the same (trivial) experimental result for any kind of semiconductor sample. The reason for this obviously undesirable fact is that all the semiconductor materials of interest here (Si, Ge, III-V's and some II-VI's) are known to be optically bulk isotropic due to their cubic crystal structure. Therefore in this somewhat strange version of nulling ellipsometry, the reflectances measured at normal incidence for two perpendicular orientations of polarization are identical.

RAS is taking advantage of the optical isotropy of the crystal bulk. During semiconductor growth at elevated temperatures the crystal surfaces are naturally bare of any oxides and the orbitals of the uppermost atomic layer are still not completely integrated into the crystal lattice. To minimize the surface energy, in most cases surface dimer bonds are formed with defined binding directions forming an ordered surface reconstruction: the surface is becoming anisotropic! Because normal incidence ellipsometry gives a zero signal from the semiconductor bulk, it is, as the null ellipsometers [12] of the early times of semiconductor oxide measurements, a highly sensitive tool to

measure the tiny anisotropy of the uppermost atomic layers of a semiconductor structure.

Starting from this fundamental idea D.E. Aspnes and co-workers published in 1985 the basic design of a RAS setup [13] which up to now is the prototype of most of the RAS systems in a number of laboratories world-wide. A typical configuration of the optical components is given in Fig. 1a. To measure the optical anisotropy between the [110] and [1̄10] direction of semiconductors (001) surface, the components are arranged as follows: The transmission axis of the polarizer prism is oriented 45° towards [110] which gives equal incoming intensities in the [110] and [1̄10] directions. If the surface is anisotropic the amplitude and phase of the reflected light will be slightly different for each direction. The photoelastic modulator subsequently modulates the phase difference between both light components. After this phase modulation is converted into amplitude modulation by the analyzer prism, the real and imaginary part of the complex reflectance anisotropy ratio

$$\frac{\Delta r}{r} = 2\frac{r_{[\bar{1}10]} - r_{[110]}}{r_{[\bar{1}10]} + r_{[110]}} \tag{1}$$

are gained from the analysis of the detected signal. Using a monochromator with a focal length of 0.1 m and a holographic UV grating, a photomultiplier as detector, and Rochon polarising prisms the reflectance anisotropy can be measured in the spectral range from 1.5 eV to 5.5 eV.

2.2 Coupling to the Growth Chamber

The RAS optical system can be arranged fairly compact ($20 \times 30 \times 15$ cm^3) and as a normal incidence technique it is easily adapted to epitaxial growth chambers (Fig. 1b) using for example the pyrometer port of a MBE chamber. Because it is a phase sensitive UV optical technique the window to the process has to be UV transparent and strain free (e.g. Bomco Inc. [14, 15]). Remaining effects of window strain can be separated either by sample rotation, reference samples (e.g. Si(001) or a-As cap [16]) or a Kramers-Kronig-consistency check [17] of the real and imaginary part of the RAS signal. In cases where standard glass windows have to be used a purposely strained quartz plate can be placed in the optical path to compensate the unwanted phase shifts induced by the window [18]. If the growth chamber provides only oblique incidence optical access, the sample can be rotated to obtain RAS spectra in an ellipsometric setup [19].

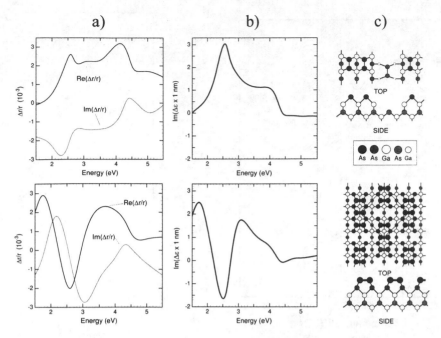

Figure 2: (a) Real and imaginary part of the RAS spectra for a c(4×4) and a β(2×4) reconstructed GaAs(001) surface with (b) the related surface dielectric function anisotropy and (c) the dimer configuration as determined by STM

3 RAS DATA BASE

As discussed in Sec. 2.1 RAS is highly sensitive to the status of the uppermost monolayers of a semiconductor structure. In order to relate the measured RAS spectra to the actual status of the semiconductor surface in a growth chamber it is necessary to compare the measured anisotropy features with those of an RAS database. In general the spectra in the database should cover the complete range of temperatures, compositions, and surface reconstructions relevant for the process to be monitored by RAS. While for ellipsometry the bulk dielectric function database for elevated temperatures at least for the most important semiconductors (GaAs [20], AlAs [21], and Si [22]) has recently been established, the RAS database even for the best studied GaAs (001) surface [7] is still incomplete and a subject of present experimental work [23]. From both the real and imaginary part of the RAS signal (Eq. 1) and the bulk dielectric function of the underlying semiconductor the anisotropic surface dielectric function $\Delta\epsilon$ can be calculated according to [19] as

$$\Delta\epsilon = \frac{\lambda(\epsilon_b - 1)}{4\pi i d} \cdot \frac{\Delta r}{r} \qquad (2)$$

with i the imaginary unit, d the effective thickness of the surface layer, l the wavelength of light and ϵ_b the bulk dielectric function. In order to assign the surface status to the measured RAS spectrum, sensitive reference measurements have to be applied. Mostly RHEED is used to assign a long range ordered surface net to the optical spectrum even if the latter is related only to the short range ordered dimer bond directions. This assignment is justified when the semiconductor surface is completely and homogeneously reconstructed. In some cases, however, the surface is dominated by domains of different reconstructions contributing to the RAS spectrum by superposition. Here sophisticated quenching procedures have to be applied to study the related surface status by STM to yield a well defined reference for the RAS spectra [23].

For MOCVD growth where both RHEED and quasi-in-situ STM cannot be applied, in-situ studies with grazing incidence X-ray scattering (GIXS) using synchrotron radiation paved the way for establishing an RAS database for precursor modified MOCVD surfaces [24]. These examples may illustrate the present efforts to establish a well characterized and completed database making this technique feasible for routineous growth monitoring in the future.

Fig. 2 shows typical RAS spectra of the c(4×4) and β(2×4) reconstructed GaAs(001) surface at 500° C together with the related anisotropic surface dielectric functions and a stick and ball model of the surface unit meshes. The surface dimers have obviously oscillator like anisotropic absorption resonances with characteristic resonance energies (about 2.5 eV

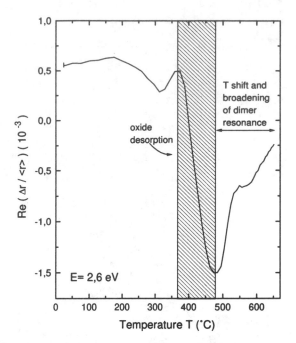

Figure 3: GaAs oxide desorption in MOCVD monitored by RAS

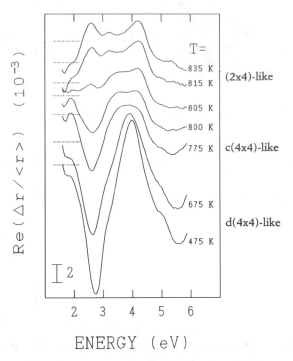

Figure 4: RAS spectra of GaAs(001) surfaces in MOCVD at different temperatures

at $500°$ C for the As dimers in Fig. 2) which can be utilized to sensitively monitor changes of the surface conditions by RAS one-wavelength transients. Thus for a bare semiconductor substrate the status of the surface can be derived directly from the RAS spectra. For layered structures where Fabry-Perot-interferences modify the RAS response the anisotropic surface dielectric function has to be implemented into an appropriate multi-layer model to calculate the RAS spectra. [25].

4 EXAMPLES OF RAS REAL TIME DIAGNOSTICS

4.1 Oxide Desorption

Under MBE conditions the thin oxide that remains on the GaAs surface after a standard etching procedure is known to desorb at about $580°$ C. In MOCVD the oxide desorption temperature depends on the H_2 carrier gas flow rate, the arsine partial pressure and the reactor total pressure and thus in-situ oxide desorption monitoring is essential. The RAS spectrum of an oxidized GaAs(001) wafer is related to a weak anisotropy in the refractive index of the thin oxide. If the oxide desorbs at about 350 to $400°$ C the bare semiconductor surface features an optical anisotropy typical for a c(4×4) surface reconstruction. This process can easily be monitored as a sharp step in the

RAS signal with a maximum sensitivity at 2.6 eV were the As dimers in the upper layer of the As terminated surface have their optical absorption resonance (Fig. 3).

4.2 Pre-Growth Surfaces and Temperature Measurement

Once the oxide is desorbed the surface evolves a surface reconstruction defined by the minimum total surface energy and the balance between the thermally induced As desorption and the arsenic adsorption from the arsine/As_2/As_4 flow. Because in MOCVD both processes depend on temperature, surface morphology, surface composition, arsine and TMG partial pressure, an in-situ monitoring of the actual status of the surface prior to growth is highly desirable. Fig. 4 gives as an example a set of RAS spectra indicating the decreasing As coverage of a GaAs(001) surface with increasing susceptor temperature. The status of a MOCVD system can be checked by measuring a complete surface phase diagram of the As coverage derived from the RAS spectra at different substrate temperatures and TMG or TEG partial pressures [26].

4.3 Composition of Ternary Alloys

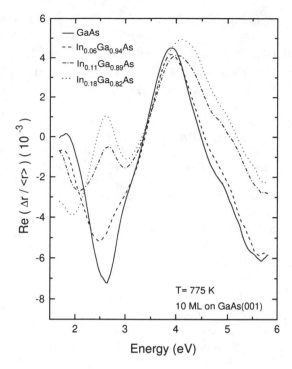

Figure 5: In-situ RAS spectra of InGaAs changing with composition

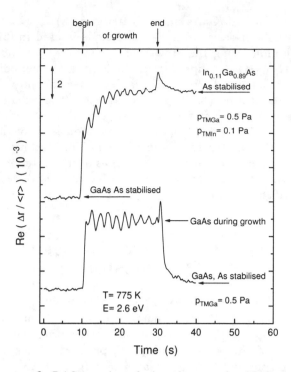

Figure 6: RAS transient during the growth of InGaAs on GaAs

If a ternary semiconductor is grown epitactically on a binary substrate, e.g. AlGaAs on GaAs or In-GaAs on GaAs, both the surface dimer resonance energy and the absorption/desorption equilibrium are shifted. Because the RAS spectra stem from both the surface dielectric function (which is dominated by the surface dimer resonance) and the semiconductor bulk dielectric function (Eqn. 2) the in-situ RAS spectra can be directly related to the composition of the grown ternary alloy. As an example Fig. 5 gives RAS spectra of InGaAs layers of varying composition illustrating the sensitivity of the method [27].

4.4 Determination of Growth Rate and Layer Thickness

Similar to RHEED in MBE, RAS can measure the growth rate in island growth mode via monolayer oscillations [28, 26, 10]. The origin of the oscillations is the modulation of the monolayer step density with island growth and island coalescence and the resulting modulation of the adsorption/desorption equilibrium. Fig. 6 gives the RAS transient at a photon energy of 2.6 eV for the growth of an InGaAs layer on GaAs in MOCVD. RAS oscillations with monolayer periodicity can clearly be identified while the mean RAS level shifts from its steady state level for a GaAs surface (before growth) to its steady state

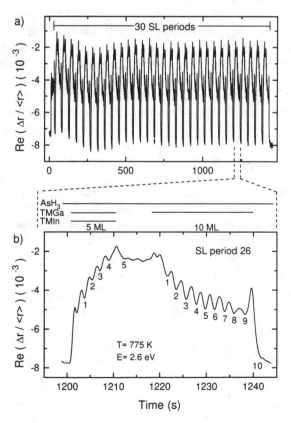

Figure 7: (a) Growth of a 30 period InGaAs/GaAs superlattice monitored by RAS. (b) Enlargement of the 26th quantum well period. Every single monolayer of the 30×15 ML structure is monitored

level for an InGaAs surface (after growth). This sub-monolayer sensitive in-situ technique is used in [27] for monitoring the growth of a complete 30 period InGaAs/GaAs superlattice. In Fig. 7a the periodic shift between barrier material level (GaAs) and well material level (InGaAs) can be seen. The enlargement of an arbitrarily chosen quantum well period in Fig. 7b shows that every single monolayer of the 30×15 ML structure is monitored. The thicknesses of well and barrier layers as well as abrupt interfaces of this structure grown under RAS control have been verified by subsequent ex-situ X-ray double crystal scattering.

5 OUTLOOK AND CRITICAL DISCUSSION

In Section 4 some examples have been given to underline that RAS is an outstanding surface sensitive technique applicable to growth processes in environments from UHV to atmospheric pressures. Thus the opportunity of surface status sensing by RAS in a complex device growth process is evident. For closed loop control by RAS however some systematic research still remains to be done. To target this aim the influence of the complete set of growth parameters to the RAS spectra has to be separated in order to feed back from the monitoring optical system to the growth machine. Layer thickness effects we assume relatively easy to separate either via the growth oscillation or via the Fabry-Perot-interference pattern in the conventional reflectance spectra which are, although not explicitly discussed here, a part of the spectral information gained with an RAS system.

As for in-situ spectroscopic ellipsometry the still missing bulk and surface dielectric function database which should cover the complete range of growth relevant temperatures and at least the most important binary and ternary semiconductor alloys is delaying the application of RAS for in-line and in-situ characterization in device production lines.

ACKNOWLEDGEMENTS

Part of this work was supported by funds of the Federal Ministry of Science and Technology (BMFT) under the project 01 BT 310/835.

REFERENCES

[1] S. Perkovitz, D. G. Seiler, and W. M. Duncan. *J. Res. Natl. Inst. Stand. Technol.* 99, 605 (1994).

[2] S. Perkovitz, D. G. Seiler, and W. M. Bullis. *(this volume)*.

[3] D. E. Aspnes. *Thin Solid Films* 233, 1 (1993).

[4] W. Gillmore and D. E. Aspnes. *(this volume)*.

[5] T. Yasuda, L. Mantese, and D.E. Aspnes. *Phys. Rev. Lett. (to be published)* (1995).

[6] M. E. Pemble, N. Shukla, A. R. Turner, J. M. Fernandesz, B. A. Joyce, J. Zhang, A. C. Taylor, T. Bitzer, B. G. Frederick, K. J. Kitching, and N. V. Richardson. *Proc. of the 1. US-European Workshop on Optical Characterization of Electronic Materials, Halle, Oct. 1994 (to be published in Phys. Stat. Sol. (a))* (1995).

[7] I. Kamiya, D. E. Aspnes, L. T. Florez, and J. P. Harbison. *Phys. Rev. B* 46(24), 15894 (1992).

[8] J.-T. Zettler, H. Wenisch, K. Stahrenberg, B. Jobs, D. Hommel, and W. Richter. *(to be published)* (1995).

[9] D. E. Aspnes. *J. El. Mater. (to be published)* (1994).

[10] J. Rumberg, P. Schützendübe, J.-T. Zettler, L. Däweritz, K. Stahrenberg, M. Wassermeier, K. Ploska, and W. Richter. *Surf. Sci. (submitted)* (1995).

[11] J. Rumberg et. al. *(to be published)*.

[12] R. M. A. Azzam and N. M. Bashara. *Ellipsometry and Polarized Light*. North-Holland, Amsterdam, (1977).

[13] D. E. Aspnes, J. P. Harbison, A. A. Studna, and L. T. Florez. *J. Vac. Sci. Technol. A* 6(3), 1327 (1988).

[14] A. A. Studna, D. E. Aspnes, L. T. Florez, B. J. Wilkens, J. P. Harbison, and R. E. Ryan. *J. Vac. Sci. Technol. A* 7, 3291 (1989).

[15] BOMCO Inc., Glouchester, MA (USA).

[16] U. Resch, S. M. Scholz, U. Rossow, A. B. Müller, and W. Richter. *Appl. Surf. Sci.* 63, 106 (1993).

[17] S. M. Scholz et. al. *to be published)* (1995).

[18] I. Kamiya, H. Tanaka, D. E. Aspnes, L.T. Florez, E. Colas, J. P. Harbison, and R. Bhat. *J. Vac. Sci. Technol. B* 10(4), 1716 (1992).

[19] K. Hingerl, D. E. Aspnes, and L. T. Florez. *Appl. Phys. Lett.* 63, 885 (1993).

[20] C. H. Kuo, S. Anand, R. Droopad, K. Y. Choi, and G. N. Maracas. *J. Vac. Sci. Technol. B* 12(2), 1214 (1994).

[21] G. Maracas. *(this volume)*.

[22] G. Vuye, S. Fisson, V. Nguyen Van, Y. Wang, J. Rivory, and F. Abeles. *Thin Solid Films* 233, 166 (1993).

[23] M. Wassermeier et. al. *(to be published)* (1995).

[24] I. Kamiya, L. Mantese, D. E. Aspnes, D. W. Kisker, P. H. Fuoss, G. B. Stephenson, and S. Brennan. *Proc. of the 1. US-European Workshop on Optical Characterization of Electronic Materials, Halle, Oct. 1994 (to be published in Phys. Stat. Sol. (a))* (1995).

[25] S. J. Morris, J.-T. Zettler, K. C. Rose, D. I. Westwood, D. A. Woolf, R. H. Williams, and W. Richter. *J. Appl. Phys. (accepted)* (1995).

[26] K. Ploska, J.-Th. Zettler, W. Richter, J. Jönsson, F. Reinhardt, J. Rumberg, M. Pristovsek, M. Zorn, D. Westwood, and R. H. Williams. *J. Cryst. Growth* 145, 44 (1994).

[27] M. Zorn, Jönsson, A. Krost, W. Richter, J.-T. Zettler, K. Ploska, and F. Reinhardt. *J. Cryst. Growth* 145, 53 (1994).

[28] J. P. Harbison, D. E. Aspnes, A. A. Studna, L. T. Florez, and M. K. Kelly. *Appl. Phys. Lett.* 52(24), 2046 (1988).

REAL-TIME MEASUREMENT OF FILM THICKNESS, COMPOSITION, AND TEMPERATURE BY FT-IR EMISSION AND REFLECTION SPECTROSCOPY

Peter R. Solomon

On-Line Technologies, Inc.
87 Church Street
East Hartford, CT 06108

Shaohua Liu, Peter A. Rosenthal
and Stuart Farquharson

Advanced Fuel Research, Inc.
87 Church Street
East Hartford, CT 06108

This paper discusses the methodology, hardware, and software to perform on-line or at-line monitoring of thin films: thickness (0.02 to 50 μm); temperature (50 to 3000°C ± 1°C); composition (wavelength dependent dielectric function, doping density, impurities, etc). Measurements of combined thickness and composition are made using Fourier transformed infrared (FT-IR) reflection spectroscopy. Determinations of temperature involves the measurement of FT-IR reflection and emission made simultaneously with the same instrument. This paper will describe the measurement methods and present several illustrative examples.

INTRODUCTION

The continued increase of device densities in integrated circuits requires smaller dimensions and thus tighter process control. In current practice, control is provided by off-line metrology measurements of film thickness, composition, uniformity, defect analysis, etc. Significant reductions in waste and faster introduction of new IC's with lower manpower could be achieved by developing on-line sensors, and accompanying real-time machine control technology. Experiments on a variety of samples made off-line, in-line and in-situ have demonstrated that a combination of infrared emission, transmission, and reflection spectroscopy yield the following process information: 1) film temperature, thickness, refractive index, and composition, and 2) gas temperature and composition. All of the measurements can be performed in-situ, during processing using a single Fourier transform infrared (FT-IR) spectrometer. In this paper, we will consider opaque samples for which measurements of transmission are unnecessary.

METHODOLOGY

These measurements are made using the geometry shown in Fig. 1. The illustration is for a single thin film on a substrate in a gas filled reactor. The thin film and substrate are characterized by their frequency dependent complex dielectric function real and imaginary parts, $\varepsilon_i(v)$ and $\varepsilon_r(v)$, thickness, z, and temperature, T. The gas is characterized by its composition and temperature. As shown in the figure, a single FT-IR spectrometer can measure the reflectance, $\rho(v)$ with detector D_ρ, transmittance, $\tau(v)$ with detector D_τ, and emission, or radiance, $R(v,T)$ with detector D_R.

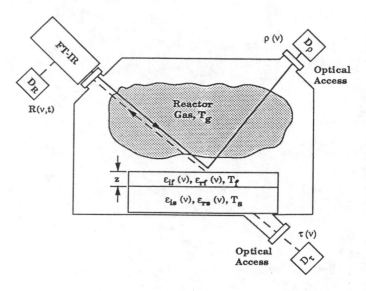

Figure 1. FT-IR Measurements During Film Processing.

Measurements of combined thickness and composition are made using FT-IR reflection spectroscopy. If the complex dielectric functions for the thin film and substrate, and the thicknesses of the film are known, then $\rho(v)$ and $\tau(v)$ can be predicted. Here, however, the inverse analysis is required, and it is not as straightforward. To perform the inverse analysis, we employ a model for the reflectance from a multi-layer film stack as in Ref. 1, and a parametric model for the complex dielectric function as in Ref. (2). An automated solution routine simultaneously solves for both the thickness of the film and its dielectric function by varying the model parameters to obtain a match between measured and theoretical reflectance. In addition, information can be obtained on the thickness of the film-substrate interface layer and the free carrier concentration. Models for multi-layered and patterned surfaces have also been successfully employed (3).

The method for temperature measurement involves the simultaneous measurement of infrared reflection and radiance spectra from a sample surface. For opaque specular samples, conservation of energy requires that reflectivity plus absorptivity equals unity. Since absorptivity equals emissivity, the measurement of reflectivity determines the emissivity, which combined with the radiance gives the temperature. Both measurements can be made simultaneously with a single spectrometer, with the use of multiple detectors (4), or using an external source and chopper and a single detector (5). This paper presents several illustrative examples of the technique.

RESULTS

Epitaxial Silicon (Epi) Thickness/Off-Line

The Semiconductor Industry Associations (SIA) Roadmap for epitaxial silicon calls for 3.0, 1.4, and 0.6 μm thickness layers in the years 1995, 2001, and 2010, respectively. Accuracies required for these measurements are 10 nm, 1.9 nm, and 0.8 nm, respectively. FT-IR spectroscopy has traditionally been employed for measurements of epitaxial silicon film thicknesses. Unfortunately, current methods (which analyze for thickness assuming a constant dielectric function using 'the interferogram') do not provide the accuracy required to meet the SIA Roadmap after 1995. In addition, current methods do not provide any doping density information or allow characterization of the epi-substrate interface layer.

Our method uses the Fourier transformed data in the reflectance spectrum. A typical measurement is presented in Fig. 2. The measured reflectance is fit using a three layer model for the thin film, epi-substrate interface layer and the substrate, and a parametric model of the silicon dielectric function which depends on the doping density. The figure shows the comparison of theory and experiment, which determines the thickness to be 4.2 μm and the interface (diffusion) layer to be 0.13 μm. The accuracy of the method has been checked using

gauge studies and comparisons to SIMS data (6). Accuracies for the 1 μm epi layers are on the order of ± 2 nm.

Submicron Epi/Off-Line

Current FT-IR methods allow measurement of epi layer thickness down to 1 μm on a heavily doped substrate. The method employed here allows measurements down to 0.1 μm and 1 μm epi on lightly doped substrates (10^{17} cm^{-3}). We have performed tests on five submicron thick epi silicon films which were provided by SEMATECH. The FT-IR thickness measurement results (1.34 ± 0.002 μm, 0.54 ± 0.003 μm, 0.22 ± 0.005 μm, 0.17 ± 0.004 μm, and 0.12 ± 0.007 μm) are compared in Fig. 3 to SIMS results supplied by SEMATECH. The data show that film thicknesses down to 0.1 μm can be measured with reasonable accuracy. More detail can be found in Fowler et al. (7).

Epi Film Thickness in a Cluster Tool

We have performed real-time measurements of epitaxial silicon films in the cool-down chamber of an Applied Materials' Centura 5200 cluster tool at Texas Instruments. The first series of tests employed a Bomem spectrometer and DTGS detector to obtain 16 cm^{-1} resolution spectra with a 60 s measurement. A 30 measurement gauge study for reproducibility and repeatability demonstrated epi thickness measurements of 1.15 μm with a standard deviation of 3 nm. A second series of tests used an On-Line Technologies, Inc. model 2100 FT-IR with a nitrogen cooled MCT detector. The spectrometer installed in the Cluster Tool is shown in Fig. 4. Good, low noise and low drift spectra were obtained using 4.5 s measurement at 16 cm^{-1} resolution. The spectra, compared to the Bomem spectra had 3 times lower noise with lower long term drift. Based on these experiments, thickness measurements with an accuracy of ± 2 nm can be obtained in under 1 s.

In-Situ Monitoring of a Pulsed Laser Deposition (PLD) Process

A Bomem MB 155 FT-IR equipped with a MCT detector was modified for sequential emission and transmission measurements and installed for in-situ diagnostics of pulsed laser deposition (PLD) of high T_c superconductor and ferro-electric thin films. The PLD system was run under conditions to produce multi-layer thin-film

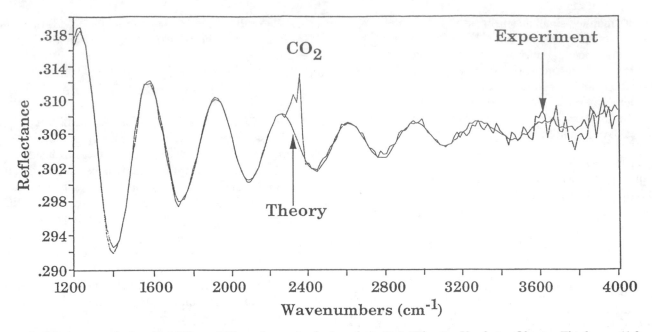

Figure 2. *Measurement of an Epi Film. A Three layer Analysis of an Epi Film is Used to Obtain Thickness (4.2 µm) and the Film Substrate Transition Width (0.13 µm).*

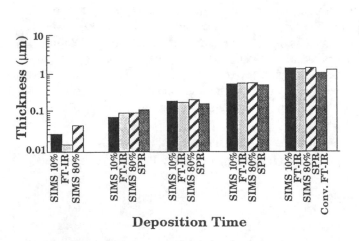

Figure 3. *Thickness Measurements of Thin Epi Silicon. SIMS 10% and 80% are the Thickness Measured to the Point where the Carrier Concentrations are 80% and 10% of the Substrate Values. SRP is the Thickness Determined by Spreading Resistance Probe.*

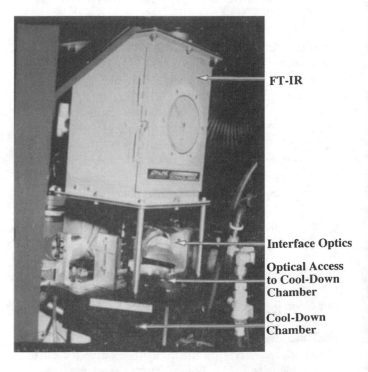

Figure 4. *Photograph of On-Line Technologies, Inc. Model 2100 FT-IR with Sample Interface Optics Mounted in the Cool Down Chamber of an Applied Materials, Inc. Centura 5200 Cluster Tool.*

materials, including: 1) epitaxial YBCO films on YSZ buffered Si(100) substrates, 2) epitaxial YBCO films on MgO buffered GaAs(100) substrates, and 3) ferroelectric BaTiO$_3$ (BTO) films on Si(100) substrates. Before each film deposition, the reflection spectra of a gold coating on silicon, a single-side polished Si(100) substrate, or a single-side polished GaAs(100) substrate were measured. These spectra served as the reference for the analysis of the reflectance spectra from the thin-film materials. Figure 5 shows reflectance spectra taken every 20 s during BTO thin-film growth on YSZ buffered Si(100) substrates. These spectra can be fitted with a multi-layer model which includes the BTO and YSZ. The thicknesses determined by our analysis are in good agreement with the expected deposition rate. Repeatability of the thickness measurement is ± 10 Å.

In-Situ Monitoring of p/n Junction Anneal

p/n junctions formed by ion implantation require annealing to activate carriers. Experiments were performed to monitor the surface by reflection and emission spectroscopy during the annealing of a n+/p shallow junction. Both reflectance and radiance spectra were obtained every 2 s. The results in Fig. 6 show the reflectance and temperature during annealing. The reflectance spectra (Fig. 6a) show that activation (indicated by an increase in the reflectance below 2000 cm^{-1}) was initiated at 667°C in 8 s and leveled off at 788°C in 12 s. The shape of the reflectance spectra above 667°C match those of reference samples of annealed junctions. The increase of reflectance at long wavelengths is due to free carriers, and can be modeled using a Drude term in the dielectric function. Further time and temperature seem to reduce the free carrier contribution to the reflectance. This may be due to diffusion of carriers into the bulk. The curves in Fig. 6b are the radiance divided by (1 - reflectance). Adjusting overlayed blackbody curves to match the measured curve determines the temperature. Measurements of both sheet resistance and temperature were performed easily at temperatures up to 950°C on a 2 s time scale.

CONCLUSIONS

New instrumentation and analytical methods have been illustrated for determining film thickness, composition and temperature during the processing of silicon wafers, and ferroelectrics including:

- silicon epitaxial ("Epi") layers
- p/n junctions
- barium titanium oxide (BTO)

Other samples which have been successfully characterized by the same techniques (3,6) include:

- silicon polycrystalline ("Poly") layers
- silicon oxide
- silicon nitride
- photo resist
- patterned silicon wafers
- multi-layer films
- mercury cadmium telluride (MCT)
- magnesium oxide (MgO)
- yttrium stabilized zirconia (YSZ)
- yttrium barium copper oxide (YBCO)
- boron nitride

ACKNOWLEDGMENT

This work was supported by the U.S. Department of Defense/Air Force Contract No. F33615–92-C-1132. Jeff Brown was the Technical Monitor. The authors wish to thank Dr. Qi. Li of AFR for his help in the PLD tests, Jason Wang of Texas Instruments and John Haigis of AFR for their work on the cluster tool tests, Burt Fowler and Ron Carpio of SEMATECH for their collaboration on the submicron epi measurements, and Dr. Mehrdad Moslehi of Texas Instruments and Karen Kinsella of AFR for their collaboration on the p/n junction anneal study.

REFERENCES

1. Yamamoto, K. and Ishida, H., *Applied Spectroscopy*, **48** (7), 775, (1994).
2. Grosse, P., "FT-IR Spectroscopy of Layered Structures - thin solid films, coated substrates, profiles, multi-layers", SPIE Vol. **1575**, 169-179, (1991).
3. Liu, S., Haigis, J.R., DiTaranto, M.B., Kinsella, K., Markham, J.R., Li, Q., Fenner., D.B., Solomon, P.R., Farquharson, S., and Morrison, P.W., Jr., "Process Monitoring and Control of Integrated Circuit Manufacturing Using FT-IR Spectroscopy", Proceedings of AWMA/SPIE Meeting, McLean, VA, (Nov. 7-10, 1994).
4. Morrison, P.W., Jr., Solomon, P.R., Serio, M.A., Carangelo, R.M., and Markham, J.R., Sensors Magazine, **8**, (12&13), (1991).
5. Markham, J.R, Kinsella, K., Carangelo, R.M., Brouillette, C.R., Carangelo, M.D., Best, P.E., and Solomon, P.R., Rev. Sci. Instrum., **64**, (9), 2515-2522, (1993).
6. Solomon, P.R., Liu, S., Haigis, J.R., Rosenthal, P.A., and Farquharson, S., "Process Monitoring and Control During Plasma and Other Processing of Semiconductors", Final Report for U.S. DoD/Air Force Contract No. F33516-92-C-1132, (1995).
7. Fowler, B.W., Simmons, D.G., Carpio, R.A., Liu, S., Solomon, P.R., and Nishikida, K., "The Measurement of Sub-Micron Epitaxial Layer Thickness and Free Carrier Concentration by Infrared Reflectance Spectroscopy", in <u>Diagnostic Techniques for Semiconductor Materials and Devices/1994</u>, PV 94-33, Perrington, NJ, The Electro-Chemical Society, 1994, pp 254-265.

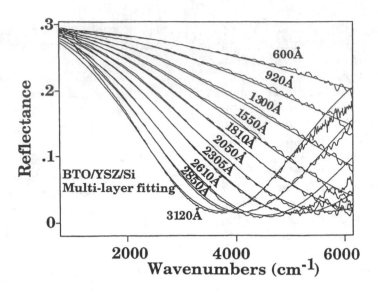

Figure 5. *FT-IR Monitoring in a Pulsed Laser Deposition Reactor. Spectra are Taken Every 20 s. The Fitted IR Reflectance Spectra of a BaTiO₃ Thin Films Deposited on a 500 Å YSZ Buffered Si(100) Substrate.*

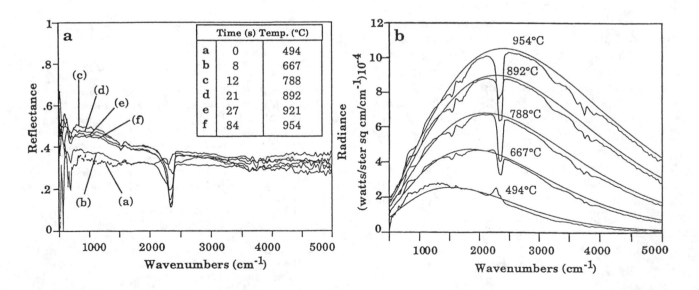

Figure 6. *Measurement of p/n Junction Anneal. a) Reflectance Spectra of n+/p Junction Taken During a Rapid Thermal Anneal; b) Surface Temperature Determination by Overlaying (Radiance/Emissivity) with a Theoretical Blackbody Temperature Curve.*

In-Situ Neutron Reflectivity of MBE Grown and Chemically Processed Surfaces and Interfaces

J. A. Dura and C. F. Majkrzak

Reactor Radiation Division
Materials Science and Engineering Laboratory
National Institute of Standards and Technology
Gaithersburg MD, 20899

Several properties make neutron reflectivity particularly suited to investigations of semiconductor structures produced by both molecular beam epitaxy (MBE) and chemical means. Reflectivity can provide quantitative information about composition, layer thickness and interface width. Since the scattering length for neutrons depends upon nuclear interactions, in contrast to x-rays, a large difference in scattering length can occur between systems of similar atomic number and one can employ isotope tagging at a particular fabrication step to determine its effect in the finished structure. Also, since neutrons are only weakly attenuated in many materials, one can also study wet chemical processes *in-situ* by passing the incident and reflected beam through a single crystal substrate. In this paper we describe a new facility for *in-situ* neutron reflectivity and related measurements within a MBE chamber, review the principles underlying reflectivity and grazing angle diffraction, discuss the unique capabilities of neutrons in comparison with x-rays, and give some examples of applications in which neutron reflectivity can be uniquely utilized.

INTRODUCTION

In neutron specular reflectivity, one measures the relative intensity of neutrons elastically scattered from a flat sample surface as a function of the glancing incident angle. A fit to these measurements yields a depth profile of the "scattering power", or scattering length density of a material. This provides information such as layer composition, with sensitivities on the order of a percent, and both layer thicknesses and interfacial widths, with Å resolution. By studying the off-specular intensity one can also derive information about the scale of in-plane structures. In grazing angle diffraction, in-plane peaks are directly measured, allowing one to determine the in-plane atomic structure of surfaces and interfaces.

To date, x-ray reflectivity and grazing angle diffraction have contributed significantly to our understanding of surface and interface structure in a wide variety of materials including various semiconducting materials (1). These studies must often be done *in-situ* to ensure that the surface being investigated is not perturbed. However, due to a lower incident beam flux relative to x-rays, neutron surface scattering has undergone a slower development. To our knowledge, the apparatus described here is the first UHV environment for surface neutron scattering to incorporate complete MBE fabrication capabilities. In addition we describe the means by which chemical processing of semiconductor surfaces can be investigated *in-situ* by neutron reflectivity.

SPECULAR REFLECTIVITY THEORY

In elastic specular reflectivity the incident and scattered beams are described by plane waves with wavevectors $\vec{k}_i$ and $\vec{k}_f$ respectively. Since the

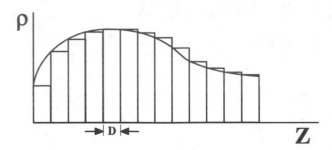

FIGURE 1. An arbitrary scattering density profile represented by slabs of uniform potential.

scattering is elastic, the magnitudes of the incident and scattered wavevector are equal, $|\vec{k}_i| = |\vec{k}_f| \equiv k$, and since it is specular the transverse components parallel to the surface are constants of the motion. Thus, the magnitude of the wavevector transfer $|\vec{Q}| = |\vec{k}_i - \vec{k}_f|$ equals $2k\sin\theta = 2k_z$, where the angles of incidence and reflection are both θ, and k_z is the component normal to the surface. Although neutrons are typically weak scatterers, at low values of Q the magnitude of the reflectivity is relatively high and the neutron wavefunction in the material can be significantly distorted from its free space plane wave form. The Born approximation is therefore not valid and a dynamical treatment must be used.

We first consider the case of neutron reflection from a single flat surface. Because the in-plane component of the scattering vector is conserved, the appropriate equation of motion is the one dimensional Schrodinger equation

$$\psi''(z) + k_z^2(z)\psi(z) = 0, \qquad (1)$$

where ψ is the neutron wavefunction, which in free space is $\exp(i k_{oz} z)$ where k_{oz} is the z-component of the neutron wavevector in vacuum. Conservation of energy requires that

$$k_{oz}^2 = k_z^2 + \frac{2mV}{\hbar^2} , \qquad (2)$$

where $\hbar$ is Planck's constant divided by 2π and m is the neutron mass. For most reflectivity measurements Q is sufficiently smaller than 2π divided by the interatomic spacing that one can treat the potential, V(z), as a constant within an isotropic medium. For neutrons (see, e.g., Ref. 2):

$$V = \frac{2\pi\hbar^2 bN}{m}, \qquad (3)$$

where b is the sum of the nuclear coherent scattering lengths of the atoms within a unit cell or molecule of number density N (unit cells/volume). With the effective scattering density $\rho = Nb$ we rewrite Eq. (2) as

$$k_z^2 = k_{oz}^2 - 4\pi\rho(z) . \qquad (4)$$

The reflection R and transmission T amplitudes are calculated by applying the boundary conditions imposed by conservation of momentum and particle number. The wave function on the vacuum side of the interface is:

$$\psi(z) = e^{ik_{zf}z} + Re^{-ik_{zf}z} , \qquad (5)$$

whereas in the material, behind the interface, it is

$$\psi(z) = Te^{ik_{zb}z}. \qquad (6)$$

For the semi-infinite barrier we obtain

$$|R(k_{oz})|^2 = \left| \frac{1 - \left[1 - (4\pi\rho/k_{oz}^2)\right]^{1/2}}{1 + \left[1 - (4\pi\rho/k_{oz}^2)\right]^{1/2}} \right|^2 \qquad (7)$$
$$\equiv \left| \frac{1 - n_z}{1 + n_z} \right|^2 .$$

Note that if $4\pi\rho > 0$, then for $k_{oz}^2 < 4\pi\rho$, the reflectivity, $|R|^2 = 1$, and there is total external reflection from the barrier. This is analogous to the total internal reflection of light inside optical fibers.

In general, for a rectangular slab of width D and constant scattering density ρ, the following matrix relation relating T and R results (see, e.g., Ref. 3):

$$\begin{pmatrix} T \\ i n_b T \end{pmatrix} = \begin{pmatrix} \cos\delta_m & \dfrac{1}{n_m}\sin\delta \\ -n_m\sin\delta_m & \cos\delta_m \end{pmatrix} \begin{pmatrix} 1+R \\ i n_f(1-R) \end{pmatrix}, \qquad (8)$$

where $n_b = k_{zb}/k_{oz}$, $n_m = k_z/k_{oz}$, $n_f = k_{zf}/k_{oz}$, $\delta = k_{oz}n_m D$, and n_m is a function of the scattering density ρ (see Eq. (4)).

Solving for R and T, one can determine the reflection and transmission coefficients, $|R|^2$ and $|T|^2$, which can be measured experimentally. The procedure outlined above can be applied in piecewise continuous fashion to arbitrary potentials (scattering density profiles) which are approximated to any desired degree of accuracy by an appropriate number of consecutive rectangular slabs, each having its own uniform scattering density ρ_j and thickness D_j as depicted in Fig. 1. For an arbitrary potential, the matrix M of Eq. (8) becomes a product matrix in which the jth matrix corresponds to the jth rectangular slab:

$$M = M_M M_{M-1} \cdots M_2 M_1 , \qquad (9)$$

where M_1 is the slab adjacent to the incident medium.

Thus one may calculate the reflectivity as a function of Q for a known profile. Obtaining an unknown density profile from measured reflectivity becomes an exercise in determining the profile that produces a calculated reflectivity which best fits the data. Typically, a model dependent approach(4) has been used, wherein one adjusts a model scattering density profile based upon knowledge of the sample's structure to best fit the data. The model may be adjusted at first by interpretation of the differences between the calculated pattern and the data, then as the fit improves by employing a least squares fitting routine. Other, model-independent, approachs have recently been introduced, which make no *a priori* assumptions about the profile (5). As an example of actual neutron data, and the quality of the fit, Fig. 2a shows the neutron reflectivity obtained by Wiesler *et al.* (6) *in-situ* during successive stages of electrochemical oxidation of a Ti film deposited on Si. The scattering length density profiles shown in Fig. 3b were obtained by least squares fits of the respective data sets.

DISTINCTIONS BETWEEN NEUTRON AND X-RAY REFLECTIVITY

While greater beam fluxes are available for x-ray reflectivity at synchrotron sources, the unique properties of neutrons make neutron reflectivity a very useful technique to complement some of the more standard methods of semiconductor characterization.

Neutron Scattering Length Density

In both x-ray and electron scattering the scattering length is a monotonically increasing function of atomic number. Therefore there is a very small contrast, or difference in scattering length density, between two elements of similar number density and Z. Neutron scattering lengths depend upon nuclear interactions the strength of which do not increase monotonically with Z and can therefore often provide a suitable contrast in cases where x-rays cannot. Neutrons also can have a considerable scattering length for light elements. Similarly, two compounds with the same average Z and number density have low x-ray contrast, since reflectivity probes the average composition of the unit

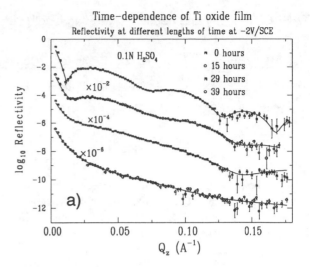

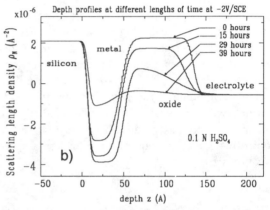

FIGURE 2. (a) Neutron reflectivity from electrochemical oxidation of Ti on Si, and least squares fits (solid lines). (b) Scattering length density profile corresponding to the fits in part a (after Ref.6).

cell. For example, in heterostructures of GaSb and InAs (which are finding applications in superlattices for infrared devices) the average atomic numbers of these compounds are the same. The x-ray contrast between these materials is small, whereas the neutron contrast is very large, as shown in Table 1. The x-ray contrast is also small in compounds that include light elements in which the number density of the common heavier element is similar, e.g., Si and SiO_2 also compared in Table 1. In some cases the scattering length for neutrons can be negative, thus causing dramatic differences when alloyed with particular elements. An example of this is hydrogen, with a neutron scattering length of -3.7406×10^{-15}m, which is virtually invisible to x-rays. In fact, using x-rays one can usually only infer the presence of H by the changes it causes in a material's lattice parameter.

TABLE 1. - X-ray and neutron scattering length density (SLD) contrast for several material systems.

Material	GaSb	InAs	Contrast	Si	SiO$_2$	Contrast	70Ge	74Ge	Contrast
X-ray SLD 10^{-5} Å^{-2}	3.964	4.064	2.5%	1.998	1.856	7.6%	3.817	3.817	0%
Neutron SLD 10^{-6} Å^{-2}	2.269	2.019	12.4%	2.072	3.425	65%	4.412	3.344	32%

Isotope effect

In the previous section the neutron scattering length densities quoted (except in the case of hydrogen) were the weighted average for the isotopic distribution of the element. Because of the often large variations in the scattering lengths of individual isotopes of a given element, further enhancements in contrast can be obtained by using isotopically pure sources. For example there is no contrast for x-ray scattering in systems in which only the isotope of the same element is different, such as ^{70}Ge/^{74}Ge(7), whereas the neutron contrast may be considerable (see Table 1). This effect is often utilized in materials containing hydrogen, by substitution with deuterium (with a scattering length of $+6.671 \times 10^{-15}$ m), to enhance the contrast relative to other materials or to isotopically tag a structural unit produced by a particular fabrication step.

Magnetism & Polarized Neutron Scattering

Another difference between neutron and x-ray scattering is the greater sensitivity of neutrons to the magnetic state of a material. While at present magnetic materials do not play a major role in the semiconductor industry, several magnetic devices are under investigation(8). The interaction between neutron and atomic magnetic moments is dependent upon their relative orientation. Two important yet simple selection rules apply in the case where the neutron polarization axis (defined by an applied magnetic field at the sample position) is perpendicular to $\vec{Q}$ (i.e., in the sample plane): the component of the in-plane magnetization parallel to this quantization axis gives rise to non-spin-flip (NSF) scattering (of the neutron) which adds coherently to the nuclear scattering, whereas the magnetic component perpendicular to the quantization axis of the neutron creates spin-flip (SF) scattering (which is purely magnetic). Consequently, the in-plane components of the magnetization can be inferred by measuring the two NSF (++, and --) and two SF (+-, and -+) reflectivities (where the pairs of signs refer to the polarizations of the incident and reflected neutrons respectively).

RELATED TECHNIQUES

Off Specular Scattering

The discussion up to this point has assumed a specular geometry, and thus the information obtained is averaged over a plane parallel to the surface of the sample. By offsetting the incident and reflected angle one probes reciprocal space away from the sample's surface normal direction and picks up an in-plane component of the wavevector transfer $\vec{Q}$. By doing so one gains sensitivity to the in-plane structure of the material, for example surface or interfacial roughness. However, one pays the cost of significantly lower intensities, and much more complicated analysis to obtain quantitative information. As such this is a developing field whose complexity is beyond the scope of this paper. The reader is referred to Ref. (9).

It is important to realize that if the coherence of the incident neutron wave (which can extend over macroscopic distances) is of sufficient spatial extent relative to the length scale of any in-plane density fluctuations, then in specular reflection these fluctuations are effectively averaged over, in-plane. Interfacial roughness is effectively manifest as a gradation in scattering density along the surface normal. Off specular scattering can, in principle, distinguish between these different effects.

Grazing Angle Diffraction

In grazing angle diffraction the incidence angle is kept below or near the critical angle of a material to allow the neutron wave to penetrate the surface only evanescently, to a depth of ~100-10,000Å, depending on the material and angle of incidence. Unlike reflectivity, the wavevector transfer $\vec{Q}$ is in the plane of the sample. This allows one to monitor, as a function of depth in the material, an in-plane lattice parameter, coherence length, and peak intensity changes during a process. One can also use it for surface crystallography, as, for example, in the x-ray determination of reconstructions of GaAs(10).

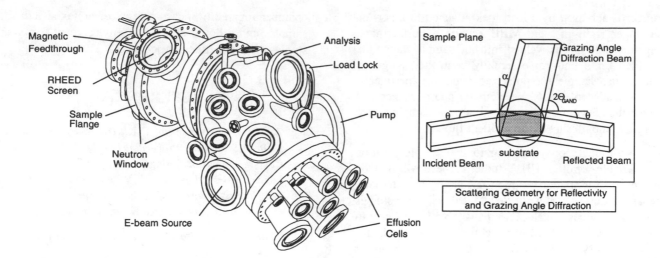

FIGURE 3. The MBE/Reflectivity chamber at NIST and scattering geometry for reflectivity and grazing angle diffraction.

Neutron grazing angle diffraction has been applied to both bulk(11) and MBE prepared thin films(12).

Prompt Gamma Activation Analysis

Analogous to x-ray fluorescence, upon exposure to neutrons, the nuclei of the sample emit gamma rays with a spectrum characteristic of the elements in the material and with sensitivities enabling trace analysis. By utilizing grazing angle techniques, surface sensitivity with adjustable skin depth can be obtained. Furthermore, with proper sample fabrication, a neutron standing wave can be created in which the beam flux can be enhanced at particular depths in the sample (13), as can be done with x-rays. For further discussion of prompt gamma activation analysis and its capabilities see the chapter in this volume by R. L. Paul and R. M. Lindstrom.

IN-SITU MBE/REFLECTOMETER

In order to be compatible with *in-situ* neutron reflectivity and grazing angle diffraction experiments several design features must be met. Due to the low flux and the grazing scattering geometry, large samples are required. The NIST MBE-reflectometer (Fig. 3) will use 3" wafers with an estimated 1% flux uniformity provided by effusion cells with conical crucibles or crucible inserts, a >10.5" crucible to sample distance, and sample rotation. On the source flange two 4.5" flanges and four 2.75" flanges are available for effusion cells with shutters attached to a 2" thick LN_2 cooled cryopanel, and actuated through pneumatic pushpull feedthroughs. In addition, there is a 2.75" flange centrally located on the source flange for either an effusion cell or device such as optical pyrometer. An additional shuttered effusion cell location is provided above the source flange, exactly perpendicular to the sample, to provide uniform flux without sample rotation for simultaneous deposition and neutron reflectivity measurements. The effusion cells on the 4.5" flanges can be replaced by electron beam hearth deposition sources.

Due to constraints both on the weight and balance (because the entire chamber will be mounted on a goniometer) and on stray magnetic fields, an ion pump will be avoided, and the pumping will be accomplished by a 2600 L/s cryopump attached by a gate valve to a port that has direct connection to the MBE section of the chamber both in front of and behind the source cryopanel.

A neutron window, provided by a cylindrical Al wall concentric with the sample will allow grazing angle diffraction at any azimuthal angle. Ports for RHEED electron gun and phosphor screen are also included in this section. These ports are placed away from the incident and reflected beam and avoid most of the area which could be used for grazing angle diffracted beams. However should the need arise the neutron window section can be rotated to allow access to other azimuthal directions. By modifying the sample holder to allow 90° tilt, this chamber could be used for high angle diffraction as well.

Because the wavevector transfer of the NIST NG1 Reflectometer lies in the horizontal plane, the surface normal of the sample must also lie in the horizontal plane. The sample holder is designed for a continuous temperature range of ~8-1200K, with appropriate uniformity across a 3" diameter substrate. Sample transfer is achieved by a load lock attached to a port on the upper right of the MBE section. Furthermore, magnetic pole pieces welded into the sample flange are provided for a uniform externally controlled magnetic field at the sample for polarized neutron experiments. Both a quadrupole mass spectrometer and quartz crystal microbalence will be incorporated into the sample flange for *in-situ* monitoring of the incident flux.

Additional ports, pointing to the sample position are available for other UHV studies. These will include an ion gun for sputter cleaning substrates or externally prepared samples, and a nozzle and leak valve for studies involving gaseous species. Also available is a 6" port to which a variety of devices could be attached, for example reverse view LEED, or electron energy analyzer for Auger or photoemission studies, etc. In addition several 2.75" flanges are present for excitation sources or other applications.

IN-SITU CHEMICAL PROCESSING CELL

While the MBE reflectometry chamber described above is a new and unique facility, another more conventional advantage of neutron reflectivity can be utilized for *in-situ* investigations of semiconductor processes that require a liquid, such as etching, oxidation, and surface passivation. Unlike x-rays, neutrons are only weakly attenuated in most materials, therefore if one avoids incoherent scattering by using the single crystal substrate as the incident medium one can do reflectivity measurements under an aqueous solution. Sample cells are made by pressing cup shaped plastic container materials directly against the substrate to form the seal. They can operate as a closed system, or with dynamic fluid flow. In this type of sample cell, electrochemical manipulations of surfaces (as in Fig. 2) have been extensively studied by neutron reflectivity. Furthermore, a second neutron reflectometer at NIST, NG7, is configured with a vertical wavevector transfer (and horizontal sample surface). This allows containerless studies of fluids on surfaces, held in place by surface tension alone. In addition, the capabilities for both horizontal and vertical sample surfaces allow for studies of *in-situ* wet chemical processes with either wafer geometry to repeat that used in production.

SUMMARY

We have briefly introduced neutron reflectivity and related techniques and described both a unique MBE chamber for *in-situ* measurements of epitaxial thin film samples and a liquid containing environment for *in-situ* studies of wet chemical processing.

REFERENCES

1. Zabel, H. and Robinson, I. K., eds., *Surface X-ray and Neutron Scattering, Proceedings of the 2nd International Conference*, Berlin: Springer-Verlag, 1992 and Lauter, H. J., and Pasyuk, V. V., eds., *Proceedings of the 3rd International Conference on Surface X-ray and Neutron Scattering*, Physica B **198**, 1-265 (1994).

2. Sears, V. F., *Neutron Optics*, Oxford, Oxford University Press, 1989.

3. Yamada, S., Ebisawa, T., Achiwa, N., Akiyoshi, T., and Okamoto, S., Annual Rep. Res. React. Inst. Kyoto Univ. 11, 8 (1978) or for a more recent review see, Majkrzak, C. F., "Fundamentals of Specular Neutron Reflectivity" in Proceedings of the Materials Research Society, Boston, 1994 (in press).

4. Ankner, J. A., and Majkrzak, C. F., "Subsurface Profile Refinement for Specular Neutron Reflectivity" in *Neutron Optical Devices and Applications*, 1992, pp. 260-269.

5. Berk, N. F. and Majkrzak, C. F., *Phys. Rev. B*, (in press).

6. Wiesler, D. G., Majkrzak, C. F., *Physica B.*, **198**, 181-186, 1994.

7. Spitzer, J., Ruf, T., Cardona, M., Dondl, W., Schorer, R., Abstreiter, G., Haller, E.E., *Phys. Rev. Lett.*, **72**, pp. 1565-1568, 1994

8. Prinz, G. A., *Science*, 250, 1092-1097, (1990)

9. Sinha, S. K., Sirota, E. B., Garoff, S., and Stanley, H. B., *Phys. Rev. B*, **38**, 2297-2311, 1988.

10. Sauvage-Simkin, M., Pinchaux, R., Massies, J., Claverie, P., Bonnet, J., Jedrecy, N., Robinson, I.K., Surf. Sci., **211-212**, p. 39-47, (1989)

11. Docsh, H., Al Usta, KJ., Lied, A., Drexel W., and Peisl, J., Rev. Sci. Instrum., **63**, 5533-5542, 1992.

12. Zhang, H, Satija, S. K., Gallagher, P. D., Dura, J. A., Ritley, K., Flynn, C. P., and Ankner, J. F., (submitted to Phys. Rev. B.)

13. Zhang, H, Gallagher, P. D., Satija, S. K., Lindstrom, R. M., Paul, R. L., Russell, T. P., Lambooy, P., and Kramer, E. J., Phys. Rev. B, **72**, 3044-3047, 1994.

FRONTIERS IN COMPOUND SEMICONDUCTORS

Frontiers in Compound Semiconductors: Bandgap Engineering and Novel Structures

T. C. McGill, D. A. Collins, M. W. Wang, J. F. Swenberg and R. W. Grant

T. J. Watson, Sr. Laboratory of Applied Physics
California Institute of Technology
Pasadena, CA 91125

Compound semiconductors have a very bright future in filling some of the deficiencies inherent in standard silicon-based electronics. They emit and detect light much more effectively and at widely different wavelengths, all the way from the infrared to the ultraviolet. Further, compound semiconductors form heterojunctions that increase dramatically the number and types of devices that can be formed. A method is presented that is particularly useful in examining the possibilities for bandgap engineering and the production of novel structures. To illustrate the points and to highlight the opportunities for characterization, we examine a number of heterojunctions and applications for new devices, new infrared materials and visible light emitters.

INTRODUCTION

If one had to pick a single semiconductor that has dominated this age, it would be silicon. Yet silicon is lacking in a number of crucial ways: it is very inefficient in emitting and absorbing light; it has a very fixed range of bandgaps; and it has very limited heterojunction possibilities. For these reasons, silicon is not the answer to all of the semiconductor industry needs and compound semiconductors are extremely important.

In this paper we will discuss some of the most recent, novel applications of compound semiconductors. The examples are selected so as to highlight unique successes of compound semiconductors produced by engineering the heterojunctions to produce novel electronic devices, light sources and detectors.

GUIDE TO BAND GAP ENGINEERING

Before we discuss a number of the exciting applications of bandgap engineering, it is useful to introduce a rather new and important way to assess material combinations for bandgap engineering. Historically, researchers have worked with diagrams showing bandgaps of materials plotted versus lattice constant. Yet this kind of plot fails to give information on band offsets and dopability, two very important properties in bandgap engineering for successful device production. J. O. McCaldin has devised a very useful

way of plotting the information that solves all of these problems.[1]

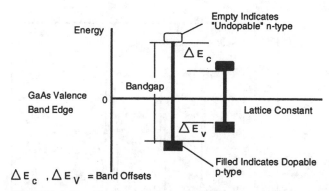

FIGURE 1. A diagram of energy versus lattice constant that allows one to simultaneously present the lattice constant, dopability, band gap and band offsets for various materials. Each material is represented by a vertical line of energy length equal to the band gap of the material located at the lattice constant for the material. The bottom of the vertical line is positioned in energy to give the value of the band offset relative to valence band edge of GaAs, the arbitrarily selected energy zero.

A schematic of such a diagram is shown in Fig. 1. The abscissa is the lattice constant of the semiconductor. The ordinate is the energy scale. Each semiconductor is represented by a line whose length is given by the bandgap and has its endpoints referenced to the valence band edge of GaAs (abitrarily taken to be the zero of energy). Dopability is indicated by whether the symbol at the end of the line is filled or empty, with materials that can be doped, the appropriate type being indicated by filled symbols.

In Fig. 2 (and in Fig. 7), we show McCaldin diagrams for a wide range of potentially useful semiconductors.[2] Such diagrams can be key components in band structure engineering. For example, they illustrate the tendency of semiconductors to group in lattice constant. In these groupings, we have the candidates for lattice matched structures. The diagrams also indicate some of the classic difficulties in getting high p-type doping in those materials with deep valence bands (e. g. ZnS, MgS, MgSe) and high n-type doping in those materials with high conduction bands (e. g. ZnTe, MgSe). Further, the diagram gives a quick guide to band offsets, indicating the approximate relative positions of the conduction band and valence band edges.

GaSb/AlSb/InAs HETEROJUNCTIONS

One of the very interesting cases that Fig. 2 shows is the case of GaSb/AlSb/InAs heterojunctions.

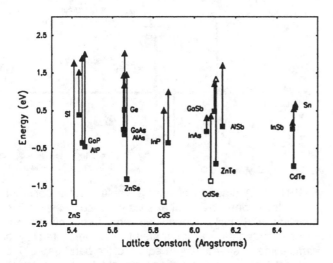

FIGURE 2. The McCaldin Diagram for a large number of semiconductor compounds.

This system is very near lattice match. The band offsets offer some unique possibilities. The conduction band of InAs is below the valence band of GaSb. These unique properties are the basis for a number of very exciting devices both for electronics and opto-electronics. We shall now look at several types of devices in more detail.

Quantum Devices

In the postshrink era quantum devices may become the workhorse for the electronics revolution.[3] A number of new electronic devices have been fabricated based on the GaSb/AlSb/InAs heterojunction system.[4] In Fig. 3 we have illustrated schematically some of these device structures, which have been the basis for a number of developments in quantum devices.

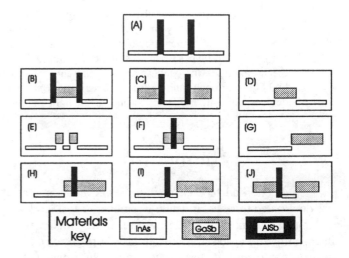

FIGURE 3. Some of the devices that have been fabricated from the InAs/GaSb/AlSb heterojunction. The devices as shown can be viewed as two-terminal devices with connections made to the layers on the far left and far right. Some of the device structures can be made into three-terminal devices with a contact made to another layer.

These devices have produced a number of success stories. The InAs/AlSb device in Fig. 3A holds the high frequency oscillator record at 720 GHz.[5] Tunnel devices based on Fig. 3B and Fig. 3C, so called resonant interband tunneling devices are the basis for artifical retinas[6] and high-speed digital signal processing circuits.[7] Stacking of these vertical devices could lead to high-density, multivalue-logic digital processing systems. The device in Fig. 3D is a barrierless-resonant-interband-tunneling device (BRIT). It shows a negative resistance in spite of the fact that the there is no barrier to produce the standard resonance. However, discontinuities in the band structure do produce a resonance confinement. This has been exploited to produce the "double barrier" structures in Figs. 3E and 3F. The standard Esaki diode is shown in Fig. 3G, with modifications thereof in Figs. 3H and 3I. A version of the so-called "Stark Effect" transistor is shown in Fig.3J where a contact is made to the center InAs layer.[8]

While we are in the earliest phase of exploration of quantum devices, they are likely to play a major role in the electronics revolution beyond 2010.

New Infrared Materials

One of the first applications of a superlattice as a tailored material was to provide new infrared structures. Building on HgTe/CdTe[9] superlattice research and then on strained layer InSb/InAsSb[10] superlattice research, Mailhiot and Smith proposed the infrared superlattice based on InAs/GaInSb.[11] The basic concept of this structure is shown in Fig. 4.

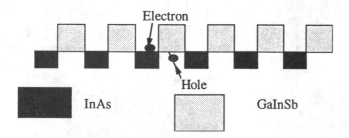

FIGURE 4. A schematic of an infrared superlattice, based on InAs/GaInSb. The InAs conduction band is below the valence band edge of GaSb. With the effects of confinement, alloying and strain, the band gap is non-zero. However the electron is primarily in the InAs layer and the hole in the GaInSb layer. The layers are quite thin (~few nm), hence the optical matrix element is comparable to the alloy systems with the same band gap.

Research by D. H. Chow and R. H. Miles [12] has brought this superlattice to the point of a high level of demonstrated interest. Infrared detectors and lasers have been fabricated that perform near or at the state of the art for a given wavelength. Single element superlattice detectors have demonstrated 12 μm performance that is equivalent to current HgCdTe alloy arrays.[13]. Infrared lasers operating at 3.47 μm have also been realized by D. Chow and R. Miles at Hughes.[14]

These superlattices have properties that could make them much more suitable for opto-electronic devices than HgCdTe alloys because of the suppression of the Auger processes. Henry Ehrenreich and co-workers[15] have found in their theoretical studies that the band structure of the infrared superlattices are such that

certain key Auger processes can be suppressed, leading to enhanced lifetimes for higher gain in lasers and better performance in detectors. The results of their calculations and the current experimental situation are illustrated in Fig. 5.[16]

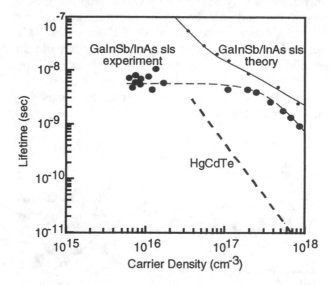

FIGURE 5. The lifetime as a function of carrier concentration for HgCdTe alloys and InAs/GaInSb superlattices (sls). The experimentally measured values from the Naval Research Laboratory are also included. The Auger processes in p-type material are used to estimate the lifetime theoretically.

Issues for GaSb/AlSb/InAs Heterojunctions

The very successes that have been briefly mentioned above have generated interest in producing high quality structures. Some of the major issues include: characterization and control of growth; the role of interface roughness and alloy induced fluctuations in small structures; control of the anion switch between Sb and As; and the role of defects and dopants.

One of the more interesting of these issues is illustrated in Fig. 6, where we have illustrated the two possible interfaces that can be obtained when growing InAs/GaSb structures. The first is the so-called InSb-like interface in which the top most layer of GaSb is terminated with Sb which is then bonded to In. The second is the so-called GaAs-like interface in which the top most layer of the GaSb is Ga-like and the growth of the InAs begins with GaAs-like bonds. The growth of these structures has been studied extensively by D. A. Collins, M. W. Wang, R. W. Grant and R. M. Feenstra.[17]

These authors have demonstrated a number of

very important properties of these interfaces and have attempted to characterize them under a wide range of growth conditions. In summary, antimonide on arsenide interfaces are relatively abrupt but arsenide on antimonide interfaces are more extended. The interdiffusion may be controllable by appropriate growth conditions, and the condition of the interface can influence many of the critical device properties. The band offsets are a function of the growth order, and the superlattice bandgap depends on the interface type. Interdiffusion of Sb into the As layers could produce some of the deep levels that are responsible for the less than optimum lifetimes observed to date.

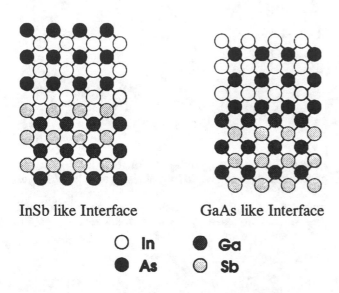

InSb like Interface   GaAs like Interface

○ In   ● Ga
● As   ◉ Sb

FIGURE 6. An illustration of the so-called anion switch problem. In the growth of GaSb/InAs structures one can either have an InSb like interface or a GaAs like interface as indicated in the left and right panels.

VISIBLE LIGHT EMITTERS

Another major development is the recent success in fabricating visible light emitters. These present some of the most challenging problems in compound semiconductor characterization. There are three major approaches for making green and blue lasers and light emitting diodes: ZnSe based light emitters, graded injectors, and GaN based light emitters.

II-VI Semiconductor Based Light Emitters

To illustrate many of the issues of the II-VI light emitters, we present in Fig. 7 a McCaldin diagram for the II-VI compound semiconductors. This diagram shows that we have relatively broad lattice constant groupings. Of the two compound II-VI light emitter approaches described in the following sections, one is based on ZnSe, which has been historically difficult to make p-type, and another is based on ZnTe, which has been difficult to make n-type. The ZnSe-based approach makes use of improved p-type doping techniques, while the ZnTe-based approach uses a heterojunction structure to solve the doping problems.

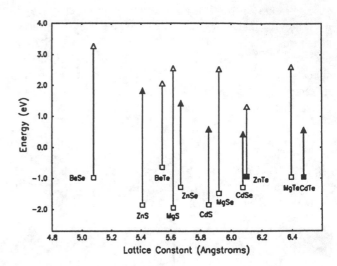

FIGURE 7. The McCaldin diagram for II-VI semiconductors that could be important in the production of visible light emitters.

Typical ZnSe Based Devices

The first significant wide bandgap light emitter success was in producing green and blue light-emitting diodes[18] and lasers[19] based on doping ZnSe p-type with nitrogen.[20] A typical structure is shown in Fig. 8. This figure demonstrates the very complex character of the laser structure, which includes ZnMgSSe cladding layers and ZnTe/ZnSe superlattice contacts. Although there are efforts underway to develop ZnSe substrates, most of the current structures are grown on GaAs substrates.

Issues for ZnSe Devices

Light emitting devices based on ZnSe present a number of very important characterization challenges. For example, the lifetime of many of the current devices is unacceptably short. Hence, the precise degradation mechanisms and the role of substrates in determining the overall quality of the heteroepitaxy layers are essential. Also, the McCaldin diagram for the II-VI materials in Fig. 7 shows that with the addition of Mg and S to the ZnMgSSe alloy, it should become even more difficult to dope p-type than ZnSe.

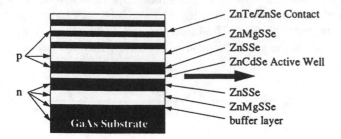

FIGURE 8. The typical ZnSe based laser structure.

Graded Injector

Another very different approach avoids the non-equilibrium doping of ZnSe by using a heterojunction approach. Study of the II-VI McCaldin diagram shows that the only wide bandgap bulk p-type II-VI material is ZnTe. The nearest lattice match to ZnTe is CdSe, which is easily made heavily n-type. However, the band offsets of this system are such that there are barriers to the electrons coming from the CdSe into the ZnTe and barriers for the holes coming from the ZnTe into the CdSe. Hence, a device consisting of n-CdSe/p-ZnTe

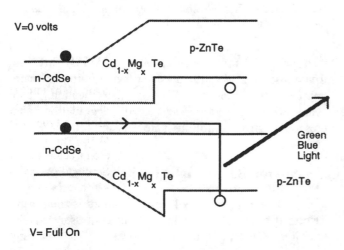

FIGURE 9. The graded injector approach to visible light emitters.

would have recombination occurring at the interface, which would be largely non-radiative.

The problems of this approach have been fixed by the addition of a special heterojunction structure that

has been termed a "graded injector".[21] The graded injector promotes the electrons from the conduction band in the CdSe into the conduction band of the ZnTe, where they recombine with holes to produce green light. We would like to preserve the hole blocking band offset between the CdSe and ZnTe. Hence, we have selected Mg as the alloying agent with CdSe to open its bandgap, with the primary change occuring in the conduction band. These points are illustrated in Fig. 9 where we have plotted a schematic band diagram for this structure at both zero bias (top) and at a voltage that produces flattening of the conduction band (bottom). By adding

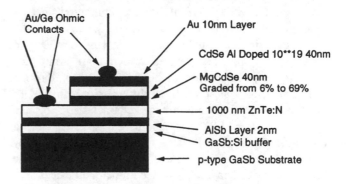

FIGURE 10. A typical graded injector device structure.

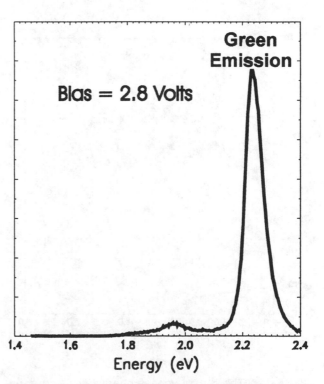

FIGURE 11. The electroluminscence spectrum from a graded injector device. The emission is mainly in the green well above the band gap of CdSe or the energy that characterizes recombination at the CdSe/ZnTe interface.

Mg to the ZnTe layer one can construct quantum wells and fabricate devices emitting far into the blue. The device has been demonstrated experimentally.[21] A typical device structure is shown in Fig. 10.

The electroluminescence spectrum, shown in Fig. 11 indicates that the graded injector approach does in fact work as predicted in Fig. 9. The light emission from the electrically stimulated devices are in the green part of the spectrum, far from the red that would characterize the CdSe or the even longer wavelength that would characterize emission by electrons recombining with holes at the interface.

One of the more important properties of a visible light emitter is its current voltage characteristic. A nearly ideal behavior of the current voltage characteristic is critical, since it usually indicates a lack of electrical inadequacies in the device, which can limit the performance, either through heating effects or through sources of degradation for the light emission. In Fig. 12 we display the current-voltage characteristic of the graded injector LED as compared to that for a commercially available green, GaP LED. As shown by this data the current-voltage characteristic of these devices compares favorably with commercial devices.

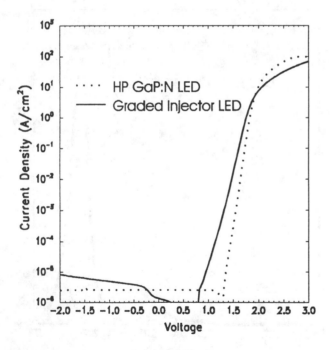

FIGURE 12. The current voltage characteristic on a log scale for the graded injector (CIT) light emitting diode compared with that from a commercially available III-V light emitter.

Issues for the Graded Injector

The graded injector provides significant challenges to characterization. The very thin MgCdSe layer must be controlled both structurally and electrically. The current structures are almost always fabricated on GaSb substrates, which leads to a number of issues with regard to the III-V/II-VI interface and its influence on the rest of the device structure. Finally, the degradation

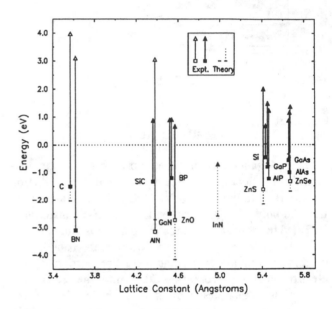

FIGURE 13. The McCaldin diagram for nitride based semiconductors.

mechanism for this device appears to be very different from those for ZnSe structures, yet their lifetimes remain unacceptably short. Thus, identification and characterization of the source of degradation and prescriptions for controlling it are very important.

Nitride Based Light Emitting Systems

The biggest story in light emitters is the recent success at Nichia in producing high brightness, relatively long-lived GaN light emitting diodes.[22] The relevant McCaldin diagram for nitride-based semiconductors is shown in Fig. 13.[23] This diagram shows many of the challenges provided by the nitrides. There are no easy lattice matches between the nitrides and readily available substrates. Further, the heterojunctions involving either InN or AlN are not well lattice matched.

A typical GaN light emitting device structure is shown in Fig. 14. This figure shows that the GaN

562

structures are grown on Al_2O_3 substrates, which have a very large lattice mismatch to GaN. The dislocation densities in these devices is very high, being greater than 10^{10} cm^{-2}, which is well beyond what one might expect for a successful light emitting device.[24] Thus, despite the remarkable success of GaN-based light emitters, many issues remain unresolved.

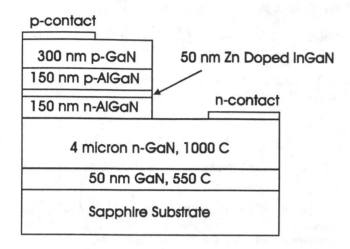

FIGURE 14. A typical GaN light emitting diode structure.

GaN Research Directions

With the promise of green, blue and ultra-violet light emitting diodes and lasers, one has the possibility of making a technical revolution. Yet nitride research is still at a very primitive stage. Very few things are known about the material system. Characterization will play a key role in mapping out the properties and providing the understanding that may lead to the realization of successful devices based on the nitrides. Some of the major issues include substrates, the properties of heterojunctions, the properties of defects and the development of new device designs.

The current substrates are not optimum. Yet the production of bulk GaN substrates seems very difficult at best. Attempts are being made to produce substrates that have more nearly ideal characteristics, for example growing GaN on Si substrates and then removing the substrate.[25] These approaches place heavy demands on the characterization tools. At present the dislocation densities are well beyond what one might expect for a successful light emitting device. Currently, the role of these dislocations is unknown and remains to be studied. For example, do they act as efficient non-radiative recombination centers? Do they migrate and lead to degradation of the overall light emission performance? These are a few of the many fundamental questions still to be answered.

HIGH TEMPERATURE ELECTRONICS

Compound semiconductors are the candidates for producing high temperature electronics.[26] Silicon, with its 1 eV bandgap, is not suitable for truly high temperature electronics. While diamond, with a bandgap greater than 5 eV, is in principle the nearly ideal candidate, difficulty in growth and doping have made it a fond hope but a distant reality at best.

The two primary candidates for high temperature electronics are SiC and GaN. Both have very large bandgaps, substantially greater than 2 eV, and are rugged enough to stand up to high temperatures. Recent successes have been reported both for SiC[27] and GaN.[28] Yet these remain major research areas with large numbers of unsolved problems in materials preparation and characterization.

FUTURE FOR COMPOUND SEMICONDUCTORS

Compound semiconductors now find and are likely to continue to find new applications where silicon-based semiconductors simply cannot do the job. However, these materials are not nearly at the same state of refinement as silicon, and their unique properties and different applications provide unique research and development challenges. Since many of the structures are fabricated by epitaxial growth methods such as molecular beam epitaxy (MBE), metal-organic chemical vapor deposition (MOCVD), or even liquid phase epitaxy (LPE), unique characterization issues are presented.

In situ characterization can be very important. Techniques such as reflection high energy electron diffraction (RHEED),[29] ellipsometry [30] and ESCA[31] are being employed heavily. Methods such as cross-sectional STM are also extremely important in characterizing the growth interfaces.[32]

Heterojunction characterization is extremely important as well. Measurement of band offsets,[33] interdiffusion at the interfaces,[34] roughness[35] and studies of growth phenomenon[36] are all critical to the future development of compound semiconductors.

Since many of the new materials, e.g. GaN,[37] have not been carefully studied in the past, even items such as defects and impurities, band structure and optical properties are not well known.

SUMMARY

Compound semiconductors have a very bright future indeed. New and technically important applications for them are constantly being discovered. Their versatility, based on their wide range of properties and ability to form heterojunctions, makes them the one most fruitful set of materials from which to develop new electronic and electro-optic applications. Since many of these materials are being used in very complicated structures, where the properties of interest can vary widely, characterization is of vital importance to the development of these devices. While it is possible to define the rather limited roadmap of silicon-based technologies, in the realm of compound semiconductors we are like explorers on the frontier, where roadmaps do not exist and trails are yet to be blazed.

ACKNOWLEDGEMENTS

The authors wish to acknowledge the constant guidance of Professor J. O. McCaldin during the development of the engineered heterojunctions discussed here. The authors have profitted from frequent candid discussions with Richard Miles and David Chow of Hughes Research Laboratories, Darryl Smith of Los Alamos National Laboratories and Randy Feenstra of IBM T. J. Watson Research Center. Mark Phillips, now of Intel, was a major participant in the early development of the graded injector. Bob Hauenstein of Okaholma State Unviersity and Tom Kuech of the University of Wisconsin were major educators on the fine art of GaN.

REFERENCES

[1] Yu, E. T., McCaldin, J. O., and McGill, T. C.,"Band Offsets in Semiconductor Heterojunctions", *Solid State Physics* Vol. **46** (Academic Press, 1992).

[2] Band lineups primarily from Harrison, W. A., and Tersoff, J., *J. Vac. Sci. Tech. B* **4**, 1068-1073 (1986).

[3] *The National Technology Roadmap for Semiconductors* (SIA Industry Association, San Jose, California, 1994) p. 62.

[4] Collins, D. A., Chow, D. H., Yu, E.T., Ting, D. Z.-Y., Soderstrom, J. R., Rajakarunanayake, Y., and McGill, T. C., "InAs/GaSb/AlSb: The Material System of Choice for Novel Tunneling Devices", in *Resonant Tunneling in Semiconductors* edited by L. L. Chang et al. pp. 515-528 (Plenum Press, New York, 1991.)

[5] Brown, E. R., Soderstrom, J. R., Parker, C. D., Mahoney, L. J., Molvar, K. M., and McGill, T. C., *Appl. Phys. Lett.* **58**, 2291-2293 (1991).

[6] Levy, H. R., Collins, D. A., and McGill, T. C., *Proc. 1992 IEEE Int'l Symp. on Circuits and Systems*, 2041-2044 (1992).

[7] Chow, D. H., Dunlap, H. L., Williamson, W., Enquist, S., Gilbert, B. K., Subramaniam, S., Lei, P. M., and Burnstein, G. H., submitted to *IEEE Electron Device Letters* (1995).

[8] Bonnefoi, A. R., Chow, D. H., and McGill, T. C., *Appl. Phys. Lett.* **47**, 888-891 (1985).

[9] Schulman, J. N., and McGill, T. C., *Appl. Phys. Lett.* **34**, 663-665 (1979); Smith, D. L., Schulman, J. N., and McGill, T. C., *Appl. Phys. Lett.* **43**, 180-182 (1983).

[10] Osbourn, G.C., *Semic. Sci. Tech.* **5**, S5-S11 (1990).

[11] Smith, D. L., and Mailhiot, C., *J. Appl. Phys.* **62**, 2545-2548 (1987).

[12] Miles, R. H., and Chow, D. H., "Infrared Detector Based GaInSb/InAs Superlattices" in *Long Wavelength Infrared Detectors*, edited by M. Razeghi (Gordon & Breach, New York, 1995).

[13] Miles, R. H., (private communication)

[14] Hasenberg, T. C., Chow, D. H., Kost, A. R., Miles, R. H., and West, L., *Electronics Letters* **31**, 275 (1995); Chow, D. H., Miles, R. H., Hasenberg, T. C., Kost, A. R., Zhang, Y. H., Dunlap, H. L., and West, L., submitted to *Appl. Physics Letters* (1995).

[15] Grein, C. H., Young, P. M., Ehrenreich, H., *J. Appl. Phys.* **76**, 1940-1942 (1994).

[16] Youngdale, E. R., Meyer, J. R., Hoffman, C., Bartoli, F. J., Grein, C. H., Young, P. M., Ehrenreich, H., Miles, R. H., and Chow, D. H., *Appl. Phys. Lett.* **64**, 3160-3162 (1994).

[17] Collins, D. A., Wang, M. W., Grant, R. W., and McGill, T. C., *J. Vac. Sci. Tech.* B **12**, 1125-1128 (1994); Feenstra, R. M., Collins, D. A., Ting, D. Z.-Y., Wang, M. W., and McGill, T. C., *Phys. Rev. Lett.* **72**, 2749-2752 (1994); Wang, M. W., Collins, D. A., Grant, R. W., Feenstra, R. M., McGill, T. C., to be published in *J. Vac. Sci. Tech. B* (1995).

[18] Yu, Z., Eason, D. B., Boney, C., Ren, J., Hughes, W. C., Rowland, W. H., Cook, J. W., Schetzina, J. F., Cantwell, G., et al., *J. Vac. Sci. Tech. B* **13**, 711-715 (1995).

[19] Haase, M. A., Qiu, J., Depuydt, J. M., Cheng, H., *Appl. Phys. Lett.* **59**, 1272-1274 (1991).

[20] Park, R. M., Troffer, M. B., Rouleau, C. M., Depuydt, J. M., Haase, M. A., *Appl. Phys. Lett.* **57**, 2127-2129 (1990); Ohkawa, K., Karasawa, T., Mitsuyu, T., *J. Cryst. Gr.* **111**,

797-801 (1991)

[21] Phillips, M. C., Wang, M. W., Swenberg, J. F., McCaldin, J. O., and McGill, T. C., *Appl. Phy. Lett.* **61**, 1962-1964 (1992).

[22] Nakamura, S., Mukai, T., Senoh, M., *Jpn. J. Appl. Phys.-2* **30,** L1998-L2001 (1991).

[23] Wang, M.W., McCaldin, J.O., Swenberg, J.F., Hauenstein, R.J., and McGill, T.C., *Appl. Phys. Lett.* **66**, 1974-1976 (1995).

[24] Lester, S. D., Ponce, F. A., Craford, M. G., Steigerwald, D. A., *Appl. Phys. Lett.* **66**, 1249-1251 (1995).

[25] T. F. Kuech (private communication).

[26] Dreike, P. L., Fleetwood, D. M., King, D. B., Sprauer, D. C., Zipperian, T. E., *IEEE Com. A.* **17**, 594-609 (1994).

[27] Neudeck, P. G., J. Elect. Mat. **24**, 283-288 (1995).

[28] Morkoc, H., Strite, S., Gao, G. B., Lin, M. E., Sverdlov, B., Burns, M., J. Appl. Phys. **76**, 1363-1398 (1994).

[29] Collins, D. A., Papa, G. O., and McGill, T. C., submitted to *J. Vac. Sci. Tech. B* (1994).

[30] See for example, Kuo, C. H., Anand, S., Fatollahnejad, H., Ramamurti, R., Droopad, R., Maracas, G. N., *J. Vac. Sci. Tech. B* **13**, 681-684 (1995) and the references contained therein.

[31] See for example, Wang, M.W., Collins, D.A., Grant, R.W., Feenstra, R.M., and McGill, T.C., *Appl. Phys. Lett.* **66,** 2981-2983 (1995).

[32] Feenstra, R.M., Collins, D.A., and McGill, T.C., *Superlattices and Microstructures* **15**, 215-220 (1994).

[33] See for example, Ref. 1 and the references contained therein.

[34] Wang, M.W., Collins, D.A., Grant, R.W., Feenstra, R.M., and McGill, T.C., to be published in *J. Vac. Sci. Tech. B* (1995).

[35] Feenstra, R. M., Collins, D. A., Ting, D. Z.-Y., Wang, M.W., and McGill, T. C., *Phys. Rev. Lett.* **72**, 2749-2752 (1994).

[36] Hauenstein, R.J., Collins, D.A., Cai, X.P., O'Steen, M.L., and McGill, T.C., submitted to *Appl. Phys. Lett.*, (1994).

[37] Strite, S., and Morkoc, H., *J. Vac. Sci. Tech. B* **10**, 1237-1266 (1992).

565

Measurement Needs and Critical Drivers
for Future High Speed Devices

Herb Goronkin
Motorola
2100 East Elliot Road
EL508
Tempe, Arizona 85284

J.R. Tucker
Department of Electrical and Computer Engineering
Beckman Institute
University of Illinois
Urbana, Illinois 61801

The scaling of devices to nanoscale dimensions in order to improve speed, power and complexity requires the invention and development of new and improved material and device technologies. Barriers to traditional scaling in silicon technology, primarily gate tunneling leakage current and drain-induced barrier lowering, will emerge with the 1 Gb DRAM in about the year 2000. Further scaling will incur performance degradation as critical dimensions are relaxed in order to reduce leakage currents. Compound semiconductor HFETs will suffer similar scaling penalties. Thin film SOI may extend silicon scaling by two or three generations. However, by the year 2000, a new technology must be in development to continue the performance trend. If the scaling/complexity trend continues, the new technology will be based on quantum devices having small numbers of charge carriers. These structures will pose new problems for design, architecture, fabrication, characterization and testing.

INTRODUCTION

The complexity of integrated circuits has increased by miniaturization according to scaling rules that preserve chip speed and power dissipation. It is anticipated that by the year 2000, critical dimensions of 0.1-0.2 μm, as required by the 1Gb DRAM, will be needed. At that time we will face a critical situation in which the normal scaling of conventional MOSFETs will no longer provide improved performance unless the operating temperature is significantly reduced.

The trend in gate length reduction has followed an exponential rule since before 1980 and continues through 1994. If we continue to follow this rule, feature size will shrink to about 100 nm in the year 2003, as shown in Figure 1. Gate oxide thickness scales with the gate length as shown in the same figure. For example, when the gate length is reduced to 150 nm the corresponding oxide thickness scales to about 3 nm. Such scaling maintains roughly constant power dissipation and chip speed over successive generations. The number of electrons in a 1 Gb DRAM transistor scales to aproximately 100 and to between 10-15 in a 16 Gb transistor. Subsequent generations of transistors that follow the same scaling trend will have current-voltage characteristics that are uniquely different from conventional saturating characteristics.

When devices and interconnects in integrated circuits are shrunk in order to increase chip complexity, the capability of the scaled device to drive the scaled interconnect must remain intact. This can be expressed by the ratio of transconductance to wiring capacitance, g_m/C_w. Traditional scaling rules lead to diminished g_m values at 0.1 μm because the combination of gate tunneling current and source-drain leakage current arising from barrier lowering exceeds the tolerable leakage current levels. This situation prevails for both silicon MOSFETs and III-V HFETs. As shown in Figure 2, experimental values of transconductance approach values that are roughly independent of material as the gate length is scaled below 0.1 μm. This reflects the inability to scale vertically as well as laterally when the gate insulator bandgap is of insufficient magnitude to minimize gate tunnel current.

Post-MOSFET ULSI technology based on semiconductors will likely be one in which the basic MOSFET building blocks are replaced with quantum-based devices of equivalent scaled size. There are currently two general classes of approach under investigation. One involves devices that perform complex functions normally requiring several or many

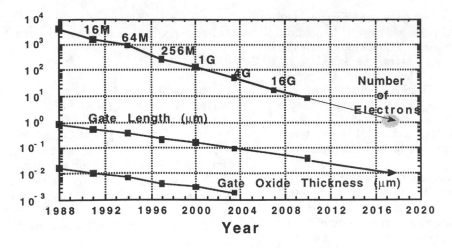

Figure 1. Scaling trends for gate length, oxide thickness and number of electrons for MOSFETs.

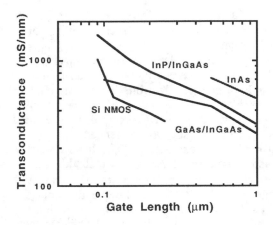

Figure 2. Transconductance versus gate length.

conventional transistors. This approach utilizes as its mode of operation one of the same physical phenomena that hinders the operation of scaled MOSFETs, namely tunneling. This new class of transistors has highly nonlinear I-V characteristics which can be utilized to create complex circuit functions with a small number of devices. The second approach is based on devices having a very small number of electrons. In this case, stepwise changes in current with voltage arise from the transport of single electrons.

In this paper we discuss some of the unique problems associated with the materials and fabrication process of quantum functional devices that consist of resonant tunnel diodes and transistors. Such device structures have not been sucessfully fabricated in silicon due to the present sparcity of materials that form suitable heterointerfaces with silicon and also have good lattice matching. Finally, we briefly examine some of the technology associated with single electron structures.

QUANTUM FUNCTIONAL DEVICES

In Quantum Functional Devices (QFD) that utilize tunneling through quantum wells to produce negative differential conductivity, two necessary conditions for tunneling must be met. The initial and final states must have the same electronic energy and the same transverse crystal momentum. These conditions are readily met in III-V *intraband* tunneling structures and in *interband* structures (Figure 3) in which the conduction band minimum and valence band maximum are located at the same point in E-k space.

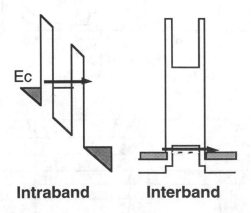

Figure 3. Intraband tunneling in the conduction band and Interband tunneling between conduction bands of InAs through the valence band of GaSb.

The resonant tunnel transistor (RTT) has I-V characteristics with regions of negative differential resistance (NDR). In an ideal situation, the RTT will occupy the same area as a conventional transistor, and have the same operating speed and power dissipation. Thus, if the critical path of the functional block in which the RTT is used has n transistors, the speed-power savings can be as high as n^2. The principle of increased functionality rests on the utilization of NDR which in a single resonant tunnel diode (RTD), for example, can support two stable voltage states as illustrated in Figure 4 which shows the I-V characteristics of a resonant interband tunnel diode fabricated in the InAs/GaSb/AlSb material system.

In order to be incorporated into transistors for future ULSICs, tunnel structures must have two important properties. They must have sufficient peak current to drive subsequent devices, and low valley current to provide low power dissipation. For example, the off-state current must be comparable to that of MOSFETs used in DRAMs. In present DRAMs the allowable leakage current density is approaching 2 pA/μm^2 while in existing RTDs the valley current density is greater than 100 nA/μm^2, irrespective of the material system.

Figure 4. I-V data of an interband tunnel diode and resistive load line showing two stable states.

The As-Sb-based tunnel structures are grown by MBE. A typical structure is shown in Figure 5. Before growth the wafer was heated to 610 °C for 15 minutes to desorb the oxide. Then 0.2 μm of GaAs was deposited at T_{sub} = 580 °C to smooth out the crystal surface. Next, a 0.5 μm superlattice buffer region was grown (T_{sub} at 530 °C) to accommodate the large lattice mismatch ($\approx$ 7%) between the semi-insulating GaAs substrate and the active region of the structure. The buffer region consists of a GaSb/AlSb superlattice followed by 0.4 μm of AlGaSb. Then T_{sub} was reduced to $\approx$ 500°C for the growth of the remaining layers. These consisted of 500 n m of heavily doped (n = 2 x 10^{18}/cm^3) and 50 nm of lightly doped InAs (n = 5 x 10^{17}/cm^3), the double barrier layer sequence (10 nm of InAs, 2.5 nm of AlSb, 6.5 nm of GaSb, 2.5 nm of AlSb, and 10 nm of InAs), 50 nm of InAs (n = 5 x 10^{17}/cm^3), and finally a cap layer of 250 nm of InAs (n = 2 x 10^{18}/cm^3).

The growth rates of GaSb, AlSb, and InAs were 1.0, 1.0 and 0.8 μm/h, respectively as determined from reflection high-energy electron diffraction (RHEED) intensity oscillation measurements. The GaSb and AlSb layers were grown using V/III flux ratios that resulted in an Sb-stabilized surface, as evidenced by a (1 x 3) RHEED reconstruction pattern. The InAs layers were grown using a minimal As$_4$ flux since this has been reported to be the optimal growth condition.[1] Furthermore, the switching technique as reported by Tuttle et al.[2] was used for the formation of an "InSb-like" interface at each of the two AlSb/InAs heterointerfaces to improve the material quality.

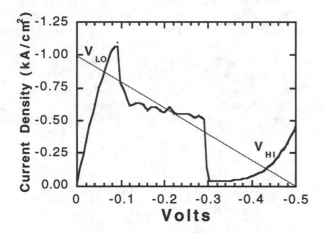

InAs	2500Å	2X10^{18}/cm^3
InAs	500Å	5X10^{17}/cm^3
InAs	100Å	nid
AlSb	25Å	nid
GaSb	65Å	nid
AlSb	25Å	nid
InAs	100Å	nid
InAs	500Å	5X10^{17}/cm^3
InAs	5000Å	2X10^{18}/cm^3

InGaAs step graded buffer

GaAs Substrate

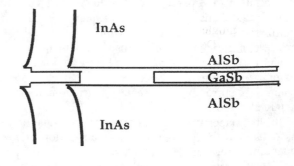

Figure 5. Layer sequence for the resonant interband tunnel diode and the corresponding band diagram.

Tunnel current is exponentially dependent on the thickness of the tunnel barrier. The barrier thickness in the above structure is 8 monolayers. A variation of ±1 monolayer can change the current by one or more orders of magnitude depending on the amount of interface over which the variation occurs.

Interface roughness alters local energy state levels according to the local quantum well width. This produces increased electron scattering and randomization of momentum compared to the incident magnitude and crystal direction. These reduce peak currents, increase valley current by phonon assisted tunneling and reduces the corresponding peak-to-valley current ratio.

Following Tuttle et al.[2], the interface between InAs and AlSb can be InSb-like, AlAs-like or an admixture of the two. The bandgap of InSb is about 0.17eV and the direct bandgap of AlAs is over 2.0eV. The small bandgap of InSb will attract electrons to accumulate at the interface while the large bandgap of AlAs will promote depletion. Accumulation is favored on the emitter side in order to maximize current density.

The following table presents an overview of the technology drivers and measurement needs for HFETs, Quantum Functional Devices and Single Electron Devices. To be candid, some of the entries under the heading, requirements are best guesses based on current technology.

Because many of the measurements needed to analyze device operation are destructive, it is not always possible to exactly correlate device performance with structure and composition. In quantum devices, this becomes especially serious because local irregularities can create large changes in electronic properties. There is a need for microscopic measurements that can be made directly on the actual device; this has always been true and remains the case for future electronic devices.

Recently, the analysis of quantum well structures has been advanced by the use of cross-sectional STM. The state-of-the-art is represented by the groups at IBM Zurich (Salemink), University of Texas at Austin (Shih), IBM Yorktown (Feenstra), and Illinois (Lyding/Tucker). The IBM Zurich group, led by H.W.M. Salemink, was the first to publish atomic resolution UHV-STM images[3] and spectroscopic measurements[4] of GaAs/AlGaAs super-lattices. More recently, they have published atomic resolution cross-sectional STM images of quantum wires fabricated by regrowth of a AlAs/GaAs superlattice on a V-grooved pre-patterned GaAs substrate.[5] In this work they also correlated the STM imaging with STM-excited luminescence (STMEL) from the quantum wires in the notches of the grooves in order to estimate the effect of surface recombination velocity uniformity. The STMEL

Technology	Technology Drivers	Measurements Needed	Requirement
Quantum Functional Devices	Epitaxy	Interface composition	1ML
		Composition	< 1 atomic %
		Band offsets	<20meV
		Layer thickness	1ML over 6"
		Carrier concentration	Non-destructive ±10%
	Etch and regrowth	Thickness	1ML
		Form factor	
	Surface stability	Surface species and concentration	$<10^{10}$ (1 ML)
	Passivation	Interface state density	$<10^{10}$
Single Electron Devices	Epitaxy	Impurity distribution	1 impurity
		Layer thickness	≈ 1ML
		Impurity species and location	<0.1nm on location
		Edge definition	0.1nm
Molecular Self Assembly		Molecule orientation	<1nm (AFM OK)
		Surface coverage	< 0.1%
		Chemisorption energy	depends on application
		Molecular structure	absolute
		Molecular energy band structure	10mV

Table 1. Technology Drivers and Measurement Needs for Future High Speed Devices

technique was developed in their laboratory. The UT Austin group has successfully imaged GaAs/AlGaAs superlattices in an attempt to determine the Al/Ga alloy distribution.[6] R.M. Feenstra at IBM Yorktown has studied the interface roughness and band offsets of InAs/InSb superlattice structures.[7] Lyding and Tucker succeeded in obtaining atomic resolution for both Al-bearing and non Al-bearing superlattices. The presence of Al greatly complicates the experiments due to rapid contamination even under UHV conditions. They recently developed a technique in which Ti is evaporated over the sample and sample holder prior to cleaving. This Ti layer effectively getters physisorbed species before they migrate to the Al-bearing layers and react. Another distinguishing feature of their cross-sectional STM program is the ability to search in two dimensions for areas of interest. The design incorporates full 2D coarse translation of the atomic resolution scan window over distances of 3 mm in any direction[8] in order to locate the interfaces as well as particular regions of interest in processed devices.

Figure 6 shows an atomic resolution STM image of a GaAs/AlGaAs superlattice grown with 100Å wells and barriers. In this occupied-states image, the As atoms on the surface are seen in fine detail. The large bright spots are individual donor impurities which happen to be located within 4 monolayers of the surface. Their patterns in these images can be analyzed to determine the depth of each. A few point defects are seen in the AlGaAs barriers due to As atoms which are missing as a result of the cleave. The slightly mottled appearance of the AlGaAs results from fluctuations in the local concentration ratio of Al to Ga, producing localized modulations in the apparent energy gap.[9]

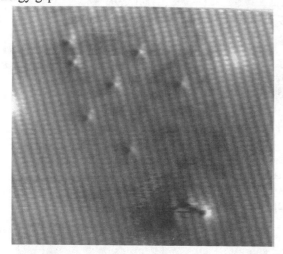

Figure 6. STM atomic resolution cross-section image of AlGaAs barrier layer and GaAs surrounding regions. GaAs monolayer = 2.8 Å

Figure 7 shows an atomic resolution image of an InGaAs/InP resonant tunneling diode.[10] Several properties which impact its electronic performance are immediately apparent. First, the inverted interfaces where the binary InP barriers were grown on top of the ternary InGaAs appear much rougher and less well defined than the normal interfaces, in which the binary InP provides a more uniform substrate for over-growth by the ternary. Electrons which encounter the rough barrier first as they tunnel from right to left in this image have many higher Fourier components at large transverse momenta when they attempt to enter the resonant well state, and this results in a higher valley current and lower peak-to-valley ratio for transport in this direction. In Fig. 7, one can also see a slightly mottled appearance within the ternary InGaAs regions, again indicating local changes in band gap due to alloy fluctuations. Detailed characterizations of this effect can be correlated with growth conditions in order to improve material quality. In some images, the tunnel barriers also appear to be "bright" at one edge and "dark" at the opposite interface. These features reflect the conditions of MBE growth terminations employed at the various interfaces, in which GaP-like and InAs-like monolayers were formed with higher and lower apparent "band gaps", respectively. Since STM injects or pulls electrons from individual atoms and bonds, electronic signatures of interface formation are to be expected; and these features can also be used to characterize and to potentially improve the growth conditions.

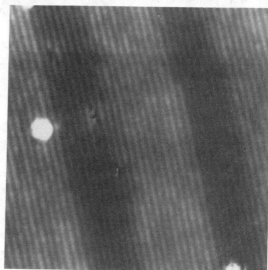

Figure 7. STM cross-section image of InGaAs/InP . Growth from right to left. Dark regions are InP. InP monolayer = 2.93 Å

The band offsets are determined by the composition of the materials on either side of the heterointerface. Global and microscopic variations in composition of the barrier and well alter the energy of quantum well states and this changes the tunneling transmission probability. The offsets can be measured directly by STM using local I-V spectroscopy which shows the band gaps of the two materials in their relative position. It is possible to follow the penetration of valence band electrons into the tunnel barriers from the surrounding regions on either side of the barrier as a function of atomic-scale position. Detailed spectroscopic data of this type will be of value in the design and characterization of future heterolayer devices.

SINGLE ELECTRON STRUCTURES

Device structures having small numbers of electrons have received significant attention in recent years.[11] In a sufficiently small capacitor, the transfer of a single electron changes the voltage across the barrier by an amount q/C where C is the capacitance. Room temperature operation requires the energy, $q^2/2C$ to be much larger than kT. This puts an upper limit of about 10^{-18} F on the capacitance. For a capacitor with dielectric constant of 12 and a thickness of 3 nm, the lateral dimension of the capacitor must be less than 10 nm. Figure 8, based on a paper by Yano et al.[12], illustrates the principle of coulomb blockade. The transfer of a single electron into the quantum well causes the energy level of the well to move upwards by an amount q/2C. In order to move another electron into the well, an additional voltage of q/2C must be applied to move the energy level downwards into alignment with the conduction band of the emitter.

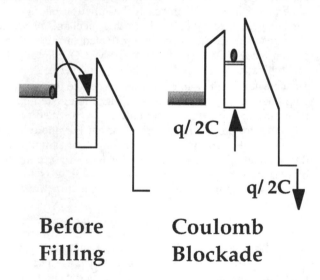

Before Filling **Coulomb Blockade**

Figure 8. Charging of quantum well by transfer of electron.

Aside from the immense difficulty in fabricating single electron devices, the characterization of structures which are on the order of 10 nm and which contain very few electrons requires precision on the order of atomic scale. For example, a 5 nm silicon quantum well which is 5 nm square contains about 1000 unit cells. The quantum well barriers are about 5-10 monomolecular layers thick. As in the case of much larger resonant

tunnel structures, small fluctuations in the composition and thickness of the well and barrier layers will cause large changes in the barrier transmission and quantum well energies.

CONCLUSION

As described in the introduction, the scaling of transistors according to historical trends leads to structures that contain a small number of electrons. In less than 10 years, the number of electrons scales to less than 20 per DRAM transistor. Long before the development of the single electron transistor, the semiconductor industry will require atom-scale characterization. The fundamental parameters that are needed to describe the physics of device structures: composition, dimension, defect location and identity, lattice parameter, interface quality will continue to be developed and refined as we approach nanoscale technology.

REFERENCES

1. S. Tehrani, J. Shen, H. Goronkin, G. Kramer, M. Hoogstra, and X. T. Zhu, *Inst. Phys. Conf.* Ser. No. 136, pp. 209-214 (1993).
2. G. Tuttle, H. Kroemer and J. H. English, *J. Appl. Phys.* **67**, 3032, (1990)
3. O. Albrektsen, D. J. Arent, H. P. Meier and H. W. M. Salemink, *Appl. Phys. Lett.* **57**, 31 (1990).
4. H. W. M. Salemink, O. Albrektsen and P. Loenraad, *Phys. Rev.* B **45**, 6946 (1992).
5. M. Phister, M. B. Hohnson, S. F. Alvarado, H. W. M. Salemink, U. Marti, D. Martin, F. Morier-Genoud, and F. K. Reinhart, *Appl. Phys. Lett.* **65**, 1168 (1994).
6. S. Gwo, K.-J. Chao and C. K. Shih, *Appl. Phys. Lett.* **64**, 493 (1994).
7. R. M. Feenstra, D. A. Collins, Z.-Y. Ting, M. W. Wang and T. C. McGill, *Phys. Rev. Lett.* **72**, 2749 (1994).
8. R. T. Brockenbrough and J. W. Lyding, *Rev. Sci. Instrum.* **64**, 2225 (1993).
9. This superlattice was grown by Professor K.-Y. Cheng.
10. RTD grown at Texas Instruments and supplied to J. Tucker by Alan Seabaugh.
11. For example, see D. V. Averin and K. K. Likharev, "Coulomb Blockade of the Single Electron Tunneling and Coherent Oscillations in Small Tunnel Junctions," *J. Low Temp. Phys.* **62**, 345 (1990).
12. K. Yano, T. Ishii, T. Hashimoto, T. Kobayashi, F. Murai, K. Seki, *IEEE Transactions on Electron Devices,* **41**, 1628 (1993).

Role of Diagnostics, Characterization, and Modeling In IR Detector Technology Development: HgCdTe–A Paradigm

W. E. Tennant

Rockwell Science Center
1049 Camino Dos Rios
Thousand Oaks, CA 91360

HgCdTe photodiodes are simple devices in a complex materials system. Many characterization methods have contributed to their development and production. Most laboratories use a relatively small subset of techniques, which combine interpretability, completeness, and cost efficiency. Modeling makes these diagnostic results more useful. For temperature-cut-off-wavelength products above 500 – 600 μm-K, HgCdTe diode models can give reasonable quantitative predictions of measured characteristics. At lower temperature wavelength products, extrinsic defects dominate, and behavior becomes much less uniform and predictable. Here diagnostics must reveal spatially dependent information about the defects. The Focal Plane Array (FPA) provides a useful tool to map defects spatially. The main objective of diagnostics in defect dominated regimes is to provide understanding to reduce defects and thus improve predictability and yield.

INTRODUCTION

For over twenty years HgCdTe has been the principal detector material for advanced IR systems.[1]

Infrared systems increasingly demand four characteristics of the technology:

> Background Limited Performance (BLIP) operation
> High operability (>98%)
> Low cost (implying high end-to-end yield)
> Custom designs on demand

HgCdTe has three features which make it highly desirable for detecting IR radiation – its tunable direct bandgap offers high operating temperature for a given level of sensitivity, its passivation is well suited to excellent photodiodes, and its thermal expansion is sufficiently close to that of silicon to make large hybrid focal plane arrays using silicon as an intimately coupled readout device. Two recent examples which illustrate the current limits of this technology have been given by Kozlowski and his co-workers.[2,3]

The photodiode on which the IR focal planes are based is a simple device. Its detection process is illustrated in Figure 1. Light incident through the CdZnTe substrate is absorbed by the n-type active layer, creating electron-hole pairs. The holes diffuse (and may drift, depending on whether there is a significant compositional gradient) until they reach the edge of the depletion region of the junction and are collected or until they recombine in the bulk or at an interface of the absorbing layer. Those optically-generated carriers which reach the junction, along with carriers generated by thermal or other mechanisms which have reached the junction, are collected in the external silicon readout chip (multiplexer or MUX). Photodiodes are typically fabricated into large arrays (as great as 10^6 detectors) and coupled device-by-device through flip-chip bonding (hybridization) to CMOS integrated circuit multiplexers comparable in circuit density to state-of-the-art computer chips. In this array format the photodiodes (located well within a diffusion length of neighboring diodes) can affect each other like parts of a complex bipolar transistor, and part of the challenge is to find operational modes in which the device interaction is beneficial.

The currents which constitute this carrier flow contain white and potentially also 1/f noise which can obscure the charges generated by the optical signal. The ratio of the signal current to the noise currents is the fundamental performance figure of merit of the photodiode.

Ideally the diode operation is described by the diode equation including a term for the photocurrent:

$$I = I_0(\exp(qV/nkT)-1)-I_{ph}$$

where $I_0=qA(n_i^2/N_D)d/\tau$ in the large diode, thin base limit. Here q is the electron charge, V is the applied bias, k is Boltzmann's constant, T is the absolute temperature, I_{ph} is the photocurrent ($=qA\eta\varphi$ where η is the quantum efficiency, also a function of several materials parameters, and φ is the incident photon flux), A is the diode area, n_i is the intrinsic carrier concentration, N_D is the donor density (a one-sided p^+/n junction is assumed with negligible contribution from the negligibly thin p-region), d is the active layer thickness (assuming no interface recombination and that the thickness is much less than a diffusion length), n is the ideality factor (= 1 for a diffusion-limited "ideal" diode), and τ is the hole lifetime in the n-type material. Real HgCdTe diodes approximate this ideal diode to a greater or lesser degree depending on several materials parameters. The most important are indicated in Figure 2.

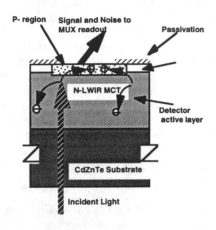

FIGURE 1. Schematic of the HgCdTe heterostructure photodiode detection process.

It is obvious from the above that, while the device structure is simple, the compound semiconductor itself is complex. The goal of technology is to push HgCdTe array performance to the fundamental physical limits and to extract this performance uniformly, predictably, and reliably. The fundamental physical limits represent the upper bound of performance; these bounds can be mitigated, depending on application, by adjusting materials properties or geometries to suit the need. More importantly, HgCdTe must overcome its nonfundamental limits associated with imperfections of material and process (such as dislocations, inclusions, surface recombination, variations in composition and doping). Characterization is key to identifying, controlling, and removing these imperfections.

HgCdTe history with respect to characterization has been varied. Inspection of the proceedings of the U.S. Workshop on the Physics and Chemistry of Mercury Cadmium Telluride reveals over 50 techniques, not to mention different experimental approaches. Seiler[4] has conducted a survey of 72 characterization methods, most of which had been applied by some workers to HgCdTe. This paper will address the significance of some of these characterization methods, the significance of device modeling where applicable, and the impact of characterization and modeling on HgCdTe photodiode technology as it matures.

CHARACTERIZATION METHODS

Table 1 shows that the most used and most popular HgCdTe characterization methods reveal information about all the key device parameters in Figure 2. The first two columns rank the importance of use to, and the frequency of use among, the respondents.

Inspecting the table shows that these most useful techniques tend to share these common characteristics: they are simple to use; give information on many parameters; can be interpreted through models, and corroborated by other techniques. Figure 3 shows the transmission curve of a LWIR HgCdTe epitaxial layer fit to a model[5] which predicted correctly the spectral response of a diode made in the layer.

For most laboratories, the ideal characterization set for HgCdTe is a combination of techniques involving both materials and test device measurements. The following is a typical set:

Materials Characterization Methods:

Optical Microscopy – visual defects (Macrodefect Density or MDD), growth characteristics
IR Transmission – composition, layer thickness
Cleaved Cross section – layer thickness, growth characteristics
Hall Effect – carrier concentration
Etch Pit Density (EPD) – dislocations, growth characteristics
IR Microscopy – substrate precipitates
Photoconductive lifetime
Profiling ellipsometry – Heterostructure compositional profile
Ellipsometry – surface prepassivation condition, passivation quality
Secondary Ion Mass Spectroscopy (SIMS) – compositional, doping, and impurity profiles and images
Double Crystal S-Ray Rocking Curve – substrate/film crystal quality
Scanning Electron Microscopy and/or Scanning Laser Microscopy–physical/electrical defect observation/ determination

Diagnostic Devices:

Diode I-V Curve (Including Temperature Dependence) – Forward bias IV gives ideality (indicating diffusion, G-R, or tunneling limit)

Statistical Sample I-V Properties vs Area (e.g. R_0A or Zero-Bias Resistance-Area Product) – should be about constant for ideal diodes; surface limited device R_0A increases with area, widely dispersed defects cause decreasing operability with increased area

Spectral Quantum Efficiency–Reveals short diode diffusion length and/or substrate interface recombination, and/or insufficient optical depth

Optical Spot Scan–shows diffusion length, anomalies in passivation or interface

Electron-Beam Induced Current (EBIC) – on cleaved cross section shows junction depth, recombining interfaces; over area gives similar information to optical spot scan

MIS Capacitors–show dielectric properties of passivation, interface state density, semiconductor carrier type near surface.

Other Devices (e.g.,Contact Chains, Gated Diodes, FET's) – a variety of occasional uses

Figure 4 shows the Current-Voltage, Differential Resistance-Voltage, and Temperature Dependence of R_0A for LWIR HgCdTe diodes. The slightly higher than ideal "ideality factor" of 1.38 (vs 1 for ideal) taken from the voltage dependence of the log of the resistance and the slightly higher than band-gap temperature dependence together suggest that the junction location on these devices is slightly into the heterojunction, creating a slight barrier which is overcome in reverse bias.

To get a complete picture of a HgCdTe processed wafer requires both materials and device analysis used together with appropriate physical models. The degree to which the characterization must be done depends on the maturity of the process. For relatively immature processes, characterization is often used to determine whether to proceed, rework, or abort a part. Mature processes may survive with minimal characterization of wafers (e.g.,. Process Evaluation Chips or PECs may be sufficient) sufficient only to monitor whether things are done as required. Diagnostics that used to be needed for screening become used for occasional troubleshooting, if at all.

THE MODELABILITY FRONTIER

At higher temperatures photodiodes tend to behave more like the ideal diode. This is simply because the band gap is typically the highest energy in the system and the densities of states associated with band-to-band transitions are by far the highest. Thus at some temperature, excitations across the band gap will dominate lower energy, lower density defect-related behavior. Technology maturity determines at how low a temperature or how high a wavelength this ideal behavior appears. For HgCdTe at its present state of maturity, the frontier appears to be roughly about 500-600 μm-K. At higher temperature-wavelength products, I-V characteristics tend to be dominated at low-moderate biases by radiative or Auger bulk materials limits. Below this temperature, defects tend to dominate. Figure 5 shows the median R_0A values of many test samples of photodiodes in MBE-grown MWIR-LWIR HgCdTe epitaxial layers. The theoretical line used for comparison is an Auger model assuming a reasonable 3×10^{13} cm^{-3} donor concentration. The shorter wavelength diodes are beginning to deviate from the theory.

The highly nonideal behavior characteristic of LWIR diodes at lower temperatures appears in Figure 6. Here the differential resistance peaks at zero bias – characteristic of tunneling behavior. The R_0A of these devices typically shows a gradual increase as temperature is decreased, indicating trap-assisted tunneling rather than band-to-band (which would produce the opposite temperature dependence due to the narrowing of the bandgap with decreasing temperature). Note—trap assisted tunneling may also cause reverse bias leakage currents to increase with decreasing temperature, depending on the nature of the trap distribution. Forward bias excess currents, on the other hand, provide unambiguous indicators of trap tunneling. In wider band-gap materials the same traps would typically produce G-R currents rather than tunneling.

Once diodes are dominated by defects, especially tunneling, their behavior becomes less uniform. Figure 7 contrasts the same set of devices measured on both sides of the modelability frontier. At 80 K the devices are exceptionally uniform while at 40 K, although much higher in performance, they vary by several orders of magnitude.

Because devices are small, they provide inherently a measure of the local materials properties which are difficult to assess by many materials techniques (like Hall effect measurements) with relatively poor spatial resolution. The importance of spatially resolved measurements is great when trying to understand device behavior in materials with significant variation. The ultimate in statistical and mapping characterization comes from characterizing the final product – the focal plane array.

Figure 8 shows the relatively high uniformity of a 128x128 element LWIR FPA operated at 80 K (in the modelable region above the threshold). The figure of merit plotted, NET or NETD (for noise equivalent temperature difference) is the temperature difference of a scene required to produce a signal equal to the measured noise in the device. Most of the NETD histograms show devices which are excellent and uniform near the background limit (BLIP). The small tail of outliers indicates the degree of imperfection due to the materials and processes which produced the device.

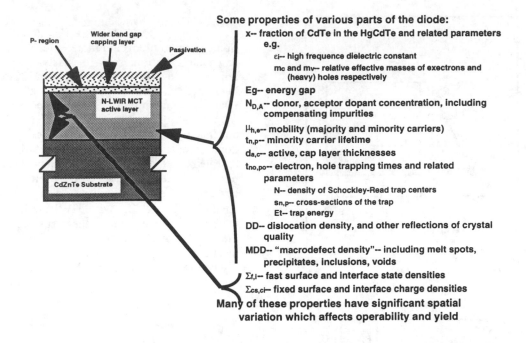

FIGURE 2. Some of the key parameters determining HgCdTe photodiode performance.

TABLE 1. Characterization methods and the parameters on which they provide information.

Meas Technique	Imp.	Freq.	x	Eg	N	μ	τ	d	τo	DD	MDD	Σ
FTIR	1	2	x	x				x				
Hall	2	3			x	x						
Current-Voltage	3	5		x			x	x				x
Capacitance Voltage	4	8		x		x						x
Defect Etch	5	1								x		
Optical Microscopy	6	4					x				x	
SIMS	7	19	x	x	x			x			x	x
DC X-ray RC	8	9								x	x	
Photoconductivity	9	12			x		x		x	x		x
Resistivity	10	6			x	x						
SEM	11	14	x	x				x			x	x
Ellipsometry	12	7	x	x				x				x
Breakdown V	13	10	x	x	x					x	x	x
MOS Capacitance	14	11	x	x		x		x	x			x
Surface Topography	15	13									x	
AES	16	23	x					x		x	x	x
DC X-ray Top.	17	27								x	x	
EBIC	18	31		x	x	x	x	x			x	x
XPS	19	22										x
Photoluminescence	20	35	x	x			x		x	x	x	x
Laser Ind. Current	24	15	x	x	x	x	x			x	x	x
μ-wave impedance	48	16			x	x						
Energy Disp. X-Ray	23	17	x	x							x	x
Opt. Mod. Absorp.	27	18					x					
Electron Diffraction	34	20								x	x	
Spectral QE*			x	x	x	x	x	x		x		x
Noise*			x	x	x		x		x	x	x	x

*Not mentioned in study

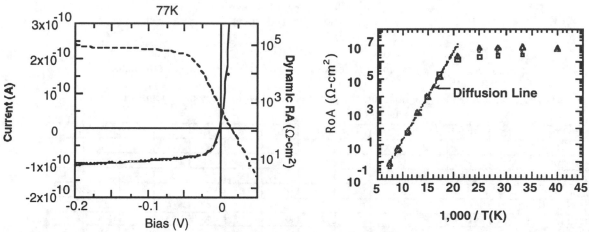

FIGURE 3. Measured and modeled room temperature transmission of a HgCdTe epitaxial layer.

TWO OBSERVATIONS

Sometimes problems are more amenable to solution than to characterization; surface passivation may be an example. For nearly fifteen years the HgCdTe surface has been studied intensely by many methods (XPS, angle resolved photoemission, Auger Electron Spectroscopy, SIMS, electrolyte electroreflectance, cyclic voltammetry, and many other techniques). Device analysis has been done extensively as well. Most of the measurements revealed a delicate, unstable surface stoichiometry and a native or natural oxide with "pretty good" electrical characteristics and stability. ZnS, SiO_2, anodic oxide, anodic sulfide, CdS, and CdTe have been deposited by various techniques to coat these surface oxides. Most of the coatings worked reasonably well up to a point, but the most exacting applications proved to be too severe. These "coated oxides" and their relatives often fell down with regard to extremes of heat, moisture, or ionizing radiation.

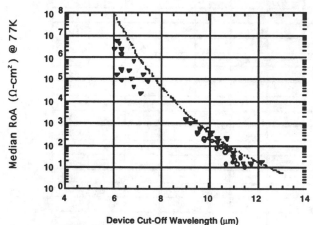

FIGURE 4. Current - and Resistance -Voltage, and Temperature Dependence of R_oA for LWIR HgCdTe Diode. R_oA value is 860 Ω-cm^2 for a cut-off wavelength of 10 µm.

FIGURE 5. R_oA vs Wavelength for MWIR-LWIR Photodiodes Made in MBE-Grown HgCdTe on CdZnTe. Dashed line gives theoretical value of R_oA for Auger-limited diodes with 10 µm layer thickness and 3×10^{15}.

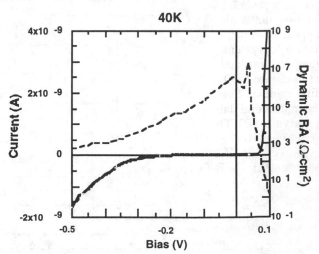

FIGURE 6. I-V Characteristic of LWIR HgCdTe at Low Temperatures Showing Trap-tunneling Dominated Behavior. R_oA for this device was 2.07×10^6 Ω-cm^2 for a cut-off wavelength of 11 µm.

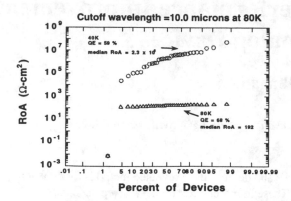

FIGURE 7. Comparison of the R_0A Values of LWIR HgCdTe diodes measured at 80 K and 40 K.

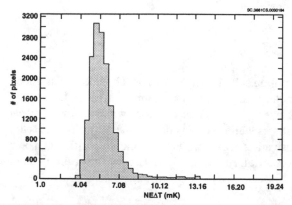

FIGURE 8. Histogram of the Noise Equivalent Temperature Difference of a 128x128 HgCdTe Focal Plane Array measured at 80 K. $\lambda T = 800$ µm-K for this device which is well-behaved, with predictable performance and few anomalous defective detectors.

Recently several laboratories have begun to take advantage of two materials characteristics which, in retrospect, appear obvious; namely, 1) that CdTe, a native to the HgCdTe system, is an insulator at typical operating temperatures, and 2) that CdTe and HgCdTe interdiffuse at moderate (>300°C) temperatures. By coating surfaces, first made as clean as possible, with CdTe and annealing them, several laboratories have established improved passivations. The hypothesis advanced for these improvements is that the CdTe interdiffuses with the HgCdTe and raises the band gap at the surface, thus rendering less significant the traps formed by the partially oxidized, poor-stoichiometry surface. Thus, the extensive characterization of the last several years, while useful to understanding the seriousness of the HgCdTe surface issue, in the end has analyzed a surface which will likely not be used much in future devices.

A second observation is that diagnostics applied to solving yield issues bring more benefit than diagnostics applied to screening materials. A low yield process will be prohibitively expensive if the entire process sequence is gone through with all parts and only in the final test are parts rejected. Screening parts along the process chain so that the part is rejected as soon as it receives a failing process step, can greatly lower cost without affecting end-to-end yield. However, diagnostics which help to raise yield to a high level will lower cost even further, since there will be few rejections and the need for screening will be minimal. Thus, when starting to develop a new process, one should concentrate on developing the understanding and control to improve yield. This will lower the cost most effectively.

SUMMARY

In conclusion, HgCdTe photodiodes provide an example of a simple structure in a difficult materials system. The definitive characterization methods have typically been simple, versatile, interpretable, and correlatable. A satisfactory characterization set combines materials and device measurements to supply a model of the device behavior where possible. Technology aims to push the modelable region downward to lower temperature-wavelength products from its current level of about 500-600 µm-K, below which defects tend to dominate and performance varies more. As this trend continues, we will be able to build what we design with higher yields, larger sizes, and higher operabilities even at higher performance levels.

REFERENCES

[1] The history of HgCdTe materials and characterization has been best documented in the proceedings of the U. S. Workshop on the Physics and Chemistry of Mercury Cadmium Telluride, which typically have been published since 1982 in the Journal of Vacuum Science and Technology, and occasionally in AIP conference proceedings.

[2] L.J. Kozlowski, J.M. Arias, G.M. Williams, K. Vural, D.E. Cooper and S.C. Cabelli, "Recent Advances in Staring Hybrid Focal Plane Arrays: Comparison of HgCdTe, InGaAs, and GaAs/AlGaAs Detector Technologies," SPIE 2274, 93 (July 1994).

[3] L.J. Kozlowski, K. Vural, S.C. Cabelli, A. Chen, D.E. Cooper, D.M. Stephenson and W.E. Kleinhans, "2.5 µm PACE-I HgCdTe 1024x1024 FPA for Infrared Astronomy," *SPIE* 2268, 353 (July 1994).

[4] D. G. Seiler, S. Mayo, and J. R. Lowney, "$Hg_{1-x}Cd_xTe$ characterization measurements: current practice and future needs," Semicond. Sci. Technol. 8, 753 (1993).

[5] G. M. Williams and R. E. DeWames, "Numerical Simulation of HgCdTe Detector Characteristics," Proceedings of the 1994 Workshop on the Physics and Chemistry of HgCdTe, to be published in June - *J. Elec. Materials*, June 1995.

Critical Materials Issues for Performance Improvement of GaAs-Based Analog Devices

James M. Ballingall

Martin Marietta Laboratories • Syracuse, Electronics Park, Building 3, Syracuse, NY 13221

Recent investigations in four areas focused on GaAs-based materials and devices have revealed methods for enhancing the performance of GaAs microwave and millimeter-wave transistors. These four areas are: InGaP as a replacement for AlGaAs in HEMT and HBT devices; carbon doping of GaAs for high hole concentrations; low-temperature-grown GaAs for insulating layers; and source materials for gas source molecular beam epitaxy (GSMBE) and chemical beam epitaxy (CBE) that improve purity levels and reduce hazards. Some observations and conclusions regarding these four areas are generally applicable to other materials systems and devices as well, such as optoelectronics.

INTRODUCTION

Our recent materials research and development in four specific areas has provided new insight for enhancing the performance of GaAs-based microwave and millimeter-wave transistors. These four areas are:

- InGaP as a replacement for AlGaAs in HEMT and HBT devices.
- Carbon doping of GaAs for high hole concentrations.
- Low temperature grown GaAs for insulating layers.
- Source materials for gas source molecular beam epitaxy (GSMBE) and chemical beam epitaxy (CBE) which improve purity levels and reduce hazards.

Each of these areas has multiple issues which need to be addressed, as briefly summarized below.

Mixed anion heterojunctions formed with InGaP and the arsenides common to HEMTs and HBTs are apparently controlled electrically by the detailed bonding arrangement at the heterointerface and the degree of alloy ordering of the InGaP. Growth technology strongly influences both of these factors. For a given epitaxial reactor, the recipe for the interfacial transition can alter the interfacial composition, but control is often difficult since chemical exchange reactions of P and As surface adatoms can dominate. High resolution x-ray diffraction (HRXRD) data coupled with a variety of other characterization techniques indicates that "parasitic" monolayers of material can be unintentionally grown into a device structure at interfaces or growth interruptions.

Carbon is an appealing p-type dopant for GaAs because of its low diffusivity and high solubility, i.e. lack of self-compensation at high concentrations. The selection is underway for the most appropriate source material for carbon, to achieve high hole concentrations with long minority carrier lifetimes while avoiding dopant memory effects. Hydrogen passivation of carbon, as with other impurities, can be quite substantial, reducing the hole concentration by as much as 100%.

Low temperature grown GaAs (LTG) is attractive as a buffer layer and for other insulating layers for GaAs-based electronic devices because of the high resistivity the material exhibits. Defects responsible for this behavior are not completely characterized or understood. An important issue is the confinement of these defects to the LTG layers, preventing their diffusion into adjacent layers. Also desired is the capability to grow high quality epilayers on top of LTG layers without an intermediary buffer on top of the LTG layer.

Source materials for GSMBE and CBE in many cases continue to limit the range of application of these growth techniques. While there are some examples reported of ultrahigh purity layers, generally these techniques produce materials with higher impurity concentrations than normally found in MBE and metal-organic chemical vapor deposition (MOCVD). Organometallic sources and low pressure hydride packages are steadily improving layer purity beyond their previous levels while substantially reducing the hazards of conventional high pressure packages of arsine and phosphine.

MIXED ANION INTERFACES

InGaP has been studied as a replacement for AlGaAs in high electron mobility transistors (HEMTs) [1-2] and heterojunction bipolar transistors (HBTs) [3-5], and appears to offer some advantages in both performance and fabrication yield. Those familiar with these results generally agree that there are a number of materials and process issues that must be addressed to fully realize these advantages. This discussion will be limited to the growth and characterization of the heterointerface. The InGaP-GaAs interface is a model system to evaluate issues associated with mixed anion heterojunctions. Figure 1 shows a schematic representation of how the transition across the heterointerface can lead to two very different bonding arrangements with a GaP-like interface or a InGaAs-like interface. As expected intuitively, both theory and experiment support significant variations in the resulting conduction and valence band discontinuities. [6-8] The discontinuities are strong determinants of device performance, and are in fact often the key feature in the designs of electronic and optoelectronic devices. For example, the AlGaAs-GaAs npn heterojunction bipolar transistor (HBT) would be a superior device if more of the band gap discontinuity spanned the valence band since this would provide a larger barrier to confine holes in the p-type GaAs base. The InGaP-GaAs system appears to possess a more advantageous band gap alignment for the HBT, though reports of the conduction band discontinuity have varied from 30 to 240 meV, apparently due to problems with reproducibility of the bulk material (alloy ordering) as well as the interface. [9-14] Growth technology strongly influences both of these factors. A significant degree of ordering is rarely observed in GSMBE material. In MOCVD InGaP, a

significant degree of ordering is typical and can be controlled with growth conditions. [15] For a given epitaxial reactor, the recipe for the interfacial transition (gas switching sequences and growth interruption times) can alter the interfacial composition and the composition of the layers near the interface. [16-20], but control is often difficult and may be complicated by chemical exchange reactions of P and As surface adatoms as well as a brief memory effect due to carryover of the Group V specie most recently switched off [19-23]. Hollinger, et. al., using photoelectron spectroscopy showed clearly that an InP surface exposed to an arsenic flux will convert the first few monolayers to InAs [23]. The converse of a phosphorus flux converting a GaAs surface to GaP was indicated by reflection high energy electron diffraction (RHEED) and x-ray diffraction data by Freundlich, et. al. [24]. Additional HRXRD and luminescence studies have supported the surface exchange phenomena, indicating that "parasitic" monolayers of material can be unintentionally grown into a device structure at interfaces or growth interruptions. [19-22]

Our own results support these findings, although we found that distinguishing between anion exchange at the interface and anion flux carryover and incorporation into the next layer can often be ambiguous and generally requires extensive characterization. Our experiments utilized GSMBE with the low pressure packages for hydrides discussed in section IV below. Substrate temperatures of 500°C and growth rates of 1 micrometer/h were used. For our first test structure, a GaAs layer was interrupted during growth every 150Å for approximately 2 seconds while the arsine flow was exchanged with the phosphine flow. After the interruption, the arsine flow was resumed, gallium shutter reopened and growth resumed. This was repeated 20 times. The modeled HRXRD scan is excellently fit by assuming that a monolayer of $GaAs_{0.425}P_{0.575}$ occurs at each of the interrupted interfaces, indicating that indeed a significant amount of P is incorporated, forming a true superlattice structure. Our second test structure is the InGaP analog of the GaAs structure, i.e., the InGaP is interrupted, and the phosphine and arsine flows are exchanged. Here again, a superlattice is formed by the growth interruptions, modeled by the inclusion of a monolayer of $In_{0.506}Ga_{0.494}As$ at each interface.

Several more GaAs multilayers were grown to investigate the presence of interfacial layers related to switching to and from arsine and phosphine during film growth. In order to determine the source of the interface layers, samples were grown with intentional delays of up to one minute built into the growth sequence. In one case, a one-minute delay was inserted between closing the gallium shutter and switching from pre-cracked arsine to pre-cracked phosphine. In the other case, the wafer was held under a pre-cracked arsine flux for one minute after pre-cracked phosphine exposure

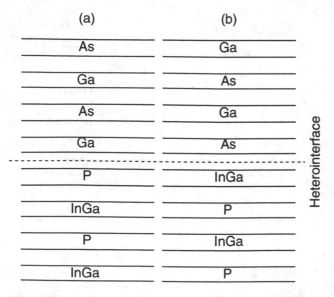

Figure 1. Depiction of hyper-abrupt heterointerfaces of GaAs-InGaP with (a) GaP-like interface and (b) InGaAs-like interface.

before opening the gallium shutter for the growth of the next layer. These two approaches yielded very different results when analyzed with HRXRD, as shown in Figure 2. The sample grown with the delay between closing the gallium and opening the phosphine was very similar to the test structures described above which were grown with no intentional delay in the growth sequence. A strong, distinct peak was apparent in the vicinity of the GaAs substrate peak. Furthermore, the +1 and -1 satellite peaks were also observed at angles corresponding to the 150Å period used for the GaAs layer growth. In contrast, the sample grown with the delay between the phosphorus exposure and the opening of the gallium source exhibited no satellite peaks and only a slight broadening on one side of the main GaAs diffraction peak. These results suggest that the interface layers observed in earlier samples may be due to a relatively high residual level of phosphorus present in the chamber after switching to arsine from phosphine. By waiting before reinitiating GaAs growth, time is provided to allow the excess phosphorus to pump from the system. If this is the case, the assumption of a distinct GaAsP interfacial layer to model the x-ray results may not be valid. The ambiguity here is that a graded GaAsP region bounded by the interface can simulate the observed diffraction pattern equally well.

In order elucidate these phenomena and to assess the impact of the growth interruptions described above on material quality, InGaP/GaAs/InGaP single quantum well structures were grown with one minute delays included on both sides of the GaAs single quantum well. Photoluminescence (PL) measurements were made and compared to assess changes in the energy, intensity, or width of the quantum well PL peak as a function of the interfacial transition recipe.

Figure 3 is a schematic diagram of the quantum well layer structure. Figure 4 is a 2K PL scan of the structure grown without growth pauses, showing excellent quality for the InGaP cladding layers and the GaAs quantum well, with state-of-the-art values for spectral linewidth (4.9 meV for InGaP and 2.4 meV for the quantum well), which to our knowledge are the best values ever achieved respectively for InGaP and InGaP/GaAs/InGaP quantum wells. [25-26] Indeed, the FWHM for these 7nm quantum wells are among the most narrow ever observed for any materials system. [27] This spectrum also shows an excellent example of monolayer splitting of the quantum well peak, a result of the exciton size being less than the average size of the monolayer ledges along the interfaces. The narrow linewidths are indicative of low defect content in the films. The InGaP

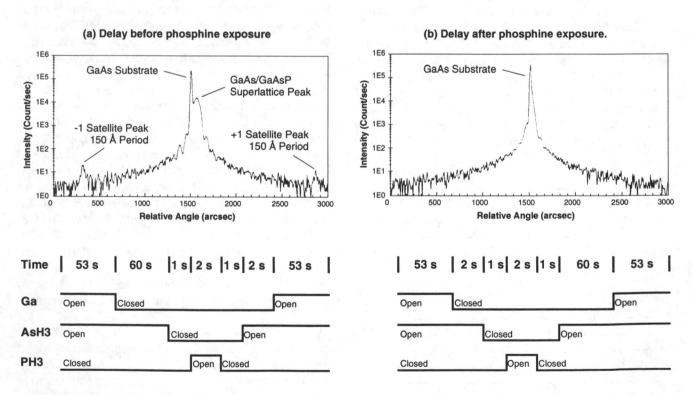

Figure 2. Effect of phosphine exposure on GaAs interrupted growth as a function of intentional delays in the gas switching sequences (a) a 3 second delay after phosphorus exposure before resuming growth provides sufficient phosphorus to create superlattice peaks in HRXRD scan. The scan can be excellently fit by assuming a discrete interfacial monolayer of GaAs$_{0.425}$P$_{0.575}$ is formed at each growth interruption, or alternatively by a graded compositional region of GaAsP bounded by each interrupted interface. (b) for a 61 second delay after phosphorus exposure before resuming growth, all traces of residual phosphorus are absent in the HRXRD scan.

GaAs:C -- 0.085 μm
InGaP -- 0.23 μm
GaAs SQW -- 70 Å
InGaP -- 0.23 μm
GaAs Buffer -- 0.12 μm
GaAs Substrate

Figure 3. Layer structure for InGaP-GaAs-InGaP single quantum well test structure

bandgap energy measured by PL indicates that minimal alloy ordering exists in the films we grew. Figure 5 compares single quantum wells grown with and without growth pauses. The sample without pauses has a similar peak position and linewidth as the sample shown in Figure 4, while the sample with 60 second pauses is shifted to a lower energy and is broadened significantly. Given that these are fairly thick (7nm) wells, the influence of the interface on the transition energy is expected to be smaller than carryover of phosphorus into the GaAs well, and thus we believe this is responsible for the difference in transition energies, while the substantial increase in broadening is most likely due to impurity uptake during the long pauses.

CARBON DOPING OF GaAs

Carbon is an important dopant for HBTs and optoelectronic devices, since it has been shown to be the most stable in terms of a low diffusion coefficient during growth and subsequent thermal treatments. [28-29] This stability is a key requirement for device reliability. Carbon filaments have been used successfully as sources of carbon for MBE-grown lasers [30], planar doped barriers diodes [31], and HBTs [32], but have shown limitations at the higher doping concentrations, presumably due to the formation of carbon-carbon pairs [30]. Trimethlgallium performs better as a dopant [33], contributing carbon from the methyl radicals (CH) available upon dissociation of the molecule. The disadvantage of this source is the growth rate dependent nature since it also contributes significant gallium at the high doping levels needed for HBTs. Carbon tetrachloride has been used with great success at the high concentrations (above $1 \times 10^{19} cm^{-3}$) in MOCVD [34,35] and the various MBE approaches [36], though the incorporation efficiency can be rather low for the MBE techniques. [37] This has led to the use of carbon tetrabromide for MBE applications. [37-39] Another factor is that while the chloride is less toxic, it is considered a greater environmental risk than the bromide, and in the USA will be legislated out of use by the year 2000. Table 1 shows growth data for chloride doping. A problem we encountered with the chloride was a large memory effect, exhibited by the data for layer Z-455, an undoped film which if grown before the use of the chloride would have yielded a low n-type residual concentration. SIMS profiling confirmed a substantial residual concentra-

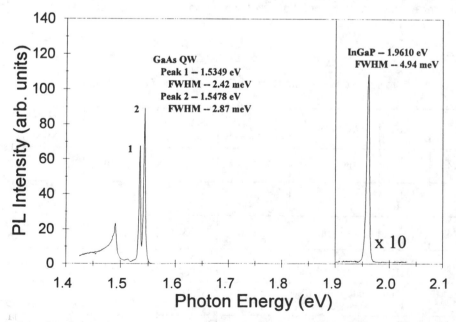

Figure 4. 2K photoluminesence of single quantum well layer depicted in Figure 3. Full width at half maxima are state-of-the-art for the quantum well and cladding layer transitions. Monolayer splitting of the quantum well transitions is clearly observed due to relative sizes of exciton wavelength and monolayer step ledges along the interfaces.

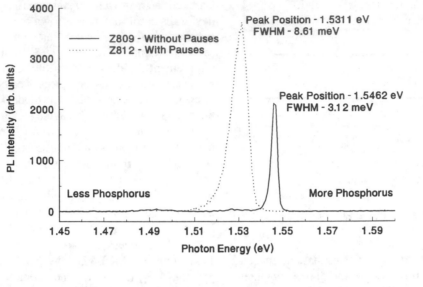

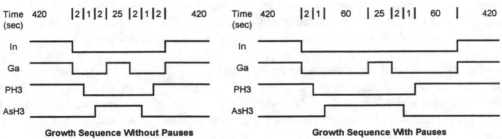

Figure 5. Effect of growth pauses on single quantum well photoluminescence transition energy and width. Layer grown with minimal pauses exhibits similar spectrum to that of Figure 4 (also grown with minimal pauses), whereas layer with long pauses is red shifted and broadened.

Table 1. Growth and Hall Effect Data for Carbon-Doped GaAs Using CCl_4.

Wafer No.	Sources	CCl_4 Flow	$|N_A-N_D|$ (300K) (cm^{-3})	μ (300K) $(cm^2/V\text{-}s)$	$|N_A-N_D|$ (77K) (cm^{-3})	μ (77K) $(cm^2/V\text{-}s)$
Z-446	Ga, AsH_3	0.5 sccm	2.6E19	88	2.6E19	120
Z-447	Ga, AsH_3	1.0 sccm	4.2E19	81	4.0E19	115
Z-442	Ga, AsH_3	3 sccm	1.1E20	55	1.0E20	86
Z-455	Ga, AsH_3	undoped	2.3E17	250	1.0E17	730
Z-458	TEGa, AsH_3	undoped, MBE furnaces cold	5.4E15	400	3.3E15	3300
Z-459	TEGa, AsH_3	undoped, MBE furnaces hot	7.9E15	340	4.5E15	3030

tion of $2 \times 10^{17} \text{cm}^{-3}$, in agreement with the Hall effect data (Figure 6). Layers Z-458 and 459 were grown to test the hypothesis that the carbon tetrachloride had doped the gallium furnace. As can be seen from the data, when the system was operated in the CBE mode, the residual doping concentration dropped considerably and was influenced by the gallium furnace temperature. Figure 6 contrasts SIMS profiles of carbon-doped GaAs layers grown with the chloride and bromide. The baseline recovers to the detection limit with the bromide, as opposed to the case with the chloride. Figures 7-8 are plots of carrier concentration vs. mobility at 300K and 77K for films doped with both compounds. There was no discernible difference between the two as far as the mobilities were concerned.

Carbon passivation by hydrogen was evaluated and found to be comparable to that observed in CBE [40], which is much less of a problem than encountered in MOCVD, where the passivation can be as large as 100% [41-42]. The passivation is due to formation of neutral dopant-hydrogen complexes, with hydrogen occupying a bond-centered position in p-type semiconductors and an anti-bonding site in n-type materials, and with carbon exhibiting the largest dissociation energy of the shallow acceptors [43]. Figure 9 shows hole concentration data for as-grown and annealed carbon doped layers using carbon tetrabromide grown by GSMBE. Layers were annealed in nitrogen at 400°C for 5 minutes. The trend is for the passivation to increase with the arsine cracker temperature employed for the growth, presumably

due to the increased atomic hydrogen content of the arsenic beam with elevated temperatures. Abernathy, et. al. have suggested an in-situ anneal to drive out hydrogen and measured a three-fold decrease in hydrogen concentration in carbon-doped HBTs after a 15 minute 600°C anneal under pre-cracked arsine [40]. Stockman, et. al. found that the bases of InGaP-GaAs HBTs grown by MOCVD exhibited minimal hydrogen passivation, while the bases of InP-InGaAs HBTs exhibited substantial passivation [41]. They also reported a strong effect on post-growth cool-down cycles. Thus, as with the mixed anion interface issues discussed above, this problem is also highly dependent on growth technology and growth recipes.

LOW TEMPERATURE GROWN GaAs

MBE GaAs is normally grown at a rate of 1 μm/h with a nominal substrate temperature of 600°C and nominal arsenic to gallium beam equivalent pressure ratios of 10-20. Under these conditions, undoped material will be conductive, typically due to shallow donor/acceptor impurity concentrations in the low $1 \times 10^{14} \text{ cm}^{-3}$ range. If on the other hand, the substrate temperature is dropped below about 450°C, then the material becomes resistive [44-46], even if heavily doped, because of deep level traps [47]. In the 1970s, Muratoni, et. al.[44], were apparently the first to ascribe some benefit to this behavior, fabricating a power FET with a resistive buffer layer that was unaffected by impurities. Later,

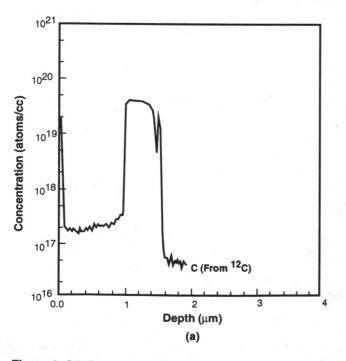

(a)

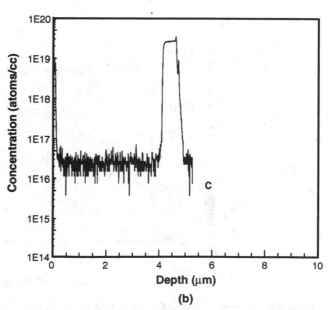

(b)

Figure 6. SIMS profiles of carbon doped pulses in GaAs using (a) CCl₄ and (b) CBr₄. A strong memory effect is observed for the chloride which correlates with Hall effect data.

583

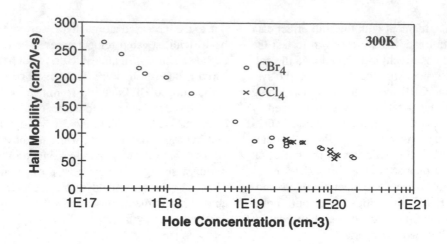

Figure 7. Hole Hall mobility vs. concentration at 300K for carbon doped GaAs using CCl₄ and CBr₄

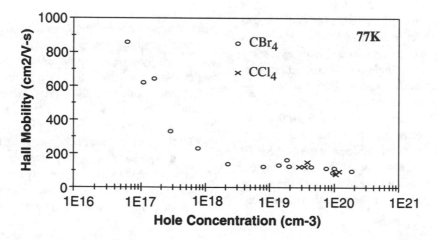

Figure 8. Hole Hall mobility vs. concentration at 77K for carbon doped GaAs using CCl₄ and CBr₄

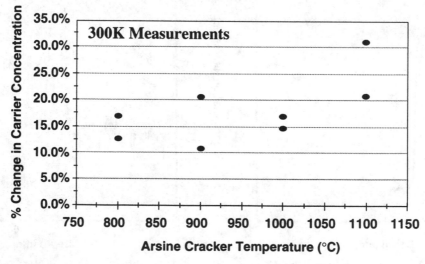

Figure 9. Hydrogen passivation in carbon doped GaAs using CBr$_4$ for films nominally doped at $5 \times 10^{19} cm^{-3}$. Two films were grown at each cracker temperature. The plotted points are the percentage increase in hole concentration after a post growth anneal at 400°C in nitrogen.

in a paper that sparked renewed interest in this material, it was shown by Smith, et. al[48] that buffers grown at 200°C could dramatically reduce FET sidegating. In 1990, the existence of arsenic precipitates in material grown below 300°C and subsequently annealed at 600°C was reported simultaneously by ourselves and two other groups.[49-51] Since then there has been considerable activity in this area, recently reviewed by Look[52], researching the physics and materials science as well as developing electronic and optoelectronic applications. As for the microwave device interests, most work has focused on utilizing insulating layers both above and below the active layer to increase breakdown voltages for higher power levels [53-54]. Much has been learned about the actual defects responsible for the resistive behavior, though there is some controversy regarding the respective roles of the macroscopic arsenic precipitates [55] and arsenic point defects[56-57]. The confinement of both of these entities to the LT layer, i.e., preventing their diffusion into contiguous layers is an important matter for devices, which was dealt with in some of the early devices by utilizing relatively thick layers of material grown at 600°C between the LT layer and the channel [58-59] to isolate the active layer from the defects that diffused from the LT layer. Figure 10 are TEM cross-sections comparing the diffusion of excess arsenic from the LT buffer and into the FET active layer for devices without an intermediate 600°C buffer. For the case where the LT layer is not heat treated before the active layer growth, arsenic precipitates extend a few

100Å into the active layer, presumably because during the heat treatment the arsenic interstitials in the near surface region can diffuse to the surface and desorb. It is interesting to compare the electrical characteristics of these layers with control layers. Layers without heat treatment and 600°C buffers exhibit lower active layer free electron concentrations and mobilities, with free electron profiles that decrease in concentration with depth, presumably due to arsenic interstitials and defects which originated with the LT buffer.[59] Yin, et. al. showed that confinement of a high percentage of the defects to the LT layer could be realized by the use of a 200Å AlAs barrier layer [54]. Since then others have confirmed this effect. [60-61] The unambiguous identification and the diffusion of these defects remains a concern today as it pertains to the reliability of devices and thus their widespread acceptance and use in the future; and thus, this is critical area for research in LTG materials.

There also has been some controversy regarding the actual benefits of LTG buffer layers for microwave power devices, with at least two different studies showing that LTG buffers provide either no RF benefit [59] or even inferior performance [62] for GaAs FETs, though both showed improved DC characteristics. On the other hand, new benchmarks for 20 GHz power performance of PHEMTs by using LTG buffers were established by Actis et al., demonstrating a significant performance margin over PHEMTs with conventional superlattice buffers in a side by side comparison. [63]

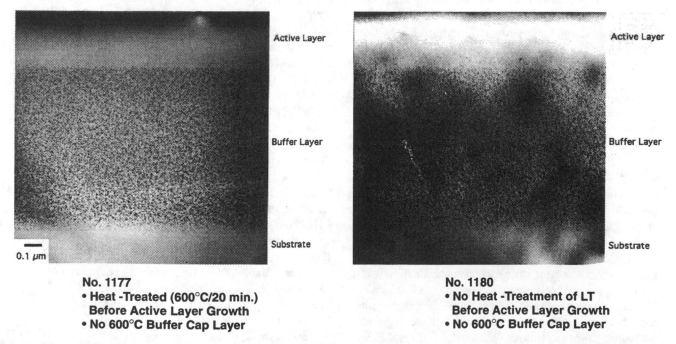

No. 1177
• Heat -Treated (600°C/20 min.)
  Before Active Layer Growth
• No 600°C Buffer Cap Layer

No. 1180
• No Heat -Treatment of LT
  Before Active Layer Growth
• No 600°C Buffer Cap Layer

Figure 10. TEM (110) bright field cross-sections of GaAs FETs with LT buffers, comparing arsenic precipitate distribution as a function of heat treatment. The micrograph of No. 1180 shows significant diffusion of arsenic into the active layer.

GSMBE WITH HYDRIDES ADSORBED ON ZEOLITE

The production of GaAs and InP epitaxial wafers requires the use of the hazardous gases arsine and phosphine. Alternative materials to arsine and phosphine, as well as alternative packages for arsine and phosphine have been proposed. Principally these alternatives are organometallic arsenic and phosphorus compounds, in-situ generators for arsine and phosphine, and low pressure packages for arsine and phosphine. With the low pressure package approach, the arsine or phosphine are stored and delivered from a vessel filled with zeolite, upon which the adsorbed hydrides exhibit substantially reduced vapor pressures relative to the liquids. The adsorption isotherms are such that practical quantities of phosphine or arsine can be delivered for GSMBE or CBE applications from vessels with starting pressures below one atmosphere (15 psi) at 300K, offering obvious safety and environmental enhancements over the conventional packages, cylinders filled with pressurized liquids (205 psi for arsine and 592 psi for phosphine).

There are several benefits to this approach compared to the alternatives, and essentially one problem which needs to be solved. First of all, the benefits include the following:

- The package is extremely safe and "environmentally-friendly". Gas evolution from a full open valve is extremely slow, less than 1 sccm. For comparison, a flow limiting orifice in the neck of a high pressure cylinder of hydride will generally flow about 10 liters per minute. Thus in the case of a release, the exposure threat and capacity for polluting the environment is diminished by four orders of magnitude.

- Waste is eliminated with this approach. Used cylinders do not need to be "recommisioned" or disposed (except after 5 years by regulation of DOT). There is no hazardous waste as with in-situ generators or organometallic bubblers. Used cylinders are simply refilled with hydride.

- Both the chemistry and hardware are simple. There are no moving parts as with some in-situ generators, and no parasitic reactions as with some organometallics. So reliability and cost are the most competitive of the approaches we are aware.

- Purity may be superior since carbon contamination is intrinsically reduced relative to the organometalics. Indeed, for InP, with one exception noted below, superior results have been reported for the zeolite/phosphine approach compared to the organometallic sources TBP and BPE, but, the ultimate purity is inferior to the best results achieved with conventional cylinders of arsine or phosphine.

Purity is the problem that needs to be addressed. In its current state of development, the zeolite/hydride approach is adequate in purity for some production applications. But in order to be a true replacement for the conventional high pressure cylinders of arsine and phosphine, and to be acceptable for the whole range of production applications, an order of magnitude of improvement is needed for purity.

A 7-liter cylinder filled with zeolite 5A can adsorb approximately 1.5 lbs of phosphine at 300K. The amount of material that can be utilized for growth will depend on the temperature of the cylinder and the pressure/flow characteristics of the delivery system. In our case, the output of the cylinder flows through a mass flow controller and into the reactor or vent line. The mass flow controller requires a pressure differential of about 50 torr to operate predictably. Thus the delivery capacity can be calculated from the difference in the starting pressure and 50 torr using the adsorption isotherms, which at 300K, works out to be approximately 50% of the adsorbed material. More hydride can be released by heating the cylnder.

Phosphide growth utilized substrate temperatures of 500-540°C, with growth rates of 1 micrometer/hour, and phosphine flow rates of 8 sccm. The low pressure cracker temperature was varied from 700-1000°C. Figure 11 shows Hall effect data for InP layers grown by GSMBE at Martin Marietta (with zeolite/phosphine) and layers grown by a variety of techniques at U of Illinois. [64] For the most part, layers fall on the line for high quality material with normal compensation ratios[65-66]. The three layers were grown with three different cracker temperatures-900, 800, and 700°C. The free electron concentration fell as the cracker temperature was reduced, a relatively common result with hydrides observed in GSMBE and CBE. [67] As a comparison, these results can be contrasted with our CBE results for InP using bisphosphinoethane (BPE) where we achieved a relatively low 77K electron mobility of 22,000 cm^2/V-s at a concentration of 4×10^{15} cm^{-3}, due to compensation.[68] Also, using tertiarybutylphosphine (TBP), we obtained a 77K mobility of 26,400 cm^2/V-s at a concentration of 5×10^{15} cm^{-3}. Similar results with BPE and TBP have been reported by other groups. [69-70]. The best reported results which we are aware of using TBP had a 77K carrier concentration of 3×10^{14} cm^{-3} with a mobility of 76,000 cm^2/V-s [71], apparently compensated, since mobilities well in excess of 100,000 cm^2/V-s have been reported at that concentration using phosphine. [72-73] In our work, the photoluminescence spectra also improved considerably with the zeolite sources. The 2K PL spectrum of a GSMBE InP film, exhibited a FWHM of 0.22 meV for the D^+X transition. With the organometallic sources, our most narrow FWHM was 0.84meV, again indicating the higher compensation of those films by residual impurities. SIMS analysis determined that sulfur

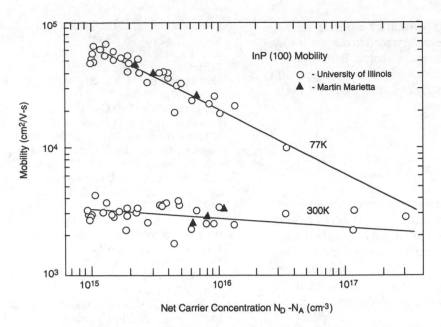

Figure 11. Hall effect data for InP grown by GSMBE using phosphine adsorbed on zeolite at Martin Marietta. University of Illinois data courtesy of Prof. Greg Stillman is shown for comparison.

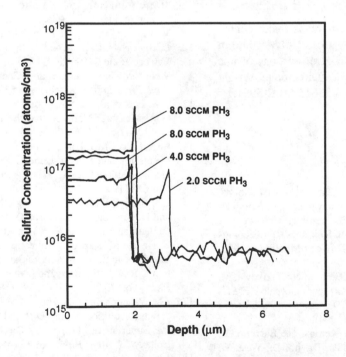

Figure 12. SIMS profile of sulfur for four different InP films grown by GSMBE as a function of phosphine flow rate. Cracker temperature is elevated above typical values to enhance the sulfur concentration.

was the dominant residual impurity. Figure 12 shows SIMS profiles for 4 different films grown with a relatively high cracker temperature of 1000°C to enhance the impurity concentration. As the figure indicates, the sulfur concentration is linearly proportional to the phosphine flow rate, implicating the source material as the source of sulfur. Hydrides are well known for being contaminated by sulfur and other group VI elements at the sub-ppm level.[67] This is the primary reason for suspecting that the original hydride source material is the limiting factor for purity. We presented similar results for GaAs with arsine adsorbed on zeolite at the 1993 North American Conference on MBE, where again sulfur was determined by SIMS profiling to be the dominant impurity.[74]

Results of the low pressure approach are encouraging, indicating that contamination may be traced to the starting hydride materials, and not due to the zeolite. Further studies with different starting hydrides will likely determine the impurity origin unambiguously as well as leading to the realization of higher purity films.

SUMMARY

This paper has briefly reviewed four areas currently under development for the performance improvement of GaAs-based devices, specifically from an R&D perspective concentrating on microwave and millimeter wave devices. There is an equally strong interest in these areas, as well as in some of these same specific issues, vis a vis optoelectronic devices, though space limitations precluded discussion of those topics here.

ACKNOWLEDGMENTS

Several colleagues at Martin Marietta made major contributions to this work, notably Bob Yanka, Paul Martin, Pin Ho, John Mazurowski, and Tom Rogers. Portions of this work were supported by the Materials Directorate, Wright Laboratory (AFSC) on contract no. F33615-90-C-5921.

REFERENCES

1. Kuroda, S.; Suehiro, H.; Miyata, T.; Asai, S.; Hanyu, I.; Shima, M.; Hara, N.; Takikawa, M." 0.25 micrometer gate length N-InGaP/InGaAs/GaAs HEMT DCFL circuit with lower power dissipation than high-speed Si CMOS circuits", Proceedings of IEEE International Electron Devices Meeting, p1022, San Francisco, 1992.

2. Suehiro, H.; Miyata, T.; Kuroda, S.; Hara, N.; Takikawa, M. "Highly doped InGaP/InGaAs/GaAs pseudomorphic HEMT's with 0.35 micrometer gates" *IEEE Transactions on Electron Devices*, vol.41, no.10,p. 1742-6 (1994).

3. Mack, M.P.; Bayraktaroglu, B.; Kehias, L.; Barrette, J.; Neidhard, R.; Fitch, R.; Scherer, R.; Davito, D.; West, W. "Microwave operation of high power InGaP/GaAs heterojunction bipolar transistors" *Electronics Letters*, vol. 29, no.12, p.1068-9 (1993).

4. Kren, D.E.; Rezazadeh, A.A.; Rees, P.K.; Tothill, J "High C-doped base InGaP/GaAs HBTs with improved characteristics grown by MOCVD" *Electronics Letters*, vol. 29, no.11, p.961-3 (1993).

5. Ren, F.; Lothian, J.R.; Pearton, S.J.; Abernathy, C.R.; Wisk, P.W.; Fullowan, T.R.; Tseng, B.; Chu, S.N.G.; Chen, Y.K.; Yang, L.W.; Fu, S.T.; Brozovich, R.S.; Lin, H.H.; Henning, C.L.; Henry, T. "Fabrication of self-aligned GaAs/AlGaAs and GaAs/InGaP microwave power heterojunction bipolar transistors" *Journal of Vacuum Science & Technology B*, vol. 12, no. 5, p.2916-28 (1994).

6. R.G. Dandrea, C.B. Duke, A. Zunger, "Interfacial Atomic Structure and Band Offsets at Semiconductor Heterojunctions," *J. Vac. Sci. Technol.B* 10(4), p. 1744, Jul/Aug (1992).

7. J.P. Landesman, J.C. Garcia, J. Massies, G. Jezequel, P. Maurel, J.P. Hirtz and P. Alnot, "GaInAs/InP and GaInP/GaAs (100) Interfaces: An Ultraviolet Photoelectron Spectroscopy Study," *J. Vac. Sci. Technol. B* 10(4), p. 1761, Jul/Aug (1992).

8. L.A. Hemstreet, C.Y. Fong, J.S. Nelson, "Effect of Interfacial Bond Type on the Electronic and Structural Properties of GaSb/InAs Superlattices," *J. Vac. Sci Technol. B* 11(4), p. 1693, Jul/Aug (1993).

9. J.S. Nelson, S.R. Kurtz, L.R. Dawson and J.A. Lott, "Demonstration of the Effects of Interface Strain on Band Offsets in Lattice-Matched III-V Semiconductor Superlattices," *Appl. Phys. Lett.*, Vol. 57 (6), pp. 578-580 (1990).

10. T. Kobayashi, K. Taira, F. Nakamura, and H. Lawai, "Band Lineup for a GaInP/GaAs Heterojunction Measured by a High-Gain Npn Heterojunction Bipolar Transistor Grown by Metalorganic Chemical Vapor Deposition," *J. of Appl. Phys.* V65 (N12), p. 4898-4902 (1989).

11. M.A. Rao, E.J. Caine, H. Kroemer, S.I. Long, and D.I. Babic, "Determination of Valence and Conduction-Band Discontinuities at the (Ga, In) P/GaAs Heterojunction by C-V Profiling," *J. of Appl. Phys.* V61 (N2), pp. 643-649 (1987).

12. D. Biswas, N. Debbar, and P. Bhattacharya, "Conduction and Valence-Band Offsets in GaAs/$Ga_{0.51}In_{0.49}P$ Single Quantum Wells Grown by Metalorganic Chemical Vapor Deposition," *Appl. Phys. Lett.* V56 (N9), pp. 833-835 (1990).

13. T.W. Lee, et al., "Conduction-Band Discontinuity in InGaP/GaAs Measured Using Both Current-Voltage and Photoemission Methods," *Appl. Phys.* V60 (4), pp. 474-476 (1992).

14. M.A. Haase, M.J. Hafich, and G.Y. Robinson, "Internal Photoemission and Energy-Band Offsets in GaAs-GaInP p-i-N Heterojunction Photodiodes," *Appl. Phys. Lett.* 58 (6), pp. 616-618 (1991).

15. L.C. Su, S.T. Pu, and G.B. Stringfellow, J. Christen, H. Selber, and D. Bimberg, "Control of Ordering in GaInP and Effect on Bandgap Energy," *J. Elec. Mat.*, Vol. 23 2, pp. 125-133 (1994).

16. X. He and M. Razeghi, "Investigation of the Heteroepitaxial Interfaces in the GaInP/GaAs Superlattices by High-Resolution X-Ray Diffractions and Dynamical Simulations," *J. App. Phys.*, Vol. 73 7, pp. 3284-3290 (1993).

17. J.M. Vandenberg, R.A. Hamm, M.B. Panish and H. Temkin, "High-Resolution X-Ray Diffraction Studies of InGaAs(P)/InP Superlattices Grown by Gas-Source Molecular-Beam Epitaxy," *J. Appl. Phys*, Vol 62 (4) pp. 1278-1281 (1987).

18. J.M. Vandenberg, A.T. Macrander, R.A. Hamm, and M.B. Panish, "Evidence for Intrinsic Interfacial Strain in Lattice-Matched $In_xGa_{1-x}As$/InP Heterostructures," *Phys. Rev. B*, Vol. 44 (8), pp. 3991-3994 (1991).

19. A.M. Moy, A.C. Chen, S.L. Jackson, X. Liu, K.Y. Cheng, G.E. Stillman and S.G. Bishop, "Optimization of Interfaces in Arsenide-Phosphide Compounds Grown by Gas Source Molecular-Beam Epitaxy," *J. Vac. Sci. Technol. B*, Vol. 11 (3), pp. 826-829 (1993).

20. T.Y. Wang, E.H. Reihlen, H.R. Jen and G.B. Stringfellow, "Systematic Studies on the Effect of Growth Interruptions for GaInAs/InP Quantum Wells Grown by Atmospheric Pres-

sure Organometallic Vapor-Phase Epitaxy," *J. Appl. Phys.* Vol. 66 (11), pp. 5376-5383 (1989).

21. A. Bensaoula, V. Rossignol, A.H. Bensaoula, and A. Freundlich, "Chemical Beam Epitaxy of $In_xAs_{1-x}P/InP$ Strained Single and Multiquantum Well Structures," *J. Vac. Sci. Technol. B*, Vol. 11 3, pp. 851-853 (1993).

22. A. Antolini, P.J. Bradley, C. Cacciatore, D. Campi, L. Gastaldi, F. Genova, M. Iori, C. Lamberti, C. Papuzza, and C. Rigo, "Investigations on the Interface Abruptness in CBE-Grown InGaAs/InP QW Structures," *J. Elec. Mater.*, Vol. 21 2, pp. 233-238 (1992).

23. G. Hollinger, D. Gallet, M. Gendry, C. Santinelli and P. Viktorovitch, "Structural and Chemical Properties of InAs Layers Grown on InP(100) Surfaces by Arsenic Stabilization," *J. Vac. Sci. Technol. B*, Vol. 8 4, pp. 832-837 (1990).

24. A. Freundlich, A. Bensaoula, A.H. Bensaoula, and V. Rossignal, "Interface and Relaxation Properties of Chemical Beam Epitaxy Grown GaP/GaAs Structures," *J. Vac. Sci. Technol. B*, Vol. 11 (3), pp. 843-846 (1993).

25. P.R. Hageman, A. van Geelen, W. Gabrielse, G.J. Bauhuis and L.J. Giling, "Optical and Electrical Quality of InGaP Grown on GaAs with Low Pressure Metalorganic Chemical Vapour Deposition," *J. Crys. Grwth*, Vol. 125, pp. 336-346 (1992).

26. F. Omnes and M. Razeghi, "Optical Investigations of GaAs-GaInP Quantum Wells and Superlattices Grown by Metalorganic Chemical Vapor Deposition," *Appl. Phys. Lett.* Vol. 59 (9), pp. 1034-1036 (1991).

27. W.T. Tsang and E.F. Schubert, "Extremely High Quality Ga0.47In0.53As/InP Quantum Wells Grown by Chemical Beam Epitaxy," *Appl. Phys. Lett.*, Vol. 49 (4), pp. 220-222 (1986).

28. N. Kobayashi, T. Makimoto, and Y. Horikoshi, "Abrupt p-Type Doping Profile of Carbon Atomic Layer Doped GaAs Grown by Flow-Rate Modulation Epitaxy," *Appl. Phys. Lett.* Vol. 50 (20), pp. 1435-1437 (1987).

29. B.T. Cunningham, L.J. Guido, J.E. Baker, J.S. Major, Jr., N. Holonyak, Jr.,, and G.E. Stillman, "Carbon Diffusion in Undoped, n-Type, and p-Type GaAs," *Appl. Phys. Lett.*, Vol. 55 (7), pp. 687-689 (1989).

30. M. Micovic, P. Evaidsson, M. Geva, G.W. Taylor, T. Vang and R.J. Malik, "Quantum Well Lasers with Carbon Doped Cladding Layers Grown by Solid Source Molecular Beam Epitaxy," *Appl. Phys. Lett.*, Vol. 64 (4), pp. 411-413 (1994).

31. R.J. Malik,, Y. Anand, M. Micovic, M. Geva, and R.W. Ryan, "Performance and Reliability Characteristics of GaAs Planar Doped Barrier Detector Diodes Using Carbon-Doped Acceptor Spikes Grown by Molecular Beam Epitaxy," Proc. 14th Bien. IEEE/Cornell U. Conf. on Adv. Conc. in High Speed Semicond. Dev. and Circ., Aug (1993).

32. H. Ito, O. Nakajima and T. Ishibashi, "Carbon Doping for AlGaAs/GaAs Heterojunction Bipolar Transistors by Molecular-Beam Epitaxy," *App. Phys. Lett.* Vol. 62 (17), pp. 2099-2101 (1993).

33. M. Meyers, N. Putz, H. Heinecke, M. Heyen, H. Luth and P. Balk, "Intentional p-Type Doping by Carbon in Metalorganic MBE of GaAs," *J. Electron. Mater.* 15, p. 57 (1985).

34. W.S. Hobson, S.J. Pearton, D.M. Kozuch and M. Stavola, "Comparison of Gallium and Arsenic Precursors for GaAs Carbon Doping by Organometallic Vapor Phase Epitaxy using CCl4," *Appl. Phys. Lett.* Vol. 60 (26), pp. 3259-3261 (1992).

35. B.T. Cunningham, G.E. Stillman and G.S. Jackson, "Carbon-Doped Base GaAs/AlGaAs Heterojunction Bipolar Transistor Grown by Metalorganic Chemical Vapor Deposition Using Carbon Tetrachloride as a Dopant Source," *App. Phys. Lett*, Vol. 56 (4), pp. 361-363 (1990).

36. T.P. Chin, P.D. Kirchner, J.M. Woodall and C.W. Tu, "Highly Carbon-Doped p-Type $Ga_{0.5}In_{0.5}As$ and $Ga_{0.5}In_{0.5}P$ by Carbon Tetrachloride in Gas-Source Molecular Beam Epitaxy," *Appl. Phys. Lett*, Vol. 59 (22), pp. 2865-2867 (1991).

37. T.J. de Lyon, N.I. Buchan, P.D. Kirchner, J.M. Woodall, G.J. Scilla, and F. Cardone, 'High Carbon Doping Efficiency of Bromomethanes in Gas Source Molecular Beam Epitaxial Growth of GaAs," *Appl. Phys. Lett.*, Vol. 58 (5), pp. 517-519 (1991).

38. Y-M. Houng, S.D. Lester, D.E. Mars and J.N. Miller, "Growth of High-Quality p-Type GaAs Epitaxial Layers Using Carbon Tetrabromide by Gas Source Molecular-Beam Epitaxy and Molecular-Beam Epitaxy," *J. Vac. Sci. Technol. B*, Vol. 11 (3), pp. 915-918 (1993).

39. P.J. Lemonias, W.E. Hoke, D.G. Weir and H.T. Hendriks, "Carbon p+ Doping of Molecular-Beam Epitaxial GaAs Films Using Carbon Tetrabromide," *J. Vac. Sci. Technol. B*, Vol. 12 (2), pp. 1190-1192 (1994).

40. C.R. Abernathy, F. Ren, S.J. Pearton, P.W. Wisk, D.A. Bohling, G.T. Muhr, A.C. Jones, M. Stavola and D.M. Kozuch, "The Impact on Impurity Incorporation on Heterojunction Bipolar Transistors Grown by Metalorganic Molecular Beam Epitaxy," *J. Crys. Grwth*, Vol. 136, pp. 11-17 (1994).

41. S.A. Stockman, A.W. Hanson, S.L. Jackson, J.E. Baker, and G.E. Stillman, "Effect of Post-Growth Cooling Ambient on Acceptor Passivation in Carbon-Doped GaAs Grown by Metalorganic Chemical Vapor Deposition," *Appl. Phys. Lett*, Vol. 62 (11), pp. 1248-1250 (1993).

42. S.A. Stockman, A.W. Hanson, S.M. Lichtenthal, M.T. Fresina, G.E. Hofler, K.C. Hsieh and G.E. Stillman, "Passivation of Carbon Acceptors During Growth of Carbon-Doped GaAs, InGaAs, and HBTs by MOCVD," *J. Elec. Mater.*, Vol. 21 (12), pp. 1111-1118 (1992).

43. S. J. Pearton, C.R. Abernathy and J. Lopata, "Thermal Stability of Dopant-Hydrogen Pairs in GaAs," *Appl. Phys. Lett.*, Vol. 59 (27), pp. 3571-3573 (1991).

44. T. Murotani, T. Shimanoe and S. Mitsui, *J. Crys. Growth*, Vol. 45, p. 302 (1978).

45. H. Kunzel, A. Fischer, K. Ploog, *Appl. Phys.* Vol. 22, p. 23 (1980).

46. C.E.C. Wood, J. Woodcock and J.J. Harris, Inst. Phys. Conf. Ser., Vol. 45, p. 29 (1979).

47. R.A. Stall, E.E.C. Wood, P.D. Kirchner and L.F. Eastman, "Growth Parameter Dependence of Deep Levels in MBE GaAs," *Electron Lett.*, Vol. 16, p. 171 (1980).

48. F.W. Smith, A.R., Calawa, C.-L. Chen, M.J. Manfra and L.J. Mahoney, "New MBE Buffer Used to Eliminate Backgating in GaAs MESFETs," *IEEE Electron Device Lett.*, Vol. 9, p. 77 (1988).

49. Z. Liliental-Weber, Workshop on Low Temp. GaAs Buffer Layers, San Francisco, CA, (1990).

50. M.R. Melloch, Workshop on Low Temp. GaAs Buffer Layers, San Francisco, CA, (1990).

51. J.M. Ballingall, Workshop on Low Temp. GaAs Buffer Layers, San Francisco, CA, (1990).

52. D.C. Look, "Molecular Beam Epitaxial GaAs Grown at Low Temperatures," *Thin Solid Films*, Vol. 231 (1-2) pp. 61-73 (1993).

53. C.-L. Chen, F.W. Smith, B.J. Clifton, L.J. Mahoney, M.J. Manfra and A.R. Calawa, "High Power Density GaAs MESFETs With a LTG Epitaxial Layer as the Insulator," *IEEE Elec. Dev. Lett.*, Vol. 12, p. 306 (1991).

54. L-W. Yin, Y. Hwang, J.H. Lee, R.M. Kolbas, R.J. Trew, U. Mishra, "Improved Breakdown Voltage in GaAs MESFETs Utilizing Surface Layers of GaAs Grown at a Low Temperature by MBE," *IEEE Elec. Dev. Lett.*, Vol 11 (12), pp. 561-563 (1990).

55. N. Antique, E.S. Harmon, J.C.P. Chang, J.M. Woodall, M.R. Melloch, and N. Otsuka, "Electrical and Structural Properties of Be- and Si-Doped Low-Temperature-Grown GaAs," *J. Appl. Phys.*, Vol. 77 (4), pp. 1-6 (1995).

56. D.C. Look, G.D. Robinson, J.R. Sizelove and C.E. Stutz, "Donor and Acceptor Concentrations in Molecular Beam Epitaxial GaAs Grown at 300 and 400°C," *Appl. Phys. Lett*, Vol. 62 (23), pp. 3004-3006 (1993).

57. D.C. Look, D.C. Walters, G.D. Robinson, J.R. Sizelove, M.G. Mier, and C.E. Stutz, "Annealing Dynamics of Molecular-Beam Epitaxial GaAs Grown at 200°C," *J. Appl. Phys.*, Vol. 74 (1) pp. 306-310 (1993).

58. B. J-F. Lin, C.P. Kocot, D.E. Mars and R. Jaeger, "Anomalies in MODFETs with a Low-Temperature Buffer," *IEEE Trans. on Elec. Dev.*, Vol 37 (1), pp. 46-49 (1990).

59. J.M. Ballingall, P. Ho, R.P. Smith, S. Mith, G. Tessmer, T. Yu, E.L. Hall, and G. Hutchins, "Material and Device Characteristics of MBE Microwave Power FETs with Buffer Layers Grown at Low Temperature (300°C)," *Mat. Res. Soc. Symp. Proc.*, Vol. 241, pp. 171-179 (1992).

60. M.R. Melloch, J.M. Woodall, N. Otsuka, K. Mahalingam, C. Chang and D.D. Nolte, "GaAs, AlGaAs, and InGaAs Epilayers Containing As Clusters: Semimetal/semiconductore Composites," *Mat. Sci. and Eng*, B22 pp. 31-36 (1993).

61. M.R. Melloch, C.L. Chang, N. Otsuka, K. Mahalingam, J.M. Woodall and P.D. Kirchner, "Two-Dimensional Arsenic-Precipitate Structures in GaAs," *J. Crys. Grwth*, 127, pp. 499-502 (1993).

62. D.C. Streit, M.M. Hoppe, C-H. Chen, J.K. Liu and K-H. Yen, "Are Low Temperature Molecular-Beam Epitaxy GaAs Buffers Good for Microwave Applications?," *J. Vac. Sci. Technol. B.*, Vol. 10 (2), pp. 819-821 (1992).

63. R. Actis, K. Nichols, W. Kopp, T. Rogers, F. Smith, " High Performance 0.15 micrometer Gate Length PHEMTs Enhanced with a Low Temperature Grown GaAs Buffer" Microwave Theory and Techniques Symposium, Orlando, FL (May 1995).

64. J. M. Ballingall, J. S. Mazurowski, P. A. Martin, J. McManus "GSMBE of InP Using Phosphine Adsorbed on Zeolite", 36th Electronics Materials Conference, Boulder CO, June, 1994.

65. T.R. Lepkowski, "Automated Variable Temperature Hall Effect Measurements and Analysis of N-type GaAs, InP, and Their Lattice-Matched Alloy Semiconductors," Ph.D. Thesis, University of Illinois, 1985.

66. E. Kuphal and D. Fritzsche, "LPE Growth of High Purity InP and N and P InGaAs," *J. Electron. Mater.* 12, 743 (1983).

67. A.P. Roth, T. Sudersena Rao, R. Benzaquen, C. Lacelle, and S. Rolfe, "Nature and Origin of Residual Impurities in High-Purity GaAs and InP Grown by Chemical Beam Epitaxy," *J. Vac. Sci, Technol. B*, Vol. 11 (3), pp. 836-839 (1993).

68. A. Chin, P. Martin, U. Das, J. Mazurowsli, J. Ballingall, "CBE Growth of InP, InGaP, and InAs Heterojunctions Using Triethylindium and Bisphosphinoethane", *J. Vac. Sci. Technol.B* 11(3), 847 (1993).

69. D. Ritter, M.B. Parish, R.A. Hamm, D. Gershoni, and I. Brener, "Metalorganic Molecular Beam Epitaxy of InP, $Ga_{0.47}In_{0.53}As$, and GaAs with Tertiarybutylarsine and Tertiarybutylphosphine," *App. Phys. Lett.* 56 (15) pp. 1448-1450 (1990).

70. H. H. Ryu, C. W. Kim, L. P. Sadwick, G. B. Stringfellow, R. W. Gedridge, A. C. Jones, "Growth of InP, GaP, and GaInP by CBE Using Alternative Sources", 21st International Symposium on Compound Semiconductors, San Diego, CA, Sept. 1994.

71. E.A. Beam III, T.S. Henderson, A.C. Seabaugh and J.Y. Yang, "The Use of Tertiarybutylphosphine and Tertiarybutylarsine for the Metalorganic Molecular Beam Epitaxy of the $In_{0.53}Ga_{0.47}As/InP$ and $In_{0.48}Ga_{0.52}P/GaAs$ Materials System," *J. Crys. Grwth*, Vol. 116, pp. 436-446 (1992).

72. T. Sudersena Rao, C. Lacelle, and A.P. Roth, "Growth of High-Purity InP by Chemical Beam Epitaxy," *J Vac. Sci, Technol. B*, Vol. 11 (3) pp. 840 - 842 (1993).

73. T. Sudersena Rao, C. Lacelle, and A.P. Roth, "3.1 x 10e5 Peak Electron Mobility in InP Grown by CBE", *Electronics Lett.*, 29(4). 373 (1993).

74. J. M. Ballingall, J.S. Mazurowski, P.A. Martin, J. McManus, G. Tom "MBE, GSMBE, and CBE of GaAs Using Arsine Adsorbed on Zeolite", 13th North American Conference on MBE, Stanford, CA, October, 1993.

Magnetoluminescence Characterization of Quantum Well Structures

E. D. Jones and S. R. Kurtz

Sandia National Laboratories
Albuquerque, NM 87185-0350

Three applications of magnetic field effects upon the photoluminescence spectrum which provide unique information about semiconductor quantum well structures are presented. The first example shows data which provide a quantitative measure of both the conduction- and valence-band the energy dispersion curves for an InGaAs/GaAs single-strained-quantum well and a GaAs/AlGaAs lattice-matched single quantum well. The second subject discusses magnetoluminescence data which provides a clear demonstration for the existence of spectral shifts related to ionized-impurity scattering and the third study involves infrared magnetoluminescence measurements on narrow bandgap semiconductor alloys and heterostructures.

INTRODUCTION

Photoluminescence (PL) characterization of semiconductor materials has been a standard nondestructive diagnostic tool for a number of years. With the advent of quantum well structures, the photoluminescence technique has been used to study exciton and impurity states, interface roughness, impurity (dopant) diffusion, quantum-well energy levels, valence-band offset energies, etc. A recent review by Herman et. al.[1] presents an excellent summary for these kinds of studies. In particular, the use of 2D-excitons to study interface roughness is discussed in detail. However, magnetic field effects on the photoluminescence spectrum were not treated. With the current availability of high-field superconductor magnet systems and the advent of novel semiconductor systems based upon epitaxy growth techniques, magnetoluminescence is becoming a valuable diagnostic tool. In order to demonstrate the utility of the magnetoluminescence technique, we present three examples of magnetic-field dependent PL spectra, each of which provide unique information about the quantum-well.

The first application of magnetic fields concerns the electronic band structures of doped quantum-well structures. From an interpretation of the magnetic field dependent PL spectrum, it will be shown how a simultaneous determination of both the conduction- and valence-band energy dispersion curves can be made from a single sample measurement. Specifically, data for a single-strained-quantum well (SSQW) structure based on InGaAs/GaAs and a GaAs/AlGaAs lattice-matched-single-quantum-well (LMSQW) structure will be presented. It is found that while the valence-bands are highly nonparabolic, the conduction-band energy dispersion curves are nearly parabolic.

The second subject of this paper is the spectral shift of the PL peak resulting from ionized-impurity scattering in highly doped semiconductor quantum wells. Many researchers in the past have used the PL peak as a measure of bandgap energy, or quantum well energies, ground and excited states. Magnetoluminescence data will be presented that graphically demonstrate the importance of these spectral shifts.

The third and final example extends magnetoluminescence measurements to the long, 2 to 5 μm, wavelength range where magneto-excitons are studied in narrow bandgap materials and heterostructures. Analysis of the linear portion of the magnetic-field induced photoluminescence energy shift provides estimates of effective masses and exciton binding energies.

BAND-STRUCTURE STUDIES

The semiconductor quantum-well laser is the principal component for optoelectronic applications. The need for custom laser wavelengths, higher laser powers, and better beam quality or control has led to many different laser diode designs and configurations. The most common combination for semiconductor laser materials for these devices have been GaAs and AlGaAs. The desirability for light-hole valence-band masses for optoelectronic devices, lasers, or high-speed electronic devices has been well documented.[2-4]

Light-hole valence-band masses have been mainly achieved in layered semiconductor structures by the introduction of compressive biaxial strain in the active quantum layers. The biaxial strain is achieved by growing layered structures from materials with differing lattice constants, e.g., layers of $In_xGa_{1-x}As$ and GaAs. However, because of level crossing repulsion between the in-plane heavy and light-hole valence-band states, the resulting in-plane valence-band energy dispersion curves (and masses) are nonparabolic. For lattice-matched quantum well devices, the same conditions hold because of the quantum confinement splitting of the heavy and light-hole valence bands. For these structures, the in-plane heavy-hole light-hole mixing and hence valence-band nonparabolicity occurs at small values of the wave vector. For the devices based upon GaAs and AlGaAs, quantum confinement can be an effective method to alter the energy difference ΔE_{HL} between the heavy and light-hole valence bands. However, for wide (~15 nm) GaAs/AlGaAs structures, ΔE_{HL} is small ($\lesssim$ 5meV) and thus the amount of heavy-hole light-hole mixing is large. For these wide quantum well structures, the in-plane valence-band ground state energy is also "heavy," i.e., for Fermi energies $E_f > 2$ meV, the valence-band mass $m_v \approx 0.35m_0$. By reducing the quantum-well width, the energy difference ΔE_{HL} can be increased by quantum confinement. For GaAs/AlGaAs lattice-matched quantum wells, a maximum $\Delta E_{HL} \approx 30$ meV occurs near a quantum-well width of about 4.5 nm.

Valence-band nonparabolicity, for lattice-matched or strained-layer optoelectronic devices, means that with increasing carrier densities the valence-band mass also increases, changing laser threshold currents, modulation frequency response, magneto-optic parameters, device transconductance, etc. Thus a device which performs satisfactory at low power levels, i.e, low-carrier densities, may not perform as expected, under high power or high-current operation. Furthermore, device modeling codes predicting the optical or carrier response function should take into account any effects which can be attributed to valence-band mixing. As digital communication speeds increase, a knowledge of the energy-band dispersion curves is mandatory if these design criteria are to be modeled or achieved.

Recently, it was demonstrated that both the conduction-band and valence-band energy dispersion curves of modulation doped SSQW[4-7] and LMSQW[8] structures can be simultaneously determined by magnetoluminescence measurements from a single sample. These papers discuss the importance of the energy difference ΔE_{HL} between the heavy and light-hole valence-bands in determining the degree of mixing between these two valence bands. Large values for ΔE_{HL} not only give rise to the smallest ground-state in-plane light-hole masses, but also help to reduce the valence-band nonparabolicity.

The structures discussed here were prepared using molecular beam epitaxy. The quantum well barrier material was silicon-doped, with a spacing of about 8 nm between the ~3-nm-wide silicon-modulation layer and the quantum well. The SSQW structure (#BC042) consisted of a single 8-nm-wide $In_{0.20}Ga_{0.80}As$ strained quantum-well and unstrained GaAs barriers. The energy difference ΔE_{HL} between the heavy-hole and light-hole valence-bands, which includes contributions from both strain effects and quantum confinement, is about 60 meV. The 4-K two-dimensional carrier concentration N_{2d} and mobility μ are respectively 5×10^{11} cm^{-2} and 1.2×10^4 cm^2/Vsec. The LMSQW GaAs/AlGaAs structure (#G0260) has a 4.5-nm-wide GaAs quantum well with lattice-matched $Al_{0.25}Ga_{0.75}As$ barriers, and as previously mentioned, the heavy-hole and light-hole valence-band energy difference $\Delta E_{HL} \approx 30$ meV. For the LMSQW and T = 4 K, $N_{2d} = 6.6 \times 10^{11}$ cm^{-2} and $\mu = 2.2 \times 10^4$ cm^2/Vsec. The magneto-luminescence measurements were made in the temperature range of 1.4 and 76 K, and the magnetic fields varied between 0 and 14 T. The luminescence measurements were made with an Argon-ion laser operating at 514.5 nm and CAMAC-based data acquisition system.[9] The direction of the applied magnetic field is parallel to the growth direction, i.e., the resulting Landau orbits are in the plane of the quantum well. With this geometry, all measurements concerning the conduction and valence-band dispersion curves and masses refer to their *in-plane* values.

A free particle, with mass m and charge e, moving in a magnetic field B forms quantized states, Landau levels, with an energy E = (n + 1/2)(e$\hbar$B/mc) $\equiv$ (n + 1/2)$\hbar\omega$, in cgs units, where n is the Landau index, $\hbar$ is Planck's constant over 2π, c is the velocity of light, and $\hbar\omega$ is the cyclotron energy. The distribution function for a degenerate two-dimensional electron gas (conduction-band states for a n-type material) is determined by the Fermi-Dirac statistics, but because of the very small number of photo-induced two-dimensional hole states, the distribution for the valence-band holes are governed by Maxwell-Boltzmann statistics. At high temperatures, where kT is much larger than $\hbar\omega_v$, the n_v = 0, 1, 2, 3, ... valence-band Landau levels are populated and all magnetoluminescence transitions between the n_c and n_v Landau levels are *allowed*, obeying the $\delta n_{cv} \equiv (n_c - n_v) = 0$ selection rule. For these high temperatures, the interband luminescence transition energy E is given by

$$E(n) = E_{gap} + \left(n + \frac{1}{2}\right)\left(\frac{e\hbar B}{\mu c}\right), \qquad (1)$$

where E_{gap} is the bandgap energy, μ is the reduced mass

$(\mu^{-1} = m_c^{-1} + m_v^{-1})$ where m_c and m_v are respectively the conduction or valence-band effective masses expressed in terms of the free electron mass m_0. In this paper we assume that the sample temperature is always less than the Fermi temperature, $kT_f = E_f$, of the degenerate two-dimensional electron gas.

A schematic showing these *allowed* transitions for a n-type structure is shown in the right side of Fig. 1. The Fermi energy E_f, the bandgap energy E_{gap}, and the Landau level indices n are also indicated in the figure. For large magnetic fields and low temperatures ($\hbar\omega_v \gg kT$) only the $n_v = 0$ valence-band Landau level is populated. Here, the PL transition between the $n_c = 0$ and $n_v = 0$ Landau level is *allowed* while transitions between the higher energy conduction-band Landau levels $n_c = 1, 2, 3, \ldots$ and the $n_v = 0$ ground state valence-band Landau level are zeroth-order forbidden, but are observable due to higher-order ionized-

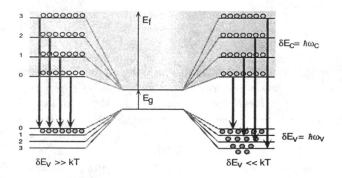

Figure 1. A schematic representation for the Landau levels in a n-type quantum well. The left side of the figure is the low temperature case, $\delta E_V \gg kT$, and the right side is the high temperature condition $\delta E_V \ll kT$.

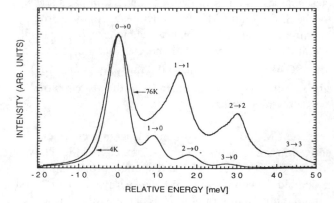

Figure 2. Magnetoluminescence spectra (B = 5.75T) at two temperatures, 4 and 76 K, for an n-type InGaAs/GaAs SSQW showing the *allowed*, $\delta n_{cv} = 0$, and the $\delta n_{cv} \neq 0$ zeroth-order forbidden transitions. The energy axes of the two spectra have been shifted in order to compare them.

impurity scattering processes.[10,11] The left side of Fig. 1 shows the energy level diagram for this case.

Two magnetoluminescence spectra (B = 5.75 T) at 4 and 76 K at are shown in Fig. 2. The origins of the energy axes of the two spectra have been adjusted in order to allow a comparison of the spacing between the magnetoluminescence transitions. For 4 K, $E_{gap} = 1330.1$ meV while at 76 K, $E_{gap} = 1321.1$ meV. The indices $n_c \rightarrow n_v$ for each peak are labeled in the figure. As can be seen in Fig. 2, all observed transitions for the 76 K data are *allowed*, i.e, $\delta n_{cv} = 0$, while for the 4.2 K data, the only zeroth-order *allowed* peak is the $0 \rightarrow 0$ transition. Because of the Maxwell-Boltzmann statistics for the holes, the energy dependence of the 76 K magnetoluminescence peak-amplitudes are also governed by Maxwell-Boltzmann distribution function. An analysis of the energy dependence of the peak-amplitudes for this spectrum yields a temperature of about 80 K, which is in good agreement with the expected temperature of liquid nitrogen. A theoretical treatment of the energy dependence of the amplitudes of the zeroth-order forbidden transitions has been performed by Lyo[11] and data shown in Fig. 2 is in good agreement with his calculation.

As is obvious from the left-hand side of Fig. 1, where $\hbar\omega_v \gg kT$, the energy difference δE between the $E(n_c)$ and $E(n_c-1)$ magnetoluminescence peaks depends only on the conduction-band cyclotron energy $\hbar\omega_c$. Setting $n_v = 0$ and using (1), the magnetoluminescence transition energy $E(n_c)$ is given by in terms of the respective conduction and valence-band cyclotron energies as

$$E(n_c) = E_{gap} + \left(n_c + \frac{1}{2}\right)\hbar\omega_c + \frac{1}{2}\hbar\omega_v, \quad (2)$$

where $n_c = 0, 1, 2, 3, \ldots$ Thus, utilizing low temperatures and measuring the energy differences as a function of magnetic field, (2) provides a method for obtaining all the information about the conduction band energy dispersion curves

Figure 3 shows the derived low-temperature-derived conduction-band dispersion curve for the InGaAs/GaAs SSQW. The method (and justification) used relating the magnetic field B to the wavevector **k** has been adequately discussed.[12,13] The maximum value of the wavevector **k** of about 3% of the Brillouin zone is determined by the conduction-band Fermi energy E_f of about 35 meV. The minimum wavevector determinations are limited by our ability to distinguish magnetoluminescence peaks at low magnetic fields. The conduction-band dispersion curve is

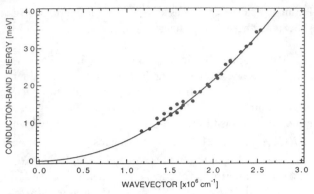

Figure 3. Conduction-band energy dispersion curve for SSQW BC042. The parabolic curve drawn line through the data is a best-fit curve with $m_c = 0.067 m_0$.

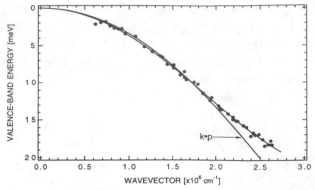

Figure 4. Valence-band energy dispersion curve for SSQW BC042. The line drawn through the data is a best-fit of the nonparabolic curve given by (3). A **k•p** calculation is also indicated in the figure.

found to have a small nonparabolic correction with the zone center conduction-band mass $m_c \sim 0.067 m_0$ and at the Fermi energy $m_c \sim 0.069 m_0$. Nonparabolic effects upon the cyclotron resonance measured conduction-band masses have been discussed in detail[14] and the magnetoluminescence results presented here are in agreement with that derived by conduction-band cyclotron resonance.

With a knowledge of the conduction-band dispersion curve, we can, using (1), derive the valence-band dispersion curve by performing 76-K magnetoluminescence measurements and analyzing the zeroth-order *allowed* transition peaks as a function of magnetic field. The valence-band data is shown in Fig. 4 and it is evident that the dispersion curve is nonparabolic. For the range of data shown in the figure, the valence-band effective mass m_v for the SSQW structure BC042 varies between a zone-center $0.11 m_0$ to about $0.3 m_0$ at $E_v(k) \sim 20$ meV. The solid line through the valence-band data is a best-fit nonparabolic curve to the data of the form

$$E(k) = \left(\frac{\hbar^2 k^2}{2 m_v^0} \right) \left[1 - \frac{K}{E_{gap}} \left(\frac{\hbar^2 k^2}{2 m_v^0} \right) \right], \qquad (3)$$

where $m_v^0 = 0.11 m_0$ is the zone center valence-band mass and $K \sim 16$ which is about an order of magnitude larger than measured for the conduction band.[14] A comparison with a **k•p** calculation for the valence-band dispersion, which includes strain and valence-band mixing, is also shown in Fig. 4. The agreement between the **k•p** calculation and the data is reasonable in view of the fact that the **k•p** calculation did not include the contributions from the conduction band or split-off valence band. Magnetoluminescence determined valence-band dispersion curves for other samples with varying indium concentration and hence, with differing values for the heavy-hole light-hole

valence band splitting ΔE_{HL}, confirm that as ΔE_{HL} increases the valence-bands become more parabolic and that the ground state in-plane valence-band mass remains relatively constant.[5]

The conduction and valence-band dispersion curves for the GaAs/AlGaAs LMSQW sample were also determined by both low and high temperature magnetoluminescence measurements. The results of the magnetoluminescence data analyses are shown in Figs. 5 and 6, where comparisons with the SSQW data are presented. The conduction-band dispersion curve (Fig. 5) for the LMSQW structure is again nearly parabolic with respective zone-center conduction-band masses of $m_c \approx 0.085 m_0$. The increase to m_c for the LMSQW conduction-band mass m_c between the bulk GaAs $m_c \approx 0.065 m_0$ to $0.085 m_0$ is a result[14,15] of the quantum confinement energy, i.e., increased bandgap energy. In order to verify these masses on our samples, the conduction-band masses for the SSQW and LMSQW structures were measured by far infrared cyclotron resonance techniques with the result that at their respective Fermi energies, $m_c \approx 0.069 m_0$ for the SSQW sample and $m_c \approx 0.082 m_0$ for the LMSQW structure in excellent agreement with the aforementioned magnetoluminescence data.

The LMSQW valence-band dispersion curve can be derived in the same manner as before, e.g., with a knowledge of the conduction-band dispersion curve the magnetic field dependent valence-band energy can be inferred from the high temperature (76 K) magnetoluminescence data using (1). The resulting valence-band dispersion curve for LMSQW sample is shown in Fig. 6. For the range of data shown in the figure, the valence-band mass m_v for the SSQW structure varied between $0.11 m_0$ and

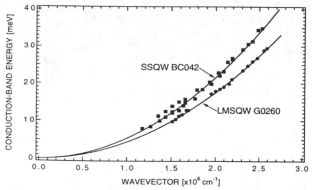

Figure 5. Conduction-band energy dispersion curves for the SSQW and LMSQW samples. The solid lines drawn through the data are best-fit parabolas with masses $0.067m_0$ (SSQW) and $0.085m_0$ (LMSQW.)

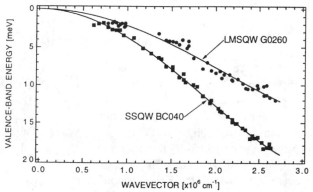

Figure 6. Comparison of the valence-band energy dispersion curves for SSQW and LMSQW samples. The two lines through the data are best-fit of the non-parabolic curve given by (3).

$0.22m_0$. For the LMSQW sample at small wavevectors, $m_v \approx 0.15m_0$ and for larger wavevectors ($E_v \approx 10$ meV) $m_v \approx 0.35m_0$. As previously mentioned, the heavy-hole light-hole energy separation ΔE_{HL} is about 60 meV for the SSQW structure and about 30 meV for the LMSQW sample. It is obvious from Fig. 6, that the valence-band dispersion curve for the SSQW structure is more parabolic than that of the LMSQW sample and this result is due to the increased heavy-hole light-hole energy separation ΔE_{LH} in the SSQW sample.[5]

SPECTRAL SHIFTS

The PL line shape function in degenerate (doped) quantum wells has recently received attention as a measure of the 2D-carrier concentration.[16,17] Recently two PL peaks were observed[18] in heavily doped n-type quantum wells and the energy difference between the two peaks was used to measure the difference between the quantum-

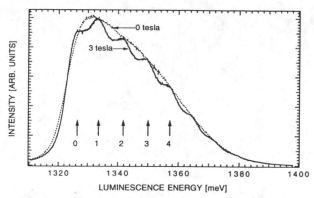

Figure 7. Magnetoluminescence spectra at 76 K for 0 and 3 tesla. The energy of the *allowed* n→n transitions are indicated. The amplitude of the 0→0 transition is smaller than the 1→1 transition. The *true* bandgap energy E_{gap} is slightly smaller than the 0→0 energy.

well's ground state and the first excited state energies. Care must be taken in interpreting the meaning the PL peak energy if this kind of information is vital to the analysis of the transport data as presented in reference 18. For example, spectral shift contributions to the PL line shape could be large and the PL peak energy may be shifted from the actual bandgap energy.

Most calculations[1] for the PL spectral line shape function in doped quantum wells ignore the second-order contributions arising from carrier-impurity scattering. For modulation doped structures, the majority of ionized-impurities are restricted to the modulation layer, located some distance from the quantum well. A microscopic theory for the PL line shape for degenerate semiconductor quantum wells which includes the effects due to ionized-impurity scattering of the carriers is treated in detail.[19] This paper shows that the experimental PL line shape is nearly accounted for by ionized-impurity scattering of the carriers. Also discussed were PL line shape changes as a function of temperature and also distance between the modulation-doped region and the quantum well.

An important result of the line shape calculation[19] is the spectral shift of the line shape, i.e., the peak intensity of the PL spectrum is shifted by from the actual bandgap energy. A complete description of the origins of the spectral shift can be found in reference 19 where it is shown how ionized impurity scattering couples the large number of conduction-band carriers with energies greater than kT with those electrons within kT of the bottom of the band. The photo-generated holes have an energy of ~ kT which is much less than the conduction-band Fermi energy E_f ~ 35 meV. The spectral shift is found to be strongly dependent upon the distance between the dopant-layer in the

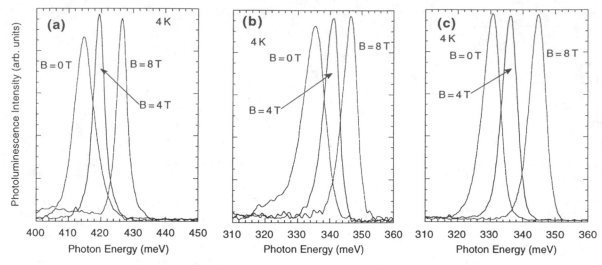

Figure 8. Magneto-photoluminescence spectra of (a) InAs, (b) InAs$_{0.93}$Sb$_{0.07}$ alloy, and (c) InAs$_{0.91}$Sb$_{0.09}$ / In$_{0.87}$Ga$_{0.13}$As SLS.

barrier and the quantum well and also upon the carrier densities. For the SSQW structure discussed earlier, the spectral shift of the PL peak intensity is about 7 meV. However, because of the many contributions to the PL line shape, it is difficult to clearly demonstrate the importance, or size, of the spectral shift by merely presenting agreements between calculated and experimental line shapes.

Referring to the 76-K *allowed* PL spectrum shown in Fig. 2, the amplitude of the 0→0 peak is the larger than the other transitions. At very high magnetic fields, where the energy spacing between the peaks is large, the line shapes of the *allowed* transitions are almost gaussian[20] and the 0→0 transition is dominant. At low magnetic fields, the amplitudes of the Landau level transitions is a convolution of the inherent gaussian line shape with the zero-field line shape function. Thus if there are large spectral shifts, there may be situations where after convolving the two line shapes (gaussian and zero-field), the peak amplitude of the 1→1 transition is larger than the 0→0 transition. Figure 7 is a graphic demonstration for this situation. This data was part of the high temperature spectra used to obtain the valence-band dispersion curves for the SSQW structure. At slightly higher magnetic fields, the spectrum appears normal. The spectral shift is again estimated from Fig. 7 to be of the order of 7 meV.

Thus if zero-field PL spectra are being used to measure bandgap energies in heavily doped semiconductor quantum wells, correct interpretation of the PL line shape must be made to obtain meaningful numbers.

NARROW BANDGAP STRUCTURES

Mid-wave infrared photoluminescence was measured by operating a Fourier transform infrared (FTIR) spectrometer in a double-modulation mode. In the magnetoluminescence experiments, a fluoride optical fiber was used to transmit infrared light in the magnet cryostat. The PL photons were collected by the fiber and analyzed with the FTIR equipped with an InSb photodiode. All measurements were made in the Faraday configuration with the magnetic field parallel to the growth, (001), direction of the sample.

In narrow gap semiconductors, magnetoluminescence measurements are used to estimate effective masses and to identify the ground state of holes in heterostructures. Figure 8 shows spectra for a 9 nm/13 nm-layer-thicknesses InAs$_{0.91}$Sb$_{0.09}$/In$_{0.87}$Ga$_{0.13}$As strained-layer superlattice (SLS) (Fig. 8(c)) are compared with those for unstrained InAs$_{0.93}$Sb$_{0.07}$ (Fig. 8(b)) and InAs (Fig. 8(a)) alloys. Photoluminescence spectra for the SLS and alloys consist of a single peak in the 3-4 μm range, with linewidth ≤ 10 meV. In these unintentionally doped samples, excitonic behavior is revealed in the magneto-photoluminescence results. For all samples, the photoluminescence peak energy is insensitive to magnetic field for B < 2T, characteristic of a diamagnetic exciton, (see Fig. 9) and in the linear region observed at higher fields, the reduced mass values obtained from the free carrier approximation are consistently too large, due in part to the binding energy of the exciton. For each photoluminescence line, the reduced mass, μ, obtained from the free electron-hole approximation for the 0→0 transition is indicated in the figure. The reduced mass of the SLS was less than that of the alloys

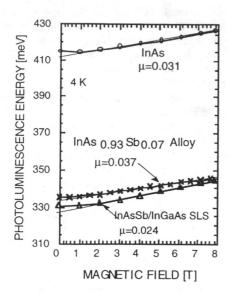

Figure 9. Magnetic field-induced shift of the photoluminescence energy for the InAsSb/InGaAs strained-layer superlattice and unstrained InAsSb and InAs alloys

because the in-plane hole mass is significantly decreased in the biaxially compressed InAsSb layers of the SLS.

Using measured parameters for InAs and semi-empirical expressions for high-field line shifts and magneto-exciton energies, we can estimate exciton binding energies and correct the reduced mass values. Including nonparabolicity and magneto-exciton contributions, the "linear" shift of the photoluminescence line at high magnetic field, B, is described by [21]

$$\Delta E = \frac{\mu_\beta B}{\mu}\left(1 - \frac{\mu_\beta K B}{E_{gap}\mu}\right) - E_{ex}(B), \qquad (4)$$

where K again is the nonparabolicity factor, E_{gap} is the bandgap energy of the alloy or quantum well material, and $E_{ex}(B)$ is the magneto-exciton binding energy.

For the three-dimensional case[22],

$$E_{ex}(B) = 1.6R\left(\frac{\mu_\beta B}{\mu R}\right)^{1/3}, \qquad (5)$$

where R is the exciton Rydberg or zero-field binding energy. Using R = 1.8 meV and K = 1.6 for InAs[23,24], a fit of the data yields $\mu = 0.023m_0$ and $\mu = 0.026m_0$ for InAs and InAs$_{0.93}$Sb$_{0.07}$ respectively. The InAs result is in good agreement with the predicted value for the InAs heavy hole exciton, $\mu = 0.022$. Again using the three-dimensional expression, the data for the SLS are described

by a reduced mass $\mu = (0.017 \pm 0.001)m_0$, with an exciton Rydberg R = (2.5 ± 1.0) meV.

CONCLUSIONS

We have shown that magnetoluminescence measurement techniques is a powerful tool for performing single-sample simultaneous measurements of both the conduction and valence-band dispersion curves (and masses). We showed that for quantum-well devices the amount of heavy-hole light-hole splitting is the determining factor for valence-band nonparabolicities and that the larger the heavy-hole light-hole energy difference, the smaller the valence-band nonparabolicity. The experimental agreement with *k•p* calculations for the valence-band energies are good and with these dispersion curves, it is possible to model not only optoelectronic behavior, but also predict electronic phenomena which rely upon these dispersion curves. In particular, the amount of valence-band nonparabolicity can now be experimentally quantified and predictions for new material systems may be possible.

We also provided a graphical example of large spectral shifts by examining a high-temperature low-magnetic field spectrum of the SSQW data used for the first example.

For characterization of narrow bandgap semiconductors, magnetoluminescence was demonstrated at long wavelength (< 5 μm). "Free carrier-like" behavior is observed at moderate magnetic fields (2 - 10 T) in these low effective mass materials, and electron-hole effective masses can be easily measured. In our example, we demonstrate that the holes in an InAsSb/InGaAs SLS are confined to the |3/2, ±3/2> state of the InAsSb layer.

Thus in conclusion, magnetoluminescence experiments provide unique quantum well information over a wide range of semiconductor bandgap parameters, and are essential to characterize novel heterostructures containing narrow bandgap semiconductors.

ACKNOWLEDGMENTS

The authors wish to thank Dr. R. M. Biefeld, Dr. J. J. Klem, and Dr. S. K. Lyo for many collaborations involving this subject. This work was performed at Sandia National Laboratories and supported by the Division of Material Science, Office of Basic Energy Science, U. S. DOE, No. DE-AC04-94AL8500.

REFERENCES

1. Herman, M. A., Bimberg, D., and Christen, J., *J. Appl. Phys.* **70**, R1 (1991).

2. Osbourn, G. C., Gourley, P. L., Fritz, I. J., Biefeld, R. M., Dawson, L. R.,and Zipperian, T. E., "Principles and applications of semiconductor strained-layer superlattices,"*Semiconductors and Semimetals*, Edited by R. Dingle, Vol. **24**, Academic Press, NY, 1987, pp. 459-503

3. T. P. Pearsall, "Strained-layer superlattices," *Semiconductors and Semimetals*, Edited by T. P. Pearsall, Vol. **32**, Academic Press, NY, 1990, pp. 1-15.

4. Morkoc, H., Sverdlov, B., and Gao, Guang-Bo, *Proceedings of the IEEE* **81**, 493 (1993).

5. Jones, E. D., Lyo, S. K., Fritz, I. J., Klem, J. F., Schirber, J. E., Tigges, C. P., and Drummond, T. J., *Appl. Phys. Lett.* **54**, 2227 (1989).

6. Jones, E. D., Biefeld, R. M., Klem, J. F., and Lyo, S. K., "Strain and density dependent valence-band masses in InGaAs/GaAs and GaAs/GaAsP strained-layer structures," *Proceedings Int. Symp. GaAs and Related Compounds, Karuizawa, Japan, 1989,* Inst. Phys. Conf. Ser. No. **106**, 435 (1990).

7. Jones, E. D., Dawson, L. R., Klem, J. F., Lyo, S. K., Heiman, D., and Liu, X. C., "Magnetic-field dependent photoluminescence studies of InGaAs/GaAs strained-single-quantum wells," Proceedings, *The Applications of High Magnetic Fields in Semiconductor Physics (SEMIMAG-94)*, August 8-12, 1994, Boston, MA.

6. Jones, E. D., L. R. Dawson, Klem, J. F., Lyo, S. K., D. Heiman, and Liu, X. C., "Optical Determination of conduction and valence-band dispersion curves in strained-single-quantum wells," *Proceedings of the 22nd International Conference of the Physics of Semiconductors*, Vancouver, Canada, August 15-19, 1994.

8. Jones, E. D., Lyo, S. K., Klem, J. F., Schirber, J. E. and Lin, S. Y., "Ground-state in-plane light-holes in GaAs/AlGaAs structures," *Proceedings. Int. Symp. GaAs and Related Compounds, Seattle, USA, 1991*, Inst. Phys. Conf. Ser. No. **120**, 407 (1992).

9. Jones, E. D. and Wickstrom, G. L., "Laser measurement techniques-laser spectroscopy in semiconductors," *Proceedings Southwest Conference on Optics*, SPIE **540**, 362 (1985)

10. Lyo, S. K., Jones, E. D., and Klem, J. F., *Phys. Rev. Lett.* **61**, 2265 (1988).

11. Lyo, S. K., *Phys. Rev.* B **40**, 8418 (1989).

12. Lyo, S. K. and Jones, E. D., "Light-heavy-hole-mixing-induced selection rules for magneto-luminescence and valence-band energy dispersion in semiconductor quantum wells," *Electronic, Optical, and Device Properties of Layered Structures*, Edited by J. R. Hayes, M. S. Hybertsen, and C. R. Weber, Fall 1990 Meeting of the Materials Research Society, Boston, MA, pp. 271-274.

13. Lyo S. K., and Jones, E. D., "Valence-band Energy dispersion in modulation-doped quantum wells: Effect of strain and confinement on heavy- and light-hole mixing," *Proceedings Int. Symp. GaAs and Related Compounds, Seattle, 1991*, Inst. Phys. Conf. Ser. No. **120**, 583 (1992).

14. Singleton, J., Nicholas, R. J., Rogers, D. C., and Foxen, C. T. B., *Surface Science* **196**, 429 (1989).

15. Osório, F. A. P., Degani, M. H., and Hipólito, O., *Superlattices and Microstructures* **6**, 107 (1989).

16. Fritz, I. J., Schirber, J. E., Jones, E. D., Drummond, T. E., and Osbourn, G. C., "Hole transport and charge transfer in GaAs/InGaAs/GaAs single strained quantum well structures," *Proceedings of the Int. Symp. GaAs and Related Compounds, Las Vegas, Nevada, 1986*, Inst. Phys. Conf. Ser. No. **83**, 233 (1986).

17. Brugger, H., Müssig, H., Wölk, C., Berlec, F. J., Sauer, R., Kern, K., and Heitman, D., "Two-dimensional electron gas analysis on pseudomorphic heterojunction field-effect transistor structures by photoluminescence," *Proceedings Int. Symp. GaAs and Related Compounds, Seattle, 1991*, Inst. Phys. Conf. Ser. No. **120**, 149 (1992).

18. Lovejoy, M.L., Simmons, J. A., Ho, P., and Martin, P. A., *Appl. Phys. Lett.* 64, 3634 (1994).

19. Lyo, S. K. and Jones, E. D., *Phys. Rev.* B **38**, 4113 (1988).

20. Tigges, C. P., Jones, E. D., and Klem, J. F., "Line shape analysis of magneto-luminescence spectra of strained-layer single quantum well InGaAs/GaAs structures," *Bulletin Amer.Phys. Soc.* **34**, 830 (1989).

21. Kurtz, S. R. and Biefeld, R. M., *Appl. Phys. Lett.* **66**, 364 (1995).

22. Johnson, E. A., "Absorption near the fundamental edge," *Semiconductors and Semimetals*, Edited by R. K. Willardson and A. C. Beer, Academic Press, NY, 1967, Vol. **3**, pp. 154-253

23. Sundaram, C. G. M., Warburton, R. J., Nicholas, R. J., Summers, G. M., Mason, N. J., and Walker, P. J., *Semi. Sci. Tech.* **7**, 985 (1992

24. Varfolomeev, D. A. V., Seisyan, R. P., and Yakimova, R. N., *Soviet Phys. Semi.* **9**, 530 (1970).

598

GaAs/Si Optoelectronic Design and Development at Hiroshima University

S. Yokoyama, K. Miyake, T. Nagata, H. Sakaue*, S. Miyazaki*, Y. Horiike**,
A. Iwata*, T. Ae*, M. Koyanagi*** and M. Hirose

Research Center for Integrated Systems, Hiroshima University
* Department of Electrical Engineering, Hiroshima University
1-4-2 Kagamiyama, Higashi-Hiroshima 724, Japan

One of the major projects at the Research Center for Integrated Systems (RCIS), Hiroshima University, is a three-dimensional, optically-coupled common memory (3D-OCC memory), which can offer parallel data processing capability for a computer while keeping the merit of the conventional Neumann system. A second project is intelligent LSI with two-dimensional optical interconnection. In this paper, these subjects will be discussed in conjunction with the GaAs/Si hybrid integration technique.

1. INTRODUCTION

The RC time constant for the metal interconnects per unit length increases as the size is scaled down. Edge capacitance, which is not decreased with scaling, enhances this trend [1]. Furthermore, the currently increasing chip size makes the problem even more serious. However, optical interconnection has the potential to overcome these problems. Signal delay time in optical interconnection is at least one order of magnitude less than that in a metal interconnection of 1 μm design rule for signal transfer length of 1 cm, even when the delay times of the optical/electrical conversion devices (total 200 ps) are considered [2]. Other high functional aspects of the optical interconnection, for example, the signal delay time independent of the fan-out, signal transfer with electrical isolation, no interaction between propagating light beams, etc., are also attractive [3].

We have been studying a three-dimensional, optically-coupled common memory (3D-OCC memory) [4-7] and two dimensional intelligent LSI with optical interconnection [8-10]. In this paper, these studies are reviewed and discussed in conjunction with the GaAs/Si hybrid integration technique.

2. THREE-DIMENSIONAL, OPTICALLY-COUPLED COMMON MEMORY (3D-OCC MEMORY)

2-1. Operation Principle

A real-time parallel processing computer system using 3D-OCC memory is shown in Fig. 1 [4]. The 3D-OCC memory is composed of multilayered two-dimensional

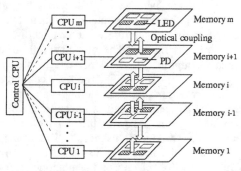

FIGURE 1. Parallel processing system with 3D-OCC memory.

LSI memory chips equipped with GaAs LEDs and photodetectors. The memory layers are optically coupled by an LED and a photodetector pair in each of the memory cells. Data is transferred in the vertical direction by optical coupling, with conventional memory operation in the horizontal planes. Each memory layer in the 3D-OCC memory is connected to its respective CPU. Therefore, data stored in the memory can be simultaneously

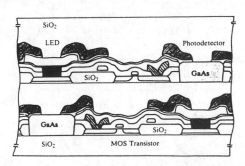

FIGURE 2. Cross sectional view of a 3D-OCC memory.

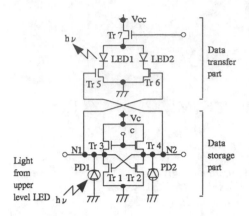

FIGURE 3. Memory cell circuit.

accessed by many CPUs without conflict. Even during data rewrite by the CPUs, simultaneous access is basically free from conflict. Consequently, very fast real-time parallel processing can be achieved in this system.

In Fig. 2, the cross sectional view of a proposed 3D-OCC memory is shown. In this figure, the GaAs LEDs and photodetectors are fabricated by heteroepitaxy on Si substrate, and then the Si substrate is thinned by chem-mechacical polishing. Therefore, you can hardly observe the Si substrate in the figure. The memory cell circuit, by which optical coupling in the 3D-OCC memory is performed, is shown in Fig. 3. The memory cell circuit can be divided into two parts; a data storage part and a data transfer part. The data storage part is the static RAM which includes two photodiodes. The data transfer part consists of two LEDs and their driving transistors. The data transfer part is connected to the nodes of the data storage flip-flop circuit. After irradiation of one of the photodetectors from the other layer, the potential of the node, where the irradiated photodetector is connected, is assured to be "low"

irrespective of the initial potential. Then, the LED connected to the "high" node of the data storage flip-flop emits light to the upper and lower memory layers. The emitted light is detected by the corresponding photodiodes of the other memory layers to achieve vertical data transfer. Computer simulation of the 3D-OCC memory using this circuit has indicated that the block data of 512 bits can be simultaneously transferred within 16 ns through four layers (see Fig. 4), which corresponds to a data transfer speed of 128 Gbits/sec [5].

2-2. Fabrication Procedure

A 3D-OCC memory test chip, on which GaAs LEDs are integrated using microbonding techniques, was fabricated using a 2 μm CMOS technology. Figure 5(a) shows the structure of the fabricated test chip [6]. A more recently developed test chip structure, which has two Si layers, is shown in Fig. 5(b). The Si LSI chip before and after hybrid integration of GaAs LEDs is shown in Figs. 6(a) and 6(b), respectively. In this chip, memory cell circuits, sense amplifiers, read/write circuits, precharge/equalize circuits and input/output circuits are integrated. The basic fabrication procedure is shown in Fig. 7. The GaAs LED wafers, n^+ substrate (N_D: ~10^{18} cm^{-3}, 350 μm in thickness)-n(50 μm)-p^+(N_A: ~10^{19} cm^{-3}, 100 μm) were first thinned to about 120 μm by polishing the n^+ substrate. Next, ring electrodes (1 μm-thick Au on 0.1 μm-thick In) for microbonding and a mesh electrode (0.25 μm-thick Au on 0.05 μm-thick Cr) for back electrodes of LEDs are formed using the lift-off technique. In the last step, LEDs are isolated by mesa etching in a NH$_4$OH+H$_2$O$_2$ (1:25) solution. The size of the GaAs LED is 50 μm x 70 μm. On the other hand, the Si LSI, on which the 2 μm CMOS

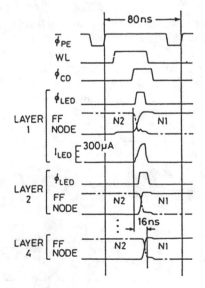

FIGURE 4. Simulated waveforms for the multilayer optical data transfer operation.

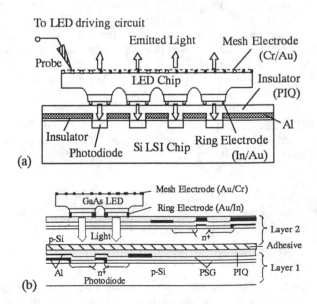

FIGURE 5. Structures of (a) prototype and (b) advanced 3D-OCC memory test chips.

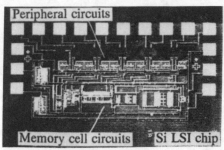

(a)

⌐ ⌐ 200μm

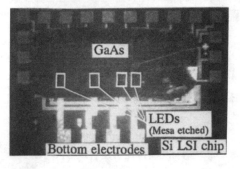

(b)

FIGURE 6. Micrographs of fabricated 3D-OCC memory test chip, (a) Si LSI chip and (b) after hybrid integration of GaAs LEDs.

circuits are fabricated, is covered by an aluminum mask (0.3 μm in thickness) except in the photodiode regions (the size of the photodiode is 20 μm x 40 μm) in order to reduce stray light. A polyimide layer (2 μm in thickness) is then spin coated on the aluminum mask for electrical insulation and surface flattening. The ring electrode (0.1 μm-thick Au on 0.9 μm-thick In) is deposited on the polyimide layer by the lift-off technique. Finally, the GaAs LED wafer is bonded on the Si LSI wafer using a newly developed three dimensional wafer aligner shown in Fig. 8 [7].

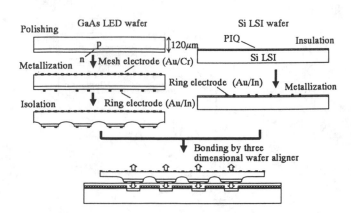

FIGURE 7. Process sequence for hybrid integration of LEDs on Si chips.

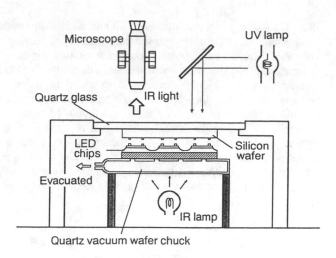

FIGURE 8. Three-dimensional wafer aligner.

2-3. Operation Experiment

The optical writing and electrical reading operations of both test chips shown in Fig. 5 (a) and (b) have been verified. Waveforms applied to the test chip and the predicted output signal are shown in Fig. 9. The test circuit consists of the memory cell circuit shown in

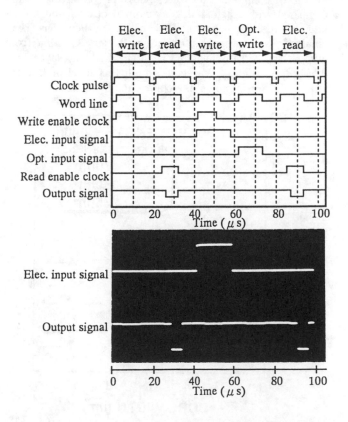

FIGURE 9. Operation waveforms in 3D-OCC memory test chip shown in Fig. 5 (b).

601

Fig. 3 and the peripheral circuits. A relatively long cycle time is used because of the limited performance of the logic analyzer. The electrical writing/reading operation for "low" is executed in the first two cycles, then the optical writing/electrical reading operation is executed in the next two cycles. In the optical writing/electrical reading operation, the electrically written "high" should be optically changed to "low" in the second write cycle as shown in the figure. Photographs of the observed oscilloscope waveforms are shown in the bottom figure. The waveforms are consistent with the predicted operation waveforms, confirming the basic function of a 3D-OCC memory test chip.

3. TWO-DIMENSIONAL OPTICAL INTERCONNECTION

3-1. Micron-Size Optical Waveguides

Optical waveguides consist of a SiN (refractive index of 2) core and SiO2 (refractive index of 1.45) cladding layers fabricated by conventional Si LSI technologies. Core width and thickness dependence of the propagation loss is shown in Fig. 10. The light propagation loss rapidly increases for core widths less than 3 μm and core thicknesses less than 0.47 μm (~wavelength in SiN). The light propagation loss for these micron-size waveguides is relatively large compared with that of the conventional optical fibers. So far, the best propagation loss of 0.6 dB/cm has been obtained for waveguides capped with an Al film (core thickness: 0.5 μm, width: 10 μm, cladding layers are 0.5~0.8 μm-thick). Waveguides with this level of loss are still useful for intrachip interconnection.

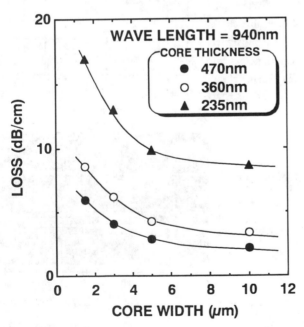

FIGURE 10. Core width and thickness dependence of the propagation loss.

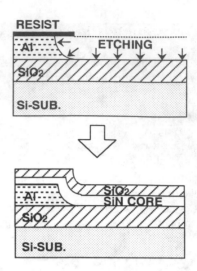

FIGURE 11. Fabrication process of Al micromirror used in vertical plane.

3-2. Micromirrors

Micromirrors which reflect the light from vertical to horizontal or reverse direction were fabricated from substrate Si, poly-Si and Al. For example, the fabrication procedure of the Al micromirrors is shown in Fig. 11. Al mirrors which change the light propagation direction from horizontal to downward have been also fabricated using a deformed photoresist mask [9]. The reflectivities of these micromirrors range from 60 to 80 %. Al corner micromirrors which are used in the horizontal plane have been also fabricated with various shapes (45° slanted, quarter circle, and right angle corner). The structure of the 45° slanted mirror is shown in Fig. 12. This mirror has the largest reflectivity of 50% and reflectivities of the right angle corner mirror and the quarter circle mirror were 10 and 25 %, respectively. Branched waveguides were also fabricated by using the Al corner mirrors [10].

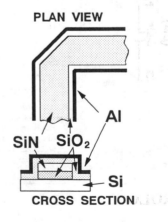

FIGURE 12. Al corner mirror used in horizontal plane.

3-3. Single-Chip Integration

GaAs LEDs, micromirrors, waveguides and photodetectors (the LED itself is used as photodetector) have been integrated on a single chip and the signal transfer from the LED to the photodetector has been verified. Figure 13 shows the structure of the fabricated test chip. First, substrate Si micromirrors are fabricated by wet-chemical etching and then the optical waveguides are fabricated. The sample surface is then planarized by a spin coated polyimide film. Finally LEDs and photodetectors are mounted on the chip by the microbonding technique [7]. Signal transfer chracteristics are shown in Fig. 14, where a good linear relationship between LED current and photodiode current is observed. The ratio of input to output currents is about 5×10^{-4} which is of the same order as the calculated transmission efficiency of 2.8×10^{-4}. The bottle neck for this small coupling efficiency is the low light incident efficiency into the waveguide from the LED (1/80), because the size of the LED ($1600\ \mu m^2$) is larger than that of the Si mirror ($20\ \mu m^2$). Therefore, the coupling efficiency can be significantly improved by shrinking the size of the LED.

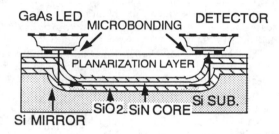

FIGURE 13. Structure of the fabricated optoelectronic test chip. Length of the waveguide is 4 mm.

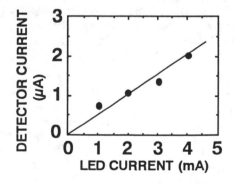

FIGURE 14. Output photodetector current versus input LED current.

4. METROLOGY USED IN THIS STUDY

4-1. Metrology for the 3-D Wafer Aligner

The bottom stage of the 3-D wafer aligner is controlled by the piezoelectric actuator, which has a minimum step of 5 nm. The alignment acuracy of this 3-D wafer aligner is, however, determined by the wavelength of the monitor light (940 nm), and is found to be $\sim\pm1\ \mu m$. Three capacitance gap sensors are used to adjust two substrates in parallel. Also three pressure gauges are set to measure the squeezing pressure between the samples to be stacked. The measurable force region is from 0.00 to 10.00 kg per each sensor and the resolution is 0.01 kg. The control of the pressure is very important, since too small pressure results in imperfect bonding and too large pressure induces the damage in the Si devices or GaAs LED's or generates cracks in the Si or GaAs wafers.

4-2. Metrology for the Characterization of Optical Waveguides

The characterization system for the optical waveguides is shown in Fig. 15. The light from the waveguides is monitored by the CCD camera attached to the optical microscope. The output electrical signal from the CCD camera is analyzed by the image processor installed in a personal computer. The performance of the measurement system is checked by using a small lamp. In Fig. 16 the output signal of the image processor is plotted as a function of the reciprocal of the square of the distance between the light source and the photodetector. A good linear relation between them is observed. It should be noticed that the lamp can be regarded as a point source, since the size of the lamp is small enough compared to the distance between the lamp and the detector. The measurement of the propagation loss of the waveguide was carried out in this linear region.

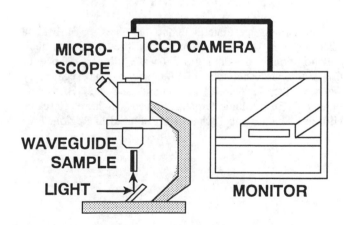

FIGURE 15. Characterization system for the optical waveguides.

603

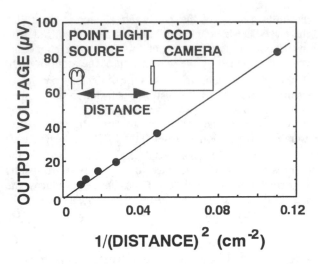

FIGURE 16. Output signal of the image processor as a function of reciprocal of the square of the distance between the light source and the photodetector.

5. CONCLUSIONS

The studies of three-dimensional, optically-coupled common memory (3D-OCC memory) and two dimensional optical interconnection in Si chips have been reviewed. The basic operation of the fabricated 3D-OCC memory test chip has been confirmed. The signal transfer in two dimensional optical interconnection has been also verified. Microbonding of GaAs devices on Si chips has been achieved using a newly developed 3D wafer aligner. Also, the metrologies for the 3-D wafer aligner and for the characterization of 2-D optical waveguide have been discussed.

ACKNOWLEDGMENTS

The authors are grateful to the members in Hiroshima R&D Center, Mitsubishi Heavy Industries LTD, for their collaboration in developing a 3D wafer aligner. A part of this study was supported by the Science and Technology Agency, the Japanese Government and the Hiroshima Industrial Technology Organization as a Joint Research Utilizing Science and Technology Potential in Region.

REFERENCES

** Present affiliation : Toyo University
*** Present affiliation : Tohoku University

1. Sze S.M., *Physics of Semiconductor Devices*, 2nd ed., John Wiley & Sons, Canada, 1981, Ch 12, pp. 501-503.
2. Iwata A., "Optical Interconnection for ULSI Technology Innovation", *Optoelectronics-Devices and Technologies*, **9**, 39-54 (1994).
3. Hayashi I., "Optoelectronic Devices and Material Technologies for Photo-Electronic Integrated Systems", *Jpn. J. Appl. Phys.* **32**, 266 (1993).
4. Koyanagi M., Takata H., Maemoto T. and Hirose M., "Optically Coupled Three-Dimensional Common Memory", *Optoelectronics-Devices and Technologies*, **3**, 83-98 (1988).
5. Koyanagi M., Takata H., Mori H. and Iba J., "Design of 4-kbitX4-Layer Optically Coupled Three-Dimensional Common Memory for Parallel Processor System", *IEEE J. of Solid-State Circuits*, **25**, 109-116 (1990).
6. Miyake K. Namba T., Hashimoto K., Sakaue H., Miyazaki S., Horiike Y., Yokoyama S., Koyanagi M. and M. Hirose, "Fabrication and Evaluation of Three-Dimensional Optically Coupled Common Memory", *Jpn. J. Appl. Phys.* **34**, No. 2B, in press, (1995).
7. Miyake K., Tanaka T., Etoh T., Tsuno M., Yokoyama S. and Koyanagi M., "New Ram-Bus Memory System with Interchip Optical Interconnection", *Jpn. J. Appl. Phys.* **33**, 848-851 (1994).
8. Nagata T., Tanaka T., Miyake K., Kurotaki H., Yokoyama S. and Koyanagi M., "Micron-Size Optical Waveguide for Optoelectronic Integrated Circuit", *Jpn. J. Appl. Phys.* **33**, 822-826. (1994)
9. Nagata T., Namba T., Kuroda Y., Miyake K., Miyamoto T., Yokoyama S., Miyazaki S., Koyanagi M. and Hirose M., "Single-Chip Integration of Light-Emitting Diode, Waveguide and Micromirrors", *Jpn. J. Appl. Phys.* **34**, No. 2B, in press (1995).
10. Yokoyama S., Nagata T., Namba T., Kuroda Y., Doi T., Miyake K., Miyazaki S., Iwata A., Ae T., Koyanagi M. and Hirose M., "Optical interconnection on silicon LSI chips", *Proceedings of Photonics West '95, San Jose*, to be published (1995).

Rapid Non-destructive Scanning of Compound Semiconductor Wafers and Epitaxial Layers

Carla J. Miner

Bell Northern Research, P.O. Box 3511, Station C, Ottawa, ON Canada K1Y 4H7

This review describes the rapid whole wafer assessment tools needed by the compound semiconductor industry to control substrates and epitaxial layers to the exacting standards set by advanced device designs. Appropriate sampling strategies and general measurement requirements for these applications are discussed. A comparison of the major non-destructive scanning techniques in use today demonstrates the great progress made in the field thus far and outlines the remaining challenges.

INTRODUCTION

The performance required of many of the advanced semiconductor devices under active development today can only be realized if the materials from which the devices are fabricated can meet increasingly stringent specifications. As growth technology advances to meet the new demands, the characterization methods that support its development are under continuous pressure to be faster, more accurate, less expensive and more adept at accessing information from complex structures. To set the stage for a comparison of non-destructive scanning techniques, this review first discusses the different characterization needs involved in the fabrication of advanced compound semiconductor devices, the attributes of methods best suited to meeting those needs and the parameters that need to be measured.

CHARACTERIZATION NEEDS

Often, the product specification and therefore the device design requires that the properties of the substrate or epitaxial layers be controlled to a degree that is at the very limit of the current growth capabilities. When this is the case, frequent characterization provides the growers with the necessary feedback to correct the gradual run-to-run changes in the grown material properties before they accumulate and exceed the design tolerances. The tighter the specifications become, the more frequently the material properties must be checked, until in the extreme, *in situ* monitoring is required to correct slight deviations even while the material is being grown. As yet, however, *in situ* monitoring can only measure a limited number of material properties on a restricted portion of the wafer. Post growth feedback can measure more parameters and also assess uniformity.

Post growth characterization also plays an important feedforward role. Because processing and device testing costs are high and the cycle times lengthy, it is cost effective to discard wafers whose properties are out of specification early in the process sequence. For this purpose, it is only necessary to establish whether the wafers are similar enough to the last ones processed. If these metrics can be calibrated to those used in the fabrication line, the results can also be used to tailor processing conditions such as etch depths or diffusion times.

Materials characterization is also critical to the designers. After fabrication and testing, knowledge of the material properties of specific wafers can be combined with the measured device results to improve the device design models. To date, most models have relied upon relatively simple expressions to describe the dependence of material properties on factors such as composition and quantum well thickness. Typically these expressions have been derived from data with large measurement uncertainties[1,2]. The designers need precise relative measurements to overcome the model inaccuracies.

To satisfy these three requirements, the assessment techniques must be accurate (or at least reproducible), precise and easily calibrated. The degree of precision required is determined by the sensitivity of the device performance to the property measured. Ideally, assessment tools should directly measure the parameters most relevant to the device performance. In contrast, the most practical techniques often produce indirect results that must be correlated with device data. In addition, characterization techniques should give statistically significant results: i.e. the data should be quantitative and be measured at enough locations on the wafer to give an accurate description of the whole wafer[3]. A final consideration is cost, both in terms of the initial investment and the on-going resources needed to generate the data.

A non-destructive measurement, made right on the product wafer, has always been recognized as being a worthy goal. Given the current tight specifications, it is now essential. In the past, devices were fabricated on part wafers, a portion of which could be set aside for destructive testing. Full wafer processing is now needed to control critical dimensions, so that this approach is no longer appropriate. Another approach is the periodic use of test structures or destructive sampling to establish the characteristics of a batch. This is successful only if the specification window is many times larger than the run-to-run reproducibility. Consider Figure 1 where the photoluminescence peak wavelength data is plotted as a function of run number for two different device structures grown by MOCVD. For structure A, corrections to the growth conditions are not made for several growth runs because the specification window is relatively large (± 15 nm). Batch monitoring using calibration wafers would suffice for this structure. For structure B, the specification window is one half as large (± 7.5 nm), and corrections have to be made after each growth run on the basis of the analysis results from the preceding run. Without non-destructive testing, a calibration wafer would have to be grown after each product run. This is clearly an expensive option.

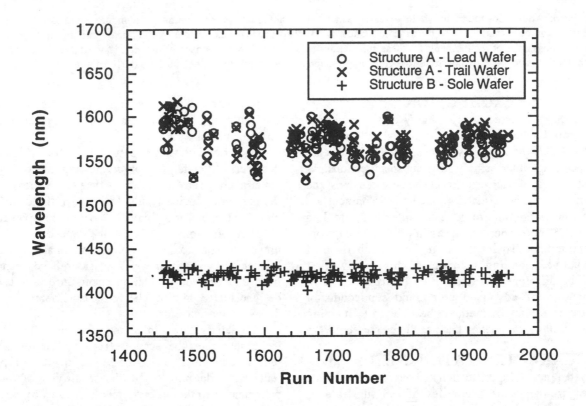

FIGURE 1. The photoluminescence peak wavelength plotted as a function of growth run number for two InGaAsP MQW structures grown in the same MOCVD reactor. Two structure A wafers are grown in the same run, while only a single wafer is used for the growth of structure B. The wavelength tolerance is much tighter for structure B, which makes it necessary to adjust the growth conditions every run. For structure A, the adjustments are made only when the measurements indicate that the wavelength is out of specification.

The use of small and potentially less costly witness pieces in epitaxial reactors is less than satisfactory. It is nearly impossible to ensure that the witness piece and product wafer experience the same gas flow or beam flux, thermal environment and degree of susceptor and reactor wall interactions: all of these affect layer properties. Moreover, the differences cannot be taken into account by the use of a tooling factor, since they continuously change depending on the details of the previous growth runs. This can be seen in the data of Figure 1 where the difference in the wavelength of the two wafers grown simultaneously for structure A varies significantly. Finally, the use of small witness pieces is problematic in growth environments such as MOCVD, where the edge enhanced cracking of PH_3 can perturb layer thickness and alloy composition over distances of 5 to 10 mm[4].

The need for a statistically significant number of measurements across the wafer constrains the choice of methods to whole wafer imaging or scanning techniques. It has been argued that since the run-to-run changes strongly affect the average value of the material parameters but affect their spatial distribution to a lesser degree, a well characterized calibration wafer and a single point on the product wafer are enough. While this is true in many cases, significant exceptions to the rule do occur. For instance, Ga or In desorption during growth of AlGaAs or InGaAs by

MBE or CBE at high temperatures affects affect growth rate and composition[5,6]. Because the dependence on temperature is exponential, small local variations in temperature cause layer non-uniformity to vary from run to run. As for MOCVD, we have observed that the patterns of thickness and composition change subtly from run to run due to the build up of susceptor deposits. Finally, if the defect density in the epitaxial layers is the parameter of interest, each wafer must be mapped: defects rarely appear in the same place in consecutive runs.

These considerations lead to the conclusion that non-destructive whole wafer testing is a necessity for advanced device fabrication. A short list of properties that need to be determined is presented in Table 1. These were identified through a SEMI sponsored survey of industry users[3]. While many users identified properties not on this list that significantly impact the performance of their devices, these are common to the majority. The second column lists related properties that may be more easily measured. They are more likely to be used for feedback and feedforward purposes. The third column lists those situations where the interpretation of the measured data in terms of the primary property is made more difficult by confounding effects. For example, while the transition energy or lattice mismatch may be sufficient to pin down the composition of an undoped bulk ternary, or the combination of the two can identify an undoped bulk quaternary, the interpretation is not straight-forward if the layer is heavily doped, very thin or strained. In real device structures any of these conditions may be present.

TABLE 1. Minimum Set of Epitaxial Layer Properties to be Measured

Basic Property	Related Properties	Special Consideration
alloy composition	band gap transition energy lattice mismatch	quantum wells, strained layers and heavily doped materials
thickness	optical thickness quantum well confinement energy	graded or rough interfaces
doping profile	mobility and carrier concentration sheet resistance or resistivity dopant profile	multi-layer structures semi-insulating materials dopant activation less than 100%
defect density	particulate density surface or interface roughness crystallographic defect density impurity levels carrier lifetime killers	

CHARACTERIZATION METHODS

In this section, the most widely used non-destructive whole wafer mapping techniques will be discussed. They are divided into categories based upon the nature of probe: optical, x-ray or electrical.

Optical Characterization Methods

Scatterometry

Visible light scatterometry is probably the most frequently used qualitative assessment method. In the simplest implementation, wafers are observed by the unaided eye or with a low power microscope under strong oblique lighting. Surface roughness or particulates stand out as bright points on a dark background. In commercial particle counters, a laser is rapidly scanned over the sample and all but the specularly reflected beam is collected. Since particles in real applications have various reflectivities and irregular shapes, the output of these instruments is in terms of equivalent scattering centers[7]. The amount of light lost to scattering from uniformly dispersed features too small to be resolved as individual particles is expressed as haze[8].

Surface roughness, film residues and the presence of very small particles all contribute to the haze signal. At the moment there are significant discrepancies between haze measurements made with different wafer scanning instruments, especially for III-V materials. This is partially due to the limitations of a single wavelength, single angle of incidence instrument. By considering the variation of scattered light intensity with angle, wavelength and polarization, more reliable information about the nature of the scatter centers, including haze, can be extracted[9].

Absorption Spectroscopy

Infrared absorption mapping and spectroscopy are mature characterization tools which have found ready application within the III-V materials community. Variations in the transmission of sub-band gap light are measured as a function of wavelength and position on the wafer and are used to identify dopants or deep levels and estimate their concentration. Infrared absorption mapping has been used to map the concentration of free carriers[10], specific dopants (e.g. carbon in GaAs[11], Fe in InP[12]) and deep levels (e.g. EL2 in semi-insulating GaAs[13]) and the distribution of implant damage[14]. The measurements are often done at low temperature to take advantage of reduced spectral broadening or to change the occupancy of the levels taking part in the transition. Because the absorption cross section for many of these transitions is small and their concentration is usually low, absorption mapping is better suited to the analysis of substrates than of epitaxial layers. Care must be taken to avoid errors due to scattering from surface roughness and subsurface defects. Changes in transmission due to interference with light reflected from the back of the wafer can lead to artifacts in a measured absorption map.

Birefringence mapping is a related technique where the polarization state of the transmitted light is used to measure strain induced birefringence within the wafer[15]. Unlike a simple polariscope measurement, where the variation in the light transmitted is measured at a fixed orientation of the crossed polarizers with respect to the crystal axes, the variation in transmitted light is measured as the polarizer and analyzer are rotated. Both the principal axis of birefringence and the phase retardation is then calculated for each point. Two dimensional maps of residual strain can be calculated from this data. This has been used to monitor improvements in substrate preparation methods, especially post growth annealing steps. A modification of the technique has also been used to image local fields around crystal defects[16].

Reflectance Spectroscopy

Reflectance spectroscopy and mapping make use of optical interference to infer the thickness of heteroepitaxial layers. In a typical implementation, the relative intensity of light reflected at normal incidence from a sample consisting of at least two different compositions is measured as a function of wavelength and position on the wafer[6]. If the indices of refraction of the layers are sufficiently different, then the total reflected light will be modified by optical interference between light reflected at the top and bottom surfaces of the layers. For the simplest structures, analysis of the spectra is straight forward[17]; however, when there are many layers or compositionally graded ones, more sophisticated data reduction approaches are required[18]. Since the change in index of refraction due to composition or doping variation across the wafer is usually a hundred times or more smaller than the variation in epitaxial layer thickness, contours of equal reflectance can be used to indicate lines of equal layer thicknesses. A combination of analyzed reflectance spectra at selected points and a reflectance map can be used to generate thickness maps with high spatial density and a precision of 1 part in 1000 (See Figure 2).

Data extraction is greatly simplified if the reflected light intensity in the maps and spectra corresponding to the extrema can be identified. Thus the method is most suitable for mapping epitaxial layers with thicknesses greater than ~1 μm and an absolute thickness variation greater than 200 nm to ensure that at least one pair of extrema are present.

The main limitation to the accuracy of the technique is the uncertainties in the indices of refraction. The index of refraction for a particular III-V compound is typically estimated from empirical data having uncertainties on the order of one part in three hundred[2]. Additional limitations in the translation of fixed wavelength reflectance maps into thickness maps are the assumptions that the thickness varies slowly on the scale of the spatial sampling grid and that the scattering due to surface or interface roughness is minimal, or at least is spatially uniform. These assumptions are not valid in patterned samples or those with poor growth morphology.

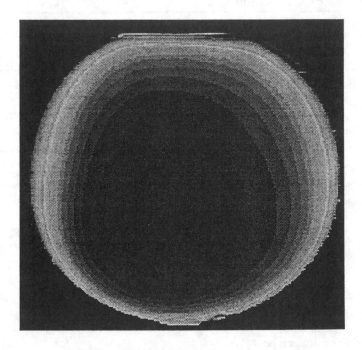

FIGURE 2. Thickness map of a nominally 1.1 μm thick layer of InP on 3.5 μm lattice matched InGaAs layer on a 50 mm diameter InP substrate. The data displayed correspond to grid points spaced every 200 μm (a total of 40,000 points) and cover a range of thickness from 0.80 μm to 1.15 μm.

Ellipsometry

Ellipsometry is a powerful optical assessment method where the polarization state of light reflected from the sample is analyzed to extract both the real and imaginary component of the complex index of refraction[19]. The wavelength or the angle of incidence may be varied to extract more complete data. From this data and detailed models, layer composition and thickness can be deduced. Very thin layers, with thicknesses on the order of the native oxides, can be readily measured. Spectroscopic ellipsometry has been used to measure the variations in the complex dielectric function associated with specific transitions such as higher energy states in quantum wells and thereby assess interface quality[20].

With the advent of computer controlled instruments, data acquisition and data reduction can be executed rapidly enough for the simpler form of this technique to be considered as one of the routine characterization methods. Commercial systems for mapping thin film dielectrics on whole wafers already exist. The method is best suited to the analysis of single dielectric or semiconductor layers grown on well characterized substrates. The extension to multi-layer structures is more problematic, given the need to know the optical constants of everything below the layer being analyzed, which is treated as an effective substrate. Approaches based on modeling to derive the effective substrate optical constants have been notoriously susceptible to error due the presence of thin surface or interfacial layers. *In situ* spectroscopic ellipsometry overcomes this limitation by making continuous measurements as the layers are being deposited, thereby establishing the optical constants for the effective substrate as growth proceeds[21]. As yet, these *in situ* techniques do not typically analyze more than one point per wafer.

Photoluminescence

Photoluminescence (PL) is a method with widespread application to III-V materials analysis. Above band gap light, usually from a low power laser, is directed onto the sample and causes electron hole pairs to be generated. After rapidly thermalizing to the lowest energy states, these pairs recombine, either non-radiatively at defects, surfaces and interfaces, or radiatively across the band gap or via inter gap levels associated with dopants or defects. At room temperature, under steady state conditions, the PL spectrum peak occurs very close to the band gap energy and its intensity can be a good relative measure of defect density[22]. Scanning PL maps of peak wavelength are routinely used to infer the variation in layer composition[6,23], quantum well thickness[6] and the expected device emission wavelength[6]. Figure 3 shows PL peak wavelength data for structure A described in Figure 1 and is typical of what may be used for *in line* characterization of a wafer prior to optoelectronic device processing.

Maps of PL intensity have been used to study defect distributions in substrates and epitaxial layers[24,25]. Often the distribution of areas with reduced intensity will lead to the identification of their origin, e.g. polishing damage, cleaning residues, misfit dislocations and substrate related defects. In the example shown in Figure 4, an isolated particle, that was present on the wafer prior to epitaxial growth, has acted as a source of the contamination which increased non-radiative recombination over an area much larger than the particle itself. The image also shows that the contamination has been transported preferentially in the direction of the gas flow.

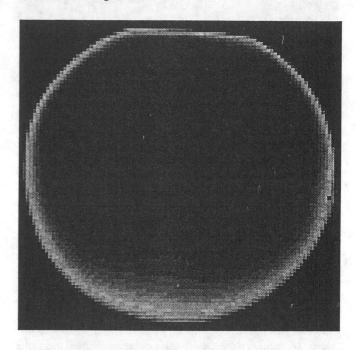

FIGURE 3. Photoluminescence peak wavelength map for one of the structure A wafers described in Figure 1. The total range of wavelengths depicted is 100 nm, which corresponds to a standard deviation of 7.5 nm over the center 40 mm of this 50 mm diameter wafer. A total of 9000 spectra were acquired to create this map. The pattern is characteristic of the gas flow distribution in the MOCVD reactor used.

Under low excitation measurement conditions, changes in PL intensity can be quantitatively can be related to such device properties as p-i-n photodiode leakage currents[25] and field effect transistor backgating[26]. High excitation conditions reduce the sensitivity of the technique to the density of non-radiative recombination centers. The non-linear dependence of the PL intensity of double hetereojunctions on excitation power can also be used to characterize the junction in a manner that is relevant to optoelectronic device performance[27].

The effective spatial resolution of the PL technique is controlled by the laser wavelength, and the excitation and collection optics. For commercial PL systems, it can range from ~1 to >250 μm. Spatial resolution may be traded off to increase the detected signal strength and thereby reduce the data acquisition time. The 38,000 valid data points shown in Figure 4 were acquired in under 35 minutes. The spot size was 25 μm. Optimized scanning monochromators and parallel detection have both been used to reduce spectral acquisition times. Each of the 200 point spectra used to compose Figure 3 were recorded in ~ 3.5 seconds.

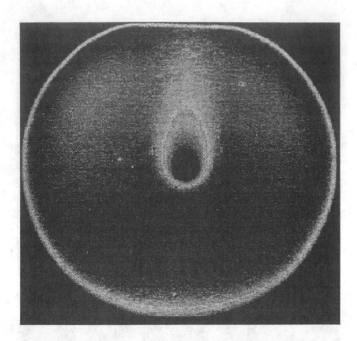

FIGURE 4. Photoluminescence map of intensity at a fixed wavelength of a InGaAsP MQW structure grown by MOCVD. The gas flow direction is from bottom to top of this figure. The area of reduced PL intensity at the center of this 50 mm diameter wafer is associated with contamination on the wafer prior to layer growth.

Although PL provides valuable information at room temperature, the samples must be cooled to cryogenic temperatures to resolve and identify separate dopant and defect related transitions[28]. The lower temperatures also increase the total PL intensity by freezing out non-radiative recombination paths. Often, this is the only feasible way to examine intrinsically dim samples such as semi-insulating GaAs and InP substrates. Except when the information cannot be obtained in any other way, the time and handling needed for low temperature PL makes the technique less suitable for routine assessment applications. Recent advances using load-locked cryostats reduce these concerns[29].

Photoreflectance

Photoreflectance (PR) is one of the modulation spectroscopies, where the reflectance spectrum of the sample is measured as it undergoes a periodic, band-structure altering perturbation[30,31]. In the case of PR, a chopped, low power, laser generates varying densities of electron hole pairs which in turn modulate the internal electric fields. The reflectance signal then contains both DC and AC components, related to the dielectric function and its derivative respectively. Details of the band structure can be calculated from the modulated reflectance spectrum, but for the purposes of rapid material assessment, the main interest

is the wavelength of the features associated with specific optical transitions. Maps of epitaxial layer composition have been inferred from peak shifts determined by PR[32]. One of the advantages of PR over PL is that even higher energy transitions and non-radiative ones can be clearly identified. In the PR spectrum of a heterojunction bipolar transistor structure shown in Figure 5 both GaAs and AlGaAs layers give rise to separate features, whereas in PL only the lowest energy transition would have been seen.

PR has another advantage that makes it unique. When a large electric field is applied to a semiconductor, either externally or through internal junctions, it causes predictable variations in the dielectric function at energies above band gap. These are known as the Franz Keldysh Oscillations[33] and show up in PR spectra as a series of extra extrema to the high energy side of main transition. From their positions, the internal electric field of the sample can be assessed non-destructively. The electric field itself may be the parameter of interest but often it is used as a sensitive monitor of doping levels, interface abruptness and the surface pinning voltage. Care must be taken, however, to ensure that the optical power in the pump or probe beam is not so great as to significantly alter the sample field by photo-voltaic effects[33].

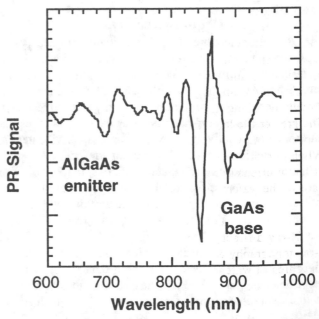

FIGURE 5. PR spectrum of a heterojunction bipolar transistor structure. The composition of the AlGaAs emitter can be determined by the position of the feature near 700 nm while the base-collector electric field may be estimated from the positions of the Franz Keldysh Oscillations immediately to the left of the GaAs transition at 870 nm.

The main limitations of PR are the relatively weak signals which necessitate long scan times (> 30 minutes) or large spot sizes (> 5 mm^2) to give acceptable signal to noise ratios and the complexity of the spectra which are fitted to standard shapes for interpretation. There is always a concern that the fit parameters may not be unique.

X-Ray Characterization Methods

X-ray Topography

X-ray topography makes use of x-ray diffraction to detect departures in either sample lattice constant (e.g. due to dopant incorporation or local strain) or lattice orientation (e.g. that associated with crystal defects). Single crystal techniques have largely given way to double crystal methods where an asymmetric reflection from a suitable first crystal produces a large non-divergent beam with which to expose the test wafer[34]. The low divergence heightens sensitivity to small lattice constant or orientation changes and the wide beam eliminates mechanical scanning and the associated image artifacts. The spatial resolution of the method is limited by the grain size in the photographic emulsion used (typically less than 1 μm), and is usually traded off against film speed.

Alternatively known as double crystal diffraction (DCD) topography or asymmetric crystal topography (ACT), these methods have been routinely used to image substrate defects such as low angle grain boundaries, inclusions and grown-in dislocations[35], and induced defects such as misfit dislocations associated with epitaxial layer mismatch[36] and slip dislocations caused by thermal processing[37]. Lattice plane curvature induced during boule growth, polishing or other handling steps is clearly revealed. The greater the lattice distortion, the more restricted is the portion of the wafer that satisfies the diffraction condition. When curvature is present, separate exposures at several different orientations are needed to reveal the defects in all parts of the wafer.

X-ray Diffractometry

X-ray diffractometry is the quantitative complement to x-ray topography where the x-ray intensity diffracted from a local area of wafer is measured as a function of the angle of incidence (rocking angle). Double crystal diffraction (DCD) is the most frequently used form for rapid substrate and epitaxial layer evaluation[38]. If two layers of distinct lattice constants are present within the sampling volume, then the x–ray rocking curve will consist of two peaks whose separation is a sensitive measure of the relative lattice mismatch (1 part in 10^5). Thus DCD mapping can be used to infer variations in ternary layer composition (See Figure 6). For quaternaries, the DCD peak separation map must be analyzed in conjunction with a map of another parameter that also depends on composition (e.g. PL wavelength).

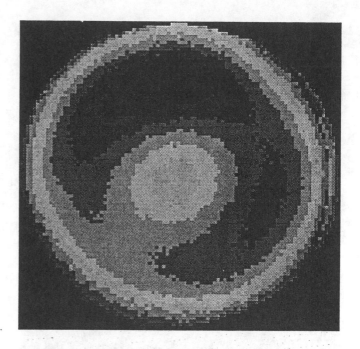

FIGURE 6. Map of InGaAs composition derived from DCD peak separation. The full data range covers In mole fraction from 0.545 to 0.555. The layer was grown on a 50 mm diameter InP substrate by chemical beam epitaxy. The departure from the circular symmetry expected for a rotated wafer was caused by enhanced indium desorption at a local hot spot.

For substrate evaluation, the peak width has been used to infer defect density, both of grown-in dislocations and polishing related defects. The rocking curves also broaden when the x-ray spot covers regions of different orientation[24,39]. Thus DCD maps of peak width or peak intensity can be used to quantitatively assess the percentage of the wafer affected by low angle grain boundaries. The severity of the low angle grain boundaries is given by the degree of peak broadening or equivalently, the shift in peak angle measured on either side of the defect. Unlike the case of an x-ray topograph, the rocking curve map is not limited by curvature: the whole wafer is displayed. Lattice distortion perpendicular to the rocking axis is quantified by the shift in the rocking curve peak angle. By combining the peak angle maps for two orthogonal rocking axes directions, the two dimensional distribution of lattice warp may be calculated.

Spatial resolution in x-ray diffractometry is limited by the divergence of the beam diffracted from the first crystal and the width of the slits used. With care, a spot under a 1 mm square can be produced on commercial equipment. The $K_{\alpha 2}$ diffracted beam must be blocked to attain a single spot. For rapid mapping applications, high speed silicon detectors are used which enable 100 point rocking curves to be acquired in only a few seconds, but they typically suffer from limited dynamic range (<500:1). To analyze rocking curves with peaks of widely differing maximum intensities it is necessary to use alternative detectors, such as scintillation counters, which have, at best, only 20% of the acquisition speed of a silicon detector.

Electrical Characterization Methods

R-F and Microwave Absorption

Eddy current mapping is a widely used method for the non-contact assessment of semiconductor conductivity, based upon the absorption of radio frequency power by free carriers[40]. For the typical conductances of substrates and epitaxial layers (0.1 $\Omega/\square$ to 3 k$\Omega/\square$), the workable compromise between spatial resolution and sensitivity is made with r-f frequencies between 500 kHz and 10 MHz and coil transducer diameters of 10-15 mm. Several commercial instruments based on this principle have been marketed.

At microwave frequencies, 5-30 GHz, a waveguide is used to transfer energy to the sample and the reflected power is used to determine the amount absorbed[41]. A waveguide probe can be much smaller than that used in eddy current mapping (typical waveguide diameter: 500 μm). By applying a magnetic field during the measurement, the mobility may also be assessed. This method is restricted to samples with mobility greater than 2000 cm^2/Vs. Photo-induced microwave absorption mapping combines the use of above band gap and sub band gap light to map the technologically significant deep levels in semi-insulating materials[42].

Since both technique are sensitive only to the a-c conductivity of carriers within the field, they can be used for routine monitoring of the most conductive layers of complete device structures as well as substrates and single layers.

Time Domain Charge Measurement

Semi-insulating III-V materials present a special characterization challenge. There are too few carriers for the application of the infrared, r-f or microwave absorption techniques. Time domain charge measurement (TDCM) has recently been developed to measure the resistivity of semi-insulating material non-destructively[43]. In TDCM, the wafer is placed between a small shielded probe and a large conductive chuck and a small voltage pulse is applied. This physical configuration corresponds to an equivalent circuit containing two capacitors: one with air and the other with semiconductor as dielectrics. The semiconductor capacitor is shunted with a large, but not infinite resistor, which bleeds off the charge on this capacitor after the initial voltage step. The resulting charge transient monitored at the probe is

analyzed to deduce the resistivity of the sample from the decay constant. One of the advantages of the technique is that no adjustable parameters are required to extract the resistivity value other than the static dielectric constant of the material. The technique has been applied to a variety of samples with resistivity greater than 10^5 Ω-cm[39,44]. Figure 7 shows typical data for semi-insulating InP.

The spatial resolution is limited by the probe size and detection electronics to approximately 1 mm. The data acquisition speed is not limited by the decay time constant (<200 μs), but rather the time required to fit the transient (~1 s). Since the activation energy associated with semi-insulating materials is 1 to 2 eV, the temperature of the sample must be precisely controlled and be uniform to better than 0.1°C.

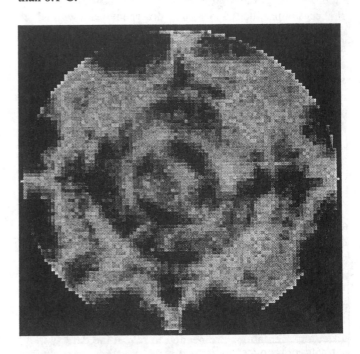

FIGURE 7. Map of the resistivity of semi-insulating Fe doped InP substrate. The full data range shown is 5 to 25 MΩ-cm. The ring like features are related to fluctuations in the Fe concentration during the growth of the boule.

CONCLUSIONS

Various forms of non-destructive mapping, either used alone or in combination, are currently satisfying a large portion of the base characterization needs of the compound semiconductor industry: namely composition, thickness, doping profile and defect density. Alloy composition can be non-destructively monitored using ellipsometry, PL, PR and x-ray diffractometry. However, the relationship between the properties measured by these techniques and composition is indirect and the accuracy of the correlations needs to be improved. Layer thickness can be measured with the required precision by reflectance spectroscopy and ellipsometry. Here, the accuracy of both methods is limited by incomplete knowledge of the indices of refraction. Doping levels can be monitored by infrared absorption in substrates and by r-f and microwave absorption in both substrates and epitaxial layers. Resistance can be measured by the latter two as well as by the time domain charge measurement method. However, doping profiles still cannot be mapped in any practical manner. The situation with measuring defect density appears to be under control. Defects are now routinely monitored by a range of techniques including scatterometry, infrared absorption, photoluminescence and x-ray methods. Continued study is required to make sure that the defects that are being measured are the ones that matter to device performance or reliability. Table 2 summarizes the techniques, their applications and limitations.

Table 2 also comments on the relative speed, accuracy and precision of the different methods. The accuracy and precision is judged with respect to the characterization needs for the particular parameter measured. Data acquisition speed is categorized as being excellent for high resolution wafer maps that take only a few minutes, good to very good if they take less than an hour, but only fair if they can be completed automatically overnight. An overnight scan is still acceptable when a high resolution map of a critical parameter is worth the delay.

It is clear that much progress has been made towards meeting the characterization needs of the compound semiconductor industry. Non-destructive mapping has become widely used and has enabled considerable improvement in the quality, reproducibility and uniformity of substrates and epitaxial layers. Many of the current methods still have significant limitations to address, so that technique development will undoubtedly continue.

TABLE 2. Comparison on Characterization Techniques for Scanning Applications

Technique	Application	Accuracy / Precision	Speed	Comments and Limitations
scatterometry	defects particles	good to excellent	excellent	haze not well understood, especially for III-Vs
absorption spectroscopy	defects, dopants impurities, strain	good	good	not sensitive enough for use on epitaxial layers
reflectance spectroscopy	layer thickness (>500 nm)	excellent	very good	limited by the accuracy of the layer indices of refraction
ellipsometry	index, thickness transition energy	excellent	fair to good	limited by the accuracy of the substrate index of refraction fit may not be unique
photoluminescence	transition energy defects	excellent	very good	limited to direct band gap materials
photoreflectance	transition energy electric field	excellent	fair	needs to be faster fit may not be unique
x-ray topography	defects, curvature doping variations	very good	very good	only qualitative measure of curvature and doping
x-ray diffractometry	lattice mismatch curvature, defects	fair to very good	fair to good	need better dynamic range and speed at same time
r-f absorption	resistance ($\rho < 10^4$ Ω-cm)	very good	fair to good	spatial resolution needs to improved
microwave absorption	mobility, resistance trap concentration	good	fair	limited to samples with mobility > 2000 cm^2/Vs
time domain charge measurement	resistance ($\rho > 10^5$ Ω-cm)	very good	good	good temperature control and uniformity essential

REFERENCES

1   Pollack F., *Properties of Aluminum Gallium Arsenide*, EMIS Data Review Series #7, London: INSPEC IEE, 1993, ch. 4.
2.   Adachi S., *J. Appl. Phys.*, **53**, 5863-5869 (1982).
3.   Miner C.J., *Inst. Phys. Conf. Ser.*, **135**, 147-156 (1993).
4.   Knight D.G., Miner C.J. and Watt B., J. Crystal Growth, **107**, 221-225 (1991).
5.   Miner C.J., Watt B., Moore W.T., Majeed A. and SpringThorpe A.J., *J. Vac. Sci. Technol.*, **B11**, 998-1002 (1993).

6. Miner C.J., *Semicond. Sci. Technol.*, **7**, A10-A15 (1992).

7. ASTM standard under development by the Committee F-1 on Electronics, draft document 94D1.

8. Specification M1-95, *Book of SEMI Standards 1995*, **Materials Volume**, 3-6 (1995).

9. Stover J.C., *in this volume.*

10. Mier M.G., Look D.C., Sizelove J.R., Walters D.C. and Beasely D.L., *in this volume.*

11. Homma Y., Ishii Y., Kobayashi T. and Osaka J., *J. Appl. Phys.*, **57**, 2931-2935 (1985).

12. Jantz W., Stibal R. and Windsheif J., *Proc. 7th Conf. on Semi-Insulating III-V Materials,* 171-176 (1992).

13. Dobrilla P. and Blakemore J., *J. Appl. Phys.*, **61**, 1442-1448 (1987).

14. Windsheif J. and Wetting W., *Defect Recognition and Image Processing in III-V Compounds II*, 195-206 (1987).

15. Yamada M., Fukuzawa M., Kimura N., Kaminaka K. and Yokogawa M., *Proc. 7th Conf. on Semi-Insulating III-V Materials,* 201-210 (1992).

16. Clayton R.D., Bassignana I.C., Macquistan D.A and Miner C.J., *Proc. 7th Conf. on Semi-Insulating III-V Materials,* 211-216 (1992).

17. Tarof L.E., Miner C.J. and Blaauw C., *J. Appl. Phys.*, **68**, 2927-2983 (1990).

18. Swart P. L. and Lacquet B.M., *J. Appl. Phys.*, **70**, 1069-1071 (1991).

19. Tompkins H., *A User's Guide to Ellipsometry*, Academic Press, 1993.

20. Nguyen N.V., Pellergino J.G., Amirtharaj P.M., Seiler D.G. and Qadri S.B., J. Appl. Phys., **71**, 7739-7746 (1993).

21. Celii F.G., Kao Y.-C., Moise T.S., Katz A.J., Harton T.B. amd Woolsey M., *in this volume.*

22. Hovel H., *Semicond. Sci. Technol.*, **7**, A1-A9 (1992).

23. Moore C.J.L. and Hennessy J., *Semicond. Sci. Technol.*, **7**, A69-A72 (1992).

24. Miner C.J., Knight D.G. and Zorzi J.M., *Inst. Phys. Conf. Ser.*, **135**, 181-186 (1993).

25. Knight D.G., Miner C.J. and SpringThorpe A.J., *MRS Symposium Proceedings*, **184,** 157-162 (1990).

26. Miner C.J., Harrison A. and Clayton R., *Proc. 7th Conf. on Semi-Insulating III-V Materials,* 189-194 (1992).

27. Komiya S., Yamaguchi A. and Umebu I., *Solid State Electronics*, **29**, 235-240 (1986).

28. Dean P.J., *Prog. Crystal Growth Charact.*, **5**, 89-174 (1989).

29. Steiner T.W. and Thewalt M.L.W., *Semicond. Sci. Technol.*, **7**, A16-A21 (1992).

30. Aspnes D.E., *Handbook on Semiconductors*, North-Holland, 1981, ch.4A, pp. 109-154.

31. Glembocki O.J., *in this volume.*

32. Pollack F.H., Okeke C.E., Vanier P.E., Raccah, P.M., *J. Appl. Phys.*, **49**, 4216-4220 (1978).

33. Shen H., Dutta M., Fotiadis L., Newman P.G., Moerkirk R.P., Chang W.H. and Sacks R.N., *Appl. Physi. Lett.*, **57**, 2118-2120 (1990).

34. Boettinger W.J., Burdette H.E., Kuriyama M. and Green R.E., *Rev. Sci. Inst.*, **47**, 906-911 (1976).

35. Bassignana I.C. and Macquistan D.A., *Proc. 7th Conf. on Semi-Insulating III-V Materials,* 183-188 (1992).

36. Bassignana I.C. and Macquistan D.A., *Advances in X-Ray Analysis*, **34**, 103-108 (1991).

37. Kitana T., Ishikawa T., Ono H. and Matsui J., *Proc. 4th Conf. on Semi-Insulating III-V Materials,* 91-96 (1986).

38. Halliwell M.A.G, *J. Crystal Growth*, **68**, 523-531 (1984).

39. Miner C.J., Knight D.G., Zorzi J.M., Macquistan D.A. and Mallard R., *Proc. 8th Conf. on Semi-Insulating III-V Materials,* 151-154 (1994).

40. Miller G.L., Robinson A.H. and Wiley J.D., *Rev. Sci. Instr.*, **47**, 799-805 (1976).

41. Jantz W., Frey Th., and Bachem K.H., *Appl.Phys. A*, **45**, 225-232 (1988).

42. Heimlich M.C., Gutmann R.J., Seielstad D. and Hou D., *Proc. 7th Conf. on Semi-Insulating III-V Materials,* 367-372 (1992).

43. Stibal R., Windsheif J. and Jantz W., *Semicond. Sci. Technol.*, **6**, 995-1001 (1991).

44. Jantz W., Stibal R., Windsheif J., Mosel F. and Müller G., *Proc. 7th Conf. on Semi-Insulating III-V Materials,* 171-176 (1992).

Minority and Majority Carrier Transport Characterization in Compound Semiconductors

Michael L. Lovejoy

Sandia National Laboratories
Albuquerque, NM 87185

A review of minority and majority carrier transport characterization techniques for compound semiconductor material is presented. Minority carrier transport is discussed in the context of base transport in HBTs where the base transit time can be an important component of the total transit time through the transistor. Characterization techniques to measure minority carrier mobilities in heavily-doped compound semiconductors and theoretical results are reviewed. Majority carrier transport in high-performance PHEMT structures is discussed. Parallel conduction of the heavily-doped contact layer and the 2-D electron gas makes measurements of the 2-D electron concentration difficult. Techniques to measure the 2-D concentration are reviewed. It is shown that recent applications of these techniques have yielded new data for GaAs that have important implications for device design and that additional measurements of compound semiconductor materials and devices are needed to realize accurate device design and optimization.

INTRODUCTION

Compound semiconductor devices can no longer rely on inherent material transport properties for high-performance device applications due to the performance gains of silicon based technologies. As compound semiconductor (CS) technology is pushed to higher performance, improved device design and modeling are necessary to bring products to the market on-time, in sufficient quantity, and at an acceptable cost. To realize accurate device modeling, accurate material transport parameters are required. Most transport properties are easily measured at low carrier concentrations and in special device structures that accentuate a particular property; however, at concentrations important for high-performance devices and in actual high-performance device structures transport parameters are difficult to measure. This paper reviews techniques for transport parameter characterization of CSs that are important for heterojunction bipolar transistors (HBTs) and pseudo-morphic high electron mobility transistors (PHEMTs).

The performance of HBTs is directly related to the minority-carrier mobility in the degenerately doped base. Minority carrier mobility data for most CS materials with the exception of GaAs is lacking; consequently, in device modeling minority-carrier mobility must be approximated by the majority-carrier mobility in comparably doped material. A review of theoretical work for GaAs is presented to show that the minority and the majority carrier mobilities are expected to exhibit very different temperature- and doping-dependencies; consequently, measurements of these parameters are needed for accurate device modeling. Techniques to measure the minority carrier mobility in heavily-doped CS materials including

conventional time-of-flight, zero-field time-of-flight and unity gain cutoff frequency (f_T) techniques are reviewed. Results of applications of these techniques to measurements of minority electron and hole mobilities in heavily-doped GaAs are surveyed and the impact on HBT device design is discussed.

The performance of unipolar devices such as the technologically important PHEMTs is largely governed by the majority-carrier concentration and mobility of the two-dimensional (2-D) carriers in the strained quantum well (SQW). Compound semiconductor PHEMTs, both GaAs- and InP-based devices, are exclusively used for microwave and millimeter-wave low-noise and power amplifiers. As the frequency spectrum is becoming more crowded, the push to higher frequencies is stronger. To achieve higher performance both device design and material structure must be optimized. Important performance figures of merit such as transconductance and f_T are largely dependent on carrier concentration (n_{2D}) and mobility in the 2-D SQW. Measurement of n_{2D} is complicated by two factors. First, high-performance PHEMTs require heavily-doped cap layers to realize low-resistance ohmic contacts. This layer presents a conduction path parallel to the 2-D gas conduction channel in the SQW and distorts the signal of standard electrical characterization methods. Second, high performance also requires high carrier concentrations that result in multiple subbands being occupied which make measurement and data analysis difficult. Techniques to deduce the 2-D gas density from these convolved signals including Hall techniques, Shubnikov-de Haas (SdH), photoluminescence (PL) and a new hybrid SdH-PL technique are reviewed.

MINORITY CARRIER MOBILITY THEORY AND CHARACTERIZATION

Minority carrier mobility determines the performance of bipolar junction transistors (BJTs), HBTs, solar cells and semiconductor lasers. The dc current gain and short circuit small signal current gain of BJTs and HBTs have a linear dependence on minority carrier diffusivity, or equivalently mobility, and the efficiency of solar cells depends on the square root of the minority carrier mobility. Mobility data, however, is sparse for minority carriers in CS materials which precludes accurate modeling and optimization of the devices. Compounding the problem of insufficient measured minority carrier mobility data is limited theoretical work on minority mobility in heavily-doped CSs because scattering mechanisms that are believed to be important for minority carrier transport can not be easily treated with current mathematical techniques [1]. Clearly, measured mobility data, for heavily-doped CSs is needed for device optimization and to seed theoretical work in developing a fundamental understanding of minority carrier transport in CS materials.

Techniques to measure minority carrier transport properties in heavily-doped CS materials must overcome formidable obstacles due to fundamental properties of CSs. Minority carrier lifetimes, *i.e.* the characteristic decay time of excess concentrations by recombination, is of the order of nanoseconds, or shorter, for technologically important heavily-doped CS materials. Such short lifetimes necessitate high-speed techniques for the study of CS transport properties. Since recombination is mainly radiative in most important CS materials, photon recycling, which is the reabsorption of an emitted photon, presents another transport mechanism which must be properly treated in the measurement technique. Another fundamental phenomenon, common to all material systems, which can make interpretation of many minority carrier transport measurements difficult, is majority carrier drag. This drag results in the measurement of an effective minority carrier mobility under high-field conditions that is lower than the true low-field value. Techniques to overcome these difficulties in measurements of minority carrier transport properties are presented after reviewing the minority carrier transport theory in CSs and reported theoretical results for GaAs.

Compound Semiconductor Transport Theory

Considerable work has been performed to understand minority carrier transport in III-V semiconductors. Iterative techniques and complicated variational techniques have been applied to the important problem of understanding carrier transport in GaAs and other III-V semiconductors [1-4]. It is common in device related fields to express theoretical formulas for important bipolar characteristics in terms of drift mobility (μ), but many of the results evolve from diffusion terms, hence inherently contain diffusivity (D). Common examples include dc current gain or small-signal current gain of BJTs and HBTs that are both ratios of diffusion currents. For the case of nondegenerate carrier systems the Einstein relation is used to relate diffusivity to mobility.

$$De = \mu kT, \tag{1}$$

where k is Boltzmann constant, T is temperature, and *e* is the electronic charge. This relationship is valid for nondegenerate minority carrier systems even in degenerately doped CS. Throughout this paper, minority diffusivity and mobility will be used interchangeably.

Scattering Mechanisms and Mobility

The scattering mechanisms that are present in a material system determine the minority and majority carrier mobilities. Many mechanisms have been characterized within the framework of Fermi's Golden Rule and the calculated scattering rates have been used in investigations of each mechanism's influence on carrier mobility. Important mechanisms for majority carriers in CSs include ionized impurity scattering, polar and acoustic lattice scattering and carrier-carrier scattering. Scattering rates for most of these mechanisms are documented in many textbooks [5].

Additional scattering mechanisms must be considered for minority carriers in degenerately doped CSs. The first mechanism to be discussed is scattering of minority carriers via a Coulomb interaction with a single majority carrier. The second is the interaction of a minority carrier and the collective charge oscillations (plasmon oscillations) that occur in a degenerate majority carrier gas. Theory for a gaseous plasma has been used to describe these oscillations.

A rigorous treatment of minority carrier-single majority carrier scattering includes the majority carrier distribution function and results in an untractable formulation. Walukiewicz *et al.* [3] used an approximation in calculations of mobilities in GaAs that was suggested by Ehrenreich [6] where, due to the greater effective mass of holes relative to electrons, the electron scattering was described by a screened impurity scattering rate. For uncompensated material, this essentially doubled the screened impurity scattering rate for minority carrier electrons from that experienced by majority carrier electrons, hence, minority carrier electron mobility was less than that of majority electrons in comparably doped material.

Lowney and Bennett recently extended the work of Walukiewicz *et al.* They considered different formulations of screened impurity scattering, included plasmon scattering, and included a correction to the approximation of treating minority-single majority carriers scattering as an elastic scattering event with all

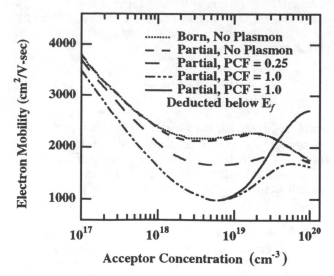

Figure 1. Theoretical calculation of minority-electron mobility in p-GaAs vs. doping for different dominant scattering mechanisms showing an increase in minority carrier mobility can occur with heavy doping. From Lowney and Bennett, *J. Appl. Phys.* **69**, 7102 (1991).

majority carriers [1]. The screened impurity scattering rate calculated with the Born approximation, which was used by Walukiewicz *et al.* as well, was compared with results using partial wave analysis, which unlike the Born treatment is quantitatively different for scattering by like charged carriers and scattering by oppositely charged carriers. Their calculations show little difference in mobility for the two treatments (see Figure 1), which indicates that scattering differences due to charge signs are probably not important.

When plasmon scattering was included in the mobility calculations a significant reduction in mobility occurred. Plasmon oscillations occur in a two-component system due to coupling between the constituent systems in the form of weak restoring forces that force small perturbations in concentration back to a uniform distributions of carriers [7]. In the context of semiconductors, the two-component system is that of majority carriers and ions on the lattice sites. As a direct consequence, plasmon oscillations of majority carriers and lattice oscillations are coupled. However, tractable coupled plasmon-phonon scattering rate expressions are not available; hence, Lowney and Bennett employed uncoupled plasmon and phonon scattering rate formulas.

Plasmon oscillations exhibit a minimum wavelength for which majority carrier collective oscillations are possible before shorter wavelength oscillations are damped out. This wavelength is of the order of the screening length but is not well defined. Lowney and Bennett defined a plasmon cutoff factor (PCF) as the square of the product of the plasmon wavenumber and the screening length; they considered a range of PCF where a larger PCF included more plasmon waves in the calculation. As shown in Figure 1, for minority electrons

in p-GaAs doped as high as $\sim 10^{19}$ cm^{-3}, plasmon scattering significantly reduced the minority electron mobility below that which would be realized if minority-majority carrier scattering was approximated by only ionized impurity scattering. Including more modes lowered the mobility.

For the calculation including more plasmon modes (*i.e.* shorter wavelengths were allowed), as the doping increased above 10^{19} cm^{-3} the strength of plasmon scattering actually decreased which resulted in minority electron mobility significantly increasing with increased hole concentration. The reason for decreased plasmon scattering was explained by the greater plasma frequency which increases with doping. Presumably, the greater energy exchange with greater plasma frequency is more important for greater wavenumber plasmon oscillations. As the hole concentration approached 10^{20} cm^{-3}, minority electron mobility approached the value found considering only single carrier scattering, as shown in Figure 1. Scattering of minority holes with plasmon oscillations of majority electron n-GaAs was found to be negligible as compared to single carrier majority-minority scattering.

Lowney and Bennett also included a correction to the treatment of the minority-single majority carrier scattering mechanism as ionized impurity scattering with each majority carrier. Their correction is an approximation to a rigorous application of the Pauli exclusion principle, while approximating minority-single majority carrier scattering by elastic scattering. However, in addition to the momentum transfer between the carrier systems, a small energy exchanges occurs since the process is an inelastic scattering event. Although the energy exchange does not quantitatively change their results, it is included in the following discussion for completeness.

In degenerately doped material where the Fermi level is deep in the majority band, as diagrammed in Figure 2, all states below the Fermi energy are filled. Because the majority carriers in states below the Fermi level do not have states within a small energy range in which to scatter

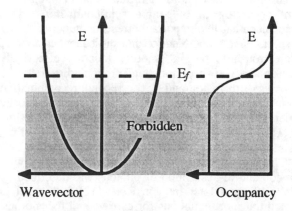

Figure 2. Dispersion curve and occupancy diagram showing that majority carriers below the Fermi level in degenerately doped CS are forbidden to participate in first order inelastic scattering events since all states within a small energy range are fully occupied.

they are forbidden to participate in minority-majority scattering. As shown in Figure 1, when this correction is made, the electron minority mobility increases greatly as the material becomes more degenerate since more majority carriers are prohibited from participating in minority carrier scattering.

This treatment of single minority-majority carrier interaction is a first approximation of the mechanism's effect on mobility and will be shown to yield good agreement with experimental data for low-field mobility. However, in highly non-equilibrium systems with a large applied field, this treatment does not account for the highly energetic majority carriers. For these conditions, the average drift momentum of majority carriers is directed diametric to that of minority carrier drift. Majority-minority scattering results in a net momentum transfer to minority carriers that will impede the minority carrier drift; consequently, minority carrier mobilities that are lower than the zero-field mobility can be observed.

Majority carrier drag is expected theoretically and has been observed experimentally. Dumke derived an expression for the minority electron mobility in terms of mobilities resulting from individual scattering mechanisms [8]. He estimated that, in the case of high-fields, majority hole drag can reduce the electron mobility by up to 20%. In low temperature measurements of the 2-D minority electron mobility in GaAs, the drag effect has been reported to be strong enough to produce negative minority electron mobilities [9]. In the review of experimental data for minority electrons in GaAs that follows the discussion of measurement techniques, we show evidence that suggests majority hole drag is important in heavily-doped p-GaAs.

Effects of Photon Recycling on Mobility

Measurements of minority carrier mobilities in CSs are complicated by photon recycling. Photon recycling is the absorption of a photon emitted by a recombination event occurring in the same semiconductor. This phenomenon has two important effects on minority carrier transport. First, recombination and subsequent creation of an electron-hole pair at another location effectively represents a transport mechanism in addition to the processes of diffusion and drift. Second, recombination and subsequent absorption effectively enhances minority carrier lifetime beyond that of the radiative lifetime [10,11]; this effect is noted for completeness but will not be discussed further.

The steady-state flow of minority carriers with emission and re-absorption of photons was described by Dumke [8], and later by von Roos [12] with a set of coupled transport equations for carriers and for photons. Separating the set of equations yields an additional term in the uncoupled minority carrier diffusion equation (MCDE) which characterizes the effects of photon recycling. The extra term is a diffusion-like term, hence, it

is an effective diffusion coefficient that appears in the uncoupled MCDE. The effective diffusivity is given by the sum of the actual diffusivity and the term due to photon recycling,

$$D_{ph} = \frac{1}{3\tau_{rad}} \int \frac{P(E)}{\alpha^2(E)} dE = \frac{1}{3\tau_{rad}} \left\langle \alpha^{-2} \right\rangle, \qquad (2)$$

where τ_{rad} is the radiative lifetime, P(E) is the probability density of photons with energy E and α(E) is the absorption coefficient. For heavily-doped, direct bandgap CS, which have radiative lifetimes of nanoseconds or less, this enhancement can be important.

Measurement Techniques for Minority Carrier Mobility Measurements

In the last few years, several techniques have been employed to measure minority carrier mobility in heavily-doped CS. Techniques such as one-sided diode I-V characterization [13,14] and the photo-Hall effect technique [15] that have been used for minority carrier mobility measurements in moderately-doped GaAs are not suitable for measurements in heavily-doped CS due to concerns associated with multiple carrier transport, high-level injection, thermal heating and majority carrier drag. Techniques that are suitable for measurements of heavily-doped CS are the conventional time-of-flight technique, the unity gain transistor cutoff-frequency technique (f_T), and the zero-field time-of-flight technique (ZFTOF). These techniques will be described and compared in the following sections. It will be shown that the f_T and the ZFTOF techniques involve measurements that are related by Fourier transformations; that is, the first is applied in the frequency domain and the latter in the time domain.

The Conventional Time-of-Flight Technique

The conventional time-of-flight (ToF) technique is conceptually the simplest of all techniques. Mobility is determined from a measurement of the average time for a carrier to drift in a uniform field across a well defined distance. From the measured time and distance, the average velocity is calculated, and the mobility is simply the average velocity divided by the field strength. The simplicity of this analysis is very attractive; however, performing measurements of mobilities in a region with an applied field does not simulate the typical environment, or yield the mobility that characterizes minority carrier transport, in many bipolar devices. Instead of regions of high fields, just the opposite are often the important regions in which minority carrier transport determines the device performance. One example is the base of a HBT, which is typically a zero-field region.

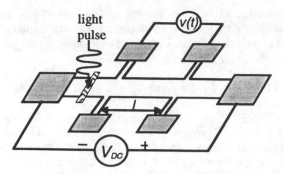

Figure 3. Time-of-flight technique for measuring minority electron mobility in heavily-doped p-GaAs. After Furuta *et al. Appl. Phys. Lett.* **55** (1989).

The device structure and experimental ToF setup is shown in Figure 3 as performed by Furuta *et al.* [16]. As shown, a device with Hall-bar geometry and a picosecond laser system to inject a high level of minority carriers along a band near the end of the Hall-bar are employed with this technique. The applied field along the channel sweeps the minority carriers into the region between the side contacts of the device. As a result of the additional carriers, the conductivity increases which decreases the voltage that is measured between the side contacts. Since the lifetime of minority carriers in heavily-doped CS is of the order of nanoseconds or less, a large field (≥ 2.5 kV/cm) is required to achieve a transit time comparable to the lifetime. This insures that photon recycling mobility enhancements do not occur and that a significant number of minority carriers traverse the region between the side contacts to provide good signal response. The high-fields, however, suggest that majority carrier drag may lower the minority carrier mobility below that of the low-field value.

Furuta *et al.* applied this technique to measurements of minority carrier mobility in p$^+$-GaAs [16-18]. Shigekawa *et al.* applied the technique to p$^+$-InGaAs [19]. A comparison of minority electron mobility in p-GaAs measured by the high-field ToF technique and low-field mobility results is presented at the end of this section. The comparison suggests that minority hole drag is important above hole concentrations of $\sim 10^{19}$ cm^{-3}.

The Short Circuit Unity Gain Cutoff-Frequency Technique

Short circuit unity gain cutoff-frequency, or f_T, measurements have been applied by several authors to measure mobility in moderately to heavily-doped p-type GaAs [20-24] and heavily-doped n-type GaAs [25]. f_T and the total transit time (τ_{ec}) through an HBT are related by

$$f_T = \frac{1}{2\pi\tau_{ec}},\qquad(3)$$

where τ_{ec} is the sum of junction capacitor charging times and of transit times through individual layers. The charging times, τ_e and τ_{cc}, are charging times for the base-emitter and base-collector junctions respectively and are current (I) dependent, as shown in the simplified expressions below.

$$\tau_e = \frac{kT}{eI}C_e \text{ and}\qquad(4)$$

$$\tau_{cc} = \frac{kT}{eI}C_c,\qquad(5)$$

where C_e is the base-emitter junction capacitance and C_c is the base-collector junction capacitance. The transit times, τ_b and τ_c, are transit times for the base-layer and base-collector depletion layer respectively and are current independent, as shown in the simplified expressions below.

$$\tau_b = \frac{W^2}{2D} \text{ and}\qquad(6)$$

$$\tau_c = \frac{W_{bc}}{2v_{sat}},\qquad(7)$$

where W is the base thickness, D is the minority carrier diffusivity (related to the mobility by the Einstein relation), W_{bc} is the depletion layer thickness and v_{sat} is the high-field saturation velocity.

Application of the technique requires transistors with good gain at low frequency, hence, homojunction BJTs in

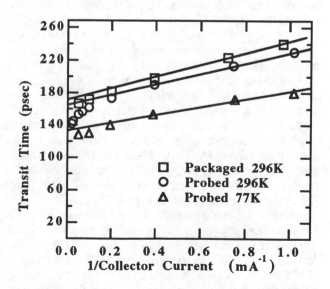

Figure 4. Total transit time *vs.* inverse current which permits extracting base transit time from f_T measurements at both room-temperature and 77K. From Nathan *et al. Appl. Phys. Lett.* **52** (1988).

619

many CS material systems are not well suited for f_T-measurements due to poor gain. More complicated HBTs are required, however, abrupt heterojunctions present the problem of spikes in the band edges at the metallurgical junction that can complicate data interpretation, therefore, junction grading is desirable. In application of this technique, f_T is not measured directly because making a true short at microwave frequencies is difficult, consequently, S-parameters are measured, converted to hybrid parameters and f_T is extrapolated from lower frequency measurements.

Inspection of equations 4 to 7 shows that measurement of f_T vs. current permits extrapolation to infinite current which yields the current independent transit times, as shown in Figure 4. These data are from Nathan et al. who employed a thick base transistor in which the base transit time dominated the total device transit time [20]. This is possible with moderately doped bases where the lifetime is sufficiently long to maintain a high base transport factor, hence, good gain; however, at heavier doping levels the base must be made thinner to maintain gain.

Applying this technique to heavier doping levels is more challenging because, with thinner bases, f_T is higher and parasitics become more important. Characterizing these parasitics and techniques for de-embedding parasitics in the context of minority electron mobility measurements in p^+-GaAs have been reported [22]. Additional characterization methods such as resistance and capacitance measurements are required to reduce the number of fitting parameters in f_T data analysis [26].

This technique has been applied to measurements of minority carrier diffusivity in both p- and n-type GaAs. Nathan et al. first applied the technique to CS and measured the electron mobility in moderately doped p-type GaAs (3.6×10^{18} cm^{-3}) at 77K and 300K [20]. Beyzavi et al. confirmed their work and extended the f_T technique to temperature dependent measurements of minority electron mobility [21]. Tiwari and Wright applied the technique for electron mobility measurements in p^+-GaAs [23]. Lee et al. reported techniques to address parasitics and measured electron mobility in p^+-GaAs [22]. Most recently Kim et al. measured the doping dependence of minority electron mobility in heavily-doped p-GaAs [24]. Measurements of n^+-GaAs have also been reported; Slater et al. measured the minority hole mobility in n^+-GaAs doped to 4×10^{18} cm^{-3} [27]. These data are shown in Figures 7 and 8.

Unlike the conventional time-of-flight technique, the f_T technique measures the low-field mobility which determines transit times through the bases of HBTs. The fact that the technique measures transit times directly in HBTs makes the data very valuable for HBT modeling. Application of this technique to CS-material systems with less mature processing technologies where HBT fabrication is difficult is a weakness of the technique, as well as difficulties associated with de-embedding parasitic contributions. The ZFTOF technique which will be discussed next will be shown to be a conjugate pair technique with the f_T technique in the sense that the ZFTOF technique is applied in the time domain and the f_T technique in the frequency domain.

The Zero-Field Time-of-Flight Technique

The ZFTOF technique was introduced by Ahrenkiel et al. [28,29]. Lovejoy et al. extended the technique to make it suitable for measurement of the doping- and temperature-dependence of minority carriers in heavily-doped CSs [30-34]. The ZFTOF experimental setup for measuring minority electrons in n^+-GaAs is shown in Figure 5. A picosecond laser pulse photoexcites minority electrons near the passivated surface of a photodiode designed with the junction ~1-4 μm from the illuminated surface. Electron-hole pairs diffuse through a quasi-neutral n^+-GaAs layer to the junction where the minority electrons are swept across by the field. Phenomenologically, this charges the pn-junction capacitor and generates a transient voltage response which characterizes the transport process in the n^+-GaAs layer. Low-level injection is maintained to insure that transport is governed by the minority electron diffusivity. From the measured voltage response, the mobility is extracted by appropriate fits to measured responses.

A digitizing sampling oscilloscope is used to record the ZFTOF transient responses. Figure 6 shows the response of a ZFTOF diode for measuring the minority hole mobility in heavily-doped n-GaAs. The slower initial voltage response is due to the required time for carriers to diffuse from near the surface to the junction. As more carriers diffuse to the junction, the signal rises quickly until the excess carriers are collected or have recombined in the quasi-neutral region. On a time scale greater than that shown in Figure 6, the response decays with an RC-decay rate given by the junction capacitance and the oscilloscope input impedance.

The rapid rise in voltage occurs when the photocurrent is large which is when the diffusion process is

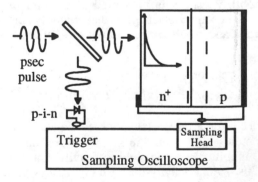

Figure 5. Schematic diagram of the ZFTOF technique for measuring minority hole mobility in n^+-GaAs. The photodiode transient voltage response characterizes minority carrier diffusion in the n^+-GaAs. From Lovejoy et al. Solid-State Electr. **35** (1992).

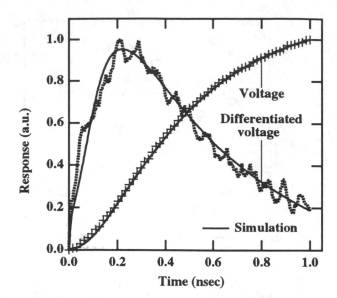

Figure 6. ZFTOF voltage response of a photodiode for measuring minority hole diffusivity in n⁺-GaAs. The differentiated voltage response which is proportional to the diffusion-driven photocurrent and fits to both voltage and differentiated voltage response are also shown. From Lovejoy et al. *Solid-State Electr.* **35** (1992).

dominating the response. For properly designed diodes, the photocurrent (i_{ph}) at the junction edge is related to the measured voltage (v_m) by

$$i_{ph} = C_{junc} \frac{dv_m}{dt}, \qquad (8)$$

where C_{junc} is the junction capacitance. To extract the low-field mobility, the minority carrier transport in the charge-neutral region is modeled by the minority carrier diffusion equation. The time-dependent carrier concentration is differentiated to yield the photocurrent at the junction edge which is fit to the differentiated voltage response with low-field mobility as a fitting parameter. As shown in Figure 6, excellent fits are obtained [32].

A critical analysis of the technique has been presented [32]. Devices can be designed so that the fit is sensitive to only the low-field mobility. In addition, techniques for device designs to minimize photon recycling effects, lateral charge transport parallel to the junction, and other parasitics have been developed [32]. The response risetimes are of the order of tens to hundreds of picoseconds which demands microwave packaging techniques to minimize parasitic signal distortion. A simple high-speed package that eliminates parasitic distortion and is suitable for low-temperature measurements has been presented [32,34].

This technique, as developed for measurements in heavily-doped CS, has been applied to minority mobility measurements in both n- and p-GaAs and p-InGaAs.

During the technique development, Lovejoy et al. measured the doping dependence of mobilities in p⁺-GaAs doped to 10^{19} cm⁻³ [30,31,33,35]. Colomb et al. employed the technique to measure minority electron mobility in p-GaAs doped to mid-10^{19} cm⁻³ at room-temperature and at 77K by submersing the sample in liquid nitrogen [36]. Lovejoy et al. compared minority electron mobilities in both Be- and C-doped GaAs [35] and measured the doping dependence of minority hole mobilities in heavily-doped n-GaAs [37]. Harmon et al. confirmed ZFTOF measurements of p-GaAs doped ≤10^{19} cm⁻³ and utilized a femtosecond laser system to measure doping dependence above 10^{19} cm⁻³ [38]. Minority electron mobility in p-InGaAs doped to ~10^{19} cm⁻³ was measured by Lovejoy [33] and was confirmed with a femtosecond laser system by Harmon et al. [39]. These data are shown in Figures 7 and 8. Lovejoy et al. continued development of the technique to permit continuously variable temperature dependent mobility measurements [33,34].

In common with the f_T technique, this technique also measures the quantity of interest to bipolar design and analysis—the low-field minority carrier mobility. Circuit parasitics are also important in ZFTOF measurements; however, unlike the f_T technique, they can be made negligibly small with the use of microwave packaging. In contrast to the f_T technique, the ZFTOF technique requires devices that are simple to fabricate. This offers an advantage for measurements of minority carrier mobility in developing CS material systems.

Relating the Unity Gain Cutoff-Frequency Technique and the ZFTOF Technique

The ZFTOF and the f_T techniques have been shown to differ mainly in the domain in which they are implemented [32,33]. The ZFTOF technique is implemented in the time domain and the f_T technique is implemented in the frequency domain; however, the same physical quantity is measured by both techniques.

In the f_T technique, the transistor is designed so that the base transit time dominates the ac common emitter current gain. The small signal short circuit gain (h_{fe}) is given by

$$h_{fe}(\omega) = \frac{\gamma(\omega)\alpha(\omega)}{1 - \gamma(\omega)\alpha(\omega)} \cong \frac{\alpha_T(\omega)}{1 - \alpha_T(\omega)}, \qquad (9)$$

where the base transport factor ($\alpha_T(\omega)$) is given by

$$\alpha_T(\omega) = \frac{1}{\cosh\left\{\frac{W_b}{L_n}\sqrt{1 + j\omega\tau_n}\right\}}, \qquad (10)$$

where $L_n = \sqrt{D_n \tau_n}$. In practice, h_{fe} is calculated from S-parameter measurements, but α_T could just as easily be calculated. It is the base transport factor that is the Fourier transform of the normalized photocurrent in the ZFTOF technique.

In an ideal implementation of the ZFTOF technique, photoexcitation occurs as a sheet of carriers at the illuminated surface and the surface recombination velocity is negligible. For these conditions, the generation can be included as a boundary condition at the illuminated surface given by

$$-D_n \frac{\partial \Delta n(0,t)}{\partial z} = N_0 \delta(t), \qquad (11)$$

where N_0 is the number of electrons photogenerated per unit area. With this boundary condition, the governing diffusion equation for minority carriers in the zero-field region can be readily solve by Fourier transformation to yield the transform of the photocurrent,

$$\mathcal{F}\{i_{ph}(t)\} = \frac{N_0 e}{\cosh\left\{\dfrac{W_b}{L_n}\sqrt{1 + j\omega\tau_n}\right\}}. \qquad (12)$$

Inspection of equations 10 and 12 relates the ZFTOF and the f_T techniques. The Fourier transformation of the ZFTOF photocurrent is proportional to the base transport factor deduced from f_T-measurements. It is interesting to note that if N_0 were accurately known, then cutoff-frequencies for transistors that were dominated by base transit could be predicted from a Fourier analysis of the ZFTOF photocurrent. Unfortunately, N_0 is not accurately known, therefore, any f_T prediction or quantitative comparison of independent data from measurements with the two techniques is not possible.

Minority Carrier Mobility in GaAs

In this section, minority carrier mobilities in moderately and heavily-doped GaAs are reviewed and compared to both minority carrier theoretical predictions and to majority Hall mobilities in comparably doped material. In Figure 7, hole mobilities *vs.* doping are shown. The minority hole mobilities measured with the f_T [27] and the ZFTOF [37] techniques are in good agreement. The theory [1] agrees qualitatively with the measured data showing the same trend but underestimates the mobility over the doping range for which measured data are available. Of particular interest is that at doping levels above 10^{18} cm^{-3} the minority carrier mobility can be as great as twice the majority carrier mobility in comparably doped material This is extremely important for Pnp-HBT design. Prior to the availability of these measured minority hole data, predictions of Pnp-HBT

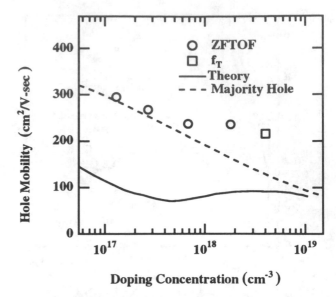

Figure 7. Hole mobility *vs.* doping. Markers are measurements made by ZFTOF [37] and f_T techniques [27]. Also shown is typical majority Hall mobility [40] and theory [1]. At high doping levels, the minority hole mobility is nearly twice the majority hole mobility.

performance based on majority hole mobility predicted poor performance; however, these data show that higher performance is possible.

In Figure 8, electron mobilities *vs.* doping are shown. As found for minority hole mobility, the minority electron mobilities measured with the f_T [20-24] and the ZFTOF [30,33,35,36,38] techniques are in good agreement. The important point for bipolar device design is that the mobility actually increases with increased doping from $\sim 10^{19}$ cm^{-3} to $\sim 10^{20}$ cm^{-3}. As shown, this behavior is very different from majority electron mobility behavior and must be included in device modeling routines for accurate device optimization. Also shown is the theory of Lowney and Bennett which includes plasmon scattering (PCF=1.0) and the correction for single majority carrier scattering being approximated by ionized impurity scattering with all majority carriers [1]. The excellent agreement of theoretical and measured data indicates that the important scattering mechanisms for minority electrons are understood and that for degenerately doped material when approximating minority-single carrier majority scattering by ionized impurity scattering a correction for degeneracy effects must be included.

Also shown in Figure 8 is minority electron mobility data measured with a high-field time-of-flight technique [17]. Comparing these data with the low-field data from f_T and ZFTOF measurements shows that for concentrations above $\sim 10^{19}$ cm^{-3} the high-field data is considerably lower which suggests that majority carrier drag is important at room temperature for heavily doped p-GaAs under high-field conditions.

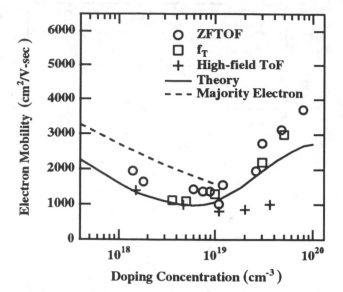

Figure 8. Electron mobility *vs.* doping. Markers are measurements made by ZFTOF [30,33,35,36,38], f_T techniques [20-24] and high-field ToF [17]. Also shown is minority theoretical predictions [1]and typical majority Hall mobility[41]. At high doping levels, the minority electron mobility increases with doping and becomes over twice the majority electron mobility.

The temperature dependencies of minority holes and minority electrons in heavily doped GaAs have been measured by the ZFTOF [34,42] and f_T [21] techniques, respectively. For both minority holes and electrons, the mobilities exhibited a roughly 1/T dependence. A slightly weaker dependence was predicted theoretically for minority electrons by Walukiewicz *et al.* [3]. This behavior is very different from majority carrier mobility temperature dependence which is roughly temperature independent in heavily doped CS. For accurate device design, this behavior must also be included in low-temperature device modeling routines.

This completes the review of minority carrier mobility theory, characterization techniques and applications of the techniques to measurements of GaAs. The theoretical discussion showed that very different doping dependence is expected for minority carriers as compared to majority carriers due to the very different scattering environments. Two techniques, the f_T and the ZFTOF techniques, for measurement of minority carrier mobility in heavily-doped CSs were reviewed. Results of applications of the techniques to measurements of GaAs verified the very different doping dependencies of minority and majority carrier mobilities. The impact on GaAs bipolar device design was discussed to exemplify the need for minority carrier transport parameter measurements of the technologically important CS materials; these data are absolutely necessary for accurate bipolar device design and optimization.

MAJORITY CARRIER TRANSPORT CHARACTERIZATION

Majority carrier transport is in general easily characterized at low doping levels or in special structures that accentuates the property being investigated. However, measurements in actual high-performance device structures can be difficult. One example is measurement of the two-dimensional (2-D) electron density in the quantum well of strained quantum well FETs (SQWFETs), or equivalently pseudomorphic HEMTs (PHEMTs). This parameter is an important parameter for device performance because figures of merit such as the drain-source current and transconductance are largely proportional to the carrier concentration (n_{2D}) in the well, hence, accurate n_{2D} data are needed for structure optimization and reliable device modeling.

Measurement of n_{2D} is complicated by two factors. First, high-performance PHEMTs require heavily-doped cap layers to realize low-resistance ohmic contacts [43-46]. This layer presents a conduction path parallel to the 2-D gas conduction channel in the SQW. Hall effect and Shubnikov-de Haas (SdH) responses of the two layers are different; consequently, the measured Hall and longitudinal voltages are non-linear, intermediate values which do not accurately characterize either layer. Second, high performance also requires high carrier concentrations which result in multiple subbands being occupied [44,46]. SdH oscillations from individual subbands are superimposed which makes measurement and data analysis difficult. Techniques to deduce the 2-D gas density from these convolved signals have been developed [44,47], but there is a need for a simple, accurate technique to extract the concentration. In this section, we review techniques to measure n_{2D} of PHEMT structures with severe parallel conduction and multiple subband occupation by various techniques including standard Hall analysis, SdH analysis, steady-state photoluminescence (PL) and a new hybrid analysis technique (SdH-PL) to extract the 2-D gas density.

Hall Effect Analyses

As noted earlier, the parallel combination of Hall effect responses of the two high-conductivity layers in a high-performance PHEMT structure results in non-linear, intermediate low-field values which do not accurately characterize either layer [46]. This is shown schematically in Figure 9 where the low-field Hall voltage is linear in field strength for measurements of devices with conduction through either a cap layer or SQW, but is highly non-linear for a nominally identical SQW structure with a heavily-doped contact layer. Variations of standard Hall effect analysis is have been developed to analyze the non-linear data or eliminate the parallel conduction channel.

623

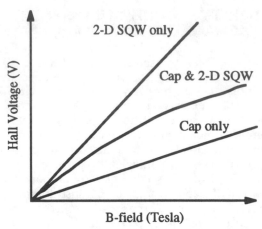

Figure 9. Schematic diagram of Hall voltages for devices with a) only a typical contact cap layer of a high-performance PHEMT, b) only a 2-D SQW of a PHEMT and c) a high-performance PHEMT with both heavily-doped cap and SQW which exhibits non-linear behavior.

The iterative-etch/Hall effect measurement technique, where the cap layer is sequentially etched to eliminate the parallel conduction path and Hall measurements are taken at each etch step [46,48], has the problem that as the cap is removed, the carrier concentration varies due to changes in band-bending. Consequently, it is difficult to identify the correct etch depth and corresponding carrier concentration that characterizes the device. A promising technique is the variable magnetic field Hall effect analysis which fits the non-linear Hall voltage and/or the magnetoresistance with a multi-parameter model [45]. The technique should prove to be very valuable for manufacturing process control of standardized device structures.

High-field measurements are used to determine n_{2D} in SQW structures but high-field data are difficult to analyze in high-performance PHEMT structures due to the parallel conduction and multiple subband occupancy. Transverse resistance ($R_{xy} \propto$ Hall voltage) for samples with both an undoped cap and a doped cap are shown in Figures 10 (see reference [46] for structure details). The effect of parallel conduction is clearly shown for these samples with nominally identical SQWs. While measured data for the sample with an undoped cap show a linear low-field dependence and the characteristic behavior of a 2-D gas with plateaus in R_{xy}, the doped-cap device does not exhibit either behavior. Parallel conduction smears out the plateaus in R_{xy}, significantly changes the low-field slope of R_{xy}, and causes a "rollover" of R_{xy} as the field strength increases.

The quantized Hall plateaus in R_{xy} allow very accurate determination of $n_{2D} = (l_0 + l_1)2e\,B/h$ and the plateaus are given by $h/\nu e^2$ where $\nu = 2(l_0 + l_1)$; the 2 is due to spin degeneracy [49]. The field strength, B, is at the corresponding minimum in magnetoresistance and l_0 and l_1 are the number of Landau levels filled in the N=0

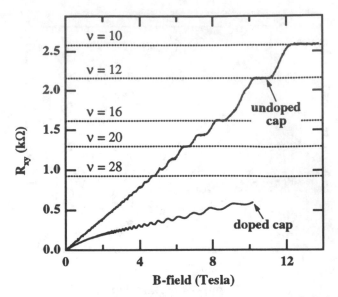

Figure 10. Transverse resistance ($R_{xy} \propto$ Hall Voltage) for samples with nominally identical SQWs but different contact layers showing the distorted ideal 2-D gas signal due to a heavily-doped contact layer. The dependence on filling factor $\nu = 2(l_0 + l_1)$ where $R_{xy} = h/\nu e^2$ is also shown. From Lovejoy et al., Appl. Phys. Lett. **64** (1994).

and n=1 subbands respectively. The effect of multiple subband occupation is evident in the R_{xy} characteristic where integer steps are skipped when the Fermi level passes through mid-points of Landau levels for both subbands simultaneously at ~6 and ~9 tesla and steps of $h/4e^2$ occur due to the combined responses of two occupied subbands. Obviously from Fig. 10 this technique can not be applied when severe parallel conduction distorts the plateaus.

Shubnikov-de Haas Effect Analysis

The Shubnikov-de Haas (SdH) effect, or magnetoresistance, analysis is another technique to measure n_{2D}. Measurements can be performed with the same Hall-bar structure used for standard Hall effect measurements with the difference that the voltage drop along the direction of current flow is monitored as a function of magnetic field. As with the standard Hall voltage measurement, the ideal 2-D gas magnetoresistance response is distorted and the analysis is more complicated for high-performance PHEMT structures as compared to simple SQW structures. Magnetoresistances (R_{xx}) for the samples discussed earlier with either doped or undoped contact layers are shown in Figure 11.

The R_{xx} response of the sample with the undoped cap exhibits classic 2-D gas response where the oscillations with field strength are observed relative to the low-field resistance value. Rapid oscillations corresponding to the n=0 subband are clearly evident and a much slower, weaker oscillation envelope due to the n=1 subband is hinted. As expected for a 2-D system, the resistivity vanishes at high field strengths. Also shown in Figure 11 is the response of a nominally identical SQW structure with a heavily-doped cap which does not exhibit either 2-D characteristic that was observed for the undoped cap sample. The strong 2-D oscillations are damped and a field-dependent background is imposed on R_{xx}.

The minima of SdH oscillations occur when the Fermi energy is midway between Landau levels of the 2-D system. To show this phenomenon, consider the density of states of an ideal 2-D gas which is given by

$$D(E) = \frac{2eB}{h} \sum_N \sum_j \delta\left(E - E_N\right), \tag{13}$$

where E_N is the energy of the N_j Landau level and the summations are over subbands N and Landau levels j of subband N [44]. The longitudinal conductivity σ_{xx} depends on the density of states as

$$\sigma_{xx} \propto \int \frac{\partial f}{\partial E}\left(N + \tfrac{1}{2}\right) D(E)\, dE, \tag{14}$$

where f is the Fermi distribution function [50]. The tensorial relation of conductivity and resistivity is $\rho_{xx} = \sigma_{xx}/\left(\sigma_{xx}^2 + \sigma_{xy}^2\right)$ which has the property that both longitudinal resistivity and conductivity vanish simultaneously. Since the derivative of f is sharply peaked at low temperatures and from equations 13 and 14, one sees that the longitudinal resistance vanishes when the Fermi level is midway between Landau levels. In a real semiconductor scattering broadens the ideal δ-function distribution and the density of states can be approximated by a summation of Gaussians as given by

$$D(E) = \frac{2e}{h} \cdot \frac{1}{\sqrt{2\pi}} \cdot \frac{B}{\Gamma} \times$$

$$\sum_N \sum_j \exp\left\{-\frac{\left(E - \left(N + \tfrac{1}{2}\right)\hbar\omega_c\right)^2}{2\Gamma^2}\right\}, \tag{15}$$

where Γ is the broadening parameter [51]. Inspection of equations 14 and 15 show that the minimum of the magnetoresistance still occurs when the Fermi level is midway between Landau levels, but the minimum in resistance is non-zero except at high B-field, as shown in Figure 11.

For a system with 2 subbands occupied, the minima will occur with 2 periods of oscillations which are periodic with inverse field strength. The oscillations associated with the n=0 subband will occur more rapidly in field strength than the n=1 subband associated oscilla-

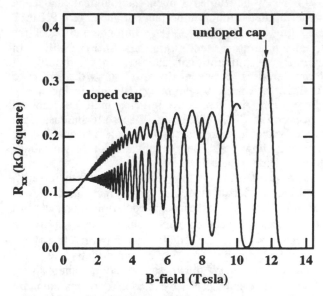

Figure 11. Longitudinal resistance (R_{xx}) for samples with nominally identical SQWs but different contact layers showing the distorted ideal 2-D gas signal due to a heavily-doped contact layer. From Lovejoy *et al.*, *Appl. Phys. Lett.* **64**, 3634 (1994).

tions. Analysis of high-performance PHEMTs requires identifying multiple frequencies of oscillations due to the multiple quantum well subbands occupancy. In principle the frequencies can be resolved by taking the Fourier power spectrum [47] or twice differentiating [44] the magnetoresistance. The Fourier power spectrum technique can require large field sweeps as high as 20 tesla or low noise measurements at lower fields; the second derivative analysis requires very low levels of noise to resolve both frequencies as well.

Photoluminescence Line Shape Analyses

Photoluminescence line shape analysis is a very attractive technique for rapid material characterization and for pre-processing wafer screening because it is a non-destructive technique. Several different models have been presented. Lyo and Jones presented a comprehensive basic physics theory and a study of line shape in degenerate SQWs which showed the importance of second order transitions due to perturbations of impurity levels [52]. Jones *et al.* extended the work to correlate PL-data to the performance of SQW field effect transistors [53]. This pioneering work showed the utility of PL-characterization of wafers by non-destructive means prior to device fabrication. Brugger *et al.* used an empirical relation for determining n_{2D} in high-performance PHEMT structures with the 2-D density of states function and found good agreement with more complicated electrical measurements [43].

Brierley developed a phenomenological line-shape model for determining important parameters for PHEMT modeling including n_{2D} [54]. Implementation of the room-temperature technique has shown utility in identifying unintentional changes in quantum well characteristics between different growths prior to fabrications which is extremely valuable in increasing circuit yield by screening inferior material prior to processing [55]. Applications of these techniques have proven valuable in process monitoring and standard wafer screening; however, general application of the techniques are not clearly established.

Martin et al. showed that application of their semi-empirical line shape model exhibited an excitation power dependence which influenced n_{2D} determination and must be considered for accurate determination of n_{2D} [48]. In agreement with Brierley, their semi-empirical line-shape model was demonstrated to be an effective tool for deducing n_{2D}, SQW thickness and indium composition. Magneto-photoluminescence can be used to measure more electronic properties of SQW materials than is possible with steady-state PL. With this technique, PL is measured as a function of magnetic field perpendicular to the SQW. The field quantizes energy states in k-space parallel to the plane of the SQW which places carriers in one-dimensional, Landau level energy states. Analysis of the field dependent line shape yields subband energy separations and Landau level energy separation ($\hbar\omega_c$).

Using $\omega_c = eB/m^*$ yields the reduced cyclotron effective (CR) masses for individual subbands. A typical spectra is shown in Figure 12. Jones et al. pioneered work in determining CR-effective masses for the techno-logically important InGaAs/GaAs SQW PHEMT structure where both electron and hole effective masses were determined [56]. Material parameters determined from magneto-PL analysis are used in both the phenomenological and semi-empirical models discussed earlier.

Hybrid Shubnikov-de Haas-Photoluminescence Analysis

The hybrid Shubnikov-de Haas—photoluminescence (SdH-PL) analysis is a new analysis technique that combines SdH and steady-state PL data to extract n_{2D} in high-performance PHEMT structures with multiple subband occupancy and severe parallel conduction of a heavily-doped contact layer [46]. The technique is based on the fact that local minima in the longitudinal magneto-resistance occur when the Fermi energy is midway between Landau levels due to the suppression of scattering between the current carrying extended states of the Landau levels (LLs).

For a system in an applied perpendicular magnetic field with two occupied subbands separated by energy E_{01}, which is diagrammed in Fig. 13, the minima occur with two periods of oscillations. The energy spacing of LLs is given by $\hbar\omega_c = \hbar eB/m^*$ where e is the electronic charge, B is the magnetic field strength and m^* is the cyclotron resonance effective mass (CR-mass). As $\hbar\omega_c$ changes with B and because more LLs are occupied in the n=0 subband (see Fig. 13), the Fermi level passes through LLs more often for n=0; therefore, oscillations associated with the n=0 subband occur more rapidly than for the n=1 subband. The rapid oscillations are periodic in 1/B [44] with minima in the magnetoresistance (see Fig. 11) occurring at field strengths denoted as B_{min}.

In general, the 2-D carrier concentration is given by the integral over energy (E) of the density of states $(D(E))$ times the Fermi-Dirac distribution $(f(E))$. At low temperatures $kT << \hbar\omega_c$, one can approximate $f(E)$ by a step function and $D(E)$ for each subband as a summation of

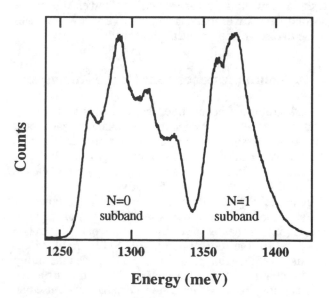

Figure 12. Magneto-photoluminescence spectrum of PHEMT structure with multiple subband occupancy showing Landau level fine-structure. Recorded with an 8 tesla field and at 77K. Courtesy of E. D. Jones, Sandia National Laboratories.

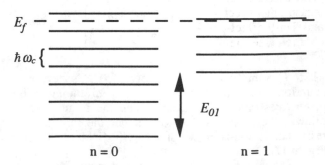

Figure 13. Spin-degenerate Landau level energy diagram showing the conditions for a local minimum in the longitudinal resistance with two subbands occupied. From Lovejoy et al., Appl. Phys. Lett. **64**, 3634 (1994).

δ-functions with LL degeneracies of $2eB/h$ where the factor of 2 is for spin-degeneracy (see equation 13). Thus $n_{2D} = (l_0 + l_1)2e\ B/h$ where, for the ith subband, l_i is the filling factor of LLs and the ith subband carrier concentration is $l_i 2e\ B/h$.

For the n=0 subband, at each applied B_{min} where minima in the longitudinal magnetoresistance occur, l_0 takes integer values. In the SdH-PL technique l_0 at each B_{min} is determined from the low-frequency oscillation in the longitudinal resistance [46] and l_1 from $l_1 = l_0 - \dfrac{E_{01}}{\hbar\omega_c}$ which follows from Fig. 13. The 2-D carrier concentrations is given by $n_{2D} = (l_0 + l_1)2e\ B_{min}/h$. Both $\hbar\omega_c$ and E_{01} can be determined in a number of ways. Cyclotron resonance measurements can be used to determine $\hbar\omega_c$ and E_{01} can be extracted from low temperature PL spectra which show two peaks when two subbands are occupied [43,46]. Of course, magneto-PL can be used to determine both $\hbar\omega_c$, and E_{01}. Another technique would be to employ nominally identical SQWs with undoped caps which can be analyzed by quantized Hall analysis to yield accurate n_{2D} and then back out $\hbar\omega_c$ with the SdH—PL expression and E_{01} determined by PL.

In actuality, disorder broadens the LLs from the δ-function expression assumed earlier and introduces localized states between them. However, since many LLs are occupied in PHEMT material at fields of several tesla, the above analysis is insensitive to the actual form of LL broadening. A sophisticated model such as one incorporating a Gaussian density of states could be used [51]; however, the δ-functions density of states model to yield excellent results [46].

The SdH-PL analysis is a hybrid analysis technique that utilizes SdH- and PL-data to accurately determine the 2-D gas concentration in PHEMT structures with multiple subband occupation and with severe parallel conduction due to the heavily-doped contact layer. The technique requires magnetic field sweeps of only a few tesla, is noise tolerant and is much faster than iterative-etch techniques. Results of the new technique are shown to yield excellent agreement with the results of established measurement techniques. Because low temperature PL and Hall effect measurements are routinely used to characterize device material, no additional device processing or testing is required to implement the hybrid SdH-PL analysis technique. Furthermore, it is easily implemented for automated computer analysis and should prove useful in both research and industrial environments.

CONCLUSION

A review of minority and majority carrier transport characterization techniques was presented. Minority transport was discussed in the context of base transport in HBTs where the base transit time can be an important component of the total transit time through the transistor. Characterization techniques to measure minority carrier mobilities in heavily-doped CS and theoretical results were reviewed. Majority carrier transport was discussed in the context of high-performance PHEMT structures where parallel conduction and multiple subband occupancy makes the measurement of 2-D electron concentration (n_{2D}) difficult. Techniques to measure n_{2D} were reviewed.

A review of techniques to measure the minority carrier mobility in heavily-doped CS was presented. Application of two techniques, the ZFTOF and f_T techniques, to measurements of GaAs and InGaAs were summarized. The recent result that minority carrier mobilities are as great as twice the majority carrier mobility in comparable doping levels in GaAs shows that minority carrier mobility measurements in technologically important CS are necessary for accurate device modeling. Theoretical results for minority carrier mobility in heavily-doped GaAs were reviewed and shown to be in good agreement with measured minority carrier mobilities. The theoretical discussion showed that very different behaviors for minority and majority carriers are expected due to the different scattering environments for minority and majority carriers.

Characterization techniques to measure the 2-D carrier concentration in high-performance PHEMT structures were reviewed. The parallel conduction of heavily-doped contact layers and the 2-D conduction channel was shown to distort standard electrical characterization methods. Electrical techniques including both standard and variable magnetic field Hall effect and Shubnikov-de Haas (SdH) techniques were reviewed. Non-destructive techniques, photoluminescence (PL) and magneto-PL, were described. In addition, a new hybrid SdH-PL was presented.

Compound semiconductor technology must include accurate design tools for device optimization. To realize good models, accurate material parameters are required. Application of the techniques reviewed in this paper to measure transport parameters of the technologically important CS materials and device structures should help improve device performance and expand applications of CS technology.

ACKNOWLEDGMENT

This work was supported by the Department of Energy under contract #DE-AC04-94AL85000.

REFERENCES

1. J. R. Lowney and H. S. Bennett, *J. Appl. Phys.* **69**, 7102 (1991).
2. D. L. Rode, *Phys. Rev. B* **3**, 3287 (1971).
3. W. Walukiewicz, J. Lagowski, L. Jastrzebski and H. C. Gatos, *J. Appl. Phys.* **50**, 5040 (1979).
4. W. Walukiewicz, J. Lagowski, L. Jastrzebski, M. Lichtensteiger and H. C. Gatos, *J. Appl. Phys.* **50**, 899 (1979).
5. M. S. Lundstrom (Ed.). (1990). *Fundamentals of Carrier Transport.* Addison-Wesley Publishing Company.
6. H. Ehrenreich, *J. Phys. Chem. Solids* **8**, 130 (1959).
7. J. D. Jackson (1975). *Classical Electrodynamics* (Second Edition ed.). John Wiley & Sons Ltd.
8. W. P. Dumke, *Solid-State Electron.* **28**, 183 (1985).
9. R. A. Höpfel, J. Shah, P. A. Wolff and A. C. Gossard, *Phys. Rev. Lett.* **56**, 2736 (1986).
10. W. van Roosbroeck and W. Shockley, *Phys. Rev* **94**, 1558 (1954).
11. H. B. Bebb and E. W. Williams (Eds.). (1972). *Photoluminescence I: Theory.* New York: Academic Press.
12. O. von Roos, *J. Appl. Phys.* **54**, 1390 (1983).
13. H. L. Chuang, P. D. Demoulin, M. E. Klausmeier-Brown, M. R. Melloch and M. S. Lundstrom, *J. Appl. Phys.* **64**, 6361 (1988).
14. H. L. Chuang, M. E. Klausmeier-Brown, M. R. Melloch and M. S. Lundstrom, *J. Appl. Phys.* **66**, 273 (1989).
15. H. Ito and T. Ishibashi, *J. Appl. Phys.* **65**, 5197 (1989).
16. T. Furuta, N. Shigekawa, T. Mizutani and A. Yoshii, *Appl. Phys. Lett.* **55**, 2310 (1989).
17. T. Furuta and M. Tomizawa, *Appl. Phys. Lett.* **56**, 824 (1990).
18. T. Furuta, M. Taniyama and M. Tomizawa, *J. Appl. Phys.* **67**, 293 (1990).
19. N. Shigekawa, T. Furuta and K. Arai, *Appl. Phys. Lett.* **57**, 67 (1990).
20. M. I. Nathan, W. P. Dumke, K. Wrenner, S. Tiwari and S. L. Wright, *Appl. Phys. Lett.* **52**, 654 (1988).
21. K. Beyzavi, K. Lee, D. M. Kim, M. I. Nathan, K. Wrenner and S. L. Wright, *Appl. Phys. Lett.* **58**, 1268 (1991).
22. S. Lee, A. Gopinath and S. J. Pachuta, *Electronics Letters* **27**, 1551 (1991).
23. S. Tiwari and S. Wright, *Appl. Phys. Lett.* **56**, 563 (1990).
24. D. M. Kim, S. Lee, M. I. Nathan, A. Gopinath, F. Williamson, K. Beyzavi and A. Ghiasti, *Appl. Phys. Lett.* **62**, 861 (1993).
25. J. Slater D. B., P. M. Enquist, J. A. Hutchby, A. S. Morris and R. J. Trew, *IEEE Electron Device Lett.* **15**, 91 (1994).
26. I. E. Getreu (1978). *Modeling the Bipolar Transistor.* Elsevier Scientific Publishing Company. .
27. J. Slater D. B., P. M. Enquist, F. E. Najjar, M. Y. Chen, J. A. Hutchby, A. S. Morris and R. J. Trew, *IEEE Electron. Dev. Lett.* **12**, 54 (1991).
28. R. K. Ahrenkiel, D. J. Dunlavy, H. C. Hamaker, R. T. Green, R. E. Lewis, R. E. Hayes and H. Fardi, *Appl. Phys. Lett.* **49**, 725 (1986).
29. R. K. Ahrenkiel, D. J. Dunlavy, D. Greenberg, J. Schlupmann, H. C. Hamaker and H. F. MacMillian, *Appl. Phys. Lett.* **51**, 776 (1987).
30. M. L. Lovejoy, B. M. Keyes, M. E. Klausmeier-Brown, M. R. Melloch, R. K. Ahrenkiel and M. S. Lundstrom, *Extended Abstracts of the 22nd (1990 International)* *Conference on Solid State Devices and Materials*, 613 (1990).
31. M. L. Lovejoy, B. M. Keyes, M. E. Klausmeier-Brown, M. R. Melloch, R. K. Ahrenkiel and M. S. Lundstrom, *Jap. J. Appl. Phys.* **30**, L135 (1991).
32. M. L. Lovejoy, M. R. Melloch, R. K. Ahrenkiel and M. S. Lundstrom, *Solid-State Electr.* **35**, 251 (1992).
33. M. L. Lovejoy (1992). *Minority Carrier Diffusivity Measurements in III-V Semiconductors by the Zero-Field Time-of-Flight Technique.* Ph. D. thesis, Purdue University.
34. M. L. Lovejoy, M. R. Melloch, M. S. Lundstrom, B. M. Keyes and R. K. Ahrenkiel, *J. Electron. Mater.* **23**, 669 (1994).
35. M. L. Lovejoy, M. R. Melloch, M. S. Lundstrom, B. M. Keyes, R. K. Ahrenkiel, T. J. de Lyon and J. M. Woodall, *Appl. Phys. Lett.* **61**, 822 (1992).
36. C. M. Colomb, S. A. Stockton, S. Varadarajan and G. E. Stillman, *Appl. Phys. Lett.* **60**, 65 (1992).
37. M. L. Lovejoy, M. R. Melloch, M. S. Lundstrom and R. K. Ahrenkiel, *Appl. Phys. Lett.* **61**, 2683 (1992).
38. E. S. Harmon, M. L. Lovejoy, M. R. Melloch, M. S. Lundstrom, T. J. de Lyon and J. M. J. M. Woodall, *Appl. Phys. Lett.* **63**, 536 (1993).
39. E. S. Harmon, M. L. Lovejoy, M. R. Melloch, M. S. Lundstrom, D. Ritter and R. A. Hamm, *Appl. Phys. Lett.* **63**, 636 (1993).
40. M. Ilegems (Ed.). (1985). *The Technology and Physics of Molecular Beam Epitaxy.* New York: Plenum.
41. D. A. Anderson, N. Apsley, P. Davies and P. L. Giles, *J. Appl. Phys.* **58**, 3059 (1985).
42. M. L. Lovejoy, M. R. Melloch, M. S. Lundstrom, B. M. Keyes and R. K. Ahrenkiel (1993). Electronic Materials Conference, Santa Barbara, CA: The Minerals, Metals and Materials Society.
43. H. Brugger, H. Müssig, C. Wölk, K. Kern and D. Heitmann, *Appl. Phys. Lett.* **59**, 2739 (1991).
44. C. S. Chang, H. R. Fetterman and C. R. Viswanathan, *J. Appl. Phys.* **66**, 928 (1989).
45. J. S. Kim, D. G. Seiler and W. F. Tseng, *J. Appl. Phys.* **73**, 8324 (1993).
46. M. L. Lovejoy, J. A. Simmons, P. Ho and P. A. Martin, *Appl. Phys. Lett.* **64**, 3634 (1994).
47. M. Santos, T. Sajoto, A. Zrenner and M. Shayegan, *Appl. Phys. Lett.* **53**, 2504 (1988).
48. P. A. Martin, J. M. Ballingall, P. Ho and T. J. Rogers, *J. Electon. Mater.* **23**, 1303 (1994).
49. s. For a comprehensive review (1987). *The Quantum Hall Effect.* Berlin: Springer-Verlag. .
50. T. Englert, D. C. Tsui, A. C. Gossard and C. Uihlein, *Surface Science* **113**, 295 (1982).
51. D. Weiss and K. v. Klitzing (Eds.). (1986). *High Magnetic Fields in Semiconductor Physics.* Berlin: Springer-Verlag.
52. S. K. Lyo and E. D. Jones, *Phys. Rev. B* **38**, 4113 (1988).
53. E. D. Jones, T. E. Zipperian, S. K. Lyo, J. E. Schirber and L. R. Dawson, *J. Electron. Mater.* **19**, 533 (1990).
54. S. K. Brierley, *J. Appl. Phys.* **74**, 2760 (1993).
55. S. K. Brierley and H. T. Hendriks (1995). *International Workshop on Semiconductor Characterization: Present Status and Future Needs,* (pp. PP 58). Gaithersburg, MD, USA: National Institute of Standards and Technology.
56. E. D. Jones, S. K. Lyo, I. J. Fritz, J. F. Klem, J. E. Schirber, C. P. Tigges and T. J. Drummond, *Appl. Phys. Lett.* **54**, 2227 (1989).

628

Room Temperature Photoluminescence Characterization of HEMT Quantum Well Structures

Steven K. Brierley and Henry T. Hendriks

Raytheon Advanced Device Center Research Laboratories
358 Lowell Street, Andover, MA 01810

Room temperature photoluminescence is presented as a non-destructive technique for characterizing High Electron Mobility Transistor epitaxial structures. Fitting a phenomenological line shape model to the photoluminescence spectra allows the extraction of quantitative parameters. It is shown how these can be used to track the reproducibility of epitaxial growth and to optimize the layer structure for improved device performance.

INTRODUCTION

High Electron Mobility Transistors (HEMTs) have assumed increasing importance in recent years for microwave and millimeter-wave electronics applications. To fabricate these devices, a complex set of process steps is imposed upon a relatively complex active layer structure: an InGaAs channel (perhaps strained) sandwiched between barrier layers of AlGaAs or AlInAs. Achievement of maximum performance requires that both the epitaxial layer structure and the processing be of high quality. Assuring the correctness of the quantum well structure prior to incurring the expense of wafer processing is thus of great importance.

Conventional characterization of heterostructures (by Hall, C-V profiling, TEM, etc.) is generally either destructive or provides limited information about the quality of the quantum well region. These difficulties can be avoided by fitting room temperature photoluminescence (PL) spectra to a phenomenological line shape model. The parameters extracted from the fit (sub-band energies, Fermi level, etc.) provide quantitative information related to key characteristics of the quantum well, such as indium mole fraction, well width, symmetry, etc. and permit calculation of the channel sheet density.

LINE SHAPE MODEL

A detailed derivation of the line shape model used for room temperature PL spectra of pseudomorphic HEMTs has been given previously.[1] This section summarizes some of the key concepts and assumptions in the model.

The intensity of the direct optical transition between an electron and a hole sub-band at a particular wave-vector depends upon the density of states and the respective occupation probabilities of the two sub-bands at the given wave-vector. For a two-dimensional system, the theoretical joint density of states is a step function in photon energy centered on the band gap. However, scattering due to carrier-carrier interactions and carrier-impurity interactions broadens the step function. Following Cingolani, et al,[2] the broadened density of states step function is represented phenomenologically by

$$D(\hbar\omega) \propto 1/\{1 + \exp[-(\hbar\omega - E_g)/\Gamma]\} \qquad (1)$$

where E_g is the energy gap separating the two sub-bands, and Γ is the broadening parameter. For room temperature spectra, Γ typically is on the order of 4 - 10 meV.

The occupancies of the electron and hole sub-bands (as a function of photon energy) are given by the Fermi distribution. Electrons are assumed to be in equilibrium with the lattice. Empirically, however, it is found that in order to fit properly the room temperature PL spectra, holes must be available over a substantial energy (or wave-vector) range, implying a hole temperature significantly in excess of the lattice temperature. This is not unreasonable, since the photogenerated holes from the barrier layers are injected into the InGaAs quantum well with energies on the order of 0.1 to 0.2 eV above the lowest energy available hole states. Recognizing that a certain amount of arbitrariness is involved, 700K is taken to be a reasonable hole temperature.

The total PL intensity is just the sum of intensities for the transitions between pairs of electron and hole sub-bands:

$$I(\hbar\omega) = \Sigma \; A_{ij}D(\hbar\omega)f_{ei}(\hbar\omega)f_{hj}(\hbar\omega) \; , \qquad (2)$$

where the A_{ij} are the numerical coefficients (which include the interband matrix elements) and the summation is over the various electron and heavy hole sub-band indices (i and j, respectively) which contribute to the optical transitions. The reduced symmetry due to quantum confinement (and, in the case of pseudomorphic HEMTs, the compressive stress of the InGaAs in the quantum well) serves to lift the degeneracy of the light and heavy hole bands, with the light hole sub-bands being shifted higher in energy. This shift of the light hole bands is greater than the separation of heavy hole bands due to quantum confinement.[3-5] Thus, only transitions to the heavy hole sub-bands are considered in the model. (Note that in this discussion, "heavy" refers to the effective hole mass perpendicular to the plane of the quantum well. The reduced symmetry due to quantum confinement and strain also mixes the character of the light and heavy hole sub-bands, resulting in an in-plane heavy hole effective mass which is expected to be less than the perpendicular mass.[6]) All sub-bands are assumed to be parabolic.

Only two electron and two hole sub-bands are sufficiently populated with carriers to make a significant contribution to the PL line shape. The model therefore contains eleven adjustable parameters to fit the four possible optical transitions: four peak amplitudes (A_{ij}), two electron sub-band energies (E_i), the Fermi energy, the hole quasi-Fermi energy, two phenomenological broadening parameters (Γ_i), and a baseline offset. The parameters for a given PL spectrum are evaluated by fitting it to Equation (2) using the Levenberg-Marquardt method for non-linear curve-fitting.[7]

RESULTS

The PL spectra were obtained using the 514.5 nm line of an Ar$^+$ laser with peak power densities in the vicinity of a few W/cm^2. The signal was dispersed through a 1-meter double monochromator, with a spectral resolution of 0.7 meV, and sensed with a cooled high purity Ge detector using phase sensitive detection. All spectra were corrected for the system response. The responsivity of Ge detectors cooled to 77 K falls sharply for energies below about 0.75 eV. Correcting for the detector response extends the useful range down to 0.73 eV which is sufficient (barely) to observe the room temperature spectra of lattice matched InGaAs/InP HEMTs.

The device structures were standard pseudomorphic or lattice-matched HEMTs. The pseudomorphic HEMT structure is double-pulse doped, with Si doping pulses on either side of the strained InGaAs channel, whereas the lattice-matched structure is single-pulse doped, containing only one Si doping pulse in the barrier layer above the lattice-matched InGaAs channel.

For pseudomorphic HEMTs, magnetoluminescence[8] as well as cyclotron resonance[9] measurements have been made of the in-plane electron effective mass in the n=1 sub-band. Jones, et al, [8] found good agreement with their data, assuming a parabolic sub-band, for an effective mass of 0.071. Also assuming a parabolic sub-band, the n=1 hole effective mass was found to be 0.15 for indium mole fractions typical of pseudomorphic HEMTs.[8] In the absence of any experimental data on the effective masses of higher electron and hole sub-bands, these values ($m_e = 0.071$ and $m_{hh} = 0.15$) are used for the n=2 sub-band effective masses as well.

Figure 1 shows the measured room temperature PL spectrum and the corresponding line shape fit for a typical double-pulse doped pseudomorphic HEMT. The distinguishable peaks are identified by the sub-bands between which the transitions occur. The energies of the first two electron sub-bands (relative to the lowest heavy

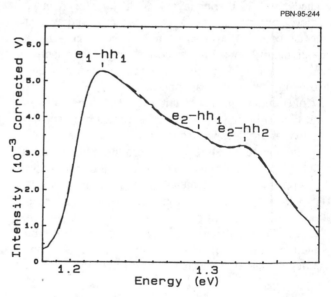

Figure 1. Room temperature photoluminescence spectrum of a typical double-pulse doped pseudomorphic HEMT. The solid curve is the experimental spectrum; the dashed line is the fit to the line shape model. The peaks are labeled by the electron and hole sub-bands between which the transitions occur.

hole sub-band) are found to be $E_1 = 1.204$ eV, $E_2 = 1.286$ eV, and the Fermi energy is $E_f = 1.262$ eV. Even for this case of a nominally symmetric double-pulse doped structure, small asymmetries due to the spacings or concentrations of the doping pulses above and below the channel partially mix the sub-band states, allowing otherwise forbidden transitions.[10] This is manifested as the small peak labeled "e_2-hh_1" around 1.29 eV.

From these energy levels, several important parameters can be derived. For instance, the amount by which the band gap of the quantum well differs from GaAs is roughly proportional to the composition and therefore can be used as a surrogate for In mole fraction. The true quantum well band gap is not simply the energy between the lowest electron and hole sub-bands, since quantum confinement raises them above the bottom of their respective wells by an amount dependent upon both the well width and depth. It is possible, however, to correct for the well width effect on the sub-band shift. For an infinite square well, the energy separation between the second and first energy level is three times that of the first level and the bottom of the well. Thus, knowledge of the sub-band separation allows the absolute energy of the first sub-band to be determined. Double-pulse doped quantum wells are reasonably symmetric, suggesting that a similar relationship between the energy levels exists. Empirically, it is found that the ratio between the sub-band separations is not three but 2.85. So the well

depth (relative to GaAs) can be approximated by $\Delta E_g = 1.424 - \{E_1 - (E_2 - E_1)/2.85\}$ eV. The term in brackets is the "effective" band gap of the quantum well if there were no quantum confinement effects.

One key parameter related to the output power of a HEMT is the gate-drain reverse breakdown voltage, defined as the reverse voltage between the gate and the drain at which the gate leakage current is 1 mA/mm of gate periphery. In Figure 2, the average gate-drain breakdown voltage of test FETs on processed wafers is plotted vs ΔE_g determined from the PL spectra on those same wafers prior to processing. In spite of the scatter in the data (other factors also affect breakdown), it is clear that as the well depth (In mole fraction) increases beyond a threshold, the breakdown voltage drops sharply.

Another example relating device performance to quantum well characteristics involves the well symmetry. The A_{ij} coefficients in Equation (2) are proportional to the interband matrix elements for the various transitions. Since transitions for which $i \neq j$ are forbidden in a perfectly symmetric well, the value of the A_{21}/A_{11} ratio is an indicator for the quantum well asymmetry. In Figure 3, the maximum source-drain current with the gate forward biased at 1 v. for standard test FETs is plotted as a function of the asymmetry of the quantum well in the underlying pseudomorphic HEMT structure. It is apparent that more symmetric wells yield higher currents.

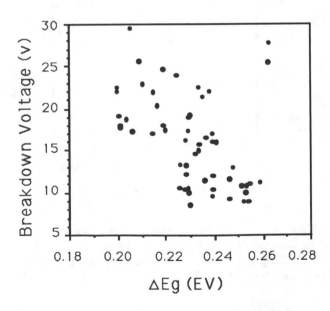

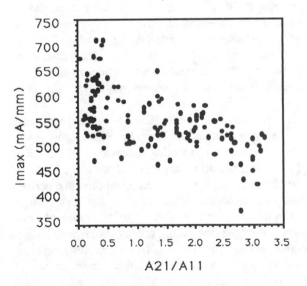

Figure 2. Dependence of the gate-drain breakdown voltage of pseudomorphic HEMTs on the the effective band gap of the quantum well, ΔE_g, which is proportional to the indium content of the quantum well.

Figure 3. Maximum open channel current of pseudomorphic HEMTs as a function of the symmetry of the quantum well, expressed by the normalized intensity of the nominally forbidden transition between the second electron sub-band and the first heavy hole sub-band.

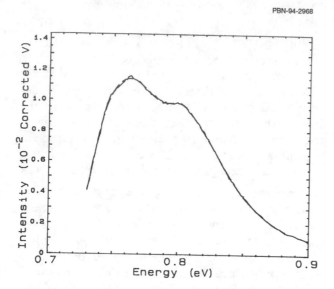

PBN-94-2968

Figure 4. Room temperature photoluminescence spectrum of a typical single-pulse doped lattice-matched In-GaAs/InP HEMT. The solid curve is the experimental spectrum; the dashed line is the fit to the line shape model.

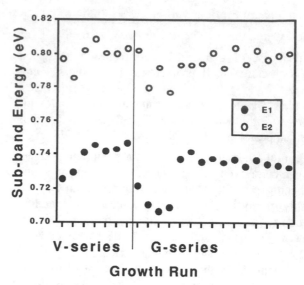

Figure 5. Trend in the first two electron sub-band energies for a series of lattice-matched HEMT structures. Note the deviation from" nominal" when the growth was shifted to a different chamber.

Turning now to lattice-matched structures, Figure 4 shows the room temperature PL spectrum of a lattice-matched InGaAs/InP HEMT along with the corresponding fit to the line shape model. The transition energies from the two lowest electron sub-bands are found to be 0.735 and 0.798 eV. For this material system, the in-plane electron sub-band effective mass is taken to be 0.049.[11] Although one expects the in-plane "heavy" hole mass to be be smaller than the bulk value, there have been no experimental determinations of that effective mass, so it is taken to be the same as the bulk effective mass, 0.47.[12]

These lattice-matched HEMTs are single-pulse doped, and the quantum wells are therefore highly asymmetric. The line shape model, however, is phenomenological: it makes no assumptions about the underlying shape of the quantum well. Thus, even though a derived parameter such as the "effective band gap" is no longer meaningful, it is still possible to to determine electron sub-band energies, the Fermi energy, and transition amplitudes. Figure 5 tracks the two electron sub-band energies over a series of wafers, encompassing a transfer of the growth process from one MBE chamber ('V' series) to another ('G' series). Note that it took several runs to calibrate the new chamber to grow the same material as had been growing in the old one, a fact which was quickly apparent from the PL spectra whereas standard post-growth electrical tests gave no indication of the problem. At a minimum, therefore, the technique can be used as a

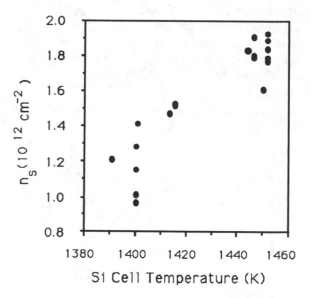

Figure 6. Correlation between the sheet density in the quantum well of lattice-matched HEMTs (from Equation (3)) and the temperature of the Si cell during the growth of the doping pulse.

quality control monitor.

From the Fermi level and the two electron sub-band energies, the channel sheet density can be determined by summing the Fermi-Dirac integrals for the two sub-bands and multiplying by the constant density of states:

$$n_s = (m_e/\hbar^2\pi)\, kT \sum_i \mathcal{F}_0(\eta_i). \qquad (3)$$

For a two-dimensional system, $\mathcal{F}_0(\eta_i)$ has the analytic form $\ln[1+\exp(\eta_i)]$, where $\eta_i=(E_f-E_i)/kT$ is the reduced Fermi energy for the ith electron sub-band.[13] Figure 6 plots the sheet density calculated from the PL energies as a function of the temperature of the Si dopant cell in the chamber during the growth of the doping pulse. For fixed growth time, the total doping in the pulse is proportional to the Si flux, which is set by the temperature of the cell. The figure shows that as the Si temperature (flux) is increased, the sheet density in the quantum well increases also, as expected.

CONCLUSION

The data presented here have demonstrated the validity of room temperature photoluminescence as a non-destructive characterization tool for HEMT structures. By fitting spectra with a phenomenological line shape model, quantitative parameters (energy levels, transition amplitudes) can be used to characterize the quantum well region of HEMTs. These quantitative parameters are useful both for tracking growth reproducibility of the heterostructure and for correlating the quantum well structure with device performance, thus providing a basis for optimizing the epitaxial structure.

ACKNOWLEDGMENT

This work was partially supported by ARPA on the MIMIC Phase II program, contract N00019-91-C-0210.

1. Brierley, S.K., *J. Appl. Phys.* **74**, 2760–2767 (1993).
2. Cingolani, R., Stolz, W., and Ploog, K., *Phys. Rev. B* **40**, 2950–2955 (1989).
3. Ridley, B.K., *J. Appl. Phys.* **68**, 4667–4673 (1990).
4. Jogai, B. and Yu, P.W., *Phys Rev. B* **41**, 12650–12658 (1990).
5. Reithmaier, J.-P., Hoger, R., Riechert,H., Hiergeist, P., and Ab-streiter, G., *Appl. Phys. Lett.* **57**, 957–959 (1990).
6. Osbourn, G.C., *Superlattices and Microstructures* **1**, 223–226 (1985).
7. Press, W.H., Flannery, B.P., Teukolsky, S.A., and Vetterling, W.D., *Numerical Recipes* New York: Cambridge University Press, 1989, ch. 14 pp. 523–528.
8. Jones, E.D., Lyo, S.K., Frist, I.J., Klem, J.F., Schirber, J.E., Tiggs, C.P., and Drummond, T.J., *Appl. Phys. Lett.* **54**, 2227–2229 (1989).
9. Liu, C.T., Lin, S.Y., Tsui, D.C., Lee, H., and Ackley, D., *Appl. Phys. Lett.* **53**, 2510–2512 (1988).
10. Brierley, S.K., Hoke, W.E., Lyman, P.S., and Hendriks, H.T., *Appl Phys Lett.* **59**, 3306-3308(1991).
11. Kopf, R. F., Wei, H. P., Perley, A. P., and Livescu, G., *Appl. Phys. Lett.* **60**, 2386–2388 (1992).
12. Alavi, K., Aggarwal, R. L., and Groves ,S. H., *Phys. Rev. B* **21**, 1311–1315 (1980).
13. Blakemore, J.S., *Semiconductor Statistics* New York: Dover, 1987 App. B, p. 361.

Use of Pressure for Quantum-Well Band-Structure Characterization

J. H. Burnett* and P. M. Amirtharaj
Semiconductor Electronics Division, NIST, Gaithersburg, MD 20899, U.S.A.

H. M. Cheong and W. Paul
Department of Physics, Harvard University, Cambridge, MA 02138, U.S.A.

E. S. Koteles
Institute for Microstructural Sciences, NRC of Canada, Ottawa, Ontario K1A OR6 Canada

B. Elman
GTE Laboratories, 40 Sylvan Road, Waltham, MA 02254, U.S.A.

We discuss the use of externally applied pressure for determining the origin of the peaks in absorption spectra (and related spectra) of semiconductor heterostructures. This technique depends on the different effective mass dependences of the various types of heterostructure energy levels involved in absorption transitions, such as electron, heavy-hole, light-hole, exciton, and defect levels, and the dissimilar pressure dependences of the various masses. Measurements of the hydrostatic or uniaxial pressure dependences of the spectral peaks thus distinguish peaks associated with these different types of energy levels. The approach is demonstrated for GaAs/AlGaAs quantum wells and HgTe/HgCdTe superlattices using hydrostatic pressure. We also briefly discuss the use of externally applied uniaxial pressure for distinguishing heavy- and light-hole-related peaks, and using the heavy-hole/light-hole splitting for determining quantitatively the amount of built-in uniaxial strain in the heterostructure layers.

INTRODUCTION

Practical applications of any semiconductor quantum-well (QW) system require quantitative knowledge of the energy-level positions for particular material structures; e.g., the operation of resonant-tunneling devices depends critically on the precise energy-level structures in various regions of the device. These are determined most accurately by measurements of absorption spectra, or the related photoluminescence excitation (PLE) or photoconductivity excitation spectra. These spectra usually consist of numerous peaks, associated primarily with heavy-hole (HH) subband-to-conduction subband ground-state-exciton transitions, light-hole (LH) subband-to-conduction subband ground-state-exciton transitions, HH and LH excited-state-exciton transitions, and defect transitions. From a comparison of the energies and shapes of the peaks with quantum-well band-structure models, one can infer important QW parameters, such as well width and width uniformity, band offsets, exciton binding energies, alloy composition, and defect composition. However, all this requires correct assignments of the peaks to particular transitions, which is not wholly reliable, since the assignments are made through comparison with the same models one is trying to infer from the peak structure.

However, the peaks can be distinguished on the basis of the different parametric dependences on the effective masses of the different energy levels associated with the various types of transitions. Thus, changes in the masses result in different and distinguishable changes in the various energy levels, and thus distinguishable changes in the transition-peak energies. In this work we have examined the technique of using externally applied pressure to continuously modify the masses and band gaps while monitoring the resulting change in the peak energies, enabling establishment of the origin of the peaks in QW optical spectra. We give two examples of the technique using externally applied hydrostatic pressure. Then we discuss how the different uniaxial pressure coefficients of the heavy- and light-hole levels can be used to distinguish these levels, and can be used to determine the amount of built-in strain in strained-layer systems.

---

*National Research Council - NIST Research Associate

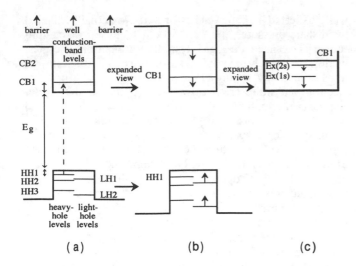

(a)   (b)   (c)

FIGURE 1. (a) Schematic diagram of a general quantum well, showing conduction-band levels, heavy-hole levels, and light-hole levels (separated for clarity). The vertical arrows above indicate the pressure dependences of the conduction-band edges. (b) Expanded view indicating the pressure dependences of the quantum-well levels. (c) Expanded view indicating the pressure dependences of the exciton levels.

GENERAL METHOD

Figure 1(a) is a schematic of the energy levels of a general semiconductor type I QW structure (electrons and holes confined in the same layers), indicating several subbands in the conduction-band (CB) QW, and several HH and LH subbands (shown separately for clarity) in the valence-band (VB) QW. An absorption transition from a HH subband to a CB subband is indicated by the dashed arrow. The energy of the absorption transition consists of four parts, which are considered to be independent in simple, envelope-function models for this system: 1) the well band gap (Eg in Fig. 1(a)), 2) the electron QW energy level (CB1 in Fig. 1(a)), 3) the heavy-hole QW energy level (HH1 Fig. 1(a)), and 4) a correction (decrease) to the sum of these three terms given by the exciton binding energy (not shown in Fig. 1(a), Ex(1s) in Fig. 1(c)). Each of these parts has an effective-mass dependence (or equivalently a band-gap dependence) and consequently, as is discussed, a pressure dependence. The pressure dependence of each part is discussed below.

1) It is observed empirically that the direct (Γ) band gap, Eg, of all zinc-blende semiconductors increases with hydrostatic pressure, at a rate of 100 to 150 meV/GPa [1]; e.g., for GaAs (with an atmospheric pressure band gap of ~1.52 eV at low temperature), an externally applied hydrostatic pressure of 1.4 GPa increases the band gap by ~10%. The arrows at the top of Fig. 1(a) indicate an increase in energy of the CB well and barrier band edges with pressure, with respect to the VB band edges.

2) In the Kane model $\mathbf{k} \cdot \mathbf{p}$ theory, the effective masses of the zone center CB electron and VB light hole increase

approximately linearly with band gap [2]. For example, for GaAs, an applied hydrostatic pressure of 1.4 GPa increases the electron and light-hole effective masses by ~10%, a result well verified experimentally [3]. In a simple one-band, infinite-barrier-height model, the quantum-well energy levels have a dependence on the effective mass of $1/m^*$ (which approximately holds in more sophisticated multiple-band, finite-barrier-height models). Thus, e.g., for GaAs, an applied pressure of 1.4 GPa *decreases* the CB quantum-well energy levels, with respect to the band edge, by ~10%. This energy level decrease with pressure is indicated by the arrows in the expanded view of the conduction band QW in Fig. 1(b).

3) In the Kane model, the heavy-hole mass is determined by higher energy bands, and so its fractional hydrostatic pressure dependence is expected to be much weaker. Since the QW energy levels have roughly a $1/m^*$ effective mass dependence, and the heavy hole is much heavier than the electron and light hole ($m_{hh}^* \cong 0.51$ versus $m_e^* \cong 0.067$ and $m_{lh}^* \cong 0.082$ for GaAs), the heavy-hole QW levels are much more closely spaced than the electron and light-hole levels, as indicated in Fig. 1(b). Since the fractional pressure dependence of the heavy-hole levels is smaller than that of the electron or light-hole levels, the actual pressure dependence of the heavy-hole levels is much smaller, and can be neglected. The light-hole levels behave similarly to the electron levels in energy and hydrostatic-pressure dependence, except that since the levels decrease in energy with pressure, the levels move *up* relative to the VB edge, as indicated by the arrows in the expanded view of the VB QW in Fig. 1(b).

4). The energies of the allowed optical transitions between VB and CB levels are decreased by excitonic effects. For bulk material, the excitonic levels are given approximately by the hydrogenic formula $Ex = \mu e^4/2\hbar^2\varepsilon^2 n^2$, where e is the electron charge, $\hbar$ is Planck's constant divided by 2π, ε is the dielectric constant, n is the exciton quantum number, and μ is the reduced effective mass of the electron and an appropriately averaged hole, $\mu^{-1} = m_e^{*-1} + m_h^{*-1}$. Note that Ex is proportional to the reduced mass. The pressure dependence of the reduced mass is approximately the same as the pressure dependence of the electron mass. Thus, e.g., for GaAs an applied external hydrostatic pressure of 1.4 GPa *increases* the exciton binding energy by ~10% for $n = 1$, the exciton ground state. (There is also a small effect due to the pressure dependence of the dielectric constant.) From the hydrogenic formula above, for $n = 2$, $Ex(n = 2) = Ex(n = 1)/4$. Thus for the $n = 2$ exciton state, the increase in the exciton energy with pressure is 1/4 that for the $n = 1$ state.

For confined excitons in heterostructures, the hydrogenic model is not valid and the situation is more complicated. Excitons are associated with each subband, and there are separate sets of HH and LH exciton levels, with more complicated dependences on the masses. However, we have calculated the pressure dependences of the HH and LH exciton ground and first excited states, using a variational method, similar to that of Greene, Bajaj, and Phelps [4], with non-separable wavefunctions, two variational parameters, and assuming finite barrier heights. The calculations show that, though the exciton energies differ significantly from the bulk values, the energies are still approximately proportional to the appropriate pressure-dependent reduced masses, with the HH exciton ground-state binding energy increasing with pressure more rapidly than the first-excited-state energy, as in the bulk [5]. These results are indicated schematically in Fig. 1(c), which shows an expanded view of the lowest energy CB QW level and some of its associated exciton states.

Other transitions are possible, for example, transitions to or from localized defect states. In these cases one or the other of the subband level pressure dependences (Fig. 1(b)) is absent. Also, if a hydrogenic impurity is involved, the pressure dependence of the impurity energy, approximately proportional to the carrier mass, must be considered in a manner similar to the exciton effects discussed above.

In summary, the HH → CB ground-state-exciton transitions, the LH → CB ground-state-exciton transitions, the corresponding first-excited-state-exciton transitions, and some defect transitions, can in principle be distinguished by the hydrostatic-pressure dependences of the various levels involved, summarized schematically in Fig. 1. Two examples follow: a III-V GaAs/AlGaAs QW and a II-VI narrow-band-gap HgTe/HgCdTe superlattice.

GaAs/AlGaAs 70-Å QUANTUM WELL

Figure 2 shows photoluminescence excitation (PLE) spectra at 2 K of a 70-Å GaAs/Al$_{0.22}$Ga$_{0.78}$As quantum

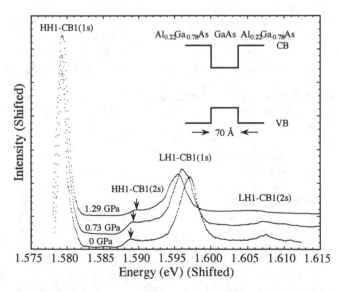

FIGURE 2. PLE spectra of a 70-Å wide GaAs/AlGaAs quantum well, taken at the hydrostatic pressures 0, 0.73, and 1.29 GPa, at 2 K. The two higher pressure spectra have been shifted vertically and shifted down in energy (78.3 meV for the middle spectrum and 141.0 meV for the top spectrum).

well, indicated in the inset, for three externally applied hydrostatic pressures. In PLE spectroscopy, an excitation source is scanned through absorption transition energies, while a monochromator, set to detect at the lowest transition energy, monitors the luminescence resulting from electrons and holes created from absorptions at the higher energies, thermalizing down to the lowest levels and recombining. Thus, in a plot of photoluminescence (PL) intensity versus excitation energy, PL intensity peaks correspond to absorption peaks; i.e., PLE spectra are essentially absorption spectra (with some modifications due to recombination dynamics). For this sample, the hydrostatic pressure was applied using a diamond anvil cell, with He as a pressure medium [6]. The energy scale of the spectra is correct for the bottom spectrum, taken at atmospheric pressure. The other two spectra, taken at 0.73 GPa and 1.29 GPa, have been shifted vertically and shifted down in energy (78.3 meV for the middle spectrum and 141.0 meV for the top spectrum) to line up the lowest energy peaks for comparison purposes.

We begin with tentative assignments of the peaks to various transitions, made from comparison of peak positions with envelope-function calculations of energy levels, with the assignments labeled above the peaks. According to these assignments, the lowest energy peak of each spectrum corresponds to the lowest HH subband-to-lowest CB subband ground-state-exciton (1s) transition, HH1-CB1(1s). The next small peak above this in energy corresponds to the lowest HH subband-to-lowest CB subband first-excited-state-exciton (2s) transition, HH1-CB1(2s). The next higher energy peaks correspond to the lowest LH subband-to-lowest CB subband exciton 1s and 2s transitions LH1-CB1(1s),(2s). Higher energy HH and LH transitions are observed, but we chose to display a scale expanded near low energy to show clearly the pressure shifts of the low-energy peaks. The spectra show that the LH1-CB1(1s) transition energy shifts *down* with pressure with respect to the HH1-CB1(1s) transition energy, and that the HH1-CB1(2s) transition energy shifts *up* with pressure with respect to the HH1-CB1(1s) transition energy.

A plot of the energies of these two peaks, with the energy of the HH1-CB1(1s) transition subtracted out, as a function of hydrostatic pressure is shown in Fig. 3. The lower data represent the pressure dependence of the energy difference between the ground- and first-excited state of the lowest-energy HH exciton. The upper data represent the pressure dependence of the HH/LH splitting. Three-band-model envelope-function calculations of the pressure dependences of these levels, incorporating the pressure effects discussed in the previous section, are shown by the solid lines. The excitonic energies were calculated using a variational method, as discussed. The absolute energies of the data and calculations differ somewhat for the upper data, due presumably to uncertainties in input parameters, such as the well width, the barrier x value, and the effective masses. However, the calculated pressure dependences, which are much less sensitive to these parameters, are quite close to the observed behavior for both sets of data. The fit

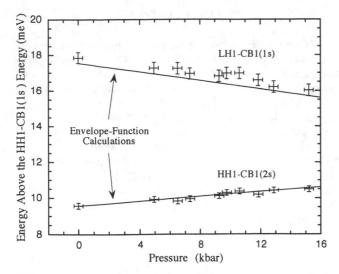

FIGURE 3. Hydrostatic-pressure dependences of the HH1-CB1(2s) transition (lower data) and the LH1-CB1(1s) transition (upper data), with the ground-state HH1-CB1(1s) transition energy subtracted out. The solid curves are envelope-function calculations, as discussed in the text.

demonstrates unambiguously the correctness of the assignments, a demonstration available by no other technique we are aware of. The calculated pressure dependences of several other higher energy peaks, such as the LH1-CB1(2s) peak, are also consistent with the measured dependences, confirming these other assignments.

HgTe/HgCdTe SUPERLATTICE

Figure 4(a) shows the photoluminescence spectrum of a HgTe/Hg$_{0.15}$Cd$_{0.85}$Te superlattice (45-Å HgTe layers/61-Å HgCdTe layers) at 2.04-GPa hydrostatic pressure [7]. The lowest energy peak has an energy of ~130 meV, which is approximately the energy of the superlattice band gap calculated for this system, in the envelope-function model. This peak, and peaks at this energy for other HgTe/HgCdTe samples with similar structures, have thus been assigned to transitions across the superlattice band gap. These peaks have, in fact, been used routinely to characterize the superlattice band gap for these structures, considered for use in high-efficiency infrared detectors.

However, Fig. 4(b) shows the pressure dependence of the peak, along with the pressure dependence of the superlattice band gap calculated with an envelope-function approach based on the eight-band Kane model. The calculated pressure dependence is clearly inconsistent with the data. In this case the pressure dependence strongly indicates that this widely accepted assignment is incorrect. In fact, the weak pressure dependence suggests that the transition is defect or interfaced related. Other explanations, preserving the superlattice-band-gap-transition assignment, are in principle possible, such as the envelope-function approximation being inappropriate for this system, or the band offset having an enormous pressure dependence. However, these possibilities are generally regarded as

improbable, or else extremely interesting if true [7]. These pressure measurements constitute the only rigorous test, to our knowledge, of the generally accepted peak assignment, which appears to be incorrect.

DISCUSSION

We have demonstrated the usefulness of the hydrostatic-pressure technique for two very different systems, the III-V GaAs/AlGaAs system and the II-VI narrow-band-gap HgTe/HgCdTe system. In fact, it can be used in general for any semiconductor heterostructure system for which the effective-mass model is appropriate, assuming the pressure dependences of the masses are known. However, even if the pressure dependences of the masses have not been measured, if the effective-mass model is valid, then the pressure dependences of the masses are easily calculable, with the accuracy needed, from the pressure dependences of the band gaps. Other approaches, such as PLE polarization analysis, have been used for distinguishing the low-energy QW HH and LH transitions [8]. However, the pressure technique has the advantage of being very reliable for the higher-energy QW levels, and distinguishes the other important transitions, such as those associated with exciton excited states and defect states, as well.

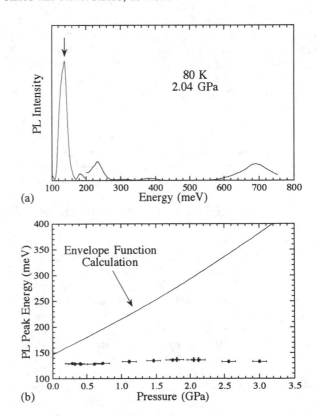

FIGURE 4. (a) Photoluminescence spectrum of a HgTe/Hg$_{0.15}$Cd$_{0.85}$Te superlattice with layer thicknesses of 45Å/61Å, taken at 80 K and 2.04 GaP hydrostatic pressure. (b) Pressure dependence of the lowest energy peak indicated by the arrow in (a), along with the calculated pressure dependence of the superlattice band gap.

The technique, however, is relatively difficult to implement on a routine basis. First, it requires the use of a hydrostatic pressure cell, operated at low temperatures. Though we have used a diamond anvil cell for the measurements shown here, a much simpler piston and cylinder pressure apparatus could be used for the limited pressure range (~ 0 to 1 GPa) needed. Still, although these cells are not expensive, they require specialized techniques and are not routine laboratory facilities. Second, the differences in hydrostatic pressure shifts of the various types of peaks are relatively small, typically being no more than several meV over the pressure range (0 to 1 GPa). Thus, the technique requires high-resolution spectra. The samples must be of high quality to exhibit sharp exciton peaks and the spectrometer must be capable of meV resolution. Nevertheless, if confident peak assignment is essential, we believe this is one of the most reliable methods available.

A related technique is the use of externally applied *uniaxial* pressure to distinguish HH and LH energy levels [9-11]. For bulk semiconductors, application of uniaxial compressive stress splits the HH- and LH-exciton degeneracy at the zone center due to symmetry breaking, the LH-exciton transition energy increasing more rapidly with pressure than the HH-exciton transition energy. For heterostructures, the HH- and LH-exciton degeneracy is already split by the quantum-well confinement. Application of externally applied uniaxial compressive stress in the plane of the QW layers results in a more rapid increase in exciton transition energies associated with the LH sublevels, than for exciton transition energies associated with the HH sublevels. Since application of uniaxial pressure in this direction is relatively easy to implement, and the effects are large, this can be the basis of a convenient method to distinguish HH and LH levels in heterostructures. Uniaxial-pressure measurements, however, are more difficult to interpret, and have not been used to distinguish excited exciton states. The technique can be considered as complementary to hydrostatic-pressure measurements.

Uniaxial-pressure dependences of the HH/LH splittings have been determined for a number of important bulk semiconductors, including GaAs, GaP, InAs, and InP [12]. Since application of external uniaxial pressure mimics the effects of built-in strain in QW's due to lattice mismatch [13], HH/LH-splitting measurements can be used as probes of the built-in strain in the heterostructure layers [14]. The amount of built-in strain in the various layers in heterostructures can be controlled by appropriate choice of constituents (and alloy composition in ternary alloys). In some circumstances, structures are designed to minimize built-in strain. In other circumstances, a specific amount of strain is deliberately incorporated, e.g. quantum-well heterostructure laser structures are often designed to incorporate strain because of the beneficial effects on laser performance characteristics [15]. In all these cases, monitoring the HH/LH splittings, and using the measured uniaxial-pressure coefficients, provides an accurate and convenient measure of the strain built into the structures.

CONCLUSIONS

We have discussed, for semiconductor quantum wells, the usefulness of using measurements of the hydrostatic- and uniaxial-pressure dependence of the quantum-well energy levels to establish the origin of peaks in absorption (and related) spectra. The technique can distinguish unambiguously transitions associated with heavy-hole, light-hole, excited-exciton, and defect states. We demonstrated the technique for the III-V GaAs/AlGaAs system and the II-VI narrow-band-gap HgTe/HgCdTe system, and considered its general applications. The difference in the uniaxial-pressure dependences of heavy-hole and light-hole related transitions can be used not only to distinguish these peaks, but can be used also as a probe of the built-in strain in strained-layer heterostructure systems.

ACKNOWLEDGMENTS

We thank P. M. Young and H. Ehrenreich for helpful discussions. The work at Harvard was supported in part by the NSF under Grant No. DMR-91-23829 and by the JSEP under Contract No. N00014-89-J1023.

1. Paul, W., *J. Appl. Phys.* **32**, 2082 (1961).
2. Kane, E. O., *J. Phys. Chem. Sol.* **1**, 249 (1957).
3. DeMeis, Walter M., Ph.D. thesis, Harvard University, 1965.
4. Greene, R. L., Bajaj, K. K., and Phelps, D. E., *Phys. Rev. B* **29**, 1807 (1984).
5. Burnett, J. H., Cheong, H. M., Paul, W., Koteles, E. S., and Elman, B., to be published.
6. Burnett, J. H., Cheong, H. M., and Paul, W., *Rev. Sci. Instrum.* **61**, 3904 (1990).
7. Cheong, H. M., Burnett, J. H., Paul, W., Young, P. M., Lansari, Y., and Schetzina, J. F., *Phys. Rev. B* **48**, 4460 (1993).
8. Miller, R. C., Kleinman, D. A., Nordland, Jr., W. A., and Gossard, A. C., *Phys. Rev. B* **22**, 863 (1980).
9. Sanders, G. D. and Chang, Yia-Chung, *Phys. Rev. B* **32**, 4282 (1985).
10. Jagannath, C., Koteles, E. S., Lee, J., Chen, Y. J., Elman, B. S., and Chi, J. Y., *Phys. Rev. B* **34**, 7027 (1986).
11. Trzeciakowski, W. and Sosin, T. P., to be published in *J. Phys. Chem. Solids*, 1995.
12. Adachi, S., *Physical Properties of III-V Semiconductor Compounds* (Wiley-Interscience, New York, 1992).
13. Pollak, F. H., in *Strained-Layer Superlattices: Physics*, Semiconductors and Semimetals Vol. 32, edited by T. P. Pearsall (Academic Press, Boston, 1990).
14. Koteles, Emil S., Elman, B., Melman, P., Bertolet, Daniel C., Hsu, Jung-Kuei, and Lau, Kei May, 20th International Conference on the Physics of Semiconductors, Thessaloniki, Greece, edited by E. M. Anastassakis and J. D. Joannopoulos (World Scientific Publishing, Singapore, 1990), p. 965.
15. Dunstan, D. J. and Adams, A. R., *Semicond. Sci. Technol.* **5**, 1202 (1990).

Double-Modulation and Selective Excitation Photoreflectance for Wafer-Level Characterization of Quantum-Well Laser Structures

D. Chandler-Horowitz, D.W. Berning, J.G. Pellegrino, J.H. Burnett[*],
and P.M. Amirtharaj

Semiconductor Electronics Division, NIST, Gaithersburg, MD

and

D.P. Bour and D.W. Treat

Xerox-PARC, Palo Alto, CA

A double-modulation photoreflectance (PR) procedure is presented, where both the probe and pump beams are modulated, and the photoreflectance signal can be isolated from the luminescence and the scattered pump beam signals. The PR signal is separated from the other two signals through detection at the sum frequency. A careful choice of frequencies and specially designed filters and tuned amplifiers were needed to achieve optimum operation.. A complete system, along with the necessary circuits, is presented and applied to the characterization of a highly luminescent quantum-well laser structure. The freedom allowed by such a system to easily accommodate any pump wavelength is an important feature. We have exploited this added versatility, and the ordering of the bandgap of the multiple layers required in complex laser structures, to extract the bandgap and alloy composition of each of the constituent regions as well as the built-in strain in the pseudomorphic quantum-well.

INTRODUCTION

Photoreflectance (PR) is a powerful, contactless and nondestructive technique capable of probing interband electronic transitions and built-in electric fields in the surface and interface regions in semiconductor materials and microstructures, and has been widely used as a characterization tool (1). However, its application to highly luminescent systems, such as quantum-well laser structures, has been limited because of the difficulty in minimizing the interference from the luminescence on the PR spectrum. Several techniques have been suggested to overcome this difficulty, but none, so far, is as simple and as robust as the double-modulation technique.

We demonstrate the applicability of the procedure by characterizing a AlGaInP-based quantum-well (QW) laser structure designed for visible emission (2). Multiple pump wavelengths were employed to selectively excite the cladding region, the barrier layers, and the QW. The observed transition energies were used to determine the alloy composition in the active QW region. In addition, the accurate measurement of the splitting between the light-hole and heavy-hole states helped deduce the degree of tetragonal strain present in the pseudomorphic QW. The intense Franz-Keldysh oscillations from the barrier regions yielded the interface electric field strengths.

EXPERIMENTAL DETAILS

Photoreflectance is a contactless form of electro-modulation spectroscopy that utilizes an intense pump beam chopped at a frequency, f_{pump}, to create photo-excited electron-hole (e-h) pairs that periodically change the built-in electric field in the near-surface, space-charge region of the semiconductor. This results in a modulated reflectivity, ΔR (at f_{pump}), which is detected by observing the intensity of a weak reflected-probe beam. The normalized PR signal, $\Delta R/R$, is measured as a function of photon energy. The magnitude of this signal is typically quite small, i.e., $\Delta R/R < 10^{-3}$. An analysis of the oscillatory line shapes yields the optical transition energy, E_i, at the critical points in the electronic band structure. The alloy composition, strain, electric field strength, and a qualitative measure of the crystalline quality, can be obtained from the PR spectra.

The conventional PR system (1) utilizes a probe beam, originating from a monochromator, reflecting off a sample, and striking a detector. The perturbation in the measured signal is caused by a chopped pump beam. In such a configuration, the detector receives not only the reflected monochromatic probe beam, but also the scattered laser pump beam and any luminescence caused by the laser excitation. Since the scattered light and luminescence are also modulated

---

[*] National Research Council - NIST Research Associate

at the chopping frequency, f_{pump}, they can create a large spurious background in the measured spectrum. Structures optimized for light emission, such as a QW laser structure, are particularly prone to such behavior. In many cases, the scattered pump intensity can be filtered with appropriate optical filters, but the removal of the luminescence is more problematic since its wavelength usually coincides with the most interesting spectral region, i.e., the near-bandgap energy range.

The modulated signal, as processed by a lock-in amplifier, can result in signal offsets due to scattering and luminescence that are much higher than the photoreflectance signal itself, and hence, will significantly reduce the sensitivity with which the

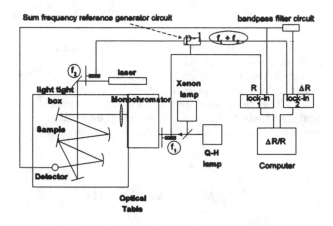

Figure 1 Double Modulation Photoreflectance Apparatus

measurements can be performed. A double-modulation apparatus was constructed where the probe beam was also chopped at f_{probe}, with the intention of isolating the PR signal at the sum $f_{pump}+f_{probe}$ or difference frequency $f_{pump}-f_{probe}$, thereby

discriminating against the other unwanted signals occurring at f_{pump}. Although this method has been used before (3), we are not aware of a detailed description of the hardware or procedure in the literature.

Compared to the conventional system, the double-modulation apparatus, as seen in Fig. 1, has only one additional mechanical chopper and lock-in amplifier associated with the probe beam, along with two specialized circuits. In the conventional system the optical filter used to block the pump beam is not needed. The detector output is split between the two lock-in amplifiers. The first amplifier samples the strong

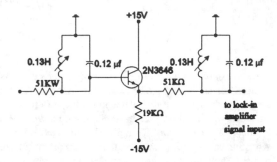

Figure 2 Sum frequency (1250 Hz) bandpass filter

reflected signal, R, at f_{probe}. The reference signal for this lock-in amplifier is obtained directly from the probe chopper. The second lock-in amplifier, with the lock-in reference signal input at the sum frequency, $f_{probe}+f_{pump}$, measures the PR signal. The sum frequency is preferable over the difference frequency to avoid low frequency measurement problems. Convenient values of 750 and 500 Hz were chosen for f_{pump} and f_{probe}, respectively. Two specialized circuits are required in this configuration. The first circuit filters out the strong

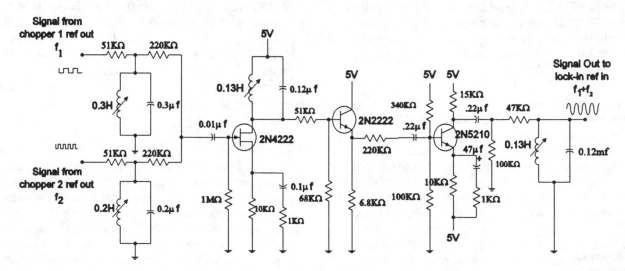

Figure 3 Mixer-amplifier circuit to generate sum frequency (1250 Hz) lock-in reference input from two chopper signals.

component in the reflected beam modulated at frequency f_{probe} that will otherwise overload the second lock-in amplifier's input. The second circuit generates the sum frequency reference signal from the two mechanical chopper reference outputs. Figure 2 shows the two-stage band-pass filter tuned to the desired PR modulation frequency, $f_{probe}+f_{pump}$, which rejects all other frequencies. Typically, the rejected signals are reduced by $\sim\times10^{-6}$, i.e., from a few volts to a few microvolts. Figure 3 shows the circuit that derives the sum frequency of 1250 Hz of the pump frequency (750 Hz) plus the probe frequency (500 Hz). This derived signal is used as the reference signal input for the second lock-in that measures the PR signal. This circuit has inputs from the reference output signals of the two mechanical choppers. The shapes of these signals are nearly square waves which are first fed through bandpass filters tuned to their respective fundamentals to suppress harmonics. The resulting sinewave signals are added and applied to the gate of a JFET 2N4222 mixing transistor. The load for the transistor is a tank network tuned to the sum frequency which is an intermodulation product of the non-linear amplification in the transistor. This tank suppresses all other frequencies, including the original pump and probe frequencies. The second transistor is configured as an emitter-follower buffer, and the third transistor provides some amplification. The last tuned stage further suppresses other signal harmonics. The inductors in both Figs. 1 and 2 are made with adjustable pot cores.

The signal, , observed by the detector is given by:

$$\Sigma = R \cos (2\pi f_{probe}t) [1+ \alpha\cos (2\pi f_{pump}t)] + \beta\cos (2\pi f_{pump}t + \Phi), \quad (1)$$

where α signifies the intensity of the PR signal, i.e, $\alpha = \Delta R/R$; β, the intensity of the background PL and scattered pump radiation; and Φ the phase difference between the PR and background components. When $\alpha \geq \beta$, a conventional, single modulation system (i.e., $f_{probe} = 0$) is adequate. It is when $\beta \gg \alpha$ that the double-modulation procedure is required. It is straight forward to show from Eq. (1) that the signal detected at the sum frequency is given by:

$$\Sigma_{sum} = (\alpha R/2) \cos [2\pi (f_{probe}+f_{pump})t]. \quad (2)$$

Because the output signal is split between the sum and difference frequency, the output (sum frequency component) will be reduced by one half from that in the single-modulation case. The modulation is produced with mechanical light chopper blades and hence is not sinusoidal and contains higher harmonics that can reduce the amplitude slightly. Furthermore, since the two channels used in detecting R and ΔR are not identical and may possess different amplifications, an additional multiplicative factor may be required.

RESULTS AND DISCUSSIONS

This double-modulation PR method was tested on a sample that had a large luminescence emission. A sample of a QW laser structure, optimized for high-power cw red emission (2),

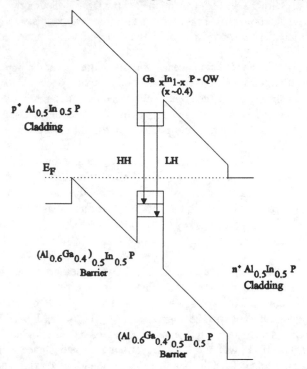

Figure 4 Band edge schematic for the QW laser structure.

comprised of $p^+Al_{0.5}In_{0.5}P/(Al_{0.6}Ga_{0.4})_{0.5}In_{0.5}P/Ga_xIn_{1-x}P\text{-}QW/(Al_{0.6}Ga_{0.4})_{0.5}In_{0.5}P/n^+Al_{0.5}In_{0.5}P$ as shown in Fig. 4, was modulated with a 488-nm Ar^+ laser beam. The intense red,

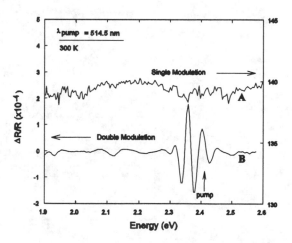

Figure 5 Single-versus-double modulation for PR on sample with large luminescence. Note the change in signal scale of ~100X between case A, single modulation, and case B, double modulation.

visible glow from the luminescence (room temperature) was easily observable by eye. A single-modulation measurement was performed, but the strong luminescence and some scattering produced a signal offset that was more than 100 times larger than the PR signal. The PR signal obtained with the use of the double modulation is shown in Fig. 5 along with a single modulation spectrum. A vast improvement is clearly evident: the large background signal is completely suppressed.

In addition to the luminescence, surface scattering of the pump radiation is a concern in samples with poor surface morphology. The conventional procedure for removing the unwanted light is to use an appropriate spectral band-pass filter that removes the pump radiation but allows the wavelength band of interest to reach the detector. In the double modulation system, as long as the linear dynamic range of the detector is not exceeded the spectral filter is not required. The versatility offered by this modification leads to a complete freedom in the choice of the pump wavelength. We have exploited this key feature also in the characterization of the complex multilayer structure.

The PR spectra, presented in Fig. 6, were measured over an energy range of 1.5 to 2.8 eV that spans the direct gaps of all the constituent layers, i.e., the doped $Al_{0.5}In_{0.5}P$ cladding regions, the undoped 250-nm-thick $(Al_{0.6}Ga_{0.4})_{0.5}In_{0.5}P$ barrier regions that sandwich a ~8 nm-thick active QW region made of $Ga_xIn_{1-x}P$, with $x \approx 0.4$. The structure was designed with the expectation that the bandgaps of the cladding, barrier, and active QW regions would be ~2.5 eV, 2.25 eV, and 1.9 eV, respectively. The ordering of the bandgaps in light emitting devices, i.e., reducing bandgap with depth between the front surface and the active QW region, and the small thickness of the QW, makes it possible to optically probe most of the layers through the front surface. The only part of the structure that is not easily accessible is the n^+ cladding layer that is closest to the substrate. In addition, the ability to vary the pump wavelength may offer a procedure for selective excitation of various regions. Such versatility, inherent in PR (4), can be exploited to untangle the rich spectra observed from a complex multilayer device structure.

The measured PR spectra, displayed in Fig. 6, were recorded with three different pump wavelengths. The first, labeled A, was obtained with λ_{pump} = 351.1 to 363.8 nm (~3.5 eV) Ar^+-laser ultraviolet lines, chosen to exceed E_c, the bandgap of the cladding layer. Clearly, the single transition observed originates in the $p^+Al_{0.5}In_{0.5}P$ front-cladding region. The simple line shapes were fit to yield a value of 2.609 eV for E_c. The absence of any features near the bandgap of the barrier regions or the QW strongly suggest that the field perturbations are confined to the near-surface region. The large carrier density of ~6 x 10^{17} cm^{-3} in the p^+ region and the consequent short screening length is most likely the mechanism responsible for such behavior. Reducing λ_{pump} to below E_c assures the excitation of deeper layers, as shown by curve B, obtained with a pump wavelength of 514.5 nm (2.54 eV). The intense oscillatory feature at ~2.4 eV and the weaker, but

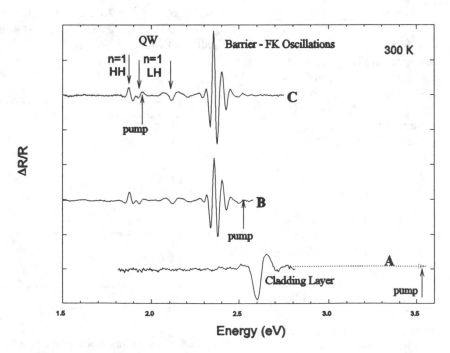

Figure 6 Photoreflectance signals of AlGaInP QW laser structure using different pump energies- A,3.5 eV; B,2.5 eV; and C,1.96 eV.

642

distinct doublet at ~1.9 eV, originate from the barrier and QW regions, respectively. A broad feature, present at 2.108 eV may also originate from the QW, most likely from the spin-orbit split component. Excitation with a $\lambda_{pump} = 632.8$ nm (1.96 eV), which is below the bandgap of the barrier regions, does not affect the spectrum significantly as shown by curve C. This suggests that the photoexcited e-h pairs are not localized in the QW region. The incomplete confinement of the electrons in these structures may be responsible for such a behavior (2).

Further examination of the line shapes strongly suggests that the oscillatory features in the ~2.4 eV region are the result of Franz-Keldysh (F-K) oscillations from the large built-in field in the barrier region. Use of the large-field, asymptotic form of the F-K solution and the widely used graphical technique of plotting $(E_m - E_b)^{3/2}$ versus m, where m is the extrema index, E_m, its energy position, and E_b, the bandgap of the barrier region, displays two regions dominated by electric field strengths of 3.8×10^4 V/cm and 8.4×10^4 V/cm and a bandgap E_b of 2.293 eV. The effective masses for this calculation were obtained from Ref. 2. The built-in field in the barrier region can be estimated from the known structure shown in Fig. 4: a drop of 2.609 eV over 500 nm, the thickness of the barrier region, yields 5.2×10^4 V/cm. The observations suggest that the field is not evenly distributed in the barrier region, but the magnitude of their strengths are what would be expected.

The doublet at ~1.9 eV was fitted using simple excitonic line shapes which yielded transition energies of 1.880 and 1.931 eV. These are most likely due to interband transitions originating from the n = 1 heavy-hole (HH) and light-hole (LH) to the n = 1 conduction band state in the QW. The intensity ratios are consistent with such an assignment. All of the energy positions can be determined to a precision of better that ±0.0005 eV.

Armed with accurate values for the barrier bandgap and the LH and HH transition energies in the QW, one may attempt to determine the alloy composition, x, in the QW. A simple model calculation, similar to that presented in Ref. 5 that includes the effects of the biaxial strain, was used to interpret the QW interband transition energies and parameter values available in the literature (2) for the alloy composition of the bandgap. Linear interpolation between the end point binaries for the lattice parameter, band offsets, deformation potential and elastic constants, and alloy composition and thicknesses expected from the growth details were employed as inputs into the calculation. A comparison of the measured 1HH and 1LH transition energies are consistent with an x-value of 0.47. Such a value is higher than expected from the growth conditions. The source of the discrepancy is most likely the simple model used that does not account for the significant perturbation of the QW energy levels as result of the large built-in field.

CONCLUSIONS

Double-modulation photoreflectance is an easy-to-implement modification to a single-modulation apparatus. The method can extract the PR signal from samples and setups that have large background signals due to scattering and luminescence that could otherwise mask the desired PR signal. The recently available dual-channel choppers might simplify the system still further, and may yield satisfactory results.

An accurate measure of the bandgap of the cladding and barrier regions and the electronic transition energies in the QW, as well as the built-in barrier electric field strengths, provides a detailed picture of the structure. The sensitive dependence of the HH-LH splitting on the biaxial strain and the dependence of the energies of the 1HH and the 1LH transition on the confining barrier potential can be used to elucidate the state of the active region. Such a detailed picture can be used to optimize laser design during the developmental phase. The procedure can also be employed in a manufacturing environment as a contactless and nondestructive wafer probe, prior to expensive processing. The simple intuitive model used and the approximate values for some parameters can be clearly improved

ACKNOWLEDGMENT

The authors wish to express their appreciation of the helpful discussions with Prof. J.R. Anderson of the Physics Department of the University of Maryland during the initial phase of this work. The support from the NIST Advanced Technology Projects office, Dr. T.F. Leedy in particular, is gratefully acknowledged.

REFERENCES

1. O.J. Glembocki, this proceedings; see also F.H. Pollak and H. Shen, *Materials Science and Engineering: R Reports* **R10**, 275 (1993).
2. D.P. Bour, R.S. Geels, D.W. Treat, T.L. Paoli, F. Ponce, P.L. Thornton, B.S. Krusor, R.D. Bringans and D.F. Welch, *IEEE J. of Quantum Electronics* **30**, 593 (1994).
3. C.R. Lu, J.R. Anderson, W.T. Beard, and R.A. Wilson, *Superlatt. Microstruct.* **8**, 155 (1990); C.R. Lu, J.R. Anderson, et al., *Phys. Rev. B* **43**, 791 (1990).
4. H. Shen, X.C. Shen and F.H. Pollak, *Phys. Rev. B* **36**, 3487 (1987).
5. S.H. Pan, H. Shen, Z. Hang, F.H. Pollak, W. Zhuang, A.P. Roth, R.A. Masut, C. Lacelle and D. Morris, *Phys. Rev. B* **38**, 3375 (1988)

Optical Characterization of MBE-Grown GaAs-AlGaAs Superlattices for Infrared Detectors

Z. C. Feng, S. Perkowitz, J. Cen and K. K. Bajaj

Department of Physics, Emory University, Atlanta, GA 30322 - 2430

D. K. Kinell and R. L. Whitney

Lockheed Research Laboratories, Palo Alto, CA 94304

We present a combined optical investigation of GaAs-AlGaAs superlattices grown by molecular beam epitaxy for use as 8-20 μm infrared detectors, using photoluminescence, Raman and Fourier transform infrared spectroscopies. Various structural and physical parameters, such as well size, barrier composition, carrier concentration, mobility, and intersubband transition energy, of these superlattices were obtained by theoretical analysis of the optical results. The use of multiple optical techniques offers comprehensive characterization and further understanding of the physics of long wavelength infrared detectors.

INTRODUCTION

Infrared (IR) methods are important in modern science and technology. Recent interest has focused on the wavelength range 8 - 20 μm, where an atmospheric transmission window makes possible long-range applications [1]. Appropriate detectors are made from GaAs-AlGaAs microstructures, where intersubband transitions provide the detection mechanism [1-3 and references therein]. Careful control of layer thickness, composition, and other parameters is needed to tune these complex structures to the desired wavelengths.

In this study, a combination of optical methods -- photoluminescence (PL), Raman scattering and Fourier transform infrared (FTIR) spectroscopies -- characterizes GaAs-AlGaAs superlattices (SL) grown by molecular beam epitaxy (MBE) for use as IR detectors over 8-20 μm. PL data at 2K, 77K and 300K give intersubband transition energies more conveniently than the usual IR absorption method. PL and Raman data give the composition x(Al) in the barrier layers. Well sizes are deduced by comparing PL data with our calculations for exciton binding energy and PL transition energy *vs.* well width. Widths were also found from FTIR data, which also gave carrier density and other parameters. The widths of the PL and Raman lines indicate SL quality.

EXPERIMENT

The samples were fabricated by MBE, using a solid-source Varian Gen II machine with an arsenic cracker to provide As_2. They were grown on substrates composed of commercial semi-insulating (100) GaAs wafers, 50 mm in diameter and about 450 μm thick. A 1 μm-thick GaAs buffer layer n^+-doped with Si to a carrier density of $\sim 2 \times 10^{18}$ cm^{-3} was grown on the substrate, followed by a 4.0 nm undoped GaAs layer, followed by the SL structure. This consisted of 50 periods of 28 nm undoped $Al_xGa_{1-x}As$ (x ~ 22%) alternating with 4 nm GaAs. A 200 nm GaAs layer, n^+-doped with Si to $\sim 2 \times 10^{18}$ cm^{-3}, formed a cap layer for the structure. Four samples, M14, M15, M41 and M42 were made in this format. Figure 1 shows their structures.

To enhance intersubband transitions within the conduction band, the GaAs wells were delta doped with Si up to $\sim 2 \times 10^{18}$ cm^{-3} within the central 3.2 nm of the wells. Undoped regions 0.4 nm were left at each side of GaAs wells, to prevent diffusion of Si dopants into the barriers. In some samples, the δ-doping regions are up to 3.8 nm wide, occupying most of well regions. For the purpose of comparative study, there is no GaAs cap on the top of one sample, M42.

PL spectra were excited by the 632.8 nm line from a 50 mW He-Ne laser. The laser beam was focused to a spot ~200 μm across on the sample surface. Each sample was freely suspended by a special holder that kept it immersed in pumped liquid helium, in liquid nitrogen, or in vacuum for measurements at 2K, 77K and 300K, respectively. The signals were dispersed by a 1 m spectrometer, detected by a photomultiplier tube, and amplified by a lock-in amplifier under microcomputer control, with a resolution of 1 Å and a sensitivity range between 10 μeV and 1 eV. Reference lines from a pencil Ar lamp were used for calibrations.

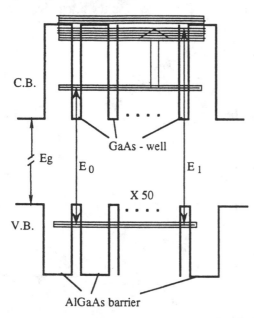

Figure 1 Structure of GaAs-AlGaAs superlattice with 50 periods for 8-20 μm infrared detection.

Raman scattering was performed in the near-backscattering geometry. Samples were mounted on a cold finger attached to a liquid nitrogen dewar, and excited by the 488 nm line from an Ar^+ laser. The scattered light was dispersed by a triple spectrometer, detected by an optical multichannel analyzer with a Si array of 1024 pixels and accumulated by computer, to give a high signal-to-noise ratio, at 2-cm^{-1} resolution.

Fourier transform infrared (FTIR) measurements were made at room temperature (RT) using a Michelson interferometer under computer control. A polished coin-silver mirror provided a reference for absolute reflectivity measurements over the range 20 - 500 cm^{-1} at 2- or 4-cm^{-1} resolution. The incoming radiation encountered the sample surface at near-normal incidence. For each sample, multiple spectra were measured and averaged to enhance the signal-to-noise ratio.

RESULTS

Raman spectra

Figure 2 shows two Raman spectra excited at 488 nm from a typical AlGaAs-GaAs MQW, sample M41, at 80 K. In 2(a), the bands at 270 cm^{-1} and 292 cm^{-1} are due, respectively, to the transverse optical (TO) and longitudinal optical (LO) phonons from the 200 nm-thick GaAs cap. Curve 2(b) shows the results as the GaAs cap layer is etched away. The spectra display LO phonon modes from the $Al_xGa_{1-x}As$ barriers, a GaAs-like one denoted LO_{b1} at 283 cm^{-1}, and an AlAs-like one denoted LO_{b2} at 378 cm^{-1}, and from the GaAs wells, denoted LO_W at 290 cm^{-1}. The LO_W mode was spatially confined inside the GaAs well and therefore shifted to a lower energy than the bulk value of 292 cm^{-1} from the unconfined GaAs layer in (a).

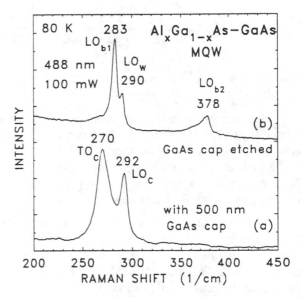

Figure 2 Raman spectra of a GaAs-AlGaAs SL M41, at 80 K.

Photoluminescence spectra

Figure 3 shows a room temperature (RT) PL spectrum of GaAs-AlGaAs sample M41 excited with the 633 nm He-Ne line operating at 20 mW focused on the sample surface. A very strong PL band, peaked at 1.5578 eV, comes from the fundamental recombination E_0 between the ground states of the electron in the conduction subband and the heavy hole in the valence subband of the GaAs well. The energy of E_0 is much higher than that of the band gap for pure GaAs, 1.424 eV at RT [4]. This broad band possesses a full width at half maximum (FWHM) of 66 meV. The intensity of the E_0

transition was 3 to 4 orders of magnitude stronger than the luminescence from bulk and film GaAs or AlGaAs, which is due to the carrier confinement in the multiple quantum wells. This strength indicates good quality for the SL structures.

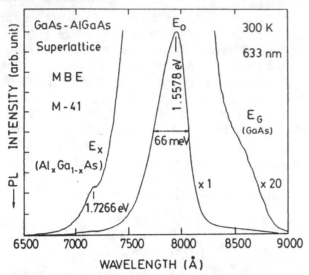

Figure 3   RT PL spectrum of a GaAs-AlGaAs MQW, M41, excited by 633 nm.

When the spectrum was amplified by a factor of 20, a band E_X at 1.7266 eV, due to the recombination from $Al_xGa_{1-x}As$ barriers, was seen on the high energy side of the basic E_0 transition. On the low energy side of the E_0 band, there is a shoulder near 1.4 eV, due to the recombination from the GaAs cap.

Figure 4 exhibits a PL spectrum excited at 633 nm from sample M15 held at 2K, and also shows the spectrum amplified by a factor of 30. The fundamental E_0 peak lies at 761 nm or 1.6288 eV, up-shifted by 71 meV from the value at RT (see Fig. 3.)

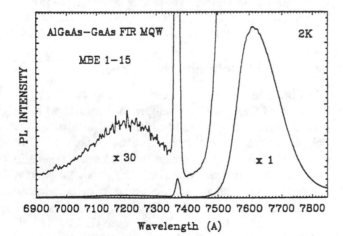

Figure 4   2 K PL spectrum of a GaAs-AlGaAs MQW, M15, excited under 633 nm.

Infrared spectra

Figure 5 shows a RT FTIR reflectivity spectrum from sample M41. Open symbols represent data. There are two high- reflectivity reststrahlen bands in the region of 100 - 450 cm^{-1}. The structure between 250 and 350 cm^{-1} is mainly due to the GaAs and GaAs-like TO phonon vibrations from the SL structure and the GaAs cap. The spectrum between 350 and 450 cm^{-1} is due to the AlAs-like TO phonon modes from the AlGaAs barrier layers. The increase in reflectivity at very low frequencies comes from the free carriers. The theoretical fit is presented by a solid line with details given in the discussion section.

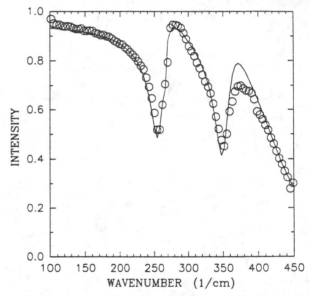

Figure 5   RT FTIR reflectivity of a GaAs-AlGaAs SL, M41. The solid line is from theoretical fit.

DISCUSSION

Deduced parameters of SLs

The parameters derived from our measurements and analysis are listed in Table I.

Raman and PL determination of x

To determine the Al composition x in the barrier layers more precisely, we performed a comprehensive Raman scattering study, including theoretical fits by the modified random-element isodisplacement (MREI) model, for $Al_xGa_{1-x}As$ over the entire x range with the x interval about 10%, grown by the same MBE machine [5]. We derived a precise relationship for the optical

TABLE I   DEDUCED PARAMETERS OF SLs

Sample	M14	M15	M41	M42
{ GaAs cap	200 nm	200 nm	200 nm	none }
PL, 300-77-2 K				
E_0 (eV, 300K)	1.5534	1.5406	1.5578	1.5600
FWHM (meV)	41	57	66	68
E_X(AlGaAs, eV)	1.7312	1.7619	1.7266	1.7266
x (from PL)	21.2%	23.3%	20.9%	20.9%
E_0 (eV, 77K)	1.6408	1.6171	1.6450	1.6485
FWHM (meV)	17	37	40	40
E_1 (eV)		1.6952		
$E_1 - E_0$ (meV)		78.1		
$E_1 - E_0$ (μm)		15.9		
E_0 (eV, 2K)	1.6471	1.6289	1.6506	1.6500
FWHM (meV)	9	31	35	37
E_1 (eV)		1.7215	1.7191	1.7254
$E_1 - E_0$ (meV)		92.9	68.5	75.4
$E_1 - E_0$ (μm)		13.3	18	16.4
calculated, 2K				
E_g(AlGaAs) (eV)	1.821	1.853	1.818	1.818
L_Z (nm, theory)	3.1	3.9	3.0	3.0
FTIR fits, 300K				
ω_{TO1} (cm^{-1})			268.7	268.7
ω_{TO2} (cm^{-1})			361.0	361.0
L_Z (nm)			3.2	3.2
μ (cm^2/V·s)			3900	5000
n (10^{18} cm^{-3})			0.39	0.87

phonon frequencies versus x, which were used to determine the x values in the $Al_xGa_{1-x}As$ barriers. However, this requires us to remove or thinnen the GaAs cap. The sample M42 with no GaAs cap showed x=0.21, which is consistent with the result determined below by PL method.

PL is a sensitive and powerful tool for determining compositions in $Al_xGa_{1-x}As$, especially using recently-derived precise dependences of E_g on x for direct-gap $Al_xGa_{1-x}As$ between 2 and 300 K [6-7]. We determined x in the barrier layers, as listed in Table I, with the following equation for RT PL:

$$E = 1.424 + 1.455x \quad (eV). \tag{1}$$

At 2 and 77 K, the carriers are more likely to occupy lower energy levels. Electrons and holes in the AlGaAs barriers may quickly tunnel into the wells to fill the ground or excited states. Therefore, no near band edge transitions from the AlGaAs barriers can be detected by

low temperature PL in Fig. 4. At RT, due to the Boltzmann distribution, the carriers are more likely to occupy higher energy levels in both the wells and barriers. Then the percentage of electrons and holes tunneling from barriers into wells is greatly decreased, leading to the observation of E_X at RT PL in Fig. 3.

Computer fits of FTIR spectra

From a classical model, the FIR dielectric response function can be expressed [8]:

$$\varepsilon(\omega) = \varepsilon_{inf} + S_j \, \omega_{Tj}^2/(\omega_{Tj}^2 - \omega^2 - i\Gamma_j\omega)$$
$$- \omega_p^2/\omega(\omega + i/\tau) \tag{2}$$

where ε_{inf} is the high frequency dielectric constant, ω_{Tj}, S_j and Γ_j are the frequency, strength and damping constant of the j-th TO mode, respectively. The last term in Eq. (2) represents the free carrier contribution with the carrier scattering time τ and the plasma frequency

$$\omega_p = (4\pi ne^2/m^*)^{1/2}, \tag{3}$$

where n is carrier concentration and m^* is effective mass. It has also been established that a SL composed of alternating layers of two types of semiconductors may have an effective dielectric function in the far IR region without strong absorption [8]:

$$\varepsilon_{SL}(\omega) = [\varepsilon_1(\omega)d_1 + \varepsilon_2(\omega)d_2]/(d_1 + d_2), \tag{4}$$

where ε_1, ε_2, d_1 and d_2 are the dielectric functions and thicknesses of the two types of layers, respectively.

Computer fits were performed for our data, based upon eqs. (2)-(4). The results, seen in Fig. 5, gave a series of physical parameters, some listed in Table I. FIR fits lead to not only the lattice vibration properties including TO phonon mode frequencies, strengths and damping constants, but also important structural parameters such as well size, and transport properties such as carrier concentration and mobility.

Theoretical calculation

Theoretical calculations of the binding energies of the heavy and light hole excitons as a function of well size with different Al compositions for the finite square potential GaAs wells have been made since 1983 [9] and used to accurately determine the well sizes from the QW PL peak energy [10]. Using this model and recent values for the band off-sets of GaAs-$Al_xGa_{1-x}As$, we have performed an extensive calculation for the dependence of heavy hole exciton binding energy E_B and PL transition energy E_T on well width L_W for different values of x in the $Al_xGa_{1-x}As$ barriers. Details of these calculations will be published elsewhere. Typical results for E_B and E_T versus L_W with x = 0.22 and 0.24 are shown in Fig. 6. The calculation uses the ratio of the conduction band off-set to the valence band off-set $\Delta E_c:\Delta E_v$ as 0.60:0.40; the band gap dependence of $Al_xGa_{1-x}As$ on x was taken as $E_g(Al_xGa_{1-x}As) - E_g(GaAs) = 1.366x + 0.22x^2$ [11]; and the effective mass of $Al_xGa_{1-x}As$ was linearly interpolated between the values for GaAs and AlAs [12]. In this work, the x values of $Al_xGa_{1-x}As$ barriers from four different samples range between 20% and 24%. By insertions between two set of data in Fig. 6, we obtained the calculated well widths listed in Table I. The value of 3.0 nm for M41-M42 agrees well with the result 3.2 nm obtained from the FTIR fits.

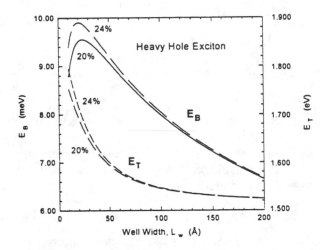

Figure 6   Calculated dependence of heavy hole exciton binding energy E_B and recombination transition energy E_T versus L_z with x=0.20 and 0.24, respectively.

PL determination of the inter-subband transition energy

Figure 3 shows a weak and broad band, E_1, at 720 nm or 1.7215 eV. It cannot come from the near-band edge emission of the $Al_xGa_{1-x}As$ barrier because its band gap is 1.853 eV, as shown in Table I. Rather, E_1 is due to the recombination between the first excited states in the conduction and valence bands, which obey the selection rule $\Delta n = 0$ [13] (Fig. 1). The difference between the excited and ground states for holes in the valence well is much less than that for electrons in the conduction band. So E_1-E_0 measures mainly the intersubband transition in the conduction band, i.e. the response to 8-20 μm radiaton of these SLs designed as detectors.

CONCLUSION

As has recently been emphasized by one of us (SP), optical techniques are powerful means to characterize semiconductors and their microstructures [14]. The combination here of several spectroscopic methods with theoretical analyses gives comprehensive characterization of complex SLs, leading to further understanding of the physics of long wavelength IR detection. Of special importance, we have presented a new PL approach to determine the intersubband transitions of GaAs-AlGaAs structures in the 8-20 μm range. It is more convenient than other methods, such as IR absorption.

ACKNOWLEDGMENT

The authors are grateful to Prof. W. J. Choyke for use of the PL facility, Dr. L.L. Clemen, M. Yoganathan and Dr. M. Macler for their technical assistance.

REFERENCES

[1] E. Rosencher, B. Vinter and B. Levine edited, *Intersubband Transitions in Quantum Wells*, Plenum, 1992, New York.

[2] Z. C. Feng ed. *Semiconductor Interfaces and Microstructures*, World Scientific, 1992, Singapore.

[3] Z. C. Feng ed. *Semiconductor Interfaces, Microstructures and Devices: Properties and Applications*, Institute of Physics, 1993, Bristol.

[4] H. C. Casey, Jr. and M. B. Panish, *Heterostructure Lasers*, Academic Press, New York, 1978, Part A, p. 193.

[5] Z. C. Feng, S. Perkowitz, D. K. Kinell, R. L. Whitney and D. N. Talwar, Phys. Rev. **B47**, 13466 (1993).

[6] G. Oelgart, R. Schwabe, M. Heider and B. Jacobs, Semicond. Sci. Technol. **2**, 468 (1987).

[7] M. Guzzi, E. Grilli, S. Oggioni, J. L. Staehli, C. Bosio and L. Pavesi, Phys. Rev. **B45**, 10951 (1992).

[8] S. Perkowitz, Solid State Commun. **84**, 19 (1992).

[9] R. L. Greene and K. K. Bajaj, Solid State Commun. **45**, 825 (1983).

[10] T. Hayakawa, T. Suyama, K. Takahashi, M. Kondo, S. Yamamoto, S. Yano and T. Hijikata, Appl. Phys. Lett. **47**, 952 (1985).

[11] C. Bosio, J. L. Staehli, M. Guzzi, G. Burri and R. A. Logan, Phys. Rev. **B38**, 3263 (1988).

[12] A. Adachi, J. Appl. Phys. **58**, R1 (1985).

[13] R. Dingle, in *Festkorperprobleme (Advances in Solid State Physics)*, Vol. 15, H. J. Queisser ed., Pergamon, 1975, Vieweg, p. 21.

[14] S. Perkowitz, *Optical Characterization of Semiconductors* (Academic press, London, 1993).

Identification of Shallow Acceptors in GaAs Using Time-Delayed Photoluminescence Spectroscopy.

A.M.Gilinsky and K.S.Zhuravlev

Institute of Semiconductor Physics,
Siberian Branch of Russian Academy of Sciences,
pr.Lavrentieva, 13, 630090 Novosibirsk,
RUSSIA
E-mail: gilinsky@ispht.nsk.su and zhuravlev@ispht.nsk.su

We describe a new technique that facilitates discrimination among the near-bandgap acceptor-related luminescence bands of undoped or lightly doped GaAs or close material, and provides observation of acceptors which are present in small amounts. It can be used for studying shallow acceptors' contents of samples prepared by different growth technologies in a wide range of sample quality. The technique exploits narrowing and modification of the bands observed in the time-delayed transient PL spectra taken on the microsecond time scale.

INTRODUCTION

Characterization of residual doping is one of the most important issues for optimization of semiconductor growth technology. One of the most sensitive and widely used tools for studying impurity contents of different semiconducting materials is photoluminescence (PL) spectroscopy [1,2,3]. For identification and quantification of dopants by the PL, transition energies and intensities of respective bands should be derived from the PL spectra. In practically important case of GaAs and related compound materials, however, characterization of shallow acceptors from PL data often becomes complicated because of spectral overlap of band-to-acceptor and donor-to-acceptor groups of bands and/or small spectral separation of the bands corresponding to different dopants [4,5]. Therefore it is often desirable to use some approach that could enhance the resolving power of PL analysis.

In this work we propose a new spectroscopic technique that facilitates discrimination among the multitude of near-bandgap acceptor-related PL bands of undoped or lightly doped GaAs or close compound material, and provides observation of acceptors which are present in small amounts simultaneously with the dominating ones. This technique can be used in the course of growth technology optimization and is applicable to samples prepared by different growth techniques in a wide range of growth conditions variation. The technique is based on our recent observation of unexpectedly long decay of transient acceptor-related PL of undoped GaAs at low temperatures (T<10K) [6]. As we have shown, all the conduction-band-to-acceptor (c,A) and pair donor-to-acceptor (D,A) PL bands of undoped GaAs samples exhibit similar prolonged nonexponential decay curves after pulsed excitation, the curves' tails being still clearly observable even at delays t≈100 μsec after the excitation pulse. Such a behavior is established by domination of indirect-in-real-space (D,A) recombination mechanism at delay t>0,5 μsec. This result disagrees with previous conclusions about dominating mechanism of recombination after interband excitation of pure GaAs, usually assumed to be due to band-to-band and/or band-to-impurity transitions in direct-gap material (see for example thorough paper by Bimberg et al [7]). Due to thermal reexcitation of electrons from donors to the conduction band, (c,A) PL bands appear in time-resolved spectra. As time after the excitation pulse goes on, sufficient narrowing of the bands is observed in the delayed time-resolved spectra due to depletion of nearby donor-acceptor pairs. As a result, the bands that appear smeared in the spectrum taken under continuous excitation conditions become clearly separated in time-delayed spectra, while red shift of some of the bands with time points unambiguously to the (D,A) recombination mechanism.

Therefore, examining evolution of time-resolved low-temperature PL spectra with time after the excitation pulse on a microsecond time scale enables one to enhance separation of the acceptor-related bands and clarify their nature. In the following sections we present the experimental results exemplifying the proposed technique, discuss its application range and describe the measurement procedure.

MEASUREMENT PROCEDURE

As it was stated before, the proposed technique requires accumulating a set of time-resolved PL spectra taken at progressively increased delays. To perform time-resolved measurements, we used experimental setup schematically shown on Fig.1. Excitation of the sample luminescence was accomplished by either a frequency-doubled Q-switched YAG or a YAG-pumped pulsed Ti:Sapphire lasers operating at a repetition frequency of 3-10 kHz (excitation wavelength of 532 nm or 760 nm, respectively), with peak power density kept in the range of 0,5-5 kW/cm². After passing diffraction grating and a diaphragm for elimination of laser pump lamps light and lasing crystal luminescence background, the laser beam was impinged on the sample placed into a variable-temperature liquid helium cryostat. Off-axis sample illumination scheme was used in order to exclude detection of luminescence of optical elements and cryostat windows. Luminescence of the sample was dispersed by a double grating monochromator and detected by a cooled S-1-photocathode photomultiplier. Time-correlated single-photon counting was used in the setup in order to achieve the maximum dynamic range of the detection system. Together with optimization of the optical path mentioned above, this results in dynamic range of up to 6-7 orders of magnitude, depending on decay curve accumulation time. We usually used accumulation time corresponding to 0,5-1 million of excitation pulses. To obtain a set of time-resolved spectra, a set of PL decay curves was taken first while scanning the monochromator across the desired spectral region; afterwards, the time-resolved spectra were recomputed from the decay curves set. This approach to time-resolved spectra acquisition enables greater flexibility of temporal window parameters' selection and

subsequent data handling (as compared with predefined temporal windows parameters used in [7] for example) at the cost of greater raw data volume. Throughout the measurements reported in this work the temporal resolution of time-to-digital converter was set to 20 nsec per channel.

MEASUREMENT RESULTS AND DISCUSSION

In this section we describe results of application of the proposed technique to characterization of residual acceptors, and discuss its application range and limitations. In the present work, we will describe data taken on samples of epitaxial GaAs grown on (100) GaAs substrates by different growth techniques. Measurements reported in this work were carried out with samples kept at 4,2K.

Figure 2 shows evolution of time-resolved spectra of acceptor-related PL of two samples grown by liquid phase epitaxy (LPE) with time after the excitation. pulse. The PL intensity is shown as photon count rate divided by excitation pulses frequency. For comparison, PL spectra taken with continuous excitation are also shown by dashed lines. Figure 2,a shows data taken on high-purity sample having $N_D-N_A=2\cdot10^{14}cm^{-3}$, mobility $\mu_{77K}=157.000$ cm²/V·sec and compensation ratio of $\approx0,23$ [8]. At zero delay with respect to excitation pulse, a broad featureless band with maximum intensity around 8310 Å wavelength dominates the time-resolved spectrum. Its high-energy tail which originates from recombination of free electrons in the conduction band shows increased electrons temperature $T_e\approx30K$. Also, an additional broad band is observed around 8380 Å (1.479 eV) at small delay values. After some 0,3 μsec T_e decreases, and small red shift of the dominating band is observed. Afterwards, at delays of 0,7 μsec and higher the dominating band begins to show some fine structure; at this moment, the spectrum is close to the one taken under CW excitation. With time, narrowing of partial components which form the band is observed, resulting in splitting of the band into four separate bands 5 μsec after the excitation. Simultaneously intensity maximum shifts to longer-wavelength band. Two of these four bands shift with time towards low-energy side, while other two remain fixed around 8312 Å (1.491 eV) and 8325 Å (1.489 eV). At large delays the bands show nearly synchronous change of intensities with time. This behavior implies that the bands are related to (c,A) and (D,A)·transitions involving Mg (or Be) and Zn [2]. The band at 8380 Å that disappears quickly after the end of excitation pulse is therefore attributed to LO-phonon replica of excitonic-related transitions but not to the presence of Ge acceptor, as could be infered from the band position alone.

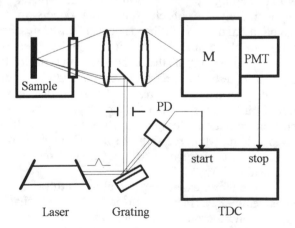

Figure 1. Schematic drawing of the measurement setup. M · monochromator, PMT · photomultiplier, PD · photodiode, TDC · time-to-digital converter.

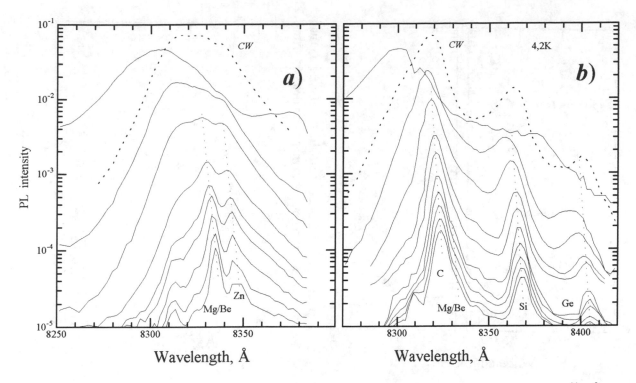

Figure 2. Evolution of the time-resolved PL spectra of two samples grown by LPE, having $N_D-N_A=2\cdot10^{14}$cm^{-3} (*a*) and $2\cdot10^{15}$cm^{-3} (*b*), with time after the excitation pulse (solid lines). The spectra were taken at delays of 0, 0.4, 0.7, 2.5, 4.5, 6, 12, 24, and 45 µsec (*a*), and 0, 0.4, 1.5, 4, 7, 12, 18, 25, 35, and 45 µsec (*b*) with respect to the excitation pulse (from upper to lower). Also shown are spectra taken under CW excitation conditions (dotted lines). Donor-to-acceptor transitions corresponding to different acceptors are marked.

On Fig.2,b the evolution of time-resolved PL spectra of another sample grown by LPE which had an order of magnitude higher impurities concentration is shown ($N_D-N_A=2\cdot10^{15}$cm^{-3}, compensation ratio ≈0,27). Its CW spectrum shows broad bands attributed to carbon, silicon, and germanium acceptors. Despite higher impurity concentration, the decay manner of acceptor-related bands is similar to that shown in Fig.2,a. Some 0,3 µsec after the excitation pulse a carbon-related (c,A$_C$) band that dominates the spectrum during the pump pulse falls in intensity below the neighboring (D,A$_C$) band and stabilizes at approximately 1/40 of its intensity. Similar change of intensities with time is seen for Si- and Ge-related bands. Because of higher donors' concentration and overlap with neighboring bands, (c,A) transitions are seen poorer in this sample than in that shown on Fig.2,a. With time, dominating (D,A) bands shift to lower energy side and become narrowed. Starting from t≈5 µsec, the shoulder is seen on the long-wavelength side of (D,A$_C$) peak that also shifts with time. This shoulder is caused by appearance of additional (D,A) band due to presence of Mg (or Be) acceptor in concentration of approximately 1/10 of that of carbon. This acceptor could not be detected using common CW PL procedures (pump intensity and/or sample temperature change).

Examples of evolution of time-resolved spectra shown on Fig.2 are not specific for the case of samples grown by LPE, which is known to produce low point defects concentration and hence low concentration of capture and nonradiative recombination centers. As indirect-in-real-space (D,A) recombination mechanism dominates decay from the excited state, spatial separation of nonequilibrium electrons and holes captured by donors and acceptors leads to isolation of carriers from defect-induced recombination channels that could be present for example because of sufficient deviation of growth conditions from the quasiequilibrium ones. Therefore, the same temporal behavior as that described previously is seen even in the case of high point-defects concentration. This is exemplified on Fig.3 that shows evolution of the time-resolved PL spectra of the samples grown by molecular beam epitaxy (MBE), the technique that utilizes growth conditions far from equilibrium to produce extremely sharp interfaces. The samples were grown under the same conditions which differed only in amount of intentional doping with silicon. The PL efficiency of all the three samples was at least 10 times less than that of the LPE-grown ones. The sample corresponding to Fig.3,a had no intentional doping and showed p-type conductivity with $N_A-N_D=2\cdot10^{14}$cm^{-3}, while sample of Fig.3,c was

651

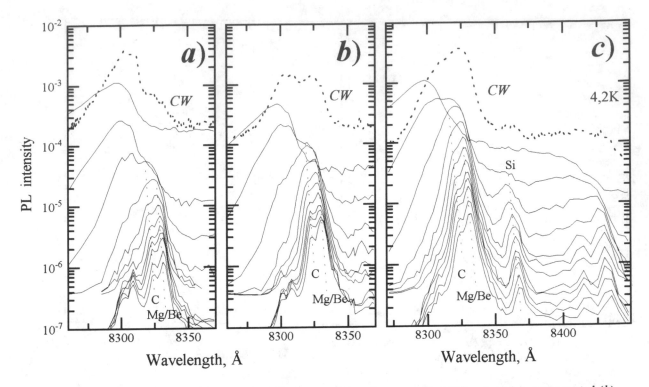

Figure 3. Evolution of the time-resolved PL spectra of samples grown by MBE: p-type (*a*), compensated (*b*), and n-type *(c)*, with time after the excitation pulse (solid lines). The spectra were taken at delays of 0, 0.4, 0.7, 1, 2, 3, 6, 9, 12, 18, 30, and 45 μsec with respect to the excitation pulse (from upper to lower). Spectra taken under CW excitation conditions are shown by dotted lines. Donor-to-acceptor transitions corresponding to different acceptors are marked.

doped to n-type with $N_D-N_A=2 \cdot 10^{14} cm^{-3}$; sample of Fig.3,b had intermediate doping level and was nearly compensated. The behavior similar to that described previously is observed on Fig.3. After initial relaxation of high-density carriers' population created by the laser pulse, two (D,A) bands begin to dominate in the spectra, indicating presence of C and Mg (or more likely Be in MBE-grown samples). It is seen that at the end of measurement time window intensities of (c,A) bands depend on the amount of donors present in the sample. In p-type sample the intensities ratio $I_{c,A}/I_{D,A}$ equals approximately 1/3, in slightly doped sample it decreases to below 1/10, and in the most doped one it falls further down to ≈1/20 with simultaneous switching to shoulder-like appearance of (c,A) bands. The dependence of $I_{c,A}/I_{D,A}$ ratio on donor density provides a method for nondestructive local evaluation of sample conductivity type and estimation of donors concentration. With Si doping level increased, new (D,A) band corresponding to Si_{As} (Si on As sites) acceptor appears in the spectra. Also, the bands that are attributed to defect-induced acceptors [9,10] are seen in the wavelength region up to 8450 Å (shown only on Fig.3,c), with two acceptors manifesting themselves in this case. Similar behavior was observed during study of samples grown by MBE at low substrate temperature, which is used to produce high-defect-density

high-resistivity material for isolation purposes.

Relaxation behavior described above enables one to use low-temperature microsecond-time-resolved PL as a tool for characterization of residual acceptors and clarification of the nature of PL bands that appear because of acceptor-related transitions. The technique enables discrimination between (c,A) and (D,A) recombination mechanisms without the need to conduct multiple spectra measurements at different pump intensity levels and/or sample temperatures. As the technique is based on spatial separation of carriers by capture to impurities, its application range covers materials having impurity density lower than that corresponding to emergence of donors degeneracy due to overlap of wavefunctions of neighboring donors, which occurs around $N_D=2 \cdot 10^{16} cm^{-3}$. Due to its essentially nonequilibrium nature, the technique can be applied for studying p-type material as well as for n-type, even though electrons are absent on donors in equilibrium in p-type material. Up to now we have observed the prolonged decay behavior similar to that described above in p-type samples with holes concentrations up to $N_h≈10^{16} cm^{-3}$, however sharp bands were not observed in the time-resolved spectra in this case.

CONCLUSIONS

We have demonstrated that all the (D,A) and (c,A) PL bands of undoped or lightly doped GaAs exhibit similar prolonged decay after pulsed interband excitation. Sufficient narrowing of these bands is observed in time-resolved spectra taken on microsecond time scale, while red shift of some of the bands with time evidences readily theirs (D,A) nature. This behavior forms a basis for acceptors characterization technique that facilitates discrimination among acceptor-related PL bands of undoped or lightly doped material, enables nondestructive estimation of local donors concentration and evaluation of sample conductivity type and provides observation of acceptors present in small amounts. The technique can be used for studying residual shallow chemical and defect-related acceptors' contents of the samples prepared by different growth technologies in a wide range of growth conditions variation and sample quality.

ACKNOWLEDGMENTS

We would like to express our gratitude to the International Science Foundation for the support of visit of one of us (A.M.G.) to the Workshop through its travel grant program for scientists from the countries of the former Soviet Union. We are also grateful to the groups at our Institute headed by Drs. N.A.Yakusheva, N.S.Rudaya, Yu.B.Bolkhovityanov, N.T.Moshegov, A.I.Toropov, V.P.Migal, and V.V.Preobrazhenskii for supplying us with samples of epitaxial structures.

REFERENCES

1. Dean P.J., *Prog. Crystal Growth Charact.* **5**, 89-114 (1982).

2. Ashen D.J., Dean P.J., Hurle D.T.J., Mullin J.B., and White A.M., *J.Phys.Chem.Solids*, **36**, 1041-1053 (1975).

3. Lu Z.H., Hanna M.C., Szmyd D.M., Oh E.G., and Majerfeld A., *Appl.Phys.Lett.*, **56,** 177-179 (1990).

4. Rao E.V.K., Alexandre F., Masson J.M., Allovon M., and Goldstein L., *J.Appl.Phys.*, **57**, 503-508 (1985).

5. Zemon S., Norris P., Koteles E.S., and Lambert G., *J.Appl.Phys.*, **59**, 2828-2832 (1986).

6. Zhuravlev K.S., Gilinsky A.M., submitted to *Phys.Rev. B*.

7. Bimberg, D., M·nzel H., Steckenborn A., and Christen J., *Phys.Rev. B*, **31**, 7788-7799 (1985).

8. Yakusheva N.A., Zhuravlev K.S., Chickichev S.I., and Shegai O.A., *Cryst.Res.Techn.*, **24**, 235 (1989).

9. Briones F., and Collins Douglas M., *J.Electron.Mater.*, **11,** 847-866 (1982).

10. Szafranek I., Plano M.A., McCollum M.J., Stockman S.A., Jacksom S.L., Cheng K.Y., and Stillman G.E., *J.Appl.Phys.*, **68**, 741-754 (1990).

653

Atomic Force Microscopy as a Process Characterization Tool for GaAs-based Integrated Circuit Fabrication

A.J. Howard, A.G. Baca, R.J. Shul, J.C. Zolper,
M.E. Sherwin, and D.J. Rieger

Sandia National Laboratories
Compound Semiconductor Research Laboratory
Albuquerque, NM 87185-0603

Numerous situations have been encountered during the fabrication of GaAs-based integrated circuits where an in-line and post-process characterization tool which operates in a non-destructive manner was required. We report several examples which demonstrate that the atomic force microscope (AFM) fills this characterization void in our laboratory. The AFM is extremely useful for measuring the unintentional removal of small amounts of GaAs which can be detrimental in the fabrication of devices with very thin active layers, e.g., FET channels within 30 nm of the GaAs surface. We have also characterized the use of AFM as a non-destructive tool to determine complete etching of 1.25 μm Si_3N_4 via holes for a two-level interconnect process. The AFM has also been useful in optimizing the ion implant activation anneal conditions of $Al_{0.75}Ga_{0.25}Sb$ epitaxial layers by comparing the morphology through RMS roughness measurements. Finally we have observed differences in wet etch results based on prior process history, such as 1-5 nm differences in etch depth and morphological differences of ion implanted vs. non-implanted areas. The results and implications from these various AFM processing studies will be presented in this paper.

INTRODUCTION

GaAs-based integrated circuits (ICs) such as complementary heterostructure field effect transistors (CHFETs) (1) and JFETs (2) are currently being developed in our research laboratory. While fabricating these devices we have encountered numerous situations where an in-line and post-process characterization tool which operates in a non-destructive manner is required. The atomic force microscope (AFM) (3), operating without sample preparation in normal clean room environment, has filled this semiconductor processing characterization void.

A number of instances occur during GaAs-based device processing where the unintentional removal of small amounts of GaAs can be detrimental to the performance of devices with very thin active layers, e.g., FET channels within 30 nm of the GaAs surface. Prior to using AFM, these processes were investigated with a conventional surface profilometer and SEM with unsatisfactory results. For example, the selectivity of refractory metal and dielectric plasma etches to GaAs has been previously estimated with very long exposures. The AFM can be used to very accurately determine GaAs

removal rates for short plasma exposures which occur during overetch.

Other uses of the AFM in GaAs IC fabrication can improve on the imaging capabilities of conventional microscopy (scanning electron and optical) and profilometry. These include in-line, non-destructive, process monitoring of micron-sized vias in a two-level interconnect process, surface morphology changes during implant activation annealing of $Al_{0.75}Ga_{0.25}Sb$, and differences in wet etch depth and shallow wet etch characterization. In this paper we report a number of examples where the AFM, with its near atomic-level z- and ~1 nm x-y-resolution has been extremely useful in characterizing and improving GaAs IC fabrication steps. In this paper the AFM is shown to be a highly effective tool for in-line process control monitoring.

RESULTS AND DISCUSSION

The specific tasks to be discussed which demonstrate the utility of the AFM as an important new characterization tool for semiconductor device fabrication are failure analysis, in-line monitoring, and process development. All AFM imaging was performed using a Digital Instruments Dimension 3000 system operating in tapping mode with Si tips in a class 100 cleanroom.

The first example of using the AFM in GaAs IC fabrication falls into the general field of failure analysis. While performing process development for GaAs-based FETs using a self-aligned refractory gate technology (see ref. 1 for more details), the AFM was used to analyze the GaAs surface at the gate region. After all steps through the gate and sidewall were completed (4000 Å W sputter deposition, gate-level optical lithography and SF_6/Ar-based plasma etch (RIE), ~2000 Å SiO_2 PECVD, and CHF_3/O_2-based plasma (RIE) sidewall spacer etch) the W gate was chemically removed supposedly without damage to the GaAs surface. The previously uncovered GaAs surface "field" was apparently etched resulting in a step where the GaAs was protected by W. This step was barely visible when viewed under an optical microscope. Further inspection by plan view SEM showed a "ghost" image pattern of the gate etched into the GaAs surface epitaxial layer, but the etched depth was not quantifiable when viewed by cross-sectional SEM due to resolution limitations. Subsequently, these same devices were imaged using AFM and a representative 3-D AFM image is shown in Figure 1. Note the trenches which appear on both sides of what was supposed to be the gate metal. These AFM images were further analyzed by averaging several line scans through the 0.8 µm gate. An averaged line scan trace for the device region shown in Figure 1 is displayed in Figure 2. Cursors mark the region of greatest interest (and concern) showing a height difference of 21.1 nm in a region where the GaAs field ideally would not be noticeably attacked during the refractory or sidewall etch.

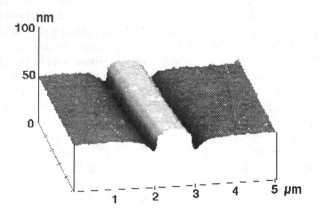

Figure 1. 3-D AFM image of a GaAs surface where a patterned W gate with SiO_2 sidewalls was chemically removed. Note the trenches at the edges of where the gate was before removal.

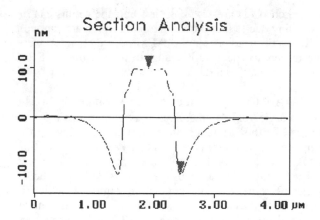

Figure 2. AFM cross-sectional analysis on the image shown in Figure 1. The cursor placement indicates that the trenches are etched 21.1 nm deep into the GaAs surface.

This material removal is especially critical for these devices since a nominally 30 nm thick GaAs contact layer separates the FET's active channel from the GaAs surface. This material loss causes an increase in source resistance and consequently a severe degradation of device performance at high GHz operating frequencies. The source of this trench was subsequently identified as a PECVD SiO_2 cusp formed at the gate sidewall due to undercutting of the refractory gate. This SiO_2 cusp was transferred into the GaAs surface during the SiO_2 reactive ion etch (RIE), hence causing the trench. Using AFM, we determined that the GaAs removal rate was ~20 Å/min during standard RIE etch conditions for both SiO_2 and refractory metal which is consistent with the measured trench depth. This severe problem, which we were unaware of before performing AFM analysis, is currently being addressed in our laboratory.

The AFM is also being used in our two-level interconnect metal process to determine if vias are completely opened. During standard FET process flow, after the ~3000 Å thick GeAuNiAu n-, and BeAu p-metal patterning steps, 4000 Å of blanket Si_3N_4 is PECVD deposited and then 1.25 µm diameter via holes are patterned through the Si_3N_4 film by optical lithography and RIE. Endpoint is determined during the plasma etch by performing in-situ optical interferometry on large open areas. If the via hole resist mask is underexposed or underdeveloped, the remaining resist scum can cause the via to be underetched causing an "open" in the circuit. Attempts at inspecting individual 1.25 µm via holes ex-situ for full etch completion by optical fluorescence microscopy and Nanospec have been unsuccessful due to the rather small size of the via holes. SEM analysis has also been unsuccessful unless operated in the destructive cross section mode. Using Dektak measurements on larger etched features is an option but, the results can be misleading due to microloading effects which are common in RIE. Prior to using the AFM, via etch rates were determined by destructive and time consuming SEM analysis on sacrificial devices.

An AFM image of four etched via holes in Si_3N_4 over two GeAuNiAu metal pads is shown in Figure 3. Images like this are taken after the via patterning step is completed. The images are then analyzed in cross section by taking line traces through the vias as shown in Figure 4. The average etch depth of the vias is 3710 Å, as shown by the placement of the cursors in Figure 4, which is correlated to previously recorded nitride thickness values measured by ellipsometry and confirmed by Nanospec. We have determined by AFM that the deposited nitride thickness on top of the metal ohmic pads is actually thinner by a consistent correction factor of 0.85 than the larger field areas from which thicknesses are measured optically. For this sample the average nitride thickness in the field was 4300 Å and this correlates, using our correction factor, to a nitride thickness on metal pads of 3650 Å. When this 3650 Å value is compared to the AFM measured via etch depth of 3710 Å, we conclude that the vias are etched to completion. Once the via holes have been etched to completion to expose the underlying GeAuNiAu, subsequent overetching does not significantly attack the exposed Au. Therefore, by comparing the AFM measured via hole etch depth with the optically measured nitride thickness at several points on the wafer, we can quickly and non-destructively determine if the vias are etched to completion.

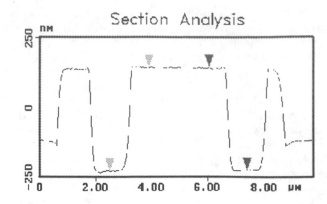

Figure 4. Cross sectional analysis on two of the 1.25 μm via holes shown in Figure 3. The cursor placement indicates that the vias are etched 3696 and 3724 Å deep.

While this technique has obvious measurement uncertainties which limit its ability to detect if the last 100 to 200 Å of Si_3N_4 are removed, it is a great improvement in detecting a commonly observed failure mechanism. A small amount of photoresist scum leads to grossly underetched features that are not imaged well by conventional methods in a non-destructive manner. We are currently investigating ways to improve upon these limitations by using properly designed on-wafer inspection sites for in-line AFM measurements. We have found that by performing this type of AFM characterization in-line, we can identify probable failures at a point in the process where rework is still a feasible option.

The final examples use the AFM to optimize a post ion implantation activation anneal process and characterize an ion implantation/wet etch sequence. In our laboratory we are exploring the feasibility of using ion implantation to fabricate Sb-based FETs. After ion implantation into III-V semiconductors, implant damage is typically removed during activation anneals (4). Increased activation typically occurs at higher temperatures; however, thermal treatments can cause the material's surface morphology to degrade causing subsequent processing and device performance problems (4). Thirty second activation anneals were performed in an RTA (SiC coated graphite crucible in flowing Ar) using a 15 °C/sec ramp rate on MBE grown 1.0 μm thick $Al_{0.75}Ga_{0.25}Sb$ on GaAs samples protected with a 50 Å thick GaSb cap layer at temperatures ranging from 500 to 750 °C. AFM images of annealed samples were quantified by RMS roughness measurements and compared to an as-grown unannealed sample. These RMS roughness results are reported in Table I. It appears that at ~650 °C the material starts to decompose, presumably due to incongruent evaporation of the group V Sb species. This effect has not yet been confirmed by surface analysis. The melting point of GaSb is ~710 °C. AFM images of representative smooth (600

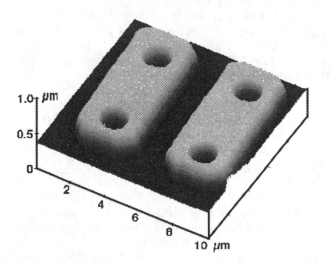

Figure 3. 3-D AFM image of four 1.25 μm via holes etched into 3650 Å thick Si_3N_4 which is conformally covering a Au-based ohmic metallization.

TABLE I.
The effect of anneal temperature (30 sec anneals) on the AFM measured RMS roughness of $Al_{0.25}Ga_{0.75}$ Sb on GaAs surfaces. The "As Grown" sample was not annealed for comparison. The RMS roughness averages were calculated using data from 25 µm x 25 µm image areas.

Anneal Temperature (°C)	Average RMS Roughness (nm)
As Grown	0.611
500	0.625
600	0.919
650	2.063
700	11.310
750	22.375

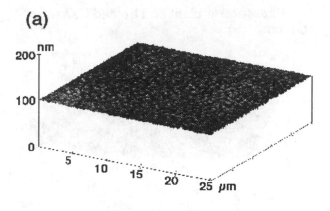

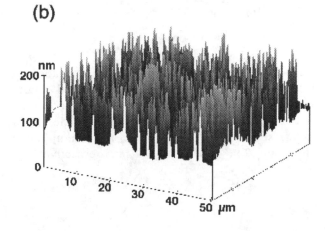

Figure 5. 3-D AFM images of an $Al_{0.25}Ga_{0.75}$ Sb on GaAs surfaces which were annealed at (a) 600 °C for 30 sec and (b) 750 °C for 30 sec. The AFM measured RMS roughness values are: (a) 0.909 nm (600 °C sample), and (b) 23.76 nm (750 °C sample).

°C) and rough (750 °C) surfaces are shown in Figure 5. Clearly, due to its rough nature, the 750 °C surface is not suitable for device processing. The 650 °C anneal was chosen as the optimum process based on these AFM surface morphology and electrical activation results.

AFM imaging was also performed on GaAs samples exposed to various steps in a patterned implant/wet etch processing sequence. The purpose of this sequence was to reposition the GaAs surface relative to the implanted dopant profile. It was observed that after wet etching in a dilute $NH_4OH:H_2O_2$ based etch, the 200 keV Si implanted etched regions of the GaAs were much rougher (RMS roughness of 2.96 nm) than the unimplanted etched field areas (RMS roughness of 1.18 nm - see Figure 6). Also, the implanted areas etched deeper by 1 - 5 nm than the unimplanted areas. These conditions caused problems during subsequent FET fabrication steps. Consequently, based on the AFM results, experiments are in progress to improve this device processing scheme.

CONCLUSIONS

Several AFM process characterization results are reported as in-line process control monitoring steps for GaAs ICs. These results are especially significant due to the apparent non-destructive nature of tapping mode AFM measurements, such that the analyses can be performed on active devices. We have not observed any adverse effects due to AFM-related contamination or physical damage on the performance of our devices. This type of AFM characterization has been extremely useful in quantitatively analyzing the unintentional loss of GaAs during several FET fabrication steps. With the AFM, we determined that a SiO_2 cusp at the gate edge was being transferred into the GaAs surface during SiO_2 etch potentially degrading high

frequency FET performance. We have also used the AFM as a non-destructive, in-line tool to determine if 1.25 µm Si_3N_4 interconnect via holes have been etched to completion. This characterization is performed at a point in the FET process where rework is still a feasible option.

AFM measured RMS roughness measurements have also aided in determining optimum ion implant activation anneal conditions for $Al_{0.75}Ga_{0.25}Sb$ epitaxial layers. Finally, with the AFM, we have observed morphological and height differences in GaAs after an ion implant/wet etch process sequence which can be correlated to prior process history. Ion implanted areas were etched deeper and were much rougher than non-implanted areas. These AFM results have helped to optimize several critical process steps and, by working in a failure analysis mode, identified process steps which are in need of improvement in our GaAs-based FET development program.

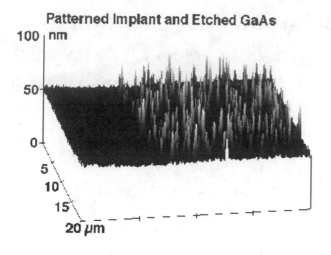

Patterned Implant and Etched GaAs

Figure 6. 3-D AFM image of a resist patterned, 200 keV Si implanted and wet etched GaAs surface. The region to the left was protected by resist only during the Si implant. The RMS roughnesses of the implanted and unimplanted regions are 2.96 nm and 1.18 nm, respectively.

ACKNOWLEDGMENTS

The authors would like to thank L. Griego and A.T. Ongstad for their expert technical assistance on this project. This work performed at Sandia National Laboratories is supported by the U.S. Department of Energy under contract DE-AC04-94AL85000.

REFERENCES

1. A.G. Baca, J.C. Zolper, M.E. Sherwin, P.J. Robertson, R.J. Shul, A.J. Howard, D.J. Rieger, and J.F. Klem, *GaAs IC Symposium Tech. Digest*, p. 59-62 (1994).

2. M.E. Sherwin, J.C. Zolper, A.G. Baca, R.J. Shul, A.J. Howard, D.J. Rieger, J.F. Klem, and V.M. Hietala, *IEEE Electron Dev. Lett.*, **15**, 242-244 (1994).

3. G. Binnig, C.F. Quate, and C. Gerber, *Phys. Rev. Lett.*, **56**, 930-933 (1986); T. Ohmi, M. Miyashita, M. Itano, T. Imaoka, and I. Kawanabe, *IEEE Trans. Electron Dev.*, **39**, 537 (1992); D. Pramanik, M. Weling, L. Zhou, *Solid State Tech.*, **July**, 79 (1994).

4. S.J. Pearton, A. Katz, and M. Geva, *J. Appl. Phys.*, **68**, 2482 (1990).

Infrared Transmission Topography: Application to Nondestructive Measurement of Dislocation Density and Carrier Concentration in Silicon-Doped Gallium Arsenide Wafers

M.G. Mier[1], D.C. Look[2], J.R. Sizelove[1], D.C. Walters[2], and D.L. Beasley[1]

1. Solid-State Electronics Directorate, Wright Laboratories, WL/ELDM, Wright-Patterson Air Force Base, OH, 45433-7323
2. University Research Center, Wright State University, Dayton OH 45435

Devices made on bulk GaAs:Si wafers may be critically dependent on both low dislocation density and high free carrier concentration in the material. Standard techniques for measuring these quantities are destructive; e.g., carrier concentration measurement by Hall effect generally requires small pieces and ohmic contacts, while dislocation counting requires etching the wafer to delineate etch pits where dislocations intersect the surface, and this process destroys the wafer surface finish. We report here a nondestructive method for measuring these quantities. For GaAs:Si with $n_{avg}=1.7 \times 10^{18}$ cm^{-3}, we find that at a wavelength of 0.9 μm we get ~2% transmission, sufficient for a high-contrast topographic map of dislocation density as well as background carrier concentration.. A similar measurement at a slightly longer wavelength 1.0 - 1.5 μm gives 20 - 40% transmission and provides even better data for a topographic map of free carrier concentration. The latter data have been calibrated by Hall effect measurements and the former by etching the wafer in molten potassium hydroxide (450°C for ~45 minutes) and counting the resulting etch pits. We attribute the unprocessed wafer transmission variations to free carrier variations for $\lambda > 1.0$ μm and free-carrier-induced bandgap variations for $\lambda < 1.0$ μm. Our correlation plot at $\lambda = 1.5$ μm gives $n = 1.91 \times 10^{17}\alpha + 2.35 \times 10^{17}$, where α is the optical absorption in cm^{-1} and n is the free carrier concentration in cm^{-3}. At the wavelength $\lambda = 0.9$ μm, we correlate the optical absorption with the counted etch pit density, giving $\rho_D = -4.60 \times 10^2\alpha + 2.24 \times 10^4$, where α is again the optical absorption and ρ_D is the etch pit density in cm^{-2}.

INTRODUCTION

Bulk n$^+$ GaAs:Si wafers are used to fabricate a variety of devices such as lasers and solar cells. Up to this time, however, very little attention has been given to GaAs:Si whole-wafer materials characterization, in contrast to the case of semi-insulating (SI) GaAs, where a variety of whole-wafer characterization techniques have been developed. For the SI case, two of the important whole-wafer parameters are EL2 and dislocation density. For bulk n$^+$ GaAs:Si, comparable attention must be given to variations in free carrier concentration (high free carrier concentration is needed for low series resistance and good ohmic contacts) and dislocation density (dislocations lead to dark-line defects and other recombination centers which limit carrier lifetime).

We have found that infrared transmission at a wavelength just above the band edge can nondestructively detect regions that have dislocations (1). Further, at slightly longer wavelengths, variations in free carrier concentration can also be detected (1). We find it useful to map the transmission of a GaAs:Si n$^+$ wafer with half-millimeter resolution. The high-dislocation regions are readily separated from the background, most of which is dislocation-free.

APPARATUS

The measurement apparatus is the same as has been described previously for EL2 measurements in SI GaAs (2). A beam of light from a tungsten-halogen source is focussed through a 1/4m Bausch and Lomb monochromator, then apertured and refocussed through a mechanical chopper into a square spot on the sample wafer. The size of this spot is adjusted according to the spatial resolution. The wafer is positioned by dc motor-driven actuators under resolution. The wafer was positioned by dc motor-driven actuators under computer control. Resolution of the position encoders is 4μm or better. Spot size was 500 μm x 500 μm for a total of 16,597 masurements on a three-inch wafer. Monochromatic light which passes through the sample is detected by a thermoelectrically-cooled photovoltaic germanium detector diode operating in the low-noise zero-bias mode. The output is lock-in detected for stability. A schematic of the apparatus is shown in Fig.1.

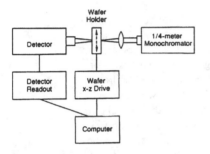

FIGURE 1. Schematic of apparatus.

To calibrate the values for dislocation density, a wafer was etched in molten KOH (450°C, ~45 minutes) which produces a hexagonal etch pit at each location where a dislocation intersects the wafer surface. Both the unetched and etched wafers were analyzed by the following formula (3):

$$T = \frac{(1-R)^2(1-S)^2 e^{-\alpha d}}{1 - R^2(1-S)^2 e^{-2\alpha d}} \qquad (1)$$

where T is the relative transmission, S is the scattering factor ($S = 0$ for an unetched wafer), α is the absorption coefficient, d is the sample thickness, and R is the reflection coefficient. The reflection coefficient is given approximately by $R=(\eta-1)^2/(\eta+1)^2$ where η, the real part of the GaAs refractive index, can be determined (4) from:

$$\eta = \left[7.10 + \frac{3.78}{1 - 0.18\,(h\nu)^2} \right]^{1/2} \qquad (2)$$

at 300 K, where $h\nu$ is the light energy in eV.

RESULTS

To check the widely accepted relationship that α is proportional to the carrier concentration n in GaAs:Si or $\lambda > 1$ μm and $n > 5 \times 10^{17}$ cm^{-3}, transmission was measured and absorption calculated from Eq. (1) at $\lambda = 1.5$ μm wavelength. The wavelength resolution was 0.0125 μm. Twenty-four 6 mm square pieces were sawn from representative areas of the wafer and a value for n determined for each piece by Hall-effect measurement. The values of n were compared with values of absorption calculated by averaging over the 144 positions (1/2 mm x 1/2 mm) corresponding to each 6 mm square piece. The map of a(1.5 μm) is shown in Fig. 2 and the correlation plot in Fig. 3.

1.5 μm Absorption

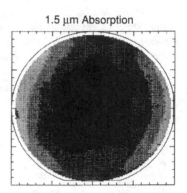

FIGURE 2. Half-millimeter resolution whole-wafer map of $\lambda = 1.5$ μm absorption for a wafer of GaAs:Si with average carrier concentration ~1.7x10^{18}cm^{-3}.

The correlation of α(1.5 μm) versus n obeys

$$n = 1.91 \times 10^{17}\alpha(1.5\ \mu m) + 2.35 \times 10^{17} \qquad (3)$$

with a correlation coefficient of 0.957. A similar map of absorption measured at a wavelength of 0.90 μm is shown in Fig. 4. The histogram of values for α(0.90 μm) is shown in Fig 5.

An adjacent wafer from the same boule was etched in molten KOH (450°C, ~45 minutes) to delineate etch pits where disloctions intersect the surface. The etch pits were manually counted under a microscope and an etch pit density ρ_D assigned to a number of locations. The whole-wafer transmission of the etched wafer at 1.45 μm was used to determine the whole-wafer dislocation density (3). Fig. 6 is the map of this whole-wafer dislocation density determined by correlating the etch pit count with the etched wafer transmission(3). Correlating the low-absorption distribution (approximately 57.3 - 58.1 cm^{-1}) of the 0.90 µm data with the whole-wafer dislocation density (Fig. 4 with Fig. 6) we get Fig. 7. The equation relating the absorption at 0.90 µm in this region with the dislocation density is

$$\rho_D = -4.60 \times 10^2\, \alpha(0.90\ \mu m) + 2.24 \times 10^4 \qquad (4)$$

with a correlation coefficient of -0.621.

DISCUSSION

In general, as the excitation energy E increases (wavelength λ decreases), the absorption coefficient α in an n-type GaAs sample will be influenced by three mechanisms: [1]interband absorption, for which α increases as the carrier concentration n increases (5); [2]interband absorption in the band tailing (Urbach) regime, for which α increases as n increases(6); and [3]interband absorption in the band-enhancement (Moss-Burstein) regime, for which α decreases as n increases(7). Lowney(8) has clearly delineated the latter two regimes, with the change in "contrast" (from α increasing to α decreasing as n increases) occurring at about 1.368 eV (0.906 µm) as n increasews from 1 to 2×10^{16} cm^{-3}. However, in our case, we believe that mechanism [1] swamps mechanism [2] (note that they both have α increasing as n increases), and we see the α vs n mechanism change directly from [1] to [3].

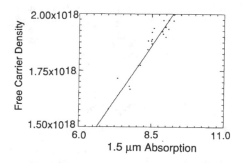

FIGURE 3. Correlation of 1.5 µm absorption with free carrier concentration measured by Hall effect. The least-squares-fit gives $n = 1.91 \times 10^{17}\alpha + 2.35 \times 10^{17}$.

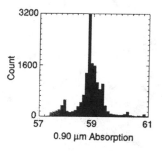

FIGURE 5. Histogram of the map of 0.90 µm absorption in Fig. 4.

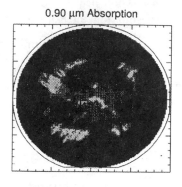

0.90 µm Absorption

FIGURE 4. Half-millimeter resolution whole-wafer map of 0.90 µm absorption for a wafer of GaAs:Si with carrier concentration ~1.7x10^{18} cm^{-3}

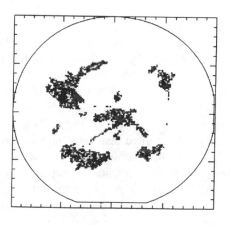

FIGURE 6. Whole-wafer dislocation density of wafer adjacent in the boule to the wafer measured in previous figures.

661

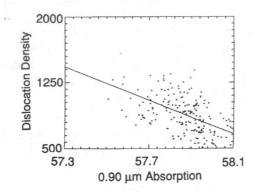

FIGURE 7. Correlation of dislocation density with low absorption distribution of the 0.90 μm absorption data. The least-squares fit gives $\rho_D = -4.60 \times 10^2 \alpha + 2.24 \times 10^4$.

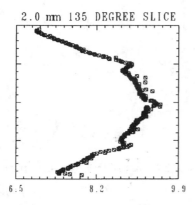

2.0 mm 135 DEGREE SLICE

6.5   8.2   9.9

FIGURE 8. Diagonal trace across the map shown in Fig. 2

The dislocated regions have a slightly higher n than the non-dislocated regions. This fact is directly seen from Hall-effect measurements, and indirectly seen from the absorption map at, say, 1.5 μm, for which α increases as n increases. For example, consider the diagonal trace (Fig. 8) across the map shown in Fig. 2; clearly, slight peaks are visible at the three dislocated regions crossed by the trace. However, the contrast is not high, as is apparent from the poor delineation of the dislocated region in the map itself. At 0.90 μm (Fig. 4), on the other hand, the contrast is very high, clearly delineating the dislocated regions. The difference is in the relationship α versus n in dislocated (D) and non-dislocated (ND) regions. For mechanism [1], intraband absorption, α versus n follows the same relationship in both (D) and (ND) regions. Thus, a slight increase in n in the (D) region results in a slight increase in α. However, for mechanism [3], interband absorption is not the same function of n in the two regions.

To prove this assertion, note in Fig. 5 that the (D) regions have an entirely separate distribution in the histogram at $\lambda=0.90$ μm. This is because the α versus n contrast change occurs at $\lambda=0.96$ μm for the (D) regions, and at $\lambda=0.93$ μm for the (ND) regions. It is this clear distinction between the behavior of α as a function of n in the interband absorption region that produces the high contrast in the region $\lambda=0.90$ - 0.93 μm. That is, at $\lambda=0.90$ μm the background (ND) regions have a weakly inverse α versus n relationship while the (D) regions have a strongly inverse α versus n relationship. For $\lambda<0.90$ μm, this distinction is gradually lost, and the contrast is accordingly reduced. The reason for the two different λs at which α versus n contrast reversal takes place is not clear. It may be that the (D) regions have a different density of conduction-band states than the (ND) regions, and thus a different Fermi level (effective bandgap) for a given n. Or it may be that the (D) regions have a nonuniform n, with very high n close to the dislocation cores. (It should be remembered that the Hall measurements are averaged over 6mm x 6mm pieces.) With a nonuniform n, the high-n regions would transmit more light than the low-n regions, and thus dominate the signal. Some of these questions could be answered by microscopic light probing near the dislocations.

CONCLUSIONS AND RECOMMENDATIONS

We have shown that free carrier concentration and dislocation density may be determined non-destructively in polished GaAs:Si wafers by measuring infrared transmission at two wavelengths. The free carrier concentration is straightforwardly determined from the absorption coefficient at a wavelength of 1.5 μm. We also observe that infrared transmission near the 0.87 μm band edge (to get sufficient transmission intensity, we use the range 0.90 - 0.93 μm) correlates with the free carrier concentration, but moreover shows features that we observe are related to dislocation density. The range of carrier concentrations for which this technique will work needs to be determined, as does the applicability to other dopants in conducting GaAs.

ACKNOWLEDGEMENTS

We thank T.A. Cooper for electrical measurements and L.V. Callahan for technical assistance. D.C.L. and D.C.W. were supported under USAF Contract

No. F33615-91-C-1765, and all of the work was performed at the Solid State Electronics Directorate, Wright Laboratories.

REFERENCES

1. D.C. Look, D.C. Walters, M.G. Mier, and J.R. Sizelove, Appl. Phys. Lett. **65**, 2188 (1994)

2. See, for example, P. Dobrilla, J.S. Blakemore, A.J. Camant, K.R. Gleason, and R.Y. Koyama, Appl. Phys. Lett. **47**, 602 (1985); F.X. Zach and A. Winnaker, Jpn. J. Appl. Phys. **28**, 957 (1989); S.K. Brierley and D.S. Lehr, Appl. Phys. Lett. **55**, 2426 (1986)

3. See, for example, J.S. Sewell, S.C. Dudley, M.G. Mier, D.C. Look, and D.C. Walters, J. Electron. Mat. **18**, 191 (1989); D.C. Look, D.C. Walters, J.S. Sewell, S.C. Dudley, M.G. Mier, and J.R. Sizelove, J. Appl. Phys. **65**, 1375 (1989); and M.G. Mier, D.C. Look, D.C. Walters, and D.L. Beasley, Solid State Electronics **35**, 319 (1992)

4. J.S. Blakemore, J. Appl. Phys. **53**, R123 (1982)

5. A.S. Jordan, J. Appl. Phys. **51**, 2218 (1980)

6. H.S. Bennett, J. Appl. Phys. **60**, 2866 (1986)

7. T.S. Moss, Proc. Roy. Soc. London, Sect B **67**, 775 (1954) and E. Burstein, Phys. Rev. **93**, 632 (1954)

8. J.M. Lowney, private communication

New Direct RF Parameter Extraction Technique for Heterojunction Bipolar Transistors Using an Extended Equivalent Circuit

D. Peters, <u>W. Daumann</u>, W. Brockerhoff, R. Reuter, E. Koenig [1], F.J. Tegude

Duisburg University, SFB 254, Solid-State Electronics Department, Kommandantenstr. 60, D-47057 Duisburg

[1] *TH Darmstadt, Department for High Frequency Technology, Germany*

A new algorithm is presented for the direct determination of the extrinsic and intrinsic small signal equivalent circuit elements for Heterojunction Bipolar Transistors (HBT). The new method is based on two scattering parameter measurements using an on-wafer measurement setup in the frequency range from 45MHz to 40GHz. The extrinsic elements were determined at zero bias condition ($V_{BE} = V_{CE} = V_{CB} = 0$V) and were kept constant. With the knowledge of these elements the intrinsic elements -**for the first time including the feedback capacitance C_{fb}**- can be directly derived at various active bias conditions.

INTRODUCTION

The extensive investigation of the high frequency performance using an equivalent circuit model and an appropriate parameter extraction technique is a powerful tool for device characterization and optimization. Different methods for the determination of the small-signal equivalent circuit were presented in the past mainly based on numerical optimization (1, 2) or direct calculation using a simplified equivalent circuit (3).

However, a simple equivalent circuit is not exactly valid for a complex triple-mesa-structure such as the AlGaAs/GaAs Heterojunction Bipolar Transistor (HBT) shown in figure 1. The feedback capacitance C_{fb}, which represents the extrinsic base-collector junction, influences strongly the device performance and can not be neglected (2). However, this capacitance is that parameter which prevents a direct and analytical determination of the equivalent circuit using the conventional *hot-cold methods* (3, 6).

In this paper we present for the first time a new and improved method for the direct determination of the complete equivalent circuit for AlGaAs/GaAs Heterojunction Bipolar Transistors. This method works without any simplifications of the corresponding equations and without any restrictions with respect to a certain frequency range.

HBT-STRUCTURE

Self aligned n-p-n AlGaAs/GaAs HBTs with an Al mole fraction of about 30% and an abrupt emitter-base junction were fabricated. Table 1 depicts the epitaxial layers of the HBT with the highly doped base and a spacer between base and emitter to prevent the outdiffusion of beryllium into the emitter, which strongly deteriorates the HBT performance.

Table 1 : Layer structure of the investigated HBT

Layer	Doping [cm^{-3}]	Thickness [nm]
GaAs emitter cap	n= 7.3·10^{18}	100
graded AlGaAs	N= 2.0·10^{18}	30
Al$_{0.3}$ Ga$_{0.7}$ As emitter	N= 2.0·10^{17}	100
GaAs spacer	undoped	10
GaAs base	p= 8.4·10^{19}	90
GaAs collector	n= 2.0·10^{16}	1500
GaAs subcollector	n= 5.0·10^{18}	1500
substrate	s.i.	

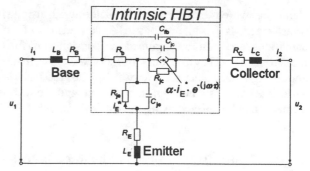

Fig. 2: Small-signal equivalent circuit of a HBT divided into an extrinsic and intrinsic part

Figure 1 schematically shows the triple-mesa-structure of the investigated HBT and the corresponding small-signal equivalent circuit elements. The extrinsic region of this device structure is modeled by the three inductances L_B, L_C and L_E and the parasitic resistances R_B, R_C and R_E. The active region underneath the emitter contact is represented by the equivalent circuit of two p-n junctions, one forward and one reverse biased, with the dynamic resistances (R_{jc}, R_{je}) and the capacitances C_{jc} and C_{je} including the depletion and diffusion parts. The base-spreading resistance R_b connects the intrinsic and the extrinsic base. The bias dependent base-collector junction is represented by the feedback capacitance C_{fb} (2). The active components are represented by the current source with the current gain α and the transit time τ.

The elements of both parts can be determined separately. Figure 3 shows, in principle, the application of the cold-hot-method.

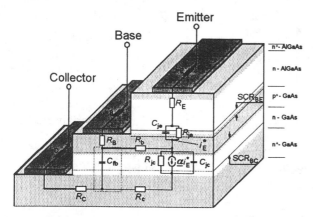

Fig. 1: Triple mesa structure of the investigated HBT including the feedback capacitance C_{fb}

NEW METHOD

The new extraction method described in this contribution is based on the known *cold/hot* measurement technique (3, 4, 6). In this case the equivalent circuit of the HBT is divided into an extrinsic and intrinsic part as suggested in (2, 3, 6) (fig. 2).

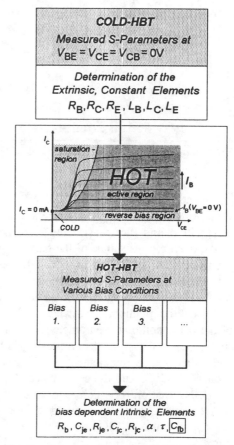

Fig. 3: Application of the *cold-hot-method* to directly determine the small-signal equivalent circuit elements of HBT

The *COLD-HBT*, which is measured at zero bias condition ($V_{BE} = V_{CE} = 0$V), is used for the determination of the extrinsic elements $R_{B...C}$ and $L_{B...C}$ of the HBT. These elements, which can be simply determined using the transformed z-parameters, are independent of bias conditions and can be assumed as constant for further measurements. They are only influenced by the material and design parameters. Afterwards, the *HOT-HBT* at various active bias conditions is measured and the intrinsic elements can be derived using the extrinsic values.

Determination of the Extrinsic Elements

For the determination of the extrinsic elements of the *COLD-HBT* the following small-signal equivalent circuit has to be analyzed :

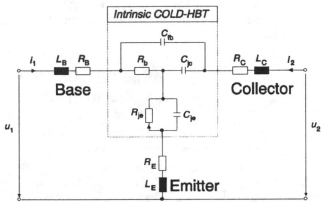

Fig. 4: Small signal equivalent circuit of the *COLD-HBT*

RF-measurements indicate that the *COLD-HBT* represents a passive network at ($V_{BE} = V_{CE} = 0$V). Hence, in this case the current gain α and the transit time τ are negligible. Moreover the dynamic resistance of the base-collector p-n junction is to be assumed as $R_{jc} \rightarrow \infty$ which was experimentally verified for the devices under test. A detailed discussion about the small-signal equivalent circuit of the *COLD-HBT* will be published elsewhere. The equation system of the corresponding z-parameters (fig. 4) directly and unambiguously allows to determine the inductances L_B, L_C and L_E as well as the elements of the base emitter p-n junction, R_{je} and C_{je} and the capacitances C_{fb} and C_{jc}. After the determination of these elements there still remains an equation system depending on the parasitic resistances R_B, R_C and R_E as well as the intrinsic base resitance R_b:

$$\text{Re}\{\underline{Z}_1\} = R_B + \frac{R_b}{2} \qquad (1)$$

$$\text{Re}\{\underline{Z}_0\} = R_E + \frac{R_b}{2} \qquad (2)$$

$$\text{Re}\{\underline{Z}_2\} = R_C - \frac{R_b}{4} \qquad (3)$$

From equation 1 and 2 it is now possible to predict the minimum and maximum value of the base resistance R_b, as follows:

$$0\Omega \le R_b \le 2 \cdot \left\{ \text{Min}\left(\text{Re}\{\underline{Z}_1\}, \text{Re}\{\underline{Z}_0\} \right) \right\} \qquad (4)$$

A short optimization algorithm referring to the equations 1, 2 and 3 directly leads to the determination of the unknown resistances R_B, R_C, R_E and R_b.

Determination of the Intrinsic Elements

For the determination of the eight small-signal equivalent circuit elements of the intrinsic *HOT-HBT* (C_{fb}, R_b, R_{je}, C_{je}, R_{jc}, C_{jc}, α and τ) the measured s-parameters have to be transformed into z-parameters. They can be obtained at each different active bias condition by extracting the known and constant extrinsic elements from the measured s-parameters of the *COLD-HBT* and transforming the reduced scattering parameters into the appropriate intrinsic z-parameters. Therefore, the intrinsic *HOT-HBT* is represented by the small-signal equivalent circuit illustrated in figure 5.

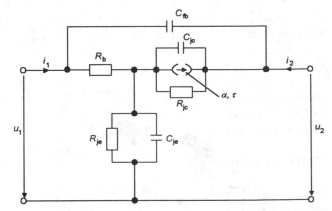

Fig. 5: Small signal equivalent circuit of the intrinsic *HOT-HBT*

The solution of the equation system of the corresponding z-parameters can be directly solved without any restrictions to a certain frequency range. Additionally and for the first time the feedback capacitance C_{fb} is included. By first performing special transformations of the corresponding equations to the intrinsic z-parameters and then dividing them into real and imaginary parts, the remaining unknown equivalent circuit elements can be found unambiguously. As an example figure 6 shows the determination of the feedback capacitance C_{fb}. Its corresponding value can be directly obtained using a linear regression of the investigated function.

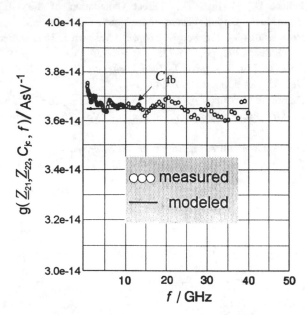

Fig. 6: Determination of the feedback capacitance C_{fb} : $g(\underline{Z}_{21}, \underline{Z}_{22}, C_{jc}, f) = \text{constant} = C_{fb}$

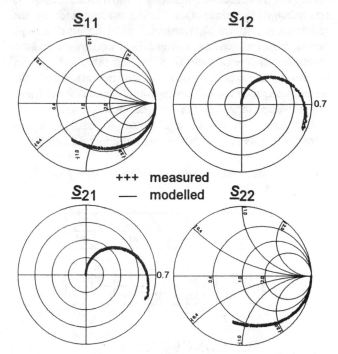

Fig. 7: Measured and modeled s-parameters of the *COLD-HBT*

RESULTS

Scattering parameters were measured in the frequency range from 45 MHz to 40 GHz for both, the *COLD-HBT* and the *HOT-HBT* using an on-wafer measurement setup (2). At first the s-parameters of the *COLD-HBT* are modeled by using the new developed technique described before. The resulting values of the extrinsic elements are summarized in table 2:

Table 2 : Extrinsic elements of the *COLD-HBT*

R_B/Ω	R_C/Ω	R_E/Ω
0.36	7.16	1.80

L_B/pH	L_C/pH	L_E/pH
42.6	68.1	33.0

The corresponding s-parameters are shown in figure 7. An excellent agreement between both the measured s-parameters and the modeled data can be obtained in the whole frequency range from 45 MHz to 40 GHz. Furthermore, the similarity between the forward and reverse transmission coefficients ($\underline{s}_{12}$, $\underline{s}_{21}$) indicates the passive character of the network of the *COLD-HBT* at $V_{BE} = V_{CE} = 0V$.

The results of the analysis of the complete *HOT-HBT* are illustrated in figure 8.

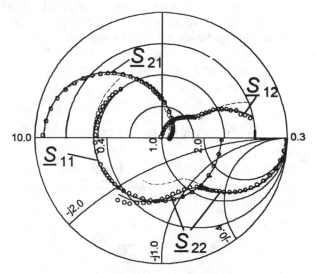

Fig. 8: Measured and modeled s-parameters of the *HOT-HBT* at $I_B = 500\mu A$, $V_{CE} = 3V$:

ooo   measured

———   new method considering C_{fb}

- - -   conventional method without C_{fb}

Moreover, this figure demonstrates the results using the conventional technique (3, 4) without considering the feedback capacitance C_{fb} (dashed lines) leading to a large error at high frequencies. The excellent agreement and the extremely small error (less than 2%) on the one hand and the large discrepancy without consideration of the feedback capacitance C_{fb} on the other hand demonstrates the relevance and validity of the new technique. All corresponding intrinsic equivalent circuit elements extracted by this new algorithm are summarized in table 4:

Table 4 : Intrinsic element values of the *HOT-HBT* at I_B = 500μA, V_{CE} = 3V.

R_{je}/Ω	C_{je}/fF	$R_{jc}/k\Omega$	C_{jc}/fF
11.72	209.0	79.71	17.88

R_b/Ω	C_{fb}/fF	α	τ/ps
16.30	36.61	0.9396	2.54

CONCLUSION

In conclusion, a new algorithm based on the full small-signal equivalent circuit has been presented. For the first time, a method has been demonstrated for the direct determination of all extrinsic and intrinsic elements including the bias dependant feedback capacitance C_{fb}, which must not be neglected. The extracted equivalent circuit of the HBT excellently fits the measured s-parameters and provides parameters with physical relevance. This powerful tool allows fast and efficient device characterization supporting the optimization procedure for the HBT development.

ACKNOWLEDGEMENTS

The author would like to thank Dr. Ulrich Seiler from the Daimler-Benz AG Research Center, Ulm (Germany), for supporting the AlGaAs/GaAs HBTs.

REFERENCES

1. Kondoh A., "An Accurate FET Modeling From Measured S-Parameters", in *Proceedings of the Conference on Mircowave Theory and Technique Symposium*, 1988, pp. 377-380
2. Peters D., Brockerhoff W., Reuter R., Meschede H., Wiersch A., Becker B., Daumann W., Seiler U., Koenig E., Tegude F.J., "RF-Characterization of AlGaAs/GaAs HBT Down to 20K", in *Proceedings of the Conference on Bipolar/BiCMOS Ciruits and Technology*, 1993, pp. 32-35
3. Pehlke D., Pavlidis D., "Direct Calculation of the HBT Equivalent Circuit From Measured S-Parameters", in *Proceedings of the conference on Microwave Theory and Technique*, 1992, pp. 735-738
4. Costa D. Liu. W. Harris J. *IEEE Trans. on Electron Devices*, Vol. 38, No. 9, pp. 2018-2024 (1991)
5. Dambrine G., Cappy A., Heliodore F., Playez E., *IEEE Trans. on Microwave Theory and Technique*, Vol. 36, No. 7, pp. 1151-1158 (1988)
6. Lee. S, Gopinath A., *IEEE Trans. on Microwave Theory and Technique*, Vol. 40, No. 3, pp. 574-577 (1992)

Contactless Room Temperature Analysis of Heterojunction Bipolar Transistor Wafer Structures using Photoreflectance

Fred H. Pollak, Wojciech Krystek, M. Leibovitch and H. Qiang

Physics Department and New York State Center for Advanced Technology in Ultrafast Photonic Materials and Applications
Brooklyn College of the City University of New York
Brooklyn, NY 11210 USA
and
Dwight C. Streit and Michael Wojtowicz

TRW Electronics and Technology Division
Redondo Beach, CA 90278 USA

In this paper we review previous work on the characterization/qualification of GaAs/GaAlAs HBT structures using photoreflectance (PR) including the correlation of certain PR features with actual device performance and present new results on the illumination dependence of the electric fields deduced from the PR spectra. We find that doping levels in the emitter/base and collector/base regions can be obtained from the illumination dependence of the observed fields, while the fields themselves can be used to evaluate variations in the doping levels.

INTRODUCTION

To improve the quality and yield of semiconductor devices one would like to perform the characterization procedure at room temperature on entire wafers, possibly even before the sample is removed from the growth chamber. The contactless electromodulation (EM) method of photoreflectance (PR) is ideally suited for this purpose. This technique gives rise to sharp, differential-like spectra in the region of interband (intersubband) transitions. Therefore, EM emphasizes relevant spectral features and suppresses uninteresting background effects (1,2).

One of the most important advantages of modulation spectroscopy is the ability to perform detailed lineshape fits to extract important parameters such as interband (intersubband) energies, broadening parameters, electric fields, Fermi energies, etc. Even at 300K it is possible to evaluate energy gaps to within a few meV. EM is particularly useful since (a) it is sensitive to surface/interface electric fields, (b) can be performed in contactless modes that require no special mounting of the sample and (c) can be employed as an optical impedance spectroscopy. Results on topographical scans with a resolution of $100\mu m$ have been published using a high quality camera lens (1) while some preliminary results indicate a resolution of $10\mu m$ can be attained using a microscope objective (3). These features make it possible to characterize wafer-sized material without altering the sample. The apparatus is simple, compact (about 3'x3'), easy to use and relatively inexpensive.

The general consensus is that the most critical region for determining overall device performance of an HBT is the emitter/base region. Small differences in the placement of the emitter/base p-n junction relative to the heterojunction can have considerable impact on the electrical characteristics of the device.

It has been demonstrated that PR can be an effective, nondestructive screening method for HBT structures (4,5). From an analysis of certain spectral features [Franz-Keldysh oscillations (FKOs)] in the room temperature spectra it has been possible to evaluate the dc electric fields in the GaAlAs emitter (F^{emit}) as well as in the n-GaAs collector (F^{coll}) region. The behavior of F^{emit} has been found to have a direct relation to actual device performance; there was a sudden drop in the dc gain when $F^{emit} > 200$ kV/cm. The explanation of this effect is the redistribution of the Be dopant in the p-region in these MBE samples. When the redistribution moves the p-n junction into the emitter, there is an increase in the electric field in this region (4).

PR also can yield important information about additional HBTs (InP/InGaAs, InAlAs/InGaAs, InGaP/GaAs and InGaAs/GaAs) as well as other device structures such as pseudomorphic high electron mobility transistors, quantum well and vertical cavity surface emitting lasers, multiple quantum well infrared detectors, etc. (6).

In this paper we report a study of the PR spectra from a GaAs/GaAlAs HBT structure as a function of the intensity of both the probe and pump beams. Our results show that there is a significant photovoltaic contribution to the PR signals. We present new results on the illumination dependence of F^{emit} and F^{coll} deduced from the PR spectra. This experiment demonstrates that (a) doping levels in the emitter/base and collector/base regions can be obtained from the illumination dependence of the observed fields while (b) the fields themselves can be used to evaluate variations in the doping levels.

EXPERIMENTAL DETAILS

The GaAs/GaAlAs HBT sample used in this study was fabricated by MBE with the following characteristics. The collector contact layer, 6000Å of n^+-GaAs:Si ($\approx 5\times 10^{18}$ cm^{-3}), was grown on an undoped liquid encapsulated Czochralski <001> semi-insulating substrate. This was followed by the collector layer of 7000Å of n^--GaAs:Si ($N^{coll} \approx 7.5\times 10^{15}$ cm^{-3}) and the 1400Å p^+-GaAs:Be base ($P^{base} \approx 1\times 10^{19}$ cm^{-3}). The 1800Å n-GaAlAs:Si emitter ($N^{emit} \approx 5\times 10^{17}$ cm^{-3}) emitter was graded in Al composition at the top and bottom over 300Å. The n^+-GaAs:Si ($\approx 7\times 10^{18}$ cm^{-3}) emitter contact layer was about 750Å thick. The PR apparatus has been described in the literature (1,2). The pump beams were the 670 nm line of a laser diode (GaAs spectra) or the 543 nm line of a green HeNe laser (GaAlAs region) modulated at 200 Hz.

EXPERIMENTAL RESULTS

Plotted in Fig. 1 is the PR spectrum at 300 K from the collector and emitter regions of the sample. The intensities of the pump and probe beams were 600 μW/cm^2 and 100 μW/cm^2, respectively. The signals occur in the region of the direct band gaps of GaAs and GaAlAs, respectively. From the position of the latter we can deduce an Al composition of 28%. Both spectra exhibit pronounced FKOs.

The positions of the mth extrema in the FKOs are given by (1,2):

$$m\pi = (4/3)[(2\mu_\parallel)^{1/2}(E_m - E_g)^{3/2}/q\hbar F] + \chi \qquad (1)$$

where E_m is the photon energy of the mth extrema, E_g is the band gap, F is the electric field, $\mu_\parallel$ is the reduced interband effective mass in the direction of $\overrightarrow{F}$ and χ is an arbitrary phase factor. Therefore, if $\mu_\parallel$ is known F can be evaluated from Eq. (1). Using $\mu_\parallel$ (in units of the free electron mass) = 0.055 (GaAs) and 0.073 (GaAlAs) we find $F^{coll} = 30$ kV/cm and $F^{emit} = 190$ kV/cm.

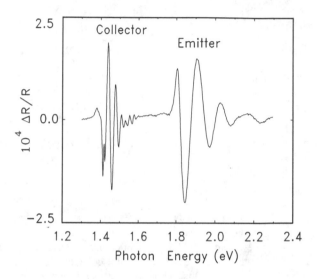

Fig. 1 Photoreflectance spectrum in the region of the direct gaps of GaAs (collector) and GaAlAs (emitter) from a GaAs/GaAlAs HBT at 300K.

It has been shown that for nonuniform fields, such as the ones that exist in p-n junctions, the FKOs are a measure of the maximum field in the structure (1).

One of the most important aspects of Fig. 2 are the FKOs associated with the emitter region, i.e., GaAlAs band gap. Values of F^{emit}, as deduced from the GaAlAs FKOs, were compared with device parameters of fabricated HBT MBE samples. Below electric field values of about 2 x 10^5 V/cm high current gains were obtained (4). Shown in Fig. 2 is F^{emit} as a function of dc current gain at 1 mA. Note that there is a sudden drop when $F^{emit} > 2$ x 10^5 V/cm. The explanation of this effect is the redistribution of the Be dopant in the p-region in these MBE samples. When the redistribution moves the p-n junction into the emitter, there is an increase in the electric field in this region, i.e. the value of F^{emit} becomes greater. The movement of the Be has been verified by secondary ion mass spectroscopy. When the p-n junction and the GaAs/GaAlAs heterojunction are not coincident, carrier recombination occurs, reducing the current and the performance of fabricated HBTs. These

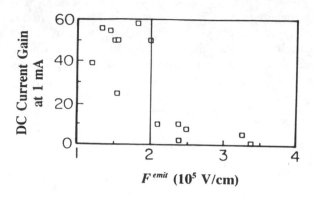

Fig. 2 Current gain at 1 mA for GaAs/GaAlAs HBT fabricated devices vs F^{emit} as determined from the GaAlAs FKOs before processing.

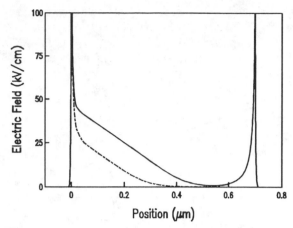

Fig. 3 Calculated field profile in the collector/base region in the dark (solid line) and under illumination (dashed line).

observations have made it possible to use PR as a screening technique to eliminate wafers with unwanted characteristics before the costly fabrication step (4).

However, the field deduced from the FKOs may not necessarily represent the actual field in the emitter/base or collector/base regions of the structure. The built-in electric fields can be reduced by the illumination (pump and/or probe beams) due to the photovoltaic effect. In order to compare the fields deduced from the FKOs with the actual fields that exist in the emitter/base and collector/base regions we have performed a computer simulation of the field profiles using a comprehensive, self-consistent model (7) with the structure parameters listed above.

Shown by the solid lines in Figs. 3 and 4 are the results of this calculation for the collector/base and emitter/base regions, respectively. The zero of the spatial ordinate has been taken to be the metallurgical *p-n* homo- or heterojunction, respectively. Note that both fields are approximately linear over a large spatial region and that the maximum values of the linear portions are about 50 kV/cm and 500 kV/cm, respectively.

To account for this apparent discrepancy in the electric field strengths we also have simulated the field profiles under conditions of illumination. Shown by the dashed lines in Figs. 3 and 4 are these results for the collector/base and emitter/base regions, respectively, for a photon flux of 2×10^{14} cm^{-2}-sec^{-1}. Note that the maximum fields of the linear portions are now in relatively good agreement with the observed values.

To explore further the influence of the photovoltaic effect we have evaluated the illumination dependence of the fields deduced from the FKOs. Shown by the squares in Figs. 5 are the variations in measured values of F^{coll} as a function of probe beam intensity (I_{pr}) in the range 20 -

1000 μW/cm^2 for a pump beam intensity of 150 μW/cm^2. Note that there is an approximately linear dependence of F^{coll} on the log of I_{pr}.

Using our computer model we also have calculated the intensity dependence of the field. The dashed, solid and dot-dashed lines in Fig. 5 show the results of such a simulation for the maximum field in the linear portion of the collector region using the $N^{coll} = 10$, 7.5 and 5.5 (in units of 10^{15} cm^{-3}), respectively, and P^{base} given above. There is very good agreement between experiment and the intended value of N^{coll}. Thus, our experiment and model show that the "slope" of F^{coll} vs log I_{pr} (Fig. 5) is sensitive to the doping level. Note that for a given illumination level variations in the doping concentrations can be evaluated from the changes in F^{coll}.

We have carried out a similar experiment/computer

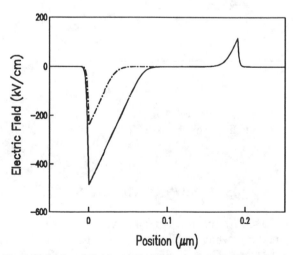

Fig.4 Calculated field profile in the emitter/base region in the dark (solid line) and under illumination (dashed line).

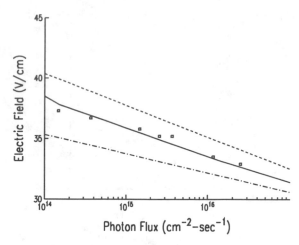

Fig. 5 Experimental (square) value and simulation (dashed, solid and dot-dashed lines) of the probe beam intensity dependence of the collector field.

simulation for the emitter/base region. Shown by the solid line in Fig. 6 are the values of the "slope" of the illumination dependence of the emitter field, in units of kV/cm-decade (of I_{pr}), as a function of N^{emit} obtained from the computer simulation. The triangle is the experimental determined "slope", yielding $N^{emit} \approx 3.5 \times 10^{17}$ cm^{-3}. This value is in reasonable agreement with the intended quantity.

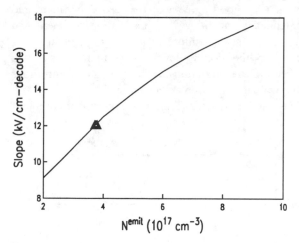

Fig.6 "Slope" of F^{emit} vs I_{pr} as a function of N^{emit}. Simulation-solid line, experiment-triangle.

DISCUSSION OF RESULTS

The HBT structure contains various hetero- and homojunctions, which govern the distribution of the fields in the device. These fields play a key role in the redistribution of the excess carriers generated by both the probe and pump beams of the PR method. The photogenerated carriers are separated at the junctions and accumulate in the quasi-neutral regions (see Figs. 2 and 3) leading to a photo-induced forward bias of the junctions, i.e., photovoltaic effect. Under such conditions the F^{emit} and F^{coll} become illumination-dependent parameters and therefore one cannot simply obtain the "actual" fields in the structure from these quantities alone.

However, as we have demonstrated both experimentally and by computer simulation, the dependence of these fields on illumination intensity is sensitive to the structural parameters of the device. Therefore, the "slope" of F^{emit} or F^{coll} vs the log of intensity can be used to deduce information about doping levels in these regions (see Figs. 4 and 5).

Furthermore, for a given light intensity F^{emit} or F^{coll} can still be used to detect changes in the doping profile from sample to sample or over the sample surface. Thus, the relationship found between F^{emit} and the dc current gain in GaAs/GaAlAs HBTs (see Fig. 2) is still valid (4).

ACKNOWLEDGEMENTS

The authors FHP, WK, ML and HQ acknowledge the support of NSF grant DMR-9120363, US Army Research Office contract DAAL03-92-G-0189 and the New York State Science and Technology Foundation through its Centers for Advanced Technology program.

REFERENCES

1. Pollak, F.H. and Shen, H., *Materials Science and Engineering* **R10**, 275-374 (1993) and references therein.
2. Glembocki, O.J. and Shanabrook, B.V., *Semiconductors and Semimetals*, Vol. **67**, ed. D.G. Seiler and C.L. Littler (Academic, New York, 1992) pp. 222-292 and references therein.
3. Boccio, V.T. and Pollak, F.H., private communication.
4. Yin, X., Pollak, F.H., Pawlowicz, L., O'Neill, T.J. and Hafizi, M., *Appl. Phys. Lett.* **56**, 1278-1280 (1990); also, *Proc. Soc. Photo-Optical Instrum. Engineers* (SPIE, Bellingham, 1990) **1286**, 404-413 (1990).
5. Bottka, N., Gaskill, D.K., Wright, P.D., Kaliski, R.W. and Williams, D.A., *J. Cryst. Growth* **107**, 893-897 (1991).
6. Qiang, H., Yan, D., Yin, Y. and Pollak, F.H., *Asia-Pacific Engineering Journal, Part A: Electrical Engineering* **3**, 167-198 (1993); Pollak, F.H., Qiang, H., Yan, D., Yin, Y. and Krystek, W., *Journal of Metals* **46**, 55-59 (1994).
7. M. Leibovitch, L. Kronik and Y. Shapira, *Phys. Rev.* **B50**, 1739-1745 (1994).

Novel Magnetic Field Characterization Techniques for Compound Semiconductor Materials and Devices†

C.A. Richter‡, D.G. Seiler, J.G. Pellegrino, W.F. Tseng, and W.R. Thurber

Semiconductor Electronics Division
National Institute of Standards and Technology
Gaithersburg, MD 20899 USA

Quantum-mechanical effects observed in the magnetoresistance of semiconductor devices and materials give important information such as carrier density, scattering rates, and band structure parameters. However, the small size of these effects often limits their observation and subsequently their practical use. The purpose of this paper is to show how these quantum mechanical effects can be more easily observed and used as characterization tools by modulating the applied magnetic field, which can increase the sensitivity of magnetotransport measurements.

INTRODUCTION

Quantum-mechanical (QM) effects observed in the magnetoresistance (MR) of semiconductor devices and materials can be used as powerful characterization tools for assessing the quality of materials, processing effects, and other device-related properties. Information such as carrier density, scattering rates, and fundamental band structure parameters can be obtained by studying these effects. However, while QM effects are widely studied for scientific purposes (1), they are often neglected as meaningful industrial characterization tools. One reason for this neglect is that QM effects are small in many instances and therefore difficult to observe. By modulating the applied magnetic field, the sensitivity of magnetotransport measurements can be increased (2,3), allowing QM effects to be more easily measured. In this paper, we give examples of this enhanced sensitivity in HgCdTe infrared detectors and GaAs/AlGaAs modulation-doped FETs, and review how the observed quantum effects can be used to characterize various device properties. The sensitivity attainable is illustrated for three two-dimensional quantum effects: Shubnikov-de Haas oscillations, the magnetophonon

effect, and universal conductance fluctuations. These examples demonstrate that device properties can be measured by using QM effects that would be difficult to obtain by using other techniques.

Conventional dc-magnetic-field transport measurements are commonly used to characterize semiconductor materials. The Hall effect gives the carrier type and density, and by combining Hall measurements with van der Pauw or bridge resistor measurements (which measure resistivity), the carrier-drift mobility is derived (4). Implicit in the analysis of these measurements is the fact that the conduction path is uniform, and only a single type of carrier is present. The dc techniques are being extended to study multicarrier systems (5), but in general, a complex correlation with theory is necessary to obtain multicarrier information. Furthermore, these measurements do not give any information concerning the fundamental band structure of semiconductor materials. By studying QM effects, band structure parameters (such as the carrier mass) are measured in addition to carrier density and mobility. Also, in many cases, information can be derived concerning the various conduction paths in multicarrier or non-uniform systems.

Experimentally, in addition to the dc magnetic field, B_{dc}, we apply an ac magnetic field of frequency ω with a constant amplitude B_0 (2,3). (Typically, $B_0 \approx 10$ mT.) As a dc current is passed through a device, lock-in

---

† Contribution of the National Institute of Standards and Technology; not subject to copyright.
‡ National Research Council Research Associate.

amplifier techniques are used to measure the ac response at ω or 2ω. This ac response at ω (2ω) has the appearance of the first (second) derivative of the device resistance with respect to magnetic field. This technique allows changes in the resistance as a function of magnetic field to be measured with unparalleled sensitivity.

SHUBNIKOV-DE HAAS OSCILLATIONS

The Shubnikov-de Haas (SdH) effect (3,6) is a quantum mechanical phenomenon that arises in electronic systems, having either electrons or holes, in the presence of a magnetic field, and leads to oscillations in the MR that are periodic in $1/B_{dc}$ for carriers with a constant effective mass. Analysis of these oscillations gives the carrier density and scattering rates of these electronic systems. SdH oscillations result from the crossing of the Fermi level by the Landau levels that are resolved at low temperatures and moderate magnetic fields ($\mu B_{dc} \geq 1$). In general, SdH oscillations are more easily observed in two-dimensional (2D) systems than in three-dimensional (3D) or bulk systems; therefore, studying these oscillations is an extremely useful technique to characterize multicarrier systems that contain both a 2D and a 3D conduction path.

In devices containing a 2D system with only a single populated subband such as GaAs/AlGaAs modulation-doped structures, SdH oscillations dominate the MR at low temperatures. By taking a Fourier transform (FT) of the resistance versus $1/B_{dc}$, a single, sharp peak is observed indicating the oscillation period, $\Delta[1/B_{dc}]$. The period gives the carrier concentration, n_s, of the 2D electron gas

$$n_s = (2e/h)(1/\Delta[1/B_{dc}]) \qquad (1)$$

where the factor of 2 accounts for spin, and a single valley is assumed for GaAs. Comparison of this density with the total carrier density obtained from Hall measurements indicates whether additional conduction paths exist in the bulk of the device. This information is valuable when determining growth and process parameters for fabricating modulation-doped FET's in either Si/Ge, III-V, or II-VI systems (7). It should be noted that in systems where many 2D subbands are populated, each subband gives rise to a different oscillation period due to the different carrier populations. These oscillations "beat" with one another in the measured MR, and therefore, this FT technique is also capable of extracting the subband populations within a 2D electronic system.

In device structures that have both 2D and bulk conduction paths, the SdH oscillations can be swamped by the

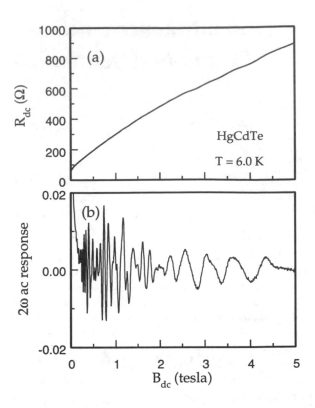

Figure 1. Shubnikov-de Haas oscillations in a commercial photoconductive HgCdTe IR detector at 6 K. (a) the dc magnetoresistance; (b) the ac response at 2ω.

conduction in the 3D carrier system. Using the modulated magnetic field technique, the large, but slowly varying, background resistance of the 3D conduction path is eliminated, and the SdH oscillations can be readily studied. The effects of modulated magnetic fields have been explicitly established and used in extensive studies of commercially available photoconductive infrared detectors made from passivated HgCdTe (8,9). In addition to the bulk conduction path, these devices have a 2D electron gas in an accumulation layer at the passivated surface. Figure 1 shows data that are typical of our results for one type of HgCdTe IR detector. The MR (Figure 1a) shows small perturbations due to the SdH oscillations. By using the modulated-magnetic-field technique, these oscillations are easily observed (Figure 1b) and can be analyzed. Fourier analysis of these data indicates that this particular device has three populated 2D subbands which lead to the observed convoluted oscillations. The carrier density in each of the subbands is derived and found to be 7.6×10^{15} m^{-2}, 2.6×10^{15} m^{-2}, and 1.1×10^{15} m^{-2}, respectively. The importance of this technique is that it allows the characterization of the passivation-layer properties on a completed commercial device. These properties, which control the ultimate performance of these detectors, cannot be obtained by other conventional means.

MAGNETOPHONON EFFECT

The magnetophonon effect (6,10) is another quantum effect that arises due to the quantization of Landau levels (LLs). The magnetophonon effect results from a resonant scattering of electrons between LLs due to the absorption or emission of a phonon. This resonance occurs when the energy difference of an integral number, n, of LLs is commensurate with the energy of a quantized, generally LO, phonon,

$$nh\omega_c = h\omega_0 \qquad (2)$$

where $\omega_c = eB_{dc}/m^*$ is the cyclotron frequency, ω_0 is the phonon frequency, and h is Planck's constant. At this resonance condition, the resistance increases, giving rise to an oscillatory MR that is periodic in $1/B_{dc}$. Unlike SdH oscillations which occur only at low temperatures, the magnetophonon effect is observed at moderate temperatures (e.g., 40 K to 300 K) where a significant phonon population exists. Also, these oscillations are independent of electron density and depend only upon the fundamental material properties. In general, magnetophonon oscillations are small in amplitude and difficult to observe without resorting to extremely large magnetic fields. The increased sensitivity of the modulated-magnetic-field techniques allows these small changes in the MR to be observable at readily available magnetic fields (i.e., < 9 T) (11).

Our results, shown in Figure 2, illustrate the 2D magnetophonon effect in a GaAs/AlGaAs modulation-doped heterostructure where the 2D electron gas resides in GaAs. (This device had a 2D carrier density of 0.8×10^{15} m^{-2} and an electron mobility equal to 100 m^2/V·s at 4.2 K.) Due to the polar nature of III-V compounds, the LO phonons couple strongly to the electrons in the system. The magnetophonon effect appears as small perturbations of the dc MR as seen in Figure 2a. By using the modulated-magnetic-field technique, these oscillations are clearly resolved in the ac response at ω (Figure 2b). Because this ω response is proportional to the first derivative, the peaks in the resistance are shifted to the inflection points of the measured oscillations. From the position of these oscillations and the GaAs LO phonon energy (36 meV), the effective electron mass is derived by using Equation 2 and found to be m$^* = 0.07$m$_e$. By studying the period of magnetophonon oscillations, information concerning the effective mass of the carriers in the system and the energy of the dominant phonon scattering mechanism is derived. Furthermore, the amplitude of these oscillations gives insight into the mobility of the carriers and electron-phonon coupling. In particular, the magnetophonon effect is often used to study hot-electron effects (10).

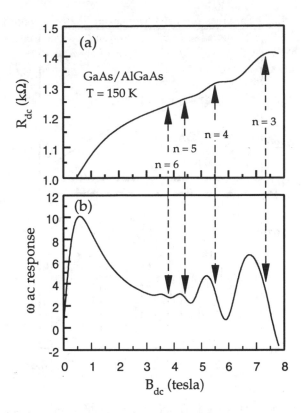

Figure 2. The magnetophonon effect in a GaAs/AlGaAs modulation-doped device at T = 150 K. (a) the dc magnetoresistance; (b) the ac response at 2ω. Arrows indicate the resonance positions (Equation 2).

UNIVERSAL CONDUCTANCE FLUCTUATIONS

Experimental studies of "universal" conductance fluctuations (UCF) are another example that illustrates the remarkable sensitivity of the modulated-magnetic-field technique. UCF are reproducible aperiodic fluctuations which are due to the quantum interference of electron waves passing through a device (1,12). Because electrons must maintain their quantum phase coherence over the entire device in order to contribute to UCF, the statistical variance of UCF is inversely proportional to the device area: i.e., the larger the device, the smaller the amplitude of the fluctuations. Because of this property, it has been assumed that at finite temperatures, UCF could not be observed in macroscopic, or large area devices.

Figure 3 shows observations of UCF in a 0.5 mm × 0.05 mm GaAs/AlGaAs Hall bar device (n$_s$ = 3.5×10^{15} m^{-2} and μ = 4.5 m^2/V·s at 4.2 K). The 2ω response clearly shows reproducible conductance fluctuations due to UCF (Figure 3b) in the magnetic field range below the onset of

675

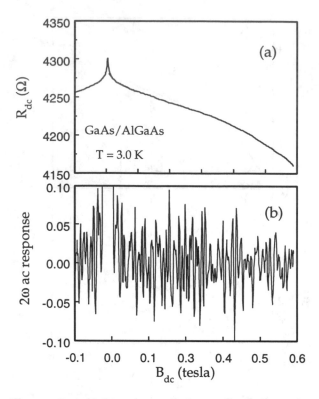

Figure 3. Universal conductance fluctuations in a GaAs/AlGaAs modulation-doped device at T = 3.0 K. (a) the dc magnetoresistance; the large peak at $B_{dc} = 0$ T is due to weak-localization (1). (b) The ac response at 2ω.

SdH oscillations. We have taken great care to ensure that these small perturbations are not a measurement artifact (13). The fluctuations are reproducible, much larger than the experimental noise level, and show all the attributes of UCF. During these studies, we have measured fluctuations corresponding to changes in the device resistance as small as 1 part in 10^7. The conventional MR (Figure 3a) shows no indication of UCF within the resolution of the measurements, illustrating that the fluctuations are too small to be observed by using more traditional low noise measurement techniques where the magnetic field is fixed and the device current is modulated. Thus, the additional sensitivity of the magnetic-field-modulation technique has permitted us to observe and study UCF in a new large-size-scale regime where to the best of our knowledge it has not been previously observed. Proper analysis of these fluctuations gives information concerning quantum scattering lengths. For this low-mobility device, the quantum phase breaking length was found to be 0.7 μm at 2.3 K. These quantum length scales are necessary design parameters for novel quantum electron devices (14).

CONCLUSIONS

We have demonstrated that the modulated-magnetic-field technique is an extremely effective way to measure small changes in magnetoresistance. The enhanced measurement sensitivity of this technique allows quantum-mechanical effects to be easily observed and studied. The combination of this measurement technique and an understanding of quantum-mechanical effects can be used to characterize the transport properties of a wide range of semiconductor materials and devices that would be difficult or impossible to characterize by using more traditional techniques.

REFERENCES

1. For a review of quantum effects, see Beenakker, C.W.J., and van Houten, H. in *Solid State Physics*, Vol. **44,** Ehrenreich and Turnbull, eds. New York: Academic Press, 1991, p. 1.

2. Goldstein, A., Williamson, S.J., and Foner, S., *Rev. Sci. Instr.* **36**, 1356 (1965).

3. Seiler, D.G., and Stephens, A.E., in *Landau Level Spectroscopy*, Landwehr and Rashba, eds. Amsterdam: North Holland, 1991, p. 1031.

4. *Standard Test Methods for Measuring Resistivity and Hall Coefficient and Determining Hall Mobility in Single-Crystal Semiconductors,* ASTM, F 76 (1991).

5. Kim, J.S., Seiler, D.G., and Tseng, W.F., *J. Appl. Phys.* **73**, 8324 (1993).

6. See, for example, Ridley,B.K., *Quantum Processes in Semiconductors,* Oxford: Clarendon Press, 1988, pp. 282–317.

7. Tobben, D. Schaffler, F., Zrenner, A., and Abstrieter, G. *Phys. Rev. B,* **46** 4344 (1992). Wilks, S.P., Cornish, A.E., Elliot, M., Woolf, D.A., Westwood, D.I., and Williams, D.I. *J. Appl Phys.* **76**, 3583 (1994). Jaeger, A., Hoerstel, W., Thiede, M., Schafer, P., and Elsing, J. *Semi. Sci. Tech.* **9**, 54 (1994), among others.

8. Seiler, D.G., Lowney, J.R., Thurber, W.R., Kopanski, J.J., and Harmon, G.G., *Semiconductor Measurement Technology:* Improved Characterization and Evaluation Measurements for HgCdTe Detector Materials, Processes, and Devices Used on the GOES and TIROS Satellites, NIST Special Publication 400-94 (April 1994).

9. Lowney, J.R., Seiler, D.G., Thurber, W.R., Yu, Z., Song, X.N., and Littler, C.L., *J. Elec. Mat.* **22**, 985 (1993).

10. Nicholas, R.J., in *Landau Level Spectroscopy*, Landwehr and Rashba, eds. Amsterdam: North Holland, 1991, p. 777.

11. Hazama, H., Sugimasa, T., Imachi, T., and Hamaguchi, C., *J. Phys. Soc. Japan* **54**, 3488 (1985).

12. See, for example, Stone, A.D., *Coherent Quantum Transport,* in *Physics of Nanostructures*, Davies and Long, eds. Bristol: IOP Publishing, 1992, p. 65.

13. Richter, C.A., Seiler, D.G., and Pellegrino, J.G., "Mesoscopic conductance fluctuations in large devices," to appear in *The Proceedings of the 22nd International Conference on the Physics of Semiconductors*, August 15–19 1994.

14. Capasso, F., and Datta, S., *Phys. Today*, Feb 1990, p.74

Nondestructive Characterization of GaAs-based Multilayer Epitaxial Structures

Patricia B. Smith, Andrew A. Allerman* and Walter M. Duncan

Corporate Research and Development
*Defense Systems and Electronics Group
Texas Instruments Incorporated
Dallas, Texas 75265

Nondestructive characterization tools are needed to screen epitaxial materials. Structural variations during epitaxial growth can lead to subsequent device yield loss. Present, destructive evaluation methods such as capacitance-voltage (C-V) measurements, limit wafer yields and are labor intensive. We report the nondestructive, optical characterization of a heterojunction bipolar transistor (HBT) structure using photoreflectance spectroscopy (PR) and spectral ellipsometry (SE). The PR results show good agreement with C-V measurements for the dopant concentration in the n-GaAs collector layer. In contrast to the C-V measurement, we find that PR provides a reasonable N_D for the emitter layer; C-V analysis of the thin emitter layer provides only an upper-limit to the dopant concentration. Thus, PR provides a useful means of obtaining emitter and collector dopant concentrations quickly, accurately and nondestructively. We have developed a detailed SE model of the HBT structure and applied this model to obtain alloy compositions and layer thicknesses for the upper five of the nine HBT layers. Although SE probes only as deeply as the emitter layer due to absorption of the incident light (250 - 900 nm), the detailed structural data obtained from the calculated fits provides valuable insight into variations which occur during HBT growth. Combining the data obtained from the PR and SE analyses, we can determine the structure and dopant concentrations in critical HBT layers, progressing toward the goal of establishing a reliable, nondestructive wafer qualification scheme.

INTRODUCTION

Nondestructive characterization techniques are receiving more attention recently in applications involving multilayer epitaxial structures due to the potential impact on semiconductor device production. The structures are, by their nature, complex, comprising binary, ternary and even quaternary alloys, superlattices, compositional grading and other features representative of today's advanced semiconductor growth technology.

The current practice of building test structures and obtaining dopant concentrations, and device characteristics such as gain and breakdown voltage is a costly and time-consuming activity. These tests are necessary in order to assess the quality and suitability of the material for device processing. It is at this point that a nondestructive qualification procedure could have the most significant impact on the cost and yield of devices.

We are progressing toward the establishment of such a procedure, which we believe would lead to a more efficient processing and device fabrication operation. We have focused our efforts on two optical characterization techniques, spectral ellipsometry (SE) and photoreflectance spectroscopy (PR). We discuss our data and its application to the characterization of heterojunction bipolar transistor (HBT) structures grown by metalorganic chemical vapor deposition (MOCVD).

EXPERIMENTAL

HBT epitaxial structures were grown at Texas Instruments using three different MOCVD reactors. Two of the reactors are multiple wafer, production reactors. The third reactor is a single wafer, rotating disk reactor. Substrates were GaAs oriented 2° off (100), toward (110).

Room temperature PR was performed using a He-Ne laser (633 nm) as the pump beam with a nominal power of 2 mW at the sample. The pump beam was chopped at 1 kHz using a mechanical chopper. A 100 W quartz tungsten halogen lamp scanned by a 0.50 m Czerny-Turner monochromator was employed for the probe beam. The spot size of the pump and probe beams at the sample are about 4 mm diameter and (1x2) mm^2, respectively. The PR signal, detected using a Si photodiode, is coupled into subsequent analysis electronics through a transimpedance amplifier. The amplified signal is analyzed separately for ΔR and R components. The in phase AC signal component, which is proportional to the modulated reflectance (ΔR), is measured using a lock-in amplifier. The DC component, which is proportional to the nominal reflectance (R), is obtained by direct digitization of the transimpedance amplifier output. Spurious AC background signals due to photoluminescence and pump beam scattering are minimized using polarization and spatial filtering. Any spurious AC background remaining after optical filtering is subtracted by zeroing the lock-in with the probe beam off. Both AC and DC signals are read into a PC and stored as the AC/DC or $\Delta R/R$ ratio versus wavelength.

Spectral ellipsometry data was obtained using a SOPRA MOSS Model ES4G instrument. The SOPRA apparatus uses a 75 W Xe white light source. The signal is scanned by a 0.75 m (grating and prism) monochromator. A PMT detector is used for photon counting. Data is obtained at a 75° incident angle. Reference dielectric functions for InGaAs(1), GaAs(2) and AlGaAs(3) were used in calculating film thickness and alloy composition from the raw data.

RESULTS and DISCUSSION

Photoreflectance

The PR technique provides dopant concentrations and alloy composition through a measurement of the change in the reflectance of a sample.(4-6) A scanning probe beam and a chopped pump beam are co-incident upon the material under investigation. The modulated pump beam causes a perturbation of the reflectance of the sample. The reflectivity change results from a change in the electric field of the sample due to the photoinjection of charge

carriers.(7-9) The oscillation period of the PR lineshape (ΔE) has been shown to be proportional to the dopant concentration.(10,11) Relationships between the electric field, the Franz-Keldysh oscillation (FKO) energy spacing and the dopant level are shown in equations (1) and (2). The surface electric field is ε_{max}, V is the potential barrier height resulting from surface states below the Fermi level, N_D is the dopant density, ε_0 is the static dielectric constant, q is the electron charge, and ΔE is the energy spacing between lobes of the FKOs occurring above the band gap energy.(10) From equations (1) and (2), we see that $\Delta E \propto N_D^{1/3}$. The empirically determined relationship we have used to determine the dopant concentrations in the AlGaAs emitter layer and the GaAs collector layer of the HBT structure is shown in equation (3). (11)

$$\varepsilon_{max} = \left[\frac{2qN_DV}{\varepsilon_0} \right]^{1/2} \qquad (1)$$

$$\Delta E \propto \varepsilon_{max}^{2/3} \qquad (2)$$

$$N_D = 10^{\,3.31(\log(\Delta E)) + 11.34} \qquad (3)$$

PR and C-V determined collector dopant concentrations generally agree within the experimental error of the techniques, estimated at 10-20%.(12) The estimated error in the C-V determined emitter dopant concentration is assumed to be much higher than 20% due to the limitations of the C-V measurement for very thin layers. Figure 1 shows a comparison of the PR and CV derived dopant data for several HBTs. We observe an (expected), large deviation between C-V and PR measurements of the emitter layer dopant level. Since the emitter layer is only on the order of 40 nm thick, C-V analysis can give, at best, an upper limit to the actual dopant level. The PR measurement should be considerably more accurate since the measurement is not dependent upon the thickness of the AlGaAs layer.

The accuracy of the PR measurement is dependent only upon the accuracy of the ΔE measurement. Generally, the accuracy of the ΔE measurement is about 1 meV. This corresponds to measurement accuracy of the emitter and collector dopant concentrations of roughly 5% and 10%, respectively. Deviations substantially greater than 10% are associated with rough InGaAs capping layers. Measurement accuracy is significantly reduced for these

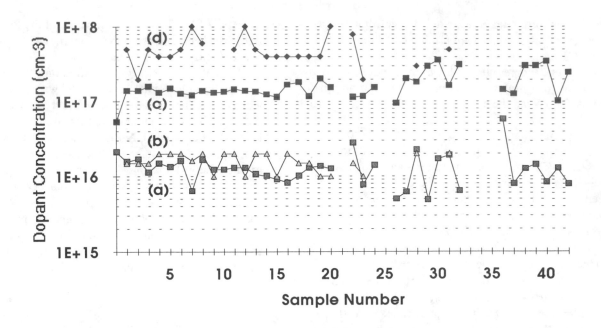

FIGURE 1. PR determined N_D for the HBT collector layer **(a)** (squares) are in good agreement with C-V values **(b)** . Emitter layer N_D values determined by PR **(c)** are more accurate than those determined by C-V analyses **(d).**

samples due to suppression of the modulation at the junctions of interest.

Spectral Ellipsometry

A structural model was developed to extract film thickness and alloy composition information from raw SE data, which is expressed in terms of the phase (delta) and amplitude (psi) changes in the complex reflectance. The model must be physically representative of the measured sample. Reference dielectric data as a function of wavelength over the measurement range is required for all materials contributing to the reflected signal. Regression analysis is carried out until the difference between the calculated and the experimental SE data is minimized. A schematic diagram of the HBT structure is shown in Figure 2. We grew several test structures, terminating with successive layers of the HBT structure, to determine the depth of the reflected light signal. From analysis of the test structure data, we determined that the reflected signal contained information up to the 80 nm thick GaAs base layer. In the model, the base and all deeper layers are treated as a GaAs substrate. SE data obtained from a test structure grown with the standard structure from the AlGaAs emitter to the InGaAs cap layer on GaAs was equivalent

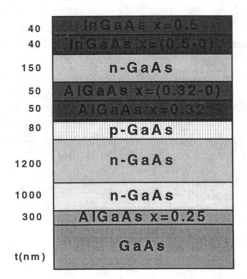

t(nm)	
40	InGaAs x=0.5
40	InGaAs x=(0.5-0)
150	n-GaAs
50	AlGaAs x=(0.32-0)
50	AlGaAs x=0.32
80	p-GaAs
1200	n-GaAs
1000	n-GaAs
300	AlGaAs x=0.25
	GaAs

FIGURE 2. Schematic diagram of the heterojunction bipolar transistor structure.

to that obtained from a standard HBT structure, validating this model assumption.

InGaAs and AlGaAs alloy compositions in the calculated spectra were allowed to vary approximately over a 50% range in fitting the two graded-composition layers, and the emitter layer of the structure. The accuracy of the InGaAs and AlGaAs alloy concentrations of test structures determined by SE were within 5% of the values determined using photoluminescence or x-ray diffraction data.

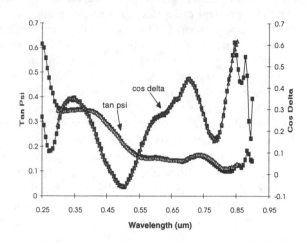

FIGURE 4. Comparison of experimental and calculated SE spectra for an HBT structure showing an excellent fit.

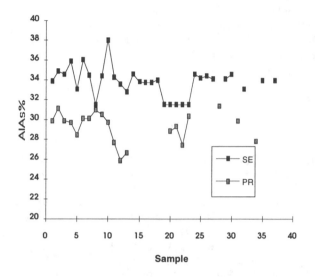

FIGURE 3. Comparison of AlGaAs emitter layer composition determined using PR and SE for several HBT structures.

Agreement between PR and SE determined AlGaAs emitter layer composition was generally within about 5% (AlAs). Equation (4) was used to calculate the AlAs mole fraction from the wavelength of the E_0 peak in the PR data.(13) Equation (4) is valid for AlAs mole fraction less than 0.45.(13)

$$Eg \ (eV) = 1.424 + 1.247 \ x \qquad (4)$$

A comparison of PR and SE derived AlGaAs composition is shown in Figure 3 for several HBTs. The SE calculation results in fits corresponding to +/- 2% uncertainty in AlAs%. PR uncertainty in the AlAs% calculation is less than +/- 0.5%. The larger uncertainty in the SE determination of the AlGaAs composition is largely a result of the reduced contribution of the emitter layer to the reflected light signal. From the surface layer to the deeper layers in a multilayer structure such as the HBT, the contribution of the layer to the reflected light signal diminishes.

We used the figure of merit of the SE fit, sigma, to assess the reliability of the SE data. Sigma (σ) is the

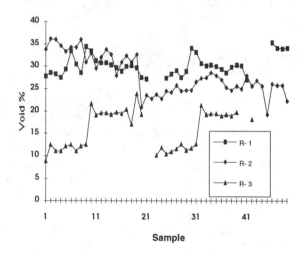

FIGURE 5. Comparison of surface roughness data determined using SE from HBTs grown using different MOCVD reactors (R-1, R-2 and R-3).

standard deviation of the calculated spectrum from the experimental spectrum. One of the best SE fits we have obtained for an HBT structure is shown along with the experimental data in Figure 4. Good fits result in sigma values less than about 0.015. The SE fit shown in Figure 4 had a sigma of 0.0066.

Surface roughness differences are dominated by variations in the InGaAs layer thickness and the InAs mole fraction. To account for the surface roughness in our SE model, we used the Bruggeman effective medium approximation (EMA)(14), mixing the dielectric functions of InGaAs and voids. We found significant, reactor-dependent differences in surface roughness. SE data obtained from HBTs grown in three different MOCVD reactors is shown in Figure 5. Thicker InGaAs layers and higher InAs mole fractions result in greater surface roughness. Figure 5 shows similar surface roughness for Reactors 1 and 2 with 25-35% voids. Reactor 3 shows substantially lower surface roughness, with about 10-20% voids. SE fits of the samples from Reactor 3 indicated 10-30 nm lower InGaAs layer thickness, compared with Reactors 1 and 2, accounting for the roughness difference shown. Additionally, the graded InGaAs composition was lower in In content for Reactor 3 compared with the other reactors. The InAs% in the graded composition layer is overestimated as a result of the SE structural model we use. The thickness of layer one corresponds with the rough portion of the InGaAs layer. Generally, this amounts to about 20 nm of the 40 nm thickness of the $In_{0.5}Ga_{0.5}As$ layer. Consequently, the InGaAs composition calculated for the "graded" layer always includes about 20 nm of $In_{0.5}Ga_{0.5}As$, leading to larger apparent In content in this layer.

CONCLUSIONS

We have shown that dopant concentrations and structural data can be obtained nondestructively from multilayer, epitaxial structures. Collector layer dopant concentrations determined nondestructively using PR are equivalent to destructively determined values within the measurement uncertainties of both techniques. Emitter layer dopant concentrations determined using PR are considerably more accurate than those obtained using the C-V technique. We find that the AlGaAs (emitter layer) composition determined using SE and PR agree within about 5% (AlAs). Good quality SE fits were obtained for numerous HBT structures grown in three different MOCVD reactors. Differences in the surface roughness were found from the SE fits and correlated with variations in InGaAs film thickness and In content in the InGaAs layers.

The quality of the InGaAs layer plays a dominant role in determining the quality of optically based measurements. Rough InGaAs layers lead to degradation in the SE fit quality, as well as suppression of the desired junction modulation observed with PR.

Optically obtained data provides a useful complement to device measurements such as gain and breakdown voltage. The structural and dopant concentration data obtained with PR and SE provides information pertaining to growth reproducibility and is useful for screening wafers in this capacity. SE and PR are both relatively easy and inexpensive techniques to implement and both techniques provide considerable information about complex device structures.

ACKNOWLEDGMENTS

Tom Brandon, Kathy Rice, David Koch, Claudia Delcampo and Ryan Keller acquired the PR and SE data. Jeff Large generously allowed us the use of the SOPRA SE instrument. Vance Ley, Tae Kim and Monica Chadwick grew the HBT structures. Tim Henderson provided the C-V data.

REFERENCES

1. Pickering, C., Carline, R.T., Emeny, M. T., Garawal, N. S., and Howard, L. K., *Appl. Phys. Lett.* **60,** 2412-2414 (1992).
2. Palik, E. D., Ed., *Handbook of Optical Constants of Solids*, Academic Press, 1985; and Jellison, G. E. Jr., *Optical Materials* **1,** 151-160 (1992).
3. Aspnes, D. E., Kelso, S. M., Logan, R. A., and Bhat, R., *J. Appl. Phys.* **60,** 754-767 (1986).
4. Pollak, F. H., and Shen, H., *Materials Sci. and Engineering* **R10**, 275-374 (1993).
5. Pollak, F. H., and Glembocki, O. J., *SPIE* **946,** 2-35 (1988).
6. Pollak, F. H., *SPIE* **276**, 142-156 (1981).
7. Aspnes, D. E., *Solid State Commun.*, **8,** 267- 270 (1970).
8. Nahory, R. E., and Shay, J. L., *Phys. Rev. Lett.*, **21,** 1569-1571 (1968).
9. Risch, L., *Phys. Status Solidi B*, **88,** 111- 116 (1978).
10. Duncan, W. M., and Schreiner, A. F., *Solid State Commun.* **31,** 457-459 (1979).
11. Sydor, M., Angelo, J., Mitchel, W., Haas, T. W., and Yen, M.-Y., *J. Appl. Phys.* **66,** 156-160 (1989).
12. Kim, T. S., and Henderson, T. S., private communication.
13. Casey, H. C., and Panish, M. B., *Heterostructure Lasers, Part A: Fundamental Principles,* New York: Academic Press, 1978, p. 193.
14. Bruggeman, D. A. G., *Ann. Physik,* **24**, 636-664 (1935).

Electrochemical Capacitance-Voltage, Magneto-Hall and Theoretical Analysis of Heterostructure Transistor Material

C.E. Stutz

Solid State Electronics Directorate, Wright Laboratory, WL/ELR, Wright Patterson AFB, Ohio 45433

B. Jogai and D.C. Look

University Research Center, Wright State University, Dayton, Ohio 45435

This work describes advanced characterization techniques for measuring AlGaAs/GaAs and AlGaAs/InGaAs heterosturcture interfaces. Electrochemical capacitance-voltage (EC-V) and magneto-Hall (M-Hall) measurements in conjunction with theoretical analysis of heterostructure transistor material are employed. The limitations of EC-V and M-Hall are explored and the measurements are compared to a self-consistent Poisson/Schrodinger k•p calculation of the electron concentration. This three tiered approach allows for a deeper understanding of the distribution of charge in the heterojunction region of device material.

INTRODUCTION

Methods to determine if heterostructure interfaces are of sufficient quality for the fabrication of heterojunction transistors have recently become very important (1-3). This is due to the usefulness of the AlGaAs/GaAs and AlGaAs/InGaAs heterointerfaces for microwave and digital device applications (4-5). Previous investigations have used conventional voltage-stepped, capacitance-voltage (C-V) profiles in conjunction with Hall measurements to study the heterointerfacial material properties of such heterojunction device structures. Unfortunately, the characterization of transistor materials is complicated by the addition of a highly doped GaAs cap layer used for ohmic contacts. This layer causes a high field breakdown in C-V measurements, due to the incrementally increased potential applied at the sample surface, so that the information desired at the heterointerface is beyond the capability of the measurement. The cap layer also causes problems for conventional Hall measurements, since this method measures a weighted average mobility and concentration of all the layers present in a multi-layered sample. Efforts to overcome this problem, such as etching off part or all of the cap layer in order to measure only the two-dimensional electron gas (2-DEG) at the heterostructure interface, have generally led to inaccurate results. Besides electrical measurments, Brierly (6) used photoluminescence linewidth measurements on heterostructure material. While useful, these measurements are restricted to the sheet carrier density only in the well of transistor material, and do not give any dopant concentration information in regard to the highly doped cap or barrier layers.

In this work we describe advanced characterization techniques for measuring AlGaAs/GaAs and AlGaAs/InGaAs heterostructure device material, namely electrochemical capacitance-voltage (EC-V) and magnetic-field-dependent Hall (M-Hall). It is shown that EC-V measurements, which employ a stepped surface etching technique, are useful for delineating the cap, delta-dopant, and 2-dimensional electron gas (2-DEG) in delta-doped high electron mobility transistor (HEMT) and pseudomorphic HEMT (pHEMT) material. The depth resolution of EC-V is ultimately limited by the spatial extent of the groundstate wavefunction, but can practically be limited by the roughness of the etchant. These limitations become a problem when two reconstructed features are separated by approximately 50 Å or less. Also, M-Hall measurements can separate two parallel conductive layers by making measurements at several different magnetic fields. This allows for the accurate determination of the cap doping level and the sheet charge in the 2-DEG. (The conductance in the AlGaAs barrier layer can generally be ignored in this case due to its low mobility.) These measurements are then compared and found to be in excellent agreement with state-of-the-art self-consistent Poisson/Schrodinger k•p theory (7). Our implementation of this theory includes effects due to strain, surface states, and interface states if desired, although the latter do not seem to be important for pHEMT structures. By using the combined approach of EC-V, M-Hall and theoretical modeling a more complete understanding of heterostructure transistor material is obtained. Therfore, we can now show that it is possible to determine, relatively accurately, how much charge is actually at the heterointerface of device material.

THEORY

The conduction and valence band edges and free charge distribution were calculated self-consistently from the Poisson and Schrodinger equations. The theory, reported in detail elsewhere (7), is described here briefly. The Poisson equation was solved for the band edges. Since the free charge is quantum-confined along the growth-axis, it is necessary to calculate the charge quantum mechanically, hence the need for the Schrodinger equation. Our Schrodinger equation is formulated within the effective mass approximation using the **k·p** perturbation theory. The basis set therein consists of electrons, heavy holes, light holes, and split-off holes.

This scheme allows for the rigorous inclusion of strain. Strain shifts the energies of the electrons and holes relative to their positions prior to strain. These shifts, however, are not uniform within the Brillouin zone. Rather, they depend on k. The present model includes this k-dependent strain. Additionally, strain-induced changes to the spin-orbit interaction are included. Since the crystal symmetry under strain is no longer zinc-blende, it is incorrect to employ the usual zinc-blende formulation for the spin-orbit interaction. The spin-orbit interaction in the |J,m_J> representation is no longer diagonal. The strain-induced off-diagonal terms are included perturbatively in the **k·p** Hamiltonian.

Also included in the model are surface states. Surface states are known to pin the Fermi level at the free surface of GaAs at about 0.7 eV at 300 K. In previous models, this pinning was implemented by simply manipulating the boundary conditions for the Poisson equation. This approach does not account for an important charge-neutrality effect. The surface states are predominantly deep-level acceptors about 0.68 eV above the valence band edge. They trap electrons which, because of charge neutrality, must come from the shallow donors. If the Poisson equation is simply manipulated to produce band bending, the resulting charge distribution can be non-physical: the excess electrons from the depletion region in the cap accumulate elsewhere in the structure, possibly in the channel or the interior, and may even form an artificial well in the interior buffer layer. The present work treats the surface states rigorously, incorporating them as a layer of sheet charge at the surface in the Poisson equation and calculating their occupancy through Fermi-Dirac statistics.

The model is also capable of incorporating interface states, usually present at the inverted GaAs/AlGaAs interface. These states, which may also involve deep acceptors, were not included in the present results.

The parameters entering into the calculation are the bulk physical properties for each constituent layer. They include the effective masses for electrons, heavy holes, light holes, and split-off holes, the bulk band gaps, spin-orbit splitting, elastic constants, deformation potentials, and dielectric constants. Also needed are the surface state density and activation energy. Wherever possible, published data for the ternary compounds are used. If unavailable, as in the case of the effective masses, the values for the ternary compounds are obtained by a linear interpolation between the values for the binary compounds.

The method of solution is based on the finite-difference technique. Poisson's equation is discretized into a series of difference equations, each representing a segment of the growth axis. The resulting tri-diagonal matrix is solved numerically. The change in dielectric constant across material boundaries is included. The **k·p** Hamiltonian must be solved in real space because of the material discontinuities along the growth axis. This means that the component of the wave vector along the growth axis becomes a differential operator with respect to real space position along the axis. The resulting coupled differential equations are solved by the finite-difference technique. If the growth axis is divided into N segments, one is required to diagonalize a 4N x 4N matrix to obtain the wave functions and energies. The charge distribution is calculated from the wave functions and inserted into Poisson's equation, from which the new electrostatic potential is obtained. This process is repeated until a convergent solution is obtained. At each step, the Fermi energy is required. This is calculated iteratively via the Newton-Raphson method by requiring the entire structure to be electrically neutral.

ELECTROCHEMICAL CAPACITANCE-VOLTAGE

The EC-V measurements were made on a PN4200 Bio-Rad system with 0.1 M Tiron (4,5-dyhydroxy-1,3-benzene-disulfonic acid) or a mixture of 0.2M NaOH/0.1M EDTA (ethylenediaminetetraacetic acid disodium salt) as the electrolytes. The electrolyte solution acted as both a Schottky contact and etchant. The system is effectively described by Blood (8) who used a carbon counterelectrode, saturated calomel electrode for voltage reference, and a platinum wire placed close to the electrolyte/semiconductor interface for AC signal access. A 3-mm or 1-mm-diameter sealing ring defined the appropiate contact area A which was calibrated with conventional Hall measurements using thick (1 μm) uniformly doped GaAs:Si epitaxial layers doped at 3×10^{18} cm^{-3}, so as to minimize errors due to depletion. The thickness of the layers was also calibrated against reflection high-energy electron diffraction (RHEED) oscillations carried out in a molecular beam epitaxy system. The bias conditions were carefully chosen so as to minimize leakage currents for accurate measurements. The capacitance C was measured with an AC signal of 0.1 V (pk-pk) at 1 kHz for the 3-mm sealing ring and 3 kHz for the 1-mm sealing ring. The differential capacitance dC/dV was measured by adding a modulated voltage of 0.2 V (pk-pk) at 10 Hz for the 3-mm sealing ring and 40 Hz for the 1-mm sealing ring. The concentration was obtained with the usual differentiated Schottky equation of $N_{CV}=C^3/[\epsilon\epsilon A^2(dC/dV)]$. However, the measured depth is

684

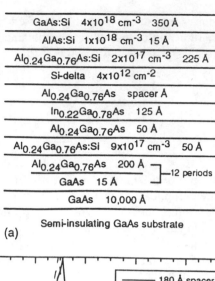

GaAs:Si	4×10^{18} cm^{-3}	350 Å
AlAs:Si	1×10^{18} cm^{-3}	15 Å
Al$_{0.24}$Ga$_{0.76}$As:Si	2×10^{17} cm^{-3}	225 Å
Si-delta	4×10^{12} cm^{-2}	
Al$_{0.24}$Ga$_{0.76}$As	spacer Å	
In$_{0.22}$Ga$_{0.78}$As	125 Å	
Al$_{0.24}$Ga$_{0.76}$As	50 Å	
Al$_{0.24}$Ga$_{0.76}$As:Si	9×10^{17} cm^{-3}	50 Å
Al$_{0.24}$Ga$_{0.76}$As	200 Å	⎤ 12 periods
GaAs	15 Å	⎦
GaAs	10,000 Å	

Semi-insulating GaAs substrate

(a)

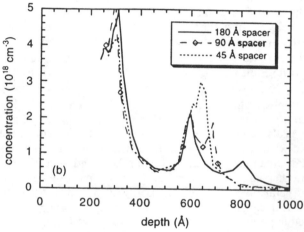

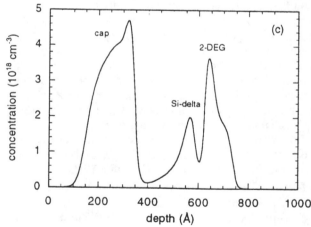

FIGURE 1. (a) pHEMT layer structure; (b) EC-V measurements for pHEMTs with 45, 90, and 180 Å spacers; (c) theoretical calculation showing the carrier concentration for a pHEMT with a 45 Å spacer.

determined by the sum of the depletion depth, using the capacitor equation, $x_d = \varepsilon A/C$, which assumes that no traps are present, and the etch depth, using Faraday's equation, $x_e = MQ/zFDA$. M and D are the molecular weight and density of the semiconductor, F is the Faraday constant, z

is the charge transferred per molecule dissolved and Q=I dt where I is the measured current and dt is the duration of the data collection.

Figure 1(a) shows our basic delta-doped pHEMT structure. Figure 1(b) shows the delineation of the cap, delta-doped region and the 2-DEG region of three different delta-doped pHEMT samples with 45Å, 90Å, and 180Å spacers. The delta-doped and 2-DEG regions are clearly defined for the 90 Å and 180 Å spacers, respectively. The 45 Å spacer material looks more like a single peak, but with a large shoulder. Using the above theory a simulated curve of what the true carrier profile should look like is shown in Figure 1(c) for the 45 Å layer. Ideally, two well defined peaks, one in the delta-doped region (depth=565 Å) and another in the 2-DEG region (depth=645 Å) should be seen. The similarities between the theoretical and EC-V plots show that the EC-V measures the free charge and not the fixed dopants. The broadening of the carrier peaks is not due to the classical Debye length, but to the groundstate wavefunction extent explained by the following quantum- mechanical relation (9),

$$\Delta x_0 = 2 \sqrt{\frac{7}{5}} \left(\frac{4}{9} \right)^{1/3} \left(\frac{\varepsilon \hbar^2}{e^2 N_{2D} m^*} \right)^{1/3} \qquad (1)$$

where N_{2D} is the delta-dopant concentration. Δx_0 can be thought of as the resolution limit for the C-V measurement technique. As can be seen by Figure 2 the resolution improves as N_{2D} increases. However, for EC-V another resolution limitation arises due to the smoothness of the etching capability of the electrolyte. For the above 45 Å spacer sample (as determined by TEM) a typical mean etch depth of about 250 Å, as determined by x_e, occurs when the measurement depth ($x_d + x_e$) reaches the delta-doped/2DEG region. If the amplitudes of the etchant

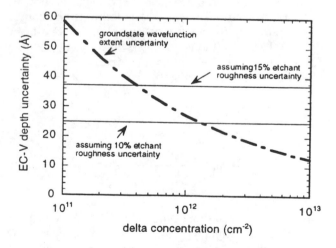

FIGURE 2. A comparison of the C-V resolution limit of a delta-doped layer and the resolution limit due to the EC-V etchant.

685

undulations are about 10% of the mean depth etched, as suggested by Blood (8), our resolution in this case is limited by the etchant and not the extent of the groundstate wavefunction. As shown in Figure 2 our etchant roughness may be closer to 15% of the mean etch depth to explain why the peaks at 565 and 645 Å cannot be well defined.

Even though EC-V measurements give a distorted view of the charge distribution they can still be useful for determining the amount of sheet charge at the heterostructure interface. This is true since the apparent charge differential $dQ=eN_{cv}dx_{cv}$ is equal to the true charge differential $dQ=CdV=eN_{true}dx_{true}$ (1). Therefore, the integrated charge of the measured EC-V profile is equal to the true integrated charge. For instance when we integrate the area under the delta-doped and 2-DEG EC-V concentration profile for the 45 Å spacer sample a value of $4.5 \pm 0.3 \times 10^{12}$ cm^{-2} is calculated. The above theory calculates the sheet charge to be 4.6×10^{12} cm^{-2}, which is in excellent agreement with EC-V. In order to properly compare the EC-V and theoretical sheet charge the integration was arbitrarily chosen to begin at the minimum between the cap and delta-doped layers and effectively go out to infinity.

MAGNETO-FIELD-DEPENDENT HALL

As is seen in Figure 1(c), there are two strongly conductive layers in a pHEMT structure, the n$^+$-GaAs cap, and the In$_x$Ga$_{1-x}$As well. (The delta-doped layer has inconsequential conductance compared to the others because of low mobility.) Another example of a two-layer problem is a GaAs n$^+$/n structure used for MESFET devices. A simple Hall-effect/conductance analysis determines only the overall conductance σ and Hall coefficient R, and these two measured parameters allow us to determine two unknowns, the mobility μ and carrier concentration n (in an n-type sample). A two-layer problem has four unknowns, μ_1, n_1, μ_2, n_2, and a Hall/conductance measurement mixes them in the following, well-known way (10):

$$\sigma_0 = e\left(n_1\mu_1 + n_2\mu_2\right) \qquad (2)$$

$$R_0 = -\frac{r_1 n_1\mu_1^2 + r_2 n_2\mu_2^2}{e\left(n_1\mu_1 + n_2\mu_2\right)^2} \qquad (3)$$

where the "0" denotes magnetic field B=0. (Note that the Hall scattering factors, r_1 and r_2, are unity for degenerate electrons, which is the case for both the cap and well electrons in a pHEMT structure. In any case, the r factors are usually near unity for electrons.) Obviously, the four unknowns in Eqns. 2 and 3 cannot be uniquely determined

from the measurement of only two parameters, σ_0 and R_0. However, futher information can be obtained from the magnetic-field dependence of σ and R (3,9):

$$\frac{1}{B^2} + C_0 = S_\rho \frac{\rho_0}{\Delta\rho} \qquad (4)$$

$$\frac{1}{B^2} + C_0 = -S_R \frac{R_0}{\Delta R} \qquad (5)$$

where $\rho=1/\sigma$, $\Delta\rho=\rho(B)-\rho(0)$, $\Delta R=R(B)-R(0)$, and C_0, S_ρ, and S_R are complicated functions of μ_1, n_1, μ_2, and n_2. We now have four measured parameters, ρ_0, R_0, S_ρ, and S_R, written in terms of four unknowns, μ_1, n_1, μ_2, and n_2. (Note that the measurement of C_0 does not give any additional information.) The unknowns can be determined uniquely by an algebraic inversion of the four equations $\rho_0(\mu_1, n_1, \mu_2, n_2)$, etc. The uniqueness holds even though there is a choice of sign on a square-root term, because the opposite choice of sign simply interchanges the μ_1, n_1 and μ_2, n_2 solutions. In the end, physical considerations must be implemented to choose which layer has the higher mobility. The final formulas can be found in Ref. 3 and are easily programmed on a computer. For example, we have written a short program in BASIC which computes the desired quantities in well under 1 second on a PC. On the other hand, several other groups have implemented four-parameter, least-squares (LSQ) fits of Eqns. 2-5, or of identical equations in somewhat different form (11-13). Such approaches are much more computer-intensive and do not necessarily lead to unique solutions of μ_1, n_1, μ_2, and n_2. It should also be mentioned that formulas for σ and R as functions of B can also be set up for situations involving more than two conductive layers (10). In these cases, a unique algebraic inversion cannot be performed, and a multi-parameter LSQ computer fit is the only way to solve the problem. However, from a practical point of view, it is still very difficult to obtain accurate solutions of the μ's and n's for more than two layers.

In our analysis of pHEMT structures we have chosen to use four magnetic fields, 0, 2, 10, and 16 kG for resistivity, and three fields, 2, 10, and 16 kG for Hall coefficient. We then carry out a linear LSQ fit of $1/B^2$ vs. $\rho_0/\Delta\rho$ (Eq. 4), with ρ_0 being the value of ρ at B=0, and $\Delta\rho=\rho(B)-\rho_0$ for B=2, 10, and 16 kG. These fits are normally excellent, with correlation coefficients (CC's) above 0.999, and give precise values of S_ρ and C_0. (A linear LSQ fit of course gives unique values of the slope and intercept because it is an analytical solution; i.e., no "fitting" is necessary.) Next, for small B ($B^2 << 1/C_0$), Eq. 5 shows that $R \propto B^2$; thus we use the 2 kG (weighted twice) and 10 kG values of R to determine R_0, which will not be too different from the value at B=2 kG. Finally, this value of R_0 is used along with the value of C_0 found from Eq. 4 and the values of $\Delta R=R(B)-R_0$ at B=10 and 16 kG to

TABLE I. Comparison of M-Hall measurements with theory.

| Sample | T (K) | Cap conc. (10^{12} cm^{-2}) | | Channel conc. (10^{12} cm^{-2}) | | Channel mob. (cm^2/V-s)) |
		M-Hall	Theory	M-Hall	Theory	M-Hall
1	296	5.4	7.6	3.1	3.0	6.600
	77	5.7	7.6	3.2	3.2	17,000
2	296	6.6	7.6	3.2	3.0	6,200
	77	8.0	7.6	3.2	3.2	14,400

determine S_R by a LSQ fit of Eq. 5. Although this sequence for solving Eqns. 4 and 5 seems somewhat arbitrary, and more data at other magnetic fields would of course be helpful, we have found that the recipe outlined here gives good, reproducible values for pHEMT structures. Furthermore, the total time for the experiment is less than twice that used for a simple Hall measurement of only one magnetic field, and could be reduced even more by having a means of rapidly changing the magnetic field.

To illustrate the use of M-Hall analysis we consider two pHEMT structrues grown at different times (several months apart) but with nominally identical structures using a 45Å spacer, see Figure 1(a). The results are shown in Table I. In our analysis we have assumed that $\mu=1500$ cm^2/V-s for the cap, as measured on separate structures, because it turns out that although the cap sheet resistance is well determined, the separation into μ_{cap} and n_{cap} is not, in this case. However, the channel μ and n are accurately determined individually and agree well with theory. Sample 2 has a somewhat lower mobility, perhaps due to a rougher channel/spacer interface. When the cap on sample 1 was etched off the channel concentration was reduced to 1.9×10^{12} cm^{-2}, while the mobility was relatively unchanged. Thus, the common methods of either etching off the cap, or growing a separate structure without a cap, give a totally different answer for the channel concentration. This fact underscores the need for implementation of a technique such as M-Hall analysis for pHEMT evaluation.

CONCLUSION

We have shown that a state-of-the-art Poisson/Schrodinger $\mathbf{k \cdot p}$ formulation along with EC-V and M-Hall measurements make possible a detailed understanding of heterostructure device material. The EC-V can delineate the cap, delta-doped, and 2-DEG regions of heterostructure device material. Measurement agrees with theory when integrating the concentration profile for the delta-doped and 2-DEG regions. EC-V can sometimes only show qualitatively the amount of charge transferred from the delta-doped layer into the quantum well region for small spacers, due to the roughness of the electrolyte/semiconductor interface. For material with spacer layers > 90 Å a quantitative amount of charge transfer can be obtained, since the layers can be clearly

defined. M-Hall, by using several magnetic fields, is able to quantitatively measure the cap and 2-DEG regions, with no etching necessary and no limit on the size of the spacer layer.

ACKNOWLEDGMENTS

The authors would like to thank T. Cooper for M-Hall measurements, Z. Liliental-Weber for TEM support, and J.M. Ballingall for material growth. This work was partially supported by the Air Force Office of Scientific Research and U.S. Air Force contract F33615-91-C-1765.

REFERENCES

1. Stutz, C.E., Jogai, B., Look, D.C, Ballingall, J.M., and Rogers, T.J., *Appl. Phys. Lett.* **64**, 2703-2705 (1994).
2. Look, D.C., Jogai, B., Stutz, C.E., Sherriff, R.E, DeSalvo, G.C., Rogers, T.J., and Ballingall, J.M., *J. Appl. Phys.* **76**, 328-331 (1994).
3. Look, D.C., Stutz, C.E., and Bozada, C.A., *J. Appl. Phys.* **74**, 311-314 (1993).
4. Beneking, H., [Ed. Malik, R.J], *III-V Semiconductor Materials and Devices,* Amsredam, North Holland, 1989, Chap. 8.
5. Ballingall, J.M. Martin, P.A., Mazurowski, J., Ho, P., Chao, P.C., Smith, P.M., and Duh, K.H.G., *Thin Solid Films* **231**, 95 (1993).
6. Brierly, S.K., J. Appl. Phys. **74**, 2760 (1993).
7. Jogai, B., J. Appl. Phys. **76**, 2316-2323 (1994).
8. Blood, P., Semicond. Sci. Technol. **1**, 7-27 (1986).
9. Schubert, E.F., Kuo, J.M., and Kopf, R.F., J. of Electronic Materials **19**, 521-531 (1990).
10. See, e.g., Look, D.C., Electrical Characterization of GaAs Materials and Devices, New York, Wiley, 1989, 227ff.
11. Aina, L. Mattingly, M., and Pande, K., Appl. Phys. Lett. **49**, 865-867 (1986).
12. Schacham, S.E., Mena, R.E., Haugland, E.J., and Alterovitz, S.A., Appl. Phys. Lett. **62**, 1283-1285 (1993).
13. Kim, J.S., Seiler, D.G. and Tseng, W.F., J. Appl. Phys. **73**, 8324-8329 (1993).

Atomic Scale Characterization of InGaAs/GaAs/Si Heterostructure by HRTEM, Atomistic Modeling and Multislice Image Simulation

T. Zheleva, M. Ichimura*, S. Oktyabrsky and J. Narayan

Department of Materials Science & Engineering, North Carolina State University, Raleigh, NC 27695-7916., USA

*Department of Electrical & Computer Engineering, Nagoya Institute of Technology, Nagoya 466, Japan

Detailed atomic structure of epitaxial layers and their interfaces is important for the devices like quantum-well-confined optoelectronic devices, high mobility modulation doped structures etc. Atomic scale characterization of InGaAs/GaAs/Si(100) heterostructure grown by MBE technique has been performed using High Resolution Electron Microscopy, Atomistic Modeling and Multislice Image Simulation. A novel feature of this study is that the calculated atomic positions associated with defects and interfaces are used to simulate HRTEM images which are compared with experimental images. HRTEM analysis shows high density of threading segments and misfit dislocations. Most of the misfit at the GaAs/Si interface is accommodated by perfect edge Lomer type - 90° misfit dislocations almost uniformly spread across the interface with average spacing 95Å corresponding to a complete misfit relaxation. Both films are continuous and perfectly epitaxial without interfacial contamination or amorphous phases. Atomistic modeling based on a Keating potential has been used to study structure and stability of interfaces. At the interface averaging of parameters (force constants and bond lengths) is done to calculate the relaxed structures. The strain energies for the two interfaces are calculated and are found to be in good agreement with similar results for structures with a similar mismatch like Ge/Si. The critical thicknesses of the GaAs is estimated and compared with that for Si. Multislice simulation analysis of HRTEM through-focus series, based on the output of the atomistic modeling have been performed to analyze the detailed coherent and dislocated interface structures. Four cases are studied: coherent interface without relaxation, relaxed coherent interface and 90° and 60° perfect dislocations which are considered to accommodate the ~4% misfit strain at the GaAs/Si interface and the 2% mismatch at $In_{0.3}Ga_{0.7}As$/GaAs interface. The variations in the interface contrast at different experimental conditions are analyzed and the simulated images are compared with the experimental images to deduce the atomic structure of defects and interfaces.

INTRODUCTION

To obtain the optimum performance from electronic and optoelectronic devices based on heterojunctions and superlattices, it is necessary to grow defect-free layers with morphologically and compositionally sharp interfaces [1]. The coherently strained structures can be controlled to obtain electronic properties for applications in high speed transistors and magnetic recording. The advantages of using dissimilar materials in terms of physical properties are often perturbed by difficulties leading to poor structure of the multilayers and hence degraded electrical and optical properties. Since the properties of the multilayer systems depend upon the structural and compositional variations at the interfaces,

they have to be characterized and controlled. The structural features, i.e. characteristics of the local displacements at and around the interfaces, requiring measurements can be divided into three classes [2]: (1) the magnitude of the rigid body displacement across the interface, (2) the measurement of the strain associated with the mismatch of the lattices on both sides of the interface, and (3) characterization of any local strain fields caused by intrinsic or interfacial defects.

TEM studies have shown that the defects in epitaxial heterostructures such as $In_xGa_{1-x}As$/GaAs/Si have different origins: (1) line or planar defects from the substrate extended during the growth into the epilayer, (2) defects introduced in the epilayer due to a loss of symmetry relative to the substrate, such as inversion boundaries in GaAs/Si, induced by the surface morphology (steps), (3) single growth mistakes, leading to microtwins, stacking faults, impurities or polycrystalline or amorphous growth, (4) clustering of excess point defects - vacancies, interstitials, impurities, and (5) stress induced by the lattice mismatch across the coherent interface and in case of systems with a large mismatch resulting in formation of misfit dislocations [3,4].

The film is coherently strained to match the substrate when the thickness of the epilayer is sufficiently small. However with the increase of the film thickness the homogeneous misfit strain energy increases and at a critical thickness the nucleation of dislocation becomes energetically favorable. The dislocations can introduce traps and recombination centers for charge carriers, thus affecting their mobility and lifetime. Also the positions of atoms in a core of a dislocation determine the electron distribution and hence their electronic properties. In addition, the atomic structure and the energetics of dislocations can be used to predict reactions and growth directions of dislocations.

The atomic scale characterization of the interfaces and defects at the interfaces in $In_{0.3}Ga_{0.7}As/GaAs/Si$ heterostructure has been performed by HRTEM, Atomistic modeling and Multislice Image Simulation. Two important structural aspects of epitaxial interfaces are considered in this paper: (i) the atomic structure at the coherent interfaces, and (ii) the atomic structure of dislocations, in particular 90° and 60° perfect dislocations.

EXPERIMENTAL

Our studies were carried out on $In_{0.3}Ga_{0.7}As/GaAs$ grown on Si(001) wafers by molecular-beam epitaxy (MBE). Cross-sectional samples were prepared for TEM observation by conventional "sandwich" technique using mechanical polishing and Ar ion-milling in the final stages. High-resolution transmission electron microscopy studies were performed using Topcon 002B microscope operated at 200 kV, with point-to-point resolution of 0.18nm and spherical aberration coefficient C_S=0.4mm.

ATOMISTIC MODEL

The atomic cells used in the simulation are two dimensional cells with a width of one unit length (3.84Å for Si) along the [110] dislocation line (z-direction). The cell length in the perpendicular direction (x-direction), w, is taken to be 25 unit lengths or 96Å. We use periodic boundary conditions for these two directions. Thus, when the cell has one dislocation in it, the cell length w corresponds to the spacing between dislocations in a periodic array of parallel dislocations. For GaAs/Si interface, the width of 96Å is almost equal to the average spacing of 90° dislocations in a fully relaxed GaAs/Si

interface. Thus the misfit strain in x-direction is completely relaxed when the cell has one 90° dislocation. A 90° dislocation is twice as effective for relaxation as a 60° dislocation, and thus only half of the misfit strain is relaxed in the cell with a 60° dislocation.

The misfit of $In_{0.3}Ga_{0.7}As/GaAs$ is 2%, about half of that of GaAs/Si. The same cell size is used to simulate a 60° dislocation at the InGaAs/GaAs interface. In this case, the misfit strain is almost completely relaxed. It should be noted that since our atomic cell is two dimensional, an array of dislocations is considered in one direction, not a square network of dislocations. Thus, the misfit strain is relaxed in only one direction and the cell is always strained in the other direction.

The thickness of the substrate used in the simulation studies is about 35Å (24 monatomic layers for Si and 12 monolayers for GaAs), and the atoms at the bottom of the substrate are fixed to simulate a semiinfinite substrate. The thickness of the layer is about 30Å (20 monatomic layers for Si and 10 monolayers for InGaAs).

The energy of the cell is calculated using the Keating potentials. The Keating potentials are known for all of the materials included in the structures studied here. The Keating potential works well for small atomic displacements, and it is not strictly valid in a presence of a dangling bond. However, its use for dislocation leads to overestimate of the strain energy near the core [5]. Therefore, the energy of the structure with a dislocation will be overestimated in our simulation. The potentials for the GaAs/Si interface bonds are not known, and even the type of the interface bond (Ga-Si or As-Si) has not been identified. Arithmetic averaging of parameters (bond length and force constants) of Si and GaAs for the interface bonds is used in the simulation procedure. For the $In_{0.3}Ga_{0.7}As$ alloy, averaging of parameters of GaAs and InAs weighted by composition is applied. The energy of the cell is minimized by moving every atom sequentially in the direction of force calculated by the interatomic potentials [6].

IMAGE SIMULATION

Detailed determination of the atomic structure from HRTEM images is usually based on a comparison with simulated images of a known structure model. The reason is that even if the microscope is modeled perfectly the dynamical scattering of the electrons, non-linear contribution to the image and phase distortions introduced by the lenses eliminate the direct relationship between the image contrast and the structure of interest. Image simulations in the [110] projection for the two interfaces - GaAs/Si and $In_{0.3}Ga_{0.7}As/GaAs$, were carried with the program MACTEMPAS for the super-cells utilized in the atomistic model and containing the coherent and defect interface structures with Lomer and 60° dislocations. First the data about the specimen - unit cell dimensions, symmetries, atom positions, occupancies and temperature factors are used to generate part of the crystal potential that produces electron scattering. Next step is generation of the electron wave function at the specimen exit surface, which uses data from step one and information about acceleration voltage of the microscope and the specimen thickness and the tilt. The generation of the image intensity at the microscope image plane is performed at the third step of the simulations as the effects of the electron wave phase variations are included via parameters: defocus, spherical aberration, incident beam convergence, spread of focus, position and size of the objective aperture. It should be noted that there are number of limitations in the image simulation procedure due to uncertainties over the experimental conditions - thickness, defocus, beam tilt, etc. which make the accurate interpretation of the images difficult.

RESULTS AND DISCUSSION

High density of threading segments and misfit dislocations is observed at GaAs/Si interface and to a much lesser extent at $In_{0.3}Ga_{0.7}As$/GaAs interface. High resolution images of GaAs/Si and $In_{0.3}Ga_{0.7}As$/GaAs interfaces in [110] orientation of the Si substrate are shown in Fig.1(a) and (b), respectively. Most of the misfit at GaAs/Si interface is accommodated by perfect edge Lomer type - 90° misfit dislocations and 60° dislocations. One can easily distinguish between the two types of dislocations by the fact that 90° dislocations terminate at two {111} planes, while the 60° dislocations terminate at one of them. It should be noted that only projection of the Burgers vector in the (110) plane of view can be determined, and hence the screw component of the 60° dislocation can not be defined by HRTEM image.

Calculated by the atomistic model strain energies of the cell (eV/Å along the dislocation line) are shown in Table 1. The energies of the dangling bond and the surface step are not included in the results for GaAs/Si. Owing to the elastic relaxation (tetragonal distortion), the energy of the coherent GaAs/Si structure is about one third of the energy of the rigid structure where all Ga and As atoms are fixed at lattice sites of a Si lattice. The energy of the cell with a 60° dislocation is larger than that of the coherent cell, while the cell with a 90° dislocation has the lowest energy. This means that the relaxation by 90° dislocations is energetically more favorable than the relaxation by 60° dislocations, and that the GaAs thickness (10ML) is larger than the critical thickness h_c for 90° dislocation but smaller than hc for 60° dislocations. For comparison, we also show results of similar simulation for Ge/Si. The Stillinger-Weber potential, which is more suitable for large distortion than the Keating potentials, is used for Ge/Si.

The coherent state energy of Ge/Si is larger than GaAs/Si because of larger elastic constants of Ge, but the difference in the energy is small for the cells with dislocations. This is because the core energy is overestimated for GaAs/Si. The energy of $In_{0.3}Ga_{0.7}As$/GaAs structure is also shown in Table 1. The energy of the cell with 60° dislocations is about twice as large as the energy of the coherent cell. Thus, the critical thickness is much larger than the layer thickness.

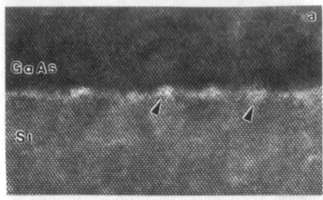
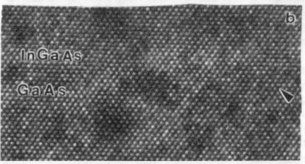

Figure 1. High-resolution lattice image of (a) GaAs(up)/Si(down) interface. Most of the dislocations are of mixed type (60°) near complete association to form 90° dislocation. (b) High-resolution lattice image of coherent $In_{0.3}Ga_{0.7}As$(up)/GaAs(down) interface

Table 1. Energies* of the atomic cells of GaAs/Si, $In_{0.3}Ga_{0.7}As$/GaAs and Ge/Si structures.

	rigid	coherent	60° dislocation	90° dislocation
GaAs/Si	11.2	3.80	4.3	3.1
InGaAs /GaAs	2.8	0.93	1.9	
Ge/Si	10.3	4.45	4.45	3.2

* The values are given in eV per unit length (Å) along the dislocation line.

Table 2. Critical thicknesses (in Å) of GaAs/Si and Ge/Si heterostructures.

	60° dislocation	90° dislocation
GaAs/Si	39	20
Ge/Si	28	9

The critical thickness h_c is estimated for GaAs/Si by calculating energy for cells with various GaAs layer thicknesses. Here the critical thickness is defined as a thickness at which the energy of the coherent cell is equal to that of a cell with a dislocation. This criterion is similar to that of Matthews or van der Merwe. [7]. The obtained values of h_c is 39Å for 60° dislocations and 20Å for 90° dislocations. The higher value for h_c in 60° dislocation is almost twice larger than for 90° dislocation because the core energy of a 60° dislocation is twice as large as that of 90° dislocation and also, the strain field generated by 60° dislocation does not compensate the biaxial misfit strain field. However, the core energy of the dislocation is not accurately calculated in our simulation, and thus the values of the critical thickness will be overestimated.

The local atomic arrangement at the interfaces established with the atomistic modeling before and after the relaxation for the coherent interfaces, and those with 90° and 60° dislocations have been employed to simulate high resolution TEM images. Here we present examples of only few of them. Figure 2 shows the experimental image (a), calculated energy minimized structure (b), and simulated high-resolution TEM image (c) of a 90° dislocation at GaAs/Si[-110] interface. The Burgers vector is along [110] direction -x-axis, and y- axis is along [001]Si, normal to the plane that contains the Burgers vector and the dislocation line. The core of this dislocation contains five-fold (pentaring) and seven-fold (septaring) rings of atoms. Thus this core does not have dangling bonds and therefore is electrically neutral. The characteristic six-fold rings in the diamond cubic lattice are present away from the dislocation line. Good agreement between the experimental, calculated and simulated images is obtained. Similar core geometry with pairs of pentarings and septarings is observed for 90° dislocations in Ge/Si interface, and also in bulk Si and Ge [4]. Example for a 60° dislocation in GaAs/Si interface is presented in Figure 3. The core of a 60° dislocation consists of an eight-fold ring (octaring). From the calculated and in simulated image it is seen that one of the

atoms has only three nearest neighbors instead of four, characteristic for the tetrahedral symmetry.

Figure 2. Experimental image (a), calculated energy-minimized atomic structure (b), and simulated HRTEM image (c) of a 90° dislocation at GaAs(up)/Si(down) interface. At the simulated image white dots represent atomic column positions.

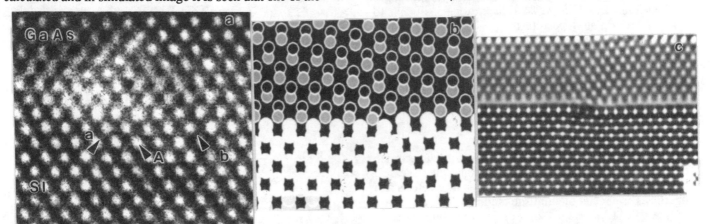

Figure 3. Experimental image (a), calculated atomic configuration after minimization (b) and simulated image of a 60° dislocation at GaAs(up)/Si(down) interface. The characteristic octaring of atoms are seen at a dislocation core of a 60° dislocation (A) pointed by arrow. Note: There are two not completely reacted 60° dislocations (a) and (b) in the vicinity of (A).

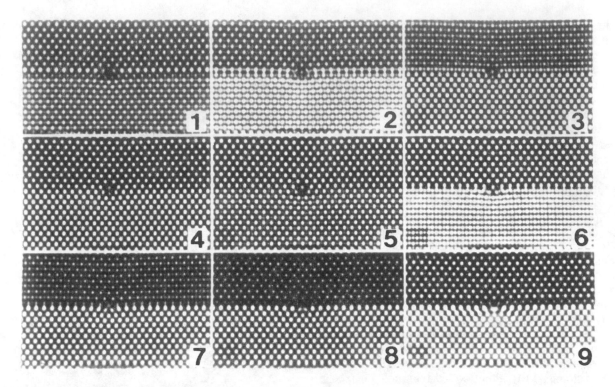

Figure 4. Simulated images for TOPCON (200kV) of a 90° dislocation at GaAs (up)/Si (down) interface in [110] orientation in through-focus series by multislice simulation. The thickness of the sample is 6nm, and the defocus step is Δf=10nm, staring from -60nm (image #1) to 20nm (image #9).

To understand the lattice image visibility of the interfaces of interest simulated images were computed from the cells in atomistic model over defocus range from -60 to 20 nm at a step of 10 nm. Figure 4 represents the simulated images of a 90° dislocation at GaAs/Si interface in through-focus series at a thickness of the sample 6nm. All images are 13-beam images computed for Cs=0.5mm at 200kV acceleration voltage. Half-period fringes in Si - #2 and #6, and #3 and #7 in GaAs part of the interface, are clearly seen when simulated images are overlapped with the atom positions. Also, half-Fourier period images are observed from image #4 (black atom positions) to image #9 (white atom positions).

The through-focal series is an example of the power of the image simulation method to extract additional information for the object, in our case interface structures, if certain experimental conditions are known.

CONCLUSION

The following procedure was established to characterize the detailed atomic structure of defects and interfaces: (i) calculation of the local atomic arrangement of defects and interfaces as a result of atomistic modeling; (ii) simulation of the HRTEM images of defects and interfaces at different conditions (mainly defocus and specimen thickness) using the calculated atomic positions; (iii) comparison of the experimental and simulated images to verify the suggested in the atomistic model defect structure. Examples for the interfaces in InGaAs/GaAs/Si heterostructures that confirm the atomistic models are given. Utilizing the established procedures we have determined the atomic structures of 60° and 90° dislocations, and those of the coherent interfaces as well. Future work should include developing of quantitative techniques for matching experimental and simulated HRTEM images.

ACKNOWLEDGMENT

This work was supported by the NSF.

REFERENCES

1. Washburn J., Kvam E., and Liliental-Weber Z., *J. Electron. Mater.* **20**, 155-161 (1991).
2. Boothroyd C., Baxter C., Bithell E., Hytch M., Ross F., SatoK., and Stobbs W., *Ultramicroscopy* **29**, 18-30 (1989).
3. Zheleva T., Jagannadham K., and Narayan J., *J. Appl. Phys.* **75**, 860-871 (1994).
4. Oktyabrsky S., Hong W., Vispute R.D., and Narayan J., *Phil. Mag.* (in press).
5. Nandedkar A., and Narayan J., *Phil. Mag.* **A61**, 873-891 (1990).
6. Ichimura M., and Narayan J., *Phil. Mag.* (in press).
7. van der Merwe, J.H., *Crit. Rev. Solid St. Mater. Sci.* **17**, 187-209 (1991).

Extraction of Dynamical Characteristics of 2-D Electron Gas in Heterostructure Field-Effect Transistors

Roberto Sung, Jau-Wen Chen, and Mukunda B. Das

Department of Electrical Engineering, and
Electronic Material and Processing Research Laboratory
The Pennsylvania State University
University Park, Pa 16802

This paper is concerned with the methodology and techniques of extraction of 2-D electron gas (2-DEG) characteristics in ultra-submicrometer heterostructure field-effect transistors (HFET's) operated in the passive and active amplifying modes. Parameters extracted from passive mode are based on magnetotransconductance and channel resistance measurements at 300 K and 77 K, and those extracted from amplifying mode of operation are based on microwave S-parameter co-planar wafer-probe measurements. These parameters provide a basis for the extraction and understanding of the intrinsic high-speed limitations of the HFET's.

INTRODUCTION

Ultra-submicron HFET's with unity-current-gain frequency above 100 GHz have been demonstrated by several research groups[1-3] in recent years. The overall high-speed performances of these HFET's depend both on the intrinsic and parasitic device network elements, and they must be delineated for a better understanding of the HFET's high-speed performance limits. In this paper we present characterization techniques to carry out this task.

Small-signal network models of high-speed devices are useful not only for understanding the connection that exists between the device structure and its electrical characteristics, but also for improving the structural designs and predicting circuit performance at frequencies where measurements are impractical. This is particularly true for ultra-submicrometer gate-length HFET's whose unity-current-gain frequency (f_τ) values usually extend beyond the limits of existing measurement equipment. (*e.g.* Network Analyzer Model HP 8510 A&B&C)

For accurate determination of equivalent network model the techniques of direct extraction of various network elements have been suggested and implemented by several groups for FET's[4,5]. In ref. [5] it was concluded that while it is possible to determine $\mathrm{Re}[Y_{21}]$ and $\mathrm{Im}[Y_{11}]$ of the intrinsic HFET's with a high degree of accuracy, it is difficult to accurately determine the related $\mathrm{Im}[Y_{21}]$ and $\mathrm{Re}[Y_{11}]$.

In this paper we re-examine this issue of accurate extraction of the intrinsic Y-parameters of ultra-submicrometer gate-length (0.15 μm) HFET's. The test devices for this work were obtained from Martin-Marietta Electronics Laboratory (formerly GE) under a collaborative arrangement.

PASSIVE MODE MEASUREMENTS

Measurement techniques have been developed for the extraction of the 2-D electron gas (2-DEG) sheet carrier concentration ($n_{2\mathrm{DEG}}$) and the related capacitance ($c_{2\mathrm{DEG}}$) in heterostructure field-effect transistors (HFET's) at room temperature and cryogenic temperatures. The extraction process depends on an accurate determination of the 2-DEG channel conductance (g_{ch}), its derivative ($\partial g_{ch}/\partial V_{GS}$) with respect to the gate voltage, and the average and differential carrier mobilities (μ_{av} & μ_{diff}) through low-frequency magneto-transconductance measurements[6]. The key equations are as follows:

$$n_{2DEG} = \frac{\ell_g}{W} \frac{g_{ch}}{\mu_{av}}$$

$$c_{2DEG} = \frac{\ell_g}{W} \frac{1}{\mu_{diff}} \frac{\partial g_{ch}}{\partial V_{GS}} \qquad (1)$$

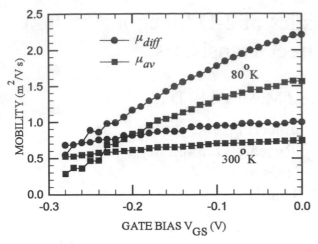

Figure 1. The average and differential mobilities measured at 80 K and 300 K.

where $g_{ch} = [r_{ds} - (r_s + r_d)]^{-1}$ and $\partial g_{ch}/\partial V_{GS} = -g_{ch}^2 \partial r_{ds}/\partial V_{GS}$, r_{ds} is the drain to source resistance when $V_{DS} \rightarrow 0$ and $(r_s + r_d)$ is the total source and drain series resistance. Typical data for mobilities, and n_{2DEG} and c_{2DEG} are shown in Figures 1 and 2, respectively. The significance of the measurement technique lies in the fact that it enables one to determine the characteristics of the 2-DEG channel carriers in ultra-submicron field-effect transistors, and these when combined with the microwave S-parameter data lead to a detailed understanding of the transimpedance phase delay and the gate charging processes of the intrinsic transistor. For the purpose of separating the total source and drain parasitic series resistance associated with channel we used the gate probe method[7] involving variation of drain current, near zero drain bias voltage (V_{DS}).

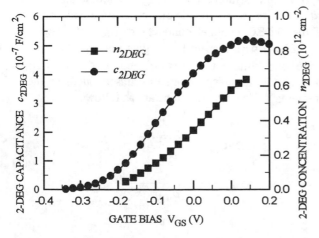

Figure 2. Dependence of 2-DEG capacitance and sheet concentration on the gate voltage.

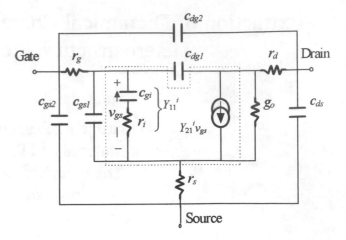

Figure 3. HF equivalent network representation of HFET's.

ACTIVE MODE MEASUREMENTS

Equivalent Network Model Parameters

By carefully examining the structural details of the lattice-matched test HFET's presented elsewhere[8], an equivalent network for small-signal operation in the common-source configuration can be developed as shown in Figure 3. The equivalent network representing the intrinsic device is shown within the dotted boundary, and the relevant input and transfer-admittance parameters can be expressed as follows[5].

$$Y_{11}^i = \frac{j\omega c_{gi}}{1 + j\omega c_{gi} r_i} \qquad (2)$$

$$Y_{21}^i = g_{mi} \frac{e^{-j0.6\omega\tau_o/4}}{(1 + j\omega\tau_o/4)} \frac{e^{-j\omega\tau_d}}{(1 + j\omega\tau_{21})} \qquad (3)$$

where $c_{gi} r_i$ represents the intrinsic gate-to-channel charging time constant, g_{mi} represents the intrinsic transconductance, τ_o represents the channel transit time due to the excess channel charge near the source above the carrier velocity saturation charge, τ_d represents the transit time associated with the effective gate-length due to carrier saturation velocity (v_{sat}), and τ_{21} is due to the distributed RC-network associated with the velocity saturation charge distribution.

The parameters c_{gi}, τ_o, and τ_d can be expressed as follows[5,9].

$$c_{gi} = c_{2DEG}[1 - \frac{\gamma}{3} + (1 - \gamma)\frac{\ell_{dg}}{\ell_g}] \qquad (4)$$

$$\tau_o = \frac{2}{3}\left(\frac{\gamma}{1-\gamma}\right)\frac{\ell_g}{v_{sat}} \qquad (5)$$

$$\text{and} \qquad \tau_d = \frac{\ell_g + \ell_{dg}}{v_{sat}} \qquad (6)$$

where c_{2DEG} represents the 2-D electron-gas gate capacitance under the gate electrode when $V_{DS}=0$, ℓ_g is the gate-length and ℓ_{dg} is the depleted separation region between the gate and the drain under drain current saturation condition of operation, and γ represents a modulation factor of the drain saturation voltage with respect to the gate voltage. For gate-length values of 0.25μm and 0.15μm typical values of $\gamma \approx 0.25$[10].

Determination of Extrinsic Network Elements

The lattice-matched AlInAs/InGaAs HFET's on InP substrate supplied by Martin Marietta Electronics Laboratory, Syracuse, NY, have their electrode layout exactly matching the footprint of the Tektronix wafer probe type TMP9610. Before carrying out any measurements on a given device, a calibration routine through LRM (Line-Reflect-Match) technique[11] using Tektronix standards was performed to minimize errors due to possible probe misalignment caused by mechanical and thermal effects. Our results of LRM open method of calibration on a 50 Ω load indicated an error range better than ±0.25 Ω for frequencies up to 26 GHz for the network analyzer used (HP8510C).

The extraction of intrinsic Y-parameters follows the deembedding procedure suggested in ref.[4] which requires a knowledge of the various parasitic elements. When the device is in its cut-off state (i.e. $V_{GS}<V_{TH}$) with the drain voltage $V_{DS}=0$, $Y_{11}^i = 0$ and $Y_{21}^i = 0$ in the

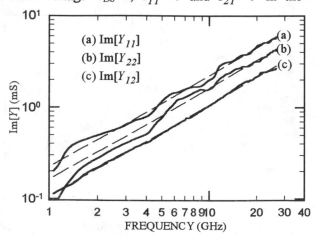

(a) Im[Y_{11}]
(b) Im[Y_{22}]
(c) Im[Y_{12}]

Figure 4. Dependence of imaginary Y-parameters on frequency measured in the cutoff region.

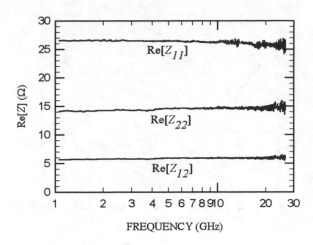

Figure 5. Dependence of real Z-parameters on frequency.

equivalent network shown in Figure 3. Thus for lower frequency extrapolation we can express the imaginary parts of the Y-parameters of the passive device as follows:

$$\text{Im}[Y_{11p}] \cong \omega(c_{gs1} + c_{gs2} + c_{dg1} + c_{dg2})$$

$$\text{Im}[Y_{22p}] \cong \omega(c_{ds} + c_{dg1} + c_{dg2}) \qquad (7)$$

$$\text{Im}[Y_{21}] = \text{Im}[Y_{12}] \cong \omega(c_{dg1} + c_{dg2})$$

Because of $V_{DS}=0$, the capacitance c_{gs1} and c_{dg1} at the two edges of the gate electrode can be expected to be nearly identical, and the value of this capacitance can be approximated[12] to $\pi \varepsilon W/2 \approx 9$ fF for $W= 50$ μm. When V_{DS} is in the normal bias range c_{dg1} can be much different than c_{gs1}. The experimental results representing the above parasitic Y-parameters are shown in Figure 4.

Next we heavily forward bias the gate while maintaining $V_{DS}=0$, and determine the following impedance parameters of this bias modified passive device[4].

$$Z_{11p} \cong r_s + r_g + \frac{r_{ch}}{3} + \frac{nkT}{ql_g} + j\omega(L_s + L_g)$$

$$Z_{22p} \cong r_s + r_d + r_{ch} + j\omega(L_s + L_d) \qquad (8)$$

$$Z_{12p} = Z_{21p} = r_s + \frac{r_{ch}}{2} + j\omega L_s$$

From the imaginary parts we extracted the values of L_s, L_g, and L_d, and they are listed in Table I. The source and drain resistances were determined following the gate probe method published elsewhere[7]. This method yielded $r_s=5$ Ω and $r_d=7$ Ω for this particular test device. The gate resistance is best determined from the linear resistance of the gate metallization pattern and its equivalent RC transmission line model. This yielded $r_g \cong 1$ Ω. The

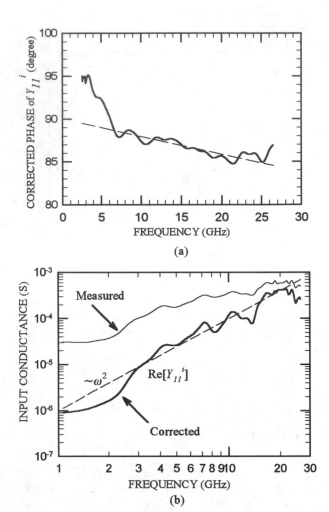

(a)

(b)

Figure 6. Dependences of Y_{11}^i on frequency: (a) phase angle and (b) real Y_{11}^i.

extracted data showing the real parts of Z_{22p}, Z_{12p} and Z_{11p} in Figure 5 do indicate consistency with a rather high Schottky diode differential resistance (~15.7 Ω).

Extraction of Intrinsic Network Parameters

From eqn (2) the phase angle of Y_{11}^i and its real part, $Re[Y_{11}^i]$, can be expressed as follows when $\omega c_{gi} r_i \ll 1$,

$$\angle Y_{11}^i = \frac{\pi}{2} - \omega c_{gi} r_i$$
$$Re[Y_{11}^i] \cong \omega^2 c_{gi}^2 r_i \qquad (9)$$

The graphical data representing $\angle Y_{11}^i$ and $Re[Y_{11}^i]$ are fitted in Figures 6(a) and 6(b), respectively. The data presented in Figure 6(b) were corrected for the leakage

Table I. Parasitic Parameters

V_DS=0, V_GS<V_TH		
Resistance (Ω)	Inductance (pH)	Capacitance (fF)
$r_s = 5$ $r_d = 7$ $r_g = 1$	$L_s = 2$ $L_d = 36$ $L_g = 18$	$c_{gs1} = 9^*$ $c_{dg1} = 9^*$ $c_{gs2} = 9.5$ $c_{dg2} = 7.5$ $c_{ds} = 12.5$

*Calculated

Schottky gate input conductance, the corrected phase angle versus frequency is shown in Figure 6(a). The data presented in Figures 6(a) and 6(b) were consistently iterated for the extraction of the r_i.

From eqn (3), the phase angle $\angle Y_{21}^i$ can be approximated as follows when $\omega \tau_{21} \ll 1$, and $\omega \tau_o/4 \ll 1$,

$$\angle Y_{21}^i \cong -\omega(0.4\tau_o + \tau_d + \tau_{21}) \qquad (10)$$

By combining equations (2) and (3) the intrinsic current gain can be approximated as

$$\left| h_{21}^i \right| = \left| \frac{Y_{21}^i}{Y_{11}^i} \right| = \left| \frac{1}{j\omega\tau_i} \right| \qquad (11)$$

The approximations that $\omega c_{gi} r_i \ll 1$, $\omega \tau_{21} \ll 1$, and $\omega \tau_o/4 \ll 1$ also apply here. Experimental data for representing $\angle Y_{21}^i$ and $|h_{21}^i|$ are presented in Figures 7 and 8, respectively, and they were used to determine $\tau_d' = (0.4\tau_o + \tau_d + \tau_{21})$ and $\tau_i = (\tau_o + \tau_d)$. We next calculated τ_o from eqn (5) for the assumed value of $\gamma \cong 0.25$, and thus obtained τ_d and τ_{21} using the known values of τ_i and τ_d'. Since τ_o is much smaller than τ_d any

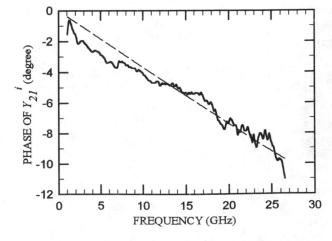

Figure 7. Dependence of phase angle of Y_{21}^i on frequency.

696

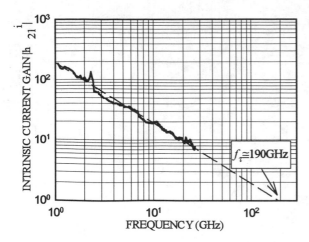

Figure 8. Magnitude of the intrinsic current gain, h_{21}, versus frequency.

Table II. Extracted and Calculated Small-signal Equivalent Network Parameters.

Parameters	V_{GS}=−0.2V V_{DS}= 0.5V	V_{GS}=0.0V V_{DS}=0.6V	V_{GS}=0.2V V_{DS}=0.7V
c_{gi} (fF)	29	44	52
r_i (Ω)	31	13.1	12
$c_{gi}r_i$ (ps)	0.899	0.576	0.624
g_{mi} (mS)	17	48.5	57
$(g_{mi}r_i)^{-1}$	1.89	1.57	1.46
c_{gd} (fF)	14.0	13.0	11.5
g_o (mS)	2.5	6.1	6.4
τ'_d (ps)	1.6	1.1	1.1
τ_i (ps)	1.706	0.907	0.912
τ_o (ps)	0.25	0.14	0.14
τ_d (ps)	1.45	0.75	0.75
τ_{21} (ps)	0.05	0.28	0.29
$f_{\tau i}$ (GHz)	93.2	175	174
l_{dg}/l_g	0.26	0.20	0.164
d (Å)	346	225	180
v_{sat}(10^7cm/sec)	1.31	2.34	2.28
$g_{mi}d/W\varepsilon(1-\gamma)$ (10^7cm/sec)	1.40	2.60	3.00
v_{av} (10^7 cm/sec)	1.1	1.97	1.91

possible variation of γ from its assumed value would have negligible impact on the determination of τ_d.

Finally we present all the extracted and calculated parameter values at 300 K in Table II for three different operating bias conditions and assuming nominal values of $l_g \cong 0.15$ μm and W=50 μm.

CONCLUSION

A methodology and measurement techniques have been presented in this paper for the extraction of HF intrinsic network elements of HFET's which are characterized by the dynamical behavior of velocity saturated 2-DEG. The measurements exploit both the passive and active mode device operations for the separation of extrinsic and intrinsic network parameters. Experimental data have clearly indicated linear phase delay in Y_{21}^i and Y_{11}^i with increasing frequency, and ω^2 dependence of Re[Y_{11}^i]. These frequency dependent characteristics provide the basic ingredients for a clear understanding of the high-speed limitations of HFET's from material and structural viewpoints.

ACKNOWLEDGMENTS

The earlier phase of this work was supported by NSF Grant No. ECS-8921694. The test devices were obtained from Martin-Marietta (formerly GE), Syracuse, NY.

REFERENCES

1. P. Ho et al., *Electronics Lett.* **27**, 325-327 (1991)

2. D.C. Streit et al., *Proc. IEEE/Cornell Conf. Adv. Conc. High. Speed Dev. & Circuit*, 455-460 (1991)

3. L.D. Nguyen et al., *IEEE Trans. Electron Dev.* **39**, 2007-2014 (1992)

4. G. Dambrine, *IEEE Trans. MTT* **36**, 1151-1159 (1988)

5. S.T. Fu et al., *IEEE Trans. Electron Dev.* **37**, 888-901 (1990)

6. S.T. Fu et al., and M.B. Das, *IEEE Trans. Electron Dev.* **38**, 1719-1729 (1991)

7. S.M. Liu et al., *IEEE Electron Dev. Lett* **10**, 85-87 (1989)

8. P.C. Chao et al., *IEEE Electron Dev. Lett.* **11**, 59-61 (1990)

9. M.B. Das, "HEMT's device physics and models", Ch.2 in *HEMT's & HBT's: Device, Fabrication and Circuits*, Eds. F. Ali and A. Gupta, Artech House, Boston, 11-63 (1991)

10. M.B. Das, *IEEE Trans. Electron Dev.* **34**, 1429-1440 (1987)

11. S. Lautzenhiser, *Microwaves & RF Journal*, 105 (1990)

APPENDICES

Appendix 1: Panel Discussions

Panels of representatives of the equipment industry concluded the afternoon session on each of the first three days of the Workshop. These panel sessions were organized by Alan Jung and Murray Bullis on behalf of the Semiconductor Equipment and Materials International (SEMI) to provide for multiple inputs and interactive discussion on important issues related to the topics of the invited paper sessions. In particular, they were intended to allow the metrology equipment suppliers and other industry representatives to provide additional inupt on these topics. Each panel discussion began with brief opening statements by the panelists. These were followed by a half-hour of open discussion on topics of interest to the audience.

NOTE — This appendix was assembled by Murray Bullis based on notes and the visual aids used by the panelists in their opening statements. It was not peer reviewed and is presented for information only.

Panel 1: Meeting the Metrology Requirements of the National Technology Roadmap for Semiconductors

Moderator: Michael E. Fossey, *ADE Corporation*

Participants: Marjorie Balazs, *Balazs Labs*; Rebecca S. Howland, *Tencor Instruments*; W. Lee Smith, *ThermaWave Corporation*; and Richard F. Spanier, *Rudolph Research*

OPENING STATEMENTS

FOSSEY: The keys to the continuous development and growth of the semiconductor industry over the last 35 years are *better, cheaper, faster*. There are no magic formulas to get to where the industry must be to achieve 0.25 μm and 0.18 μm technologies; a concerted effort will be required.

BALAZS: The wet chemical analysis laboratory is capital equipment intensive even at the present time and accuracy and sensitivity requirements are continuing to increase. The services provided by such laboratories can answer real problems for the semiconductor industry. Table 1 lists ways in which the services of one such laboratory are used by semiconductor manufacturers and suppliers. However, the advent of the 300 mm wafer is expected to introduce a number of problems for the analyst. These revolve around wafer cost (many analytical procedures require breaking the wafer so it cannot be reclaimed), lack of 300 mm stages for vacuum analytical equipment (even if available they would be slow and awkward), and wafer transport. The time to consider these issues is now if 300 mm wafer technology is to be in place by 1998–2000.

HOWLAND: Detection of particles as small as 0.08 μm on bare wafers, as required by the NTRS in the year 2000, is

Table 1. How Wet Chemical Lab Services are Used by Semiconductor Manufacturers and Suppliers.

Area of Effort	Application
Contamination Free Manufacturing	Wet benches Contamination on wafer Contamination in air Contamination from reactors Contamination from liquids Leachable contamination from gowns, wafer carriers, etc.
Quality Control	BPSG, PSG Si/Cu in Al Ti/W ratios, etc. Other thin films Cleaning, etching, processing chemicals High purity water, etc.
Special activities; new development	Cleanroom air sampler TOC in chemicals Particle monitor Consulting
Education	SPWCC In-house seminars

within today's capabilities, but detection of still smaller particles will require development of reliable, shorter wavelength lasers. Detection of particles on films depends

Table 2. Film Metrology Challenges.

Area	Challenges
Technology	More parameters simultaneously on more complex stacks • 3 → 4 → 5 unknowns • ONO, OPO, ONPO, PONPO, ... Greater repeatability • <0.1 Å on a 50-Å gate oxide film Production conditions • Gate oxide under poly in cluster tool Increasing robustnessactors • CMP process control
Productivity	Reduce metrology cost • Capital, maintenance, spare parts, cleanroom space, training, test wafer Increse fab efficiency • Cross training, standard-ization, uptime and MTBF Reduce process excursions • Enable measurements on product wafers
Migration to in-situ	Core competencies Partnering • may lead to conflicts Respecting metrology output • higher level of scrutiny

on suppression of the surface scatter by techniques such as oblique incidence. There is also a general issue with respect to true sizing of particles in that it is necessary to know what the particle is before its physical size can be established. Film thickness measurements are intensively model-based. One trend is toward a combination of spectroscopic ellipsometry with reflectometry.

SMITH: Challenges for the future include technology challenges, productivity challenges, and challenges associated with the migration of metrology from in-line to in-situ. These challenges are summarized in Table 2. In response to these challenges, many companies (including ThermaWave) are combining multiple technologies and capabilities into a single tool.

SPANIER: The goal is to provide the needed production worthy metrology tools when they are needed at the *lowest CoO to a user* that yields a *reasonable profit to the supplier*. In film thickness measurements increasing levels of precision are being demanded, but it is essential to realize that the product of the standard deviation and the measurement area is a constant. That is, a reduction in

standard deviation can be achieved only by averaging over larger areas. Stable film thickness standard artifacts are needed to qualify tools to 0.015 nm, 3-sigma levels. Finally, cooperation with users and process equipment manufacturers is essential beginning in the early instrument design phases, but this can bring its own problems of communication and understanding.

GENERAL DISCUSSION

The first issue raised related to the need for ellipsometry in the cooling chamber of a cluster tool. It was pointed out that a large test site was essential unless a microscope can be placed very close to the wafer position. It was also noted that business problems were as much an issue as technical problems in incorporating in-situ metrology into process tools.

More modeling efforts were also identified as a key to improved thin film metrology; more effort was called for in development of analytical approaches to the modelling problem. Also the need to better understand the physics of the problem was emphasized.

Discussion turned to the specifications of film thickness; several attendees deplored the lack of "sense" in many specifications. However, it was noted that Raman-based tools were available for fractional Å determinations, but the distance over which this precision could be achieved was not really clear. The necessity for understanding the meaning of the specification was emphasized. Also it was brought out that overspecification adds unnecessary cost to production, but this is a difficult issue to discuss with semiconductor manufacturers.

The increasing recognition of metrology in device production was viewed both as good and as not so good. It is clear that cost reduction is a major metrology issue. Introduction of "respectable," meaning precise but not necessarily accurate, tools into production areas was suggested as one way to reduce costs. Such an approach would not be acceptable, however, for development applications. Another avenue could be the incorporation of more standardized common elements. Value engineering was also identified as another path. In any event, the need to do more in cost reduction was widely recognized.

The changing nature of the metrology equipment industry was a recurring theme in the discussion. It was questioned whether existing instrument companies could be as successful in supplying high volume in-situ measurement tools, but it was noted that the market for reference instruments would also grow with the industry.

Panel 2: Choosing the Proper Instruments — Benchmarking, Cost of Ownership, Standardization

Moderator: Robert I. Scace, *NIST*

Participants: Daren Dance, *SEMATECH*; James J. Greed, Jr., *VLSI Standards*; Richard S. Hockett, *Charles Evans and Associates*; and Robert Mazur, *Solid State Measurements*

OPENING STATEMENTS

DANCE: The model for cost of ownership (CoO) was developed to address the economic and productivity performance for wafer fab tools. This model is well accepted for process tools, but a CoO model for metrology equipment is also required. For metrology tools, the CoO analysis is carried out in two parts. One is an estimate of the cost per measurement. The second is an estimate of the impact of the measurement on the process or product. If the metrology information improves the process, then metrology is a value-added step.

GREED: On its first day, the workshop came up with several attention grabbers. Both Rose's (1) economic assessment of metrology in high volume fabs and Hosch's (2) metrology hierarchy should be factored into the planning of instrument manufacturers. In considering the structure of future instruments, there are several issues that should be included in the design. First, the operating protocol should both emulate a standard method promulgated by ASTM, SEMI, or other consensus standards setting bodies and support traceability paths. Second, the data evaluation routines should use the ISO/NIST methodology (3) to evaluate and express the measurement uncertainty.

HOCKETT: Many different physical analysis techniques are finding application in the semiconductor industry in both development and production areas. Standards for these techniques are being developed by groups in ASTM, ISO and other bodies. Nevertheless, a double message is being received from the end users of these methods. On the one hand they are saying "We love you! Just make it more reliable (95% uptime), sooner (1 month delivery), and of course cheaper (15% less than the last one.)" On the other hand, they say "We will abandon you as soon as possible. You have no value; your are more than worthless — you drain our profitability and keep us from growing. We can become something great if you leave."

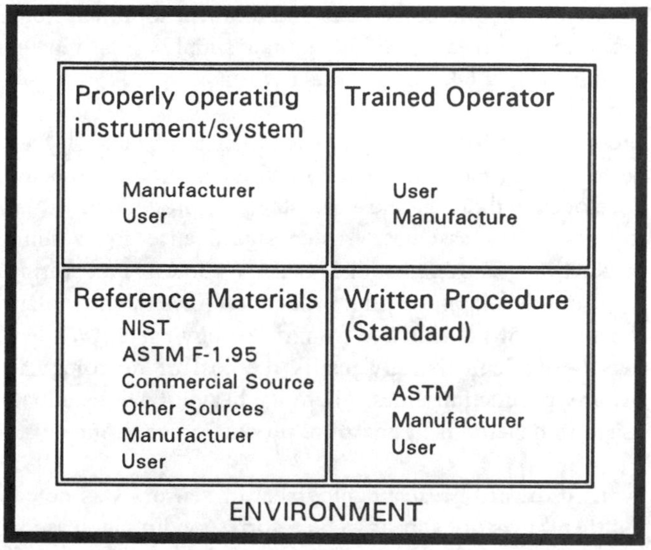

Figure 1. The elements of a good measurement system, surrounded (and influenced) by the environment in which it is being used.

MAZUR: The elements of a good measurement instrument or system involve the four areas schematically depicted in Figure 1. Responsibility for the proper operation of the system lies first with the instrument manufacturer (design and construction) and then with the user (environment and maintenance). The user, and secondarily the instrument manufacturer, are responsible for ensuring operator training. Reference materials, such as SRMs from NIST, consensus reference materials from ASTM interlaboratory activity, or commercially supplied reference standards, are needed to monitor and verify instrument performance. And finally the instrument should be operating in conformance with a written standard procedure, such as a consensus standard from ASTM or an internally generated procedure. The instrument must be operated in a proper environment and should be benchmarked by a procedure, perhaps one similar to the "Ironman" procedure used for process equipment.

SCACE: There is a need for improved communication among instrument manufacturers, metrology users, and developers of test methods and metrology to increase awareness of the capabilities and limitations of various test procedures.

GENERAL DISCUSSION

The discussion opened with a suggestion that metrology should be introduced in well-thought out stages, not in one single jump. This is because commitment to a metrology program involves much more than simply installing the equipment and taking the measurements.

Metrology programs can pay back their costs through fault detection and fault classification while processes are being developed and during the early stages of production. Run-to-run control assumes greater significance for volume production. This distinction between volume manufacturing lines and development-pilot production lines might explain the duality of the message from metrology users. While it may be difficult to fully justify the cost of metrology in mature production lines, there is heavy (and essential) usage in development and pilot production environments.

A trend toward greater deployment of sensors was noted, but there does not appear to be a corresponding increase in the development of models for the larger in-situ sensors that are being introduced. Further discussion of the application of sensors led to the concept of calibrating sensors through the use of designed experiments. In addition, it was noted that sensor response can be related to process models.

Processes can be modeled using TCAD (except for plasma etching); such model-based analysis can determine the sensitivity of the process to the "knobs" or process parameters. Process parameters can be related to product parameters using product data taken with a minimum number of sensors. It was noted that every sensor must earn its place in the process. Data should be collected and used judiciously; neither too much time spent nor too much data collected adds value.

REFERENCES

1. Rose, Don, "Cost Benefit of Process Metrology," *this volume*.

2. Hosch, J. W., "In-Situ Sensors for Monitoring and Control of Process and Product Parameters," *this volume*.

3. ISO/TAG 4/WG 3, "Guide to the Expression of Uncertainty in Measurement," June 1992; NIST Technical Note 1297, "Guidelines for Evaluating and Expressing the Uncertainty of NIST Measurement Results," January 1993.

Panel 3: In-situ Measurements — Promise and Barriers

Moderator: John C. Bean, *AT&T Bell Labs*

Participants: Norman E. Schumaker, *EMCORE Corp.*; John C. Stover, *TMA Technologies*; and John A. Woollam, *J. A. Woollam Co.*

OPENING STATEMENTS

SCHUMAKER: Critical technology relationships exist between process engineering, equipment development, materials science, and device production. In-situ metrology can play a key role in these relationships. Metrology should be:

- Used to design and maintain status of equipment;
- Symbiotic rather than parasitic with equipment; and
- Used to maintain process yield and product performance.

Sensors should monitor system status while metrology should validate system status for process conditions.

STOVER: Based on experience with other industries, up-front communication between suppliers of in-situ instrumentation and both process equipment manufacturers and end users is critical in defining the functional specifications the instruments are intended to test. One such attribute, surface roughness is becoming more important as line widths continue to shrink. Light scattering is a useful, non-invasive technique for determining the magnitude of the surface roughness.

WOOLLAM: The objective of in-situ spectroscopic ellipsometry is control of the process from various perspectives including thickness, composition, source or substrate temperature, growth rate, roughness, and void fraction. This technique has been used to control a wide range of processes including MBE growth, CVD growth, ECR etching, RIE, magnetron sputtering, e-beam reactive evaporation, oxygen plasma ashing, electrochemical deposition and organic films from vapor. In some cases, chamber retrofits were necessary to install the capability for spectroscopic ellipsometry on the system. There are several issues in connection with implementing process control by in-situ spectroscopic ellipsometry. One major issue is how to motivate process equipment vendors to put optical ports on process chambers. Technical issues include focusing, spot size, window effects, and rotating substrates. The community needs greater understanding of

correlated variables and what various instruments can and can NOT do. Finally, there is the ever critical issue of software integration between the tool and the process.

GENERAL DISCUSSION

One important application of in-situ metrology is to maintain yields. But a major question remains as to whether yields can be improved enough to justify the installation and use of in-situ sensors. It was noted that wafer state sensors were the most important, but the issue of how to monitor patterned wafers must be resolved. Later it was noted that areas for test structures must be made available on patterned wafers. The probe area needs to be scaled to the material variability.

The principal use of sensors now is for monitoring process and equipment variables. One critical application area noted was the diagnosing of problems associated with mass-flow controllers. Others suggested that there is also a need to monitor hydrogen purifiers, pressure transducers, and temperature. Because of the increasing interest in reducing equipment down time, equipment state sensors are assuming a greater importance.

Discussion turned to practices used in Japan. There processes appear to be "more manufacturable." Operator training is more comprehensive and intensive, equipment documentation is more complete, and detailed performance records are kept to assure equipment operation. It was noted that Japanese customers do not insist on modifications and customization to the extent that is practiced in the US, but that they do conduct exhaustive design reviews to assure that the equipment performs as expected. Then the equipment is scrupulously maintained to assure that it continues to perform as it was designed.

It was opined that we do not now take enough advantage of the information that is already available from sensors. Sensors are available for a number of applications, but they are not being used. Perhaps process equipment manufacturers are not ready for more widespread implementation of in-situ sensors There are issues to be resolved such as assumption of system responsibility and reliance on equipment designed and made by others. In any event, users must drive the use of in-situ sensors, but it was concluded that wafer state sensors are not needed to assure business success at the present time. On the other hand, fast, short term payback can sometimes be obtained from the use of equipment and process state sensors. Looking ahead to the NTRS requirements of the near future, it is possible to predict that in-situ monitoring of product parameters as well as process parameters will become more necessary.

Appendix 2: Rump Session
Roadmapping Synchrotron X-Ray Metrology

This rump session was organized by Alain Diebold of SEMATECH and Bob McDonald of Intel to discuss an area that represents a good example of how partnerships between industry and the national labs can be developed. Such partnerships would appear to work most satisfactorily when there is a convergence between the interests and capabilities of a national lab and an industry need that cannot be met using standard laboratory tools. In the present case, several synchrotron facilities have capabilities to provide analysis services that would not otherwise be possible. A roadmap for determining the feasibility of establishing a commercial facility to provide these services has been established and is being implemented.

NOTE — This appendix was prepared by Murray Bullis based on notes and the roadmap visual aid used in the discussions. It was not peer reviewed and is presented for information only.

Bob McDonald opened the rump session with a statement regarding the interest of the industry in cooperative arrangements that could utilize the facilities and expertise of the various national laboratories. Alain Diebold then reviewed the Roadmap for Synchrotron X-Ray Metrology, see Figure 1. He noted that the plan is based on the use of existing facilities and does not require the construction of any new facility. In general, cooperative arrangements, such as the ones considered here, thrive best when they involve activities that cannot be done in-house with standard laboratory instruments. In this case, although the usage level is expected to be relatively low, the information that can be obtained is essential for solving certain critical problems.

Steve Laderman of Hewlett-Packard Labs described his early efforts in developing this concept. The interest began with a convergence of an industry need for TXRF analysis using a synchrotron source for increased sensitivity with the interests and capabilities of several synchrotron laboratories in the Bay Area. Nevertheless, it was not a simple task to couple the need with the capability. The process Laderman used involved the seven Cs:

- campaign to illuminate the need;
- communicate the need to an organization with the capability;
- cross calibrate with others in the industry;
- collaborate with others to promote the concept;
- commit resources to demonstrate usefulness;
- collaborate with others in demonstration projects; and
- create a neutral organization to coordinate the development of a plan (roadmap) and the conduct of experiments.

Dick Brundle, consultant to Applied Materials, described

other activities that could usefully employ synchrotron x-ray sources. These include x-ray lithography (which only needs the intense illumination from the source) and particle analysis by spectromicroscopy using μ-XPS or μ-XANES to provide information on both composition and chemical state. He noted that the June 30, 1995, decision date in the roadmap is too aggressive for development of μ-XANES capabilities.[*]

Dick Hockett of Charles Evans Associates discussed the conditions under which a viable commercial business might be developed to exploit the synergy in synchrotron x-ray capabilities. Most important is guaranteed access on a timely basis in order to meet commercial schedules. This would almost certainly require the cooperation of two or more synchrotron facilities so that access could always be maintained.

There was general discussion of ways that such cooperative activities might be conducted. An issue yet to be resolved is the question of proprietary work which conflicts with the generally open policies of the synchrotron facilities. Skepticism was expressed concerning the viability of such partnerships. In response, it was noted that they would not be expected to function in a production environment, but that they could answer questions during development of processes that could not be answered any other way. The session closed with general appreciation both for the potential of the concept and the difficulty of developing specific arrangements of this type.

---

[*] NOTE ADDED IN PROOF: Subsequent to the Workshop, it was decided to postpone the Go/No Go decision on TXRF until August 31, 1995 and the decision on μ-XANES and related technologies for several years.

Synchrotron X-Ray Metrology

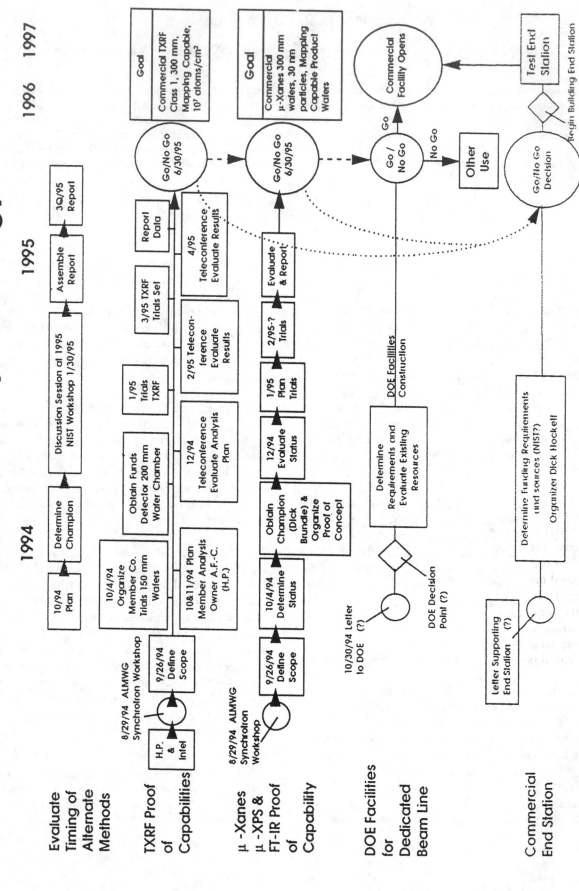

Figure 1. Roadmap for implementing synchrotron x-ray metrology facilities.

708

Historical Perspective on Semiconductor Characterization

W. Murray Bullis

Materials & Metrology
1477 Enderby Way
Sunnyvale, CA 94087-4015

Today's sleek, high speed semiconductor characterization instruments with automated wafer loading and unloading are a far cry from the relatively crude instruments used to establish the fundamental semiconductor characteristics in the 50s and 60s. These early instruments were constructed out of components, some of which had to be home-built, and computations were laborious. The development of the methodology and its standardization is traced through a series of anecdotes together with some commentary on what we have gained and lost as a result of this evolution.

NOTE — The remarks in this appendix have not been peer reviewed and are presented for information only.

INTRODUCTION

Semiconductor characterization tools have changed dramatically since the early days of semiconductor research and development. Today's sleek, high speed instruments with computer controls and automated wafer loading and unloading are a far cry from the relatively crude instruments used to establish the fundamental semiconductor characteristics in the 50s and 60s. The field has benefitted greatly from the evolution of semiconductor devices and circuits over the last 50 years or so. But this evolution has also imposed more severe demands on the characterization technologies. It is rather like a perpetual motion machine: No sooner does someone come up with a brilliant scheme for measuring some elusive characteristic, often based on the cutting edge of semiconductor technology, than the requirement tightens and new ideas and techniques are needed.

This paper provides an anecdotal rather than exhaustive history of semiconductor characterization beginning with the period around the time of the invention of the transistor in 1947. It is based largely on personal recollections that illustrate in a schematic way how this field evolved. No attempt has been made to include a complete list of references that recognize the contributions of the many dedicated scientists and engineers that contributed to the development of the field. This task is left for another time and place.

EARLY DAYS OF THE TRANSISTOR

There had been semiconductor characterization activity for more than 20 years before the invention of the transistor, but it was only after reproducible crystals could be grown that semiconductor characterization could be considered a science rather than a hobby. In the early days of the transistor, the researchers at Bell Labs and a few other places around the country were beginning to unravel the mysteries of germanium and silicon by finding out how to synthesize them in a controllable way and to tailor their properties so they did what was desired in a predictable fashion. At the same time, they also developed many of the basic tools and techniques for semiconductor characterization. These developments and others were recorded in many scientific papers in the *Proceedings of the IRE*, the *Review of Scientific Instruments*, *The Physical Review*, the *Journal of Applied Physics*, and the *Bell System Technical Journal*. Some of these techniques were based on research that had taken place decades before. Papers that describe the initial development of some of the most important semiconductor characterization techniques have been reprinted in a collection of pioneering semiconductor papers (1).

As a student, I was not at all involved in these very early activities. In fact, we were quite incredulous that people had reported electrical conduction with holes (which we interpreted, erroneously of course, as *nothing*). When I started graduate school in 1951, vacuum tubes were still the dominant active electronic components.

I first became involved with semiconductor characterization during my thesis research. For this I chose to study the electrical transport properties of germanium. Since there were no on-campus facilities for semiconductor measurements, I actually worked in the semiconductor physics group at Lincoln Laboratory. At this point the die was cast; one way or another, I have been concerned with semiconductor characterization during most of the intervening years.

The group at Lincoln Laboratory at that time had skills in materials preparation as well as in characterization. This was essential, because one of the aspects of semiconductor characterization in those days was that you could not ring up your favorite silicon supplier and ask for some 200 mm diameter circular wafers with known characteristics. In fact, silicon itself was not very popular in the early 50s, even though the crystal detectors for WWII radar had been made of silicon, because it was very difficult to obtain single crystals of this material. In fact, it was not even easy to obtain uniformly doped samples of germanium; at one time we seriously considered using high a-c fields in heated samples in order to try to homogenize the dopant distribution. This scheme would not have worked out, and it is well that we did not spend much time on it.

The methodology for collection of the data for this research illustrates the state of the art of measurement at the time. We were interested in the potential drops both along and across a rectangular bar of germanium as a function of magnetic field, temperature, and crystal orientation. The samples were cut from crystals grown by the Teal-Little (modified Czochralski) technique using an outside diameter, glass cutting saw. The crystals were really large for the time, nearly 1 inch in diameter. The surfaces were oriented by means of x-ray diffraction techniques, commonly used in metallurgy; these were reasonably precise ($\pm 1°$), but additional degradation occurred because the saw could not be aligned so accurately.

Our electrical test equipment was quite standard. A battery in series with a large resistor served as the current source. The potential drop across a smaller precision resistor was a measure of the magnitude of the current. Large end contacts to the bar were composed of indium solder applied using a sharpened soldering iron and a special flux, a much dirtier combination than today's evaporated films. Potential contacts on the sides of the bar were formed from dots of the same solder. Five-mil copper bell wire was used to complete the circuit. A manually operated, multi-pole switch in the circuit provided the means to select the appropriate terminals for potential measurement.

A potentiometer-galvanometer combination was used to measure the potential drops. You may remember that a potentiometer works on a null principle; a voltage opposing the one being measured is varied until there is zero current in the galvanometer circuit. The voltage is then read off the precision vernier dial. As an aside, I should note that many ingenious types of potentiometers were developed here at the National Bureaus of Standards (NBS), the predecessor to NIST. Some, I expect, can still be seen in the museum next to the Library entrance.

We also needed several other systems for the experiments. The magnetic field was obtained using a very large water-cooled, d-c electromagnet. Similar equipment is still in use today, although superconducting magnets are more frequently used to obtain the higher fields. Temperatures were controlled by water baths or by immersion in boiling liquids. Fancy Dewar systems of non-magnetic stainless steel or glass were in common use, and we spent some time designing our own stainless steel dewar system for use between the poles of the electromagnet. Finally, we occasionally measured the carrier lifetime by the photoconductive decay method which involved a spark source and an oscilloscope.

The data recording system was equally archaic, consisting of an eyeball to read the dials or scales, a pen, and a notebook. The data collection process was facilitated by using two operators, one to set the dials and read the indicator and the other to record the results. We had many discussions as to whether it was more efficient to collect data in the morning or afternoon. Once the data were recorded, they had to be analyzed. What better way to do this than with a slide rule, or if the task required greater precision, as this one did, a mechanical calculator? Graphs were prepared manually, usually using "Leroy" lettering templates, which explains the general uniformity of graphs in publications of the day.

Reports were typed; it was a major step forward when IBM brought out its line of electric typewriters with proportionally spaced letters and the possibility of typing mathematical symbols and Greek letters with the use of special keys. A small number of such keys could be inserted into the machine and operated from the keyboard, but more symbols could be obtained by using loose keys, manually placed in a special frame near the paper so that they could be struck against the ribbon by one of the regular keys. As might be suspected, all of this was very labor intensive, from the assembly of the apparatus, through data collection, to reporting, leading to the characterization of graduate students as "slave labor."

While I was preoccupied with this activity, other interesting things were taking place in the immediate vicinity. One of these involved a complex apparatus, lovingly called the "kluge" by its mentors, Don Stevenson and Bob Keyes, who were also the developers of the photoconductive decay method for carrier lifetime. This apparatus was made up to study the effect of moisture on the surface recombination and surface states of germanium and silicon. Moisture levels were established by using carefully controlled solutions of various salts, and the moisture was transported to the surface of the sample by means of plastic tubing. The whole thing was encased in a big metal water bath in order to control the temperature of the system. This apparatus was not very elegant, but it did its job well even though it was very temperamental at times.

Another group at Lincoln was making pioneering cyclotron resonance measurements. These measurements provided direct evidence of the shapes of the band edges near their maxima and minima. These results could then be used to predict the anisotropies in other properties of these semiconductors, such as the galvanomagnetic properties I was studying in my thesis research.

REFINEMENTS AND EXTENSIONS

By the end of the 50s, the basic tools for characterization of semiconductor materials were well in hand. Silicon was well established as the semiconductor of choice, and so much had been learned about this material that a degree of complacency had come over the industry. New fields lay ahead in the conquering of III-V and other compound semiconductors. About this time, I joined the Central Research Laboratory at Texas Instruments; on arrival I was informed that because everything essential was already known about silicon, we should concentrate on the III-V compounds with particular attention to gallium arsenide which was of interest at the time because of the possibility, later proven false, of producing bipolar transistors that would operate at even higher temperatures than silicon.

The characterization techniques developed for germanium and silicon also worked well for the compounds. We set up a Hall effect system to study the properties of both dopant and deep lying impurities. The basic equipment and techniques for these measurements had not much changed since my thesis research, but electronic constant current sources were beginning to become available, and stable, precise digital voltmeters were on the horizon.

Other activity in the group involved studies of the photomagnetoelectric effect, an interesting but not too useful property. We also attempted to measure spreading resistance on indium antimonide and gallium arsenide without much success. John Stamper was more successful in his study of thermomagnetic effects; he developed an ingenious thermocouple probe that enabled these tiny effects to be determined without introducing heat losses through the probe, but no one really cared because these effects were too small to be of practical use.

Characterization tools were also being introduced in the industry. At RCA, Hildebrand and Gold extended the capacitance-voltage method to the measurement of dopant profiles of arbitrary shape in p-n junctions in 1960. Nearly 20 years before, Schottky had pointed out that impurity distributions could be derived from the $C(V)$ relationship, and others had applied this technique to measurement of dopant distributions of known shape. But the RCA work introduced a more general relationship and demonstrated that sufficiently precise instrumentation (based on the venerable 1-MHz Boonton Capacitance Bridge) was available to enable the measurements to be made on a practical level. This extremely important technique has been refined extensively over the years by many workers, and it is still in widespread use today. It has proved to be especially useful for characterizing epitaxial films.

At Westinghouse Research Labs, Mazur and Dickey successfully developed the spreading resistance technique for measuring resistivity profiles with very high spatial resolution. By making the measurement on an angle beveled surface, depth profiles could also be determined. This technique has been further refined by a number of workers over the years and various correction factors and calibration procedures have been developed.

One very important set of these correction factors was developed by Schumann and Gardner at IBM. They were part of an active metrology group that was also instrumental in developing phase shift corrections for the infrared reflectance technique of measurement of the thickness of epitaxial layers that had been developed earlier both at the Bell Telephone Laboratories and at Monsanto. This method was somewhat tedious when carried out with dispersive spectrometers, but nevertheless it was widely used. Later on, Fourier Transform Infrared Spectrometers supplanted the dispersive types and provided a much faster, automated method for the determination.

711

Another major development was the application of ellipsometry to the characterization of thin films of SiO_2 and other insulators on silicon. A symposium on applications of ellipsometry for characterizing surfaces and thin films was held at NBS in 1964 (2). At TI, Ray Hilton began to develop this technique in the early 60s; however, he did not report on his work until much later. Ellipsometry was very tedious because of the extensive calculations that were involved. As a result, this technique was one of the first to benefit from the increased computational power of semiconductor based computers. As can be seen from the many papers at this conference, it has since been greatly extended and is routinely employed for both laboratory and production measurements.

Also at this time, trace analysis of unwanted impurities at the ppb levels was becoming critical. Techniques that were available, including spark-source mass spectrometry, neutron activation analysis, and autoradiography, were emphasized in the Analysis and Characterization section of the first Silicon Symposium in 1969 (3). Although some of these techniques, especially neutron activation analysis, are still employed, they are not used extensively both because of their limited precision and their inconvenience.

Several other analytical chemistry techniques, such as Auger electron spectroscopy, x-ray photoelectron spectroscopy (ESCA), and flame emission spectroscopy, were also beginning to be used for trace analysis of semiconductor materials, but at this time these were generally not sensitive enough for many applications. These techniques were discussed extensively, along with many other trace analysis methods, in a symposium at NBS (4) in 1966. A few years later, Kane and Larrabee (5) published a now-classic volume that detailed the application of these techniques to semiconductor characterization.

In the mid-60s I switched my attention from the III-Vs to silicon. I had the good fortune to work with Walt Runyan during the period he was writing his classic volume, *Silicon Semiconductor Technology* (6) which, unfortunately, is now out of print. We set up a Hall effect system for detailed studies of 1) the Hall factor r (the ratio of the Hall mobility to the conductivity mobility) as a function of crystal orientation and magnetic field and 2) the electrical activity of the gold centers in both n- and p-type silicon.

We also did an interesting experiment in which we measured the Hall effect in the wide, high resistivity base regions of power n-p-n transistors after each major process step. Both the conductivity type and the resistivity of this region changed dramatically at various stages of the process, eventually ending up very close to the starting condition of about 50 $\Omega\cdot$cm p-type after the final phosphorus diffusion to form the emitter. We never published the results because we were unable to explain them, but with the benefit of hindsight, it is quite clear that we were observing the influence of iron impurities, and that these impurities were being gettered out during the phosphorus diffusion.

BEGINNINGS OF STANDARDIZATION

The 60s also brought the beginnings of test method standardization. The semiconductor industry was expanding and the variability of measurement results was affecting the requirements of commercial transactions. Even in such "simple" measurements as resistivity of silicon wafers, disagreements between supplier and user were commonplace. This was not because people were making errors in measurement, but rather because of imprecise definition of measurement conditions, with the result that each location employed its own, usually not well controlled, procedures. This is a classic example of the fact that the marketplace makes demands on definitions and metrology that are much more severe than those made by the research environment. In the latter, it is generally sufficient to demonstrate a scientific principle, but the marketplace demands that nominally identical products be replicated with high precision.

Standardization of materials for the vacuum tube industry had been well established during the late 50s. In ASTM, Committee F-1 on Electronics had evolved from a subcommittee on non-ferrous materials for vacuum tubes in Committee B-5 on Copper and Copper Alloys. Its scope covered development of standards for materials for all kinds of electron devices. By 1960, through the visionary leadership of its early chairmen, Sid Standing of Raytheon and Frank Biondi of Bell Telephone Laboratories, the committee had made a rapid transition to include semiconductor materials in its activities.

The first two semiconductor related documents involved tests for properties of GeO_2, a source material for germanium crystals which could be made by hydrogen reduction of very pure GeO_2. By 1964, the semiconductor subcommittee had developed standards to cover all of the basic test procedures: crystal orientation by x-ray diffraction and optical figures, "minority" carrier recombination lifetime by the decay of photoconductivity, conductivity type by thermoelectric and rectification tests, resistivity by the four-probe and two-probe methods, oxygen content by infrared absorption, and crystallographic perfection by preferential etch techniques. Except for the last two, all these tests could be applied to both silicon and germanium. In addition, standard procedures were developed for preparation of germanium and silicon test ingots.

At the same time, another F-1 subcommittee was working on test methods for determining particulate contamination in air, in fluids, and on surfaces. This subcommittee conducted a series of landmark symposia that defined industry requirements and led to the development of almost 20 standards. Many of these early test procedures were later incorporated into Federal Standard 209; still others supplement this venerable document which has just recently been revised for the fifth time to accommodate the increasingly demanding contamination control requirements of the microelectronics and other industries.

During the last half of the decade, other standard test procedures were developed. These covered Hall effect on rectangular and van der Pauw samples, four-point probe measurement of radial resistivity variation, crystallographic perfection of silicon epitaxial layers, and thickness of epitaxial layers by infrared reflectance. At the same time, the industry was encountering difficulties with the initial standardized test method for resistivity because of the increasing requirements as the technology developed. The solution to these problems involved the contributions of many organizations, including NBS.

EARLY NBS ACTIVITIES

The NBS Electron Devices Section had begun to investigate test methods for semiconductor materials in 1960, at the request of Committee F-1. Resistivity, probably the single most important characteristic to be controlled for commercial silicon wafers, was the subject of one of the first projects undertaken. This work was directed first toward establishment of reference standards for the four-probe method. Because the prevailing wisdom was that more accurate measurements could be obtained with the use of rectangular test samples measured by the two-probe method, factors were developed to correct four-probe measurements made on rectangular samples. It was soon found that agreement between the two methods could not be achieved because of non-uniformities in the resistivity of the rectangular samples.

Consequently, the work was redirected to improving the four-probe method. By improving the control of factors such as probe spacing, probe wobble, surface and probe condition, and measurement temperature, it was possible to improve the 3-sigma reproducibility attainable with the four-probe method to about $\pm 2\%$, significantly better than had been possible previously. The demonstration of reproducibility was made by means of a round robin experiment conducted by Committee F-1. This experiment, and much of the work that preceded it, involved a great deal of cooperation from several companies, including Bell Labs, IBM, RCA, and Monsanto. The project was considered to be an outstanding success and became a model for future projects, both in Committee F-1 and at NBS.

It is worth noting that this method, published as an ASTM standard in 1968, was developed with old fashioned apparatus. The current was supplied by a battery in series with a large series resistance, and potential drops were determined with a potentiometer-galvanometer combination. Later on, of course, the system was upgraded to include an electronic constant current supply and a high-precision digital voltmeter.

During the 60s, another project at NBS was concerned with the issue of second breakdown in n-p-n power devices. This work, which was carried out in cooperation with the Power Device Committee of the Joint Electron Device Engineering Councils, led to the development of a test instrument for nondestructive determination of the reverse-bias safe operating area and an extensive program in thermal characterization of devices and device packaging.

Another useful output from the group at NBS was a series of bibliographies on various measurement techniques including resistivity (7), second breakdown (8), doping inhomogeneities (9), and carrier lifetime (10). These compilations provide comprehensive listings of the papers that report early work in these areas of metrology.

OTHER TEST PROCEDURES

Other government agencies were also working to develop improved standards for semiconductor processing and devices. A group at Rome Air Development Center developed a revolutionary procedure for qualification of integrated circuits. When first introduced in 1968, the rigid testing procedures of MIL-STD-883 were widely considered to be too difficult or even unnecessary, so many vendors sought and received waivers. Although this standard continued the philosophy of "testing-in" rather than "building-in" quality, the discipline imposed by the testing hierarchy provided a methodology for control and improvement of processes, especially when it was combined with the determination of the causes of testing failures. It should be noted that this testing philosophy was widely adopted by the fledgling Japanese electronics industry, and contributed significantly to its success at producing high quality products.

NASA took a complementary approach. The Line Certification Program, developed at Marshall Space Flight Center, focussed on the control of incoming materials (including silicon wafers, chemicals, gases, and parts), the clean room environment, and fabrication processes. Where they were available, ASTM test methods were cited in the specifications. This approach helped several early IC manufacturers significantly improve their processing capabilities.

A two-day symposium on semiconductor characterization cosponsored by NBS and ASTM Committee F-1 in 1970 provided a summary of the metrology developments during the 60s (11). The principal emphasis was on techniques that had been refined or standardized during this period, especially those relating to characterization of epitaxial and diffused regions. However, several other analysis techniques such as Rutherford back scattering, neutron activation analysis, Auger spectroscopy, and ellipsometry were also discussed, and one paper discussed an early form of process modeling.

SILICON WAFER STANDARDS

In the early 70s, the rapid expansion of the industry caused a shortage of silicon wafers. At the time, wafers were being manufactured on a custom basis with a wide range of dimensions to meet the differing requirements of various device lines. These lines had grown up using equipment designed and built in-house with little regard for what others in the industry were doing. The materials suppliers felt that the impending shortage could be averted if the dimensions could be standardized, thus reducing the need to inventory similar, but not interchangeable, material. A limited number of values for wafer diameter, thickness and thickness variation, and flat length was proposed. Although a system of flats on the wafer periphery to differentiate wafer type and orientation was included in the proposal, no attempt was made to standardize electrical and crystallographic properties. This distinction between dimensional and other properties in silicon wafer specifications has endured to the present.

These specifications were formalized by the newly formed Standards Program of the Semiconductor Equipment and Materials Institute, and they were soon generally accepted throughout the free world. They were instrumental in increasing industry awareness of the ASTM test method standards and led directly to the standardization of new test methods for dimensional characteristics, including thickness, thickness variation, bow, diameter, and flat length. The first of these methods were based on contacting probes, but after a short while, non-contacting techniques, based on a capacitance method for sensing the distance from a probe to the wafer surface, were introduced. A non-contacting technique was also introduced for resistivity measurement. In addition to the dimensional tests, a procedure for visual inspection of wafers for surface defects was also developed.

A second major impact of the existence of standardized wafer geometry was the ability of the equipment industry to design and construct wafer handlers, transport systems, carriers, and processing equipment on an interchangeable basis. This, together with the almost universal adoption of the planar process for device and circuit fabrication, fostered the orderly growth of the merchant semiconductor process and measurement equipment industry.

The wafer specifications have evolved over time. They were introduced at about the time the leading edge of the industry was shifting to from 2 in. to 3 in. wafers, but they have since been extended to cover larger wafers including 100 mm, 125 mm, 150 mm, and 200 mm. The industry is now seriously considering the next transition to 300 mm wafers, and it is expected that a similar format will be adapted for specifying these wafers.

NEW MEASUREMENT TECHNOLOGY

During the 70s, new requirements for device reliability and performance caused the expansion of semiconductor characterization activity to increase at NBS and elsewhere in the industry. At NBS, George Harman developed an ingenious capacitor microphone technique to detect potentially weak bonds while the bond was being made.

A major ARPA-funded program at NBS significantly expanded the development of semiconductor measurement technology (12). The program had three major elements: in-house projects; workshops and symposia; and contract research. One of the principal in-house projects was the development of optical techniques for line-width measurement. This system was one of the first to use microprocessor control.

Microelectronic test structures were studied in detail. These structures, fabricated by advanced device processes, were used as test vehicles for determination of a wide range of material and device properties. One of the first applications of these structures was in the redetermination of the resistivity-dopant density relationship for silicon, but later they were applied to characterization of a variety of material and processing properties including pattern alignment, dopant profiling, process uniformity, and metallization integrity.

Also during this period, NBS issued the first standard reference materials, for calibration of four-probe and spreading resistance measurements. These were later followed by line-width standards for photomasks, oxide thickness standards, and, just recently, standards for oxygen content in silicon. There is a constant clamor for more such reference materials; however, it must be recognized that highly reproducible test methods and stable test specimens with adequate uniformity are necessary prerequisites for their development. These critical elements, unfortunately, frequently do not exist.

Symposia and workshops were conducted to provide a forum for the industry to discuss IC metrology issues. Topics covered included a general review of measurement problems in IC processing (13), hermeticity testing (14), spreading resistance (15), test patterns (16), and surface analysis techniques (17).

Also as part of the program, many different measurement techniques were developed in industry and academia under contract to NBS. Among the techniques developed were micrometer-scale imaging for acoustic microscopy, improved SEM analysis procedures, a scanning photometric microscope for optical linewidth measurements, and surface quality tests for sapphire substrates. Auger and x-ray photoelectron spectroscopies were applied to detailed investigations of the Si-SiO$_2$ interface. Specialized electrical test structures were developed for a variety of applications. Electrical test methods and theoretical models were combined to obtain better data on carrier mobilities as a function of dopant density and temperature. Very preliminary data on the transfer of metal ions from solutions to wafer surfaces were developed by combining advanced chemical analysis with theoretical models. Some of these techniques and data were applied directly by the industry, but others did not meet with wide acceptance at the time because they were either too sketchy or too advanced.

Elsewhere in the industry, established techniques such as spreading resistance, four-point probe, and capacitance measurements were refined and extended. For example, improved correction factors and surface preparation techniques were developed for spreading resistance measurements. Perloff and co-workers extended the four-probe method by the use of configuration switching techniques and made mapping of the resistivity distribution over the wafer feasible. Sah and his students developed a variety of capacitance spectroscopy techniques for investigating deep levels in semiconductors; these were further refined by Lang as deep level transient spectroscopy (DLTS). The procedures introduced earlier by Zerbst for generation lifetime determinations by measuring the time dependence of the capacitance were refined and extended. As time went on, many of these extensions and improvements were incorporated into new or revised standard test methods by ASTM Committee F-1.

There were significant developments in the application of various chemical techniques both for surface analysis and for dopant profiling. Secondary ion mass spectrometry was of particular importance for depth profiling because of its high spatial resolution. In Japan, photoluminescence was introduced as a very sensitive test for bulk impurities. The techniques for x-ray topography were refined and the speed of this measurement was increased. Reflectivity measurements in the visible and near-UV were applied to the measurement of dielectric film thickness. Although reflectivity had been used for film thickness measurement in the laboratory from the mid-50s, the later automation of this technique led to its widespread use in production environments.

Many of these developments were reported in the Characterization Sections of the Silicon Symposia held in 1973 (18) and 1977 (19) and in the first stand-alone symposium on semiconductor characterization held by The Electrochemical Society in 1978 (20). Carrier lifetime measurements were the subject of a one-day workshop conducted by ASTM Committee F-1 in 1979 (21).

Widespread interest in these techniques and in new analytical methods continues to grow. Significant improvements in sensitivity and spatial resolution are still being made in many areas as the demands of the industry become more severe. Although many of the techniques and standards were developed for the silicon industry, the methodology employed and the techniques themselves can frequently be applied to the characterization of other semiconductor materials and devices made from them (22).

RESPONSE TO INCREASING DEMANDS

Late in the 70s, computer assisted measurement equipment began to proliferate. Use of such equipment permitted collection of much more data than had been possible previously. This raised the question of how to present the data for easy comprehension and comparison. Although it is relatively easy to visualize contour and density maps of data distributions, such maps do not lend themselves to quantitative comparison; thus, the characteristics of two samples cannot easily be compared by means of such maps. As a result, the data were frequently reduced to single numbers, an average or the peak-to-valley excursion, for these comparisons. This made comparison easy, but at the expense of discarding a great deal of the information from the data set. Furthermore, it was usually not recognized that the average of such a data set does not have the same significance as the average of a data set resulting from random deviations about a central value, as is obtained, for example, when machining a part to a target dimension.

The second consequence of the introduction of computerized equipment was that the instruments were most often of the "black-box" type; that is, the user did not have much information about how the system actually performed; at the same time, the measurement could be performed without such knowledge. This had the advantage that the instrument could be used by less highly trained operators. On the other hand, there was no satisfactory way to ensure that the instrument was performing correctly. The first example of this problem occurred in connection with the measurement of epitaxial film thickness by infrared interference techniques. The standard test method was based on direct analysis of the reflectance spectrum; however, soon after it was adopted, Fourier transform instruments, based on the analysis of the interferogram were introduced. These instruments contained proprietary software and each treated the data somewhat differently from the others. Consequently, it was not feasible to standardize this technique.

In the early 80s, automated instruments for wafer surface inspection began to appear. These instruments involved scanning of a beam across the surface and detecting the scattered light. Depending on the design, these instruments were more or less sensitive to particular types of defects on the surface. Again, standardization has proved to be an elusive goal; just now is the standards community beginning to make some headway in this area.

The 80s placed continually increasing demands on semiconductor characterization. Tighter tolerances on wafer dimensions and especially flatness led to the development of new measurement tools. For use in production lines, automated wafer handling and data analysis by computer were essential. Early microprocessors had to be programmed in assembly language; the ubiquitous personal computer and "user friendly" applications packages had not yet appeared on the scene.

One of the areas in which confusion reigned was in the determination of the interstitial oxygen content of silicon. Control of this impurity assumed a greater importance because of the conflicting requirements between wafer strength and gettering capability. There were two problems: the dependence of the measurement on the surface condition of the test sample and the calibration of the measurement. Many calibration factors had been reported in the literature with a factor of about 2½ between the largest and the smallest. This issue was resolved by means of a world-wide interlaboratory experiment which showed that the infrared absorption technique, used in production, had much better precision than the absolute techniques needed for its calibration. The issue of sample surface influence on the measurement has largely been solved, but standardization has been slow to develop (23).

There was also an increasing interest in microscopic studies of both surface contamination and crystal defects in silicon. Transmission electron microscopy studies revealed many details concerning stacking faults and precipitates of various kinds. Analysis of tiny particulate contamination found on surfaces by SEM and energy dispersive x-ray analysis gave information on particle composition and clues as to their origin.

Point defects took on a greater measure of significance because of their influence on precipitation kinetics and diffusion of impurities. Studies of diffusion under various surface conditions allowed the testing of various models for point defect interactions. Positron annihilation and x-ray studies provided further information on point defect types and densities.

For detection of surface metals contamination, analytical techniques such as total reflection x-ray fluorescence (TXRF), graphite-furnace atomic absorption spectrometry (GF-AAS), and inductively-coupled plasma—mass spectrometry (ICP-MS) came into use. In some cases, these analysis methods were preceded by vapor phase dissolution of the native oxide to capture and concentrate the surface metal contamination. Standard procedures for these methods are beginning to appear. Standards for secondary ion mass spectrometry are also under development.

Surface topography is being investigated by atomic force microscopy, liyght scatter, and interferometry. This characteristic is playing an important role in development of cleaning processes and in the observation of sub-0.1 μm particles on wafer surfaces.

Most of these techniques have been described in a number of symposia sponsored by The Electrochemical Society (24), the Materials Research Society (25), ASTM (26), and others. In addition, specialized symposia and workshops have been held on a number of topics including profiling of shallow junctions (27) and mercury cadmium telluride (28).

Several text books and encyclopedias on semiconductor characterization have been published in the past several years. Examples are the book by Schroder (29) which emphasizes electrical and optical measurements and the encyclopedia by Brundle et al. (30) which emphasizes chemical and physical analysis methods.

A PARADIGM SHIFT

There is currently an increasing interest in metrology for controlling high density integrated circuit fabrication processes. This topic was discussed extensively in a recent workshop sponsored by ASTM and other organizations (31), in a report by Robert Scace of NIST (32), and in the recently updated National Roadmap for Semiconductor Technology (33) and associated Metrology supplement (34). A major paradigm shift toward *in situ* and in-line measurement of product characteristics is projected for future manufacturing environments. This will require significant alteration of the traditional view that measurements are made on product variables for diagnostic or analysis purposes and in-line measurements are made on process variables. Up-time, repeatability, and reliability requirements for in-line and *in situ* measurement instruments are significantly greater than for off-line diagnostic and analysis tools. Improvements in measurement sensitivity and ease-of-operation will be needed. Relationships between suppliers of measurement instruments, sensors, and process equipment will need to be developed and strengthened so that mutual confidence can be established. The challenges ahead for the metrology community are indeed as formidable as they have ever been. The present Workshop provides a comprehensive guide along the path to solution of these challenges.

REFERENCES

1. Sze, S. M., ed., *Semiconductor Devices: Pioneering Papers*, Singapore: World Scientific Publishing Pte (1991), pp. 153–207.

2. Passaglia, E., Stromberg, R. R., and Krueger, J., eds., *Ellipsometry in the Measurement of Surfaces and Thin Films*, NBS Misc. Publ. 256 (1964).

3. Haberecht, R. R., and Kern, E. L., eds., *Semiconductor Silicon*, New York: The Electrochemical Society (1969) Section 6, pp. 516–584.

4. Meinke, W. W., and Scribner, B. F., eds., *Trace Characterization, Chemical and Physical*, NBS Monograph 100 (1967).

5. Kane, P. F., and Larrabee, G. B., *Characterization of Semiconductor Materials*, New York: McGraw Hill (1970).

6. Runyan, W. R., *Semiconductor Silicon Technology*, New York: McGraw-Hill (1965).

7. French, J. C., "Bibliography on the Measurement of Bulk Resistivity of Semiconductor Materials for Electron Devices," NBS Technical Note 232 (October 21, 1964).

8. Schafft, H. A., "Second Breakdown in Semiconductor Devices — A Bibliography," NBS Technical Note 431 (October 1967).

9. Schafft, H. A., and Needham, S. G., "A Bibliography on Methods for the Measurement of Inhomogeneities in Semiconductors (1953–1967)," NBS Technical Note 445 (May 1968).

10. Bullis, W. M., "Measurement of Carrier Lifetime in Semiconductors — An Annotated Bibliography Covering the Period 1949–1967," NBS Technical Note 465 (November 1968).

11. Marsden, C. P., ed., *Silicon Device Processing*, NBS Spec. Publ. 337 (Nov. 1970).

12. Bullis, W. M., "Advancement of Reliability, Processing and Automation for Integrated Circuits with the National Bureau of Standards (ARPA/IC/NBS)," NBSIR 81-2224 (March 1981).

13. Schafft, H. A., "ARPA/NBS Workshop I. Measurement Problems in Integrated Circuit Processing and Assembly," NBS Special Publication 400-3 (February 1974).

14. Schafft, H. A., "ARPA/NBS Workshop II. Hermeticity Testing for Integrated Circuits," NBS Special Publication 400-9 (December 1974).

15. Ehrstein, J. R., ed., "Spreading Resistance Symposium," NBS Special Publication 400-10 (December 1974).

16. Schafft, H. A., ed., "ARPA/NBS Workshop III. Test Patterns for Integrated Circuits," NBS Special Publication 400-15 (January 1976).

17. Lieberman, A. G., ed., "ARPA/NBS Workshop IV. Surface Analysis for Silicon Devices," NBS Spec. Publ. 400-23 (March 1976).

18. Huff, H. R., and Burgess, R. R., eds., *Semiconductor Silicon 1973*, Princeton: The Electrochemical Society (1973), Chapter 3, pp 416–623.

19. Huff, H. R., and Sirtl, E., eds., *Semiconductor Silicon 1977*, Proceedings Volume 77-2, Princeton: The Electrochemical Society (1973), Chapters 6–8, pp 359–501.

20. Barnes, P. A., and Rozgonyi, G. R., eds., *Semiconductor Characterization Techniques*, Proceedings Volume 78-3, Princeton: The Electrochemical Society (1978).

21. Westbrook, R. D., *Lifetime Factors in Silicon*, ASTM Special Publication 712, Philadelphia: American Society for Testing and Materials (1980).

22. Bullis, W. M., "Evolution of Silicon Materials Characterization: Lessons Learned for Improved Manufacturing," NIST Special Publication 400-92 (July 1993); see also —, "Evolution of Silicon Materials Characterization: Lessons Learned for Improved Manufacturing," Ref. 28, pp 777–787.

23. Bullis, W. M., "Oxygen Concentration Measurement," in *Oxygen in Silicon*. Shimura, F., ed., Boston: Academic Press, Inc. (1994), Chapter 4.

24. Schaffner, T. J., and Schroder, D. K., eds., *Diagnostic Techniques for Semiconductor Materials and Devices*, Proceedings Volume 88-20, Pennington, NJ: The Electrochemical Society (1988); Kolbesen, B. J., McCaughan, D. V., and Vandervorst, W., eds., *Analytical Techniques for Semiconductor Materials and Process Characterization*, Proceedings Volume 90-11, Pennington, NJ: The Electrochemical Society (1990); Benton, J. L., Maracas, G. N., and Rai-Choudhury, P., eds., *Diagnostic*

Techniques for Semiconductor Materials and Devices, Proceedings Volume 92-2, Pennington, NJ: The Electrochemical Society (1992); and Schroder, D. K., Benton, J. L., and Rai-Choudhury, P., eds., *Diagnostic Techniques for Semiconductor Materials and Devices*, Proceedings Volume 94-33, Pennington, NJ: The Electrochemical Society (1994).

25. Cheung, N., and Nicolet, M-A., eds., *Materials Characterization*, MRS Symposia Proceedings Volume 69, Pittsburgh: Materials Research Society (1986) and Glembocki, O. J., Pang, S. W., Pollak, F. H., Crean, G. M., and Larrabee, G., eds., *Diagnostic Techniques for Semiconductor Materials Processing*, MRS Symposia Proceedings Volume 324, Pittsburgh: Materials Research Society (1994).

26. Gupta, D. C., ed., *Silicon Processing*, ASTM STP 804, Philadelphia: American Society for Testing and Materials (1983); Gupta, D. C., ed., *Semiconductor Processing*, ASTM STP 850, Philadelphia: American Society for Testing and Materials (1984); Gupta, D. C., and Langer, P. L., ed., *Emerging Semiconductor Technology*, ASTM STP 960, Philadelphia: American Society for Testing and Materials (1987); and Gupta, D. C., ed., *Semiconductor Fabrication—Technology and Metrology*, ASTM STP 990, Philadelphia: American Society for Testing and Materials (1989).

27. International Workshops on the Measurement and Characterization of Ultra-Shallow Doping Profiles in Semiconductors, Microelectronics Center of North Carolina, March 1991 [*J. Vac. Sci. Tech.* B 10, 286−549 (1992)], March 1993 [*ibid.* 12, 165− 406 (1994)], and March 1995 [*ibid.*, to be published].

28. International Workshop on Mercury Cadmium Telluride Characterization, October 1992 [*Semicond. Sci. Tech.* 8, 753−957 (1993)].

29. Schroder, D. K., *Semiconductor Material and Device Characterization*, New York: John Wiley & Sons (1990).

30. Brundle, C. R., Evans, C. A., and Wilson, S., eds., *Encyclopedia of Materials Characterization*, Boston: Butterworth-Heinemann (1992).

31. Langer, P. H., and Bullis, W. M., eds., *Process Control Measurements for Advanced IC Manufacturing*, Mountain View: Semiconductor Equipment and Materials International (1993).

32. Scace, R. I., "Metrology for the Semiconductor Industry," NISTIR 4653 (September 1991).

33. *The National Technology Roadmap for Semiconductors*, San Jose: Semiconductor Industry Association (1994).

34. Diebold, A. C., "Metrology Roadmap: A Supplement to the National Technology Roadmap for Semiconductors," SEMATECH Technology Transfer # 94102578A−TR (January 25, 1995); see also —, "Critical Metrology and Analytical Technology Based on the Process and Materials Requirements of the 1994 National Technology Roadmap for Semiconductors," this volume.

Author Index

Subject Index

723